Functional

and

Biotechnology

FOOD SCIENCE AND TECHNOLOGY

Functional Foods
and
Biotechnology

edited by
Kalidas Shetty
Gopinadhan Paliyath
Anthony L. Pometto
Robert E. Levin

CRC Press
Taylor & Francis Group
Boca Raton London New York

CRC Press is an imprint of the
Taylor & Francis Group, an **informa** business

A TAYLOR & FRANCIS BOOK

The material was previously published in *Food Biotechnology, Second Edition.* © CRC Press LLC 2005.

CRC Press
Taylor & Francis Group
6000 Broken Sound Parkway NW, Suite 300
Boca Raton, FL 33487-2742

First issued in paperback 2019

© 2007 by Taylor & Francis Group, LLC
CRC Press is an imprint of Taylor & Francis Group, an Informa business

No claim to original U.S. Government works

ISBN-13: 978-0-8493-7527-9 (hbk)
ISBN-13: 978-0-367-39029-7 (pbk)

Library of Congress Cataloging-in-Publication Data

Functional foods and biotechnology / editors, Kalidas Shetty ... [et al.].
    p. cm. -- (Food science and technology)
Includes bibliographical references and index.
ISBN-13: 978-0-8493-7527-9 (alk. paper)
ISBN-10: 0-8493-7527-4 (alk. paper)
ISBN-10: 1-4200-0772-6
1. Food--Biotechnology. 2. Functional foods. I. Shetty, Kalidas. II. Series.

TP248.65.F66F84 2006
664--dc22
    2006012676

**Visit the Taylor & Francis Web site at
http://www.taylorandfrancis.com**

**and the CRC Press Web site at
http://www.crcpress.com**

Preface

The World Health Organization (WHO) has highlighted that the disease profile of the world is changing, and this is more so in low- and middle-income countries where there is a double burden of diet-related chronic disease along with infectious diseases. WHO further states that 80% of chronic disease deaths now occur in less developed countries. Globally there are more than 1 billion overweight and obese adults, and since 2001 this figure is higher than the number of people (0.8 billion) on this planet who are malnourished. Obesity-related diseases seriously contribute to chronic disease and disability. Therefore, major challenges facing the world today are not just of food production and quality for meeting protein, calorie, vitamin, and mineral needs but also of better health once the basic nutrient needs are met, for which additional protective food ingredients are essential. Clearly, significant challenges are from major oxidation-linked chronic disease epidemics from calorie sufficiency and excess calories in the developed world, and particularly in the newly industrialized countries such as China, Brazil, Mexico, and India, which have the most rapidly growing diet-related chronic disease problems in the world. Chronic disease such as diabetes — which is linked to other oxidation-linked diseases such as CVD (cardiovascular diseases) — along with cancer, will place a tremendous burden on the current healthcare systems in both developing and developed countries. In developing countries, this will further strain the existing challenges of infectious diseases such as acquired immune deficiency (AIDS), tuberculosis, and food-borne illness among the lower income population. In the more developed countries, the continuous and steady development of obesity and its associated complications of diabetes, CVD, and perhaps cancer is already posing more challenges. All the major health challenges, whether excess calorie-linked chronic diseases or undernutrition-linked infectious diseases, are directly or indirectly diet- and environmental-linked diseases. Therefore technologies for chemoprevention through diet (reduced calories with more fruit and vegetables and novel ingredients from other food-grade biological/microbial systems) will be very important to help manage the current and emerging healthcare challenges.

With these critical issues in mind, a more focused edition of the book *Functional Foods and Biotechnology* has been developed from the recently published volume of *Food Biotechnology* (CRC Press, Boca Raton, FL, 2005). This book focuses on those chapters (25 of the original 70 chapters) related to food biotechnology concepts that have the potential to contribute to advances in the areas of functional foods. Functional foods refers to the improvement of conventional foods with added health benefits. Biotechnology concepts related to advances in functional foods will be significant at a time when diet will play a major role in a global population that is projected to increase to 9 billion by 2050. The topics in the book focus on molecular, biochemical, cellular, and bioprocessing concepts for designing ingredients for functional foods and cover major nutrients such as starch, lipids, minerals, and vitamins to specialty ingredients and their disease preventive role as in several phenolic metabolites of several well-known food botanical species. Many chapters are focused on ingredient role in oxidation-linked disease, which is the basis of major chronic diseases. Several specialty topics such as phytochemicals and breast cancer, nonnutritive sweeteners, immune factors from eggs, passive immunity improvement through probiotics and role of prebiotics, phytochemicals as antimicrobials, and various potentials of microbial processing of ingredients have been highlighted. These concepts are by no means exhaustive but give a good conceptual insight to this emerging area of functional foods and point to the role of biotechnology in the development of this rapidly growing research area. Biotechnology has become an important tool in recent years and scientists around the world are investigating advanced and novel whole tissue, cellular, molecular and biochemical strategies for improving food production and processing, enhancing food safety and quality, and improving the organoleptic to the functional aspects of food and food ingredients for better human health. The strength of this book is the conceptual insights it provides in some key emerging areas of functional foods, where biotechnology principles will be key to new advances.

The editors thank all the contributors for their outstanding efforts to document and present their research and conceptual information on their current understanding of food biotechnology. Their efforts have particularly advanced the conceptual knowledge with regard to the use of biotechnology concepts and tools to develop functional food ingredients for better health.

The editors also thank the staff of Taylor & Francis for their help and support in the timely publication of this volume and in targeting an audience focused on research in functional foods. All these efforts have advanced the frontiers of both functional foods and biotechnology, and therefore the title of the book.

The Editorial Board

Editors

Kalidas Shetty, Ph.D. is a professor of food biotechnology in the Department of Food Science at the University of Massachusetts-Amherst. He earned his B.S. from the University of Agricultural Sciences, Bangalore, India, majoring in applied microbiology, and his M.S./Ph.D. from the University of Idaho, Moscow, in microbiology. He then pursued postdoctoral studies in plant biotechnology at the National Institute of Agro-Biological Sciences, Tsukuba Science City, Japan and at the University of Guelph, Canada, prior to joining the University of Massachusetts in 1993.

Dr. Shetty's research interests focus on Proline and Redox pathway-linked biochemical regulation of phenolic phytochemicals in food botanicals using novel tissue culture, seed sprout, and fermentation systems. This focus contributes to innovative advances in the areas of nutraceuticals, functional foods, and food antimicrobial strategies. The susceptibility of bacterial food pathogens to phenolic phytochemicals at low pH and the role of proline metabolism through redox-linked pathways for chemoprevention of diabetes and cardiovascular disease are his major interests in developing new food safety and disease chemoprevention strategies. He has published over 120 manuscripts in peer-reviewed journals and over 25 as invited reviews and in conference proceedings. He holds four U.S. patents.

Dr. Shetty is the editor of the journal *Food Biotechnology* (Taylor & Francis). He is also on the editorial boards of three other journals in the areas of food and environmental sciences.

In 2004, Professor Shetty was selected by U.S. State Department as a Jefferson Science Fellow to advise the Bureau of Economic and Business Affairs on scientific issues as they relate to international diplomacy and international development. This program, administered by the U.S. National Academies, allowed Dr. Shetty to serve as science advisor at the U.S. State Department for one year in 2004–2005, and he will continue to serve in this position for 5 more years following his return to the University of Massachusetts. Dr. Shetty has traveled widely and has been invited to present numerous lectures and seminars in the areas of food biotechnology, functional foods and dietary phytochemicals, and food safety in over 20 countries in Asia, Europe, and the Americas. In 1998 he was awarded the Asia-Pacific Clinical Nutrition Society Award for his contributions to the area of phytochemicals, functional foods, and human health based on his understanding of Asian food traditions. At the University of Massachusetts he has won the College of Food and Natural Resources Outstanding Teaching Award and the Certificate of Achievement for Outstanding Outreach Contributions.

Gopinadhan Paliyath, Ph.D. is an associate professor with the Department of Plant Agriculture, University of Guelph, Ontario, Canada. Dr. Paliyath is a plant biochemist and has an interest in various aspects of plant development. He obtained his B.Sc.Ed. (botany and chemistry) from the University of Mysore, M.Sc. (Botany) from the University of Calicut, and Ph.D. (biochemistry) from the Indian Institute of Science, Bangalore. Subsequently, he did postdoctoral work at Washington State University, University of Waterloo, and the University of Guelph.

The focus of Dr. Paliyath's current research is in the area of post-harvest biology and technology of fruits, vegetables, and flowers. He is investigating signal transduction events in response to ethylene and the role of phospholipase D in such events. Various aspects dealing with improvement in the shelf life and quality of highly perishable commodities are being investigated. Technologies and products have been developed for enhancing the shelf life and quality of fruits, vegetables, and flowers based on phospholipase D inhibition (US Patent #6,514,914). He is also investigating the nutraceutical properties of fruit polyphenols and their role in disease prevention. In this context, the mechanism behind the selectively cytotoxic action of grape polyphenols on breast cancer cell lines is being investigated.

Dr. Paliyath is the author of several peer-reviewed articles, book chapters, books, and research reports. He has also actively participated in conference presentations and several media events promoting fruit and vegetable consumption and its beneficial effects. He is currently serving as associate editor of physiology and molecular biology of plants and is on the editorial board of *Food Biotechnology*.

Anthony L. Pometto III, Ph.D. is a professor of industrial microbiology in the Department of Food Science and Human Nutrition at Iowa State University. He earned his B.S. from George Mason University, Fairfax, Virginia, in biology, and his M.S./Ph.D. from the University of Idaho, Moscow, Idaho in bacteriology. Dr. Pometto worked as a full time scientific aide in the Department of Bacteriology and Biochemistry at the University of Idaho for twelve years. He joined the faculty at Iowa State University in 1988.

Dr. Pometto's research focus is on microbial degradation of degradable plastics, bioconversion of agricultural commodities into value-added products via fermentation, development of novel bioreactors, production of enzymes for the food industry, and the utilization of food industrial wastes. He has co-authored over 60 peer-reviewed journal articles and over 25 articles as invited reviews, book chapters, and conference proceedings. He is a coinventor on three U.S. patents. He is also a member of the editorial board of the journal *Food Biotechnology* (Taylor & Francis), and coeditor of *Food Biotechnology, Second Edition* K. Shetty, G. Paliyath, A.L. Pometto, and R.E. Levin, Eds. (Taylor & Francis, Boca Raton, 2005).

Dr. Pometto became director of the NASA Food Technology Commercial Space Center at Iowa State University in 2000. The center is associated with NASA Johnson Space Center, Houston, Texas, which manages all the food systems for the shuttle, International Space Center, and planetary exploration missions. The NASA Food Technology Commercial Space Center at Iowa State University was founded in August 1999 and completed February 2006. Its mission was to engage industry and academia to develop food products and processes which would benefit NASA and the public. The specific objectives were to develop food products that meet the shelf life requirements for shuttle, ISS, and planetary outposts (which are 9 months, 1 year, and 5 years, respectively); to develop equipment and process technologies to convert the proposed over 15 crops grown on planetary outposts, Moon, or Mars, into safe, edible foods; and to build partnerships with food companies to develop these new food products and processes to make them available for NASA utilization. The space food challenges addressed by the center's commercial partners and affiliate faculty were development of new food products, development of new food processing equipment, extending the shelf life of foods, improving and monitoring food safety, packaging of foods, development of food waste management systems, and development of disinfection systems for space travel. For more information please see www.ag.iastate.edu/centers/ftcsc/.

In January 2006, Dr. Pometto was named the associate director of the Iowa State University Institute for Food Safety and Security, which was created in 2002 as one of six presidential academic initiatives. Dr. Pometto works with the institute's director, Dr. Manjit Misra, to bring together the research, education, and outreach components of food safety and security at Iowa State University into one umbrella institute for the purpose of efficient teamwork that is well-positioned among government, industry, and producers. For more information please see www.ifss.iastate.edu/.

Robert E. Levin, Ph.D. is a professor of food microbiology in the Department of Food Science at the University of Massachusetts-Amherst. He earned his B.S. degree in biology from the California State University, Los Angeles, his M.S. degree in bacteriology from the University of Southern California, and his Ph.D. from the University of California, Davis. Dr. Levin's research interests involve toxicology, dietary modulators of mutagenesis, industrial fermentations, enzymology, and molecular methods of rapid detection and enumeration of bacterial pathogens in foods. He has authored over 150 peer-reviewed research publications and has served on the editorial boards of several journals dealing with food safety and food biochemistry and on U.S.D.A. and N.S.F study groups.

Introduction

Many parts of the world, especially low-income countries, are facing the double burden of diet-related chronic disease along with infectious diseases. Analysis by the World Health Organization states that 80% of chronic disease deaths now occur in low income and less developed countries. As indicated in the preface of this book, it is evident that globally there are more overweight and obese adults (1 billion) than those who are malnourished (0.8 billion), and obesity-related diseases seriously contribute to chronic disease and disability. As a result, significant challenges are linked to major oxidation-linked chronic disease from calorie sufficiency and excess calories in developing and developed countries, with a higher dual burden on less developed countries where people still have to deal with the higher burden of infectious diseases. Technologies for low-cost chemoprevention strategies and dietary means through design of "functional foods" will therefore be very important to help manage the emerging health care challenges, and in this regard tools of biotechnology will be important.

In light of the urgency of the health challenges linked to diet and chronic disease, *Functional Foods and Biotechnology* has been developed from the recently published volume of *Food Biotechnology* (CRC Press, Boca Raton, 2005) in order to highlight some of the challenges. This book focuses on those chapters (25 of the original 70 chapters) related to food biotechnology concepts that have the potential to contribute to advances in the area of functional foods. Functional foods refers to the improvement of conventional foods with added health benefits. The topics focus on biochemical and bioprocessing concepts for designing ingredients for functional foods, and cover improvement of major nutrient sources such as starch, lipids, minerals, and vitamins to specialty ingredients and their disease prevention role, as in several phenolic metabolites of some well-known food botanical species. Many chapters focus on ingredient role in oxidation-linked disease, which is the core basis of major chronic diseases. Several specialty topics such as phytochemicals and breast cancer, nonnutritive sweetners, immune factors from eggs, passive immunity improvement through probiotics and role of prebiotics, phytochemicals as antimicrobials, and various potentials of microbial processing of ingredients have been highlighted. These concepts are by no means exhaustive but provide a good conceptual insight to this emerging area of functional foods and point to the role of biotechnology in the development of this rapidly growing research area.

Chapters 1 through 15 focus on various biotechnological aspects of design of functional ingredients in plants. Chapter 1 focuses on concepts related to the use of clonal screening and sprout-based bioprocessing for designing functional phenolic phytochemicals, and has a section on its relevance to functional foods. This chapter's introduction to the general aspects of regulatory issues related to functional foods is relevant to all chapters that follow. Chapters 2 through 5 focus on concepts of ingredient modification in the context of starch, plant oils and lipids, soybean proteins, and mineral and vitamin enrichment. These chapters provide excellent conceptual insights and ideas also relevant to other photosynthetic plant species that are sources of the above major nutrients. Chapters 6 through 9 focus on phenolic ingredient development in specific food plant species such a soybean (Chapter 6), cranberry (Chapter 7), fava bean (Chapter 9), and the phenolic metabolite rosmarinic acid-enriched culinary herb family, Lamiaceae (Chapter 8). Good conceptual insights are also provided for several health benefits, and there is excellent discussion of potential mechanisms of action of key metabolites with critical chemoprotective roles for enzymatic-based redox regulation in cellular systems. Chapter 10 covers some basic concepts in antioxidant mechanism and provides some additional perspectives to the better understanding of Chapters 6 through 9. Chapter 11, on chemo-prevention of breast cancer by phytochemicals, is a good example of diet-influenced cancer, and these concepts could be relevant for other diet-influenced cancers such as colon and stomach cancers. The influence of dietary antimicrobial phytochemicals is discussed in Chapter 12 in the context of chronic infections such as *Helicobacter pylori*, and in the context of food-borne pathogens. New control strategies are essential in light of emerging challenges with antibiotic resistance, and the role of dietary phenolic phyto-chemicals is promising. In view of the established benefits of wine (Chapter 13) in car-diovascular health, and the need for non-nutritive sweeteners (Chapter 14) in the context of diabetes, these concepts have been separately discussed. The rationale for Chapter 15, on biotechnological strategies to improve nutrients, is that fruits and vegetables are major sources of protective phytochemicals against chronic disease, and many of these com-pounds are produced during post-harvest development. Therefore, protective phyto-chemical synthesis during post-harvest stages has implications for both preservation and enhancement of protective phytochemical factors for managing chronic human disease that is oxidation linked. Chapters 16 through 19 focus on protective immune modulating factors from eggs (Chapter 15) and immune-modulating lactic acid bacteria (Chapter 16) which could be delivered through dairy and soy-milk fermented products, with their functionality enhanced through the use of prebiotics (Chapters 18 and 19). Chapters 20 through 25 cover various microbial and biochemical concepts, specific metabolite types, and bioprocessing systems to develop functional food ingredients. The concepts from these chapters can be extended to many microbial ingredients of relevance to human health. Chapter 25, on solid-state bioprocessing, is relevant for ingredient production and mobilization from chemically bound forms from food biomass and food waste byproducts using aerobic fungal and anaerobic yeast and bacterial systems.

The strength of this book is the conceptual insights it provides on some key emerg-ing areas of functional foods where biotechnology principles will be key to new advances. These novel conceptual ideas and tools could be adapted and applied to develop diverse food-relevant biological systems and biochemical processes for ingredient and whole foods production in order to manage both diet-linked chronic and infectious diseases in an economically feasible manner.

Kalidas Shetty
Professor of Food Biotechnology
Chenoweth Laboratory
University of Massachusetts
Amherst, Massachusetts

Contributors

Motoyasu Adachi
Laboratory of Food Quality Design and
  Development
Graduate School of Agriculture
Kyoto University
Uji, Japan

Tamara Casci
School of Food Biosciences
The University of Reading
Whiteknights, UK

Hanne Risager Christensen
BioCentrum-DTU
Biochemistry and Nutrition
The Technical University of Denmark
Kgs. Lyngby, Denmark

Fergus M. Clydesdale
Department of Food Science
Chenoweth Laboratory
University of Massachusetts
Amherst, Massachusetts, USA

Ali Demirci
Deptartment of Agricultural and
  Biological Engineering
The Hucks Institute of Life Sciences
Pennsylvania State University
University Park, Pennsylvania, USA

K. Helen Fisher
Department of Plant Agriculture
University of Guelph
Vineland, Ontario, Canada

Hanne Frøkiær
BioCentrum-DTU
Biochemistry and Nutrition
The Technical University of Denmark
Kgs. Lyngby, Denmark

Glenn R. Gibson
School of Food Biosciences
The University of Reading
Whiteknights, UK

Ramon Gonzalez
Departments of Chemical
  Engineering and Food Science
  and Human Nutrition
Iowa State University
Ames, Iowa, USA

Peter L. Keeling
BASF Plant Science
Ames Research
Ames, Iowa, USA

Anthony J. Kinney
Crop Genetics Research and Development
DuPont Experimental Station
Wilmington, Delware, USA

Jeffrey D. Klucinec
BASF Plant Science
Ames Research
Ames, Iowa, USA

Jennifer Kovacs-Nolan
Department of Food Science
University of Guelph
Guelph, Ontario, Canada

Reinhard Krämer
Institute of Biochemistry
University of Köln
Zülpicher, Germany

Yuan-Tong Lin
Department of Food Science
Chenoweth Laboratory
University of Massachusetts
Amherst, Massachusetts, USA

V. Maitin
School of Food Biosciences
The University of Reading
Whiteknights, Reading, UK

Nobuyuki Maruyama
Graduate School of Agriculture
Kyoto University
Uji, Japan

Yukie Maruyama
Graduate School of Agriculture
Kyoto University
Uji, Japan

Patrick P. McCue
Program in Molecular and Cellular
  Biology
University of Massachusetts
Amherst, Massachusetts, USA

Yoshinori Mine
Department of Food Science
University of Guelph
Guelph, Ontario, Canada

Moustapha Oke
Ontario Ministry of Agriculture
  and Food
Guelph, Ontario, Canada

Juan Alberto Osuna-Castro
Laboratorio de Biotecnología
Facultad de Ciencias Biológicas y
  Agropecuarias
Universidad de Colima
Tecoman, Colima, México

Gopinadhan Paliyath
Department of Plant Agriculture
University of Guelph
Guelph, Ontario, Canada

Octavio Paredes-López
Centro de Investigación y de Estudios
Avanzados del IPN
Apdo, Gto., México

Reena Grittle Pinhero
Department of Food Science
University of Guelph
Guelph, Ontario, Canada

Anthony L. Pometto III
Department of Food Science and Human
  Nutrition
NASA Food Technology Commercial
  Space Center
Iowa State University
Ames, Iowa, USA

Reena Randhir
Department of Food Science
Chenoweth Laboratory
University of Massachusetts
Amherst, Massachusetts, USA

Robert A. Rastall
School of Food Biosciences
The University of Reading
Whiteknights, UK

Louis A. Roberts
Pioneer Valley Life Sciences Institute
Springfield, Massachusetts, USA

Gerhard Sandmann
Botanical Institute
J. W. Goethe Universität
Frankfurt, Germany

Kalidas Shetty
Department of Food Science
  University of Massachusetts
  Amherst, Massachusetts, USA

Preethi Shetty
Department of Food Science
Chenoweth Laboratory
University of Massachusetts
Amherst, Massachusetts, USA

Sallie Smith-Schneider
Pioneer Valley Life Sciences Institute
Springfield, Massachusetts, USA

Ian W. Sutherland
Institute of Cell and Molecular
Biology Edinburgh University
Edinburgh, UK

Evelyn Mae Tecson-Mendoza
Institute of Plant Breeding
College of Agriculture
University of the Philippines
Los Banos
Laguna, Philippines

Shigeru Utsumi
Graduate School of Agriculture
Kyoto University
Uji, Japan

Dhiraj A. Vattem
Nutritional Biomedicine and
  Biotechnology
FCS Department
Texas State University-San Marcos
San Marcos, Texas, USA

Dhiraj A. Vattem
Nutritional Biomedicine and Biotechnology
FCS Department
Texas State University-San Marcos
San Marcos, Texas, USA

Contents

Chapter 1 Clonal Screening and Sprout Based Bioprocessing of Phenolic
Phytochemicals for Functional Foods   1
Kalidas Shetty, Fergus M. Clydesdale, and Dhiraj A. Vattem

Chapter 2 Molecular Design of Soybean Proteins for Enhanced
Food Quality   25
*Nobuyuki Maruyama, Evelyn Mae Tecson-Mendoza, Yukie
Maruyama, Motoyasu Adachi, and Shigeru Utsumi*

Chapter 3 Genetic Modification of Plant Starches for Food Applications   51
Jeffrey D. Klucinec and Peter L. Keeling

Chapter 4 Genetic Modification of Plant Oils for Food Uses   85
Anthony J. Kinney

Chapter 5 Molecular Biotechnology for Nutraceutical Enrichment
of Food Crops: The Case of Minerals and Vitamins   97
Octavio Paredes-López and Juan Alberto Osuna-Castro

Chapter 6 Potential Health Benefits of Soybean Isoflavonoids and
Related Phenolic Antioxidants   133
Patrick P. McCue and Kalidas Shetty

Chapter 7 Functional Phytochemicals from Cranberries: Their Mechanism
of Action and Strategies to Improve Functionality   151
Dhiraj A. Vattem and Kalidas Shetty

Chapter 8 Rosmarinic Acid Biosynthesis and Mechanism of Action   187
Kalidas Shetty

Chapter 9 Bioprocessing Strategies to Enhance L-DOPA and Phenolic
Antioxidants in the Fava Bean *(Vicia faba)*   209
Kalidas Shetty, Reena Randhir, and Preethi Shetty

Chapter 10   Biochemical Markers for Antioxidant Functionality   229
Dhiraj A. Vattem and Kalidas Shetty

Chapter 11   Phytochemicals and Breast Cancer Chemoprevention   253
Sallie Smith-Schneider, Louis A. Roberts, and Kalidas Shetty

Chapter 12   Phenolic Antimicrobials from Plants for Control
of Bacterial Pathogens   285
Kalidas Shetty and Yuan-Tong Lin

Chapter 13   Biotechnology in Wine Industry   311
Moustapha Oke, Gopinadhan Paliyath, and K. Helen Fisher

Chapter 14   Biotechnology of Nonnutritive Sweeteners   327
Reena Randhir and Kalidas Shetty

Chapter 15   Biotechnological Approaches to Improve Nutritional Quality
and Shelf Life of Fruits and Vegetables   345
Reena Grittle Pinhero and Gopinadhan Paliyath

Chapter 16   Egg Yolk Antibody Farming for Passive Immunotherapy   381
Jennifer Kovacs-Nolan and Yoshinori Mine

Chapter 17   Human Gut Microflora in Health and Disease:
Focus on Prebiotics   401
Tamara Casci, Robert A. Rastall, and Glenn R. Gibson

Chapter 18   Immunomodulating Effects of Lactic Acid Bacteria   435
Hanne Risager Christensen and Hanne Frøkiær

Chapter 19   Enzymatic Synthesis of Oligosaccharides: Progress
and Recent Trends   473
V. Maitin and Robert A. Rastall

Chapter 20   Metabolic Engineering of Bacteria for Food Ingredients   501
Ramon Gonzalez

Chapter 21   Technologies Used for Microbial Production of Food Ingredients   521
Anthony L. Pometto III and Ali Demirci

Chapter 22   Production of Carotenoids by Gene Combination in
Escherichia coli   533
Gerhard Sandmann

Chapter 23   Production of Amino Acids: Physiological and
Genetic Approaches   545
Reinhard Krämer

Chapter 24   Biotechnology of Microbial Polysaccharides in Food   583
Ian W. Sutherland

Chapter 25   Solid-State Bioprocessing for Functional Food Ingredients
and Food Waste Remediation   611
Kalidas Shetty

Index   625

1

Clonal Screening and Sprout Based Bioprocessing of Phenolic Phytochemicals for Functional Foods

Kalidas Shetty, Fergus M. Clydesdale, and Dhiraj A. Vattem

CONTENTS

1.1 Phenolic Phytochemical Ingredients and Benefits..1
1.2 Biological Function of Phenolic Phytochemicals...2
1.3 Phytochemicals and Functional Food 4
1.4 Relevance of Phenolic Antioxidants for Functional Food and Comparative
     Metabolic Biology Considerations..9
1.5 Clonal Screening of Phenolic Synthesis in Herb Shoot Culture Systems..............11
1.6 Phenolic Synthesis in Seed Sprouts 14
1.7 Summary of Strategies and Implications 15
References..16

1.1 PHENOLIC PHYTOCHEMICAL INGREDIENTS AND BENEFITS

Phenolic phytochemicals are secondary metabolites synthesized by plants to protect themselves against biological and environmental stresses such as pathogen attack or high energy radiation exposure (1,2). These compounds involved in the plant defense response are one of the most abundant classes of phytochemicals and are also invariably important components of our diets (3,4,5). Commonly consumed fruits such as apples, bananas, grapes, and several types of berries and their beverages are examples of plant foods as sufficiently rich sources of phenolic phytochemicals. Similar phytochemicals in our diet are also obtained from diverse commonly consumed vegetables such as tomato, cabbage, and onions to grains such as cereals and millets as well as legumes such as soybean, common beans, mung beans, fava beans, and peas, depending on the specific regions of the world (4,5,6). In addition many different types of herbs and spices containing phenolic

1

phytochemicals are widely consumed through the diet. Therefore, there are many different types of phenolic phytochemicals from the diet that are mediators of different biological functions for health and wellbeing. The profile of phenolic phytochemicals is often a characteristic of that plant species and is a result of the evolutionary pressures experienced by that species (5,6,7). The most abundant phenolic compounds in fruits are flavonols and flavonoids. Flavonoids, isoflavonoids, phenolic acids, and tannins are important phenolic phytochemicals in many legumes (5,6,7). Biphenyls such as rosmarinic acid are common in many herbs (1).

Chemically phenolic phytochemicals refers to a wide range of chemical compounds containing at least one aromatic ring and a hydroxyl substituent. Metabolic processing in plants after their synthesis results in chemical variations in basic phenolic structure (8). More than 8000 different phenolic structures, categorized into 10 subclasses, have been identified and are a result of differences in substituent groups and linkages. Structurally different phenolic phytochemicals having distinct properties range from simple molecules (e.g., phenolic acids with a single ring structure) to biphenyls and flavanoids having two or three phenolic rings (9,10). Polyphenols that contain >10 phenolic groups are another abundant group of phenolic phytochemicals in fruits and vegetables. Proanthocyanidins, tannins, and their derivatives phlorotannins are often referred to as polyphenols (9,10).

Phenolic phytochemicals are synthesized in plants via a common biosynthetic pathway and derive precursors from the shikimate–phenylpropanoid and the acetate–malonate (polyketide) pathways (1,11). Cinnamic acid, coumaric acid, and caffeic acid, and their derivatives are widespread in fruits, vegetables, and herbs and are derived primarily from the shikimate/phenylpropanoid pathway (1). Oxidative modifications of side chains produce benzoic acid derivatives, which include protocatechuic acid and its positional isomer gentisic acid. Fruits and legumes are especially rich sources of another group of phenolic phytochemicals called flavonoids and iso-lavonoids, which constitute the most abundant group of phenolic phytochemicals derived from the phenylpropanoid–acetate–malonate pathway (2,11).

Flavonoids are subdivided into several families such as flavonols, flavones, flavanols, isoflavones, and antocyanidins, which are formed as a result of hydroxylation, methylation, isoprenylation, dimerization, and glycosylation of the substituents in the aromatic rings (2,11). Phenolic phytochemicals are often esterified with sugars and other chemicals such as quinic acid to increase their solubility and to prevent their enzymatic and chemical degradation. Esterification also helps to target the phenolics to specific parts of the plant (11). Phenolic phytochemicals esterified via their hydroxyl groups to sugars are called glycosides. The sugar most commonly involved in esterification is glucose. However, the glycosides of phenolics with galactose, sucrose, and rhamnose are also found in some plant species (11).

1.2   BIOLOGICAL FUNCTION OF PHENOLIC PHYTOCHEMICALS

Reactive oxygen species (ROS) are now associated with manifestation of several oxidation-linked diseases such as cancer, cardiovascular diseases (CVD) and diabetes; epidemiological studies indicate that diets rich in fruits and vegetables are associated with lower incidences of such oxidation linked diseases. These disease protective effects of fruits and vegetables are now linked to the presence of antioxidant vitamins and phenolic phytochemicals having antioxidant activity, which support the body's antioxidant defense system (12–15). This has led to an interest in the use of diet as a potential tool for the control of these oxidative diseases (5,16–18). This is further supported by recent *in vitro* and clinical studies which have shown that lack of physical activity, exposure to environmental toxins, and consumption of diets rich in carbohydrate and fats induced oxidative stress, which was decreased by consuming fruits, vegetables, and their products (19–26).

Phenolic phytochemicals with antioxidant properties are now widely thought to be the principle components in fruits, vegetables, and herbs that have these beneficial effects. Most phenolic phytochemicals that have positive effect on health are believed to be functioning by countering the effects of reactive oxygen species (ROS) species generated during cellular metabolism. Consumption of natural dietary antioxidants from fruits, vegetables, and herbs has been shown to directly enhance scavenging of ROS, prevent the formation of ROS, and enhance the function of the antioxidant defense response mediated by Glutathione, Ascorbate, superoxide dismutase (SOD), catalase (CAT), and glutathione–s–transferase (GST) interface (15–20).

Specifically, oxidation of biological macromolecules as a result of free radical damage has now been strongly associated with development of many physiological conditions that can manifest into disease (27–31). The first and widely accepted mode of action of these phenolic phytochemicals in managing oxidation stress related diseases is due to the direct involvement of the phenolic phytochemicals in quenching the free radicals from biological systems. Phenolic phytochemicals, due to their phenolic ring and hydroxyl substituents, can function as effective antioxidants due to their ability to quench free electrons and chelate metal ions that are responsible for generating free radicals (32). Phenolic antioxidants can therefore scavenge the harmful free radicals and thus inhibit their oxidative reactions with vital biological molecules (32).

Emerging research into the biological functionality of phenolic phytochemicals also strongly suggests their ability to modulate cellular physiology both at the biochemical/ physiological and at molecular level. Structural similarities of phenolic phytochemicals with several key biological effectors and signal molecules have been suggested to be involved in induction and repression of gene expression or activation and deactivation of proteins, enzymes, and transcription factors of key metabolic pathways (27,33–35). They are believed to be able to critically modulate cellular homeostasis as a result of their physiochemical properties such as size, molecular weight, partial hydrophobicity, and ability to modulate acidity at biological pH through enzyme (dehydrogenases) coupled reactions. As a consequence of many modes of action of phenolic phytochemicals they have been shown to have several different functions. Potential anticarcinogenic and antimutagenic properties of phenolic phytochemicals such as gallic acid, caffeic acid, ferulic acid, catechin, quercetin, and resveratrol have been described in several studies (36–38). It is believed that phenolic phytochemicals might interfere in several of the steps that lead to the development of malignant tumors, including, inactivating carcinogens, inhibiting the expression of mutant genes (39). Many studies have also shown that these phenolic phytochemicals can repress the activity of enzymes such as cytochrome P 450 (CYP) class of enzymes involved in the activation of procarcinogens. Their protective functions in liver against carbon tetrachloride toxicity (40) have shown that phenolic phytochemicals also decrease the carcinogenic potential of a mutagen and can activate enzymatic systems (Phase II) involved in the detoxification of xenobiotics (41). Antioxidant properties of the phenolic phytochemicals linked to their ability to quench free radicals has been shown to prevent oxidative damage to the DNA which has been shown to be important in the age related development of some cancers (42). Phenolic phytochemicals have been shown to inhibit the formation of skin tumors induced by 7, 12-dimethyl-benz(a) anthracene in mice (43). Skin tumors in mice, and development of preneoplastic lesions in rat mammary gland tissue in cultures in the presence of carcinogens, were inhibited by resveratrol which is an important biphenyl found in wine (44,45).

Ability of phenolic phytochemicals is preventing of cardiovascular diseases (CVD) has been well described by epidemiological studies. The "French paradox" describes a famous study linking the lower incidences of CVD in the population consuming wine as part of their regular diet (46). Recent research has revealed that these beneficial effects of wine are due to the presence of a biologically active phenolic phytochemical "resveratrol." Inhibiting of LDL oxidation (47) and preventing platelet aggregation (48) are now believed to be the mechanisms by which resveratrol and other phenolic antioxidants prevent development of CVD. Phenolic phytochemicals have also been able to reduce blood pressure and have antithrombotic and antiinflammatory effects (48,49). Phenolic phytochemicals have also been shown to inhibit the activity of alpha-amylase and alpha-glucosidase which are responsible for postprandial increase in blood glucose level, which has been implicated in the manifestation of type-II diabetes and associated cardiovascular diseases (50,51).

In addition to managing oxidation linked diseases, immune modulatory activities of phenolic phytochemicals such as antiallergic (52) properties as a result of suppressing the hypersensitive immune response have also been defined. Antiinflammatory responses mediated by suppression of the TNF-alpha mediated proinflammatory pathways were also shown to be mediated by phenolic phytochemicals (53). Several studies have shown phenolic phytochemicals to have antibacterial, antiulcer, antiviral, and antifungal properties (54–57) and therefore are being implicated in management of infectious diseases.

1.3 PHYTOCHEMICALS AND FUNCTIONAL FOOD

The use of foods and food components such as phenolic phytochemicals as medicine has had an extraordinarily long history in the East and is still practiced successfully in many countries in Asia with Ayurvedic Medicine in India being one of many examples. However, the origin of the modern concept of "functional foods" as a separate and government regulated category of foods is quite new and was first developed in Japan in the 1980s. Faced with inflationary health care costs, the Japanese government instituted a regulatory system to approve certain foods with documented health benefits in order to improve health. These foods have a special seal and are known as "Foods for Specified Health Use" (FOSHU) (1). According to a recent report, as of September 2001, 271 food products had FOSHU status in Japan (58).

Interestingly, with this exception, there is no international, nor in most cases national, agreement on how functional foods should be defined nor is there a legal or regulatory definition. Functional foods seem to be regarded simply as foods that have physiological and psychological effects beyond traditional nutritional effects (59). In spite of the lack of legal or regulatory status consumers are extremely interested in the potential "medicinal" effects of food because they are becoming more and more convinced that their health and quality of life is, in part, a controllable gift, and they want to be an active participant in the process of this control. Added to this desire is a health care system, or lack thereof, which is increasingly perceived as distant, cold, and uncaring along with an explosion in emerging science which points to nontraditional health benefits of foods, food components, and plants (botanicals). This does not imply that consumers wish to give up traditional food, medicine, or health care but means that other options are being offered to the consumer, that form a continuum of treatment. These options are attractive to the consumer but they are also extremely attractive to governments who are being overwhelmed by the cost of health care and to growers and the food industry as a source of new value added products, which fulfill consumer demand and have the potential to create a healthier and higher quality of life.

The concept of functional foods is thus a key component in the new paradigm for health care. This paradigm is unique in that it appeals to not only consumers but governments and industries as well. The paradigm is illustrated graphically in Table 1.1, Table 1.2, and Table 1.3, which have been modified from Clydesdale (60). Each table shows some of the disease states that may be linked to diet and an approximation of the percentage of the population at risk. For these disease states three traditional options are shown "Established Public Health Programs," "Emerging/Controversial Public Health Recommendations," and "Clinical Research." In Table 1.1 the scope of therapy to achieve these treatments is shown as a continuum from individualized treatment of the disease for critical care patients to reduction of risk of the disease for those responding to public health programs. In Table 1.2 the level of health professional involvement in order to

Table 1.1
Disease States Linked to Diet and Current Scope of Therapy

Reported Problem	Nutrient Deficiency	Dental Caries	Obesity	Cardiovascular Disease	Cancer	Osteoporosis	Critical Care
% of Population At risk	100%	25%	33%	25%	20%	8%	<5%

ESTABLISHED PUBLIC HEALTH PROGRAM	EMERGING/CONTROVERSIAL PUBLIC HEALTH RECOMMENDATIONS		CLINICAL RESEARCH
Scope of Therapy	← REDUCTION OF RISK	TREATMENT OF DISEASE →	

Table 1.2
Disease States Linked to Diet with Health Professional Involvement

Reported Problem	Nutrient Deficiency	Dental Caries	Obesity	Cardiovascular Disease	Cancer	Osteoporosis	Critical Care
% of Population At risk	100%	25%	33%	25%	20%	8%	<5%

ESTABLISHED PUBLIC HEALTH PROGRAM	EMERGING/CONTROVERSIAL PUBLIC HEALTH RECOMMENDATIONS		CLINICAL RESEARCH
Scope of Therapy	REDUCTION OF RISK ←	TREATMENT OF DISEASE →	
Level of Health Professional Involvement	← LOW	HIGH →	

Table 1.3
Disease States Linked to Diet and Scope of Functional Foods

Reported Problem	Nutrient Deficiency	Dental Caries	Obesity	Cardiovascular Disease	Cancer	Osteoporosis	Critical Care
% of Population At risk	100%	25%	33%	25%	20%	8%	<5%

ESTABLISHED PUBLIC HEALTH PROGRAM	EMERGING/CONTROVERSIAL PUBLIC HEALTH RECOMMENDATIONS	CLINICAL RESEARCH

Scope of Therapy

⟵ REDUCTION OF RISK   TREATMENT OF DISEASE ⟶

Level of Health Professional Involvement

⟵ LOW   HIGH ⟶

Scope of Delivery Options

FUNCTIONAL FOODS AND BEVERAGES   HERBS   SUPPLEMENTS   MEDICAL FOODS   DRUGS

Individual Participation

⟵ HIGH   LOW ⟶

Cost

⟵ LOW   HIGH ⟶

achieve these therapies is also shown as a continuum from a high level for individual treatment to a low level for reduction of risk utilizing established public health programs. In the latter case there is an important educational component even though health professional involvement is low compared to the rest of the continuum. In Table 1.3 several other continua are overlapped on those in Table 1.2. With the advent of functional foods it can be seen that the scope of delivery options has been enlarged. The options continuum begins with drugs at the high cost, low consumer participation end, and then moves through medical foods, supplements, herbs, and finally functional foods and beverages at the low cost, high consumer participation preventative end of the spectrum. Functional foods and beverages are becoming, therefore, a part of established public health programs to reduce the risk of specific diseases.

The ultimate success of functional foods will depend on the ability to deliver bioactive healthful components from plants in a predictable and assured manner after they have been proven scientifically to be efficacious in reducing the risk of disease. In addition to the availability of bioactives such as phenolic phytochemicals, we will also need to ensure that the foods we create with these components will provide a stable environment for them so they can deliver their physiological benefits as well as being sensorially pleasing, convenient, safe, and affordable.

It has been proposed that the effectiveness of a public health measure can be defined in the following equation:

$$\text{Effectiveness} = \text{Efficacy} \times \text{Compliance} \tag{1.1}$$

Therefore it is essential that functional foods do not compromise flavor or convenience to insure continued use (compliance) by consumers.

Functional foods are intended to be eaten as part of a healthy diet and in so doing reduce the risk of disease in the future along with the symptoms of disease which destroy our quality of life.

The other part of the equation, efficacy, requires sound science in order to validate a food's effectiveness. The question always arises as to the type of science and how much is necessary to validate effectiveness and thus make a claim of some sort for the food. The Keystone National Policy Dialogue on Food, Nutrition, and Health (61) noted that three types of evidence were considered by the FDA in assessing the validity of health claims:

Epidemiology: clinical data derived from observational epoidemiologic studies assessing associations between food substances and disease

Biologic mechanisms: data derived from chemical, cellular, or animal models investigating plausible mechanisms of action of food substances

Intervention trials: controlled assessment of clinical food substance interventions in the human population

The FDA felt that these combinations of data met their "significant scientific agreement" standard of proof. It should be mentioned that the "significant scientific agreement" standard has been called into question and FDA is facing legal challenges in the landmark "Pearson vs. Shalala" dietary supplement claims case. A Federal appeals court directed FDA, based on first amendment rulings, to allow certain health claims accompanied by FDA approved label disclaimers, even though they had failed to meet FDA's "significant scientific agreement" standard (62). This issue has not yet been fully resolved.

A slightly different approach to validate efficacy has been suggested in a consensus document for Europe (63): "The design and development of functional foods is a key issue, as well as a scientific challenge, which should rely on basic scientific knowledge relevant to target functions and their possible modulation by food components. Functional foods themselves are not universal, and a food based approach would have to be influenced by local considerations. In contrast, a science based approach to functional food is universal and, because of this, is very suitable for a panEuropean approach. The function driven approach has the science base as its foundation — in order to gain a broader understanding of the interactions between diet and health. Emphasis is then put on the importance of the effects of food components on well identified and well characterized target functions in the body that are relevant to health issues, rather than solely on reduction of disease risk."

In this case the European position suggests validation of the efficacy of functional foods on functions in the body rather than disease states as is the case with the FDA. In both cases the need for government accepted biomarkers (biological markers of disease) is implicit and unfortunately there is no standard procedure to follow to have biomarkers approved. Hopefully this situation will be rectified in the not to distance future.

One might argue that the three types of evidence suggested by the FDA would be valid for proving the efficacy of a food to reduce the risk of disease or to affect a function in the body that is relevant to health. Thus, in the U.S. evidence from epidemiology, biological mechanisms and clinical trials could be used to validate both health claims and structure-function claims with the difference being the end point in question. In health claims the end point would be disease or an accepted biomarker of disease and in structure function claims the end point would be the specific structure or function in question.

It is probably fair to say that the majority of countries who have policies on functional foods would be within the parameters of evidence set by the U.S., Europe, and Japan (59). Functional foods, however, may be viewed more broadly than the preceding discussion indicates. For instance the American Dietetic Association (ADA) takes a very broad view in a position statement issued in 1999 (58). ADA stated that functional foods are "any potentially healthful food or food ingredient that may provide a health benefit beyond the traditional nutrients it contains." In addition they added that functional foods may be whole, fortified, enriched, or enhanced foods, to have a beneficial effect on health. A functional food would have to be consumed as part of a varied diet on a regular basis, at effective levels, and it is likely that all foods are functional at some physiological level.

Indeed, Milner (64) has asked "What is a functional food? I have never identified a nonfunctional food. I do not think anyone can truly describe one. All foods should be functional under the right circumstances." He goes on to use diet colas as an example of being functional for one who wishes to avoid calories, to increase palatability or to use the caffeine in it as a stimulant. All this simply means is that both consumers and those working on functional foods should have a definition in mind because most governments do not currently have one. In fact as noted by Clydesdale (59) it might be more productive to focus on what might ethically and scientifically be said about foods and what health claims might be made, rather than on the concept of functional foods themselves. If the discussion becomes involved with terminology, one can easily become lost in a morass of terms, such as functional foods, nutraceuticals, phytochemicals, natural remedies, bio-chemo-preventatives, medical foods, dietary supplements, and foods for special medical purposes that may or may not be regulated, depending on the country.

Further, we should be clear that our focus is on foods and not on drugs and that we make a clear difference between the two. In the U.S., the terms prevention and cure are generally ascribed, by statute, to drugs. Indeed, one of the U.S. statutory drug definitions relies on intended use whereby a drug is any article intended for use in the diagnosis, cure, mitigation, treatment, or prevention of disease in man (65). Therefore, potential health claims for foods or beverages that ascribe prevention, mitigation, or cure relationship between the compound and the disease would therefore, be drug claims. Thus food and beverage claims should be limited to a level that ascribes risk reduction or improved biological function (60). Foods, in general, will have a future benefit rather than an immediate benefit like drugs. Because of this they rarely have side effects as their level of bioactivity is lower and therefore long term rather than high and short term like drugs. Foods therefore have a presumption of safety for the entire population, at every age, gender, and ethnic group, which drugs do not. The benefit/risk of drugs is used as part of their regulatory approval in the U.S. whereas the regulation of foods cannot consider benefit/risk assessments.

If the bioactive component of a food botanical was found to be drug-like in its function then a functional food utilizing the bioactive would become a drug if a claim for prevention or cure of a disease was made. This would mean that there would not be a presumption of safety and testing under the drug statutes would have to be conducted.

As noted previously the European Commission has taken a different approach (63) in which they select and discuss key target functions in detail (66). Based on this they concluded that key target functions should (63):

play a major role in maintaining an improved state of health and wellbeing or reduction of risk of disease
appropriate markers should be available and feasible
potential opportunities should exist for modulation by candidate food components

Such reasoning would lend scientific credibility to structure function claims and this along with the scientific rigor required for health claims in the U.S. would provide for a meaningful measure of efficacy for use in the equation:

$$\text{Effectiveness} = \text{Efficacy} \times \text{Compliance} \tag{1.2}$$

Food is a wonderful vehicle for bioactive ingredients such as phenolic phytochemicals because the consumer finds little difficulty with compliance as long as the food is appealing; tastes, smells, and looks good; has good storage stability; is convenient; and is reasonably priced. However, one should not gloss over the importance of food as a vehicle because the technology to create an optimum vehicle for bioactive ingredients is as critical as the science and technology to create physiologically important bioactive compounds. Food, the vehicle which will carry these bioactive compounds must provide an environment which maintains their stability, bioactivity, and bioavailability during preparation and storage such that the bioactive ingredients are consumed at a known and efficacious level by the consumer.

It is important that consumers understand that claims for functional foods are based on sound science. This is at times made difficult by the different types of claims, differences in the laws which govern foods and supplements, and the wide diversity of terminology mentioned previously. However, as society becomes more accustomed to foods with specific biological properties these issues should and must be resolved. The science to provide proof of quality, safety, and efficacy is critical to the future success of functional foods. As Gruenwald (67) notes: the key market for herbal medicines is Germany, followed by France, because these countries have the highest sales but also the most rigid requirements regarding quality, safety, and efficacy.

Understanding the physiological benefits of foods and their ingredients has the potential to provide significant advances in public health, not only from the benefits provided by the bioactive ingredients but also from the impetus provided to consumers to take some responsibility for their own health. Certainly sales of supplements botanicals and functional foods have been growing with functional foods showing the largest rate of growth in the U.S. since 2000 (68). This, however, may be somewhat deceptive as the greatest growth has been in sports bars, beverages, and meal supplements, many of which are legally classified as supplements in the U.S. rather than functional foods. Nevertheless, the future has great potential and new foods will be developed as science allows us to uncover the bioactivity of the foods and plants that we eat.

1.4 RELEVANCE OF PHENOLIC ANTIOXIDANTS FOR FUNCTIONAL FOOD AND COMPARATIVE METABOLIC BIOLOGY CONSIDERATIONS

It is clear that food plants are excellent sources of phenolic phytochemicals, especially as bioactives with antioxidant property. As is evident, phenolic antioxidants from dietary sources have a history of use in food preservation, however, many increasingly have therapeutic and disease prevention applications (69–72). Therefore, understanding the nutritional and the disease protective role of dietary phytochemicals and particularly phenolic antioxidants is an important scientific agenda well into the foreseeable future (73). This disease protective role pf phytochemicals is becoming more significant at a time when the importance of in the prevention of oxidation linked chronic diseases is gaining rapid recognition globally. Therefore, disease prevention and management through the diet can be considered an effective tool to improve health and reduce the increasing health care costs for these oxidation linked chronic diseases, especially in low income countries.

As discussed earlier, phenolic phytochemicals have been associated with antioxidative action in biological systems, acting as scavengers of singlet oxygen and free radicals (74–77). Recent studies have indicated a role for phenolics from food plants in human health and, in particular, cancer (76,78). Phenolic phytochemicals (i.e., phenylpropanoids) serve as effective antioxidants due to their ability to donate hydrogen from hydroxyl groups positioned along the aromatic ring to terminate free radical oxidation of lipids and other biomolecules (79). Phenolic antioxidants, therefore, short circuit destructive chain reactions that ultimately degrade cellular membranes. Examples of food based plant phenolics that are used as antioxidant and antiinflammatory compounds are curcumin from *Curcuma longa* (80–82), *Curcuma mangga* (83), and *Zingiber cassumunar* (84), and rosmarinic acid from *Rosmarinus officinalis* (72,85). Examples of phenolics with cancer chemopreventive potential are curcumin from *Curcuma longa* (80,86–89), isoflavonoids from *Glycine max* (90–92), and galanigin from *Origanum vulgare* (93). Other examples of plant phenolics with medicinal uses include lithospermic acid from *Lithospermum* sp. as antigonadotropic agent (94), salvianolic acid from *Salvia miltiorrhiza* as an antiulcer agent (95), proanthocyanidins from cranberry to combat urinary tract infections (96,97), thymol from *Thymus vulgaris* for anti-caries (98), and anethole from *Pimpinella anisum* as an antifungal agent (99).

We have targeted enhanced production of oxidation disease relevant plant biphenyl metabolites such as rosmarinic acid, resveratrol, ellagic acid, and curcumin using novel tissue culture and bioprocessing approaches. Other phenolic phytochemicals also targeted are flavonoids, quercetin, myrcetin, scopoletin, and isoflavonoids. Among simple phenolics, there is major interest in the overexpression of L-tyrosine and L-DOPA from legumes in a high phenolic antioxidant background (100,101). Rosmarinic acid has been targeted from clonal herbs (1,69) for its antiinflammatory and antioxidant properties (85,102,103). Resveratrol has shown antioxidant and cancer chemopreventive properties (104,105) and its overproduction has been targeted from several fruits using solid-state bioprocessing (106,107). Ellagic acid has been targeted for antioxidant and cancer chemopreventive properties (108,109) and has been similarly targeted via solid-state bioprocessing from fruits and fruit processing byproducts (106). Extensive studies have shown cancer chemopreventive and antioxidant properties for *Curcuma longa* and its major active compound, curcumin, (81,86,110) and the developmental and elicitor stress mediated overexpression of curcumin is being investigated.

The emergence of dietary and medicinal applications for phenolic phytochemicals, harnessing their antioxidant and antimicrobial properties in human health and wellness has sound rationale. As stress damage on the cellular level appears similar among eukaryotes, it is logical to suspect that there may be similarities in the mechanism for cellular stress mediation between eukaryotic species. Plant adaptation to biotic and abiotic stress involves the stimulation of protective secondary metabolite pathways (111–113) that results in the biosynthesis of phenolic antioxidants. Studies indicate that plants exposed to ozone responded with increased transcript levels of enzymes in the phenylpropanoid and lignin pathways (114). Increase in plant heat tolerance is related to the accumulation of phenolic metabolites and heat shock proteins that act as chaperones during hyperthermia (115).

Phenolics and specific phenolic-like salicylic acid levels increase in response to infection, acting as defense compounds, or to serve as precursors for the synthesis of lignin, suberin, and other polyphenolic barriers (116). Antimicrobial phenolics called phytoalexins are synthesized around the site of infection during pathogen attack and, along with other simple phenolic metabolites, are believed to be part of a signaling process that results in systemic acquired resistance (111–113). Many phenylpropanoid compounds such as flavonoids, isoflavonoids, anthocyanins, and polyphenols are induced in response to wounding (117), nutritional stress (118), cold stress (119), and high visible light (120). UV irradiation induces light-absorbing flavonoids and sinapate esters in *Arabidopsis* to block radiation and protect DNA from dimerization or cleavage (121). In general, the initiation of the stress response arises from certain changes in the intracellular medium (122) that transmit the stress induced signal to cellular modulating systems and results in changes in cytosolic calcium levels, proton potential as a long distance signal (123), and low molecular weight proteins (124). Stress can also initiate free radical generating processes and shift the cellular equilibrium toward lipid peroxidation (125). It is believed that the shift in prooxidant–antioxidant equilibrium is a primary nonspecific event in the development of the general stress response (126). Therefore, protective phenolic antioxidants involved in such secondary metabolite linked stress responses in food plant species has potential as a source of therapeutic and disease-preventing functional ingredients for oxidation disease linked diet (high carbohydrate and high fat diets) and environment (physical, chemical, and biological) influenced chronic disease problems (69).

1.5 CLONAL SCREENING OF PHENOLIC SYNTHESIS IN HERB SHOOT CULTURE SYSTEMS

The hypothesis that the biosynthesis of plant phenolic metabolites is linked to the proline linked pentose–phosphate pathway (1,69; Figure 1.1) was developed based on the role of the proline linked pentose–phosphate pathway in regulation of purine metabolism in mammalian systems (127). Proline is synthesized by a series of reduction reactions from glutamate (69). In this sequence, P5C and proline known to be metabolic regulators function as a redox couple (127,128). During respiration, oxidation reactions produce hydride ions, which augment reduction of P5C to proline in the cytosol. Proline can then enter mitochondria through proline dehydrogenase (129) and support oxidative phosphorylation (alternative to NADH from Krebs/TCA cycle). This is important because shunting the TCA cycle toward proline synthesis likely deregulates normal NADH synthesis. The reduction of P5C in the cytosol provides $NADP^+$, which is the cofactor for glucose-6-phosphate dehydrogenase (G6PDH), an enzyme that catalyzes the rate-limiting step of the pentose–phosphate pathway. Proline synthesis is therefore hypothesized and has been partly shown to both regulate and stimulate pentose–phosphate pathway activity in erythrocytes (130) and cultured fibroblasts (131) when P5C is converted to proline. This was shown to stimulate purine metabolism via ribose-5-phosphate, which affects cellular physiology and therefore function (127,132).

Based on several studies, Shetty (1,69) proposed a model that the proline linked pentose–phosphate pathway (PLPPP) could be the critical control point (CCP) of shikimate and phenylpropanoid pathways and hypothesized that stress linked modulation of PLPPP could lead to the stimulation of phenolic phytochemicals (1,69; Figure 1.1). Using this model, proline, proline precursors, and proline analogs were effectively utilized to stimulate total phenolic content and a specific phenolic metabolite, rosmarinic acid (133,134). Further, it was shown that proline, proline precursors, and proline analogs

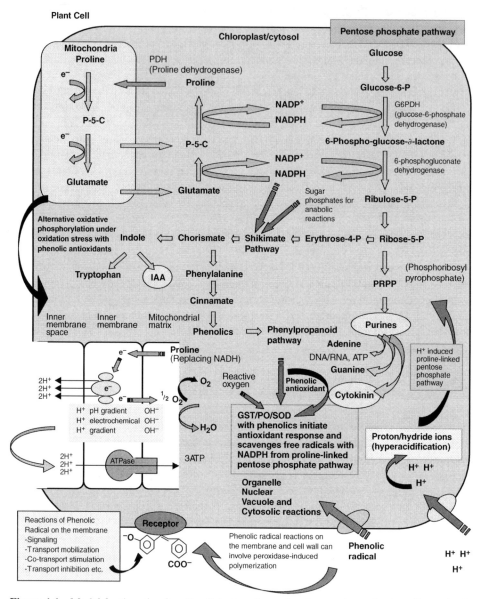

Figure 1.1   Model for the role of proline linked pentose–phosphate pathway in phenolic synthesis and its utilization for screening high phenolic clonal lines of single seed genetic origin.

stimulated somatic embryogenesis in anise, which correlated with increased phenolic content (135). It was also established that during *Pseudomonas* mediated stimulation of total phenolic and rosmarinic acid (RA), proline content was stimulated in oregano clonal shoot cultures (136). Therefore, it was proposed that NADPH demand for proline synthesis during response to microbial interaction and proline analog treatment (1,69), may reduce the cytosolic NADPH/NADP$^+$ ratio, which should activate G6PDH (137,138). Therefore, deregulation of the pentose–phosphate pathway by proline analog and microbial induced proline synthesis may provide the excess erythrose-4-phosphate (E4P) for shikimate and, therefore, the phenylpropanoid pathways (69). At the same time, proline and P5C could serve as superior reducing equivalents (RE), alternative to NADH (from Krebs/TCA cycle) to support increased oxidative phosphorylation (ATP synthesis) in the mitochondria during the stress response (69,127,128).

The proline analog, azetidine-2-carboxylate (A2C), is an inhibitor of proline dehydrogenase (139). It is also known to inhibit differentiation in Leydig cells of rat fetal testis, which can be overcome by exogenous proline addition (140). Another analog, hydroxyproline, is a competitive inhibitor of proline for incorporation into proteins. According to the model of Shetty (1,69), either analog, at low levels, should deregulate proline synthesis from feedback inhibition and stimulate proline synthesis (1,69). This would then allow the proline linked pentose–phosphate pathway to be activated for NADPH synthesis, and concomitantly drive metabolic flux toward E4P for biosynthesis of shikimate and phenylpropanoid metabolites, including RA. Proline could also serve as a RE for ATP synthesis via mitochondrial membrane associated proline dehydrogenase (129,69).

Using this rationale for the mode of action of proline analogs and links to the pentose–phosphate pathway, high RA-producing, shoot based clonal lines originating from a single heterozygous seed among a heterogeneous bulk seed population of lavender, spearmint, and thyme have been screened and isolated based on tolerance to the proline analog, A2C, and a novel *Pseudomonas* sp. isolated from oregano (Figure 1.2; 141–143). This strategy for screening and selection of high RA clonal lines is also based on the model that proline linked pentose phosphate pathway is critical for driving metabolic flux (i.e., E4P) toward shikimate and phenylpropanoid pathways (Figure 1.1). Any clonal line with a deregulated proline synthesis pathway should have an overexpressed pentose phosphate pathway which allows excess metabolic flux to drive shikimate and phenylpropanoid pathway toward total phenolic and RA synthesis (69). Similarly, such proline overexpressing clonal lines should be more tolerant to proline analog, A-2-C. If the metabolic flux to RA is overexpressed, it is likely to be stimulated in response to *Pseudomonas* sp (69). Therefore,

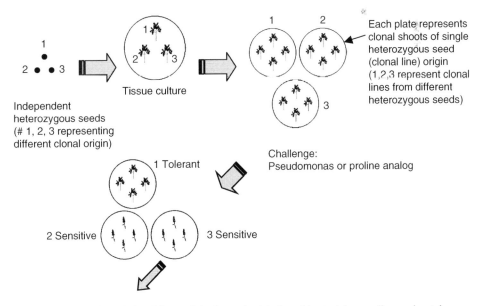

Figure 1.2 Isolation of clonal lines of single seed origin based bacterial or proline analog tolerance. Clonal lines can be used for investigations on Rosmarinic acid inducibility and endogenous antioxidant enzyme response in various clonal lines under various nutritional conditions. Further, high phenolic clonal lines can be used for development of ingredient profiles of single seed clonal origin.

such a clonal line is equally likely to be tolerant to *Pseudomonas* sp. Further, such a clonal line should also exhibit high proline oxidation and RA content in response to A2C and *Pseudomonas* sp. The rationale for this model is based on the role of the pentose–phosphate pathway in driving ribose-5-phosphate toward purine metabolism in cancer cells (127), differentiating animal tissues (140), and plant tissues (144). The hypothesis of this model is that the same metabolic flux from over expression of proline linked pentose–phosphate pathway regulates the interconversion of ribose-5-phosphate to E4P driving shikimate pathway (69). Shikimate pathway flux is critical for both auxin and phenylpropanoid biosynthesis, including RA (69). This hypothesis has been strengthened by preliminary results in which RA biosynthesis in several oregano clonal lines was significantly stimulated by exogenous addition of proline analog (e.g., A2C) and ornithine (133,134). The same clonal lines are also tolerant to *Pseudomonas* sp. and respond to the bacterium by increasing RA and proline biosynthesis (136,145). High RA-producing clonal lines selected by approaches based on this model (141–143,145–147) are being targeted for preliminary characterization of the key enzymes in phenolic synthesis and large scale production of such phenolics using tissue culture generated clonal lines in large scale field production systems. This strategy for investigation and stimulation of phenolic biosynthesis in clonal plant systems using PLPPP as the CCP can be the foundation for designing dietary phenolic phytochemicals from cross-pollinating, heterogeneous species for functional foods (1,69).

1.6 PHENOLIC SYNTHESIS IN SEED SPROUTS

Preliminary results (1,69,72,133,134,148) have provided empirical evidence for a link between proline biosynthesis and oxidation, as well as stimulation of G6PDH. In light mediated sprout studies in pea (*Pisum sativum*), acetylsalicylic acid in combination with fish protein hydrolysates (a potential source of proline precursors) stimulated phenolic content and guaiacol peroxidase (GPX) activity during early germination with corresponding higher levels of proline and G6PDH activity (149). In parallel light mediated studies in pea, low pH and salicylic acid treatments stimulated increased phenolic content and tissue rigidity. Similarly, there was concomitant stimulation of G6PDH and proline (148). This work supported the hypothesis that pentose–phosphate pathway stimulation may be linked to proline biosynthesis and that modulation of a proton linked redox cycle may also be operating through proline linked pentose–phosphate pathway (148). In dark germinated studies in pea, high cytokinin-containing anise root extracts stimulated phenolic content and antioxidant activity, which correlated with increased proline content but inversely with G6PDH activity (150). In further dark germination studies in mung bean (*Vigna radiata*), dietary grade microbial polysaccharide treatments stimulated phenolic content and enzyme activity, G6PDH and GPX compared to controls (151), with concomitant stimulation of proline content. In addition, specific elicitors xanthan gum, yeast extract, and yeast glucan stimulated antioxidant activity. In additional studies, oregano phenolic extracts were used as elicitors to stimulate phenolic content during dark germination of mung bean. Again, increased phenolic content corresponded to an increase in activity of G6PDH and GPX and phenolic related antioxidant activity were also stimulated (152). In studies with dark germinated fava bean, support for the hypothesis that stimulation of proline linked pentose–phosphate pathway would stimulate phenolic metabolism under elicitor and stress response was probed (69). In polysaccharide elicitor studies, gellan gum stimulated fava bean total phenolic content ninefold in late stages of germination with a corresponding increase in proline content and GPX activity, although the effect on antioxidant and G6PDH activity was inconclusive (100). In the same fava bean system, UV mediated stimulation of phenolic content in dark germinated fava bean

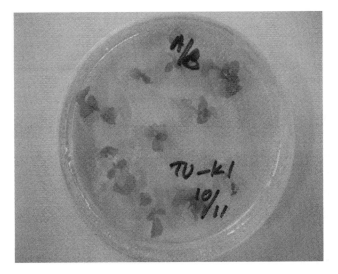

Figure 1.3 High phenolic clonal shoot culture of oregano (*Origanum vulgare*) with genetic origin from single heterozygous seed.

sprouts indicated a positive correlation to G6PDH and GPX activities with a concomitant increase in proline content (153). It was further confirmed that proline analog, A2C also stimulated phenolic content in fava bean with positive correlation to G6PDH and GPX activities as well as proline content (154). Similar to studies in clonal shoot cultures of thyme (133) and oregano (134), the proline analog mediated studies in fava bean confirmed that proline overexpression was not only possible, but involved stimulating G6PDH and therefore likely diverted the pentose–phosphate pathway toward phenylpropanoid biosynthesis. The late stage stimulation of phenolic content and GPX activity in response to microwave mediated thermal stress in dark germinated fava bean strongly correlated with stimulation of free radical scavenging activity of free phenolics as measured by the quenching of 1,1-diphenyl-2-picrylhydrazyl (DPPH) free radical assay and stimulation of superoxide dismutase (SOD) activity (155). Therefore, using elicitor and physical stress in early stages of seed sprouting could be effectively used to stimulate key phenolic phytochemicals for design of functional foods. In designing appropriate phenolic phytochemical extracts diverting carbohydrate and protein metabolic flux through the proposed CCP, proline linked pentose–phosphate pathway (PLPPP) has substantial potential. This CCP link to PLPPP could be the basis for a more effective Systems Biology approach for metabolic control of functional phenolic phytochemicals and appropriate design of phytochemicals from different seed sprout systems for different oxidation linked disease targets.

1.7 SUMMARY OF STRATEGIES AND IMPLICATIONS

In an age where diet and environmental related oxidation and infectious diseases are on the rise, the role of dietary phenolic phytochemicals in preventive management of diseases is becoming important. Strategies to improve functional phytochemicals that addresses the problem of consistency have to be developed. Within this framework we have developed strategies for isolating clonal dietary herbs of single seed genetic origin with a specific profile of phenolic phytochemicals and also stress elicitation based enhancement of phenolic phytochemicals during legume seed sprouting. The rationale for development of these techniques for phenolic phytochemical enhancement for functional food design is based on harnessing the potential of proline linked pentose–phosphate pathway (PLPPP) as the critical control point (CCP) in clonal shoots of single seed genetic origin such as herbs from the family Lamiaceae and seed sprouts in self-pollinating species such as various legumes. This strategy can be extended to develop foods with better phenolic phytochemical profile and functionality. Further it can be extended to develop functional foods and supplements with consistent ingredient profiles targeted against a disease condition. This concept is now being extended to specifically isolate antioxidants for diverse disease conditions, antimicrobials against bacterial pathogens, phytochemicals for diabetes management, angiotensin converting enzyme inhibitors for hypertension management, L-DOPA for Parkinson's management, dietary cyclooxygenase (COX-2 inhibitors) for inflammatory diseases and isoflavones for women's health.

REFERENCES

1. Shetty, K. Biotechnology to harness the benefits of dietary phenolics: focus on *Lamiaceae*. *Asia Pac. J. Clin. Nutr.* 6:162–171, 1997.
2. Briskin, D.P. Medicinal plants and phytomedicines. Linking plant biochemistry and physiology to human health. *Plant Physiol.* 124:507–14, 2000.
3. Mann, J. *Oxford Chemistry Series: Secondary Metabolism.* Oxford: Clarendon Press, 1978.
4. Bravo, L. Phenolic phytochemicals: chemistry, dietary sources, metabolism, and nutritional significance. *Nutr. Rev.* 56:317–333, 1998.
5. Crozier, A., J. Burns, A.A. Aziz, A.J. Stewart, H.S. Rabiasz, G.I. Jenkins, C.A. Edwards, M.E.J. Lean. Antioxidant flavonols from fruits, vegetables and beverages: measurements and bioavailability. *Biol. Res.* 33:79–88, 2000.
6. Urquiaga, I., F. Leighton. Plant Polyphenol Antioxidants and Oxidative Stress. *Biol. Res.* 33:55–64, 2000.
7. Moure, A., J.M. Cruz, D. Franco, J.M. Domínguez, J. Sineiro, H. Domínguez, M.H. Núñez, J.C. Parajó. Natural antioxidants from residual sources. *Food Chem.* 72(2):145–171, 2001.
8. Bors, W., C. Michel. Chemistry of the antioxidant effect of polyphenols. *Ann. NY Acad. Sci.* 957:57–69, 2002.
9. Harborne, J.B. Plant phenolics. In: *Encyclopedia of Plant Physiology*, Bella, E.A., B.V. Charlwood, eds, Heidelburg: Springer-Verlag, 8:329–395, 1980.
10. Haslam, E. *Practical Polyphenolics: From Structure to Molecular Recognition and Physiological Action.* Cambridge: Cambridge University Press, 1998.
11. Strack, D. Phenolic metabolism. In: *Plant Biochemistry,* Dey, P.M., J.B. Harborne, eds., San Diego: Academic Press, 1997, pp 387–416.
12. Jakus, V. The role of free radicals, oxidative stress and antioxidant systems in diabetic vascular disease. *Bratisl. Lek. Listy.* 101(10):541–551, 2000.
13. Barbaste, M., B. Berke, M. Dumas, S. Soulet, J.C. Delaunay, C. Castagnino, V. Arnaudinaud, C. Cheze, J. Vercauteren. Dietary antioxidants, peroxidation and cardiovascular risks. *J. Nutr. Health Aging.* 6(3):209–23, 2002.
14. Gerber, M., C. Astre, C. Segala, M. Saintot, J. Scali, J. Simony-Lafontaine, J. Grenier, H. Pujol. Tumor progression and oxidant-antioxidant status. *Cancer Lett.* 19:114(1–2): 211–214, 1997.
15. Block, G., B. Patterson, A. Subar. Fruit, vegetables, and cancer prevention: a review of the epidemiological evidence. *Nutr. Cancer* 18:1–29, 1992.
16. Serdula, M.K., M.A.H. Byers, E. Simoes, J.M. Mendlein, R.J. Coates. The association between fruit and vegetable intake and chronic disease risk factors. *Epidemiology* 7:161–165, 1996.

17. Tapiero, H., K.D. Tew, G.N. Ba, G. Mathe. Polyphenols: do they play a role in the prevention of human pathologies? *Biomed. Pharmacother.* 56(4):200–207, 2002.

18. Duthie, G.G., P.T. Gardner, J.A. Kyle. Plant polyphenols: are they the new magic bullet? *Proc. Nutr. Soc.* 62(3):599–603, 2003.

19. Lundberg, A.C., A. Åkesson, B. Åkesson. Dietary intake and nutritional status in patients with systemic sclerosis. *Ann. Rheum. Dis.* 51:1143–1148, 1992.

20. Yoshioka, A., Y. Miyachi, S. Imamura, Y. Niwa. Anti-oxidant effects of retinoids on inflammatory skin diseases. *Arch. Dermatol. Res.* 278:177–183, 1986.

21. Wilks, R., F. Bennett, T. Forrester, N. Mcfarlane-Anderson. Chronic diseases: the new epidemic. *West Ind. Med. J.* 47(4):40–44, 1998.

22. Leighton, F., A. Cuevas, V. Guasch, D.D. Perez, P. Strobel, A. San Martin, U. Urzua, M.S. Diez, R. Foncea, O. Castillo, C. Mizon, M.A. Espinoza, I. Urquiaga, J. Rozowski, A. Maiz, A. Germain. Plasma polyphenols and antioxidants, oxidative DNA damage and endothelial function in a diet and wine intervention study in humans. *Drugs Exp. Clin. Res.* 25(2–3):133–41, 1999.

23. Gillman, M.W., L.A. Cupples, D. Gagnon, B.M. Posner, R.C. Ellison, W.P. Castelli, P.A. Wolf. Protective effect of fruits and vegetables on development of stroke in men. *J. Am. Med. Assoc.* 273:1113–1117, 1995.

24. Joshipura, K.J., A. Ascherio, J.E. Manson, M.J. Stampfer, E.B. Rimm, F.E. Speizer, C.H. Hennekens, D. Spiegelman, W.C. Willett. Fruit and vegetable intake in relation to risk of ischemic stroke. *J. Amer. Med. Asso.* 282:1233–1239, 1999.

25. Cox, B.D., M.J. Whichelow, A.T. Prevost. Seasonal consumption of salad vegetables and fresh fruit in relation to the development of cardiovascular disease and cancer. *Public Health Nutr.* 3:19–29, 2000.

26. Strandhagen, E., P.O. Hansson, I. Bosaeus, B. Isaksson, H. Eriksson. High fruit intake may reduce mortality among middle-aged and elderly men: the study of men born in 1913. *Eur. J. Clin. Nutr.* 54:337–341, 2000.

27. Droge, W. Free radicals in the physiological control of cell function. *Physiol. Rev.* 82(1): 47–95, 2002.

28. Morel, Y., R., Barouki Repression of gene expression by oxidative stress. *Biochem J.* 342(3):481–96, 1999.

29. Parke, A.L., C. Ioannides, D.F.V. Lewis, D.V. Parke. Molecular pathology of drugs: disease interaction in chronic autoimmune inflammatory diseases. *Inflammopharmacology* 1:3–36, 1991.

30. Schwarz, K.B. Oxidative stress during viral infection: a review. *Free Radic. Biol. Med.* 21(5):641–9, 1996.

31. Offen, D., P.M. Beart, N.S. Cheung, C.J. Pascoe, A. Hochman, S. Gorodin, E. Melamed, R. Bernard, O. Bernard. Transgenic mice expressing human Bcl-2 in their neurons are resistant to 6-hydroxydopamine and 1-methyl-4-phenyl-1,2,3,6-tetrahydropyridine neurotoxicity. *Proc. Natl. Acad. Sci.* 95:5789–5794, 1998.

32. Rice-Evans, C.A., N.J. Miller, G. Paganga. Antioxidant properties of phenolic compounds. *Trends Plant Sci.* 2:152–159, 1997.

33. Lusini, L., S.A. Tripodi, R. Rossi, F. Giannerini, D. Giustarini, M.T. del Vecchio, G. Barbanti, M. Cintorino, P. Tosi, P. Di Simplicio. Altered glutathione anti-oxidant metabolism during tumor progression in human renal-cell carcinoma. *Int. J. Cancer* 91(1):55–59, 2001.

34. Mates, J.M., F. Sanchez-Jimenez. Antioxidant enzymes and their implications in pathophysiologic processes. *Front. Biosci.* 4D:339–345, 1999.

35. Mates, J.M., C. Perez-Gomez, I. Nunez de Castro. Antioxidant enzymes and human diseases. *Clin. Biochem.* 32(8):595–603, 1999.

36. Yamada, J., Y. Tomita. Antimutagenic activity of caffeic acid and related compounds. *Biosci. Biotechnol. Biochem.* 60(2):328–329, 1996.

37. Mitscher, L.A., H. Telikepalli, E. McGhee, D.M. Shankel. Natural antimutagenic agents. *Mutat. Res.* 350:143–152, 1996.

38. Uenobe, F., S. Nakamura, M. Miyazawa. Antimutagenic effect of resveratrol against Trp-P-1. *Mutat. Res.* 373:197–200, 1997.

39. Kuroda, Y., T. Inoue. Antimutagenesis by factors affecting DNA repair in bacteria. *Mutat. Res.* 202(2):387–391, 1988.

40. Kanai, S., H. Okano. Mechanism of protective effects of sumac gall extract and gallic acid on progression of CC14-induced acute liver injury in rats. *Am. J. Chin. Med.* 26:333–341, 1998.

41. Bravo, L. Phenolic phytochemicals: chemistry, dietary sources, metabolism, and nutritional significance. *Nutr. Rev.* 56:317–333, 1998.

42. Halliwell, B. Establishing the significance and optimal intake of dietary antioxidants: the biomarker concept. *Nutr. Rev.* 57:104–113, 1999.

43. Kaul, A., K.I. Khanduja. Polyphenols inhibit promotional phase of tumorigenesis: relevance of superoxide of superoxide radicals. *Nurt. Cancer* 32:81–85, 1998.

44. Clifford, A.J., S.E. Ebeler, J.D. Ebeler, N.D. Bills, S.H. Hinrichs, P.-L. Teissedre, A.L. Waterhouse. Delayed tumor onset in transgenic mice fed an amino acid-based diet supplemented with red wine solids. *Am. J. Clin. Nutr.* 64:748–756, 1996.

45. Jang, M., L. Cai, G.O. Udeani, K.V. Slowing, C. Thomas, C.W.W. Beecher, H.H.S. Fong, N.R. Farnsworth, A.D. Kinghorn, R.G. Mehta, R.C. Moon, J.M. Pezzuto. Cancer chemopreventive activity of resveratrol, a natural product derived from grapes. *Science* 275:218–220, 1997.

46. Ferrieres, J. The French paradox: lessons for other countries. *Heart* 90(1):107–111, 2004.

47. Frankel, E.N, J. Kanner, J.B. German, E. Parks, J.E. Kinsella. Inhibition of oxidation of human low-density lipoprotein by phenolic substances in red wine. *Lancet* 341:454–457, 1993.

48. Gerritsen, M.E., W.W. Carley, G.E. Ranges, C.-P. Shen, S.A. Phan, G.F. Ligon, C.A. Perry. Flavonoids inhibit cytokine-induced endothelial cell adhesion protein gene expression. *Am. J. Pathol.* 147:278–292, 1995.

49. Muldoon, M.F., S.B. Kritchevsky. Flavonoids and heart disease. *BMJ* 312(7029):458–459, 1996.

50. McCue, P., K. Shetty. Inhibitory effects of rosmarinic acid extracts on porcine pancreatic amylase *in vitro. Asia Pac. J. Clin. Nutr.* 13(1):101–106, 2004.

51. Andlauer, W., P. Furst. Special characteristics of non-nutrient food constituents of plants: phytochemicals: introductory lecture. *Int. J. Vitam. Nutr. Res.* 73(2):55–62, 2003.

52. Noguchi, Y., K. Fukuda, A. Matsushima, D. Haishi, M. Hiroto, Y. Kodera, H. Nishimura, Y. Inada. Inhibition of Df-protease associated with allergic diseases by polyphenol. *J. Agric. Food Chem.* 47:2969–2972, 1999.

53. Ma, Q., K. Kinneer. Chemoprotection by phenolic antioxidants. Inhibition of tumor necrosis factor alpha induction in macrophages. *J. Biol. Chem.* 277(4):2477–2484, 2002.

54. Chung, K.T., T.Y. Wong, C.I. Wei, Y.W. Huang, Y. Lin. Tannins and human health: a review. *Crit. Rev. Food Sci. Nutr.* 38:421–464, 1998.

55. Ikken, Y., P. Morales, A. Martínez, M.L. Marín, A.I. Haza, M.I. Cambero. Antimutagenic effect of fruit and vegetable ethanolic extracts against N-nitrosamines evaluated by the Ames test. *J. Agric. Food Chem.* 47:3257–3264, 1999.

56. Vilegas, W., M. Sanomimiya, L. Rastrelli, C. Pizza. Isolation and structure elucidation of two new flavonoid glycosides from the infusion of *Maytenus aquifolium* leaves. Evaluation of the antiulcer activity of the infusion. *J. Agric. Food Chem.* 47:403–406, 1999.

57. Abram, V., M. Donko. Tentative identification of polyphenols in *Sempervivum tectorum* and assessment of the antimicrobial activity of *Sempervivum* L. *J. Agric. Food Chem.* 47:485–489, 1999.

58. Meister, K. *Facts About Functional Foods*. New York: A report by the American Council on Science and Health, 2002, pp 5–29.

59. Clydesdale, F. M. A proposal for the establishment of scientific criteria for health claims for functional foods. *Nutr. Rev.* 55:413–422, 1997.

60. Clydesdale, F. M. Science, education technology: new frontiers for health. *Crit. Rev. Fd. Sci. Nutr.* 38:397–419, 1998.

61. Keystone Center. *The Final Report of the Keystone National Policy Dialogue on Food, Nutrition and Health,* Keystone, CO and Washington, DC, 1996, pp 33–56.

62. FDA seeks public comment after first amendment rulings. *Food Chemical News,* Washington, DC. 2002:44, pp. 1, 26.

63. Diplock, A.T., P.J. Aggett, M. Ashwell, F. Bornet, E.B. Fern, M.B. Roberfroid. Scientific concepts of functional foods in Europe: consensus document. *Brit. J. Nutr.* 81(1):S1–S27, 1999.

64. Milner, J.A. Expanded definition of nutrients. *Proceedings of the Ceres Forum on What is a Nutrient?: Defining the Food – Drug Continuum.* Georgetown University Center for Food and Nutrition Policy, Washington, DC. 1999, pp 50–55.

65. 21 U.S.C. 1994, 321 (g).

66. Bellisle, F., A.T. Diplock, G. Hornstra, B. Koletzko, M. Roberfroid, S. Salminen, W.H.M. Saris. Functional food science in Europe. *Brit. J. Nutr.* 80(1):S1–S193, 1998.

67. Gruenwald, J. The future of herbal medicines in Europe. *Nutraceuticals World* 5(6):26, 2002.

68. Gruenwald, J., F. Herzberg. The situation of major dietary supplement markets. *Nutracos.* 1(3):18021, 2002.

69. Shetty, K., M.L. Wahlqvist. A model for the role of proline-linked pentose phosphate pathway in phenolic phytochemical biosynthesis and mechanism of action for human health and environmental applications; a Review. *Asia Pac. J. Clin. Nutr.* 13:1–24, 2004.

70. Shetty, K., Labbe, R.L. Food-borne pathogens, health and role dietary phytochemicals. *Asia Pac. J. Clin. Nutr.* 7:270–276, 1998.

71. Shetty, K. Phytochemicals: biotechnology of phenolic phytochemicals for food preservatives and functional food applications. In: *Wiley Encyclopedia of Food Science and Technology,* 2nd ed., Francis, F.J., ed., New York: Wiley Interscience, 1999, pp 1901–1909.

72. Shetty, K. Biosynthesis of rosmarinic acid and applications in medicine. *J. Herbs Spices Med. Plants* 8:161–181, 2001.

73. Shetty, K., P. McCue. Phenolic antioxidant biosynthesis in plants for functional food application: integration of systems biology and biotechnological approaches. *Food Biotechnol.* 17:67–97, 2003.

74. Rice-Evans, C.A., N.J. Miller, P.G. Bolwell, P.M. Bramley, J.B. Pridham. The relative antioxidant activities of plant-derived polyphenolic flavonoids. *Free Rad. Res.* 22:375–383, 1995.

75. Hertog, M.G.L., P.C.H. Hollman, M.B. Katan. Content of potentially anticarcinogenic flavonoids of 28 vegetables and 9 fruits commonly consumed in the Netherlands. *J. Agric. Food Chem.* 40:2379–2383, 1992.

76. Hertog, M.G.L., D. Kromhout, C. Aravanis, H. Blackburn, R. Buzina, F. Fidanza, S. Giampaoli, A. Jansen, A. Menotti, S. Nedeljkovic, M. Pekkarinen, B.S. Simic, H. Toshima, E.J.M. Feskens, P.C.H. Hollman, M.B. Kattan. Flavonoid intake and long-term risk of coronary heart disease and cancer in the seven countries study. *Arch. Intern. Med.* 155:381–386, 1995.

77. Jorgensen, L.V., H.L. Madsen, M.K. Thomsen, L.O. Dragsted, L.H. Skibsted. Regulation of phenolic antioxidants from phenoxyl radicals: an ESR and electrochemical study of antioxidant hierarchy. *Free Rad. Res.* 30:207–220, 1999.

78. Paganga, G., N. Miller, C.A. Rice-Evans. The polyphenolic contents of fruits and vegetables and their antioxidant activities: what does a serving constitute? *Free Rad. Res.* 30:153–162, 1999.

79. Foti, M., M. Piattelli, V. Amico, G. Ruberto. Antioxidant activity of phenolic meroditerpenoids from marine algae. *J. Photochem. Photobiolo.* 26:159–164, 1994.

80. Huang, M.T., T. Lysz, T. Ferraro, A.H. Conney. Inhibitory effects of curcumin on tumor promotion and arachidonic acid metabolism in mouse epidermis. In: *Cancer chemoprevention*, Wattenberg, L., M. Lipkin, C.W. Boone, G.J. Kellof, eds., Boca Raton, FL: CRC Press, 1992, pp 375–391.

81. Osawa, T., Y. Sugiyama, M. Inayoshi, S. Kawakishi. Antioxidant activity of tetrahydrocurcuminoids. *Biosci. Biotechnol. Biochem.* 59:1609–1612, 1995.

82. Lim, G.P., T. Chu, F. Yang, W. Beech, S.A. Frauschy, G.M. Cole. The curry spice curcumin reducces oxidative damage and amyloid pathology in an Alzheimer transgenic mouse. *J. Neurosci.* 21:8370–8377, 2001.

83. Jitoe, A., T. Masuda, G.P. Tengah, D.N. Suprapta, I.W. Gara, N. Nakatani. Antioxidant activity of tropical ginger extracts and analysis of the contained curcuminoids. *J. Agric. Food Chem.* 40:1337–1340, 1992.

84. Masuda, T., A. Jitoe. Antioxidative and anti-inflammatory compounds from tropical gingers: isolation, structure determination, and activities of cassumunins A, B and C, new complex curcuminoids from *Zingiber cassumunar*. *J. Agric. Food Chem.* 42:1850–1856, 1994.

85. Peake, P.W., B.A. Pussel, P. Martyn, V. Timmermans, J.A. Charlesworth. The inhibitory effect of rosmarinic acid on complement involves the C5 convertase. *Int. J. Immunopharmac.* 13:853–857, 1991.

86. Ruby, A.J., G. Kuttan, K.D. Babu, K.N. Rajasekharan, R. Kuttan. Anti-tumour and antioxidant activity of natural curcuminoids. *Cancer Lett.* 94:79–83, 1995.

87. Singh, S.V., X. Hu, S.K. Srivastava, M. Singh, H. Xia, J.L. Orchard, H.A. Zaren. Mechanism of inhibition of benzo[a] pyrene-induced forestomach cancer in mice by dietary curcumin. *Cacinogenesis* 19:1357–1360, 1998.

88. Khar, A., A.M. Ali, B.V. Pardhasaradhi, Z. Begum, R. Anjum. Antitumor activity of curcumin is mediated through the induction of apoptosis in AK-5 tumor cells. *FEBS Lett.* 445:165–168, 1999.

89. Inano, H., M. Onoda, N. Inafuku, M. Kubota, Y. Kamada, T. Osawa, H. Kobayashi, K. Wakabayashi. Potent preventive action of curcumin on radiation-induced initiation of mammary tumorigenesis in rats. *Carcinogenesis* 21:1835–1841, 2000.

90. Hutchins, A.M., J.L. Slavin, J.W. Lampe. Urinary isoflavonoid phytoestrogen and lignan excretion after consumption of fermented and unfermented soy products. *J. Am. Dietetic Assoc.* 95:545–551, 1995.

91. Messina, M.J. Legumes and soybeans: overview of their nutritional and health effects. *Am. J. Clin. Nutr.* 70:439S–450S, 1999.

92. Nagata, C., N. Takatsuka, N. Kawakami, H. Shimizu. A prospective cohort study of soy product intake and stomach cancer death. *Br. J. Cancer* 87:31–36, 2002.

93. Kanazawa, K., H. Kawasaki, K. Samejima, H. Ashida, G. Danno. Specific desmutagens and (antimutagens) in oregano against a dietary carcinogen, Trp-P-2 are galagin and quercetin. *J. Agric. Food Chem.* 43:404–409, 1995.

94. Winterhoff, H., H.G. Gumbinger, H. Sourgens. On the antiogonadotropic activity of *Lithospermum* and *Lycopus* species and some of their phenolic constituents. *Planta. Medica.* 54:101–106, 1988.

95. Harbone, J.B., H. Baxter. Phenylpropanoids. In: *Phytochenical Dictionary: A Handbook of Bioactive Compounds from Plants*, Harbone, J.B., H. Baxter, eds., London: Taylor and Francis, 1993, pp 472–488.

96. Howell, A.B., N. Vorsa, A. Der Marderosian, L.Y. Foo. Inhibition of adherence of P-fimbriated *Escherichia coli* to uroepithelial-cell surfaces by proanthocyanidin extracts from cranberries. *N. Eng. J. Med.* 339(15):1085–1086, 1998.

97. Howell, A.B., B. Foxman. Cranberry juice and adhesion of antibiotic-resistant uropathogens. *JAMA* 287(23):3082–3083, 2002.

98. Guggenheim, S., S. Shapiro. The action of thymol on oral bacteria. *Oral Microbiol. Immunol.* 10:241–246, 1995.

99. Himejima, M., I. Kubo. Fungicidal activity of polygodial in combination with anethol and indole against *Candida albicans*. *J. Agric. Food Chem.* 41:1776–1779, 1993.

100. Shetty, P., M.T. Atallah, K. Shetty. Enhancement of total phenolic, L-DOPA and proline contents in germinating fava bean (*Vicia faba*) in response to bacterial elicitors. *Food Biotechnol.* 15:47–67, 2001.

101. Randhir, R., P. Shetty, K. Shetty. L-DOPA and total phenolic stimulation in dark geriminated fava bean in response to peptide and phytochemical elicitors. *Process Biochem.* 37:1247–1256, 2002a.

102. Engleberger, W., U. Hadding, E. Etschenberg, E. Graf, S. Leyck, J. Winkelmann, M.J. Parnham. Rosmarinic acid: A new inhibitor of complement C3 – convertase with anti-inflammatory activity. *Intl. J. Immunopharmac.* 10:729–737, 1988.

103. Kuhnt, M., A. Probstle, H. Rimpler, A. Bauer, M. Heinrich. Biological and pharmacological activities and further constituents of *Hyptis verticillata*. *Planta. Medica.* 61:227–232, 1995.

104. Fremont, L. Biological effects of resveratrol. *Life Sci.* 66:663–673, 2000.

105. Pinto, M.C., J.A. Garcia-Barrado, P. Macias. Resveratrol is a potent inhibitor of the dioxygenase activity of lipoxygenase. *J. Agric. Food Chem.* 47:4842–4846, 1999.

106. Vattem, D.A., K. Shetty. Solid-state production of phenolic antioxidants from cranberry pomace by *Rhizopus oligosporus*. *Food Biotechnol.* 16:189–210, 2002.

107. Vattem, D.A., K. Shetty. Ellagic acid production and phenolic antioxidant activity in cranberry pomace mediated by *Lentinus edodes* using solid-state system. *Process Biochem.* 39:367–379, 2003.

108. Singh, K., A.K. Khanna, R. Chander. Protective effect of ellagic acid on t-butyl hydroperoxide induced lipid peroxidation in isolated rat hepatocytes. *Ind. J. Exp. Biol.* 37:939–943, 1999.

109. Narayanan, B.A., G.G. Re. IGF-II down regulation associated cell cycle arrest in colon cancer cells exposed to phenolic antioxidant ellagic acid. *Anticancer Res.* 21:359–364, 2001.

110. Quiles, J.L., M.D. Mesa, C.L. Ramirez-Tortosa, C.M. Aguilera, M. Battino, M.C. Ramirez-Tortosa. *Curcuma longa* extract supplementation reduces oxidative stress and attenuates aortic fatty streak development in rabbits. *Arterioscler. Thromb. Vasc. Biol.* 22:1225–1231, 2002.

111. Dixon, R.A., M.J. Harrison, C.J. Lamb. Early events in the activation of plant defense responses. *Ann. Rev. Phytopath.* 32:479–501, 1994.

112. Dixon, R.A., N. Paiva. Stress-induced phenylpropanoid metabolism. *Plant Cell.* 7:1085–1097, 1995.

113. Rhodes, J.M., L.S.C. Wooltorton. The biosynthesis of phenolic compounds in wounded plant storage tissues. In: *Biochemistry of wounded plant tissues,* Kahl, G., ed., Berlin: W de Gruyter, 1978, pp –286.

114. Brooker, F.L., J.E. Miller. Phenylpropanoid metabolism and phenolic composition of soybean [Glycine max (L) Merr.] leaves following exposure to ozone. *J. Exp. Bot.* 49:1191–1202, 1998.

115. Zimmerman, Y.Z., P.R. Cohill. Heat shock and thermotolerance in plant and animal embryogenesis. *New Biol.* 3:641–650, 1991.

116. Yalpani, N., A.J. Enyedi, J. Leon, I. Raskin. Ultraviolet light and ozone stimulate accumulation of salicylic acid, pathogenesis related proteins and virus resistance in tobacco. *Planta* 193:372–376, 1994.

117. Hahlbrock, K., D. Scheel. Physiology and molecular biology of phenylpropanoid metabolism. *Plant Mol. Biol.* 40:347–369, 1989.

118. Graham, T.L. Flavanoid and isoflavanoid distribution in developing soybean seedling tissue and in seed and root exudates. *Plant Physiol.* 95:594–603, 1991.

119. Christie, P.J., M.R. Alfenito, V. Walbot. Impact of low temperature stress on general phenylpropanoid and anthocyanin pathways: enhancement of transcript abundance and anthocyanin pigmentation in maize seedlings *Planta* 194:541–549, 1994.

120. Beggs, C.J., K. Kuhn, R. Bocker, E. Wellmann. Phytochrome induced flavonoid biosynthesis in mustard *Sinapsis alba* L cotyledons: enzymatic control and differential regulation of anthocyanin and quercitin formation. *Planta* 172:121–126, 1987.

121. Lois, R., B.B. Buchanan. Severe sensitivity to ultraviolet light radiation in *Arabidopsis* mutant deficient in flavanoid accumulation: mechanisms of UV-resistance in *Arabidopsis*. *Planta* 194:504–509, 1994.

122. Kurganova, L.N., A.P. Veselov, T.A. Goncharova, Y.V. Sinitsyna. Lipid peroxidation and antioxidant system protection against heat shock in pea (*Pisum sativum* L.) chloroplasts. *Fiziol. Rast. Russ. J. Plant Physiol., Engl. Transl.* 44:725–730, 1997.

123. Retivin, V., V. Opritov, S.B. Fedulina. Generation of action potential induces preadaptation of *Cucurbita pepo* L. stem tissues to freezing injury. *Fiziol. Rast. Russ. J. Plant Physiol. Engl. Transl.* 44:499–510, 1997.

124. Kuznetsov, V.V., N.V. Veststenko. Synthesis of heat shock proteins and their contribution to the survival of intact cucumber plants exposed to hyperthermia. *Fiziol. Rast. Russ. J. Plant Physiol. Engl. Transl.* 41:374–380, 1994.

125. Baraboi, V.A. Mechanisms of stress and lipid peroxidation. *Usp. Sovr. Biol.* 11:923–933, 1991.

126. Kurganova, L.N., A.P. Veselov, Y.V. Sinitsina, E.A. Elikova. Lipid peroxidation products as possible mediators of heat stress response in plants. *Exp. J. Plant Physiol.* 46:181–185, 1999.

127. Phang, J.M. The regulatory functions of proline and pyrroline-5-carboxylic acid. *Curr. Topics Cell. Reg.* 25:91–132, 1985.

128. Hagedorn, C.H., J.M. Phang. Transfer of reducing equivalents into mitochondria by the interconversions of proline and δ-pyrroline-5-carboxylate. *Arch. Biochem. Biphys.* 225:95–101, 1983.

129. Rayapati, J.P., C.R. Stewart. Solubilization of a proline dehydrogenase from maize (*Zea mays* L.) mitochondria. *Plant Physiol.* 95:787–791, 1991.

130. Yeh, G.C., J.M. Phang. The function of pyrroline-5-carboxylate reductase in human erythrocytes. *Biochem. Biophys. Res. Commun.* 94:450–457, 1980.

131. Phang, J.M., S.J. Downing, G.C. Yeh, R.J. Smith, J.A. Williams. Stimulation of hexosemonophosphate-pentose pathway by δ-pyrroline-5-carboxylic acid in human fibroblasts. *Biochem. Biophys. Res. Commun.* 87:363–370, 1979.

132. Hagedorn, C.H., J.M. Phang. Catalytic transfer of hydride ions from NADPH to oxygen by the interconversions of proline and δ-pyrroline-5-carboxylate. *Arch. Biochem. Biphys.* 248:166–174, 1986.

133. Kwok, D., K. Shetty. Effect of proline and proline analogs on total phenolic and rosmarinic acid levels in shoot clones of thyme (*Thymus vulgaris* L.). *J. Food Biochem.* 22:37–51, 1998.

134. Yang, R., K. Shetty. Stimulation of rosmarinic acid in shoot cultures of oregano (*Origanum vulgare*) clonal line in response to proline, proline analog and proline precursors. *J. Agric. Food Chem.* 46:2888–2893, 1998.

135. Bela, J., K. Shetty. Somatic embryogenesis in anise (*Pimpinella anisum* L.): the effect of proline on embryogenic callus formation and ABA on advanced embryo development. *J. Food Biochem.* 23:17–32.

136. Perry, P.L., K. Shetty. A model for involvement of proline during *Pseudomonas*-mediated stimulation of rosmarinic acid. *Food Biotechnol.* 13:137–154, 1999.

137. Lendzian, K.J. Modulation of glucose-6-phosphate dehydrogenase by NADPH, NADP+ and dithiothreitol at variable NADPH/NADP+ ratios in an illuminated reconstituted spinach (*Spinacia oleracea* L.) choroplast system. *Planta* 148:1–6, 1980.

138. Copeland, L., J.F. Turner. The regulation of glycolysis and the pentose-phosphate pathway. In: *The Biochemistry of Plants*, Vol. 11, Stumpf, F., E.E. Conn, eds., New York: Academic Press, 1987, pp 107–125.

139. Elthon, T.E., C.R. Stewart. Effects of proline analog L-thiazolidine-4-carboxylic acid on proline metabolism. *Plant Physiol.* 74:213–218, 1984.

140. Jost, A., S. Perlman, O. Valentino, M. Castinier, R. Scholler, S. Magre. Experimental control of the differentiation of Leydig cells in the rat fetal testis. *Proc. Natl. Acad. Sci. USA.* 85:8094–8097, 1988.

141. Al-Amier, H., B.M.M. Mansour, N. Toaima, R.A. Korus, K. Shetty. Tissue culture-based screening for selection of high biomass and phenolic-producing clonal lines of lavender using *Pseudomonas* and azetidine-2-carboxylate. *J. Agric. Food Chem.* 47:2937–2943, 1999.

142. Al-Amier, H., B.M.M. Mansour, N. Toaima, R.A. Korus, K. Shetty. Screening of high biomass and phenolic-producing clonal lines of spearmint in tissue culture using *Pseudomnas* and azetidine-2-carboxylate. *Food Biotechnol.* 13:227–253, 1999.

143. Al-Amier, H.A., B.M.M. Mansour, N. Toaima, L. Craker, K. Shetty. Tissue culture for phenolics and rosmarinic acid in thyme. *J. Herbs Spices Med. Plant* 8:31–42, 2001.

144. Kohl, D.H., K.R. Schubert, M.B. Carter, C.H. Hagdorn, G. Shearer. Proline metabolism in N$_2$-fixing root nodules: energy transfer and regulation of purine synthesis. *Proc. Natl. Acad. Sci. USA* 85:2036–2040, 1988.

145. Eguchi, Y., O.F. Curtis, K. Shetty. Interaction of hyperhydricity-preventing *Pseudomonas* spp. with oregano *(Origanum vulgare)* and selection of high rosmarinic acid-producing clones. *Food Biotechnol.* 10:191–202, 1996.

146. Shetty, K., T.L. Carpenter, D. Kwok, O.F. Curtis, T.L. Potter. Selection of high phenolics-containing clones of thyme (*Thymus vulgaris* L.) using *Pseudomonas* spp. *J. Agric. Food Chem.* 44:3408–3411, 1996.

147. Yang, R., O.F. Curtis, K. Shetty. Selection of high rosmarinic acid-producing clonal lines of rosemary (*Rosmarinus officinalis*) via tissue culture using *Pseudomonas* sp. *Food Biotechnol.* 11:73–88, 1997.

148. McCue, P., Z. Zheng, J.L. Pinkham, K. Shetty. A model for enhanced pea seedling vigor following low pH and salicylic acid treatments. *Process Biochem.* 35:603–613, 2000.

149. Andarwulan, N., K. Shetty. Improvement of pea (*Pisum sativum*) seed vigor by fish protein hydrolysates in combination with acetyl salicylic acid. *Process Biochem.* 35:159–165, 1999.

150. Duval, B., K. Shetty. The stimulation of phenolics and antioxidant activity in pea (*Pisum sativum*) elicited by genetically transformed anise root extract. *J. Food Biochem.* 25:361–377, 2001.

151. McCue, P., K. Shetty. A biochemical analysis of mungbean (*Vigna radiata*) response to microbial polysaccharides and potential phenolic-enhnacing effects for nutraceutical applications. *Food Biotechnol.* 6:57–79, 2002.

152. McCue, P., K. Shetty. Clonal herbal extracts as elicitors of phenolic synthesis in dark-germinated mungbeans for improving nutritional value with implications for food safety. *J. Food Biochem.* 26:209–232, 2002.

153. Shetty, P., M.T. Atallah, K. Shetty. Effects of UV treatment on the proline-linked pentose phosphate pathway for phenolics and L-DOPA synthesis in dark germinated *Vicia faba*. *Process Biochem.* 37:1285–1295, 2002.

154. Shetty, P., M.T. Atallah, K. Shetty. Stimulation of total phenolics, L-DOPA and antioxidant activity through proline-linked pentose phosphate pathway in response to proline and its analog in germinating fava beans (*Vicia faba*). *Process Biochem.* 38:1707–1717, 2003.

155. Randhir, R., K. Shetty. Microwave-induced stimulation of L-DOPA, phenolics and antioxidant activity in fava bean (*Vicia faba*) for Parkinson's diet. *Process Biochem.* 39:1775–1785, 2004.

2

Molecular Design of Soybean Proteins for Enhanced Food Quality

Nobuyuki Maruyama, Evelyn Mae Tecson-Mendoza, Yukie Maruyama, Motoyasu Adachi, and Shigeru Utsumi

CONTENTS

2.1 Introduction 26
2.2 Soybean Proteins 27
    2.2.1 β-Conglycinin..27
    2.2.2 Glycinin...27
2.3 Three Dimensional Structures of Soybean Proteins 28
    2.3.1 Crystallization 28
    2.3.2 X-ray Crystallography...29
        2.3.2.1 β-Conglycinin 29
        2.3.2.2 Glycinin...31
    2.3.3 Cupin Superfamily 33
2.4 Structural Features of Soybean Proteins 34
    2.4.1 β-Conglycinin..34
    2.4.2 Glycinin...35
2.5 Improvement of Nutritional Quality by Protein Engineering.....................36
    2.5.1 Increasing Sulfur Containing Amino Acids 36
    2.5.2 Increasing Digestibility 37
2.6 Improvement in Physicochemical Functions by Protein Engineering....................38
    2.6.1 Structure to Physicochemical Function Relationships of
           Soybean Proteins..38
    2.6.2 Improvement in the Physicochemical Functions of Glycinin................39
2.7 Improvement and Addition of Physiological Properties of Soybean Proteins 42
    2.7.1 Fortifying the Bile Acid Binding Ability 42
    2.7.2 Addition of Phagocytosis Stimulating Activity 43
    2.7.3 Antimitotic Activity of Lunasin 45

2.8   Future Trends 46
References..46

2.1   INTRODUCTION

Generally, the protein contents of seeds are high; those of legume seeds and cereal seeds range from ~20% to 40% and from ~7% to 15%, respectively (1). Protein contents of soybean seeds are especially high, and those of some cultivars are more than 40%. Thus, they play an important role as a protein resource for human and domestic animals. In addition, soybean seeds are oil seeds, with oil content comprising ~20% of total weight. The annual production of soybean seeds in the world is ~180 million tons with more than 80% being used for oil expression. The amount of proteins in the residues after oil expression reaches around 58 million tons. The residues after oil expression have a history of safe use as food material for many years. Most of them are used as feed for domestic animals and as fertilizer. They are very cheap (U.S. $1.2/10kg). Even isolated soybean proteins extracted from the residues are cheap (700–850 yen/kg or U.S. $6–8/kg).

Fuji Oil in Japan and Dupont in U.S.A. have endeavored to develop new foods for which soybean proteins are utilized, but their applications have been limited due to inadequate physicochemical properties of soybean proteins. Food proteins should have specific nutritional quality, palatability, physicochemical functions, safety, economical efficiency, and physiological properties as food material (1). Nutritional quality is a property determined by the amino acids, especially the essential amino acids, of the protein. Soybean proteins are deficient in sulfur containing amino acids. Nutritional quality is also influenced by the digestibility of the protein. On the other hand, palatability is determined by taste and sensory quality (texture and hardness). Generally, animal proteins are superior to plant proteins in terms of palatability.

Physicochemical properties are properties such as gelation, emulsification, foaming, and water absorption properties and are determined by structural features of soybean proteins. Thus, the physicochemical properties unique to soybean make it possible to produce traditional foods such as tofu, koori-tofu, and ganmodoki from soybeans, and to use soybean proteins as a modifier for chilled food and sausage.

The physiological property of a protein can help maintain and enhance human health. It is well known that ingestion of 6 g of soybean proteins per day lowers serum cholesterol levels in humans (2). Recently, it has been reported that soybean proteins also lower serum triglyceride levels in humans (3). Soybean proteins, therefore, are capable of improving lipid metabolism. The soybean Bowman-Birk protease inhibitor has been reported to be capable of suppressing cancer of the colon (4), prostrate cancer (5,6), oral cancers, and cancer of the head and neck (7). A soybean peptide called *lunasin* has been isolated and shown to have antimitotic activity (8,9) and is a possible chemopreventive agent (10). While some soybean proteins may have excellent physiological or biological activities, in general, the nutritional quality and physicochemical functions of soybean proteins are inferior to those of animal proteins.

Recently, in Japan and in many developed countries, health problems related to obesity, such as hypertension and heart disease, have increased. Aging of the population has also progressed. In many developing countries, the problems of lack of food and protein malnutrition exist, and legumes like soybeans are the major source of proteins. Therefore, it is desirable to develop foods that will address these problems. It is estimated that world population will be about 8.4 billion around 2070, and food production at the present rate will be insufficient to feed the growing population. Therefore, this situation demands an increase of food production at a worldwide scale.

Soybeans could serve as a major source of proteins for the world's growing population. Soybean proteins have excellent physiological properties, are cheap, and can be produced in large quantities. Further, their physicochemical and functional properties can be improved. Protein engineering provides a means to improve the properties of soybean proteins by modifying their primary structures through gene manipulation. The ultimate objective, of course, is to develop crops in which soybean proteins with enhanced food quality accumulate. As a requisite, modification by protein engineering for enhanced quality must not disturb the folding ability of proteins. If the modified proteins cannot fold correctly, they may be degraded during the processes of biosynthesis, transportation and accumulation, and may therefore not accumulate in seeds (11,12). Thus, it is most important to understand the structures of soybean proteins.

This review will discuss the structures and improvement of nutritional quality, physicochemical functions, and physiological properties of soybean proteins, with emphasis on seed storage proteins, by protein engineering.

2.2 SOYBEAN PROTEINS

Most soybean proteins are seed storage proteins. Globulins, salt soluble proteins, are the primary storage proteins. The globulins of legume seeds are classified into two types according to their sedimentation coefficients; the 7S globulin and 11S globulin (13). Both account for approximately 80% of soybean proteins. The ratio of 11S and 7S globulins varies from 0.5 to 1.7 among soybean cultivars (14). This ratio determines the nutritional quality and physicochemical functions of soybean proteins. Aside from the 7S and 11S globulins, γ-conglycinin and basic 7S globulin also exist in soybean proteins.

2.2.1 β-Conglycinin

The 7S globulin of soybean is called β-conglycinin. β-conglycinin is a glycoprotein with a trimeric structure composed of three subunits and has a molecular mass of 150–200 kDa. The constituent subunits of β-conglycinin are α, α', and β. The α and α' subunits are composed of an extension region and a core region, while the β subunit consists of only the core region (15). The core regions of three subunits exhibit high absolute homologies (90.4%, 76.2%, and 75.5% between α and α', between α and β, and between α' and β, respectively). The extension regions of the α and α' subunits exhibit lower absolute homologies (57.3%) and a highly acidic property. The α and α' subunits are glycosylated at two sites (α, Asn199 and Asn455; α', Asn215 and Asn471), and the β subunit is glycosylated at one site (Asn328) (16). The N-glycosylation site of the β subunit corresponds to the latter site of the α and α' subunits. The α and α' subunits and the β subunit are synthesized on polysomes as preproform and preform, respectively. Their signal peptides are removed cotranslationaly in the endoplasmic reticulum, and the resultant subunits assemble into trimers. The trimers are targeted from the endoplasmic reticulum to protein storage vacuoles. Finally, the propeptides of the α and α' subunits are removed, resulting in the mature forms. Molecular heterogeneity of a subunit composition is present (17,18). In other words, ten molecular species having different subunit compositions with random combinations exist (homotrimers; α3, α'3, and β3; heterotrimers; α2β1, α1β2, α'2β1, α'1β2, α2α'1, α1α'2, and α1β1α'1).

2.2.2 Glycinin

The 11S globulin is called glycinin. Glycinin is a simple protein composed of six subunits, with a molecular mass of 300–400 kDa. Some of them have a potential glycosylation site, but none of them is glycosylated. The constituent subunits are A1aB1b, A1bB2, A2B1a,

A3B4, and A5A4B3. Five subunits are classified into two groups according to their amino acid sequences: group I (A1aB1b, A1bB2 and A2B1a) and group II (A3B4 and A5A4B3). The sequence identity in each group and between groups is about 80% and 45%, respectively. Each subunit is composed of an acidic polypeptide with acidic pI and a molecular mass of 30–40 kDa, and a basic polypeptide with basic pI and a molecular mass of 18–22 kDa. The acidic and basic polypeptides are linked together by a disulfide bond. It is considered that many molecular species having different subunit compositions with random combinations exist (19). The processes of synthesis, assembly, and accumulation of glycinin are shown in Figure 2.1. The constituent subunits are synthesized as a single polypeptide precursor, the preproglycinin. The signal peptide is removed cotranslationaly in the endoplasmic reticulum and the resultant proglycinin assembles into a trimer. Proglycinins are sorted into a protein storage vacuole and a specific posttranslational cleavage occurs. The cleavage results in a mature form of the glycinin. In contrast to β-conglycinin, none of the peptides is removed by this processing.

2.3 THREE DIMENSIONAL STRUCTURES OF SOYBEAN PROTEINS

2.3.1 Crystallization

X-ray crystallography is a suitable technique to elucidate three dimensional structures of proteins of high molecular masses. X-ray crystallography needs high quality crystals. However, it is difficult to obtain crystals of samples prepared from normal soybean seeds because of the molecular heterogeneity of both β-conglycinin and glycinin. Both proteins from soybean seeds had not been crystallized. To exclude the influence of molecular

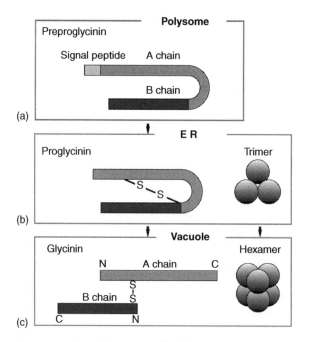

Figure 2.1 Schematic representation of processes of (a) biosynthesis, (b) assembly, and (c) transportation of soybean glycinin.

heterogeneity, we purified the storage proteins by constructing an *Escherichia coli* expression system and by using soybean seeds of mutant cultivars containing β-conglycinin or glycinin composed of a single or limited subunits.

Molecular species composed of a single subunit can be prepared easily by means of an *E. coli* expression system, because the cDNA of a single subunit is expressed in *E. coli*. Thus, we constructed an *E. coli* expression system for deletion mutants of the α and α′ subunits lacking the extension regions (α$_{core}$ and α′$_{core}$) in addition to individual normal subunits of β-conglycinin (α, α′ and β) and A1aB1b and A3B4 subunits of glycinin. The pET-21d was used as an expression vector. *Escherichia coli* does not contain enzymes for the processing of the pro regions of the α and α′ subunits and a specific posttranslational cleavage of the proglycinins. Therefore, we expressed the α and α′ subunits without the pro regions. As a result, mature forms of the α and α′ subunits and proglycinin A1aB1b and A3B4 were prepared by means of the *E. coli* expression system. Meanwhile, researchers of the Ministry of Agriculture, Forestry, and Fisheries of Japan have developed mutant soybean cultivars containing β-conglycinin lacking the α subunit, the α′ subunit or both subunits (20,21). Further, mutant soybean cultivars containing glycinin composed of only the A3B4 subunit were developed (22,23). As a collaborative effort, we prepared native homotrimers of individual subunits of β-conglycinin (α3, α′3 and β3) and A3B4 homohexamer (glycinin A3B4) from seeds of these mutant cultivars.

Crystallization of the purified glycinin and β-conglycinin from the mutant cultivars was done by means of vapor diffusion, dialysis, and a batch method under various conditions such as pH, protein concentration, and temperature using polyethyleneglycol, 2-methyl-2,4-pentanediole and ammonium sulfate as a precipitant. The crystals of the recombinant and native β homotrimer (β3), the recombinant α′$_{core}$ homotrimer (α′$_{core}$3), the recombinant A1aB1b homotrimer (proglycinin A1aB1b), the recombinant A3B4 homotrimer (proglycinin A3B4), and the native glycinin A3B4 were suitable for X-ray crystallography for analysis of their three dimensional structures.

2.3.2 X-ray Crystallography

When we started X-ray crystallography of soybean proteins, the crystal structures of the 11S globulin had not been reported; only those of French bean phaseolin and jack bean canavalin, 7S globulins, had been determined (24–26).

2.3.2.1 β-Conglycinin

We tried to determine the crystal structures of the native and recombinant β3 by the molecular replacement method using the protein structure models of phaseolin and canavalin as a search model. As a result, we determined the three dimensional structure of β3 by molecular replacement using canavalin as a search model (27) [Figure 2.2]. The β monomers consist of aminoterminal and carboxyterminal modules which are very similar to each other and are related by a pseudodiad axis. Each module of the β monomer is subdivided into a β-barrel and an α-helix domain. The superposition of the models of the native and recombinant β monomers shows a root mean square deviation (RMSD) of 0.43–0.51 Å for 343 common Cα atoms within 2.0 Å. This result indicates that the N-linked glycan does not influence the final structure of β3. Four regions including N- and C-termini were not included in the models of all the recombinant and native β monomers, because electron density maps in these regions were too thin to trace the correct sequence. These are likely to be disordered on the molecular surface [Figure 2.3(a)]. Comparison of sequences between 7S globulins by Wright (28) showed conserved regions (amino acid identity is high) and variable regions (amino acid identity is low), and it was noted that five variable

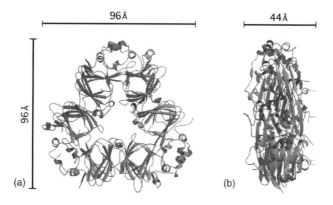

Figure 2.2 Ribbon diagrams of the three dimensional structures of (a) β-conglycinin and (b) β-homotrimer. Numbers indicate dimensions. (From: Maruyama, N., M. Adachi, K. Takahashi, K. Yagasaki, M. Kohno, Y. Takenaka, E. Okuda, S. Nakagawa, B. Mikami, S. Utsumi, *Eur. J. Biochem.* 268:3595–3604, 2001.)

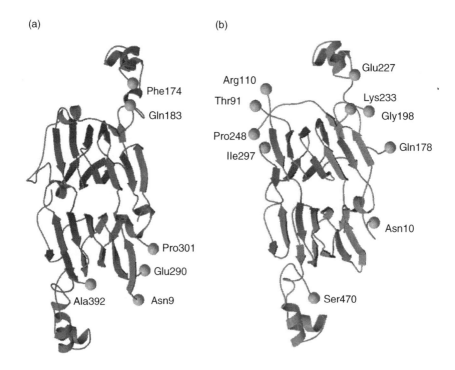

Figure 2.3 Disordered regions of soybean proteins. Disordered regions of β-conglycinin β (a) and of glycinin A1aB1b (b). The termini of the disordered regions are indicated by balls.

regions exist. In fact, there are only four variable regions, because one of them was identified by using an incorrect sequence of the α′ subunit (15). Therefore, the disordered regions in the β subunit are consistent with the four variable regions. An amino acid replacement is liable to occur in variable regions, because they are disordered regions.

α3 and α′3 among the three homotrimers of β-conglycinin formed only small crystals, but α$_{core}$3 and α′$_{core}$3 with no extension region formed crystals suitable for X-ray crystallography. Therefore, the presence of the extension region probably perturbs the formation of crystals of good quality. We determined the three dimensional structure of α′$_{core}$3 by the molecular replacement using the model of α3 (unpublished data). The scaffold of α′core3 was identical to that of β3. All α atoms of the α′core monomer and homotrimer could be superimposed on corresponding atoms of the β monomer and homotrimer with a small RMSD of 0.6 and 0.7 Å, respectively. These values indicate that scaffolds of these two proteins are very similar to each other.

2.3.2.2 *Glycinin*

The amino acid identity between 7S and 11S globulins is only 15%, but they do exhibit partially significant identity. Therefore, it is considered that these two globulins are evolutionarily related. The 11S globulin has a hexameric structure, but proglycinin has a trimeric structure. In other words, there is a possibility that pro forms of 11S globulins have a structure analogous to the 7S globulin. Thus, we tried to determine the structure of proglycinin A1aB1b by molecular replacement using the model of β-conglycinin β3, canavalin, and phaseolin but failed. Finally, we successfully determined the structure of proglycinin A1aB1b by the multiple isomorphous replacement method using heavy atom derivatives (29). Our results showed that the structure of proglycinin A1aB1b is similar to that of β-conglycinin β (Figure 2.4). Superposition of the structure of proglycinin A1aB1b on that of β-conglycinin β exposes the high similarity between the proteins (Figure 2.5). A least squares fit of a protomer between them produced an RMSD of 1.35 Å, and that of β-barrel in a protomer between them was 1.2 Å. Further, RMSDs of the comparable Cα atoms of β-barrel domains in both N- and C-terminal modules between proglycinin A1aB1b and β-conglycinin β were approximately 0.9 Å, whereas those of α-helix domains in N- and C-terminal modules were 1.35 and 1.39 Å, respectively. This indicates that the scaffold of proglycinin A1aB1b is very similar to that of β-conglycinin β, but that the configurations between a β-barrel and an α-helix domain in protomers are slightly different.

All six disordered regions existing in proglycinin A1aB1b protomer are located on the molecular surface similarly to β-conglycinin [Figure 2.3(b)]. Amino acid sequences have been previously aligned among various 11S globulins from legumes and nonlegumes, and five variable regions have been proposed (14). Five of the six disordered regions correspond to the five variable regions. The only disordered region that does not correspond to the variable region is the shortest one. The second disordered region from the C-terminus is the

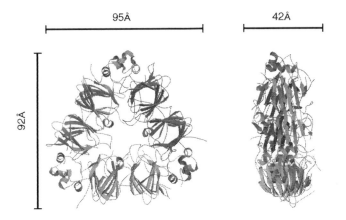

Figure 2.4 Ribbon diagram of the three dimensional structure of A1aB1b homotrimer. The view in the left diagram is depicted with the threefold symmetry axis running perpendicular to the paper, whereas the depiction on the right is related to the view on the left by a rotation of 90°. (From: Adachi, M., Y. Takenaka, A.B. Gidamis, B. Mikami, S. Utsumi., *J. Mol. Biol.* 305:291–305, 2001.)

longest, and the region of A1aB1b is composed of 48 residues. This region is called the hypervariable region, because both an amino acid sequence and a number of an amino acid in this region are the most variable among the five.

The three dimensional structure of the native glycinin A3B4 homohexamer is determined by the molecular replacement using the model of proglycinin A1aB1b (Figure 2.6) (30). As a result, a trimer of a proglycinin assembles into a hexamer by stacking of faces in which the processing site for a mature form exists.

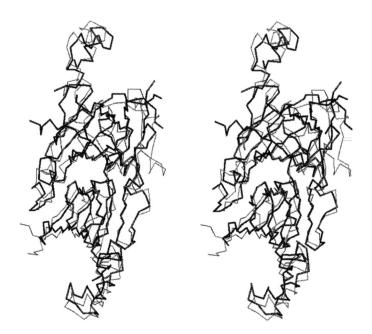

Figure 2.5 Superposition of the Cα trace of the β-conglycinin β and proglycinin A1aB1b. The Cα traces of β-conglycinin β and proglycinin A1aB1b are shown in gray and black lines, respectively.

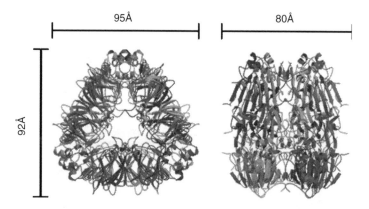

Figure 2.6 Ribbon diagram of the three dimensional structure of mature A3B4 homohexamer. The view in the left diagram is depicted with the threefold symmetry axis running perpendicular to the paper. The depiction on the right is related to the view on the left by a rotation of 90°. Numbers indicate dimensions. (From: Adachi, M., J. Kanamori, T. Masuda, K. Yagasaki, K. Kitamura, M. Bunzo, S. Utsumi. *Proc. Natl. Acad. Sci. USA*, 100:7395–7400, 2003.)

2.3.3 Cupin Superfamily

The 7S and 11S globulins are classified into the cupin superfamily (31,32).

Cupin comes from "cupa," the Latin term for a small barrel. The characteristic cupin domain comprises two conserved motifs, G(x)5HxH(x)3,4E(x)6G and G(x)5PxG(x)3N (31,32). Based on structural features, cupin proteins are classified as either a monocupin having one cupin motif per one monomer, and a bicupin having two cupin motifs per one monomer. The 7S and 11S globulins are bicupins. Cupin proteins are also known to have diverse functions. For example, in addition to seed storage proteins, an oxalate oxidase and a sucrose binding protein belong to this group (33,34).

In the three dimensional structure of the cupin protein phaseolin, which was determined at the early stage of research of the cupin superfamily, the cupin motifs are located at the C and D strands and at the G and H strands of the β-barrel (Figure 2.7). The length between motifs varies largely among species of proteins. Later, the three dimensional structures of many cupin proteins were shown to have a jellyroll type β-barrel structure. Further, it was elucidated that in addition to a tertiary structure, a quaternary structure is common among cupin proteins. Three dimensional structures of many cupin proteins indicate that an arabinose binding domain of AraC transcription factor from *E. coli*, a dimeric dTDP-4-dehydrorhamnose 3,5-epimerase from *Salmonella enterica*, a hexameric germin from barley and an oxalate decarboxylase from *Bacillus subtilis* are very similar to the domain, protomer, and trimer of proglycinin and β-conglycinin, and the hexamer of glycinin, respectively (27,29,33,34) (Figure 2.7).

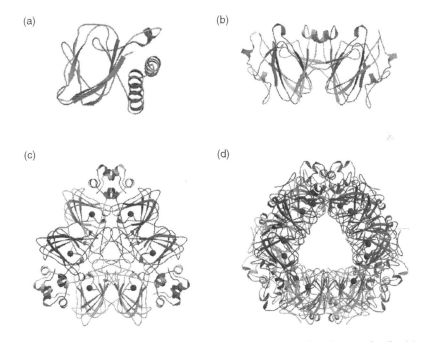

Figure 2.7 Ribbon diagrams of the three dimensional structures of cupin superfamily. (a) arabinose-binding domain of AraC transcription factor from *E. coli* (PDB code: 2ARA); (b) dimeric dTDP-4-dehydrorhamnose 3,5-epimerase from *Salmonella enterica* (1DZR); (c) hexameric germin from barley (1FI2); (d) oxalate decarboxylase from *Bacillus subtilis* (IJ58). The cupin motifs in (b) are shown in dark gray. Manganese ions are indicated by balls in (c) and (d). (From: Dunwell, J.M., A. Culham, C.E. Carter, C.R. Sosa-Aguirre, P.W. Goodenough. *Trends Biochem. Sci.* 26:740–746, 2001; and Anand, R., P.C. Dorrestein, C. Kinsland, T.P. Begley, S.E. Ealick. *Biochemistry* 41:7659–7669, 2002.)

The cupin fold is evolutionally conserved, although amino acid sequences are not conserved except the cupin motifs. This indicates that the cupin motif is essential for the formation and maintenance of the structures of glycinin and β-conglycinin.

2.4 STRUCTURAL FEATURES OF SOYBEAN PROTEINS

2.4.1 β-Conglycinin

We analyzed the structural features of the native homotrimers (α3, α'3, and β3) of the individual subunits of β-conglycinin prepared from seeds of mutant cultivars containing β-conglycinin with limited subunit compositions, the recombinant homotrimers (α3, α'3, β3, α core3, and α' core3) of individual subunits, and the α core and α' core of β-conglycinin by means of *E. coli* expression system (35–37).

On a sucrose density gradient centrifugation, three recombinant homotrimers sedimented at similar positions. However, the recombinant α3 and α'3 eluted faster than did the recombinant β3 on a gel filtration chromatography, similarly to the native glycinin hexamer. These indicate that the extension regions largely contribute to the dimensions of α3 and α'3. Therefore, the structural models indicated in Figure 2.8 can be proposed. The order of thermal stability of the recombinant and native homotrimers was α3 < α'3 < β3, and β3 was the most stable among the three subunits. On the other hand, the α core3 and α' core3 exhibited similar thermal stability to those of α3 and α'3, respectively. These results indicate that the extension regions and the glycans do not affect the thermal stability, and that the core regions determine the thermal stability of the individual subunits of β-conglycinin. Further, an analysis of the thermal stability of the heterotrimers of β-conglycinin indicated that thermal stabilities of most heterotrimers are determined by that of the constituent subunit which has the lower thermal stability (38). On the other hand, the order of surface hydrophobicity was α3 > α'3 ≥ β3, and dependent on that of the core regions. Surface hydrophobicities of the heterotrimers were the arithmetic mean of those of constituent subunits. The extension regions and the glycans largely affect solubility under lower ionic strength, but not under higher ionic strength. The solubility of the trimer containing even one α or α' subunit is much higher than that of the β3, α core3 and α' core3, and close to that of the α3 and α'3, indicating that the extension regions contribute largely to the solubility.

Because the thermal stability of the core region determines that of the subunit, we tried to analyze the structural factors related to thermal stability by comparing the three dimensional structure of the α' core3 with that of β3 in detail. Thus, five structural factors that can account for the result that β3 is more thermostable than α' core3 were elucidated (39).

First, protein packs tightly but some cavities are often observed inside the molecule, resulting in a decrease in its thermal stability (40). We calculated the total cavity volume

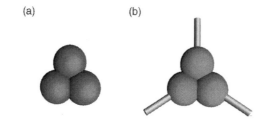

(a)   (b)

Figure 2.8 Structure models of soybean β-conglycinin. (a) β3, (b) α3 or α3. The core and the extension regions are shown in ball and stick, respectively.

of the $\alpha'_{core}3$ (172 Å^3) to be larger than that of $\beta3$ (79 Å^3), indicating that $\beta3$ is more tightly packed than $\alpha'_{core}3$. The main cavities are located in the β-barrels of the N- and C-terminal modules. The cavities of the C-terminal β-barrels of $\beta3$ and $\alpha'_{core}3$ exhibit a prominent difference between them. Twelve residues among 15 residues lining the cavity were common between the β and α'_{core}, and three residues were different. In other words, Phe233, Leu262, and Phe325 of β were substituted with Leu376, Phe405, and Val468 of α'_{core}, respectively. Because of the compensating substitutions of Leu376 and Phe405 in the α'_{core} to Phe and Leu in β, respectively, a substitution of Val468 in the α'_{core} to Phe in β contributes mainly to the difference in the cavity volume of the C-terminal β-barrel.

Second, because long ion pair networks have been found in several enzymes from hyperthermophilic bacterium (41,42), they are thought to be a common mechanism to stabilize the protein of hyperthermophiles (43). Sequence alignment of β and α'_{core} indicates that some of the charged residues in β are substituted with noncharged residues in the α'_{core}. These residues exist in the α-helix domain of the C-terminal module, and form a cluster at the intermonomer interface. The length of an ion pair network of $\beta3$ was longer than that of the $\alpha'_{core}3$, because E353, R359, E362, R363, R379, K383, and R385 of β were replaced with S496, S502, P505, S506, N522, S526, and S528 of the α'_{core}, respectively.

Third, the hydration of nonpolar groups apparently destabilizes proteins (44). Experimental analysis of surface hydrophobicity using a hydrophobic column demonstrated that the molecular surface of $\alpha'_{core}3$ was more hydrophobic than that of $\beta3$. Further, a comparison of three dimensional structures between $\beta3$ and $\alpha'_{core}3$ indicated that the ratio of hydrophobic atoms (carbon and sulfur) to all atoms of solvent accessible surface of $\alpha'_{core}3$ was higher than that of $\beta3$.

Fourth, the number of proline residues, which stabilize the protein structure by decreasing the entropy of denatured structure, is 57 in the $\alpha'_{core}3$ compared to 63 in $\beta3$. According to Matthews et al. (45), this difference contributes 4.8 kcal/mol to the ΔG. Therefore, the difference in the number of proline residues may affect the thermal stability.

Fifth, shorter loops are one of stabilizing factors in thermophilic bacteria (46). The loop between helix 3 and strand J' is five residues shorter in β than that in α'_{core}. The electron densities of this loop region were broken in all monomers of the $\alpha'_{core}3$ and two of $\beta3$, but appeared in one monomer of $\beta3$, giving the clearest density map among the three monomers. Because the resolution of $\beta3$ is lower, this may suggest relatively rigid conformation of this loop region in the β subunit. The length of the loop is also considered one of the important factors related to the structural stability of the cupin superfamily to which β-conglycinin belongs.

As described, we investigated possible factors for the difference in the thermal stability between the $\alpha'_{core}3$ and $\beta3$ through the comparison of three dimensional structures. More hydrogen bonds were observed in each module of $\alpha'_{core}3$, which suggests more stable packing of $\alpha'_{core}3$, and is not consistent with the experimental data. This difference should be more than compensated by the accumulation of the effects of many factors which account for less thermal stability of the $\alpha'_{core}3$, such as larger cavity volumes, lack of short ion pair networks, higher surface hydrophobicity, lower number of proline residues, and a longer loop. We think that $\beta3$ would be stabilized more than the $\alpha'_{core}3$ by the sum of the contributions of each of these factors.

2.4.2 Glycinin

Researchers of the Ministry of Agriculture, Forestry, and Fisheries of Japan have developed mutant soybean cultivars containing glycinin of limited subunit compositions, such as only group I, only group II, and only A5A4B3, in addition to only A3B4 (22,23). In collaboration with them, we prepared glycinin (group I, group II, A3B4, and A5A4B3) from these mutant cultivars and analyzed their structural features (unpublished data). On gel filtration chromatography, subunits having the longer hypervariable region eluted faster. In other words, A5A4B3 and group II eluted faster than A3B4 and group I did, respectively. These results indicate that the variable regions influence the dimensions of glycinin similarly to the extension region of β-conglycinin. The thermal stability of A3B4 (87.2°C) was slightly lower than those of the other subunits (92.9–95.0°C), but group II containing A3B4 exhibited thermal stability similar to the others. Therefore, the effect of subunit compositions on the thermal stability is not significant. The order of surface hydrophobicity on a hydrophobic column chromatography was group I < 11S < A3B4 < group II < A5A4B3. Thus, the surface hydrophobicity of glycinin comprised of several subunits is the arithmetic mean of those of individual subunits. Further, we confirmed that the molecular surfaces of the three dimensional structures of recombinant proglycinin A1aB1b and native glycinin A3B4 corresponded to the difference of surface hydrophobicity on a hydrophobic column chromatography. However, it is difficult to discuss the difference of the thermal stability in detail, because we have only determined the three dimensional structure of A3B4 among native glycinins.

2.5 IMPROVEMENT OF NUTRITIONAL QUALITY BY PROTEIN ENGINEERING

There are two approaches to improve the nutritional quality of soybean proteins: fortify the content of a limiting amino acid such as amino acids containing sulfur, and increase the digestibility of the proteins.

2.5.1 Increasing Sulfur Containing Amino Acids

In grain legumes such as soybean, methionine is the primary limiting essential amino acid. The biological value of legume protein is limited to 55–75% of that of animal protein because of this limitation or imbalance in amino acid composition (47). Several ways to increase the content of sulfur containing amino acids have been cited (48): (1) the insertion and replacement of amino acid residues at appropriate sites in a storage protein; (2) increasing the level of endogenous methionine rich protein; (3) introduction of heterologous genes for methionine rich protein; and (4) manipulating key enzymes in the biosynthetic pathway. A description of the work done in our laboratory using the first strategy, protein engineering, follows.

Both glycinin and β-conglycinin have variable regions which are less conservative. Thus, it may be possible to replace and insert amino acids in the variable regions to enhance nutritional quality.

At the start of our research to improve glycinin by protein engineering, no data for the three dimensional structures of glycinin were available. Therefore, we attempted to modify the variable regions based on the alignment of amino acids. Five variable regions were named I, II, III, IV, and V from the N-terminus [Figure 2.9(a)]. We designed modified glycinins with five contiguous Met residues inserted into variable region IV or V (IV+4Met and V+4Met). Further, we also designed a modified glycinin whose five contiguous Glu in the variable region IV were substituted with five Met and one Leu [IV(Met)]. These modified glycinins were produced by means of an *E. coli* expression system and their folding ability was analyzed. All of them were accumulated as soluble proteins in *E. coli* at a high level, and each one self assembled. Because the modified glycinins exhibit solubility as a globulin, we concluded that the modified glycinins fold correctly. Moreover, the three modified glycinins can be crystallized, and IV+4Met and V+4Met can be accumulated in tobacco seeds, rice seeds, and potato tubers in a way similar to the intact molecule (49–53). In other words, an insertion and a replacement in variable regions could be useful in improving the nutritional quality of glycinin.

It is difficult to insert amino acids in conserved regions, but it is possible to replace amino acids based on three dimensional structures. For example, Figure 2.10(a) shows one β-strand in the β-barrel of C-terminal module of A1aB1b. This strand contains a hydrophobic peptide, Val-Ile-Leu-Val. Based on molecular modeling, we considered that it is possible to replace these four amino acids with four Met residues [Figure 2.10(b)]. We have not examined the folding ability of this modified glycinin yet, but we think that it probably can fold correctly.

2.5.2 Increasing Digestibility

There are two approaches to improving digestibility: by introducing a cleavable site for digestive enzymes, and by destabilizing a structure. If aromatic residues recognized by chymotrypsin are introduced inside the molecule, however, its structures may be stabilized

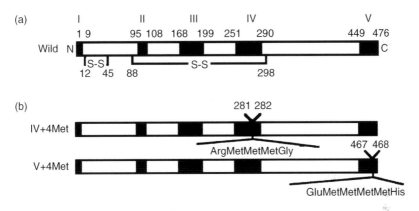

Figure 2.9 Schematic representation of the wild and modified proglycinin A1aB1b. (a) wild type, (b) tetramethionine-inserted proglycinin. White and black areas indicate the conserved and variables regions, respectively. The number of the residues from the N-terminus for the variable regions (I–V) are shown above the alignment. S-S indicates a disulfide bond.

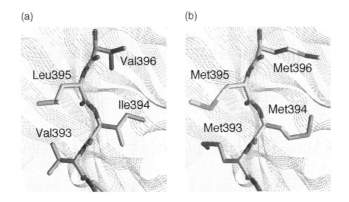

Figure 2.10 Fortification of Met into β-barrel of the proglycinin A1aB1b. (a) wild type; (b) methionine-introduced proglycinin. Backbone and side chains are indicated by bond models.

by the increased hydrophobicity inside the molecules, resulting in the lowering of digestibility. On the other hand, introduction of lysine and arginine residues recognized by trypsin inside the molecule may probably have a large influence on digestibility, because they not only add the site for cleavage, but also destabilize the structure due to the introduction of a charged residue inside the molecule. Most of the salt bridges inside molecules contribute largely to structural stability. Therefore, a substitution of an acidic residue which forms a salt bridge inside the molecule with another amino acid destabilizes a structure while keeping the site for trypsin. In glycinin A1aB1b, a salt bridge between Asp157-Arg161 exists inside the molecule. We replaced Asp157 with Ala (D157A) (54). This mutant, produced in *E. coli*, was found to fold correctly. Furthermore, we confirmed that D157A formed crystals suitable for X-ray crystallography. The T_m value of D157A was 15°C lower than that of the wild type and its sensitivity to chymotrypsin increased.

2.6   IMPROVEMENT IN PHYSICOCHEMICAL FUNCTIONS BY PROTEIN ENGINEERING

2.6.1   Structure to Physicochemical Function Relationships of Soybean Proteins

The physicochemical functions of soybean proteins are determined by solubility, degree of hydrophobicity, and hydrophilicity of a molecular surface. Their distribution and balance, structural stability, exposure of hydrophobic regions under denaturation, and interactions between the same and the different molecular species also influence physiochemical function. Therefore, elucidation of the structure to physicochemical function relationships of soybean proteins is necessary to improve rationally the physicochemical functions of soybean proteins. Such studies have mostly been done in our laboratory using mutant cultivars, and by means of protein engineering.

We prepared homotrimers of the individual subunits of β-conglycinin and analyzed their physicochemical functions (35–38). The order of emulsifying ability was α3 > α'3 > β3 in the cases of both the recombinant and native homotrimers. In comparison with the homotrimers lacking the extension region, the order of emulsifying ability was α$_{core}$3 > α'$_{core}$3 > β3, although the emulsifying abilities of α'$_{core}$3 and α$_{core}$3 were lower than those of α'3 and α3, respectively. This order was related to those of thermal stability and surface hydrophobicity. The heterotrimers containing two α or α' subunits (α2β1, α'2β1) exhibited emulsifying ability similar to those of α3 or α'3, respectively. The heterotrimers containing one α or α' subunit (α1β2, α'1β2) exhibited emulsifying ability similar to that of β3. Therefore, the number of the extension regions, structural stability, and surface hydrophobicity contribute to the emulsifying abilities, but the glycans do not. The native and recombinant α3 and α'3 formed soluble aggregates without accompanying insoluble aggregates by heating at greater than their T_m values. In contrast, the native and recombinant β3 did not form the soluble aggregates at all at greater than their T_m values, but formed insoluble aggregates. The size and the amount of the soluble aggregates of the native α3 and α'3 were smaller and fewer than those of the recombinant α3 and α'3, respectively, but α'$_{core}$3 and α$_{core}$3 formed insoluble aggregates without soluble aggregates similarly to that of β3. The heterotrimers containing two α or α' subunit (α2β1 and α'2β1) exhibited a similar tendency about heat induced associations to α3 and α'3, but the behaviors of heterotrimers containing two β subunits (α1β2 and α'1β2) resembled that of β3 rather than those of α3 and β3. These results indicate that the presence and the number of the extension regions contribute to the heat induced association, but the glycans prevent such heat induced association.

Similarly, we measured the emulsifying ability of glycinin prepared from mutant cultivars having subunit compositions different from a normal one (unpublished data). Group II exhibited higher emulsifying ability than that of group I, and that of glycinin containing all subunits (normal glycinin) was intermediate between those of group I and II. A5A4B3 exhibited slightly higher ability than A3B4 did, and the ability of group II was intermediate between those of A3B4 and A5A4B3. These results indicated that A5A4B3 has the highest emulsifying ability among all subunits of glycinin, and that the heterohexamer exhibits the arithmetic mean of the abilities of constituent subunits. Variable region IV (the hypervariable region), the longest in five variable regions of glycinin, is rich in negative charges. The length of this region is different among subunits of glycinin, and that of group II, especially A5A4B3, is longer than that of group I. Therefore, it can be considered that the length of the hypervariable region is largely related to the emulsifying ability of glycinin. This resembles the effect of the extension region of β-conglycinin on the emulsifying ability. However, the order of the emulsifying ability of glycinins did not correlate with those of the surface hydrophobicity and the structural stability, in contrast to β-conglycinin. In other words, the degree of the contribution by each factor is different among the species of proteins and subunits.

To exhibit excellent emulsifying ability, proteins need the affinity for both oil and water at their interface. Therefore, flexibility and amphiphatic properties are important factors for emulsifying ability. In fact, it is pointed out that in many proteins surface hydrophobicity and structural stability are related to emulsifying ability. The contribution of surface hydrophobicity and structural stability of glycinin to its emulsifying ability is different from that of β-conglycinin, although the presence of large hydrophilic regions contributes to the emulsifying ability of both glycinin and β-conglycinin. Elucidation of the three dimensional structures of all subunits of glycinin and β-conglycinin can probably shed light on this difference. It is expected that destabilization of a structure and strengthening of hydrophilicity or hydrophobicity can improve emulsifying ability. However, the effect is probably different between glycinin and β-conglycinin.

The subunit compositions of glycinins vary among cultivars (55). Mori et al. analyzed heat induced gel forming ability of glycinins prepared from five cultivars containing different subunit compositions (56). As a result, glycinins which have larger amounts of A3B4 formed harder gel. Pseudoglycinins were reconstituted from A1aB1b, A2B1a, and A3B4 isolated on a chromatography in the presence of urea, and their heat induced gel forming abilities were analyzed (57,58). The results showed that the hardness of the gels is directly proportional to the amount of A3B4. The number of cysteine residues in A3B4 is less than those of A1aB1b and A2B1a. The topology of free sulfhydryl groups of A3B4 is different from those of A1aB1b and A2B1a. Further, A3B4 has lower structural stability than A1aB1b and A2B1a. Changes in conformation and SH/S-S interchange reactions are essential for heat induced gel forming ability of glycinin. Therefore, it can be considered that the reactivity of SH/S-S interchange of A3B4 is higher than those of A1aB1b and A2B1a due to the difference in the topology of the free sulfhydryl group and lower structural stability. As a result, A3B4 can form hard gels with fine network structures. Therefore, we can consider that a destabilization of a structure and modification of the topology of free sulfhydryl groups are an approach to improving gel forming ability of glycinin. In terms of fineness of network, the addition of a number of SH and S-S could be a strategy. We cannot describe the mechanism of gel forming ability of β-conglycinin, because analysis is still insufficient.

2.6.2 Improvement in the Physicochemical Functions of Glycinin
Variable regions of glycinin are rich in hydrophilic amino acids. Therefore, deletions of these regions can change the balance between hydrophilicity and hydrophobicity of glycinin.

There is a possibility that the deletions destabilize the molecule, because the regions exist in the intact molecule. So we designed mutants of glycinin lacking each of the variable regions (ΔI, ΔII, ΔIII, ΔIV, ΔV, ΔV36 and ΔV8) [Figure 2.9(a)] (49). Existing restriction enzyme sites were used to construct the mutants because the use of PCR was not yet popular during the time of these experiments. Therefore, some parts of conserved regions were also deleted, and glycinin lacking C-terminal 8 or 36 residues (ΔV8, ΔV36) has two extra amino acids, Leu and Asn, at its C-terminus.

If a peptide composed of four Met residues is introduced into the variable region [Figure 2.9(b)], a small hydrophobic patch is formed. This may result in structure destabilization because a small hydrophobic patch is introduced into a hydrophilic region. Therefore, it is expected that this mutation improves the protein's emulsifying ability and gel forming ability in addition to nutritional quality.

Glycinin has two disulfide bonds in each constituent subunit: one is an intraacidic chain bond (Cys12-Cys45 in A1aB1b) and another links the acidic and basic chains (Cys88-Cys298 in A1aB1b). These disulfide bonds can be confirmed in the three dimensional structure of A1aB1b (29). Because the corresponding cysteine residues are conserved in all 11S globulins whose primary structures have been determined, it is considered that these two disulfide bonds are also conserved in all 11S globulins (1). In case of the subunits of group I, it is assumed that one more disulfide bond (C271-C278 in A1aB1b) exists. However, we cannot confirm this disulfide bond in the three dimensional structure, because these residues are located in the disordered region. If one of the cysteine residues forming a disulfide bond is replaced by other amino acids, a free sulfhydryl group arises. The lack of a disulfide bond probably induces a destabilization of a structure. It is therefore expected that the emulsifying ability and gel forming ability are improved by a disruption of a disulfide bond. To test this hypothesis, mutants lacking one disulfide bond (C12G and C88S) and both disulfide bonds (C12G/C88S) were designed [Figure 2.11(b)] (59).

On the other hand, Cys53 and Cys377 in A1aB1b are free sulfhydryl groups. If an amino acid near the site of Cys53 or Cys377 is replaced by a cysteine residue, a new disulfide bond can be rationally introduced, considering the directions of the side chains of the replaced, the cysteine residues, and the distance of Cα between them shown by a simulated model. On the other hand, if two amino acids are replaced by cysteine residues by a similar method, a new disulfide bond can be introduced. Two and three mutants containing a new sulfhydryl residue and a new disulfide bond were designed, respectively (60).

These mutants were expressed in *E. coli*. Solubilities, self assemblies, and stabilities of expressed proteins indicated that ΔI, ΔV8, C88S, C12G/C88S, R161C, P248C, N116C, P248C, and N116C/P248C, in addition to IV+4Met, V+4Met. and D157A can fold correctly. In other words, modifications of the variable regions and the conserved regions based on three dimensional structure can be acceptable. Further, it was elucidated that the two disulfide bonds conserved among 11S globulins are not essential for the formation and the maintenance of the conformation of glycinin. However, it was difficult to purify C12G/C88S, because it was susceptible to attack of proteinase of *E. coli* during purification under a low ionic strength. This indicates that the removal of both disulfide bonds makes the conformation of C12G/C88S less compact than that of normal glycinin under low ionic strength.

We confirmed the expected introduction of a new free sulfhydryl group and a new disulfide bond by measuring the content of the free sulfhydryl group of the modified glycinins after blocking a free sulfhydryl group (60).

We analyzed the physicochemical functions of the modified glycinins, which can be prepared at a large scale. We compared the heat induced gel forming ability of ΔI, ΔV8, IV+4Met, V+4Met, C12G, C88S, R161C, F163C, N116C, P248C, and N116C/P248C with those of native glycinin and normal proglycinin A1aB1b. Proglycinin A1aB1b exhibits

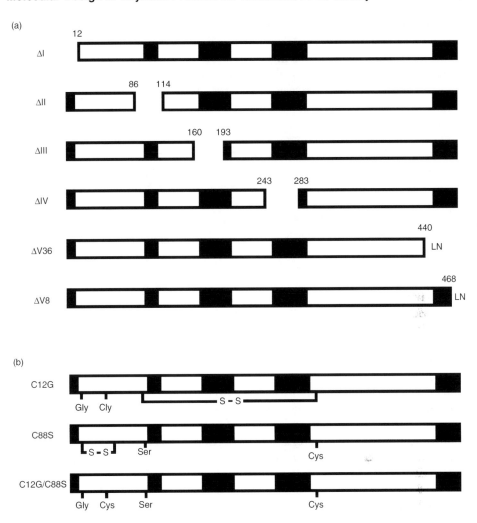

Figure 2.11 Schematic representation of modified proglycinin A1aB1b. (a) variable region-deleted mutants; (b) disulfide bond-deleted mutants. Numbers and S-S indicate the residue number from N-terminus and a disulfide bond, respectively. (From: Kim, C.S., S. Kamiya, T. Sato, S. Utsumi, M. Kito, *Protein Eng.* 3:725–731, 1990.)

gel forming ability and gel hardness similar to those of the native glycinin. On the other hand, ΔI, IV+4Met, V+4Met, C88S, F163C, N116C, P248C, and N116C/P248C formed harder gels than the normal one did. C88S and N116C/P248C showed the most prominent results. On the other hand, R161C formed harder gels than normal at protein concentrations higher than 7%, but formed gels with gel hardness similar to that of normal to one at protein concentration lower than 7%. C12G formed gels with gel hardness similar to that of normal glycinin at higher protein concentration, but could not form gel at a protein concentration lower than 5.6%. These results indicate: (1) the disulfide bond between Cys12 and Cys45 plays an important role in the initiation of a disulfide exchange reaction for a gelation; (2) introduction of a new sulfhydryl group and a disulfide bond is useful for improvement of heat induced gel forming ability and gel hardness; (3) the reactivity of the disulfide bond between Cys12 and Cys45 of ΔI, where its N-terminus is Cys12, increases; and (4) enhancement of surface hydrophobicity may improve gel forming ability.

We compared the emulsifying ability of ΔI, ΔV8, IV+4Met, V+4Met, C12G, C88S, and D157A with that of proglycinin A1aB1b. ΔI, IV+4Met, C12G, and C88S exhibited similar emulsifying ability to that of proglycinin A1aB1b. D157A exhibited worse emulsifying ability than proglycinin A1aB1b did (the details will be described elsewhere). In contrast, ΔV8 and V+4Met exhibited better emulsifying ability. These results indicate: (1) the presence of a hydrophobic region at the C-terminus increases emulsifying ability; (2) the effect of increasing hydrophobicity is different among the mutants; and (3) a destabilization of a structure does not have an effect on the improvement of the emulsifying ability of proglycinin A1aB1b. These observations correspond to the results of the analyses of glycinin having different subunit compositions described in the previous section.

Protein engineering is useful not only for the improvement of physicochemical functions, but also to elucidate the relationship between structure and physicochemical function. Therefore, the elucidation of the relationship between structure and physicochemical function at a molecular level, by protein engineering, will enable us to improve the physicochemical functions of soybean proteins effectively and rationally.

2.7   IMPROVEMENT AND ADDITION OF PHYSIOLOGICAL PROPERTIES OF SOYBEAN PROTEINS

Many kinds of bioactive peptides have been isolated from enzymatic digests of food proteins (61–64). Some proteins contain such peptides in a noncleavable form, and some contain peptides with a weak activity in a cleavable form. In these cases, physiological properties can be introduced or changed by replacement of a few amino acids. It is easy to replace a few amino acids in variable regions. Further, simulation models based on three dimensional structures enable introductions of bioactive peptides into even conserved regions. In case of the variable regions, if there is no sequence similar to that of the bioactive peptide, it is possible to introduce bioactive peptides that are not long and highly hydrophobic by replacements and insertions.

The peptide lunasin, which is the small subunit of the soybean MRP GM2S-1, has antimitotic properties and is a potential chemopreventive agent (8,10). As earlier mentioned, the soybean Bowman-Birk protease inhibitor has been shown to suppress various types of cancer: colon cancer (4), prostrate cancer (5,6), oral cancers, and cancers of the head and neck (7).

In this section, we review the research status of the improvement of physiological properties of soybean protein.

2.7.1 Fortifying the Bile Acid Binding Ability

It is well known that soybean proteins have a cholesterol lowering effect in human serum (65,66). Several mechanisms for the regulation of serum cholesterol have been proposed. As a major one, an undigested insoluble fraction of soybean protein interacts with cholesterols and bile acids in the intestine, and they are excreted into an easing (67,68). Most probably, hydrophobic interactions play an important role in these interactions. Makino et al. found that the 114–161 residues in A1aB1b, one of the peptides generated by trypsin digestion of glycinin, have a bile acid binding ability (69). On the other hand, Iwami et al. confirmed that peptides generated by hydrolysis of mild acids contain residues having highly hydrophobic properties derived from β-conglycinin, and these peptides actually bind bile acids (70).

The sequence of 114–161 residues in A1aB1b found by Makino et al. is very similar to those of the corresponding regions of the other glycinin subunits, because the region is located in the conserved region. Further, the cleavage sites for trypsin exist at both sides of the corresponding regions of the other subunits. Makino et al. confirmed that a peptide generated from A2B1a also has bile acid binding activity. However, the activity of peptides generated from any of the subunits except A1aB1b and A2B1a has not been identified. There is a possibility that the peptide derived from A1bB2 was not detected because the sequence of the corresponding region of A1bB2 is very similar to that of A1aB1b, and the amount of A1bB2 in seeds is smaller than those of A1aB1b and A2B1a. All normal cultivars examined so far contain A3B4, whereas some normal cultivars of soybean do not contain A5A4B3. Therefore, Makino et al. probably used a cultivar containing no A5A4B3, and the activity of the corresponding region in A3B4 is weaker than those of A1aB1b and A2B1a.

Our experiments using an affinity column conjugated with cholic acid, one of the bile acids, supports this idea (71). By replacing the corresponding region in A3B4 with the region of 114–161 residues in A1aB1b, the bile acid binding ability can be increased in A3B4. Recently, we constructed deletion mutants of A1a and evaluated their bile acid binding ability by affinity column and their dissociation constants by fluorescence measurement (71). As a result, VAWWMY, the most hydrophobic sequence in the region of 114–161 residues of A1aB1b, was found to be very important in binding bile acids. We can easily fortify the bile acid binding ability of food proteins by using this peptide. For example, the sequences of A3B4 and A5A4B3 corresponding to VAWWMY of A1aB1b are VPYWTY and VPYWTY, respectively. Therefore, a replacement of three amino acids will probably enable the addition a bile acid binding ability into them. However, the folding ability of the mutants in these cases must also be considered, because the amino acids to be replaced are located inside the β-barrel. As described in the section on improvement of nutritional quality, the folding ability of the mutants can be evaluated by simulations before constructing and preparing them. On the other hand, the bile acid binding sequence (VAWWMY) can be introduced into variable regions easily, because it is easy to insert and replace amino acids in these regions. Thus, we introduced the bile acid binding sequence into two sites (M1 and M2) in the variable region IV (hypervariable region) of A1aB1b. The dissociation constants of M1 and M2 for sodium cholate were calculated by fluorescence measurements to be 30 mM and 37 mM, respectively, and that of A1a was 49 mM. These results indicate that the bile acid binding activity of A1a can be fortified as expected.

The bile acid binding peptide sequence can probably be introduced into β-conglycinin by a similar method, because β-conglycinin also has variable regions. Both glycinin and β-conglycinin have β-barrels rich in hydrophobic amino acids. Therefore, the bile acid binding ability of soybean proteins may be fortified by introduction of the bile acid binding sequence (VAWWMY) into the β-barrel based on simulation modeling.

2.7.2 Addition of Phagocytosis Stimulating Activity

Recently, Yoshikawa et al. isolated a phagocytosis stimulating peptide from trypsin digests of soybean proteins (72). This peptide, named soymetide, is derived from the α′ subunit of β-conglycinin, and its primary sequence is MITLAIPVNKPGR. Met at its N-terminus is essential for the activity and MITL, the four residues from N-terminus, is the shortest peptide exhibiting phagocytosis stimulation. When the Thr, third residue from the N-terminus of the soymetide, was replaced by Phe or Trp, the activities of the modified peptides greatly increased (Thr < Phe < Trp) (72). Although the regions of the α and β subunits corresponding to the soymetide are highly conserved among the three subunits, and the cleavage sites for trypsin exist at both sides of the regions, their digests by trypsin do not exhibit the activity. The reason is that the residues of the α and β subunits corresponding to the N-terminal Met in the soymetide, which is essential for the activity, are Leu and Ile, respectively. It was also found that Lys124 in the β subunit was not favorable for phagocytosis stimulation (72).

We attempted to introduce the phagocytosis stimulating peptide sequence into the β subunit by replacing Ile122 and Lys124 of the wild type with Met and Thr/Phe/Trp (I122M/K124T, I122M/K124F and I122M/K124W), respectively (73). These residues are located in the β barrel of the core region and their side chains are inside the molecule. Therefore, there is a possibility that the introduction of the phagocytosis stimulating peptide sequence into the β subunit might induce the loss of the folding ability by causing unfavorable contacts, especially in the case of the replacement of Lys with Phe or Trp, which have side chains bigger than that of Lys.

At first, we simulated the models of the three mutants using the three dimensional structure of the β subunit by the programs Insight II and Discover to confirm whether these mutants could fold correctly or not. The RMSD for the entire Cα atoms between the starting and simulated monomer models was around 0.67 Å for all the mutants. The distance of Cα atoms between the wild type and the simulated models at the positions 122 and 124 were 0.47–0.49 and 0.29–0.50 Å in all the mutants, respectively. These values mean that the position of the backbone scarcely moved. Furthermore, no unfavorable van der Waal's interactions between the side chains of the replaced residues and those of their neighboring residues were observed. Therefore, we concluded that all the mutants could fold correctly. To confirm our assumption that all the mutants could fold correctly based on the molecular modeling, we characterized the structural features of the mutants expressed in *E. coli* by circular dichroism measurement, differential scanning calorimetry, and gel filtration column chromatography. No significant difference in circular dichroism spectra was observed between the wild type and the mutants. This result indicates that the mutations for introducing the phagocytosis stimulating peptide sequence into the β subunit have little effect on the secondary structure. T_m values of all the mutants measured by differential scanning calorimetry were 1.9–3.1°C lower than that of the wild type. Although there is a hydrogen bond between Lys124 and Tyr109 in the β barrel of the wild type, all the mutants lost this hydrogen bond by the replacement of Lys124. Therefore, the loss of the hydrogen bond of the mutants might induce a slight decrease in T_m values. In gel filtration chromatography, all the mutants eluted similarly to the wild type. The results of circular dichroism measurement, differential scanning calorimetry, and gel filtration chromatography indicate that all the mutants folded correctly similar to the wild type. Further, we determined the crystal structure of I122M/K124W to investigate the effect of mutations in detail, because Trp was biggest among the introduced residues (Figure 2.12). The distances of Cα between Ile and Met (residue number 122) and between Lys and Trp (residue number 124) were 0.48 Å and 0.17 Å, respectively. Furthermore, no unfavorable van der Waal's interactions between the side chains of the replaced residues and those of their neighboring residues were observed. These results indicate that the replacement has little influence on the backbone structures and that the assumption about the conformation of I122M/K124W from the simulated model is correct. Further, all the mutants exhibited phagocytosis stimulating activity in the order of I122M/K124T < I122M/K124F < I122M/K124W as expected, and the wild type did not (Figure 2.13). The results confirm that we could introduce the phagocytosis stimulating peptide sequence into the β subunit with the correct folding.

When bioactive peptides are introduced into food proteins, a simulation based on three dimensional structures is a powerful tool to estimate the folding ability of the mutants. In the future, the accumulation of data on three dimensional structures of higher resolutions and on the folding abilities of mutants will make it possible to introduce bioactive peptide sequences exactly into food proteins by protein engineering.

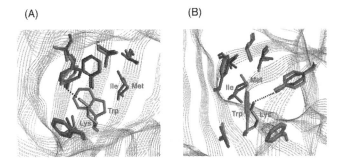

Figure 2.12 Structural comparison of the mutation site of I122M/K124W with corresponding site of the wild type. The Cα traces of wild type and I122M/K124W are represented by lines, respectively. The view in panel (b) is related to that depicted in panel (a) by a rotation of 90°. Dotted line indicates a hydrogen bond. (From: Maruyama, N., Y. Maruyama, T. Tsuruki, E. Okuda, M. Yoshikawa, S. Utsumi, *Biochim. Biophys. Acta* 1648:99–104, 2003.)

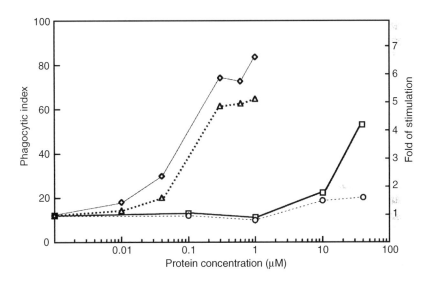

Figure 2.13 Phagocytic index and fold of stimulation of tryptic digests of the wild type and mutant β subunits. Dashed line with open circles, wild type; thick solid line with open squares, I122M/K124T; dashed line with open triangle, I122M/K124F; thin solid line with open lozenges, I122M/K124W. (From: Maruyama, N., Y. Maruyama, T. Tsuruki, E. Okuda, M. Yoshikawa, S. Utsumi, *Biochim. Biophys. Acta* 1648:99–104, 2003.)

2.7.3 Antimitotic Activity of Lunasin

Lunasin is the small subunit peptide of a seed specific methionine rich 2S albumin (Gm2S-1) (74). It consists of 43 amino acids, and contains the cell adhesion motif Arg-Gly-Asp (RGD) followed by eight aspartic acid residues at its C-terminal end. Lunasin was found to arrest mitosis, leading to cell death, when the lunasin gene was transfected and expressed in mammalian cells (8). In addition to arresting cell division, the lunasin peptide caused abnormal spindle fiber elongation, chromosomal fragmentation, and cell lysis when transiently transfected into murine embryo fibroblast, murine hepatoma, and human breast cancer cells. Deletion of the polyaspartyl tail abolished the biological activity of lunasin. The antimitotic activity of lunasin is therefore ascribed to the binding of its polyaspartyl tail carboxyl end to regions of hypoacetylated chromatin. Galvez et al. (10) have also shown that exogenous application of the peptide inhibits chemical carcinogen induced transformation of murine fibroblast cells to cancerous foci. When applied on the skin, lunasin (250 µg/week) reduces skin tumor incidence by 70%, tumor yield per mouse significantly, and delays tumor appearance by 2 weeks compared to the positive control.

It was estimated that commercial soy products contain about 5.48 mg lunasin/g of protein (defatted soy flour) to 16.52 mg of lunasin/g protein (soy concentrate) (10). However, it is not known whether ingestion of soy products at the FDA recommended daily consumption of 25 g of soy protein (75) will be adequate for chemoprevention. Bioavailability studies of natural and recombinant lunasin are needed to determine the physiological doses of lunasin that will be effective for preventing cancer. At present, the biological role of the methionine rich protein GM2S-1 or lunasin in soybean seeds is not known. Its allergenic potential has also not been established. Lunasin with identical structure and anti mitotic properties has also been isolated from barley (76).

Lunasin is presently produced either by recombinant DNA technology or chemical synthesis (10). It may be possible to add this biologically active peptide to one of the storage proteins of soybean based on three dimensional structure to prevent or minimize changes in structural conformation as described earlier. This can possibly increase the amount of lunasin in the soybean seeds.

2.8 FUTURE TRENDS

Understanding and use of three dimensional structures have made it possible for us to increase or add specific nutritional properties, physicochemical functions, and physiological properties to soybean proteins more accurately and rationally by protein engineering. The nutritional quality of soybean proteins can be improved by adding sulfur containing amino acids, while physiological properties can be enhanced or added by the introduction of bioactive peptide sequences. However, to improve physicochemical function, which is most important for usage as a food material, requires the elucidation of the relationship between structure and physicochemical function in detail. This can be done to a certain extent by analysis of a molecular species consisting of a single subunit prepared by means of an *E. coli* expression system and a limited number of subunits by using the mutant cultivars. However, this is still insufficient, as the relationship between the structure and the physicochemical function are different among a species of proteins. It still remains to be elucidated whether there is a general rule on structure and physicochemical function.

Protein engineering is a powerful tool for describing these subjects. If we can elucidate the relationship between structures and physicochemical functions of soybean proteins at a molecular level by protein engineering in the near future, it is expected that we can develop soybean proteins with excellent physicochemical functions in addition to nutritional and physiological properties. This will contribute to solving the food and health problems of the twenty-first century.

REFERENCES

1. Utsumi, S. Plant food protein engineering. In: *Advances in Food and Nutrition Research*, Kinsella, J.E., ed., San Diego, CA: Academic Press, 1992, Vol. 36, pp 89–208.
2. Kito, M., T. Moriyama, Y. Kimura, H. Kambara. Changes in plasma lipid levels in young healthy volunteers by adding an extruder-cooked soy protein to conventional meals. *Biosci. Biotechnol. Biochem.* 57:354–355, 1993.

3. Aoyama, T., M. Kohno, T. Saito, K. Fukui, K. Takamatsu, T. Yamamoto, Y. Hashimoto, M. Hirotsuka, M. Kito. Reduction by phytate-reduced soybean beta-conglycinin of plasma triglyceride level of young and adult rats. *Biosci. Biotechnol. Biochem.* 65:1071–1075, 2001.

4. Kennedy, A.R., P.C. Billings, X.S. Wan, P.M. Newberne. Effects of Bowman-Birk inhibitor on rat colon carcinogenesis. *Nutr. Cancer* 43:174–186, 2002.

5. Kennedy, A.R., X.S. Wan. Effects of the Bowman-Birk inhibitor on growth, invasion, and clonogenic survival of human prostate epithelial cells and prostate cancer cells. *Prostate* 50:125–133, 2002.

6. Malkowicz, S.B., W.G. McKenna, D.J. Vaughn, X.S. Wan, K.J. Propert, K. Rockwell, S.H. Marks, A.J. Wein, A.R. Kennedy. Effects of Bowman-Birk inhibitor concentrate (BBIC) in patients with benign prostatic hyperplasia. *Prostate* 48:16–28, 2001.

7. Meyskens, F.L. Development of Bowman-Birk inhibitor for chemoprevention of oral head and neck cancer. *Ann. N.Y. Acad. Sci.* 952:116–123, 2001.

8. Galvez, A.F., B.O. de Lumen. A soybean cDNA encoding a chromatin-binding peptide inhibits mitosis of mammalian cells. *Nat. Biotechnol.* 17:495–500, 1999.

9. de Lumen, B.O., A.F. Galvez, M.J. Revilleza, D.C. Krenz. Molecular strategies to improve the nutritional quality of legume proteins. *Adv. Exp. Med. Biol.* 464:117–126, 1999.

10. Galvez, A.F., N. Chen, J. Macasieb, B.O. de Lumen. Chemopreventive property of a soybean peptide (lunasin) that binds to deacetylated histones and inhibits acetylation. *Cancer Res.* 61:7473–7478, 2001.

11. Utsumi, S., T. Katsube, T. Ishige, F. Takaiwa. Molecular design of soybean glycinins with enhanced food qualities and development of crops producing such glycinins. In: *Food Proteins and Lipids*, Damodaran, S., ed., New York: Plenum Press, 1997, pp 1–15.

12. Katsube, T., N. Maruyama, F. Takaiwa, S. Utsumi. Food protein engineering of soybean proteins and the development of soy-rice. In: *Engineering Crop Plants for Industrial End Uses*, Shewry, P.R., J.A. Napier, P. Davis, eds., London: Portland Press, 1998, pp 65–76.

13. Derbyshire, E., D.J. Wright, D. Boulter. Legumin and vicilin, storage proteins of legume seeds. *Phytochemistry* 15:3–24, 1976.

14. Wright, D.J. The seed globulins: part II. In: *Developments in Food Proteins: 5*, Hudson, B.J.F., ed., London: Elsevier, 1987, pp 81–157.

15. Maruyama, N., T. Katsube, Y. Wada, M.H. Oh, A.P. Barba de la Rosa, E. Okuda, S. Nakagawa, S. Utsumi. The roles of the N-linked glycans and extension regions of soybean β-conglycinin in folding, assembly and structural features. *Eur. J. Biochem.* 258:854–862, 1998.

16. Utsumi, S., Y. Matsumura, T. Mori. Structure-function relationships of soy proteins. In: *Food Proteins and Their Applications*, Damodaran, S., A. Paraf, eds., New York: Marcel Dekker, 1997, pp 257–291.

17. Thanh, V.H., K. Shibasaki. Heterogeneity of beta-conglycinin. *Biochim. Biophys. Acta* 439:326–338, 1976.

18. Thanh, V.H., K. Shibasaki. Major proteins of soybean seeds: subunit structure of β-conglycinin. *J. Agric. Food Chem.* 26:692–695, 1978.

19. Utsumi, S., H. Inaba, T. Mori. Heterogeneity of soybean glycinin. *Phytochemistry* 20:585–589, 1981.

20. Takahashi, K., H. Banba, A. Kikuchi, M. Ito, S. Nakamura. An induced mutant line lacking the α subunit of β-conglycinin in soybean [*Glycine max* (L.) Merril]. *Breed. Sci.* 44:65–66, 1994.

21. Takahashi, K., Y. Mizuno, S. Yumoto, K. Kitamura, S. Nakamura. Inheritance of the α-subunit deficiency of β-conglycinin in soybean [*Glycine max* (L.) Merril] line induced by γ-ray irradiation. *Breed. Sci.* 46:251–255, 1996.

22. Yagasaki, K., N. Kaizuma, K. Kitamura. Inheritance of glycinin subunits and characterization of glycinin molecules lacking the subunits in soybean [*Glycine max* (L.) Merril]. *Breed. Sci.* 46:11–15, 1996.

23. Yagasaki, K., T. Takagi, M. Sasaki, K. Kitamura. Biochemical characterization of soybean protein consisting of different subunits of glycinin. *J. Agric. Food Chem.* 45:656–660, 1997.

24. Lawrence, M.C., T. Izard, M. Beuchat, R.J. Blagrobe, P.M. Colman. Structure of phaseolin at 2.2 Å resolution: implications for a common vicilin/legumin structure and the genetic engineering of seed storage proteins. *J. Mol. Biol.* 238:748–776, 1994.

25. Ko, T.-P., J.D. Ng, A. McPherson. The three-dimensional structure of canavalin from jack bean (*Canavalia ensiformis*). *Plant Physiol.* 101:729–744, 1993.

26. Lawrence, M.C., E. Suzuki, J.N. Varghese, P.C. Davis, A. Van Donkelaar, P.A. Tulloch, P.M. Colman. The three-dimensional structure of the seed storage protein phaseolin at 3 Å resolution. *EMBO J.* 9:9–15, 1990.

27. Maruyama, N., M. Adachi, K. Takahashi, K. Yagasaki, M. Kohno, Y. Takenaka, E. Okuda, S. Nakagawa, B. Mikami, S. Utsumi. Crystal structures of recombinant and native soybean β-conglycinin β homotrimers. *Eur. J. Biochem.* 268:3595–3604, 2001.

28. Wright, D.J. The seed globulins: part II. In: *Developments in Food Proteins: 6*. London: Elsevier, 1987, pp 119–178.

29. Adachi, M., Y. Takenaka, A.B. Gidamis, B. Mikami, S. Utsumi. Crystal structure of soybean proglycinin A1aB1b homotrimer. *J. Mol. Biol.* 305:291–305, 2001.

30. Adachi, M., J. Kanamori, T. Masuda, K. Yagasaki, K. Kitamura, M. Bunzo, S. Utsumi. Crystal structure of soybean 11S globulin: glycinin A3B4 homohexamer. *Proc. Natl. Acad. Sci. USA*, 100:7395–7400, 2003.

31. Dunwell, J.M. Cupins: a new superfamily of functionally diverse proteins that include germins and plant seed storage proteins. *Biotechnol. Genet. Eng. Rev.* 15:1–32, 1998.

32. Dunwell, J.M., P.J. Gane. Microbial relatives of seed storage proteins: conservation of motifs in a functionally diverse superfamily of enzymes. *J. Mol. Evol.* 46:147–154, 1998.

33. Dunwell, J.M., A. Culham, C.E. Carter, C.R. Sosa-Aguirre, P.W. Goodenough. Evolution of functional diversity in the cupin superfamily. *Trends Biochem. Sci.* 26:740–746, 2001.

34. Anand, R., P.C. Dorrestein, C. Kinsland, T.P. Begley, S.E. Ealick. Structure of oxalate decarboxylase from *Bacillus subtilis* at 1.75 Å resolution. *Biochemistry* 41:7659–7669, 2002.

35. Maruyama, N., R. Sato, Y. Wada, Y. Matsumura, H. Goto, E. Okuda, S. Nakagawa, S. Utsumi. Structure-physicochemical function relationships of soybean β-conglycinin constituent subunits. *J. Agric. Food Chem.* 47:5278–5284, 1999.

36. Utsumi, S., N. Maruyama, R. Satoh, M. Adachi. Structure-function relationships of soybean proteins revealed by using recombinant systems. *Enzyme Microb. Tech.* 30:284–288, 2002.

37. Maruyama, N., M.S. Mohamad Ramlan, K. Takahashi, K. Yagasaki, H. Goto, N. Hontani, S. Nakagawa, S. Utsumi. The effect of the N-linked glycans on structural features and physicochemical functions of soybean β-conglycinin homotrimers. *J. Am. Oil Chem. Soc.* 79:139–144, 2002.

38. Maruyama, N., M.S. Mohamad Ramlan, K. Takahashi, K. Yagasaki, H. Goto, N. Hontani, S. Nakagawa, S. Utsumi. Structure-physicochemical function relationships of soybean β-conglycinin heterotrimers. *J. Agric. Food. Chem.* 50:4323–4326, 2002.

39. Maruyama, Y., N. Maruyama, B. Mikami, S. Utsumi. Structure of the core region of the soybean β-conglycinin α' subunit. *Acta Cryst.* D60:289–297, 2004.

40. Pace, C.N. Contribution of the hydrophobic effect to globular protein stability. *J. Mol. Biol.* 226:29–35, 1992.

41. Yip, K.S., T.J. Stillman, K.L. Britton, P.J. Artymiuk, P.J. Baker, S.E. Sedelnikova, P.C. Engel, A. Pasquo, R. Chiaraluce, V. Consalvi. The structure of *Pyrococcus furiosus* glutamate dehydrogenase reveals a key role for ion-pair networks in maintaining enzyme stability at extreme temperatures. *Structure* 3:1147–1158, 1995.

42. Lim, J.H., Y.G. Yu, Y.S. Han, S. Cho, B.Y. Ahn, S.H. Kim, Y. Cho. The crystal structure of an Fe-superoxide dismutase from the hyperthermophile *Aquifex pyrophilus* at 1.9 Å resolution: structural basis for thermostability. *J. Mol. Biol.* 270:259–274, 1997.

43. Karshikoff, A., R. Ladenstein. Ion pairs and the thermotolerance of proteins from hyperthermophiles: a "traffic rule" for hot roads. *Trends Biochem. Sci.* 26:550–556, 2001.

44. Spassov, V.Z., A.D. Karshikoff, R. Ladensten. The optimization of protein-solvent interactions: thermostability and the role of hydrophobic and electrostatic interactions. *Protein Sci.* 4:1516–1527, 1995.

45. Matthews, B.W., H. Nicholson, W.J. Becktel. Enhanced protein thermostability from site-directed mutations that decrease the entropy of unfolding. *Proc. Natl. Acad. Sci. USA* 84:6663–6667, 1987.

46. Gerhard, V., W. Stefanie, A. Patrick. Protein thermal stability, hydrogen bonds, and ion pairs. *J. Mol. Biol.* 269:631–643, 1997.

47. Müntz, K., V. Christov, G. Saalbach, I. Saalbach, D. Waddell, T. Pickardt, O. Schieder, T. Wustenhagen. Genetic engineering for high methionine grain legumes. *Nahrung* 42:125–127, 1998.

48. de Lumen, B.O., H. Uchimiya. Molecular strategies to enhance the nutritional quality of legume protein: an update. *AgBiotech. News Info.* 9:53N–58N, 1997.

49. Kim, C.S., S. Kamiya, T. Sato, S. Utsumi, M. Kito. Improvement of nutritional value and functional properties of soybean glycinin by protein engineering. *Protein Eng.* 3:725–731, 1990.

50. Utsumi, S., S. Kitagawa, T. Katsube, I.J. Kang, A.B. Gidamis, F. Takaiwa, M. Kito. Synthesis, processing and accumulation of modified glycinins of soybean in the seeds, leaves and stems of transgenic tobacco. *Plant Sci.* 92:191–202, 1993.

51. Takaiwa, F., T. Katsube, S. Kitagawa, T. Hisaga, M. Kito, S. Utsumi. High level accumulation of soybean glycinin in vacuole-derived protein bodies in the endosperm tissue of transgenic tobacco seed. *Plant Sci.* 111:39–49, 1995.

52. Utsumi, S., S. Kitagawa, T. Katsube, T. Higasa, M. Kito, F. Takaiwa, T. Ishige. Expression and accumulation of normal and modified soybean glycinins in potato tubers. *Plant Sci.* 102:181–188, 1994.

53. Katsube, T., N. Kurisaka, M. Ogawa, N. Maruyama, R. Ohtsuka, S. Utsumi, F. Takaiwa. Accumulation of soybean glycinin and its assembly with the glutelins in rice. *Plant Physiol.* 120:1063–1074, 1999.

54. Okuda, E., M. Adachi, B. Mikami, S. Utsumi. Significance of the salt bridge Asp157-Arg161 for structural stability of soybean proglycinin A1aB1b homotrimer. [In preparation]

55. Mori, T., S. Utsumi, H. Inaba, K. Kitamura, K. Harada. Differences in subunit composition of glycinin among soybean cultivars. *J. Agric. Food. Chem.* 29:20–23, 1981.

56. Nakamura, T., S. Utsumi, K. Kitamura, K. Harada, T. Mori. Cultivar differences in gelling characteristics of soybean glycinin. *J. Agric. Food Chem.* 32:647–651, 1984.

57. Mori, T., T. Nakamura, S. Utsumi. Formation of pseudoglycinins and their gel hardness. *J. Agric. Food Chem.* 30:828–831, 1982.

58. Nakamura, T., S. Utsumi, T. Mori. Formation of pseudoglycinins from intermediary subunits of glycinin and their gel properties and network structure. *Agric. Biol. Chem.* 49:2733–2740, 1985.

59. Utsumi, S., A.B. Gidamis, J. Kanamori, I.J. Kang, M. Kito. Effect of deletion of disulfide bonds by protein engineering on the conformation and functional properties of soybean proglycinin. *J. Agric. Food Chem.* 41:687–691, 1993.

60. Adachi, M., E. Okuda, Y. Kaneda, A. Hashimoto, A.D. Shutov, C. Becker, K. M ntz, S. Utsumi. Crystal structures and structural stabilities of the disulfide bond-deficient soybean proglycinin mutants C12G and C88S. *J. Agric. Food Chem.* 51:4633–4639, 2003.

61. Fujita, H., R. Sasaki, M. Yoshikawa. Potentiation of the antihypertensive activity of orally administered ovokinin, a vasorelaxing peptide derived from ovalbumin, by emulsification in egg phosphatidylcholine. *Biosci. Biotechnol. Biochem.* 59:2344–2345, 1995.

62. Takahashi, M., S. Moriguchi, M. Ikeno, S. Kono, K. Ohata, H. Usui, K. Kurahashi, R. Sasaki, M. Yoshikawa. Studies on the ileum-contracting mechanism and identification as a complement C3a receptor agonist of oryzatensin, a bioactive peptide derived from rice albumin. *Peptides* 17:5–12, 1996.

63. Fujita, H., H. Usui, K. Kurahashi, M. Yoshikawa. Isolation and characterization of ovokinin, a bradykinin B1 agonist peptide derived from ovalbumin. *Peptides* 16:785–790, 1995.

64. Fujita, H., M. Yoshikawa. LKPNM: a prodrug type ACE-inhibitory peptide derived from fish protein. *Immunopharmacology* 44:123–127, 1999.

65. Nagata, C., N. Takatsuka, Y. Kurisu, H. Shimizu. Decreased serum total cholesterol concentration is associated with high intake of soy products in Japanese men and women. *J. Nutr.* 128:209–213, 1998.

66. Wong, W.W., E.O. Smith, J.E. Stuff, D.L. Hachey, W.C. Heird, H.J. Pownell. Cholesterol-lowering effect of soy protein in normocholesterolemic and hypercholesterolemic men. *Am. J. Clin. Nutr.* 68:1385S–1389S, 1998.

67. Sugano, M., Y. Yamada, K. Yoshida, Y. Hashimoto, T. Matsuo, M. Kishimoto. The hypocholesteroletic action of the undigested fraction of soybean protein in rats. *Atherosclerosis* 72:115–122, 1988.

68. Sugano, M., S. Goto, Y. Yamada, K. Yoshida, Y. Hashimoto, T. Matsuo, M. Kimoto. Cholesterol-lowering activity of various undigested fractions of soybean protein in rats. *J. Nutr.* 120:977–985, 1990.

69. Makino, S., H. Nakashima, K. Minami, R. Moriyama, S. Takano. Bile acid-binding protein from soybean seed: isolation, partial characterization and insulin-stimulating activity. *Agric. Biol. Chem.* 52:803–809, 1988.

70. Kato, N., K. Iwami. Resistant protein: its existence and function beneficial to health. *J. Nutr. Sci. Vitaminol. (Tokyo)* 48:1–5, 2002.

71. Choi, S.K., M. Adachi, S. Utsumi. Identification of the bile acid-binding region in the soy glycinin A1aB1b subunit. *Biosci. Biotechnol. Biochem.* 66:2395–2401, 2002.

72. Tsuruki, T., K. Kishi, M. Takahashi, M. Tanaka, T. Matsukawa, M. Yoshikawa. Soymetide, an immunostimulating peptide derived from soybean β-conglycinin, is an fMLP agonist. *FEBS Lett.* 540:206–210, 2003.

73. Maruyama, N., Y. Maruyama, T. Tsuruki, E. Okuda, M. Yoshikawa, S. Utsumi. Creation of soybean β-conglycinin β with strong phagocytosis-stimulating activity. *Biochim. Biophys. Acta* 1648:99–104, 2003.

74. Galvez, A.F., M.J.R. Revilleza, B.O. de Lumen. A novel methionine-rich protein from soybean cotyledon: cloning and characterization of cDNA (Accession No. AF005030) (PGR97-103). *Plant Physiol.* 114:1567, 1997.

75. FDA Talk Paper. *FDA approves new health claim for soy protein and coronary heart disease: FDA, United States Department of Heath and Human Services, October 26, 1999.* Washington DC: United States Government Printing Office, 1999.

76. Jeong, H.J., Y. Lam, B.O. de Lumen. Barley lunasin suppresses ras-induced colony formation and inhibits core histone acetylation in mammalian cells. *J. Agric. Food Chem.* 50(21):5903–5908, 2002.

3

Genetic Modification of Plant Starches for Food Applications

Jeffrey D. Klucinec and Peter L. Keeling

CONTENTS

3.1 Introduction 52
3.2 Structure 52
3.3 Synthesis..56
    3.3.1 Granule Bound Starch Synthase 57
    3.3.2 Soluble Starch Synthase 58
    3.3.3 Branching Enzyme 58
    3.3.4 Debranching Enzyme 58
3.4 Modifications 59
    3.4.1 Modification of GBSSI Activity 60
    3.4.2 Modification of SS Activity 61
    3.4.3 Modification of SBE Activity 62
    3.4.4 Modification of DeBE Activity 63
    3.4.5 Modification of Multiple Pathway Enzymes 63
3.5 Functionality...64
    3.5.1 Amylopectin Retrogradation 65
    3.5.2 Amylose Gelation...66
    3.5.3 Gelation of Mixtures of Amylose and Amylopectin....................67
3.6 Applications 68
    3.6.1 Amylose Free Starch..68
    3.6.2 Low Amylose Starch 69
    3.6.3 High Amylose Starch 70
    3.6.4 Amylopectin Chain Length 71
    3.6.5 Novel Starches..72
Acknowledgments...73
References..73

3.1 INTRODUCTION

Starch is a unique natural material, valued for its uses in food, feed, and industry. It is found in higher plants, mosses, ferns, and some microorganisms where it serves as an important store of energy. In higher plants, starch is deposited as transitory starch in leaves and as storage starch in specialized storage organs such as seeds or tubers. Starch is also an important component of many fruit crops such as apple, pear, melon, banana, and tomato. Storage starch is one of the main components of cereal grain (seeds) harvested from crops like wheat, maize, oats, barley, sorghum, and rice as well as of tubers harvested from crops like cassava, yam, and potato. Whether in its native state in grain or tubers, or in isolated granular form, starch is a convenient stable material, cheap to produce, suitable for long term storage without spoilage, convenient for high volume transport, and an important source of calories, retaining functional properties for use in many potential product applications. Grain and tubers are often used directly for animal feed or human food, with little or no processing, such as cooked whole cereal grain or potatoes. Cereal grain is also ground or milled to make flour or meal, which is subsequently mixed with other ingredients and cooked to make breads and pastries. Alternatively, starch may be extracted from the storage organs and the purified starch used as a key functional ingredient added to foodstuffs such as pie fillings, puddings, soups, sauces, gravies, coatings, candies, confectionary products, yogurts, and other dairy products. Extracted starch also has many nonfood industrial uses, such as paper sizing aids, textile sizing aids, molded plastics, ceramics, dye carriers, or suspension aids. Globally, starch is an essential commodity providing most (~80%) of the worlds calories. This vital commodity supply comes from just six different plant species: three cereal crops (rice, maize, and wheat) and three tuber crops (potato, yam, and cassava).

As a result of advances in genetics and biochemistry, we have discovered much about how starch is synthesized in crop plants. Furthermore we have also unraveled the biochemical and genetic basis of some useful natural genetic variations that affect starch synthesis and consequently starch structure and functionality. Some of these variants are already commercially exploited. Examples include variants that accumulate less starch and more sugar (e.g., sweet peas, sweet corn, sweet potato) and others that cook to form clear sols rather than opaque gels (e.g., waxy corn, waxy rice, waxy wheat) and yet others that are useful industrially (e.g., amylose extender corn), and finally others valued for imparting stickiness when cooked (e.g., indica vs. japonica rice). Further progress in this area depends upon improvements in our understanding of the relationship between starch synthetic genes and enzymes, starch structure and functionality. Thus, by linking these findings with further advances in our understanding of the genes required for starch synthesis, an opportunity has appeared for us to make starches with increased usefulness and value.

This paper seeks to pull together the disciplines of biochemistry, genetics, biotechnology, and food technology of plant starches. First we will review current knowledge of starch structure and how starch is synthesized in plants. The primary focus of this review will be storage starch because this provides the main source of food starch today. Next we will summarize the effects on starch composition, physical properties, and functionality due to genetic modifications that cause changes in starch biosynthetic enzymes. Finally we will focus on food applications that might benefit from genetic modifications of crop plants and discuss future opportunities coming from traditional plant breeding and modern biotechnology.

3.2 STRUCTURE

Physically, after extraction and drying, normal starch is a white powder consisting of a mixture of amylose and amylopectin in semicrystalline granules. Starch granules are microscopic structures approximately 0.5 to 100 μm in diameter. In shape, they are spherical, elliptical, or polyhedral. The size and morphology of starch granules is characteristic of the organ and species in which they are produced (102). Starch granules appear rather similar in size and morphology with and without amylose. Under most environmental conditions, starch granules can be considered moderately inert with little capacity to hold water. These characteristics of starch granules make them ideal vessels for storage and shipping, whether in grain or tubers or from processed isolated starch.

Chemically, starch is classified as a complex carbohydrate and is a mixture of two polymers of glucose: amylose and amylopectin. Amylose is a generally linear α-1,4 glucan which is sometimes lightly branched with α-1,6-glycosidic linkages. Amylopectin is normally higher in molecular weight than amylose. It is also an α-1,4 glucan, but is highly branched with α-1,6-glycosidic linkages. The proportions on a dry weight basis of amylose and amylopectin in starches isolated from storage tissues like potato tubers or cereal grain is normally between 20 to 30 percent amylose and 70 to 80 percent amylopectin. In addition to amylose and amylopectin, granules contain small quantities of protein and lipid. Between species there is variation in the structure of amylopectin (104), the size and structure of amylose (84,207,209,211,213), and the nature and amounts of proteins (77,78,155) and lipids (154,214). Because starch physical behavior is dependent on all of these components (55,70,104,116–119) there are specific uses of starches from different species. In addition, within a given species, rare examples have been found of grains, tubers, or roots producing starches that deviate from the typical amylose to amylopectin ratio or have altered amylopectin structure. These plants have been selected because of their unique cooking behavior due to their unusual starch composition that confers unique properties to the crop storage organ. Some of these natural variants are now cultivated on a commercial scale.

Examination of starch structure began over 60 years ago (191). The first widely accepted model shows starch as a branched structure with alternating regions of higher branching density separated by more lightly branched regions (110). A more widely accepted model shows amylopectin arranged in alternating clusters (178,179). Based on the chain length profile of debranched amylopectin and a refinement of the cluster model, the amylopectin chains were categorized into type A, B1, B2, B3, B4 and a single C chain (82). Recently, three refinements for the different modes of interconnection of the amylopectin clusters were presented (82,216) (see Figure 3.1). In starch granules, some of the chains of amylopectin are believed to be associated with one another through hydrogen bonding, forming double helices. The double helices either form higher ordered crystalline structures or may exist independently of crystalline order. The double helices are oriented radially within the granule, with the reducing ends of the chains oriented toward the center or hilum of each granule. Within the granule, crystalline regions, often referred to as growth rings, are separated in a radial fashion from each other by amorphous regions. The crystalline regions are further subdivided into amorphous and crystalline lamellae, which have a periodicity between clusters of approximately 9 nm (105). The branch points in amylopectin are believed to be the primary component of the amorphous lamellae, with the ordered amylopectin side chain double helices clustered in the crystalline lamellae. Differences in the internal chain lengths of amylopectin affect starch crystallinity (163). Important new insights into how amylopectin chain architecture may affect packing have been advocated based on small angle x-ray scattering studies and analogies with liquid crystals (230–233). Using these models it is possible to discuss the mechanisms and kinetics of interchain associations in the context of visualizing starch as a liquid crystalline polymer having different degrees of crystalline order depending on physical conditions.

Amylose contributes to the overall crystallinity of normal starch through the formation of crystalline complexes of amylose with lipids and, it is believed, through participa-

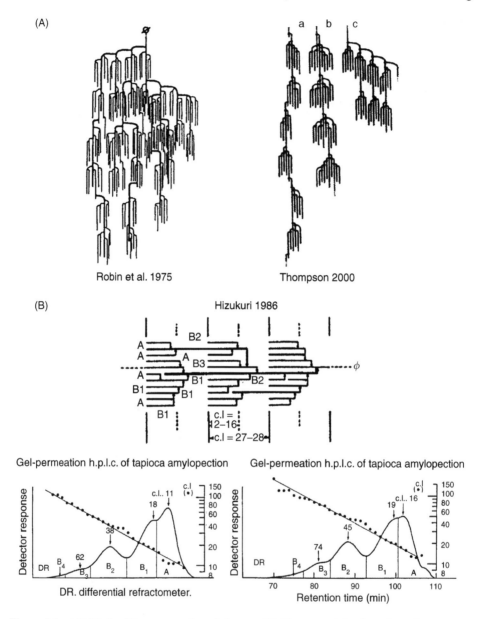

Figure 3.1   (A) Mode of Interconnection of clusters. (B) Cluster models of amylopectin.

tion in some starch double helices. However, although amylose readily forms double helices in solution, it readily leaches out of hydrated starch at granule gelatinization temperatures (68). This observation shows that amylose in starch granules does not associate in the same manner inside granules as it does outside of the granule. This observation additionally raises some intriguing questions about the nature of amylose synthesis and the kinetic trapping of amylose within starch granules.

While normal starch can be readily separated into two components, amylose and amylopectin, studies of high amylose starches have revealed what some authors describe as a third starch component, sometimes termed intermediate starch. For normal starches containing ~25% amylose, the results from quantification by iodine inclusion complex formation

(11,139,181,127), or precipitation in the presence of complexing agents (191,192,126,240,241) are in general agreement. However, for high amylose starches, these methods give different results. Thus, from detailed observations, high amylose starches were initially considered to contain either normal amylopectin and short chain linear amylose (1,2,75) or longer chain amylopectin compared with amylopectin from normal starch (148,149). Liquid chromatographic studies (140) have confirmed the latter. Additional studies of high amylose starches by differential precipitation with 1-butanol combined with Sepharose 2B chromatography (245) revealed an inability to clearly separate amylose from amylopectin. Further fractionation revealed high molecular weight material in what traditionally was the amylose fraction (8,9,209). Later studies showed that this material can be removed by repeated precipitation with 1-butanol and was most likely contaminating amylopectin (209,213), while removal of the low molecular weight material in the amylopectin fraction using differential precipitation techniques has not been successful. In summary, starch from high amylose mutants appears to contain a significant amount of an amylopectin-like component having an altered architecture. This intermediate starch component is characterized by:

1. An inability to precipitate with 1-butanol
2. An ability to eluete within the same molecular size range as amylose
3. An ability to bind iodine and having a lambda maximum between amylose and amylopectin

Estimates of intermediate material defined in this way have exceeded 55% (w/w) of the total starch of high amylose mutants (10).

Other workers (116,209,239) have defined intermediate starch to be the material that has the ability to precipitate in the presence of some complexing agents (e.g., 2-nitropropane) or mixtures of complexing agents (1-butanol with isoamyl alcohol) but not others (e.g., 1-butanol, 1-nitropropane). This type of intermediate material obtained by differential precipitation is a relatively small proportion of normal starches (<10%) (116,209,220,239) irrespective of the amylose content of the high amylose starch. The molecules have been considered by some as amylopectin molecules with long external chains and a limited capacity to form clathrate complexes and precipitate (116,209,220,239). It has also been suggested (209) that this type of intermediate material could be a mixture of amylopectin and a small amount of contaminating amylose, which might be the case if the amylopectin from high amylose starch has an overall molecular size similar to that of amylose from normal starch (116).

Amylopectin molecules within granules are believed to be organized radially, with the long C-chain innermost (84). As a result of high magnification microscopy studies it was proposed that the radially oriented amylopectin clusters are organized into super helices, which may relate to the blocklets seen in microscopy studies (166). In turn the super helices may be responsible for the formation of concentric spherical rings. Although these growth rings are a characteristic of all starch granules, the mechanisms determining their formation are still not well understood (60,170). Recently, Bertoft proposed a bidirectional backbone model, where the super helix could be organized so that the longer amylopectin chains are oriented in line with the super helix, while the amylopectin clusters may be oriented radially (18) (Figure 3.2).

X-ray crystallography has shown that there are three distinct types of crystalline order in starch: A-type, B-type and V-type. V-type crystallinity is often associated with crystalline packing of amylose lipid complexes. The A-type and B-type starches differ in the organization of helices: A-type crystals are densely packed hexagonal arrays of double helices, B-type crystals, though also double helices packed in a hexagonal array, are unlike A-type

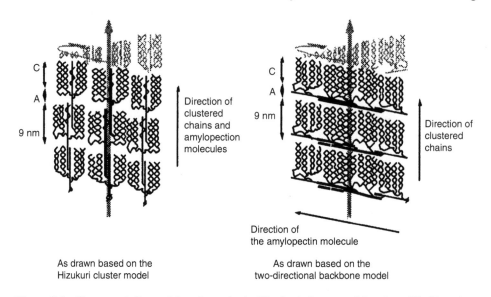

Figure 3.2   The super helix model as drawn in the Hizukuri cluster model and modified based on the two-directional backbone model.

crystals because they have an open central cavity which results in an increased water content (92). C-type starches are a mixture of A-type and B-type crystallinity (22,27,69).

3.3   SYNTHESIS

For an overview of starch synthesis, the reader is referred to recent reviews (99,151,218). In general terms, it is important to note that storage organs are composed of individual starch storing cells and each cell contains several subcellular compartments: including particularly the cytosol and amyloplast compartments. Sucrose, made in the leaves, is transported to the storage organ where it is imported into the cytosolic compartment of each cell. A well characterized pathway (see Figure 3.3) of starch synthesis achieves the enzymatic conversion of sucrose to starch. The first part of this pathway is localized in the cytosol, while the final steps of starch synthesis are located in a specialized subcellular compartment called the amyloplast.

In the cytosol, the glucosyl and fructosyl moieties of sucrose are converted into sugar phosphates. One of these types of hexose phosphates, glucose-1-phosphate is converted into ADP-glucose by the enzyme ADP-glucose pyrophosphorylase (AGPase), the first committed step in starch synthesis. In the endosperm of monocot crops, AGPase is located predominantly in the cytosol, whereas in the storage organs of dicot crops, AGPase is located in the amyloplast. Thus in monocot crops, ADP-glucose is transported into the amyloplast, while in dicot crops hexose phosphate is imported to support ADP-glucose synthesis inside the amyloplast. The conversion of ADP-glucose to starch is performed by several enzymes, which include, but may not be limited to, soluble starch synthase (SS), granule bound starch synthase (GBSS), starch branching enzymes (SBE), and isoamylase (ISA). Other enzymes that also may be involved in starch synthesis include phosphorylase (PHO), pullulanase (PU) and disproportionating enzyme (DisPE). GBSS is involved primarily in amylose biosynthesis and SSs, SBEs, and ISAs are involved in the biosynthesis of amylopectin.

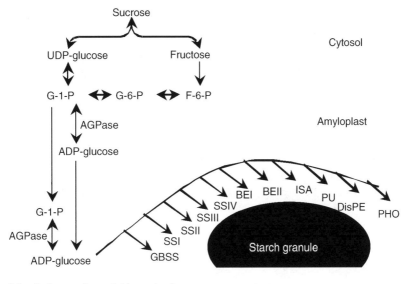

Figure 3.3 Pathway of starch biosynthesis.

Across species, well conserved families of genes encode the enzymes in the pathway. Thus, diverse plant species possess similar sets of multiple forms (isoforms) of each enzyme. The conservation of isoforms suggests that each has a unique role in the process of starch synthesis. This idea has been largely confirmed in several plant species by studies of genetic modifications in which particular isoforms have been overexpressed or reduced.

3.3.1 Granule Bound Starch Synthase

Granule bound starch synthase (GBSS) is a member of the glycosyltransferase class of enzymes more generally known as starch synthases. GBSS is more formally named ADPglucose:1,4-α-D-glucan-4-α-D-glucosyltransferase (E.C. 2.4.1.21). In storage organs there are several starch synthase proteins associated with starch. However, only one isoform (GBSSI) is found predominantly associated with starch, and based on studies of amylose free mutants which lack the GBSSI protein, this isoform appears to be exclusively responsible for amylose synthesis in maize (200) and other plant species (44,88). There is a similar isoform in leaves (229). GBSSI has been identified, mapped and cloned from several species (218). The activity of GBSSI is correlated with the product of the waxy, lam, and amf locus of cereals, pea, and potato respectively (44,88,89,160,200).

As well as associating with starch granules, another key attribute of GBSSI is its ability to elongate a growing glucan chain processively. This means that the enzyme does not necessarily dissociate from the glucan chain after the addition of the first glucose but can remain associated with it to add further glucose units before finally dissociating. Soluble starch synthases enzymes in comparison add single glucose units per encounter with the glucan chain (distributive elongation). This processive mechanism of elongation by GBSSI is consistent with the proposed role of GBSSI in amylose synthesis (45,46). One of the more intriguing aspects of the function of GBSSI is that in isolated granules, its amylose synthesizing activity appears to be dependent on the presence of low concentrations of maltooligosaccharides (45,46). These experiments with isolated granules also showed that GBSSI can elongate glucan chains within amylopectin, a finding that may explain why mutants lacking GBSSI have altered amylopectin structure as well as lacking amylase (83,172,251).

3.3.2 Soluble Starch Synthase

Soluble starch synthase (SS) isoforms are all members of the glycosyltransferase class of enzymes. SS is more formally named ADPglucose:1,4-α-D-glucan-4-α-D-glucosyltransferase (E.C. 2.4.1.21). Like GBSSI, the SS enzymes are also associated with starch granules, but are also present in the stroma of the amyloplast, leading to their classification as soluble starch synthase. SS catalyses the transfer of glucose from ADP-glucose to an acceptor glucan chain, by a mechanism that has been described as distributive (46).

Four SS forms have been identified, mapped and cloned from several species (218). Although the various isoforms are believed to be present in all starch synthesizing cells, they appear to have different relative activities in different species and tissues. SS is believed to be primarily responsible for amylopectin synthesis, and there is evidence that each isoform is responsible for the synthesis of different chain lengths. SSI has been cloned and mapped from several species, but the only known mutant type described so far is in rice (159). SSI appears to be primarily responsible for synthesizing shorter chains of amylopectin based on biochemical studies of the isolated enzyme (38) and structural studies of the rice mutant (159). The SSII and SSIII isoforms appear to be involved in synthesizing the intermediate chains of amylopectin, as evidenced by changes observed with mutant and transgenic rice (40,63,221). The precise role of SSIV remains to be elucidated, and it may even be revealed that SSIV is a subclass of the SSIII isoform. All of these classifications are consistent with evidence obtained from starch isolated from potatoes transformed with antisense constructs to the potato enzymes (51,68).

3.3.3 Branching Enzyme

Starch branching enzyme (SBE) is a member of the glycosyltransferase class of enzymes. The branching enzymes are more formally and collectively named 1,4-α-D-glucan:1,4-α-D-glucan 6-α-D-(1,4-α-D-glucano)-transferase (E.C. 2.4.1.18). Like the SS enzymes, the SBEs appear to have a weak association with the starch granule as evidenced by the fact that these enzymes may be found entrapped inside starch granules. SBEs catalyze the interchain cleavage of a glucan chain with subsequent bonding of the cleaved portion to the parent glucan chain. SBEs act by cutting an α-1,4 linkage forming a new linkage between this cleaved glucan chain and an adjacent glucan chain via an α-1,6 linkage. There appear to be at least two major classes of highly conserved SBEs across plant species (SBEI and SBEII). Thus, like the soluble starch synthases, SBEs appear to have a distinct role in the mechanism of amylopectin structure development. The SBEs must also play some role in amylose synthesis, as amylose is lightly branched. All known SBEs have been mapped, cloned, and sequenced from several plant species. SBEIIb appears to be primarily responsible for transferring longer chains of amylopectin based on biochemical studies of the isolated enzyme (76,212) and structural studies of the maize and rice mutant (162,205,212). The role of SBEI and SBEIIa is less clear as mutants of SBEI in rice have reduced intermediate and long chains (159) while in maize the chain lengths are unaffected in mutants of SBEI and SBEIIa (20,21).

3.3.4 Debranching Enzyme

Debranching enzymes are members of the amylase class of enzymes. There appear to be at least four debranching enzymes in the storage organs: three isoforms of Isoamylase (ISA) and one Pullulanase (PU). Although ISA is known to catalyze the selective cleavage of α-1,6 linkages in branched glucans, their precise catalytic role in starch granule biosynthesis remains unresolved. Plants lacking ISAI accumulate higher levels of sucrose, more starch granules and phytoglycogen and have reduced amounts of starch, strongly indicating that ISAI is required for normal starch biosynthesis (28,29,59,122). Some studies have highlighted the involvement of ISA in starch granule initiation, probably mediated by degradation of soluble glucans which would otherwise initiate granule formation (28,29,122). The role of ISAII and ISAIII in starch biosynthesis remains unresolved at the present time, as no mutants are known to exist. What little is known about their catalytic properties has led to the proposal that each isoform has retained a specific role (90). Studies in maize of the limit dextrinase/pullulanase (PUI) class of debranching enzymes has successfully identified a mutant, which on its own appears to have minimal effects on starch deposition (13,47).

3.4 MODIFICATIONS

Most of the maize starch mutants affecting starch structure were discovered before the genes and enzymes responsible for starch synthesis were known. Their names are therefore based upon the phenotypic changes observed in that storage organ or on the properties of the starch, rather than that of the gene or enzyme affected by the mutation. For example, numerous mutant phenotypes have been reported for maize (73,36) and several phenotypes (e.g., waxy, amylose extender, dull, shrunken, sugary-2, and sugary) have been described extensively with regard to their effects on carbohydrate composition and response to genetic background, allelic dosage, or interaction with other mutations (42,64,65,87). Examination of maize kernels with differing starch phenotypes has been instrumental in determining which enzymes are required for starch synthesis in this storage organ (24,194). Mutations that are responsible for most of the abnormal starch phenotypes have been located in genes encoding starch synthetic enzymes. Related isoforms, for which there are as yet no mutants available, have also been identified and characterized. Table 3.1 summarizes the enzymes

Table 3.1
Summary of individual isoforms of enzymes linked with specific mutants and the effects of these mutations on starch structure (similar genetic modifications affecting these starch biosynthetic enzymes exist in other species).

Enzyme and Specific Isoform	Mutant (Maize)	Effect on Starch
Granule Bound Starch Synthase		
GBSSI	Waxy (*wx*)	Low amylose content.
Soluble Starch Synthase		
SSI	None known	Not known
SSII	Sugary2 (*su2*)	Lacks intermediate glucan chains in amylopectin
SSIII	Dull1 (*du*)	Lacks longer glucan chains in amylopectin
SSIV	None known	Not known
Branching Enzyme		
BEI	BEI (*be1*)	None/Minimal
BEIIb	Amylose extender (*ae*)	High amylose content.
Debranching Enzyme		
ISAI	Sugary1 (*su1*)	Forms compound granules and phytoglycogen
ISAII	None known	Not known
ISAIII	None known	Not known
PUI	Pullulanase1 (*pu1*)	None/Minimal

required for normal starch synthesis in maize kernels, the associated mutant loci and the starch phenotypes of these mutants.

As a result of the great progress made in genetics in the last decade, in addition to the more traditional modifications used in plant breeding, it has become possible to modify or engineer starch synthesis enzymes using modern biotechnology. Collectively this is opening an opportunity to consider using all the tools of breeding, genetics and biotechnology to genetically control starch deposition in plants so as to enhance starch yield and starch quality. Thus, instead of crudely controlling enzyme activity by selecting certain mutants or selecting among different alleles of starch synthesis enzymes using plant breeding, it is possible to more precisely reduce the expression or activity of certain enzymes (such as by using antisense or RNAi technology). Alternatively the enzymes may be overexpressed: an opportunity that is unavailable using conventional plant breeding. What is additionally exciting about enzyme overexpression is that novel enzyme isoforms can be selected in order to act in concert with the existing starch pathway enzymes. With full sequencing of many homologous genes from different plants, we have begun to be able to identify domains within enzymes, which impart particular catalytic properties. This reclassification of enzymes into domain classes is an important new facet of our time, and is generating new understandings of the origins, mode of action and potential for genetic modification of different enzymes. The starch created from such work has the potential to be extremely novel and valuable. However, while the opportunities seem limitless, we are constrained by our limited understanding of the links between starch structural changes and functionality. Furthermore we are significantly constrained by our understanding of what determines the specific catalytic properties of the enzymes.

In the following sections, we will expand upon the properties and roles of the best studied of the enzymes of starch synthesis and describe the effects on the properties of starch of genetic modifications that eliminate or modify the genes encoding them.

3.4.1 Modification of GBSSI Activity

Modifications in granule bound starch synthase activity result in changes in amylose content or amylose structure. In maize, the GBSSI mutant phenotype was named waxy because the intact seed has a waxy phenotypic appearance, unlike normal seed that has a shiny and glossy appearance (194,236). No other enzyme appears to be involved in determining amylose content, as the waxy mutants of cereals (96,157), the amf mutants of potato (88), the lam mutants of pea (44), and GBSSI antisense lines of potato (4,228) show either a reduction or elimination of GBSS activity and a specific reduction of amylose in starch from tubers. Some studies have attempted to restore the production of amylose in amylose free potato plants by transforming the plants with genes for GBSSI enzymes produced by other plants. Between 3.5% and 13% amylose was restored to amylose free mutants of potato by transformation with the cassava GBSSI enzyme (183). Similarly, amylose free potatoes transformed with pea GBSSI isoforms resulted in potatoes with amylose contents of between 0.8% and 1%. Like, the other low amylose potatoes and pea starch, heterogeneity in amylose and amylopectin content was observed within the granules: the granules stained with iodine revealed amylose in concentric rings or having blue-staining granule cores (52).

Extensive work, initially in Japan, has identified wheat mutants lacking GBSSI activity (143,144,145,156,158,255). Miura and Sugawara (144) showed that substitution of genes producing functional GBSSI enzyme with the null alleles could result in starches with a 22 to 23% amylose content rather than the 25.5% amylose content of the normal control. Likewise, Miura et al. have shown that elimination of an active GBSSI enzyme at 2 of the 3 loci in wheat endosperm results in a wheat starch that has an amylose content of 16% to 21% (143). Thus, the presence of a functional GBSSI enzyme from a single locus pair is sufficient to produce a starch with an amylose content of at least 16%. Others (164) have shown that low amylose wheat starches having amylose content between 14.1 and 16.7% can be created through ethyl methanesulphonate (EMS) mutagenesis of the seeds. Amylose free wheat starches were created using triple null combinations of GBSSI mutants (255) and using mutagenesis of a double null wheat known as Ike to generate a nonnull wheat (WO09815621) which stained red when stained with iodine.

Two functional Wx alleles of rice exist: Wx^a, which produces a large amount of amylose, and Wx^b, which produces a smaller amount of amylose. Studies of the effects of the two alleles on the gene expression at the waxy locus in rice (187) showed that the Wx^b allele resulted in an ineffective production of GBSSI enzyme and amylose in japonica rice, while the Wx^a allele produced larger quantities of GBSSI enzyme and amylose in indica rice (186). On a specific activity basis, other authors have shown that the Wx^a allele was less effective in the production of amylose than the Wx^b allele based on analysis of 40 rice varieties (226). It was observed that for two wild-type rice alleles, Wx^a and Wx^b, Wx^b had a GBSS expression tenfold lower than Wx^a at the protein and mRNA levels (97). The decrease in the expression of Wx^b compared to Wx^a was the result of a point mutation within the genetic sequence for the normal rice enzyme (Wx^a allele). The Wx^b allele resulted in the synthesis of a 3.4 kilobase pair mRNA transcript compared to a 2.3 kilobase pair mRNA transcript for Wx^a as a result of the inclusion of an intron into the mRNA sequence as a result of the point mutation (97). Amylose produced from rice plants was related to the ability of the plant to excise the intron 1 from the mRNA sequence (235). Plants expressing high levels of mature mRNA (without intron 1) and no pre-mRNA (containing intron 1) produced the highest levels of GBSSI protein and the highest levels of amylose (20.0 to 27.8% amylose). These were all indica species. With more balanced expression of mature and pre-mRNA, lower levels of GBSSI protein and amylose were observed (6.7 to 16.0% amylose). Both indica and japonica species were within this group. When all of the mRNA contained intron 1 and no mature mRNA was observed, no GBSSI protein was observed and no amylose was detected (235). This pattern relating amylose content to mature mRNA with properly excised intron 1 could be applied across 31 different rice cultivars (235). Thus, based on extensive studies (97,198,235), low amylose rice appears to be the result of a decrease in the amount of normal GBSSI through a mutation which results in problems with mRNA processing rather than due to a mutation in the mature mRNA sequence. However, some differences in the behavior of Wx^a and Wx^b may be present in different rice species.

In recent years a number of patents have been filed covering genetic modifications of granule bound starch synthase in plants, utilizing knowledge gained after cloning, sequencing and transforming plants with the GBSSIs. Examples include WO9211376, US5365016, WO09827212, WO028052, WO02018606, and WO04078983.

3.4.2 Modification of SS Activity

Modifications in starch synthase activity result in changes in amylopectin content or amylopectin structure. Starch synthase may have either a subtle or profound impact on the starch, depending on the activity or inactivity of a specific isoform on the structure and composition of the starch. Although SSI is a relatively minor isoform in potato, it is the predominant isoform of SS in the cereals. Of the starch synthases, the effects of SSIIa and SSIII on starch structure and composition are the best elucidated, especially in maize. Mutants lacking the putative SSIV isoform have not yet been reported. The first reported SSI mutant was in rice (159) and no mutants have yet been reported in other plant species. In rice, the SSI mutant has only a minor effect on starch content and quality. In maize, SSIIa maps to a locus known as Sugary-2 which when mutated produces the sugary-2 mutant. Starch from mutants lacking SSIIa lack certain intermediate amylopectin chains in maize, pea and rice (31,40,221). Wheat, null for the SSIIa enzyme in each of the A, B, and D genomes, has an increase in starch chains with a DP below 10, an elevated amylose content, and a poor x-ray diffraction pattern (246). Barley null for the SSIIa enzyme produces a starch with an apparent amylose content between 50 and 70% of the dry weight of the starch, with a concurrent increase in the proportion of short chains through a DP of 35 (150). The structural gene for SSIII in maize is known to be the Dull1 locus which when mutated gives the dull1 mutant. Starch from the dull1 mutant is relatively lacking in longer amylopectin chains (62). In potato the simultaneous antisense inhibition of SSII and SSIII resulted in a grossly modified amylopectin (51,136), with yet further changes in structure if GBSSI as well as SSII and SSIII are inhibited (108).

In recent years a number of patents have been filed covering genetic modifications of starch synthase in plants, utilizing knowledge gained after cloning, sequencing and transforming plants with the starch synthase enzymes. Examples include WO9409144, US6013861, US5824790, JP06070779, US6130367, WO9720936, EP779363, WO09726362, WO09744472, WO09745545, WO9844780, WO9924575, WO9966050, WO006755, WO031274, and WO112826.

3.4.3 Modification of SBE Activity

Modifications in branching enzyme activity result in changes in apparent amylose content which are believed to be due to an increase in amylose and to changes in amylopectin structure. The first naturally high amylose starch was reported 50 years ago when Vinyard and Bear successfully found a maize endosperm mutant termed *amylose extender* (*ae*) (12,227). Despite extensive research, kernels containing exclusively amylose as the reserve starch polysaccharide have never been found (197,201). The relationship between the Ae mutation and SBE activity was proposed nearly 30 years ago (25). We now know that the structural gene for SBEIIb is the *Ae* locus (205), which when mutated produces *ae* mutant starch having an amylose content of up to 50%. Although the biochemical basis of maize starches having amylose contents above 50% clearly requires the *ae* mutant (SBEIIb), the biochemical basis of the additional increases over 50% amylose is not clear at the present time (201). Furthermore, it does not appear as though SBI is involved, as SBI mutants do not have increased amylose alone or in combination with SBEIIb mutants (20,250). Similarly, SBEIIa mutants (21) had no detectable effects on seed starch structure.

Irrespective of the means of defining or partitioning *ae* starch into amylose and amylopectin components, it is clear that the relatively clear demarcation between amylose and amylopectin existing for normal starch is blurred for *ae* starch, giving rise to a third type of starch termed intermediate starch. Additional evidence for this comes from material collected as amylopectin from *wxae* intermediate starch which has no GBSSI activity and hence produces no true amylose, and has reduced amylopectin branching (61,104,116,135,196,254). Further, the beta-amylolysis limits for *wx* starch, *wxae* starch, and the amylopectins from normal maize and *ae* starches are all between 55 and 61% (259, 260). Thus, both the average exterior chain length and average interior chains length are proportionally longer for the amylopectin from *ae*-containing starches and the double mutant *wxae* starch than for *wx* starch and the amylopectin for normal maize starch. The combination of the absent GBSSI and BEIIb activity in the *wxae* double mutant produces an amylopectin which has sufficient linearity to have apparent amylose content approximately 20% as measured by iodine binding (25).

SBEIIb mutants also exist in rice (162) but there appear to be pleiotropic effects that complicate directly linking SBE activity with increased amylose. Rice SBEI mutants have also been found (189) which, like maize, do not have increased amylose content and do not further increase amylose when combined with SBEIIb mutants. High amylose rice starches appear to have amylose contents between 30 and 50% of the weight of the starch (226). However, with the variability of amylose content of what may be considered normal rice varieties (19,173,210) and the recent implication that other starch biosynthetic enzymes affect the amylose content of such normal rice starch (221) with a few exceptions (146,147,225), it is difficult to know whether deficiencies in SBE are explicitly responsible for some of these high amylose rice starches. In such instances, the rice starch granules have the characteristic changes in starch properties, amylose properties, and amylopectin structure seen with high amylose maize starches (6,225,249). Similar to rice, high amylose barley starches with an amylose content between 30 and 45% of the weight of the starch clearly exist (184,203,252), although it is unknown whether these starches are a result of down regulation of SBE activity or are a result of changes in the expression and activity of other enzymes.

Reports of high amylose potato starch obtained through transgenic down regulation of multiple starch branching enzymes have been published recently (85,107,182,193). Work on development of a high amylose potato starch has been occurring for at least the past 10 years. Amylose contents as high as commercially available maize starches have been obtained in potato with a decrease in the overall molecular weight distribution.

In recent years a number of patents have been filed covering genetic modifications of branching enzymes in plants, utilizing knowledge gained after cloning, sequencing and transforming plants with the SBEs. Examples include US6013861, WO9211375, WO9214827, WO09507355, WO9634968, WO9722703, WO9720040, WO9820145, WO9837214, WO9837213, WO9914314, WO9964562, WO015810, WO031282, WO022140, JP0621767, JP06098656, JP05317057, and JP04088987.

3.4.4 Modification of DeBE Activity

Modifications in debranching enzyme activity can result in significant changes in starch granule structure. ISAI mutants accumulate starch in compound instead of simple starch granules (compound refers to amyloplasts containing many small granules, while simple refers to amyloplasts containing one major granule) and sometimes also accumulate phytoglycogen (a highly branched nongranular storage product). Mutants known in ISAI include the sugary1 locus in maize and sugary of rice (100,122) and in barley by lines named *Riso17* and *Notch-2* (28). Similar results were observed using antisense technology to reduce ISAI activity in rice (59). In potato where antisense constructs for ISAI and ISAII were combined, the tubers accumulated large numbers of small granules (28,29,122). Mutations in PUI have been identified (47), but effects on starch content are minimal. Modifications in the other isoamylase (ISAII and ISAIII) have not yet been identified, although there is some evidence that they each play distinct roles in starch synthesis (90). The maize sugary mutants are important because they are one of the main sources of producing sweet corn.

In recent years a number of patents have been filed covering genetic modifications of debranching enzymes in plants, utilizing knowledge gained after cloning, sequencing and transforming plants with the ISA and PU enzymes. Examples include WO9202614, WO09504826, WO09619581, US5750876, WO9603513, US5912413, WO09732985, WO09742328, WO9850562, WO09906575, WO9912950, WO09958690, and WO0001796.

3.4.5 Modification of Multiple Pathway Enzymes

By eliminating multiple starch biosynthesis enzymes, other alterations of the starch biosynthetic pathway can occur resulting in even more novel starches. Several patents exist on the creation and use of such starches (US4428972, 4615888, 4767849, 4789557, 4789738, 4801470, 5009911, and 5482560). More recently, several patents and published applications have described the production and utilization of heterozygous combinations of mutations in the starch biosynthetic pathway to obtain commercially useful starches (WO9535026, US5356655, US5502270, and 5516939). The production of many of these starches involves the use of double or triple mutant plants. Due to the number of mutations required to sufficiently alter the starch (at least 2 or 3 within a single plant) many of these starches are difficult and costly to produce commercially, so many of these starches from plants with mutations in the starch biosynthetic pathway are uncompetitive with chemically modified starches. Further, these combinations of two or more mutations, whether they are combined homozygously or heterozygously in the plant endosperm, rely on the alteration of the structure of amylopectin from normal or waxy starch.

3.5 FUNCTIONALITY

The origin of wide angle x-ray scattering (WAXS) diffraction patterns from amylose fibers was determined to be due to the crystalline packing of double helices of starch chains (242,243). This work was further developed to explain the crystallinity of starch granules. When starch granules are heated sufficiently in excess water, the WAXS diffraction pattern is lost (257) and an endothermic transition is observed (39). Double helical order is additionally lost during this endothermic event, and it is this loss and not the loss of crystalline order that is believed to be responsible for the endothermic event (39). This loss of granular, crystalline, and double helical order is called gelatinization, and it is irreversible with respect to the three dimensional granule organization (7). Subsequent to gelatinization, starch chains of both amylose and amylopectin may organize into new double helical and crystalline structures in a process called retrogradation (7,70,176).

In its native granular form mixed with water, starch makes a high solids mixture, useful in making batters and doughs. During heating, starch granules swell, associating with water up to 30-fold its dry weight, to form a hydrated starch. This process is reversible provided temperatures and pressures are lower than those that lead to disruption of the organized intramolecular glucan chain associations within the granule. Above these temperatures and pressures, the granules will undergo irreversible changes and glucan chains will dissociate, permitting further swelling. With additional heating and applied shear forces, swollen hydrated granules will eventually collapse to form an unorganized paste of starch molecules. Such hydration, with heating, results in a thickening effect, imparting texture and structure. In functional terms, this process of starch granule swelling and dissociation is known as gelatinization (7,18,214,238). The glucan polymers of a gelatinized starch are able to associate with each other or other components of food to impart additional character to the food system including stickiness, tackiness or a rubbery texture (258). In addition, amylose readily leaches out of the granule during gelatinization and interacts with the other food components (17,53,165). Under mild cooking conditions, amylopectin does not so readily leach from the granule and hence is enriched inside the gelatinized granule. However, in more extreme cooking conditions, the amylopectin is much more dispersed, resulting in products with a tacky or sticky character.

Once a cooked starch is allowed to cool, the glucan chains begin to reassociate in a process resulting in a change in the functional characteristics, including a decrease in paste clarity and gelation of the paste. In functional terms, this process of reorganization is known as retrogradation (7,70,176). Normal starches are generally recognized for their ability to retrograde within hours (175) and usually form opaque gels (41). Retrogradation of amylopectin gels occurs during days or weeks of storage. As a consequence of their different molecular weights and chain length profiles, the rates of retrogradation of amylose and amylopectin are not the same. The rapid setting of the structure of breads is believed to be due to rapid (within seconds or minutes) amylose retrogradation to form a network structure. Starch functionality is therefore a consequence of the degree of gelatinization and is influenced by retrogradation, time, temperature, concentration, and the presence of other food components or additives. In addition, modifying starch using chemical, enzymatic, or physical treatments alters and extends its functional properties.

Measuring starch functionality and applying it to food applications is problematic because the results are subject to extrapolation to systems and processes which are far more complex than laboratory testing is able to emulate. Analytical instruments (e.g., Differential Scanning Calorimetry, DSC) are frequently used to quantify the temperature range and amount of energy needed to melt crystalline starch. The amounts and molecular size of amylose and amylopectin may be measured using gel permeation or size exclusion chromatography. Granule size and shape are measured microscopically or by light diffraction techniques, and granule viscosity is measured using various rheometers. Rheological measurements may include various temperature and shear programs that attempt to mimic thermal treatments, pumping, and shearing forces that occur during food processing. Such measurements of texture provide information on adhesiveness, cohesiveness, yield stress, viscous flow, and rigidity of starch sols and gels.

3.5.1 Amylopectin Retrogradation

Because retrogradation manifests itself in many ways and over many different spatial scales, establishing relationships between the structural orders and the physical properties of the starch are complicated. Thus, a number of techniques are required for such an assessment. NMR has been used to probe the most fundamental physical order of the starch: double helical order (71), and x-ray diffraction has been used to examine the crystallinity of starches (114). However, neither of these two measures of starch order adequately represents the rheological properties of a cooked starch paste because rheology is dependent on not just the order of the starch but also the larger scale interrelationships of this order with the covalent structure of the starch molecules themselves. There are additional problems that make relationships between low level molecular order and larger scale order difficult to establish: residual granular order (41,141,153) or polymer incompatibility (49,66,112) may result in macromolecular heterogeneity in the system. To avoid these complications, many investigators interested in starch retrogradation have examined the behavior of waxy-type starches (16,30,57,134,135,144,154,175,176,195,253). Likewise, the properties of amylopectin free amylose have been studied either by preparing synthetic amylose using glucose-1-phosphate and the enzyme phosphorylase (35,68,70,180), or by studying highly pure amylose fractionated from the native starch mixtures of amylose and amylopectin (116,126,209,239). Amylopectin retrogradation has been examined by multiple techniques including DSC (57,95,137,169,196,254), WAXS (175,176,217), turbidimetry (98,176), and rheology (16,30,55,111,117,176,253).

The length of a double helix, whether in a starch granule or in a retrograded starch, is limited by the length of the external chains participating in the double helix. Linear chain lengths of amylopectin are generally shorter than the DP 40 to DP 70 lengths observed for amylose double helices (103,129,138). Amylopectin double helices are generally observed to dissociate below 70°C by DSC (39) when the starch is in excess water (<30% starch). The DSC endotherm of retrograded amylopectins containing a higher proportion of longer chains than normal amylopectin (i.e., amylopectin from some mutant starches) may extend above 100°C. The endotherms from these mutant starches are also broader than an endotherm observed for retrograded amylopectin from normal starch, perhaps indicating that amylopectin with longer external chains has an ability to form a broader distribution of double helical lengths than amylopectin with shorter chains. The absolute lower limit for the length of a stable double helix appears to be six anhydroglucose residues (69), the length required to complete one turn of a chain in a double helix (242,243). Practically, the lower limit appears to be closer to ten anhydroglucose units, given that in systems of a pure oligosaccharide, chains with a length (or degree of polymerization, DP) less than ten anhydroglucose units could not self associate into stable double helices and crystallize (69). However, this minimum appears to be somewhat dependent on the lengths of other chains available for double helix formation, because in mixed systems of short chain oligosaccharides chains longer than a DP of 6 may form double helices with longer chains.

As with double helical and crystalline formation with oligosaccharides, the ability for amylopectin chains to retrograde is closely related to the average external chain length (ECL) of the molecules. With stubs of DP 1, 2, or 3 anhydroglucose residues as external chains, β-limit dextrins of amylopectin do not retrograde after one month of storage at 4°C (176). Shortening the average ECL by just three anhydroglucose units (from 14 to 11) by alpha-amylase was sufficient to reduce the enthalpy and solid character (assessed by DSC and pulsed NMR, respectively) to 10% of those of the native starch (12,244). Chromatography of the modified starches was instructive, because it showed that the component chains of the starch do not all need to be equally shortened to have an effect on starch retrogradation: some chains were reduce to a DP less than 5, with the remainder remaining largely unchanged (12,244).

Examination of starches with intact chains (135,196,254) has indicated that starches with high proportions of chains below a DP of 11 to 12 have a slower rate of retrogradation than those starches with fewer short chains, irrespective of concentration. The retrogradation rate of starches with higher proportions of shorter chains is also more concentration dependent than for those starches with fewer short chains (57,135). However, conclusions from these studies must be tempered by any possible confounding effects of additional changes in the fine structure of the amylopectin (e.g., branching pattern).

3.5.2 Amylose Gelation

Amylose retrogradation has been monitored by the development of supramolecular organization using turbidimetric and rheological methods (35,48,56,70,141,142). Gels of amylose have been suggested to develop within minutes through the retrogradation of amylose into double helices (35,70) or through an initial phase separation of amorphous amylose, followed by retrogradation into double helices in these concentrated amylose regions. The aggregation event is followed by an additional retrogradation event: the development of double helical crystallites within hours (91,128,141,142). Increasing the chain length and concentration of the amylose both have a destabilizing effect on amorphous amylose in solution (70).

Destabilization of amylose may lead to retrogradation in the form of gelation, precipitation, or both. The relatively long, stable double helices of amylose result in more thermally stable gels than gels of amylopectin. In the formation of amylose gels, one outcome of amylose retrogradation, gel development is understood to be related to the ability of a single molecule to interact in double helical junction zones which bind molecules together into a three dimensional network structure, and the subsequent ability for the double helices to either orient or aggregate once formed (70,128). For synthetic, completely linear amylose, the network formation is possible for amylose molecules with a DP above 100 because these molecules are able to interact with more than one additional molecule to form a network (70). Precipitation, the other outcome of amylose retrogradation, is favored over gelation for amylose molecules below a DP of 250, gelation is favored over precipitation for amylose molecules above a DP of 1100, and both network forming (gelation) and breaking (precipitation) retrogradation phenomena are similarly favored for amylose molecules with a DP of 440 and 660 (35). The concentration of amylose also influences whether amylose will gel or precipitate. Higher concentrations of amylose result in an increased frequency for chain to chain interactions, which would both increase the rate of gel development and decrease the time required for the onset of a three dimensional network.

Amylose gels prepared in water have been described as phase separated systems consisting of a solvent rich phase and a filamentous polymer rich phase consisting of bundles of amylose chains (74,91,129,222). Portions of amylose chains that participate in the double helices residing in the crystalline regions of amylose gels have been estimated to range in length between DP 40 and DP 70 (103,128,137,138). As measured by DSC, the double helices of amylose appear to dissociate over a very broad temperature range (~50°C) with an endothermic peak temperature between 120°C and 160°C (23,37,117,129,202).

3.5.3　Gelation of Mixtures of Amylose and Amylopectin

Understanding the retrogradation behavior of normal starches, considered as mixtures of amylose and amylopectin, is even more complex than understanding amylose or amylopectin retrogradation in isolation. This is due to the possibilities for phase separation of the two types of molecules and for interactions between amylose and amylopectin to develop.

Cooked normal starches have been considered phase separated systems of swollen granules consisting of amylopectin trapped within a continuous gel matrix of leached amylose (141,153,167). In these systems, the amylose and amylopectin behaviors appear to be relatively independent of each other: the amylose forms a thermally irreversible (at <100°C) gel network within hours, followed by a thermally reversible increase in gel elastic modulus attributable to an increase in granule rigidity due to amylopectin retrogradation (113,141,167). In the absence of granular order, amylose and amylopectin mixtures have been shown to phase separate over time (66,111). Examination of gels of artificial mixed systems of amylose and amylopectin showed that over a small increase (10–15% w/w) in the amylose content of gels (from 25 to 40% amylose on a total starch basis), the gels developed the behavior of pure amylose gels in their rate of turbidity development (49), their elastic modulus, and their susceptibility to enzymatic and acid hydrolysis (128). Both groups suggested that a phase inversion from an amylopectin to an amylose continuous phase at an amylose to amylopectin ratio of about 30:70 or about 17:83 was the cause of the observed differences in gel characteristics at high and low amylose to amylopectin ratios. The differences may have been with the starches and concentrations used, with the methods of preparation of both the amylose and amylopectin prior to mixing, and the method of mixing the amylose and amylopectin. From examination of the fracture of gels of jet cooked high amylose maize starches prepared between temperatures of 121°C and 166°C, it was hypothesized (33) that higher processing temperatures resulted in more homogeneous mixtures of amylose and amylopectin than processing at lower temperatures, and that the more homogeneous mixtures would take longer to phase separate and gel during cooling at a constant rate. The delay in phase separation during cooling was thought to result in phase separation and gelation at lower temperatures, resulting in weaker gels due to the more rapid formation of a less perfect gel network.

Others have suggested that amylose and amylopectin interact more extensively than suggested by the relatively independent behavior of mixtures of phase separated amylose and amylopectin (101,117,168,190) and that the short chains of branched molecules interfere with the molecular association of amylose molecules (23,117). In gels prepared from reconstituted mixtures of amylose and amylopectin, the chain length distribution of the amylopectin, the size of the amylopectin, and the amylose content appear to influence the formation and properties of starch gels. From examination of gels formed by waxy maize starch (considered amylopectin) and oligosaccharide chains with a DP of 21 or 35, linear chains as short as DP 21 appear sufficient to form junction zones with multiple amylopectin chains resulting in a gel (190). Acetylation, substitution of the hydroxyl groups of the starch with structure-inhibiting moieties, also appear to destroy the gel-forming ability of the amylopectin, and the addition of amylose instead of the oligosaccharides appears to result in stronger gels. Substitution was suggested to limit associations preventing gelation, but was not sufficiently high to prevent gelation in the presence of amylose or short, linear chains (190). Using a variety of amylopectin sources with different chain lengths and a consistent source of amylose and ratio of amylose to amylopectin ratio (20:80), the strongest gels appear to form from the mixtures containing the amylopectin with the longest chains (101). Using 25:75 or 50:50 (w/w) mixtures of amylose with either *wx* starch or small, medium, or large dextrins prepared from waxy maize starch, gels prepared with the medium sized dextrins, irrespective of amylose content, formed weaker gels than gels prepared from the other dextrins or the *wx* starch (168). Examination of starches by DSC shows that amylose appears to retrograde independently when the starch is heated to temperatures below 140°C, but does not do so when heated to temperatures above 160°C in an excess water environment (23). Further, dynamic shear rheology of these same starches indicated that the length of these double helices formed between amylose and amylopectin was dictated by the length of the external chains of the amylopectin: when gels prepared from mixtures of completely dispersed starches were heated, the decrease in the elastic modulus paralleled the decrease in the elastic modulus for the amylopectin isolated from the starch (117).

3.6   APPLICATIONS

As well as providing essential calories, starches from different crops play an important role in foods, such as improving processing, shelf life, consistency and appearance by providing texture and thickening capability for suspending solids. There are myriads of applications of starches in foods and these uses have been extended due to application work by food technologists using physical and chemical modifications (designated by E-numbers in the EU). These include oxidation (E1404, E1451), monostarch phosphorylation (E1410), distarch phosphorylation (E1412), acetylation (E1414, E1420, E1422), hydroxypropylation (E1440, E1442), and octenylsuccination (E1450). In this section, we will focus on starch applications of various genetic modifications of the major crop plants and how these modifications impact the uses of these starches in foods. In particular we will emphasize modifications in amylose and amylopectin content and molecular weight, amylopectin chain length, starch granule size and morphology, crystal structure and phosphate content. We will conclude with a discussion of new opportunities for enhanced starches resulting from new genetic modifications of plants.

3.6.1 Amylose Free Starch

The most widely used amylose free starches are obtained from waxy maize and rice and in all cases are considered non-GM natural variants of normal maize and rice. Staining starch with iodine readily identifies amylose free types: normal starch will stain blue or purple whereas the starch produced from amylose free plants will stain red or brown or brownish red in color. Amylose free maize starches are generally considered to contain zero or almost zero amylose. Amylose free rice starches have been shown to contain between 0 and 3% amylose, though collectively these starches are referred to as amylose free or waxy rice starches (109,173,185). With these amylose free rice starches, it has been assumed that the differing cooking and paste properties are due to differences in the structure of the amylopectin of the starch rather than variations in the low levels of amylose of the starch (234). The effects of amylose and other molecular and compositional characteristics of rice starches on rice (19,34) or rice starch properties remain unclear (125).

Amylose free starches have useful functionality that has encouraged their commercial development. They are considered useful as water binders, viscosity builders, and texturizers in food as well as in industrial applications (171). However, these starches are less resistant to shear, acid, and high temperatures than are normal starches, and extended cooking results in stringy, cohesive pastes. Amylose free starches are generally recognized for their improved transparency after processing compared to normal starches (41) and have better freeze and thaw stability compared to normal starches once cooked (171,237). They are also recognized for their improved long term storage capability as they require weeks to gel if they could be considered to gel at all (15,253). To correct for some of the negative paste attributes of amylose free starch, such as poor stability to temperature, shear and acid and undesirable paste quality, most amylose free starches are chemically modified by substitution, cross linking, or both (171,238,256).

In recent years there has been significant interest in developing amylose free starches in other crops so as to take further advantage of any species specific qualities of the starch produced from that species. However, although mutants are readily found in some plant species (such as maize, rice, and barley) this is more difficult in other plant species where there are multiple copies of each gene (such as the polyploid species like potato, oats, and wheat). In the case of wheat, the advances have been made by screening for mutants (non-GM), while in potato the waxy types were made using biotechnology (GM). Applications of amylose free wheat and potato are at present still being developed. However, the most likely applications for amylose free potato include the paper, adhesive, textile, and packing industries. In the EU, certain GM varieties of modified amylose free (high amylopectin) potato have not been approved thus far (http://europa.eu.int/comm/food/fs/sc/scp/out24_en.html), while others appear to be in the approval process (http://europa.eu.int/comm/food/fs/sc/scp/out129_gmo_en.pdf). Amylose free wheats are finding applications in foods such as noodles and baked goods including breads (5,247).

3.6.2 Low Amylose Starch

Although staining starch with iodine readily identifies amylose free starches (161), care must be taken when using this quantitatively to adequately account for the iodine binding capacity of the amylopectin (120). For example, amylopectin from an amylose free plant having an inactive GBSSI enzyme might appear to contain 5% amylose based on iodine binding, blue value, measurement (116). After carefully considering these potential problems certain low amylose starches have been identified. For example, in the early 1940s, a waxy maize mutant (wx^a) was discovered in two exotic Argentinean small seeded flint varieties that contained a starch that had an amylose content of 2.4% and stained a pale violet color with iodine (26). Additionally, the amylose content of the starch increased from 0% (full waxy) to 0.65% to 1.3% to 2.4% (full wx^a) with increasing dose of the wx^a allele (26,204). With these same crosses, the viscosity of starch pastes decreased with increasing dose of the wx^a allele. The wx^a allele was described as resulting in a 95% reduction in the amount of GBSSI protein produced and a starch with a low amylose content (174,204,261).

Wheat starches have been produced with amylose contents of about 7.5% and 13.5% by crossing normal wheat with amylose free wheat (188). Peak viscosities of all starches differed by less than 20% of the peak viscosity of the amylose free wheat starch, with the low amylose starches having a higher peak viscosity than both normal and waxy wheat starch. The gelatinization temperatures and enthalpy were highest for waxy wheats and decreased in the order waxy > 13.5%, amylose wheat > 7.5%, and amylose wheat > normal wheat starch. The retrogradation temperatures and enthalpy were insignificantly different for amylose free wheat, normal wheat, or any of the low amylose wheat starches. Starch granules extracted from a wheat strain derived from mutagenized Tanikei A6099 had an apparent amylose content of 1.6% and stained dark brown with dark cores compared to red-staining waxy wheat starch with 0.4% apparent amylose (115). The Tanikei A6099 mutant wheat starch had a higher initial pasting stability than an amylose free wheat starch (0.4% amylose). However, the viscosity of the low amylose starch paste decreased dramatically, to the same viscosity as the amylose free wheat, during continued cooking and remained at the same viscosity as amylose free wheat after cooking. The mutagenized Tanikei A6099 wheat is known to produce a mutant GBSSI enzyme (115,248).

By screening microtubers from plants exposed to x-ray radiation an amylose free mutant of potato was identified (88). Potato starches considered amylose free have been shown to have an amylose content varying from 0 to 7.9% (183,204,223,224). The amylose free potatoes were null for the GBSSI enzyme. Staining with iodine gave varied results: sometimes the starch stained red and other times reddish brown and blue. These results are indicative of a heterogeneous mixture of amylose free starch and amylose containing starch of unknown quality within the potato tuber. In further attempts to understand the link between function and activity of GBSSI in potato, antisense transgenic plants having amylose contents between 3.0 and 8% were produced (123,224,228). These tubers had both blue and red brown staining portions (228) again indicative of heterogeneous mixtures of amylose free starch and amylose containing starch of unknown quality. Others (123) observed additional heterogeneity at a granule level, with starch granules having blue cores surrounded by a red brown shell of starch. The size of the blue core appeared to be correlated with the amylose content of the starch. Starch extracted from plants produced from crosses between an amylose free potato and a normal potato had no linear correlation between GBSS activity and amylose content (58). Additionally it was observed that the swelling and rheological properties of the granules could not be clearly linked with amylose content (183).

The waxy barley starches have been shown to contain up to approximately 5% apparent amylose (215). However, this apparent amylose is due to a mixture of starch granules within the barley seeds. The amylose content of the granules typically ranges from an undetectable level up to approximately 10%, with the granules closest to the surface of the seed having the highest amylose content (3). Recent work with waxy barley starch (with amylose contents up to 6.44%) shows delayed peak viscosity development and viscosity varying during cooking under shear (132). Additionally, all of the waxy barley starches began to develop viscosity at a similar time and temperature in the cooking process.

Low amylose rice starches have been shown to have amylose contents between 7 and 15% (124). Shimada et al. (199) produced several antisense rice plants with starch having an amylose contents between 6 and 13%. The iodine staining qualities of these starch granules were not reported. Further, any cooking properties of the starches, the elastic properties and gelling abilities of pastes and the gel properties of gels produced from these low amylose rice starches produced by transgenic rice plants are unknown.

3.6.3 High Amylose Starch

Commercialized high amylose maize starch is a result of down regulation of the maize BEIIb enzyme, utilizing the *amylose extender* mutation. High amylose maize starches bring important differentiated properties for food applications, having amylose contents between 30 and 90% of the weight of the starch. The starch imparts gelling ability to the food system, improving adhesion to water impermeable surfaces and altering product texture. These starches also have improved film forming ability and improved fat impermeability compared to the normal starch counterparts. High amylose starches provide firmness, extend cooking times and increase the crispiness of coatings.

Because of their high resistance to processing and subsequent digestion, in addition to their rapid retrogradation if gelatinization occurs, high amylose maize starches are also being utilized as a source of resistant starch in a number of food products. Reviews (43,79,93) provide an overview of resistant starch and its valued physiological benefits in foods. Commercialized resistant starches include unprocessed starches from different botanical sources (e.g., green banana, legumes, potato) as well as more highly processed forms of high amylose starches. High amylose starches from other sources are also being advocated for their resistant starch properties.

The most widely available high amylose starches originate from maize using the *amylose extender* mutation that results in a starch having 40–50% amylose. As a result of extensive breeding and selection work higher amylose contents (up to 90%) have been achieved (201). High amylose starches are also available in barley and rice.

3.6.4 Amylopectin Chain Length

As already stated, care must be taken when using the iodine binding assay for quantifying amylose content. In a more extreme example of this problem, the combination of the absent GBSSI activity and the *ae* mutation produces an amylose free maize starch that stains blue or purple and appears to have an amylose content of 15–26% (194). This is because the *ae* mutation causes a decrease in starch branching enzyme activity (24), which results in the formation of long chain amylopectin (104), which will itself stain blue with iodine. Other examples of this come from genetic modifications that affect amylopectin structure.

Loss of SSIIa activity in maize results in a starch with an amylose content near 40% of the total starch weight. This starch develops viscosity very slowly at temperatures above 90°C and forms stable gels that strengthen only very slowly compared to normal starch gels (31). The loss of SSIIa activity resulting in elevated amylose contents highlights the often observed disconnect between the viscosity development of the starch and the thermally detected gelatinization of the starch. Despite the resistance of the starch to develop viscosity, likely a consequence of the elevated amylose content, loss of SSIIa activity results in a decrease in the gelatinization temperature range of the starch from about 70°C to 80°C for normal starch to approximately 55°C to 65°C (31,21). Examination of the chain length distribution of the maize starch in combination with the absence of amylose, as a result of the inactivity of GBSSI, indicates that the starch has a elevated proportion of short component chains below a DP of 30 compared to normal starch (212) and additionally an elevated proportion of very short component chains below a DP of 10 compared to normal starch (94,134). This high proportion of very short chains imparts a decreased tendency to retrograde compared to normal starch (134,135).

Recently, potatoes have been engineered to eliminate both SSIIa and GBSSI activity, resulting in an increase in short chains below a DP of 14 compared with normal starch (108). This had the benefit of a decrease in the tendency for the starch to retrograde after cooking which may have implications for improved freeze and thaw tolerance. Potatoes engineered to have reduced SSIII activity had decreased amounts of chains longer than DP 17 (60).

The SSIIa enzyme of rice has been implicated as one of the major enzymes that affects whether the grain is of the indica-type vs. the japonica-type (221). Indica-type rices have been long known to have higher kernel integrity and higher granule stability than japonica-type rices (131). The properties of indica rice starches have been attributed to their higher proportion of longer chains than japonica-type rice starches.

It has been known for some time that the inactivation of SSIII in the dull mutant of maize results in an elevation in the apparent amylose content to 30 to 40% (81). Beyond this, a number of chromatographic studies on dull starches, including those additionally lacking GBSSI enzyme activity, indicate that the chain distribution of maize starch is only slightly different from normal starch, with some elevation in the chains with a DP less than 30 to 50 (94,196,254). The beta-limit dextrins, starches with the exterior chains digested to stubs of maltose to maltotriose, provide a more complete picture of the changes observed with elimination of SSIII in maize. In this case the lengths of the residual chains containing all of the branching are considerably shorter when SSIII is absent during synthesis. It has been suggested (135) that this change has implications for the retrogradation of starch produced in the absence of SSIII. Thus, during gelatinization and then continued heating after gelatinization, the starch chains are less able to randomly orient themselves compared with starches having longer spacing between branch points (i.e., normal or waxy starch). Thus, the retrogradation rate of *du wx* starch is less concentration dependent than *wx* starch despite the similar chain length distributions of the two starches after debranching.

3.6.5 Novel Starches

Recent developments in biotechnology are opening ways of making novel starches in plants. Thus starches obtained from different crops having varied functionalities due to differences in composition may now theoretically be readily transferred into crops from one or another different plant species. In recent years a number of patents have been filed covering novel genetic modifications of starch in plants. Examples include WO09711188, WO008184, WO09961580, WO9814601, WO9404692, WO09601904, WO09729186, WO014246, WO9924593, WO9839460, WO047727, WO011192, WO9929875, WO9747806, WO9747807, WO9747808, and WO9822604. In general we can consider making changes in any aspect of amylose/amylopectin, phosphate, protein, phospholipids, crystallinity, gelation, and pasting characteristics, flavor and starch granule morphology. One merely has to understand the genetic basis of the differences.

One interesting example of a cross species difference that may be exploited is phosphate content in potato. Recent work (177) has shown that starch phosphate content may be attributed to a novel glucan water dikinase enzyme. This discovery opens up the possibility of creating novel genetically modified starches having varied phosphate contents. Although it is too early to conclude whether there would be useful applications in food, possibilities include changing starch digestibility or starch viscosity after cooking. Another interesting character in one species that is not available in others is associated with the genes responsible for endosperm texture in some cereal crops. In this case the puroindoline genes, described as PinA and PinB are interesting candidates (14,72,86,152). Transforming the wheat genes into rice successfully created the softness trait in the seed (121). At this time it is still too early to say how useful this character will be in food applications.

Although starch granule morphology varies extensively across species (102) and starch from different species is highly valued for certain applications, there are few specific reports that granule morphology is the attribute desired for valuable food applications. Furthermore, although there have been some reports of progress in understanding the mechanisms that control and influence starch granule size (28,29,122), for granule morphology there is little that can be defined as clear enough to arouse interest in a genetic approach. Similarly, flavor can be considered to vary extensively across species and a bland flavor of (e.g., tapioca starch) is valued over certain cereal starches that impart a mealy character. However, as with granule morphology, our understanding of the genetic components determining this character is rather poor, making flavor a difficult target for genetic modification at the present time. Thus in these cases it is difficult to see a way of using such differences in granule morphology and flavor to exploit that character in another species.

Another novel concept has come from bifunctional domains involving cellulose and starch, which may enable developments with novel biomaterials (130). Another possibility is to explore incorporation of inulins with starch (80). Even more novel possibilities to consider include adding new functionality to starch using gene constructs containing fusion proteins (106). The fusion proteins can be proteins, which might affect starch properties or enzymes, or other bioactive proteins.

ACKNOWLEDGMENTS

The authors wish to thank Dr Kay Denyer for comments and constructive criticism of the manuscript. All starch structure diagrams kindly supplied by Dr Eric Bertoft (with permission).

REFERENCES

1. Adkins, G.K., C.T. Greenwood. Studies on starches of high amylose-content, part VIII: the effect of low temperature on the interaction of amylomaize starch with iodine: a unique characterization. *Carbohydr. Res.* 3:152–156, 1966.
2. Adkins, G.K., C.T. Greenwood. Studies on starches of high amylose content, part X: an improved method for fractionation of maize and amylomaize starches by complex formation from aqueous dispersion after pretreatment with methulsuphoxide. *Carbohydr. Res.* 11:217–224, 1969.
3. Anderson, J.M. Characterisation of starch from inner and peripheral parts of normal and waxy barley kernels. *J. Cereal Sci.* 30:165–171, 1999.
4. Andersson, M., A. Trifonova, A. Andersson, M. Johansson, L. Bulow, P. Hofvander. A novel selection system for potato transformation using a mutated AHAS gene. *Plant Cell Rep.* 22:261–267, 2004.
5. Araki, E., H. Miura, S. Sawada. Differential effects of the null alleles at the three *Wx* loci on the starch-pasting properties of wheat. *Theor. Appl. Gen.* 100:1113–1120, 2000.
6. Asaoka, M., K. Okuno, Y. Sugimoto, M. Yano, T. Omura, H. Fuwa. Characterization of endosperm starch from high-amylose mutants of rice (*Oryza sativa* L.). *Starch* 38:114–117, 1986.
7. Atwell, W., L.F. Hood, D.R. Lineback, E. Varriano-Martson, H.F. Zobel. The terminology and methodology associated with basic starch phenomena. *Cereal Foods World* 33:306–311, 1988.
8. Baba, T., M. Arai, T. Yamamoto, T. Itoh. Some structural features of amylomaize starch. *Phytochemistry* 21:2291–2296, 1982.
9. Baba T, Y. Arai. Structural characterization of amylopectin and intermediate material in amylomaize starch granules. *Agric. Biol. Chem.* 48:1763–1775, 1984.
10. Baba, T., R. Uemura, M. Hiroto, Y. Arai. Structural features of amylomaize starch: components of amylon 70 starch. *J. Jpn. Soc. Starch Sci.* 34:213–217, 1987.
11. Bates, F.L., D. French, R.E. Rundle. Amylose and amylopectin content of starches determined by their iodine complex formation. *J. Am. Chem. Soc.* 65:142–148, 1943.
12. Bear, R., M.L. Vineyard, M.M. MacMasters, W.L. Deatherage. Development of "amylomaize": corn hybrids with high amylose starch, II: results of breeding efforts. *Agron. J.* 50:598–602, 1958.
13. Beatty, M.K., A. Rahman, H.P. Cao, W. Woodman, M. Lee, A. Myers, M. James. Purification and molecular genetic characterization of ZPU1, a pullulanase-type starch-debranching enzyme from maize1. *Plant Physiol.* 119:255–266, 1999.

14. Beecher, B., A.D. Bettge, E. Smidansky, M. Giroux. Expression of wild-type pinB sequence in transgenic wheat complements a hard phenotype. *Theor. Appl. Genet.* 105:870–877, 2002.

15. Beliaderis, C.G. Characterization of starch networks by small strain dynamic oscillatory rheometry. In: *Developments in Carbohydrate Chemistry,* Alexander, R.J., H.F. Zobel, eds., St Paul: American Association of Cereal Chemists, 1992, p 103.

16. Bello-Pérez, L.A., O. Paredez-López. Starch and amylopectin: rheological behavior of gels. *Starch/Stärke* 46:411–413, 1994.

17. Bello, A.B., R.D. Waniska, M.H. Gomez, L.W. Rooney. Starch solubilization and retrogradation during preparation of To (a food jell) from different sorghum varieties. *Cereal Chem.* 72:80–84, 1995.

18. Bertoft, E. On the nature of categories of chains in amylopectin and their connection to the super helix model. *Carbohydr. Polym.* 57:211–234, 2004.

19. Bett-Garber, K.L., E.T. Champagne, A.M. McClung, K.A. Moldenhauer, S.D. Linscombe, and K.S. McKenzie. Categorizing rice cultivars based on cluster analysis of amylose content, protein content, and sensory attributes. *Cereal Chem.* 78:551–558, 2001.

20. Blauth, S.L., K.N. Kim, J.D. Klucinec, J.C. Shannon, D.B. Thompson, M.J. Guilitinan. Identification of *Mutator* insertional mutants of starch-branching enzyme 1 (SBEI) in *Zea mays* L. *Plant Mol. Biol.* 48:287–297, 2002.

21. Blauth, S.L., Y. Yao, J.D. Klucinec, J.C. Shannon, D.B. Thompson, M.J. Guilitinan. Identification of *Mutator* insertional mutants of starch-branching enzyme 2a in corn. *Plant Physiol.* 125:1396–1405, 2001.

22. Bogracheva, T.Y., T. Wang, V.J. Morris, S. Ring, C. Hedley. The granular structure of C-type pea starch and its role in gelatinisation. *Biopolymers* 45:323–332, 1998.

23. Boltz, K.W., D.B. Thompson. Initial heating temperature and native lipid affects ordering of amylose during cooling of high-amylose starches. *Cereal Chem.* 76:204–211, 1999.

24. Boyer, C.D. Soluble starch synthases and starch branching enzymes from developing seeds of sorghum. *Phytochemistry* 24:15–18, 1985.

25. Boyer, C., D.L. Garwood, J. Shannon. Interaction of the amylose-extender and waxy mutants of maize. *J. Heredity* 67:209–214, 1976.

26. Brimhall, B., G.F. Sprague, and J.E. Sass. A new waxy allel in corn and its effect on the properties of endosperm starch. *J. Am. Soc. Agron.* 37:937–944, 1945.

27. Buleon, A., C. Gerard, C. Riekel, R. Vuong, H. Chanzy. Details of the ultrastructure of C-starch granules revealed by synchrotron microfocus mapping. *Macromolecules* 31:6605–6610, 1998.

28. Burton, R.A., H. Jenner, L. Carrangis, B. Fahy, G.B. Fincher, C. Hylton, D.A. Laurie, M. Parker, D. Waite, S. Van Wegen, T. Verhoeven, K. Denyer. Starch granule initiation and growth are altered in barley mutants that lack isoamylase activity. *Plant J.* 31:97–112, 2002.

29. Bustos, R.,B. Fahy, C. Hylton, R. Seale, N.M. Nebane, A. Edwards, C. Martin, A.M. Smith. Starch granule initiation is controlled by a heteromultimeric isoamylase in potato tubers. *Proc. Nat. Acad. Sci. USA* 101:2215–2220, 2004.

30. Cameron, R.E., C.M. Durrani, A.M. Donald. Gelation of amylopectin without long range order. *Starch/Stärke* 46:285–287, 1994.

31. Campbell, M.R., L.M. Pollak, P.J. White. Dosage effect at the sugary-2 locus on maize starch structure and function. *Cereal Chem.* 71:464–468, 1994.

32. Campbell, M.R., L.M. Pollak, P.J. White. Interaction of two sugary-1 alleles (su1 and su1st) with sugary-2 (su2) on characteristics of maize starch. *Starch/Stärke* 48:391–395, 1996.

33. Case, S.E., T. Capitani, J.K. Whaley, Y.C. Shi, P. Trzasko, R. Jeffcoat, H.B. Goldfarb. Influence of amylose-amylopectin ratio on gel properties. *J. Cereal Sci.* 27:301–314, 1998.

34. Champagne, E.T., K.L. Bett, B.T. Vinyard, A.M. McClung, F.E. Barton, K. Moldenhauer, S. Linscombe, K. McKenzie. Correlation between cooked rice texture and rapid visco analyser measurements. *Cereal Chem.* 76:764–771, 1999.

35. Clark, A.H., M.J. Gidley, R.K. Richardson, S.B. Ross-Murphy. Rheological studies of aqueous amylose gels: the effect of chain length and concentration on gel modulus. *Macromolecules* 22:346–351, 1989.

36. Coe, E. The genetics of corn. In: *Corn and Corn Improvement*, Sprague, G.W., J.W. Dudley, eds., Madison: American Society of Agronomy, 1988.

37. Colonna, P., C.F. Morris. Behavior of amylose in binary DMSO-water mixtures. *Food Hydrocolloids* 1:573–574, 1987.

38. Commuri, P.D., P.L. Keeling. Chain-length specificities of maize starch synthase I enzyme: studies of glucan affinity and catalytic properties. *Plant J.* 25:475–486, 2001.

39. Cooke, D., M. Gidley. Loss of crystalline and molecular order during starch gelatinization: origin of the enthalpic transition. *Carbohydr. Res.* 227:103–112, 1992.

40. Craig, J., J.R. Lloyd, K. Tomlinson, L. Barber, A. Edwards, T.L. Wang, C. Martin, C.L. Hedley, A.M. Smith. Mutations in the gene encoding starch synthase II profoundly alter amylopectin structure in pea embryos. *Plant Cell.* 10:413–426, 1998.

41. Craig, S.A.S., C.C. Maningat, P. Seib, R.C. Hoseney. Starch paste clarity. *Cereal Chem.* 66:173–182, 1989.

42. Creech, R.G. Genetic control of carbohydrate synthesis in maize endosperm. *Genetics* 52:1175–1186, 1965.

43. Cummings, J.H., E.R. Beatty, S.M. Kingman, S.A. Bingham, H.N. Englyst. Digestion and physiological properties of resistant starch in the human large bowel. *Br. J. Nutr.* 75:733–747, 1995.

44. Denyer, K., C.M. Hylton, C.F. Jenner, A.M. Smith. Identification of multiple isoforms of soluble and granule bound starch synthase in developing wheat endosperm. *Planta* 196:256–265, 1995.

45. Denyer, K., D. Waite, A. Edwards, C. Martin, A.M. Smith. Interaction with amylopectin influences the ability of granule-bound starch synthase I to elongate malto-oligosaccharides. *Biochem. J.* 342:647–653, 1999.

46. Denyer, K., D. Waite, S. Motawia, B.L. Moller, A.M. Smith. Granule-bound starch synthase I in isolated starch granules elongates malto-oligosaccharides processively. *Biochem. J.* 340:183–191, 1999.

47. Dinges, J.R., C. Colleoni, M. James, A. Myers. Purification and Molecular Genetic Characterization of ZPU1, a Pullulanase-Type Starch-Debranching Enzyme from Maize1. *Plant Cell* 15:666–680, 2003.

48. Doublier, J.L., I. Coté, G. Llamas, M.L. Meur. Effect of thermal history on amylose gelation. *Prog. Coll. Polymer Sci.* 90:61–65, 1992.

49. Doublier, J.L., G. Llamas. A rheological description of amylose-amylopectin mixtures. In: *Food Colloids and Polymers: Stability and Mechanical Properties*, Doublier, J.L., G. Llamas, eds., Cambridge: The Royal Society of Chemistry, 1993, pp 138–146.

50. Echt, C., D. Schwartz. The *wx* locus is the structural gene for the *Wx* protein. *Maize Genetics Cooperation Newsletter* 55:8–9, 1981.

51. Edwards, A., D.C. Fulton, C.M. Hylton, S.A. Jobling, M. Gidley, U. Rössner, C. Martin, A.M. Smith. A combined reduction in activity of starch synthases II and III of potato has novel effects on the starch of tubers. *Plant J.* 17:251–261, 1999.

52. Edwards, A., J.P. Vincken, L.C.J.M. Suurs, R.G.F. Visser, S. Zeeman, A. Smith, C. Martin. Discrete forms of amylose are synthesized by isoforms of GBSSl in pea. *Plant Cell* 14:1767–1785, 2002.

53. Eerlingen, R.C., H. Jacobs, K. Block, J.A. Delcour. Effects of hydrothermal treatments on the rheological properties of potato starch. *Carbohydr. Res.* 297:347–356, 1997.

54. Eliasson, A.C., H.R. Kim. Changes in rheological properties of hydroxypropyl potato starch pastes during freeze-thaw treatments, I: a rheological approach for evaluation of freeze-thaw stability. *J. Texture Studies* 23:279–295, 1992.

55. Eliasson, A.C., H.R. Kim. A dynamic rheological method to study the interaction between starch and lipids. *J. Rheol.* 39:1519–1534, 1995.

56. Ellis, H.S., S. Ring. A study of some factors influencing amylose gelation. *Carbohydr. Polymers* 5:201–213, 1985.

57. Fisher, D.K., D.B. Thompson. Retrogradation of maize starch after thermal treatment within and above the gelatinization temperature range. *Cereal Chem.* 74:344–351, 1957.

58. Flipse, E., C.J.M. Keetels, E. Jacobsen, R.G.F. Visser. The dosage effect of the wild type GBSS allele is linear for GBSS activity but not for amylose content: Absence of amylose has a distinct influence on the physico-chemical properties of starch. *Theor. Appl. Genet.* 92:121–127, 1996.

59. Fujita, N., A. Kubo, D. Suh, K.S. Wong, J.-L. Jane, K. Ozawa, F. Takaiwa, Y. Inaba, Y. Nakamura. Antisense inhibition of isoamylase alters the structure of amylopectin and the physicochemical properties of starch in rice endosperm. *Plant Cell Physiol.* 44:607–618, 2003.

60. Fulton, D., A. Edwards, E. Pilling, H.L. Robinson, B. Fahy, R. Seale, L. Kato, A.M. Donald, P. Geigenberger, C. Martin, A.M. Smith. Role of granule bound starch synthase in determination of amylopectin structure and starch granule morphology in potato. *J. Biol. Chem.* 277:10834–10841, 2002.

61. Fuwa, H., D.V. Glover, K. Miyaura, N. Inouchi, Y. Konishi, Y. Sugimoto. Chain length distribution of amylopectins of double- and triple-mutants containing the waxy gene in the inbred Oh43 maize background. *Starch/Stärke* 39:295–298, 1987.

62. Gao, M., D.K. Fisher, K.N. Kim, J.C. Shannon, M.J. Guiltinan. Independent genetic control of maize starch-branching enzymes IIa and IIb: isolation and characterization of a *Sbe2a* cDNA. *Plant Physiol.* 114:69–78, 1997.

63. Gao, M., J. Wanat, P.S. Stinard, M.G. James, A.M. Myers. Characterization of *dull1*, a maize gene coding for a novel starch synthase. *Plant Cell* 10:399–412, 1998.

64. Garwood, D.L., J.C. Shannon, R.G. Creech. Starches of endosperms possessing different alleles at the *amylose-extender* locus in *Zea mays* L. *Cereal Chem.* 53:355–364, 1976.

65. Garwood, D.L., S.F. Vanderslice. Carbohydrate composition of alleles at the sugary locus in maize. *Crop Sci.* 22:367–371, 1982.

66. German, M.L., A.L. Blumenfeld, Y.V. Guenin, V.P. Yuryev, V.B. Tolstoguzov. Structure formation in systems containing amylose. *Carbohydr. Polym.* 18:27–34, 1992.

67. Gidley, M. Factors affecting the crystalline type (A-C) of native starches and model compounds: a rationalisation of observed effects in terms of polymorphic structures. *Carbohydr. Res.* 161:301–304, 1987.

68. Gidley, M. Molecular mechanisms underlying amylose aggregation and gelation. *Macromolecules* 22:351–358, 1989.

69. Gidley, M., C. Bulpin. Crystallisation of malto-oogisaccharides as models of the crystalline forms of starch: minimum chain-length requirement for the formation of double helices. *Carbohydr. Res.* 161:291–300, 1987.

70. Gidley, M., C. Bulpin. Aggregation of amylose in aqueous systems: the effect of chain length on phase behavior and aggregation kinetics. *Macromolecules* 22:341–346, 1989.

71. Gidley, M.J., S.M. Bociek. Molecular organization in starches: a 13C CP/MAS NMR study. *J. Am. Chem. Soc.* 107:7040–7044, 1985.

72. Giroux, M., C.F. Morris. Wheat grain hardness results from highly conserved mutations in the friabilin components puroindoline a and b. *Proc. Natl. Acad. Sci. USA* 95:6262–6266, 1998.

73. Glover, D.V., E.T. Mertz. Corn. In: Nutritional quality of cereal grains. Genetic and agronomic improvement (Olson, R.A., Frey, K.J., eds). Agronomy No. 28. American Society of Agronomy, Madison, pp 183–336, 1987.

74. Goodfellow, B.J., R.H. Wilson. A fourier transform IR study of the gelation of amylose and amylopectin. *Biopolymers* 30:1183–1189, 1990.

75. Greenwood, C.T., S. McKenzie. Studies on starches of high amylose-content, Part IV. The fractionation of amylomaize starch; a study of the branched component. *Carbohydr. Res.* 3:7–13, 1966.

76. Guan, H.P., J. Preiss. Differentiation of the properties of the branching isozymes from maize (*Zea mays*). *Plant Physiol.* 102:1269–1273, 1993.

77. Han, X.-Z., O.H. Campanella, H. Guan, P.L. Keeling, B.R. Hamaker. Influence of maize starch granule-associated protein on the rheological properties of starch pastes, I: large deformation measurements of paste properties. *Carbohydr. Polym.* 49:323–330, 2002.

78. Han, X.-Z., O.H. Campanella, H. Guan, P.L. Keeling, B.R. Hamaker. Influence of maize starch granule-associated protein on the rheological properties of starch pastes, II: dynamic measurements of viscoelastic properties of starch pastes. *Carbohydr. Polym.* 49:315–321, 2002.

79. Haralampu, S.G. Resistant starch: a review of the physical properties and biological impact of RS3. *Carbohydr. Polym.* 41:285–292, 2000.

80. Hellwege, E.M., S. Czapla, A. Jahnke, L. Willmitzer, A.G. Heyer. Transgenic potato (*Solanum tuberosum*) tubers synthesize the full spectrum of inulin molecules naturally occurring in globe artichoke (*Cynara scolymus*) roots. *Proc. Natl. Acad. Sci. USA* 97:8699–8704, 2000.

81. Helm, J.L., V.L. Ferguson, M.S. Zuber. Interaction of dosage effects on amylose content of corn at the *Du* and *Wx* loci. *J. Heredity* 60:259–260, 1969.

82. Hizukuri, S. Polymodal distribution of the chain lengths of amylopectins and its significance. *Carbohydr. Res.* 147:342–347, 1986.

83. Hizukuri, S., Y. Takeda, N. Maruta. Molecular structure of rice starches. *Carbohydr. Res.* 189:227–235, 1989.

84. Hizukuri, S., Y. Takeda, M. Yasuda, A. Suzuki. Multi-branched nature of amylose and the action of de-branching enzymes. *Carbohydr. Res.* 94:205–213, 1981.

85. Hofvander, P., M. Andersson, C. Larsson, H. Larsson. Field performance and starch characteristics of high-amylose potatoes obtained by antisense gene targeting of two branching enzymes. *Plant Biotechnol. J.* 2:311–320, 2004.

86. Hogg, A.C., T. Sripo, B. Beecher, J.M. Martin, M.J. Giroux. Wheat puroindolines interact to form friabilin and control wheat grain hardness. *Theor. Appl. Genet.* 108:1089–1097, 2004.

87. Holder, D.G., D.V. Glover, J.C. Shannon. Interaction of shrunken-2 with five other carbohydrate genes in corn endosperm. *Crop Sci.* 14:643–646, 1997.

88. Hovenkamp-Hermelink, J.H.M., E. Jacobsen, A.S. Ponstein, R. Visser, G.H. Vos-Scheperkeuter, B. Witholt, J.N. de Vries, E.W. Bijmolt, W.J. Feenstra. Isolation of an amylose-free starch mutant of the potato (*Solanum tuberosum* L.). *Theor. Appl. Genet.* 75:217–221, 1987.

89. Hseih, J.S. Genetic studies of the Wx gene of sorghum (*Sorghum bicolor* [L.] Moench). *Crop Sci.* 14:643–646, 1988.

90. Hussain, H., A. Mant, R. Seale, S. Zeeman, E. Hinchliffe, A. Edwards, C. Hylton, S. Bornemann, A.M. Smith, C. Martin, R. Bustos. Three isoforms of isoamylase contribute different catalytic properties for the debranching of potato glucans. *Plant Cell* 15:133–149, 2003.

91. I'Anson, K.J., M.J. Miles, V.J. Morris, S. Ring. A study of amylose gelation using a synchrotron x-ray source. *Carbohydr. Polym.* 8:45–53, 1988.

92. Imberty, A., A. Buleon, V. Tran, S. Perez. Recent advances in knowledge of starch structure. *Starch/Stärke* 43:375–384, 1991.

93. Björck, I. Starch: Nutritional Aspects. In: *Carbohydrates in Food*, Eliasson, A.C., ed., New York: Marcel Dekker, 1996.

94. Inouchi, N., D.V. Glover, H. Fuwa. Chain length distribution of amylopectins of several single mutants and the normal counterpart, and sugary-1 phytoglycogen in maize (*Zea mays* L.). *Starch/Stärke* 39:259–266, 1987.

95. Inouchi, N., D.V. Glover, H. Fuwa. Structure and physicochemical properties of endosperm starches of a waxy allelic series and their respective normal counterparts in the inbred Oh43 maize background. *Starch/Stärke* 47:421–426, 1995.

96. Ishikawa, N., J. Ishihara, M. Itoh. Artificial induction and characterization of amylose-free mutants of barley. *Barley Genet. News* 24:4953, 1994.

97. Isshiki, M., K. Morino, M. Nakajima, R. Okagaki, S.R. Wessler, T. Isawa, K. Shimamoto. A naturally occurring functional allele of the rice *waxy* locus has a GT to TT mutation at the 5' splice site of the first intron. *Plant J.* 15:133–138, 1998.

98. Jacobson, M.R., J.N. BeMiller. Method for determining the rate and extent of accelerated starch retrogradation. *Cereal Chem.* 75:22–29, 1998.

99. James, M., K. Denyer, A. Myers. Starch synthesis in the cereal endosperm. *Curr. Opin. in Plant Biol.* 6:215-222, 2003.

100. James, M.G., D.S. Robertson, A.M. Myers. Characterization of the maize gene sugary1, a determinant of starch composition in kernels. *Plant Cell* 7:417–429, 1995.

101. Jane, J.-L., J.F. Chen. Effect of amylose molecular size and amylopectin branch chain length on paste properties of starch. *Cereal Chem.* 69:60–65, 1992.

102. Jane, J.-L., T. Kasemsuwan, S. Leas, I. Ames, H.F. Zobel, I.L. Darien, J.F. Robyt. Anthology of starch granule morphology by scanning electron microscopy. *Starch/Stärke* 46:121–129, 1994.

103. Jane, J.-L., J.F. Robyt. Structure studies of amylose-V complexes and retrogradaded amylose by action of alpha amylases, and a new method for preparing amylodextrins. *Carbohydr. Res.* 132:105–118, 1984.

104. Jane, J.L., Y.Y. Chen, L.F. Lee, A.E. McPherson, K.-S. Wong, M. Radosavljevic, T. Kasemsuwan. Effects of amylopectin branch chain length and amylose content on the gelatinization and pasting properties of starch. *Cereal Chem.* 76:629–637, 1999.

105. Jenkins, P.J., R.E. Cameron, A.M. Donald. A universal feature in the structure of starch granules from different botanical sources. *Starch/Stärke* 45:417–420, 1993.

106. Ji, Q., J.P. Vincken, L.C. Suurs, R.G.F. Visser. Microbial starch-binding domains as a tool for targeting proteins to granules during starch biosynthesis. *Plant Mol. Biol.* 51:789–801, 2003.

107. Jobling, S.A., G.P. Schwall, R.J. Westcott, C.M. Sidebottom, M. Debet, M.J. Gidley, R. Jeffcoat, R. Safford. A minor form of starch branching enzyme in potato (*Solanum tuberosum* L.) tubers has a major effect on starch structure: cloning and characterisation of multiple forms of *SBE A*. *Plant J.* 18:163–171, 1999.

108. Jobling, S.A., R.J. Westcott, A. Tayal, R. Jeffcoat, G.P. Schwall. Production of freeze-thaw stable potato starch by antisense inhibition of three starch synthase genes. *Nat. Biotechnol.* 20:295–299, 2002.

109. Juliano, B.O., M.B. Nazareno, N.B. Ramos. Properties of waxy and isogenic nonwaxy rices differing in starch gelatinization temperature. *J. Agric. Food Chem.* 17:1364–1369, 1969.

110. Kainuma, K., D. French. Nägeli amylodextrin and its relationship to starch granule structure, I: preparation and properties of amylodextrins from various starch types. *Biopolymers* 10:1673–1680, 1971.

111. Kalichevski, M.T., P.D. Orford, S. Ring. The retrogradation and gelation of amylopectins from various botanical sources. *Carbohydr. Res.* 198:49–55, 1990.

112. Kalichevski, M.T., S. Ring. Incompatability of amylose and amylopectin in aqueous solution. *Carbohyd. Res.* 162:323–328, 1987.

113. Keetels, C.J.A.M., T.V. Vilet, P. Walstra. Gelation and retrogradation of concentrated starch systems, 2: retrogradation. *Food Hydrocolloids* 10:355–362, 1996.

114. Kim, J.O., W.S. Kim, M.S. Shin. A comparative study on retrogradation of rice starch gels by DSC, X-ray and α-amylase methods. *Starch/Stärke* 49:71–75, 1997.

115. Kiribuchi-Otobe. *Cereal Chem.* 75:671–672, 1998.

116. Klucinec, J.D., D.B. Thompson. Fractionation of high-amylose maize starches by differential alcohol precipitation and chromatography of the fractions. *Cereal Chem.* 75:887–896, 1998.

117. Klucinec, J.D., D.B. Thompson. Amylose and amylopectin interact in retrogradation of dispersed high-amylose starches. *Cereal Chem.* 76:282–291, 1999.

118. Klucinec, J.D., D.B. Thompson. Amylopectin nature and amylose-to-amylopectin ratio as influences on the behaviours of gels of dispersed starch. *Cereal Chem.* 79:24–35, 2002.

119. Klucinec, J.D., D.B. Thompson. Structure of amylopectins from ae-containing maize starches. *Cereal Chem.* 79:19–23, 2002.

120. Knutson, C.A., M.J. Grive. Rapid method for estimation of amylose in maize starches. *Cereal Chem.* 71:469–471, 1994.

121. Krishnamurthy, K., M.J. Giroux. Expression of wheat puroindoline genes in transgenic rice enhances grain softness. *Nat. Biotechnol.* 19:162–166, 2001.

122. Kubo, A., N. Fujita, K. Harada, T. Matsuda, H. Satoh, Y. Nakamura. The starch-debranching enzymes isoamylase and pullulanase are both involved in amylopectin biosynthesis in rice endosperm. *Plant Physiol.* 121:399–409, 1999.

123. Kuipers, A.G.J., E. Jacobsen, R.G.F. Visser. Formation and deposition of amylose in the potato tuber starch granule are affected by the reduction of granule-bound starch synthase gene expression. *Plant Cell* 6:43–52, 1994.

124. Kumar, I., G.S. Khush. Inheritance of amylose content in rice (*Oryza sativa* L.). *Euphytica* 38:261–269, 1988.

125. Lai, V.M.-F., S. Lu, C.Y. Lii. Molecular characteristics influencing retrogradation kinetics of rice amylopectins. *Cereal Chem.* 77:272–278, 2000.

126. Lansky, S., M. Kooi, T.J. Schoch. Properties of the fractions and linear subfractions from various starches. *J. Am. Chem. Soc.* 71:4066–4075, 1949.

127. Larson, B.L., K.A. Gilles, R. Jenness. Amperometric method for determining the sorption of iodine by starch. *Anal. Chem.* 25:802–804, 1953.

128. Leloup, V.M., P. Colonna, A. Buleon. Influence of amylose-amylopectin ratio on gel properties. *J. Cereal Sci.* 13:1–13, 1991.

129. Leloup, V.M., P. Colonna, S.G. Ring, K. Roberts, B. Wells. Microstructure of amylose gels. *Carbohydr. Polym.* 18:189–197, 1992.

130. Levy, I., T. Paldi, O. Shoseyov. Engineering a bifunctional starch-cellulose cross-bridge protein. *Biopolymers* 25:1841–1849, 2004.

131. Lii, C.-Y., Y.-Y. Shao, K.-H. Tseng. Gelation mechanism and rheological properties of rice starch. *Cereal Chem.* 72:393–400, 1995.

132. Li, J.H., T. Vasanthan, B. Rossnagel, R. Hoover. Starch from hull-less barley: II. Thermal, rheological, and acid hydrolysis characteristics. *Food Chem.* 74:407–415, 2001.

133. Liu, H., L. Ramsden, H. Corke. Physical properties and enzymatic digestibility of acetylated *ae*, *wx*, and normal maize starch. *Carbohydr. Polym.* 34:283–287, 1997.

134. Liu, Q., D.B. Thompson. Effects of moisture content and different gelatinization heating temperatures on retrogradation of waxy-type maize starches. *Carbohydr. Res.* 314:221–235, 1998.

135. Liu Q., D.B. Thompson. Retrogradation of *du wx* and *su2 wx* maize starches after different gelatinization heat treatments. *Cereal Chem.* 75:868–874, 1998.

136. Lloyd, J.R., V. Landschütze, J. Kossmann. Simultaneous antisense inhibition of two starch-synthase isoforms in potato tubers leads to accumulation of grossly modified amylopectin. *Biochem. J.* 338:515–521, 1999.

137. Lu, S., L.N. Chen, C.-Y. Lii. Correlations between the fine structure, physicochemical properties, and retrogradation of amylopectins from Taiwain rice varieties. *Cereal. Chem.* 74:34–39, 1997.

138. Lu, T.-J., J.-L. Jane, P.L. Keeling Temperature effect on retrogradation rate and crystalline structure of amylose. *Carbohydr. Polymers* 33:19–26, 1997.

139. McCready, R.M., W.Z. Hassid. The separation and quantitative estimation of amylose and amylopectin in potato starch. *J. Am. Chem. Soc.* 65:1154–1157, 1943.

140. Mercier, R. The fine structure of corn starches of various amylose-percentage: waxy, normal, and amylomaize. *Starch/Stärke* 25:78–83, 1973.

141. Miles, M.J., V.J. Morris, P.D. Orford, S. Ring. The roles of amylose and amylopectin in the gelation and retrogradation of starch. *Carbohydr. Res.* 135:271–281, 1985.

142. Miles, M.J., V.J. Morris, S. Ring. Some recent observations on the retrogradation of amylose. *Carbohydr. Polym.* 4:73–77, 1984.

143. Miura, H., E. Araki, S. Tarui. Amylose synthesis capacity of the three *Wx* genes of wheat cv. Chinese Spring. *Euphytica* 108:91–95, 1999.

144. Miura, H., A. Sugawara. Dosage effects of the three Wx genes on amylose synthesis in wheat endosperm. *Theor. Appl. Genet.* 93:1066–1070, 1996.

145. Miura, H., S. Tanii, T. Nakamura, N. Watanabe. Genetic control of amylose content in wheat endosperm starch and differential effects of three Wx genes. *Theor. Appl. Genet.* 89:276-280, 1994.

146. Mizuno, K., T. Kawasaki, H. Shimada, H. Satoh, E. Kobayashi, S. Okumura, Y. Arai, T. Baba. Alteration of the structural properties of starch components by the lack of an isoform of starch branching enzyme in rice seeds. *J. Biol. Chem.* 268:19084–19091, 1993.

147. Mizuno, K., K. Kimura, Y. Arai, T. Kawasaki, H. Shimada, T. Baba. Starch branching enzymes from immature rice seeds. *J. Biochem.* 112:643–651, 1992.

148. Montgomery, E.M., K.R. Sexon, F.R. Senti. High-amylose corn starch fractions. *Starch/Stärke* 13:215–222, 1961.

149. Montgomery, E.M., K.R. Sexon, F.R. Senti. Physical properties and chemical structure of high-amylose corn starch fractions. *Starch/Stärke* 16:345–351, 1964.

150. Morell, M.K., B. Kosar-Hashemi, M. Cmiel, M.S. Samuel, P. Chandler, S. Rahman, A. Buleon, I.L. Batey, Z. Li. Barley *sex6* mutants lack starch synthase IIa activity and contain a starch with novel properties. *Plant J.* 34:173–185, 2003.

151. Morell, M.K., Z. Li, S. Rahman. Starch biosynthesis in the small grained cereals: wheat and barley. In: *Starch: Advances in Structure and Function,* Barsby, T., A.M. Donald, P.J. Frazier, eds., Cambridge: The Royal Society of Chemistry, 2001, pp 129–137.

152. Morris, C.F. Puroindolines: the molecular genetic basis of wheat grain hardness. *Plant Mol. Biol.* 48:633–647, 2002.

153. Morris. V.J. Starch gelation and retrogradation. *Trends Food Sci. Technol.* 1:2–6, 1990.

154. Morrison, W.R. Lipids in cereal starches: a review. *J. Cereal Sci.* 8:1–15, 1988.

155. Mu-Forster, C., B.P. Wasserman. Surface localization of zein storage proteins in starch granules from maize endosperm: proteolytic removal by thermolysin and *in vitro* cross-linking of granule-associated polypeptides. *Plant Physiol.* 116:1563–1571, 1996.

156. Murai, J., T. Taira, D. Ohta. Isolation and characterization of the three *Waxy* genes encoding the granule-bound starch synthase in hexaploid wheat. *Gene* 234:71–79, 1999.

157. Murata, K., T. Sagiyama, T. Akazawa. Enzymic mechanism of starch synthesis in glutinous rice grains. *Biochem. Biophys. Res. Commun.* 18:371–376, 1965.

158. Nakamura, T., M. Yamamori, H. Hirano, S. Hidaka. Identification of 3 Wx proteins in wheat (*Triticum aestivum* L). *Biochem. Genet.* 31:75–86, 1993.

159. Nakamura, Y. Towards a better understanding of the metabolic system for amylopectin biosynthesis in plants: rice endosperm as a model tissue. *Plant Cell Physiol.* 43:718–725, 2002.

160. Nakamura, Y., M. Yamamori, H. Hirano, S. Hidaka, Y. Nagamura. Production of waxy (amylose-free) wheats. *Mol. Gen. Genet.* 248:253–259, 1995.

161. Neuffer, G. *In Mutants of Maize.* Cold Spring Harbor, NY: Cold Spring Harbor Laboratory Press, 1997, p 298.

162. Nishi, A., Y. Nakamura, N. Tanaka, H. Satoh. Biochemical and genetic analysis of the effects of *amylose-extender* mutation in rice endosperm. *Plant Physiol.* 127:459–472, 2001.

163. O'Sullivan, A.C., S. Perez. The relationship between internal chain length of amylopectin and crystallininty in starch. *Biopolymers* 50:381–390, 1999.

164. Oda, S., C. Kiribuchi, H. Seko. A bread wheat mutant with low amylose content induced by ethyl methanesulphonate. *Jpn. J. Breed.* 42:151–154, 1992.

165. Ong, M.H., J.M.V. Blanshard. Texture determinations of cooked, parboiled rice, II: physicochemical properties and leaching behaviour of rice. *J. Cereal Sci.* 21:261–269, 1995.

166. Oostergetel, G.T., E.F.J. Vanbruggen. The crystalline domains in potato starch granules are arranged in a helical fashion. *Carbohydr. Polym.* 21:7–12, 1993.

167. Orford, P.D., S. Ring, V. Carroll, M.J. Miles, V.J. Morris. The effect of concentration and botanical source on the gelation and retrogradation of starch. *J. Sci. Food Agric.* 39:177, 1987.

168. Paravouri, P., R. Manelius, T. Suortti, E. Bertoft, K. Autio. Effects of enzymatically modified amylopectin on the rheological properties of amylose-amylopectin mixed gels. *Food Hydrocolloids* 11:471–477, 1997.

169. Paredez-López, O., L.A. Bello-Pérez, M.G. López. Amylopectin: structural, gelatinization and retrogradation studies. *Food Chem.* 50:411–417, 1994.

170. Pilling, E., A.M. Smith. Growth ring formation in the starch granules of potato tubers. *Plant Physiol.* 1323:365–371, 2003.

171. Reddy, I., P.A. Seib. Modified waxy wheat starch compared to modified waxy corn starch. *J. Cereal Sci.* 31:25–39, 2000.

172. Reddy, R.K., Z.S. Ali, K.R. Bhattacharya. The fine structure of rice-starch amylopectin and its relation to the texture of cooked rice. *Carbohydr. Polym.* 22:267–275, 1993.

173. Reyes, A.C., E.L. Albano, V.P. Briones, B.O. Juliano. Varietal differences in physicochemical properties of rice starch and its fractions. *J. Agric. Food Chem.* 13:438–442, 1965.

174. Echt, C., D. Schwartz. The *wx* locus is the structural gene for the *Wx* protein. *Maize Genetics Cooperation Newsletter* 55:8-9, 1981.

175. Ring, S. Some studies on starch gelation. *Starch/Stärke* 37:80–83, 1985.

176. Ring, S., P. Colonna, K.J. I' Anson, M.T. Kalichevski, M.J. Miles, V.J. Morris, P.D. Orford. The gelation and crystallization of amylopectin. *Carbohydr. Res.* 162:277–293, 1987.

177. Ritte, G., J.R. Lloyd, N. Eckermann, A. Rottmann, J. Kossmann, M. Steup. The starch-related R1 protein is an α-glucan, water dikinase. *Proc. Natl. Acad. Sci. USA* 99:7166–7171, 2002.

178. Robin, J.P., C. Mercier, R. Charbonierre, A. Guilbot. Lintnerized starches: gel filtration and enzymatic studies of insoluble residues from prolonged acid treatment of potato starch. *Cereal Chem.* 51:389–406, 1974.

179. Robin, J.P., C. Mercier, F. Duprat, R. Charbonierre, A. Guilbot. Amidons lintnerises: etudes chromatographique et enzymatique des residus insolubles provenant de l'hydrolyse chlorhydrique d'amidons de cereales, en particulier de mais cireux. *Starch/Stärke* 27:36–45, 1975.

180. Roger, P., P. Colonna. The influence of chain length on the hydrodynamic behaviour of amylose. *Carbohydr. Res.* 227:73–83, 1992.

181. Rundle, R.E., R.R. Baldwin. The configuration of starch and the starch-iodine complex, I: The dichroism of flow of starch-iodine solutions. *J. Am. Chem. Soc.* 65:554–558, 1943.

182. Safford, R., S.A. Jobling, C. Sidebottom, R.J. Westcott, D. Cooke, K.J. Tober, B. Strongitharm, A.L. Russell, M. Gidley. Consequences of antisense RNA inhibition of starch branching enzyme activity on properties of potato starch. *Carbohydr. Polym.* 35:155–168, 1998.

183. Salehuzzaman, S.N.I.M., J.P. Vincken, M. Van de Wal, I. Straatman-Engelen, E. Jacobsen, R.G.F. Visser. Expression of a cassava granule-bound starch synthase gene in the amylose-free potato only partially restores amylose content. *Plant Cell Environ.* 22:1311–1318, 1999.

184. Salomonsson, A.C., B. Sundberg. Amylose content and chain profile of amylopectin from normal, high amylose and waxy barleys. *Starch/Stärke* 46:325–328, 1994.

185. Sanchez, P.C., B.O. Juliano, V.T. Laude, C.M. Perez. Nonwaxy rice for tapuy (rice wine) production. *Cereal Chem.* 65:240–243, 1988.

186. Sano, Y. Differential regulation of waxy gene expression in rice endosperm. *Theor. Appl. Genet.* 68:467–471, 1984.

187. Sano, Y., M. Katsumata, K. Okuno. Genetic studies of speciation in cultivated rice. 5. Inter- and intraspecific differentiation in the waxy gene expression of rice. *Euphytica* 35:1–9, 1986.

188. Sasaki, T., T. Yasui, J. Matsuki. Effect of amylose content on gelatinization, retrogradation, and pasting of starches from waxy and nonwaxy wheat and their F1 seeds. *Cereal Chem.* 77:58–63, 2000.

189. Satoh, H., A. Nishi, K. Yamishita, Y. Takemoto, Y. Tanaka, Y. Hosaka, Y. Sakurai, N. Fujita, Y. Nakamura. Starch-branching enzyme I-deficient mutation specifically affects the structure and properties of starch in rice endosperm. *Plant Physiol.* 133:1111–1121, 2004.

190. Schierbaum, F., W. Vorwerg, B. Kettlitz, F. Reuther. Interaction of linear and branched polysaccharides in starch gelling. *Die Nahrung* 30:1047–1049, 1986.

191. Schoch, T.J. Fractionation of starch by selective precipitation with butanol. *J. Am. Chem. Soc.* 64:2957–2961, 1942.

192. Schoch, T.J. Preparation of starch and the starch fractions. In: *Methods in Enzymology III*, Colowick, S.P., N.O. Kaplan, eds., New York: Academic Press, 1954, pp 5–17.

193. Schwall, G.P., R. Safford, R.J. Westcott, R. Jeffcoat, A. Tayal, Y.C. Shi, M.J. Gidley, S.A. Jobling. Production of very-high-amylose potato starch by inhibition of SBE A and B. *Nat. Biotechnol.* 18:551–554, 2000.

194. Shannon, J.C., D.L. Garwood. Genetics and physiology of starch development. In: *Starch: Chemistry and Technology*, ed. 2, Whistler, R.L., J.N. BeMiller, E.F. Paschall, eds., New York: Academic Press, 1984, pp 25–86.

195. Shi, Y.C., P. Seib. The structure of four waxy starches related to gelatinization and retrogradation. *Carbohydr. Res.* 227:131–145, 1992.

196. Shi, Y.C., P. Seib. Fine structure of maize starches from four *wx*-containing genotypes of the W64A inbred line in relation to gelatinization and retrogradation. *Carbohydr. Polym.* 26:141–147, 1995.

197. Shi Y.C., P. Seib. Molecular structure of a low-amylopectin starch and other high-amylose maize starches. *J. Cereal Sci.* 27:289–299, 1998.

198. Shimada, H., Y. Tada, T. Kawasaki, T. Fugimura. Antisense regulation of the rice waxy gene expression using a PCR-amplified fragment of the rice genome reduces the amylose content in grain starch. *Theor. Appl. Genet.* 86:665–672, 1993.

199. Shimada, H., Y. Tada, T. Kawasaki, T. Fujimura. Antisense regulation of the rice waxy gene expression using a PCR-amplified fragment of the rice genome reduces the amylose content in grain starch. *Theor. Appl. Genet.* 86:665–672, 1993.

200. Shure, M., S.R. Wessler, N. Federoff. Molecular identification and isolation of the waxy locus in maize. *Cell* 225–233, 1983.

201. Sidebottom, C., M. Kirkland, B. Strongitharm, R. Jeffcoat. Characterization of the difference of starch branching enzyme activities in normal and low-amylopectin maize during kernel development. *J. Cereal Sci.* 27:279–287, 1998.

202. Sievert, D., P. Wursch. Thermal behavior of potato amylose and enzyme-resistant starch from maize. *Cereal Chem.* 70:333–338, 1993.

203. Song, Y., J.-L. Jane. Characterization of barley starches from waxy, normal and high amylose varieties. *Carbohydr. Polym.* 41:365–377, 2000.

204. Sprague, G.F., M.T. Jenkins. The development of waxy corn for industrial use. *Iowa State Coll. J. Sci.* 22:205–213, 1948.

205. Stinard, P.S., D.S. Robertson, P.S. Schnable. Genetic isolation, cloning and analysis of a mutator-induced, dominant antimorph of the maize amylose extender1 locus. *Plant Cell* 5:1555–1566, 1993.

206. Sun, C.X., P. Sathish, S. Ahlandsberg, C. Jansson. The two genes encoding starch-branching enzymes IIa and IIb are differentially expressed in barley. *Plant Physiol.* 118:37–49, 1998.

207. Takeda, C., Y. Takeda, S. Hizukuri. Structure of amylomaize amylose. *Cereal Chem.* 66:22–25, 1989.

208. Takeda, Y., H.P. Guan, J. Preiss. Branching of amylose by the branching isoenzymes of maize endosperm. *Carbohydr. Res.* 240:253–263, 1993.

209. Takeda, Y., S. Hizukuri, B.O. Juliano. Purification and structure of amylose from rice starch. *Carbohydr. Res.* 148:299–308, 1986.

210. Takeda, Y., S. Hizukuri, B.O. Juliano. Structures of rice amylopectins with low and high affinities for iodine. *Carbohydr. Res.* 168:79–88, 1987.

211. Takeda, Y., S. Hizukuri, C. Takeda, A. Suzuki. Structures of branched molecules of amyloses of various origins, and molar fractions of branched and unbranched molecules. *Carbohydr. Res.* 165:139–145, 1987.

212. Takeda, Y., Preiss, J. Structures of B90 (sugary) and W64A (normal) maize starches. *Carbohydr. Res.* 240:265–275, 1993.

213. Takeda, Y., K. Shirakawa, S. Hizukuri. Examination of the purity and structure of amylose by gel-permeation chromatography. *Carbohydr. Res.* 132:83–92, 1984.

214. Tester, R.F., W.R. Morrison. Swelling and gelatinization of cereal starches, I: effects of amylopectin, amylose, and lipids. *Cereal Chem.* 67:551–557, 1990.

215. Tester, R.F., W.R. Morrison. Swelling and gelatinization of cereal starches, 3: some properties of waxy and normal nonwaxy barley starches. *Cereal Chem.* 69:654–658, 1992.

216. Thompson, D.B. On the non-random nature of amylopectin branching. *Carbohydr. Polym.* 43:223–239, 2000.

217. Thompson, D.B., J.M.V. Blanshard. Retrogradation of selected *wx*-containing maize starches. *Cereal Foods World* 40:670, 1995.

218. Tomlinson, K., K. Denyer. Starch synthesis in cereal grain. *Adv. Bot. Res.* 40:1–61, 2003.

219. Topping, D.L., M.K. Morell, R.A. King, Z. Li, A.R. Bird, M. Noakes. Resistant starch and health – Himalaya 292, a novel barley cultivar to deliver benefits to consumers. *Starch/ Stärke* 55:539–545, 2003.

220. Tziotis, A., K. Seetharaman, K.S. Wong, J.D. Klucinec, J.-L. Jane, P. White. Structural properties of starch fractions isolated from normal and mutant corn genotypes by using different methods. *Cereal Chem.* 2004.

221. Umemoto, T., M. Yano, H. Satoh, J.M. Short, A. Shamura, Y. Nakamura. Mapping of a gene responsible for the difference in amylopectin structure between japonica-type and indica-type rice varieties. *Theor. Appl. Genet.* 104:1–8, 2002.

222. Vallêra, A.M., M.M. Cruz, S.G. Ring, F. Boue. The structure of amylose gels. *J. Phys.* 6:301–320, 1994.

223. Hovenkamp, J.H.M., E. Jacobsen, A.S. Ponstein, R.G.F. Visser, G.H. Vos-Scheperkeuter, E.W. Bijmolt, J.N. de Vries, B. Witholt, W.J. Feenstra. Isolation of an amylose-free starch mutant of the potato (*Solanum tuberosum* L.). *Theor. Appl. Genet.* 75:217–221, 1991.

224. Van der Leij, F.R., R.G.F. Visser, K. Oosterhaven, D.A.M. van der Kop, E. Jacobsen, W.J. Feenstra. Complementation of the amylose-free starch mutant of potato (*Solanum tuberosum* L.) by the gene encoding granule-bound starch synthase. *Theor. Appl. Genet.* 82:289–295, 1991.

225. Villareal, C.P., S. Hizukuri, B.O. Juliano. Amylopectin staling of cooked milled rices and properties of amylopectin and amylose. *Cereal Chem.* 74:163–167, 1997.

226. Villareal, C.P., B.O. Juliano. Comparative levels of waxy gene product of endosperm starch granules of different rice ecotypes. *Starch/Stärke* 41:369–371, 1989.

227. Vinyard, M.L., R. Bear. Amylose content. *Maize Genet. Coop. Newsl.* 26:5, 1952.

228. Visser R.G.F., I. Somhorst, G.J. Kuipers, N.J. Ruys, W.J. Feenstra, E. Jacobsen. Inhibition of the expression of the gene for granule-bound starch synthase in potato by antisense constructs. *Mol. Gen. Genet.* 225:289–296, 1991.

229. Vrinten, P.L., T. Nakamura. Wheat granule-bound starch synthase I and II are encoded by separate genes that are expressed in different tissues. *Plant Physiol.* 122:255–263, 2000.

230. Waigh, T.A., A.M. Donald, F. Heidelbach, C. Riekel, M.J. Gidley. Analysis of the native structure of starch granules with small angle x-ray microfocus scattering. *Biopolymers* 49:91–105, 1999.

231. Waigh, T.A., M. Gidley, B.U. Komanshek, A.M. Donald. The phase transformations in starch during gelatinisation:a liquid crystalline approach. *Carbohydr. Res.* 328:165–176, 2000.

232. Waigh, T.A., K.L. Kato, A.M. Donald, M. Gidley, C. Riekel. Side chain liquid crystalline models for starch. *Starch/Stärke* 52:450–460, 2000.

233. Waigh, T.A., P.A. Perry, C. Riekel, M. Gidley, A.M. Donald. Chrial side chain liquid crystalline properties of starch. *Macromolecules* 31:7980–7904, 1998.

234. Wang, Y.J., L.F. Wang. Structures of four waxy rice starches in relation to thermal, pasting and textural properties. *Cereal Chem.* 79:252–256, 2002.

235. Wang, Z.Y., F.Q. Zheng, G.Z. Shen, J.P. Gao, D.P. Snustad, M.G. Li, J.L. Zhang, M.M. Hong. The amylose content in rice endosperm is related to the post-transcriptional regulation of the *waxy* gene. *Plant J.* 7:613–622, 1995.

236. Weatherwax, P. A rare carbohydrate in waxy maize. *Genetics* 7:568–572, 1922.

237. Whistler, R.L., J.N. BeMiller. Starch. In: *Carbohydrate Chemistry for Food Scientists,* Whistler, R.L., J.N. BeMiller, eds., St. Paul: Eagan Press, 1997, p 146.

238. Whistler, R.L., R.M. Daniel. Carbohydrates in feed chemistry. In: *Food Chemistry,* Fennema, O.R., ed., New York: Marcel Dekker, 1985, pp 118–120.

239. Whistler, R.L., W.M. Doane. Characterization of intermediary fractions of high-amylose corn starches. *Cereal Chem.* 38:251–255, 1961.

240. Whistler, R.L., G.E. Hilbert. Separation of amylose and amylopectin by certain nitroparaffins. *J. Am. Chem. Soc.* 67:1161–1165, 1945.

241. Wilson, E.J., T.J. Schoch, C.S. Hudson. The action of *macerans* amylase on the fractions from starch. *J. Am. Chem. Soc.* 65:1380–1383, 1943.

242. Wu, H.C., A. Sarko. The double-helical molecular structure of crystalline A-amylose. *Carbohydr. Res.* 61:7–25, 1978.

243. Wu, H.C., A. Sarko. The double-helical molecular structure of crystalline B-amylose. *Carbohydr Res* 61:27–40, 1978.

244. Würsch, P, D. Gumy. Inhibition of amylopectin retrogradation by partial beta-amylolysis. *Carbohydr Res* 256:129–137, 1994.

245. Yamada, T., M. Taki. Fractionation of maize starch by gel-chromatography. *Starch/Stärke* 28:374–377, 1976.

246. Yamamori, M., S. Fujita, K. Hayakawa, J. Matsuki, T. Yasui. Genetic eliminiation of a starch granule protein, SGP-1, of wheat generates an altered starch with apparent high amylose. *Theor. Appl. Genet.* 101:21–29, 2000.

247. Yamamori, M., N.T. Quynh. Differential effects of Wx-A1,-B1 and-D1 protein deficiencies on apparent amylose content and starch pasting properties in common wheat. *Theor. Appl. Genet.* 100:32–38, 2000.

248. Yanagisawa, T., C., Kiribuchi-Otobe, H. Yoshida. An alanine to threonine change in the Wx-D1 protein reduces GBSS I activity in waxy mutant wheat. *Euphytica* 121:209–214, 2001.

249. Yano, M., K. Okuno, J. Kawakami, H. Satoh, T. Omura. High amylose mutants of rice, *Oryza sativa* L. *Theor. Appl. Genet.* 69:253–257, 1985.

250. Yao, Y., D.B. Thompson, M.J. Guilitinan. Maize starch-branching enzyme isoforms and amylopectin structure: in the absence of starch-branching enzyme IIb, the further absence of starch-branching enzyme Ia leads to increased branching. *Plant Physiol.* 136:3515–3523, 2004.

251. Yeh, J.Y., D.L. Garwood, J. Shannon. Characterisation of starch from maize endosperm mutants. *Starch/Stärke* 33:222–230, 1981.

252. Yoshimoto Y., J. Tashiro, T. Takenouchi, Y. Takeda. Molecular structure and some physico-chemical properties of high-amylose barley starches. *Cereal Chem.* 77:279–285, 2000.

253. Yuan, R.C., D.B. Thompson. Rheological and thermal properties of aged starch pastes from three waxy maize genotypes. *Cereal Chem.* 75:117–123, 1998.

254. Yuan, R.C., D.B. Thompson, C.D. Boyer. Fine structure of amylopectin in relation to gela-tinization and retrogradation behavior of maize starches from 3 wx containing genotypes in 2 inbred lines. *Cereal Chem.* 70:81–89, 1993.

255. Zhao, X.C., P.J. Sharp. Production of all eight genotypes of null alleles at 'waxy' loci in bread wheat, *Triticum aestivum* L. *Plant Breed.* 117:488–490. 1998.

256. Zheng, G.H., H.L. Han, R.S. Bhatty. Functional properties of cross-linked and hydroxypro-pylated waxy hull-less barley starches. *Cereal Chem.* 76:182–188, 1999.

257. Zobel, H.F., S.N. Young, L.A. Rocca. Starch gelatinization: an x-ray diffraction study. *Cereal Chem.* 65:443–446, 1988.

258. Slade, L., H. Levine. Glass transitions and water-food structure interactions. In: *Advances in Food and Nutrition Research,* Kinsella, J.E., S.L. Taylor, eds., San Diego, CA: Academic Press, 1995, pp 103–269.

259. Yun, S.-H. N.K. Matheson. Structures of the amylopectins of waxy, normal, amylose-extender, and wx:ae genotypes and of the phytoglycogen of maize. *Carbohydr. Res.* 243:307–321, 1993.

260. Hizukuri, S. Starch:analytical aspects. Pgs. 347–429 in: *Carbohydrates in Food.* Eliasson, A.-C., ed., Marcel Dekker, New York, 1996.

261. Sprague, G.F., M.T. Jenkins. The development of waxy corn for industrial use. *Iowa State College Journal of Science* 22:205-213, 1948.

4

Genetic Modification of Plant Oils for Food Uses

Anthony J. Kinney

CONTENTS

4.1   Introduction 85
4.2   Frying Oils..86
4.3   Oils for Baking and Confectionary Applications....................................87
4.4   Fatty Acids for Food Ingredients 89
References..92

4.1   INTRODUCTION

Almost 100 million metric tons of vegetable oil are produced annually, and most of it is used for food purposes. The majority of edible vegetable oil in the world (70 million metric tons) is derived from four crops: soybean, canola, palm and sunflower. About half finds its way (after refining, bleaching, and deodorizing) directly to the retail market in the form of bottled oil for cooking and salads (1). These oils tend to be rich in linoleic acid, an omega-6 polyunsaturated fatty acid, and for this reason the ratio of omega-6 to omega-3 fatty acids in the Western diet is greater than 10 moles of omega-6 to one of omega-3. Most medical experts now believe that this ratio should be about three to one, and suggest a reduction in omega-6 fatty acid consumption or an increased intake of omega-3 fatty acids (2,3).

The half of edible vegetable oil production that is not bottled directly is used by the food industry for the manufacture of shortening, confections, and margarine, and as such is usually hydrogenated to a greater or lesser degree depending upon the application (1,4). Oils are hydrogenated to remove oxidatively unstable polyunsaturated fatty acids, and thus increase the shelf life of oil containing food products. Oils are also hydrogenated to achieve solid fat functionality, by increasing the saturated fatty acid content and by the introduction of *trans* fats, which have a much higher melting point than their *cis* isomers. The relative abundance of saturated and monounsaturated (*cis* and *trans*) fatty acids in the oil will determine the solid fat content (SFC) curve of the oil and thus its food application. A developing understanding of the negative health effects associated with the consumption of *trans* fatty acids is leading food manufacturers to consider alternatives to hydrogenated oils, such as fractionation and transesterification, for many food applications (5).

In recent years the importance in the diet of relatively modest concentrations of bioactive fatty acids, such as long chain omega-3 and conjugated fatty acids, especially conjugated linoleic acid (CLA), has been recognized. Currently the main dietary sources of these fatty acids are dairy products (for CLA) and fish oils (for long chain omega-3 fatty acids). Consumption of these fatty acids has been strongly correlated with a reduced risk for coronary disease and a number of other positive health benefits (2,3). There is a potential for including these bioactive fatty acids in the diet as food ingredients, although their widespread use is currently limited by the high cost of purifying and refining them to a point where they will not negatively impact the flavor of the food to which they are added. Food ingredient quality refined fish oil, for example, can cost as much as $50 per kilogram.

Biotechnology has the potential to impact both the low added value, high volume commodity oil market and the low volume, high added value specialty oil market. Changing the relative abundance of saturated, monounsaturated, and polyunsaturated fatty acids in seed oils by manipulating existing biosynthetic pathways could result in oils that have desired health and functional properties without hydrogenation, fractionation, or other modifications (6). Adding new pathways for the synthesis of conjugated or long chain polyunsaturated fatty acids represents a new direction for food oil biotechnology, and could result in lower cost, higher quality replacements for fish and microbial oils as health promoting food ingredients.

Many of the technical challenges of both these areas have been met, and the feasibility of producing a broad range of novel plant oils in the major oilseed crops has been demonstrated in both model plants and oilseed crops. Some of these oils (such as lauric acid rich canola and high oleic acid soybean oils) have even been commercialized in limited quantities. It is not expected, however, that any of these new oils will make a commercial impact until the end of the current decade. It is hard to imagine any of the major edible plant oils (with the exception of gourmet, organic, and artisanal oils) that will not have been improved, by the middle of this century, by the use of biotechnology.

4.2 FRYING OILS

The most important property of frying oils is resistance to oxidation and polymerization (4). Oxidative stability is usually measured by either the active oxygen method (AOM) or the oxidative stability index (OSI). In both cases the higher the number, the more stable the oil to oxidation and polymerization, and the two measurements normally correlate with each other. A typical refined, bleached, deodorized (RBD) vegetable oil will have an AOM in the 15–25 hour range (4). This is fine for home frying that is done in small batches and often discarded after a single use. For the kind of heavy duty, continuous frying done in the food industry, however, far more stable oils are needed, with AOMs of around 100–200 hours (4,6,7). This stability is usually achieved by either partial hydrogenation or a combination of partial hydrogenation, transesterification, and fractionation. Partial hydrogenation will remove most of the polyunsaturated fatty acids and replace them with *cis* and *trans* monounsaturated fatty acids. There is now a substantial body of evidence linking the consumption of *trans* fatty acids to coronary heart disease, and food manufacturers will soon be required to label the *trans* content of packaged foods (8). Thus there is an incentive to remove polyunsaturated fatty acids from oils by means other than hydrogenation.

The polyunsaturated fatty acids linoleic and linolenic acid are synthesized in the developing oilseed by the action of two types of desaturase enzyme. A delta-12 desaturase inserts a second double bond into oleic acid while it is esterified to the membrane lipid phosphatidylcholine (9). This enzyme is encoded by the Fad 2 gene in plants. Linolenic acid is synthesized by the insertion of a third double bond into linoleic acid by an omega-3 desaturase, encoded by the Fad 3 gene. Again, this reaction occurs while the fatty acid is esterified to a membrane lipid. Because this synthesis of linolenic acid from oleic acid is sequential, blocking the first step by silencing the Fad 2 gene should theoretically result in blocking the formation of polyunsaturated fatty acids and result in an increase in oleic acid. This generally proves to be the case, although most oilseed plants have two or three Fad 2 genes. To achieve the maximum oleic acid content all three may have to be silenced in a seed specific manner. A small amount of polyunsaturated found in seed oils is actually synthesized in a separate pathway located in the plastid. Nevertheless, silencing of Fad 2 genes has resulted in oils with an oleic acid content up to 90%. High oleic soybean oil (HOS), which has an oleic acid content of about 85% and less than 5% total polyunsaturated fatty acids, has an AOM of 150 hours. Polymer formation in HOS oil during frying is very low, equivalent to heavy duty, hydrogenated shortening (7).

Interestingly, high oleic soybean oil has a higher AOM than other modified vegetable oils with a similar oleic and polyunsaturate content. This is most likely because soybean oil is naturally richer in tocopherols than other plant oils, and the form is predominantly *gamma*-tocopherol. When, for example, high oleic sunflower oil is stripped of tocopherols, and tocopherols are added back to the final content and composition of soybean oil, the AOM of the sunflower increases close to that of high oleic soy (7). Most of the genes that control tocopherol formation and the degree of tocol methylation have been cloned (10–12). It is now theoretically possible, therefore, to combine genetic modifications in fatty acid desaturation with modifications of tocopherol metabolism to produce very high stability frying oils from most oilseed species.

Because of concerns surrounding the consumption of genetically enhanced foods and the advent of novel, cheaper transesterification techniques, the commercial production of transgenic high oleic acid oils, even in the US, has been limited. The increasing demand for *trans* free oils created by the new labeling requirements and increased public awareness of *trans* related health issues is likely to reverse this situation. Consequently, in the near future genetically enhanced high oleic oils should become more widely available.

4.3 OILS FOR BAKING AND CONFECTIONARY APPLICATIONS

The bulk of the hydrogenated vegetable oil produced in the US is used as shortening for baking applications (4). These oils were originally developed to replace animal fats such as lard and butter. They typically need a high oxidative stability (AOM between 100 and 200 hours), a high melting point (35–60°C) and usually a SFC curve that is basically flat or very gently sloping between 10 and 40°C. The actual SFC curve requirement depends upon the application, although SFC curve requirements for piecrusts, puff pastry, and baker's margarine would be very similar (4). Consumer margarine applications, which represent about 5% of hydrogenated vegetable oil produced in the US, require very similar SFC curves and properties. Hydrogenated vegetable oils are also used for confectionary coatings, cookie fillers, and wafer fillers. These particular oils, however, need to be solid at room temperature but liquid at body temperature, and thus have a very different SFC requirement from baking and margarine oils (4).

Producing modified plant oils with these two very different types of SFC curves can be done with similar technical approaches. Margarine type baking oils may be replicated by oils with a higher stearic acid and higher oleic acid content than regular plant oils. The coating, confectionary, and filler-type fats can be mimicked by fractionating these same oils to enrich for particular triacylglycerol species. Oils with a confectionary-type SFC curve can also be made by engineering seeds to produce structured triacylglycerols that are also rich in certain medium chain fatty acids.

High oleic, high stearic acid oils can be made by combining the Fad 2 silencing effect described with a gene for increased stearic acid. There are a number of possible approaches to increase stearic acid in oilseeds. All of these approaches involve providing more favorable flux conditions for 18:0-ACP leaving the cytoplasm (where it becomes a substrate for triacylglycerol [TAG] biosynthesis) than for it being desaturated to 18:1-ACP.

Stearic acid is the predominant fatty acid (45–55%) in the seed oil of *Garcinia mangostana* (mangosteen). This is result of the activity of an unusual FatA-type thioesterase, *GarmFatA1*, which has low activity toward 16:0-ACP but high activity toward 18:0 and 18:1-ACPs (13). When a cDNA encoding this FatA was expressed in seeds of canola, it was able to siphon away large amounts of plastidal 18:0-ACP from the acyl-ACP desaturase and transfer it to the cytoplasm as stearoyl-CoA. The resultant stearate accumulation in the seed oil was increased from 1% to 20% of the total fatty acids (13).

It is also possible to slow flux through the acyl-ACP desaturation step to allow the endogenous FatB-type thioesterase to release more stearic acid from 18:0-ACP into cytoplasm. This can be achieved by silencing the gene for one of the number of SAD genes in oilseeds that encode 18:0-ACP desaturases. Knocking out a single gene of the acyl-ACP desaturase gene family usually results in a stearic acid content of around 25% of total fatty acids (6,14).

Because, in both these cases, flux into 18:1 has been reduced, the final oleic acid content of high stearate, high oleic combinations is usually in the 55–60% range (6,9). Vegetable oils which contain 20–25% stearic acid and 55–60% oleic acid have a unique SFC functionality that makes them suitable for baking and margarine applications. This functionality may be further extended by fractionation of these oils.

It has been possible to make oils suitable for chocolate substitutes, confectionary coatings, and cookie fillings by modifying fatty acid synthesis to produce shorter chain fatty acids, from 8:0 to 12:0 carbons in length. A 12:0-ACP FatB-type thioesterase was isolated from California bay laurel. Expressing this FatB-type thioesterase in canola resulted in seed oil rich in lauric acid, up to 50% of the total fatty acids (15). The initial intent of producing this oil (which has the trade name Laurical ®) was for nonfood uses, such as soaps. Once the oil had been produced and its functionality measured, it was also found to be suitable for various food applications (16). Commercial production of this oil has been limited, however, by the availability of relatively inexpensive coconut oil. Nevertheless, the invention of Laurical was a major milestone along the road to more exotic food oils (16).

Subsequently, cDNAs encoding 8:0-ACP, 10:ACP, and 14:0-ACP FatB-type thioesterases have been isolated from *Cuphea* species with oils rich in the corresponding acyl chains. Expression of these cDNAs in canola resulted in oil rich in these medium chain fatty acids (18).

The functionality of all these medium chain oils is limited by the positions of the fatty acids on the triaclyglycerol molecule. In the bay thioesterase oil, for example, the lauric acid is found exclusively at the sn-1 and sn-3 positions. Coexpression in canola seeds of a coconut 12:0-CoA Lysophosphatidic acid acyltransferase, along with the bay thioesterase, facilitated efficient laurate deposition at the sn-2 position resulting in the accumulation of lauric acid at all three positions of the seed triacylglycerol (19).

The cloning and expression of acyl transferases has paved the way for the production of triacylglycerols (TAGs) structured with specific fatty acids at different positions on the glycerol backbone. The utility of these structured triacylglycerols (STAGs) extends beyond the functional properties related to their SFC curve and their nutritional importance is more profound than their bulk fatty acid profile might suggest (20,21). For example, STAGS with medium chain fatty acids at positions sn-1 and sn-3 of the glycerol backbone and an essential fatty acid at the sn-2 position have been useful in the diets of patients suffering from malabsorbtion because they provide both essential fatty acids and a source of energy equivalent to soluble sugars. The medium chain fatty acids are absorbed as free fatty acids, independently of acylcarnitine, and hence faster and more efficiently than longer chain fatty acids. They thus provide a rapid source of energy without having the side effects of high sucrose supplements (21).

In another example, most of the palmitic acid in human milk is at the sn-2 position of triacylglycerol. Enriching infant formula with TAGs having palmitic acid at this position results in a food that more closely resembles natural sources and results in a number of health benefits for the infant, including reduced calcium loss (21,22).

The utility of foods containing STAGs is currently limited by the very high cost of producing the structured molecule itself. By combining the fatty acid pathway manipulation techniques described with the expression of specific acyltransferases it should be possible to directly produce relatively inexpensive plant oils enriched in specific STAGs for various food and medical purposes.

4.4 FATTY ACIDS FOR FOOD INGREDIENTS

There is an increasing body of evidence supporting the conclusion that the fatty acids in the food we eat have a major influence on our physical and mental health (23). This is particularly the case with the essential omega-6 and omega-3 polyunsaturated fatty acids (PUFAs) linoleic and linolenic acids, which are the precursors of the long chain PUFAs arachidonic acid (ARA) and eicosopentaenoic acid (EPA) in humans. In turn these long chain PUFAS are the direct precursors of bioactive eicosanoids (prostaglandins, prostacyclins, thromboxanes, and leukotrienes) and the bioactive omega-3 fatty acid docosohexaenoic acid (DHA). The eicosanoids mediate many functions in the human body including the inflammatory response, the induction of blood clotting, the regulation of mental functions such as the sleep/wake cycle, and the regulation of blood pressure (24). The balance of omega-6 and omega-3 derived eicosanoids is important in the body because those derived from omega-6 ARA are proinflammatory and those derived from omega-3 EPA are antiinflammatory (24). DHA is an important component of mammalian brain membranes and has been shown to play a role in the cognitive development of infants and the mental health of adults (25–27).

Because of dietary changes over the last century many populations, particularly those in the U.S. and Europe, now consume a high proportion of omega-6 fatty acids, mostly in the form of linoleic acid. The main source of omega-3 fatty acids in human diets is fish, the consumption of which is not consistent across any population (23). Thus many people are thought to have suboptimal concentrations of long chain omega-3 fatty acids in their bodies (28). It is no surprise, therefore, that consumption of omega-3 fatty acids, particularly the long chain PUFAs EPA and DHA, has been associated with many positive health benefits such as the prevention of cardiovascular disease in adults, improved cognitive development in infants (25–27), and a greatly reduced risk of sudden cardiac arrest, one of the major causes of death in the U.S. (29).

It has been calculated that adding between 800mg and 1000mg of EPA per day to the human diet would redress the omega-6:omega-3 balance in the body and lead to substantially improved health (2). This is equivalent to eating a 4–12 oz serving of fresh tuna every day, depending on the EPA/DHA content of the fish itself (2). It is very difficult to persuade people to change drastically their dietary habits, particularly if the change involves eating more of a strongly flavored food such as fish. In addition, the world supply of fish (either wild or farmed, because farmed fish need fish oil from the wild to sustain them) is limited, environmentally sensitive, and subject to periodic issues of contamination with mercury and other pollutants (30).

Biotechnology has the potential to provide a sustainable source of high quality and relatively inexpensive long chain omega-3 PUFAs that could be added to many different foods, from salad dressings to high protein nutrition bars. This can be achieved by adding a long chain PUFA biosynthetic pathway to the existing fatty acid biosynthetic pathway in developing oilseeds and thus producing a PUFA enriched plant oil that could be used as a food ingredient.

There are a number of potential pathways that could be added, and a number of potential sources for the pathway genes (31). Fish make few if any long chain PUFAs, obtaining all they need from marine plankton. Over the past few years a wide range of PUFA pathways have been discovered in microbes. Certain marine prokaryotes, such as *Shewanella* species, *Moritella* species, and *Photobacter profundum*, contain significant quantities of either EPA or DHA which they synthesize via a polyketide synthase (PKS) enzyme complex (32). Some eukaryotic microbes, such as many thraustochytrids, also contain a PUFA synthesizing PKS (31,32). It is likely that PUFA PKSs make EPA or DHA from malonyl-CoA by a cyclical extension of the fatty acid chain with isomerization and enoyl reduction occurring in some of the cycles.

Many other eukaryotic microbes synthesize EPA and DHA using an aerobic pathway similar to that of humans, involving desaturases and fatty acid elongating enzymes. In these pathways EPA or DHA is made by the elongation and aerobic desaturation of linoleic and linolenic acids (31).

The PUFA synthesizing PKS was first identified in the EPA rich prokaryote *Shewanella*. The five open reading frames necessary for reconstituting this system in *E. coli* had homology to type I and II PKS components and to type II fatty acid synthase enzymes. A similar PKS from the DHA rich eukaryotic microbe *Schizochytrium* was found to contain three open reading frames (ORFs) with domains homologous to those in *Shewanella* (32). When expressed with a 4′-phosphopantetheinyl transferase, to activate the PKS ACP, the *Schizochytrium* PKS was able to synthesize DHA in *E. coli*. Current efforts are underway to engineer *Schizochytrium* PKS into oilseeds crops such as safflower and thus provide a renewable source of DHA in the seed oil (33).

In humans EPA is synthesized by the delta-6 desaturation of linolenic acid, and delta-6 elongation of the resulting stearidonic acid, followed by delta-5 desaturation to EPA. The same enzymes also convert linoleic acid to ARA (31). This pathway is also used by a wide range of microbes including fungi (e.g., *Mortierella alpina* and *Saprolegnia diclina*), mosses (e.g., *Physcomitrella patens*) and microalgae (e.g., *Phaeodactylum tricornutum*). A number of other freshwater and marine microbes (*Isochrysis galbana*, *Euglena gracillus*) have been shown to use an alternative aerobic pathway of EPA biosynthesis that begins with the delta-9 elongation of linoleic acid. This is followed by delta-8 and delta-6 desaturations to yield ARA, and then an omega-3 (delta-17) desaturation to EPA (34).

The synthesis of DHA from EPA in humans is complex, involving two elongation steps, a delta-6 desaturation, and β-oxidation in the peroxisome to DHA (31). Microbial DHA synthesis follows a simpler route. EPA is elongated to docosopentaenoic acid (DPA) which is then desaturated by a delta-4 desaturase to DHA.

There has also been some success in engineering both the delta-8 and the delta-6 aerobic pathways for EPA biosynthesis into plants. The delta-8 pathway was reconstructed in the model oilseed plant *Arabidopsis*, by the heterologous expression of an *I. galbana* delta-9 elongase, a *Euglena* delta-8 desaturase, and an *M. alpina* delta-5 desaturase. Significant accumulation of both ARA and EPA (7% and 3% respectively) was observed in *Arabidopsis* leaves, although the seed content of these long chain PUFAs was not reported (35). A similar EPA content was observed in flax seeds containing a delta-6 pathway transferred from *Physcomitrella patens* (36). Flax was chosen because it has a naturally abundant linolenic acid content, thus encouraging the formation of long chain omega-3, rather than omega-6, fatty acids (37). Another approach to omega-3 enrichment is to add extra omega-3 desaturase genes along with the long chain PUFA pathway (38,39). Transforming soybeans with a delta-6 desaturase and delta-6 elongase from *M. alpina*, a delta-5 desaturase from *S. diclina* and two extra omega-3 desaturases (under the control of various seed specific promoters) resulted in soybean oil containing up to 20% EPA and no ARA (40). It has also been possible to include two more genes, encoding a C20-elongase and a delta-4 desaturase, to the set for this aerobic pathway in transgenic plants, resulting in the production of small quantities of DHA in seed oils (36,40).

All of these results are very encouraging regarding future plant derived sources of long chain omega-3 fatty acids although considerable progress still needs to be made in optimizing pathway expression, producing crops with good agronomics and in stabilizing the new, highly unsaturated oil during production (to prevent oxidation).

Another bioactive fatty acid that is attracting some interest from the biotechnology world is conjugated linoleic acid (CLA). This fatty acid is normally found in dairy products, and has attracted this interest because of its reported effects on reducing body fat mass and increasing lean body mass (41). Other beneficial health related effects of CLA that have been reported in the literature include anticarcinogenic effects, and beneficial effects on the prevention of coronary heart disease, as well as modulation of the immune system (42). Most of the research on CLA, however, has been done with animals and there is not yet a scientific consensus about the human benefits of CLA consumption (42). Part of the problem may be that CLA is really a collective term signifying a group of different geometrical and positional isomers of linoleic acid and it is still unclear which isomers are the most active forms (43,44). Most natural food sources of CLA consist predominantly of the *cis-9 trans-11* isomer, while commercial CLA preparations are usually a mixture of *cis-9 trans-11* and *trans-10 cis-12* isomers. Consequently, most research has focused on these two isomers, although it is quite possible that other isomers and other conjugated fatty acids may also have some biological activity. Only very recently have researchers begun to examine the effect of purified CLA isomers in humans, and in these studies both *cis-9 trans-11* and *trans-10 cis-12* isomers have been shown to have some biological effects on blood lipids and immune function (44,45).

Attempts have been made to provide new food sources of CLA by enriching seed oils with conjugated fatty acids. One approach to this is expressing a polyenoic *cis-trans* fatty acid isomerase in developing oilseeds to convert newly synthesized linoleic acid to CLA before it is incorporated into oil (46). Isomerase genes have been cloned from algal (*Ptilota filicina*) and bacterial (*Propionium, Bacterium acnes*) sources, and attempts made to express these genes in plant cells (46,47). The substrates for these isomerases, however, are free fatty acids, which are only present in trace amounts in plant cells. Thus only trace amounts of CLA have been observed when isomerases were expressed in plants (47). Because most fatty acids in plant cells are part of complex lipids such as phospholipids, galactolipids and triacylglycerol, it seems likely that coexpressing an isomerase in developing oil seeds with some kind of lipase might be necessary to observe useful quantities of CLA in plant oils.

Another approach to making significant quantities of conjugated fatty acids in plant oils is to use plant enzymes that operate on esterified fatty acids. Dienoic conjugated fatty acids, such as CLA, are only present in plants in trace amounts although trienoic fatty acids are abundant in some plant species. The trienoic isomer of linolenic acid, calendic acid (*trans-8, trans-10, cis-12* CLnA), is found in high relative abundance (over 65% of total fatty acids) in the seeds of the pot marigold *Calendula officinalis* (48). This conjugated linolenic acid isomer contains the *trans-10 cis-12* double bonds thought to be involved in the biological activity of commercial CLA. The *trans-8, trans-10, cis-12* CLnA is synthesized in the developing marigold seed from linoleic acid, while the fatty acid is esterified to the membrane lipid phosphatidylcholine (48). The synthesis is catalyzed by a delta-12 oleic acid desaturase related enzyme encoded by a member of the Fad 2 gene family. When the marigold Fad 2 conjugase was expressed in soybean seeds, significant concentrations of calendic acid (15–20%) were observed in the seed oil (48).

Similar delta-12 oleic acid desaturase related enzymes catalyze the synthesis of other linolenic acid isomers, such as α-eleostearic acid (*cis-9, trans-11, trans-13* CLnA) in Chinese bitter melon (*Momordica charantia*) seeds (49) and trichosanic acid (*cis-9, trans-11, cis-13* CLnA) in *Trichosanthes kirilowii* and *Punica granatum* seeds (50). Again, it is possible to produce these other CLnA isomers in plant oils by expressing the Fad 2-related gene encoding the specific linoleic acid conjugase in the developing seeds of an oilseed plant (49,50).

It is not yet known how the biological activities of various CLnA isomers compare with those of CLA isomers, although there have been a number of provocative recent studies that have even raised the suggestion that dietary CLnA may be more biologically potent than CLA (51–54). For example, it has been shown that eleostearic acid prepared from both pomegranate seed and bitter melon seeds had very potent anticarcinogenic effects in rats (51,52). In another study, chemically synthesized eleostearic acid was found to be substantially more effective at suppressing tumors in rats than was CLA (53). Other studies in rats have found calendic acid to be more effective than CLA at increasing lean body mass and reducing adipose tissue weight (54).

The biotechnology of bioactive food lipids such as long chain omega-3 PUFAs and conjugated fatty acids is still in its infancy. Nevertheless, the ability to produce these fatty acids in plant oils has now been demonstrated and has the potential to impact the long term physical and mental health of human populations in a significant and positive way.

REFERENCES

1. *USDA Economic Research Service Oil Crops Situation and Outlook Yearbook*, U.S. Department of Agriculture, Springfield, VA, October, 2003.
2. Kris-Etherton, P.M., W.S. Harris, L.J. Appel. American Heart Association Nutrition Committee: Fish consumption, fish oil, omega-3 fatty acids, and cardiovascular disease. *Circulation* 106:2747–2757, 2002.
3. Wijendran, V., Hayes, K.C. Dietary n-6 and n-3 fatty acid balance and cardiovascular health. *Annu. Rev. Nutr.* 24:597–615, 2004.
4. Stauffer, C. *Fats and Oil*. St. Paul: Eagen Press, 1996.

5. Ascherio, A. Epidemiologic studies on dietary fats and coronary heart disease *Am. J. Med.* 113(9)B:9S–12S, 2002.

6. Kinney, A.J. Development of genetically engineered soybean oils for food applications. *J. Food Lipids* 3:209–273, 1996.

7. Kinney, A.J., Knowlton, S. Designer oils: the high oleic soybean. In: *Genetic Modification in the Food Industry*, Roller, S., S. Harlander, eds. Glasgow: Blackie & Son, 1998.

8. Food and Drug Administration, HHS. Food labeling: trans fatty acids in nutrition labeling, nutrient content claims, and health claims, final rule. *Fed. Regist.* 68:1433–1506, 2003.

9. Voelker, T., A.J. Kinney. Variations in the biosynthesis of seed-storage lipids. *Annu. Rev. Plant Physiol. Plant. Mol. Biol.* 52:335–361, 2001.

10. Van Eenennaam, A.L., K. Lincoln, T.P. Durrett, H.E. Valentin, C.K. Shewmaker, G.M. Thorne, J. Jiang, S.R. Baszis, C.K. Levering, E.D. Aasen, M. Hao, J.C. Stein, S.R. Norris, R.L. Last. Engineering vitamin E content: from Arabidopsis mutant to soy oil. *Plant Cell* 15:3007–3019, 2003.

11. Cahoon, E.B., S.E. Hall, K.G. Ripp, T.S. Ganzke, W.D. Hitz, S.J. Coughlan. Metabolic redesign of vitamin E biosynthesis in plants for tocotrienol production and increased antioxidant content. *Nat. Biotechnol.* 21:1082–1087, 2003.

12. Cheng, Z., S. Sattler, H. Maeda, Y. Sakuragi, D.A. Bryant, D. DellaPenna. Highly divergent methyltransferases catalyze a conserved reaction in tocopherol and plastoquinone synthesis in cyanobacteria and photosynthetic eukaryotes. *Plant Cell* 15:2343–2356, 2003.

13. Hawkins, D.J., J.C. Kridl. Characterization of acyl-ACP thioesterases of mangosteen (*Garcinia mangostana*) seed and high levels of stearate production in transgenic canola. *Plant J.* 13:743–752, 1998.

14. Knutzon, D.S., G.A. Thompson, S.E. Radke, W.B. Johnson, V.C. Knauf, J.C. Kridl. Modification of Brassica seed oil by antisense expression of a stearoyl-acyl carrier protein desaturase gene. *Proc. Natl. Acad. Sci. USA* 89:2624–2628, 1992.

15. Voelker, T.A., A.C. Worrell, L. Anderson, J. Bleibaum, C. Fan, D.J. Hawkins, S.E. Radke, H.M. Davies. Fatty acid biosynthesis redirected to medium chains in transgenic oilseed plants. *Science* 257:72–74, 1992.

16. Del Vecchio, A.J. High laurate canola. *Inform* 7:230–243, 1996.

17. Dehesh, K., A. Jones, D.S. Knutzon, T.A. Voelker. Production of high levels of 8:0 and 10:0 fatty acids in transgenic canola by overexpression of Ch FatB2, a thioesterase cDNA from *Cuphea hookeriana*. *Plant J.* 9:167–172, 1996.

18. Voelker, T.A., A. Jones, A.M. Cranmer, H.M. Davies, D.S. Knutzon. Broad-range and binary-range acyl-acyl-carrier protein thioesterases suggest an alternative mechanism for medium-chain production in seeds. *Plant Physiol.* 114:669–677, 1997.

19. Knutzon, D.S., T.R. Hayes, A. Wyrick, H. Xiong, H. Maelor Davies, T.A. Voelker. Lysophosphatidic acid acyltransferase from coconut endosperm mediates the insertion of laurate at the sn-2 position of triacylglycerols in lauric rapeseed oil and can increase total laurate levels. *Plant Physiol.* 120:739–746, 1999.

20. Mu, H., C.E. Hoy. The digestion of dietary triacylglycerols. *Prog. Lipid Res.* 43:105–133, 2003.

21. Mu, H., C.E. Hoy. Intestinal absorbtion of specific structured triacylglycerols. *J. Lipid Res.* 42:792–798, 2001.

22. Tomarelli, R.M., B.J. Meyer, J.R. Weaber, F.W. Bernhart. Effect of positional distribution on the absorption of the fatty acids of human milk and infant formulas. *J Nutr.* 95:583–590, 1968.

23. Lands, W.E. Essential fatty acids in the foods we eat have a subtle but powerful influence on hundreds of different processes in the life and death of humans *Nutr. Metab. Cardiovasc. Dis.* 13:154–164, 2003.

24. Yaqoob, P. Fatty acids and the immune system: from basic science to clinical applications. *Proc. Nutr. Soc.* 63:89–104, 2003.

25. Willatts, P., Forsyth, J.S. The role of long-chain polyunsaturated fatty acids in infant cognitive development. *Prostaglandins Leukot. Essent. Fatty Acids* 63:95–100, 2000.

26. Iribarren, C., J.H. Markovitz, D.R. Jacobs, Jr., P.J. Schreiner, M. Daviglus, J.R. Hibbeln. Dietary intake of n-3, n-6 fatty acids and fish: relationship with hostility in young adults – the CARDIA study. *Eur. J. Clin. Nutr.* 58:24–31, 2004.

27. Stoll, A.L., K.E. Damico, B.P. Daly, W.E. Severus, L.B. Marangell. Methodological considerations in clinical studies of omega 3 fatty acids in major depression and bipolar disorder. *World Rev. Nutr. Diet.* 88:58-67, 2001.

28. Harris, W.S., C. Von Schacky. The Omega-3 Index: a new risk factor for death from coronary heart disease? *Prev. Med.* 39:212–220, 2004.

29. Richter, W.O. Long-chain omega-3 fatty acids from fish reduce sudden cardiac death in patients with coronary heart disease. *Eur. J. Med. Res.* 8:332–336, 2003.

30. Naylor, R.L., R.J. Goldburg, J.H. Primavera, N. Kautsky, M.C. Beveridge, J. Clay, C. Folke, J. Lubchenco, H. Mooney, M. Troell. Effect of aquaculture on world fish supplies. *Nature* 405:1017–1024, 2000.

31. Wallis, J.G., J.L. Watts, J. Browse. Polyunsaturated fatty acid synthesis: what will they think of next? *Trends Biochem. Sci.* 27:467–473, 2002.

32. Metz, J.G., P. Roessler, D. Facciotti, C. Levering, F. Dittrich, M. Lassner, R. Valentine, K. Lardizabal, F. Domergue, A. Yamada, K. Yazawa, V. Knauf, J. Browse. Production of polyunsaturated fatty acids by polyketide synthases in both prokaryotes and eukaryotes. *Science* 293:290–293, 2001.

33. Metz, J.G. PUFA synthases: characterization, distribution and prospects for PUFA production in heterologous systems. *6th Congress of the International Society for the Study of Fatty Acids and Lipids, Brighton, UK, Abstract book,* 2004, p 41.

34. Sayanova, O.V., J.A. Napier. Eicosapentaenoic acid: biosynthetic routes and the potential for synthesis in transgenic plants. *Phytochemistry* 65:147–158, 2004.

35. Qi, B., T. Fraser, S. Mugford, G. Dobson, O. Sayanova, J. Butler, J.A. Napier, A.K. Stobart, C.M. Lazarus. *Nat. Biotechnol.* 22:739–745, 2004.

36. Heinz, E. Biosynthesis of VLCPUFAs in transgenic oilseeds. *16th International Plant Lipid Symposium, Budapest, Hungary, Abstract book,* 2004, p 8.

37. Drexler, H., P. Spiekermann, A. Meyer, F. Domergue, T. Zank, P. Sperling, A. Abbadi, E. Heinz. Metabolic engineering of fatty acids for breeding of new oilseed crops: strategies, problems and first results. *J. Plant Physiol.* 160:779–802, 2003.

38. Spychalla, J.P., A.J. Kinney, J. Browse. Identification of an animal omega-3 fatty acid desaturase by heterologous expression in *Arabidopsis. Proc. Natl. Acad. Sci. USA* 94:1142–1147, 1997.

39. Pereira, S.L., Y.S. Huang, E.G. Bobik, A.J. Kinney, K.L. Stecca, J.C. Packer, P. Mukerji. A novel omega3-fatty acid desaturase involved in the biosynthesis of eicosapentaenoic acid. *Biochem. J.* 378:665–671, 2004.

40. Cahoon, E.B., H.G. Damude, W.D. Hitz, A.J. Kinney. Production of very long chain polyunsaturated fatty acids in oilseed plants. U.S. Patent Application US20040049813, 2004.

41. Jeukendrup, A.E., S. Aldred. Fat supplementation, health, and endurance performance. *Nutrition* 20:678–688, 2004.

42. Rainer, L., C.J. Heiss. Conjugated linoleic acid: health implications and effects on body composition *J. Am. Diet Assoc.* 104:963–968, 2004.

43. Wang, Y., P.J. Jones. Dietary conjugated linoleic acid and body composition. *Am. J. Clin. Nutr.* 79(6):1153S–1158S, 2004.

44. O'Shea, M., J. Bassaganya-Riera, I.C. Mohede. Immunomodulatory properties of conjugated linoleic acid. *Am. J. Clin. Nutr.* 79(6):1199S–1206S, 2004.

45. Burdge, G.C., B. Lupoli, J.J. Russell, S. Tricon, S. Kew, T. Banerjee, K.J. Shingfield, D.E. Beever, R.F. Grimble, C.M. Williams, P. Yaqoob, P.C. Calder. Incorporation of cis-9,trans-11 or trans-10,cis-12 conjugated linoleic acid into plasma and cellular lipids in healthy men. *J. Lipid Res.* 45:736–771, 2004.

46. Zhen, W., L. Yuan, J. Metz. Nucleic acid sequences encoding polyenoic fatty acid isomerase and uses thereof, International Patent Application WO2001009296, 2002.

47. Hornung, E. Production of conjugated fatty acids in plants with bacterial double bond isomerases: biosynthesis of VLCPUFAs in transgenic oilseeds, *16th International Plant Lipid Symposium, Budapest, Hungary,* Abstract book, 2004, p 27.

48. Cahoon, E.B., K.G. Ripp, S.E. Hall, A.J. Kinney. Formation of conjugated delta8,delta10-double bonds by delta12-oleic-acid desaturase-related enzymes: biosynthetic origin of calendic acid. *J. Biol. Chem.* 276:2637–2643, 2001.

49. Cahoon, E.B., T.J. Carlson, K.G. Ripp, B.J. Schweiger, G.A. Cook, S.E. Hall, A.J. Kinney. Biosynthetic origin of conjugated double bonds: production of fatty acid components of high-value drying oils in transgenic soybean embryos. *Proc. Natl. Acad. Sci. USA* 96:12935–12940, 1999.

50. Iwabuchi, M., J. Kohno-Murase, J. Imamura. Delta 12-oleate desaturase-related enzymes associated with formation of conjugated trans-delta 11, cis-delta 13 double bonds. *J. Biol. Chem.* 278:4603–4610, 2003.

51. Kohno, H., R. Suzuki, Y. Yasui, M. Hosokawa, K. Miyashita, T. Tanaka. Pomegranate seed oil rich in conjugated linolenic acid suppresses chemically induced colon carcinogenesis in rats. *Cancer Sci.* 95:481–486, 2004.

52. Kohno, H., Y. Yasui, R. Suzuki, M. Hosokawa, K. Miyashita, T. Tanaka Dietary seed oil rich in conjugated linolenic acid from bitter melon inhibits azoxymethane-induced rat colon carcinogenesis through elevation of colonic PPARgamma expression and alteration of lipid composition. *Int. J. Cancer* 20(110):896–901, 2004.

53. Tsuzuki, T., Y. Tokuyama, M. Igarashi, T. Miyazawa. Tumor growth suppression by alpha-eleostearic acid, a linolenic acid isomer with a conjugated triene system, via lipid peroxidation. *Carcinogenesis* 25:1417–1425, 2004.

54. Koba, K., A. Akahoshi, M. Yamasaki, K. Tanaka, K. Yamada, T. Iwata, T. Kamegai, K. Tsutsumi, M. Sugano. Dietary conjugated linolenic acid in relation to CLA differently modifies body fat mass and serum and liver lipid levels in rats. *Lipids* 37:330–343, 2002.

5

Molecular Biotechnology for Nutraceutical Enrichment of Food Crops: The Case of Minerals and Vitamins

Octavio Paredes-López and Juan Alberto Osuna-Castro

CONTENTS

5.1 Micronutrients: Alleviating Nutritional Disorders by Nutraceuticals.....................97
5.2 Micronutrient Bioavailability and Approaches to Reduce Micronutrient
    Malnutrition..108
5.3 Minerals...112
    5.3.1 Iron...112
    5.3.2 Zinc...116
    5.3.3 Calcium...116
    5.3.4 Phytic Acid...116
5.4 Vitamins and Nutraceuticals..117
    5.4.1 Carotenoids as Food Pigment and Provitamin A......................................117
        5.4.1.1 Carotenogenesis in Tomato..120
        5.4.1.2 Carotenogenesis in Other Food Crops.......................................122
    5.4.2 Vitamin E...124
    5.4.3 Vitamin C..125
Acknowledgments...128
References..128

5.1 MICRONUTRIENTS: ALLEVIATING NUTRITIONAL DISORDERS BY NUTRACEUTICALS

Ever since the discovery of vitamins, there has been a great interest in deciphering the effects of food constituents on human health. The requirements for a healthy life with a balanced metabolism (metabolic homeostasis) and preservation of the body cell mass require a protein balance having all the essential amino acids, fatty acids, carbohydrates, and micronutrients (1,2).

Approximately 40 micronutrients (the vitamins, essential minerals, and other compounds needed in small amounts for normal metabolism) are required in human diet (Tables 5.1, 5.2). Obtaining maximum health and life span requires metabolic harmony. For each micronutrient, metabolic harmony requires an optimal intake (i.e., to give maximal life span); deficiency distorts metabolism in numerous and complicated ways, many of which may lead to DNA damage (3–7). Many micronutrient minerals and vitamins act as substrates or cofactors in key DNA maintenance reactions, and the exact concentration of these in the cell may be critical. Suboptimal levels of key micronutrients will thus lead to impaired activity of enzymes needed for genomic stability in man, producing effects similar to inherited genetic disorders or exposure to carcinogens (Tables 5.1, 5.2) (3–7). In fact, a deficiency of some micronutrients (folic acid, vitamin B12, vitamin B6, niacin, vitamin C, vitamin E, selenium, iron, or zinc) appears to mimic exposure to radiation or chemicals by causing single and double strand breaks in DNA, oxidative lesions, or both (4–12). Chromosomal aberrations such as double strand breaks are a strong predictive factor for human cancer.

In adults, superoxide, hydrogen peroxide, and hydroxy radicals (produced by some 10^{10} free radicals per cell each day) can potentially cause in the order of 10^6 mutational alterations of DNA per cell per day. Fortunately, the activities of these dangerous mutagens are countered by antioxidants, DNA repair, the removal of persistent alterations by apoptosis (normally a cellular program is activated, that causes the cell to self destruct), differentiation, necrosis, and activation of the immune system so that only about one mutation per cell per day persists (13). By old age, many mutations have accumulated because the repair system is not working properly, and as a consequence, cancer may occur. However, the increased consumption of dietary antioxidants such as vitamins C and E, quercetin (flavonoid), and carotenoids (zeaxanthin and lycopene) can diminish DNA oxidation and therefore, less cancer incidence is presented (3,13–18).

The U.S. Recommended Dietary Allowances (RDAs) for micronutrients refer to daily levels of intake of essential nutrients, at or above defined minimum values, judged by the Food and Nutrition Board to be adequate to meet the known nutrients needed for practically all healthy persons and to prevent deficiency diseases. The level of daily requirement for each nutrient varies with age, sex, and physiological status (e.g., pregnancy, lactation, disease related stress) (Tables 5.1, 5.2) (19). RDAs are established by estimating the requirement for the absorbed nutrient, adjusting for incomplete utilization of the ingested nutrient, and incorporating a safety factor to account for variability among individuals (19).

Also, now it is known that vitamins, minerals, and other dietary components consumed at varying levels, higher than RDAs, are significant contributors to the reduction of risks of chronic diseases such as cancers, cardiovascular diseases, and degenerative diseases associated with aging (Alzheimer's, premature aging). Their role in maintaining human genomic stability is likely to be critical to alleviating the classical nutritional deficiency diseases which include scurvy (vitamin C), anemia (folic acid, iron), and pellagra (niacin) (Tables 5.1, 5.2) (4–7,17,20,21).

While deficiencies of dietary energy (i.e., calories) and protein currently affect more than 800 million people in food insecure regions, incredibly, micronutrient malnutrition, known as "hidden hunger," now afflicts over 40% of the world population, especially the most vulnerable being in many developing nations (resource-poor women, infants, and children), and a surprisingly large amount of people in the advanced countries, where food diversity, abundance, and supply are excellent, but also poor food eating habits are common and in both cases the numbers are rising (4,22–25). Even though micronutrients are needed in minute quantities (i.e., micrograms to milligrams per day), they have tremendous impact on human health and well-being (Tables 5.1, 5.2) (22,26,27). Insufficient dietary intake of these microcomponents impairs the functions of the brain, the immune and reproductive systems, and energy metabolism (26,27).

At present, deficiencies of iron, vitamin A and iodine are of the most concern to the community and healthcare systems. Unfortunately, for zinc there is a lack of a simple, quick, and cheap clinical screening test for determining marginal zinc deficiency in humans. Specialists in zinc nutrition suggest that its deficiency presents similar problems as caused by a deficiency of iron (10,22,24,25,28). Iron, zinc, and vitamin A all play important roles in brain development, and when deficiencies of these micronutrients are manifest during pregnancy, or even for up to two years postpartum, permanent damage to offspring is possible. Thereafter, further loss of cognitive ability is found (10,22,24–31).

Iodine deficiency is the greatest single cause of preventable brain damage and mental retardation in the world today. More than 2 billion people around world live in iodine deficiency environments. Deficiency in iodine occurring in late infancy and childhood have been demonstrated to produce mental retardation, delayed motor development, and stunted growth, occurrence of neuromuscular disorders, and speech and hearing defects. Even mild iodine deficiency has been reported to decrease intelligence quotients by 10–15 points (22,31).

Rapid growth during fetal development, infancy, childhood, and adolescence demand greater amount of micronutrients, when their deficiency can cause developmental abnormalities (22,25). In addition to multiple micronutrient deficiencies (those of vitamin A and C, decreased iron absorption), interactions between micronutrient deficient conditions and infectious diseases may become significant. These effects can complicate health care efforts to control various diseases such as vitamin A deficiency and measles, and increased severity of measles leading to vitamin A deficiency blindness and death. Malaria and hookworm infections are also associated with Fe deficiency anemia (22,25,30,32).

Plants synthesize and accumulate an astonishingly diverse array of vitamins, and nutraceuticals that have health-promoting properties (33,34). Many naturally occurring compounds have health promoting or disease preventing properties beyond the mere provision of nutrients for basic nutrition (1,34–36). In this way, plants not only constitute the base of the human food chain but they are also important means to improve human health and well-being, with the exception of vitamin B12 and D. A diverse and well-balanced plant-based diet, that includes mixed sources of grains, fruits, and vegetables, can ensure the proper micronutrient nutrition and health at all stages of the life cycle (Tables 5.1, 5.2) (18,20,21,37–39). It is known that seeds are good sources of lipid-soluble vitamins, but tend to have low levels of bioavailable iron, zinc, and calcium, whereas leafy vegetables can supply most minerals and vitamins. Fruits provide water-soluble vitamins and several types of carotenoids but generally are minor sources of certain minerals. Thus, concentrations of very few individual plant foods are able to supply all the daily recommended intake of any micronutrient in an average or reasonable serving size (20–22,37–39). Unfortunately, many people do not consume a sufficiently diverse diet. In fact, many low-income families from the developing world subsist with a simple diet formed mainly of a staple food (i.e., maize, soybean, beans, wheat, tubers, root) that are poor sources of many micronutrients (2,16,20–25,27,37–39). Heavy and monotonous consumption of cereal-based foods with low concentrations and diminished bioavailability of iron and zinc has been considered a major reason for the current widespread deficiency of both minerals in developing countries (16,20–25,27,37–39). Furthermore, even in developed nations, the average intake of fruits and vegetables may fall below official recommendations (19,21).

Table 5.1
Minerals: Dietary allowances per day and safe upper intake limit values, sources, deficiency and importance on human genomic stability.

Minerals	Maximum Adult RDA$^\phi$	Safe Upper Intake Limits (Relative to RDA)$^\theta$	Predominant Food Sources	Deficiency Some Classical Nutritional Disorders and Diseases	Deficiency DNA Damage (Negative Effect on Genomic Stability)	Deficiency New Health Effects	Functions on Genomic Stability
Selenium	70 μg	13X	Seeds	Keshan disease, an endemic cardiomyopathy in China	DNA Oxidation (radiation mimic)	Prostate cancer, increased risk for heart diseases and other type of cancer	Retardation of oxidative damage to DNA, proteins and lipids, modulation of cellular events critical in cell growth inhibition and multi-step carcinogenesis process, cofactor of glutathione peroxidase
Iodine	150 μg	13X	Iodized salt, sea foods, plants and animal grown in areas where soil iodine is not depleted	Goiter, mental retardation, brain damage and reproductive failure	—	—	—
Chromium	50–200 μg	1X	Various	Rare	—	—	—
Molybdenum	75–250 μg	1X	Seeds	Rare	—	—	—
Copper	1.5–3 mg	1X	Organ meats, sea foods, nuts, seeds	Rare	—	—	Cofactor of cuproenzymes of antioxidant system
Fluoride	1.5–4 mg	1X	Aerial tissues	Problems with bone and teeth	—	—	—

Manganese	2.–5 mg	1X	Whole seeds, fruit	Extremely rare	—	—	—
Zinc	15 mg	1X	Meat, eggs, nut, seeds	Growth retardation, delayed skeletal and sexual maturity, dermatitis, diarrhea, alopecia and defects in immune system function with decreased resistance to infections	Elevated DNA Oxidation (radiation mimic) DNA breaks and increased chromosome damage rate	Immune and brain dysfunction, cancer	Co-factor of enzymes: Cu/Zn superoxide dismutase, endonuclease IV, zinc fingers
Iron	13–15 mg	5X	Meat, seeds, leafy vegetables	Nutritional anemia, problem pregnancies, stunted growth, tiredness and diminished food-energy conversion, poor work performance	DNA breaks, radiation mimics	Immune and brain dysfunction with lower resistance to infections and long-term impairment of neural motor development, and mental function, cancer	Fe-containing enzymes that are particularly important for gene regulation
Magnesium	350 mg	1X	Whole grains, nuts, green leafy vegetables	Rare, but when it occurs there are some problems with bone structures	—	—	—
Sodium	500 mg	5X	Common salt	Rare, but excesive intakes may lead to hypertension	—	—	—
Chloride	750 mg	5X	Various	Rare	—	—	—

(Continued)

Table 5.1 (*Continued*)

Minerals	Maximum Adult RDA^φ	Safe Upper Intake Limits (Relative to RDA)^θ	Predominant Food Sources	Deficiency Some Classical Nutritional Disorders and Diseases	Deficiency DNA Damage (Negative Effect on Genomic Stability)	Deficiency New Health Effects	Functions on Genomic Stability
Phosphorus	1200 mg	2X	Ubiquitous, animal products tend to be good sources	Rare due to presence in virtually all foods	—	—	—
Calcium	1200 mg	2X	Milk products, green leafy vegetables, tofu, fish bones	Osteoporosis in elderly, rickets	—	—	—
Potassium	2000 mg	9X	Fruits, vegetables, meats	Rare	—	—	—

φ Recommended daily dietary allowances. RDA values given mean that highest either for male or female adult, but not for pregnant or lactating women. θ The concept assumes that there is individual variation in both requirement for the minerals and tolerance for elevated intake.

Sources: References 3–7, 9–11, 16, 19–22, 24–29, 31, 37, 43, 46.

Table 5.2

Vitamins: Dietary allowances per day and safe upper intake limit values, sources, deficiency and importance on human genomic stability.

Vitamins	Maximum Adult RDA$^\phi$	Safe Upper Intake Limits (Relative to RDA)$^\theta$	Predominant Food Sources	Deficiency/Traditional Functions — Classical Nutritional Disorders and Diseases	Deficiency — DNA Damage (Negative Effects on Genomic Stability)	Deficiency — New Health Effects	Role on Genomic Stability
Fat Soluble							
Vitamin D (Chole-calciferol)	10 μg	4X	Milk and dairy products	Rickets, whose common symptoms include weak and misshapen bone, bowlegged and poor muscle tone; Required to incorporate calcium, phosphate and magnesium into bones	—	Colorectal cancer and adenoma, increase cell proliferation	—
Vitamin K	80 μg	375X	Leafy vegetables (spinach, cauliflower, cabbage), tomato and some vegetable oils	Elevated hemorrhage due to reduced prothrombin levels, particularly in premature or anoxic infants	—	—	—
Vitamin A	1 mg RE$^\psi$	5X (retinol); 100X (β-carotene)	Some pigmented vegetables and fruits	Poor night vision, eye lesions, and in severe cases, permanent blindness; elevated illness and death from infections	—	—	

(Continued)

Table 5.2 (*Continued*)

Vitamins	Maximum Adult RDA[◊]	Safe Upper Intake Limits (Relative to RDA)[θ]	Predominant Food Sources	Deficiency/Traditional Functions	Deficiency	Deficiency	Role on Genomic Stability
				Classical Nutritional Disorders and Diseases	DNA Damage (Negative Effects on Genomic Stability)	New Health Effects	
Fat Soluble							
Vitamin E	10 mg α-TE [θ]	100X	Oilseeds, leafy vegetables	Increased fragility in red cells and hemolysis, as well as impaired sensation and neuromuscular activity	Increased baseline level of DNA strand breaks, chromosome breaks and oxidative DNA damage and lipid radicals adduct on DNA	Heart disease, colon cancer, immune system impairment	Prevention of lipid and DNA (radiation mimic) oxidation
Water Soluble							
Vitamin B12 (Cyano-cobalamin)	2 µg	500X	Fish, dairy products, meat	Megaloblastic anemia where proper maturation of red blood cells is hampered, among other symptoms such as anorexia, intestinal discomfort, depression and some neurological problems	Chromosome breaks	Neuronal damage also see folate	Maintaining methylation level in DNA

Biotin	30–100 µg	300X	Seeds, animal products	Rare	—	—	—
Folate	200 µg	50X	Fruits, leafy vegetables, legumes including dark and dried beans	Megaloblastic anemia, weight loss, nausea, confusion, irritability and dementia are typical.	Uracil misincorporation into DNA provoking DNA breakage (DNA hypomethylation), chromosome breaks (radiation mimic)	Colorectal cancer, most heart diseases, altered risks of birth defects (neural tube), brain dysfunction (Down's syndrome), tumorigenesis, acute lymphoblastic leukemia in children	Maintaining methylation level in DNA, efficient recycling of folate
Vitamin B1 (Thiamine)	1.5 mg	67X	Crops	Beriberi, symptoms include weak muscles, fatigue, depression, irritability, weight loss, with frequency cardiovascular complications and paralysis	—	—	—
Vitamin B2 (Riboflavin)	1.7 mg	—	Leafy vegetables (i.e., broccoli), cereals, liver, beef and cheese	Nouse dermatitis, altered coloration and texture of lips and tongue, fatigue and sensitive and vascularized eyes, bloodshot	—	—	—

(Continued)

Table 5.2 (Continued)

Vitamins	Maximum Adult RDA[θ]	Safe Upper Intake Limits (Relative to RDA)[θ]	Predominant Food Sources	Deficiency/Traditional Functions — Classical Nutritional Disorders and Diseases	Deficiency — DNA Damage (Negative Effects on Genomic Stability)	Deficiency — New Health Effects	Role on Genomic Stability
Vitamin B6 (Pyridoxin)	2 mg	125X	Whole grains, meat	Vomiting, irritability, weakness, ataxia	Chromosome breaks	Cancer of prostate, and lung, See folate	See folate
Pantothenic acid	4–7 mg	150X	Cereal seeds, eggs, meat, milk, fresh vegetables	—	—	—	—
Vitamin B3 (Niacin)	19 mg NE*	150X	Leafy vegetables, crops	Pellagra, characterized by skin lesions, diarrhea, and mental apathy	Disables DNA repair, (augmented percentage of unrepaired nicks in DNA); High levels of chromosome fractures, mutagen sensitivity	Neurological problems; memory loss	Substrate of poly (ADP-ribose) polymerase enzyme, that cleaves DNA and rejoins cut ends, and telomere length maintenance, component of DNA involved in poly ADP-ribose

| Vitamin C | 60 mg | 16X | Citrus fruits, vegetables | Irritability, growth delay, anemia, poor wound healing, increased tendency to bleed. and susceptibility to infections; weak cartilages and tenderness in the legs are typical symptoms of scurvy | See vitamin E | Cataract, See vitamin E | defense against DNA strand fractures See vitamin E |

$^{\phi}$Recommended daily dietary allowances. RDA values given mean that highest either for male or female adult, but not for pregnant or lactating women. $^{\theta}$The concept assumes that there is individual variation in both requirement for the vitamin and tolerance for elevated intake. $^{\psi}$ Vitamin A activity is expressed in retinol equivalent (RE). One RE represents 1 mg of all-*trans*-retinol or 6 mg of all-*trans*-β-carotene. $^{\vartheta}$ one α-tocopherol equivalent (TE) is similar to 1 mg of (R, R, R)-α-tocopherol. $^{\circ}$ One mg NE (niacin equivalent) represents 1 mg of dietary tryptophan.

Sources: References 3–8,12–14, 16–22, 24, 25, 31, 37, 38, 42, 45, 92.

The world population is expected to grow from 6 billion to around 10 billion by 2050. Virtually all of the anticipated increase in population will occur in Africa, Latin America, and Asia (23–25,30,31). These regions already face serious problems of food production and access, water scarcity, nutritional disorders, and malnutrition of macronutrients and micronutrients (23–25,30,31). Among the undernourished, children under five years of age are particularly at risk. A high percentage of children are underweight, and low birth weight becomes an important factor in child malnutrition and premature death (23–25,30). Notably, countries that have dramatically reduced the incidence of malnutrition have also concurrently reduced their birth rates dramatically over the last century.

Therefore, to ensure an adequate dietary intake of all essential micronutrients and to augment the consumption of various nutraceuticals by people around the world, great and intelligent efforts are needed to improve the nutritional and nutraceutical quality of agricultural produce, with regard to their micronutrient composition, concentration as well as their biological availability (16,17,27,38–41).

5.2 MICRONUTRIENT BIOAVAILABILITY AND APPROACHES TO REDUCE MICRONUTRIENT MALNUTRITION

The mineral and vitamin content of food crops should be considered at the same time with its bioavailability (20,42,43). The total content, or absolute concentration, of a given nutrient in a food is not always a reliable indicator of its useful nutritional quality, because not all of the nutrients in food are absorbed (20,42,43).

The human nutritionists employ the concept of bioavailability of micronutrients, which refers to the percentage of a given nutrient that is potentially available for absorption from an ingested food, and once absorbed, utilizable for normal physiological functions and even for storage in the body (25,40,44). Bioconversion is defined as the fraction of a bioavailable micronutrient (i.e., absorbed provitamin A carotenoids) that is converted to the active form of a nutrient (retinol) (44,45).

The digestive processes help to release and solubilize nutrients so they can diffuse out of the bulk food matrix into the enterocytes of the human intestine. Micronutrients can be present in several chemical species in plant foods and their quantities may vary according to growth environment, plant species and genotype, and cultural practices utilized to grow the plant (21,22,24,43,46,47). These chemical forms may have characteristically distinct solubilities and reactivities related to other plant components and other meal constituents; absorption then is a function of physicochemical properties and form of micronutrients (i.e., free, complexed, charged, dispersible). The absorption and transport of nutrients may be influenced by the presence of inhibitory (i.e., antinutrients) and promotive (i.e., promoters that can enhance absorption and utilization) substances in the food matrix (Table 5.3) (21,22,24,32,39,43,46,47).

As a result, for a complete and realistic description of the nutritional adequacy of food from dietary micronutrient viewpoint, three factors should be assessed: (1) the concentration of the micronutrients at the time of consumption; (2) the nature of chemical species of the micronutrients present; and (3) the bioavailability of these forms of the micronutrient as they exist in the meal consumed (21,22,24,32,39,42,43,46,47).

The bioavailability of normal micronutrients from plant sources ranges from less than 5% for some minerals such as calcium and iron, over 90% for sodium and potassium (20,21,43). The three most common and widely recognized strategies for reducing micronutrient mineral and vitamin malnutrition include: (1) supplementation with pharmaceutical preparations, (2) fortification, (3) and dietary food diversification (16,21,22,31,39–43,48–50).

For instance, iron and zinc supplementation has been useful in developing countries for rapid improvement of the status of both minerals in deficient individuals, but this approach is relatively expensive and often has poor compliance, particularly with medicinal Fe because of unpleasant side effects (22,43).

Commercial fortification of foods is familiar to most of us. Minerals and vitamins are added to a particular food vehicle during its processing, well after the foods have left the farm and before they are distributed through various marketing channels for consumer purchase and consumption (31). Thus, fortification refers to incorporation of nutrients at superior levels than those existing in the original (i.e., unprocessed) (42,48,49). This may include adding nutrients not normally associated with the food. Food fortification is usually the best long term strategy, but its effectiveness depends on the compatibility of fortified foods with local culinary habits. Iron is the most difficult mineral to incorporate into foods (due to technical problems related to the choice of suitable iron compounds) and to ensure adequate absorption (22,43,48,49). For example, those with relatively high bioavailability, which are water-soluble iron compounds such as ferrous sulfate, usually provoke unacceptable color and flavor changes in food. When water-soluble iron compounds are incorporated into cereal flour, they often cause rancidity; and in low grade salt, they rapidly lead to undesirable color formation. Furthermore, other compounds which are organoleptically inert, such as elemental iron powders, are so poorly absorbed as to be of little or no nutritional value (22,43,48,49).

Both supplementation and fortification have treated the symptoms of micronutrient malnutrition rather than the underlying causes. While many of these interventions have been successful in the short term, these strategies have proved to be prohibitively expensive in marginal economies, unsustainable and incapable of reaching all the individuals affected. Ironically, those segments of population at high risk (poor women, infants, and children) usually live in remote places either from a clinic and health care professionals or do not have ready access to processed and fortified foods (22,24,30,43,48,49). Finally, food diversification may also be quite difficult, but not impossible, in developing countries for economic, social, or traditional reasons, and people who avoid milk products in developed nations normally do so because they dislike or are intolerant to dairy products (16,20,38).

Within the agricultural community, conventional plant breeding efforts during the green revolution were focused on productivity and efficiency of cropping systems (mostly cereal crops), being greatly successful in supplying enough calories and macronutrients to prevent the threatening global starvation and shortage of foods predicted at the beginning of 1960s (37,51). Remarkably, the nutrition community has never embraced agriculture as a key tool to be used to fighting "hidden hunger"; paradoxically the green revolution may have contributed to some unforeseen negative consequences on human nutrition and health referent to micronutrient malnutrition (22,25,37). Thus, in some developing countries the rise in micronutrient deficiencies may be associated to changes in the patterns in cultivation toward cereals, which was paralleled by a significant decreased per capita production of traditional edible legumes that are a much richer source of micronutrients than cereals, especially after cereals have been milled and polished before consumption (22,23,25).

Plant mineral and vitamin concentrations vary among plant sources (i.e., species, cultivars) and within plant tissues (i.e., leafy structures against seeds); thereby demonstrating the existence of genetic variability, which can contribute to the plant's ability to acquire, sequester, synthesize, and accumulate micronutrients (16,21,22,33,34,37,38,52). In contrast to macronutrients which can represent up to 30 to 50% in dry weight in some tissues (g/100 g portion of food), individually, minerals and vitamins when present constitute a very tiny quantity ranging from a microgram to a milligram per 100 g of food (less than 0.1%) (2,16,21,22,33,34,37,38,52). Therefore, significant quantitative and qualitative

Table 5.3

Antinutritional and promoter substances that reduce and enhance, respectively, the bioavailability of some important micronutrients for human nutrition and health, and even genomic stability, and their food distribution.

Antinutrient(s)	Micronutrient(s) Negatively Affected (Less Bioavailability)	Common Dietary Food Sources of Antinutrient(s)	Promoter(s) [Enhancer(s)]	Micronutrient(s) Positively Affected (Improved Bioavailability)	Common Dietary Food Sources of Promoter(s)
Some tannins and other polyphenols	Iron, zinc	Pigmented beans, sorghum, coffee, tea	Meat factors (cysteine-rich polypeptides)	Iron, zinc	Animal meats (i.e., pork, beef, fish)
Hemagglutinins (i.e., lectins)	Iron, zinc	Wheat and most legumes	Some free and essential amino acids (i.e., lysine, histidine, sulfur-containing)	Iron and/or zinc	Animal meats (i.e., pork, beef, fish)
Fiber (i.e., cellulose, cutin, hemicellullose, lignin)	Iron, zinc	Whole-cereal-based products	Some organic acids and/or their salts (i.e., ascorbic acid, citrate, fumarate, malate)	Iron and/or zinc	Several fresh fruits and vegetables
Toxic heavy metals (i.e., Cd, Hg, Pb, Ag)	Iron, zinc	Contained leafy vegetables, tubers and grains (i.e., Cd in rice) from metal-polluted soils	Plant ferritin	Iron	Legume crops, leafy vegetables

Phytic acid or phytin	Iron, zinc, magnesium, calcium	Whole legume and cereal crops	Hemoglobin	Iron	Animal red meats
Oxalic acid	Calcium	Spinach	β-Carotene, vitamin A	Iron	Green and orange vegetables, red palm oil, yellow maize
—	—	—	Iron, zinc	Vitamin A	Animal meats
—	—	—	Selenium	Iodine	Sea foods, organ meats, tropical nuts and cereals, amounts vary depending on soil levels
—	—	—	Vitamin E (α-tocopherol)	Vitamin A	Vegetable oils, green leafy vegetables

Sources: Adapted from References 22, 24, 27.

changes are feasible for minor plant components. Genetic manipulation to augment the levels of these components would require minimal diversion of precursors and some modifications in the plant's ability to store or sequester the target micronutrients (16,33,34,52,53). Interestingly, the tolerable upper intake levels in humans for minerals fluctuate between 1- and 13-fold in their RDA values, whereas those for vitamins are higher than those for minerals, and, as it has been reviewed earlier, micronutrient intakes superior to RDAs, but not higher than those tolerable levels for humans, provide health benefits including genomic stability (Tables 5.1, 5.2) (3–7).

All this raises the opportunity for the molecular biotechnology as an emerging and powerful approach with the potential for improving nutritional quality of food plants, altering the composition, content, and bioavailability of the existing micronutrients [i.e., modifying chemical forms of the stored micronutrient, removing (or reducing the level of) antinutritional compounds, or elevating the amount of promoter substances] (Table 5.3), or accumulating novel and bioavailable minerals and vitamins in edible parts (i.e., the endosperm of cereals), which usually lack these components (18,33,34,41,49,50).

5.3   MINERALS

In developing nations, cereal grains such as wheat, maize, rice, and sorghum, and some legumes such as common bean and soybean, are the primary and cheap sources of essential minerals as iron, zinc, and calcium (20,37,43). Minerals with chemical similarities can compete for transport proteins or other uptake mechanisms, as well as for chelating organic substances, hindering absorption (Table 5.3) (32,54,55). In fact, it has also been suggested that antinutritional factors that interfere with proper nutrient absorption and bioavailability account for a large proportion of world wide micronutrient deficiencies (40,55).

Thus, increasing the amount of bioavailable micronutrients in plant foods for human consumption by molecular biotechnology is a challenge that is not only important for developing countries, but also for many industrialized countries. Theoretically, it could be achieved by increasing the total level of micronutrients in the edible part of staple crops, such as cereals and pulses, while simultaneously increasing the concentration of compounds which promote their uptake, for example ascorbic acid, and by decreasing the concentration of chemicals that inhibit their absorption, such as phytic acid or some phenolic compounds (Table 5.3) (22,41,43,50).

5.3.1   Iron

Iron is both an essential micronutrient and a potential toxicant to cells; as such, it requires a highly sophisticated, coordinated, and complex set of regulatory mechanisms to meet the demands of cells as well as prevent excess accumulation (29). The human body requires Fe for the synthesis of the oxygen transport proteins hemoglobin and myoglobin and for the formation of heme enzymes and other Fe-containing enzymes that are particularly important for energy production, gene regulation, immune defense, regulation of cell growth and differentiation and thyroid function (Table 5.1) (28,29,43). The body normally regulates Fe absorption so as to replace the obligatory iron losses of about 1–1.5 mg per day. Thus, the body must be economical in its handling of iron, for example, when a red blood cell dies, its iron is reutilized, and excess level of iron can be stored by a specially designed protein, ferritin, which is used at times of increased iron metabolic requirements (28). In spite of these ingenious physiological approaches, iron deficiency is estimated to affect around 30% of the world population, making iron by far the most deficient nutrient worldwide. In general, the etiology of iron deficiency can be viewed as a negative balance between iron intake and iron loss. Whenever there is a rapid growth, as occurs during infancy, early childhood, adolescence, and pregnancy, positive iron balance is difficult to maintain (9,28). The blood volume expands in parallel with growth, with a corresponding increase in iron requirement (9).

Dietary iron is constituted by heme iron (animal origin) and nonheme iron (inorganic salts mainly from vegetal sources); the first one is absorbed by a distinct route and more efficiently than nonheme iron (Table 5.1). Usually, nonheme iron bioavailability is very low (less than 5–10%) due to its poor solubility and interaction with other diet components known as antinutrients (28,43). Inadequate absorption of this mineral will first lead to the mobilization of storage iron, and finally to lower hemoglobin levels or anemia (29). Iron deficiency is the most common cause of anemia and is usually due to inadequate dietary intake of bioavailable iron and/or excessive loss due to physiological conditions of parasitic infections (30,32). For example, dietary iron sources in developing countries consist mainly of nonheme iron. Because cereal and legume staples are rich in phytic acid, a potent inhibitor of mineral absorption, and in addition, the intake of foods that enhance nonheme iron absorption such as fruits, vegetables, and animal muscle tissues is often limited, these conditions may serve as major factors responsible for the anemia (Table 5.3) (21,22,43,47,49,55). The major consequences of this deficiency are poor pregnancy outcome, including increased mortality of mothers and children, reduced psychomotor and mental development in infants, decreased immune functions, tiredness, and poor work performance (25,26,30–32).

Increasing ferritin, the natural iron store protein, in food crops, has been suggested as an approach to raise iron levels and bioavailability. Ferritin is a multimeric iron storage protein, composed of 24 subunits, and has a molecular structure highly conserved among plants, animals, and bacteria (56,57). This protein is capable of storing up to 4500 Fe atoms in its central cavity, which are nontoxic, biologically available, and releases them when iron is required for metabolic functions. In fact, recent studies have demonstrated that iron from animal and plant ferritin can successfully be utilized by anemic rats and humans (56,58,59).

While staple food, such as corn and wheat flours are usually fortified with iron, rice grains present much hard problems and challenges. In addition, whole brown rice is barely consumed, and its commercial milling (polishing) produces considerable loss of micronutrients, up to 30% and 67% for zinc and iron, respectively, by eliminating its outer layers where these metals are accumulated (27).

With the aim of increasing the iron content and its bioavailability in rice, two different research groups have overexpressed a ferritin gene into its endosperm isolated from either common bean or soybean (Table 5.4) (60–62). In both cases the plant ferritin is produced at high levels and correctly accumulated in the cereal endosperm. Notably, the iron content of bean ferritin rice is 22.1 µg/g dry weight whereas soybean ferritin rice stores up to 31.8 µg/g dry weight, resulting in two- and threefold greater levels, respectively, than that of the corresponding untransformed crop (10–11 µg/g dry weight). A two to three times extra iron enrichment in ferritin in transgenic grains would appear to be of nutritional significance. In fact, a daily consumption of about 300 g of the iron-rich rice by an adult would be sufficient to provide 50–75% of the daily adult requirements for this mineral, which is about 13–15 mg (Table 5.1).

Recently, Vansconcelos et al. (63), a third distinct research group, also reported the expression of soybean ferritin gene, driven by the endosperm-specific glutelin promoter, leading to higher iron accumulation in transgenic *indica* rice seed than control grains, even after commercial milling. They selected as target the *indica* rice line IR68144-3B-2-2-3, an elite line, which presents high tolerance for tungro virus and an excellent grain quality, good yield and resistance for growing in mineral-poor soils, and high iron level in the crop (15–17 µg/g untransformed brown rice). Transgenic rice lines were obtained containing as much as 71 µg iron/g dry weight unpolished rice. This accounts for a 4.4-fold increase in iron compared with that of the control; a two- to threefold extra iron content of transformed rice with plant ferritins would already be of nutritional relevance as noted earlier (Tables 5.1, 5.4) (60–62). But when the iron levels of the rice grains (untransformed and transgenic) were assessed after the seeds were polished, they indicated that the highest iron content of transgenic lines ranged from 19–37 µg/g milled rice, versus control material of only 10 µg/g milled rice. This is the first report which shows that after commercial milling the iron concentration remains higher than that of that milled negative control, and even that of untransformed brown crop. These results with transgenic rice expressing a ferritin from either soybean or bean would imply that low iron concentration in food seeds may not result from low iron availability for transport, but rather from a lack of sequestering and storing capacity in the seeds.

In order to explore and test the potential benefit of iron-improved transgenic rice incorporating soybean ferritin in its edible tissue (60), a standard hemoglobin depletion bioassay was employed with anemic rats followed by complete diets having equivalent quantities of either iron as $FeSO_4$ (a popular compound used in anemic human beings in medical treatments) or bioengineered ferritin rice. Iron-rich rice diet was as effective as the diet containing $FeSO_4$, and it was shown that full recovery of anemia in rats occurred after 28 days of treatment with any of the iron sources (Table 5.4) (64).

It is generally agreed that nutrients are effectively utilized from breast milk and that breast-fed infants possess a lower prevalence of infections than those fed with commercial formula. Breast milk not only provides the infant with a well balanced supply of nutrients, but also several unique components that facilitate nutrient digestion and absorption, protection against pathogenic microorganisms, and promotion of healthy growth and development. It is believed that those benefits are due in part to milk proteins (41,65,66). One of such bioactive proteins is lactoferrin. Lactoferrin is an 80 kDa iron-binding glycoprotein belonging to the transferrin family and is found in elevated levels (1–2 g/l) in human milk. Proposed biological activities for this protein include antimicrobial properties, regulation and facilitating iron absorption, immune system modulation, cellular growth activity, and antivirus and anticancer activities (41).

Rice was used as a useful bioreactor to produce, in its edible endosperm, recombinant human lactoferrin to infant food because it presents a low allergenicity, and is likely a vehicle safer than transgenic microorganisms or animals. Therefore, a human milk lactoferrin linked to a rice glutelin 1 promoter was inserted into rice cells and a very high expression level was reached in a large scale field trial (5 g of recombinant human lactoferrin per kilogram of dehusked transgenic rice), being stable for four generations. In fact, the boosting expression of lactoferrin in rice endosperm turned this cereal grains pink, as a consequence of iron bound to lactoferrin (65). The gross nutrient composition of transgenic cereal was similar to that of nontransformed rice, except for a twofold increase in iron content (negative control, 5.7 µg/g dehusked rice; transgenic rice, 19.3 µg/g dehusked rice) probably because each molecule of lactoferrin is able to bind two Fe^{3+} ions (Table 5.4) (65).

Additionally, the lactoferrin purified from transgenic rice exhibited similar pI, antimicrobial activity against a human pathogen (i.e., inhibition of growth of enteropathogenic *Escherichia coli*, one of the most common causes of diarrhea in infants and children) and bind and release iron capacity at acidic gastric pH as those of native human lactoferrin (Table 5.4) (65,66). Lactoferrin-rich rice crops (as lactoferrin is bioactive and therefore, has the ability to store iron in a bioavailable manner) can be incorporated directly into infant formula or baby foods, and even to be consumed by people at any age; without purification

Table 5.4

Manipulation of selected micronutrients for human nutrition and nutraceutical uses by molecular biotechnology.

Micronutrient	Molecular Approach	Improved Plant	Results and Comments	Ref.
Mineral				
Iron	Expression of bean-ferritin- or soybean-ferritin	Rice	Both bean-ferritin- and soybean-ferritin-rice presented high levels of stored iron, in their edible endosperm being of nutritional and nutraceutical significance for fighting against anemia disease and other iron-related deficiencies. In fact, anemic rat fed with iron-rich transgenic rice had a full recovery from that disease, and thus a food with enhanced iron bioavailability.	60–63.
	Production of human milk lactoferrin	Rice	Overaccumulation of lactoferrin in rice endosperm, which turned pink; transgenic protein showed similar antimicrobial activity against a human pathogen and both bind capacity and release iron at acidic gastric pH as native human lactoferrin; also immune system modulation, cellular growth activity, antivirus and anticancer properties are expected	65, 66.
Phosphorus	Expression of microbial or plant phytase	Canola, soybean, rice, maize, wheat	Phytic acid reduced amount in crops; significant improvement in essential mineral bioavailability such as iron, calcium and zinc and in protein digestibility are expected. Low-phytic-acid maize enhanced iron absorption in humans that consumed tortillas whereas pigs fed with this cereal exhibited a higher phosphorus absorption	43, 47, 61, 68–71.
Vitamin				
Vitamin A	Production of carotenoid biosynthetic enzymes	Canola, rice, tomato	Enhanced content of provitamin A and other nutraceutical carotenoids such zeaxanthin	79–82.
Vitamin E	γ-Tocopherol methyltransferase expression	Arabidopsis	Increased levels of α-tocopherol, higher vitamin E activity, nutraceutical	90.
Vitamin C	Expression of rat L–gulono–γ–lactone oxidase or strawberry D-galacturonic acid reductase	Lettuce, Arabidopsis	Improved antioxidant and nutraceutical properties and iron bioavailability; reduction, or elimination of bisulfite utilization to prevent browning in leaf lettuce	94, 95.

this milk protein provides a convenient advantage over other heterologous protein expression systems (65,66).

These achievements have demonstrated the feasibility of producing ferritin or lactoferrin in a very important food crop as rice, as a cheap and good source of bioavailable iron (Table 5.4). However, further investigations are now needed to show biological usefulness in the human diet and their contribution to a solution to global problems of iron deficiency.

5.3.2 Zinc

Most of the zinc in the human body is in the bones and muscles. This mineral acts as a stabilizer of the structures of membranes and cellular components. Its biochemical function is as an essential component of a large number of Zn-dependent enzymes, especially in the synthesis and degradation of biomacromolecules such as carbohydrates, proteins, lipids, and nucleic acids, as well as wound healing (43). Because of a significant portion of cellular zinc is found in the nucleus, it seems that mechanistically this metal is participating in processes which include stabilization of chromatin structure, DNA replication and transcription by the activity of transcription factors and DNA and RNA polymerases, as well as playing a key role in DNA repair and programmed cell death (Table 5.1) (10).

These features give Zn an essential and unique role for healthy growth and development of human beings. In fact, its deficiency reduces appetite, growth, sexual maturity, and the immune defense system (22,30,43). Meat and seafood are good sources of Zn for people in industrialized nations, providing up to 70% of their requirements. However, its deficiency has just recently taken dimensions as a serious public health problem (10).

5.3.3 Calcium

Calcium is required for the normal growth and development of the bones. It accumulates at the rate of about 150 mg per day during human skeletal growth until genetically predetermined peak bone mass is reached in the early twenties. Bone mass is then stable until about 50 years of age in men or before menopause in women (43). After that time, Ca balance becomes negative and bone is lost from all skeletal sites, which is related with a marked rise in fracture rate in both sexes, but particularly in women. So, adequate Ca intake during adolescence is critical. Osteoporosis in old age is characterized by a microarchitectural deterioration of bone tissue, leading to increased bone fragility as well as to an increased risk of fractures (Table 5.1) (22,46). Also, during early childhood this mineral is of a significant concern as its deficiency can cause rickets (22).

Besides its structural role in humans, Ca plays major regulatory functions in several biochemical and physiological processes such as blood clotting, muscle contraction, cell division, transmission of nerve impulses, enzyme activity, cell membrane function and hormone secretion (37,46). Milk and dairy products are the most important sources of calcium, but again these foods are scarce in less developed countries; in Mexico maize tortillas are a good and important Ca source for great part of its population (20,39).

5.3.4 Phytic Acid

Seeds normally accumulate severalfold more phosphorus than that which is needed to support basic cellular functions. In the normal seed this excess P is incorporated into a single small molecule referred to as phytic acid (*myo*-inositol-1,2,3,4,5,6-hexa*kis*phosphate). Grain crops typically contain about 10 mg of phytic acid per gram of dry seed weight, representing about 60–85% of total P (47). Once synthesized, most phytic acid is deposited as a mixed phytate or phytin salt of K and Mg, although phytates also contain other mineral cations such as Fe and Zn. During germination, phytate salts are broken down by the action of phytases, releasing P and *myo*-inositol for use by the growing seedling. Phytic acid is a polyanion at physiological pH and an effective chelator of nutritionally important mineral cations (Ca, Zn, Mg, and Fe) (Table 5.3) (22,32,40,47,54,55,67). Once consumed in human foods or animal feeds, phytic acid binds to these minerals in the intestinal tract to form mixed and unavailable salts that are largely excreted. This phenomenon contributes to mineral deficiency in human populations (40). In addition, diverse investigations suggest that phytic acid may also react with proteins making them partially unavailable for human absorption (22,67). On the other hand, phytic acid can readily be degraded in cereal and legume foods by the addition of exogenous phytases either during food processing or during digestion, increasing mineral absorption dramatically (22,43,47,50). Other interesting approach that has been proposed is for an *in vivo* decrease of phytic acid levels through raising phytase activity in crops using genetic engineering. Also, it is important that the enzyme should be able to withstand the cooking temperatures that occur during food preparation. Thus, highly important agronomic crops such as canola, soybean, rice, wheat, and even tobacco seeds have been successfully transformed with fungal phytase genes, driving their expression by either constitutive or seed-specific promoters (Table 5.4) (43,50,61,68,69). Significant enhancement on mineral bioavailability is expected in transgenic crops synthesizing and accumulating functional microbial phytases. On the other hand, consumption of natural low phytic-acid mutant maize improved iron absorption in humans fed with maize-based diets, for example tortillas prepared with this type of maize. The phytic acid level was reduced significantly representing only 35% of that found in wild-type maize and outstandingly, absorption of iron from transgenic tortillas was nearly 50% greater than from normal tortillas (Table 5.4) (70). Moreover, when pigs were fed with low phytic-acid maize, a higher P bioavailability of up to 62% than that obtained from the nonmutant, wild-type grain, was observed (71).

5.4 VITAMINS AND NUTRACEUTICALS

Vitamins are defined as a diverse group of food-based essential organic substances (relatively small molecules but comparable in size to amino acids or sugars) that are not synthesized by the human body, but by plants and microorganisms. Therefore, vitamins are nutritionally essential micronutrient for humans and function *in vivo* in several ways, including: (1) as coenzymes or their precursors (niacin, thiamin, riboflavin, biotin, pantothenic acid, vitamin B6, vitamin B12, and foliate); (2) in specialized functions such as vitamin A in vision and ascorbate in distinct hydroxylation reactions; and (3) as components of the antioxidative defense systems (vitamins C and E and some carotenoids), and as factors involved in human genetic regulation and genomic stability (folic acid, vitamin B12, vitamin B6, niacin, vitamin C, vitamin E, and vitamin D) (Table 5.2) (5,6,42).

Vitamins present in a food source, once taken up by the body, are dissolved either in water or fat (37,42). As a consequence vitamins are classified on the basis of their solubility as water-soluble or fat-soluble vitamins. Vitamins and the chemical structures of each group are presented in Figures 5.1 and 5.2.

5.4.1 Carotenoids as Food Pigment and Provitamin A

The term carotenoids summarizes a class of structurally related compounds, which are mainly found in plants, algae, and several lower organisms, bacteria, and fungi. At present, more than 600 different carotenoids have been identified (72). Saffron, pepper, leaves, and red palm oil possessing carotenoids as their main color components, have been exploited as food colors for several centuries. The color of carotenoids, together with beneficial properties such as vitamin A precursor and antioxidant activity, has led to their wide application in the food industry. They have been used for pigmentation of margarine, butter, fruit juices and

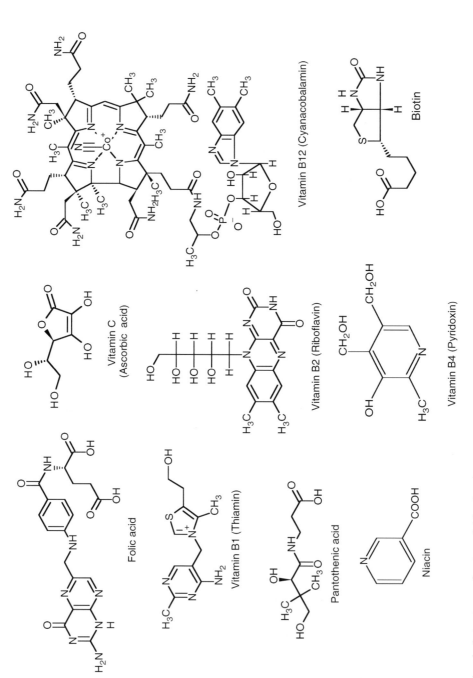

Figure 5.1 Chemical structures of water soluble vitamins.

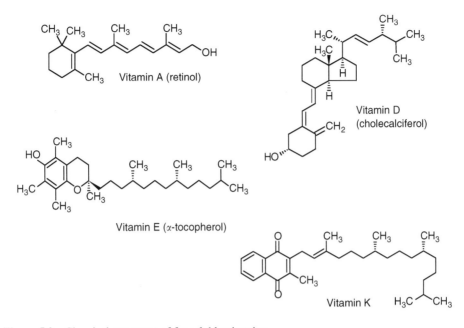

Figure 5.2   Chemical structures of fat soluble vitamins.

beverages, canned soups, dairy and related products, desserts and mixes, preserves and syrups, sugar and flour confectionery, salad dressings, meat, pasta and egg products (17,73).

Carotenoids form one of the most important classes of plant pigments and play a crucial role in defining the quality parameters of fruits and vegetables. Their function as antioxidants in plants in the prevention of photo oxidative damage shows interesting parallels with their potential roles as antioxidants in foods and humans (72). Carotenoids as food colorants take advantage of their very high stability to pH and reducing agents (ascorbic acid); their colors fluctuate from yellow to red, including orange, with variations of brown and purple (17).

Epidemiological studies suggest that the onset of chronic disease states such as coronary heart disease, certain cancers and macular degeneration can be reduced by high dietary intakes of carotenoid-rich foods (72–74). Lycopene has been shown to prevent the incidence of prostate cancer, while zeaxanthin and lutein offer protection against the occurrence of age-related macular degeneration. In addition, β-carotene is the most potent provitamin A carotenoid; α-carotene and β-cryptoxanthin also possess provitamin A activity, but to a lesser extent than β-carotene. Their deficiency can result in blindness and premature death (Table 5.2) (17,74).

Marigold (*Tagetes erecta* L.) varieties range in petal color from white to dark orange. This pigmentation of flowers is due to the massive synthesis and accumulation of carotenoids during petal development. The dark orange varieties possess concentrations of carotenoids that are 20-fold greater than those found in ripe tomato fruit (75). For this reason, marigolds are grown commercially as an important source of carotenoids and used as an animal feed supplement (73).

In our laboratory, Delgado-Vargas et al. (76) found that sunlight illumination produces a favorable equilibrium toward all *trans*-lutein isomers. Also, *trans*-isomers showed increased red hues, and the poultry fed with sunlight-illuminated marigold meal (lutein as its main carotenoid) showed a better egg yolk-pigmentation than controls. These authors also suggested that other components different from carotenoids may participate in pigmentation efficiency.

Also, the antimutagenicity activity of carotenoids from Aztec marigold has been evaluated. It was concluded that lutein was the compound with a higher activity in marigold extracts; however, the mixture of carotenoids in the marigold extract had higher antimutagenicity activity. In addition, it was suggested that lutein and 1-nitropyrene (mutagen) formed an extracellular complex that limits the bioavailability of 1-nitropyrene and consequently its mutagenicity (77).

5.4.1.1 Carotenogenesis in Tomato

In general, carotenoids are compounds comprised of eight isoprenoid units whose order is inverted at the center of the molecule. Formally, all carotenoids can be considered as lycopene ($C_{40}H_{56}$) derivatives by reactions involving hydrogenation, dehydrogenation, cyclization of the ends of the molecules, or oxidation (Figure 5.3A), which in addition to the number and location of conjugated double bonds inside their structure, influence biological action (antioxidant properties) and pigmentation efficiencies of colorants in food systems (17,74,76,78).

Based on their chemical composition, carotenoids are subdivided into two groups. Those which contain only carbon and hydrogen atoms are collectively assigned as carotene (β-carotene, α-carotene, and lycopene). The majorities of natural carotenoids have at least one oxygen function, such as keto, hydroxy, or epoxy groups and are referred to as xanthophylls (oxy-carotenoids) (17,72,74,76,78). The first step in the carotenoid biosynthetic pathway is the head-to-head condensation of two molecules of geranylgeranyl pyrophosphate (GGPP) to produce the colorless intermediate phytoene (Figure 5.3A) (17,74,75,78,79). This reaction is catalyzed by the enzyme phytoene synthase (PSY). Introduction of four double bonds convert phytoene to red lycopene. In plants, two enzymes carry out those transformations, phytoene desaturase (PDS), and ζ-carotene desaturase (ZDS). Each desaturase catalyzes two symmetric dehydrogenation steps; thus, the first enzyme catalyzes the conversion of ζ-carotene via phytofluene (phytoene to phytofluene to ζ-carotene) whereas the second transforms ζ-carotene to lycopene through neurosporene. In contrast, the bacterial phytoene desaturase (ctrl) is capable of introducing the four double bonds converting phytoene to lycopene. Subsequently, the ends of the linear lycopene can be cyclized by lycopene cyclases, resulting in the formation of β-carotene and α-carotene. Hydroxylation reactions carried out by hydroxylases [β-carotene hydroxylase (β-*Chy*) and ε-hydroxylase (ε-*Chy*)] produce the xanthophylls β-cryptoxanthin, zeaxanthin, and lutein (Figure 5.3A) (17,74,75,78,79).

Vitamin A refers to a family of essential, fat-soluble dietary compounds required for vision, growth, reproduction, cell proliferation, cell differentiation, and the integrity of the immune system (Table 5.2) (45). The most potent vitamin A compound, all-*trans*-retinol, is able to reverse signs and symptoms of vitamin A deficiency (44,45).

All vitamin A is originally derived from carotenoids, and all green plants produce carotenoids as β-carotene in their tissues as essential photosynthetic pigments. But for many staple crops, to have nutritional value from a micronutrient viewpoint, the carotenoids must also be produced in the nonphotosynthetic edible tissues consumed by humans; however traditional breeding methods have had little success in producing staple crops containing a high vitamin A concentration (17,22).

Research on carotenoid metabolic engineering in plants has focused on increasing nutritional quality, and it includes: (1) introducing variation in carotenoid products in tomato; (2) production of higher levels of preexisting carotenoids in canola; and (3) accumulation of carotenoids in normally carotenoid-free tissues, such as rice endosperm (Table 5.4) (78–83).

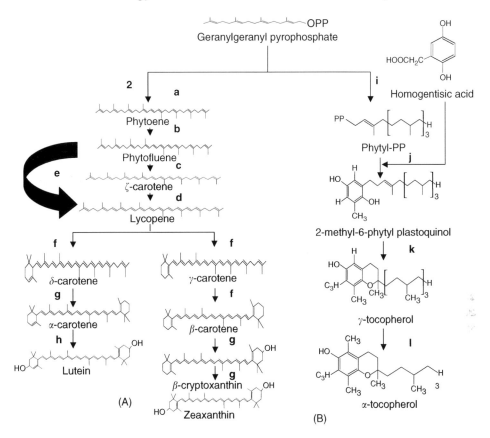

Figure 5.3 Summary of carotenoid (A) and vitamin E (B) biosynthesis pathway in plants. Enzymes carrying out specific reactions (a-h, for carotenoids and i-l, for vitamin E); are named as: a, phytoene synthase (PSY); b and c, phytoene desaturase (PDS); d, ζ-carotene desaturase (ZDS); e, the formation of lycopene through a bacterial phytoene desaturase (ctrl) is indicated with a black thick arrow; f, lycopene ε-cyclase, g, β-carotene hydroxylase (β-Chy); h, ε-hydroxylase (ε-Chy); i, hydrogenation enzyme; j, phytyl/prenyl transferase; k, methyl transferase 1 and cyclase; l, tocopherol γ-methyltransferases (γ-TMT). Adapted from Ref. 17, 21, 74, 75, 78, 79, 90.

Tomato fruit and its processed products are the principal dietary sources of lycopene, and also useful for their β-carotene content in minor amounts in the red varieties. At the green stage, tomato fruit has a carotenoid content similar to leaves (essentially β-carotene, lutein, and violaxanthin). But during fruit ripening, the genes mediating lycopene synthesis are up regulated whereas those controlling its cyclization are down regulated leading to over accumulation of lycopene in ripe fruit (up to 90% of total carotenoid, followed by β-carotene and only traces of lutein) (Figure 5.3A) (79).

Because the activity of the enzyme phytoene synthase (PSY) increases during tomato fruit ripening and when lycopene deposition occurs, it has been a preferred target to modify carotenoid profiles and levels in the fruit (Figure 5.3A). The constitutive expression of tomato PSY-1 in transgenic tomato plants led to dwarfism (plants with reduced size), due to redirecting GGPP from the gibberellins biosynthetic pathway into the carotenoid pathway, which reduced the amount of those hormones in the plants. Transgenic fruits produced lycopene earlier in development, but the final levels in the ripe fruits were lower than those in control samples, due to silencing of the endogenous gene (Table 5.4) (84). On the other hand,

Romer and coworkers (81) achieved a significant increment in the levels of β-carotene (provitamin A) in transgenic tomatoes by manipulating their desaturation activity. They produced transgenic lines overexpressing a PDS from *Erwinia uredovora* bacterium using a constitutive promoter; an enzyme capable of producing lycopene from phytoene (Figure 5.3A). Although total carotenoid concentration was not affected, ripe tomato fruits from transformed plants showed a threefold increase in β-carotene levels, accounting for up to 45% of total carotenoids. Tomato with improved provitamin A quality possessed about 5 mg/g fresh weight of all-*trans*-β-carotene (800 retinal equivalents), which satisfy 42% of the RDA, in contrast to normal fruit satisfying only 23% (Tables 5.2, 5.4).

Tomato is one of most productive based on carotenoid production per unit of cultivated area. By metabolic engineering, the large pool of red lycopene could be converted into high value added downstream carotenoids such as xanthophylls, which are an important type of target nutraceuticals, because of their antioxidant properties, their chemical stability and the difficulty in their chemical synthesis (79).

Recently, Dharmapuri et al. (79) reported the hyperexpression in tomato of an *Arabidopsis* β-lycopene cyclase (β-*Lcy*) gene with and without pepper β-carotene hydroxylase (β-*Chy*) gene under the control of the fruit-specific PDS promoter (Figure 5.3A). They found that the color of the ripe fruit varied from complete red as in wild-type tomato (by natural lycopene accumulation) to red-orange or to complete orange for transgenic tomatoes expressing only recombinant β-lycopene cyclase, or both recombinant β-lycopene cyclase and β-carotene hydroxylase enzymes, suggesting significant changes in carotenoid composition. The transformed fruits showed up to a 12-fold increase in β-carotene content with respect to their untransformed parental line. The transgenic tomato (containing genes β-*Lcy* + β-*Chy*) accumulated β-carotene as much as 63 µg/g fresh weight as compared to 5 µg/g fresh weight produced by the untransformed fruit. Notably, modified tomato also stored very good levels of both nutraceutical xanthophylls; β-cryptoxanthin (11 µg/g fresh weight) and zeaxanthin (13 µg/g fresh weight), but they are not detectable in control parent fruit or transformed tomato with only β-*Lcy* gene (Table 5.4). These results demonstrate that β-carotene pool can be converted into xanthophylls by the overexpression of pepper β-*Chy* and thus adding nutraceutical and commercial value to tomato.

5.4.1.2 *Carotenogenesis in Other Food Crops*

Another successful genetic manipulation of carotenogenesis has been reported in canola (*Brassica napus*). The bacterial (*Erwinia uredovora*) phytoene synthase with a plastid targeting sequence was overexpressed in a seed-specific manner using a napin promoter from rapeseed (Figure 5.3A) (80). Transgenic embryos were visibly orange, as compared with those green from control seeds, and the mature seed exhibited up to 50-fold more carotenoids, mainly α- and β-carotene, both presenting activities of provitamin A. Carotenoid-rich transgenic canola seed reached up to 1.6 mg per gram of fresh weight and produced an oil with 2 mg of carotenoids per gram of oil (Tables 5.2, 5.4).

Worldwide, vitamin A deficiency causes visible eye damage in around 3 million preschool children and up to 500 thousand of those children become partially or totally blind each year because of this deficiency, and approximately two thirds of them die within months of going blind. The estimates of the subclinical prevalence of vitamin A deficiency range between 100 and 200 million across the world. In developing countries several clinical trials have shown that vitamin A capsules can reduce mortality rates among preschool children by 23%, whereas improved vitamin A nutrition could prevent 1.2 million deaths annually among children aged 1–4 years (31). In addition, because vitamin A deficiency is common among vast populations of Asia, Africa, and South America, whose principal source of food is rice, engineering these crops to produce provitamin A-type carotenoids is of great importance (72,74).

Rice is generally consumed in its polished version, as commercial milling removes the oil-rich aleurone layer which becomes rancid during storage mainly in tropical and subtropical regions. Mature rice (*Oryza sativa*) endosperm is capable of synthesizing and accumulating GGPP but completely lacks carotenoids (Figure 5.3A) (85). In rice endosperm, the additional enzymatic activities needed to produce β-carotene was genetically engineered (82). To achieve this, four plant enzymes are necessary, but alternatively the number may be reduced to three as a bacterial carotene desaturase catalyzes the introduction of four double bonds required to produce lycopene (Figure 5.3A). Therefore three new genes (a plant phytoene synthase gene from daffodil, a bacterial phytoene desaturase gene from *Erwinia uredovora* and a lycopene β-cyclase also from daffodil plant) were transferred to rice. The first and third genes were driven by the endosperm-specific promoter of the rice glutelin gene, whereas a constitutive CaMV 35S promoter was used for the phytoene desaturase gene. Interestingly, transgenic rice seeds exhibited a beautiful golden yellow color after milling and increased accumulation of β-carotene in edible endosperm; as well as the xanthophylls, zeaxanthin, and lutein were formed to some extent, resulting in a carotenoid qualitative profile somehow analogous to that of green leaves. However, golden rice showed a higher proportion of β-carotene in their endosperm than that of the other two carotenoids, with a maximum amount of 1.6 µg/g dry weight; however, this quantity only represents 1–2% of carotenoid concentration in transgenic rapeseed (80) (Table 5.4). Nevertheless, it is noteworthy that in a typical Asian diet (about 300 g of rice per day), provitamin A-rich golden rice could provide nearly the full daily vitamin A requirement (85) (Table 5.2). It has been suggested that β-carotene accumulated in golden rice endosperm can be converted to retinol easier than β-carotene in vegetables, where provitamin A is converted to retinol at a rate equivalent to 26 to 1, due to the physicochemical properties of endosperm matrix (31,85). This significant accomplishment in plant biotechnology of improving the nutritional value of rice with beneficial carotenoids can benefit human nutrition and health (83). A future project aims to join iron-rich rice lines expressing ferritin with golden rice lines because it is known that provitamin A improves the iron bioavailability (Table 5.3) (85).

Nopal cactus (its young cladodes are called nopalitos) is low in calories and since Mesoamerican times has been eaten as a vegetable and fruit source in Mexico. At present, the economic and social importance of nopal is not only because large areas are covered with wild and commercial species in all arid and semiarid Mexican regions, but also because of its remarkable nutritional and nutraceutical qualities. For example, the intake of broiled nopalitos improved glucose control in adult people with noninsulin-dependent diabetes mellitus; it is also known that nopal diminishes human cholesterol levels (35,86). On the other hand, nopal is a good source of dietary fiber, calcium, iron, zinc, and vitamin C, and thereby can also be used in treatments against scurvy. However, although β-carotene is accumulated in nopal, its amount is very low; in fact its provitamin A level depends on nopalito development stage and nopal variety (35,86).

In our laboratory, Paredes-López et al. (87) have developed a genetic transformation system for nopal cactus (*Opuntia* sp.) through *Agrobacterium tumefaciens*, obtaining regenerated transgenic nopal plants. Thus, it has been proposed to use the nopal plant as a bioreactor for improving the production and storage of provitamin A by introducing additional genes for carotenoid biosynthesis to obtain β-carotene (P. Garcia-Saucedo, O. Paredes-López, personal communication, 2004). Nopal could become an important, cheap, and accessible source of vitamin A, and other nutraceuticals such as lutein or zeaxanthin for a great part of the population in Mexico.

5.4.2 Vitamin E

The intense research efforts which have surrounded vitamin E, a lipid-soluble antioxidant, support the hypothesis that preventing free radical-mediated tissue damage, for example to cellular lipids, proteins, or DNA, may play a key role in decreasing or delaying the pathogenesis of a variety of degenerative diseases such as cardiovascular disease, cancer, inflammatory diseases, neurological disorders, cataract, and age-related macular degeneration, and a decline in the immune system function (Table 5.2) (8,21). It has been suggested that vitamin E supplementation of 100 to 400 international units (IU) or around 250 mg of α-tocopherol per day may help reduce the magnitude of occurrence of such health disorders. Vitamin E is represented by a family of structurally related compounds, eight of which are known to occur in nature, being isolated from vegetable oils and other plant materials. The eight naturally occurring compounds are α-, β-, γ-, and δ-tocopherol (which differ only in the number and position of methyl substituents on the aromatic ring), and α- β-, γ-, and δ-tocotrienols (8,33). Tocotrienols differ from the corresponding tocopherols in that the isoprenoid side chain is unsaturated at $C3'$, $C7'$, and $C11'$. The phenolic hydroxyl group is key for the antioxidant activity of vitamin E, as donation of hydrogen from this group stabilizes free radicals (Figure 5.2). The presence of at least one methyl group on the aromatic ring is also critical. α-tocopherol, with three methyl groups, is the most biologically active of all homologues occurring in nature as a single isomer, followed by β-tocopherol, γ-tocopherol, and δ-tocopherol (8,88). Changes in the isoprenoid side chain also influence vitamin E activity; this biological activity is defined in terms of equivalents of α-tocopherol (α-TE). (R, R, R)-α-tocopherol has an activity of 1 α-tocopherol (α-TE) equivalent per milligram of compound. The activities of (R, R, R)-β-, (R, R, R)-γ-, and (R, R, R)-δ-tocopherols are 0.5, 0.1 and 0.03 per mg of compound, respectively (8,21). Of the tocotrienols, only α-tocotrienol has significant biological activity (0.3 mg α-TE/mg). Lengthening or shortening the side chain results in a progressive loss of vitamin E activity. The previous information is based on all tocopherols which are absorbed to similar extents during digestion, however single (R, R, R)-α-tocopherol is successfully stored and distributed in the entire body, while the other species are not processed with the same efficiency. It has been estimated that one α-TE molecule is capable of protecting 2000 phospholipids (8,88).

The physiological role of vitamin E centers on its ability to react with and quench free radicals in cell membranes and other lipid environments, thereby preventing polyunsaturated fatty acids (PUFAs) from being damaged by lipid oxidation. An imbalance in the production of free radicals and the natural protective system of antioxidants may lead to oxidized products, able to harm tissue; in fact tissue damage due to free radicals has been associated to several human chronic diseases (Table 5.2) (4–8).

Refined and processed foods are usually exposed to light, heat, or metal ions that can cause structural degradation of their constituent lipids by triggering the process of lipid oxidation (42). The rate of lipid oxidation in a food depends on the concentration and type of PUFAs it contains, the amount and effectiveness of the antioxidants present in the food and the heating, processing and storage conditions to which it is subjected to. Tocopherols and tocotrienols are the most important natural antioxidants in fats and oils, acting as primary or chain breaking antioxidants by converting lipid radicals to more stable products (8,42). At normal oxygen pressure, the major lipid radical is the peroxyl radical (ROO°) which can be converted to a hydroperoxide (ROOH) by proton donors such as tocopherols and tocotrienols. The hydrogen is donated from their phenolic groups, stabilizing the radicals and stopping the propagation phase of the oxidative chain reaction.

Supplementation of α-tocopherol has been demonstrated to positively affect sensorial quality of meat as well as saving money (i.e., in the U.S. beef industry meat, color degradation provokes losses up to $1 billion each year). In poultry, the high tocopherol level increased the stability of its meat; whereas in the case of pork and beef, vitamin E protects against rancid flavor, odor, and discoloration improving shelf life of packaged meat (89).

However, traditional plant breeding and food processing technologies have not concerned themselves with maximizing the levels of tocopherols in the human diets and even in diets for domestic animals used for meat, and the supplementation is necessary both for nutritional reasons and for the protection of fat-rich foods against oxidative rancidity (8,42). Unfortunately, synthetic α-tocopherol used as supplement is a complex mixture of stereo-isomers with less biological activity than natural single (R, R, R)-α-tocopherol (16,21,88). Significant changes in the α-tocopherol levels of major edible crops are necessary because there is a growing body of evidence to suggest that the dietary intake of vitamin E is insufficient to protect against the long term health risks associated with oxidative stress (8,16,21,88,89). Normally, the tocopherol composition of cultivated sunflower (*Helianthus annuus* L.) seed is primarily α-tocopherol, 95–100% of the total tocopherol pool. However, two mutant sunflower lines have been identified with tocopherol compositions of 95% γ-tocopherol/5% α-tocopherol, and 50% β-tocopherol/50% α-tocopherol. Although these presumed tocopherol methylation mutants showed severe alterations in their tocopherol profiles in seeds, their overall levels do not differ significantly from those of wild-type sunflower (8,89). These results suggest that it should be possible to alter the tocopherol profile of different crop species by manipulating the expression of one or both tocopherol methyltransferases (TMT), without having a detrimental effect on the total tocopherol pool size (Figure 5.3B). An exquisite and noteworthy research in the context of increasing the overall level of vitamin E activity available to consumers from plant foods was carried out by Shintani and DellaPenna (90). By overexpressing γ-TMT (Figure 5.3B), it was possible to increase α-tocopherol content in the *Arabidopsis* seed to about 85–95% of the total tocopherol, as compared with levels of 1.1% α-tocopherol and 97% γ-tocopherol in the untransformed seeds. Transgenic *Arabidopsis* showed a vitamin E activity about nine times greater than the negative control (Table 5.4). The authors speculate that if γ-TMT activity is limiting in commercially important oilseed crops such as soybean, corn, and canola, all of which have low γ-tocopherol to α-tocopherol ratios, overexpressing the γ-TMT gene in these crops should also elevate α-tocopherol amounts and improve their nutraceutical value (8,16,21,33,34,89).

5.4.3 Vitamin C

Vitamin C is used in large scale as an antioxidant in food, animal feed, beverages, pharmaceutical formulations, and cosmetic applications (91). This water-soluble vitamin, defined as L-ascorbic acid (L-AA), structurally is one of the simplest vitamins (Figure 5.1); its oxidation product is termed dehydroascorbate. It is related to the C6 sugars, being the aldono-1,4-lactone of a hexonic acid (L-galactonic or L-gulonic acid) and contains an enediol group on carbons 2 and 3 (92). In animal metabolism, including that of humans, the biological functions of L-AA are centered on its antioxidant properties and on its role to modulate a number of important enzymatic reactions. Thus, generally it acts as an enzyme cofactor, free radical scavenger and donor and acceptor in electron transfer reactions. For example, L-AA is required for collagen synthesis, and consequently in the formation and maintenance of cartilage, bones, gums, skin, teeth, and wound healing. In fact, the Fe-dioxygenases involved in collagen biosynthesis need L-AA for maximal activity, where the function of L-AA is to keep the transition metal ion centers of these enzymes in a reduced form. In the disease scurvy, which is known to be the result of vitamin C deficiency, its symptoms are directly related with the inadequate collagen formation (Table 5.2) (12,92). This micronutrient is also crucial for the normal, and enhanced, functioning of immune system, and is required for carnitine synthesis. There is now strong evidence to link high intake dietary vitamin C with reduced risk for several oxidative stress-associated diseases such as cardio-vascular diseases, various types of cancers, aging, neurodegenerative diseases, and cataract formation. Cataracts appear to be due to the oxidation of lens protein, and antioxidants such as vitamin C and E, and carotenoids seem to protect against cataracts and macular degeneration of eye in rodents and humans. On the other hand, increased oxidative damage from low vitamin C intake, chronic inflammation, smoking, or radiation, together with elevated levels of uracil in DNA, would be expected to lead to more double strand (chromosome) breaks in individuals who are deficient in both folate and antioxidant (Table 5.2) (4–7,12).

However, there is also large body literature on supplementation studies with vitamin C in humans using biomarkers of oxidative damage to DNA, lipids (its oxidation releases mutagenic aldehydes), and protein. Some studies suggest that blood cell saturation occurs at about 100 mg vitamin C/day and the evidence suggests that this level minimizes DNA damage. Both experimental and epidemiological data support that vitamin C provides protection against stomach cancer, a result that is plausible because of the role of oxidative damage from inflammation by *Helicobacter pylori* infection, which is the main risk factor for stomach cancer (4–7,12). Unfortunately, the plasma levels of L-AA in large sections of the population around the world are suboptimal for those health benefic effects of this vitamin; in fact about 15% of the population consumes less than half the RDA (60 mg/day) of ascorbate (Table 5.2).

It is thought that L-AA secreted in gastric juices in animals enhances the absorption of iron from plant foods through two mechanisms: by forming Fe(III) complexes and by reducing the less soluble Fe^{3+} to the more soluble and bioavailable Fe^{2+} valence state (Table 5.3) (12,43).

Plants and most animals (i.e., rats, dogs, cats) can synthesize their own vitamin C, but a few mammalian species, including primates, human beings, and guinea pigs have lost this capability, and thus entirely depend upon dietary sources to meet needs for this vital micronutrient. This deficiency has been localized to a lack of the terminal flavo-enzyme L-gulono-1,4-lactone oxidase (L-gulono-γ-lactone oxidase, [GuLO]); the gene encoding it was found in the human genome, but was not expressed due to the accumulation of various mutations. In vitamin C-producing animals, GuLO catalyzes the final reaction in the L-AA route corresponding to the oxidation of L-gulono-1,4-lactone, whereas in plants the enzyme L-galactono-1,4-lactone dehydrogenase employs L-galactono-1,4-lactone as a substrate for carrying out the terminal step in the vitamin C production (Figure 5.4) (92–95).

Vitamin C is the single most important specialty chemical manufactured in the world. The current world market of ascorbic acid is 60,000 to 70,000 metric tons each year and generates annual revenues in excess of US$ 500 million (91). But its industrial production is a lengthy procedure involving microbial fermentation and diverse chemical steps (38,91). Until quite recently, little focus has been given to improving the L-AA content of plant foods, either in terms of the amounts present in commercial crop varieties or in minimizing losses prior to consumption. Notably, plants and animal possess different pathway for synthesizing L-AA; the expression in transgenic lettuce plants of an animal cDNA encoding a rat GuLO under the control of CaMV 35S promoter led to accumulation up to seven times more L-AA than untransformed crops (the basal levels of L-AA varied among the three unmodified lettuce cultivars from 0.36–0.58 μmol/g fresh weight) (Table 5.4) (95). In food science and technology, vitamin C as well as bisulfites are used to prevent oxidation in peaches, potato chips, apples, potatoes, peanut butter, beer, fat, and oils (20,42). Therefore, in the future, L-AA-rich transgenic lettuce may diminish the commercial application of bisulfite to avoid browning of its leaf, as well as enhancing the nutritional and nutraceutical value of this food vegetable. A recent report also showed that the L-AA content of *Arabidopsis thaliana* (untransformed plants have a vitamin C level of about 2 μmol/g fresh weight) was increased two- and threefold by hyperexpression of a D-galacturonic acid reductase gene from strawberry

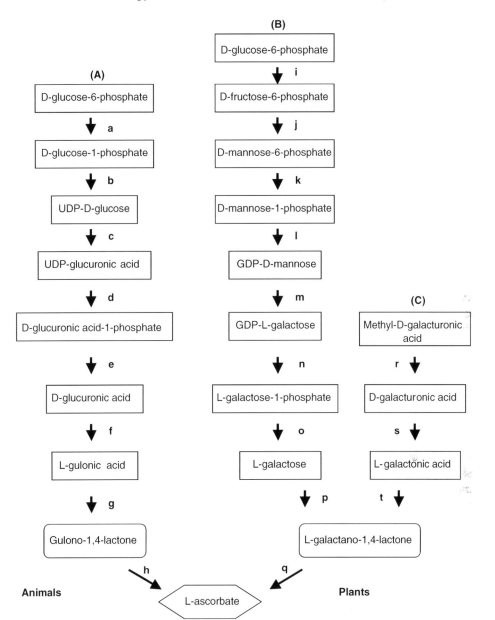

Figure 5.4 Proposed pathways for L-ascorbic acid biosynthesis in animals (A) and plants (B and C). Enzymes catalyzing the individuals reactions (a-h, for animals; i-t, for plants) are given next: a, phosphoglucomutase; b, UDP-glucose pyrophosphorylase; c, UDP-glucose dehydrogenase; d, glucuronate-1-phosphate uridylyltransferase; e, glucurono kinase; f, glucuronate reductase; g, aldono-lactonase; h, L-gulono-1,4-lactone oxidase, GuLO; i, glucose-6-phosphate isomerase; j, mannose-6-phosphate isomerase; k, phosphomannomutase; l, GDP-mannose pyrophosphorylase; m, GDP-mannose-3,5-epimerase; n, phosphodiesterase; o, sugar phosphatase; p, L-galactose-1- dehydrogenase; q, L-galactono-1,4-lactone dehydrogenase, r, methylesterase, s, D-galacturonate reductase; t, aldono-lactonase. Adapted from Ref. 92–95.

(Table 5.4) (94). This gene encodes the enzyme D-galacturonate reductase that converts D-galacturonic acid into L-galactonic acid, which is readily transformed to L-galactono-1,4-lactone, the immediate precursor of L-AA (Figure 5.4C).

The previous works demonstrate the possibility that the basal content of minerals and vitamins in important food crops can be augmented by metabolic engineering, and therefore raise the realistic possibility that such increases may substantially benefit vulnerable populations in their daily dietary intakes, without the need for fortification or for a change in dietary habits as a whole.

ACKNOWLEDGMENTS

We acknowledge partial financial support from National Council of Science and Technology (CONACYT).

REFERENCES

1. Blackburn, G.L. Pasteur's quadrant and malnutrition. *Nature* 409:397–401, 2001.
2. Willet, W.C., M.J. Stampfer. Rebuilding the food pyramid. *Sci. Am.* 288:64–71, 2003.
3. Lachance, P.A. Overview of key nutrients: micronutrient aspects. *Nutr. Rev.* 56:S34–S39, 1998.
4. Ames, B.N. DNA damage from micronutrient deficiencies is likely to be a major cause of cancer. *Mut. Res.* 475:7–20, 2001.
5. Ames, B.N., P. Wakimoto. Are vitamin and mineral deficiencies a major cancer risk? *Nat. Rev.* 2:694–704, 2002.
6. Fenech, M., L.R. Ferguson. Vitamins/minerals and genomic stability in humans. *Mut. Res.* 475:1–6, 2001.
7. Fenech, M. Micronutrients and genomic stability: a new paradigm for recommended dietary allowances (RDAs). *Food Chem. Toxicol.* 40:1113–1117, 2002.
8. Bramley, P.M., I. Elmadfa, A. Kafatos, F.J. Kelly, Y. Manios, H.E. Roxborough, W. Shuch, P.J. Sheehy, K.H. Wagner. Vitamin E. *J. Sci. Food Agric.* 80:913–938, 2000.
9. De Freitas, J.M., R. Meneghini. Iron and its sensitive balance in the cell. *Mut. Res.* 475: 153–159, 2001.
10. Dreosti, I.E. Zinc and gene. *Mut. Res.* 475:161–167, 2001.
11. El-Bayoumy, K. The protective role of selenium on genetic damage and on cancer. *Mut. Res.* 475:123–139, 2001.
12. Halliwell, B. Vitamin C and genomic stability. *Mut. Res.* 475:29–35, 2001.
13. Trewavas, A., D. Stewart. Paradoxical effects of chemicals in the diet on health. *Curr. Opin. Plant Biol.* 6:185–190, 2003.
14. Collins, A.R. Carotenoids and genomic stability. *Mut. Res.* 475:21–28, 2001.
15. Kaur, C., H.C. Kapoor. Antioxidants in fruits and vegetables: the millennium's health. *Int. J. Food Sci. Technol.* 36:703–725, 2001.
16. DellaPenna, D. Nutritional genomics: manipulating plant micronutrients to improve human health. *Science* 285:375–379, 1999.
17. Delgado-Vargas, F., O. Paredes-López. *Natural Pigments for Food and Nutraceutical Uses.* Boca Raton, FL: CRC Press, 2003.
18. Smil, V. Magic beans. *Nature* 407:567, 2000.
19. National Research Council (U.S.), Food and Nutrition Board. *Recommended Dietary Allowances*, 10th ed. Washington: National Academy Press, 1989.
20. Guzmán-Maldonado, S.H., O. Paredes-López. Biotechnology for the improvement of nutritional quality of food crop plants. In: *Molecular Biotechnology for Plant Food Production*, Paredes-López, O., ed., Boca Raton, FL: CRC Press, 1999, pp 553–620.

21. Grusak, M.A., D. DellaPenna. Improving the nutrient composition of plants to enhance human nutrition and health. *Annu. Rev. Plant Physiol. Plant Mol. Biol.* 50:133–161, 1999.

22. Graham, R.D., R.M. Welch, H.E. Bouis. Addressing micronutrient malnutrition through enhancing the nutritional quality of staple foods: principles, perspectives and knowledge gaps. *Adv. Agron.* 70:77–142, 2001.

23. Cakmak, I. Plant nutrition research: priorities to meet human needs for food in sustainable ways. *Plant Soil* 247:3–24, 2002.

24. Welch, R.M. Breeding strategies for biofortified staple plant foods to reduce micronutrient malnutrition globally. *J. Nutr.* 132:495S–499S, 2002.

25. Welch, R.M., R.D. Graham. Breeding crops for enhanced micronutrient content. *Plant Soil* 245:205–214, 2002.

26. Grantham-McGregor, D.M., C.C. Ani. The role of micronutrients in psychomotor and cognitive development. *Brit. Med. Bull.* 55:511–527, 1999.

27. Welch, R.M. The impact of mineral nutrients in food crops on global human health. *Plant Soil* 247:83–90, 2002.

28. Hallberg, I. Perspectives of nutritional iron deficiency. *Annu. Rev. Nutr.* 21:1–21, 2001.

29. Beard, J.L. Iron biology in immune function, muscle metabolism and neuronal functioning. *J. Nutr.* 131:568S–580S, 2001.

30. Bhan, M.J., H. Sommerfelt, T. Strand. Micronutrient deficiency in children. *Brit. J. Nutr.* 85:199S–203S, 2001.

31. Bouis, H.E. Plant breeding: a new tool for fighting micronutrient malnutrition. *J. Nutr.* 132:491S–494S, 2002.

32. Sandström, B. Micronutrient interactions: effects on absorption and bioavailability. *Brit. J. Nutr.* 85:181S–185S, 2001.

33. MA Grusak, M.A. Genomics-assisted plant improvement to benefit human nutrition and health. *Trends Plant. Sci.* 4:164–166, 1999.

34. Grusak, M.A. Phytochemicals in plants: genomics-assisted plant improvement for nutritional and health benefits. *Curr. Opin. Biotechnol.* 13:508–511, 2002.

35. Guzmán-Maldonado, S.H., O. Paredes-López. Functional products of indigenous plants to Latin America: amaranth, quinoa, common beans and botanicals. In: *Functional Foods: Biochemical and Processing Aspects,* G. Mazza, ed., Boca Raton, FL: CRC Press 1978, pp 293–328.

36. Elliot, R., T.J. Ong. Nutritional genomics. *BMJ* 324:1438–1442, 2002.

37. Chrispeels, M.J., D.E. Sadava. *Plants, Genes, and Agriculture.* Sudbury, MA: Jones & Bartlett Publishers, Inc, 1994.

38. Osuna-Castro, J.A., O. Paredes-López. Introduction to molecular food biotechnology. In: *Food Science and Food Biotechnology,* Gutiérrez-López, G.F., G.V. Barbosa-Canovas, eds, Boca Raton FL: CRC Press, 2003, pp 1–62.

39. Osuna-Castro, J.A., O. Paredes-López. Mejoramiento de características y calidad alimentarias y nutracéuticas de plantas mediante biotecnología molecular; algunos ejemplos. In: *Fundamentos y Casos Exitosos de la Biotecnología,* Bolívar, F., ed. México: El Colegio Nacional, 2004, pp 451–503.

40. King, J.A. Biotechnology: a solution for improving nutrient bioavailability. *Int. J. Vitam. Nutr. Res.* 72:7–12, 2002.

41. Lönnerdal, B. Genetically modified plants for improved trace element nutrition. *J. Nutr.* 133:1490S–1493S, 2003.

42. Gregory, J.F. III. Vitamins, 3rd ed. In: *Food Chemistry*, Fennema, O.R., ed., New York: Marcel Dekker, Inc., 1996, pp 531–616.

43. Frossard, E., M. Bucher, M. Machler, A. Mozafar, R. Hurrell. Potential for increasing the content and bioavailability of Fe, Zn and Ca in plants for human nutrition. *J. Sci. Food Agric.* 80:861–879, 2000.

44. Van Lieshout, M., C.E. West, R.B. Van Breemen. Isotopic tracer techniques for studying the bioavailability and bioefficacy of dietary carotenoids, particularly β-carotene, in humans. *Am. J. Clin. Nutr.* 77:12–28, 2003.

45. Dawson, M.I. The importance of vitamin A in nutrition. *Curr. Pharm. Des.* 6:311–325, 2000.
46. Miller, D.D. Minerals, 3rd ed. In: *Food Chemistry*, Fennema, O.R., ed. New York: Marcel Dekker, Inc., 1996, pp 617–649.
47. Raboy, V. Seeds for a better future: 'low phytate' grains help to overcome malnutrition and reduce pollution. *Trends Plant. Sci.* 6:458–462, 2001.
48. Hurrell, R.F. Fortification: overcoming technical and practical barriers. *J. Nutr.* 132:806S–812S, 2002.
49. King, J.A. Evaluating the impact of plant biofortification on human nutrition. *J. Nutr.* 132:511S–513S, 2002.
50. Lucca, P., R. Hurrell, I. Potrykus. Approaches to improving the bioavailability and level of iron in rice seeds. *J. Sci. Food Agric.* 81:828–834, 2001.
51. NE Borlaug, N.E. Ending world hunger: the promise of biotechnology and the threat of antiscience zealotry. *Plant Physiol.* 124:487–490, 2000.
52. Grusak, M.A. Enhancing mineral content in plant food products. *J. Am. Coll. Nutr.* 21:178S–183S, 2002.
53. Kochian, L.V., D.F. Garvin. Agricultural approaches to improving phytonutrient content in plants: an overview. *Nutr. Rev.* 2:S13–S18, 1999.
54. Carbonaro, M., G. Grant, M. Mattera, A. Aguzzi, A. Pusztai. Investigation of the mechanisms affecting Cu and Fe bioavailability from legumes. *Biol. Trace Elem. Res.* 84:181–196, 2001.
55. Sandberg, A.-S. Bioavailability of minerals in legumes. *Brit. J. Nutr.* 88:281S–285S, 2002.
56. Masuda, F., F. Goto, T. Yoshihara. A novel plant ferritin subunit from soybean that is related to a mechanism in iron release. *J. Biol. Chem.* 276:19575–19579, 2001.
57. Theil, E.C. Ferritin: at the crossroads of iron and oxygen metabolism. *J. Nutr.* 133:1549S–1553S, 2003.
58. Skikne, B., D. Fonz, S.R. Lynch, J.D. Cook. Bovine ferritin iron bioavailability in man. *Eur. J. Clin. Invest.* 27:228–233, 1997.
59. Murray-Kolb, L., R. Welch, E.C. Theil, J.L. Beard. Women with low iron stores absorb iron from soybeans. *Am. J. Clin. Nutr.* 77:180–184, 2003.
60. Goto, F., T. Yoshihara, N. Shigemoto, S. Toki, F. Takaiwa. Iron fortification of rice seed by the soybean ferritin gene. *Nat. Biotechnol.* 17:282–286, 1999.
61. Lucca, P., R. Hurrell, I. Potrykus. Genetic engineering approaches to improve the bioavailability and the level of iron in rice grains. *Theor. Appl. Genet.* 102:392–397, 2001.
62. Lucca, P., R. Hurrell, I. Potrykus. Fighting iron deficiency anemia with iron-rich rice. *J. Am. Coll. Nutr.* 21:184S–190S, 2002.
63. Vasconcelos, M., K. Datta, N. Oliva, M. Khalekuzzaman, L. Torrizo, S. Krishnan, M. Oliveira, F. Goto, S.K. Datta. Enhanced iron and zinc accumulation in transgenic rice with the *ferritin* gene. *Plant Sci.* 164:371–378, 2003.
64. Murray-Kolb, L., F. Takaiwa, F. Goto, T. Yoshihara, E. Theil, J.L. Beard. Transgenic rice is a source of iron for iron-depleted rats. *J. Nutr.* 132:957–960, 2002.
65. Nandi, S., Y.A. Suzuki, J. Huang, D. Yalda, P. Pham, L. Wu, G. Bartley, N. Huang, B. Lönnerdal. Expression of human lactoferrin in transgenic rice grains for the application in infant formula. *Plant Sci.* 163:713–722, 2002.
66. Suzuki, Y.A., S.L. Kelleher, D. Yalda, L. Wu, J. Huang, N. Huang, B. Lönnerdal. Expression, characterization and biologic activity of recombinant human lactoferrin in rice. *J. Pediatr. Gastroenterol. Nutr.*, 36:190–199, 2003.
67. Guzmán-Maldonado, S.H., J. Acosta-Gallegos, O. Paredes-López. Protein and mineral content of a novel collection of wild and weedy common bean (*Phaseolus vulgaris* L). *J. Sci. Food Agric.* 80:1874–1881, 2000.
68. Tramper, J. Modern biotechnology: food for thought. In: *Food Biotechnology*, Bielecki, S., J. Tramper, J. Polak, eds., Amsterdam: Elsevier Science BV, 2000, pp 3–12.
69. Ponstein, A.S., J.B. Bade, T.C. Verwoerd, L. Molendijk, J. Storms, R.F. Beudeker, J. Pen. Stable expression of phytase (*phyA*) in canola (*Brassica napus*) seeds: towards a commercial product. *Mol. Breed.* 10:31–44, 2002.

70. Mendoza, C., F.E. Viteri, B. Lonnerdal, K.A. Young, V. Raboy, K.H. Brown. Effect of genetically modified, low-phytic acid maize on absorption of iron from tortillas. *Am. J. Clin. Nutr.* 68:1123–1127, 1998.

71. Golovan, S.P., R.G. Meidinger, A. Ajakaiye, M. Cottrill, M.Z. Wiederkehr, D.J. Barney, C. Plante, J.W. Pollard, M.Z. Fan, M.A. Hayes, J. Laursen, J.P. Hjorth, R.R. Hacker, J.P. Phillips, C.W. Forsberg. Pigs expressing salivary phytase produce low-phosphorus manure. *Nat. Biotechnol.* 19:741-745, 2001.

72. Van den Berg, H., R. Faulks, H.F. Granado, J. Hirschberg, B. Olmedilla, G. Sandmann, S. Southon, W. Stahl. The potential for the improvement of carotenoid levels in foods and the likely systemic effects. *J. Sci. Food Agric.* 80:880–912, 2000.

73. Delgado-Vargas, F., A.R. Jiménez, O. Paredes-López. Natural pigments: carotenoids, anthocyanins, and betalains: characteristics, biosynthesis, processing, and stability. *Crit. Rev. Food Sci. Nutr.* 40:173–289, 2000.

74. Fraser, P.D., S. Romer, J.W. Kiano, C.A. Shipton, P.A. Mills, R. Drake, W. Schuch, P.M. Bramley. Elevation of carotenoids in tomato by genetic manipulation. *J. Sci. Food Agric.* 81:822–827, 2001.

75. Moehs, C.P., L. Tian, K.W. Osteryoung, D. DellaPenna. Analysis of carotenoid biosynthetic gene expression during marigold petal development. *Plant Mol. Biol.* 45:281–293, 2001.

76. Delgado-Vargas, F., O. Paredes-López, E. Avila-González. Effects of sunlight illumination of marigold flower meals on egg yolk pigmentation. *J. Agric. Food Chem.* 46:698–706, 1998.

77. González-de-Mejía, E., G. Loarca-Piña, M. Ramos-Gómez. Antimutagenicity of xanthophylls present in Aztec marigold (*Tagetes erecta*) against 1-nitropyrene. *Mut. Res.* 389: 219–226, 1997.

78. Sandmann, G. Genetic manipulation of carotenoid biosynthesis: Strategies, problems and achievements. *Trends Plant Sci.* 6:14–17, 2001.

79. Dharmapuri, S., C. Rosati, P. Pallara, R. Aquilani, F. Bouvier, B. Camara, G. Guiliano. Metabolic engineering of xanthophyll content in tomato fruits. *FEBS Lett.* 519:30–34, 2002.

80. Shewmaker, C.K., J.A. Sheehy, M. Daley, S. Colburn, D.Y. Ke. Seed specific overexpression of phytoene synthase: increase in carotenoids and other metabolic effects. *Plant J.* 20:401–412, 1999.

81. Romer, S., P.D. Fraser, J.W. Kiano, C.A. Shipton, N. Misawa, W. Schuch, P.M. Bramley. Elevation of the provitamin A content of transgenic tomato plants. *Nat. Biotechnol.* 18: 666–669, 2000.

82. Ye, X., S. Al-Babili, A. Klöti, J. Zhang, P. Lucca, P. Beyer, I. Potrykus. Engineering provitamin A (β-carotene) biosynthetic pathway into (carotenoid-free) rice endosperm. *Science* 287:303–305, 2000.

83. Potrykus, I. Golden rice and beyond. *Plant Physiol.* 125:1157–1161, 2001.

84. Fray, R.G., A. Wallace, P.D. Fraser, D. Valero, P. Hedden, P.M. Bramley, D. Grierson. Constitutive expression of a fruit phytoene synthase gene in transgenic tomatoes causes dwarfism by redirecting metabolites from the gibberellin pathway. *Plant J.* 8:693–701, 1995.

85. Beyer, P., S. Al-Babili, X. Ye, P. Lucca, P. Schaub, R. Welch, I. Potrykus. Golden rice: introducing the β-carotene biosynthesis pathway into rice endosperm by genetic engineering to defeat vitamin A deficiency. *J. Nutr.* 132:506S–510S, 2002.

86. Mizrahi, Y., A. Nerd, P. Nobel. Cacti as crops. *Horticul. Rev.* 18:291–321, 1997.

87. Paredes-López, O., H. Silos, J.L. Cabrera, Q. Rascón. Método para la transformación genética y regeneración de plantas transgénicas de nopal. Patente en trámite. México, D.F.: Agosto 9, 2000.

88. Hirschberg, J. Production of high-value compounds: carotenoids and vitamin E. *Curr. Opin. Biotechnol.* 10:186–191, 1999.

89. Rocheford, T.R., J.C. Wong, C.O. Egesel, R.J. Lambert. Enhancement of vitamin E levels in corn. *J. Am. Coll. Nutr.* 21:191S–198S, 2002.

90. Shintani, D., D. DellaPenna. Elevating the vitamin E content of plants through metabolic engineering. *Science*, 282:2098–2100, 1998.

91. Chotani, G., T. Dodge, A. Hsu, M. Kumar, R. LaDuca, D. Trimbur, W. Weyler, K. Sanford. The commercial production of chemicals using pathway engineering. *Biochim. Biophys. Acta*, 1543:434–455, 2000.

92. Davey, M.W., M. Van Montagu, D. Inze, M. Sanmartin, A. Kanellis, N. Smirnoff, I.J. Benzei, J.J. Strain, D. Favell, J. Fletcher. Plant L-ascorbic acid: chemistry, function, metabolism, bioavailabilty and effect of processing. *J. Sci. Food Agric.* 80:825–860, 2000.

93. Wheeler, G.L., M.A. Jones, N. Smirnoff. The biosynthetic pathway of vitamin C in higher plants. *Nature* 393:365–369, 1998.

94. Agius, F., R. González-Lamothe, J.L. Caballero, J. Muñoz-Blanco, M.A. Botella, V. Valpuesta. Engineering increased vitamin C levels in plants by overexpression of a D-galacturonic acid reductase. *Nat. Biotechnol.* 21:177–181, 2003.

95. Jain, A.K., C.L. Nessler. Metabolic engineering of an alternative pathway for ascorbic acid biosynthesis in plants. *Mol. Breed.* 6:73–78, 2000.

6

Potential Health Benefits of Soybean Isoflavonoids and Related Phenolic Antioxidants

Patrick P. McCue and Kalidas Shetty

CONTENTS

6.1 Introduction 134
6.2 Consumption of Soybean and Reduced Incidence of Disease 134
  6.2.1 Soybean Isoflavonoids 134
  6.2.2 Major Bioactivities of Soybean Isoflavonoids...135
    6.2.2.1 Phytoestrogenic and Postmenopausal Activity...........................135
    6.2.2.2 Cancer Chemoprevention 135
    6.2.2.3 Prevention of Cardiovascular Disease 136
  6.2.3 Approaches Toward Isoflavone Enrichment of Soybean..........................136
    6.2.3.1 Genetic Modulation of Soybean..136
    6.2.3.2 Enrichment of Soybean Isoflavone Content
            via Nongenetic Approaches 137
  6.2.4 Toward a Model Mechanism for Action of Soybean Isoflavonoids
        and Related Phenolic Antioxidants against Cancer 137
    6.2.4.1 Metabolism of Dietary Isoflavonoids.......................................137
    6.2.4.2 Control of Energy Metabolism and Oxidative Stress
            in a Healthy Cell..138
    6.2.4.3 Control of Energy Metabolism and Oxidative Stress in a
            Tumorigenic Cell 138
  6.2.5 Stress Response and the Proline Linked Pentose–Phosphate Pathway 138
6.3 A Hypothetical Model for the Cancer Chemopreventive Action
    of the Soybean Isoflavone Genistein...139
6.4 Summary and Implications 141
References...142

6.1 INTRODUCTION

Soybeans are a well-known dietary staple of Asian countries such as Japan and are now consumed worldwide (1). Unlike string beans or snap peas, soybeans cannot be eaten raw, because they contain trypsin inhibitors, which can disrupt digestion activities in the stomach, leading to cramping and associated discomfort. Soybeans are usually fermented to produce distinct cultural or ethnic foods or food ingredients, or sometimes are sprouted for use in salads, as are mungbean and alfalfa sprouts. In Japan, soybeans are sometimes blanched (i.e., boiled for about five minutes), cooled, and then consumed readily with beer as a snack. The Japanese also process soybeans to produce a protein rich meat substitute, tofu. In Indonesia, soybeans are fermented by a food grade fungus of the *Rhizopus* species to produce an alternative protein rich meat substitute called tempeh (1).

6.2 CONSUMPTION OF SOYBEAN AND REDUCED INCIDENCE OF DISEASE

East Asian populations that regularly consume soybeans as a part of their dietary intake seem to have lower incidences of cancers and oxidation linked diseases of old age than are prevalent in Western populations. Numerous epidemiological studies have demonstrated an association between the consumption of soybeans and improved health, particularly as a reduced risk for cancers or diseases, such as breast cancer, cardiovascular disease, and atherosclerosis (2–8). Consumption of soy foods has also been associated with a reduced risk of prostate cancer (9,10).

Although soybean protein was first suspected to potentiate the health-promoting benefits of soybean consumption, these properties have more recently been linked to the biological activities of a specific group of phenolic compounds, found mainly in soybeans, known as isoflavonoids (5). While the chemopreventive properties of purified and synthetic isoflavonoids have been heavily investigated, a fermented soybean extract was recently shown to perform better at reducing the incidence of mammary tumor risk than a similar mixture of its constituent isoflavonoids, suggesting that the food background may play a positive role in the chemopreventive actions associated with soybean consumption, in addition to that of isoflavonoids (11–16).

6.2.1 Soybean Isoflavonoids

Isoflavonoids are a unique subgroup of the flavanoids, one of the largest classes of plant phenolics, with more than 5000 compounds currently identified. Isoflavonoids are found mainly in soybeans, and possess a chemical structure that is similar to the hormone estrogen (1). The chief isoflavonoids found in soybeans are genistein and daidzein. Because their structures resemble estrogen and they can interact with the estrogen receptor, soybean isoflavonoids are sometimes referred to as phytoestrogens (17).

Isoflavonoids are flavanoid variants in which the location of one of the phenolic rings is shifted. As diphenolic secondary metabolites, isoflavonoids are synthesized from products of the shikimic acid and malonyl pathways in the fusion of a phenylpropanoid with three malonyl CoA residues (1). Isoflavonoid content in soybeans ranges from 0.14 to 1.53 mg/g, and in soy flour from 1.3 to 1.98 mg/g (18). The Japanese are estimated to consume 25–100 mg of isoflavonoids per day (19). Chinese women are estimated to consume 39 mg of isoflavonoids per day (20). The consumption of isoflavonoids in Western diets is much lower, at less than 1 mg/day in the U.S. and U.K. (18,21).

Phenolic compounds normally occur as glucoside bound moieties called glycones (22,23). However, it is the aglycone (glucoside free) form that is metabolically active (24). After consumption, probiotic enzymes in the intestine cleave the glycoside moieties from glycone isoflavonoids and release the biologically active health-promoting aglycone isoflavonoids. Aglycone phenolic compounds possess higher antioxidant activity and are absorbed faster in the intestines than glucoside bound forms (23,25,26). Interestingly, fermented soy foods are rich in phenolic aglycones due to microbial bioprocessing during fermentation (27,28). However, once inside the bloodstream, biologically active aglycone genistein travels to the liver were it is converted back into an inactive glycone (β-glucuronide) (24). Cellular glucuronidases must remove the glycone moiety before genistein can exert its biological activity (24).

Isoflavonoids have been well studied and possess numerous biological activities (1). For example, genistein possesses inhibitory activity against topoisomerase II, tyrosine kinase, NF-κB, cancer cell proliferation, and nonoxidative pentose–phosphate pathway ribose synthesis in cancer cells (29–33). Many of the health-promoting benefits of isoflavonoids have been linked to the ability of phenolics to serve as antioxidants (34–37).

6.2.2 Major Bioactivities of Soybean Isoflavonoids

6.2.2.1 *Phytoestrogenic and Postmenopausal Activity*

The structure of soybean isoflavonoids is uniquely similar to that of estrogen (17) and may account for their weak ability to act as agonists at estrogen receptors (38). Many have speculated that soybean isoflavonoids may be useful for the treatment of somatic, mood, and cognitive disturbances associated with the onset of menopause (39). Diet supplementation with soybean phytoestrogens has been reported to ameliorate hot flashes and other symptoms of menopause (40–43).

Soybean isoflavonoids may also have potential in natural chemoprevention therapies against long term health problems associated with menopause, particularly for osteoporosis (44–47). After menopause, the ovaries stop producing estrogen. Because estrogen positively affects the metabolism of calcium, lack of sufficient estrogen can lead to bone loss and osteoporosis (48). Hormone replacement therapy (HRT) can reduce bone loss and the risk of osteoporosis in postmenopausal women, but unfortunately appears to also increase the risk for certain estrogen linked cancers (49–51).

Current osteoporosis prevention research is focused on the development of estrogen-like compounds (selective estrogen receptor modulators, or SERMs) that can selectively act against bone loss without causing negative estrogenic action against the uterus (52). The soybean isoflavonoid genistein has shown SERM activity in ovariectomized mice (53). When provided at optimal dosages, soybean isoflavonoids (especially genistein and daidzein) have been shown to improve bone mass and reduce bone resorption (54,55).

6.2.2.2 *Cancer Chemoprevention*

Soybean isoflavonoids also possess various biological activities that may help to explain the cancer chemopreventive properties associated with the consumption of soybean foods (3,8,49,56). In *in vitro* studies, daidzein was reported to activate the catalase promoter, to stimulate caspase-3 and apoptosis, and to down regulate the activities of Bcl-2 and Bcl-xL (57,58). Genistein can stimulate p53, antioxidant enzyme activities, BRCA2, caspase-3 and apoptosis, and chloride efflux (59–63). Genistein has also been reported to suppress activation of NF-κB, matrix metalloproteinases, lipogenesis, and COX-2 (31,64–66).

The exact mechanism by which these compounds exert their chemopreventive properties is not yet clear.

6.2.2.3 *Prevention of Cardiovascular Disease*

Soybean consumption has also been linked to a reduced risk for cardiovascular disease (47). Addition of soybean to foods has been shown to result in reduced cholesterol (67). In 1999, the US Food and Drug Administration reported that the consumption of soy protein as part of a healthy diet could help reduce the risk of coronary heart disease by lowering blood cholesterol levels (68). Soy protein isolates typically contain soybean isoflavonoids, which are believed to be largely responsible for the health benefits assigned to soy protein. Related herbal flavanoids prevented *in vitro* platelet aggregation and *in vivo* thrombogenesis in mouse arteries (69). Inclusion of isoflavonoid rich soybean in diets was also reported to protect against coronary heart disease by causing reductions in blood lipids, oxidized LDL, homocysteine, and blood pressure (7).

6.2.3 Approaches Toward Isoflavone Enrichment of Soybean

6.2.3.1 *Genetic Modulation of Soybean*

Throughout recorded history, man has used conventional breeding and selection techniques to improve crop species for desired traits. When modern genetic engineering techniques became available, agricultural scientists sought to improve crop species' phenolic content through the use of genetic technologies. At first, progress was slow, as knowledge of the biosynthetic pathways responsible for producing beneficial phenolic phytochemicals was limited.

One of the most studied pathways is the anthocyanin biosynthesis pathway, as phenotypic changes in flower color aided genetic analysis and metabolic understanding. Study of anthocyanin biosynthesis has also aided in the understanding of isoflavonoid biosynthesis as both pathways share a dependence on substrate flux through the flavanoid biosynthetic pathway. The isoflavonoid biosynthetic pathway is now almost completely characterized and genetic manipulation techniques have matured enough that it is now possible to alter synthesis at many different stages (70).

Knowledge of key enzymes involved in isoflavonoid biosynthesis in legumes led to attempts to engineer isoflavonoid biosynthesis in nonlegumes through genetic manipulation, in order to expand the delivery of dietary isoflavonoids as well as to develop new sources for their isolation (1). Unfortunately, initial attempts to incorporate key enzymes involved in isoflavonoid synthesis, such as soybean isoflavone synthase (IFS) and alfalfa chalcone isomerase in *Arabidopsis*, corn, and tobacco resulted in little to no formation of genistein or daidzein, the major isoflavonoids produced by soybean (71–73). Significant accumulation of isoflavonoids in nonlegumes was thought to be hindered by limited activity of the introduced IFS enzyme, by precursor pool limitations, and by competition (flux partitioning) between IFS and other enzymes that use the flavanoid naringenin as a substrate (73,74). More recent attempts to engineer flavanoids in bacteria and increased isoflavonoid content in soybean have been more successful, with the latter largely by coengineering the suppression of the naringenin-utilizing enzyme flavanone-3-hydroxylase to block competing pathways (72,75).

Dietary safety of genetically engineered foods remains a major concern of potential consumers. Although genetically modified (GM) foods appear no more harmful than conventionally produced foods, concerns remain as to the safety of the newly added DNA, its gene product, the overall safety of the rest of the food, the potential toxicology of the expressed protein, potential changes in allergenicity, changes in nutrient composition, unintended effects that could give rise to allergens or cause toxicity, and the safety of antibiotic resistance marker encoded proteins included with the transgene (76).

6.2.3.2 Enrichment of Soybean Isoflavone Content via Nongenetic Approaches

The technological challenge of stably introducing a foreign gene into a food crop and having that gene product function as desired, the problems of controlling substrate flux partitioning to drive a desired biosynthetic pathway, and the potential risks posed by transgenic food crops are troublesome issues that have underscored the need for continued research on and development of nongenetic approaches for the enrichment of isoflavonoids in soybean foods and food ingredients (1). Major nongenetic approaches for increasing phenolic content in dietary plants include bioprocessing of soybean substrates and stimulation of the plant defense responses, which is known to result in the stimulation of phenolic synthesis. Both of these approaches could potentially be used to increase isoflavonoid content in soybeans.

Bioprocessing of various plant based foods by dietary fungus is a technology that has been used throughout history in the context of producing fermented foods, such as tempeh. Currently, this technology is being utilized in conjunction with specific dietary fungi to produce certain desired products. In this context, fungal bioprocessing (also known as solid-state fermentation) has been employed to enrich various solid food substrates such as grape pomace, cornmeal, mango, date, wheat bran, and wine for products such as protein, C/N ratio, β-glucan, and pectinase (77–81). The use of dietary fungal bioprocessing of fruit and legume food substrates for enrichment of aglycone phenolic antioxidants such as ellagic acid in cranberry and isoflavonoids from soybeans has been reported (27,28,82–84). Enhanced isoflavonoid content in soybeans and soybean meal following bioprocessing by *Aspergillus* species has also been reported (85,86). The number of different microbial species and substrates available for isoflavonoid and other phenolic enrichment by fungal bioprocessing is likely to grow as knowledge of dietary microbial species increases.

One of the results of elicitor mediated activation of the plant defense system is an increase in phenolic secondary metabolite biosynthesis (87,88). Exogenous application of salicylic acid, a phenolic metabolite thought to act as a chemical signal in the defense response system, can stimulate phenolic content in peas (89). Similarly, pure dietary phenolics and phenolic rich extracts have been reported to stimulate phenolic content in legumes (90–96). Further, application of bacteria, bacterial polysaccharides, and UV and microwave radiation (all of which are known inducers of plant defense responses) have been shown to stimulate plant phenolic content (96–99). The type of phenolics elicited by activation of the defense response is largely determined by the nature of the treated plant (i.e., phenolic profiles vary from plant to plant). Therefore, application of plant defense response elicitors to soybean may potentially stimulate higher isoflavonoid content without the need for genetic manipulation.

6.2.4 Toward a Model Mechanism for Action of Soybean Isoflavonoids and Related Phenolic Antioxidants against Cancer

6.2.4.1 Metabolism of Dietary Isoflavonoids

As stated earlier, genistein and daidzein occur in soybeans as glucoside conjugates that must be converted into aglycones to be metabolically active (22,24). In humans, this can be performed by intestinal flora. Interestingly, fermented soy foods are rich in phenolic aglycones due to microbial bioprocessing during fermentation, which may allow for rapid intestinal absorption upon consumption (27,28).

6.2.4.2 Control of Energy Metabolism and Oxidative Stress in a Healthy Cell

In cells, the production of energy adenosine triphosphate (ATP) occurs in mitochondria by reduced nicotinamide adenine dinucleotide (NADH)-mediated oxidative phosphorylation (oxPHOS) (100). The tricarboxylic acid (TCA) cycle produces NADH to support mitochondrial ATP synthesis. Although some reactive oxygen species (ROS) are generated during mitochondrial oxPHOS, cells possess an extensive antioxidant response system which operates to scavenge ROS and protect cellular components from oxidative damage (100,101).

Antioxidant enzymes play a key role in cellular antioxidant response (100). Chief among these enzymes are superoxide dismutase (SOD) and catalase. SOD converts superoxide (O_2^-) into hydrogen peroxide (H_2O_2), which is less reactive. Catalase converts H_2O_2 into water (H_2O) and oxygen (O_2). Manganese superoxide dismutase (MnSOD) occurs within mitochondria to protect the organelle from oxidative damage, while copper–zinc SOD (CuZnSOD) and catalase occur in the cytosol, both to protect cytosolic bound cellular components from oxidative damage and to maintain a proper redox environment, because redox imbalances can activate certain cellular activities (102).

It has been proposed that phenolic antioxidants may have chemopreventive potential through modulation of the antioxidant enzyme response through the proline linked pentose–phosphate pathway (103). Exogenous antioxidants, such as dietary plant phenolic compounds, have been shown to scavenge ROS in cells *in vitro* and may help protect cells against oxidative damage *in vivo* (103,104).

6.2.4.3 Control of Energy Metabolism and Oxidative Stress in a Tumorigenic Cell

Many of the diseases for which a reduced risk of incidence has been associated with soy food consumption are oxidation linked diseases, such as cancer and cardiovascular disease (100). Oxidation linked diseases have been linked to a general breakdown in the regulation of cellular activities (such as growth or energy production), an accumulation of ROS such as superoxide and hydrogen peroxide, a cellular redox imbalance, and accumulated oxidative damage in normal cells (106).

Evidence indicates that, similar to healthy cells, tumor cells obtain much of their ATP for energy requirements via NADH linked mitochondrial oxPHOS (107). However, in tumorigenic cells, energy generation is inefficient as mitochondrial respiration activities are defective compared to healthy cells (108–110). Interestingly, in many cancer cells the activity of glucose-6-phosphate dehydrogenase (G6PDH), the key regulatory enzyme of the oxPPP, is reduced by up to 90%, while the nonoxidative pentose–phosphate pathway (nonoxPPP) flux is increased, possibly to support higher glycolytic flux toward the TCA cycle to support additional demand for NADH (111,112).

In addition to an abnormal energy metabolism, tumor cells possess a reduced antioxidant response system. Catalase and CuZnSOD activities, important for controlling cytosolic ROS levels, are decreased in numerous cancer cell lines (113,114). Low activity of antioxidant enzymes leaves cancer cells particularly susceptible to increased oxidative damage upon ROS accumulation, and eventually apoptosis (cell death) or necrosis (113,114).

6.2.5 Stress Response and the Proline Linked Pentose–Phosphate Pathway

Healthy cells possess a mechanism that couples increased mitochondrial ATP synthesis to increased glucose flux through the oxPPP for biosynthetic substrates (glucose phosphates, reduced nicotinamide adenine dinucleotide phosphate (NADPH) that support cellular energy demands during times of stress (100). In this mechanism, mitochondrial ATP generation and oxPPP activity are coupled through the biosynthesis of proline (115). Proline biosynthesis provides a mechanism for the transfer of reducing equivalents from NADPH into mitochondria (via proline oxidation) and is linked to glucose oxidation in the oxPPP by NADPH turnover, a coupling known as the proline linked pentose–phosphate pathway (PL-PPP) (116). Notably, in this mechanism mitochondrial oxPHOS switches its dependence on NADH to proline as a source of reducing equivalents to support ATP synthesis while maintaining cellular NADPH biosynthesis for anabolic reactions. Increased proline metabolism has been shown to stimulate oxPPP activity via NADP mediated redox regulation (117).

As the activity of several antioxidant enzymes depends on the availability of NADPH, activity of the cellular antioxidant response system in stressed cells may depend directly upon flux through the oxPPP and, therefore, indirectly upon the activity of a functional proline cycling mechanism (100). Increased oxPPP activity has been shown to protect cells against H_2O_2 and NO stresses (118,119). Similarly, a phagocyte derived ROS increase occurs during the immune response and is followed by an increase in G6PDH activity (120). G6PDH and oxPPP activity protected cells from oxidant and radiation induced apoptosis (121). Recently, antioxidant enzymes were found to be essential for protecting cells against ROS mediated damage (122).

6.3 A HYPOTHETICAL MODEL FOR THE CANCER CHEMOPREVENTIVE ACTION OF THE SOYBEAN ISOFLAVONE GENISTEIN

The mechanistic coupling of mitochondrial ATP generation and oxPPP through proline metabolism that supports normal cells during stress and which appears to be dysfunctional in many tumorigenic cells provides a foundation for a possible model mechanism for the chemopreventive action of phenolic antioxidants, such as genistein, against certain cancers (100).

When soybean is consumed, its main phenolic phytochemical, the isoflavone genistein, is converted by intestinal flora bioprocessing and subsequent hepatic activities into a β-glucuronide conjugate. In an interesting twist of fate, the expression of β-glucuronidase is low or not detected in normal tissues, but is high in tumors (123). Therefore, in tissues that take up the genistein-β-glucuronide from the bloodstream, free and biologically active genistein is most likely to occur in a higher amount in tumor cells than in healthy cells due to the selective nature of β-glucuronidase expression in these tissues.

Once activated by glucuronidase, free genistein may stimulate proline metabolism through the activation of proline dehydrogenase (PDH) via p53. Genistein can induce p53 expression in colorectal cancer cells (124). The mechanism for stimulation of p53 by genistein is not clear, but other dietary antioxidants can activate p53 by a ref1 dependent redox mechanism (125,126). The transcription factor p53 can activate PDH, as well as ROS and apoptosis in cancer cells (127–129).

Stimulation of mitochondrial PDH in the tumor cell by phenolic antioxidants such as genistein should cause a demand for proline (117) that would have several major metabolic effects: (1) increased mitochondrial oxPHOS supported by proline oxidation would generate increased ROS that could leak into the cytosol and damage essential cellular components; (2) proline would be shunted to the mitochondria and away from collagen biosynthesis, potentially crippling tumor growth, expansion, and proliferation activities; and (3) energy metabolism would be redirected toward proline mediated mitochondrial ATP synthesis and away from TCA linked, NADH mediated oxPHOS via activity of the PL-PPP. An occurrence of these metabolic effects would be uniquely detrimental to the functioning of a tumor cell because, by all reported indications, normal cellular metabolism, including the antioxidant response system, in tumor cells is dysfunctional at mitochondrial and cytosolic levels.

First, increasing the activity of PDH would drive mitochondrial oxPHOS toward ATP synthesis, and produce increased ROS levels in the process. In a normal cell, increased ROS production by increased mitochondrial activity would likely be countered by activation of the cellular antioxidant enzyme response system, but in cancer cells antioxidant enzymes appear to be less active and may be less able to adequately protect tumor cells from oxidative damage, and potentially promote apoptosis (109,110,122,130). Because tumor cell mitochondria are already dysfunctional, increased respiration activities activity driven by genistein stimulated PDH activity should produce increased amounts of cytosolic ROS into the cytosol, endangering numerous essential cytosolic components such as proteins and organelles. MnSOD is expressed at low levels in normal cells, but at high levels in tumors, perhaps in response to high ROS levels (131). However, expression of catalase, glutathione peroxidase (GPx), and CuZnSOD is greatly diminished, which may facilitate accumulation of excessive oxygen radicals and oxidative damage in tumor cells (109,110). Although dietary antioxidants, such as curcumin, ascorbic acid, and flavanoids, have been shown to stimulate antioxidants and important phase II antioxidant enzymes such as SOD and catalase, genistein did not stimulate catalase or SOD in prostate cancer (132–135). In fact, genistein and soy isoflavone extracts have been shown to stimulate caspase-3 and apoptosis in cancer cells (58,62). Thus, phenolic antioxidants such as genistein may stimulate apoptosis linked activities that support the generation and accumulation of oxygen radicals and oxidative damage in tumor cells. The action of genistein could be further enhanced synergistically by other soluble phenolics from a food system. This is further supported by recent evidence that fermented soymilk and whole soy extracts inhibit tumor growth better than genistein alone (16,136,137).

Second, the stimulation of PDH activity by genistein and synergistic phenolic profiles would cause a metabolic demand for proline as a reductant to support mitochondrial oxPHOS (energy production) and may derail tumor growth supporting collagen biosynthesis by redirecting a necessary substrate (e.g., proline). Genistein has been shown to inhibit the growth and proliferation of cancer cells *in vitro* (32). If genistein activates of mitochondrial PDH and the proline cycle as part of a key stress response mechanism, proline cycle demand for proline may have higher cellular priority than collagen biosynthesis during times of stress (117).

Finally, and perhaps most importantly, induction of mitochondrial PDH activity and proline cycling by genistein would shift cellular energy metabolism in the tumor cell away from NADH linked mitochondrial oxPHOS to a proline linked mitochondrial oxPHOS system (100). Dysfunctional tumor cell mitochondria may produce increased ROS generation via increased mitochondrial respiration that may lead to mitochondrial membrane damage, caspase activation, and eventually apoptosis. In support of this idea, genistein has been shown previously to inhibit nonoxPPP ribose synthesis and cell proliferation in cancer cells (32,33). In healthy cells, stress induced stimulation of the proline cycle also drives the oxPPP via G6PDH recycling of NADPH, and G6PDH (and thereby the stress induced PL-PPP) can be inhibited by high NADPH levels. In contrast, in tumorigenic cells G6PDH is dysfunctional (with activity decreased by up to 90%) and is no longer inhibited by high NADPH (105,117,38). Therefore, while genistein stimulated PDH activity may be supported by NADPH cycling between proline biosynthesis and G6PDH activity, a dysfunctional G6PDH enzyme may not allow the tumor cell to disengage the stress response mechanism in the presence of high levels of NADPH. Because the dysfunctional enzyme operates at such a low efficiency (~10%) and likely does not produce high NADPH levels, there may not be enough NADPH produced to support the anabolic demands of the antioxidant enzyme response system which would further hinder the tumor cell to defend itself against oxPHOS derived ROS.

Further, genistein possesses other biological activities that may help to promote apoptosis in tumor cells. Aside from the stimulation of caspase-3 which could result from mitochondrial PDH activation, genistein can inhibit NF-κB (13), which can block apoptosis (139).

6.4 SUMMARY AND IMPLICATIONS

Starvation is an efficient way to kill a living organism, even a diseased cell. An effective strategy to starve a tumor cell could be to potentiate a switch in energy metabolism from a dysfunctional and inefficient TCA/NADH linked mechanism to an even more dysfunctional alternative pathway (100). In tumor cells, mitochondrial ATP-generating activities are known to be negatively altered and to function inefficiently (103,104,140). If the energy metabolism could somehow be forced to revert back to the dysfunctional mitochondria, the inefficiency of the system may starve the tumor cell of chemical energy (ATP) for cellular activities (100). For a tumor cell to knowingly switch a core metabolism to favor a pathway or mechanism that would be detrimental to its survival seems unlikely, but it is possible that just such an action may occur through the activation of an underlying key stress response mechanism that functions in normal, healthy cells, for which activation triggers may still remain even after the normal cell transitions into a tumorigenic cell (100).

Here we describe a potential mechanism by which phenolic antioxidants such as the isoflavone genistein may act to promote tumor cell death by inducing the diseased cell to switch its energy metabolism from growth-promoting TCA cycle/ NADH linked system to a dysfunctional proline linked system through the activation of a key stress response mechanism involving the PL-PPP, whereby the cell is essentially starved of chemical energy (100). As the PL-PPP in plants can be stimulated by dietary (89,92,93,41), the same may be true in animal (e.g., human) cells. We have described how p53 mediated activation of PDH by genistein may create a metabolic demand for proline, therein activating the PL-PPP and causing a shift in energy metabolism to facilitate mitochondrial ATP generation, even though components of the stress response mechanism in tumor cells may be dysfunctional. A mitochondrial demand for proline might divert proline away from collagen biosynthesis, which may explain the interrupted tumor growth and proliferation activities observed in cancer cells treated with phenolic compounds. It is unknown whether or not tumor cell PDH is dysfunctional (100).

ROS accumulation precipitated by increased activity of leaky mitochondria should stimulate activity of antioxidant enzymes and the oxPPP for NADPH to support reductant cycling systems (100). However, in tumor cells G6PDH is also dysfunctional and inefficient, such that the generation of adequate NADPH levels to support the demands of both proline synthesis and the antioxidant enzyme response system is unlikely. Further, as the dysfunctional G6PDH is no longer inactivated by high NADPH, it is likely that once the PL-PPP mediated stress response mechanism is activated, the tumor cell may not be able to disengage it, and may become locked in its fate and thus prevented from returning to TCA cycle/ NADH linked energy metabolism (100).

Further, our hypothetical model helps to explain how phenolic antioxidants such as genistein could have both detrimental effects against tumor cells and beneficial effects on healthy cells. In healthy cells, addition of a phenolic antioxidant that can stimulate mitochondrial activity and the antioxidant enzyme response system (i.e., SOD, catalase, GPx) may aid in protection against oxidative damage and thus promote healthy cellular conditions (102,142). Furthermore, as normal G6PDH can be inactivated by high NADPH levels, the PL-PPP-mediated stress response mechanism can be turned off as needed, something that a tumor cell may be unable to do and which may, ultimately, lead to the death of the diseased cell (100).

Numerous dietary phenolic antioxidants have been shown to inhibit cancer cell growth and proliferation, but until now no overall cellular mechanism that integrates the many observed phenolic linked activities has been put forth. We believe our hypothetical model for the chemopreventive actions of the soybean isoflavonoid genistein as described here offers a logical mode of action for dietary phenolic antioxidants based on key cellular metabolism linked to alternative energy and redox management. This model merits further experimental investigation in conjunction with genetic and signal transduction mechanisms associated with cancer biology (107). As various dietary phenolic compounds have been reported to possess activities that may facilitate apoptosis, our model provides a core mechanism by which many of these compounds could act, and a mechanism that could be supported by other such anticancer properties, such as NF-κB inhibition and related signal pathways.

REFERENCES

1. McCue, P., K. Shetty. Health benefits of soy isoflavonoids and strategies for enhancement: a review. *Crit. Rev. Food Sci. Nutr.* 44:1–7, 2004.
2. Wu, A.H., R.G. Ziegler, P.L. Horn-Ross, A.M. Nomura, D.W. West, L.N. Kolonel, J.F. Rosenthal, R.N. Hoover, M.C. Pike. Tofu and risk of breast cancer in Asian-Americans. *Cancer Epidemiol. Biomarkers Prev.* 5:901–906, 1996.
3. Zheng, W., Q. Dai, L.J. Custer, X.O. Shu, W.Q. Wen, F. Jin, A.A. Franke. Urinary excretion of isoflavonoids and the risk of breast cancer. *Cancer Epidemiol. Biomarkers Prev.* 8:35–40, 1999.
4. Anderson, J.W., B.M. Smith, C.S. Washnock. Cardiovascular and renal benefits of dry bean and soybean intake. *Am. J. Clin. Nutr.* 70(Suppl):464S–474S, 1999.
5. Yamakoshi, J., M.K. Piskula, T. Izumi, K. Tobe, M. Saito, S. Kataoka, A. Obata, M. Kikuchi. Isoflavone aglycone-rich extract without soy protein attenuates atherosclerosis development in cholesterol-fed rabbits. *J. Nutr.* 130:1887–1893, 2000.
6. Dai, Q., A.A. Franke, F. Jin, X.O. Shu, J.R. Hebert, L.J. Custer, J. Cheng, Y.T. Gao, W. Zheng. Urinary excretion of phytoestrogens and risk of breast cancer among Chinese women in Shanghai. *Cancer Epidemiol. Biomarkers Prev.* 11:815–821, 2002.
7. Jenkins, D.J., C.W. Kendall, C.J. Jackson, P.W. Connelly, T. Parker, D. Faulkner, E. Vidgen, S.C. Cunnane, L.A. Leiter, R.G. Josse. Effects of high- and low-isoflavone soyfoods on blood lipids, oxidized LDL, homocysteine, and blood pressure in hyperlipidemic men and women. *Am. J. Clin. Nutr.* 76:365–372, 2002.
8. Yamamoto, S., T. Sobue, M. Kobayashi, S. Sasaki, S. Tsugane. Soy, isoflavones, and breast cancer risk in Japan. *J. Natl. Cancer Inst.* 95:906–913, 2003.
9. Jacobsen, B.K., S.F. Knutsen, G.E. Fraser. Does high soy milk intake reduce prostate cancer incidence?: the Adventist health study (United States). *Cancer Causes Control* 9:553–557, 1998.
10. Lee, M.M., S.L. Gomez, J.S. Chang, M. Wey, R.T. Wang, A.W. Hsing. Soy and isoflavone consumption in relation to prostate cancer risk in China. *Cancer Epidemiol. Biomarkers Prev.* 12:665–668, 2003.
11. Darbon, J.M., M. Penary, N. Escalas, F. Casagrande, F. Goubin-Gramatica, C. Baudouin, B. Ducommun. Distinct Chk2 activation pathways are triggered by genistein and DNA-damaging agents in human melanoma cells. *J. Biol. Chem.* 275:15363–15369, 2000.
12. Lamartiniere, C.A. Protection against breast cancer with genistein: a component of soy. *Am. J. Clin. Nutr.* 71:1705S–1707S, 2000.

13. Xu, J., G. Loo. Different effects of genistein on molecular markers related to apoptosis in two phenotypically dissimilar breast cancer cell lines. *J. Cell. Biochem.* 82(1):78–88, 2001.

14. Lamartiniere, C.A., M.S. Cotroneo, W.A. Fritz, J. Wang, R. Mentor-Marcel, E. Elgavish. Genistein chemoprevention: timing and mechanisms of action in murine mammary and prostate. *J. Nutr.* 132:552S–558S, 2002.

15. Tanos, V., A. Brzezinski, O. Drize, N. Strauss, T. Peretz. Synergistic inhibitory effects of genistein and tamoxifen on human dysplastic and malignant epithelial breast cancer cells *in vitro. Eur. J. Obstet. Gynecol. Reprod. Biol.* 102:188–194, 2002.

16. Ohta, T., S. Nakatsugi, K. Watanabe, T. Kawamori, F. Ishikawa, M. Morotomi, S. Sugie, T. Toda, T. Sugimura, K. Wakabayashi. Inhibitory effects of *Bifidobacterium*-fermented soy milk on 2-amino-1-methyl-6-phenylimidazo- [4, 5-*b*]-pyridine-induced rat mammary carcinogenesis, with a partial contribution of its component isoflavones. *Carcinogenesis* 21: 937–941, 2000.

17. Wuttke, W., H. Jarry, T. Becker, A. Schultens, V. Christoffel, C. Gorkow, D. Seidlova-Wuttke. Phytoestrogens: endrocrine disrupters or replacement for hormone replacement therapy? *Maturitas* 44(1):S9-S20, 2003.

18. Safford, B., A. Dickens, N. Halleron, D. Briggs, P. Carthew, V. Baker. A model to estimate the oestrogen receptor mediated effects from exposure to soy isoflavones in food. *Regul. Toxicol. Pharmacol.* 38:196–209, 2003.

19. Coward, L., N. Barnes, K.D. Setchell, S. Barnes. Genistein, daidzein, and the β-glycoside conjugates: anti-tumor isoflavones in soybean foods from American and Asian diets. *J. Agric. Food Chem.* 41:1961–1967, 1993.

20. Chen, Z., W. Zheng, L.J. Custer, Q. Dai, X.O. Shu, F. Jin, A.A. Franke. Usual dietary consumption of soy foods and its correlation with the excretion rate of isoflavonoids in overnight urine samples among Chinese women in Shanghai. *Nutr. Cancer* 33:82–87, 1999.

21. De Kleijn, M.J., Y.T. van der Schouw, P.W. Wilson, H. Adlercreutz, W. Mazur, D.E. Grobbee, P.F. Jacques. Intake of dietary phytoestrogens is low in postmenopausal women in the United States: the Framingham study. *J. Nutr.* 131:1826–1832, 2001.

22. Peterson, T.G., G.P. Ji, M. Kirk, L. Coward, C.N. Falany, S. Barnes. Metabolism of the isoflavones genistein and biochanin A in human breast cancer cell lines. *Am. J. Clin. Nutr.* 68:1505S–1511S, 1998.

23. Rao, M., G. Muralikrishna. Evaluation of the antioxidant properties of free and bound phenolic acids from native and malted finger millet (Ragi, Eleusine coracana Indaf-15). *J. Agric. Food Chem.* 50:889–892, 2002.

24. Yuan, L., C. Wagatsuma, M. Yoshida, T. Miura, T. Mukoda, H. Fujii, B. Sun, J.H. Kim, Y.J. Surh. Inhibition of human breast cancer growth by GCP™ (genistein combined polysaccharide) in xenogeneic athymic mice: involvement of genistein biotransformation by β-glucuronidase from tumor tissues. *Mutat. Res.* 523-524:55–62, 2003.

25. Murota, K., S. Shimizu, S. Miyamoto, T. Izumi, A. Obata, M. Kikuchi, J. Terao. Unique uptake and transport of isoflavone aglycones by human intestinal caco-2 cells: comparison of isoflavonoids and flavanoids. *J. Nutr.* 132:1956–1961, 2002.

26. Setchell, K.D., N.M. Brown, L. Zimmer-Nechemias, W.T. Brashear, B.E. Wolfe, A.S. Kirshner, J.E. Heubi. Evidence for lack of absorption of soy isoflavone glycosides in humans, supporting the crucial role of intestinal metabolism for bioavailability. *Am. J. Clin. Nutr.* 76:447–453, 2002.

27. McCue, P., A. Horii, K. Shetty. Solid-state bioconversion of phenolic antioxidants from defatted soybean powders by *Rhizopus oligosporus*: role of carbohydrate cleaving enzymes. *J. Food. Biochem.* 27(6):501–514, 2003.

28. McCue, P., K. Shetty. Role of carbohydrate-cleaving enzymes in phenolic antioxidant mobilization from whole soybean fermented with *Rhizopus oligosporus. Food Biotechnol.* 17:27–37, 2003.

29. Okura, A., H. Arakawa, H. Oka, T. Yoshinari, Y. Monden. Effect of genistein on topoisomerase activity and on the growth of [Val 12]Ha-ras-transformed NIH 3T3 cells. *Biochem. Biophys. Res. Commun.* 157:183–189, 1988.

30. Akiyama, T., J. Ishida, S. Nakagawa, H. Ogawara, S. Watanabe, N. Itoh, M. Shibuya, Y. Fukami. Genistein, a specific inhibitor of tyrosine-specific protein kinases. *J. Biol. Chem.* 262:5592–5595, 1987.

31. Gong, L., Y. Li, A. Nedeljkovic-Kurepa, F.H. Sarkar. Inactivation of NF-κB by genistein is mediated via Akt signaling pathway in breast cancer cells. *Oncogene* 22:4702–4709, 2003.

32. Wang, S.Y., K.W. Yang, Y.T. Hsu, C.L. Chang, Y.C. Yang. The differential inhibitory effects of genistein on the growth of cervical cancer cells *in vitro*. *Neoplasma* 48:227–233, 2001.

33. Boros, L.G., S. Bassilian, S. Lim, W.N.P. Lee. Genistein inhibits non-oxidative ribose synthesis in MIA pancreatic adenocarcinoma cells: a new mechanism of controlling cancer growth. *Pancreas* 22:1–7, 2001.

34. Arora, A., M.G. Nair, G.M. Strasburg. Antioxidant activities of isoflavones and their biological metabolites in a liposomal system. *Arch. Biochem. Biophys.* 356(2):133–141, 1998.

35. Hollman, P.C., M.B. Katan. Health effects and bioavailability of dietary flavonols. *Free Radic. Res.* 31:S75-S80, 1999.

36. Barnes, S., B. Boersma, R. Patel, M. Kirk, V.M. Darley-Usmar, H. Kim, J. Xu. Isoflavonoids and chronic disease: mechanisms of action. *Biofactors* 12(1-4):209-215, 2000.

37. Cos, P., M. Calomme, J.B. Sindambiwe, T. De Bruyne, K. Cimanga, L. Pieters, A.J. Vlietinck, D. Van den Berghe. Cytotoxicity and lipid peroxidation-inhibiting activity of flavanoids. *Planta. Med.* 67(6):515–519, 2001.

38. Pike, A.C., A.M. Brzozowski, R.E. Hubbard, T. Bonn, A.G. Thorsell, O. Engstrom, J. Ljunggren, J.A. Gustafsson, M. Carlquist. Structure of the ligand-binding domain of oestrogen receptor beta in the presence of a partial agonist and a full antagonist. *EMBO J.* 18(17):4608–4618, 1999.

39. Hochanadel, G., J. Shifren, I. Zhdanova, T. Maher, P.A. Spiers. Soy isoflavones (phytoestrogens) in the treatment of the cognitive and somatic symptoms of menopause. *PCRS Abtracts* 71(4,1):20S, 1999.

40. Murkies, A.L., C. Lombard, B.J. Strauss, G. Wilcox, H.G. Burger, M.S. Morton. Dietary flour supplementation decreases post-menopausal hot flushes: effect of soy and wheat. *Maturitas* 21(3):189–195, 1995.

41. Dalais, F.S., G.E. Rice, A.L. Murkies, R.J. Bell, M.L. Wahlqvist. Effects of dietary phytoestrogens in postmenopausal women. *Maturitas* 27(1):214, 1997.

42. Albertazzi, P., F. Pansini, G. Bonaccorsi, L. Zanotti, E. Forini, D. De Aloysio. The effect of dietary soy supplementation on hot flushes. *Obstet Gynecol* 91(1):6–11, 1998.

43. Carusi, D. Phytoestrogens as hormone replacement therapy: an evidence-based approach. *Primary Care Update OB/GYNS* 7(6):253–259, 2000.

44. Arjmandi, B.H., R. Birnbaum, N.V. Goyal, M.J. Getlinger, S. Juma, L. Alekel, C.M. Hasler, M.L. Drum, B.W. Hollis, S.C. Kukreja. Bone-sparing effect of soy protein in ovarian hormone-deficient rats is related to its isoflavone content. *Am. J. Clin. Nutr.* 68(6):1364S–1368S, 1998.

45. S.M. Potter, J.A. Baum, H. Teng, R.J. Stillman, N.F. Shay, J.W. Erdman. Soy protein and isoflavones: their effects on blood lipids and bone density in postmenopausal women. *Am. J. Clin. Nutr.* 68(6):1375S–1379S, 1998.

46. J.J. Anderson, M.S. Anthony, J.M. Cline, S.A. Washburn, S.C. Garner. Health potential of soy isoflavones for menopausal women. *Public Health Nutr.* 2(4):489–504, 1999.

47. Scheiber, M.D., J.H. Liu, M.T. Subbiah, R.W. Rebar, K.D. Setchell. Dietary inclusion of whole soy foods results in significant reductions in clinical risk factors for osteoporosis and cardiovascular disease in normal postmenopausal women. *Menopause* 8(5):384–392, 2001.

48. Gallagher, J.C. Role of estrogens in the management of postmenopausal bone loss. *Rheum. Dis. Clin. North. Am.* 27:143–162, 2001.

49. Goodman, M.T., L.R. Wilkens, J.H. Hankin, L.C. Lyu, A.H. Wu, L.N. Kolonel. Association of soy and fiber consumption with the risk of endometrial cancer. *Am. J. Epidemiol.* 146(4):294–306, 1997.

50. Morishige, K., K. Matsumoto, M. Ohmichi, Y. Nishio, K. Adachi, J. Hayakawa, K. Nukui, K. Tasaka, H. Kurachi, Y. Murata. Clinical features affecting the results of estrogen

replacement therapy on bone density in Japanese postmenopausal women. *Gynecol. Obstet. Invest.* 52(4):223–226, 2001.

51. Lacey, J.V., P.J. Mink, J.H. Lubin, M.E. Sherman, R. Troisi, P. Hartge, A. Schatzkin, C. Schairer. Menopausal hormone replacement therapy and risk of ovarian cancer. *JAMA* 288(3):334–341, 2002.

52. Black, L.J., M. Sato, E.R. Rowley, D.E. Magee, A. Bekele, D.C. Williams, G.J. Cullinan, R. Bendele, R.F. Kauffman, W.R. Bensch, C.A. Frolik, J.D. Termine, H.U. Bryant. Raloxifene (LY139481 HCI) prevents bone loss and reduces serum cholesterol without causing uterine hypertrophy in ovariectomized rats. *J. Clin. Invest.* 93(1):63–69, 1994.

53. Ishimi, Y., C. Miyaura, M. Ohmura, Y. Onoe, T. Sato, Y. Uchiyama, M. Ito, X. Wang, T. Suda, S. Ikegami. Selective effects of genistein, a soybean isoflavone, on B-lymphopoiesis and bone loss caused by estrogen deficiency. *Endocrinology* 140(4):1893–1900, 1999.

54. Anderson, J.J., S.C. Garner. The effects of phytoestrogens on bone. *Nutr. Res.* 17(10):1617–1632, 1997.

55. Mühlbauer, R.C., F. Li. Frequency of food intake and natural dietary components are potent modulators of bone resorption and bone mass in rats. *Biomed. Pharmacother.* 51(8):360–363, 1997.

56. Barnes, S., T.G. Peterson, L. Coward. Rationale for the use of genistein-containing soy matrices in chemoprevention trials for breast and prostate cancer. *J. Cell. Biochem. Suppl.* 22:181–187, 1995.

57. Röhrdanz, E., S. Ohler, Q.H. Tran-Thi, R. Kahl. The phytoestrogen daidzein affects the antioxidant enzyme system of rat hepatoma H4IIE cells. *J. Nutr.* 132:370–375, 2002.

58. Su, S.J., N.H. Chow, M.L. Kung, T.C. Hung, K.L. Chang. Effects of soy isoflavones on apoptosis induction and G2-M arrest in human hepatoma cells involvement of caspase-3 activation, Bcl-2 and Bcl-xL down-regulation, and Cdc2 kinase activity. *Nutr. Cancer* 45:113–123, 2003.

59. Wilson, L.C., S.J. Baek, A. Call, T.E. Eling. Nonsteroidal anti-inflammatory drug-activated gene (NAG-1) is induced by genistein through the expression of p53 in colorectal cancer cells. *Int. J. Cancer* 105:747–753, 2003.

60. Cai, Q., H. Wei. Effect of dietary genistein on antioxidant enzyme activities in SENCAR mice. *Nutr. Cancer* 25:1–7, 1996.

61. Vissac-Sabatier, C., Y.J. Bignon, D.J. Bernard-Gallon. Effects of the phytoestrogens genistein and daidzein on BRCA2 tumor suppressor gene expression in breast cell lines. *Nutr. Cancer* 45:247–255, 2003.

62. Song, D., X. Na, Y. Liu, X. Chi. Study on mechanisms of human gastric carcinoma cells apoptosis induced by genistein. *Wei Sheng Yan Jiu* 32:128-130, 2003.

63. Andersson, C., Z. Servetnyk, G.M. Roomans. Activation of CFTR by genistein in human airway epithelial cell lines. *Biochem. Biophys. Res. Commun.* 308:518–522, 2003.

64. Yan, C., R. Han. Effects of genistein on invasion and matrix metalloproteinase activities of HT1080 human fibrosarcoma cells. *Chin. Med. Sci. J.* 14:129–133, 1999.

65. Naaz, A., S. Yellayi, M.A. Zakroczymski, D. Bunick, D.R. Doerge, D.B. Lubahn, W.G. Helferich, P.S. Cooke. The soy isoflavone genistein decreases adipose deposition in mice. *Endocrinology* 144:3315–3320, 2003.

66. Murakami, A., D. Takahashi, K. Hagihara, K. Koshimizu, H. Ohigashi. Combinatorial effects of nonsteroidal anti-inflammatory drugs and food constituents on production of prostaglandin E2 and tumor necrosis factor-alpha in RAW264.7 murine macrophages. *Biosci. Biotechnol. Biochem.* 67:1056–1062, 2003.

67. Ridges, L., R. Sunderland, K. Moernlan, B. Meyer, L. Astheimer, P. Howe. Cholesterol lowering benefits of soy and linseed enriched foods. *Asia. Pac. J Clin. Nutr.* 10:204–211, 2001.

68. Department of Health and Human Services. Food labeling: health claims, soy protein and coronary heart disease, final rule. Federal registers, Food and Drug Administration, 64:57700–57733, 1999.

69. Cheng, J., K. Kondo, Y. Suzuki, Y. Ikeda, X. Meng, K. Umemura. Inhibitory effects of total flavones of *Hippophae Rhamnoides* L. on thrombosis in mouse femoral artery and *in vitro* platelet aggregation. *Life Sci.* 72:2263–2271, 2003.

70. Dixon, R.A., C.L. Steele. Flavanoids and isoflavonoids – a gold mine for metabolic engineering. *Trends Plant Sci* 4(10):394–400, 1999.

71. Jung, W., O. Yu, S.M. Lau, D.P. O'Keefe, J. Odell, G. Fader, B. McGonigle. Identification and expression of isoflavone synthase, the key enzyme for biosynthesis of isoflavones in legumes. *Nat. Biotechnol.* 18(2):208–212, 2000.

72. Yu, O., W. Jung, J. Shi, R.A. Croes, G.M. Fader, B. McGonigle, J.T. Odell. Production of the isoflavones genistein and daidzein in non-legume dicot and monocot tissues. *Plant Physiol.* 124(2):781–794, 2000.

73. Liu, C.J., J.W. Blount, C.L. Steele, R.A. Dixon. Bottlenecks for metabolic engineering of isoflavone glycoconjugates in Arabidopsis. *Proc. Natl. Acad. Sci. USA* 99(22):14578–14583, 2002.

74. Dixon, R.A., L.W. Sumner. Legume natural products: understanding and manipulating complex pathways for human and animal health. *Plant Physiol.* 131:878–885, 2003.

75. Hwang, E.I., M. Kaneko, Y. Ohnishi, S. Horinouchi. Production of plant-specific flavanones by *Escherichia coli* containing an artificial gene cluster. *Appl. Environ. Microbiol.* 69(5):2699–2706, 2003.

76. Chassy, B.M. Food safety evaluation of crops produced through biotechnology. *J. Am. Coll. Nutr.* 21(3, Suppl):166S–173S, 2002.

77. Jwanny, E.W., M.M. Rashad, H.M. Abdu. Solid-state fermentation of agricultural wastes into food through *Pleurotus* cultivation. *Appl. Biochem. Biotechnol.* 50(1):71–78, 1995.

78. Okamura, T., T. Ogata, N. Minamimoto, T. Takeno, H. Noda, S. Fukuda, M. Ohsugi. Characteristics of wine produced by mushroom fermentation. *Biosci. Biotechnol. Biochem.* 65(7):1596–1600, 2001.

79. Sanchez, A., F. Ysunza, M.J. Beltran-Garcia, M. Esqueda. Biodegradation of viticulture wastes by *Pleurotus*: a source of microbial and human food and its potential use in animal feeding. *J. Agric. Food Chem.* 50:2537–2542, 2002.

80. Han, J. Solid-state fermentation of cornmeal with the basidiomycete *Hericium erinaceum* for degrading starch and upgrading nutritional value. *Int. J. Food Microbiol.* 80:61–66, 2003.

81. Kashyap, D.R., S.K. Soni, R. Tewari. Enhanced production of pectinase by *Bacillus* sp. DT7 using solid state fermentation. *Bioresource Technol.* 88:251–254, 2003.

82. Zheng, Z., K. Shetty. Solid-state bioconversion of phenolics from cranberry pomace and role of *Lentinus edodes* beta-glucosidase. *J. Agric. Food Chem.* 48(3):895–900, 2000.

83. Vattem, D.A., K. Shetty. Solid-state production of phenolic antioxidants from cranberry pomace by *Rhizopus oligosporus*. *Food Biotechnol.* 16:189–210, 2002.

84. Vattem, D.A., K. Shetty. Ellagic acid production and phenolic antioxidant activity in cranberry pomace mediated by *Lentinus edodes* using solid-state system. *Process. Biochem.* 39:367–379, 2003.

85. Esaki, H., R. Watanabe, H. Onozaki, S. Kawakishi, T. Osawa. Formation mechanism for potent antioxidative o-dihydroxyisoflavones in soybeans fermented with *Aspergillus saitoi*. *Biosci. Biotechnol. Biochem.* 63(5):851–858, 1999.

86. Kishida, T., H. Ataki, M. Takebe, K. Ebihara. Soybean meal fermented by *Aspergillus awamori* increases the cytochrome P-450 content of the liver microsomes of mice. *J. Agric. Food Chem.* 48:1367–1372, 2000.

87. Ohlsson, A.B., T. Berglund, P. Komlos, J. Rydstrom. Plant defense metabolism is increased by the free radical-generating compound AAPH. *Free Radic. Biol. Med.* 19(3):319–327, 1995.

88. Cantos, E., J.C. Espin, F.A. Tomas-Barberan. Effect of wounding on phenolic enzymes in six minimally processed lettuce cultivars upon storage. *J. Agric. Food Chem.* 49:322–330, 2001.

89. McCue, P., Z. Zheng, J.L. Pinkham, K. Shetty. A model for enhanced pea seedling vigor following low pH and salicylic acid treatments. *Process. Biochem.* 35:603–613, 2000.

90. Andarwulan, N., K. Shetty. Improvement of pea (*Pisum sativum*) seed vigor by fish protein hydrolysates in combination with acetyl salicylic acid. *Process. Biochem.* 35:159–165, 1999.

91. Zheng, Z., K. Shetty. Enhancement of pea (*Pisum sativum*) seedling vigor and associated phenolic content by extracts of apple pomace fermented with *Trichoderma* spp. *Process. Biochem.* 36:79–84, 2000.

92. Duval, B., K. Shetty. Stimulation of phenolics and antioxidant activity in pea (*Pisum sativum*) elicited by genetically transformed anise root extract. *J. Food Biochem.* 25:361–377, 2001.

93. McCue, P., K. Shetty. Clonal herbal extracts as elicitors of phenolic synthesis in dark-germinated mungbean for improving nutritional value with implications for food safety. *J. Food Biochem.* 26:209–232, 2002.

94. Randhir, R., K. Shetty. Light-mediated fava bean (*Vicia faba*) response to phytochemical and protein elicitors and consequences on nutraceutical enhancement and seed vigor. *Process. Biochem.* 38:945–952, 2003.

95. Randhir, R., Y.T. Lin, K. Shetty. Stimulation of phenolics, antioxidant and antimicrobial activity in dark-germinated mungbean sprouts in response to peptide and phytochemical elicitors. *Process. Biochem.* 39:637–646, 2004.

96. Randhir, R., Y.T. Lin, K. Shetty. Phenolics, antioxidant and antimicrobial activity in dark-germinated fenugreek sprouts in response to peptide and phytochemical elicitors. *Asia Pac. J. Clin. Nutr.* 2003.

97. McCue, P., K. Shetty. A biochemical analysis of mungbean (*Vigna radiata*) response to microbial polysaccharides and potential phenolic-enhancing effects for nutraceutical applications. *Food Biotechnol.* 16:57–79, 2002.

98. Shetty, P., M.T. Atallah, K. Shetty. Effects of UV treatment on the proline-linked pentose-phosphate pathway for phenolics and L-DOPA synthesis in dark-germinated *Vicia faba*. *Process. Biochem.* 37:1285–1295, 2002.

99. Strycharz, S., K. Shetty. Effect of *Agrobacterium rhizogenes* on phenolic content of *Mentha pulegium* elite clonal line for phytoremediation applications. *Process. Biochem.* 38:287–293, 2002.

100. McCue, P., K. Shetty. A hypothetical model for action of soybean isoflavonoids against cancer involving a shift to proline-linked energy metabolism through activation of the pentose-phosphate pathway. *Food Biotechnol.* 18:19–37, 2004.

101. Benzie, I.F. Evolution of antioxidant defense mechanisms. *Eur. J. Nutr.* 39:53–61, 2000.

102. Shi, D.Y., Y.R. Deng, S.L. Liu, Y.D. Zhang, L. Wei. Redox stress regulates cell proliferation and apoptosis of human hepatoma through Akt protein phosphorylation. *FEBS Lett.* 542:60–64, 2003.

103. Shetty, K., P. McCue. Phenolic antioxidant biosynthesis in plants for functional food application: integration of systems biology and biotechnological approaches. *Food Biotechnol.* 17(2):67–97, 2004.

104. Sang, S., K. Lapsley, W.S. Jeong, P.A. Lachance, C.T. Ho, R.T. Rosen. Antioxidative phenolic compounds isolated from almond skins (*Prunus amygdalus* Batsch). *J. Agric. Food Chem.* 50:2459–2463, 2002.

105. Middleton, E., C. Kandaswami, T.C. Theoharides. The effects of plant flavanoids on mammalian cells: implications for inflammation, heart dizease, and cancer. *Pharmacol. Rev.* 52:673–751, 2000.

106. Squier, T.C. Oxidative stress and protein aggregation during biological aging. *Exp. Gerontol.* 36:1539–1550, 2001.

107. Spitz, D.R., J.E. Sim, L.A. Ridnour, S.S. Galoforo, Y.J. Lee. Glucose deprivation-induced oxidative stress in human tumor cells. *Ann. NY Acad. Sci.* 899:349–362, 2000.

108. Simonnet, H., N. Alazard, K. Pfeiffer, C. Gallou, C. Beroud, J. Demont, R. Bouvier, H. Schagger, C. Godinot. Low mitochondrial respiratory chain content correlates with tumor aggressiveness in renal cell carcinoma. *Carcinogenesis* 23(5):759–768, 2002.

109. Savagner, F., B. Franc, S. Guyetant, P. Rodien, P. Reynier, Y. Malthiery. Defective mitochondrial ATP synthesis in oxyphilic thyroid tumors. *J. Clin. Endocrinol. Metab.* 86:4920–4925, 2001.

110. Dey, R., C.T. Moraes. Lack of oxidative phosphorylation and low mitochondrial membrane potential decrease susceptibility to apoptosis and do not modulate the protective effect of Bcl-x_L in osteosarcoma cells. *J. Biol. Chem.* 275(10):7087–7094, 2000.

111. Dominguez, J.E., J.F. Graham, C.J. Cummins, D.J. Loreck, J. Galarraga, J. Van der Feen, R. De La Paz, B.H. Smith. Enzymes of glucose metabolism in cultured human gliomas: neoplasia is accompanied by altered hexokinase, phosphofructokinase, and glucose-6-phosphate dehydrogenase levels. *Metab. Brain Dis.* 2(1):17–30, 1987.

112. Chesney, J., R. Mitchell, F. Benigni, M. Bacher, L. Spiegel, Y. Al-Abed, J.H. Han, C. Metz, R. Bucala. An inducible gene product for 6-phosphofructo-2-kinase with an AU-rich instability element: role in tumor cell glycolysis and the Warburg effect. *Proc. Natl. Acad. Sci. USA* 96(6):3047–3052, 1999.

113. Hasegawa, Y., T. Takano, A. Miyauchi, F. Matsuzuka, H. Yoshida, K. Kuma, N. Amino. Decreased expression of catalase mRNA in thyroid anaplastic carcinoma. *Jpn. J. Clin. Oncol.* 33(1):6–9, 2003.

114. Sander, C.S., F. Hamm, P. Elsner, J.J. Thiele. Oxidative stress in malignant melanoma and non-melanoma skin cancer. *Br. J. Dermatol.* 148:913–922, 2003.

115. Phang, J.M., S.J. Downing, G.C. Yeh. Linkage of the HMP pathway to ATP generation by the proline cycle. *Biochem. Biophys. Res. Commun.* 93:462–470, 1980.

116. Hagedorn, C.H., J.M. Phang. Transfer of reducing equivalents into mitochondria by the interconversions of proline and Δ^1-pyrroline-5-carboxylate. *Arch. Biochem. Biophys.* 225:95–101, 1983.

117. Phang, J.M. The regulatory functions of proline and pyrroline-5-carboxylic acid. *Curr. Top. Cell. Regul.* 25:91–132, 1985.

118. Nissler, K., H. Petermann, I. Wenz, D. Brox. Fructose 2,6-bisphosphate metabolism in Ehrlich ascites tumour cells. *Cancer Res. Clin. Oncol.* 121:739–45, 1995.

119. Le Goffe, C., G. Vallette, L. Charrier, T. Candelon, C. Bou-Hanna, J.F. Bouhours, C.L. Laboisse. Metabolic control of resistance of human epithelial cells to H_2O_2 and NO stresses. *Biochem. J.* 364(2):349–359, 2002.

120. Spolarics, Z. Endotoxemia, pentose cycle, and the oxidant/antioxidant balance in the hepatic sinusoid. *J. Leukoc. Biol.* 63:534–541, 1998.

121. Tuttle, S., T. Stamato, M.L. Perez, J. Biaglow. Glucose-6-phosphate dehydrogenase and the oxidative pentose phosphate cycle protect cells against apoptosis induced by low doses of ionizing radiation. *Radiat. Res.* 153:781–787, 2000.

122. Neumann, C.A., D.S. Krause, C.V. Carman, S. Das, D.P. Dubey, J.L. Abraham, R.T. Bronson, Y. Fujiwara, S.H. Orkin, R.A. Van Etten. Essential role for the peroxiredoxin Prdx1 in erythrocyte antioxidant defense and tumor suppression. *Nature* 424:561–565, 2003.

123. Friedmann, Y., I. Vlodavsky, H. Aingorn, A. Aviv, T. Peretz, I. Pecker, O. Pappo. Expression of heparanase in normal, dysplastic, and neoplastic human colonic mucosa and stroma. Evidence for its role in colonic tumorigenesis. *Am. J. Pathol.* 157:1167–1175, 2000.

124. Wilson, L.C., S.J. Baek, A. Call, T.E. Eling. Nonsteroidal anti-inflammatory drug-activated gene (NAG-1) is induced by genistein through the expression of p53 in colorectal cancer cells. *Int. J. Cancer* 105:747–753, 2003.

125. Seo, Y.R., M.R. Kelley, M.L. Smith. Selenomethionine regulation of p53 by a ref1-dependent redox mechanism. *Proc. Natl. Acad. Sci. USA* 99:14548–14553, 2002.

126. Brash, D.E., P.A. Havre. New careers for antioxidants. *Proc. Natl. Acad. Sci. USA* 99:13969–13971, 2002.

127. Donald, S.P., X.Y. Sun, C.A. Hu, J. Yu, J.M. Mei, D. Valle, J.M. Phang. Proline oxidase, encoded by p53-induced gene-6, catalyzes the generation of proline-dependent reactive oxygen species. *Cancer Res.* 61(5):1810–1815, 2001.

128. Maxwell, S.A., G.E. Davis. Differential gene expression in p53-mediated apoptosis-resistant vs. apoptosis-sensitive tumor cell lines. *Proc. Natl. Acad. Sci. USA* 97:13009–13014, 2000.

129. Maxwell, S.A., A. Rivera. Proline oxidase induces apoptosis in tumor cells, and its expression is frequently absent or reduced in renal carcinomas. *J. Biol. Chem.* 278:9784–9789, 2003.

130. Sun, Y. Free radicals, antioxidant enzymes, and carcinogenesis. *Free Radic. Biol. Med.* 8:583–599, 1990.
131. Cobb, C.S., D.S. Levi, K. Aldape, M.A. Israel. Manganese superoxide dismutase expression in human central nervous system tumors. *Cancer Res.* 56:3192-3195, 1996.
132. Iqbal, M., S.D. Sharma, Y. Okazaki, M. Fujisawa, S. Okada. Dietary supplementation of curcumin enhances antioxidant and phase II metabolizing enzymes in ddY male mice: possible role in protection against chemical carcinogenesis and toxicity. *Pharmacol. Toxicol.* 92:33–38, 2003.
133. Zheng, Q.S., Zhang, Y.T., Zheng, R.L. Ascorbic acid induces redifferentiation and growth inhibition in human hepatoma cells by increasing endogenous hydrogen peroxide. *Pharmazie.* 57 (11):753–757, 2002.
134. Fahey, J.W., K.K. Stephenson. Pinostrobin from honey and Thai ginger (*Boesenbergia pandurata*): a potent flavanoid inducer of mammalian phase 2 chemoprotective and antioxidant enzymes. *J. Agric. Food. Chem.* 50:7472–7476, 2002.
135. Suzuki, K., H. Koike, H. Matsui, Y. Ono, M. Hasumi, H. Nakazato, H. Okugi, Y. Sekine, K. Oki, K. Ito, T. Yamamoto, Y. Fukabori, K. Kurokawa, K. Yamanaka. Genistein, a soy isoflavone, induces glutathione peroxidase in the human prostate cancer cell lines LNCaP and PC-3. *Int. J. Cancer* 99(6):846–852, 2002.
136. Chang, W.H., J.J. Liu, C.H. Chen, T.S. Huang, F.J. Lu. Growth inhibition and induction of apoptosis in MCF-7 breast cancer cells by fermented soy milk. *Nutr. Cancer* 43(2):214–226, 2002.
137. Hewitt, A.L., K.W. Singletary. Soy extract inhibits mammary adenocarcinoma growth in a syngeneic mouse model. *Cancer Lett.* 192(2):133–143, 2003.
138. Loreck, D.J., J. Galarraga, J. Van der Feen, J.M. Phang, B.H. Smith, C.J. Cummins. Regulation of the pentose phosphate pathway in human astrocytes and gliomas. *Metab. Brain Dis.* 2:31–46, 1987.
139. Javelaud, D., F. Besancon. NF-κB activation results in rapid inactivation of JNK in TNF α-treated Ewing sarcoma cells: a mechanism for the anti-apoptotic effect of NF-κB. *Oncogene* 20:4365–4372, 2001.
140. Stadtman, E.R., B.S. Berlett. Reactive oxygen-mediated protein oxidation in aging and dizease. *Drug Metab. Rev.* 30:225–243, 1998.
141. Andarwulan, N., K. Shetty. Stimulation of novel phenolic metabolite, epoxy-Psuedoisoeugenol-(2-Methylbutyrate) [EPB], in transformed anize (*Pimpinella anisum* L.) root cultures by fish protein hydrolysates. *Food Biotechnol.* 14:1–20, 2000.
142. Shetty, K., M.L. Wahlqvist. A model for the role of proline-linked pentose-phosphate pathway in phenolic phytochemical biosynthesis and mechanism of action for human health and environmental applications. *Asia Pac. J. Clin. Nutr.* 13(1):1–24, 2004.

7

Functional Phytochemicals from Cranberries: Their Mechanism of Action and Strategies to Improve Functionality

Dhiraj A. Vattem and Kalidas Shetty

CONTENTS

7.1 Introduction: Phenolic Phytochemicals...152
7.2 Chemical Nature and Biosynthesis of Phenolic Phytochemicals 152
    7.2.1 Phenolic Phytochemicals from Berries 153
        7.2.1.1 Cranberry 156
        7.2.1.2 Cranberry Pomace...156
    7.2.2 Oxidative Stress Mediated Pathogenesis 157
        7.2.2.1 Antioxidant Defense Systems..159
        7.2.2.2 Phenolic Phytochemicals in Nutritional Management of
                     Antioxidant Defense 161
    7.2.3 Biological Function of Phenolic Phytochemicals 162
        7.2.3.1 Biological Functionality of Cranberry Phenolics 163
        7.2.3.2 Ellagic Acid...165
        7.2.3.3 Alternate Model for the Function of Ellagic Acid and
                     Related Phenolic Phytochemicals from Cranberries 166
        7.2.3.4 A Model for Antimicrobial Activity of Ellagic Acid and
                     Related Phenolics Against Bacterial Pathogens 169
7.3 Innovative Strategies to Enrich Fruits with Phenolic Antioxidants......................171
7.4 Solid-State Bioprocessing 172
7.5 Cranberry Synergies with Functional Phytochemicals and Other Fruit Extracts.....173
7.6 Conclusions 174
References..175

7.1   INTRODUCTION: PHENOLIC PHYTOCHEMICALS

Phenolic compounds or phenolic phytochemicals are secondary metabolites of plant origin which constitute one of the most abundant groups of natural compounds and form an important component of both human and animal diets (1,2,3). These phenolic metabolites function to protect the plant against biological and environmental stresses and are therefore synthesized in response to pathogenic attack, such as fungal or bacterial infection, or high energy radiation exposure, such as prolonged UV exposure (4,5). Because of their important biological functions, phenolic phytochemicals are ubiquitous in plants and therefore find their place in almost all food groups. Common fruits such as apples, cranberries, grapes, raspberries, and strawberries, and fruit beverages like red wine and apple and orange juices, are rich sources of phenolic phytochemicals. In addition to fruits, vegetables such as cabbage and onion, and food grains such as sorghum, millet, barley, peas, and other legumes (6) are also described as important sources of phenolic phytochemicals. Varied biological functions of phenolic phytochemicals in plants have led to the evolution of diverse types of phenolic compounds. Depending on the evolutionary pressures experienced, plants have evolved a constituent profile of inducible phenolic phytochemicals, usually characteristic of a particular species of fruit or vegetable (6,7). For example, the most abundant phenolic compound in fruits and their products are flavonols. Cereals and legumes are rich in flavonoids, phenolic acids, and tannins (6,7). The major phenolic phytochemicals in wine include phenolic acids, anthocyanins, tannins, and other flavonoids (6,7).

7.2   CHEMICAL NATURE AND BIOSYNTHESIS OF PHENOLIC PHYTOCHEMICALS

The International Union for pure and applied chemists (IUPAC) defines phenol as hydroxybenzene. The term phenolic compounds refers to a relatively wide range of chemical compounds that contain at least one aromatic ring and usually one or more hydroxyl substituents (8). There are numerous different types of these phenolic phytochemicals, classified according to their ring structure and the number of carbon atoms substituting the ring and linking them together (Table 7.1).

Metabolic processing of phenolic phytochemicals in plants for their final biological function has led to chemical variations in basic phenolic structure. Differences in substituent groups and linkages have resulted in a wide variety of chemical structures having distinct properties. More than 8000 different phenolic structures, categorized into 10 classes, have been identified. They vary structurally from simple molecules (e.g., phenolic acids with a single ring structure), to biphenyls and flavonoids having 2–3 phenolic rings (9,10) (Figure 7.1). Another abundant group of phenolic phytochemicals in fruits and vegetables, often referred to as polyphenols, contain 12–16 phenolic groups and approximately 5–7 aromatic rings per 1000 relative molecular mass (Figure 7.1). These polyphenols are classified as condensed proanthocyanidins, tannins which include galloyl and hexahydroxy-diphenoyl (or ellagoyl) esters and their derivatives, or phlorotannins (9,10) (Figure 7.1).

All the phenolic phytochemicals are derived from a common biosynthetic pathway, incorporating precursors from the shikimate or the acetate–malonate pathways or both (1,11) (Figure 7.2). Simple phenolic acids such as cinnamic acid and its derivatives are widespread in fruits and vegetables. They are derived primarily from the shikimate pathway via phenylalanine or tyrosine (1). Other phenolics of biological importance that are formed in this manner include coumaric acid and caffeic acid. Chemical modification of side chains by processes such as oxidation produce another group of compounds, such as protocatechuic

Table 7.1

The major classes of phenolic compounds in plants

Number of Carbon Atoms	Basic Skeleton	Class
6	C_6	Simple phenols
		Benzoquinones
7	C_6-C_1	Phenolic acids
8	C_6-C_2	Acetophenones
		Tyrosine derivatives
		Phenylacetic acids
9	C_6-C_3	Hydroxycinnamic acids
		Phenylpropenes
		Coumarins
		Isocoumarins
		Chromones
10	C_6-C_4	Naphthoquinones
13	C_6-C_1-C_6	Xanthones
14	C_6-C_2-C_6	Stilbenes
		Anthraquinones
15	C_6-C_3-C_6	Flavonoids
		Isoflavonoids
18	$(C_6$-$C_3)_2$	Lignans
		Neolignans
30	$(C_6$-C_3-$C_6)_2$	Biflavonoids
N	$(C_6$-$C_3)_n$	Lignins
	$(C_6)_n$	Catechol melanins
	$(C_6$-C_3-$C_6)_n$	Flavolans (Condensed Tannins)

acid and its positional isomer gentisic acid, which are referred to as benzoic acid derivatives (Figure 7.2). To increase the solubility and target the phenolics to specific parts of the plant, and to prevent their enzymatic and chemical degradation, phenolic phytochemicals are often esterified with sugars and other chemical components such as quinic acid through the hydroxyl groups of the phenolic ring. The esters are sugars and phenolic phytochemicals are called glycosides. A majority of glycosides contain glucose, but the glycosides of phenolics with galactose, sucrose, and rhamnose have also been described (11).

Flavonoids and biphenyls are another important class of phenolic phytochemicals that are especially rich in fruits and legumes. Flavonoids like quercetin constitute the most abundant group of phenolic phytochemicals. Their common structure consists of two aromatic rings linked through three carbons that usually form an oxygenated heterocycle; this type of chemical arrangement is often described as diphenylpropanes (C_6-C_3-C_6) (2,11). Structural variations within the rings resulting in an alteration in the extent of hydroxylation, methylation, isoprenylation, dimerization, and glycosylation (producing O- or C-glycosides) subdivide the flavonoids into several families: flavonols, flavones, isoflavones, antocyanidins and others (Figure 7.2).

7.2.1   Phenolic Phytochemicals from Berries

Phenolic phytochemicals are ubiquitous in nature and are especially rich in fruits (12,13). Fruits such as berries (14,15) are known to be good sources of phenolic antioxidants having protective health benefits. Fruit beverages such as wines have been shown to

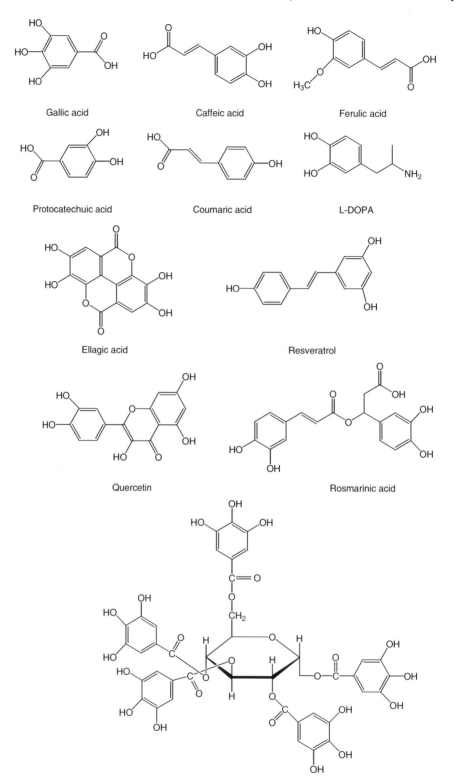

Figure 7.1 Common simple phenol, biphenyls, flavonoids, and tannins in plants.

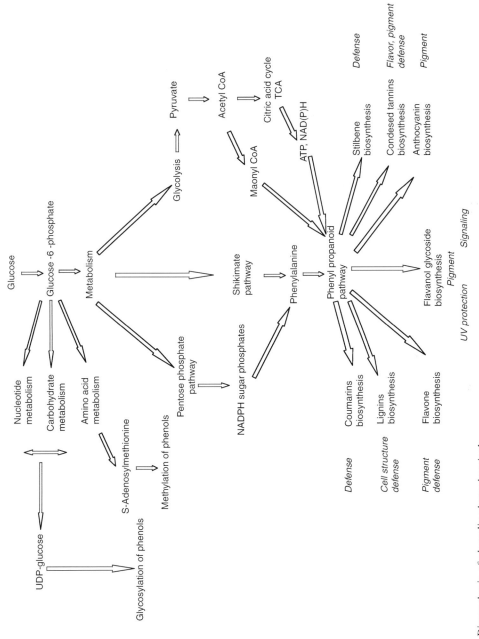

Figure 7.2 Biosynthesis of phenolic phytochemicals.

contain a variety of polyphenolic compounds, the most abundant being flavonoids such as anthocyanins (16,17,18).

Berries, like many other fruits, are rich in phenolic compounds, including diphenyls, flavonoids and phenolic acids, which exhibit a wide range of biological effects, including antioxidant (19,20,21) and anticarcinogenic properties (22). A high free radical scavenging activity of berry extracts toward chemically generated active oxygen species has been described in several studies (19–21). Increased fruit consumption in daily diets has shown to significantly reduce the incidence and mortality rates of cancer, cardiovascular disorders, and other degenerative diseases caused by oxidative stress (22–25). Epidemiological evidence suggests that high consumption of flavonoids, which are an important component of berries, may provide protection against coronary heart disease, cardiac stroke (26–28), lung cancer (22,28,29), and stomach cancer (30).

Berries are typically rich in flavonoids and diphenyls such as ellagic acid (Figure 7.1), and represent a large group of secondary plant metabolites (31). The diversity and complexity of the flavonoids found in berries depends on at least two factors: different variety of aglycones and the high number of glycosides, sometimes in acylated form; and condensation into complex molecules. Recent research has determined the antioxidant activity of different phenolic compounds (32–34) and attempted to define the structural characteristics which contribute to their activity (33,35). Phenolic acids present in berries are hydroxylated derivatives of benzoic acid and cinnamic acid (36). The other simple phenolics in berries include caffeic, chlorogenic, ferulic, sinapic, and *p*-coumaric, acids (37).

7.2.1.1   Cranberry

The cranberry *(Vaccinium macrocarpon),* also known as American cranberry, belongs to the family Ericaceae, which also includes blueberry (*V. angustifolium*) and bilberry (*V. myrtillus*). Even though cranberries have been historically associated with positive health benefits, scientific investigation into the positive health benefits of the cranberry has received little attention (38,39). Many studies have demonstrated high phenolic content in fruits such as grapes, apples, oranges, prunes, and berries (31). The American cranberry is a prominent agricultural food crop produced in Massachusetts, Wisconsin, Michigan, Canada, New Jersey, Oregon, and Washington. The crop size is approximately 500 million pounds annually and is processed into three basic categories: fresh (5%); sauce products, concentrate, and various applications (35%); and juice drinks (60%) (40).

7.2.1.2   Cranberry Pomace

Cranberry pomace is the byproduct of the cranberry juice processing industry with limited applications. Once the juice is extracted from the fruit, the remaining product is called cranberry pomace. Pomace is mainly composed of the skin, flesh, and seed of the fruit. It is rich in fiber and has relatively small amounts of protein and carbohydrates (41). Traditionally it has been used as an ingredient in animal feed, however, due to its low protein and carbohydrate content, it has little nutritive value. Its disposal into the soil or in a landfill poses considerable economic loss, and causes potential environmental problems due to its low pH (41).

Agricultural and industrial residues are attractive sources of natural antioxidants. Potato peel waste (42,43), olive peel (44), grape seeds, and grape pomace peels (45,46) have been studied for their use as cheap sources of antioxidants. Increased antioxidant activity in rat plasma after oral administration of grape seed extracts was reported recently (47). Identification of polyphenolic compounds from apple pomace (48) has also been reported. Phenolics are ubiquitous in plants, but seeds and skins are especially rich sources of phenolics (42,47,49–51), probably because of the role they play in protecting the fruit and the seed to ensure healthy propagation of the species. Pomace, which mainly consists of fruit skins and seeds, is also a rich source of phenolic compounds (47,48). However, several phenolics that are found in pomace and other plant products exist in conjugated forms either with sugars (primarily glucose), as glycosides, or as other moieties. This conjugation occurs via the hydroxyl groups of the phenolics, which reduces their ability to function as good antioxidants, because availability of free hydroxyl groups on the phenolic rings is important for resonance stabilization of free radicals. Lowered antioxidant capacity has direct implications on decreasing health functionality when these phenolics are ingested via food or as nutraceuticals. Therefore, if free phenolics are released from their glycosides or other conjugates, then the antioxidant, and thus the health functionality of these phytochemicals could be improved. Enzymatic hydrolysis of these phenolic glycosides appears to be an attractive means of increasing the concentration of free phenolic acids in fruit juice and wines to enrich taste, flavor, and aroma, also potentially increasing nutraceutical value (52–54).

7.2.2 Oxidative Stress Mediated Pathogenesis

Emerging evidence suggests that free radical mediated oxidative stress is responsible for the induction of pathogenicity in biological systems. The primary manner in which oxidative stress is mediated in cellular systems is by the generation of reactive oxygen species (ROS). Oxygen in its ground state is often described as a biradical, as it contains two electrons in its outer shell having the same spin. When one of the electron changes its spin, the oxygen is transformed into a singlet, and becomes a powerful oxidant and undergoes reduction. However, in biological systems, incomplete reduction of molecular oxygen occurs due to errors in metabolism, or due to controlled reduction of oxygen, resulting in the formation of what are now known as ROS. These ROS are highly reactive molecules, and include the hydroxyl radicals (HO$^-$), superoxide anions (O$_2^-$) and hydrogen peroxide (H$_2$O$_2$) (Figure 7.3). The superoxide anion is the precursor of most ROS in biological systems, and is formed by the one electron reduction of molecular oxygen. Spontaneous or enzymatic dismutation of superoxide anion produces the peroxide radical, which can undergo a complete reduction to form water or an incomplete reduction to form the highly reactive hydroxyl radical (55). Singlet oxygen can also react with nitric oxide (NO) in a diffusion limited reaction to form peroxynitrite which is a powerful

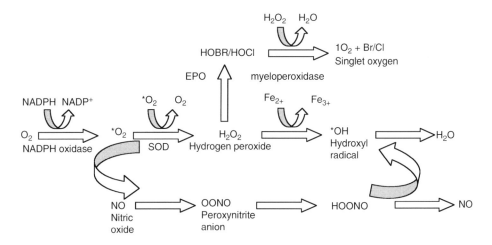

Figure 7.3 Biological mechanisms for ROS formation.

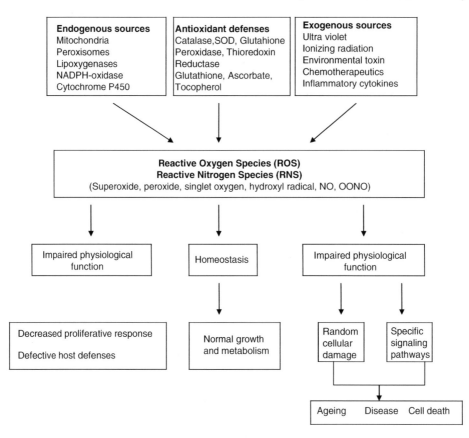

Figure 7.4   Reactive oxygen species mediated pathogenecity.

oxidant (56,57). The oxidants derived from NO are often referred to as reactive nitrogen species (RNS) (Figure 7.3).

Reactive oxygen species and reactive nitrogenspecies (RNS) are constantly produced in aerobic organisms both enzymatically and non-enzymatically. Many sources for the formation of ROS have been identified in cellular systems of living organisms (58,59). Superoxide is formed upon the one electron reduction of oxygen mediated by enzymes such as NADPH oxidase located on the cell membrane of polymorphonuclear cells, macrophages, and endothelial cells (60–62); from xanthine oxidase or from the respiratory chain (63). Superoxide radicals can also be formed from the electron cycling carried out by the cytochrome P450 dependent enzymes (64,65). Mitochondrial leakage of electrons or direct transfer of electron to oxygen via coenzymes or prosthetic groups such as flavins and Fe-S centers constitute another important source of ROS in cellular systems (58,59). The homolytic cleavage of water, or the breakdown of hydrogen peroxide in a high energy radiation (e.g., x-rays, UV) or metal catalyzed process, forms the hydroxyl radical which is the most reactive oxygen species (Figure 7.3, Figure 7.4). Another ROS, the hypochlorite ion, is formed by the macrophage myeloperoxidase, or related peroxidase activities, which catalyze the halide driven reduction of hydrogen peroxide to form the oxidant hypochlorous acid (HOCl) to kill invading microorganisms (63). Biological conversion of L-arginine to L-citrulline by nitric oxide synthase forms a cytotoxic oxygen species, nitric oxide (NO), which is involved in the cellular defense against malignant cells, invading fungi, and protozoa (56,57). Nitric oxide is also involved critically in signaling mechanisms in the vasculature controlling the vascular force in the blood vessels by vasodilation and inflammation (66).

Reactive oxygen species are ideally suited to be signaling molecules, as they are small, can diffuse small distances, and have several mechanisms of production that can be controlled and regulated. Low levels of ROS therefore have been implicated in many cellular processes, including intracellular signaling responsible for proliferation or apoptosis (67), for modulation of immune response (68), and for mounting a defense response against pathogens (69) (Figure 7.4). Even though the exact mechanism of action of ROS in effecting these physiological processes is not very well understood, there is growing evidence that ROS at some level are capable of activating or repressing many biological effector molecules (70). It has been shown to activate or repress transcription factors by directly activating them by oxidizing the sulphydryl groups present, or by modulating a complex array of kinases and phosphatases, which are important in signal transduction. Many ions that maintain the electrostatic balance of the cell required for many physiological processes such as cell growth and cell death are also now shown to be regulated by ROS. Another important mechanism by which ROS are now shown to be regulating the cell function is by altering the redox status of the cell, often by regulating the levels of oxidized and reduced glutathione (GSH/GSSG) (71).

7.2.2.1 Antioxidant Defense Systems

Redox regulated physiological processes are inevitably sensitive against excessive ROS production by any source. Such excessive levels of ROS may be generated either by overstimulation of the otherwise tightly regulated NAD(P)H oxidases or by other mechanisms that produce ROS in a nonregulated fashion, including the production of ROS by the mitochondrial electronic transport chain (ETC) or by xanthine oxidase. Diets rich in saturated fatty acids, and carbohydrates; and environmental factors such as exposure to high energy radiation, and ingestion of or exposure to toxins and carcinogens can also overstimulate metabolic systems resulting in the formation of ROS, which are beyond the cellular need for their regular functions (63–65) (Figure 7.4).

Several antioxidant systems are in place in the cell that can quickly remove the ROS from cellular systems. Biological defenses against ROS comprise a complex array of endogenous antioxidant enzymes, endogenous antioxidant factors including glutathione (GSH) and other tissue thiols, heme proteins, coenzyme Q, bilirubin and urates, and a variety of nutritional factors, primarily the antioxidant vitamins and phenolic phytochemicals from diet (72–74) (Figure 7.4, Figure 7.5).

Glutathione is the most abundant intracellular reductant in all aerobic cells. Tissue GSH and other tissue thiols exist at millimolar concentrations and are important systems to protect the cell against oxidative stress and tissue injury (71). Glutathione acts as a redox sensor and as a sulfhydryl buffer in cellular systems. Upon generation of ROS, GSH preferentially reacts with ROS in an almost sacrificial manner to form oxidized glutathione (GSSH) to prevent the oxidation of other biological molecules. Enzymes such as glutathione peroxidases (GPX) reduce soluble peroxides and membrane bound peroxides to the corresponding alcohols, at the expense of GSH, which is oxidized to GSSG and ascorbic acid, which is then oxidized to dehydroascorbic acid (71–73). Another class of proteins, called glutaredoxins (GRx), are reduced by GSH, and are capable of reducing protein disulfides from oxidative stress (Figure 7.4, Figure 7.5). However, for this antioxidant response to continue, cellular systems have to regenerate GSH constantly (74). Tissue GSH/GSSH ratios are maintained in the reduced state by the concerted action of antioxidant vitamins such as ascorbate and tocopherols using many antioxidant enzyme systems

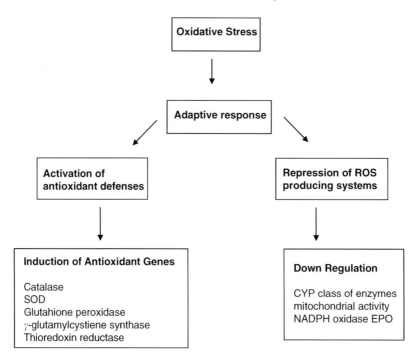

Figure 7.5 Biological adaptive responses to manage oxidative stress. (Morel, Y., R. Barouki, *Biochem. J.* 342(3): 481–496, 1999.)

involving several oxidoreductases (72–74). The cascade of endogenous antioxidant enzymes requires energy to maintain the living system in the reduced state. Glutathione reductase and ascorbate peroxidase maintain the tissue GSH in the reduced state at the expense of NADPH and $FADH_2$ (72–74) (Figure 7.6). Enzymatic reactions of the thioredoxin system contain a set of oxidoreductases with wide range substrate specificity that reduce the active site disulfide in thioredoxin (Trx) and several other substrates, directly under the consumption of NADPH. Reduced Trx is highly efficient in reducing disulfides in proteins and peptides including glutathione disulfide (GSSG).

Another biological defense against superoxide free radicals is the superoxide dismutase (SOD) enzyme, which is often considered the most effective antioxidant system.

Superoxide dismutase is an enzyme that catalyzes dismutation of two superoxide anions into hydrogen peroxide and molecular oxygen

$$2O_2^{\cdot-} + 2H^+ \rightarrow H_2O_2 + O_2 \qquad (7.1)$$

The importance of SOD is that it is a very efficient enzyme system in removing superoxide radicals from biological systems. The function of SOD is so imperative for the protection of cells that it represents a substantial proportion of the proteins manufactured by the body (72,73). One of the contributing factors for the high efficiency of SOD is that its dismutation is always coupled to another enzyme system called catalase (CAT). Catalase is involved in removing hydrogen peroxide molecules, which are byproducts of the reactions created by SOD (72,73) (Figure 7.6). Catalase is also abundantly present in the body and is integrated into all the cellular systems that operate in an oxidant environment. Red blood cells, which transport oxygen to different cellular systems, have high levels of CAT and help to remove hydrogen peroxide from tissues to prevent both cellular damage and propagation

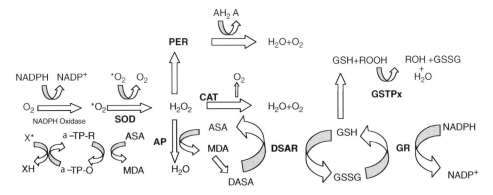

Figure 7.6 The antioxidant defense response of the cell carried out by enzymatic as well as the nonenzymatic antioxidants.

of free radicals. This natural interaction synergy between these two antioxidant enzymes constitutes the most effective system of free radical control (72,73).

Imbalance in the generation of ROS can occur, and deficiency of biological systems to control such an imbalance can result in conditions of oxidative stress, which is now believed to be a primary cause for the induction of many chronic diseases. Small fluctuations in the concentration of ROS may actually play a role in intracellular signaling (67). Uncontrolled increases in the concentrations of these oxidants lead to free radical mediated chain reactions that indiscriminately target proteins (75), lipids (76), and DNA (77,78), which can impair their normal biological functions (Figure 7.4).

Excessive generation of ROS can result in activation or repression of several key signaling pathways and transcription factors, which can impair normal physiological responses, resulting in impaired host defenses. These can lead to impaired cell growth and cell death, which have been shown to develop into cancer and diseases related to aging (67,70,79).

Pathogenesis mediated by virus and *Helicobacter pylori* are also now linked to oxidative stress and are associated with lower responses of CAT and SOD (80,81). Recent experimental findings suggest that overproduction of ROS and RNS lowered antioxidant defense; and alterations of enzymatic pathways in humans with poorly controlled diabetes mellitus can contribute to endothelial, vascular, and neurovascular dysfunction (82). Neurodegenerative disorders such as Alzheimer's disease, Parkinson's disease, and dementia have also been associated with the oxidation of carbohydrates and proteins by ROS, and reduced activities of the enzymes involved in the antioxidant defense response: GPX, SOD and CAT (83).

7.2.2.2 Phenolic Phytochemicals in Nutritional Management of Antioxidant Defense

Association of ROS with manifestation of several diseases has led to an increase in the management of oxidation linked diseases such as cancer, cerebrovascular disease, and diabetes by supporting the body's antioxidant defense system through consuming dietary antioxidants such as ascorbic acid, tocopherols, and carotenoids (82,84,85). Recent epidemiological studies have shown that diets rich in fruits and vegetables are associated with lower incidences of oxidation linked diseases such as cancer, cerebrovascular disease, and diabetes. These protective effects of fruits and vegetables are now linked to the presence of antioxidant vitamins and phenolic phytochemicals having antioxidant activity.

The ability of dietary antioxidants in managing diseases manifested by oxidative stress is not clearly understood. Most phenolic phytochemicals that have positive effects on health are believed to function by countering the effects of ROS species generated during cellular metabolism. Consumption of natural dietary antioxidants from fruits and vegetables has been shown to enhance the function of the antioxidant defense response mediated by GSH, ascorbate, SOD, CAT and GST interface (86–89). This was observed in systemic sclerosis and rheumatoid arthritis patients (90). It was observed that consumption of fruits and vegetables led to a direct increase in scavenging ROS, and prevented polymorphonuclear leukocytes from generating hydroxyl radicals (91).

7.2.3   Biological Function of Phenolic Phytochemicals

Chronic and infectious diseases have become the primary cause of mortality and are expected to become a major public health challenge. These chronic diseases such as cardiovascular disease, hypertension, diabetes mellitus, and some forms of cancer, have now been associated with changes in diet and lifestyle associated with calorie sufficiency. These include excessive dietary carbohydrate and fat intake, low intake of fruits and vegetables, smoking, lack of physical activity, and exposure to environmental toxins (92).

Compelling epidemiological and scientific evidence has led to an understanding that oxidative stress, as a consequence of an imbalance of prooxidants and antioxidants, is a key phenomenon in the manifestation of chronic diseases (90). Powerful strategies to control oxidative stress related pathogenicities are gaining prominence. Epidemiological evidence showing that populations consuming diets rich in fruits and vegetables have lower incidences of many chronic diseases such as cancer, cardiovascular diseases and diabetes has led to an interest in the use of diet as a potential tool for the control of these oxidative diseases (86–89). Recent *in vitro* and clinical studies have shown that diets rich in carbohydrates and fats induced oxidative stress, which was decreased by consuming fruits, vegetables, and their products (93). Among all the dietary components, fruits and vegetables have especially been shown to exert a protective effect (23–26). Phenolic phytochemicals with antioxidant properties are now widely thought to be the principle components in fruits and vegetables which have these beneficial effects. Phenolic phytochemicals exhibit a wide range of biological effects and can broadly be divided into two categories. The first and the most well described mode of action of these phenolic phytochemicals in managing oxidation stress related diseases is due to the direct involvement of the phenolic phytochemicals in quenching the free radicals from biological systems. It is well known that free radicals cause oxidative damage to nucleic acids, proteins, and lipids. Oxidation of biological macromolecules as a result of free radical damage has now been strongly associated with development of many physiological conditions which can develop into disease (67,70,79–83). Phenolic phytochemicals, due to their phenolic ring and hydroxyl substituents, can function as effective antioxidants due to their ability to quench free electrons. Phenolic antioxidants can therefore scavenge the harmful free radicals and inhibit their oxidative reactions with vital biological molecules (19).

The second and more significant mode of action is a consequence of their ability to modulate cellular physiology both at the biochemical or physiological level and at the molecular level. This mode of action is a result of the structure to function phenomenon linked to metabolic pathways. Because of their structural similarities with several key biological effectors and signal molecules, phenolic phytochemicals are able to participate in induction or repression of gene expression; or activation or deactivation of proteins, enzymes, and transcription factors of key metabolic pathways (67,71–73). They can critically modulate cellular homeostasis as a result of their physiochemical properties such as size, molecular weight, partial hydrophobicity, and ability to modulate acidity at biological pH through enzyme coupled reactions. As a consequence of many modes of action of phenolic phyto-chemicals they have been shown to have several different functions. Several studies have demonstrated anticarcinogenic properties of phenolic phytochemicals such as gallic acid, caffeic acid, ferulic acid, catechin, quercetin, and resveratrol (94–96). It is believed that phe-nolic phytochemicals might interfere with several of the steps that lead to the development of malignant tumors, including inactivating carcinogens and inhibiting the expression of mutant genes (97). Potential anticarcinogenic functions of phenolic phytochemicals have also been shown due to their ability to act as animutagens in the Ames test (94–96). Many studies have also shown that these phenolic phytochemicals can repress the activity of enzymes such as the CYP class of enzymes, involved in the activation of procarcinogens. The protective functions of the liver against carbon tetrachloride toxicity (98) have shown that phenolic phytochemicals also decrease the carcinogenic potential of a mutagen and can activate enzymatic systems (Phase II) involved in the detoxification of xenobiotics (2). Antioxidant properties of the phenolic phytochemicals can also prevent oxidative damage to DNA which has been shown to be important in the age related development of some cancers (99). Other phenolics such as caffeic and ferulic acids react with nitrite *in vitro* and inhibit nitrosamine formation *in vivo*. They inhibit the formation of skin tumors induced by 7,12-dimethyl-benz(a) anthracene in mice (100). Important biphenyls such as resveratrol (Figure 7.1) which is found in wine has been shown to inhibit the development of preneoplastic lesions in rat mammary gland tissue in cultures in the presence of carcinogens, and it was also shown to inhibit skin tumors in mice (101,102).

Another well described function of phenolic phytochemicals is the prevention of cardiovascular diseases (CVDs). The lower incidences of CVDs in populations consuming wine as part of their regular diet is well established, and is often referred to as the French paradox (103). Recent research into the beneficial effects of wine has led to an understand-ing that resveratrol, which is a phenolic phytochemical, is the active component in wine responsible for its beneficial effects. Resveratrol and other phenolic antioxidants have also been shown to prevent development of CVDs by inhibiting LDL oxidation *in vitro* (104) and preventing platelet aggregation. Phenolic phytochemicals have also been able to reduce blood pressure and have antithrombotic and antiinflammatory effects (105,106). Phenolic phytochemicals have also been shown to inhibit the activity of α-amylase and α-glucosidase, which are resposible for postprandial increase in blood glucose level, which has been implicated in the manifestation of type II diabetes (107,108).

In addition to managing oxidation, several studies have indicated the ability of phe-nolic phytochemicals to manage infectious diseases. Antibacterial, antiulcer, antiviral, and antifungal properties (109–112) of the phenolic extracts have been described. Immune modulatory activities of phenolic phytochemicals such as antiallergic properties as a result of suppressing the hypersensitive immune response have also been defined (113). Antiinflammatory responses mediated by suppression of the TNF-α mediated proinflam-matory pathways have also been shown to be mediated by phenolic phytochemicals (114).

7.2.3.1 Biological Functionality of Cranberry Phenolics

Cranberries and their products have been part of North American and Western European cui-sines for many centuries. Foods that contain cranberries and their products have been associ-ated historically with many positive benefits on human health. For many decades, cranberry juice has been widely used, particularly in North America, as a folk remedy to treat urinary tract infections (UTIs) in women, as well as other gastrointestinal (GI) disorders (115,116). These infections have now been shown to be caused by the infections of the GI tract by *Escherichia coli* and other pathogens. Recent clinical studies have established a positive link in prevention of urinary tract infection with the consumption of cranberry juice (117). Cranberry, like other fruits and berries is rich source of many bioactive components including phenolic phytochemicals such as phenolic acids, flavonoids, anthocyanins and their derivatives (118). p-Hydroxybenzoic acid, a phenolic acid present at high concentrations in cranberry, was believed to be the primary bioactive component in preventing urinary tract infections (38). This was believed to be due to the bacteriostatic effect of hippuric acid which is formed from metabolic conversion of p-Hydroxybenzoic acid in the liver. Hippuric acid, when excreted into the renal system, causes acidification of urine and thus prevents the growth of *Escherichia coli* on the urinary tract (38). It is now well established that adherence of the pathogen to the host tissue is also one of the most important steps required for the colonization of the bacteria and their subsequent infection. A majority of infectious diseases, including UTIs that are caused by microorganisms, have now been shown to involve the adherence of the pathogen to the host tissue (119,120). Investigations into the mechanism of adherence to host tissue has led to an understanding that these are mediated by specific glycoprotein receptors called fimbriae or lectins on the bacterial cell surface which can specifically bind to sugars present on the mucosal or intestinal cell surfaces of the host tissue (119,120). Many soluble and nondigestible sugars and oligosaccharides, such as fructose and mannooligosaccharides, can act as decoy sugars, forcing the bacteria to bind to them instead of the host cell. Inability of the pathogen to bind to the cell surface causes the microorganism to be washed away by the constant peristaltic motion in the intestine. This type of binding however, occurs only via a specific type of fimbriae called type 1 (mannose sensitive) fimbriae (119,120). Recent investigations have shown that type P fimbriae [α-Gal(1→4)β-Gal] mediated adhesion, which is mannose resistant, is also involved in bacterial adhesion. Components of fruit juices, including cranberry juice, have been proposed to inhibit bacterial adherence to the epithelial cells by competing to bind with both these fimbriae (119,120). In addition to the extensive studies done on the inhibition of the adherence of components of *Escherichia coli* to host mucus cells, recent *in vitro* studies indicate a high molecular weight component in cranberry to inhibit the siallylactose specific (S-fimbriae) adhesion of *Helicobacter pylori* strains to immobilized human mucus, erythrocytes, and cultured gastric epithelial cells. It is suspected that these high molecular weight components from cranberries can inhibit the adhesion of *Helicobacter pylori* to the stomach *in vivo* and therefore may have a potential inhibitory effect on the development of stomach ulcers (121,122). Certain high as well as low molecular weight preparations of cranberry juice were also effective in decreasing the congregation and salivary concetration of *Streptococcus mutans,* which causes tooth decay (123,124). The formation of catheter blocking *Proteus mirabilis* biofilms in recovering surgical patients was also significantly decreased by the consumption of cranberry juice (125). Adherence of *Fusobacterium nucleatum* to buccal cells was also reduced by the high molecular weight extract from cranberry juice (123,124). Low and high molecular weight components from cranberry are also suspected to have antiviral properties because of the ability of tannins and other polyphenols to form noninfectious complexes with viruses. Cranberry and its products are also known to inhibit many fungi belonging to *Candida* species, *Microsporum* species and *Trycophyton* species (126,127).

Recent studies have reported on the radical scavenging activities of the various flavonol glycosides and anthocyanins in whole cranberry fruit and their considerable ability to protect against lipoprotein oxidation *in vitro*. The flavonoid and hydrocinnamic acid derivatives in cranberry juice reduced the oxidation of LDL and LDL mobility (128). In an *in vitro* study cranberry samples significantly inhibited both H_2O_2 and TNFα induced vascular endothelial growth factor (VEGF) expression by the human keratinocytes (129). Matrigel assay using human dermal microvascular endothelial cells showed that edible cranberries impair angiogenesis (129). It is therefore believed that cranberry juice may also have beneficial effects on cardiovascular health (130,131).

Cranberry and cranberry extracts have been shown to have anticancer properties. Phenolic extracts from berries of *Vaccinium* species were able to modulate the induction and repression of ornithine decarboxylase (ODC) and quinone reductase that critically regulate tumor cell proliferation (132). Cranberry extracts showed *in vitro* antitumor activity by inhibiting the proliferation of MCF-7 and MDA-MB-435 breast cancer cells. Cranberry extracts also exhibited a selective tumor cell growth inhibition in prostate, lung, cervical, and leukemia cell lines (132,133). Solid-state bioprocessing of natural products including cranberry pomace had shown to enhance its functionality. The antioxidant activity of the cranberry pomace was improved significantly after solid-state bioprocessing with fungi. Bioprocessing of cranberry pomace was found to release phenolic aglycones and enhance the phenolic profile of the pomace with important functionally relevant diphenyls such as ellagic acid. The antimicrobial properties of the pomace extract against foodborne and human pathogens such as *Listeria monocytogenes, Eschereschia coli O157: H7, Vibrio parahemolyticus* and *Helicobacter pylori* significantly improved after solid-state bioprocessing (134).

7.2.3.2   Ellagic Acid

Ellagic acid is a naturally occurring phenolic lactone compound found in a variety of natural products (Figure 7.1). Ellagic acid is present in plants in the form of hydrolyzable tannins called ellagitannins as the structural components of the plant cell wall and cell membrane. Ellagitannins are esters of glucose with ellagic acid which, when hydrolyzed, yield ellagic acid. Ellagic acid is seen at high concentrations in many berries including strawberries, raspberries, cranberries, and grapes (38,39). Other sources of ellagic acid include walnuts, pecans (135), and distilled beverages (136). Recent studies have indicated that ellagic acid possesses antimutagenic, antioxidant, and antiinflammatory activity in bacterial and mammalian systems (137–140).

Ellagic acid has been shown to be a potent anticarcinogenic agent. One of the main mechanisms by which ellagic acid is proposed to have anticancer benefits is by modulating the metabolism of environmental toxins and therefore preventing initiation of carcinogenesis induced by these chemicals (141). It is also proposed to show antimutagenic activity by inhibiting the direct binding of these carcinogens to the DNA (142).

Ellagic acid was found to inhibit the mutagenesis induced by aflatoxin B1 in *Salmonella* tester strains TA 98 and TA 100 (143). On oral administration, ellagic acid exhibited hepatoprotective activity against carbon tetrachloride both *in vitro* and *in vivo* (144). Ellagic acid inhibited the DNA binding and DNA adduct formation of N-nitrosobenzylmethylamine (NBMA) in cultured explants of rat esophagus (145). Related studies have shown that ellagic acid inhibited both the metabolism of NBMA and the binding of NBMA metabolites to DNA (146). In human epithelial cells ellagic acid also inhibited the binding of carcinogenic benzo[a]pyrene metabolites to DNA (142), and dibenzo[a,l]pyrene-DNA adduction in human breast epithelial cell line MCF-7 (147). Smith et al. (148) also showed that ellagic acid resulted in substantially reduced (>70%) DNA binding of 7,12-dimethylbenz[a]anthracene (DMBA) and suggested that possible mechanisms for the observed adduct reduction include direct interaction of the chemopreventive agent with the carcinogen or its metabolite, inhibition of phase I enzymatic activity, or formation of adducts with DNA, thus masking binding sites to be occupied by the mutagen or carcinogen (142). Ellagic acid also significantly increased the GST enzyme activity, GST isozyme levels, and glutathione levels, and therefore is proposed to show strong chemoprotective effects by selective enhancement of members of the GST detoxification system in the different cancerous cells (142).

Ellagic acid was also found to significantly reduce the number of bone marrow cells with chromosomal aberrations and chromosomal fragments as effectively as alpha tocopherol (149). Moreover, administration of ellagic acid inhibited radiation induced DNA strand breaks in rat lymphocytes. Ellagic acid induced G1 arrest, inhibited overall cell growth, and caused apoptosis in tumor cells (150). One of the studies reported a better protection by ellagic acid than vitamin E against oxidative stress (151). Ellagic acid reduced cytogenetic damage induced by radiation, hydrogen peroxide, and mitomycin C in bone marrow cells of mice (149,152). Chen et al. (153) suggested that the antitumor promoting action of ellagic acid and other related phenolics may be mediated in part by inducing a redox modification of protein kinase C (PKC) which serves as a receptor for tumor promoters. Ellagic acid is also suggested to carry out its antimutagenic and anticarcinogenic effects through the inhibition of xenobiotic metabolizing enzymes (141), and by the induction of antioxidant responsive element (ARE) mediated induction of NAD(P)H: quinone reductase and glutathione S-transferase (GST) genes, which can detoxify carcinogens and reduce carcinogen induced mutagenesis and tumorigenesis (154,155). Wood et al. (140) showed that ellagic acid is a potent antagonist to the adverse biological effects of the ulimate carcinogenic metabolites of several polycyclic aromatic hydrocarbons, and suggested that this naturally occurring plant phenolic, normally ingested by humans, may inhibit the carcinogenicity of polycyclic aromatic hydrocarbons.

Studies have shown that ellagic acid is a potent inhibitor of DNA topoisomerases, which are involved in carcinogenesis. Structure and activity studies identified the 3,3′-hydroxyl groups and the lactone groups as the most essential elements for the topoisomerase inhibitory actions of plant phenolics (156). Some recent studies have shown that ellagic acid was found to be better than quercetin for chemoprevention (139). When the effect of both of these compounds on reduced glutathione (GSH), an important endogenous antioxidant, and on lipid peroxidation, was investigated in rats, both ellagic acid and quercetin caused a significant increase in GSH and decrease in NADPH dependent and ascorbate dependent lipid peroxidation. However, ellagic acid was found to be more effective in decreasing the lipid peroxidation and increasing the GSH (139). This suggested that it may be more effective in inducing the intracellular synthesis of GSH and may be more capable of regenerating the oxidized GSH. This may be one of the reasons for the well documented anticarcinogenic activity of ellagic acid compared to quercetin (139). When the ability of vitamin E succinate and ellagic acid to modulate 2,3,7,8-tetrachlorodibenzo-p-dioxin (TCDD)-induced developmental toxicity and oxidative damage in embryonic or fetal and placental tissues was compared in C57BL/6J mice (151), ellagic acid provided better protection than vitamin E succinate and decreased lipid peroxidation in embryonic and placental tissues.

7.2.3.3   Alternate Model for the Function of Ellagic Acid and Related Phenolic Phytochemicals from Cranberries

Recent research has shown that phenolic phytochemicals such as ellagic acid from cranberries and other fruits have several health benefits. Several studies have suggested many mechanisms as the mode of action of ellagic acid and several other related phenolics. Primarily the mechanism of action has been defined as being able to counter the negative effects of stress at late stages of pathogenecity by aiding the regeneration of cellular antioxidants such as GSH and ascorbate, and by activation or induction genes responsible for expressing enzymes such as superoxide dismutase (SOD), catalase (CAT), glutathione-S-transferase (GST), Quinone: NADPH oxidoreductase (QR), and others which are involved in managing the oxidative stress (141,155). Control of oxidative stress induced diseases is also believed to be brought about by repressing certain genes such as the cytochrome P450 dependent Phase I enzymes (157,158), inhibition of NADPH oxidase, and other systems that generate ROS (63). Though these proposed mechanisms are valid and justified with experimental findings, however, they do not explain the larger, more comprehensive, functions of cranberry phenolics such as ellagic acid and other phenolic antioxidants in maintaining specific cellular homeostasis, which contribute to its preventive mode of action. The mechanisms and models proposed so far only explain a particular specific response mediated by ellagic acid or related phytochemicals in a disease such as preventing the binding of a carcinogen to the DNA (141), or repressing the activity of Phase I enzymes in the liver (154,155). These observations often explain the beneficial effect of phenolic phytochemicals based on either one mechanism of action such as a free radical scavenging activity, or by explaining consequences on end results such as activation or repression of some genes. These models, however, do not explain several different effects mediated by a single phenolic phytochemical and the synergistic actions of phenolic phytochemicals in foods. Biological antioxidant protection is believed to occur through an adaptive response (Figure 7.5, Figure 7.6), wherein the cell shifts its functions in a manner that induces genes and transcription factors such as AP1, NFκB and *cfos* (70), which in turn stimulate the antioxidant enzyme response mediated by GSH, SOD/CAT and GST interface, and also reduces mitochondrial function to prevent ROS generation (67,70). One drawback of these models is that they are limited in explaining the several upstream metabolic processes that ultimately contribute to manifestation of the adaptive response that maintains the cellular homeostasis. The adaptive response, which comprises the cascade of antioxidant enzyme activity, is an energy intensive process and therefore requires a constant supply of ATP and reducing equivalents (NADPH) (74). For this antioxidant response to function in an efficient manner, the cell would have to constantly replenish its energy. It would, therefore, be imperative for the mitochondria to function efficiently to replenish the energy needs. Also, a specific mode of functionality of the individual phytochemical does not explain other functionality such as antimicrobial activity in preventing *Helicobacter pylori* or *Escherichia coli* infections for maintaining gastrointestinal health. It is challenging to explain, using existing models, the reason for the same phytochemical to promote survival in eukaryotes and discourage the survival of pathogens. These apparently conflicting modes of action of phytochemicals have prompted a need to understand the functionality of phenolic phytochemicals such as ellagic acid in a much broader sense. There is a need to understand the mechanism of action of these phenolic phytochemicals at the early stages of stimulating the antioxidant response mediated cellular homeostasis in the cell (159,160). Therefore, in addition to the described mechanisms of action of phenolic antioxidants, an alternative model has been proposed for the mode of action of dietary phenolic phytochemicals. The antioxidant homeostasis in the cell occurs via the functioning of a diverse array of redox processes, primarily carried out by cellular antioxidants such as glutathione, ascorbate, tocopherols, and an array of antioxidant enzymes such as SOD, CAT and GST (72). However, to maintain high efficiency of this system it is important to regenerate the oxidized substrates such as glutathione disulfides (GSSG), dehydroascorbate, and other proteins with oxidized sulfhydryl groups. The regeneration of oxidized glutathione, ascorbate and tocopherol occurs by a group of oxidoreductases which use cellular reducing equivalents such as $FADH_2$ and NADPH, and therefore are energy intensive processes (72–74). In order to meet the cellular requirement for these reducing equivalents, in this model, it is proposed that phenolic antioxidants aid the antioxidant response of the cell not only by themselves acting as redox modulators by virtue of their antioxidant (free radical scavenging) activity, but also able to stimulate pathways in the cell that can replenish the needs for reducing equivalents. One such pathway that could be up regulated by phenolic phytochemicals such as cranberry phenolics, and ellagic acid could be the pentose and phosphate pathway (PPP) (Figure 7.7). The pentose phosphate pathway is an important pathway that commits glucose to the production of ribose sugars for nucleotide synthesis and, in the process, also produces reducing equivalents (NADPH) (161–163). Phenolic phytochemicals, especially biphenyls and polyphenols, are structurally similar to many biological signaling molecules which

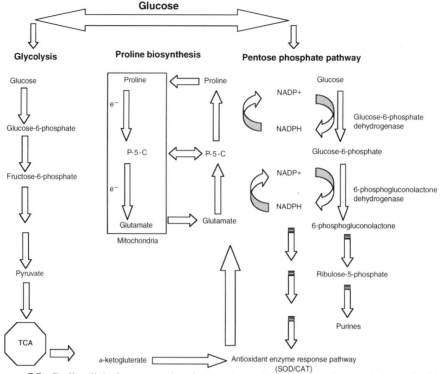

Figure 7.7   Proline linked pentose–phosphate pathway in eukaryotes for regulating antioxidant response.

interact with the receptors on the cell surface responsible activating biological signal transduction processes. Recent empirical evidence has now shown that some phenolic phytochemicals can also mimic the functions of biological signaling molecules and trigger the signal transduction (153,164,165). Phenolics from cranberry, especially biphenyls and flavonoids are large and partially hydrophobic, and can be perfect candidates for the activating signaling pathways responsible for the stimulation of PPP (153,164,165). In addition, phenolic phytochemicals such as ellagic acid and flavonoids from cranberry have been known to be good chelators; these phytochemicals can effectively chelate the ions such as calcium in the extracellular matrix (ECM) or in the cytosol and alter the net concentration of free calcium (8,166,167). Calcium in the cytosol and extracellular matrix is critical for maintaining cellular homeostasis and is an important factor in regulating cell division and cell death, and is often regulated by the calcium sensing receptor (168–172). Many cellular signaling cascades are sensitive to a calcium gradient across the cell membrane (171,172). An apparent modulation in the concentration of calcium, either by calcium binding or by the modulating the calcium sensing receptor, can activate these cellular signaling cascades which can result in changes in many physiological pathways including the stimulation of the PPP (161, 171–173). Phenolic phytochemicals such as ellagic acid and flavonoids have been shown to interact with proteins and alter their configuration. These phytochemicals can therefore directly interact with the cell surface receptors and ion pumps, and directly activate signaling cascades by inducing structural changes in the membrane receptor proteins and pumps (174–177). Phenolic phytochemicals, especially phenolic acids, are weak acids and are capable of dissociating at the cell membrane at a biological pH of 6.8–7.2 (178). This dissociation can create a proton gradient across the cell membrane which can alter the function of many proton pumps on the cell membrane (159,160). The partially hydrophobic ellagic acid and flavonoids can also directly interact with the proton pumps and modulate their function. In addition, phenolic phytochemicals, especially flavonoids, have been shown to be able to interact with the membranes (179,180). Such an interaction can alter the permeability of the cell and cause changes in the electrochemical gradient across the cell membrane, causing rapid influx of protons into the cytosol and activation of many signal cascades leading to the stimulation of PPP (Figure 7.7). The stimulation of PPP which results in the NADPH synthesis can help in decreasing the excess protonation of the cytosol by the phenolic phytochemical, by combining the protons with $NADP^+$ to make NADPH through dehydrogenases. Stimulation of the pentose phosphate pathway could be an important cellular mechanism in managing antioxidant stress response and may be closely linked to stimulation of antioxidant response pathways (160,181). The availability of reducing equivalents in the cell could now further stimulate the antioxidant stress response managed by GSH, ascorbate, SOD, CAT, and GST, which require NADPH for their effective functioning (72–74). Phenolic phytochemicals are now known to modulate gene expression either by activating or repressing transcription factors or by directly binding to the DNA (164,182,183). The phenolic phytochemicals such as ellagic acid and other cranberry flavonoids can aid in the cellular antioxidant defense response by activating the expression of enzymes involved in antioxidant defense and repressing the expression of oxidative stress producing pathways such as NADPH oxidase and cytochrom-P450 dependent phase I enzymes.

The stimulation of the pentose phosphate pathway could further be coupled to the biosynthesis of proline made from glutamic acid using NADPH (184–186). Phenolic phytochemicals can stimulate the biosynthesis of proline and subsequently create a demand for the tricarboxylic acid (TCA) cycle intermediates such as α-keto glutarate to be channeled to glutamic acid and then to NADPH requiring proline biosynthesis (Figure 7.7). This channeling of the TCA cycle intermediates to proline biosynthesis can generate a cellular demand for NADPH and can therefore stimulate the pentose phosphate pathway (160,181). This coupling of proline biosynthesis with PPP can generate more NADPH which can be used by the proline biosynthesis pathway, antioxidant response pathways, and anabolic reactions (160,181). The cellular demands for reducing equivalents are coupled to the needs for ATP, which is the source of energy in biological systems. ATP is synthesized by oxidative phosphorylation of ADP by an enzyme, ATPase, in the mitochondria by reduction of molecular oxygen to water with the help of electrons from reducing equivalents such as NADH and $FADH_2$. Excessive cellular requirement for ATP usually results in incomplete reduction of oxygen to make reactive oxygen species, which have implications in manifestation of various oxidative stress related diseases (187,188). Proline has been shown to be able to function as a reductant in cellular systems (189,190). Therefore, it has also been proposed that proline can function as an alternative reductant (instead of NADH) (Figure 7.7) for mitochondrial oxidative phosphorylation to generate ATP (159,160). This can reduce the cellular need for NADH linked ATP synthesis, which can reduce excessive mitochondrial oxidative burst limit leakage of reactive oxygen species into the cytosol during oxidative phosphorylation.

7.2.3.4   *A Model for Antimicrobial Activity of Ellagic Acid and Related Phenolics Against Bacterial Pathogens*

Ellagic acid and other related phenolics can also have a different mode of action that can be used to explain their other functionalities, such as antimicrobial activity. Phenolic phytochemicals and phenolic acids are weak acids, and therefore are capable of dissociating at the plasma membrane at biological pH (178). This weak dissociation of the phenolics is proposed to be an important mechanism by which ellagic acid and other phenolic antioxidants exert their antimicrobial effect. The dissociation of a proton from the carboxyl or the hydroxyl substituent on the phenolic phytochemical may result in hyperacidification at the plasma membrane interface of the microorganism (160,191). The hyperacidification of the plasma membrane by the phenolic antioxidant could change the resting potential of the membrane, as it results in increased positive charges outside of the membrane. Disruption in the proton and electrostatic gradient across the membrane can have several implications. Many ion pumps and channels such as Na^+/K^+ or at Ca^+ pumps are responsible for critically regulating many cellular functions such as motility, cell division, and cell death. Impairment of calcium homeostasis or calcium signaling can cause disruption in the normal cellular function of the microorganism and therefore death. Phytochemicals such as tannins, and their hydrolyzed products such as ellagic acid, can inhibit the growth of microorganisms by sequestering metal ions critical for the microbial growth and metabolism (192–195), or by inhibiting critical functions of the bacterial membrane such as ion channels and proteolytic activity (196), which are all dependent on pH and ionic strength. Many key proteins and receptors on the membrane are responsible for receptor mediated transport of nutrients and cofactors. These proteins are sensitive to the pH and ionic strength. Disruption of the electrochemical gradient and phenolic induced acidity at the membrane could cause inactivation of the receptors involved in uptake of the nutrients and thus cause death of the target bacteria.

The cell membrane is a site of electron transport and ATP generation in prokaryotes. Ellagic acid related phytochemicals are phenolic chemicals and have an ideal chemical structure to function as antioxidants. Presence of phenolic moieties confers upon them an excellent ability to quench the electrons from free radicals and delocalize them within the phenolic ring. Consequently, they can easily quench the electrons from the electron transport chain (ETC) along the bacterial membrane. This could disrupt the flow of the electrons at the level of cytochromes and inhibit the growth of bacteria by disrupting oxidative phosphorylation (160,181,191). Localized protonation and reduced plasticity of the membrane can diffuse the proton gradient essential for the functioning of H^+-ATPase required for ATP synthesis. The diffusion in the proton gradient results in lowered efficiency of the H^+-ATPase, and therefore the organism will synthesize reduced ATP, or no ATP at all. This can force the microorganism to switch to an anaerobic metabolism to generate ATP. The reduced intake of nutrients because of disruption in the membrane transport induced by ellagic acid and other related phenolics, increased demand for ATP, and disruption of the ATPase activity could prove fatal to the microorganism. For a phenolic to function effectively as an antimicrobial, it should be able to function at the lipid to water interface and therefore has to be partially hydrophobic. Biphenyls such as ellagic acid and other polymeric phenoics are partially hydrophobic, making them effective to act efficiently at the membrane to water interface of microorganisms. Molecules such as ellagic acid can possibly stack or embed themselves in the membrane. Stacking and or embedding of phenolic phytochemicals such as ellagic acid (160,181,191) at the surface of the membrane can induce a change in the membrane structure by creating an environment of hydrophobicity around the cell, and may result in inactivation of proteins by inducing protein unfolding or by the exclusion of the polar and hydrophilic component from the membrane such as water polar lipids and surface proteins (179,180). This can result in changes in the composition of the membrane and therefore can severely impair the plasticity of the membrane (179,180). A rigid membrane destabilizes the cell by weakening membrane integrity, which may result in the disruption of critical transport processes and sometimes collapsing of the bacterial membrane. This could be an important mechanism of antimicrobial activity in Gram-negative bacteria such as *Eschereschia coli* which have an external lipopolysac-charide layer and additional minor membrane components besides an intact plasma mem-brane. This gives them potentially more buffering capacity and hydrophobicity and couls, therefore, create an unfavorable environment for simple phenolics to exert their hyper-acidification effect.

7.3   INNOVATIVE STRATEGIES TO ENRICH FRUITS WITH PHENOLIC ANTIOXIDANTS

A large variety of phenolic phytochemicals that have several beneficial functions on human health are present in plants and especially in fruits. The phenolic phytochemicals are generally present in their glycosidic and nonglycosidic forms. The glycosides are mainly confined to hydrophilic regions in the cells, such as in vacuoles and apoplasts, probably because of their higher water solubility (197,198). Glycosylation of the hydroxyl groups on the phenolic ring of a phenolic phytochemical renders the molecule more water soluble and less reactive toward free radicals (6). Glucose is the most commonly involved sugar in glycoside formation, although phenolic glycosides of galactose, rhamnose, xylose, and arabinose; and disaccharides such as rutinose, have also been reported in plants (6). Polymeric phenolics such as tannins exist primarily as condensed tannins or proanthocyanidins and are formed biosynthetically by the condensation of single catechins and flavonols. They are present either as soluble tannins or bound to the cell wall. Hydrolysable tannins are esters of a sugar with either gallic acid (gallotannins) or ellagic acid (ellagitannins). Tannins, even though they have higher antioxidant properties than individual simple phenolics, are usually not bioavailable, and are to some extent antinutri-tive in their function because of their ability to bind and precipitate biological macromol-ecules such as proteins and carbohydrates (109). The total phenolic phytochemical content in plant foods also varies greatly. Their presence in plant foods is largely influenced by genetic factors and environmental conditions. Other factors, such as cultivar, variety, matu-rity, processing, and storage, also influence the content of plant phenolics (199–201). The effects of processing and storage on the changes and content of polyphenols in cranberry (202), plum (203), and grape juice (204) have been evaluated.

As a consequence of evidence that consumption of fruits and vegetables has been linked to decreased manifestation of chronic diseases, there has been a constant increase in the demand for diets rich in phenolic phytochemicals. Vast variation in the amounts of phenolic antioxidants available via diet (205) coupled with reduced bioavailability and functionality has led to an urgent need to develop innovative strategies to enrich diets with phenolics and specifically phenolic antioxidants with consistent phytochemical profile for enhanced health functionality.

Among many strategies, two are important to enrich phenolic antioxidants. The first is genetic modification of cultivars to produce plants that will yield fruits and vegetables with higher phenolic concentration. Currently, in terms of genetic improvement, breeding strategy coupled with micropropagation using tissue culture is being developed (206–208). These strategies, along with genetic modification, could be directed toward phytochemical enrich-ment and quality improvement. However, this method presents important issues, such as regulation of key metabolites by multiple genes and biochemical pathways, acceptance of genetically modified foods, and relative time and economic considerations that are involved (209). Another exciting strategy that can be used is the bioprocessing of botanicals using solid-state bioprocessing and synergies to generate phytochemical profiles with enhanced health functionality. This strategy can be used for juice and pulp as well as pomace that remains after the juice is extracted from the fruits. Fermentation of fruit juices, such as grape juice to wine, has already been shown to improve nutritive and health promoting activities (210–212). Solid-state bioprocessing done on the pulps using food grade fungi can result in enrichment of the pulps with phenolic antioxidants and functionally important phenolic phytochemicals, and also improve phytochemical profile consistency.

7.4   SOLID-STATE BIOPROCESSING

Fermented foods have been consumed by humans all over the world for centuries. Most fermentation processes are conducted with liquid nutrient broths. Well known examples in the food industry are the production of yogurt, beer, wine, lactic acid, and many food flavors (213). However, partial fermentation and aerobic microbial growth based bioprocessing has also been used for processing food and food wastes. Here, instead of a nutrient broth, moist solid nutrients with minimal water are used as a substrate for microbial growth. This process is referred to as solid-state bioprocessing. Microbial fermentation and aerobic microbial growth on foods in solid state, for preservation of food and flavor enhancement, has been done for centuries and some of the common examples for these processes include manufacture of cheese and bread (214). Other wellknown examples are the production of microbe laced cheeses such as Roquefort, and the production of fermented sausages. In Asia, solid-state bioprocessing has been used for food fermentation for over 2000 years, for the production of fermented foods such as tempeh, natto, and soy sauce (193). The preservation of fish and meat by solid-state bioprocessing has also been reported to be carried by early human civilizations (193,214). Fruit wastes have been extensively used as substrates for solid-state bioprocessing. These wastes have mostly been used for the production of fertilizers, animal feed, as a growth substrate for mushrooms, ethanol production, production of organic acids such as citric acid, tartaric acid, and lactic acid, and for the production of various kinds enzymes such as pectinases (41,213). Solid-state bioprocessing of fruit substrates has also been carried out for several decades to produce compounds like gallic acid and vinegar (193,214). Recent research has also shown that consumption of fermented foods, especially solid-state bioprocessed foods is beneficial for health (215,216). Solid-state bioprocessing of fruit wastes such as cranberry pomace using the food grade fungi *Rhizopus oligosporus* and *Lentinus edodes* has been shown to enrich phenolic antioxidants and improve phytochemical consistency. These studies have shown that during the course of solid-state growth, the antioxidant activity and phenolic content of the pomace extracts increase several fold (52,53). The process resulted in enrichment of functional phenolics to a level found usually in fresh fruits and their juice products (217). The antimicrobial activity of the extracts against pathogens such as *Escherichia coli O157:H7, Listeria monocytogenes, Vibrio parahemolyticus* and *Helicobacter pylori* of cranberry pomace was also enhanced by solid-state bioprocessing. It is suspected that the increase in phenolic and antioxidant activity could have been due to the production of various hydrolyzing enzymes by the fungi during the course of solid-state growth. These fungal hydrolases such as glucosidase and fructofuranosidases could possibly be hydrolyzing the glycosidic linkages between the phenolic moieties and sugars. A similar observation in the increase in phenolic aglycones was observed during the fermentation of soy milk for the production of tofu (218). The fungus, in adapting itself to utilize the fruit substrates, may produce various other types of hydrolases such as laccases and lignocellulases. The activity of these enzymes is suspected to responsible for the increase in the polymeric phenolics, and potentially to contribute to the enhanced bioavailability of such phenolic antioxidants. Enrichment of the solid-state bioprocessed fruit wastes such as cranberry pomace with ellagic acid after bioprocessing has been reported (52,53). This may have resulted due to the hydrolysis of ellagotannins by tannin hydrolyzing enzymes produced by the fungus. Further, it is suspected that phenolic enrichment could also occur through contribution from the growing fungal species. The endogenous phenolics present in the fruit wastes could be toxic to the growing fungus. In an attempt to adapt and utilize the substrate for growth, the fungus could be detoxifying the phenolics biochemically using a variety of enzymatic systems present in the fungus. The fungal detoxification can occur by a variety of mechanisms including methylating or demethylating the pheniolic ring, or by hydroxylation (219,220). Recent studies have shown methylated phenolic phytochemicals have excellent antibacterial properties against Gram-positive bacteria (221). Hydroxylation of the phenolic ring by the fungal system during its growth increases antioxidant properties (19) and therefore, phenolics resulting from biotransformation occurring during the solid-state bioprocessing may improve their functionality and be beneficial for human health. The advantage of this strategy is that the fungi, such as *Rhizopus oligosporus* and *Lentinus edodes* and other fungi used in this solid-state growth process are food grade and are generally recognized as safe (GRAS). This approach can easily be adapted to different substrates as well as be extended to liquid fermentation of juices to develop food and ingredients with enhanced functionality.

7.5 CRANBERRY SYNERGIES WITH FUNCTIONAL PHYTOCHEMICALS AND OTHER FRUIT EXTRACTS

Recent research has documented the evidence that whole foods, and not single compounds, have a better functionality in maintaining our health against many of the antioxidant diseases. Recent work on the effect of wine has shown that resveratrol is responsible for the decrease in atherogenesis in rats (101,102). However, when resveratrol was used as a supplement in diet such an effect was not seen. Other researchers have shown that the combination of resveratrol and quercetin exerts a synergic effect in the inhibition of growth and proliferation of human oral squamous carcinoma cells (221). Carbonneau et al. (223) during *in vivo* antioxidant assays with red wine, observed that different phenolics in wine could play a coantioxidant role, similar to that described for vitamin C, and a sparing role toward vitamin E, which increases due to supplementation with phenolics. Synergistic interactions between wine polyphenols, quercetin, and resveratrol were found to decrease the inducible nitric oxide synthase (iNOS) activity in cell culture systems (224). This suggests that the phytochemical profile in which the specific functional phenolic is present plays an important role in determining its functionality. Synergy can be defined as the ability of two or more functional components, such as antioxidants in a phytochemical background, to mutually enhance their functionalities. Typically, in a whole food background such as red wine, resveratrol is present in a background containing several simple as well as polyphenols, such as gallic acid, protocatcheuic acid and hydrolyzed tannins. Each phenolic phytochemical has its own mode of action against a particular target. These modes of actions could be due to their ability to function as classical antioxidants or because of their ability to modulate cellular physiology by disrupting membrane functions or by altering the redox balance and energy metabolism of the cell (160). However, when they are present together, their ability to function together rapidly improves the overall result of maintaining the cellular homeostasis, especially in eukaryotes, or killing the pathogen in prokaryotes. Also, it was observed that coadministration of coffee with ellagic acid enhanced the antigenotoxic effect compared with that of either coffee or the ellagic acid alone, suggesting that there is a significant synergistic interaction between coffee and the dietary constituents for antigenotoxic effects against different mutagens (225). The conditions created due to the mode of action of one phenolic significantly improves the chances of the other phenolics being able to function effectively, thereby reducing the overall dosage required to observe the desired positive effect. This may be one of the reasons why whole foods have a better functionality in maintaining human health compared to the consumption of supplements. This synergy concept can also be artificially duplicated, and this provides another way to significantly enhance the functionality of foods. Synergistic supplementation of foods such as fruit beverages with flavonoids and other functional phenolic phytochemical can drastically improve their functionality.

In a recent study the potential antimutagenic properties of cranberry phenolics, ellagic acid, rosmarinic acid, and their synergistic interactions on enhancing antimutagenic properties in *Salmonella typhimurium* tested the system against mutagens sodium azide and N-methyl-N′-nitro-N-nitrosoguanidine (MNNG) was investigated (226). Ability of these phytochemical treatments to prevent oxidative damage to DNA was also investigated, using the supercoiled DNA strand scission assay. Results showed that ellagic acid was most effective in inhibiting the mutations in *Salmonella typhimurium* system, whereas rosmarinic acid and ellagic acid were equally effective in protecting the DNA from oxidative damage. The antimutagenic functionality of cranberry powder made from juice extracts was significantly enhanced when 30% (w/w) of phenolics in cranberry powder were substituted with rosmarinic acid and ellagic acid possibly due to synergistic redox modulation which can influence mutagen function. It was suggested that the synergistic mixture of cranberry phenolics with rosmarinic acid could also be protecting the cell from mutations by modulating DNA repair systems.

Phenolic antioxidant and α-amylase inhibition activity of cranberry, wild blueberry and grape seed extracts and their synergistic mixtures were investigated in a recent study to develop an additional strategy to manage type II diabetes (227). The results indicated that all the extracts had α-amylase inhibition activity which correlated to the presence of specific phenolic phytochemicals such as chlorogenic acid, ellagic acid, and rosmarinic acid, suggesting a possible structure related inhibition of α-amylase. Among the fruit juice powders the cranberry powder had the highest α-amylase inhibition activity. A mixture containing 75% cranberry, 15% blueberry and 10% grape showed a synergistic mode of action and was the most optimal mixture to control α-amylase activity.

A similar study investigated the effect of cranberry, blueberry and grape seed extracts on inhibiting *H. pylori* (228). The results showed that the anti-*H. pylori* activity of cranberry juice extract was significantly improved by its synergistic blending with blueberry or grape seed extract. In both the studies, the lower efficacy of purified phenolics in inhibiting α-amylase or *H. pylori* compared with fruit powder at similar dosage levels suggests a synergistic mode of functionality of these individual phenolics in whole food background.

Consumption of blends of fruit juices with biologically active biphenyls or other fruit as well as herb extracts can impart unique functional attributes and could be a more effective strategy in developing diet based management of *H. pylori* infections, as well as other oxidation linked diseases, including diabetes and mutagen and DNA damage induced carcinogenesis.

7.6   CONCLUSIONS

Emerging epidemiological evidence is increasingly pointing to the beneficial effects of fruits and vegetables in managing chronic and infectious diseases. These beneficial effects are now believed to be due to the constituent phenolic phytochemicals having antioxidant activity. Cranberry, like other fruits, is also rich in phenolic phytochemicals such as phenolic acids, flavonoids and ellagic acid. Consumption of cranberry has been historically been linked to lower incidences of urinary tract infection and now has been shown to have a capacity to decrease peptic ulcer caused by *Helicobacter pylori*. Isolated compounds from cranberry have been shown to reduce the risk of CVD and cancer. Functional phenolic antioxidants from cranberry such as ellagic acid have been well documented to have antimutagenic and anticarcinogenic functionality. Even though many benefits have been associated with phytochemicals from cranberry, such as ellagic acid, their mechanism of action is still not very well understood. Emerging research exploring the mechanism of action of these phytochemicals from cranberry usually follows a reductionist approach, and is often focused on the disease or pathological target. These approaches to understanding the mechanism of action of phytochemicals have limitations as they are unable to explain the overall preventive mode of action of phenolic phytochemicals. The current proposed mechanisms of action of these phenolic phytochemicals have also overemphasized the antioxidant activity of phenolic phytochemicals associated with free radical quenching capacity. This mechanism of action is unable to explain the nonantioxidant (free radical scavenging) role, as well as sometimes contradictory modes of action of phenolic phytochemicals across different species, such as being protective in eukaryotes while being inhibitory to prokaryotes. The often promising results seen at laboratory scale have very rarely been successful at clinical levels in terms of seeing the beneficial effects of these phenolic phytochemicals. All these limitations of the present understanding strongly suggest the involvement of phenolic phytochemicals much earlier in the cellular response towards maintaining antioxidant stress response. A current line of investigation after a newly proposed integrated model suggests the involvement of phenolic phytochemicals in critically regulating energy metabolism of the cell by stimulating the PPP in order to supply the cell with reducing equivalents necessary for the efficient functioning of antioxidant response. A coupling of the proline biosynthesis with PPP has been suggested, which can further stimulate the PPP. Proline can also function as an alternative reductant to participate in the oxidative phosphorylation for ATP synthesis. This would effectively reduce the cellular needs for NADH, which could contribute to reducing oxidative stress as a result of reduced oxidative burst in the mitochondria. This model emphasizes the structure and function aspect of phenolic phytochemicals in their ability to cause hyperacidification, alter membrane permeability to ions, and interact with the membrane proteins and receptors to activate many signaling pathways which can be responsible for their beneficial mechanism of action in both prokaryotes and eukaryotes. Recent evidence suggests the ability of whole foods in being more effective in managing human health compared to individual phenolic phytochemicals. This suggests that the profile of phenolic phytochemicals determines the functionality of the phenolic phytochemical as a result of synergistic interaction of constituent phenolic phytochemicals. Solid-state bioprocessing using food grade fungi, as well as cranberry phenolic synergies with functional biphenyls such as ellagic acid and rosmarinic acid, and other fruit extracts have helped to advance these concepts. These strategies could be further explored to enhance cranberry and cranberry products with functional phytochemicals and further improve their functionality.

REFERENCES

1. Mann, J. *Secondary Metabolism: Oxford Chemistry Series*. Oxford: Clarendon Press, 1978.
2. Bravo, L. Phenolic phytochemicals: chemistry, dietary sources, metabolism, and nutritional significance. *Nutr. Rev.* 56:317–333, 1998.
3. Crozier, A., J. Burns, A.A. Aziz, A.J. Stewart, H.S. Rabiasz, G.I. Jenkins, C.A. Edwards, M.E.J. Lean. Antioxidant flavonols from fruits, vegetables and beverages: measurements and bioavailability. *Biol. Res.* 33:79–88, 2000.

4. Shetty, K. Biotechnology to harness the benefits of dietary phenolics: focus on *Lamiaceae*. *Asia Pac. J. Clin. Nutr.* 6:162–171, 1997.
5. Briskin, D.P. Medicinal plants and phytomedicines: linking plant biochemistry and physiology to human health. *Plant Physiol.* 124:507–514, 2000.
6. Urquiaga, I., F. Leighton. Plant polyphenol antioxidants and oxidative stress. *Biol. Res.* 33:55–64, 2000.
7. Moure, A., J.M. Cruz, D. Franco, J.M. Domínguez, J. Sineiro, H. Domínguez, M.H. Núñez, J.C. Parajó. Natural antioxidants from residual sources. *Food Chem.* 72(2):145–171, 2001.
8. Bors, W., C. Michel. Chemistry of the antioxidant effect of polyphenols. *Ann. NY Acad. Sci.* 957:57–69, 2002.
9. Harborne, J.B. Plant phenolics. In: *Encyclopedia of Plant Physiology*, Vol. 8, Bella, E.A., B.V. Charlwood, eds., Heidelburg: Springer–Verlag. 1980, pp 329–395.
10. Haslam, E. *Practical Polyphenolics: From Structure to Molecular Recognition and Physiological Action*. Cambridge: Cambridge University Press, 1998.
11. Strack, D. Phenolic metabolism. In: *Plant Biochemistry*, Dey, P.M., J.B. Harborne, eds., San Diego: Academic Press, 1997, pp 387–416.
12. Wang, H., G. Cao, R.L. Prior. Total antioxidant capacity of fruits. *J. Agric. Food Chem.* 44:701–705, 1996.
13. Kalt, W., C.F. Forney, A. Martin, P.L. Prior. Antioxidant capacity, vitamin C, phenolics, and anthocyanins after fresh storage of small fruits. *J. Agric. Food Chem.* 47:4638–4644, 1999.
14. Abuja, P.M., M. Murkovic, W. Pfannhauser. Antioxidant and prooxidant activities of elderberry (*Sambucus nigra*) extract in low-density-lipoprotein oxidation. *J. Agric. Food Chem.* 46:4091–4096, 1998.
15. Heinonen, M., P.J. Lehtonen, A.L. Hopia. Antioxidant activity of berry and fruit wines and liquors. *J. Agric. Food Chem.* 46:25–31, 1998.
16. Sato, M., N. Ramarathnam, Y. Suzuki, T. Ohkubo, M. Takeuchi, H. Ochi. Varietal differences in the phenolic content and superoxide radical scavenging potential of wines from different sources. *J. Agric. Food Chem.* 44:37–41, 1996.
17. Lapidot, T., S. Harel, B. Akiri, R. Granit, J. Kranner. pH-dependent forms of red-wine anthocyanins as antioxidants. *J. Agric. Food Chem.* 47:67–70, 1999.
18. Larrauri, J.A., C. Sánchez-Moreno, P. Rupérez, F. Saura-Calixto. Free radical scavenging capacity in the aging of selected red spanish wines. *J. Agric. Food Chem.* 47:1603–1606, 1999.
19. Rice–Evans, C.A., N.J. Miller, G. Paganga. Antioxidant properties of phenolic compounds. *Trends Plant Sci.* 2:152–159, 1997.
20. Wang, S.Y., H. Lin. Antioxidant activity in fruits and leaves of blackberry, raspberry, and strawberry varies with cultivar and developmental stage. *J. Agric. Food Chem.* 48:140–146, 2000.
21. Kähkönen, M.P., A.I. Hopia, H.J. Vuorela, J.-P. Rauha, K. Pihlaja, T.S. Kujala, M. Heinonen. Antioxidant activity of plant extracts containing phenolic compounds. *J. Agric. Food Chem.* 47:3954–3962, 1999.
22. Knekt, P., R. Järvinen, R. Reppänen, M. Heliövaara, L. Teppo, E. Pukkala, A. Aroma. Dietary flavonoids and the risk of lung cancer and other malignant neoplasms. *Am. J. Epidemiol.* 146:223–230, 1997.
23. Gillman, M.W., L.A. Cupples, D. Gagnon, B.M. Posner, R.C. Ellison, W.P. Castelli, P.A. Wolf. Protective effect of fruits and vegetables on development of stroke in men. *J. Am. Med. Assoc.* 273:1113–1117, 1995.
24. Joshipura, K.J, A. Ascherio, J.E. Manson, M.J. Stampfer, E.B. Rimm, F.E. Speizer, C.H. Hennekens, D. Spiegelman, W.C. Willett. Fruit and vegetable intake in relation to risk of ischemic stroke. *J. Am. Med. Assoc.* 282:1233–1239, 1999.
25. Cox, B.D., M.J. Whichelow, A.T. Prevost. Seasonal consumption of salad vegetables and fresh fruit in relation to the development of cardiovascular disease and cancer. *Public Health Nutr.* 3:19–29, 2000.
26. Strandhagen, E., P.O. Hansson, I. Bosaeus, B. Isaksson, H. Eriksson. High fruit intake may reduce mortality among middle-aged and elderly men. The study of men born in 1913. *Eur. J. Clin. Nutr.* 54:337–341, 2000.

27. Hertog, M.G., E.J. Feskens, P.C. Hollman, M.B. Katan, D. Kromhout. Dietary antioxidant flavonoids and risk of coronary heart disease: the Zetphen Elderly Study. *Lancet* 342(8878):1007–1011, 1993.

28. Keli, S.O., M.G.L. Hertog, E.J.M. Feskens, D. Kromhout. Dietary flavonoides, antioxidant vitamins, and incidence of stroke: the Zetphen study. *Arch. Intern. Med.* 156:637–642, 1996.

29. Le Marchand, L., S.P. Murphy, J.H. Hankin, L.R. Wilkens, L.N. Kolonel. Intake of flavonoids and lung cancer. *J. Natl. Cancer Inst.* 92(2):154–160, 2000.

30. Garcia-Closas, R., C.A. Gonzalez, A. Agudo, E. Riboli. Intake of specific carotenoids and flavonoids and the risk of gastric cancer in Spain. *Cancer Causes Control* 10(1):71–75, 1999.

31. Zuo, Y., C. Wang, J. Zhan. Separation, characterization, and quantitation of benzoic and phenolic antioxidants in American cranberry fruit by GC-MS. *J. Agric. Food Chem.* 50(13):3789–3794, 2002.

32. Guo, C., G. Cao, E. Sofic, R.L. Prior. High-performance liquid chromatography coupled with coulometric array detection of electroactive components in fruits and vegetables: relationship to oxygen radical absorbance capacity. *J. Agric. Food Chem.* 45:1787–1796, 1997.

33. Cao, G., E. Sofic, R.L. Prior. Antioxidant and prooxidant behavior of flavonoids: structure-activity relationships. *Free Rad. Biol. Med.* 22:749–760, 1997.

34. Foti, M., M. Piattelli, M.T. Baratta, G. Ruberto. Flavonoids, coumarins, and cinnamic acids as antioxidants in a micellar system structure-activity relationship. *J. Agric. Food Chem.* 44:497–501, 1996.

35. Velioglu, Y.S., G. Mazza, L. Gao, B.D. Oomah. Antioxidant activity and total phenolics in selected fruits, vegetables, and grain products. *J. Agric. Food Chem.* 46:4113–4117, 1998.

36. Macheix. J.J., A. Fleuriet, J. Billot. *Fruit Phenolics.* Boca Raton, FL: CRC Press, 1990.

37. Larson, R.A. The antioxidants of higher plants. *Phytochemistry* 27:969–978, 1988.

38. Marwan, A.G., C.W. Nagel. Characterization of cranberry benzoates and their antimicrobial properties. *J. Food Sci.* 51:1069–1070, 1986.

39. Chen, H., Y. Zuo, Y. Deng. Separation and determination of flavonoids and other phenolic compounds in cranberry juice by high-performance liquid chromatography. *J. Chromatogr. A.* 13:913(1,2):387–395, 2001.

40. National Agricultural Statistics Service. *"Cranberries" Annual Reports, Fr Nt 4,* Agricultural Statistics board, Washington, DC, USDA, 2001.

41. Zheng, Z., K. Shetty. Cranberry processing waste for solid-state fungal inoculant production. *Proc. Biochem.* 33:323–329, 1998.

42. Rodríguez de Sotillo, D., M. Hadley, E.T. Holm. Potato peel waste, stability and antioxidant activity of a freeze-dried extract. *J. Food Sci.* 59:1031–1033, 1994.

43. Rodríguez de Sotillo, D., M. Hadley, E.T. Holm. Phenolics in aqueous potato peel extract, extraction, identification and degradation. *J. Food Sci.* 59:649–651, 1994.

44. Sheabar, F.Z., I. Neeman. Separation and concentration of natural antioxidants from the rape of olives. *J. Am. Oil Chem. Soc.* 65:990–993, 1988.

45. Yamaguchi, F., Y. Yoshimura, H. Nakazawa, T. Ariga. Free radical scavenging activity of grape seed extract and antioxidants by electron spin resonance spectrometry in an H_2O_2/NaOH/DMSO system. *J. Agric. Food Chem.* 47:2544–2548, 1999.

46. Koga, T., K. Moro, K. Nakamori, J. Yamakoshi, H. Hosoyama, S. Kataoka, T. Ariga. Increase of antioxidative potential of rat plasma by oral administration of proanthocyanidin-rich extract from grape seeds. *J. Agric. Food Chem.* 47:1892–1897, 1999.

47. Lu, Y., L.Y. Foo. The polyphenol constituents of grape pomace. *Food Chem.* 65:1–8, 1999.

48. Lu, Y., L.Y. Foo. Antioxidant and radical scavenging activities of polyphenols from apple pomace. *Food Chem.* 68:81–85, 2000.

49. Peleg, H., M. Naim, R.L. Rouseff, U. Zehavi. Distribution of bound and free phenolic acids in oranges (*Citrus sinensis*) and grapefruits (*Citrus paradisi*). *J. Sci. Food Agric.* 57:417–426, 1991.

50. Meyer, A.S., S.M. Jepsen, N.S. Sørensen. Enzymatic release of antioxidants for human low-density lipoprotein from grape pomace. *J. Agric. Food Chem.* 46:2439–2446, 1998.

51. Bocco, A., M.E. Cuvelier, H. Richard, C. Berset. Antioxidant activity and phenolic composition of citrus peel and seed extracts. *J. Agric. Food Chem.* 46:2123–2129, 1998.

52. Zheng, Z., K. Shetty. Solid-state bioconversion of phenolics from cranberry pomace and role of *Lentinus edodes* beta-glucosidase. *J. Agric. Food Chem.* 48:895–900, 2000.

53. Vattem, D.A., K. Shetty. Solid-state production of phenolic antioxidants from cranberry pomace by *Rhizopus oligosporus*. *Food Biotechnol.* 16(3):189–210, 2002.

54. Vattem, D.A., K. Shetty. Ellagic acid production and phenolic antioxidant activity in cranberry pomace (*Vaccinium macrocarpon*) mediated by *Lentinus edodes* using solid-state system. *Proc. Biochem.* 39(3):367–379, 2003.

55. Liochev S.I., I. Fridovich. The relative importance of HO* and ONOO- in mediating the toxicity of O*-. *Free Radic. Biol. Med.* 26(5,6):777–778, 1999.

56. Beckman J.S., W.H. Koppenol. Nitric oxide, superoxide, and peroxynitrite: the good, the bad, and ugly. *Am. J. Physiol.* 271(5,1):C1424–C1437, 1996.

57. Radi, R., A. Cassina, R. Hodara, C. Quijano, L. Castro. Peroxynitrite reactions and formation in mitochondria. *Free Radic. Biol. Med.* 33(11):1451–1464, 2002.

58. Barber, D.A., S.R. Harris. Oxygen free radicals and antioxidants: a review. *Am. Pharm.* NS34:26–35, 1994.

59. Betteridge, D.J. What is oxidative stress? *Metabolism* 49:3–8, 2000.

60. Vignais, P.V. The superoxide-generating NADPH oxidase: structural aspects and activation mechanism. *Cell Mol. Life Sci.* 59(9):1428–1459, 2002.

61. Babior, B.M., J.D. Lambeth, W. Nauseef. The neutrophil NADPH oxidase. *Arch. Biochem. Biophys.* 397(2):342–344, 2002.

62. Babior, B.M. The leukocyte NADPH oxidase. *Isr. Med. Assoc. J.* 4(11):1023–1024, 2002.

63. Parke, A., D.V. Parke. The pathogenesis of inflammatory disease: surgical shock and multiple system organ failure. *Inflammopharmacology* 3:149–168, 1995.

64. Parke, D.V. The cytochromes P450 and mechanisms of chemical carcinogenesis. *Environ. Health Perspect.* 102:852–853, 1994.

65. Parke, D.V., A. Sapota. Chemical toxicity and reactive oxygen species. *Int. J. Occ. Med. Environ. Health* 9:331–340, 1996.

66. Triggle, C.R., M. Hollenberg, T.J. Anderson, H. Ding, Y. Jiang, L. Ceroni, W.B. Wiehler, E.S. Ng, A. Ellis, K. Andrews, J.J. McGuire, M. Pannirselvam. The endothelium in health and disease: a target for therapeutic intervention. *J. Smooth Muscle Res.* 39(6):249–267, 2003.

67. Droge, W. Free radicals in the physiological control of cell function. *Physiol Rev.* 82(1):47–95, 2002.

68. Niess, A.M., H.H. Dickhuth, H. Northoff, E. Fehrenbach. Free radicals and oxidative stress in exercise: immunological aspects. *Exerc. Immunol. Rev.* 5:22–56, 1999.

69. Sculley, D.V., S.C. Langley-Evans. Salivary antioxidants and periodontal disease status. *Proc. Nutr. Soc.* 61(1):137–143, 2002.

70. Morel, Y., R. Barouki. Repression of gene expression by oxidative stress. *Biochem. J.* 342(3):481–496, 1999.

71. Lusini, L., S.A. Tripodi, R. Rossi, F. Giannerini, D. Giustarini, M.T. del Vecchio, G. Barbanti, M. Cintorino, P. Tosi, P. Di Simplicio. Altered glutathione anti-oxidant metabolism during tumor progression in human renal-cell carcinoma. *Int. J. Cancer 91* (1):55–59, 2001.

72. Mates, J.M., F. Sanchez-Jimenez. Antioxidant enzymes and their implications in pathophysiologic processes. *Front. Biosci.* 4:D339–345, 1999.

73. Mates, J.M., C. Perez-Gomez, I. Nunez de Castro. Antioxidant enzymes and human diseases. *Clin. Biochem.* 32(8):595–603, 1999.

74. Nordberg, J., E.S. Arner. Reactive oxygen species, antioxidants, and the mammalian thioredoxin system. *Free Radic. Biol. Med.* 31(11):1287–1312, 2001.

75. Stadtman, E.R., R.L. Levine. Protein oxidation. *Ann. NY Acad. Sci.* 899:191–208, 2000.

76. Rubbo, H., R. Radi, M. Trujillo, R. Telleri, B. Kalyanaraman, S. Barnes, M. Kirk, B.A. Freeman. Nitric oxide regulation of superoxide and peroxynitrite-dependent lipid peroxidation: formation of novel nitrogen-containing oxidized lipid derivatives. *J. Biol. Chem.* 269(42):26066–26075, 1994.

77. Richter, C., J.W. Park, B.N. Ames. Normal oxidative damage to mitochondrial and nuclear DNA is extensive. *Proc. Natl. Acad. Sci. USA* 85(17):6465–6467, 1988.

78. LeDoux, S.P., W.J. Driggers, B.S. Hollensworth, G.L. Wilson. Repair of alkylation and oxidative damage in mitochondrial DNA. *Mutat. Res.* 434(3):149–159, 1999.

79. Parke, A.L., C. Ioannides, D.F.V. Lewis, D.V. Parke. Molecular pathology of drugs: disease interaction in chronic autoimmune inflammatory diseases. *Inflammopharmacology* 1:3–36, 1991.

80. Schwarz, K.B. Oxidative stress during viral infection: a review. *Free Radic. Biol. Med.* 21(5):641–649, 1996.

81. Götz, J., C.I. va Kan, H.W. Verspaget, I. Biemond, C.B. Lamers, R.A. Veenendaal. Gastric mucosal superoxide dismutases in *Helicobacter pylori* infection. *Gut* 38:502–506, 1996.

82. Jakus, V. The role of free radicals, oxidative stress and antioxidant systems in diabetic vascular disease. *Bratisl. Lek. Listy.* 101(10):541–551, 2000.

83. Offen, D., P.M. Beart, N.S. Cheung, C.J. Pascoe, A. Hochman, S. Gorodin, E. Melamed, R. Bernard, O. Bernard. Transgenic mice expressing human Bcl-2 in their neurons are resistant to 6-hydroxydopamine and 1-methyl-4-phenyl-1,2,3,6- tetrahydropyridine neurotoxicity. *Proc. Natl. Acad. Sci.* 95:5789–5794, 1998.

84. Barbaste, M., B. Berke, M. Dumas, S. Soulet, J.C. Delaunay, C. Castagnino, V. Arnaudinaud, C. Cheze, J. Vercauteren. Dietary antioxidants, peroxidation and cardiovascular risks. *J. Nutr. Health Aging* 6(3):209–223, 2002.

85. Gerber, M., C. Astre, C. Segala, M. Saintot, J. Scali, J. Simony-Lafontaine, J. Grenier, H. Pujol. Tumor progression and oxidant-antioxidant status. *Cancer Lett.* 19:114(1,2):211–214, 1997.

86. Block, G., B. Patterson, A. Subar. Fruit, vegetables, and cancer prevention: a review of the epidemiological evidence. *Nutr. Cancer* 18:1–29, 1992.

87. Serdula, M.K., M.A.H. Byers, E. Simoes, J.M. Mendlein, R.J. Coates. The association between fruit and vegetable intake and chronic disease risk factors. *Epidemiology* 7:161–165, 1996.

88. Tapiero, H., K.D. Tew, G.N. Ba, G. Mathe. Polyphenols: do they play a role in the prevention of human pathologies? *Biomed. Pharmacother.* 56(4):200–207, 2002.

89. Duthie, G.G., P.T. Gardner, J.A. Kyle. Plant polyphenols: are they the new magic bullet? *Proc. Nutr. Soc.* 62(3):599–603, 2003.

90. Lundberg, A.C., A. Åkesson, B. Åkesson. Dietary intake and nutritional status in patients with systemic sclerosis. *Ann. Rheum. Dis.* 51:1143–1148, 1992.

91. Yoshioka, A., Y. Miyachi, S. Imamura, Y. Niwa. Anti-oxidant effects of retinoids on inflammatory skin diseases. *Arch. Dermatol. Res.* 278:177–183, 1986.

92. Wilks, R., F. Bennett, T. Forrester, N. Mcfarlane-Anderson. Chronic diseases: the new epidemic. *West Ind. Med. J.* 47(4):40–44, 1998.

93. Leighton, F., A. Cuevas, V. Guasch, D.D. Perez, P. Strobel, A. San Martin, U. Urzua, M.S. Diez, R. Foncea, O. Castillo, C. Mizon, M.A. Espinoza, I. Urquiaga, J. Rozowski, A. Maiz, A. Germain. Plasma polyphenols and antioxidants, oxidative DNA damage and endothelial function in a diet and wine intervention study in humans. *Drugs Exp. Clin. Res.* 25(2,3):133–141, 1999.

94. Yamada, J., Y. Tomita. Antimutagenic activity of caffeic acid and related compounds. *Biosci. Biotechnol. Biochem.* 60(2):328–329, 1996.

95. Mitscher, L.A., H. Telikepalli, E. McGhee, D.M. Shankel. Natural antimutagenic agents. *Mutat. Res.* 350:143–152, 1996.

96. Uenobe, F., S. Nakamura, M. Miyazawa. Antimutagenic effect of resveratrol against Trp-P-1. *Mutat. Res.* 373:197–200, 1997.

97. Kuroda, Y., T. Inoue. Antimutagenesis by factors affecting DNA repair in bacteria. *Mutat. Res.* 202(2):387–391, 1988.

98. Kanai, S., H. Okano. Mechanism of protective effects of sumac gall extract and gallic acid on progression of CC14-induced acute liver injury in rats. *Am. J. Chin. Med.* 26:333–341, 1998.

99. Halliwell, B. Establishing the significance and optimal intake of dietary antioxidants: the biomarker concept. *Nutr. Rev.* 57:104–113, 1999.

100. Kaul, A., K.I. Khanduja. Polyphenols inhibit promotional phase of tumorigenesis: relevance of superoxide radicals. *Nurt. Cancer* 32:81–85, 1998.

101. Clifford, A.J., S.E. Ebeler, J.D. Ebeler, N.D. Bills, S.H. Hinrichs, P.-L. Teissedre, A.L. Waterhouse. Delayed tumor onset in transgenic mice fed an amino acid-based diet supplemented with red wine solids. *Am. J. Clin. Nutr.* 64:748–756, 1996.

102. Jang, M., L. Cai, G.O. Udeani, K.V. Slowing, C. Thomas, C.W.W. Beecher, H.H.S. Fong, N.R. Farnsworth, A.D. Kinghorn, R.G. Mehta, R.C. Moon, J.M. Pezzuto. Cancer chemopreventive activity of resveratrol, a natural product derived from grapes. *Science* 275:218–220, 1997.

103. Ferrieres, J. The French paradox: lessons for other countries. *Heart* 90(1):107–111, 2004.

104. Frankel, E.N, J. Kanner, J.B. German, E. Parks, J.E. Kinsella. Inhibition of oxidation of human low-density lipoprotein by phenolic substances in red wine. *Lancet* 341:454–457, 1993.

105. Gerritsen, M.E., W.W. Carley, G.E. Ranges, C.-P. Shen, S.A. Phan, G.F. Ligon, C.A. Perry. Flavonoids inhibit cytokine-induced endothelial cell adhesion protein gene expression. *Am. J. Pathol.* 147:278–292, 1995.

106. Muldoon, M.F., S.B. Kritchevsky. Flavonoids and heart disease. *BMJ* 312(7029):458–459, 1996.

107. McCue, P.P., K. Shetty. Inhibitory effects of rosmarinic acid extracts on porcine pancreatic amylase *in vitro*. *Asia Pac. J. Clin. Nutr.* 13(1):101–106, 2004.

108. Andlauer, W., P. Furst. Special characteristics of non-nutrient food constituents of plants: phytochemicals: introductory lecture. *Int. J. Vitam. Nutr. Res.* 73(2):55–62, 2003.

109. Chung, K.T., T.Y. Wong, C.I. Wei, Y.W. Huang, Y. Lin. Tannins and human health: a review. *Crit. Rev. Food Sci. Nutr.* 38:421–464, 1998.

110. Ikken, Y., P. Morales, A. Martínez, M.L. Marín, A.I. Haza, M.I. Cambero. Antimutagenic effect of fruit and vegetable ethanolic extracts against N-nitrosamines evaluated by the Ames test. *J. Agric. Food Chem.* 47:3257–3264, 1999.

111. Vilegas, W., M. Sanomimiya, L. Rastrelli, C. Pizza. Isolation and structure elucidation of two new flavonoid glycosides from the infusion of *Maytenus aquifolium* leaves: evaluation of the antiulcer activity of the infusion. *J. Agric. Food Chem.* 47:403–406, 1999.

112. Abram, V., M. Donko. Tentative identification of polyphenols in *Sempervivum tectorum* and assessment of the antimicrobial activity of *Sempervivum* L. *J. Agric. Food Chem.* 47:485–489, 1999.

113. Noguchi, Y., K. Fukuda, A. Matsushima, D. Haishi, M. Hiroto, Y. Kodera, H. Nishimura, Y. Inada. Inhibition of Df-protease associated with allergic diseases by polyphenol. *J. Agric. Food Chem.* 47:2969–2972, 1999.

114. Ma, Q., K. Kinneer. Chemoprotection by phenolic antioxidants. Inhibition of tumor necrosis factor alpha induction in macrophages. *J. Biol. Chem.* 277(4):2477–2484, 2002.

115. Mowrey, D.B. *The scientific validation of herbal medicine.* New York: McGraw-Hill, 1990, pp 255–264.

116. Yan, X., B.T. Murphy, G.B. Hammond, J.A. Vinson, C. Neto. Antioxidant activities and antitumor screening of extracts from cranberry fruit (*Vaccinium macrocarpon*) *J. Agric. Food Chem.* 50:5844–5849, 2002.

117. Griffiths, P. The role of cranberry juice in the treatment of urinary tract infections. *Br. J. Comm. Nurs.* 8(12):557–561, 2003.

118. Vvedenskaya, I.O., R.T. Rosen, J.E. Guido, D.J. Russell, K.A. Mills, N. Vorsa. Characterization of flavonols in cranberry (*Vaccinium macrocarpon*) powder. *J. Agric. Food Chem.* 52(2):188-95, 2004.

119. Zafriri, D., I. Ofek, R. Adar, M. Pocino, N. Sharon. Inhibitory activity of cranberry juice on adherence of type 1 and type P fimbriated *Escherichia coli* to eukaryotic cells. *Antimicrob. Agents Chemother.* 33(1):92–98, 1989.

120. Sharon N, I. Ofek. Fighting infectious diseases with inhibitors of microbial adhesion to host tissues. *Crit. Rev. Food Sci. Nutr.* 42(3):267–272, 2002.

121. Burger, O., I. Ofek, M. Tabak, E.I. Weiss, N. Sharon, I. Neeman. A high molecular constituent of cranberry juice inhibits *Helicobactor pylori* adhesion to human gastric mucus. *FEMS Immunol. Med. Microbiol.* 29:295–301, 2000.

122. Burger, O., E. Weiss, N. Sharon, M. Tabak, I. Neeman, I. Ofek. Inhibition of *Helicobacter pylori* adhesion to human gastric mucus by a high-molecular-weight constituent of cranberry juice. *Crit. Rev. Food Sci. Nutr.* 42(3):279–284, 2002.

123. Weiss, E.I., R. Lev-Dor, Y. Kashamn, J. Goldhar, N. Sharon, I. Ofek. Inhibiting interspecies coaggregation of plaque bacteria with a cranberry juice constituent. *J. Am. Dent. Assoc.* 129(12):1719–1723, 1998.

124. Weiss, E.L., R. Lev-Dor, N. Sharon, I. Ofek. Inhibitory effect of a high-molecular-weight constituent of cranberry on adhesion of oral bacteria. *Crit. Rev. Food Sci Nutr.* 42(3):285–292, 2002.

125. Morris, N.S., D.J. Stickler. Does drinking cranberry juice produce urine inhibitory to the development of crystalline, catheter-blocking *Proteus mirabilis* biofilms? *BJU Int.* 88:192–197, 2001.

126. Swartz, J.H., T.F. Medrek. Antifungal properties of cranberry juice. *Appl. Microbiol.* 16(10):1524–1527, 1968.

127. Cavanagh, H.M., M. Hipwell, J.M. Wilkinson. Antibacterial activity of berry fruits used for culinary purposes. *Med. Food* 6(1):57–61, 2003.

128. Wilson, T., J.P. Porcari, D. Harbin. Cranberry extract inhibits low-density lipoprotein oxidation. *Life Sci.* 62:381–386, 1998.

129. Bagchi, D., C.K. Sen, M. Bagchi, M. Atalay. Anti-angiogenic, antioxidant, and anti-carcinogenic properties of a novel anthocyanin-rich berry extract formula. *Biochemistry* 69(1):75–80, 2004.

130. Reed, J.D., Krueger, C.G., Porter, M.L. Cranberry juice powder decreases low density lipoprotein cholesterol in hypercholesterolemic swine. *FASEB J.* 15(4,5):54, 2001.

131. Reed, J. Cranberry flavonoids, atherosclerosis and cardiovascular health. *Crit. Rev. Food Sci. Nutr.* 42:301–316, 2002.

132. Bomser, J., D.L. Madhavi, K. Singletary, M.A. Smith. *In vitro* anticancer activity of fruit extracts from *Vaccinium* species. *Planta Med.* 62(3):212–216, 1996.

133. Krueger, C.G., M.L. Porter, D.A. Weibe, D.G. Cunningham, J.D. Reed. Potential of cranberry flavonoids in the prevention of copper-induced LDL oxidation. *Polyphenols Communications.* 2:447–448, 2000.

134. Vattem, D.A., Y.-T. Lin, R.G. Labbe, K. Shetty. Phenolic antioxidant mobilization in cranberry pomace by solid-state bioprocessing using food grade fungus *Lentinus edodes* and effect on antimicrobial activity against select food borne pathogens. *Innovative Food Sci. Emerg. Technol.* 5(1):81–91, 2003.

135. Daniel, E.M., A.S. Krupnick, Y.H. Heur, J.A. Blinzler, R.W. Mims, G.D. Stoner. Extraction, stability and quantitation of ellagic acid in various fruits and nuts. *J. Food Comp. Anal.* 2:385–398, 1989.

136. Goldberg, D.M., B. Hoffman, J. Yang, G.J. Soleas. Phenolic constituents, furans and total antioxidant status of distilled spirits. *J. Agric. Food. Chem.* 47:3978–3985, 1999.

137. Loarca-Pina, G., P.A. Kuzmicky, E.G. de Mejia, N.Y. Kado. Inhibitory effects of ellagic acid on the direct-acting mutagenicity of aflatoxin B1 in the *Salmonella* microsuspension assay. *Mutat. Res.* 398(1,2):183–187, 1998.

138. Kaur, S., I.S. Grover, S. Kumar. Antimutagenic potential of ellagic acid isolated from *Terminalia arjuna. Ind. J. Exp. Biol.* 35:478–482, 1997.

139. Khanduja, K.L., R.K. Gandhi, V. Pathania, N. Syal. Prevention of N-nitrosodiethylamine-induced lung tumorigenesis by ellagic acid and quercetin in mice. *Food Chem. Toxicol.* 37(4):313–318, 1999.

140. Wood, A.W., M.T. Huang, R.L. Chang, H.L. Newmark, R.E. Lehr, H. Yagi, J.M. Sayer, D.M. Jerina, A.H. Conney. Inhibition of the mutagenicity of bay-region diol epoxides of polycyclic aromatic hydrocarbons by naturally occurring plant phenols: exceptional activity of ellagic acid. *Proc. Natl. Acad. Sci. USA* 79(18):5513–5517, 1982.

141. Zhang, Z., S.M. Hamilton, C. Stewart, A. Strother, R.W. Teel. Inhibition of liver microsomal cytochrome P450 activity and metabolism of the tobacco-specific nitrosamine NNK by capsaicin and ellagic acid. *Anticancer Res.* 13(6A):2341–2346, 1993.

142. Teel, R.W., M.S. Babcock, R. Dixit, G.D. Stoner. Ellagic acid toxicity and interaction with benzo[a]pyrene and benzo[a]pyrene 7,8-dihydrodiol in human bronchial epithelial cells. *Cell Biol. Toxicol.* 2(1):53–62, 1986.

143. Soni, K.B., M. Lahiri, P. Chackradeo, S.V. Bhide, R. Kuttan. Protective effect of food additives on aflatoxin-induced mutagenicity and hepatocarcinogenicity. *Cancer Lett.* 115(2):129–133, 1997.

144. Singh, K., A.K. Khanna, R. Chander. Hepatoprotective effect of ellagic acid against carbon tetrachloride induced hepatotoxicity in rats. *Ind. J. Exp. Biol.* 37:1025–1026, 1999.

145. Mandal S., N.M. Shivapurkar, A.J. Galati, G.D. Stoner. Inhibition of N-nitrosobenzyl-methylamine metabolism and DNA binding in cultured rat esophagus by ellagic acid. *Carcinogenesis* 9(7):1313–1316, 1988.

146. Mandal, S., G.D. Stoner. Inhibition of N-nitrosobenzyl-methylamine-induced esophageal tumorigenesis in rats by ellagic acid. *Carcinogenesis* 11:55–61, 1990.

147. Smith, W.A., J.W. Freeman, R.C. Gupta. Modulation of dibenzo[a,l]pyrene-DNA adduction by chemopreventive agents in the human breast epithelial cell line MCF-7 (meeting abstract). *Proc. Annu. Meeting Am. Assoc. Cancer Res.* 38:A2422, 1997.

148. Smith, W.A., U. Devanaboyina, R.C. Gupta. Use of a microsomal activation system as a potential screening method for cancer chemopreventive agents (meeting abstract). *Proc. Annu. Meeting. Am. Assoc. Cancer Res.* 36:A3555, 1995.

149. Thresiamma, K.C., J. George, R. Kuttan. Protective effect of curcumin, ellagic acid and bixin on radiation induced genotoxicity. *J. Exp. Clin. Cancer Res.* 17(4):431–434, 1998.

150. Narayanan, B.A, O. Geoffroy, D.W. Nixon. P53/p21 (WAF1/CIP1) expression and its possible role in G1 arrest and apoptosis in ellagic acid treated cancer cells. *Cancer Lett.* 136(2):215–221, 1999.

151. Hassoun, E.A., A.C. Walter, N.Z. Alsharif, S.J. Stohs. Modulation of TCDD-induced fetotoxicity and oxidative stress in embryonic and placental tissues of C57BL/6J mice by vitamin E succinate and ellagic acid. *Toxicology* 124(1):27–37, 1997.

152. Cozzi, R., R. Ricordy, F. Bartolini, L. Ramadori, P. Perticone, R. De Salvia. Taurine and ellagic acid: two differently acting natural antioxidants. *Environ. Mol. Mutag.* 26:248–254, 1995.

153. Chen, C., R. Yu, E.D. Owuor, A.N. Kong. Activation of antioxidant-response element (ARE), mitogen-activated protein kinases (MAPKs) and caspases by major green tea polyphenol components during cell survival and death. *Arch. Pharm. Res.* 23(6):605-612, 2000.

154. Barch, D.H., L.M. Rundhaugen. Ellagic acid induces NAD(P)H:quinone reductase through activation of the antioxidant responsive element of the rat NAD(P)H:quinone reductase gene. *Carcinogenesis* 15(9):2065–2068, 1994.

155. Barch, D.H., L.M. Rundhaugen, N.S. Pillay. Ellagic acid induces transcription of the rat glutathione S-transferase-Ya gene. *Carcinogenesis* 16(3):665–668, 1995.

156. Constantinou, A., G.D. Stoner, R. Mehta, K. Rao, C. Runyan, R. Moon. The dietary anti-cancer agent ellagic acid is a potent inhibitor of DNA topoisomerases *in vitro. Nutr. Cancer* 23(2):121–130, 1995.

157. Szaefer, H., J. Jodynis-Liebert, M. Cichocki, A. Matuszewska, W. Baer-Dubowska. Effect of naturally occurring plant phenolics on the induction of drug metabolizing enzymes by o-toluidine. *Toxicology* 186(1,2):67–77, 2003.

158. Jimenez-Lopez, J.M., A.I. Cederbaum. Green tea polyphenol epigallocatechin-3-gallate protects HepG2 cells against CYP2E1-dependent toxicity. *Free Radic. Biol. Med.* 36(3):359–370, 2004.

159. Shetty, K., P. McCue. Phenolic antioxidant biosynthesis in plants for functional food application: integration of systems biology and biotechnological approaches. *Food Biotechnol.* 17:67–97, 2003.

160. Shetty, K., M.L. Wahlqvist. A model for the role of proline-linked pentose phosphate pathway in phenolic phytochemical biosynthesis and mechanism of action for human health and environmental applications. *Asia Pac. J. Clin. Nutr.* 13(1):1–24, 2004.

161. Fabregat, I., J. Vitorica, J. Satrustegui, A. Machado. The pentose phosphate cycle is regulated by NADPH/NADP ratio in rat liver. *Arch. Biochem. Biophys.* 236(1):110–118, 1985.

162. Pfeifer, R., G. Karl, R. Scholz. Does the pentose cycle play a major role for NADPH supply in the heart? *Biol. Chem. Hoppe Seyler* 367(10):1061–1068, 1986.

163. Cabezas, H., R.R. Raposo, E. Melendez-Hevia. Activity and metabolic roles of the pentose phosphate cycle in several rat tissues. *Mol. Cell Biochem.* 201(1,2):57–63, 1999.

164. Owuor, E.D., A.N. Kong. Antioxidants and oxidants regulated signal transduction pathways. *Biochem. Pharmacol.* 64(5,6):765–770, 2002.

165. Chen, C., G. Shen, V. Hebbar, R. Hu, E.D. Owuor, A.N. Kong. Epigallocatechin-3-gallate-induced stress signals in HT-29 human colon adenocarcinoma cells. *Carcinogenesis* 24(8):1369–1378, 2003.

166. Hider, R.C., Z.D. Liu, H.H. Khodr. Metal chelation of polyphenols. *Methods Enzymol.* 335:190–203, 2001.

167. Yang, C.S., J.M. Landau, M.T. Huang, H.L. Newmark. Inhibition of carcinogenesis by dietary polyphenolic compounds. *Annu. Rev. Nutr.* 21:381–406, 2001.

168. Nicholson, C. Modulation of extracellular calcium and its functional implications. *Fed. Proc.* 39(5):1519–1523, 1980.

169. Brown, E.M. Physiology and pathophysiology of the extracellular calcium-sensing receptor. *Am. J. Med.* 106(2):238–253, 1999.

170. Rizzuto, R., P. Pinton, D. Ferrari, M. Chami, G. Szabadkai, P.J. Magalhaes, F. Di Virgilio, T. Pozzan. Calcium and apoptosis: facts and hypotheses. *Oncogene* 22(53):8619–8627, 2003.

171. Stout, C., A. Charles. Modulation of intercellular calcium signaling in astrocytes by extracellular calcium and magnesium. *Glia* 43(3):265–273, 2003.

172. Cohen, J.E., R.D. Fields. Extracellular calcium depletion in synaptic transmission. *Neuroscientist* 10(1):12–17, 2004.

173. Bellomo, G., H. Thor, S. Orrenius. Increase in cytosolic Ca^{2+} concentration during t-butyl hydroperoxide metabolism by isolated hepatocytes involves NADPH oxidation and mobilization of intracellular Ca^{2+} stores. *FEBS Lett.* 168(1):38–42, 1984.

174. Hagerman, A.E., L.G. Butler. The specificity of proanthocyanidin-protein interactions. *J. Biol. Chem.* 256(9):4494–4497, 1981.

175. Pan, C.Y., Y.H. Kao, A.P. Fox. Enhancement of inward $Ca(2^+)$ currents in bovine chromaffin cells by green tea polyphenol extracts. *Neurochem. Int.* 40(2):131–137, 2002.

176. Papadopoulou, A., R.A. Frazier. Characterization of protein–polyphenol interactions. *Trends Food Sci. Technol.* 5:(3,4)186–190, 2003.

177. Kim, H.J., K.S. Yum, J.H. Sung, D.J. Rhie, M.J. Kim, S. Min do, S.J. Hahn, M.S. Kim, Y.H. Jo, S.H. Yoon. Epigallocatechin-3-gallate increases intracellular $[Ca^+]$ in U87 cells mainly by influx of extracellular Ca^{2+} and partly by release of intracellular stores. *Naunyn. Schmiedebergs Arch. Pharmacol.* 369(2):260–267, 2004.

178. Choi, S.H., M.B. Gu. Phenolic toxicity: detection and classification through the use of a recombinant bioluminescent *Escherichia coli*. *Environ. Toxicol. Chem.* 20(2):248–255, 2001.

179. Tsuchiya, H. Biphasic membrane effects of capsaicin, an active component in *Capsicum* species. *J. Ethnopharmacol.* 75(2,3):295–299, 2001.

180. Tsuchiya H., M. Sato, Y. Kameyama, N. Takagi, I. Namikawa. Effect of lidocaine on phospholipid and fatty acid composition of bacterial membranes. *Lett. Appl. Microbiol.* 4(6):141–144, 1987.

181. Shetty, K. Role of proline-linked pentose phosphate pathway in biosynthesis of plant phenolics for functional food and environmental applications: a review. *Process. Biochem.* 39:789–803, 2004.

182. Mazumder, A., N. Neamati, S. Sunder, J. Schulz, H. Pertz, E. Eich, Y. Pommier. Curcumin analogs with altered potencies against HIV-1 integrase as probes for biochemical mechanisms of drug action. *J. Med. Chem.* 40 (19):3057–3063, 1997.

183. Durant, S., P. Karran. Vanillins: a novel family of DNA-PK inhibitors. *Nucleic Acids Res.* 31(19):5501–5512, 2003.

184. Wu, G. Intestinal mucosal amino acid catabolism. *J. Nutr.* 128(8):1249–1252, 1998.

185. Brosnan, J.T. Glutamate, at the interface between amino acid and carbohydrate metabolism. *J. Nutr.* 130(4S):988S–990S, 2000.

186. Newsholme, P., J. Procopio, M.M. Lima, T.C. Pithon-Curi, R. Curi. Glutamine and glutamate: their central role in cell metabolism and function. *Cell Biochem. Funct.* 21(1):1–9, 2003.

187. Sarkela, T.M., J. Berthiaume, S. Elfering, A.A. Gybina, C. Giulivi. The modulation of oxygen radical production by nitric oxide in mitochondria. *J. Biol. Chem.* 276(10):6945–6949, 2001.

188. Cadenas, E. Mitochondrial free radical production and cell signaling. *Mol. Aspects Med.* 25(1,2):17–26, 2004.

189. Phang, J.M. The regulatory functions of proline and pyrroline-5-carboxylic acid. *Curr. Topics Cell Reg.* 25:91–132, 1985.

190. Hagedorn, C.H., J.M. Phang. Transfer of reducing equivalents into mitochondria by the interconversions of proline and α-pyrroline-5-carboxylate. *Arch. Biochem. Biophys.* 225:95–101, 1983.

191. Shetty, K., R.L. Labbe. Food-borne pathogens, health and role dietary phytochemicals. *Asia Pac. J. Clin. Nutr.* 7:270–276, 1998.

192. Acamovic, T., C.S. Stewart. Plant phenolic compounds and gastrointestinal microorganisms. *AICR Proc.* 137–139, 1992.

193. Aidoo, K.E., R. Hendry, B.J.B. Wood. Solid state fermentation. *Ad. Appl. Microbiol.* 28:201–237, 1982.

194. McDonald, M., I. Mila, A. Scalbert. Precipitation of metal ions by plant polyphenols: optimal conditions and origin of precipitation. *J. Agric. Food Chem.* 44:599–606, 1996.

195. Kainja, C., L. Bates, T. Acamovic. The chelation of trace elements by tannins. In: *Toxic Plants and Other Natural Toxicants*, Garland, T., A.C. Barr, eds., Wallingford: CAB Intl., 1998, pp 111–114.

196. Muhammed, S.A. Anti-nutrient effects of plant polyphenolic compounds. PhD Thesis, University of Aberdeen, 1999.

197. McClure, J.W. Physiology and functions of flavonoids. In: *The Flavonoids*, Harborne, J.B., ed., New York: Academic Press, 1975, pp 45–77.

198. Wollenweber, E., V.H. Dietz. Occurrence and distribution of free flavonoid aglycones in plants. *Phytochemistry* 20:869–932, 1981.

199. Kähkönen, M.P., A.I. Hopia, M. Heinonen. Berry phenolics and their antioxidant activity. *J. Agric. Food Chem.* 49:4076–4082, 2001.

200. Onyeneho, S.N., N.S. Hettiarachchy. Antioxidant activity, fatty acids and phenolic acids compositions of potato peels. *J. Sci. Food Agric.* 62:345–350, 1993.

201. Torres, A.M., T. Mau-Lastovicka, R. Rezaaiyan. Total phenolics and high-performance liquid chromatography of phenolic acids of avocado. *J. Agric. Food Chem.* 35:921–925, 1987.

202. Gil, M.I., D.M. Holcroft, A.A. Kader. Changes in strawberry anthocyanins and other polyphenols in response to carbon dioxide treatment. *J. Agric. Food Chem.* 45:1662–1667, 1997.

203. Raynal, J., M. Moutounet, J.M. Souquet. Intervention of phenolic compounds in plum technology, 1: changes during drying. *J. Agric. Food Chem.* 37:1046–1050, 1989.

204. Spanos, G.A., R.E. Wrolstad. Phenolics of ale, pear, and white grape juices and their changes with processing and storage: a review. *J. Agric. Food Chem.* 40:1478–1487, 1992.

205. Kris-Etherton, P.M., K.D. Hecker, A. Bonanome, S.M. Coval, A.E. Binkoski, K.F. Hilpert, A.E. Griel, T.D. Etherton. Bioactive compounds in foods: their role in the prevention of cardiovascular disease and cancer. *Am. J. Med.* 30(113)9B:71S–88S, 2002.

206. De Lumen, B.O., A.F. Galvez, M.J. Revilleza, D.C. Krenz. Molecular strategies to improve the nutritional quality of legume proteins. *Adv. Exp. Med. Biol.* 464:117–126, 1999.

207. Tabe, L.M., T. Wardley-Richardson, A. Ceriotti, A. Aryan, W. McNabb, A. Moore. Exceptional activity of ellagic acid. *Proc. Natl. Acad. Sci. USA* 79:5513–5517, 1982.

208. Viera Diaz, J. Genetic improvement of legumes. *Arch. Latinoam. Nutr.* 44(4,1):41S–43S, 1996.

209. Brar, D.S., T. Ohtani, H. Uchimiya. Genetically engineered plants for quality improvement. *Biotechnol. Genet. Eng. Rev.* 13:167–179, 1996.

210. Auger, C., B. Caporiccio, N. Landrault, P.L. Teissedre, C. Laurent, G. Cros, P. Besancon, J.M. Rouanet. Red wine phenolic compounds reduce plasma lipids and apolipoprotein B and prevent early aortic atherosclerosis in hypercholesterolemic golden Syrian hamsters (*Mesocricetus auratus*). *J. Nutr.* 132(6):1207–1213, 2002.

211. Mattivi, F., C. Zulian, G. Nicolini, L. Valenti. Wine, biodiversity, technology, and antioxidants. *Ann. NY Acad. Sci.* 957:37–56, 2002.

212. Waterhouse, A.L. Wine phenolics. *Ann. NY Acad. Sci.* 957:21–36, 2002.

213. Raimbault, M. General and microbiological aspects of solid substrate fermentation. *Electron. J. Biotechnol.* 1(3):1–15, 1998.

214. Pandey, R. Recent progress developments in solid-state fermentation. *Process Biochem.* 27:109–117, 1992.

215. Shekib, L.A. Nutritional improvement of lentils, chick pea, rice and wheat by natural fermentation. *Plant Foods Hum. Nutr.* 46(3):201–205, 1994.

216. Hadajini, S. Indegenous mucana tempe as functional food. *Asia Pac. J. Lin. Nutr.* 10(3):222–225, 2001.

217. Wang, S.Y., A.W. Stretch. Antioxidant capacity in cranberry is influenced by cultivar and storage temperature. *J. Agric. Food Chem.* 40:969–974, 2001.

218. Zheng, G.L., Y.G. Zhou, W.G. Gong. Isolation of soybean isoflavones from tofu wastewater. *ACTA Academiae Medicinae Zhejang* 8:23–25, 1997.

219. Morrissey, P., A.E. Osbourn. Fungal resistance to plant antibiotics as a mechanism of pathogenesis. *Microbiol. Mol. Biol. Rev.* 62(3):708–724, 1999.

220. Osbourn, A.E. Antimicrobial phytoprotectants and fungal pathogens: a commentary. *Fung. Genet. Biol.* 26:163–168, 1999.

221. Fukai, T., A. Marumo, K. Kaitou, T. Kanda, S. Terada, T. Nomura. Anti-*Helicobacter pylori* flavonoids from licorice extract. *Life Sci.* 71(12):1449–1463, 2002.

222. Elattar, T.M.A., A.A. Virji. The effect of red wine and its components on growth and proliferation of human oral squamous carcinoma cells. *Anticancer Res.* 19:5407–5414, 1999.

223. Carbonneau, M.A., C.I. Léger, B. Descomps, F. Michel, L. Monnier. Improvement in the antioxidant status of plasma and low-density lipoprotein in subjects receiving a red wine phenolics mixture. *J. Am. Oil Chem. Soc.* 75:235–240, 1998.

224. Chan, M.M., J.A. Mattiacci, H.S. Hwang, A. Shah, D. Fong. Synergy between ethanol and grape polyphenols, quercetin, and resveratrol, in the inhibition of the inducible nitric oxide synthase pathway. *Biochem. Pharmacol.* 60(10):1539–1548, 2000.

225. Abraham, S.K. Anti-genotoxic effects in mice after the interaction between coffee and dietary constituents. *Food Chem. Toxicol.* 34(1):15–20, 1996.

226. Vattem, D.A., H.-D. Jang, R. Levin, K. Shetty. Synergism of cranberry phenolics with ellagic acid and rosmarinic acid for antimutagenic and DNA-protection functions. *J. Food Biochem.*, 30:98–116, 2006.

227. Vattem, D.A., Y.-T. Lin, R. Ghaedian, K. Shetty. Cranberry synergies for dietary management of α-amylase activity for type-II diabetes. [Unpublished results]

228. Vattem, D.A., Y.-T. Lin, R. Ghaedian, K. Shetty. Cranberry synergies for dietary management of *Helicobacter pylori* infections. *Process. Biochem.* 40(5): 1583–1592, 2005.

8

Rosmarinic Acid Biosynthesis and Mechanism of Action

Kalidas Shetty

CONTENTS

8.1 Introduction 187
8.2 Rosmarinic Acid Sources and Functional Pharmacological Effects.....................188
8.3 RA Biosynthesis in Cell Cultures 190
     8.3.1 Pathways Associated with Rosmarinic Acid Biosynthesis 190
     8.3.2 RA Biosynthesis and Generation of High RA Clonal Lines....................192
     8.3.3 Role of Proline Linked Pentose–Phosphate Pathway in
           RA Biosynthesis in Clonal Systems...192
8.4 Mechanism of RA Action Through Stimulation of Host Antioxidant Response 194
8.5 Summary 200
References...201

8.1 INTRODUCTION

Rosmarinic acid (RA) (Figure 8.1) is commonly found in substantial amounts in the family Lamiaceae, which has many important species such as oregano (*Origanum vulgare*), thyme (*Thymus vulgaris*), rosemary (*Rosmarinus officinalis*), holy basil (*Ocimum sanctum*), perilla or shiso (*Perilla frutescens*), spearmint (*Mentha spicata*), and several other species that have food and medicinal applications. This chapter highlights various sources of rosmarinic acid and their functional effects, RA biosynthesis in cell cultures and pathways associated with RA biosynthesis. In addition, investigations from my own research group have been summarized, where we have developed methods for RA biosynthesis through generation of high RA clonal lines using tissue cultures for field production and regulation of RA biosynthesis through a proposed critical control point (CCP), proline linked pentose–phosphate pathway. In terms of applications we have developed

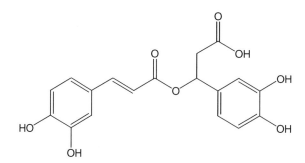

Figure 8.1 Rosmarinic acid.

new applications for RA such as for amylase inhibition, which has potential for hyperglycemia management and antimicrobial effects of high rosmarinic acid-containing herbal
extracts. Further, we have proposed a model of how RA can regulate human health relevant antioxidant response through the CCP, proline linked pentose–phosphate pathway of
human cells.

8.2   ROSMARINIC ACID SOURCES AND FUNCTIONAL
PHARMACOLOGICAL EFFECTS

Rosmarinic acid (RA) is an important caffeoyl ester (phenolic depside) with proven
medicinal properties and well characterized physiological functions. Rosmarinic acid is
found in substantial sources in the family Lamiaceae (1,2). *Salvia lavandulifolia* is used as
choleretic, antiseptic, astringent, and hypoglycemic drug in Southern Europe and contains
high levels of RA (3). Rosmarinic acid-containing *Ocimum sanctum* (holy basil) is used to
reduce fevers and against gastrointestinal disorders in India and has antioxidant properties
(2,4). In Mexico, high RA-containing *Hyptis verticillata* is widely used by Mixtec Indians
against gastrointestinal disorders and skin infections (5). In Indonesia and several countries in Southeast Asia, RA-containing *Orthosiphon aristatus* is known for diuretic properties and is also used against bacterial infections of the urinary system (6). *Salvia cavaleriei*,
a high RA-containing species is used in China for treatment of dysentery, boils, and inju-
ries (7). The antioxidative, antimicrobial, and antiviral effects of *Prunella vulgaris* indicate
its potential as a medicinal herb (8). Rosmarinic acid-containing *Origanum vulgare*
(oregano), *Thymus vulgaris* (thyme), *Ocimum basilicum* (sweet basil), and *Rosmarinus
officinalis* (rosemary) are important sources of antioxidants in food preservation (9–11)
and for stability and enhancement of anthocyanin and related pigment color in berry based
juices (12,13) and have potential health benefits as dietary amylase inhibitor in diabetes
management (14).

   Many pharmacological effects of RA are known. Rosmarinic acid inhibits several complement dependent inflammatory processes and has potential as a therapeutic
agent for control of complement activation diseases (15,16). Rosmarinic acid has been
reported to have effects on both the classical C3-convertase and on the cobra venom
factor and ovalbumin/antiovalbumin mediated passive cutaneous anaphylaxis (15).
Rosmarinic acid also inhibits prostacyclin synthesis induced by complement activation
(17,18). Rosmarinic acid is also known to have complement independent effects, such
as scavenging of oxygen free radicals and inhibiting elastase and is known to be safe
(19). Other actions of RA are antithyrotropic activity in tests with human thyroid

membrane preparations, inhibition of complement dependent components of endotoxin shock in rabbits, and the ability to react rapidly to viral coat proteins and so inactivate the virus (15). Rosmarinic acid also inhibits Forskolin induced activation of adenylate cyclase in cultured rat thyroid cells (20) and inhibits external oxidative effects of human polymorphonuclear granulocytes (21).

Recent research has indicated other benefits of RA containing *Perilla frutescens* on the reduction of lipopolysaccharide (LPS) induced liver injury in D-galactosamine sensitized mice (22). High RA *P. frutescens* also inhibited lung injury in mice induced by diesel exhaust particles (23) and also had antiallergic effect (24). The antiallergic titer of rosmarinic acid was more effective than tranilast, which is a widely used antiallergic drug (24). Rosmarinic acid also has potential antidepressive-like effect in mice based on a forced swimming test (25). Another interesting study has shown that RA inhibits calcium dependent pathways of T-cell antigen receptor mediated signaling (26). However, investigations so far on pharmacological effects of RA have not clarified the antiinflammatory effects but more evidence suggests RA's ability to block complement activation (27) and to inhibit cyclooxygenase (4).

Research findings on pharmacological potential of RA have substantially increased since 2000. RA synergistically inhibited LDL oxidation in combination with lycopene indicating its potential against atherosclerosis (28). In relation to HIV type 1, RA in addition to being an integrase inhibitor also inhibited reverse transcriptase (29,30). In mice studies, *Perilla frutescens* rich in RA reduced allergenic reactions using mice ear passive cutaneous anaphylaxis (PCA) reaction (24,31). In another mice model, RA in *Perilla* extract inhibited allergic inflammation induced by mite allergen (32). In a human clinical study related to allergy reduction, RA enriched *Perilla frutescens* proved to be an effective intervention for mild seasonal allergic rhino-conjunctivitis (SAR), partly through inhibition of polymorphonuclear leukocytes (PMNL) infiltration into the nostrils, which could contribute to reduction in treatment costs for allergic diseases (33). In mice studies, oral and intraperitoneal administration of RA had antidepressive effects and mechanism was suggested to not involve inhibition of monoamine transporters and monoamine oxidase (25,34). In relation to improvement of kidney related functions, RA has shown suppressive effects on mesangioproliferative glomerulonephritis in rats (35). Inhibitory effect of RA on the proliferation of cultured murine mesangial cells was previously reported (36). In this study, RA inhibited cytokine induced mesangial cell proliferation and suppressed PDGF and c-myc mRNA expression in PDGF mediated mesangial cells (36). Suppressive effects of RA enriched *Perilla frutescens* on IgA nephropathy in HIGA mice were also observed (37). In other pharmacological studies, RA reduced the defensive freezing behavior of mice exposed to conditioned fear stress (38). In relation to signal transduction, RA inhibited Ca2+ dependent pathways of T-cell antigen receptor mediated signaling by inhibiting the PLC-gamma 1 and Itk activity (26). RA is also known to inhibit TCR induced T cell activation and proliferation in an Lck dependent manner (39–41) and can also influence Lck dependent apoptotic activity (42). In other T cell studies, RA alone and in conjunction with currently used immunosuppressive drugs, inhibited *in vitro* splenic T-cell proliferation (43). RA enriched herb extract was also shown to have beneficial effects on suppression of collagen induced arthritis (44) and showed significant reduction in tumorigenesis in a murine two stage skin carcinogenesis model (45). Herbal extracts enriched in RA yielded higher inhibition of amylase than purified RA, suggesting RA in synergy with other phenolic compounds may contribute to amylase inhibition for potential modulation of hyperglycemia (14). Recently it was shown that oregano clonal extracts high in RA also have antimicrobial activity against *Listeria monocytogenes* (46) and *Helicobacter pylori* (47).

8.3   RA BIOSYNTHESIS IN CELL CULTURES

Rosmarinic acid has been targeted for production using undifferentiated cell suspension cultures in several species (6,48–56). Further, elicitors such as yeast extract and methyl jasmonate were used to stimulate RA content in cell cultures (6,51,57,58). The biosynthesis of RA were also evaluated and stimulated in hairy root cultures by elicitors (59). The main purpose of cell suspension cultures for production of RA is the potential for large scale production in bioreactors (60,61). Although large scale production in bioreactors is feasible for RA (62), undifferentiated cell suspension cultures are generally not practical for metabolites produced in differentiated structures (e.g., anethole in seeds of anise, curcumin in rhizomes of turmeric, eugenol in barks of cinnamon, and thymol in glandular cells of leaves). In a comparison of nodal shoot cultures and callus cultures of *Ocimum basilicum* L in airlift bioreactors, RA production in cell suspension cultures was 29 µg/g dry weight compared to 178 µg/g dry weight for shoot cultures (63). An additional disadvantage of undifferentiated callus based suspension cultures is that the DNA is more error prone and therefore more genetically unstable (64). Further, bioreactor based production requires high initial operating costs.

In terms of regulation of RA in cell cultures it is constitutively expressed in *Coleus blumei* without any medium manipulation (65). Plant cell cultures are known to accumulate 8–10% of their dry weight as RA, a content much higher than parent plants (53,66), contradicting the callus and shoot culture comparisons of the *Ocimum* study (63). The pathway of RA biosynthesis is through the aromatic amino acids, phenylalanine, and tyrosine (65). Cell suspension cultures of *C. blumei* (66), *Rosmarinus officinalis*, *Salvia officinalis* (67), and *Anchusa officinalis* (68,69) have been used to produce RA in cell suspension cultures. The influence of various macronutrients and growth regulators on growth of *A. officinalis* has been investigated (68,69). Concentrations of 3% sucrose, 15 mM nitrate, 3 mM phosphate, and 0.25 mM calcium were best for increasing both biomass and RA contents with yields in the range of 10–15% (68). Similar nitrogen, potassium, and phosphate optimizations proved useful for RA enhancement in *Lavandula vera* MM cell suspension cultures (70). When auxin effects were tested, they maintained growth and integrity of the cell suspensions, whereas cytokinins alone did not, suggesting that the culture of *A. officinalis* was auxin dependent (69). Among auxins, NAA had the most pronounced effect on RA content (69) and was confirmed in other recent studies where NAA at levels of 2 mg/l induced maximum RA level of 355 mg/l in cultures of *Zataria multiflora* (71). The kinetics of growth and RA production suggested that the increase in the final RA content and initiation of the period of biosynthesis was in the exponential, rather than the linear growth phase (69). Other studies have shown that increased sucrose concentrations stimulate RA content (72). The highest RA content reported so far was 36% of the cell dry weight in suspension cultures of *Salvia officinalis* at 5% sucrose (49), which is unusually high and never been reproduced.

8.3.1   Pathways Associated with Rosmarinic Acid Biosynthesis

The amino acids phenylalanine and tyrosine have been shown to be precursors of RA biosynthesis (52,73–75). Phenylalanine is transformed to an activated hydroxycinnamic acid by the enzymes of the general phenylpropanoid pathway, which are already well known for biosynthesis of flavonoid or lignin (75). Using radioactive phenylalanine and tyrosine, it was established that these two amino acids are incorporated into caffeic acid and 3,4-dihydroxyphenyllactic acid moieties, respectively (73). Steps in RA biosynthesis originating from phenylalanine and tyrosine have been characterized (Figure 8.2)

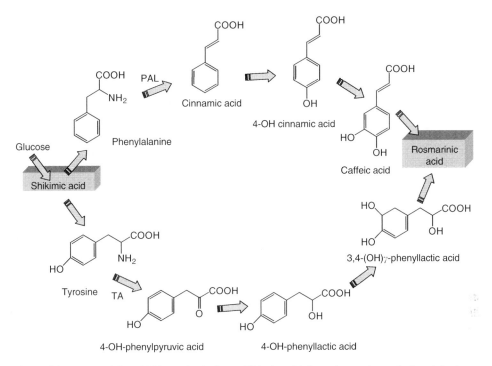

Figure 8.2 Rosmarinic acid biosynthesis from shikimic acid through tyrosine and phenylalanine.

(6,52,65,76,77). In several cell suspension cultures, activity of phenylalanine ammonia-lyase (PAL) was correlated to RA (6,77). Basic characteristics of this enzyme were determined in protein preparations from suspension cell suspension cultures of *Coleus blumei* (78). The reaction, following formation of cinnamic acid, is the hydroxylation in the position 4 to 4-coumaric acid by cytochrome P450 monooxygenase cinnamate 4-hydroxylase (74). Generally, coenzyme A thioester, or glucose or chlorogenic acid have been shown to serve as donors of hydroxycinnamic acid moieties (75). Further, using *A. officinalis* cell suspension cultures, it was reported tyrosine amino transferase catalyzes the first step of the transformation of tyrosine to 3,4-dihydroxyphenyllactic acid. Several isoforms of tyrosine aminotransferase were found to be active in cell suspension cultures of *A. officinalis* (76,77). Prephenate aminotransferase in *A. officinalis* cell suspension cultures was found to be important, and its activity was affected by 3,4-dihydroxyphenyllactic acid (79). Other enzymes of late steps in the RA biosynthesis pathway, like hydroxyphenylpyruvate reductase and RA synthase (hydroxycinnamoyl-CoA, hydroxyphenyllactate, hydroxycinnamoyl transferase), were isolated and characterized in cell suspension cultures of *C. blumei* (80–82). Under the release of coenzyme A, the ester linkage is formed between carboxyl group of 4-coumaric acid and the aliphatic hydroxyl group of 4-coumaric acid and the aliphatic hydroxyl group of 4-hydroxyphenyllactate (75). Other studies have isolated microsomal hydroxylase, later confirmed as cytochrome P450 monooxygenases (74,83), whose activities introduce hydroxyl groups at position 3 and 3′-hydroxyphenyllactate to give rise to the aromatic rings of ester 4-coumaroyl-4′-hydroxyphenyllactate to give rise to RA (52). A number of cDNAs encoding cytochrome P450s, which can hydroxylate 4-coumaric acid or 4-coumaroyl moiety in an ester, have been isolated (84–86). This enzyme isolation led to the proposal that the complete biosynthetic pathway for RA biosynthesis originates from phenylalanine and tyrosine (52,75).

8.3.2   RA Biosynthesis and Generation of High RA Clonal Lines

High RA production can be achieved at low cost by incorporating superior varieties in traditional agronomic systems. Superior varieties can be isolated using tissue culture techniques using shoot based clonal lines of single seed origin (1). The major limitation of using dietary herbs for pharmacological applications from traditional wild collections and heterogeneous seeds is the inconsistency of phenolic phytochemicals such as RA due to the heterogeneity resulting from the cross pollinating nature of their breeding characteristics, and especially species in the family Lamiaceae (1). Plants which originate from different heterozygous seeds in a given pool of extract are phenotypically variable, resulting in the substantial phytochemical inconsistency, and therefore leads to unreliable clinical effects as well as inconsistent health benefits and functional value. In order to overcome the problem of phytochemical inconsistency due to genetic heterogeneity, plant tissue culture techniques have been developed to isolate a clonal pool of plants originating from a single heterozygous seed (87,88). A single elite clonal line with superior RA and phenolic profile can then be screened and selected based on tolerance to *Pseudomonas* sp. (89–91) and proline analogs (92–94). These elite clonal lines (each clonal line originating from a different heterozygous seed), following large scale clonal propagation (micropropagation) and evaluation of functionality, can be targeted as dietary sources of phenolics (with focus on RA compared in diverse total phenolic clonal backgrounds) for diverse food and pharmaceutical applications.

8.3.3   Role of Proline Linked Pentose–Phosphate Pathway in RA Biosynthesis in Clonal Systems

The hypothesis that synthesis of plant phenolic metabolites is linked to the critical control point (CCP), proline linked pentose–phosphate pathway (PLPPP) (1,95,96) (Figure 8.3) was developed based on the role of the PLPPP in regulation of purine metabolism in mammalian systems (97). Proline is synthesized by a series of reduction reactions from glutamate. In this sequence, P5C and proline known to be metabolic regulators function as a redox couple (97,98). During respiration, oxidation reactions produce hydride ions, which augment reduction of P5C to proline in the cytosol. Proline can then enter mitochondria through proline dehydrogenase (99) and support oxidative phosphorylation (alternative to NADH from Krebs/TCA cycle). This is important because shunting the TCA cycle toward proline synthesis likely deregulates normal NADH synthesis. The reduction of P5C in the cytosol provides $NADP^+$, which is the cofactor for glucose-6-phosphate dehydrogenase (G6PDH), an enzyme that catalyzes the rate-limiting step of the pentose–phosphate pathway. Proline synthesis is therefore hypothesized, and has been partly shown to both regulate and stimulate pentose–phosphate pathway activity in erythrocytes (100) and cultured fibroblasts (101) when P5C is converted to proline. This was shown to stimulate purine metabolism via ribose-5-phosphate, which affects cellular physiology and therefore function (97,102). Therefore, understanding the CCP, PLPPP is important for designing high RA clonal extracts isolated from single seed genetic origin among a heterogeneous seed population.

From the above insights Shetty (1,95,96) first proposed a model that CCP, PLPPP could stimulate shikimate and phenylpropanoid pathways and hypothesized that stress linked modulation of this pathway can lead to the stimulation of phenolic phytochemicals, including RA (1,95,96) (Figure 8.3). Using this model, proline, proline precursors, and proline analogs were effectively utilized to stimulate total phenolic content and RA in clonal shoot cultures (103,104). Further, it was shown that proline, proline precursors, and proline analogs stimulated somatic embryogenesis in anise, which correlated with increased soluble phenolic content (105). It was also established that during *Pseudomonas* mediated stimulation of total soluble phenolics, RA and proline content was stimulated in

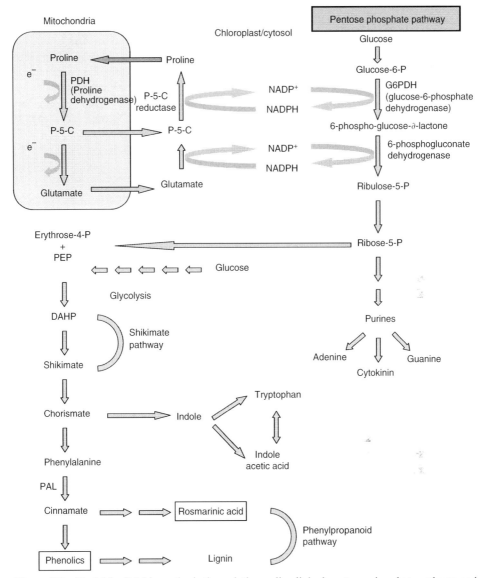

Figure 8.3 Model for RA biosynthesis through the proline linked pentose–phosphate pathway and its utilization to screen high RA and phenolic-producing clonal lines using microbial interaction and proline analogs (targeted at proline dehydrogenase-PDH).

oregano clonal shoot cultures (106). Therefore, it was proposed that NADPH demand for proline synthesis during response to microbial interaction and proline analog treatment (1) may reduce the cytosolic $NADPH/NADP^+$ ratio, which should activate G6PDH (107,108). Therefore, deregulation of the pentose–phosphate pathway by proline analog and microbial induced proline synthesis may provide the excess erythrose-4-phosphate (E4P) for shikimate and, therefore, the phenylpropanoid pathway leading to RA (95). At the same time, proline and P5C could serve as superior reducing equivalents (RE) reductant, an alternative to NADH (from Krebs/TCA cycle) to support increased oxidative phosphorylation (ATP synthesis) in the mitochondria during the stress response (97,98).

Therefore, proline analog, azetidine-2-carboxylate (A2C), and proline stimulating *Pseudomonas* interactions were used to isolate high RA clonal lines (1,92–95). A2C is an inhibitor of proline dehydrogenase (109) and is also known to inhibit differentiation in Leydig cells of rat fetal testis, which can be overcome by exogenous proline addition (110). Another analog, hydroxyproline, is a competitive inhibitor of proline for incorporation into proteins. According to the model of Shetty (1,96), either analog at low levels should deregulate proline synthesis from feedback inhibition, and stimulate proline synthesis (1). This would then allow the proline linked pentose–phosphate pathway to be activated for NADPH synthesis, and concomitantly drive metabolic flux toward E4P for biosynthesis of shikimate and phenylpropanoid metabolites, including RA (Figure 8.4). Proline could also serve as a RE reductant for ATP synthesis via mitochondrial membrane associated proline dehydrogenase (99). Therefore, any high RA clonal line with a deregulated proline synthesis pathway should have an overexpressed pentose phosphate pathway which allows excess metabolic flux to drive shikimate and phenylpropanoid pathway toward total phenolic and RA synthesis. Similarly, such proline overexpressing clonal lines should be more tolerant to proline analog, A-2-C (95,96). At the same time if the metabolic flux to RA is overexpressed, it is likely to be stimulated in response to *Pseudomonas* sp (95). Therefore, such a clonal line is equally likely to be tolerant to *Pseudomonas* sp. Further, such a clonal line should also exhibit high proline oxidation and RA content in response to A2C and *Pseudomonas* sp. In addition, in the presence of A2C or *Pseudomonas* sp., increased activity of key enzymes G6PDH (pentose–phosphate pathway), P5C reductase (proline synthesis pathway), proline dehydrogenase (proline oxidation pathway), 3-deoxy-D arabino-heptulosonate-7-phosphate synthase (shikimate pathway), and phenylalanine ammonia-lyase (phenylpropanoid pathway) should be stimulated (96). As mentioned earlier, the rationale for the PLPPP model for RA biosynthesis is based on the role of the pentose–phosphate pathway in driving ribose-5-phosphate toward purine metabolism in cancer cells (97), differentiating animal tissues (110), and plant tissues (111). The success of this CCP, PLPPP strategy (89–94) provides ready access to critical interlinking metabolic pathways associated with RA biosynthesis and will allow more detailed analyses, which could lead to large scale greenhouse and Agronomic based field production systems for efficient RA biosynthesis. This strategy for investigation and production of RA can be the foundation for designing other dietary phenolic phytochemicals from cross pollinating, heterogeneous species (1,95,96).

8.4   MECHANISM OF RA ACTION THROUGH STIMULATION OF HOST ANTIOXIDANT RESPONSE

Investigations so far in food grade clonal herb systems (1) and legume sprouts (112–114) led to the development of the model that activity of CCP, proline linked pentose–phosphate pathway is important for stress induced phenolic biosynthesis such as RA and phenolics and that this stimulation of phenolics is likely closely linked to stimulation of antioxidant response pathways (Figure 8.4) (96,112–114). Further research has indicated that the proline biosynthesis pathway coupled to stress induced antioxidant response pathways could be also stimulated in legume sprouts using exogenous treatment of phenolic extracts from clonal oregano (113,115,116). Phenolic extracts from these clonal oregano lines have high free radical scavenging activity (47). Proline linked stimulation of antioxidant enzyme response pathways may also be stimulated by low pH and salicylic acid (117). Further, exogenous seed treatment with oregano phenolic antioxidant extracts enhanced endogenous phenolic content, GPX activity, and consequently, enhanced seeding vigor during germination (116). From these initial plant studies and plant PLPPP model (Figure 8.4), a human/mammalian cell PLPPP model has been developed (Figure 8.6) wherein a proton donation by phenolic antioxidants

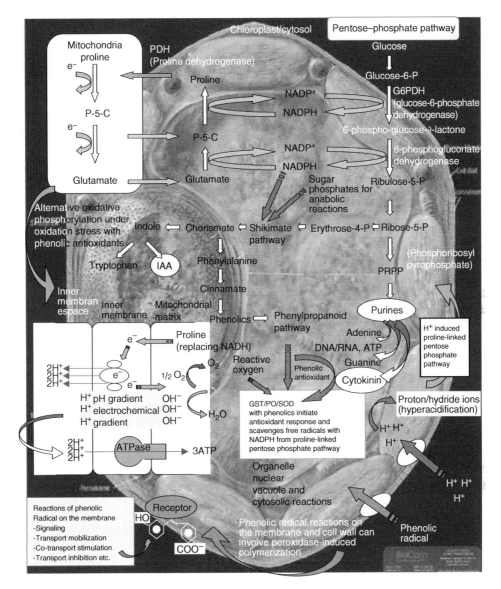

Figure 8.4 Expanded Model for the role of proline linked pentose–phosphate pathway in regulating phenolic biosynthesis, which also accommodates the mechanism of action of external phenolic phytochemicals such as RA from herbal extracts to trigger an endogenous antioxidant enzyme response. (Abbreviations: P5C: pyrroline-5-carboxylate; IAA: indole acetic acid; GST: Glutathione-s-transferase; PO: peroxidase; SOD: superoxide dismutase)

such as RA at the outer plasma membrane initiates a proton/hydride ion influx into the cytosol which activates an antioxidant response through the stimulation of the, CCP, proline linked pentose–phosphate pathway (95,96). Demand for NADPH by stimulated proline biosynthesis also drives the production of precursors for phenolic (only in plants and fungi), purine and antioxidant pathways. In this host response PLPPP model, proline can also be used as a reducing equivalent (RE) reductant to support oxidative phosphorylation for ATP synthesis. Using this approach and rationale, we first developed several RA and phenolic overexpressing plant clonal systems for functional food and agro-environmental applications. Subsequently, PLPPP linked and optimized phenolic phytochemical clonal profiles can be used as sources of antioxidants and antimicrobials in biological systems based on host PLPPP response and have implications for human health and wellness.

Human health applications have been developed with additional insights based on the animal antioxidant response model (Figure 8.7) and the plant antioxidant response model (Figure 8.5). From these insights an innovative model for the mechanism of action of phenolic antioxidants like RA for improving human health through protection

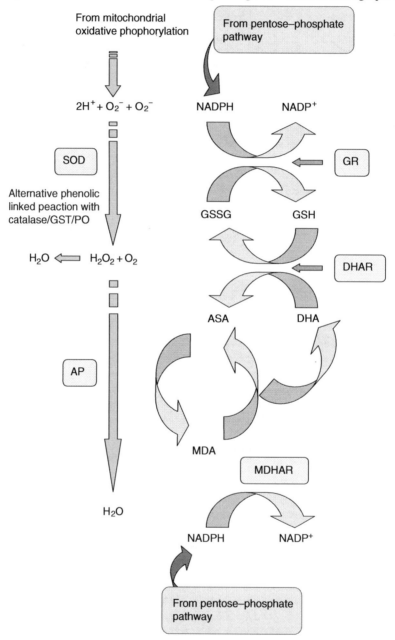

Figure 8.5   Model for specific steps in antioxidant enzyme response pathway in plants. (Abbreviations: SOD: superoxide dismutase; AP: ascorbate peroxidase; GR: glutathione reductase; GSSG: oxidized glutathione; GSH: reduced glutathione; DHAR: dehydroascorbate reductase; ASA: reduced ascorbate: DHA: dehydroascorbate; MDA: monodehydroascorbate; MDHA: monodehydroascorbate reductase)

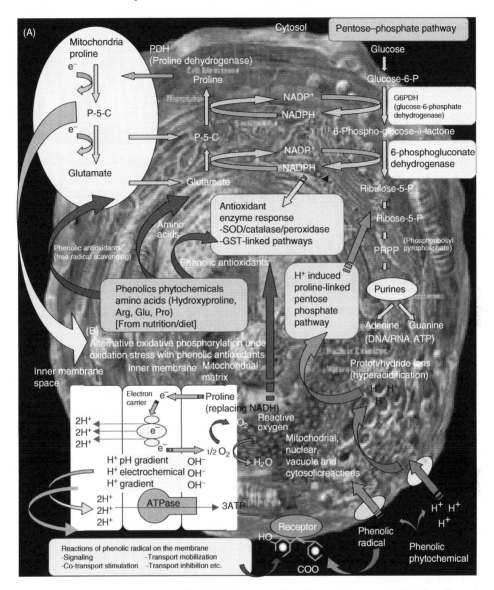

Figure 8.6 Extension of plant proline linked pentose–phosphate pathway model for the effect of external phenolic phytochemicals in human and mammalian systems. (Abbreviations: P5C: pyrroline-5-carboxylate; GST: Glutathione-s-transferase; SOD: superoxide dismutase)

against oxidation linked diseases involving the host proline linked pentose–phosphate pathway has been proposed (Figure 8.6). The major diseases afflicting humans today, related to excess calories, obesity, and environmental pollutant exposure, are oxidation linked chronic diseases such as cancer, cardiovascular disease, arthritis, cognition diseases, and diabetes. Oxidation linked immune dysfunction and the inability to fight pathogenic infection under a very low calorie and protein diet still remains a problem in several parts of the world and the challenge has to be addressed. Oxidation linked and infectious diseases involve free radical reactions. Free radicals are potential

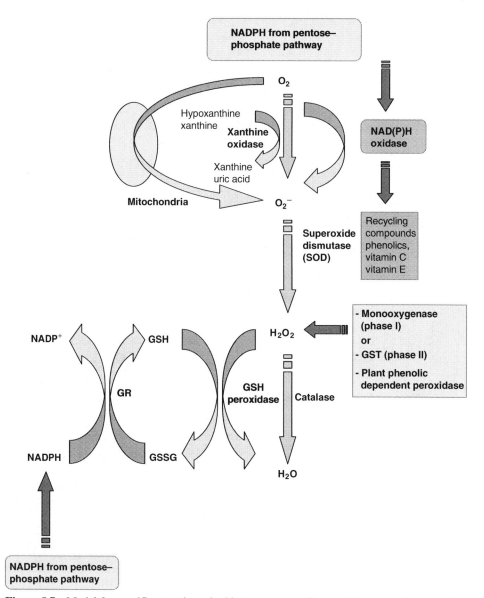

Figure 8.7 Model for specific steps in antioxidant response pathways in human and mammalian systems. (Abbreviations: SOD: superoxide dismutase; GST: glutathione-s-transferase; GR: glutathione reductase; PO: peroxidase; GP: glutathione peroxidase; GSSG: oxidized glutathione; GSH: reduced glutathione)

carcinogens because they can facilitate mutagenesis, tumor promotion and progression (118–120). For example, in the case of cardiovascular diseases, free radicals are implicated in the pathogenesis of atherosclerosis, which is characterized by the hardening of the arterial wall (118,121,122). Rheumatoid arthritis is a systemic autoimmune disease characterized by chronic joint inflammation with infiltration of macrophages and activated T cells. Production of free radicals at sites of inflammation is thought to contribute to the pathogenesis (118,123,124). Free radicals are implicated in the pathogenesis of Alzheimer disease (118,125). As significant amounts of lipid peroxidation in brain tissues have been observed and may help explain the progressive decline in cognition function and excessive neuronal loss in afflicted patients. The brain tissue shows numerous amyloid plaques. Free radicals have been implicated in diabetes mellitus (118,126,127). These oxidative stress linked disease conditions are associated with a prooxidative shift of glutathione redox state in the blood (118,128). Elevated glucose levels are also associated with increased production of free radicals by several different mechanisms (118,129,130).

Diet, environment, and lifestyle influence oxidation linked diseases and improvement of diet is an important part of the preventive management of these diseases (95,96). It is evident that a variety of plant based phenolic antioxidants have a positive impact on the prevention and modulation of various oxidation linked diseases (16,131–135) and must be considered important part of the dietary management of these diseases. Currently the modes of action and early stage effects of these phenolic antioxidants in positively modulating and preventing various diseases are not completely clear (95,96). Extensive research is underway to ascertain how free radicals modulate physiological control of cell function at the level of cell proliferation and deterioration (118) and at the level of gene expression (136). Phenolic antioxidants such as RA have been targeted to control the free radical linked cellular deterioration that can otherwise lead to major oxidation linked chronic diseases. We have proposed in this chapter that an understanding of RA biosynthesis through CCP and PLPPP could help to not only design the right RA profile through functional foods, but the host response to this RA profile could also operate effectively through an antioxidant enzyme response regulated by host (human) PLPPP.

Therefore, an important strategy to develop diet based interventions is through design of functional foods (conventional foods with clinically defined health promoting components) based on the understanding of phenolic antioxidant biosynthesis, such as RA in food plants (Figure 8.4) and the effect of RA in human and mammalian systems (Figure 8.6). In order for diet based interventions (through functional foods) to be effective, it is also important to understand the early stage modes of action of these functional compounds such as RA and how to deliver RA at consistent levels and with no toxicity problems. In the antioxidant enzyme response model for human health, proposed previously (95,96), a consistent and defined phytochemical profile of RA and in high phenolic background can be developed using clonal shoot systems using various dietary botanicals from the family Lamiaceae. In the human model (Figure 8.6), the early stage mode of action of RA in human cells has parallels to the models for plant systems (Figure 8.4), but within the scope of human cellular physiology, function, and diversity (95). By this model, plant extract enriched RA similarly initiate an inward proton flux at the outer human cell membrane, which increases the cytosolic proton/hydride ion concentration and activates the PLPPP. Some phenolic antioxidant radicals in the total phenolic background or hydrolyzed RA, depending on their size, may penetrate the plasma membrane along with the proton/hydride ion flux (cotransport) into the cytosol. The cytosolic proton/hydride ion flux then drives PLPPP generating NADPH, sugar phosphates for anabolic reactions and proline as an alternative RE reductant to generate ATP via oxidative pentose–phosphate pathway (95,96). The products of the pentose–phosphate pathway are important for purine biosynthesis and for stimulating antioxidant enzyme response pathways in conjunction with action of the dietary phenolic antioxidants (Figure 8.6). The control of free radicals that is likely associated with proline or TCA cycle generated NADH linked mitochondrial oxidative phosphorylation at this early stage could have a positive effect on any subsequent oxidation linked cellular deterioration and consequent oxidation linked chronic disease development. Other roles for RA or RA hydrolyzed phenolic radicals that penetrate the membrane could involve:

1. Stability and protection of organelle membranes and proteins from free radical damage
2. Participation in the antioxidant response pathway to quench super oxide and peroxide radicals
3. Protection of DNA and protein stability
4. Stimulation of proline linked pentose–phosphate pathway activity to satisfy demand for NADPH in reactions involving penetrating phenolic radicals (95,96)

In specific cases where RA or phenolic radicals cannot normally penetrate the outer plasma membrane, other conceivable roles could include:

1. Stability and protection of the outer membranes and membrane proteins from free radical damage
2. Modulation of membrane transport
3. Inhibition of specific membrane proteins, including those involved in PMF and electron transport chain in Prokaryotes
4. Modulation of signal transduction
5. Modulation of membrane receptors
6. Co-transport with H^+, sugars or amino acids
7. Passive membrane transport through damaged membranes

An inward proton flux to the cytosol could be created even without phenolic radical penetration, which then could stimulate the proline linked pentose–phosphate pathway and couple its action to the various reactions and roles that may be initiated and modulated through interactions of phenolic radicals at the outer plasma membrane (95,96).

8.5   SUMMARY

It is clear that RA containing herb extracts have wide potential applications for functional food and pharmacological applications. In optimizing RA biosynthesis for such applications the potential role of microbial elicitation and proline linked pentose–phosphate pathway (PLPPP) has been exploited to develop clonal tissue culture systems of single seed genetic origin that can be grown in traditional and efficient greenhouse and agronomic systems. Such clonal systems allow the screening of phenotypic specific RA enriched herbal herbs extracts that offer potential for consistency for various applications and design of health specific functional foods. Further, using clues about the role of proline linked pentose–phosphate pathway as a CCP in relation to RA biosynthesis in herb clonal systems (Figure 8.4), a model has been developed for the mode of action of RA in mammalian and human systems (95,96) (Figure 8.6). In this model for animal systems, RA has been hypothesized to stimulate host antioxidant enzyme response and enhance free radical scavenging to counter oxidation pressure. Efficient operation of this free radical scavenging system requires the proper functioning of pentose phosphate pathway. Further, pentose phosphate pathway could be optimally stimulated by coupling it to proline biosynthesis. This coupled proline linked pentose–phosphate pathway in normal human cells could be stimulated by RA enriched herb extracts, thereby providing critical precursors like NADPH for antioxidant enzyme response and therefore serving as a critical control point (CCP). Through this antioxidant response regulation and other specific RA related structure and function activity at the cellular gene and signal pathway levels, protective chemopreventive functions of RA can be potentially enhanced.

REFERENCES

1. Shetty, K. Biotechnology to harness the benefits of dietary phenolics: focus on Lamiaceae. *Asia Pac. J. Clin. Nutr.* 6:162–171, 1997.
2. Shetty, K. Biosynthesis and medical applications of rosmarinic acid. *J. Herbs Spices Med. Plant.* 8:161–181, 2001.
3. Canigueral, S., J. Iglesias, M. Hamburger, K. Hostettmann. Phenolic constituents of *Salvia lavdandulifolia* spp. lavdandulifolia. *Planta Medica* 55:92, 1989.
4. Kelm, M.A., M.G. Nair, G.M. Strasburg, D.L. Dewitt. Antioxidant and cyclooxygenase inhibitory phenolic compounds from *Ocimum sanctum* Linn. *Phytomedicine* 7:7–13, 2000.
5. Kuhnt, M.A., A. Probstle, H. Rimpler, R. Bauer, M. Heinrich. Biological and pharmacological activities and further constituents of *Hyptis verticilllata*. *Planta Medica* 61:227–232, 1995.
6. Sumaryono, W., P. Prokash, T. Hartmann, M. Nimtz, V. Wray. Induction of rosmarinic acid accumulation in cell suspension cultures of *Orthosiphon aristatus* after treatment with yeast extract. *Phytochemistry* 30:3267–3271, 1991.
7. Zhang, H.J., L.N. Li. Salvionolic acid I: a new depside from *Salvia cavaleriei*. *Plant Medica* 60:70–72, 1994.
8. Psotova, J., M. Kolar, J. Sousek, Z. Svagera, J. Vicar, J. Ulrichova. Biological activities of *Prunella vulgaris* extract. *Phytother. Res.* 17:1082–1087, 2003.
9. Kikuzaki, H., N. Nakatani. Structure and a new antioxidative phenolic acid from oregano (*Origanum vulgare L*). *Agric. Biol. Chem.* 53:519–524, 1989.
10. Madsen, H.L., G. Bertelsen. Spices as antioxidants. *Trends Food Sci. Tech.* 6:271–277, 1995.
11. Jayasinghe, C., N. Gotoh, T. Aoki, S.Wada. Phenolics composition and antioxidant activity of sweet basil (*Ocimum basilicum* L.). *J. Agric. Food Chem.* 16:4442–4449, 2003.
12. Eiro, M.J., M. Heinonen. Anthocyanin color behavior and stability during storage: effect of intermolecular co-pigmentation. *J. Agric. Food Chem.* 50:7461–7466, 2002.
13. Rein, M.J., M. Heinonen. Stability of berry juice color. *J. Agric. Food Chem.* 52:3106–3114, 2004.
14. McCue, P., K. Shetty. Inhibitory effects of rosmarinic acid extracts on Porcine amylase and implications for health. *Asia Pac. J. Clin. Nut.* 13:101–106, 2004.
15. Engleberger, W., U. Hadding, E. Etschenberg, E. Graf, S. Leyck, J. Winkelmann, M.J. Parnham. Rosmarinic acid: A new inhibitor of complement C3 – convertase with antiinflammatory activity. *Intl. J. Immunopharmac.* 10:729–737, 1988.
16. Peake, P.W., B.A. Pussel, P. Martyn, V. Timmermans, J.A. Charlesworth. The inhibitory effect of rosmarinic acid on complement involves the C5 convertase. *Int. J. Immunopharmac.* 13:853–857, 1991.
17. Bult, H., A.G. Hermann, M. Rampert. Modification of endotoxin-induced haemodynamic and haemolytical changes in rabbit by methylprednisolone. F(ab)2 fragments and rosmarinic acid. *Brit. J. Phamacol.* 84:317–327, 1985.
18. Rampart, M., J.R. Beetens, H. Bult, A.J. Herman, M.J. Parnham, J. Winklemann. Complement-dependent stimulation of prostacyclin biosynthesis: inhibition by rosmarnic acid. *Biochem. Pharmc.* 35:1397–1400, 1986.
19. Nuytinck, J.K.S., R.J.A. Goris, E.S. Kalter, P.H.M. Schillings. Inhibition of experimentally induced microvascular injury by rosmarinic acid. *Agents Actions* 17:373–374, 1985.
20. Kleemann, S., H. Winterhoff. Rosmarinic acid and freeze-dried extract of *Lycos virginicus* are able to inhibit Forskolin-induced activation of adenylate cyclase in cultured rat thyroid cells. *Planta Medica* 56:683–687, 1990.

21. Van Kessel, K.P., E.S. Kalter, J. Verhoef. Rosmarinic acid inhibits external oxidative effects of human polymorphonuclear granulocytes. *Agents Actions* 17:375–376, 1986.

22. Osakabe, N., A. Ysuda, M. Natsume, C. Sanbongi, Y. Kato, T. Osawa, T. Yoshikawa. Rosmarinic acid, a major polyphenolic component of *Perilla frutescens* reduces lipopolysaccharide (LPS)-induced liver injury in D-Galactosamine (D-GalN)-sensitized mice. *Free Rad. Biol. Med.* 33:798–806, 2002.

23. Sanbongi, C., H. Takano, N. Oskabe, N. Sasa, M. Natsume, R. Yanagisawa, K. Inoue, Y. Kato, T. Osawa, T. Yoshikawa. Rosmarinic acid inhibits lung injury by diesal exhaust particles. *Free Rad. Biol. Med.* 34:1060–1069, 2003.

24. Makino, T., Y. Furuta, H. Wakushima, H. Fujii, K. Saito, Y. Kano. Anti-allergic effect of *Perilla frutescens* and its active constituents. *Phytother. Res.* 17:240–243, 2003.

25. Takeda, H., M. Tsuji, M. Inazu, T. Egashira, T. Matusmiya. Rosmarinic acid produce anti-depressive-like effect in the forced swimming test in mice. *Eur. J. Pharmcol.* 449:261–267, 2002.

26. Kang, M-A., S.-Y. Yun, J. Won. Rosmarinic acid inhibits Ca2+- dependent pathways of T-cell antigen receptor-mediated signaling by inhibiting the PLC-γl and Itk activity. *Blood* 101:3534–3542, 2003.

27. Sahu, A., N. Rawal, M.K. Pangburn. Inhibition of complement by covalent attachment of rosmarinic acid to activated C3b. *Biochem. Pharmacol.* 57:1439–1446, 1999.

28. Fuhrman, B., N. Volkova, M. Rosenblat, M. Aviram. Lycopene synergistically inhibits LDL oxidation in combination with vitamin E, glabridin, rosmarinic acid, carnosic acid or garlic. *Antioxid. Redox. Signal.* 2:491–506, 2000.

29. Hooker, C.W., W.B. Lott, D. Harrich. Inhibitors of humandeficiency virus type 1 reverse transcriptase target distinct phases of early reverse transcription. *J. Virol.* 75:3095–3104, 2001.

30. Tewtrakul, S., H. Miyashiro, N. Nakamura, M. Hattori, T. Kawahata, T. Otake, T. Yoshinaga, T. Fujiwara, T. Supavita, S. Yuenyongsawad, P. Rattanasuwon, S. Dej-Adisai. HIV-1 integrase inhibitory substances from *Coleus parivifolius*. *Phytother. Res.* 17:232–239, 2003.

31. Makino, T., A.S. Furuta, H. Fujii, T. Nakagawa, H. Wakushima, K. Saito, Y. Kano, Y. Effect of oral treatment of *Perilla frutescens* and its constituents on type-I allergy in mice. *Biol. Pharm. Bull.* 24:1206–1209, 2001.

32. Sanbongi, C., H. Takano, N. Osakabe, N. Sasa, M. Natsume, R. Yanagisawa, K.I. Inoue, K. Sadakane, T. Ichinose, T. Yoshikawa. Rosmarinic acid in perilla extract inhibits allergic inflammation induced by mite allergen, in a mouse model. *Clin. Exp. Allerg.* 34:971–977, 2004.

33. Takano, H., N. Osakabe, C. Sanbongi, R. Yanagisawa, K. Inoue, A. Yasuda, M. Natusume, S. Baba, E. Ichiishi, T. Yoshikawa. Extract of *Perilla frutescens* enriched for rosmarinic acid, polyphenolic phytochemical inhibits seasonal allergic rhinoconjunctivitis in humans. *Exp. Biol. Med.* 229:247–254, 2004.

34. H. Takeda, M. Tsuji, J. Miyamoto, T. Matsumiya. Rosmarinic acid and caffeic acid reduce the defensive freezing behavior of mice exposed to conditioned fear stress. *Psychopharmacology* 164:233–235, 2002.

35. Makino, T., T. Ono, N. Liu, T. Nakamura, E. Muso, G. Honda. Suppressive effects of rosmarinic acid on mesangioproloferative glomerulonephritis in rats. *Nephron* 92:898–904, 2002.

36. Makino, T., T. Ono, E. Muso, H. Yoshida, G. Honda, S. Sasayama. Inhibitory effects of rosmarinic acid on proliferation of cultured murine mesangial cells. *Nephrol. Dial. Transplant.* 15:1140–1145, 2000.

37. Makino, T., T. Ono, K. Matsuyama, F. Nogaki, S. Miyawaki, G. Honda, E. Muso. Suppressive effects of *Perilla frutescens* on IgA nephropathy in HIGA mice. *Nephrol. Dial. Transplant.* 18:484–490, 2003.

38. Takeda, H., M. Tsuji, T. Matsumiya, M. Kubo. Identification of rosmarinic acid in the leaves of *Perilla frutescens* Britton var. acuta Kudo (*Perilla Herba*). *Nihon Shinkei Seishin Yakurigaku Zasshi* 22:15–22, 2002.

39. Won, J., Y.G. Hur, E.M. Hur, S.H. Park, M.A. Kang, Y. Choi, C. Park, K.H. Lee, Y. Yun. Rosmarinic acid inhibits TCR-induced T cell activation and proliferation in an Lck-dependent manner. *Eur. J. Immunol.* 33:870–879, 2003.

40. Ahn, S.C., W.K. Oh, B.Y. Kim, D.O. Kang, M.S. Kim, G.Y. Heo, J.S. Ahn. Inhibitory effects of rosmarinic acid in Lck SH2 domain binding to a synthetic phosphopeptide. *Planta Med.* 69:642–646, 2003.

41. Park, S.H., S.H. Kang, S.H. Lim, H.S. Oh, K.H. Lee. Design and synthesis of small chemical inhibitors containing different scaffolds of lck SH domain. *Biorg. Med. Chem. Lett.* 13:3455–3459, 2003.

42. Hur, Y.G., Y. Yun, J. Won. Rosmarinic acid induces p56lck-dependent apoptosis in Jurkat and peripheral T cells via mitochondrial pathway independent from Fas/Fas ligand interaction. *J. Immunol.* 172:79–87, 2004.

43. Yun, S.Y., Y.G. Hur, M.A. Kang, J. Lee, C. Ahn, J. Won. Synergistic immunosuppressive effects of rosmarinic acid and rapamycin *in vitro* and *in vivo*. *Transplantation* 75:1758–1760, 2003.

44. Youn, J., K.H. Lee, J. Won, S.J. Huh, H.S. Yun, W.G. Cho, D.J. Paik. Beneficial effects if rosmarinic acid on suppression of collagen induced arthritis. *J. Rheumatol.* 30:1203–1207, 2003.

45. Osakabe, N., A. Yasuda, M. Natsume, T. Yoshikawa. Rosmarinic acid inhibits epidermal inflammatrory responses: anticarcinogenic effect of *Perilla frutescens* extracts in the murine two-stage skin model. *Carcinogenesis* 25:549–557, 2004.

46. Seaberg, A., R.L. Labbe, K. Shetty. Inhibition of *Listeria monocytogenes* by elite clonal extracts of oregano (*Origanum vulgare*). *Food Biotechnol.* 17:129–149, 2003.

47. Chun, S.-S., D.A. Vattem, Y.-T. Lin, K. Shetty. Phenolic antioxidants from clonal oregano (*Origanum vulgare*) with antimicrobial activity against *Helicobacter pylori*. *Process. Biochem.* 40:809–816, 2005.

48. De-Eknamkul, W., B.E. Ellis. Rosmarinic acid production and growth characterization of *Anchusa officinalis* cell suspension cultures. *Planta Medica* 50:346–350, 1984.

49. Hipplolyte, I.B., J.C. Marin, J.C. Baccou, R. Jonard. Growth and rosmarinic acid production in cell suspension cultures of *Salvia officinalis*. *Plant Cell Rep.* 11:109–112, 1992.

50. Lopez-Arnoldos, T., M. Lopez-Serano, A.R. Barcelo, A.A. Calderon, M.M. Zapata. Spectrophotometric determination of rosmarinic acid in plant cell cultures by complexation with Fe^{2+} ions. *Fresenius J. Anal. Chem.* 351:311–314, 1995.

51. Mizukami, H., Y. Tabira, B.E. Ellis. Methyl jasmonate-induced rosmarinic acid biosynthesis in *Lithospermum erythrorhizon* cell suspension cultures. *Plant Cell Rep.* 12:706–709, 1993.

52. Petersen, M., B. Hausle, B. Karwatzki, J. Meinhard. Proposed biosynthetic pathway of rosmarinic acid in cell cultures of *Coleus blumei* Benth. *Plant* 189:10–14, 1993.

53. Zenk, M.H., H. El-Shagi, B. Ulbrich. Production of rosmarinic acid by cell suspension cultures of *Coleus blumei*. *Naturwissenschaften* 64:585–586, 1977.

54. Nitzsche, A., S.V. Tokalov, H.O. Gutzeit, J. Ludwig-Muller. Chemical and biological characterization of cinnamic acid derivatives from cell cultures of lavender (*Lavandula officinalis*) induced by stress and jasmonic acid. *J. Agric. Food Chem.* 52:2915–2923, 2004.

55. Santos-Gomes, P.C., R.M. Seabra, P.B. Andrade, M. Fernandes-Ferreira. Determination of phenolic antioxidant compounds produced by calli and cell suspension of sage (*Salvia officinalis* L). *J. Plant Physiol.* 160:1025–1032, 2003.

56. Kintzios, S., O. Makri, E. Panagiotopoulos, M. Scapeti. *In vitro* rosmarinic acid accumulation in sweet basil (*Ocimum basilicum* L.). *Biotechnol. Lett.* 25:405–408, 2003.

57. Mizukami, H., T. Ogawa, H. Ohashi, B.E. Ellis. Induction of rosmarinic acid biosynthesis in *Lithospermum erythrorhizon* cell suspension cultures by yeast extract. *Plant Cell Rep.* 11:480–483, 1992.

58. Szabo, E., A. Thelen, M. Petersen. Fungal elicitor preparations and methyl jasmonate enhance rosmarinic acid accumulation in suspension cultures of *Coleus blumei*. *Plant Cell Rep.* 18:485–489, 1999.

59. Chen, H., F. Chena, F.C. Chiu, C.M. Lo. The effect of yeast elicitor on the growth and secondary metabolism of hairy root cultures of *Salvia miltiorrhiza*. *Enzyme Microb. Technol.* 28:100–105, 2001.

60. Kreis, W., E. Reinhard. The production of secondary metabolites by plant cells cultivated in bioreactors. *Plant Medica* 55:409–416, 1989.

61. Ulbrich, B., W. Wiesner, H. Arens. Large-scale production of rosmarinic acid from plant cell cultures of *Coleus blumei* Benth. In: *Primary and Secondary Metabolism of Plant Cell Cultures,* Neumann, K. Ed., Heidelberg: Springer-Verlag, 1985, pp 292–303.

62. Georgiev, M., A. Pavlov, M. Ilieva. Rosmarinic acid production by *Lavandula vera* MM cell suspension: the effect of temperature. *Biotechnol. Lett.* 26:855–856, 2004.

63. Kinzios, S., H. Kollias, E. Straitouris, O. Makri. Scale-up micropropagation of sweet basil (*Ocimum basilicum* L.) in an air-lift bioreactor and accumulation of rosmarinic acid. *Biotechnol. Lett.* 26:521–523, 2004.

64. Phillips, R.L., S.M. Kaeppler, J. Olhoft. Genetic instability of plant tissue cultures: Breakdown of normal controls. *Proc. Natl. Acad. Sci. USA* 91:5222–5226, 1994.

65. De-Eknamkul, W., B.E. Ellis. Purification and characterization of tyrosine aminotransferase activities from *Anchusa officinalis* cell cultures. *Arch. Biochem. Biophys.* 257:430–438, 1987.

66. Razzaque, A., B.E. Ellis. Rosmarinic acid production in *Coleus* cell cultures. *Planta* 137:287–291, 1977.

67. Whitaker, R.J., T. Hashimoto, D.A. Evans. Production of secondary metabolites, rosmarinic acid by plant cell suspension cultures. *Ann. N.Y. Acad. Sci.* 435:363–368, 1984.

68. De-Eknamkul, W., B.E. Ellis. Effects of macronutrients on the growth and rosmarinic acid formation in cell suspension cultures of *Anchusa officinalis*. *Plant Cell Rep.* 4:46–49, 1985.

69. De-Eknamkul, W., B.E. Ellis. Effects of auxins and cytokinins on the growth and rosmarinic acid formation in cell suspension cultures of *Anchusa officinalis*. *Plant Cell Rep.* 4:50–53, 1985.

70. Pavlov, A.I., M.P. Ilieva, I.N. Panchev. Nutrient medium optimization for rosmarinic acid production by *Lavandula vera* MM cell suspension. *Biotechnol. Prog.* 16:668–670, 2000.

71. Mohagheghzadeh, A., M. Shams-Ardakani, A. Ghannadi, M. Minaeian. Rosmarinic acid from *Zataria multiflora* tops and *in vitro* cultures. *Fitoterapia* 75:315–321, 2004.

72. Gertolowski, C. M. Petersen. Influence of carbon source on growth and rosmarinic acid production in suspension cultures of *Coleus blumei*. *Plant Cell Tissue Org. Cult.* 34:183–190, 1993.

73. Ellis, B.E., G.H.N. Towers. Biogenesis of rosmarinic acid in *Mentha*. *Biochem. J.* 118:287–291, 1970.

74. Petersen, M. Cytochrome P-450-dependent hydroxylation in the biosynthesis of rosmarinic acid in *Coleus*. *Phytochemistry* 45:1165–1172, 1997.

75. Petersen, M., M.S.J. Simmonds. Molecules of interest: rosmarinic acid. *Phytochemistry* 62:121–125, 2003.

76. De-Eknamkul, W., B.E. Ellis. Tyrosine aminotransferase: the entry point enzyme of the tyrosine-derived pathway in rosmarinic acid biosynthesis. *Phytochemistry,* 26:1941–1946, 1987.

77. Mizukami, H., B.E. Ellis. Rosmarinic acid formation and differential expression of tyrosine aminotransferase isoforms in *Anchusa officinalis* cell suspension cultures. *Plant Cell Rep.* 10:321–324, 1991.

78. Petersen, M., E. Hausler, J. Meinhard, B. Karwatzki, C. Gertlowski. The biosynthesis of rosmarinic acid in suspension cultures of *Coleus blumei*. *Plant Cell Tissue Org. Cult.* 38:171–179, 1994.

79. De-Eknamkul, W., B.E. Ellis. Purification and characterization of prephenate aminotransferase from *Anchusa officinalis* cell cultures. *Arch. Biochem. Biophys.* 267:87–94, 1988.

80. Hausler, E., M. Petersen, A.W. Alfermann. Hydroxyphenylpyruvate reductase from cell suspension cultures of *Coleus blumei*. *Benth. Naturforsch.* 46C:371–376, 1991.

81. Petersen, M. Characterization of rosmarinic acid synthase from cell cultures of *Coleus blumei*. *Phytochemistry* 30:2877–2881, 1991.

82. Petersen. M., A.W. Alfermann. Two enzymes of rosmarinic acid biosynthesis from cell cultures of *Coleus blumei*: Hydroxyphenylpyruvate reductase and rosmarinic acid synthase. *Z. Naturfosch.* 43C:501–504, 1988.

83. Petersen, M. Cinnamic acid 4-hydroxylase from cell cultures of the hornwort *Anthoceros agrestis. Planta* 217:96–101, 2003.

84. Matsuno, M., A. Nagatsu, Y, Ogihara, B.E. Ellis, H. Mizukami. CYP98A6 from *Lithospermum erythrorhizon* encodes 4-coumaroyl-4'-hydroxyphenyllactic acid 3-hydroxylase involved in rosmarinic acid biosynthesis. *FEBS Lett.* 514:219–224, 2002.

85. Schoch, G., S. Goepfert, M. Morant, A. Hehn, D. Meyer, P. Ullmann, D. Werck-Reichhart. CYP98A3 from *Arabidopsis thaliana* is a 3'-hydroxylase of phenolic esters, a missing link in the phenylpropanoid pathway. *J. Biol. Chem.* 276:36566–36574, 2001.

86. Anterola, A.M., J.H. Jeon, L.B. Davin, N.G. Lewis. Transcriptional control of monolignol biosynthesis in *Pinus taeda*. Factors affecting monolignol ratios and carbon allocation in the phenylpropanoid metabolism. *J. Biol. Chem.* 277:18272–18280, 2002.

87. Shetty, K., O.F. Curtis, R.E. Levin, R. Witkowsky, W. Ang. Prevention of vitrification associated with *in vitro* shoot culture of oregano (*Origanum vulgare*) by *Pseudomonas* spp. *J. Plant Physiol.* 147:447–451, 1995.

88. Shetty, K., O.F. Curtis, R.E. Levin. Specific interaction of mucoid strains of *Pseudomonas* spp. with oregano (*Origanum vulgare*) clones and the relationship to prevention of hyperhydricity in tissue culture. *J. Plant Physiol.* 149:605–611, 1996.

89. Eguchi, Y., O.F. Curtis, K. Shetty. Interaction of hyperhydricity-preventing *Pseudomonas* spp. with oregano (*Origanum vulgare*) and selection of high rosmarinic acid-producing clones. *Food Biotechnol.* 10:191–202, 1996.

90. Shetty, K., T.L. Carpenter, D. Kwok, O.F. Curtis, T.L. Potter. Selection of high phenolics-containing clones of thyme (*Thymus vulgaris* L.) using *Pseudomonas* spp. *J. Agric. Food Chem.* 44:3408–3411, 1996.

91. Yang, R., O.F. Curtis, K. Shetty. Selection of high rosmarinic acid-producing clonal lines of rosemary (*Rosmarinus officinalis*) via tissue culture using *Pseudomonas* sp. *Food Biotechnol.* 11:73–88, 1997.

92. Al-Amier, H., B.M.M. Mansour, N. Toaima, R.A. Korus, K. Shetty. Tissue culture-based screening for selection of high biomass and phenolic-producing clonal lines of Lavender using *Pseudomonas* and azetidine-2-carboxylate. *J. Agric. Food Chem.* 47:2937–2943, 1999.

93. Al-Amier, H., B.M.M. Mansour, N. Toaima, R.A. Korus, K. Shetty. Screening of high biomass and phenolic-producing clonal lines of Spearmint in tissue culture using *Pseudomonas* and azetidine-2-carboxylate. *Food Biotechnol.* 13:227–253, 1999.

94. Al-Ameir, H., B.M.M. Mansour, N. Toaima, L. Craker, K. Shetty. Tissue culture for phenolics and rosmarinic acid in thyme. *J. Herbs Spices Med. Plant.* 8:31–42, 2001.

95. Shetty, K. Role of proline-linked pentose phosphate pathway in biosynthesis of plant phenolics for functional food and environmental applications: a review. *Process. Biochem.* 39:789–804, 2004.

96. Shetty, K., M.L. Wahlqvist. A model for the role of proline-linked pentose phosphate pathway in phenolic phytochemical biosynthesis and mechanism of action for human health and environmental applications: a review. *Asia Pac. J. Clin. Nutr.* 13:1–24, 2004.

97. Phang, J.M. The regulatory functions of proline and pyrroline-5-carboxylic acid. *Curr. Topics Cell. Reg.* 25:91–132, 1985.

98. Hagedorn, C.H., J.M. Phang. Transfer of reducing equivalents into mitochondria by the interconversions of proline and δ–pyrroline-5-carboxylate. *Arch. Biochem. Biophys.* 225:95–101, 1983.

99. Rayapati J.P., C.R. Stewart. Solubilization of a proline dehydrogenase from maize (*Zea mays* L.) mitochondria. *Plant Physiol.* 95:787–791, 1991.

100. Yeh, G.C., J.M. Phang. The function of pyrroline-5-carboxylate reductase in human erythrocytes. *Biochem. Biophys. Res. Commun.* 94:450–457, 1980.

101. Phang, J.M., S.J. Downing, G.C. Yeh, R.J. Smith, J.A. Williams. Stimulation of hexos-emonophosphate-pentose pathway by δ-pyrroline-5-carboxylic acid in human fibroblasts. *Biochem. Biophys. Res. Commun.* 87:363–370, 1979.

102. Hagedorn, C.H., J.M. Phang. Catalytic transfer of hydride ions from NADPH to oxygen by the interconversions of proline and δ-pyrroline-5-carboxylate. *Arch. Biochem. Biophys.* 248:166–174, 1986.

103. Kwok, D., K. Shetty. Effect of proline and proline analogs on total phenolic and rosmarinic acid levels in shoot clones of thyme (*Thymus vulgaris* L.). *J. Food Biochem.* 22:37–51, 1998.

104. Yang, R., K. Shetty. Stimulation of rosmarinic acid in shoot cultures of oregano (*Origanum vulgare*) clonal line in response to proline, proline analog and proline precursors. *J. Agric. Food Chem.* 46:2888–2893, 1998.

105. Bela, J., K. Shetty. Somatic embryogenesis in anise (*Pimpinella anisum* L.): The effect of proline on embryogenic callus formation and ABA on advanced embryo development. *J. Food Biochem.* 23:17–32, 1999.

106. Perry, P.L., Shetty, K. A model for involvement of proline during *Pseudomonas*-mediated stimulation of rosmarinic acid. *Food Biotechnol.* 13:137–154, 1999.

107. Lendzian, K.J. Modulation of glucose-6-phosphate dehydrogenase by NADPH, NADP+ and dithiothreitol at variable NADPH/NADP+ ratios in an illuminated reconstituted spinach (*Spinacia oleracea* L.) choroplast system. *Planta* 148:1–6, 1980.

108. Copeland, L., J.F. Turner. The regulation of glycolysis and the pentose-phosphate pathway. In: *The Biochemistry of Plants*, Vol. 11, Stumpf, F., E.E. Conn, eds., New York: Academic Press, 1987; 107–125.

109. Elthon, T.E., C.R. Stewart. Effects of proline analog L-thiazolidine-4-carboxylic acid on Proline metabolism. *Plant Physiol.* 74:213–218, 1984.

110. Jost, A., S. Perlman, O. Valentino, M. Castinier, R. Scholler, S. Magre. Experimental control of the differentiation of Leydig cells in the rat fetal testis. *Proc. Natl. Acad. Sci. USA* 85:8094–8097, 1988.

111. Kohl, D.H., K.R. Schubert, M.B. Carter, C.H. Hagdorn, G. Shearer. Proline metabolism in N_2-fixing root nodules: energy transfer and regulation of purine synthesis. *Proc. Natl. Acad. Sci. USA* 85:2036–2040, 1988.

112. McCue, P., K. Shetty. A biochemical analysis of mungbean (*Vigna radiata*) response to microbial polysaccharides and potential phenolic-enhancing effects for nutraceutical applications. *Food Biotechnol.* 6:57–79, 2002.

113. McCue, P., K. Shetty. Clonal herbal extracts as elicitors of phenolic synthesis in dark-germinated mungbeans for improving nutritional value with implications for food safety. *J. Food Biochem.* 26:209–232, 2002.

114. Randhir, R., K. Shetty. Microwave-induced stimulation of L-DOPA, phenolics and antioxidant activity in fava bean (*Vicia faba*) for Parkinson's diet. *Process Biochem.* 39:1775–1784, 2004.

115. Randhir, R., P. Shetty, K. Shetty. L-DOPA and total phenolic stimulation in dark geriminated fava bean in response to peptide and phytochemical elicitors. *Process Biochem.* 37:1247–1256, 2002.

116. Randhir, R., K. Shetty. Light-mediated fava bean (*Vicia faba*) response to phytochemical and protein elicitors and consequences on nutraceutical enhancement and seed vigor. *Process Biochem.* 38:945–952, 2003.

117. McCue, P., Z. Zheng, J.L. Pinkham, K. Shetty. A model for enhanced pea seedling vigor following low pH and salicylic acid treatments. *Process Biochem.* 35:603–613, 2000.

118. Droge, W. Free radicals in the physiological control of cell function. *Physiol. Rev.* 82:47–95, 2002.

119. Dreher, D., A.F. Junod. Role of oxygen free radicals in cancer development. *Eur. J. Cancer* 32A:30–38, 1996.

120. Ha, H.C., A. Thiagalingam, B.D. Nelkin, R.A. Casero, Jr. Reactive oxygen species are critical for the growth and differentiation of medullary throid carcinoma cells. *Clin. Cancer Res.* 6:3783–3787, 2000.

121. Alexander, R.W. Theodore Cooper Memorial Lecture: hypertension and the pathogenesis of atherosclerosis: oxidative stress and the mediation of arterial inflammatory response: a new perspective. *Hypertension* 25:155–161, 1995.

122. Griendling, K.K., C.A. Minieri, J.D. Ollerenshaw, R.W. Alexander. Angiostatin II stimulates NADH and NADPH oxidase activity in cultured vascular smooth muscle cells. *Circ. Res.* 74:1141–1148, 1994.

123. Araujo, V., C. Arnal, M. Boronat, E. Ruiz, C. Dominguez. Oxidant-antioxidant imbalance in blood of children with juvenile rheumatoid arthritis. *Biofactors* 8:155–159, 1998.

124. Mapp, P.I., M.C. Grootveld, D.R. Blake. Hypoxia, oxidative stress and rheumatoid arthritis. *Br. Med. Bull.* 51:419–436, 1995.

125. Multhaup, G., T. Ruppert, A. Schlicksupp, L. Hesse, D. Beher, C.L. Masters, K. Beyreuther. Reactive oxygen species in Alzheimer's disease. *Biochem. Pharmacol.* 54:533–539, 1997.

126. Wolff, S.P. Diabetes mellitus and free radicals: free radicals, transition metals and oxidative stress in the aetiology of diabetes mellitus and complications. *Br. Med. Bull.* 49:642–652, 1993.

127. Baynes, J.W. Role of oxidative stress in development of complications in diabetes. *Diabetes* 40:405–412, 1991.

128. De Mattia, G., M.C. Bravi, O. Laurenti, M. Cassone-Faldetta, A. Armeinto, C. Ferri, F. Balsano. Influence of reduced glutathione infusion on glucose metabolism in patients with non-insulin-dependent diabetes mellitus. *Metabolism* 47:993–997, 1998.

129. Nishikawa, T., D. Edelstein, X.L. Du, S. Yamagishi, T. Matsumura, Y. Kaneda, M.A. Yorek, D. Beebe, P.J. Oates, H.P. Hammes, I. Giardino, M. Brownlee. Normalizing mitochondrial superoxide production blocks three pathways of hyperglycemic damage. *Nature* 404:787–790, 2000.

130. Van Dam, P.S., van Asbeck, B.S., Erkelens, D.W., Marx, J.J.M., Gispen, W.H. and Bravenboer, B. The role of oxidative stress in neuropathy and other diabetic complications. *Diabetes Metab. Rev.* 11:181–192, 1995.

131. Hertog, M.G.L., D. Kromhout, C. Aravanis, H. Blackburn, R. Buzina, F. Fidanza, S. Giampaoli, A. Jansen, A. Menotti, S. Nedeljkovic, M. Pekkarinen, B.S. Simic, H. Toshima, E.J.M. Feskens, P.C.H. Hollman, M.B. Kattan. Flavonoid intake and long-term rosk of coronary heart disease and cancer in the seven countries study. *Arch. Intern. Med.* 155:381–386, 1995.

132. Masuda, T., A. Jitoe. Antioxidative and anti-inflammatory compounds from tropical gingers: isolation, structure determination, and activities of cassumunins A, B and C, new complex curcuminoids from *Zingiber cassumunar*. *J. Agric. Food Chem.* 42:1850–1856, 1994.

133. Narayanan, B.A., G.G. Re. IGF-II down regulation associated cell cycle arrest in colon cancer cells exposed to phenolic antioxidant ellagic acid. *Anticancer Res.* 21:359-364, 2001.

134. Labriola, D., R. Livingston. Possible interactions between dietary antioxidants and chemotherapy. *Oncology* 13:1003–1008, 1999.

135. Freudenheim, J.L., J.R. Marshall, J.E. Vena. Premenupausal breast cancer risk and intake of vegetables, fruits and related nutrients. *J. Natl. Cancer Inst.* 88:340–348, 1996.

136. Morel, Y., R. Barouki. Repression of gene expression by oxidative stress. *Biochem. J.* 342:481–496, 1999.

9

Bioprocessing Strategies to Enhance L-DOPA and Phenolic Antioxidants in the Fava Bean *(Vicia faba)*

Kalidas Shetty, Reena Randhir, and Preethi Shetty

CONTENTS

9.1 Introduction and Role of Phenolic Secondary Metabolites 209
9.2 Plant Phenolics in Human Health and as Antioxidants 211
9.3 Biosynthesis of Phenolic Metabolites and L-DOPA...213
9.4 L-DOPA and Parkinsonian Syndrome 215
    9.4.1 L-DOPA from Natural Sources..217
    9.4.2 L-DOPA from Fava Bean 217
9.5 Linking L-DOPA Synthesis to the Pentose Phosphate and Phenylpropanoid
    Pathways..218
9.6 Recent Progress on Elicitation Linked Bioprocessing to Enhance
    L-DOPA and Phenolics in Fava Bean Sprouts 220
9.7 Recent Progress on Solid-State Bioprocessing of the Fava Bean to Enhance
    L-DOPA and Phenolics Using Food Grade Fungal Systems 222
9.8 Implications and Summary 222
References..223

9.1 INTRODUCTION AND ROLE OF PHENOLIC SECONDARY METABOLITES

In plants, primary products such as carbohydrates, lipids, proteins, photosynthetic components, and nucleic acids are common to all; they are involved in the primary metabolic processes of building and maintaining cells. In contrast, secondary metabolites do not appear to have such a vital biochemical role, but studies have indicated a role of these chemicals in defense and stress response of plants. Some of the most abundant stress induced secondary metabolites synthesized by plants are phenolics and their derivatives. Phenolic compounds include a large array of chemical compounds possessing an aromatic ring bearing one or more hydroxyl groups, together with a number of other side groups. Plant phenolics are a chemically heterogeneous group (1–3). These phenolics usually occur in conjugated or esterified form as glycosides (1,2,4). The diverse arrays of plant phenolics have many roles in plant growth and development. Therefore, emergence of dietary and medicinal applications for phenolic phytochemicals, harnessing especially their antioxidant and antimicrobial properties, for the benefit of human health and wellness is not altogether surprising. As stress damage at the cellular level appears similar among eukrayotes, it is logical to suspect that there may be similarities in the mechanism for cellular stress mediation between eukaryotic species. Plant adaptation to biotic and abiotic stress involves the stimulation of protective secondary metabolite pathways (5–7), resulting in the biosynthesis of phenolic antioxidants. Studies indicate that plants exposed to ozone responded with increased transcript levels of enzymes in the phenylpropanoid and lignin pathways (8). Increase in plant thermotolerance is related to the accumulation of phenolic metabolites and heat shock proteins that act as chaperones during hyperthermia (9). Phenolics and specific phenolic like salicyclic acid levels increase in response to infection, acting as defense compounds or serving as precursors for the synthesis of lignin, suberin, and other polyphenolic barriers (10). Antimicrobial phenolics, called phytoalexins, are synthesized around the site of infection during pathogen attack and, along with other simple phenolic metabolites, are believed to be part of a signaling process that results in systemic acquired resistance (5–7). Many phenylpropanoid compounds such as flavonoids, isoflavonoids, anthocyanins, simple phenolics, and polyphenols are induced in response to wounding (11), nutritional stress (12), cold stress (13), and high visible light (14). Ultraviolet (UV) irradiation induces light-absorbing flavonoids and sinapate esters in *Arabidopsis* to block radiation and protect DeoxyriboNucleic Acid (DNA) from dimerization and cleavage (15). In general, the initiation of the stress response arises from certain changes in the intracellular medium (16) that transmits the stress induced signal to cellular modulating systems, resulting in changes in cytosolic calcium levels, proton potential as a long distance signal (17), and low molecular weight proteins (18). Stress can also initiate free radical generating processes and shift the cellular equilibrium toward lipid peroxidation (19). It is believed that the shift in prooxidant antioxidant equilibrium is a primary nonspecific event in the development of the general stress response (20). Therefore, phenolic compounds are ubiquitous and have important roles in all vascular plants, and as a result are integral part of the human diet (21–23). These phenolic secondary metabolites that are synthesized through the shikimic acid pathway vary from simple phenolics such as the hydroxy benzoic acids and levodopa (L-DOPA) to biphenyls such as resveratrol and rosmarinic acid to large condensed tannins and hydrolysable tannins with high molecular weights (23,24). The polymers formed from plant phenolics in the cell wall provide structural support and form barriers to prevent moisture loss diffusion and pathogen encroachment. The phenolics also function in defense mechanisms with UV protectant, antifungal, antibacterial, antifeedant, and antimitotic properties, and in morphogenesis (25,26). When exposed to air, most phenolics readily undergo oxidation to colored quinone containing products. This response is frequently observed as a browning reaction of plant tissues as a part of a healing response. The oxidation of these compounds by polyphenoloxidase (PPO) has been suggested to be the main cause of apple browning (27). Therefore, protective phenolic metabolites involved in such secondary metabolite linked stress responses in food plant species can be targeted as a source of therapeutic and disease-preventing functional ingredients, especially in oxidation disease linked diets (diets containing foods with a high glycemic index and saturated fats) and environmentally (physically, chemically, and biologically) influenced chronic disease problems (23).

9.2   PLANT PHENOLICS IN HUMAN HEALTH AND AS ANTIOXIDANTS

It is evident that plant phenolic compounds constitute one of the most numerous and widely distributed groups of substances with more than 8000 phenolic structures currently known (28). In addition to stress linked phenolics coming only from the shikimate and phenylpropanoid pathways, a number of the phenolic compounds are found in plants, including the flavonoids that contribute to the characteristic flavor and fragrance of vegetables, fruits, tea, and wine. These compounds, which come from phenylpropanoid and polyketide (acetate–malonate) pathways, also have biological properties that are beneficial to human health. Flavonoids such as quercetin and catechin and isoflavonoids, genistein for example, are being investigated for properties which may reduce the incidence of cancer (22,23). Flavonoids and isoflavonoids are a class of phenolic compounds that have appeared sequentially during plant evolution and are simple aromatic compounds generated from both the phenypropanoid and acetate–malonate (polyketide) pathways (24). From a functional health point of view, it is suggested that such phenolics, for example, through the consumption of tea, may provide protection against certain cancers; soybeans may provide protection against breast cancer and osteoporosis (22). The Japanese and Chinese frequently used plants rich in polyphenol tannins in their folk medicines for the treatment of inflammation, liver injury, kidney ailments, hypertension, and ulcers. Oregano extracts have been shown to inhibit lipid peroxidation by their flavonoid fractions such as flavone apigenin, the flavanone, eriodictoyl, dihydrokaempferol, and dihydroquercetin (29). Rosmarinic acid-containing *Ocimum sanctum* (holy basil), derived from the phenylpropanoid pathway, is commonly used in India to reduce fevers and gastrointestinal disease (30). Essential oils from thyme (*Thyme vulgaris* L.) have phenolic linked antioxidant properties, which may result from the presence of free radical scavengers in these oils (31–33). These diverse phenolic compounds in thyme are unusually derived from phenylpropanoid, polyketide, and terpenoid pathways. Many other phenolic compounds have the ability to inhibit platelet aggregation, block calcium influx, and protect low density lipoproteins (LDL) from oxidation (34). L-DOPA from the fava bean, that is the focus of this chapter, has been evaluated in Parkinson's management (35–37); it is derived from L-tyrosine of the phenylpropanoid pathway (Figure 9.1).

   Among many functional roles, the most important function assigned to phenolics is their antioxidant activity. Antioxidants may be defined as substances, which when present in low concentrations compared with those of an oxidizable substrate, such as proteins, carbohydrates, and fats, delay or prevent the oxidation of the substrate (38). Phenolic antioxidants from dietary plants can be useful to counter reactive oxygen species related to human diseases. Reactive oxygen species (ROS) are able to oxidize cellular components such as DNA, lipids, and proteins (23). Dysfunction of oxidative phosphorylation at the mitochondrion has been recognized as a major physiological source of ROS (39). Lipid peroxidation damages the structural integrity of the mitochondria, which can result in organelle swelling, resulting in increased permeability to cations, decreased membrane potential, and damage to electron transfer activities (40). This form of tissue damage can ultimately lead to some of the major diseases such as cancer, cardiovascular disease, immune dysfunction, diabetes, and neurodegeneration (23). To deal with ROS, the biological systems, including human cellular systems, have an effective defense system, which includes enzymes such as superoxide dismutase (SOD), catalase (CAT), high molecular weight antioxidants, and an array of low molecular weight antioxidants such as ascorbic acid, α-tocopherol, β-carotene, and glutathione (41). The endogenous antioxidant responses, both enzymatic and nonenzymatic could also be enhanced by dietary intake of

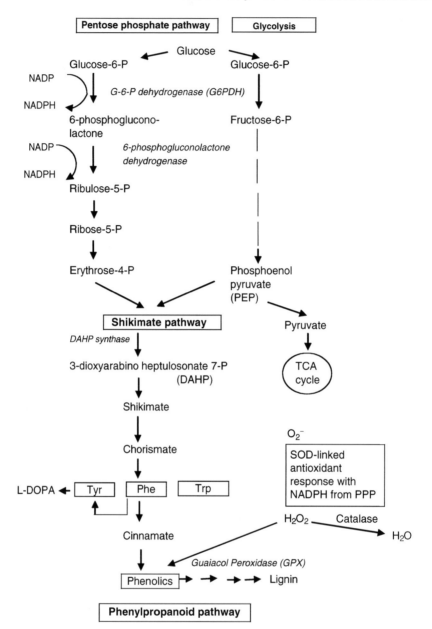

Figure 9.1 Pentose–phosphate pathway and L-DOPA biosynthesis in plants.

phenolic antioxidants from plant foods (23,42). Therefore phenolic antioxidants, through design and use of functional foods, are receiving increasing interest from consumers and food manufacturers due to their synergistic roles as antioxidants, antimutagens, and scavengers of free radicals (23,43). They have the potential to function as antioxidants by trapping free radicals generated in the oxidative chemistry, which then normally undergo coupling reaction leading to polymeric or oligomeric products (43) or by enhancing host antioxidant enzyme response through a stimulation of antioxidant enzyme response through SOD, peroxidases, and CAT (23). Epidemiological studies have also suggested associations between the consumption of phenolic antioxidant rich foods or beverages and the prevention of diseases (44–49).

9.3   BIOSYNTHESIS OF PHENOLIC METABOLITES AND L-DOPA

Plant phenolics are synthesized using several different routes. This constitutes a heterogeneous group from a metabolic point of view. Two basic pathways are the shikimic acid pathway and the acetate–malonate (polyketide) pathway. The shikimic acid pathway represents the principal mode of accumulation of plant stress related phenolic compounds. The acetate–malonate pathway is also an important source of phenolic secondary products for biosynthesis of flavonoids and isoflavonoids that have many human disease protective chemoprevention properties.

The shikimic acid pathway requires substrates such as erythrose-4-phosphate (E4P) from the pentose phosphate pathway (PPP) and phosphoenol pyruvate (PEP) from glycolysis (Figure 9.1). The oxidative pentose phosphate pathway in plants is thought of as comprising two stages (50). The first is an essential irreversible conversion of glucose 6 phosphate (G6P) to ribulose-5-phosphate (Ril5P) by the enzymes glucose-6-phosphate dehydrogenase (G6PDH), 6-phosphogluconolactonase, and 6-phospho gluconolactonate dehydrogenase (6 PGDH). This oxidative stage provides reductant in the form of NADPH for a wide range of anabolic pathways including the synthesis of fatty acids, the reduction of nitrite, and synthesis of glutamate (51,52). The second stage is the irreversible series of interconversions between phosphorylated carbon sugars. The function of this stage of the pathway is to provide a carbon skeleton for the shikimate pathway via erythrose-4-phosphate (E4P), nucleotide synthesis utilizing ribose-5-phosphate, as well as recycling of sugar phosphate intermediates for use in the glycolytic pathway (53).

The shikimate pathway is often referred to as the common aromatic biosynthetic pathway, even though in nature it does not synthesize all aromatic compounds by this route (54). The shikimic acid pathway converts the simple carbohydrate precursors to aromatic amino acids phenylalanine, tyrosine, and tryptophan. The flux from this pathway is critical for both auxin and phenylpropanoid synthesis (23,30). Auxins are plant hormones that regulate plant development (2). Up to 60% of the dry weight in some plant tissue consists of metabolites derived from the shikimate pathway. Activity of the distinct isoenzyme of 3-deoxy arabinose heptulosonate-7-phosphate (DAHP) synthase in the shikimate pathway is dependent on metabolite flux from E4P. This enzyme has been shown to be subject to feedback inhibition by L-phenylalanine, L-tyrosine, and L-tryptophan (54,55). Therefore, this enzyme controls the carbon flow into the shikimate pathway.

The most abundant class of secondary phenolic compounds in plants is derived from phenylalanine via the elimination of an ammonia molecule to form cinnamic acid. This reaction is catalyzed by phenylalanine ammonia lyase (PAL). This is the branch point between the primary (shikimate pathway) and the secondary (phenylpropanoid pathway) (6). Studies with several different species of plants have shown that the activity of PAL is increased by environmental factors such as low nutrient levels, light (through the effect of phytochrome), and fungal infection (11). Fungal invasion triggers the transcription of messenger RNA (mRNA) in the plant, which then stimulates the synthesis of phenolic compounds (6). Many phenylpropanoid compounds are induced in response to wounding or in response to microbial pathogens, insect pests, or herbivores. Anthocyanins increase in response to high visible light levels and it is thought that they attenuate the amount of light reaching the photosynthetic cells (6).

The product of PAL, *trans*-cinnamic acid, is converted to *para*-coumaric acid by the addition of a hydroxyl group on the aromatic ring in *para* position. Subsequent reactions lead to the addition of more hydroxyl groups and other substituents. These are simple phenolic compounds called phenylpropanoids because they contain a benzene ring and a three carbon side chain. Phenylpropanoids are building blocks for more complex phenolic compounds (1,3).

As previously discussed, phenolic compounds have wide ranges of functions. The synthesized phenolics can be either antioxidant in nature or they may function in lignification of the plant cells. Depending on the requirements, the type of the phenolics synthesized and their complexity vary from species to species in different environmental niches. Flavonoids, tannins, caffeic acids, curcumin, gallic acids, eugenol, rosmarinic acid, and many more have antioxidant properties (56,57). These ranges of phenolics provide plants with defense mechanisms and act as scavengers of free radicals as described earlier. Other functions of phenolics include their ability to provide structural stability to the plants by lignins and lignans. These are complex phenolics are formed from the polymerization of simple phenols (58). Figure 9.2 illustrates the origins of varied phenolic compounds from simple phenols.

Lignin is a polymer of aromatic subunits usually derived from phenylalanine. It serves as a matrix around the polysaccharide components of some plant cell walls, providing additional rigidity and compressive strength, as well as rendering the walls hydrophobic for water impermeability (59,60). The final enzymatic steps of lignin biosynthesis, the production of mesomeric phenoxy radicals from cinnamoyl alcohol, is catalyzed by peroxidase and must occur outside the cell to allow these short lived radicals to polymerize *in situ* (61). Phenol polymerization is catalyzed by the peroxidase enzyme (62) and a specific enzyme, Guaiacol peroxidase (GPX), is suggested to be important in the metabolic interconversion of phenolic antioxidants. The same phenylpropanoid pathway also supports the synthesis of L-DOPA, which is a simple phenolic compound found in many seeded legumes such as fava beans and velvet beans with relevance for Parkinson's therapy (37,63–66). This pathway

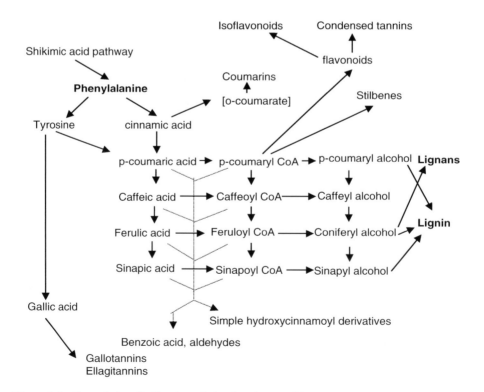

Figure 9.2   General phenolic flux through the phenylpropanoid pathway from the shikimic pathway.

provides the precursor chorismate, which oxidizes to L-phenylalanine, before going through a hydroxylation step to form L-tyrosine and than L-DOPA (Figure 9.1).

9.4   L-DOPA AND PARKINSONIAN SYNDROME

The Parkinsonian syndrome has long held the interest of psychologists, psychiatrists, and other behavioral investigators. Lately, the field of cerebral monoamines has received increased attention; first with the serotonin hypothesis of Brodie, then the noradrenaline hypothesis of Schildrant, in relation to neuron function, is receiving particular attention. (67). Dopamine is postulated to be the key neurotransmitter responsible for the origin of human intelligence (67). Dopamine, the direct precursor for noradrenaline, has a specific distribution pattern within the brain (Figure 9.3). It is concentrated mainly in the striatum and substantia nigra. It is deficient in patients with Parkinson's diseases (PD). L-DOPA has been studied as a treatment for neurological disorders such as Parkinson's disease (68,69).

Parkinson's disease is a common neurological disorder, affecting 1% of the people over the age of 60 years; it is a disease with the signs of rigidity, resting tremor, postural instability, which is due to underproduction of the neurotransmitter dopamine (DA) (70). One of the factors leading to PD is the marked loss of melanized nigrostriatal dopamine neurons, one of the principal components responsible for the normal control of movement. Signs of PD do not appear until a large majority of the nigrostriatal DA neurons have been damaged. Another factor that may mitigate the loss of nigrostriatal DA neurons in PD is a reduced rate of DA activation (71). Also, mitochondrial swelling was noticed as a result of low concentrations of DA, causing a mitochondrial damage triggering neurodegenerative process and PD (72). DA receptors are expressed not only in the central nervous system (CNS), but also in several peripheral tissues including arteries, heart, thymus, and peripheral blood lymphocytes (73). Unconjugated DA represents only 20–40% of the total DA excreted in human urine, with the remainder being mainly excreted in the form of phenolic sulfate (74). It has been suggested that a low level of DA excretion is associated with hypokinesia and normal or excess rate of excretion with hyperkinesia and associated renal dysfunction (74). Overproduction of this catecholamine appears to be associated with the psychological disorder schizophrenia (68).

Introduction of L-DOPA has improved the clinical status of patients with Parkinson's disease (75). Levodopa (L-DOPA — synonym: levo-dihydroxyphenylalanine) therapy in Parkinson's disease was initiated after the suggestion of the role of dopamine in the neuronal system (71,76–81). Dopamine is also known to have a role in renal function, through its action on renal dopaminergic receptors and improved concentration of intrarenal area is dependent on L-DOPA availability (82). The biosynthesis dopamine is considered to start from tyrosine. obtained from dietary sources (82) (Figure 9.3). Blood borne tyrosine is taken into the brain to effect the functioning of dopaminergic neurons. When tyrosine has entered the neurons, L-DOPA (dihydoxyphenylalanine) is made using the cytsolic enzyme tyrosine hydroxylase (71). Subsequently, another cytosolic enzyme aromatic amino acid decarboxylase converts L-DOPA to dopamine (71). In the neurons, striatum is one of the main components of basal ganglia that are responsible for normal control of movement and role of dopamine is essential for this action (71). In Parkinson's disease, the importance of dopamine for striatum is clear due to the motor abnormalities seen in patients, which is characterized by a marked loss of melanized nigrostriatal dopamine neurons (71). Dopamine release in the striatum modulates activity in direct and indirect circuits, which are important for voluntary control of movement (71). Signs of Parkinson's

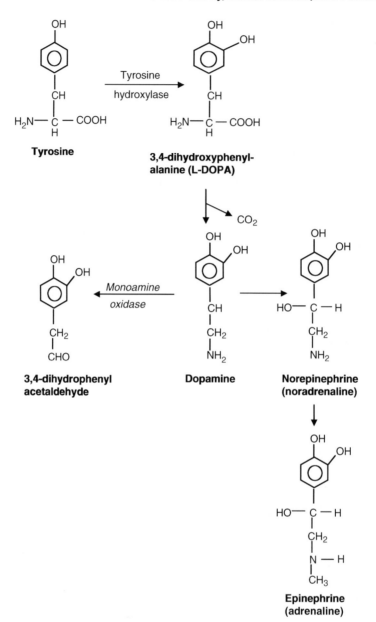

Figure 9.3 Dopamine pathway in neuronal cells.

disease are observed after damage to majority of nigrostriatal dopamine neurons when 80% of striatal dopamine loss has occurred (71).

However, declining efficacy is observed over the course of L-DOPA treatment for Parkinson's (75). In addition, the L-DOPA therapy is associated with a variety of side effects such as dyskinesias, fluctuations in motor performance, confusion, hallucinations and sensory syndrome (37,75). To reduce the side effects, L-DOPA dosage is reduced, which limits the antiParkinsonian benefits (75). The benefits of lower dosage and short absence of L-DOPA intake to reduce side effects has had limited success (75) and therefore L-DOPA from fava beans is promising as an alternative Parkinson's therapy (35,37,69).

9.4.1   L-DOPA from Natural Sources

L-DOPA is present in certain food, and L-DOPA glycoside has been extracted from the broad bean (fava bean) *Vicia faba* (83). The fava bean is one of the best plant sources of L-DOPA; clinical studies have shown that seed sprouts have antiParkinson's effects without any of the side effects seen from the pure synthetic form (74). Fava beans are a widely cultivated legume and commonly consumed in the Mediterranean region (37). L-DOPA was identified in the seedling, pods, and beans of broad bean (fava bean) initially by Guggenheim in the early 1930s (37,69,74). Fava bean ingestion correlated with a significant increase in levels of L-DOPA in plasma and with an improvement in motor performance (35). As discussed earlier, exogenously synthesized L-DOPA is not effective in all PD patients and has side effects (37). The mechanism of efficiency of the natural source of L-DOPA may be due to the amino acid milieu generated in broad fava beans that may favor the selective transport of L-DOPA across the blood brain barrier (37). Fava beans seedling ingestion is useful in the treatment of PD patients, especially those in low income societies, where limited resources do not allow the purchase or manufacture of expensive drugs. Fava beans are available cheaply in all seasons and can be easily grown. Furthermore, fava beans serve as natriuretic agents (82). Other natural sources of levodopa, besides broad bean, have also been reported in *Stizolobium deeringianum* and *Mucuna pruriens*. *Mucuna pruriens* beans have been used as an efficacious herbal drug for PD treatment in India for many years (37).

9.4.2   L-DOPA from Fava Bean

Studies have shown that ingestion of *Vicia faba* (fava bean) fruits (pods and seeds) improved substantially the clinical features in Parkinson's patients (35,36). In further studies the benefits from fava bean seedlings, which had 20-fold higher L-DOPA than fruit was even better, with higher plasma L-DOPA and substantial clinical improvement (69). In a limited clinical case study, fava beans as a source of L-DOPA, prolonged the "ON" periods in patients with Parkinson's disease who have "ON-OFF" fluctuations (37). In these studies, patients had been previously administered higher doses of L-DOPA, up to 800–100 mg per day, which failed to optimize the "ON" time and resulted in peak dose dyskinesias, but the fava bean showed beneficial effects by prolonging the "ON" time and shortened "OFF" time (37). Other studies using L-DOPA from *Mucana pruriens* beans have been used as an effective drug for Parkinson's disease in India for many years (84). In a clinical study of 60 patients over 12 weeks, an extracted powder from *M. pruriens* mixed with water significantly improved Parkinsonian motor scores, and from the studies authors speculated that the extract benefits may be a result of other antiParkinsonian compounds besides L-DOPA (Parkinson's study group for HP-200). The observation from these limited legume studies indicates that there may be additional complementary compounds that may work in conjunction with L-DOPA. Our hypothesis is that these additional complementary compounds could be other phenolics that have antioxidant benefits and amino acid cofactors. Therefore, the rationale for enhancing biosynthesis of L-DOPA and total phenolic profiles in various stages of fava bean seedlings following elicitation with physical and elicitor stress has merit. Further, because seedlings were shown to be the best source of L-DOPA (69), its biosynthesis is being investigated in light and dark germinated fava bean seedlings and correlated to total phenolic content and fava bean antioxidant activity at the metabolite and enzyme level. From this understanding of L-DOPA phenolic content and antioxidant response, the goal is to use various stages of the seedlings with different optimized levels of L-DOPA and total phenolics and to confirm that these ratios will affect the antioxidant enzyme response in a Parkinson's neuronal cell model system. This approach not only helps to optimize the best stage for L-DOPA biosynthesis in fava bean seedlings but will also confirm whether the additional factors that contribute to the benefits of L-DOPA from the fava bean is potentially the result of " phenolic antioxidant" and "amino acid" factors.

9.5 LINKING L-DOPA SYNTHESIS TO THE PENTOSE PHOSPHATE AND PHENYLPROPANOID PATHWAYS

L-DOPA in the seed is potentially derived from the phenylalanine. As mentioned earlier, phenylalanine is synthesized in the plant from the pentose phosphate pathway and the shikimate pathway, which forms the starting material of the phenylpropanoid pathway for the synthesis of phenolic acid (Figure 9.1). PAL catalyses its further conversion to cinnamate, which leads to the secondary metabolite synthesis. An alternative route for phenylalanine modification is the synthesis of tyrosine by phenylalanine hydroxylase. Tyrosine, in turn, is the precursor for L-DOPA production in the seed.

It is hypothesized that synthesis of free soluble phenolics and L-DOPA is regulated via the proline linked pentose–phosphate pathway (PLPPP), shikimate pathway, and phenylpropanoid pathway (23,30) (Figure 9.1). PPP is an alternate route for the breakdown of carbohydrates generating NADPH for use in anabolic reactions and for providing erythrose-4-phosphate for the shikimate pathway (23,30). This route is vital for the biosynthesis of phenylpropanoid secondary metabolites, including L-DOPA (Figure 9.1). Glucose-6-phosphate dehydrogenase (G6PDH) catalyses the first committed and rate-limiting step of PPP (85). A putative correlation has been observed between proline levels and total soluble phenolics in thyme, oregano, seeds of *Pangium edule*, and pea, which suggests that proline accumulation may be linked to the regulation of the phenylpropanoid pathway (86–88). The stimulation of PPP, purine, (89) and soluble phenolic (23,30) synthesis is believed to occur through a redox cycle. Cytosolic pyrroline-5-carboxylate (P5C) is reduced to proline and NADPH is oxidized to NADP$^+$ (89). The enzymes G6PDH and 6-phosphogluconate dehydrogenase utilize the generated NADP$^+$ as cofactors in PPP for their reactions. In many plants, free proline accumulates in response to the imposition of a wide range of biotic and abiotic stresses (90). The stimulation of L-DOPA through the proline linked pentose phosphate pathway (PLPPP) could also serve as the critical control point (CCP) for generating NADPH, not just for proline synthesis (as an alternative reductant for energy), but it could also meet the reductant needs of antioxidant enzyme response through superoxide dismutatse (SOD) and catalase (CAT) (Figure 9.4) (23,30).

Current investigations using fava bean sprouts indicate that under various biotic and abiotic elicitors, L-DOPA content is enhanced in the early stages of germination, which then gradually gets reduced and the total soluble phenolics steadily increases with germination, reaching highest synthesis in the later stages (63,64,66). Further, it is evident that stimulation of soluble phenolics in the late stages is regulated through the pentose phosphate pathway as reflected in the activity of glucose-6-phosphate dehydrogenase. This is likely coupled to proline synthesis (63,64). It is likely that as soon as germination takes place L-DOPA content is mobilized rapidly in early stages, and once germination leads to hypocotyl development, the phenolic flux is redirected for structural development with maximum demand for the pentose phosphate pathway because elicitor stress demand increases soluble phenolics in late stages (63,64,66). This late stage soluble phenolic stimulation in response to stress factors could be as important as

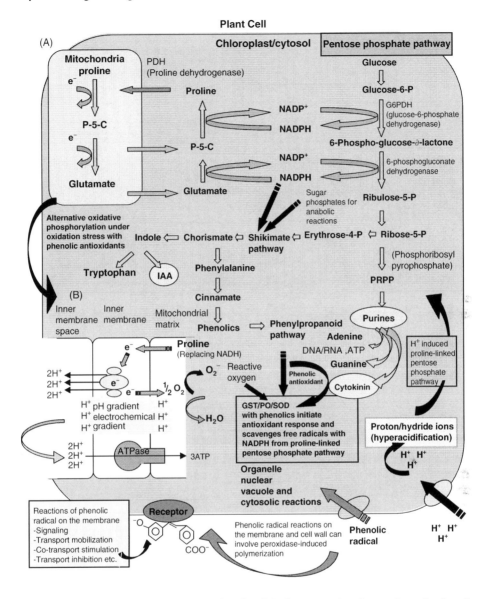

Figure 9.4 Improved model for the role of praline linked pentose–phosphate pathway in phenolic biosynthesis, which also accommodates the biosynthesis and mechanism of action of L-DOPA in a high phenolioc background from fava bean sprouts, and the effect of external elicitors like oregano phenolics. (Abbreviations: P5C: pyrroline-5-carboxylate; IAA: indole acetic acid: GST: Glutathione-s-transferase: PO: peroxidase; SOD: superoxide dismutase)

L-DOPA for Parkinson's management, as this could provide additional "phenolic anti-oxidant" and "amino acid" factors that could moderate the L-DOPA insensitivity often seen with pure synthetic L-DOPA treatments (37). These additional phenolic antioxidant and amino acid factors such as glutamic acid and proline could contribute soluble factors for proper redox management of neuronal cells as proposed for human and mammalian cells (23,91) (Figure 9.5).

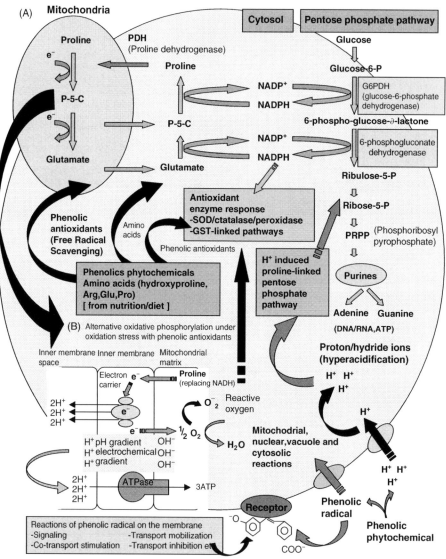

Figure 9.5 Extension of plant praline linked pentose–phosphate pathway model for the effect of external phenolic phytochemicals like L-DOPA in a high total phenolic background from fava bean sprouts or fermented extracts in human and mammalian systems. (Abbreviations: P5C: pyrroline-5-carboxylate; GST: Glutathione-s-transferase; SOD: superoxide dismutase)

9.6 RECENT PROGRESS ON ELICITATION LINKED BIOPROCESSING TO ENHANCE L-DOPA AND PHENOLICS IN FAVA BEAN SPROUTS

We have developed biotic and abiotic induced elicitation techniques to enhance L-DOPA, total phenolics, and antioxidant activity during sprouting of fava beans (63,64,66,92–94). The rationale for this approach is based on the understanding that pretreatment in the dry presoak stage or soak stage enhances critical stress responses as the seed germinates and that this will alter the stress related phenylpropanoid pathways linked to soluble phenolic and L-DOPA biosynthesis. This also likely mobilizes free amino acid from protein stored in the cotyledons. In our earliest work we investigated the stimulation of L-DOPA, total soluble phenolics, and related antioxidant activity in dark germinated fava bean seeds in response to bacterial polysaccharides from *Pseudomonas elodea* (gellan gum) and *Xanthomonas campestris* (xanthan gum) (92). Results indicated that gellan gum stimulated a ninefold increase in total soluble phenolics, compared to control. and this stimulation may be regulated via the pentose phosphate pathway based on the stimulation of glucose-6-phosphate dehydrogenase (92). The L-DOPA content was high in the initial days in the hypocotyl in all treatments and steadily reduced over the later stages. This approach allowed the development of two kinds of extracts, an early stage high L-DOPA with low soluble phenolics, and later stage medium L-DOPA with high soluble phenolics. The relative antioxidant activity based on the β-carotene assay did not change over the sprouting phase (92). Subsequent studies explored the stimulation of L-DOPA and total soluble phenolic in response to phytochemical and peptide elicitors in dark and light germinated conditions (93,94). Under dark germination conditions, the elicitors stimulated soluble phenolic content, with fish protein hydrolysates, lactoferrin, and oregano phenolic extracts stimulating the highest phenolic contents of 2.9–5.2 mg/g fresh weight (FW) during early stages of germination (93). The stimulation of primary metabolism through the pentose phosphate pathway occurred on day 5 for fish protein hydrolysates, and early stage of day 3 for lactoferrin and oregano phenolics (93). L-DOPA was stimulated from 20–40% for all elicitors immediately after treatment on day 1 and declined steadily. Under these conditions, fava extracts with any of the foregoing elicitations could be suitable for designing a dietary source of L-DOPA with high soluble phenolics between 2–5 days, depending on the elicitor. The same elicitors were tested for stimulation of light modulated seed vigor response and it was clear that L-DOPA content in seedlings was stimulated to over 80–100% by fish protein hydrolysate and oregano phenolics, which was followed by a change in flux toward higher soluble phenolics and enhanced antioxidant activity on day 20 (94).

In another study, UV treatment of seeds for 15 hours after overnight soaking stimulated highest L-DOPA content on day 1 and highest total soluble phenolic content day 6. The stimulation of soluble phenolic content was correlated to glucose-6-phosphate dehydrogenase, the first committed step in the pentose phosphate pathway. Further, proline content was slightly elevated compared non-UV treated control, indicating that soluble phenolic synthesis in late stages may be driven through the proline linked pentose–phosphate pathway (63). In another very exciting abiotic stress study, 30 seconds of microwave treatment of dry seed prior to overnight soaking and dark germination, resulted in over 50% stimulation of L-DOPA on day 1 and over 700 % stimulation of total soluble phenolics on day 7 (66). Each of these stages correlated to higher antioxidant activity both in terms of free radical scavenging assay as well as stimulation of superoxide dismutase (SOD) (66). Further, it was clear that late stage soluble phenolic stimulation correlated strongly to the stimulation of the pentose phosphate pathway (66). Additional studies using proline analog and proline combinations were used as elicitor treatments to confirm that the stimulation of soluble phenolics in the late stages of dark germination correlated with simultaneous stimulation of glucose-6-phosphate dehydrogenase and proline, further supporting the hypothesis that the proline linked pentose–phosphate pathway may be involved in the stress induced stimulation of phenolics (64). These studies support the model that PLPPP may be a critical control point (CCP) that not only stimulates the pentose phosphate pathway to support proline as an alternative reductant for energy through oxidative phosphorylations (alternative to NADH), but could also provide the critical NADPH for supporting antioxidant enzyme response through SOD and CAT (Figure 9.4). The optimization of these steps during sprouting and seedling development could help to generate fava bean extracts with optimum L-DOPA with additional soluble phenolics and amino acid factors that could better modulate neuronal antioxidant enzyme response (Figure 9.5) and help to overcome any L-DOPA insensitivity seen with extended use of pure synthetic L-DOPA.

9.7 RECENT PROGRESS ON SOLID-STATE BIOPROCESSING OF THE FAVA BEAN TO ENHANCE L-DOPA AND PHENOLICS USING FOOD GRADE FUNGAL SYSTEMS

A solid-state bioconversion system using the food grade fungus *Rhizopus oligosporus* was developed to enrich the fava bean substrate with phenolic antioxidants and L-DOPA (65). The L-DOPA content in the fungal grown fava bean increased significantly to approximately twice that of control, accompanied by moderate soluble phenolic linked antioxidant activity based on free radical scavenging assay and higher fungal SOD activity during early stages of growth. This indicated that L-DOPA can be mobilized and formed from fava bean substrates by fungal bioconversion, conributing to the antioxidant functionality of such extracts (65). A high superoxide dismutase (SOD) activity during early and late growth stages indicates the likely oxidation stress of initial fungal colonization and during later growth stages due to nutrient depletion. High levels of soluble phenolics were observed during late growth stages. During the course of solid-state growth there was an increase in β-glucosidase activity, which correlated to an increase in total soluble phenolic content during the late stages. This suggests that the enzyme may play an important role in the release of phenolic aglycones from fava bean substrate, thereby increasing the soluble phenolic content and accompanying antioxidant activity (65). The implication from this study is that solid-state bioconversion of the fava bean by *R. oligosporus* can significantly improve the phenolic antioxidant activity and Parkinson's relevant L-DOPA content.

9.8 IMPLICATIONS AND SUMMARY

It is clear that elicitor induced sprout systems and fungal based solid-state bioprocessing systems can be used for specific stimulation of L-DOPA. In the case of sprouts, the L-DOPA increase was observed immediately after elicitor and soaking treatment on day 1, but was observed over an extended growth phase when using a fungal solid-state system, The contents were consistently higher over the middle growth stage from days 4–12. In contrast, the total soluble phenolics were substantially stimulated concurrent with high proline in late stages of sprout growth with general stimulation of the pentose phosphate pathway supporting this total soluble phenolic metabolite production. In the solid-state fungal bioprocessing system, soluble phenolic content was observed in the later stages. The extent of antioxidant activity based on free radical scavenging generally coincided with higher L-DOPA content in the early stage, and with total soluble phenolics in the late stage in both elicitor stimulated sprouts and fungal bioconversion system. These bioprocessing approaches provide avenues to optimize L-DOPA content in the fava bean system with optimum "phenolic antioxidant" and "amino acid" profiles that could be more effective than the current approaches observed for pure synthetic L-DOPA drug treatments. Insights provided by studies so far also indicate that the L-DOPA content and total soluble phenolic contents are likely stimulated through an alternative proline liked pentose–phosphate pathway (Figure 9.4). In this model under elicitor stress, soluble phenolic synthesis is enhanced to counter the stress, which likely involves both direct free radical scavenging using the produced phenolics as well as antioxidant enzyme response through SOD. Further, an alternative proline linked pentose–phosphate pathway more efficiently facilitates NADPH and sugar phosphate flow for anabolic pathways, including phenolic and antioxidant response pathways, while proline meets the reductant need for oxidative phosphorylation replacing NADH from TCA Krebs cycle (Figure 9.4) (23). Extending this model into animal systems, we have hypothesized fava bean extracts with optimum total soluble phenolics, L-DOPA, and proline could better maintain the cellular homeostasis and function of neuronal cells by modulating host antioxidant response through a proline linked pentose–phosphate pathway (23) (Figure 9.5). This mode of alternative regulation could facilitate better management of cellular redox environment through dehydrogenases and antioxidant enzyme response through SOD and CAT under which L-DOPA can work effectively.

REFERENCES

1. Goodwin, T.W., E.I. Mercer. Plant phenolics. In: *Introduction to Plant Biochemistry*, New York: Pergamon Press, 1983, pp 528–566.
2. Taiz, L, E. Zeigler. Auxins. In: *Plant Physiology*, 2nd ed., Sunderland, MA: Sinauer Associates, 1998, pp 543–557.
3. Croteau, R., T.M. Kutchan, N.G. Lewis. Natural products: secondary metabolites. In: *Biochemistry and Molecular Biology of Plants*, Buchanan, B.B., W. Gruissem, R.L. Jones, eds., Rockville, MD: American Society of Plant Physiology, 2002, pp 1250–1318.
4. Walker, J.R.L. *The Biology of Plant Phenolics*, 1st ed. London: Edward Arnold, 1975.
5. Dixon, R.A., M.J. Harrison, C.J. Lamb. Early events in the activation of plant defense responses. *Ann. Rev. of Phytopath.* 32:479–501, 1994.
6. Dixon, R.A., N. Paiva. Stress-induced phenylpropanoid metabolism. *Plant Cell*, 7:1085–1097, 1995.
7. Rhodes, J.M., L.S.C. Wooltorton. The biosynthesis of phenolic compounds in wounded plant storage tissues. In: *Biochemistry of Wounded Plant Tissues*, Kahl, G., ed, Berlin: Walter deGruyter, 1978, p 286.
8. Brooker, F.L., J.E. Miller. Phenylpropanoid metabolism and phenolic composition of soybean [*Glycine max.* (L) Merr.] leaves following exposure to ozone. *J. Exp. Bot.* 49:1191–1202, 1998.
9. Zimmerman, Y.Z., P.R. Cohill. Heat shock and thermotolerance in plant and animal embryogenesis. *New Biology* 3:641–650, 1991.
10. Yalpani, N., A.J. Enyedi, J. Leon, I. Raskin. Ultraviolet light and ozone stimulate accumulation of salicylic acid, pathogenesis related proteins and virus resistance in tobacco. *Planta.* 193:372–376, 1994.
11. Hahlbrock, K., D. Scheel. Physiology and molecular biology of phenylpropanoid metabolism. *Plant Mol. Biol.* 40:347–369, 1989.
12. Graham, T.L. Flavanoid and isoflavanoid distribution in developing soybean seedling tissue and in seed root exuddates. *Plant Physiol.* 95:594–603, 1991.
13. Christie, P.J., M.R. Alfenito, V. Walbot. Impact of low temperature stress on general phenylpropanoid and anthocyanin pathways: enhancement of transcript abundance and anthocyanin pigmentatation in maize seedlings. *Planta.* 194:541–549, 1991.
14. Beggs, C.J., K. Kuhn, R. Bocker, E. Wellmann. Phytochrome induced flavanoid biosynthesis in mustard (*Sinapsis alba* L.) cotyledons: enzymatic control and differential regulation of anthocyanin and quercitin formation. *Planta.* 172:121–126, 1987.

15. Lois, R., B.B. Buchanan. Severe sensitivity to ultraviolet light radiation in *Arabidopsis* mutant deficient in flavanoid accumulation: mechanisms of UV-resistance in *Arabidopsis*. *Planta*, 194:504–509, 1994.

16. Kurganova, L.N., A.P. Veselov, T.A. Goncharova, Y.V. Sinitsyna. Lipid peroxidation and antioxidant system protection against heat shock in pea *(Pisum sativum* L.*)* chloroplasts. *Fiziol. Rast.* (Moscow), *(Russ. J. Plant Physiol., Eng. Transl.)* 44:725–730, 1997.

17. Retivin, V., V. Opritov, S.B. Fedulina. (1997) Generation of action potential induces preadaptation of *Cucurbita pepo* L. stem tissues to freezing injury. *Fiziol. Rast.* (Moscow), *(Russ. J. Plant Physiol., Engl. Transl.)* 44:499–510.

18. Kuznetsov, V.V., N.V. Veststenko. Synthesis of heat shock proteins and their contribution to the survival of intact cucumber plants exposed to hyperthermia. *Fiziol. Rast* (Moscow), *(Russ. J. Plant Physiol., Engl. Transl.)* 41:374–380, 1994.

19. Baraboi, V.A. Mechanisms of stress and lipid peroxidation. *Usp. Sovr. Biol.* 11:923–933, 1991.

20. Kurganova, L.N., A.P. Veselov, Y.V. Sinitsina, E.A. Elikova. Lipid peroxidation products as possible mediators of heat stress response in plants. *Exp. J. Plant Physio.* 46:181–185, 1999.

21. Parr, A.J., G.P. Bolwell. Phenols in the plant and in man. The potential for possible nutritional enhancement of the diet by modifying the phenols content or profile. *J. Sci. Food and Agric.* 80:985–1012, 2000.

22. Scalbert, A., G. Williamson. Dietary intake and bioavailability of polyphenols. *Am. Soc. Nutr. Sci.* 130:2073S–2085S, 2000.

23. Shetty, K., M.L. Wahlqvist. A model for the role of proline-linked pentose phosphate pathway in phenolic phytochemical biosynthesis and mechanism of action of human health and environmental applications: a review. *Asia Pacific J. Clin. Nutrition* 13:1–24, 2004.

24. Driver, G.A.C., M. Bhattacharya. Role of phenolics in plant evolution. *Phytochemistry* 49:1165–1174, 1998.

25. Cowan, M.M. Plant products as antimicrobial agents. *Clin. Micro. Rev.* 12:564–582, 1999.

26. Lewis, N.G. Antioxidants in higher plants. In: *Plant Phenolics,* Boca Raton, FL: CRC Press, 1993, pp 135–169.

27. Lu, Y., L.Y. Foo. Identification and quantification of major polyphenols in apple pomace. *Food Chem.* 59:187–194, 1997.

28. Bravo, L. Phenolic phytochemicals: chemistry, dietary sources, metabolism, and nutritional significance. *Nutr. Rev.* 56:317–333, 1998.

29. Vekiari, S.A., V. Oreopoulou, C. Taia, C.D. Thompson. Oregano flavonoids as lipid antioxidants. *J. Amer. Oil Chem. Soc.* 70:483–487, 1993.

30. Shetty, K. Biotechnology to harness the benefits of dietary phenolics: focus on *Lamiaceae*. *Asia Pac. J. Clin. Nutr.* 6:162–171, 1997.

31. Shetty, K., T.L. Carpenter, D. Kwok, O.F. Curtis, T.L. Potter. Selection of high phenolic containing clones of thyme (*Thymus vulgaris*) using *Pseudomonas* sp. *J. Agric. Food Chem.* 44:3408–3411, 1996.

32. Deighton, N., S.M. Glidewell, S.G. Deans, B.A. Goodman. Identification by EPR spectroscopy of carvacrol and thymol as the major sources of free radicals in the oxidation of plant essential oils. *J. Fd. Sci. Agric.* 63:221–225, 1993.

33. Deighton, N., S.M. Glidewell, B.A. Goodman, S.G. Deans. The chemical fate of the endogeneous plant antioxidants carvacrol and thymol during oxidative stress. *Proc. R. Soc. Edinburgh* 102B:247–252, 1994.

34. Frankel, E.N., A.L. Waterhouse, P.L. Teissedrel. Principal phenolic phytochemicals in selected California wines and their antioxidant activity in inhibiting oxidation of human low density lipoproteins. *J. Agric. Food Chem.* 43:890–894, 1995.

35. Rabey, J.M., Y. Vered, H. Shabtai, E. Graff, A.D. Korczyn. Improvement of Parkinsonian features correlate with high plasma levodopa values after broad bean (*Vicia faba*) consumption. *J. Neurosurg. Psychiatr.* 55:725–727, 1992.

36. Kempster, P.A., Z. Bogetic, J.W. Secombe, H.D. Martin, N.D.H. Balazs, M.L. Wahlqvist. Motor effects of broad beans (*Vicia faba*) in Parkinson's disease: single dose studies. *Asia Pacific J. Clin. Nutr.* 2:85–89, 1993.

37. Apaydin, H., S. Ertan, S. Ozekmekci. Broad bean (*Vicia faba*): a natural source of L-DOPA: prolongs "on" periods in patients with Parkinson's disease who have "on-off" fluctuations. *Movement Disorders.* 15:164–166, 2000.

38. Aruoma, O.I. Free radicals, antioxidants and international nutrition. *Asia Pac. J. Clin. Nutr.* 8:53–63, 1999.

39. Narayanswami, V., H. Sies. Oxidative damage to mitochondrial and protection by ebselen and other antioxidants. *Biochem. Pharm.* 40:1623–1629, 1990.

40. Bindoli, A. Lipid peroxidation in mitochondria. *Free Radical Bio. Med.* 5:247–261, 1998.

41. Whalley, C.V., S.M. Rankin, R.J. Hoult, W. Jessu, D.S. Leakes. Flavonoids inhibit the oxidative modification of low density lipoproteins by macrophages. *Biochem. Pharm.* 39:1743–1750, 1990.

42. Randhir, R., D.A. Vattem, K. Shetty. Antioxidant enzyme response studies in H_2O_2-stressed porcine muscle tissue following treatment with Oregano phenolic extracts. *Process Biochem.* 40:2123–2134, 2005.

43. Malaveille, C., A. Hautefeuille, B. Pignatelli, G. Talaska, P. Vineis, H. Bartsch. Antimutagenic dietary phenolics as antigenotoxic substances in urothelium of smokers. *Mutation Research.* 402:219–224, 1998.

44. Meyer, A.S., O.S. Yi, A. Pearson, A.L. Waterhouse, E.N. Frankel. Inhibition of human low-density lipoprotein oxidation in relation to consumption of phenolic antioxidants in grapes (*Vitis vinifera*). *J. Agric. Food Chem.* 45:1638–1643, 1997.

45. Jakus, V. The role of free radicals, oxidative stress and antioxidant systems in diabetic vascular disease. *Bratisl Lek Listy* 101:541–551, 2000.

46. Barbaste, M., B. Berke, M. Dumas, S. Soulet, J.C. Delaunay, C. Castagnino, V. Arnaudinaud, C. Cheze, J. Vercauteren. Dietary antioxidants, peroxidation and cardiovascular risks. *J. Nutr. Health Aging.* 6:209–23, 2002.

47. Block, G., B. Patterson, A. Subar. Fruit, vegetables, and cancer prevention: a review of the epidemiological evidence. *Nutr. Cancer.* 18:1–29, 1992.

48. Serdula, M.K., M.A.H. Byers, E. Simoes, J.M. Mendlein, R.J. Coates. The association between fruit and vegetable intake and chronic disease risk factors. *Epidemiology.* 7:161–165, 1996.

49. Yoshioka, A., Y. Miyachi, S. Imamura, Y. Niwa. Anti-oxidant effects of retinoids on inflammatory skin diseases. *Arch. Of Derm. Research.* 278:177–183, 1986.

50. Puskas, F., P. Gergely, K. Banki, P. Andras. Stimulation of the pentose phosphate pathway and glutathione levels by dehydroascorbate, the oxidized form of vitamin C. *FASEB J.* 14:1352–1361, 2000.

51. Appeldoorn, N.J.G., S.M. Bruijn, E.A.M. Koot Gronsveld, R.G.F. Visser, D. Vreughenhi, L.H.W. Van der Plas. Developmental changes in enzymes involved in the conversion of hexose phosphate and its subsequent metabolites during early tuberiztion of potato. *Plant Cell Environ.* 22:1085–1096, 1999.

52. Chugh, L.K., S.K. Sawhney. Effect of cadmium on activities of some enzymes of glycolysis and pentose phosphate pathway in pea. *Biologia Plantarum.* 42:401–407, 1999.

53. Debnam, P., M.J. Emes. Subcellular distribution of enzymes of the oxidative pentose phosphate pathway in root and leaf tissues. *J. Experimental Botany.* 50:1653–1661, 1999.

54. Herrmann, K.M. The shikimate pathway as an entry point to aromatic secondary metabolism. *Plant Physiol.* 107:7–12, 1995.

55. Shetty, K., D.L. Crawford, A.L. Pometto. Production of L-phenylalanine from starch by analog-resistant mutants of *Bacillus polymyxa*. *App. Environ. Micro.* 52:637–643, 1986.

56. Hollman, P.C.H. Evidence for health benefits of plant phenols: local or systemic effects? *J. Sci. Food Agric.* 81:842–852, 2001.

57. Rajalakshmi, D., S. Narashimhan. Food antioxidants: sources, and methods of evaluation. In: *Food Antioxidants: Technological, Toxicological and Health Perspectives.* Madhavi, D.L., S.S. Deshpande, D.K. Salunkhe, eds., New York: Marcel Dekker, 1995, pp 65–158.

58. Zheng, Z., U. Sheth, M. Nadiga, J.L. Pinkham, K. Shetty. A model for the role of proline-linked phenolic synthesis and peroxidase activity associated with polymeric dye tolerance in oregano. *Process. Biochem.* 36:941–946, 2001.

59. Whetten, R., R. Sederoff. Lignin biosynthesis. *The Plant Cell* 7:1001–1013, 1995.

60. Fukushima, R.S., R.D. Hatfield. Extraction and isolation of lignin for utilizing as a standard to determine lignin concentration using acetyl bromide spectrophotometric method. *J. Plant Physio.* 49:3133–3139, 2001.

61. McDougall, G.J. Cell wall associated peroxidases and lignification during growth of flax fibers. *J. Plant Physio.* 139:182–186, 1991.

62. Ebermann, R., H. Pichorner. Detection of proxidase catalyzed phenol polymerization induced by enzymatically reduced paraquat. *Phytochemistry* 28:711–714, 1989.

63. Shetty, P., M.T. Atallah, K. Shetty. Effects of UV treatment on the proline-linked pentose phosphate pathway for phenolics and L-DOPA synthesis in dark germinated *Vicia faba*. *Process Biochem.* 37:1285–1295, 2002.

64. Shetty, P., M.T. Atallah, K. Shetty. Stimulation of total phenolics, L-DOPA and antioxidant activity through proline-linked pentose phosphate pathway in response to proline and its analog in germination fava beans (*Vicia faba*). *Process Bichem.* 38:1707–1717, 2003.

65. Randhir, R., D. Vattem, K. Shetty. Solid-state bioconversion of fava bean by *Rhizopus oligosporus* for enrichment of phenolic antioxidants and L-DOPA. *Innovative Food Sci. and Emerging Technologies.* 5:235–244, 2004a.

66. Randhir, R., K. Shetty. Microwave-induced stimulation of L-DOPA, phenolics and antioxidant activity in fava bean (*Vicia faba*) for Parkinson's diet. *Process Biochem.* 39:1775–1784, 2004b.

67. Previc, F.H. Dopamine and the origin of human intelligence. *Brain and Cognition.* 41:299–350, 1999.

68. Lehninger, A.L., D.L. Nelson, M.C. Michael. *In Principles of Biochemistry.* Worth Publishers, 1993.

69. Vered Y., J.M. Rabey, D. Paleveitch, I. Grosskopf, A. Harsat, A. Yanowski, H. Shabtai, E. Graff. Bioavailability of levodopa after consumption of *Vicia faba* seedlings by Parkinsonian patients and control subjects. *Clin. Neuropharm.* 17:138–146, 1994.

70. Ahlskog, J.E. Treatment of Parkinson's disease, from theory to practice. *Postgrad. Med.* 95:55–69, 1994.

71. Elsworth, J.D., R.H. Roth. Dopamine synthesis, uptake, metabolism and receptors: relevance to gene therapy of Parkinson's Disease. *Exp. Neurol.* 144:4–9, 1997.

72. Boada , J., B. Cutillas, T. Roig, S. Bermudez, J. Ambrosio. MPP+ induced mitochondrial dysfunction is potentiated by dopamine. *Biochem. Biophys. Res. Comm.* 268:916–920, 2000.

73. Hiroi, T., S. Imaoka, Y. Funae. Dopamine formation from tyramine by CYP2D6. *Biochem. Biophys. Res. Comm.* 249:838–843, 1988.

74. Malhere, H.W., J.M. Van Buren. The excretion of dopamine metabolites in Parkinson's disease and the effect of diet thereon. *J. Lab. Clin. Med.* 74:306–318, 1969.

75. Kurlan, R., C.M. Tanner, C. Goetz, J. Sutton, D. Lichter, C. Deeley, L. Cui, C. Irvine, M.P. McDermott. Levodopa drug holiday versus drug dosage reduction in Parkinson's disease. *Clin. Neuropharm.* 17:117–127, 1994.

76. Blaschko, H. Metabolism and storage of biogenic amines. *Experientia* 13:9–12, 1959.

77. Carlsson, A. The occurrence, distribution and physiological role of catecholamines in the nervous system. *Pharmacol. Rev.* 11:490–493, 1959.

78. Ehringer, H., O. Hornykiewicz. Verteilung von Noradrenalin und Dopamin (3-Hydroxytyramin) im Gehirn des Menschen und ihr Vehalten bei Erkrankungen des extrapyramidalen Systems. *Wien. Klin. Wochenschr.* 38:1236–1239, 1960.

79. Barbeau, A. Biochemistry of Parkinson's disease. *Proc. 7ʰ Intl. Cong. Neurology.* 2:925, 1961.

80. Birkmayer, W., O. Hornykiewicz. Der L-3,4-Dioxyphenylalanin (DOPA)-Effect bei der Parkinson-Akinese. *Wien. Klin. Wochenschr.* 73:787–788, 1961.

81. Cotzias, G.C., M.H. Van Woert, L.M. Schiffer. Aromatic amino acids and modification of parkinsonism. *N. Engl. J. Med.* 276:374–379, 1967.

82. Vered Y., I. Grosskopf, D. Palevitch, A. Harsat, G. Charach, M.S. Weintraub, E. Graff. Influence of *Vicia faba* (broad bean) seedlings on urinary sodium excretion. *Planta Medica.* 63:227–249.

83. Banwart, B., T.D. Miller, J.D. Jones, G.M. Tyce. Plasma dopa and feeding. *P.S.E.B.M.* 191:357–361, 1989.

84. Manyam, B.V., K.M. Patikh. HP-200: an herbal drug for treatment of Parkinson's disease. *Parkinson's Mag. Euro. Parkinson's Assoc.* 8:10–11, 1997.

85. Garrett R.H., C.M. Grisham. Nitrogen acquisition and amino acid metabolism. In: *Biochemistry,* New York: Harcourt Brace College Publishers, 1995, pp 826–870.

86. McCue, P., Z. Zheng, J.L. Pinkham, K. Shetty. A model for enhanced pea seedling vigour following low pH and salicylic acid treatment. *Process. Biochem.* 35:603–13, 2000.

87. Perry, P.L., K. Shetty. A model for improvement of proline during *pseudomonas*-mediated stimulation of rosmarinic acid levels in oregano shoot clones. *Food Biotech.* 13:137–154, 1989.

88. Andarwulan, N., K. Shetty. Improvement of pea (*Pisum sativu*) seed vigour response by fish protein hydrolysates in combination with acetyl salicylic acid. *Process Bichem.* 35:159–165, 1999.

89. Phang, J.M. The regulatory functions of proline-5-carboxylic acid. *Cur. Top. Cell Regul.* 25:91–132, 1985.

90. Hare, P.D., W.A. Cress. Metabolic implications of stress induced proline accumulation in plants. *Plant Growth Regulation* 21:79–120, 1997.

91. Shetty, K. Role of proline-linked pentose phosphate pathway in biosynthesis of plant phenolics for functional food and environmental applications: a review. *Process. Biochem.* 39:789–804, 2004.

92. Shetty, P., M.T. Atallah, K. Shetty. Enhancement of total phenolic, L-DOPA and proline contents in germinating fava bean (*Vicia faba*) in response to bacterial elicitors. *Food Biotech.* 15:47–67, 2001.

93. Randhir, R., P. Shetty, K. Shetty. L-DOPA and total phenolic stimulation in dark germinated fava bean in response to peptide and phytochemical elicitors. *Process. Biochem.* 37:1247–1256, 2002.

94. Randhir, R., K. Shetty. Light-mediated fava bean (*Vicia faba*) response to phytochemical and protein elicitors and consequences on nutraceutical enhancement and seed vigor. *Process. Biochem.* 38:945–952, 2003.

10

Biochemical Markers for Antioxidant Functionality

Dhiraj A. Vattem and Kalidas Shetty

CONTENTS

10.1 Introduction 230
10.2 Direct Measurement of Reactive Oxygen Species 231
10.3 Antioxidant Asssays 232
    10.3.1 Antioxidant Assays Based on Free Radical Neutralization 232
    10.3.2 2,2′-Azinobis-(3-ethylbenzthiazoline-6-sulfonic acid) 233
    10.3.3 Ferric Reducing Ability of Plasma (FRAP) 234
    10.3.4 Oxygen Radical Absorbance Capacity (ORAC) Assay 234
    10.3.5 β-carotene Oxidation (Bleaching) Assay 235
10.4 Lipid Oxidation Products 235
    10.4.1 Malondialdehyde 236
    10.4.2 LDL Resistance to Oxidation 236
    10.4.3 F2 Isoprostanes 237
10.5 DNA Oxidation Products 237
    10.5.1 8-Oxo-7,8-dihydro-2′-deoxyguanosine 237
    10.5.2 Glutathione Reduced (GSH) and Oxidized (GSSG) 238
    10.5.3 Glutathione Peroxidase 240
    10.5.4 Glutathione-S-Transferases 240
    10.5.5 Superoxide Dismutase 240
    10.5.6 Catalase 241
    10.5.7 Xanthine Oxidase and Uric Acid 241
    10.5.8 NADPH Oxidoreductase 242
    10.5.9 Myeloperoxidase 242
10.6 Plant Phenolic Dependent Peroxidases 243
References 243

10.1 INTRODUCTION

An imbalance of prooxidants and antioxidants, which leads to oxidative stress, is now believed to be a key contributing factor in the manifestation of chronic diseases such as cardiovascular disease, hypertension, diabetes mellitus, and some forms of cancer (1,2). Recent research has shown that populations consuming diets rich in fruits and vegetables have lower incidences of many chronic diseases (3–6), which has encouraged the use of diet as a complementary strategy for the management of these oxidative diseases (7–10). Fruits and vegetables are rich sources of phytochemicals and highly antioxidant vitamins, which are now believed to be responsible for these beneficial effects. The function of antioxidants from fruits and vegetables has been categorized into two types depending on their mode of action. First, oxidation of biological macromolecules such as nucleic acids, proteins, and lipids as a result of free radical damage (oxidative stress) has now been strongly associated with development of many physiological conditions which can manifest as disease (11–17). The protective effect of antioxidants from fruits and vegetables, therefore, is believed to be due to their direct involvement in neutralizing free radicals in biological systems and preventing oxidative damage to cellular systems (18).

Second, reactive oxygen species (ROS) are naturally formed in the body as a result of many metabolic processes. Low levels of ROS therefore, have been implicated in many cellular processes, including intracellular signaling responsible for cell division (11,19,20), immune response (67–69), and via the activation and repression of many genes (11). However, excess formation of ROS can cause damage and therefore disturb cellular homeostasis, so cells have evolved several antioxidant enzyme coupled metabolic systems which can which can quickly remove the excess ROS from cellular systems (22,23). These antioxidant systems involved in the protection against ROS consist of many antioxidant factors including glutathione (GSH) and other tissue thiols, heme proteins, coenzyme Q, bilirubin and urates, and several antioxidant enzymes such as superoxide dismutase (SOD), catalase (CAT), glutathione peroxidase (GPX), and glutathione-S-transferase (GST) (Figure 10.1). Dietary antioxidants from fruits and vegetables and other functional foods are believed to modulate cellular physiology and participate in induction/repression of gene expression or activation/deactivation of proteins, enzymes, and transcription factors of key metabolic pathways (11,21–23) leading to activation of these biological antioxidant systems in the cell.

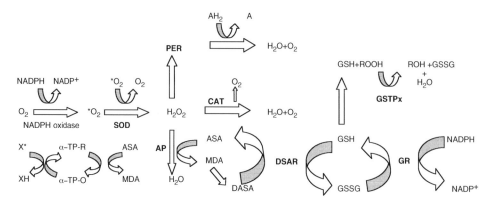

Figure 10.1 The antioxidant defense response of the cell carried out by enzymatic as well as the nonenzymatic antioxidants.

The ability of fruits and vegetables to manage oxidation linked diseases has resulted in a strong interest in the development functional foods and other dietary strategies for management of oxidation diseases (3–10). These developments have led to the need for good biomarkers and analytical methodologies to determine how these diets work. Evaluation of the functional antioxidant efficacy of new phytochemicals or functional food products requires a wide range of specific, accurate, and sensitive biomarkers appropriately associated with oxidative stress and related diseases. The biomarkers help to characterize food and food products with respect to their antioxidant activity and associated functional properties; in *in vitro* and *in vivo* models as well as in clinical studies. Some well established biochemical biomarkers currently used for studying antioxidant function in the management of oxidative stress are described here.

10.2 DIRECT MEASUREMENT OF REACTIVE OXYGEN SPECIES

Molecular oxygen in biological system is relatively unreactive and plays an important role in the proper functioning of many biological processes (11). However, metabolic pathways in the cell can reduce oxygen incompletely to produce reactive oxygen species (ROS), which are unstable compared to molecular oxygen, and can react with a number of biological macromolecules and disturb the cellular homeostasis (11–17). ROS are constantly produced in aerobic organisms both enzymatically and nonenzymatically (Figure 10.2). One electron reduction of molecular oxygen in biological systems results in the formation of the superoxide anion which is the precursor of most ROS (24). Biological dismutation of the superoxide anion produces the peroxide radical, which can undergo a complete reduction to form water, or an incomplete reduction to form the highly reactive hydroxyl radical (24) (Figure 10.2). Mitochondrial leakage of electrons or direct transfer of electrons to oxygen via the electron transport chain is another important source of ROS in cellular systems (25,26). Many other sources for the formation of ROS have been identified in cellular systems of living organisms (25,26). Superoxide is also formed upon one electron reduction of oxygen mediated by enzymes such as NADPH oxidase located on the cell membrane of polymorphonuclear cells, macrophages, and endothelial cells; from xanthine oxidase or from the respiratory chain (27–30). Superoxide radicals can also be formed from the electron cycling carried out by the

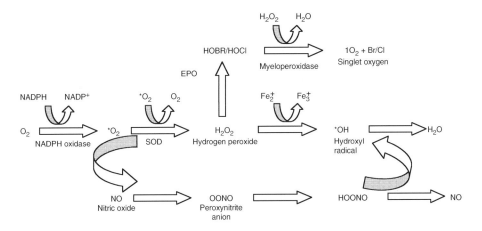

Figure 10.2 Biological mechanisms for ROS formation.

cytochrome P450-dependent enzymes in detoxification reactions (31,32). Oxidation of biological molecules has been linked to manifestation of many diseases and therefore the presence of reactive oxygen species in biological systems is often considered as a marker for oxidative stress.

Many quantitative assays are available for the detection of ROS. The majority of the methods are based on the identification of the peroxide radical, as it more stable in aqueous solution at neutral pH (33–35). H_2O_2 and other ROS are powerful oxidizing agents; therefore the most common quantification of H_2O_2 has been based on their oxidative reaction with chemiluminescent or fluorescent dyes. The neutralization of the radicals from the H_2O_2 by the dyes results in their gain or loss of fluorescence which is monitored fluorometrically (36,37). Peroxidases are used to remove the H_2O_2 radicals from the reaction mixture, and enzyme activity is monitored by the change in the fluorescence. Dependence upon peroxidases to detect H_2O_2 and lack of absolute specificity for H_2O_2 (38) are some of the limitations of this method. Other methods that are also used for direct measurement of the ROS include the use of a Clark-type oxygen electrode. Here the oxygen released from H_2O_2 in the cell supernatant is measured after the addition of catalase (39,40). Superoxide radicals are measured by electron spin resonance spectroscopy (ESR) (40). $O_2^{\cdot-}$ in solution is indirectly detected by a spin trap method by using common trapping agents such as 5,5-dimethyl-1-pyrroline N-oxide (DMPO) (41). The ESR spectrum pattern obtained from DMPO is specific to $O_2^{\cdot-}$ and therefore enables its accurate detection (41). However, due to a lower rate constant between DMPO and $O_2^{\cdot-}$, a large amount of DMPO should be added to the solution, which increases the cost per assay. Another drawback with this method is the requirement of a relatively expensive ESR instrument (41,42).

10.3 ANTIOXIDANT ASSSAYS

The total antioxidant capacity of cell extract or blood plasma can give us an indication about the functionality of the compound or food of interest (39). A number of antioxidant assays are available to study the antioxidant activity.

10.3.1 Antioxidant Assays Based on Free Radical Neutralization

Many antioxidant assays are based on their ability to neutralize or quench free radicals. The two free radicals that have been most commonly used for assessing antioxidant activity are 1,1-diphenyl-2-picrylhydrazyl (DPPH) (43,44), and 2,2′-azinobis(3-ethyl-benzothiazoline-6-sulfonic acid) (ABTS) (45,46) (Figure 10.3).

The DPPH free radical is a stable radical with one electron delocalized over the molecule (Figure 10.3). This delocalization gives a deep purple color with absorption maxima at 517 nm in an ethanol solution (43,44). When an antioxidant capable of donating a hydrogen reacts with the DPPH radical it gives rise to a nonradical reduced form of DPPH which has a yellow color. The decrease in the absorption is measured spectrophotometrically and is compared with an ethanol control to calculate the DPPH free radical scavenging activity (43,44).

$$\text{DPPH (Purple)} + \text{AO} \rightarrow \text{DPPH (Yellow)} + \text{AO} \tag{10.1}$$

This method is a very quick and simple method for the measurement of antioxidant activity. The antioxidant capacity in this assay depends on the chemical structure of the antioxidant (43,44). The reduction in the DPPH radical is dependent on the number of hydroxyl groups present in the antioxidant. This method, therefore, gives an indication of the structural dependence of the antioxidant functionality of many biological antioxidants (43,44).

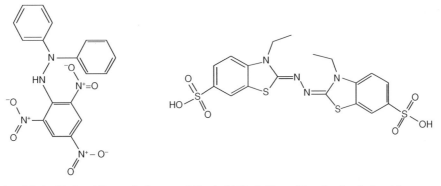

1-1-diphenyl-2-picryl-hydrazyl (free radical)   2,2'-azinobis(3-ethyl-benzothiazoline-6-sulfonic acid)

Figure 10.3 DPPH and ABTS — the two most commonly used indicators in antioxidant assays based on free radicals quenching.

The absorption maxima of the DPPH radical in ethanol solution at 517 nm is considered as one of the main drawbacks of the assay as it precipitates out most of the proteins and reduces the activity of a majority of single valance phenolic hydroxides (47,48). Dependence on the number of hydroxyl groups also results in the different kinetic rates of reaction between the antioxidant and the DPPH radical and causes fluctuation in the reaction times significantly, suggesting the need for standardization of reaction time and kinetic modeling for different samples (43,44,47,48).

10.3.2   2,2'-Azinobis-(3-ethylbenzthiazoline-6-sulfonic acid)

The antioxidant assay based on the 2,2'-Azinobis-(3-ethylbenzthiazoline-6-sulfonic acid) (ABTS) is similar to the DPPH assay (Figure 10.3). ABTS can react with hydroxyl, peroxyl, alkoxyl and inorganic radicals to form a relatively stable radical cation (ABTS) with absorption maxima at 734 nm (45,46). The antioxidant, added to the reaction mixture prior to the addition of the ROS or the radical, scavenges the free radical and prevents the formation of the ABTS radical cation. This scavenging activity can be spectrophotometrically and quantitated by comparing the antioxidant activity of the sample, extract, or biological fluid with trolox (45,46).

$$\text{ABTS} + \text{ROS} \rightarrow \text{ABTS} \qquad (10.2)$$
$$\uparrow \text{ Inhibition}$$
$$\text{ROS} + \text{AO} \nrightarrow \text{R}-\text{H}+ \text{AO}$$

The ABTS assay can determine the antioxidant contribution from different components in the system as it is capable of reacting with hydroxyl, cysteinyl, and glutathione in biological systems, and therefore has been widely used to study the antioxidant activity of various compounds, both lipophilic and hydrophilic, food products and biological samples (45–48). However, in biological samples the antioxidant enzymes such as peroxidases have been shown to be able to promote the formation of ABTS radical therefore contributing to potential interference resulting in higher absorbance values. Due to the relatively low concentrations of hydrogen peroxide, this assay cannot accurately measure the free radical quenching, as the antioxidants can react with the ABTS radical cation directly instead of the hydrogen peroxide, which can make it look as if there is more antioxidant activity than there really is (45,46). The assay also measures the antioxidant activity of the sample in a fixed time, and does not take into account the time for the completion of the reaction, which may not be suitable as it overlooks the total antioxidant capacity of the sample (47,48).

10.3.3 Ferric Reducing Ability of Plasma (FRAP)

In the ferric reducing ability of plasma (FRAP), plasma or extract is allowed to react with ferric tripyridyltriazine (Fe-III-TPTZ) complex at low pH. Reduction of ferric iron (Fe III) in the complex to ferrous iron (Fe II) results in an intense blue color. This resulting complex can be measured spectrophotometrically at 593 nm, and the intensity of the blue color developed is directly proportional to the antioxidant capacity (49).

$$Fe^{3+}(TPTZ)_2 + AO \rightarrow Fe^{2+}(TPTZ)_2 \text{ (Blue)} + AO \qquad (10.3)$$

The main advantages of the assay lie in its relative simplicity, which enables high throughput, and its cheaper cost (48,49). It is a fast assay, and reproducible results can be obtained from plasma and simple antioxidant solutions with minimal sample preparation. However, the reaction is nonspecific, as any compound with a lower redox potential than the standard redox potential of Fe(III) and Fe(II) (0.77 V) can reduce the complex and contribute to high FRAP values. The reaction also assumes that completion of the reduction is spontaneous. However, many biological antioxidants, especially phytochemicals such as caffeic acid and quercetin, continue to reduce the ferric complex even after the completion of reaction time (48,50). The antioxidant activity of any system depends on the pH of the medium in which it is present. The low pH of the reaction in the FRAP assay significantly inhibits one electron transfer from the antioxidant to the ferric ion, resulting in lower values which may not be accurate at physiologically relevant pH (pH 6.8–7.0) (47,48,50).

10.3.4 Oxygen Radical Absorbance Capacity (ORAC) Assay

The ORAC assay is very similar to the direct measurement ROS assay. In this assay, the time dependent decrease in the fluorescence of a fluorescent indicator is monitored at 37°C. The indicator often used is a protein called β-phycoerythrin (beta-PE) (50,51). 2-2′-azobis (2-amidinopropane)dihydrochloride (AAPH) is used to generate peroxyl radicals; oxidative damage of the beta-PE protein results in a decreased fluorescence, which is prevented by the antioxidant (50,51). The total area under the curve determines the antioxidant capacity of the compound or extract. This is a simple, sensitive, and reproducible method of measuring the antioxidant capacity of the natural extracts, biological fluids, and serum against peroxyl radicals (47,48). The method has also been adopted to measure antioxidant capacity against hydroxyl radicals, and the relative simplicity of the assay has resulted in the development of a microplate assay which can be used for screening large number of samples (50,51). Inconsistencies in the purity of the beta-PE protein, coupled with its interaction with many biological antioxidants, especially polyphenols, are one of the drawbacks of this method (48). Recently, fluorescein salt has been used in place of beta-PE to reduce some of the inconsistencies in the analysis (52) (Figure 10.4). One of the main limitations of this method is that it measures the ORAC capacity in comparison to a synthetic antioxidant, trolox, and the ORAC capacity is expressed as trolox equivalents (48). The assay also suggests comparison of the antioxidant activity to a synthetic antioxidant (trolox) which can give misleading information, because the antioxidant capacity of a biological sample depends on many factors such as structure, type of antioxidants, and the synergistic interaction among the constituent antioxidants (53–55). When copper is used as oxidant, trolox cannot be used, because it would act as a prooxidant in the presence of copper (47,56). In biological samples, especially plasma, interference in the scavenging of peroxyl radicals by protein thiol groups may occur; therefore protein needs to be removed from the extract before the assay (57). Nonspecificity in the reaction, especially in biological materials, is introduced by the presence of oxidizable compounds such as citric acid, which can react with the ROS and thus prevent the oxidative damage of the fluorescein and beta-PE protein, which reacts with oxygen radicals more slowly than most biological antioxidants such as

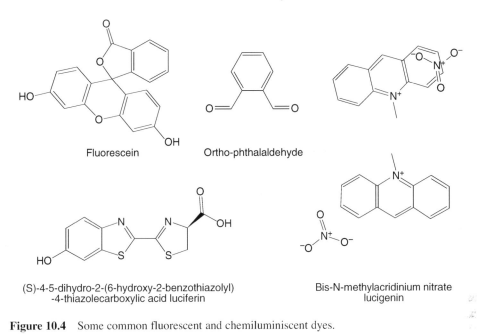

Fluorescein

Ortho-phthalaldehyde

(S)-4-5-dihydro-2-(6-hydroxy-2-benzothiazolyl)
-4-thiazolecarboxylic acid luciferin

Bis-N-methylacridinium nitrate
lucigenin

Figure 10.4 Some common fluorescent and chemiluminiscent dyes.

thiols, uric acid, and ascorbate (48,52) which can cause a falsely high value for the antioxidant capacity.

10.3.5 β-carotene Oxidation (Bleaching) Assay

The antioxidant protection factor (APF) directly measures the ability of the antioxidant extract to prevent the H_2O_2 catalyzed oxidation of β-carotene. Many antioxidant methodologies based on oxidation of an indicator dye are time consuming, as they depend on the conditions used for the reaction and the oxidizability of the reagent (58–60). However, the β-carotene bleaching assay is a relatively fast method for analysis of antioxidant activity. The reaction is accelerated by carrying out the oxidation at 50°C for 30 min, and free radicals are directly generated using hydrogen peroxide (58,60). Because β-carotene is insoluble in aqueous media, it is introduced into the reaction mixture in the form of an emulsion. Many biological sites of ROS generation and oxidation are at a lipid water interface; therefore, it is believed that an antioxidant that is capable of preventing the bleaching of the β-carotene in the emulsion would be able to act efficiently at the lipid to water interface in biological systems, so the β-carotene bleaching assay will give a measure of the true antioxidant capacity (61,62). This method is sensitive and simple to perform, however, irregularities are introduced in the analysis due to the difficulties in keeping the emulsion particle size constant. The method also suffers for its nonspecificity as it is subject to interference from other naturally occurring oxidizing and reducing agents (60,63).

10.4 LIPID OXIDATION PRODUCTS

In aerobic organisms, reactive oxygen species (ROS) are formed by various endogenous and exogenous processes, such as the mitochondrial electron transport chain, cytochrome P-450 systems, NADPH oxidases, myeloperoxidases and xanthine oxidases, ultraviolet, and ionizing irradiation (11–17,64,65). Excessive formation of ROS, can lead to oxidative stress, which can damage cellular macromolecules including nucleic acids, proteins, and lipids

(11–17,64,65). Biological membranes contain unsaturated fatty acids which are especially prone to oxidation. This oxidation of lipids results in the formation of lipid peroxidation (LPO) products which further propagate free radical reactions. Lipid hydroperoxides (LOOH) are the primary oxidation products (66), and have been shown to oxidize other lipids and cellular proteins (67), and to form cytotoxic and genotoxic compounds (68,69). LPO has been therefore been implicated in the pathogenesis of a number of degenerative diseases, such as cancer (70), atherosclerosis (71–73), and Alzheimer's disease (74,75). It is now believed that primary and secondary lipid oxidation products can serve as good markers for oxidative stress and the functionality of antioxidants (70–75).

10.4.1 Malondialdehyde

Oxidation of lipids in biological systems by reactive oxygen species results in the formation malondialdehyde (MDA) which is a metabolite of lipid hydroperoxides (74–78). MDA is a secondary oxidation product of lipids and serves as a good marker for lipid oxidation and cell membrane injury (78). Recent research has shown that elevated levels of lipid peroxides are associated with cancer, heart disease, stroke, and aging. The MDA concentration in plasma can be determined by its reaction with thiobarbituric acid (TBA) (76). The reaction of TBA with MDA in biological materials can be detected spectrophotometrically at 532–535 nm. This method is simple and easy to perform, but suffers from lack of sensitivity as many compounds, especially the products of the side reactions, also have absorption maxima in the 532–535 nm range (73). Other products of lipid peroxidation, such as hydroperoxides and conjugated aldehydes, can also react with TBA and lower the sensitivity of the method (77). Reducing sugars and aldehydes that have a carbonyl group can also react with TBA to give higher readings (77). MDA in biological samples can also react with proteins and give abnormally low values in an oxidizing system (78). The method has been improved vastly by using diaminonapthalene (DAN) instead of TBA to form a DAN–MDA complex. (78). Unlike TBA, DAN reacts very specifically with MDA; other biological compounds such as sugars, aldehydes, and proteins cannot react with DAN. The DAN-MDA complex can also be easily measured by HPLC and has analytical sensitivity and specificity (78). The HPLC modification of the MDA assay has been extensively used to study lipid peroxidation in *in vivo* studies (78). Other methods of analysis include GC using N-methylhydrazine derivatization (79) and ELISA (80).

10.4.2 LDL Resistance to Oxidation

Lipid hydroperoxides are primary product of free radical oxidation of fatty acids and serve as substrates for the reactions that produce free radicals, which in turn form more lipid hydroperoxides (81). Due to their high reactivity and instability, measurement of LOOH *in vivo* is a very challenging (81). At present, the precise measurement of LOOH in plasma remains a difficult task because of low concentration, instability during extraction, and solvent evaporation, leading to both changes in LOOH composition and production of new hydroperoxides if chain reactions are initiated (82,83). Recent research has shown that peroxidative oxidation of LDL in biological systems is responsible for initiation of plaque formation and enhanced atherogenecity (81). Ability to reduce LDL oxidation by peroxidation is used as an important biomarker for studying the antiatherogenic functionality of antioxidant compounds (82,83). The LDL oxidation is induced by prooxidant metals such as copper, and other free radical generating systems such as 2,2′-azobis(2-amidinopropane) dihydrochloride (AAPH) (82,83). The presence and absence of antioxidants, signifying the lag phase before the oxidation of LDL, can be detected by measuring the rate of oxidation and total amount of dienes at 234 nm using the molar extinction coefficient for total lipid

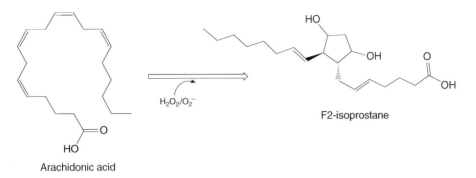

Figure 10.5 Nonenzymatic metabolism of arachidonic acid to F-2 Isoprostanes.

hydroperoxides $[\varepsilon_{234} = 29,500 \ (mol/L)^{-1} \ cm^{-1}]$ (82,83). The total oxidation in the LDL can be measured spectrofluorometrically by extracting the lipid phase and drying it under nitrogen, then resuspending it in chloroform (84–86). This chloroform extract of the lipid phase is subjected to 360 nm excitation and 430 nm emission, and the results are expressed as units of relative fluorescence per milligram of LDL (84–86). The lipid peroxide content can be estimated fluorometrically by the formation of TBARS and expressed as nanomoles of malondialdehyde equivalents per milligram of LDL (84–86).

10.4.3 F2 Isoprostanes

Prostaglandin F2-like compounds (F2-isoprostanes) are formed by nonenzymatic free radical induced peroxidation of arachidonic acid (87) (Figure 10.5). These compounds are formed in phospholipids and then cleaved and released into the circulation before excretion in the urine as free isoprostanes (87). They have been extensively used as novel markers of endogenous lipid peroxidation and potential mediators of oxidant injury (88,89). Elevated levels of F2-isoprostanes have been seen in many diseases including diabetes, myocardial infraction, atherosclerosis, Alzheimer's, and in hepatic toxicity (90–94). Detection and quantification of F2-isoprostanes in biological samples including blood serum and plasma is now used as a tool in assessing oxidative stress status (95). The most common isoprostane formed is 8-isoprostaglandin F_2, which can be detected accurately and easily by using GC/MS or by HPLC/MS (95–97). The longer analysis time required for GC/MS and HPLC/MS has resulted in the development of many immunochemical assays (91,98). However, these immunochemical methods suffer crossreactivity from biological prostaglandins due to their close structural similarity (91).

10.5 DNA OXIDATION PRODUCTS

10.5.1 8-Oxo-7,8-dihydro-2′-deoxyguanosine

Free radicals have been shown to cause damage to lipids, DNA, and proteins. DNA oxidation in biological systems can occur by its interaction with ionic radiation, trace metals, and many carcinogens (11–17,64,65). DNA damage is caused by the oxidation and degeneration of nucleotide bases resulting in the formation of 2-OH-2′deoxyadenosine, 8-Oxo-7,8-dihydro-2′-deoxyguanosine, 5-hydroxymethyl-2′-deoxyuridine (99). However, 8-OHdG, formed from oxidative damage of guanine, is considered a typical marker of DNA damage and repair because guanine is the most fragile nuclear base (99,100). 8-Oxo-7,8-dihydro-2′-deoxyguanosine (8-OHdG), is formed from deoxyguanosine after reaction with a hydroxyl radical or

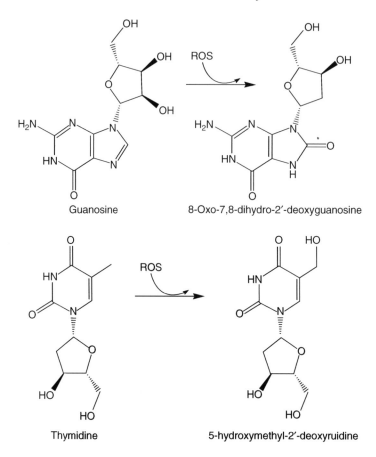

Figure 10.6   DNA oxidation by reactive oxygen species.

singlet oxygen (Figure 10.6). 8-OHdG is not metabolized and is quantitatively excreted in urine independent of other metabolic processes, enabling it to be used as a noninvasive marker of oxidative DNA damage by ROS in various physiological and nutritional studies (99,100). Oxidative damage to DNA has now been shown to be responsible for aging, cardiovascular disease, degenerative neurological disease, and various types of cancer (11–17,64,65,99,100). 8-OHdG can be analyzed by many different assays. The most common methodology is high performance liquid chromatography using an electrochemical detector (HPLC-ECD) (99,100). ^{32}P-postlabeling assay using thin layer chromatography (TLC) (101), and gas chromatography (102) are also used in detecting 8-OHdG. Immunodetection using specific antibodies against 8-OHdG. Another possible biomarker, 5-hydroxymethyl-2′-deoxyuridine (HMdU) (103,104), which is a product of thymine oxidation, is gaining popularity because its higher sensitivity, specificity, and stability (105) (Figure 10.6).

10.5.2   Glutathione Reduced (GSH) and Oxidized (GSSG)

Glutathione (GSH, χ-glutamylcysteinylglycine) and its oxidized form (GSSG) (Figure 10. 7) together form the first line of defense against the cellular protection against reactive oxygen species (22,23,106,107). They carry out their protective functions by reducing ROS via their thiol groups, and therefore are involved in a variety of processes in the cellular systems that involve generation of free radicals such as detoxification of xenobiotics, reduction of hydroperoxides, and synthesis of leukotrienes and prostaglandins (106,107). The reducing power of glutathione is also involved in the maintenance of protein and membrane structure, and regulation of numerous enzyme activities and transcription factors that could be effected by oxidation as a result of reaction with ROS (11,106,107). The thiol groups are capable of participating in a number of chemical reactions including redox transitions, exchange of thiol groups, and radical scavenging (106,107). Recent research has also shown that the redox capability of thiol groups in the GSH/GSSG couple are also involved in gene regulation and intracellular signal transduction by critically regulating the functions of transcription factors and key cellular signaling molecules through oxidation and reduction (106). Because of the many protective functions of glutathione in the body, it is now considered one of the many biomarkers important to an understanding of the oxidative status of the cell (107). Many assays are based on conjugation of GSH with a chromophore or fluorophore, such as *o*-phthalaldehyde (OPA) (Figure 10.4). OPA does not fluoresce until it reacts with a thiol group in the presence of a primary amine. The OPA method for the detection of GSH has been limited due to the presence of other fluorogenic thiol and nonthiol components that react with OPA (107,108). Many biological constituents have been shown to quench fluorescence or inhibit the reaction of OPA with GSH, causing the underestimation of GSH levels or the overestimation of GSSG levels (107,108). Recent modifications in the OPA method by using a phosphate buffer at a lower pH, higher concentrations of OPA, and dithionite as the reductant for

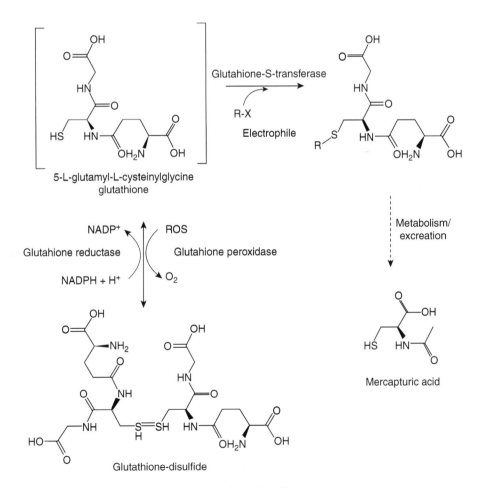

Figure 10.7 Metabolism of glutathione in eukaryotic cell.

GSSG have led to rapid analysis, lower cost, and high sensitivity. These revised protocols have been able to estimate GSH and GSSG with accuracy comparable to established HPLC methods (108).

10.5.3 Glutathione Peroxidase

Glutathione peroxidase (GPX)(EC 1.11.1.9) is a selenium containing peroxidase involved in the cellular reduction of hydroperoxides. Two main types of GPX are found in the cells: mitochondrial peroxidase and phospholipid hydroperoxide glutathione peroxidase (22,23). Both types of GPX react directly with fatty acid hydroperoxides and hydrogen peroxide, and reduce them at the expense of glutathione (Figure 10.1, Figure 10.7). Depletion of GPX has now been observed in several forms of cancer, in heart disease, and in hypersensitive pathologies, and therefore is thought to make a good marker for studying the effects of certain types of foods and bioactive compounds on reducing the symptoms of certain types of heart disease, liver disease, and cancer (22,23). The most common assay method for measuring the activity of GPX is by glutathione reductase (GR) coupled assay. The reaction is initiated in the presence of a lipid hydroperoxide or hydrogen peroxide (109,110). The glutathione in the GPX reacts with GSH and oxidizes to GSSG which is reduced by GR back to GSH using NADPH. The rate of disappearance of NADPH is monitored at 340 nm and is used to calculate the enzyme activity (109–111).

10.5.4 Glutathione-S-Transferases

The glutathione-S-transferases (GST) are important phase II detoxification enzymes found mainly in the cytosol, and are involved in the detoxification of drugs and harmful chemicals in the body (112). They are found in very high concentrations in the liver, where they conjugate glutathione to the electrophilic centers of lipophilic compounds and increase their solubility in order to facilitate their excretion (113,114) (Figure 10.7). GST posses a wide range of substrate specificities and can catalyze the reduction of breakdown products of macromolecules formed as a result of oxidative stress including reactive unsaturated carbonyls, oxidized DNA bases, and hydroperoxides, and therefore play a vital role in protecting tissues against oxidative damage and oxidative stress (112–115). As a result of the beneficial functions of GST it serves as an important biomarker for oxidative stress (115). Most of the methods to assay the GST activity involve measuring the conjugation of 1-chloro-2,4-dinitrobenzene (CDNB) with reduced glutathione (114). The conjugation of GSH with CDNB results in an increase in absorbance at 340 nm which is measured spectrophotometrically. The rate of increase is directly proportional to the GST activity in the sample. This assay can also be used to measure GST activity in plasma, cell lysates and tissue homogenates (114).

10.5.5 Superoxide Dismutase

The superoxide radicals are formed by the activity of enzymes like xanthine oxidase and NADPH-oxidase (22,23) (Figure 10.1). They are extremely reactive and have been shown to be able to oxidize biological macromolecules such as DNA, lipids and proteins (11,19,21–23). Superoxide dismutase (SOD) (EC 1.15.1.1) is involved in the dismutation of the superoxide radicals into hydrogen peroxide and molecular oxygen. In humans, there are four forms of SOD: cytosolic Cu, Zn-SOD, mitochondrial Mn-SOD, and extracellular-SOD (EC-SOD) (22,23).

The process of dismutation yields hydrogen peroxide, which is also a reactive oxygen species and causes oxidative damage to cellular systems. Therefore, in the body, activity of SOD is often coupled to another enzyme called "Catalase" (CAT) to remove hydrogen peroxide molecules which are byproducts of the reactions created by SOD (22,23). CAT is also abundant in the body and is prevents both cell damage and, more importantly, the formation of other, more toxic, free radicals (22,23).

Pathogenesis, mediated by virus and *Helicobacter pylori* (116,117), has been implicated in lowered antioxidant defense and alterations of enzymatic pathways; diabetes mellitus, endothelial, vascular, and neurovascular dysfunction. Neurodegenerative disorders such as Alzheimer's and Parkinson's disease are also now linked to oxidative stress and can be associated with lower responses of enzymes CAT and SOD (118,119). The activity of SOD and CAT is therefore considered as an important biomarker for these diseases (116–119).

The most common SOD detection method is based on spectrophotometric detection. Superoxide, which is a substrate of SOD, is generated by a xanthine to xanthine oxidase, which reacts with either cytochrome C (120) or nitroblue tetrazolium (NBT) (121), and changes color, which can be monitored spectrophotometrically. Detection of O_2^- by the cytochrome C reducing method is based on the color change to generate purple colored dye from reduced cytochrome C. The NBT method is based on the generation of water insoluble blue formazan dye (λmax: 560 nm) by a reaction with O_2^- (120,121). Recent methods developed use chemiluminescence probes such as lucigenin, and luciferin derivative (MCLA) for O_2^- detection (122,123). They react with the superoxide radical and produce a signal which can be used to quantitate SOD activity.

10.5.6 Catalase

Catalase (EC 1.11.1.6) is a ubiquitous oxidoreductase that is present in most aerobic cells. Catalase (CAT) is involved in the detoxification of hydrogen peroxide (H_2O_2), a (ROS), into molecular oxygen and two molecules of water (22,23). CAT can be assayed directly by measuring the rate of appearance of molecular oxygen using a Clark-type electrode (124). Here, the enzyme extract is allowed to react with a known concentration of hydrogen peroxide. The formation of oxygen changes the potential of the Clark-type electrode which is proportional to the enzyme activity (124). Catalase activity has also been determined spectrophotometrically at 240 nm by monitoring the disappearance of hydrogen peroxide at pH 7.0 (125). Catalase also has a peroxidative activity which has been used to assay its activity by reaction with methanol in the presence of an optimal concentration of H_2O_2 (126,127). The formaldehyde produced is measured spectrophotometrically with 4-amino-3-hydrazino-5-mercapto-1,2,4-triazole (Purpald) as the chromogen, in which low molecular weight alcohols can serve as electron donors. Fluorogenic dyes such as amplex red have also been recently used to measure CAT activity. Amplex red reacts with hydrogen peroxide to form a highly fluorescent resorufin which can be measured in a spectrofluoremeter (excitation and emission of 563/587 nm) (126,127).

10.5.7 Xanthine Oxidase and Uric acid

Xanthine oxidoreductase (XOR), a metalloflavoprotein, catalyzes the oxidation of the purine bases hypoxanthine and xanthine to uric acid. XOR is a site for ROS generation in the cell (128) (Figure 10.1). High levels of XOR activity resulting in ROS production has now been shown to play an important role in reperfusion injury; and in congestive heart failure (129). Xanthine oxidase and its reaction product uric acid are therefore now considered to be important biomarkers for oxidative stress and cardiovascular diseases (129,130). Uric acid in urine, serum, and plasma is measured using a spectrophotometric assay based on phosphotungstic acid reagent (131); however, for improved specificity and characterization, HPLC methods have been used recently (132,133). The most popular assay method for xanthine oxidoreductase involves the spectrophotometric measurement of urate production at 292 nm, using xanthine or hypoxanthine as the substrate (134). A spectrofluorometric assay utilizing pterin as the substrate for XOR, and the formation of fluorescent isoxanthopterin, is quantified for xanthine oxidase activity (135). Chemiluminescence methods are also used alternatively which assess the ROS production by xanthine oxidase which reacts with chemiluminescent dyes such as luminol or lucigenin (Figure 10.4) which are then measured and quantitated as enzyme activity (136).

10.5.8 NADPH Oxidoreductase

NADPH oxidase (NOX) is a membrane bound enzyme present in neutrophils and other phagocytic cells (138). This plasma membrane associated enzyme complex catalyzes the univalent reduction of oxygen using NADPH as an electron donor to produce superoxides, which are involved in the bactericidal activity of the neutrophils (137,138). Though NOX is involved in the generation of ROS for bactericidal activities, recent research has shown that a new family of NADPH oxidases are expressed in various cell types, including the epithelium, smooth muscle cells, and the endothelium (134,135). These enzymes produce superoxide and other ROS, including hydrogen peroxide (Figure 10.1). The activity of this enzyme in vascular cells is a main source of ROS in inflammation and is believed to be a cause for atherosclerosis and hypertension. NOX expression is also found to be high in certain forms of cancer and endocrine disorders as a result of superoxide and other ROS production (137–139). Elevated levels of ROS are also seen in patients with progressive neurological diseases like Alzheimer's, and Parkinson's disease (140). These recent discoveries about the many different functions of the nonphagocytic functions of NOX has made it an attractive target for therapeutic agents and an important biomarker for these diseases in different cell types (139,140). NOX is usually measured in cell free extracts by using them with NADPH for superoxide radical generation, which is measured by following the reduction of ferric cytochrome-C at 550 nm (141,142) or by monitoring the inhibition of the formation of formazan blue from NBT at 560 nm (143). Use of a new tetrazolium based assay to study the production of superoxide radicals has also been done, with the NOX and NADPH as a source for superoxide radical generation (143). Chemiluminescence assays based on oxidation of a chromophore such as lucigenin are also being used to measure NOX activity (144). In this assay, the chromophore becomes chemiluminescent after reacting with the superoxide radical generated from the NOX reaction. This is then measured and quantitated as NOX activity (145).

10.5.9 Myeloperoxidase

Myeloperoxidase (MPO) is a lysosomal enzyme that is found in white blood cells and neutrophils. In these cells the NADPH oxidase produces superoxide radicals which then dismutates to hydrogen peroxide. Myeloperoxidase uses hydrogen peroxide and chloride to produce another highly reactive ROX called hypochlorous acid (HOCl) (145,146). This potent oxidant then reacts with proteins, DNA, and lipids to cause cellular injury. The primary function of these enzymes is to mediate a defense response against invading pathogens. However, it has been shown recently that, due to the remarkably high oxidant potential of MPOs, they function as mediators of inflammatory pathologies, autoimmune diseases, and rheumatoid arthritis (145,146). The hypochlorous radical and tyrosyl radicals made by MPO have been shown to cause oxidation of LDL leading to atherosclerosis (148) and other cardiovascular diseases (145,146). MPO therefore has evolved into the most promising cardiac marker and a good inflammatory biomarker. The strong oxidant capacity of MPO is often exploited in assaying its activity. MPO is capable of oxidizing many different chromophores such as tyrosine (147) and 3′,5,5′-tetramethylbenzinide (TMB) (148) which are monitored spectrophotometrically or fluorometrically in cell free extracts. Possible interference from other peroxidases and ROS in the system has led to the development of many enzyme linked immunosorbent assays which are more specific to myeloperoxidases and their isoforms (149–151).

10.6 PLANT PHENOLIC DEPENDENT PEROXIDASES

Phenolic dependent peroxidases catalyze the oxidative coupling of phenolic compounds using hydrogen peroxide to make lignins, phenolic hemicellulose, and phenolic to tyrosine cross links to aid structural development in plant tissues (152,153). These phenolic dependent peroxidases, often referred to as guaiacol, peroxidases have been well characterized in plants and are known to be inducible enzymes, expressed in response to a need for structural growth and development (154,155). Plant peroxidases are induced in response to oxidative stress, exposure to environmental toxins, and high energy radiation (156–160). They protect the plants from these stresses by removing the reactive oxygen species, and by using them to oxidatively couple phenolic phytochemicals to make lignin and other cross-linked phenolics important in plant cell wall protection (154). The oxidative polymerization of phenolics also contributes to the increase in antioxidant activity which is believed to further protect the plant against oxidative stress (54,155). Recently, it has been shown that these peroxidases are also stimulated in response to exogenous phenolic elicitors in germinating fava bean sprouts and that the proxidases contributed to increased antioxidant activity observed during the course of germination (161). It has now been proposed that peroxidases similar to the phenolic dependent guaiacol peroxidase could be stimulated in human cells in response to exogenous phenolics for protection against oxidative stress (162). It is also believed that that these phenolic dependent peroxidases reduce oxidative stress by removing reactive oxygen species by oxidatively polymerizing dietary phenolics without affecting the cellular pools of glutathione, ascorbate, and other antioxidants (162). Stimulation of these peroxidases was recently observed in peroxide stressed porcine muscle models in response to treatment with phenolic extract from oregano (163).

The most common analytical method for assaying the activity of phenolic dependent peroxidases is based on oxidation of a substrate catalyzed by the enzyme in the presence of hydrogen peroxide, which can be followed spectrophotometrically (154,164). The time dependent oxidation of guaiacol to tertraguaiacol ($\varepsilon = 26.6$ mM^{-1}cm^{-1}) at 470 nm, described by Laloue et al. (165), has been extensively used in characterization of these peroxidases. Other substrates that are used for assaying the activity of peroxidases include o-dianisidine, pyrogallol, syringaldazine and 2,2′-azinobis(3-ethylbenzothiazoline-6-sulphonate) (ABTS) (154).

REFERENCES

1. Lundberg, A.C., A. Åkesson, B. Åkesson. Dietary intake and nutritional status in patients with systemic sclerosis. *Ann. Rheum. Dis.* 51:1143–1148, 1992.
2. Wilks, R., F. Bennett, T. Forrester, N. Mcfarlane-Anderson. Chronic diseases: the new epidemic. *West Ind. Med. J.* 47(4):40–44, 1998.
3. Gillman, M.W., L.A. Cupples, D. Gagnon, B.M. Posner, R.C. Ellison, W.P. Castelli, P.A. Wolf. Protective effect of fruits and vegetables on development of stroke in men. *J. Am. Med. Assoc.* 273:1113–1117, 1995.
4. Joshipura, K.J, A. Ascherio, J.E. Manson, M.J. Stampfer, E.B. Rimm, F.E. Speizer, C.H. Hennekens, D. Spiegelman, W.C. Willett. Fruit and vegetable intake in relation to risk of ischemic stroke. *J. Am. Med. Assoc.* 282:1233–1239, 1999.

5. Cox, B.D., M.J. Whichelow, A.T. Prevost. Seasonal consumption of salad vegetables and fresh fruit in relation to the development of cardiovascular disease and cancer. *Public Health Nutr.* 3:19–29, 2000.

6. Strandhagen, E., P.O. Hansson, I. Bosaeus, B. Isaksson, H. Eriksson. High fruit intake may reduce mortality among middle-aged and elderly men: the study of men born in 1913. *Eur. J. Clin. Nutr.* 54:337–341, 2000.

7. Block, G., B. Patterson, A. Subar. Fruit, vegetables, and cancer prevention: a review of the epidemiological evidence. *Nutr. Cancer* 18:1–29, 1992.

8. Serdula, M.K., M.A.H. Byers, E. Simoes, J.M. Mendlein, R.J. Coates. The association between fruit and vegetable intake and chronic disease risk factors. *Epidemiology* 7:161–165, 1996.

9. Tapiero, H., K.D. Tew, G.N. Ba, G. Mathe. Polyphenols: do they play a role in the prevention of human pathologies? *Biomed. Pharmacother.* 56(4):200–207, 2002.

10. Duthie, G.G., P.T. Gardner, J.A. Kyle. Plant polyphenols: are they the new magic bullet? *Proc. Nutr. Soc.* 62(3):599–603, 2003.

11. Droge, W. Free radicals in the physiological control of cell function. *Physiol. Rev.* 82(1):47–95, 2002.

12. Sculley, D.V., S.C. Langley-Evans. Salivary antioxidants and periodontal disease status. *Proc. Nutr. Soc.* 61(1):137–143, 2002.

13. Parke, A.L., C. Ioannides, D.F.V. Lewis, D.V. Parke. Molecular pathology of drugs: disease interaction in chronic autoimmune inflammatory diseases. *Inflammopharmacology* 1:3–36, 1991.

14. Schwarz, K.B. Oxidative stress during viral infection: a review. *Free Radic. Biol. Med.* 21(5):641–649, 1996.

15. Götz, J., C.I. va Kan, H.W. Verspaget, I. Biemond, C.B. Lamers, R.A. Veenendaal. Gastric mucosal superoxide dismutases in *Helicobacter pylori* infection. *Gut* 38:502–506, 1996.

16. Jakus, V. The role of free radicals, oxidative stress and antioxidant systems in diabetic vascular disease. *Bratisl. Lek. Listy.* 101(10):541–551, 2000.

17. Offen, D., P.M. Beart, N.S. Cheung, C.J. Pascoe, A. Hochman, S. Gorodin, E. Melamed, R. Bernard, O. Bernard. Transgenic mice expressing human Bcl-2 in their neurons are resistant to 6-hydroxydopamine and 1-methyl-4-phenyl-1,2,3,6-tetrahydropyridine neurotoxicity. *Proc. Natl. Acad. Sci.* 95:5789–5794, 1998.

18. Rice-Evans, C.A., N.J. Miller, G. Paganga. Antioxidant Properties of Phenolic Compounds. *Trends Plant Sci.* 2:152–159, 1997.

19. Niess, A.M., H.H. Dickhuth, H. Northoff, E. Fehrenbach. Free radicals and oxidative stress in exercise-immunological aspects. *Exerc. Immunol. Rev.* 5:22–56, 1999.

20. Sculley, D.V., S.C. Langley-Evans. Salivary antioxidants and periodontal disease status. *Proc. Nutr. Soc.* 61(1):137–143, 2002.

21. Lusini, L., S.A. Tripodi, R. Rossi, F. Giannerini, D. Giustarini, M.T. del Vecchio, G. Barbanti, M. Cintorino, P. Tosi, P. Di Simplicio. Altered glutathione anti-oxidant metabolism during tumor progression in human renal-cell carcinoma. *Int. J. Cancer* 91(1):55–59, 2001.

22. Mates, J.M., F. Sanchez-Jimenez. Antioxidant enzymes and their implications in pathophysiologic processes. *Front Biosci.* 4:D339–345, 1999.

23. Mates, J.M., C. Perez-Gomez, I. Nunez de Castro. Antioxidant enzymes and human diseases. *Clin. Biochem.* 32(8):595–603, 1999.

24. Liochev, S.I., I. Fridovich. The relative importance of HO* and ONOO- in mediating the toxicity of O*-. *Free Radic. Biol. Med.* 26(5–6):777–778, 1999.

25. Barber, D.A., S.R. Harris. Oxygen free radicals and antioxidants: a review. *Am. Pharm.* NS34:26–35, 1994.

26. Betteridge, D.J. What is oxidative stress? *Metabolism* 49:3–8, 2000.

27. Vignais, P.V. The superoxide-generating NADPH oxidase: structural aspects and activation mechanism. *Cell Mol. Life Sci.* 59(9):1428–1459, 2002.

28. Babior, B.M., J.D. Lambeth, W. Nauseef. The neutrophil NADPH oxidase. *Arch. Biochem. Biophys.* 397(2):342–344, 2002.

29. Babior, B.M. The leukocyte NADPH oxidase. *Isr. Med. Assoc. J.* 4(11):1023–1024, 2002.

30. Parke, A., D.V. Parke. The pathogenesis of inflammatory disease: surgical shock and multiple system organ failure. *Inflammopharmacology* 3:149–168, 1995.

31. Parke, D.V. The cytochromes P450 and mechanisms of chemical carcinogenesis. *Environ. Health Perspect.* 102:852–853, 1994.

32. Parke, D.V., A. Sapota. Chemical toxicity and reactive oxygen species. *Int. J. Occ. Med. Environ. Health.* 9:331–340, 1996.

33. Baker, C.J., G.L. Harmon, J.A. Glazener, E.W. Orlandi. A Noninvasive Technique for Monitoring Peroxidative and H2O2-Scavenging Activities during Interactions between Bacterial Plant Pathogens and Suspension Cells. *Plant Physiol.* 108(1):353–359, 1995.

34. Low, P.S., J.R. Merida. The oxidative burst in plant defense: function and signal transduction. *Physiol. Plant* 96:533–542, 1996.

35. Wojtaszek, P. Oxidative burst: an early plant response to pathogen infection. *Biochem. J.* 322(3):681–692, 1997.

36. Low, P.S., P.F. Heinstein. Elicitor stimulation of the defense response in cultured plant cells monitored by fluorescent dyes. *Arch. Biochem. Biophys.* 249(2):472–479, 1986.

37. Levine, A., R. Tenhaken, R. Dixon, C. Lamb. H_2O_2 from the oxidative burst orchestrates the plant hypersensitive disease resistance response. *Cell* 79:583–593, 1994.

38. Yoshiki, Y., T. Kahara, K. Okubo, K. Igarashi, K. Yotsuhashi. Mechanism of catechin chemiluminescence in the presence of active oxygen. *J. Biolumin. Chemilumin.* 11(3):131–136, 1996.

39. Halliwell, B., J.M.C. Gutteridge *Free Radicals In Biology And Medicine,* Third Ed., Oxford: Oxford University Press, 1999.

40. Able, A.J., D.I. Guest, M.W. Sutherland. Hydrogen peroxide yields during the incompatible interaction of tobacco suspension cells inoculated with Phytophthora nicotianae. *Plant Physiol.* 124(2):899–910, 2000.

41. Rosen, G.M., E.J. Rauckman. Spin trapping of superoxide and hydroxyl radicals. *Methods Enzymol.* 105:198–209, 1984.

42. Pou, S., G.M. Rosen, B.E. Britigan, M.S. Cohen. Intracellular spin-trapping of oxygen-centered radicals generated by human neutrophils. *Biochim. Biophys. Acta.* 991(3):459–464, 1989.

43. Bondet, V., W. Brand-Williams C. Berset. Kinetics and Mechanisms of Antioxidant Activity using the DPPH. Free Radical Method. *Lebensm.-Wiss. u.-Technol.* 30:609–615, 1997.

44. Molyneux, P. The use of the stable free radical diphenylpicrylhydrazyl (DPPH) for estimating antioxidant activity. *Songklanakarin. J. Sci. Technol.* 26(2):211–219, 2004.

45. Miller, N.J., C.A. Rice-Evans. Factors influencing the antioxidant activity determined by the ABTS.+ radical cation assay. *Free Radic. Res.* 26(3):195–199, 1997.

46. Re, R., N. Pellegrini, A. Proteggente, A. Pannala, M. Yang, C. Rice-Evans. Antioxidant activity applying an improved ABTS radical cation decolorization assay. *Free Radic. Biol. Med.* 26(9–10):1231–1237, 1999.

47. Sanchez-Moreno, C. Methods used to evaluate the free radical scavenging activity in foods and biological systems. *Food Sci. Tech. Int.* 8(3):121–137, 2002.

48. McAnalley, S., C.M. Koepke L. Le, R.A.C. Vennum, B. Bill McAnalley. *In vitro* methods for testing antioxidant potential: a review. *Glycoscience.* 4(2):1–9, 2003.

49. Benzie, I.F., J.J. Strain. The ferric reducing ability of plasma (FRAP) as a measure of "antioxidant power": the FRAP assay. *Anal. Biochem.* 239(1):70–76, 1996.

50. Cao, G., R.L. Prior. Comparison of different analytical methods for assessing total antioxidant capacity of human serum. *Clin. Chem.* 1998. 44(6 Pt 1):1309–15.

51. Cao, G., R.L. Prior. Measurement of oxygen radical absorbance capacity in biological samples. *Methods Enzymol.* 299:50–62, 1999.

52. Ou, B., M. Hampsch-Woodill, R.L. Prior. Development and validation of an improved oxygen radical absorbance capacity assay using fluorescein as the fluorescent probe. *J. Agric. Food Chem.* 49(10):4619–4626, 2001.

53. Rice-Evans, C.A., N.J. Miller, G. Paganga. Antioxidant Properties of Phenolic Compounds. *Trends Plant Sci.* 2:152–159, 1997.

54. Lotito, S.B., L. Actis Goretta, M.L. Renart, M. Caligiuri, D. Rein, H.H. Schmitz, F.M. Steinberg, C.L. Keen, C.G. Fraga. Influence of oligomer chain length on the antioxidant activity of procyanidins. *Biochem. Biophys. Res. Commun.* 276:945–951, 2000.

55. Vattem, D.A., H.-D. Jang, R. Levin, K. Shetty. Synergism of cranberry phenolics with ellagic acid and rosmarinic acid for antimutagenic and DNA-protection functions. *J. Food Biochem.* 30:98–116, 2006.

56. Frankel, E.N. A.S. Meyer. The problems of using one-dimensional methods to evaluate multifunctional food and biological antioxidants. *J. Sci. Food Agric.* 80:1925–1941, 2000.

57. Ghiselli, A., M. Serafini, G. Maiani, E. Azzini, A.A. Ferro-Luzzi. Fluorescence-based method for measuring total plasma antioxidant capability. *Free Rad. Biol. Med.* 18:29–36, 1995.

58. Marco, G.J. A rapid method for evaluation of antioxidants. *J. Am. Oil Chem. Soc.* 45:594–598, 1968.

59. Miller, H.E. A simplified method for evaluation of antioxidants. *J. Am. Oil Chem. Soc.* 48:91, 1971.

60. Lee, Y., L.R. Howard, B. Villalón. Flavonoids and antioxidant activity of fresh pepper (*Capsicum annuum*) cultivars. *J. Food Sci.* 60:473–476, 1995.

61. Vattem, D.A., K. Shetty. Solid-state production of phenolic antioxidants from cranberry pomace by *Rhizopus oligosporus*. *Food Biotechnol.* 16(3):189–210, 2002.

62. Vattem D.A., K. Shetty. Ellagic acid production and phenolic antioxidant activity in cranberry pomace (*Vaccinium macrocarpon*) mediated by *Lentinus edodes* using solid-state system. *Proc. Biochem.* 39(3):367–379, 2003.

63. Frankel, E.N., J. Kanner, J.B. German, E. Parks, J.E. Kinsella. Inhibition of oxidation of human low-density lipoprotein by phenolic substances in red wine. *Lancet* 20:341(8843):454–457, 1993.

64. Ames, B.N., M.K. Shigenaga, T.M. Hagen. Oxidants, antioxidants, and the degenerative diseases of aging. *Proc. Natl. Acad. Sci. USA* 90(17):7915–7922, 1993.

65. Beckman, J.S., W.H. Koppenol. Nitric oxide, superoxide, and peroxynitrite: the good, the bad, and ugly. *Am. J. Physiol.* 271(5,1):C1424–1437, 1996.

66. McLeod, L.L., A. Sevanian. Lipid peroxidation and modification of lipid composition in an endothelial cell model of ischemia and reperfusion. *Free Radic. Biol. Med.* 23(4):680–694, 1997.

67. Kawakami, A., A. Tanaka, T. Nakano, K. Nakajima, F. Numano. The role of remnant lipoproteins in atherosclerosis. *Ann. NY Acad. Sci.* 902:352–356, 2000.

68. Pryor, W.A., N.A. Porter. Suggested mechanisms for the production of 4-hydroxy-2-nonenal from the autoxidation of polyunsaturated fatty acids. *Free Radic. Biol. Med.* 8(6):541–543, 1990.

69. Esterbauer, H., R.J. Schaur, H. Zollner. Chemistry and biochemistry of 4-hydroxynonenal, malonaldehyde and related aldehydes. *Free Radic. Biol. Med.* 11(1):81–128, 1991.

70. Khanzode, S.S., M.G. Muddeshwar, S.D. Khanzode, G.N. Dakhale. Antioxidant enzymes and lipid peroxidation in different stages of breast cancer. *Free Radic. Res.* 38(1):81–85, 2004.

71. Yilmaz, M.I., K. Saglam, A. Sonmez, D.E. Gok, S. Basal, S. Kilic, C. Akay, I.H. Kocar. Antioxidant system activation in prostate cancer. Biol. Trace Elem. Res. 98(1):13–19, 2004.

72. Hennig, B., C.K. Chow. Lipid peroxidation and endothelial cell injury: implications in atherosclerosis. *Free Radic. Biol. Med.* 4(2):99–106, 1988.

73. Massy, Z.A., W.F. Keane. Pathogenesis of atherosclerosis. *Semin. Nephrol.* 16(1):12–20, 1996.

74. Pratico, D., M.Y. Lee V., J.Q. Trojanowski, J. Rokach, G.A. Fitzgerald. Increased F2-isoprostanes in Alzheimer's disease: evidence for enhanced lipid peroxidation *in vivo*. *FASEB J.* 12(15):1777–1783, 1998.

75. Montine, T.J., W.R. Markesbery, J.D. Morrow, L.J. Roberts, 2nd. Cerebrospinal fluid F2-isoprostane levels are increased in Alzheimer's disease. *Ann. Neurol.* 44(3):410–413, 1998.

76. Moore, K., L.J. Roberts, 2nd. Measurement of lipid peroxidation. *Free Radic. Res.* 28(6):659–671, 1998.

77. Karatas, F., M. Karatepe, A. Baysar. Determination of free malondialdehyde in human serum by high-performance liquid chromatography. *Anal. Biochem.* 1:311(1):76–79, 2002.

78. Steghens, J.P., A.L. van Kappel, I. Denis, C. Collombel. Diaminonaphtalene, a new highly specific reagent for HPLC-UV measurement of total and free malondialdehyde in human plasma or serum. *Free Radic. Biol. Med.* 31(2):242–249, 2001.

79. Wong, J.W., S.E. Ebeler, R. Rivkah-Isseroff, T. Shibamoto. Analysis of malondialdehyde in biological samples by capillary gas chromatography. *Anal. Biochem.* 220(1):73–81, 1994.

80. Khan, M.F., X. Wu, G.A. Ansari. Anti-malondialdehyde antibodies in MRL+/+ mice treated with trichloroethene and dichloroacetyl chloride: possible role of lipid peroxidation in autoimmunity. *Toxicol. Appl. Pharmacol.* 170(2):88–92, 2001.

81. Mayne, S.T. Antioxidant nutrients and chronic disease: use of biomarkers of exposure and oxidative stress status in epidemiologic research. *J. Nutr.* 133(3):933S–940S, 2003.

82. Jialal, I., C.J. Fuller. Effect of vitamin E, vitamin C and beta-carotene on LDL oxidation and atherosclerosis. *Can. J. Cardiol.* G:97G–103G, 1995.

83. Gaziano, J.M., A. Hatta, M. Flynn, E.J. Johnson, N.I. Krinsky, P.M. Ridker, C.H. Hennekens, B. Frei. Supplementation with beta-carotene *in vivo* and *in vitro* does not inhibit low density lipoprotein oxidation. *Atherosclerosis* 112(2):187–195, 1995.

84. Pierdomenico, S.D., F. Costantini, A. Bucci, D. De Cesare, F. Cuccurullo, A. Mezzetti. Low-density lipoprotein oxidation and vitamins E and C in sustained and white-coat hypertension. *Hypertension* 31(2):621–626, 1998.

85. Costantini, F., S.D. Pierdomenico, D. De Cesare, P. De Remigis, T. Bucciarelli, G. Bittolo-Bon, G. Cazzolato, G. Nubile, M.T. Guagnano, S. Sensi, F. Cuccurullo, A. Mezzetti. Effect of thyroid function on LDL oxidation. *Arterioscler. Thromb. Vasc Biol.* 18(5):732–737, 1998.

86. Princen, H.M., W. van Duyvenvoorde, R. Buytenhek, A. van der Laarse, G. van Poppel, J.A. Gevers Leuven, V.W. van Hinsbergh. Supplementation with low doses of vitamin E protects LDL from lipid peroxidation in men and women. *Arterioscler. Thromb. Vasc. Biol.* 15(3):325–333, 1995.

87. Morrow, J.D., L.J. Roberts, 2nd. The Isoprostanes: Novel Markers of Lipid Peroxidation and Potential Mediators of Oxidant Injury. *Adv. Prostaglandin Thromboxane Leukot. Res.* 23:219–224, 1995.

88. Morrow, J.D., L.J. Roberts, 2nd. The Generation and Actions of Isoprostanes. *Biochemistry* 121–135, 1997.

89. Halliwell, B. Establishing the significance and optimal intake of dietary antioxidants: The biomarker concept. *Nutr. Rev.* 57:104–113, 1999.

90. Gopaul, N.K., E.E. Anggard, A.I. Mallet, D.J. Betteridge, S.P. Wolff, J. Nourooz-Zadeh. Plasma 8-epi-PGF2a levels are elevated in individuals with non-insulin dependent diabetes mellitus. *FEBS Lett.* 368:225–229, 1995.

91. Janssen, L.J. Isoprostanes: Generation, pharmacology, and roles in free-radical-mediated effects in the lung. *Pulm. Pharmacol. Therap.* 13(4):149–155, 2000.

92. Pratico, D., Iuliano, L., A. Mauriello, L. Spagnoli, J.A. Lawson, J. Rokach, J. Maclouf, F. Violi, G.A. FitzGerald. Localization of distinct F2-isoprostanes in human atherosclerotic lesions. *J. Clin. Invest.* 100:2028–2034, 1997.

93. Mobert, J., B.F. Becker. Cyclooxygenase inhibition aggravates ischemia-reperfusion injury in the perfused guinea pig heart: involvement of isoprostanes. *J. Am. Coll. Cardiol.* 31:1687–1694, 1998.

94. Montine, T.J., M.F. Beal, M.E. Cudkowicz, H. O'Donnell, R.A. Margolin, L. McFarland, A.F. Bachrach, W.E. Zachert, L.J. Roberts, J.D. Morrow. Increased CSF F2-isoprostane concentration in probable AD. *Neurology* 52(3):562–565, 1999.

95. Morrow, J.D., T.M. Harris, L.J. Roberts, 2nd. Noncyclooxygenase oxidative formation of a series of novel prostaglandins: Analytical ramifications for measurements of eicosanoids. *Anal. Biochem.* 184:1–10, 1990.

96. Morrow, J.D., L.J. Roberts. Mass spectrometry of prostanoids: F2-isoprostanes produced by non-cyclooxygenase free radical-catalyzed mechanism. *Methods Enzymol.* 233:163–174, 1994.

97. Walter, M.F., J.B. Blumberg, G.G. Dolnikowski, G.J. Handelman. Streamlined F2-isoprostane analysis in plasma and urine with high-performance liquid chromatography and gas chromatography/mass spectroscopy. *Anal. Biochem.* 280(1):73–79, 2000.

98. Sasaki, D.M., Y. Yuan, K. Gikas, K. Kanai, D. Taber, J.D. Morrow, L.J. Roberts 2nd, D.M. Callewaert. Enzyme immunoassays for 15-F2T isoprostane-M, an urinary biomarker for oxidant stress. *Adv. Exp. Med. Biol.* 507:537–541, 2002.

99. Halliwell, B. Why and how should we measure oxidative DNA damage in nutritional studies? How far have we come? *Am. J. Clin. Nutr.* 72(5):1082–1087, 2000.

100. Kasai, H. Analysis of a form of oxidative DNA damage, 8-hydroxy-2′-deoxyguanosine, as a marker of cellular oxidative stress during carcinogenesis. *Mutat. Res.* 387:147–163, 1997.

101. Devanaboyina, U., R.C. Gupta. Sensitive detection of 8-hydroxy-2′deoxyguanosine in DNA by 32P-postlabeling assay and the basal levels in rat tissues. *Carcinogenesis* 17(5):917–924, 1996.

102. Teixeira, A.J., M.R. Ferreira, W.J. van Dijk, G. van de Werken, A.P. de Jong. Analysis of 8-hydroxy-2′-deoxyguanosine in rat urine and liver DNA by stable isotope dilution gas chromatography/mass spectrometry. *Anal. Biochem.* 226(2):307–319, 1995.

103. Santella, R.M. Immunological methods for detection of carcinogen-DNA damage in humans. *Cancer Epidemiol. Biomarkers Prev.* 8:733–739, 1999.

104. Frenkel, K., J. Karkoszka, T. Glassman, N. Dubin, P. Toniolo, E. Taioli, L.A. Mooney, I. Kato. Serum autoantibodies recognizing 5-hydroxymethyl-2′-deoxyuridine, an oxidized DNA base, as biomarkers of cancer risk in women. *Cancer Epidemiol. Biomarkers Prev.* 7:49–57, 1998.

105. Hu, J.J., C.X. Chi, K. Frenkel, B.N. Smith, J.J. Henfelt, M. Berwick, S. Mahabir, R.B. D'Agostino. Alpha-tocopherol dietary supplement decreases titers of antibody against 5-hydroxymethyl-2′-deoxyuridine (HMdU). *Cancer Epidemiol. Biomarkers Prev.* 8:693–698, 1999.

106. Dalton, T.P., H.G. Shertzer, A. Puga. Regulation of gene expression by reactive oxygen. *Annu. Rev. Pharmacol. Toxicol.* 39:67–101, 1999.

107. Senft, A.P., T.P. Dalton, H.G. Shertzer. Determining glutathione and glutathione disulfide using the fluorescence probe o-phthalaldehyde. *Anal. Biochem.* 280(1):80–86, 2000.

108. Floreani, M., M. Petrone, P. Debetto, P. Palatini. A comparison between different methods for the determination of reduced and oxidized glutathione in mammalian tissues. *Free Radic. Res.* 26(5):449–455, 1997.

109. Imai, H., K. Narashima, M. Arai, H. Sakamoto, N. Chiba Y. Nakagawa: Suppression of leukotriene formation in RBL-2H3 cells that overexpressed phospholipid hydroperoxide glutathione peroxidase. *J. Biol. Chem.* 273:1990–1997, 1998.

110. De Haan, J., C. Bladier, P. Griffiths, M. Kelner, R.D. O′Shea, N.S. Cheung, R. T. Bronson, M.J. Silvestro, S. Wild, S.S. Zheng, P.M. Beart, P.J. Herzog, I. Kola. Mice with a homozygous null mutation for the most abundant glutathione peroxidase, Gpx1, show increased susceptibility to the oxidative stress-inducing agents paraquat and hydrogen peroxide. *J. Biol. Chem.* 273:22528–22536, 1998.

111. Ding, L., Z. Liu, Z. Zhu, G. Luo, D. Zhao, J. Ni. Biochemical characterization of selenium-containing catalytic antibody as a cytosolic glutathione peroxidase mimic. *Biochem J.* 332:251–255, 1998.

112. Sheehan, D., G. Meade, V.M. Foley, C.A. Dowd. Structure, function and evolution of glutathione transferases: implications for classification of non-mammalian members of an ancient enzyme superfamily. *Biochem. J.* 360(1):1–16, 2001.

113. Smith, R.A., J.E. Curran, S.R. Weinstein, L.R. Griffiths. Investigation of glutathione S-transferase zeta and the development of sporadic breast cancer. *Breast Cancer Res.* 3(6):409–411, 2001.

114. Ito, N., S. Tamano, T. Shirai. A medium-term rat liver bioassay for rapid *in vivo* detection of carcinogenic potential of chemicals. *Cancer Sci.* 94(1):3–8, 2003.

115. Schwartz, J.L. Biomarkers and molecular epidemiology and chemoprevention of oral carcinogenesis. *Crit. Rev. Oral Biol. Med.* 11(1):92–122, 2000.

116. Schwarz, K.B. Oxidative stress during viral infection: a review. *Free Radic. Biol. Med.* 21(5):641–649, 1996.

117. Götz, J., C.I. va Kan, H.W. Verspaget, I. Biemond, C.B. Lamers, R.A. Veenendaal. Gastric mucosal superoxide dismutases in *Helicobacter pylori* infection. *Gut* 38:502–506, 1996.

118. Jakus, V. The role of free radicals, oxidative stress and antioxidant systems in diabetic vascular disease *Bratisl. Lek. Listy.* 101(10):541–551, 2000.

119. Offen, D., P.M. Beart, N.S. Cheung, C.J. Pascoe, A. Hochman, S. Gorodin, E. Melamed, R. Bernard, O. Bernard. Transgenic mice expressing human Bcl-2 in their neurons are resistant to 6-hydroxydopamine and 1-methyl-4-phenyl-1,2,3,6- tetrahydropyridine neurotoxicity. *Proc. Natl. Acad. Sci.* 95:5789–5794, 1998.

120. Oberley, L.W., D.R. Spitz. Assay of SOD activity in tumor tissue. In: *Methods in Enzymology*, Academic Press, 1984, 105:457–461.

121. Spychalla, J.P., S.L. Desborough. Superoxide dismutase, catalase and alpha tocopherol content of stored potato tubers. *Plant Physiol.* 94:1214–1218, 1990.

122. Antier, D., H.V. Carswell, M.J. Brosnan, C.A. Hamilton, I.M. Macrae, S. Groves, E. Jardine, J.L. Reid, A.E. Dominiczak. Increased levels of superoxide in brains from old female rats. *Free Radic. Res.* 38(2):177–183, 2004.

123. Yamaguchi, K., D. Uematsu, Y. Itoh, S. Watanabe, Y. Fukuuchi. *In vivo* measurement of superoxide in the cerebral cortex during anoxia-reoxygenation and ischemia-reperfusion. *Keio. J. Med.* 51(4):201–207, 2002.

124. Escobar, L., C. Salvador, M. Contreras, J.E. Escamilla. On the application of the Clark oxygen electrode to the study of enzyme kinetics in apolar solvents: the catalase reaction. *Anal. Biochem.* 184(1):139–144, 1990.

125. Beers, R., I. Sizer. A spectrophotometric method for measuring the breakdown of hydrogen peroxide by catalase. *J. Biol. Chem.* 195:133, 1952.

126. Johansson, L.H., L.A. Borg. A spectrophotometric method for determination of catalase activity in small tissue samples. *Anal. Biochem.* 174(1):331–336, 1988.

127. Wheeler, C.R., J.A. Salzman, N.M. Elsayed, S.T. Omaye, D.W. Korte Jr. Automated assays for superoxide dismutase, catalase, glutathione peroxidase, and glutathione reductase activity. *Anal. Biochem.* 184(2):193–199, 1990.

128. Fridovich, I. Quantitative aspects of the production of superoxide anion radical by milk xanthine oxidase. *J. Biol. Chem.* 245(16):4053–4057, 1970.

129. Berry, C.E., J.M. Hare. Xanthine oxidoreductase and cardiovascular disease: molecular mechanisms and pathophysiological implications. *J. Physiol.* 555(3):589–606, 2004.

130. Culleton, B.F., M.G. Larson, W.B. Kannel, D. Levy. Serum uric acid and risk for cardiovascular disease and death: the Framingham Heart Study. *Ann. Intern. Med.* 131(1):7–13, 1999.

131. Crowley, L.V. Determination of uric acid. An automated analysis based on a carbonate method. *Clin. Chem.* 10:838–844, 1964.

132. Kock, R., S. Seitz, B. Delvoux, H. Greiling. A method for the simultaneous determination of creatinine and uric acid in serum by high-performance-liquid-chromatography evaluated versus reference methods. *Eur. J. Clin. Chem. Clin. Biochem.* 33(1):23–29, 1995.

133. Czauderna, M., J. Kowalczyk. Quantification of allantoin, uric acid, xanthine and hypoxanthine in ovine urine by high-performance liquid chromatography and photodiode array detection. *J. Chromatogr. B. Biomed. Sci. Appl.* 744(1):129–138, 2000.

134. Harris, C.M., V. Massey. The reaction of reduced xanthine dehydrogenase with molecular oxygen: reaction kinetics and measurement of superoxide radical. *J Biol. Chem.* 272(13): 8370–8379, 1997.

135. Haining, J.L., J.S. Legan. Fluorometric assay for xanthine oxidase. *Anal. Biochem.* 21(3): 337–343, 1967.

136. Corbisier, P., A. Houbion, J. Remacle. A new technique for highly sensitive detection of superoxide dismutase activity by chemiluminescence. *Anal. Biochem.* 164(1):240–247, 1987.

137. Bokoch, G.M., U.G. Knaus. NADPH oxidases: not just for leukocytes anymore! *Trends Biochem. Sci.* 28(9):502–508, 2003.

138. Cheng, G., Z. Cao, X. Xu, E.G. van Meir, J.D. Lambeth. Homologs of gp91phox: cloning and tissue expression of Nox3, Nox4, and Nox5. *Gene* 269(1,2):131–134, 2001.

139. Cai, H., Griendling, K.K., Harrison, D.G. The vascular NAD(P)H oxidases as therapeutic targets in cardiovascular diseases. *Trends Pharmacol. Sci.* 24(9):471–478, 2003.

140. Shimohama, S., H. Tanino, N. Kawakami, N. Okamura, H. Kodama, T. Yamaguchi, T. Hayakawa, A. Nunomura, S. Chiba, G. Perry, M.A. Smith, S. Fujimoto. Activation of NADPH oxidase in Alzheimer's disease brains. *Biochem. Biophys. Res. Commun.* 273(1):5–9, 2000.

141. Cohen, H.J., P.E. Newburger, M.E. Chovaniec. NAD(P)H-dependent superoxide production by phagocytic vesicles from guinea pig and human granulocytes. *J. Biol. Chem.* 255(14):6584–6588, 1980.

142. Lopes, L.R., C.R. Hoyal, U.G. Knaus, B.M. Babior. Activation of the leukocyte NADPH oxidase by protein kinase C in a partially recombinant cell-free system. *J. Biol. Chem.* 274:15533–15537, 1999.

143. Able, A.J., D.I. Guest, M.W. Sutherland. Use of a new tetrazolium-based assay to study the production of superoxide radicals by tobacco cell cultures challenged with avirulent zoospores of phytophthora parasitica var nicotianae *Plant Physiol.* 117(2):491–499, 1998.

144. Schepetkin, I.A. Lucigenin as a substrate of microsomal NAD(P)H-oxidoreductases. *Biochemistry (Mosc)* 64(1):25–32, 1999.

145. Brennan, M.L., S.L. Hazen. Emerging role of myeloperoxidase and oxidant stress markers in cardiovascular risk assessment. *Curr. Opin. Lipidol.* 14(4):353–359, 2003.

146. Rutgers, A., P. Heeringa, J.W. Tervaert. The role of myeloperoxidase in the pathogenesis of systemic vasculitis. *Clin. Exp. Rheumatol.* 6(32):S55–63, 2003.

147. Savenkova, M.L., D.M. Mueller, J.W. Heinecke. Tyrosyl radical generated by myeloperoxidase is a physiological catalyst for the initiation of lipid peroxidation in low density lipoprotein. *J. Biol. Chem.* 269(32):20394–20400, 1994.

148. Haqqani, A.S., J.K. Sandhu, H.C. Birnboim. A myeloperoxidase-specific assay based upon bromide-dependent chemiluminescence of luminol. *Anal. Biochem.* 273(1):126–132, 1999.

149. Seim, S. Role of myeloperoxidase in the luminol-dependent chemiluminescence response of phagocytosing human monocytes. *Acta. Pathol. Microbiol. Immunol. Scand.* 91(2):123–128, 1983.

150. Cohen Tervaert, J.W. Association of autoantibodies to myeloperoxidase with different forms of vasculitis. *Arthritis Rheum.* 33:1264–1272, 1990.

151. Kallenberg, C.G.M. Antineutrophil cytoplasmic autoantibodies with specificity for myeloperoxidase. In: *Autoantibodies.* Peter, J.B., Y. Shoenfeld, eds., Amsterdam:Elsevier, 53–60, 1996.

152. Moorales, M., A.R. Barcelo. A basic peroxidase isoenzyme from vacuoles and cell walls of *Vitis vinifera. Phytochemistry* 45:229–232, 1997.

153. Barcelo, A.R. Peroxidase and not laccase is the enzyme responsible for cell wall lignification in the secondary thickening of xylem vessels in *Lupinus. Protoplasma* 186:41–44, 1995.

154. Wallace, G., S.C. Fry. Action of diverse peroxidases and laccases on six cell wall-related phenolic compounds. *Phytochemistry* 52:769–773, 1999.

155. Barcelo, A.R., F. Pomar. Oxidation of cinnamyl alcohols and aldehydes by a basic peroxidase from lignifying Zinnia elegans hypocotyls. *Phytochemistry* 57(7):1105–1113, 2001.

156. Kwak, S.S., S.K. Kim, I.H. Park, J.R. Liu. Enhancement of peroxidase activity by stress related chemicals in sweet potato. *Phytochemistry* 43:565–568, 1996.

157. Rao, M.V., G. Paliyath, D.P. Ormrod. Ultraviolet-B- and ozone-induced biochemical changes in antioxidant enzymes of Arabidopsis thaliana. *Plant Physiol.* 110(1):125–136, 1996.

158. Nicoli, M.C., S. Calligaris, L. Manzocco. Effect of enzymatic and chemical oxidation on the antioxidant capacity of catechin model systems and apple derivatives. *J. Agric. Food Chem.* 48(10):4576–4580, 2000.

159. Randhir, R., K. Shetty. Microwave-induced stimulation of L-DOPA phenolics and antioxidant activity in fava bean (*Vicia faba*) for Parkinson's diet. *Process Biochem.* 39(11):1775–1784, 2004.

160. Strycharz, S., K. Shetty. Effect of *Agrobacterium rhizogenes* on phenolic content of *Mentha pulegium* elite clonal line for phytoremediation applications. *Process Biochem.* 38:287–293, 2002.

161. Randhir, R., Y.-T. Lin, K. Shetty. Stimulation of phenolics, Antioxidant and antimicrobial activities in dark germinated mung bean (*Vigna radiata*) Sprouts in response to peptide and phytochemical elicitors. *Process Biochem.* 39:637–646, 2004.

162. Shetty, K. Role of proline-linked pentose phosphate pathway in biosynthesis of plant phenolics for functional food and environmental applications; A review. *Process Biochem.* 39:789–803, 2004.

163. Randhir, R., D.A. Vattem, K. Shetty. Antioxidant response studies on the effect of oregano phenolics on H_2O_2 stressed porcine muscle. *Process Biochem.* 40:2123–2134, 2005.

164. McDougall, G.J., D. Stewart, I.M. Morrison. Cell-wall-bound oxidases from tobacco (Nicotiana tabacum) xylem participate in lignin formation. Planta 194:9–14, 1994.

165. Laloue, H., F. Weber-Lofti, A. Lucau- Danila, P. Gullemat. Identification of ascorbate and guaiacol peroxidase in needle chloroplasts of spruce trees. *Plant Physiol. Biochem.* 35:341–346, 1997.

11

Phytochemicals and Breast Cancer Chemoprevention

Sallie Smith-Schneider, Louis A. Roberts, and Kalidas Shetty

CONTENTS

11.1 Introduction 254
    11.1.1 Background 254
    11.1.2 Susceptibility and Chemoprevention: Molecular Pathways 255
11.2 Molecular Pathways of Breast Cancer Chemoprevention 258
    11.2.1 p53 258
    11.2.2 Estrogen Signaling 259
        11.2.2.1 Background 259
        11.2.2.2 Estrogen Regulation 259
        11.2.2.3 Protection and Susceptibility Duality of Estrogen Signaling 260
        11.2.2.4 Estrogenic Phytochemicals 260
    11.2.3 CYP1A1 and Aryl Hydrocarbon Receptor (AhR) 261
    11.2.4 Growth Factor Signaling 262
        11.2.4.1 Background 262
        11.2.4.2 Epidermal Growth Factor (EGF) 262
        11.2.4.3 Insulin-Like Growth Factor (IGF) 263
        11.2.4.4 Transforming Growth Factor Beta (TGF-β) 264
        11.2.4.5 Vascular Endothelial Growth Factor (VEGF) 266
    11.2.5 Arachidonic Acid Metabolism 266
        11.2.5.1 Background 266
        11.2.5.2 Cyclooxygenase (COX) 266
        11.2.5.3 Lipoxygenase (LO) 268
11.3 Strategies for Phytochemical Enrichment for Chemoprevention 269
    11.3.1 Consistency of Phytochemicals 269
    11.3.2 Clonal Screening of Phytochemicals from a Heterogeneous Genetic
        Background 270
    11.3.3 Stress and Elicitor Induced Sprouting 270

11.3.4 Solid-State Bioprocessing 271
11.3.5 Liquid Fermentation 272
References 272

11.1 INTRODUCTION

11.1.1 Background

According to the National Cancer Institute, approximately 13% of American women will develop breast cancer during their lifetimes. There are approximately 216,000 newly diagnosed cases of breast cancer each year, and breast cancer is second only to lung cancer in cancer related deaths among women in the USA (American Cancer Society). Despite a gradual decline in deaths due to breast cancer (likely attributable to increased screening), there is a rise in the incidence of newly diagnosed breast cancer. Numerous pharmaceuticals exist to prevent breast cancer growth, and several drugs are used chemopreventatively to block the formation of new tumors in previously diagnosed patients. Several plant derived compounds have been identified that may offer promise as chemopreventative agents that can be used routinely by the general population. However, in order to develop improved strategies aimed at preventing breast cancer, a better understanding of the cellular and molecular mechanisms that contribute to breast cancer initiation and protect epithelial cells against initiation or progression to cancer is needed.

During normal development of the mammary gland, controlled cell growth and changes in differentiation that resemble the unchecked cell proliferation that is a hallmark of breast cancer occur in response to hormones (1,2). In particular, the signaling pathways used during development and cancer initiation or growth are the same, with the loss of regulation over normal signals representing the most significant difference in initiation of cancer. Early in fetal development an extended epidermal mammary ridge forms and subsequently regresses, producing one nipple per breast with a bud of epithelial cells that essentially remains dormant and unchanged in females until puberty. Pubertal hormones (i.e., estrogen) initiate rapid division, branching, and migration of epithelial cells to form a system of ducts which extend to the limits of the fat pad, with lobules surrounded by stroma and microvascularized connective tissue. The ends of the lobule are referred to as terminal end buds, which maintain the ability to proliferate when triggered by pregnancy hormones (including estrogen and progesterone). During pregnancy the cells of the terminal end buds rapidly divide, then differentiate during lactation into milk producing alveoli. Upon weaning, rapid apoptosis of the alveoli ensues (termed involution), and the breast is then virtually indistinguishable from those of nulliparous females. However, the epithelial terminal end buds retain the ability to proliferate, invade the stroma, and differentiate in response to hormones produced during pregnancy and lactation. Thus, mammary gland development is a highly ordered and controlled process that directs epithelial cell proliferation, differentiation, and apoptosis.

A vast majority of breast cancers initiate from the ductal and lobular epithelial cells already described. Initially, the cells that comprise ductal carcinoma *in situ* (DCIS) are still hormone responsive (express estrogen and progesterone receptors), but often display hormone independent growth upon progression to more metastatic tumors. Additionally, growth factor receptors are often overexpressed in aggressive metastatic tumors that further promote unrestrained proliferation, inhibition of apoptosis, and vascularization of the growing tumor (3). Many chemopreventative strategies are based upon promoting or restoring normal signaling mechanisms within these ductal and lobular epithelial cells to reduce the likelihood of breast cancer initiation. Interestingly, pregnancy exerts a protective effect on the breast, such that a full term pregnancy lowers breast cancer risk by ~50% (4). This effect appears to be mediated in part by sensitizing the tumor suppressor protein p53, and in part by altering growth factor signaling (5–9). Chemoprevention using dietary phytochemicals aims to mimic the protective effect of pregnancy hormones by modulating such signaling pathways (Figure 11.1).

11.1.2 Susceptibility and Chemoprevention: Molecular Pathways

There are numerous factors and molecular events that can increase or decrease the susceptibility toward developing breast cancer. Understanding the molecular basis of this disease often relies on identifying the signal transduction pathways whose activation (or inhibition) directly regulates the development of breast cancer or acts as a marker for breast cancer susceptibility or protection. Once these pathways are identified, development of chemopreventative strategies to decrease the risk of breast cancer and chemotherapeutic regimens to treat existing cancers will be greatly facilitated.

p53 is a tumor suppressor protein that when activated acts as a transcription factor to induce expression of a host of genes involved in responses to DNA damage (triggered by ultraviolet light, ionizing radiation, or carcinogenic chemicals) (6,10). The central function of p53 is to induce cell cycle arrest and decide if DNA damage can be fixed (arrest followed by resumption of the cell cycle), or is irreparable (activation of the apoptotic pathway). p53 thus plays a critical role in preventing genotoxic initiation of precancerous lesions. Mutation of the *p53* gene or loss of p53 responsivity is one of the most common events identified in human breast cancer (detected in ~50% of cases), and loss of the p53 allele in mouse models predisposes them to cancer. In fact, transformation of normal ductal epithelium to DCIS typically progresses to the formation of a malignant, invasive cancer of the breast, and is frequently linked to mutation of p53 (2). p53 can be sensitized, i.e., made more responsive, by the pregnancy hormones estrogen and progesterone (11) and many dietary phytochemicals (12). Sensitization of p53 minimizes cancer initiation by increasing the likelihood that cells with activated p53 will either repair damaged DNA, or will be removed by apoptosis.

The estrogen biosynthetic and signaling pathways have been very well studied in normal development of the breast (i.e., directing proliferation of ductal epithelial cells during puberty), as well as in breast cancer (13). The enzyme aromatase (itself a popular target of estrogen responsive breast cancer) (14,15) converts androgen into estrogen, produced distally by the ovaries and locally in the breast. Estrogen is highly mitogenic, and directs cell proliferation by binding to estrogen receptors and activating transcription. Estrogen also may ultimately lead to the release of epidermal growth factor (EGF), itself a mitogenic factor, and may be converted into hydroxylated estrogens which in turn can act to stimulate cell division. Inappropriate signaling through the estrogen response pathway can lead to unchecked cell proliferation, and breast cancer.

Upregulation of signaling through the aryl hydrocarbon receptor (AhR) has recently been closely linked with breast cancer initiation, likely by inducing hyperplasia in epithelial breast tissue. Environmental toxins such as dioxin bind to cytosolic AhR, which then acts as a transcription factor to induce the expression of several genes, including the monooxygenase CYP1A1. Protoxins are then metabolized into mutagenic substances by CYP1A1. With respect to breast cancer specifically, CYP1A1 is capable of generating hydroxylated estrogens (i.e., 16α-hydroxyestrone) with mitogenic activity, and variant or increased CYP1A1 activity is associated with an increased risk of breast cancer (16). Additionally, AhR activation leads to the upregulation of other genes that regulate the antioxidant response and the cell cycle.

Many phytochemicals have been shown to modulate AhR activation and CYP1A1 expression and activity, including polyphenols, flavonoids, and phytoalexins, and may have a role in breast cancer prevention.

Signaling by growth factors [such as insulin-like growth factor (IGF), vascular endothelial growth factor (VEGF), transforming growth factor beta (TGF-β), and EGF represents an important area of focus in cancer initiation and development. Increased signaling through IGF can inhibit apoptosis (likely by upregulating the protein kinase Akt and the transcription factor NF-κB) (17), and some correlation exists between high circulating blood levels of IGF and the risk of premenopausal breast cancer (18). Epithelial growth factor, signaling through its receptors epidermal growth factor receptor (EGFR) and HER-2/neu plays a significant role in stimulating cell proliferation in the mammary gland (19). Vascular endothelial growth factor expression is regulated by NF-κB as well as cyclooxygenase and lipoxygenase metabolites of arachidonic acid (AA), and signaling it increases the growth and migration of endothelial cells that leads to vascularization of growing tumors (20). In contrast, TGF-β acts as a negative regulator of cell growth, as disruption of TGF-β signaling is correlated with cancer development (21). Additionally, crosstalk amongst growth factor signaling networks exists. For example, both IGF and EGF are capable of regulating the expression of VEGF. Antagonistic effects are also seen. IGF activation can down regulate TGF-β signaling (22), while TGF-β can decrease the circulating levels of IGF. Modulation of growth factor signaling obviously represents a complex yet important area of research in the development of chemopreventative strategies.

The aberrant metabolism of arachidonic acid by cyclooxygenases (COX) and lipoxygenases (LO) has recently been shown to be a predictive marker for initiation and progression

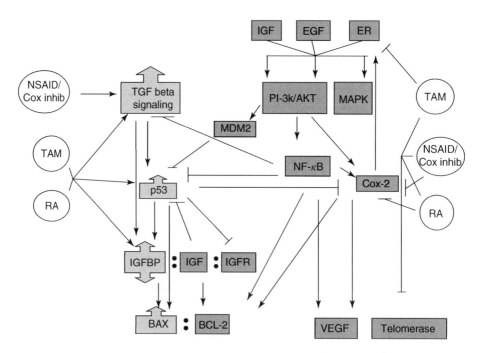

Figure 11.1 Signaling pathways which promote cell growth, survival, and metastasis and stimulate angiogenesis (dark grey) exist in a balance with signals that suppress growth, sensitize cell to apoptosis, and induce differentiation (light grey). Chemopreventive agents appear to have common overlapping roles in their ability to control growth signals through TGF beta family signaling, inhibit cyclooxygenase activity, regulate IGF, control oxidation, and sensitize or activate p53. (RA = retinoic acid, TAM = tamoxifen, NSAID = nonsteroidal antiinflammatory drug)

of a variety of cancers, and has been correlated with ultimate outcome in patients with breast cancer. COX-2 upregulation was first associated with a predisposition for colon cancer development, and COX-2 expression and activity are increased in breast cancer cells (23). The upregulation of COX-2 is indicative of, and is at least partially responsible for, increased cell growth and migration and decreased apoptotic rate. The AA metabolizing enzymes 5-lipoxygenase and 12-lipoxygenase are also overexpressed in many cancerous tissues, while decreased expression of 15-lipoxygenase is observed in many carcinomas. Recent studies have correlated increased expression of COX-2, 5-LO, and 12-LO with poor clinical outcome of breast cancer patients (24–26). Pharmacological COX inhibitors are currently in clinical trials as chemopreventative or chemotherapeutic agents (27), and the discovery and application of plant derived inhibitors of COX and LO represent an intense area of scientific and clinical research (28).

Many plant derived compounds either have been shown, or offer promise, to act as chemopreventative agents that block the development of breast cancer (29). These phytochemicals include but are not limited to the flavonoids, catechins, lignans, (poly)phenolics, and organosulfones. Interestingly, a subset of these phytochemicals act on multiple signaling pathways to coordinately suppress breast cancer initiation and progression, often by acting on a common signaling intermediate. Additionally, many of these plant derived compounds exert their protective effects by acting as antioxidants, which minimize oxidative stress and DNA damage. Several of the specific signaling pathways through which phytochemopreventative agents are known to act are detailed in the following (Figure 11.2).

	Estrogen signals	EGF signals	IGF signals	VEGF	COX2	TGF beta family signals	p53	Antioxidant enzyme response	AhR/ CYP1A1	LOX
Genistein	–		–	–		+	+		–	
Resveratrol	–					+	+		–	
Catechins	–	–	–	–	–					
Flaxseed lignans	–	–	–	–						
Ellagic acid							+			
Querciten	–		–				+		–	
Curcumin	–	–		–	–	+	+/–		–	–
Anthocyanidins	–	–			–	–				
PEITC							+			
Indole 3 carbinol	–	–							–	
Diallyl sulfide						+	+			
Perillyl alcohol						+	+			
Limonene										
Lycopene			–				–			
Baicalein					–				–	–

Figure 11.2 Numerous phytochemicals have been shown to regulate the balance of signals that promote cancer as well as the signals that control growth and sensitivity to DNA damaging agents. The (+) symbol indicates that in certain cell lines the phytochemical has been shown to activate or promote this pathway in some manner. The (−) symbol indicates that the phytochemical has been shown to have repressive or inhibitory effects on the signaling pathway at some level.

11.2 MOLECULAR PATHWAYS OF BREAST CANCER CHEMOPREVENTION

11.2.1 p53

p53 is a protein that activates transcription in response to a wide range of genotoxic stresses. Activation of p53-induced genes (i.e., p21, 14-3-3, bax, killer/DR5, GADD45, and NAG-1) results in execution of a cell cycle checkpoint important for the repair of DNA, cellular differentiation, or engagement of the apoptotic pathway. Loss of p53 activity due to mutation or viral inactivation results in immortalization of cells, and can represent one factor in the multihit process required for the progression to cancer.

The importance of proper control of p53 is underscored by its highly intricate regulation at numerous levels. The phosphorylation of p53 is complex and cooperative. Phosphorylation at specific residues are required for DNA binding, oligomerization, and transactivation. Ataxia telangiectasia mutated/ATM and Rad3 related kinase (ATM/ATR) are kinases which are believed to act as sensors of genotoxic stress. Ataxia telangiectasia mutated kinase appears to respond to double stranded DNA breaks, while ATR responds to stalled replication. These sensors transport signals of genotoxic damage through downstream targets such as Chk2 kinases, p53, and Brca. Activity of ATM/ATR is believed to be responsible for the phosphorylation of numerous N-terminal p53 residues such as serines 9, 15, and 46 as well as indirectly regulating the phosphorylation states of serines 20 and 376 (30). The phosphorylations of serines 15 and 20 control the binding of the ubiquitin ligase, MDM2, which regulates stability and cellular localization of p53 (31). Phosphorylated serine 15 (along with cooperative phosphorylations at serines 33, 37, 392) is believed to regulate the transcriptional activity of p53 to allow recruitment of general transcription factors (32–34). MAP kinase family members, in particular the stress activated members p38 and c-Jun N-terminal kinase (JNK), can also phosphorylate p53. c-Jun N-terminal kinase can phosphorylate threonine 81 to regulate p53 stability, transcriptional activity, and ability to initiate apoptosis (35), while p38 can phosphorylate serines 33 and 46 in response to UV radiation (36). The stability and cytoplasmic localization of the p53 protein is regulated by sequestration through interaction with the MDM2 protein, a ubiquitin ligase. The MDM2 protein can block p53 activity by binding to the N-terminus of p53 and either targeting p53 for degradation through ubiquitination, or by sequestering p53 in the cytoplasm. The regulation of MDM2 through the PI3k pathway is one mechanism by which deregulated signals through receptor tyrosine kinases can repress p53 activity and thus increase genetic instability (37,38). p53 can also be regulated by sumoylation and protein to protein interactions mediated through oxidative stress. The redox state of the cell determines the activity of thioredoxin reductase, which can control the activity of ref-1, which in turn can bind to and regulate transcriptional activity of p53 (39).

The mammary gland appears essentially normal in mice with a null deletion of p53, suggesting that p53 is not required for normal development. However, these mice are more prone to tumors in general and die at a relatively young age, mostly due to lymphomas. The loss of a single allele in the breast cancer susceptible mouse strain BALB/c results in 55% developing breast cancer by one year of age (40). A spectrum of p53 mutations has also been detected in cases of human breast cancer, and p53 appears to be inactivated by mutation in approximately 40% of human breast cancer cases (41,42).

A relatively new and interesting observation is that the mammary glands of virgin rodents are refractile to the activation of p53 by gamma radiation, while parous glands are sensitive (43). This offers a possible explanation for why the virgin rodent mammary gland is more susceptible to cancer induced by carcinogen exposure than the parous gland. The implications of pregnancy induced sensitization of p53 are obvious, as cells with DNA damage can be eliminated or repaired in the glands of parous mice, but may not be cleared or corrected in virgin mice. The importance of this pathway in humans remains to be determined; however, it is known that women who undergo pregnancy at an early age have a significantly reduced risk of breast cancer. Chemopreventative agents such as tamoxifen and retinoic acid have also been shown to sensitize p53 activity and the mammary gland (Tu, Schneider, and Jerry, unpublished). This suggests that the sensitization of p53 and the gland utilizes a common and important pathway in the protective response.

The ability to sensitize the mammary gland to repair and death via p53 represents an excellent target in the search for dietary phytochemicals which might act as strong chemopreventative agents. To date, numerous phytochemicals have been identified that can activate p53 and thus the apoptotic pathway in tumor cells. These agents include the organosulfur compounds phenethyl isothiocyanate (PEITC) (44) and diallyl disulfide (45); the phenolic compounds resveratrol (46), ellagic acid (47), and curcumin (48); and the plant isoflavonoids apigenin, kaempferol, luteolin (49), and genistein (50). While the mechanisms of p53 induction by most of these agents is not known, the flavonoids appear to induce p53 through the activation of ATM/ATR and the p38 stress related kinase; and some data suggest that genistein can alter Chk2 kinase activity which controls the activity of cdc25, a dual specificity kinase (49).

11.2.2 Estrogen Signaling

11.2.2.1 Background

Ovarian hormones have dichotomous effects when applied to the biology of the breast. On one hand, prolonged exposure through early menarche, belated menopause, or postmenopausal hormone replacement therapy increases the risk of breast cancer. On the other hand, an early full term pregnancy can significantly reduce breast cancer risk to humans (51), and in rodents pregnancy levels of estrogen and progesterone will impart a significant protection against carcinogen induced breast cancer (52,53). The context and duration of the signaling is likely to be the important determinant in the effects of estrogen. There is no doubt, though, that estrogen remains one of the critical hormones that regulate both normal and malignant development of the mammary gland.

11.2.2.2 Estrogen Regulation

Androgen is converted to estrogen by aromatase. Estrogen is produced primarily in the ovaries and released into the bloodstream, though aromatase activity has been identified in the stroma of the mammary gland. Estrogen is highly mitogenic, and regulates transcription of a variety of genes by activating several distinct signaling pathways within the cell. Estrogen specifically binds to estrogen receptors (ERα or ERβ) and activates transcription. Alternatively, recent data suggests estrogen may also bind a G-protein coupled plasma membrane receptor, which signals through Src, leading to matrix metalloproteinase (MMP) activation. Matrix metalloproteinase activavtion in turn cleaves heparin bound EGF such that EGF is released and can signal through EGF receptors (54). Finally, estrogen may also signal through metabolites generated by cytochrome P450 enzymes. 16a-hydroxyestrone has been shown to exert agonistic and proliferative effects upon mammary epithelial cells (55).

Estrogen is critical for the normal development of the mammary gland during puberty and pregnancy. Proper regulation of estrogen receptor (ER) signaling is important for directed proliferative events. During puberty proliferation of ductal epithelial cells increases dramatically in response to hormones, leading to extension of the ductal network through the stroma toward the limits of the fat pad. Mice with a targeted deletion of ERα exhibit a very prevalent phenotype in which the ducts are severely stunted. Estrogen is also partially required for differentiation and development of lobuloalveolar structures during pregnancy.

11.2.2.3 Protection and Susceptibility Duality of Estrogen Signaling

Ovariectomy in humans and mice (which reduces estrogen levels) substantially reduces the risk of breast cancer. A subset of potent chemopreventive and chemotherapeutic agents has the ability to inhibit certain ER activities (e.g., selective estrogen receptor modulators [SERM]) or inhibit estrogen synthesis (e.g., aromatase inhibitors). Tamoxifen (the best studied SERM) has partial ER antagonistic properties, and has been very popular as adjuvant therapy in humans to reduce the risk of recurrent breast cancer. Some of the documented successes of tamoxifen may be due to its pleiotropic effects. Tamoxifen decreases IGF signaling, increases TGF-β secretion, prevents oxidative damage, inhibits protein kinase C, and induces apoptosis. However, tamoxifen has agonistic effects on endometrial tissue, and because of its potential side effects long term use is discouraged. Raloxifene is a newer member of the SERM family which is also an ER antagonist, but lacks some of the side effects associated with tamoxifen. Preliminary studies with Raloxifene have yielded comparable preventative features. Furthermore, the aromatase inhibitor Letrozole decreases levels of estrogen and has also shown promising initial results.

Paradoxically, pregnancy or prolonged treatment with pregnancy levels of estrogen and progesterone (E+P) result in a significant reduction in the susceptibility to breast cancer. In humans, a full term pregnancy at a relatively young age results in approximately a 50% reduced risk of getting breast cancer (51,56). In rodents, a similar response is noted. Exposure to pregnancy levels of estrogen and progesterone for extended periods alters proliferative responses in untreated and carcinogen treated glands (5). Proliferation in the mammary glands of virgin estrogen and progesterone treated animals was overall lower than in untreated animals. Interestingly, the proliferation in ER positive cells was 10 times lower, suggesting that exposure to hormones that mimicked pregnancy altered important signaling pathways in a population with proliferative potential. The carcinogen n-nitroso methyl urea (NMU) induced a much more significant mitogenic response in virgin animals as compared to the hormone treated population. These hormones have also been shown to be the vital agent needed to sensitize p53 in ductal epithelial cells to make them responsive to DNA damage (5,7).

11.2.2.4 Estrogenic Phytochemicals

Epidemiological studies of populations with low versus high risk of breast cancer note that female Asian populations experience a much lower percentage of breast cancer cases (57). This protection is lost once the women relocate to the USA, suggesting that it is a difference in diet which imparts this protection. Analysis of the Asian diet indicates a higher amount of soy is consumed in comparison to a typical North American diet. Interestingly, soy contains significant amounts of estrogenic activity stemming from isoflavone components (specifically genistein and diadzein). Chemoprevention studies in rodents have indicated that soy can delay tumor development, and that administration of either diadzein or soy protein with isoflavones results in some protection (58,59). Genistein has been shown to have a chemopreventative effect on prepubertal rats (60). Besides the antiestrogenic ability of isoflavones, these compounds can enhance the antioxidant enzyme response, inhibit NF-kB and Akt activity, prevent angiogenesis, and reduce levels of IGF-I, ER-α, and PR (61,62).

Additional studies focusing on food benefits have identified a number of phytoestrogens, compounds found in plants that have both agonistic and antagonistic effects on the estrogen receptors. These phytoestrogens include the lignans found in flaxseed, the coumestans in alfalfa sprouts, and the polyphenolic catechins found in teas (63). Chemopreventative studies in rodents examining phytoestrogens indicate that at high concentrations, these agents can impart protection in the mammary gland, which can be hoped will translate into human chemoprotection.

11.2.3 CYP1A1 and Aryl Hydrocarbon Receptor (AhR)

The CYP1A1 p450 monooxygenase enzyme has recently been identified as a target of chemopreventative strategies. The protoxin dimethylbenz[a]anthracene (DMBA) is metabolized by CYP1A1 (and CYP1B1) into a DNA damaging agent that is a potent initiator of breast cancer (64). The CYP1A1 gene also catalyzes the conversion of estradiol to form hydroxylated estrogen; high estrogen levels are associated with an increased risk of breast and ovarian cancers. The CYP1A1 enzyme is capable of metabolizing arachidonic acid into 19-OH-AA (major product) and 14,15-EET (minor product) (65), but no link between these metabolites and breast cancer initiation or progression has been uncovered. Additionally, CPY1A1 expression is upregulated by a wide array of environmental toxins such as dioxins (most notably TCDD) and polychlorinated biphenyls (PCBs). The CYP1A1 gene is induced by the AhR, which is translocated to the nucleus upon ligand binding to act as a transcription factor. Dioxins act as ligands of AhR to induce expression of a class of genes (including CYP1A1) that possess a dioxin responsive element upstream of the coding sequence. In a screen of 90 vegetable derived compounds that may act as AhR antagonists, 37 were found to block dioxin mediated induction of AhR (16). The aryl hydrocarbon receptor also induces genes that regulate the cell cycle and the antioxidant response, two pathways that affect breast cancer development. Additionally, allelic variation or polymorphisms of the CYP1A1 gene may be correlated with an increased risk of breast cancer. For these reasons, identifying phytochemicals that modulate AhR/CYP1A1 signaling may be of significant value as chemopreventative agents.

The best characterized function of phytochemicals as chemopreventative agents is their role as antioxidants. Such plant derived compounds are divided into distinct classes: phenolics, flavonoids, and phytoalexins. Polyphenols are present at high levels in foods such as olives or olive oil (66) and fresh cut potatoes (67). Several polyphenolics (caffeic acid, syringic acid, and sinapic acid) have been shown to inhibit the growth of the human breast cancer cell line T47D (68). Caffeic acid was the most potent anti proliferative agent in these studies, and was capable of stimulating apoptosis of T47D cells. Induction of the apoptotic program in MCF-7 cells by the phenethyl ester of caffeic acid is mediated by inhibition of the transcription factor NF-κB and activation of Fas by a p53- and p38-dependent mechanism (69). Additionally, caffeic acid was able to bind to AhR antagonistically, and decrease CYP1A1 activity by inhibiting the activation of CYP1A1 transcription (68). Thus, polyphenols like caffeic acid are promising chemopreventative agents with respect to decreasing the likelihood of developing breast cancer.

The flavonoids have also received a great deal of attention for their potential as chemoprotective agents, especially with regards to their proestrogenic and antiestrogenic activity (70). Flavonoids are a diverse class of compounds that can act as either AhR agonists or antagonists, depending on the specific phytochemical and cell type. Focusing on the AhR responsive MCF-7 cell line, four compounds have been identified that act as AhR agonists by their induction of CYP1A1 gene expression: cantharidin (a lactone derived from insect extract); baicalein (an isoflavone that also is an inhibitor of 12-LO); chrysin (a flavonoid); and emodin (an herbal laxative) (71). Additionally, baicalein and emodin increase the protein level of CYP1A1, while baicalein and cantharidin increase the level of AhR protein. However, as dietary components it has not been shown the flavonoids would act as AhR agonists, due to their relatively low serum concentrations (16). Conversely, luteolin acts as a potent AhR antagonist in MCF-7 cells, while quercetin, kaempferol, and myricetin can act as mild AhR antagonists (71). Genistein and baicalein also are capable of inhibiting the conversion of environmental protoxins (like DMBA) into DNA damaging agents by blocking CYP1A1 activity (72).

Another class of plant derived compounds that modulate AhR/CYP1A1 functions and have potential applications as chemopreventative agents are the phytoalexins. These primarily stress induced compounds can act as antifungal agents in plants, but possess additional activities that make them attractive chemopreventative agents. For example, resveratrol (3,5,4'-trihydroxystilbene) is a polyphenolic phytoalexin found in red wine that has a wide spectrum of inhibitory properties (73). Resveratrol has been shown to induce apoptosis in established human breast cancer cell lines that have wild-type p53 tumor suppressor protein but not mutant p53 (46). In addition to its ability to inhibit cyclooxygenases, resveratrol is a competitive antagonist of AhR that causes nuclear translocation of AhR but does not induce CYP1A1 (74). Thus, resveratrol may act to prevent cancer initiation by inhibiting signaling through AhR and blocking the upregulation of CYP1A1.

In addition to these phytochemicals, other plant derived compounds affect the expression and activity of CYP1A1. Curcumin, a major component of the spice turmeric, binds to and activates AhR and can bind to the xenobiotic response element upstream of CYP1A1 (75). However, several studies suggest that curcumin partially blocks DMBA induced CYP1A1 activity, as well as the conversion of procarcinogens into DNA damaging agents in breast and squamous cell carcinoma cells (75,76), and *in vitro* (77). Curcumin also inhibits 5-HETE (78) and prostaglandin formation (79), and has antioxidant properties; thus, its role as a tumor repressor or chemopreventative agent may be due to its effects on more than one single pathway. Indole-3-carbinol (derived from cruciferous vegetables) is also antitumorigenic, likely resulting from its ability to inhibit the activity of CYP1A1, blocking the hydroxylation of estradiol and resulting in the repression of estrogen signaling.

11.2.4 Growth Factor Signaling

11.2.4.1 Background

Growth factors are very important in the normal and malignant development of the mammary gland, and include epidermal growth factor (EGF), platelet derived growth factor (PDGF), hepatocyte growth factor (HGF), amphiregulin (AR), insulin-like growth factor (IGF), vascular endothelial growth factor (VEGF), fibroblast growth factor (FGF), transforming growth factor alpha (TGF-α), and transforming growth factor beta (TGF-β). These factors have diverse effects and work on different cell types, but ultimately their activities (in concert with hormones and other factors) regulate the architecture of the mammary gland. These growth factors bind receptor kinases on the surface of the plasma membrane. The deregulated expression or activity of these receptors often results in a susceptibility to breast cancer. Among them, EGF, IGF, TGF-β, and VEGF are particularly interesting with regards to chemoprevention due to their ability to regulate homeostasis in the mammary gland.

11.2.4.2 Epidermal Growth Factor (EGF)

Epidermal growth factor is secreted from cells and can act in an autocrine or paracrine fashion. Epidermal growth factor is one of approximately ten ligands (including amphiregulin and TGF-α) that bind to and activate a family of EGF receptors (HER1, 2, 3, and 4) found on the surface of the epithelial cells. These receptors are tyrosine kinases and play a strong role in the proliferation and differentiation of the mammary gland. The EGFR and HER-2 receptors are frequently upregulated or activated in breast cancer. New advances in breast cancer treatments of ER negative tumors target this family of receptors. Herceptin (Trastuzumab) is a humanized antibody to the HER-2 receptor which has been found to be successful in clinical trials of patients with very high expression of HER-2 (80,81).

Numerous dietary phytochemicals have been identified that can inhibit EGFR or HER-2/neu signaling. Typically these agents do not block the binding of the ligand to the receptor, but instead inhibit the autophosphorylation of the receptor and downstream signaling. Flaxseed has been shown to down regulate EGF receptor levels (82). Curcumin can inhibit EGFR activity and proliferation in A431 epidermoid carcinoma cells (83). Phase 1 clinical trials indicate curcumin is nontoxic up to 8 g/day, and patients with precancerous growths often saw stability of the disease or a histological improvement (84) with this curcumin regimen. However, trials with higher numbers of patients are needed.

Black and green teas have also been shown to have constituents which will decrease EGFR activity. Epigallocatechin-3-gallate (EGCG) or theaflavin-3,3′-digallate (TF-3) in particular can inhibit the autophosphorylation of the EGF receptor and inhibit proliferation. EGCG in prolonged treatment could enhance the apoptotic response by upregulating p53 (85), while TF-3 was better at blocking the binding of EGF to its receptor (86). Fruit derived anthocyanidins are structurally similar to EGCG and can also inhibit EGF receptor activation in A431 epidermoid carcinomas cells. Cyanidin and delphinidin have been shown to inhibit growth and block MAPK phosphorylation of ELK, and in turn inhibit EGF induced transcription though the ELK transcription factor (87).

11.2.4.3 Insulin-Like Growth Factor (IGF)

Insulin-like growth factor is a mitogen and a prosurvival factor which can act through autocrine and paracrine pathways. Signaling through the IGF axis is highly regulated at numerous levels, including expression of the IGF receptor (IGFR), the regulation of IGF binding protein (IGFBP) expression (which modulate the ability of IGF to bind to and activate receptors), regulation of IGFR autophosphorylation, and transmission of downstream signals through insulin receptor substrate-1 (IRS-1). The activation of the IGF receptor results in phosphorylation and activation of IRS, which in turn signals to the nucleus. The inositol triphosphate kinase (PI3k) pathway is one of the major pathways elicited by IGF in mammary epithelial cells.

Production of IGF occurs both systemically and locally. Growth hormone regulates the levels of IGF generated by the liver, the major site of IGF production. Increased serum levels of IGF are critical for the proper maintenance of terminal end bud structures in the mammary gland. The role of IGF in prevention of or susceptibility to breast cancer stems largely from correlative evidence in human studies and direct evidence in rodent studies. Epidemiological studies analyzing possible roles for IGF in breast cancer have indicated that premenopausal women with serum IGF levels in the highest quartile have an increased incidence of breast cancer. Conversely, well established protective therapies such as tamoxifen, fenretinide, and caloric restriction are associated with a decrease in IGF serum levels. However, the most convincing evidence that high IGF levels and signaling may be related to breast cancer is obtained from genetically defined rodent models with alterations in the IGF signaling pathway. Mice overexpressing IGF-I are sensitized to TPA induced tumor development (88). Mice that have a 25% reduction in serum IGF levels have a significantly delayed onset of breast tumor occurrence and a reduction in the number of tumors induced by C3(1)/SV40-LTA or in response to DMBA treatment (89). Interestingly, the reduced levels of IGF had no effect on the levels of hyperplasia induced in these models, but a significant difference in metaplasia, suggesting that the serum levels of IGF affect the progression, rather than the initiation, of breast cancer. Serum IGF levels also appeared to affect the growth of the tumors, as differences in tumor size were noted, especially in the early weeks after tumor detection (89).

The mechanism by which IGF predisposes mammary epithelial cells to tumorigenesis is unknown, but studies on various cell types indicate that signaling through this pathway counteracts most of the pathways important for the protective response. IGF has been shown to activate the PI3k pathway, which often leads to survival by activating Akt and NF-kB, or by phosphorylating and activating MDM2, which will degrade or sequester p53 in the cytoplasm (90). The PI3k pathway also can regulate the expression and localization of β-catenin, which will activate TCF/LEF signaling. In renal mesangial cells and colon carcinoma cells, IGF enhances cyclooxygenase activity (91,92). In certain cell types IGF has also been shown to down regulate the level of the progesterone receptor and TGF-β (22), suggesting pathways by which IGF might upset the balance of homeostasis. Furthermore, IGF has been shown to upregulate telomerase activity (93). Cancer cells have higher telomerase activity, which allows a bypass of naturally occurring senescence that limits the number of cell division cycles. Agents that inhibit telomerase thus represent a new potential target for cancer therapy.

Known chemopreventative agents appear to regulate IGF signaling at multiple levels. Retinoic acid (RA) inhibits growth of the MCF-7 breast cancer cells through inhibition of IRS and PI3k signaling (94), which in turn regulates c-fos expression (83). Retinoic acid also induces the secretion of IGFBPs in a cultured cancer cell line (95), which decreases the serum level of IGF-I. Data also suggests that in normal mammary epithelial cells RA will upregulate other binding proteins, such as IGFBP7 (96). Inhibitors of cyclooxygenase activity also upregulate IGFBP3 in immortalized breast epithelial cells, and tamoxifen and TGF-β inhibit the IGFBP3 protease to stabilize IGFBP3 in MCF-7 cells. In addition to decreasing plasma levels of IGF-I and upregulating IGFBP3, tamoxifen also decreases the levels of IRS-1 and Akt signaling in MCF-7 cells (97).

Dietary phytochemicals aimed at reducing IGF signaling have only recently been utilized in human chemoprevention. Flaxseed can inhibit growth of mammary tumors in rats, and has recently found favor in humans to prevent breast cancer. In addition to its antiestrogenic activity, flaxseed can inhibit IGF-I plasma levels in rats and IGF receptor levels in breast cancer cells (82,98). In prostate cancer cells genistein, biochaninA, quercetin, and kaempferol downregulated the activity and reduced phosphorylation of IRS as well as downstream Akt (99). In a different prostate cancer cell line black tea phenolics were also shown to reduce IGF signaling by decreasing the phosphorylation of the IGFR-I and Akt (100).

11.2.4.4   Transforming Growth Factor Beta (TGF-β)

The TGF-β growth factor is a cytokine with growth inhibitory properties for many cell types, including epithelial cells. Numerous members of the TGF-β family exist, with TGF-β1, 2, and 3 the most studied. Other family members include activin, BMP, and the more distantly related NAG-1 (also known as MIC-1, PLAB, PTGF-β, or GDF15). The three main forms of TGF-β have some overlapping properties and specific roles, and are expressed in response to different stimuli. All three (TGF-β 1, 2, and 3) are upregulated during involution, but TGF-β 3 expression is increased to the highest extent (101). Furthermore, mammary gland tissue from mice null for TGF-β 3 display a significant inhibition of involution, suggesting that this form is responsible for the apoptosis that occurs after weaning (102). The TGF-β growth factor is secreted in a latent form as part of a complex with other proteins, and is retained as such in the extracellular matrix until activation. Proteolytic cleavage is required to release the active form of TGF-β, and occurs in the mammary gland in response to estrogen and progesterone, as well as radiation (103,104). The active form of TGF-β then binds to one of three types of TGF-β receptors which either heterodimerize or homodimerize. Some receptors have higher affinities for particular TGF-β subtypes and some transduce signals more efficiently. Type I and type II receptors have cytoplasmic serine or threonine kinase domains responsible for downstream signaling events.

Cytokines of TGF-β are required for the normal development of the mammary gland. They are expressed at high levels in the stroma and can act in a paracrine fashion to control the growth and branching of ductal epithelium. Animal studies have also indicated that they have a role in regulating lobular alveolar development and involution. Pregnancy has been shown to upregulate the transcription of all three TGF-β isoforms. In rats, TGF-β 2 is increased late in pregnancy, and TGF-β 3 increases with milk stasis induced by weaning. Interestingly, TGF-β 3 remains at relatively high levels even after weaning, and is one of the few differences observed between the glands of virgin and parous females (9).

The loss of the TGF-β antiproliferative response is observed in many cases of breast cancer and appears to be an early hallmark of the immortalization of cells in culture. Intriguingly, the expression of TGF-β in tumor cells increases while the levels of TGF-β receptors are decreased. However, the lack of responsivity may be due to defects at multiple levels, such as loss of SMADs, mutation of receptors, or loss of other cooperative factors needed for growth arrest. Genetic manipulation of the genes encoding TGF-β and its receptor has demonstrated that modulation of this complex pathway within the mammary gland can impact the balance between a protective or susceptible phenotype. A constitutively active form of TGF-β crossed onto a mouse which has a TGF-α oncogene reduces breast cancer development (105). In mice that overexpress a dominant negative type II TGF-β receptor in the mammary gland, there is increased susceptibility to breast tumor development (21,106). These experiments suggest that properly regulated TGF-β can function as a tumor suppressor in the mammary gland.

Numerous chemopreventative agents regulate TGF-β expression. Natural and synthetic retinoids will induce TGF-β 1 or 2 in a number of cell types (107–109). Retinoic acid and tamoxifen will increase the expression of latent TGF binding proteins (LTBPs) (110). Tamoxifen upregulates the expression of TGF-β 2 (111,112) and activates TGF-β 1 in MCF-7 cells (113–115). The upregulation of TGF-β family members may be responsible for controlling the proliferation in response to growth factors and hormones, and may play a role in regulating cell death in response to ionizing radiation. Death by ionizing radiation requires TGF-β in concert with p53. It is also known that TGF-β inhibits telomerase activity (116).

The activation of TGF-β by dietary phytochemicals (such as perillyl alcohol) has been observed in mammary carcinomas (117). In addition, NAG-1 (NSAID activated gene) is a newly identified and more distant relative of TGF-β that is currently receiving significant attention. The NAG-1 gene was first identified as a factor that is expressed at high levels in the placenta, prostate, and brain (118,119); and is synthesized as a latent proenzyme which is cleaved to generate a 28kDa active form. Like other members of the TGF-β family, NAG-1 can inhibit tumor development and activate apoptosis. The upregulation of the NAG-1 protein is controlled by at least two different pathways: a p53-independent mechanism, which stabilizes Nag-1 RNA; or a p53-dependent transcriptional increase in expression. Expression of Nag-1 is induced by known chemopreventative agents like NSAIDs (120,121) and retinoids (122), but has also been shown to be upregulated by dietary phytochemicals such as diallyl disulfide, genistein, and resveratrol (120,121,123). These agents appear to induce NAG-1 through the p53-dependent mechanism. The role of NAG-1 in protection of the mammary gland is still under investigation, and its growth inhibitory and proapoptotic properties make it worthy of further study.

11.2.4.5   Vascular Endothelial Growth Factor (VEGF)

Oxygen is required for the sustained growth of tumors, and is delivered by an induced and directed growth of vascular tissue. Angiogenesis of tumors is controlled primarily by the tumor and growth factors released from the stoma, most notably VEGF and FGF. Vascular endothelial growth factor has been found to be released by breast tumor cells, and its expression is regulated by IGF, as well as by products of cyclooxygenase and lipoxygenase. In a synergistic fashion with FGF, VEGF binds receptors on the endothelial cells and stimulates growth and neovascularization. The targeting of angiogenesis has become a major focus of cancer research. In animal models significant inhibition of tumor development is observed; however, these results have not yet translated as successfully in human studies. Inhibitors that block VEGFR signaling (Neovastat), as well as monoclonal antibodies to VEGF are being examined. Drugs are also being developed that resemble endogenous angiogenesis inhibitors such as endostatin and angiostatin.

Phytochemicals may provide a partial defense against angiogenesis. Green tea, flaxseed, and numerous berry extracts decrease the expression of inducible VEGF. Green tea (in particular epigallocatechin gallate (EGCG) has been extensively studied in numerous cell types including the MDA-MB-231 breast cancer cell line, which constitutively expresses high levels of VEGF (124,125). EGCG ultimately blocks VEGF expression through the inhibition of the EGF receptor, which regulates NF-κB and STAT-3, two signaling proteins which have been demonstrated to impact the VEGF promoter. Green tea catechins in endothelial cell lines also appear to block VEGF induced tubule formation through the regulation of VE-cadherin and the inhibition of Akt activity (126). Fruits such as *Gleditsia sinensis* effectively control VEGF in the MDA-MB-231 breast cancer cell line (127), and other berries (cranberry, raspberry, strawberry, blueberry, elderberry, and bilberry) appear to significantly reduce VEGF expression in HaCaT cells (128). While most of the studies looking at phytochemical reduction of VEGF have used cell lines, a flaxseed study corroborated these findings *in vivo*. A diet that was supplemented with flaxseed reduced tumor formation in a breast cancer xenograft model and decreased the expression of VEGF (129).

11.2.5   Arachidonic Acid Metabolism

11.2.5.1   Background

Arachidonic acid and its metabolites play key roles in cell growth, signaling, and adhesion, and are key mediators of the inflammatory and immune responses. Arachidonic acid is released from plasma membrane phospholipids (i.e., phosphatidylcholine) by phospholipase A_2, where it can serve as a signaling molecule (autocrine or paracrine), or alternatively be metabolized by the monooxygenase (MO), lipoxygenase (LO), or cyclooxygenase (COX) enzymes. Natural product and pharmacological inhibitors of AA metabolism have been successfully employed to inhibit cell proliferation, transformation, and migration. Additionally, human breast cancer tissues overexpress 12-lipoxygenase and cyclooxygenase-2 (26). Thus, phytochemicals blocking AA and its metabolites offer great promise in chemopreventative strategies against a variety of cancers, including breast cancer (130,131). The COX and LO branches of AA metabolism are discussed in the following, and plant derived agents that block each of these pathways are presented.

11.2.5.2   Cyclooxygenase (COX)

Arachidonic acid can be metabolized by cyclooxygenase (COX) to produce prostaglandins (PGs), thromboxanes (TXs), and prostacyclins. These prostanoids have physiological roles in reproduction, pain response, fever, cell growth, and inflammation (23). There is much interest in prostaglandin E_2 (PGE_2), which is a potent mediator of the inflammatory response. PG synthesis can be achieved by two functionally distinct isoforms of cyclooxygenase: COX-1 and COX-2. Cyclooxygenase-1 is generally regarded as a constitutively expressed housekeeping gene used to synthesize PGs for necessary, routine functions, and is present in most cell types. In contrast, COX-2 is normally expressed only when induced by growth factors, cytokines, or upon viral transformation as an immediate early gene (132). Upregulation of COX-2 activity increases PGE_2 levels, which acts as a signaling molecule in an autocrine or paracrine manner to stimulate cell proliferation.

Constitutive expression of COX-2 (but not COX-1) and high PGE_2 levels have been detected in a variety of epithelial derived tumors, including cancers of the breast, colon, esophagus, and skin (23), and in human breast cancer cell lines (133). Several lines of evidence suggest that COX-2 overexpression may be important in breast cancer initiation. Cyclooxygenase-2 was typically overexpressed in invasive carcinomas (134), in adjacent DCIS (134,135), and in normal epithelium surrounding DCIS (135). Overexpression of COX-2 was sufficient to cause tumorigenesis in mice (136). Additionally, COX-2 overexpression was found to correlate with markers for premalignant breast cell transformation (137), and may be of prognostic significance in patient outcome (24,25). Prostaglandin E_2 has also been shown to be required for cell migration and metastasis by signaling through cAMP dependent protein kinase (PKA) (138,139), and along with TXA_2 and PGI_2 is involved in angiogenesis to supply blood to established tumors (140,141). Many oncogenes also stimulate COX-2 expression, including HER-2/neu (142), which directs EGF signaling leading to cell proliferation. Recently it has been determined that PGE_2 increases the expression of HER-2/neu, indicating that a positive loop drives overexpression of both genes in breast tumors (143). Increased COX-2 activity also has been shown to lead to a decrease in apoptosis, primarily through upregulation of the antiapoptotic factor Bcl-2 (144). The tumor suppressor protein p53 has also been shown to down regulate COX-2 expression (145). Thus, inhibition of COX-2 activity may serve to decrease breast tumor initiation and development.

The relationship between COX inhibition and cancer was first noted when links between frequent use of nonsteroidal antiinflammatory drugs (NSAIDs) that block COX activity, and decreased colon cancer incidence were identified (146). However, many of the first generation COX inhibitors used to treat pain and inflammation (e.g., aspirin, indomethacin) are nonselective; i.e., the activity of both COX isoforms are blocked, leading to many undesirable side effects, most frequently gastrointestinal discomfort or bleeding (23). Currently, there are many pharmacological selective COX-2 inhibitors (e.g., celeCOXib, SC-236, NS-398) available that possess potent antiinflammatory properties, lower colon cancer incidence when used regularly, and have greatly reduced side effects. There are many reported studies of NSAID use and breast cancer (147), most recently linking aspirin use to a reduced risk of hormone receptor positive breast cancer (148). However, more clinical studies are required to conclusively establish the use of aspirin or other NSAIDs in breast cancer chemoprevention.

Several scientific studies have demonstrated that inhibition of COX-2 activity and PGE_2 secretion may offer promise in breast cancer prevention and treatment. Using established human breast cell lines, indomethacin (nonselective COX inhibitor) is capable of inhibiting the proliferation of MCF-7 and MDA-MB-231 cancer cells and of normal human mammary epithelial cells (HMEC) (149,150). Treatment with the selective COX-2 inhibitor NS-398 was capable of inducing apoptosis in MDA-MB-231 cells (151). Treatment with NS-398 also increased sensitivity to radiation in COX-2 expressing rat intestinal epithelial cells (152). These scientific studies indicate a potential role may exist for COX-2 inhibitors in chemoprevention of breast cancer.

In addition to these pharmacological inhibitors of COX, many phytochemicals are capable of blocking COX activity. The best studied plant derived compound that inhibits COX is the phytoalexin polyphenol resveratrol, found in red wine, peanuts, and many herbs. Resveratrol induces growth arrest and apoptosis in several human cancer cell lines, including the breast cancer line MCF-7 (153). Treatment of mouse mammary glands with resveratrol inhibited the development of preneoplastic lesions (154), thus demonstrating resveratrol has chemopreventative functionality. Most of the initial research indicated that resveratrol was selective for inhibiting COX-1; however, it has since been shown in mammary epithelial cells to inhibit the both the expression and the activity of COX-2 (155,156). Resveratrol likely blocks COX-2 expression by a protein kinase C-dependent mechanism (155).

Several other phytochemicals have demonstrated the ability to inhibit COX-2 transcription and activity. Curcumin (a significant component of turmeric) prevents lipopolysaccharide and phorbol ester mediated upregulation of COX-2 gene expression, likely by blocking signaling through extracellular signal regulated kinase (ERK) and decreasing NF-κB and activator protein 1 (AP-1) binding to the COX-2 promoter (157,158). Curcumin also directly inhibits the enzymatic activity of COX-2 (159), indicating that curcumin acts on multiple levels to decrease PGE$_2$ levels. The phytochemical baicalein (flavone isolated from the roots of *Scutellaria baicalensis* that is also a 12-lipoxygenase selective inhibitor) is capable of inhibiting both cell growth and the production of PGE$_2$ in breast and prostate cancer cell lines (160). Additionally, various flavonoids such as EGCG (green tea) and apigenin (vegetables and fruits) block COX-2 gene expression by suppressing NF-κB (161).

11.2.5.3   Lipoxygenase (LO)

Three main branches of the lipoxygenase pathway give rise to specific products of AA, including leukotrienes (LTs) and hydroxyeicosatetraenonic acids (HETEs). Arachidonic acid can be metabolized by 5-lipoxygenase (5-LO) to LTs (key mediators of asthma and anaphylactic shock) or 5-HETE. 12-LO converts AA into 12-HETE. Metabolism of AA by 15-LO generates 15-HETE, which acts as a ligand for the growth controlling nuclear receptor PPARγ (162). Both 5-HETE and 12-HETE act as a mitogen to cultured breast and prostate cancer cells (163). Interestingly, studies with colon cancer cells indicate that unlike 5-LO and 12-LO, 15-LO activity is antiproliferative (164). Tissue samples taken from breast cancer patients indeed show that 15-LO expression is decreased in tissues taken from patients with breast cancer as compared to paired normal tissues. Therefore, the application of phytochemicals that block the activities of 5-LO or 12-LO or both in combination may be useful in the treatment of several cancers, including breast cancer. While 15-LO activity can be blocked by several flavonoids such as luteolin (165), the antiproliferative activity of this enzyme makes inhibition undesirable.

Increased metabolism of AA to 5-HETE by 5-LO has been observed in a variety of cancers (166), and overexpression of 5-LO in cancerous breast tissue has been correlated with poor clinical outcomes in patients with breast cancer (26). Application of exogenous 5-HETE to the breast cancer cell lines MCF-7 and MDA-MB-231 increased cell proliferation (167). Accordingly, treatment of these cell lines with the 5-LO inhibitor Rev-9501 induced apoptotic events, such as cytochrome *c* release and decreasing levels of antiapoptotic proteins. Several researchers have attempted to identify the signal pathways utilized by which 5-HETE exerts its mitogenic effect. In pancreatic cancer cell lines, 5-HETE was shown to induce tyrosine phosphorylation, and work through Erk and Akt to induce cell proliferation (168). Many pharmacological inhibitors of 5-LO activity exist (e.g., AA-861, MK-866, and the antiallergenic drugs montelukast, zafirlukast, and zileuton), but are not used for cancer chemoprevention. Natural phytochemicals that selectively inhibit 5-LO include the COX and CYP1A1 inhibitor curcumin (78), the prenylated flavanones sigmoidins A and B (isolated from *Erythrina sigmoidea hua*) (169), the dual 5-LO/12-LO inhibitor 2-(3,4-dihydroxyphenyl)ethanol from olive extracts (170) and hyperforin, a lipophilic component of *Hypericum perforatum* (St. John's wort) extract (171).

When normal breast tissue was compared to lobular and ductal carcinomas taken from the same sets of patients, cancerous tissues exhibited increased expression of 12-LO. Additionally, prognostic value of relating clinical outcome with the degree of increased 12-LO expression existed, where higher mortality rates were directly correlated with higher overexpression of 12-LO (26). Baicalein is a potent and selective inhibitor of 12-LO activity, and inhibits proliferation and induces apoptosis in MCF-7 and MDA-MB-231 cells, two well established breast cancer cell lines (160,167). Additionally, baicalein may be antiestrogenic in ER-alpha expressing MCF-7 cells and better at inducing apoptosis than the soy isoflavone genistein (172). Expression of 12-LO and levels of 12-HETE increase with progression of prostate cancer, and a similar correlation may also exist for breast cancer. Therefore, the use of baicalein as a chemopreventative agent against the development and progression of breast cancer may be of significant value as it inhibits both 12-LO and COX-2, two enzymes upregulated in breast cancer.

11.3 STRATEGIES FOR PHYTOCHEMICAL ENRICHMENT FOR CHEMOPREVENTION

11.3.1 Consistency of Phytochemicals

The major challenge in developing and characterizing phytochemicals for chemopreventative applications lies in the inconsistency of phytochemicals from botanicals. Additionally, in many cases a whole extract or food phytochemical profile provides more effective protection than individual phytochemicals in the profile. Soybeans are an example of this complication. Epidemiological studies have associated high soy intake with a reduced risk for certain types of cancer (173). Soybeans are a rich source of phenolic antioxidants known as isoflavonoids. The functionally active isoflavonoid found in soybeans is genistein, which has weak estrogenic activity that appears to antagonize the action of estradiol at the estrogen receptor in breast cells, and thus may protect against breast cancer development (174,175). While the chemopreventative properties of purified and synthetic genistein have been demonstrated (58,61,176–178), recent research has shown that fermented soy products performed better at reducing incidence of mammary tumor risk as compared to a similar mixture of soy constituent isoflavonoids (179). This suggests that the food background may play a positive role in the chemopreventative actions of its resident isoflavonoids (particularly genistein). The challenge posed by inconsistency of phytochemicals is even greater in highly cross pollinating botanical species, such as those belonging to *Lamiaceae*. These species are the source of rosmarinic acid (180), which in proper phenolic profiles has chemopreventative potential through inhibition of COX (181). However, each extract is likely to vary in content from source to source and batch to batch (180).

Development of consistent and optimized phytochemicals and whole extract profiles is critical for consistent dosage that is clinically relevant and safe for chemoprevention. Attempts to address this need include the development of approaches such as clonal screening from heterogeneous botanical sources, elicitor and stress induced sprouting of phytochemicals from seedlings, solid-state and liquid bioprocessing and fermentation of botanical extracts. By combining various optimized dietary phytochemicals, synergistic enhancement of chemoprevention can also be attempted.

11.3.2 Clonal Screening of Phytochemicals from a Heterogeneous Genetic Background

Rosmarinic acid is a phenolic biphenyl found in several species in the family *Lamiaceae* with many pharmacological effects. Rosmarinic acid inhibits several complement dependent inflammatory processes and thus has potential as a therapeutic agent for control of complement activation diseases (182,183). Rosmarinic acid has been reported to have effects on both the classical C3-convertase and on the cobra venom factor and ovalbumin and antiovalbumn mediated passive cutaneous anaphylaxis (182). Recent research has indicated other benefits of rosmarinic acid containing *Perilla frutescens* on the reduction of lipopolysaccharide (LPS)-induced liver injury in D-galactosamine sensitized mice (184). Additional evidence suggests that rosmarinic acid inhibits COX (181), and therefore represents a good candidate for chemoprevention of oxidation linked diseases, including cancer.

High rosmarinic acid producing, shoot based clonal lines originating from a single heterozygous seed amongst a heterogeneous bulk seed population of oregano, rosemary, lavender, spearmint, and thyme have been isolated. These lines were screened based upon tolerance to the proline analog A2C and a novel *Pseudomonas* species isolated from oregano (185–187). This strategy for screening and selection of high rosmarinic acid clonal lines is additionally based on the model that the proline linked pentose phosphate pathway is critical for driving metabolic flux (i.e., E4P) toward the shikimate and phenylpropanoid pathways (188). Any clonal line with a deregulated proline synthesis pathway should have increased pentose phosphate pathway activity, which would allow excess metabolic flux to drive the shikimate and phenylpropanoid pathways toward total phenolic and rosmarinic acid synthesis. This model hypothesizes that the same metabolic flux from increased activity of the proline linked pentose phosphate pathway also regulates the interconversion of ribose-5-phosphate to E4P, driving the shikimate pathway. Shikimate pathway flux is critical for both auxin and phenylpropanoid biosynthesis, including rosmarinic acid. High rosmarinic acid producing clonal lines selected by these approaches (185,186,189–191) are being targeted to produce consistent dietary phenolic phytochemicals from cross pollinating, heterogeneous species for functional foods and for chemoprevention strategies (180,191).

11.3.3 Stress and Elicitor Induced Sprouting

In many legumes such as soy (192), pea (193), fava bean (194), mung bean (195) and fenugreek (196), sprouting in combination with biological, chemical, and physical stress linked elicitation can stimulate phytochemicals relevant in chemoprevention. Preliminary results have provided empirical evidence for a link between proline biosynthesis and oxidation, as well as stimulation of G6PDH during phenolic phytochemical synthesis (180,188,197–199). In light mediated sprout studies in peas (*Pisum sativum*), acetylsalicylic acid in combination with fish protein hydrolysates (a potential source of proline precursors) stimulated phenolic content and guaiacol peroxidase (GPX) activity during early germination, with corresponding higher levels of proline and G6PDH activity (200). In other light mediated studies in peas, low pH and salicylic acid treatments stimulated increased phenolic content and tissue rigidity, with concomitant stimulation of G6PDH and proline (199). This work supports the hypothesis that pentose phosphate pathway stimulation may be linked to proline biosynthesis, and that modulation of a proton linked redox cycle may also be operating through the proline linked pentose phosphate pathway. In dark germinated studies in peas, high cytokinin containing anise root extracts stimulated phenolic content and antioxidant activity, which correlated directly with increased proline content but inversely with G6PDH activity (193).

In dark germination studies in mung bean (*Vigna radiata*), dietary grade microbial polysaccharide, oregano phenolics and peptide elicitor treatments stimulated phenolic content and G6PDH and GPX enzyme activities compared to controls (193,195,201,202) with concomitant stimulation of proline content. In addition, specific elicitors xanthan gum, oregano phenolics, yeast extract, and yeast glucan stimulated antioxidant activity. The hypothesis that stimulation of proline linked pentose phosphate pathway would stimulate phenolic metabolism under elicitor and stress response was tested using dark germinated fava bean. In polysaccharide elicitor studies, gellan gum stimulated fava bean total phenolic content by ninefold in late stages of germination with a corresponding increase in proline content and GPX activity. However, the effect on antioxidant and G6PDH activity was inconclusive (203). In the same fava bean system, UV mediated stimulation of phenolic content in dark germinated fava bean sprouts indicated a positive correlation to G6PDH and GPX activities with a concomitant increase in proline content (204). It was further confirmed that the proline analog A2C also stimulated phenolic content in fava bean, with positive correlation to G6PDH and GPX activities as well as proline content (205). Similar to studies in clonal shoot cultures of thyme (197) and oregano (198), the proline analog mediated studies in fava bean confirmed that proline overexpression was not only possible, but involved stimulating G6PDH, and therefore likely diverted the pentose phosphate pathway toward phenylpropanoid biosynthesis. The late stage stimulation of phenolic content and GPX activity in response to microwave mediated thermal stress in dark germinated fava bean strongly correlated with stimulation of free radical scavenging activity of free phenolics as measured by the quenching of 1,1-diphenyl-2-picrylhydrazyl (DPPH) and stimulation of SOD activity (202). Recent studies have extended this concept of elicitor induced stimulation to fenugreek (196). These strategies and insights have allowed the development of a technology platform where stress induced elicitation can be used to capture and optimize the production of phytochemicals with chemopreventative potential in germinating seeds, which would not be otherwise possible. This approach could theoretically capture protective phytochemical profiles that are specifically produced under stress to protect the emerging young embryo and seedling, at a stage when respiration and oxidation stress during the sprouting process is high.

11.3.4 Solid-State Bioprocessing

Solid-state bioprocessing using dietary microbial systems under minimal water conditions has been used for over 2000 years in many parts of the world to develop specific foodstuffs such as soy based products like tempeh in Indonesia, tofu and related foods in East Asia, several legume products in India and cheese in Europe (206,207). In the context of chemoprevention, enhanced isoflavonoid content in soybean and soybean meal following bioprocessing by *Aspergillus* species has been reported (208,209). Dietary fungal bioprocessing of fruit and legume substrates to enrich aglycone phenolic antioxidants such as ellagic acid (cranberry), isoflavonoids (soybean) and L-DOPA (fava bean) have been similarly developed (210–215). Critical microbial enzymes that are involved in the optimal release of these phenolic phytochemicals have recently been identified (194,215–217). A better understanding of growth and enzyme regulation during phenolic phytochemical mobilization and release would help to optimize the desired phytochemical profile with maximum chemopreventative potential. In noncancer related chemoprevention investigations, results indicate that cranberry and soybean bioprocessing by solid-state methods not only stimulated free radical scavenging antioxidant activity, but also increased phenolic content in specific stages and inhibited foodborne pathogens (217–219) and the ulcer associated *Helicobacter pylori* (217,219). Such extracts are now being targeted for breast cancer chemoprevention and diabetes through modulation of the antioxidant enzyme response (207,220).

11.3.5 Liquid Fermentation

Recent research has shown that fermented soymilk performed better at reducing incidence of mammary tumor risk than a similar mixture of its constituent isoflavonoids, suggesting that the food background may play a positive role in the chemopreventative actions of soy based products (179). Fermented soymilk is rich in phenolic aglycones (which are more active and more readily taken up than their (β–glycosides), thus increasing the free phenolic content of soy based food through microbial bioprocessing may positively affect its medicinal and nutritional value (221–224). The role of lignin degrading enzymes in phenolic antioxidant mobilization during yogurt production from soymilk by active probiotic kefir cultures was investigated (225). Total soluble phenolic content and free radical scavenging antioxidant activity were measured every 8 hours for a 48 hour period. The activity of several enzymes (β-glucosidase, laccase, and peroxidase) associated with the microbial degradation of polymeric phenolics and lignin and previously linked to phenolic mobilization from soybean during solid-state bioprocessing by dietary fungi were also investigated. Soluble phenolic content initially increased with culture time and was strongly correlated to total peroxidase and laccase activity. However, phenolic content dropped sharply at 48 hours. Antioxidant activity increased with culture time and was strongly correlated to decreased soluble phenolic content over the same time period. This research has important implications for the optimization of functional phytochemicals in commercial soymilk based yogurts, which can be targeted for chemoprevention strategies (225).

REFERENCES

1. Wiseman, B., Z. Werb. Stromal effects on mammary gland development and breast cancer. *Science* 296:1046–1049, 2002.
2. Ronnov-Jessen, L., O. Petersin, M. Bissell. Cellular changes involved in conversion of normal to malignant breast: importance of the stromal reaction. *Physiol. Rev.* 78:69–125, 1996.
3. Beenken, S., K. Bland. Biomarkers for breast cancer. *Minerva Chirurgica* 57:437–448, 2002.
4. Rosner, B., G. Colditz, M. Willett. Reproductive risk factors in a prospective study of breast cancer: the nurses' health study. *Am. J. Epidemiol.* 139:819–835, 1994.
5. Sivaraman, L., S.G. Hilsenbeck, L. Zhong, J. Gay, O.M. Conneely, D. Medina, B.W. O'Malley. Early exposure of the rat mammary gland to estrogen and progesterone blocks co-localization of estrogen receptor expression and proliferation. *J. Endocrinol.* 171:75–83, 2001.
6. Sivaraman, L., O.M. Conneely, D. Medina, B.W. O'Malley. p53 is a potential mediator of pregnancy and hormone-induced resistance to mammary carcinogenesis. *Proc. Natl. Acad. Sci. USA* 98:12379–12384, 2001.
7. Kuperwasser, C., J. Pinkas, G.D. Hurlbut, S.P. Naber, D.J. Jerry. Cytoplasmic sequestration and functional repression of p53 in the mammary epithelium is reversed by hormonal treatment. *Cancer Res.* 60:2723–2729, 2000.
8. Medina, D. Breast cancer: the protective effect of pregnancy. *Clin. Cancer. Res.* 10:380S–384S, 2004.
9. D'Cruz, C.M., S.E. Moody, S.R. Master, J.L. Hartman, E.A. Keiper, M.B. Imielinski, J.D. COX, J.Y. Wang, S.I. Ha, B.A. Keister, L.A. Chodosh. Persistent parity-induced changes in growth factors, TGF-beta3, and differentiation in the rodent mammary gland. *Mol. Endocrinol.* 16:2034–2051, 2002.
10. Oren, M. Decision making by p53: life, death, and cancer. *Cell Death Diff.* 10:431–442, 2003.
11. Jerry, D., L. Minter, K. Becker, A. Blackburn. Hormonal control of p53 and chemoprevention. *Breast Cancer Res.* 4:91–94, 2002.
12. Singh, R., S. Dhanalakshmi, R. Agarwal. Phytochemicals as cell cycle modulators: a less toxic approach in halting human cancers. *Cell Cycle* 1:156–161, 2002.
13. Katzenellenbogen, B., J. Frasor. Therapeutic targeting in the estrogen receptor hormonal pathway. *Semin. Oncol.* 31:28–38, 2004.

14. Miller, W.R. Aromatase inhibitors: mechanism of action and role in the treatment of breast cancer. *Semin. Oncol.* 30:3–11, 2003.

15. Goss, P.E., K. Strasser-Weippl. Aromatase inhibitors for chemoprevention. *Best Pract. Res. Clin. Endocrinol. Metab.* 18:113–130, 2004.

16. Amakura, Y., T. Tsutsumi, K. Sasaki, T. Yoshida, T. Maitani. Screening of the inhibitory effect of vegetable constituents on the aryl hydrocarbon receptor-mediated activity induced by 2,3,7,8-tetrachlorodibeno-p-dioxin. *Biol. Pharmacol. Bull.* 26:1754–1760, 2003.

17. Zheng, W., S. Kar, R. Quirion. Insulin-like growth factor-1-induced phosphorylation of transcription factor FKHRL1 is mediated by phosphatidylinositol 3-kinase/Akt kinsase and role of this pathway in insulin-like growth factor-1-induced survival of cultured hippocampal neurons. *Mol. Pharmacol.* 62:225–233, 2002.

18. Sugumar, A., Y. Liu, Q. Xia, Y. Koh, K. Matsuo. Insulin-like growth factor (IGF)-I and IGF-binding protein 3 and the risk of premenopausal breast cancer: a meta-analysis of literature. *Int. J. Cancer* 111:293–297, 2004.

19. Darcy, K., A. Wohlhueter, D. Zangani, M. Vaughan, J. Russell, P. Masso-Welch, L. Varela, S. Shoemaker, E. Horn, P. Lee, R. Huang, M. Ip. Selective changes in EGF receptor expression and function during the proliferation, differentiation and apoptosis of mammary epithelial cells. *Eur. J. Cell Biol.* 78:511–523, 1999.

20. Ferrara, N. Molecular and biological properties of vascular endothelial growth factor. *J. Mol. Med.* 77:527–543, 1999.

21. Bottinger, E.P., J.L. Jakubczak, D.C. Haines, K. Bagnall, L.M. Wakefield. Transgenic mice overexpressing a dominant-negative mutant type II transforming growth factor beta receptor show enhanced tumorigenesis in the mammary gland and lung in response to the carcinogen 7,12-dimethylbenz-[a]-anthracene. *Cancer Res.* 57:5564–5570, 1997.

22. Huynh, H., W. Beamer, M. Pollak, T.W. Chan. Modulation of transforming growth factor beta1 gene expression in the mammary gland by insulin-like growth factor I and octreotide. *Int. J. Oncol.* 16:277–281, 2000.

23. Howe, L., K. Subbaramaiah, A. Brown, A. Dannenberg. Cyclooxygenase-2: a target for the prevention and treatment of breast cancer. *Endocrine Relat. Cancer* 8:97–114, 2001.

24. Ristimaki, A., A. Sivula, J. Lundin, M. Lundin, T. Salminen, C. Haglund, J. Joensuu, J. Isola. Prognostic significance of elevated cyclooxygenase-2 expression in breast cancer. *Cancer Res.* 62:632–635, 2002.

25. Tan, K., W. Yong, T. Putti. Cyclooxygenase-2 expression: a potential prognostic and predictive marker for high-grade ductal carcinoma *in situ* of the breast. *Histopathology* 44:24–28, 2004.

26. Jiang, W., A. Douglas-Jones, R. Mansel. Levels of expression of lipoxygenases and cyclooxygenase-2 in human breast cancer. *Prostaglandins Leukotrienes Essential Fatty Acids* 69:275–281, 2003.

27. Arun, B., P. Goss. The role of COX-2 inhibition in breast cancer treatment and prevention. *Semin. Oncol.* 31:22–29, 2004.

28. Cline, J., C. Hughes. Phytochemicals for the prevention of breast and endometrial cancer. *Cancer Treatment Res.* 94:107–134, 1998.

29. Surh, Y. Molecular mechanisms of chemopreventative effects of selected dietary and medicinal phenolic substances. *Mutat. Res.* 428:305–327, 1999.

30. Saito, S., A.A. Goodarzi, Y. Higashimoto, Y. Noda, S.P. Lees-Miller, E. Appella, C.W. Anderson. ATM mediates phosphorylation at multiple p53 sites, including Ser(46), in response to ionizing radiation. *J. Biol. Chem.* 277:12491–12494, 2002.

31. Vargas, D.A., S. Takahashi, Z. Ronai. Mdm2: a regulator of cell growth and death. *Adv. Cancer Res.* 89:1–34, 2003.

32. Turenne, G.A., P. Paul, L. Laflair, B.D. Price. Activation of p53 transcriptional activity requires ATM's kinase domain and multiple N-terminal serine residues of p53. *Oncogene* 20:5100–5110, 2001.

33. Pise-Masison, C.A., M. Radonovich, K. Sakaguchi, E. Appella, J.N. Brady. Phosphorylation of p53: a novel pathway for p53 inactivation in human T-cell lymphotropic virus type 1-transformed cells. *J. Virol.* 72:6348–6555, 1998.

34. Kapoor, M., R. Hamm, W. Yan, Y. Taya, G. Lozano. Cooperative phosphorylation at multiple sites is required to activate p53 in response to UV radiation. *Oncogene* 19:358–364, 2000.

35. Buschmann, T., O. Potapova, A. Bar-Shira, V.N. Ivanov, S.Y. Fuchs, S. Henderson, V.A. Fried, T. Minamoto, D. Alarcon-Vargas, M.R. Pincus, W.A. Gaarde, N.J. Holbrook, Y. Shiloh, Z. Ronai. Jun NH2-terminal kinase phosphorylation of p53 on Thr-81 is important for p53 stabilization and transcriptional activities in response to stress. *Mol. Cell Biol.* 21:2743–2754, 2001.

36. Bulavin, D.V., S. Saito, M.C. Hollander, K. Sakaguchi, C.W. Anderson, E. Appella, A.J. Fornace, Jr. Phosphorylation of human p53 by p38 kinase coordinates N-terminal phosphorylation and apoptosis in response to UV radiation. *EMBO J.* 18:6845–6854, 1999.

37. Ogawara, Y., S. Kishishita, T. Obata, Y. Isazawa, T. Suzuki, K. Tanaka, N. Masuyama, Y. Gotoh. Akt enhances Mdm2-mediated ubiquitination and degradation of p53. *J. Biol. Chem.* 277:21843–21850, 2002.

38. Mayo, L.D., D.B. Donner. A phosphatidylinositol 3-kinase/Akt pathway promotes translocation of Mdm2 from the cytoplasm to the nucleus. *Proc. Natl. Acad. Sci. USA* 98:11598–11603, 2001.

39. Ueno, M., H. Masutani, R.J. Arai, A. Yamauchi, K. Hirota, T. Sakai, T. Inamoto, Y. Yamaoka, J. Yodoi, T. Nikaido. Thioredoxin-dependent redox regulation of p53-mediated p21 activation. *J. Biol. Chem.* 274:35809–35815, 1999.

40. Kuperwasser, C., G.D. Hurlbut, F.S. Kittrell, E.S. Dickinson, R. Laucirica, D. Medina, S.P. Naber, D.J. Jerry. Development of spontaneous mammary tumors in BALB/c p53 heterozygous mice: a model for Li-Fraumeni syndrome. *Am. J. Pathol.* 157:2151–2159, 2000.

41. Coles, C., A. Condie, U. Chetty, C.M. Steel, H.J. Evans, J. Prosser. p53 mutations in breast cancer. *Cancer Res.* 52:5291–5298, 1992.

42. Moll, U.M., G. Riou, A.J. Levine. Two distinct mechanisms alter p53 in breast cancer: mutation and nuclear exclusion. *Proc. Natl. Acad. Sci. USA* 89:7262–7266, 1992.

43. Minter, L.M., E.S. Dickinson, S.P. Naber, D.J. Jerry. Epithelial cell cycling predicts p53 responsiveness to gamma-irradiation during post-natal mammary gland development. *Development* 129:2997–3008, 2002.

44. Huang, C., W.Y. Ma, J. Li, S.S. Hecht, Z. Dong. Essential role of p53 in phenethyl isothiocyanate-induced apoptosis. *Cancer Res.* 58:4102–4106, 1998.

45. Hong, Y.S., Y.A. Ham, J.H. Choi, J. Kim. Effects of allyl sulfur compounds and garlic extract on the expression of Bcl-2, Bax, and p53 in non small cell lung cancer cell lines. *Exp. Mol. Med.* 32:127–134, 2000.

46. Laux, M., M. Aregullin, J. Berry, J. Flanders, E. Rodriguez. Identification of a p53-dependent pathway in the induction of apoptosis of human breast cancer cells by the natural product, resveratrol. *J. Alt. Compl. Med.* 10:235–239, 2004.

47. Narayanan, B.A., O. Geoffroy, M.C. Willingham, G.G. Re, D.W. Nixon. p53/p21(WAF1/CIP1) expression and its possible role in G1 arrest and apoptosis in ellagic acid treated cancer cells. *Cancer Lett.* 136:215–221, 1999.

48. Choudhuri, T., S. Pal, M.L. Agwarwal, T. Das, G. Sa. Curcumin induces apoptosis in human breast cancer cells through p53-dependent Bax induction. *FEBS Lett.* 512:334–340, 2002.

49. O'Prey, J., J. Brown, J. Fleming, P.R. Harrison. Effects of dietary flavonoids on major signal transduction pathways in human epithelial cells. *Biochem. Pharmacol.* 66:2075–2088, 2003.

50. Ye, R., A. Bodero, B.B. Zhou, K.K. Khanna, M.F. Lavin, S.P. Lees-Miller. The plant isoflavenoid genistein activates p53 and Chk2 in an ATM-dependent manner. *J. Biol. Chem.* 276:4828–4833, 2001.

51. Wigle, D.T. Breast cancer and fertility trends in Canada. *Am. J. Epidemiol.* 105:428–438, 1977.

52. Grubbs, C.J., D.L. Hill, K.C. McDonough, J.C. Peckham. N-nitroso-N-methylurea-induced mammary carcinogenesis: effect of pregnancy on preneoplastic cells. *J. Natl. Cancer Inst.* 71:625–628, 1983.

53. Russo, I.H., M. Koszalka, P.A. Gimotty, J. Russo. Protective effect of chorionic gonadotropin on DMBA-induced mammary carcinogenesis. *Br. J. Cancer* 62:243–247, 1990.

54. Razandi, M., A. Pedram, S.T. Park, E.R. Levin. Proximal events in signaling by plasma membrane estrogen receptors. *J. Biol. Chem.* 278:2701–2712, 2003.

55. Telang, N.T., A. Suto, G.Y. Wong, M.P. Osborne, H.L. Bradlow. Induction by estrogen metabolite 16 alpha-hydroxyestrone of genotoxic damage and aberrant proliferation in mouse mammary epithelial cells. *J. Natl. Cancer Inst.* 84:634–638, 1992.

56. Kelsey, J.L., M.D. Gammon, E.M. John. Reproductive factors and breast cancer. *Epidemiol. Rev.* 15:36–47, 1993.

57. Ketcham, A.S., W.F. Sindelar. Risk factors in breast cancer. *Prog. Clin. Cancer.* 6:99–114, 1975.

58. Constantinou, A.I., D. Lantvit, M. Hawthorne, X. Xu, R.B. van Breemen, J.M. Pezzuto. Chemopreventive effects of soy protein and purified soy isoflavones on DMBA-induced mammary tumors in female Sprague-Dawley rats. *Nutr. Cancer.* 41:75–81, 2001.

59. Gallo, D., S. Giacomelli, F. Cantelmo, G.F. Zannoni, G. Ferrandina, E. Fruscella, A. Riva, P. Morazzoni, E. Bombardelli, S. Mancuso, G. Scambia. Chemoprevention of DMBA-induced mammary cancer in rats by dietary soy. *Breast Cancer Res. Treat.* 69:153–164, 2001.

60. Cotroneo, M.S., J. Wang, W.A. Fritz, I.E. Eltoum, C.A. Lamartiniere. Genistein action in the pre-pubertal mammary gland in a chemoprevention model. *Carcinogenesis* 23:1467–1474, 2002.

61. Lamartiniere, C.A., M.S. Cotroneo, W.A. Fritz, J. Wang, R. Mentor-Marcel, A. Elgavish. Genistein chemoprevention: timing and mechanisms of action in murine mammary and prostate. *J. Nutr.* 132:552S–558S, 2002.

62. Sarkar, F.H., Y. Li. Soy isoflavones and cancer prevention. *Cancer Invest.* 21:744–757, 2003.

63. Goodin, M.G., K.C. Fertuck, T.R. Zacharewski, R.J. Rosengren. Estrogen receptor-mediated actions of polyphenolic catechins *in vivo* and *in vitro. Toxicol. Sci.* 69:354–361, 2002.

64. Guengerich, F. Characterization of roles of human cytochrome P450 enzymes in carcinogen metabolism. *Asia Pac. J. Pharmacol.* 5:327–345, 1990.

65. Schwarz, D., P. Kisselev, S. Ericksen, G. Szklarz, A. Chernogolov, H. Honeck, W. Schunck, I. Roots. Arachidonic and eicosapentaenoic acid metabolism by human CYP1A1: highly stereoselective formation of 17(R),18(S)-epoxyeicosatetraenoic acid. *Biochem. Pharmacol.* 67:1445–1457, 2004.

66. Boskou, D., F. Visioli. Biophenols in olive oil and olives. *Ind. Res. Signpost* 2003, pp 161–169.

67. Tudela, J., E. Cantos, J. Espin, F. Tomas-Barberan, M. Gil. Induction of antioxidant flavonol biosynthesis in fresh-cut potatoes: effect of domestic cooking. *J. Agric. Food Chem.* 50:5925–5931, 2002.

68. Kampa, M., V. Alexaki, G. Notas, A. Nifli, A. Nistikaki, A. Hatzoglou, E. Bakogeorgou, E. Kouimtzoglou, G. Blekas, D. Boskou, A. Gravanis, E. Castanas. Antiproliferative and apoptotic effects of selective phenolic acids on T47D human breast cancer cells: potential mechanisms of action. *Breast Cancer Res.* 6:R63–R74, 2004.

69. Watabe, M., K. Hishikawa, A. Takayanagi, N. Shimizu, T. Nakaki. Caffeic acid phenethyl ester induces apoptosis by inhibition of NFkappaB and activation of Fas in human breast cancer MCF-7 cells. *J. Biol. Chem.* 279:6017–6026, 2004.

70. Han, D., H. Tachibana, K. Yamada. Inhibition of environmental estrogen-induced proliferation of human breast carcinoma MCF-7 cells by flavonoids. In Vitro *Cell Develop. Biol. Anim.* 37:275–282, 2001.

71. Zhang, S., C. Qin, S. Safe. Flavonoids as aryl hyrdocarbon receptor agonists/antagonists: effects of structure and cell context. *Environ. Health Perspect.* 111:1877–1882, 2003.

72. Chan, H., L. Leung. A potential protective mechanism of soya isoflavones against 7,12-dimethylbenz(a)anthracene tumour initiation. *Br. J. Nutr.* 90:457–465, 2003.

73. Granados-Soto, V. Pleiotropic effects of resveratrol. *Drug News Perspect,* 73 16(5): 299–307, 2003.

74. Casper, R., M. Quesne, I. Rogers, T. Shirota, A. Jolivet, E. Milgrom, J. Savouret. Resveratrol has antagonist activity on the aryl hydrocarbon receptor: implications for prevention of dioxin toxicity. *Mol. Pharmacol.* 56:784–790, 1999.

75. Ciolino, H., P. Daschner, T. Wang, G. Yeh. Effect of curcumin on the aryl hydrocarbon receptor and cytochrome P450 1A1 in MCF-7 human breast carcinoma cells. *Biochem. Pharmacol.* 56:187–206, 1998.

76. Rinaldi, A., M. Morse, H. Fields, D. Rothas, P. Pei, K. Rodrigo, R. Renner, S. Mallery. Curcumin activates the aryl hydrocarbon receptor yet significantly inhibits (-)-benzo(a)pyrene-7R-trans-7,8-dihydrodiol bioactivation in oral squamous cell carcinoma cells and oral mucosa. *Cancer Res*. 62:5451–5456, 2002.

77. Thapliyal, R., G. Mar. Inhibition of cytochrome P450 isozymes by curcumins *in vitro* and *in vivo*. *Food Chem. Toxicol*. 39:541–547, 2001.

78. Flynn, D., M. Rafferty, A. Boctor. Inhibition of 5-hydroxy-eicosatetraenoic acid (5-HETE) formation in intact human neutrophils by naturally-occurring diarylheptanoids: inhibitory activities of curcuminoids and yakuchinones. *Prostaglandins Leukotriene Med*. 22:357–360, 1986.

79. Conney, A. Enzyme induction and dietary chemicals as approaches to cancer chemoprevention: the seventh DeWitt S. Goodman Lecture. *Cancer Res*. 63:7005–7031, 2003.

80. Vogel, C.L., M.A. Cobleigh, D. Tripathy, J.C. Gutheil, L.N. Harris, L. Fehrenbacher, D.J. Slamon, M. Murphy, W.F. Novotny, M. Burchmore, S. Shak, S.J. Stewart, M. Press. Efficacy and safety of trastuzumab as a single agent in first-line treatment of HER2-overexpressing metastatic breast cancer. *J. Clin. Oncol*. 20:719–726, 2002.

81. Baselga, J., D. Tripathy, J. Mendelsohn, S. Baughman, C.C. Benz, L. Dantis, N.T. Sklarin, A.D. Seidman, C.A. Hudis, J. Moore, P.P. Rosen, T. Twaddell, I.C. Henderson, L. Norton. Phase II study of weekly intravenous trastuzumab (Herceptin) in patients with HER2/neu-overexpressing metastatic breast cancer. *Semin. Oncol*. 26:78–83, 1999.

82. Chen, J., P.M. Stavro, L.U. Thompson. Dietary flaxseed inhibits human breast cancer growth and metastasis and downregulates expression of insulin-like growth factor and epidermal growth factor receptor. *Nutr. Cancer* 43:187–192, 2002.

83. Korutla, L., R. Kumar. Inhibitory effect of curcumin on epidermal growth factor receptor kinase activity in A431 cells. *Biochim. Biophys. Acta* 1224:597–600, 1994.

84. Cheng, A.L., C.H. Hsu, J.K. Lin, M.M. Hsu, Y.F. Ho, T.S. Shen, J.Y. Ko, J.T. Lin, B.R. Lin, W. Ming-Shiang, H.S. Yu, S.H. Jee, G.S. Chen, T.M. Chen, C.A. Chen, M.K. Lai, Y.S. Pu, M.H. Pan, Y.J. Wang, C.C. Tsai, C.Y. Hsieh. Phase I clinical trial of curcumin, a chemopreventive agent, in patients with high-risk or pre-malignant lesions. *Anticancer Res*. 21:2895–2900, 2001.

85. Sah, J.F., S. Balasubramanian, R.L. Eckert, E.A. Rorke. Epigallocatechin-3-gallate inhibits epidermal growth factor receptor signaling pathway: evidence for direct inhibition of ERK1/2 and AKT kinases. *J. Biol. Chem*. 279:12755–12762, 2004.

86. Liang, Y.C., Y.C. Chen, Y.L. Lin, S.Y. Lin-Shiau, C.T. Ho, J.K. Lin. Suppression of extracellular signals and cell proliferation by the black tea polyphenol, theaflavin-3,3'-digallate. *Carcinogenesis* 20:733–736, 1999.

87. Meiers, S., M. Kemeny, U. Weyand, R. Gastpar, E. von Angerer, D. Marko. The anthocyanidins cyanidin and delphinidin are potent inhibitors of the epidermal growth-factor receptor. *J. Agric. Food Chem*. 49:958–962, 2001.

88. Bol, D.K., K. Kiguchi, I. Gimenez-Conti, T. Rupp, J. DiGiovanni. Overexpression of insulin-like growth factor-1 induces hyperplasia, dermal abnormalities, and spontaneous tumor formation in transgenic mice. *Oncogene* 14:1725–1734, 1997.

89. Wu, Y., K. Cui, K. Miyoshi, L. Hennighausen, J.E. Green, J. Setser, D. LeRoith, S. Yakar. Reduced circulating insulin-like growth factor I levels delay the onset of chemically and genetically induced mammary tumors. *Cancer Res*. 63:4384–4388, 2003.

90. Heron-Milhavet, L., D. LeRoith. Insulin-like growth factor I induces MDM2-dependent degradation of p53 via the p38 MAPK pathway in response to DNA damage. *J. Biol. Chem*. 277:15600–15606, 2002.

91. Guan, Z., S.Y. Buckman, L.D. Baier, A.R. Morrison. IGF-I and insulin amplify IL-1 beta-induced nitric oxide and prostaglandin biosynthesis. *Am. J. Physiol*. 274:F673–F679, 1998.

92. Di Popolo, A., A. Memoli, A. Apicella, C. Tuccillo, A. di Palma, P. Ricchi, A.M. Acquaviva, R. Zarrilli. IGF-II/IGF-I receptor pathway up-regulates COX-2 mRNA expression and PGE2 synthesis in Caco-2 human colon carcinoma cells. *Oncogene* 19:5517–5524, 2000.

93. Wetterau, L.A., M.J. Francis, L. Ma, P. Cohen. Insulin-like growth factor I stimulates telomerase activity in prostate cancer cells. *J. Clin. Endocrinol. Metab*. 88:3354–3359, 2003.

94. del Rincon, S.V., C. Rousseau, R. Samanta, W.H. Miller, Jr. Retinoic acid-induced growth arrest of MCF-7 cells involves the selective regulation of the IRS-1/PI 3-kinase/AKT pathway. *Oncogene* 22:3353–3360, 2003.

95. Adamo, M.L., Z.M. Shao, F. Lanau, J.C. Chen, D.R. Clemmons, C.T. Roberts, Jr., D. LeRoith, J.A. Fontana. Insulin-like growth factor-I (IGF-I) and retinoic acid modulation of IGF-binding proteins (IGFBPs): IGFBP-2, -3, and -4 gene expression and protein secretion in a breast cancer cell line. *Endocrinology* 131:1858–1866, 1992.

96. Swisshelm, K., K. Ryan, K. Tsuchiya, R. Sager. Enhanced expression of an insulin growth factor-like binding protein (mac25) in senescent human mammary epithelial cells and induced expression with retinoic acid. *Proc. Natl. Acad. Sci. USA* 92:4472–4476, 1995.

97. Guvakova, M.A., E. Surmacz. Tamoxifen interferes with the insulin-like growth factor I receptor (IGF-IR) signaling pathway in breast cancer cells. *Cancer Res.* 57:2606–2610, 1997.

98. Rickard, S.E., Y.V. Yuan, L.U. Thompson. Plasma insulin-like growth factor I levels in rats are reduced by dietary supplementation of flaxseed or its lignan secoisolariciresinol diglycoside. *Cancer Lett.* 161:47–55, 2000.

99. Wang, S., V.L. DeGroff, S.K. Clinton. Tomato and soy polyphenols reduce insulin-like growth factor-I-stimulated rat prostate cancer cell proliferation and apoptotic resistance *in vitro* via inhibition of intracellular signaling pathways involving tyrosine kinase. *J. Nutr.* 133:2367–2376, 2003.

100. Klein, R.D., S.M. Fischer. Black tea polyphenols inhibit IGF-I-induced signaling through Akt in normal prostate epithelial cells and Du145 prostate carcinoma cells. *Carcinogenesis* 23:217–221, 2002.

101. Faure, E., N. Heisterkamp, J. Groffen, V. Kaartinen. Differential expression of TGF-beta isoforms during postlactational mammary gland involution. *Cell Tissue Res.* 300:89–95, 2000.

102. Nguyen, A.V., J.W. Pollard. Transforming growth factor beta3 induces cell death during the first stage of mammary gland involution. *Development* 127:3107–3118, 2000.

103. Barcellos-Hoff, M.H., T.A. Dix. Redox-mediated activation of latent transforming growth factor-beta 1. *Mol. Endocrinol.* 10:1077–1083, 1996.

104. Ewan, K.B., R.L. Henshall-Powell, S.A. Ravani, M.J. Pajares, C. Arteaga, R. Warters, R.J. Akhurst, M.H. Barcellos-Hoff. Transforming growth factor-beta1 mediates cellular response to DNA damage *in situ*. *Cancer Res.* 62:5627–5631, 2002.

105. Pierce Jr., D.F., A.E. Gorska, A. Chytil, K.S. Meise, D.L. Page, R.J. Coffey, Jr., H.L. Moses. Mammary tumor suppression by transforming growth factor beta 1 transgene expression. *Proc. Natl. Acad. Sci. USA* 92:4254–4258, 1995.

106. Gorska, A.E., R.A. Jensen, Y. Shyr, M.E. Aakre, N.A. Bhowmick, H.L. Moses. Transgenic mice expressing a dominant-negative mutant type II transforming growth factor-beta receptor exhibit impaired mammary development and enhanced mammary tumor formation. *Am. J. Pathol.* 163:1539–1549, 2003.

107. Herbert, B.S., B.G. Sanders, K. Kline. N-(4-hydroxyphenyl)retinamide activation of transforming growth factor-beta and induction of apoptosis in human breast cancer cells. *Nutr. Cancer* 34:121–132, 1999.

108. Kishi, H., E. Kuroda, H.K. Mishima, U. Yamashita. Role of TGF-beta in the retinoic acid-induced inhibition of proliferation and melanin synthesis in chick retinal pigment epithelial cells *in vitro*. *Cell Biol. Int.* 25:1125–1129, 2001.

109. Jakowlew, S.B., H. Zakowicz, T.W. Moody. Retinoic acid down-regulates VPAC(1) receptors and TGF-beta 3 but up-regulates TGF-beta 2 in lung cancer cells. *Peptides* 21:1831–1837, 2000.

110. Weikkolainen, K., J. Keski-Oja, K. Koli. Expression of latent TGF-beta binding protein LTBP-1 is hormonally regulated in normal and transformed human lung fibroblasts. *Growth Factors* 21:51–60, 2003.

111. MacCallum, J., J.C. Keen, J.M. Bartlett, A.M. Thompson, J.M. Dixon, W.R. Miller. Changes in expression of transforming growth factor beta mRNA isoforms in patients undergoing tamoxifen therapy. *Br. J. Cancer* 74:474–478, 1996.

112. Kopp, A., W. Jonat, M. Schmahl, C. Knabbe. Transforming growth factor beta 2 (TGF-beta 2) levels in plasma of patients with metastatic breast cancer treated with tamoxifen. *Cancer Res.* 55:4512–4515, 1995.

113. Harpel, J.G., S. Schultz-Cherry, J.E. Murphy-Ullrich, D.B. Rifkin. Tamoxifen and estrogen effects on TGF-beta formation: role of thrombospondin-1, alphavbeta3, and integrin-associated protein. *Biochem. Biophys. Res. Commun.* 284:11–14, 2001.

114. Malet, C., F. Fibleuil, C. Mestayer, I. Mowszowicz, F. Kuttenn. Estrogen and antiestrogen actions on transforming growth factorbeta (TGFbeta) in normal human breast epithelial (HBE) cells. *Mol. Cell Endocrinol.* 174:21–30, 2001.

115. Perry, R.R., Y. Kang, B.R. Greaves. Relationship between tamoxifen-induced transforming growth factor beta 1 expression, cytostasis and apoptosis in human breast cancer cells. *Br. J. Cancer* 72:1441–1446, 1995.

116. Yang, H., S. Kyo, M. Takatura, L. Sun. Autocrine transforming growth factor beta suppresses telomerase activity and transcription of human telomerase reverse transcriptase in human cancer cells. *Cell Growth Differ.* 12:119–127, 2001.

117. Ariazi, E.A., Y. Satomi, M.J. Ellis, J.D. Haag, W. Shi, C.A. Sattler, M.N. Gould. Activation of the transforming growth factor beta signaling pathway and induction of cytostasis and apoptosis in mammary carcinomas treated with the anticancer agent perillyl alcohol. *Cancer Res.* 59:1917–1928, 1999.

118. Lawton, L.N., M.F. Bonaldo, P.C. Jelenc, L. Qiu, S.A. Baumes, R.A. Marcelino, G.M. de Jesus, S. Wellington, J.A. Knowles, D. Warburton, S. Brown, M.B. Soares. Identification of a novel member of the TGF-beta superfamily highly expressed in human placenta. *Gene* 203:17–26, 1997.

119. Strelau, J., M. Bottner, P. Lingor, C. Suter-Crazzolara, D. Galter, J. Jaszai, A. Sullivan, A. Schober, K. Krieglstein, K. Unsicker. GDF-15/MIC-1: a novel member of the TGF-beta superfamily. *J. Neural Transm. Suppl.* 273–276, 2000.

120. Baek, S.J., L.C. Wilson, T.E. Eling. Resveratrol enhances the expression of non-steroidal anti-inflammatory drug-activated gene (NAG-1) by increasing the expression of p53. *Carcinogenesis* 23:425–434, 2002.

121. Bottone Jr., F.G., S.J. Baek, J.B. Nixon, T.E. Eling. Diallyl disulfide (DADS) induces the antitumorigenic NSAID-activated gene (NAG-1) by a p53-dependent mechanism in human colorectal HCT 116 cells. *J. Nutr.* 132:773–778, 2002.

122. Newman, D., M. Sakaue, J.S. Koo, K.S. Kim, S.J. Baek, T. Eling, A.M. Jetten. Differential regulation of nonsteroidal anti-inflammatory drug-activated gene in normal human tracheobronchial epithelial and lung carcinoma cells by retinoids. *Mol. Pharmacol.* 63:557–564, 2003.

123. Wilson, L.C., S.J. Baek, A. Call, T.E. Eling. Nonsteroidal anti-inflammatory drug-activated gene (NAG-1) is induced by genistein through the expression of p53 in colorectal cancer cells. *Int. J. Cancer* 105:747–753, 2003.

124. Masuda, M., M. Suzui, J.T. Lim, A. Deguchi, J.W. Soh, I.B. Weinstein. Epigallocatechin-3-gallate decreases VEGF production in head and neck and breast carcinoma cells by inhibiting EGFR-related pathways of signal transduction. *J. Exp. Ther. Oncol.* 2:350–359, 2002.

125. Sartippour, M.R., Z.M. Shao, D. Heber, P. Beatty, L. Zhang, C. Liu, L. Ellis, W. Liu, V.L. Go, M.N. Brooks. Green tea inhibits vascular endothelial growth factor (VEGF) induction in human breast cancer cells. *J. Nutr.* 132:2307–2311, 2002.

126. Tang, F.Y., N. Nguyen, M. Meydani. Green tea catechins inhibit VEGF-induced angiogenesis *in vitro* through suppression of VE-cadherin phosphorylation and inactivation of Akt molecule. *Int. J. Cancer* 106:871–878, 2003.

127. Chow, L.M., C.H. Chui, J.C. Tang, F.Y. Lau, M.Y. Yau, G.Y. Cheng, R.S. Wong, P.B. Lai, T.W. Leung, I.T. Teo, F. Cheung, D. Guo, A.S. Chan. Anti-angiogenic potential of *Gleditsia sinensis* fruit extract. *Int. J. Mol. Med.* 12:269–273, 2003.

128. Roy, S., S. Khanna, H.M. Alessio, J. Vider, D. Bagchi, M. Bagchi, C.K. Sen. Anti-angiogenic property of edible berries. *Free Radic. Res.* 36:1023–1031, 2002.

129. Dabrosin, C., J. Chen, L. Wang, L.U. Thompson. Flaxseed inhibits metastasis and decreases extracellular vascular endothelial growth factor in human breast cancer xenografts. *Cancer Lett.* 185:31–37, 2002.

130. Cuendet, M., J. Pezzuto. The role of cyclooxygenase and lipoxygenase in cancer chemoprevention. *Drug Metab. Drug Interact.* 17:109–157, 2000.

131. Natarajan, R., J. Nadler. Role of lipoxygenases in breast cancer. *Frontiers Biosci.* 3:E81–E88, 1998.

132. Hershman, H. Prostaglandin synthase 2. *Biochem. Biophys. Acta* 1299:125–140, 1996.

133. Liu, X., D. Rose. Differential expression and regulation of cyclooxygenase-1 and -2 in two human breast cancer cell lines. *Cancer Res.* 56:5125–5127, 1996.

134. Half, E., X. Tang, K. Gwyn, A. Sahin, K. Wathen, F. Sinicrope. Cyclooxygenase-2 expression in human breast cancers and adjacent ductal carcinoma *in situ*. *Cancer Res.* 62:1676–1681, 2002.

135. Shim, V., M. Gauthier, D. Sudilovsky, K. Mantei, K. Chew, D. Moore, I. Cha, T. Tlsty, L. Esserman. Cyclooxygenase-2 expression is related to nuclear grade in ductal carcinoma *in situ* and is increased in its normal adjacent epithelium. *Cancer Res.* 63:2347–2350, 2003.

136. Liu, C., S. Chang, K. Narko, O. Trifan, M. Wu, E. Smith, C. Haudenschild, T. Lane, T. Hla. Overexpression of cyclooxygenase-2 is sufficient to induce tumorigenesis in transgenic mice. *J. Biol. Chem.* 276:18563–18569, 2001.

137. Crawford, Y., M. Gauthier, A. Joubel, K. Mantei, K. Kozakiewicz, C. Afshari, T. Tlsty. Histologically normal human mammary epithelia with silenced p16(INK4a) overexpress COX-2, promoting a premalignant program. *Cancer Cell* 5:263–273, 2004.

138. Stockton, R., B. Jacobson. Modulation of cell-substrate adhesion by arachidonic acid: lipoxygenase regulates cell spreading and ERK1/2-inducible cyclooxygenase regulates cell migration in NIH-3T3 fibroblasts. *Mol. Biol. Cell* 12:1937–1956, 2001.

139. Roberts, L., H. Glenn, R. Whitfield, B. Jacobson. Regulation of cell-substrate adhesion by the lipoxygenase and cyclooxygenase branches of arachidonic acid metabolism. *Adv. Exp. Med. Biol.* 507:525–529, 2002.

140. Gately, S., R. Kerbel. Therapeutic potential of selective cyclooxygenase-2 inhibitors in the management of tumor angiogenesis. *Progress Experim. Tumor Res.* 37:179–192, 2003.

141. Chang, S., C. Liu, R. Conway, D. Han, K. Nithipatikom, O. Trifan, T. Lane, T. Hla. Role of prostaglandin E2-dependent angiogenic switch in cyclooxygenase-2-induced breast cancer progression. *Proc. Nat. Acad. Sci. USA* 101:591–596, 2004.

142. Vadlamudi, R., M. Mandal, L. Adam, B. Steinbach, J. Mendelsohn, R. Kumar. Regulation of cyclooxygenase-2 pathway by HER2 receptor. *Oncogene* 18:305–314, 1999.

143. Benoit, V., B. Relic, X. Leval, A. Chariot, M. Merville, V. Bours. Regulation of HER-2 oncogene expression by cyclooxygenase-2 and prostaglandin E2. *Oncogene* 23:1631–1635, 2004.

144. Sheng, H., J. Shao, J. Morrow, R. Beauchamp, R. DuBois. Modulation of apoptosis and Bcl-2 expression by prostaglandin E2 in human colon cancer cells. *Cancer Res.* 58:362–366, 1998.

145. Subbaramaiah, K., N. Altorki, W. Chung, J. Mestre, A. Sampat, A. Dannenberg. Inhibition of cyclooxygenase-2 gene expression by p53. *J. Biol. Chem.* 274:10911–10915, 1999.

146. Thun, M. Aspirin use and reduced risk of fatal colon cancer. *New Eng. J. Med.* 325:1593–1596, 1991.

147. DuBois, R. Aspirin and breast cancer chemoprevention. *J. Am. Med. Assoc.* 291:2488–2489, 2004.

148. Terry, M., M. Gammon, F. Zhang, H. Tawfik, S. Teitelbaum, J. Britton, K. Subbaramaiah, A. Dannenberg, A. Neugut. Association of frequency and duration of aspirin use and hormone receptor status with breast cancer risk. *J. Am. Med. Assoc.* 291:2433–2440, 2004.

149. Earashi, M., M. Noguchi, K. Kinoshita, M. Tanaka. Effects of eicosanoid synthesis inhibitors on the *in vitro* growth and prostaglandin E and leukotriene F secretion of a human breast cancer cell line. *Oncology* 52:150–155, 1995.

150. Cunningham, D., L. Harrison, T. Shultz. Proliferative responses of normal human mammary and MCF-7 breast cancer cells to linoleic acid, conjugated linoleic acid and eicosanoid synthesis inhibitors in culture. *Anticancer Res.* 17:197–203, 1997.

151. Michael, M., M. Badr, A. Badawi. Inhibition of cyclooxygenase-2 and activation of peroxisome proliferator-activated receptor-gamma synergistically induces apoptosis and inhibits growth of human breast cancer cells. *Int. J. Molecular Med.* 11:733–736, 2003.

152. Pyo, H., H. Choy, G. Amorino, J. Kim, Q. Cao, S. Hercules, R. DuBois. A selective cyclooxy-genase-2 inhibitor, NS-398, enhances the effect of radiation *in vitro* and *in vivo* preferentially on the cells that express cyclooxygenase-2. *Clin. Cancer Res.* 7:2998–3005, 2001.

153. Joe, A., H. Liu, M. Suzui, M. Vural, D. Xiao, I. Weinstein. Resveratrol induces growth inhibition, S-phase arrest, apoptosis, and changes in biomarker expression in several human cancer cell lines. *Clin. Cancer Res.* 8:893–903, 2002.

154. Jang, M., L. Cai, G. Udeani, K. Slowing, C. Thomas, C. Beecher, H. Fong, N. Farnsworth, A. Kinghorn, R. Mehta, R. Moon, J. Pezzuto. Cancer chemopreventative activity of resve-ratrol, a natural product derived from grapes. *Science* 275:218–220, 1997.

155. Subbaramaiah, K., W. Chung, P. Michaluart, N. Telang, T. Tanabe, J. Inoue, M. Jang, J. Pezzuto, A. Dannenberg. Resveratrol inhibits cyclooxygenase-2 transcription and activity in phorbol ester-treated human mammary epithelial cells. *J. Biol. Chem.* 273:21875–21882, 1998.

156. Banerjee, S., C. Bueso-Ramos, B. Aggarwal. Suppression of 7,12-dimethylbenz(a)anthra cene-induced mammary carcinogenesis in rats by resveratrol: role of nuclear factor-KB, cyclooxygenase-2, and matrix metalloprotease-9. *Cancer Res.* 62:2002.

157. Kang, G., P. Kong, Y. Yuh, S. Lim, W. Chun, S. Kim. Curcumin suppresses lipopolysaccha-ride-induced cyclooxygenase-2 expression by activator protein 1 and nuclear factor kappaB bindings in BV2 microglial cells. *J. Pharmacol. Sci.* 94:325–328, 2004.

158. Chun, K., Y. Keum, S. Han, Y. Song, S. Kim, Y. Surh. Curcumin inhibits phorbol ester-induced expression of cyclooxygenase-2 in mouse skin through suppression of extracellular signal-regulated kinase activity and NF-kappaB activation. *Carcinogenesis* 24:1515–1524, 2003.

159. Zhang, F., N. Altorki, J. Mestre, K. Subbaramaiah, A. Dannenberg. Curcumin inhibits cyclooxygenase-2 transcription in bile acid-and phorbol ester-treated human gastrointestinal epithelial cells. *Carcinogenesis* 20:445–451, 1999.

160. Ye, F., L. Xui, J. Yi, W. Zhang, D. Zhang. Anticancer activity of *Scutellaria baicalensis* and its potential mechanism. *J. Alt. Compl. Med.* 8:567–572, 2002.

161. Liang, Y., Y. Huang, S. Tsai, S. Lin-Shiau, C. Chen, J. Lin. Suppression of inducible cyclo-oxygenase and inducible nitric oxide synthase by apigenin and related flavonoids in mouse macrophages. *Carcinogenesis* 20:1945–1952, 1999.

162. Chen, G.G., H. Xu, J.F. Lee, M. Subramaniam, K.L. Leung, S.H. Wang, U.P. Chan, T.C. Spelsberg. 15-hydroxy-eicosatetraenoic acid arrests growth of colorectal cancer cells via a peroxisome proliferator-activated receptor gamma-dependent pathway. *Int. J. Cancer* 107:837–843, 2003.

163. Nie, D., M. Che, D. Grignon, K. Tang, K. Honn. Role of eicosanoids in prostate cancer progression. *Cancer Metastasis Rev.* 20:195–206, 2001.

164. Nixon, J., K. Kim, P. Lamb, F. Bottone, T. Eling. 15-lipoxygenase-1 has anti-tumorigenic effects in colorectal cancer. *Prostaglandins Leukotrienes Essential Fatty Acids* 70:7–15, 2004.

165. Sadik, C., H. Sies, T. Schewe. Inhibition of 15-lipoxygenases by flavonoids: structure-activ-ity relations and mode of action. *Biochem. Pharmacol.* 65:773–781, 2003.

166. Ghosh, J., C. Myers. Arachidonic acid stimulates prostate cancer cell growth: critical role of 5-lipoxygenase. *Biochem. Biophys. Res. Commun.* 235:418–423, 1997.

167. Tong, W., X. Ding, T. Adrian. The mechanisms of lipoxygenase inhibitor-induced apoptosis in human breast cancer cells. *Biochem. Biophys. Res. Commun.* 296:942–948, 2002.

168. Ding, X., W. Tong, T. Adrian. Multiple signal pathways are involved in the mitogenic effect of 5(S)-HETE in human prostate cancer. *Oncology* 65f:285–294, 2003.

169. Njamen, D., J. Mbafor, Z. Fomum, A. Kamanyi, J. Mbanya, M. Recio, R. Giner, S. Manez, J. Rios. Anti-inflammatory activities of two flavanones, sigmoidin A and sigmoidin B, from *Erythrina sigmoidea*. *Planta Medica* 70:104–107, 2004.

170. Kohyama, N., T. Nagata, S. Fujimoto, K. Sekiya. Inhibition of arachidonate lipoxygenase activities by 2-(3,4-dihydroxyphenyl)ethanol, a phenolic compound from olives. *Biosci. Biotechnol. Biochem.* 61:347–350, 1997.

171. Albert, D., I. Zundorf, T. Dingermann, W. Muller, D. Steinhilber, O. Werz. Hyperforin is a dual inhibitor of cyclooxygenase-1 and 5-lipoxygenase. *Biochem. Pharmacol.* 64:1767–1765, 2002.

172. Po, L., Z. Chen, D. Tsang, L. Leung. Baicalein and genistein display differential actions on estrogen receptor (ER) transactivation and apoptosis in MCF-7 cells. *Cancer Lett.* 187:33–40, 2002.

173. Dai, Q., A.A. Franke, F. Jin, X.O. Shu, J.R. Hebert, L.J. Custer, J. Cheng, Y.T. Gao, W. Zheng. Urinary excretion of phytoestrogens and risk of breast cancer among Chinese women in Shanghai. *Cancer Epidemiol. Biomarkers Prev.* 11:815–821, 2002.

174. Adlercruetz, H., B. Goldin, S. Gorbach. Soy phytoestrogen intake and cancer risk. *J. Nutr.* 125:757S–770S, 1995.

175. Wiseman, H. *Phytochemicals: Epidemiological factors.* London: Academic Press, 1998, pp 1549–1561.

176. Darbon, J.M., M. Penary, N. Escalas, F. Casagrande, F. Goubin-Gramatica, C. Baudouin, B. Ducommun. Distinct Chk2 activation pathways are triggered by genistein and DNA-damaging agents in human melanoma cells. *J. Biol. Chem.* 275:15363–15369, 2000.

177. C.A. Lamartiniere. Protection against breast cancer with genistein: a component of soy. *Am. J. Clin. Nutr.* 71:1705S–1707S; *discussion* 1708S–1709S, 2000.

178. Tanos, V., A. Brzezinski, O. Drize, N. Strauss, T. Peretz. Synergistic inhibitory effects of genistein and tamoxifen on human dysplastic and malignant epithelial breast cells *in vitro.* *Eur. J. Obstet. Gynecol. Reprod. Biol.* 102:188–194, 2002.

179. Ohta, T., S. Nakatsugi, K. Watanabe, T. Kawamori, F. Ishikawa, M. Morotomi, S. Sugie, T. Toda, T. Sugimura, K. Wakabayashi. Inhibitory effects of Bifidobacterium-fermented soy milk on 2-amino-1-methyl-6-phenylimidazo[4,5-b]pyridine-induced rat mammary carcinogenesis, with a partial contribution of its component isoflavones. *Carcinogenesis* 21:937–941, 2000.

180. Shetty, K. Biotechnology to harness the benefits of dietary phenolics; focus on *Lamiaceae.* *Asia Pac. J. Clin. Nutr.* 6:162–171, 1997.

181. Kelm, M.A., M.G. Nair, G.M. Strasburg, D.L. DeWitt. Antioxidant and cyclooxygenase inhibitory phenolic compounds from *Ocimum sanctum* Linn. *Phytomedicine* 7:7–13, 2000.

182. Engleberger, W., U. Hadding, E. Etschenberg, E. Graf, S. Leyck, J. Winkelmann, M. Parnham. Rosmarinic acid: a new inhibitor of complement C3-convertase with anti-inflammatory activity. *Int. J. Immunopharmacol.* 10:729–737, 1988.

183. Peake, P., B. Pussel, P. Martyn, V. Timmermans, J. Charlesworth. The inhibitory effect of rosmarinic acid on complement involves the C5 convertase. *Int. J. Immunopharmacol.* 13:853–857, 1991.

184. Osakabe, N., A. Yasuda, M. Natsume, C. Sanbongi, Y. Kato, T. Osawa, T. Yoshikawa. Rosmarinic acid, a major polyphenolic component of Perilla frutescens, reduces lipopoly-saccharide (LPS)-induced liver injury in D-galactosamine (D-GalN)-sensitized mice. *Free Radic. Biol. Med.* 33:798-806, 2002.

185. Al–Amier, H., B.M. Mansour, N. Toaima, L. Craker, K. Shetty. Tissue culture for phenolics and rosmarinic acid in thyme. *J. Herbs Spices Med. Plants* 8:31–42, 2001.

186. Al-Amier, H., B.M. Mansour, N. Toaima, R.A. Korus, K. Shetty. Screening of high biomass and phenolic-producing clonal lines of spearmint in tissue culture using *Pseudomonas* and azetidine-2-carboxylate. *Food Biotechnol.* 13: 227–253, 1999.

187. Al-Amier, H., B.M. Mansour, N. Toaima, R.A. Korus, K. Shetty. Tissue culture based screening for selection of high biomass and phenolic producing clonal lines of lavender using pseudomonas and azetidine-2-carboxylate. *J. Agric. Food Chem.* 47:2937–2943, 1999.

188. Shetty, K. Role of proline-linked pentose phosphate pathway in biosynthesis of plant phenolics for functional food and environmental applications: a review. *Process Biochem.* 39:789–804, 2004.

189. Eguchi, Y., O. Curtis, K. Shetty. Interaction of hyperhydricity-preventing *Pseudomonas* spp. with oregano (*Origanum vulgare*) and selection of high rosmarinic acid-producing clones. *Food Biotechnol.* 10:191–202, 1996.

190. Yang, R., O. Curtis, K. Shetty. Selection of high rosmarinic acid-producing clonal lines of rosemary (*Rosmarinus officinalis*) via tissue culture using *Pseudomonas* sp. *Food Biotechnol.* 11:73–88, 1997.

191. Shetty, K., T. Carpenter, O. Curtis, T. Potter. Selection of high phenolics-containing clones of thyme (*Thymus vulgaris* L.) using *Pseudomonas* spp. *J. Agric. Food Chem.* 44:3408–3411, 1996.

192. McCue, P., K. Shetty. A hypothetical model for the action of soybean isoflavonoids against cancer involving a shift to proline-linked energy metabolism through activation of the pentose phosphate pathway. *Food Biotechnol.* 18:19–37, 2004.

193. Duval, B., K. Shetty. The stimulation of phenolics and antioxidant activity in pea (*Pisum sativum*) elicited by genetically transformed anise root extract. *J. Food Biochem.* 25:361–377, 2001.

194. Randhir, R., D. Vattem, K. Shetty. Solid-state bioconversion of fava bean by *Rhizopus oligosporus* for enrichment of phenolic antioxidants and L-DOPA. *Innovative Food Sci. Emerg. Technol.* 5:235–244, 2004.

195. McCue, P., K. Shetty. A biochemical analysis of mungbean (*Vigna radiata*) response to microbial polysaccharides and potential phenolic-enhancing effects for nutraceutical applications. *Food Biotechnol.* 16:57–79, 2002.

196. Randhir, R., Y.T. Lin, K. Shetty. Phenolics, their antioxidant and antimicrobial activity in dark germinated fenugreek sprouts in response to peptide and phytochemical elicitors. *Asia Pac. J. Clin. Nutr.* 13:295–307, 2004.

197. Kwok, D., K. Shetty. Effect of proline and proline analogs on total phenolic and rosmarinic acid levels in shoot clones of thyme (*Thymus vulgaris* L.). *J. Food Biochem.* 22:37–51, 1998.

198. Yang, R., K. Shetty. Stimulation of rosmarinic acid in shoot cultures of oregano (*Origanum vulgare*) clonal line in response to proline, proline analog, and proline precursors. *J. Agric. Food Chem.* 46:2888–2893, 1998.

199. McCue, P., Z. Zheng, J. Pinkham, K. Shetty. A model for enhancing pea seedling vigor following low pH and salicylic acid treatments. *Process Biochem.* 35:603–613, 2000.

200. Andarwulan, N., K. Shetty. Improvement of pea (*Pisum sativum*) seed vigor by fish protein hydrolysates in combination with acetyl salicylic acid. *Process Biochem.* 35:159–165, 1999.

201. McCue, P., K. Shetty. Clonal herbal extracts as elicitors of phenolic synthesis in dark-germinated mungbeans for improving nutritional value with implications for food safety. *J. Food Biochem.* 26:209–232, 2002.

202. Randhir, R., Y. Lin, K. Shetty. Stimulation of phenolics, antioxidant and antimicrobial activities in dark-germinated mung bean (*Vigna radiata*) sprouts in response to peptide and phytochemical elicitors. *Process Biochem.* 39:637–646, 2004.

203. Shetty, P., M. Atallah, K. Shetty. Enhancement of total phenolic, L-DOPA and proline contents in germinating fava bean (*Vicia faba*) in response to bacterial elicitors. *Food Biotechnol.* 15:47–67, 2001.

204. Shetty, P., M. Atallah, K. Shetty. Effects of UV treatment on the proline-linked pentose phosphate pathway for phenolics and L-DOPA synthesis in dark-germinated *Vicia faba*. *Process Biochem.* 37:1285–1295, 2002.

205. Shetty, P., M. Atallah, K. Shetty. Stimulation of total phenolics, L-DOPA and antioxidant activity through proline-linked pentose phosphate pathway in response to proline and its analog in germinating fava beans (*Vicia faba*). *Process Biochem.* 38:1707–1717, 2003.

206. Zheng, Z., K. Shetty. *Solid-State Fermentation and Value-Added Utilization of Fruit and Vegetable Processing By-Products.* New York: Wiley Publishers, 1999, pp 2165–2174.

207. McCue, P., K. Shetty. Health benefits of soy isoflavonoids and strategies for enhancement: a review. *Crit. Rev. Food Sci. Nutr.* 44:1–7, 2004.

208. Esaki, H., R. Watanabe, H. Onozaki, S. Kawakishi, T. Osawa. Formation mechanism for potent antioxidative o-dihydroxyisoflavones in soybeans fermented with *Aspergillus saitoi*. *Biosci. Biotechnol. Biochem.* 63:851–858, 1999.

209. Kishida, T., H. Ataki, M. Takebe, K. Ebihara. Soybean meal fermented by *Aspergillus awamori* increases the cytochrome P-450 content of the liver microsomes of mice. *J. Agric. Food Chem.* 48:1367–1372, 2000.

210. McCue, D., A. Horii, K. Shetty. Solid-state bioconversion of phenolic antioxidants from defatted powdered soybean fermented with *Rhizopus oligosprous*: role of carbohydrate-cleaving enzymes. *J. Food Biochem.* 27:501–514, 2003.

211. McCue, P., K. Shetty. Role of carbohydrate-cleaving enzymes in phenolic antioxidant mobilization from whole soybean fermented with *Rhizopus oligosprorus*. *Food Biotechnol.* 17:27–37, 2003.

212. Randhir, R., K. Shetty. Microwave-induced stimulation of L-DOPA, phenolics, and antioxidant activity in fava bean (*Vicia faba*) for Parkinson's diet. *Process Biochem.* 39:1775–1784, 2004.

213. Vattem, D., K. Shetty. Solid-state production of phenolic antioxidants from cranberry pomace by *Rhizopus oligosprous*. *Food Biotechnol.* 16:189–210, 2002.

214. Vattem, D., K. Shetty. Ellagic acid production and phenolic antioxidant activity in cranberry pomace mediated by *Lentinus edodes* using solid-state system. *Process Biochem.* 39:367–379, 2003.

215. Zheng, Z., K. Shetty. Solid-state bioconversion of phenolics from cranberry pomace and role of *Lentinus edodes* beta-glucosidase. *J. Agric. Food Chem.* 48:895–900, 2000.

216. McCue, P., A. Horii, K. Shetty. Mobilization of phenolic antioxidants from defatted soybean powders by *Lentinus edodes* during solid-state fermentation is associated with production of laccase. *Innovative Food Sci. Emerg. Technol.* 5:385–392, 2004.

217. McCue, P., Y. Lin, R. Labbe, K. Shetty. Sprouting and solid-state bioprocessing of *Rhizopus oligosporus* increase the *in vitro* antibacterial activity of aqueous soybean extacts against *Helicobacter pylori*. *Food Biotechnol.* 18:229–249, 2004.

218. Vattem, D., Y. Lin, R. Labbe, K. Shetty. Antimicrobial activity against select food-borne pathogens by phenolic antioxidants enriched cranberry pomace by solid-state bioprocessing using food-grade fungus *Rhizopus oligosporus*. *Process Biochem.* 39:1939–1946, 2004.

219. Vattem, D., Y. Lin, K. Shetty. Enrichment of phenolic antioxidants and anti-*Helicobacter pylori* properties of cranberry pomace by solid-state bioprocessing. *Food Biotechnol.* 19:51–68, 2005.

220. Shetty, K., M.L. Wahlqvist. A model for the role of the proline-linked pentose-phosphate pathway in phenolic phytochemical bio-synthesis and mechanism of action for human health and environmental applications. *Asia Pac. J. Clin. Nutr.* 13:1–24, 2004.

221. Izumi, T., M.K. Piskula, S. Osawa, A. Obata, K. Tobe, M. Saito, S. Kataoka, Y. Kubota, M. Kikuchi. Soy isoflavone aglycones are absorbed faster and in higher amounts than their glucosides in humans. *J. Nutr.* 130:1695–1699, 2000.

222. Rao, M., G. Muralikrishna. Evaluation of the antioxidant properties of free and bound phenolic acids from native and malted finger millet (Ragi, *Eleusine coracana* Indaf-15). *J. Agric. Food Chem.* 50:889–892, 2002.

223. Setchell, K.D., N.M. Brown, L. Zimmer-Nechemias, W.T. Brashear, B.E. Wolfe, A.S. Kirschner, J.E. Heubi. Evidence for lack of absorption of soy isoflavone glycosides in humans, supporting the crucial role of intestinal metabolism for bioavailability. *Am. J. Clin. Nutr.* 76:447–453, 2002.

224. Yuan, L., C. Wagatsuma, M. Yoshida, T. Miura, T. Mukoda, H. Fujii, B. Sun, J.H. Kim, Y.J. Surh. Inhibition of human breast cancer growth by GCP (genistein combined polysaccharide) in xenogeneic athymic mice: involvement of genistein biotransformation by beta-glucuronidase from tumor tissues. *Mutat. Res.* 523–524:55–62, 2003.

225. McCue, P., K. Shetty. Phenolic antioxidant mobilization during yogurt production from soymilk. *Process Biochem.* 40:1791–1797, 2005.

12

Phenolic Antimicrobials from Plants for Control of Bacterial Pathogens

Kalidas Shetty and Yuan-Tong Lin

CONTENTS

12.1 Phenolic Antimicrobials from Plants 286
12.2 Potential of *Lamiaceae* as Source of Phenolic Antimicrobials............................288
    12.2.1 Control of *Staphylococcus Aureus* by Phenolic Phytochemicals
        as a Model for Plant Based Antimicrobials 288
    12.2.2 Control of *Listeria Monocytogenes* by Oregano Clonal Extracts
        as a Model for Developing Herb Phenolics as Antimicrobials..................289
12.3 Filling Gaps in Antimicrobial Strategy with Tissue Culture Based Screening
    of Elite Phenolic Phytochemical-Producing Lines 291
    12.3.1 Novel Model for Mechanism of Action of Phenolic Phytochemicals.......292
    12.3.2 Control of *Helicobacter Pylori* by Phenolic Antimicrobials
        from Plants 296
    12.3.3 Current Treatment Strategies for *H. Pylori* and Its Limitations 297
    12.3.4 Dietary Phenolic Phytochemical Strategies for Control of *H. Pylori* 298
12.4 Dietary Soybean Approach for Control of *H. Pylori*...298
    12.4.1 Sprouting of Soybean and Phenolic Antioxidant Stimulation Methods........299
    12.4.2 Phenolic Response and Characterization of Soybean Extracts
        Inhibiting *H. Pylori* Urease...301
    12.4.3 Solid-State Bioprocessing of Soybean Using *Rhizopus Oligosporus*........301
12.5 Recent Progress and Other Strategies for Designing Phenolics
    Phytochemicals as Antimicrobials 301
12.6 Summary 302
References..302

12.1 PHENOLIC ANTIMICROBIALS FROM PLANTS

Plants are excellent sources of phenolic metabolites. In particular, phenolic antioxidants from food grade plants have potential for long term chemo preventive and therapeutic applications against oxidation linked diseases (1–4) and increasingly have antimicrobial potential (2). Such phenolic metabolites are also broadly called phytochemicals. Phenolic phytochemicals have shown to be associated with antioxidative action in biological systems, acting as scavengers of singlet oxygen and free radicals (5,6). Much of this work has emphasized the role of phenolics from higher plants in relation to human health and in particular cancer and cardiovascular health (3,7,8). Phenolic phytochemicals (phenylpropanoids) serve as effective antioxidants (phenolic antioxidants) due to their ability to donate hydrogen from hydroxyl groups positioned along the aromatic ring to terminate free radical oxidation of lipids and other bio molecules (9). Phenolic antioxidants therefore short circuit a destructive chain reaction that ultimately degrades cellular membranes and in prokaryotes such phenolic antioxidants have potential for antimicrobial activity.

Plant phenolic extracts that impart flavor and aroma also have potential for inhibiting pathogenic microorganisms (10–12). These phenolic secondary metabolites are defensive antimicrobials produced against invading pathogens and stress and therefore the methods for exploiting them have to take this into account. In certain cases the induction is associated with action of diphenolic oxidases, and resulting modified compounds can have antimicrobial activity (13). In other situations dihydroxy phenolics are oxidized to quinones, which can interact with the proteins of the invading pathogens, forming melanoid polymers (13). Quinones are aromatic rings with 2 ketone substitutions and are highly reactive (10). These compounds are responsible for the enzymatic browning reaction of cut fruits and vegetables, and are an intermediate in melanin pigment production in humans (10). The quinones are sources of stable free radicals and complex irreversibly with nucleophilic amino acids and proteins (14) leading to inactivation of protein and loss of function (14). Therefore potential antimicrobial benefits of quinones are substantial (14). The potential targets for inhibition in the bacterial cells are surface adhesions; cell wall polypeptides and membrane bound enzymes (14).

In many herb and spice species investigated so far phenolic derivatives found in essential oils have been largely investigated for antimicrobial activity. Hydrophobic phenolics derived from the terpenoid pathway, thymol, and carvacrol, present in essential oil of thyme, oregano, savory, sage, and related species, have antimicrobial activity (11). The essential oil containing thymol can inhibit *Vibrio parahaemolyticus* (15). The addition of 0.05% of alcoholic extracts of thyme can inhibit the growth of *Staphylococcus aureus (16)*. Sage extract was inhibitory to *Bacillus cereus* and *S. aureus* (17). Rosemary extract of 0.1% substantially inhibited the growth of *S. aureus* and *Salmonella typhimurium* (18). Among other types of phenolics, hydroxycinnamic acid derivatives such as caffeic acid, ferulic acid and p-coumaric acid derived from the phenylpropanoid pathway inhibit *E. coli, S. aureus* and *B. cereus* (19). Polymeric phenolics, such as tannins are inhibitory toward *Listeria monocytogenes, E. coli, S. aureus, Aeromonas hydrophila,* and *Streptococcus faecalis* (20). Hydroxylated phenols, such as catechol and pyrogallol, are known to be toxic to microorganisms (10). The site and number of hydroxyl groups is linked to the antimicrobial effect and, in some cases, more oxidized forms are more inhibitory (21,22). In other combination studies, growth of *E. coli, Salmonella spp., L. monocytogenes,* and *S. aureus* were inhibited by oregano essential oils in broth cultures and its effect in emulsion systems depended on pH, temperature, and other environmental factors (23,24). In meat studies essential oil of oregano inhibited *L. monocytogenes* and *S. typhimurium*

(25,26). Other diverse examples of antimicrobial effects of herbs and spices are documented well in Tassou et al. (27).

The mode of action of phenolics against bacterial pathogens has not been defined or clearly understood. It is suspected changes in membrane permeability through lipid compatible hydrophobic phenolics, like thymol from certain essential oils, and membrane localized hyperacidity from water and ethanol soluble phenolics like benzoic acid and rosmarinic acid, may affect proton motive force across the membrane and therefore energy depletion may take place (11,28,29). It is suggested as a result of damage to structural and functional properties of membranes, pH gradient and electrical potential of the proton motive force is disrupted (30,31). The disruption of the proton motive force and reduction of ATP pool lead to cell death (11,31,32). Further, leakage of ions, nucleic acid, and amino acids can occur (27,33–35). It is also proposed that enzyme inhibition by the oxidized compounds through reaction with sulfhydryl groups, or through nonspecific interactions with membrane proteins, may also be the reason for inhibition (10,36). Other proposed mechanisms include formation of Schiff's bases with membrane proteins by aldehyde groups of phytochemicals, which prevent cell wall biosynthesis (37,38), and interaction of ferrous iron with phenolic compounds, which can damage membranes by enhancing oxidative stress (39).

In spite of wide investigations around the world, using many diverse botanical sources, the results of antimicrobial efficacy of phytochemicals are mixed, to very poor, and mechanism of action is poorly understood. The three major reasons for this are:

1. Many phytochemicals have been derived from mixed heterogeneous genetic sources and consistency cannot be guaranteed from batch to batch and from source to source (1,2,40).
2. Many of the antimicrobial phytochemicals, and especially phenolics, are biotic or abiotic stress inducible and, if extracts are used from uninduced states, the efficacy is potentially low.
3. Investigations have largely focused on hydrophobic essential oils and focused solely on disruption of plasma membrane related functions.

Now, with the emergence of antibiotic resistance from overuse of single antibiotics from fungal and bacterial sources, new strategies using plant based sources are promising (10) and urgently needed (2,40). In line with this need, one of our major strategies focus on the hypothesis that high antioxidant phenolics from single seed origin clonal lines of herbs such as oregano, thyme, rosemary, and lavender, and genetically uniform sprouted legumes such as soybean, fava bean, fenugreek, chickpea, pea, and mung bean, which are self pollinating species would have excellent antimicrobial potential. Genetically uniform, high antioxidant phenolic profiles have been screened, evaluated for antioxidant efficacy and targeted to inhibit bacterial pathogens (41,42). In this screening strategy, both the concept of phytochemical consistency through clonal lines in herbs and stress (elicitor) based inducibility of phenolic antioxidants in self pollinating and genetically uniform legumes can be developed to explore the antimicrobial potential against bacterial pathogens. Further, we have targeted only water and ethanol extracted phenolic profiles from these systems that could work effectively and synergistically to inhibit biochemical targets at the cellular membrane and cytosol of the bacterial pathogen at the same time, thereby impacting multiple targets and offering fewer opportunities for resistance by the pathogen.

12.2 POTENTIAL OF *LAMIACEAE* AS SOURCE OF PHENOLIC ANTIMICROBIALS

Extracts from dietary herb species belonging to the family *Lamiaceae* (mint family) have been used by humanity as sources of complimentary medicine and food preservatives for over 4000 years. It is only recently that some of the bioactive components linked to medicinal and preservative function have been determined to be phenolic metabolites (4,43,44). Specific phenolic metabolites from Lamiaceae like rosmarinic acid (from rosemary, spearmint, thyme, lavender, and oregano) and thymol (from thyme and oregano) have anti inflammatory (4,45), antioxidant and antimicrobial properties (11,46–48), respectively. The antioxidant pharmacological functions from food grade dietary phytochemicals from various mint family species could contribute to long term prevention of oxidation influenced disease such as diabetes, cancer, CVD, cognition and inflammatory diseases (1,3,8). In general food processing applications oregano, rosemary, thyme, and lavender has been used widely as a source of flavors and potential food antioxidants (49–51). These herb species are native to the Mediterranean and are widely used in South European cuisine. The beneficial phenolic antioxidants that are being targeted for food and health applications in our research are biphenyl, rosmarinic acid that also has antioxidant, anti inflammatory, and antimicrobial properties (4,45,47), and several other simple soluble phenolic metabolites have antioxidant properties (49,51). The beneficial antioxidant activity from phenolic metabolites was also observed in many food preservation conditions such as in lard (50), salad dressing (52) and other food model systems. The phenolic antioxidants from these herb species also have potential antimicrobial properties (2,11,27,40,47,48) and, being partially ethanol and water soluble, could be effective at inhibiting targets on bacterial cell membrane as well as key dehydrogenase linked metabolic reactions in the cytosol, as opposed to being effective only in the hydrophobic regions as in case of wide variety of plant essential oils. These water and ethanol extracted phenolics are also less volatile and have fewer flavor related problems compared to volatile essential oils targeted to cell membranes. The water and ethanol extractable phenolics are also more compatible in a more diverse array of foods, are less volatile, and have fewer flavor problems.

12.2.1 Control of *Staphylococcus Aureus* by Phenolic Phytochemicals as a Model for Plant Based Antimicrobials

Staphylococcus aureus is an important pathogen in humans that has significant impact in the food chain (11) and high density communities and hospitals (53). The pathogen causes serious infections in open wounds, vascular tissues, bones, and joints (53,54). The importance of *S. aureus* as the etiological agent in catheter related and peritoneal dialysis patients is well known and often requires aggressive antibiotic treatment (55,56). Further it is well known that as a result of antibiotic resistance, it has been difficult to control infections ranging from abscesses, pneumonia, endocarditis, septicemia, and toxic shock syndrome. The one last line of defense is the antibiotic vancomycin (57–59). *Staphylococcus aureus* infections leading to complications from septic and toxic shock syndromes may result in organ failure (60,61). The existence of ecologically abundant hyper virulent clones has suggested that factors promoting ecological fitness of *S. aureus* also increase its virulence (62). The prospects of untreatable *S. aureus* infections have raised the need to search for alternative therapies, and development of a broadly protective vaccine based on an *in vivo* expressed antigen has some merit (63). In addition, consistent clonal profiles of plant phenolic metabolites from food grade herbs and inducible phenolics from legume sprouts can be an important complimentary therapy which can be used in conjunction with other therapies, including new microbial based antibiotics and vaccines.

Studies have indicated the antimicrobial potential of plant phenolic metabolites against *S. aureus*. The addition of 0.05% alcoholic extract of thyme can inhibit the growth of *Staphylococcus aureus* (16). Sage extract was also inhibitory to *S. aureus* (17). Rosemary extract of 0.1% substantially inhibited the growth of *S. aureus* and *Salmonella typhimurium* (18). Hydroxycinnamic acid derivatives such as caffeic acid, ferulic acid, and p-coumaric acid inhibited *S. aureus* (19). Polymeric phenolics, such as tannins, are inhibitory toward *S. aureus* (20). The hypothesis that elicitor inducible phenolic antioxidants from clonal herbs and legumes can also inhibit *S. aureus* holds much promise for improved efficacy and is being currently investigated. We also speculate that plant based phenolic antioxidants of consistent clonal origin have the potential to support human antioxidant enzyme response and therefore reduce human tissue damage from *S. aureus* toxin induced apoptosis (61) and improve host antimicrobial immune response and these investigations on mammalian cell responses are part of our future investigations.

12.2.2 Control of *Listeria Monocytogenes* by Oregano Clonal Extracts as a Model for Developing Herb Phenolics as Antimicrobials

Listeria monocytogenes is emerging to be a major food borne pathogen in post processing environments in meats and dairy foods. One of the many plants studied extensively for antimicrobial activity against *Listeria monocytogenes* is oregano (*Origanum vulgare*) (11,25,26,41,64). Oregano belongs to the family *Lamiaceae*, which also includes thyme and rosemary. Many studies indicated the antimicrobial activity of oregano and two of its major components, carvacrol and thymol, along with enhanced synergy with high rosmarinic acid extract (41). Kim et al. (65) reported that carvacrol was the most antibacterial phenolic compound they tested against five food borne pathogens (*Escherichia coli, E. coli* 0157:H7, *Salmonella typhimurium, L. monocytogenes,* and *Vibrio vulnificus*) using paper disk assays. Carvacrol was shown to have a larger zone of inhibition than did the other phenolic compounds at the same concentrations (citral, geraniol, terpineol, perillaldehyde, eugenol, linalool, and citronellal). Other studies have shown that nisin, when used in combination with carvacrol, inhibited the growth of *Bacillus cereus* and *L. monocytogenes* (66). One of the results found in this study was that at 20°C, growth of *L. monocytogenes* was completely inhibited in the presence of a concentration of nisin that was 16-fold lower (when used in combination with carvacrol) than when nisin was applied as the sole preservative. Thus, using carvacrol in combination with nisin did increase the effectiveness of the antimicrobial and allowed it to be used at much lower concentrations than it was used previously. Hurdle technology was also used in a study done by Karatzas et al. (67), which showed the combinations of carvacrol and high hydrostatic pressure inhibited *L. monocytogenes*. Other studies have also indicated that phenolic metabolites from oregano and other related herbs can be effective against *Listeria monocytogenes* (11,41).

Therefore the use of herbs and spices as a potential preservative for the food industry does have prospects for improving hurdle technology for controlling food borne pathogens. As discussed previously, due to their natural cross pollination nature, these herbs and spices are genetically heterogeneous (1,2). This heterogeneity results in a high degree of variability in the levels of phenolics in these plants (68). Furthermore, problems associated with climate and diseases can lead to even more quality variation in the plants. This produces many serious problems for the routine use of these plants as a hurdle to inhibit the growth of food borne pathogens in products developed by the food industry. Therefore tissue culture based clonal propagation was developed to develop consistent oregano extracts with antimicrobial activity against *Listeria monocytogenes* (41). Clonal propagation also allows us to obtain more detailed analysis of biosynthetic pathways of primary and secondary metabolites from uniform genetic material. This also allows us to study the effects of manipulating the pathways, such as adding specific precursors of a pathway or manipulating or deleting key enzymes (1,2).

In order to screen for high phenolic stimulated clonal lines with high antimicrobial activity, microbial elicitors, such as *Pseudomonas* sp., have been employed (69–72) (Figure 12.1). The rationale for this screening approach is that phenolic metabolites are stimulated in response to microbial elicitors (73) and, therefore, any clonal line tolerant to *Pseudomonas* must be a phenolic overexpressing phenotype (72). By using these techniques, it is now possible to obtain a reliable and genetically uniform plant source with excellent functionality for antioxidant and antimicrobial applications that can be used in the food industry. Using this approach we have shown that an elite clonal line of oregano was superior in inhibiting *Listeria monocytogenes* compared to commercial herbs extracts or individual terpenoid pathway derived phenolics normally found in essential oils of oregano (41). We are working further on obtaining *Listeria* specific inducible phenolics from specific clonal lines that could be more effective. In order to develop such *Listeria* specific herb clonal lines we have developed an innovative two layered plate assay to screen specific clonal lines that target

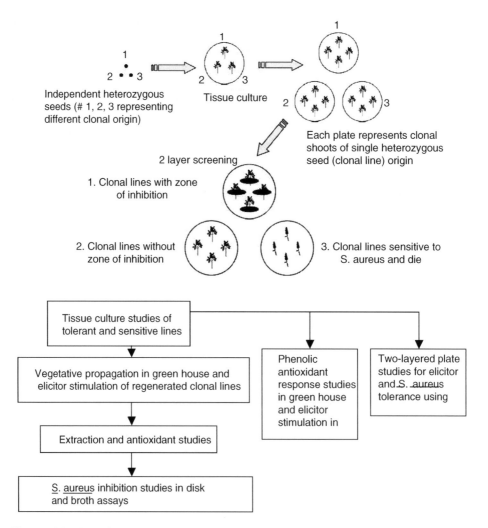

Figure 12.1 Steps for isolation of clonal lines of single seed origin heterozygous seed with potential for high phenolics with antimicrobial activity.

Listeria monocytogenes (Figure 12.2*)*. Further, the hypothesis that elicitor inducible phenolic antioxidants from low flavored clonal lines can also inhibit *L. monocytogenes* holds much promise for improved efficacy and is being investigated. In addition, in our recent studies we have also found in meat systems that ethanol extracted phenolics from clonal oregano are more effective in inhibiting *Listeria monocytogenes* than individual hydrophobic phenolics such as carvacrol that are found in essential oils of oregano (41).

12.3 FILLING GAPS IN ANTIMICROBIAL STRATEGY WITH TISSUE CULTURE BASED SCREENING OF ELITE PHENOLIC PHYTOCHEMICAL-PRODUCING LINES

As mentioned, the major limitation of using dietary food grade herbs like oregano, thyme, and lavender for antioxidant and antimicrobial applications is the inconsistency of phenolic phytochemicals due to the heterogeneity resulting from the natural cross pollinating nature of their breeding characteristics (74). Plants which originate from different heterozygous seeds in a given pool of extract are phenotypically variable, resulting in the substantial phytochemical inconsistency, and therefore leads to unreliable clinical effects as well as inconsistent health benefits and functional value.

In order to overcome the problem of phytochemical inconsistency due to genetic heterogeneity, our laboratory has developed patented plant tissue culture techniques to isolate a clonal pool of plants originating from a single heterozygous seed (75,76) (Figure 12.1). A single elite clonal line with superior phenolic profile (a combination of several bioactive phenolics such as rosmarinic acid and simple phenolic acids) can then be screened and selected based on tolerance to *Pseudomonas* sp. (70–72). This screening strategy is being further developed to select and characterize several elite high antioxidant clonal lines of food grade herbs such as oregano (*Origanum vulgare)*, thyme (*Thymus vulgaris*) and lavender (*Lavandula angustifolia*) based on response to microbial polysaccharide and bacterial pathogen targeted in a 2 layered plate assay (Figure 12.2) and based on tissue culture antioxidant response studies. These elite clonal lines (each clonal line originating from a different heterozygous seed), following large scale clonal propagation (micropropagation) in greenhouse systems are being evaluated for antioxidant profiles and functionality. High and low

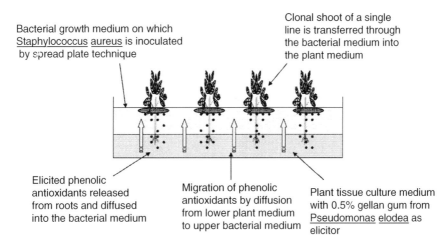

Figure 12.2   Innovative 2-layered plate assay from specific identification of clonal lines of herbs with potential of antimicrobial activity against the targeted bacterial pathogen.

antioxidant signature profiles of different clonal origin (different heterozygous seeds) are compared and targeted as dietary sources of phenolics that have high antioxidant efficacy and therefore potentially have antimicrobial benefits against food borne pathogens as evaluated *in vitro* disk (Figure 12.3) and broth assays. The advantages and importance of this approach is that it provides a rapid nontransgenic approach to obtain genetically consistent clonal lines with high rosmarinic acid as antioxidant in different total soluble phenolic and high antioxidant backgrounds from diverse heterogeneous gene pools that are generated due to natural cross pollination. Once the elite line is isolated by *in vitro* tissue culture methods it can be rapidly clonally propagated by vegetative cuttings for greenhouse and field production and therefore consistent phytochemical extracts of single seed genetic origin can obtained for potential antimicrobial applications against the pathogenic bacteria being targeted. Further, the clonal lines containing higher and lower total or specific phenolic content can be compared for synergistic contribution in different high total soluble phenolic and antioxidant clonal backgrounds. Therefore the approach envisioned in this chapter is excellent for generating consistent phytochemical antioxidant profiles for developing clinically or food safety relevant antimicrobial applications and for consistency of ingredients for food applications as a part of a hurdle technology. In addition, in future studies synergistic antimicrobial activity in combination with antimicrobials from multiple food grade sources can be evaluated against antibiotic resistant pathogens. Because antimicrobial phenolic profiles of single genetic origin are generated they offer consistency and yet have multiple phenolics that will make it difficult for pathogens to develop antimicrobial resistance. This synergy can be further enhanced in combinations with other dietary phenolic profiles. This strategy will be useful against hospital infections or food processing and packaging environments, especially infections from antibiotic resistant strains. Once a clonal profile antimicrobial efficacy is established, individual phenolic phytochemcials contributing to antimicrobial efficacy in specific clonal profiles can be investigated based on the model proposed on the mechanism of action of these phenolic phytochemicals. Such clonal screening is also ideal for screening for more ethanol and water soluble phenolic antimicrobials that can be effective in inhibiting or disrupting biochemical targets both in the plasma membrane and the cytosol.

12.3.1 Novel Model for Mechanism of Action of Phenolic Phytochemicals

Current theory and emerging data suggests that eukaryotes evolved from prokaryotes (77). Genetic evidence suggests that plant organelles like chloroplasts, mitochondria, and even vacuoles may have origins as free prokaryotes. The compartmentalized organization and

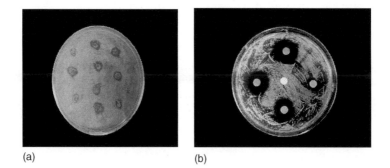

(a)   (b)

Figure 12.3   (a) Example of antimicrobial zone of inhibition by specific oregano clonal line against *S.aureus* as outlined in Figure 12.2. (b) Dose-dependent inhibition of *S. aureus* by phenolic extracts from mature clonal plants developed from clonal screening as outlined in Figure 12.1.

cellular differentiation (into tissues) of plants apparently evolved in terrestrial environments about 500 million years ago. It is likely that through the millennia, plastic plant species constantly interacted with many microorganisms and environmental stresses, especially UV radiation and oxidative stress. The speculation is that constant environmental stress in many ways may have shaped the antimicrobial and phenolic antioxidant responses of plants through the evolutionary process. Natural selection of plants under various changing environmental conditions over millions of years, in many critical ways (directly or indirectly) may have shaped the current mammalian systems, including human nutrition, health, and dietary evolution (40).

It is from these assumptions that we are exploring the mechanism of antimicrobial action of plant phenolics on prokaryotic pathogens and the activation and maintenance of plant and eukaryotic antioxidant responses, at a similar phenolic concentration needed for bacterial inhibition (40). Plant phenolics and related synthetic food grade phenolics have been suggested to inhibit bacterial pathogens (particularly at low pH) by disrupting the proton motive force (PMF) (40) and through lipid membrane stacking with hydrophobic phenolics (11). This implies that regulation of H^+-ATPases and cofactors at the external membrane is critical, and may be a logical point to study the mechanism of inhibition by plant phenolics. Additionally, plant phenolics may disrupt the electron transport chain or destabilize the plasma membrane (10,11,40). At the same time it is important to investigate the role of plant H^+-ATPases at both the external membrane and organelle (chloroplast, mitochondria, and vacuolar tonoplast) membrane levels to determine their potential involvement in modulation of a phenolic radical and proton linked redox cycle (40). This cytosolic phenolic radical proton linked redox cycle (40) is hypothesized to activate the proline linked pentose–phosphate pathway, to utilize proline as an alternative RE for mitochondrial ATP synthesis, and to generate precursors and cosubstrates (NADPH and sugar phosphates) for anabolic reactions, including a phenolic linked antioxidant response and purine synthesis (40). The genetic comparisons of H^+-ATPase alleles from bacterial and eukaryotes could provide clues toward understanding their susceptibility to plant phenolics in bacteria and the antioxidant response linked tolerance of plants and other eukaryotes (40). Such knowledge could facilitate the development of better structure function strategies for the development of better botanical phenolic profiles (i.e., elicitor induced herb and legume clonal extracts) that may provide multiple beneficial activities. A better conceptual understanding of phenolic antimicrobial effects in bacteria and antioxidant responses in eukaryotic hosts could also facilitate the development of diet based nutritional and functional food strategies to control bacterial pathogens with reduced potential for antibiotic resistance due to the concerted effect of a phenolic profile, instead of a single compound, as in case of antibiotics. Such an integrated approach involving phenolic phytochemicals has excellent potential to compliment current antimicrobial strategies (40).

Taking into account this rationale, a phenolic antimicrobial activity model (Figure 12.4) that incorporates the role of proline linked pentose–phosphate pathway provides a better perspective (40) than current putative antimicrobial models for phytochemical action (10). The phytochemical profiles that have the potential to inhibit pathogenic microorganisms (10–12) contain secondary metabolites that are defensive and inducible antimicrobials produced against both invading pathogens and stress. Therefore the methods for exploiting them must take this into account. In certain cases, the induction is associated with action of diphenolic oxidases and resulting modified compounds can have antimicrobial activity (13). In some cases, dihydroxy phenolics are oxidized to highly reactive quinones, which can interact with proteins of the invading pathogens and form melanoid polymers (10). As the compounds responsible for the enzymatic browning reaction of cut fruits and vegetables and an intermediate in melanin pigment production in humans (10), quinones are a

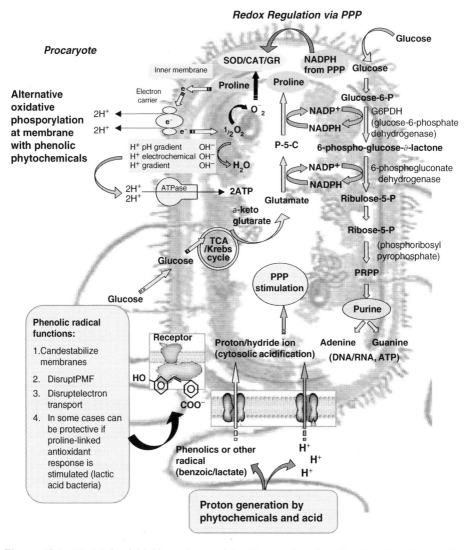

Figure 12.4 Model for inhibition of bacterial pathogens by phenolic phytochemical profiles (containing phenolics that are water soluble and effective in the cytosol as well as partially ethanol soluble phenolics that are effective at the membrane interface) based on membrane changes linked to energy (electron transport and proton flux changes) and cytosolic proton adjustment or lack of it that is linked to praline linked pathways and antioxidant response.

source of stable free radicals and complex irreversibly with nucleophilic amino acids (14) leading to inactivation of proteins and loss of function (10). Potential targets for inhibition in the bacterial cell are surface adhesions, cell wall polypeptides, and membrane bound enzymes (10). In herb species of the Lamiaceae family, phenolic derivatives are largely responsible for antimicrobial activity (2,41,42). Thymol, present in the essential oil of thyme, oregano, savory, sage, and related species, has good antimicrobial activity (11) but has flavor problems and is volatile. Other simple phenolics such as hydroxycinnamic acid derivatives such as caffeic acid, ferulic acid, and *p*-coumaric acid inhibited *E. coli*, *S. aureus* and *B. cereus* (19) could facilitate more hyper acidification in the cytosol and could be more effective with phenolics that are membrane targeted. Polymeric phenolics, such as tannins, were inhibitory toward *Listeria monocytogenes, E. coli, S. aureus, Aeromonas hydrophila,* and *Streptococcus faecalis* (20) and could be effective by stacking on the cell membrane and affecting energy function. Tannins could be made more potent in combination with cytosol targeted soluble phenolics. Hydroxylated phenols, such as catechol and pyrogallol, are known to be toxic to microorganisms (10) with a likely mechanism through membrane and cytosolic hyperacidification. The site and number of hydroxyl groups is linked to the antimicrobial effect and in some cases more oxidized forms are more inhibitory (21,22), which could be by inhibiting membrane related functions. In our laboratory we have shown that high rosmarinic acid enriched clonal profiles in ethanol and water extracts are effective in inhibiting *Listeria monocytogenes* (41) and *Helicobacter pylori* (42). We also have had similar success against the same pathogens using elicitor induced or fungal bio processed legume phenolics from soybean and other legumes.

From all the cited studies it is evident that the mode of action of phenolics against bacterial pathogens has not been clearly defined or understood and many modes of action have been suggested. Major mechanism models proposed have focused on the suspected changes in membrane permeability through hydrophobic linked lipid partitioning with certain phenolics to more membrane and cytosolic localized hyperacidity with soluble phenolics, which may affect PMF across the membrane resulting in energy depletion (11,28,29,31,32). Another model has proposed that enzyme inhibition by oxidized compounds through reaction with enzyme sulfhydryl groups or through nonspecific interactions with membrane proteins may explain the inhibition (10,27,36). Even with better models on the mechanism of action, another major limitation of using many phytochemicals is that they are derived from mixed heterogeneous genetic sources and consistency cannot be guaranteed (1,2). Now, with the emergence of antibiotic resistance from overuse of single antibiotics new strategies incorporating plant based antimicrobials are both promising (2,10,40). However, effective development of plant based antimicrobial strategies will require a better model to understand the mechanism of antimicrobial action involving consistent profiles of phenolic phytochemicals that inhibit bacterial pathogens at more than one biochemical steps. This could involve affecting cellular functions both at the membrane level with lipid or cellular interface compatible phytochemicals in conjunction with soluble phenolics that can cause cytosolic hyper acidification that will disrupt dehydrogenases and therefore disrupt the generation of proton motive force for ATP synthesis at an early cytosolic step along with reduced production of reductants for anabolic pathways (40).

An effective strategy proposed in this chapter focuses on the hypothesis that high antioxidant phenolics from single seed origin clonal lines of herbs, sprouted legumes, and fermented fruits would have excellent antimicrobial potential (40). Single seed high antioxidant phenolic profiles that are ethanol and water extracted have been screened, evaluated for antioxidant efficacy, and targeted to inhibit various food borne pathogens and chronic human infections, such as those by peptic ulcer-causing *Helicobacter pylori* (42) and *Listeria monocytogenes* (41). This strategy addresses both the concept of phytochemical consistency through the use of clonal lines and stress (elicitor) based inducible phenolic antioxidants from various developmental phases of growth and also the putative impact on antimicrobial potential. Therefore, using our strategy for developing consistent and inducible phytochemical profiles, an alternative and more robust model (Figure 12.4) for a mechanism of antimicrobial action incorporating the role of a critical control point (CCP) through proline linked pentose–phosphate pathway have been proposed. The mechanism of action of phenolic phytochemicals that may operate in prokaryotes involves the use hyperacidification (protons) from acids and phenolic metabolites and transport protons inside the cell (Figure 12.4), by more acid tolerant prokaryotes (lactic acid bacteria and some moderately acid tolerant Gram-negative pathogens like *H. pylori, Escherichia coli* and *Salmonella*) either passively or through H^+-transport membrane proteins (40). Such acid tolerant prokaryotes may stimulate redox cycling through the proline linked pentose–phosphate pathway and use proline (like the mitochondria of eukaryotes) as an alternative reductant for ATP synthesis at the single outer plasma membrane with oxygen as the terminal electron acceptor (40). As mentioned before, the process of proline biosynthesis could also be coupled to the pentose–phosphate pathway for NADPH recycling and to make sugar phosphates for all anabolic needs. By this model, acid tolerant microorganisms could efficiently manage the ATP needs from stress increased oxidative phosphorylation through coupling to pentose–phosphate pathway activity, while recycling excess proton flux from hyperacidification. Proline linked oxidative phosphorylation could excrete excess protons outside the plasma membrane and augment the generation of PMF for ATP synthesis (40). This proline linked metabolism model may occur in acid and phytochemical tolerant lactic acid bacteria (Gram-positive) and in moderately acid and phytochemical tolerant Gram-negative bacteria but is less likely in other acid and phytochemical susceptible Gram-positive bacteria (40). The ethanol, water, and organic acid extracted phenolic antimicrobials would be highly compatible for the given model.

One way to prove this model would be to determine if proline overproducing mutants are more acid tolerant and if this tolerance is associated with increased generation of NADPH. An important difference in case of prokaryotes is that they have only one outer plasma membrane and no organelles for metabolic adjustment. Depending on the type of membrane modifications that occur between various bacterial species, phenolic radicals may negatively or positively affect membrane related functions, including transport, signaling, receptor modification, and energy metabolism. Membrane related modulation of metabolism could be closely linked to the cytosolic proton linked modulation of proline linked pentose–phosphate pathway. Based on this model, it is likely that Gram-positive bacteria (excluding lactic acid bacteria) would be most susceptible, followed by Gram-negative bacteria, and then by acid tolerant lactic acid bacteria, likely being more tolerant and actually protected by phenolic phytochemicals (as antioxidants) (40).

12.3.2 Control of *Helicobacter Pylori* by Phenolic Antimicrobials from Plants

We have extended our studies on recruiting dietary plant based phenolic antimicrobials in inhibiting food borne pathogens to control and manage ulcer and gastric cancer associated *Helicobacter pylori*. *Helicobacter pylori* induced gastrointestinal disease afflicts millions worldwide, in both developed and developing countries. It was long believed that bacteria could not live in the acidic environment of the stomach, but in 1984, Marshall and Warren (78) reported success in culturing an organism from the gastric mucosal layer of a clinical specimen and proposed that this bacterium, later named *Helicobacter pylori*, was the cause of gastritis and peptic ulcer (78). This proposal has since proved to be correct, as *H. pylori* has been associated with the presence of gastric inflammation (79) and gastric non-Hodgkin's lymphomas (80), and has been identified as a major cause of peptic ulcer disease (81) and of gastric cancer (82).

Helicobacter pylori is a spiral, microaerophilic, Gram-negative bacteria that mainly infects the gastric mucosal layer, but can also enter human primary cells from gastric epithelium, possibly to evade host defenses and antimicrobial treatments (83,84). It infects up to 50% of the global human population (85). *Helicobacter pylori* is associated with gastritis, peptic ulcer and ultimately gastric cancer (80,85,86–89). The infections are mostly asymptomatic, but in all cases involve inflammation of the gastric mucosa, and in about 10% of the infected cases develop duodenal or gastric ulcers leading to cancer of stomach

(90). In developing countries, *H. pylori* infection is as high as 80–90% of the population and this incidence correlates with higher rates of stomach cancers (90). *Helicobacter pylori* infection is currently the highest prevailing bacterial infection worldwide (90). Characteristic traits of *H. pylori* infection are chronic inflammation of the stomach and development of chronic gastritis, which may proceed to gastric cancer (91). An integral part of *H. pylori* infection is the induction of gastric inflammation, which is believed to aid in colonization of the mucosal surfaces (84,92). Pathogenesis of *H. pylori* has also been linked to the presence of an external urease that is embedded on the outer surface of the bacterium (93). Urease is an important enzyme that is produced up to 5–6% of the total protein of *H. pylori* and is linked to its infectivity (90). All clinical isolates of *H. pylori* produce a urease with characteristics unlike those of any other bacterial urease (93–96). Urease negative mutants cannot colonize the gastric tissue environment as evidenced by the failure of urease negative mutants to colonize mice and gnotobiotic piglets (97,98). It is thought that hydrolysis of urea by urease generates ammonia to counterbalance gastric acidity by forming a neutral microenvironment surrounding the bacterium within the gastric lumen (99). This hypothesis is supported by results showing that *H. pylori* survive low pH *in vitro* in the presence of urea and functional urease activity (100,101). A unique feature of *H. pylori* urease is that a significant fraction is found exclusively in the outer membrane (94,102). This urease becomes associated with the surface of *H. pylori* and the urease released as a result of autolysis of a part of bacteria becomes adsorbed to the remaining intact bacteria (99,102). Further research by the same group has indicated that cytoplasmic urease activity alone, though needed, is not sufficient to allow survival of *H. pylori* in acid and the activity of surface localized urease is essential for resistance of *H. pylori* to acid conditions (99). Recent advances, however, have indicated that acidity trigger cytoplasmic urease activity and deletion of the gene *urel* prevents activation of this enzyme (103). Urel protein is a membrane protein with 6 trans membrane segments and is needed for acid stimulated urea uptake triggers the activity of the cytoplasmic urease, which is active down to pH 2.5 (103). Another feature of urease action in addition to sustaining *H. pylori* is the prospect of physical injury in the gastric lumen from ammonia (104,105). Ammonia may also lead to inflammation due to host response or due to other mucosal damage (106). Other factors that may be involved in the pathogenesis of *H. pylori* and the colonization of mucosal surfaces are the adhesion and motility of *H. pylori* (91), the induction of IL-8 and ROS in host cells (83), and the induction of COX-2 in host cells (107). High rates of *H. pylori* infection are associated with low socioeconomic status and high densities of living and prevalence of infection increases with age worldwide (108–110).

12.3.3 Current Treatment Strategies for *H. Pylori* and Its Limitations

Before the link between *H. pylori* and gastrointestinal diseases was discovered, the main course of treatment for these diseases was to assuage the symptoms with acid (proton pump) inhibitors (PPI). Now that a link to *H. pylori* has been established, antibiotics are the current standard of care for patients with gastritis and peptic ulcers, with the rationale that elimination of *H. pylori* will eliminate a major cause of gastric disease and reduce the risk for subsequent development of gastric adenocarcinomas (111). Typical therapy for *H. pylori* infection consists of a first line combination of a PPI and a bismuth compound, and two antibiotics taken for at least 7 days, followed by a second line treatment of four PPIs taken for at least 7 additional days (112). When followed religiously, the antibiotic regimen has been shown to eradicate *H. pylori* in approx. 80–90% of the cases, although there is some debate as to whether *H. pylori* is actually being eradicated or merely suppressed (112).

Difficulties with current *H. pylori* therapies contribute to treatment failure at any age. Despite the high success rate of antibiotic–PPI combination therapy against *H. pylori*, low compliance due to negative side effects (i.e., diarrhea, drug hypersensitivity, antibiotic associated colitis), not to mention the risk of nephrotoxicity when bismuth is used, continues to be a problem (84,113,114). Cost and practicality of the treatments are also concerns. Primary and acquired resistance of *H. pylori* to the antimicrobial agents is a major reason for the failure of eradication therapies (115). Due to the increasing incidence of antibiotic resistant strains of *H. pylori*, there is still a need for new compounds or alternative phenolic phytochemical strategies to combat *H. pylori* as future therapies.

12.3.4 Dietary Phenolic Phytochemical Strategies for Control of *H. Pylori*

Much recent research has focused on compounds from dietary sources as new or alternative strategies to fight *H. pylori*. Studies have indicated the antimicrobial potential of medicinal herbs and food plant metabolites against *H. pylori*. Highly selective antimicrobial activity of novel alkyl quinolone alkaloids from Chinese herbal medicine, Gosyuyu, against *H. pylori* *in vitro* has been reported (116). Eradication potential of Chinese herbal medicine, without emergence of resistant colonies, has also been reported (117). Among plant food extracts and ingredients, tea catechins have been reported to have *in vitro* and *in vivo* activities against *H. pylori* (118). Among various catechins, epigallocatechin gallate showed the strongest activity (118). In studies on synergistic antimicrobial effects of garlic, the combination of garlic based on allicin content with omeprazole showed antimicrobial benefits indicating the need for clinical studies (119,120). Susceptibility of *H. pylori* to the antimicrobial activity of manuka honey has been reported (113). Honey is a traditional remedy for dyspepsia and use of phytochemical enriched honey is a traditional practice without a known rational basis (113). Therefore, evaluating the inhibition by honey of *H. pylori* that has been suggested as the probable cause of dyspepsia has merit and results are promising (113). Plant phenolic metabolites like capsaisin from diet are known to be associated with low rate of ulcers through inhibition of *Helicobacter pylori* (121). Another interesting food based inhibition of *H. pylori* is the *in vitro* antibacterial activity of Chilean red wines (122). The main active compound that was suggested to be the antimicrobial compound was resveratrol (122). In other studies, extracts from a yeast fermented soymilk yogurt (123), bee propolis (124), licorice (125), broccoli (126), and turmeric (127), and in our laboratory, clonal oregano (42) demonstrated an ability to inhibit the *in vitro* growth of *H. pylori*. Although the mechanism of action of each extract remains largely unknown, in majority of these cases antibacterial activity was linked to the presence of phenolic compounds.

12.4 DIETARY SOYBEAN APPROACH FOR CONTROL OF *H. PYLORI*

A strong inverse association between consumption of microbially fermented (bioprocessed) soybean (natto, tofu) and *H. pylori* infection (128) and incidence of death from stomach cancer (129) has been observed in Japanese populations. Fermented soybean curd (tofu) is a common source of vegetable protein in Japan (128). Sprouts with inducible phenolics and microbially fermented or bioprocessed soybean are rich in isoflavonoids and phenolics that have antioxidant activity (130–133). Many phenolic antioxidants (in eukaryotic cells) also possess strong antimicrobial activities against prokaryotic bacteria, therefore, phenolic extracts from sprouts and microbially bioprocessed soybean that possess antioxidant activity potentially may also possess antibacterial activity, specifically anti *H. pylori* activity. Therefore the main focus of our investigations is to explore the potential of phenolic rich extracts of sprouted soybean and solid-state bioprocessed defatted soybean by the dietary fungus *Rhizopus oligosporous* (used in tempeh production in

Indonesia and the U.S.*)* to inhibit the *in vitro* growth and pathogenesis related urease of *H. pylori*. Our preliminary results suggest:

That the phenolic rich, bioprocessed soybean extracts possess strong anti *H. pylori* activity.

That sprouting linked elicitor stimulation of phenolics and bioprocessing of soybean by *R. oligosporous* increases anti *H. pylori* activity in soybean extracts in a manner dependent upon the sprouting stage, elicitor concentration, and duration of solid-state growth of the dietary fungus.

Hint at a mechanism of action which may involve diphenolic and polymeric phenolic antioxidants, which may inhibit urease.

Therefore, the hypothesis that elicitor inducible phenolic antioxidants from sprouted and fermented soybean can also inhibit *H. pylori* holds much promise for improved efficacy and support dietary therapy using soybean phenolic extracts in food systems. Other studies have postulated that antioxidants such as vitamin C and astaxanthin may constitute supportive treatment for controlling *H. pylori* infections through support of host (eukaryotic) response (134). Our preliminary studies suggest that phenolic antioxidants from soybean may be better. They can not only, potentially, provide a supportive redox environment in the stomach (eukaryotic cell) through the antioxidant improved functionality, but more importantly, can be inhibitory to prokaryotic pathogens like *H. pylori,* a prokaryote with a single membrane is susceptible to the same phenolic antioxidant*s*. Further, phenolic antioxidants could be potentially inhibitory against the pathogenesis linked urease and this will be evaluated using a novel plate assay described in this chapter. Through inhibition of urease by phenolics, hyperacidification at the membrane level can be maintained and as discussed for the general model (Figure 12.4), this proton efflux coupled to other functions of dietary phenolics can now be more effective in inhibiting *H. pylori* (Figure 12.5*)*. Because dietary phenolics are offered as phenolic profiles it is likely to reduce the emergence of antibiotic resistance.

Extracts from dietary legumes and other phenolic phytochemicals can be important source of disease preventive phytochemicals which potentially can inhibit the target prokaryotic bacterial pathogens like *H. pylori*, which have a single plasma membrane that also contains the pathogenesis associated urease, and at the same time can be antioxidant, in function, in host eukaryotic cells that have organelle compartmentalization to manage a redox response linked to phenolics, triggering a protective response. Our strategy focuses on the hypothesis that inducible high antioxidant phenolics from clonal herbs screened from heterogeneous sources, seed sprouts of single genetic origin, and phenolic aglycones with high antioxidant activity from fermented dietary botanicals would have excellent antimicrobial potential. Our strategy also addresses the concept of phytochemical consistency through the mobilization of stress (elicitor) inducible phenolic antioxidants in various developmental phases of growth of soybean sprouts and bioprocessed extracts and evaluates the antimicrobial potential against *H. pylori*.

12.4.1 Sprouting of Soybean and Phenolic Antioxidant Stimulation Methods

We are exploring the use of high antioxidant sprout extracts of soybean and other legumes for antimicrobial benefit against gastric cancer linked *H. pylori* and therefore developed a diet based bacterial management strategy. Seeds of soybean are imbibed overnight in bacterial polysaccharide elicitors normally used in foods, such as gellan gum and xanthan gum (135). In earlier work on the legume mung bean (135), there was clear evidence that stress-modulating polysaccharide elicitors will stimulate phenolic content and antioxidant activity during the sprouting stages as a likely protective mechanism for the young

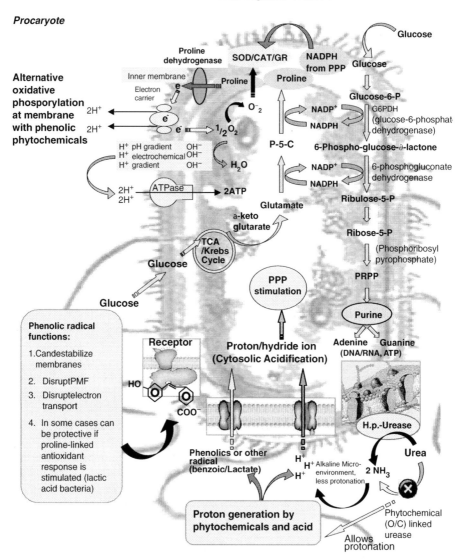

Figure 12.5 Model for inhibition of *Helicobacter pylori* by phenolic phytochemicals as in Figure 12.4 but also takes into account the role of urease.

seedling. Previously it has also been established that the percent of phenolic aglycones were the highest in soybean sprouts (136). The biological activity of the phenolics is stronger in the aglycone form than in the glycoside form. It is thought that isoflavonoids such as genistein and daidzein have a specific role during growth stages of soybean (130,136). Preliminary results from our laboratory have indicated that during dark germination, phenolic content was stimulated (inducible phenolics) by the given elicitors (137). Polysaccharide elicitor extracts from *Pseudomonas elodea* (gellan gum) are currently being tested in the concentration range of 0–500 mg/l of imbibed water. After elicitor treatment for 12 hours, sprouting through a seven day dark germination stage was done in specially designed plastic containers. Phenolic profiles, antioxidant activity and *H. pylori* inhibition studies was undertaken on extracts of various development stages. Optimized phenolic extracts was used for further *H. pylori* inhibition (137) and in the future for urease inhibition studies. Several soybean varieties from genetic stocks in U.S. and Asia are currently being tested. We have, currently, access to over 50 soy genetic accessions for testing. The basal uninduced phenolic content of soybean is around 1 to 3 mg/g of dry weight (DW) of sprouted and fermenting soybean, which is than stimulated to 4–5 mg/g DW.

12.4.2 Phenolic Response and Characterization of Soybean Extracts Inhibiting *H. Pylori* Urease

Initial preliminary efforts have so far focused on polysaccharide elicitor stimulated soybean seedling extracts, based on the rationale that if an individual clonal line (single seed origin variety soybean is self pollinating) has genetically evolved to produce high phenolics in a heterogeneous background, it may also have the genetic, biochemical and developmental mechanism to tolerate the same high level of phenolics and such high phenolics could have protective antioxidant benefits and antimicrobial effects against the bacterial pathogens or its elicitors. Further, we have also developed an innovative plate assay system to screen phenolic extracts that are inhibitory to *H. pylori* urease. In this assay system we have *H. pylori* medium at pH 5.5 with addition of 10 mM urea and bromophenol blue (0.1 g/L) as pH indicator. *Helicobacter pylori* will survive low pH conditions by using urea from the acidic environment and create a microenvironment of high pH by excreting urease mediated ammonia. When this happens, the bromophenol blue will turn blue as pH moves higher toward neutral. We have evidence that soybean extracts in a concentration dependent (100 μg (micro gram)/disk to 400 μg/disk) manner can inhibit urease activity as indicated the bromophenol blue remaining yellow at pH 5.5.

12.4.3 Solid-State Bioprocessing of Soybean Using *Rhizopus Oligosporus*

Rhizopus oligosporous used in soy Tempeh foods has been developed through solid-state bioprocessing of whole soy and would be a more suitable approach to enhance phenolic metabolites for inhibition of *Helicobacter pylori* (131,137). Using the methods developed in our laboratory (131), phenolic aglycones are being optimized from various varieties of defatted soy, and the fungal growth is carried out in Erlenmeyer flasks over a ten day period. Total soluble phenolics in water and ethanol extracts have been optimized with various specific phenolic ratios as in sprouts and using low, high, and medium phenolic categories of various soybean varieties.

12.5 RECENT PROGRESS AND OTHER STRATEGIES FOR DESIGNING PHENOLICS PHYTOCHEMICALS AS ANTIMICROBIALS

Using the clonal screening strategy we have isolated several high phenolic clonal herbs of oregano, thyme, rosemary, and spearmint of single seed origin. One superior oregano clonal line has been shown to be effective against *L. monocytogenes* in broth and meat system (41) and *H. pylori* in disk assay system (42). Further we have found that oregano extracts optimized on a phenolic basis and combined with cranberry phenolics shows synergistic effect in both broth and food systems against *L. monocytogenes* and *Vibrio parahaemolyticus* (64,138) and in disk assay system against *H. pylori* (139).

The elicitor induced spouting model in seed legumes was used to evaluate antimicrobial potential of various phenolic and antioxidant activity stages of sprouting against

H. pylori. Phenolic extracts of pea (140), chickpea (141), fenugreek (142), soy (137), and mung bean (143) were inhibitory against *H. pylori* in disk assays. Except for fenugreek, in all other sprout extract tested, the antimicrobial activity coincided with likely polymeric stage as indicated by high sprout peroxidase activity, when soluble phenolics were the lowest (137,140,141,143). In case of fenugreek, the highest antimicrobial activity coincided with the stage of highest soluble phenolics (142).

We have also developed solid-state bioprocessing systems using filamentous fungi, where under aerobic conditions phenolics from soy, cranberry, and pineapple substrates were mobilized (137,144–146). In case of soy bioprocessing with Tempeh fungus, *Rhizopus oligosporous* enhanced functional polymeric phenolics, as indicated by laccase activity at 2 days of growth, and had the highest anti *H. pylori* activity (137). In the same study, late stages, even though they had highest phenolics, had reduced antimicrobial activity. In the case of cranberry bioprocessing with *Rhizopus oligosporous* and *Lentinus edodes,* highest antimicrobial activity was found after bioprocessing, and effective antimicrobial activity varied for targeted bacterial food borne pathogens based on nitrogen sources and method of extraction (144,145). It was postulated that the type of phenolics mobilized during bioprocessing could have affect on the antimicrobial functionality (144,145). In other studies, pineapple wastes enriched with soybean and bioprocessed with *R. oligosporous* had effective anti *H. pylori* activity after 10 days of growth.

12.6 SUMMARY

Novel tissue culture techniques for screening high phenolic clonal lines and phenolic phytochemical bioprocessing approaches through sprouting from seed legumes and microbial solid-state mobilization from diverse dietary botanical substrates have been developed to isolate novel dietary and food grade phenolic metabolites as antimicrobials. In this effort only water and ethanol soluble phenolics which can likely create a localized hyper acidification at the cellular membrane interface and cytosol have been extracted and evaluated for antimicrobial effects. The goal is to design dietary phenolics that work at the cellular interface, and cytosol to disrupt the simpler prokaryotic cellular metabolism at membrane, energy, and cytosolic dehydrogenase level. In this model the functional property of dietary phenolics for localized cellular hyperacidification is coupled to a proposed critical control point (CCP), proline linked pentose–phosphate pathway (PLPPP). The presence and management of this PLPPP with membrane linked energy metabolism and cytosolic linked proton management through dehydrogenases serve as the central weak points for antimicrobial strategies. The same phenolics in eukaryotes, due to organelle compartmentalization, are likely to be managed through an antioxidant enzyme response and which in turn can positively support host response to bacterial pathogens (40).

REFERENCES

1. Shetty, K. Biotechnology to harness the benefits of dietary phenolics; focus on Lamiaceae. *Asia Pac. J. Clin. Nutr.* 6:162–171, 1997.
2. Shetty, K., R.L. Labbe. Food-borne pathogens, health and role dietary phytochemicals. *Asia Pac. J. Clin. Nutr.* 7:270–276, 1998.
3. Hertog, M.G.L., D. Kromhout, C. Aravanis, H. Blackburn, R. Buzina, F. Fidanza, S. Giampaoli, A. Jansen, A. Menotti, S. Nedeljkovic, M. Pekkarinen, B.S. Simic, H. Toshima, E.J.M. Feskens, P.C.H. Hollman, M.B. Kattan. Flavonoid intake and long-term risk of coronary heart disease and cancer in the seven countries study. *Arch. Intern. Med.* 155:381–386, 1995.

4. Peake, P.W., B.A. Pussel, P. Martyn, V. Timmermans, J.A. Charlesworth. The inhibitory effect of rosmarinic acid on complement involves the C5 convertase. *Int. J. Immunopharmac.* 13:853–857, 1991.

5. Rice-Evans, C.A., N.J. Miller, P.G. Bolwell, P.M. Bramley, J.B. Pridham. The relative antioxidant activities of plant-derived polyphenolic flavonoids. *Free Rad. Res.* 22:375–383, 1995.

6. Jorgensen, L.V., H.L. Madsen, M.K. Thomsen, L.O. Dragsted, L.H. Skibsted. Regulation of phenolic antioxidants from phenoxyl radicals: an ESR and electrochemical study of antioxidant hierarchy. *Free Rad. Res.* 30:207–220, 1999.

7. Paganga, G., N. Miller, C.A. Rice-Evans. The polyphenolic contents of fruits and vegetables and their antioxidant activities: what does a serving constitute ? *Free Rad. Res.* 30:153–162, 1999.

8. Hertog, M.G.L., P.C.H. Hollman, M.B. Katan. Content of potentially anticarcinogenic flavonoids of 28 vegetables and 9 fruits commonly consumed in the Netherlands. *J. Agric. Food Chem.* 40:2379–2383, 1992.

9. Foti, M., M. Piattelli, V. Amico, G. Ruberto. Antioxidant activity of phenolic meroditerpenoids from marine algae. *J. Photochem. Photobiol.* 26:159–164, 1994.

10. Cowan, M.M. Plant products as antimicrobial agents. *Clin. Microbiol. Rev.* 12:564–582, 1999.

11. Burt, S. Essential oils: their antibacterial properties and potential applications in foods: a review. *Int. J. Food Microbiol.* 94:223–253, 2004.

12. Shelef, L.A. Antimicrobial effects of spices. *J. Food Safety* 6:29–44, 1984.

13. Walker, J.R.L. Antimicrobial compounds in food plants. In: *Natural Antimicrobial Systems and Food Preservation,* Board, R.G., V. Dillon, eds., Wallingford: CAB International, 1994, pp 181–204.

14. Stern, J.L., A.E. Hagerman, P.D. Steinberg, P.K. Mason. Phlorotannin-protein interactions. *J. Chem. Eol.* 22:1887–1899, 1996.

15. Beuchat, L.R. Sensitivity of *Vibrio parahaemolyticus* to spices and organic acids. *J. Food Sci.* 41:899–902, 1976.

16. Aktug, S.E., M. Karapinar. Sensitivity of some common food poisoning bacteria to thyme, mint and bay leaves. *Int. J. Food Microbiol.* 3:349–354, 1986.

17. Shelef, L.A., E.K. Jyothi, M.A. Bugarelli. Growth of enteropathogenic and spoilage bacteria in sage-containing broth and foods. *J. Food Sci.* 49:737–740, 1984.

18. Farbood, M.I., J.H. McNeil, K. Ostovar. Effect of rosemary spice extractive on the growth of microorganisms in meat. *J. Milk Food Technol.* 39:675–679, 1976.

19. Herald, P.J., P.M. Davidson. Antibacterial activity of selected hydroxycinnamic acids. *J. Food Sci.* 48:1378–1379, 1983.

20. Chung, K.T., C.A. Murdock. Natural systems for preventing contamination and growth of microorganisms in foods. *Food Microstruct.* 10:361–366, 1991.

21. Scalbert, A. Antimicrobial properties of tannins. *Phytochemistry* 30:3875–3883, 1991.

22. Urs, N.V.R.R., J.M. Dunleavy. Enhancement of the bactericidal activity of peroxidase system by phenolic compounds. *Phytopathology* 65:686–690, 1975.

23. Koutsomanis, K., K. Lambropoulou, G.J.E. Nychas. A predictive model for the non-thermal inactivation of *Salmonella enteritidis* in a food model supplemented with a natural antimicrobial. *Int. J. Food Microbiol.* 49:67–74, 1999.

24. Skandamis, P.N., G.J.E. Nychas. Development and evaluation of a model predicting the survival of *Escherichia coli* 0157: H7 NCTC 12900 in homemade eggplant salad at various temperatures, pHs, and oregano essential oil concentrations. *Appl. Environ. Microbiol.* 66:1646–1653, 2000.

25. Tsigarida, E., P. Skandamis, G.J.E. Nychas. Behavior of *Listeria monocytogenes* and autochthonous flora on meat stored under aerobic, vacuum and modified atmosphere packaging conditions with or without the presence of oregano essential oil at 5 C. *J. Appl. Bacteriol.* 89:901–909, 2000.

26. Skandamis, P., E. Tsigarida, G.J.E. Nychas. The effect of oregano essential oil on survival/ death of *Salmonella typhimurium* in meat stored at 5°C under aerobic, vp/map conditions. *Food Microbiol.* 19:65–75, 2002.

27. Tassou, C.C., G.J.E. Nychas, P.N. Skandamis. Herbs and spices and antimicrobials. In: *Handbook of Herbs and Spices*, Peter, K.V., ed., Boca Raton, FL: CRC press, 2004, pp 23–40.

28. Conner, D.E., L.R. Beuchat, R.E. Worthington, D.A. Kautter. Effects of essential oils and oleoresins of plants on ethanol production, respiration and sporulation of yeasts. *Int. J. Food Microbiol.* 1:63–74, 1984.

29. Baranowski, J.D., P.M. Davidson, C.W. Nagel, A.L. Branen. Inhibition of *Saccharomyces cerevisiae* by naturally occurring hydroxycinnamates. *J. Food Sci.* 45:592–594, 1980.

30. Davidson, P.M. Chemical preservatives and natural antimicrobial compounds. In: *Food Microbiology Fundamentals and Frontiers*, Doyle, M.P., L.R. Beuchat, T.J. Montville, eds., New York: ASM Press, 1997, pp 520–526.

31. Ultee, A., M.H.J. Bennik, R. Moezellaar. The phenolic hydroxyl group of carvacrol is essential for action against the food-borne pathogen *Bacillus cereus*. *Appl. Environ. Microbiol.* 68:1561–1568, 2002.

32. Ultee, A., E.P.W. Kets, E.J. Smid. Mechanisms of action of carvacrol on the food-borne pathogen *Bacillus cereus*. *Appl. Environ. Microbiol.* 65:4606–4610, 1999.

33. Tahara, T., M. Ohsimura, C. Umezawa, K. Kanatani. Isolation and partial characterization and mode of action of acidocin J1132, a two-component bacteriocin produced by *Lactobacillus acidophilus* JCM 1132. *Appl. Environ. Microbiol.* 62:892–897, 1996.

34. Cox., S.D., J.E. Gustafson, C.M. Mann, J.L. Markham, Y.C. Liew, R.P. Hartland, H.C. Bell, J.R. Warmington, S.G. Wyllie. Tea tree oil causes K+ leakage and inhibits respiration in *Escherichia coli*. *Lett. Appl. Microbiol.* 26:355–358, 1998.

35. Tassou, C.C., K. Koutsoumanis, G.J.E. Nychas. Inhibition of *Salmonella enteritidis* and *Staphylococcus aureus* in nutrient broth by mint essential oil. *Food Res. Int.* 33:273–280, 2000.

36. Mason, T.L., B.P. Wasserman. Inactivation of red beet beta-glucan synthase by native and oxidized phenolic compounds. *Phytochemistry* 26:2197–2202, 1987.

37. Friedman, M. Chemistry nutrition and microbiology of D-amino acids. *J. Agric. Food Chem.* 47:3457–3479, 1999.

38. Patte, J. Biosynthesis of threonine and lysine. In: Escherichia coli *and* Salmonella, 2nd ed., Fredeci, M., F. Neidhardt, eds., Washington, D.C.: ASM Press, 1996, pp 528–541.

39. Friedman, M., G.A. Smith. Inactivation of quercetin mutagenicity. *Food Chem. Toxicol.* 22:535–539, 1984.

40. Shetty, K., M.L. Wahlqvist. A model for the role of proline-linked pentose phosphate pathway in phenolic phytochemical biosynthesis and mechanism of action for human health and environmental applications: a review. *Asia Pac. J. Clin. Nutr.* 13:1–24, 2004.

41. Seaberg, A., R.L. Labbe, K. Shetty. Inhibition of *Listeria monocytogenes* by elite clonal extracts of oregano (*Origanum vulgare*). *Food Biotechnol.* 17:129–149, 2003.

42. Chun, S.-S., D.A. Vattem, Y.-T. Lin, K. Shetty. Phenolic antioxidants from clonal oregano (*Origanum vulgare*) with antimicrobial activity against *Helicobacter pylori*. *Process Biochem*, 40:809–816, 2005.

43. Deighton, N., S.M. Glidewell, S.G. Deans, B.A. Goodman. Identification by EPR spectroscopy of carvacrol and thymol as the major sources of free radicals in the oxidation of plant essential oils. *J. Food Sci. Agric.* 63:221–225, 1993.

44. Deighton, N., S.M. Glidewell, B.A. Goodman, S.G. Deans. The chemical fate of the endogeneous plant antioxidants carvacrol and thymol during oxidative stress. *Proceedings of the Royal Society of Edinburgh*, 102B:247–252, 1994.

45. Engleberger, W., U. Hadding, E. Etschenberg, E. Graf, S. Leyck, J. Winkelmann, M.J. Parnham. Rosmarinic acid: a new inhibitor of complement C3 – convertase with anti-inflammatory activity. *Intl. J. Immunopharmac.* 10:729–737, 1988.

46. Frankel, E.N., S.W. Huang, R. Aeschbach, E. Prior. Antioxidant activity of a rosemary extract and its constituents carnosic acid, carnosol and rosmarinic acid in bulk oil, and oil-in-water emulsion. *J. Agric. Food Chem.* 44:131–135, 1996.

47. Kuhnt, M., A. Probstle, H. Rimpler, R. Bauer, M. Heinrich. Biological and pharmacological activities and further constituents of *Hyptis verticillata*. *Planta Medica* 61:227–232, 1995.

48. Guggenheim, S., S. Shapiro. The action of thymol on oral bacteria. *Oral Microbiol. Immunol.* 10:241–246, 1995.

49. Kikuzaki, H., N. Nakazaki. Structure of a new antioxidative phenolic acid from oregano (*Origanum vulgare*). *Agric. Biol. Chem.* 53:519–524, 1989.

50. Economou, K.D., V. Oeropoulou, C.D. Thomoupoulos. Antioxidant activity of some plant extracts of the family Labiate. *J. Am. Oil Chem. Soc.* 68:109–113, 1991.

51. Madsen, H., G. Bertelsen. Spices as antioxidants. *Trends Food Sci. Tech.* 6:271–277, 1995.

52. Chipault, J.R., G. Mizuno, W.O. Lundberg. The antioxidant properties of spices in foods. *Food Technol.* 10:209–211, 1956.

53. Emori, T.G. and Gaynes, R.P. An overview of nosocomial infections, including the role of the microbiology laboratory. *Clin. Microbiol. Rev.* 6:428–442, 1993.

54. Steinberg, J.P., C.C. Clark, B.O. Hackman. Nosocomial and community-acquired *Staphylococcus aureus* bacteremias from 1980 to 1993: impact of intravascular devices and methicillin resistance. *Clin. Infect. Dis.* 23:255–259, 1996.

55. Vychytil, A., M. Loren, B. Schneider, W.H. Horl, M. Haag-Weber. New strategies to prevent *Staphylococcus aureus* infections in peritoneal dialysis patients. *J. Am. Soc. Nephrol.* 669–670, 1998.

56. Lowy, N. *Staphyloccocus aureus* infections. *New. Engl. J. Med.* 339:520–532, 1998.

57. Tenover, F.C., M.V. Lancaster, B.C. Hill, C.D. Steward, S.A. Stocker, G.A. Hancock. Characterization of stapylococci with reduced susceptibilities to vancomycin and other glycopeptides. *J. Clin. Microbiol.* 36:1020–1027, 1998.

58. Hiramatsu, K., et al. Dissemination in Japanese hospitals of strains of *Staphylococcus aureus* heterogeneously resistant to vancomycin. *Lancet* 350:1670–1675, 1997.

59. Smith, T.L., M.L. Pearson, K.R. Wilcox. Emergence of vancomycin resistance in *Staphylococcus aureus*. *N. Engl. J. Med.* 340:493–501, 1999.

60. Marrack, P., J. Kappler. The staphylococcal enterotoxins and their relatives. *Science* 240:705–711, 1990.

61. Bantel, H., B. Sinha, W. Domschke, G. Peters, K. Schulze-Osthoff, R.U. Janicke.α-Toxin is a mediator of *Staphylococcus aureus*-induced cell death and activates caspases via the intrinsic death pathway independently of death receptor signaling. *J. Cell. Biol.* 155:637–647, 2001.

62. Day, N.P.J., C.E. Moore, M.C. Enright, A.R. Berendt, J.M. Smith, M.F. Murphy, S. Peacock, B.G. Spratt, E.J. Feil. A link between virulence and ecological abundance in natural populations of *Staphylococcus aureus*. *Science* 292:114–116, 2001.

63. McKenney, D., K.L. Pouliot, Y. Wang, V. Murthy, M. Urlich, G. Doring, J.C. Lee, D.A. Goldmann, G.B. Pier. Broadly protective vaccine for *Staphylococcus aureus* based on an *in vivo*-expressed antigen. *Science* 284:1523–1527, 1999.

64. Lin, Y.-T., R.G. Labbe, K. Shetty. Inhibition of *Listeria monocytogenes* in fish and meat systems using oregano and cranberry synergies. *Appl. Environ. Microbiol.*, 2004. [Accepted for publication]

65. Kim, J., M.R. Marshall, C. Wei. Antibacterial activity of some essential oil components against five foodborne pathogens. *J. Agric. Food Chem.* 43:2839–2845, 1995.

66. Pol, I.E., E.J. Smid. Combined action of nisin and carvacrol on *Bacillus cereus* and *Listeria monocytogenes*. *Lett. Appl. Microbiol.* 29:166–170, 1999.

67. Karatzas, A.K., E.P.W. Kets, E.J. Smid, M.H.J. Bennik. The combined action of carvacrol and high hydrostatic pressure on *Listeria monocytogenes* Scott A. *J. Appl. Microbiol.* 90:463–469, 2001.

68. Fleisher, A, N. Sneer. Oregano spices and *Origanum* chemotypes. *J. Sci. Food Agric.* 33:441–446, 1982.

69. Al-Amier, H., B.M.M. Mansour, N. Toaima, R.A. Korus, K. Shetty. Tissue culture based screening and selection of high biomass and phenolic producing clonal lines of lavender using *Pseudomonas* and azetidine-2-carboxylate. *J. Agric. Chem.* 47:2937–2943, 2000.

70. Yang, R., O.F. Curtis, K. Shetty. Tissue-culture-based selection of high rosmarinic acid-producing clonal lines of rosemary (*Rosmarinus officinalis*) using hyperhydricity-reducing *Pseudomonas*. *Food Biotechnol.* 11:73–88, 1997.

71. Eguchi, Y., O.F. Curtis, K. Shetty. Interaction of hyperhydricity-preventing *Pseudomonas* sp. with oregano (*Origanum vulgare*) and selection of high phenolics and rosmarinic acid-producing clonal lines. *Food Biotechnol.* 10:191–202, 1996.

72. Shetty, K., T.L. Carpenter, D. Kwok, O.F. Curtis, T.L. Potter. Selection of high phenolics-containing clones of thyme (*Thymus vulgaris* L.) using *Pseudomonas* sp. *J. Agric. Food Chem.*, 40:3408–3411, 1996.

73. Dixon, R.A., N. Paiva. Stress-induced phenylpropanoid metabolism. *Plant Cell* 7:1085–1097, 1995.

74. Richards, A.J. Gynodioecy. In: *Plant Breeding Systems,* Richards, A.J. ed., London: George Allen and Unwin Ltd., 1986, pp 89–331.

75. Shetty, K., O.F. Curtis, R. E. Levin, R. Witkowsky, W. Ang. Prevention of vitrification associated with *in vitro* shoot culture of oregano (*Origanum vulgare*) by *Pseudomonas* spp. *J. Plant Physiol.* 147:447–451, 1995.

76. Shetty, K. Plant clones containing elevated secondary metabolites. U.S. patent # 5,869,340, 1996.

77. Margulis, L. Symbiosis everywhere and other chapters. In: *Symbiotic Planet (A New Look at Evolution)*, New York: Basic Books, 1998.

78. Marshall, B.J., J.R. Warren. Unidentified curved bacilli in the stomach of patients with gastritis and peptic ulceration. *Lancet* 1:1311–1315, 1984.

79. Blaser, M.J. *Helicobacter pylori* and the pathogenesis of gastroduodenal inflammation. *J. Infect. Dis.* 161:626–633, 1990.

80. Parsonnet, J., S. Hansen, L. Rodriguez, A.B. Gelb, R.A. Warnke, E. Jellum, N. Orentreich, J.H. Vogelman, G.D. Friedman. *Helicobacter pylori* infection and gastric lymphoma. *N. Engl. J. Med.* 330:1267–1271, 1994.

81. NIH Consensus Conference. *Helicobacter pylori* in peptic ulcer disease. NIH Consensus Development Panel on *Helicobacter pylori* in Peptic Ulcer Disease. *JAMA* 272:65–69, 1994.

82. Uemura. N., S. Okamoto, S. Yamamoto, N. Matsumura, S. Yamaguchi, M. Yamakido, K. Taniyama, N. Sasaki, R.J. Schlemper. *Helicobacter pylori* infection and the development of gastric cancer. *N. Engl. J. Med.* 345:784–789, 2001.

83. Dunn, B.E., H. Cohen, M.J. Blaser. *Helicobacter pylori. Clin. Microbiol. Rev.* 10:720–741, 1997.

84. Rhen, M., S. Eriksson, M. Clements, S. Bergström, S.J. Normark. The basis of persistent bacterial infections. *Trends Microbiol.* 11:80–86, 2003.

85. Covacci, A., J.L. Telford, G. Del Giudice, J. Parsonnet, R. Rappuoli. *Helicobacter pylori* virulence and genetic geography. *Science* 284:1328–1333, 1999.

86. Marshall, B.J., C.S. Goodwin, J.R. Warren, R. Murray, E.D. Blincow, S.J. Blackboun, M. Phillips, T.E. Warers, C.R. Sanderson. Prospective double-blind trial of duodenal ulcer relapse after eradication of *Campylobacter pylori. Lancet* 2:1437–1442, 1988.

87. Rauws, E.A., G.N. Tytgat. Cure of duodenal ulcer associated with eradication of *Helicobacter pylori. Lancet* 335:1233–1235, 1990.

88. Parsonnet, J., G.D. Friedman, D.P. Vandersteen, Y. Chang, J.H. Vogelman, N. Orentreich, R.K. Sibley. *Helicobacter pylori* infection and the risk of gastric carcinoma. *N. Engl. J. Med. Med.* 325:1127–1131, 1991.

89. Forman, D., P. Webb, J. Parsonnet. *H.pylori* and gastric cancer. *Lancet* 34:243–244, 1994.

90. Labigne, A., H. De Reuse. Determinants of *Helicobacter pylori* pathogenicity. *Infect. Agents Dis.* 5:191–202, 1996.

91. Gerhard, M., R. Rad, C. Prinz, M. Naumann. Pathogenesis of *Helicobacter pylori* infection. *Helicobacter* 7(1):17–23, 2002.

92. Nedrud, J.G., S.S. Blanchard, S.J. Czinn. *Helicobacter pylori. Inflam. Immun. Helicobacter* 7(1):24–29, 2002.

93. Dunn, B.E., N.B. Vakil, B.G. Schneider, M.M. Miller, J.B. Zitzer, T. Puetz S.H., Phadnis. Localization of *Helicobacter pylori* urease and heat shock protein in human gastric biopsies. *Infect. Immun.* 65:1181–1188, 1997.

94. Dunn, B.E., G.P. Campbell, G.I. Perez-Perez, M.J. Blaser. Purification and characterization of urese from *Helicobacter pylori*. *J. Biol. Chem.* 265:9464–9469, 1990.

95. Mobley, H.L.T., M.J. Cortesia, L.E. Rosenthal, B.D. Jones. Chracterization of urease from *Campylobacter pylori*. *J. Clin. Microbiol.* 26:831–836, 1988.

96. Mobley, H.L.T., M.D. Island, R.P. Hausinger. Molecular biology of microbial ureases. *Microbiol. Rev.* 59:451–480, 1995.

97. Eaton, K.A., C.L. Brooks, D.R. Morgan, S. Krakowka. Essential role of urease in pathogenesis of gastritis induced by *Helicobacter pylori* in gnotobiotic piglets. *Infect. Immun.* 59:2470–2475, 1991.

98. Eaton, K.A., S. Krakowka. Effect of gastric pH on urease-dependent colonization of gnotobiotic piglets by *Helicobacter pylori*. *Infect. Immun.* 62:3604–3607, 1994.

99. Krishnamurthy, P., M. Parlow, J.B. Zitzer, N.B. Vakil, H.L.T. Mobley, M. Levy, S.H. Phadnis, B.E. Dunn. *Helicobacter pylori* containing only cytoplasmic urese is susceptible to acid. *Infect. Immun.* 66:5060–5066, 1998.

100. Marshall, B.J., L.J. Barret, C. Prakash, R.W. McCallum, R.I. Guerrant. Urea protects *Helicobacter pylori* from bactericidal effect of acid. *Gastroenterology* 99:697–702, 1990.

101. McGowan, C.C., T.L. Cover, M.J. Blaser. The proton pump inhibitor omeprazole inhibits acid survival of *Helicobacter pylori* by an urease-independent mechanism. *Gastroenterology* 107:1573–1578, 1994.

102. Phadnis, S.H., M.H. Parlow, M. Levy, D. Ilver, C.M. Caulkins, J.B. Connors, B.E. Dunn. Surface localization of *Helicobacter pylori* urease and a heat shock protein homolog requires bacterial autolysis. *Infect. Immun.* 64:905–912, 1996.

103. Weeks, D.L., S. Eskandari, D.R. Scott, G. Sachs. An H+ -Gated urea channel between *Helicobacter pylori* urease and gastric colonization. *Science* 287:482–485, 2000.

104. Desai, M.A., P.M. Vadgama. An *in vivo* study of enhanced H+ diffusion by urease action on urea: implications for *Helicobacter pylori*-associated peptic ulceration. *Scand. J. Gastroenterol.* 28:915–919, 1993.

105. Kawano, S.M., M. Tsujii, H. Fusamoto, N. Sato, T. Kamada. Chronic effect of intragastric ammonia on gastric mucosal structures in rats. *Dig. Dis. Sci.* 36:33–38, 1991.

106. Smoot, D.T. How does *Helicobacter pylori* cause mucosal damage?: direct mechanisms. *Gastroenterology* 113:S31–S34, 1997.

107. Sepulveda, A.R., L.G.V. Coelho. *Helicobacter pylori* and gastric malignancies. *Helicobacter* 7(1):37–42, 2002.

108. Mitchell, H., F. Mégraud. Epidemiology and diagnosis of *Helicobacter pylori* infection. *Helicobacter* 7(1):8–16, 2002.

109. Pilotto, A., N. Salles. *Helicobacter pylori* infection in geriatrics. *Helicobacter* 7(1):56–62, 2002.

110. Roma-Giannikou, E., P.L. Shcherbakov. *Helicobacter pylori* infection in pediatrics. *Helicobacter* 7(1):50–55, 2002.

111. Solnick, J.V., D.B. Schauer. Emergence of diverse *Helicobacter* species in the pathogenesis of gastric and enterohepatic diseases. *Clin. Microbiol. Rev.* 14(1):59–97, 2001.

112. Bazzoli, F., Pozzato, P., T. Rokkas. *Helicobacter pylori*: the challenge in therapy. *Helicobacter* 7(1):43–49, 2002.

113. Somal, N.A., K.E. Coley, P.C. Molan, B.M. Hancock. Susceptibility of *Helicobacter pylori* to the antibacterial activity of manuka honey. *J. Royal Soc. Med.* 87:9–12, 1994.

114. Tabak, M., R. Armon, I. Potasman, I. Neeman. *In vitro* inhibition of *Helicobacter pylori* by extracts of thyme. *J. Appl. Bacteriol.* 80:667–672, 1996.

115. Aldana, L.P., M. Kato, S. Nakagawa, M. Kawarasaki, T. Nagasako, T. Mizushima, H. Oda, J. Kodaira, Y. Shimizu, Y. Komatsu, R. Zheng, H. Takeda, T. Sugiyama, M. Asaka. The relationship between consumption of antimicrobial agents and the prevalence of primary *Helicobacter pylori* resistance. *Helicobacter* 7(1):306–309, 2002.

116. Hamasaki, N., E. Ishii, K. Tominaga, Y. Tezuka, T. Nagaoka, S. Kadota, T. Kuroki, I. Yano. Highly selective antibacterial activity of novel alkyl quinolone alkaloids from a Chinese

herbal medicine, Gosyuyu (Wu-Chu-Yu), against *Helicobacter pylori in vitro. Microbiol. Immunol.* 44:9–15, 2000.

117. Higuchi, K., T. Arakawa, K. Ando, Y. Fujiwara, T. Uchida, T. Kuroki. Eradication of *Helicobacter pylori* with a Chinese herbal medicine without emergence of resistant colonies. *Amer. J. Gastroenterol.* 94:119–120, 1999.

118. Mabe, K., M. Yamada, I. Oguni, T. Takahashi. *In vitro* and *in vivo* activities of tea catechins against *Helicobacter pylori. Antimicrob. Agents Chemother.* 43:1788–1791, 1999.

119. Jonkers, D., E. van den Broek, I. van Dooren, C. Thijs, G. Hageman, E. Stobberingh. Antibacterial effect of garlic and omeprazole on *Helicobacter pylori. J. Antimicrob. Chemother.* 43:837–839, 1999.

120. Canizares, P., I. Garcia, L.A. Gómez, C.M. de Argila, L. de Rafael, A. García. Optimization of *Allium sativum* solvent extraction for the inhibition of *in vitro* growth of *Helicobacter pylori. Biotechnol. Prog.* 18:1227–1232, 2002.

121. Jones, N.L., S. Shabib, P.M. Sherman. Capsaicin as an inhibitor of the growth of the gastric pathogen *Helicobacter pylori. FEMS Microbiol. Lett.* 146:223–227, 1997.

122. Daroch, F., M. Hoeneisen, C. Gonzalez, F. Kawaguchi, F. Salgado, H. Solar, A. Garcia. *In vitro* antibacterial activity of Chilean red wines against *Helicobacter pylori. Microbios* 104:79–85, 2001.

123. Oh, Y., M.S. Osato, X. Han, G. Bennett, W.K. Hong. Folk yogurt kills *Helicobacter pylori. J. Appl. Microbiol.* 93:1083–1088, 2002.

124. Boyanova, L., S. Derejian, R. Koumanova, N. Katsarov, G. Gergova, I. Mitov, R. Nikolov, Z. Krastev. Inhibition of *Helicobacter pylori* growth *in vitro* by Bulgarian propolis: preliminary report. *J. Med. Microbiol.* 52:417–419, 2003.

125. Fukai, T., A. Marumo, K. Kaitou, T. Kanda, S. Terada, T. Nomura. Anti-*Helicobacter pylori* flavonoids from licorice extract. *Life Sciences*, 71:1449–1463, 2002.

126. Fahey, J.W., X. Haristoy, P.M. Dolan, T.W. Kensler, I. Scholtus, K.K. Stephenson, P. Talalay, A. Lozniewski. Sulforaphane inhibits extracellular, intracellular, and antibiotic-resistant strains of *Helicobacter pylori* and prevents benzo[a]pyrene-induced stomach tumors. *Proc. Natl. Acad. Sci. USA* 99(11):7610–7615, 2002.

127. Mahady, G.B., S.L. Pendland, G. Yun, Z.Z. Lu. Turmeric (*Curcuma longa*) and curcumin inhibit the growth of *Helicobacter pylori*, a group 1 carcinogen. *Anticancer Res.* 22:4179–4182, 2002.

128. Shinchi, K., H. Ishii, K. Imanishi, S. Kono. Relationship of cigarette smoking, alcohol use, and dietary habits with *Helicobacter pylori* infection in japanese men. *Scand. J. Gastroenterol.* 32:651–655, 1997.

129. Nagata, C., N. Takatsuka, N. Kawakami, H. Shimizu. A prospective cohort study of soy product intake and stomach cancer death. *Br. J. Cancer* 87:31–36, 2002.

130. McCue, P., K. Shetty. A role for amylase and peroxidase-linked polymerization in phenolic antioxidant mobilization in dark-germinated soybean and implications for health. *Process. Biochem.* 39:1785–1791, 2004.

131. McCue, D.A., A. Horii, K. Shetty. Solid-state bioconversion of phenolic antioxidants from defatted powdered soybean by *Rhizopus oligosprous*: role of carbohydrate cleaving enzymes. *J. Food Biochem.* 27:501–514, 2003.

132. McCue, P., K. Shetty. Role of carbohydrate-cleaving enzymes in phenolic antioxidant mobilization from whole soybean fermented with *Rhizopus oligosprorus. Food Biotechnol.* 17:27–37, 2003.

133. Gyorgy, P., K. Murata, H. Ikehata. Antioxidants isolated from soybeans (tempeh). *Nature* 203:870–872, 1964.

134. Akyon, Y. Effet of antioxidants on the immune response of *Helicobacter pylori. Eur. Soc. Clin. Microbiol. Infect. Dis.* 8:438–441, 2002.

135. McCue, P., K. Shetty. A biochemical analysis of ming bean (*Vigna radiata*) response to microbial polysaccharides and poential phenolic-enhancing effects for nutraceutical applications. *Food Biotechnol.* 16:57–79, 2002.

136. Nakamura, Y., A. Kaihara, K. Yoshii, Y. Tsumura, S. Ishimitsu, Y. Tonogai. Content and composition of isoflavonoids in mature or immature beans and bean sprouts consumed in Japan. *J. Health Sci.* 47:394–406, 2001.

137. McCue, P., Y.-T. Lin, R.G. Labbe, K. Shetty. Sprouting and solid-state bioprocessing by Rhizopus oligosporus increase the *in vitro* antibacterial activity of aqueous soybean extracts against *Helicobacter pylori. Food Biotechnol.,* 18:229–249, 2004.

138. Lin, Y.-T., R.G. Labbe, K. Shetty. Inhibition of *Vibrio parahaemolyticus* in seafood systems using oregano and cranberry phytochemical synergies and lactic acid. *Innovative Food Science and Emerging Technologies,* 6:453–458, 2005.

139. Vattem, D.A., Y.-T. Lin, R. Ghaedian, K. Shetty. Cranberry synergies for dietary management of *Helicobacter pylori* infections. *Process Biochem,* 40:1583–1592, 2005.

140. Ho, C-Y., Y.-T. Lin, R.G. Labbe, K. Shetty. Inhibition of *Helicobacter pylori* by phenolic extracts of sprouted peas (*Pisum sativum* L.). *J. Food Biochem.,* 30:21–34, 2006.

141. Ho, C.-Y., Y.-T. Lin, R.G. Labbe, K. Shetty. Phenolic extracts from sprouted chickpeas (Cicer arietinum L.) to inhibit ulcer-associated Helicobacter pylori. *J. Food Biochem.,* In preparation.

142. Randhir, R., Y.-T. Lin, K. Shetty. Phenolics, antioxidant and antimicrobial activity in dark germinated fenugreek sprouts in response to peptide and phytochemical elicitors. *Asia Pac. J. Clin. Nutr.,* 13:295–307, 2004.

143. Randhir, R., Y.-T. Lin, K. Shetty. Stimulation of phenolics, antioxidant and antimicrobial activities in dark germinated mung bean (*Vigna radiata*) sprouts in response to peptide and phytochemical elicitors. *Process. Biochem.* 39:637–646, 2004.

144. Vattem, D.A., Y.T. Lin, R.G. Labbe, K. Shetty. Phenolic antioxidant mobilization in cranberry pomace by solid-state bioprocessing using food grade fungus *Lentinus edodes* and effect on antimicrobial activity against select food-borne pathogens. *Innovative Food Sci. Emerg. Technol.* 5:81–91, 2004.

145. Vattem, D.A., Y.T. Lin, R.G. Labbe, K. Shetty. Antimicrobial activity against select food-borne pathogens by phenolic antioxidants enriched cranberry pomace by solid-state bioprocessing using food-grade fungus *Rhizopus oligosporus. Process Biochemistry,* 39:1939–1946, 2004.

146. Correia, R.T.P., P. McCue, D.A. Vattem, M.A.M. Margarida, G.R. Macedo, K. Shetty. Amylase and *Helicobacter pylori* inhibition by phenolic extracts of pineapple wastes bioprocessed by *Rhizopus oligosporus. J. Food Biochem.,* 28:419–434, 2004.

13

Biotechnology in Wine Industry

Moustapha Oke, Gopinadhan Paliyath, and K. Helen Fisher

CONTENTS

13.1 Introduction 311
13.2 Grape Cultivars and Wine Types 312
    13.2.1 *Vitis vinifera* Cultivars...313
        13.2.1.1 White Cultivars 313
        13.2.1.2 Red Cultivars 314
    13.2.2 French–American Hybrids 315
    13.2.3 American Hybrids 316
13.3 Genetic Engineering of Wine Grapes 316
    13.3.1 Clonal Selection 317
    13.3.2 Somaclonal Selection 318
    13.3.3 Biotechnology in Viticulture 318
13.4 Genetic Engineering of Yeast for Fermentation..319
    13.4.1 Ideal Yeast 320
    13.4.2 Yeast Breeding and Wine Quality 321
    13.4.3 Wine as a Functional Food...322
13.5 Conclusions 322
References..322

13.1 INTRODUCTION

Wine is a product of the alcoholic fermentation of grape juice, which is stored in a manner to retain its wine like properties (1). Clear evidence of winemaking dates from 5000 years ago, although archaeological records on wine date back to 6000 B.C. (2,3). It is believed that winemaking technology was first developed in Caucasia (currently part of Turkey, Iraq, Azerbaijan, and Georgia). The domestication of the commercial grape cultivar *Vitis vinifera* is also assumed to have originated in the same area (4). Some researchers believe that domestication of grapes may have independently occurred in Spain (5). From Caucasia, grape growing and winemaking spread toward Palestine, Syria, Egypt, Mesopotamia, and around the Mediterranean. In southern Spain however, there is evidence that extensive grape cultivation existed before the colonization of the region by the Phoenicians (6). It is not until the seventeenth century that modern winemaking technology became established. The quality of the wine was improved by the use of sulphur in barrels, glass bottles, and corks.

The grape growing regions of the world are located between 10°C and 20°C isotherms (lines drawn around the earth connecting the average yearly temperatures). It is also possible to grow grapes in cooler and subtropical regions, but on those regions viticultural practices are limited because of less than ideal conditions. Currently, grapes are important fresh fruit crops globally, with production reaching over 60 million metric tons (second only to bananas, with a production level of 68.6 million metric tons in 2001) (7). Italy, France, and Spain produced more than 35% of the total world production of grapes. More than 80% of the grapes grown each year are fermented, and the rest are used for fresh consumption or dried for raisins. World wine production has varied from 250 to 330 million hectolitres since 1970, with the production level reaching 282 million hectolitres in 2001. Italy, France, and Spain produce about 50% of the world's wine. Other major wine producers include the USA, Argentina, South Africa, Germany, Australia, Portugal, and Chile.

Wine consumption varies in different regions of the world. Wine consumers are found primarily in European countries. For centuries, wine has been an integral part of daily regional food consumption and an important source of calories. The political conditions and religious beliefs of certain regions, and alcohol abuse in other countries, have resulted in increased restriction in alcohol consumption. It has long been accepted that excessive alcohol consumption is detrimental to human health, although recent studies support the beneficial effects of moderate wine drinking (8,9). There has been a noticeable decline in world wine consumption due possibly to the shift toward the use of less, but better quality, wine. The production of high quality wines involves the use of excellent quality grapes, and of ideal strains of yeasts called superyeasts. The objective of this chapter is to evaluate the current technologies for improving wine quality.

13.2 GRAPE CULTIVARS AND WINE TYPES

There is no unanimous classification for grapevine cultivars even though there are nearly 15,000 of them. Growers in France use ecogeographic associations based on ampelographic (structural) and physiological properties of grapes, while others use aroma profiles to assess the relationship between cultivars (10). Numerical taxonomy is another way of classifying grapevine cultivars (11). The common disadvantage of these methods is the subjectivity of data interpretation. Grouping of cultivars based on evolutionary evidence may be obtained using modern techniques such as DNA fingerprinting (restriction fragment length polymorphisms and amplified fragment length polymorphisms -RFLP and AFLP) (12). The application of microsatellite (SSR) markers is now available to *Vitis* species (13–18). Raisins and grape berries have been typed by microsatellites (19,20), and clonal line diversity has been investigated with such markers (21,22). Many authors have evaluated these markers for cultivar identification, and parentage studies (14,19,23–30). AmpeloCAD, a computer aided digitizing system for determining ampelographic measurements is another means of simplifying the identification (31).

Generally the classification of grape cultivars is based on their specific or interspecific origin. Currently there are four majors groups of grapevine cultivars that include the pure line of *Vitis vinifera*, the French–American hybrids, the American hybrids (*Vitis labrusca*), and some interspecific cultivars (1).

13.2.1 *Vitis vinifera* Cultivars

The genus *Vitis* is divided into two distinct subgenera, *Vitis* and *Muscadinia*. Except for *V. rotundifolia* and *V. popenoei*, all the rest of the species are members of the largest subgenus *Vitis*. The difference between the two genera is in the number of their chromosomes; 38 for *Vitis* against 40 for *Muscadinia* (32). Crossing between species of *Vitis* and *Muscadinia* is possible, but often result in nonfertile progenies, due to the imprecise pairing and unequal separation of the chromosomes during meiosis. The genetic imbalance creating infertility is probably caused by the synthesis of inhibitors such as quercetin glycosides in the pistil (33). However, interspecies crossings among *Vitis* and *Muscadinia* are relatively easy and often yield fertile progeny. This, on the other hand, complicates the classification of grapevine species, because the differences in shoot and leaf morphological characteristics such as hairiness are strongly dependent on environmental conditions.

Most commercial cultivars belong to the species *V. vinifera*. Their reputation comes from their fine wines with distinctive, balanced, and long aging potential. Their weakness, however, is the low productivity and demanding cultivation practices. *Vitis vinifera* cultivars are used to produce both white and red wines. There is room for improving quality attributes of *V. vinifera* cultivars using modern techniques such as genetic engineering.

13.2.1.1 White Cultivars

Chardonnay is the most prestigious white French cultivar, and does well in most wine producing regions of the world. It also yields one of the finest sparkling wines. Under optimal conditions, chardonnay produces wine with an aroma reminiscent of apple, peach, and melon. The cultivar is susceptible to powdery mildew and bunch rot.

Riesling (White Riesling or Johannesburg Riesling) is Germany's most appreciated variety, popular also in the U.S. and in Australia. Wines made from these varieties of grapes have the characteristics of a fresh, aromatic, and well aged wine, reminiscent of roses. Fruit clusters possess small to medium size berries. Although a cold hardy variety, the yield is moderate, and it is also susceptible to powdery mildew and bunch rot.

Parellada is a Spanish variety from Catalonia, producing white wine with apple to citrus like aromas, and sometimes with hints of licorice or cinnamon. Pinot Blanc and Pinot Gris are popular in France, Germany, and other cool regions of Europe, and are used mostly in the production of dry, botrytised, and sparkling wines. They yield wines with the fragrances of passionfruit and hard cheese, respectively.

Sauvignon Blanc is originally from Loire Valley in France, but is also grown in California, Eastern Europe, Italy, and New Zealand. Some clones possess a floral character, but most Sauvignon Blanc wines have an herbaceous aspect, and the aroma of green peppers. Clusters produce small size berries, resistant to bunch rot and downy mildew. Sauvignon Blanc is susceptible to powdery mildew and black rot.

Muller-Thurgau variety was developed by H. Muller-Thurgau in 1882, presumably as a cross between Riesling and Silvaner. However the DNA fingerprint showed that it was more likely a cross between Riesling and Chasselas de Courtillier (34). Muller-Thurgau is probably the best known modern *V. vinifera* in Germany, but is also extensively grown in cool regions of Europe and in New Zealand. It is a high-yielding cultivar, producing a light wine with fruity fragrance and mild acidity. The fruits are susceptible to powdery and downy mildew and bunch rot.

Traminer is an aromatic cultivar, grown throughout the cooler regions of the world. It is used to produce dry and sweet white wines. Gewürztraminer is a clone with an intense fragrance and litchi fruit aroma. Savagnin, in France, is a clone of Traminer with a mild aroma. The basic characteristic of Traminer is the relatively small size of the fruit and their tough skin. The modest clusters are susceptible to powdery mildew, bunch rot, and coulure.

Muscat Blanc is mostly used in the production of dessert wines. Muscat grapes are extensively grown throughout the world because of their marked and distinctive muscaty aroma. They also possess high levels of soluble proteins and flavanoids. Muscato Bianco is used for the production of sparkling wine in Asti, Italy.

Another well known white varietie is Viognier, from the Rhône Valley of France, which is characterized by the smaller size of the fruit and its Muscat character. The wine matures quickly and has a peach or apricot fragrance. Viura, cultivated in Rioja, Spain is another white cultivar, with excellent attributes. Generally, clusters are few and fruits are large. The wine has a floral aroma with aspects of citron when produced in cooler regions, and a rich butterscotch or banana fragrance, after aging in wood. Some *V. vinifera* cultivars that are regionally popular, but less well known are Fiano, Garganega, and Torbato, in Italy; Malvasia, in Spain; Arinto, in Portugal; Tămîoasa Romînească, in Romania; and Furmint, in Hungary.

13.2.1.2 Red Cultivars

Cabernet Sauvignon is the most well known red variety, because of its association with Bordeaux, one of Europe's finest red wine producing areas. The wine has a black currant aroma under favorable conditions or a bell pepper aroma otherwise. The berries are small, seedy, and acidic, with darkly pigmented tough skin. The microsatellite study of Cabernet Sauvignon suggests a crossing between Cabernet Franc and Sauvignon Blanc (35). In order to accelerate the maturation of wine produced by Cabernet Sauvignon, it is blended with Merlot or Cabernet Franc. The cultivar is susceptible to several fungal diseases including powdery mildew, *Eurypa* and *Esca* wood decays, and phomopsis.

Merlot is a red variety grape, similar to Cabernet Sauvignon, which has the advantage of growing in cooler regions and moist soils. The ability to mature more quickly has made Merlot a popular substitute for Cabernet Sauvignon. The clusters are susceptible to coulure.

Pinot Noir is a famous French variety from Burgundy, suitable for the production of rosé and sparkling wines. Under favorable conditions it produces a wine with a distinctive aroma, which some researchers attribute to peppermint, beets, or cherries. Pinot Noir has a large number of clones such as Pinot Gris, Pinot Blanc, Meunier, Pinotage, (crossed with Cinsaut), and the California varietal Gamay Beaujolais (12,36). The clusters are compact with small to medium size fruits susceptible to bunch rot.

Barbera, ranked third after Sangiovese and Trebbiano in Italy, is also grown in California and in Argentina. The clusters possess long green stalks, and are susceptible to leaf roll and sometimes to bunch rot. Because of its high acidity, Barbera is often blended with some low acidic cultivars, but can also be used to make fruity wines by itself.

Gamay Noir à jus Blanc (Gamay with white juice), is mostly used in the production of Beaujolais wines. It produces fruit with medium size and tough skins. Gamay is a high yielding cultivar giving light red wines with little aroma. The fruity aroma of Beaujolais comes from the carbonic maceration technique. Gamay is susceptible to most fungal diseases.

Zinfandel is mostly grown in California and produces a wide range of wines varying from ports to light blush. The predominant aroma in rosé wines is reminiscent of raspberry. The full bodied wines possess berry flavors. The main problem with Zinfandel is that the fruits ripen unevenly.

Nebbiolo is basically grown in Northwestern Italy and adapts well in a wide range of soil pH and types. The wine produced from Nebbiolo is high in tannin and is acidic with a good aging potential. This cultivar is susceptible to powdery mildew and to bunch rot.

Tempranillo (Ull de Llebre) is the famous red variety in Spain, which is also grown in Argentina and in California under the name of Valdepeñas. Under optimal conditions it yields a fine wine with a complex berry jam fragrance with nuance of citrus. Tempranillo is sensitive to powdery and downy mildew.

Another important red variety of *V. vinifera* is Graciano, a Spanish variety with relatively low yield, but resistant to most fungal diseases and drought. Sangiovese is grown in central Italy and consists of a number of clones. It produces light to full bodied wines with nuances of citrus. The clusters are small to medium in size with oval fruits sensitive to bunch rot, powdery mildew and downy mildew. Syrah is a cultivar of the Rhône Valley in France, characterized by a low yield and high tannin content, allowing a good aging potential. It is also cultivated in Australia, where it is known under the name of Shiraz. The clusters are elongated with oval or small round fruits susceptible to drought, bunch rot, and berry moths.

Touriga National, mostly grown for the production of Port wine, is one of the famous Portugaese red cultivars. The clusters contain small berries and produce wine with a rich flavor and deep color. Ramisco, in Portugal, and Corvina and Dolcetto, in Italy, are nationally popular varieties with less international recognition.

13.2.2 French–American Hybrids

French–American hybrids are the second largest group of cultivars, and are derived from crosses between *V. vinifera* and one or more of the following: *V. riparia*, *V. rupestris*, and *V. aestivalis*. Primarily developed in France, their expansion was banned in the European Economic Community (EEC) in the late 1950s, because of their high yield and their non-traditional fragrance. Extensively grown in some South American countries and Asia as well, French–American hybrids have been widely used by North American wine industries since 1960. These hybrids were developed in order to overcome the difficulties and high expenses of grafting *V. vinifera* onto *Phylloxera* resistant rootstocks. The hybrids obtained are easy to grow with high yielding characteristics, and show reduced sensitivity to most of leaf pathogens. These varieties are used in Europe by the breeding programs as sources of disease resistance. The most popular French–American hybrids are:

Baco Noir, resulting from a cross between Folle Blanche and *V. riparia,* yields wine with a specific flavor and a very good aging potential. This hybrid is unfortunately sensitive to bunch rot, and several soil borne viruses.

Marechal Foch, derived from a cross between *V. riparia* and *V. rupestris* selections, with Goldriesling (Riesling × Courtiller Musqué). This hybrid has characteristics such as early maturation, winter hardiness, high productivity, and resistance to downy mildew, which makes it popular in North America.

Chambourcin is a Johannes Seyve hybrid of unknown parentage, and is presumed to be one of the best French–American hybrids. The cultivar was popular in France in the 1960s and 1970s, and also in Australia, because of its long growing season. Chambourcin is characterized with resistance to downy and powdery mildews and produces wines with a rich flavor.

Vidal Blanc has excellent winemaking and viticultural properties including tough skin and long maturity, making it suitable for the production of high quality ice wines. Vidal Blanc yields wine with Riesling like character, under favorable conditions. It is a cold hardy cultivar, but susceptible to soil borne virus diseases, as are many other French–American hybrids (37).

De Chaunac, once extensively grown in Eastern North America, has lost its importance nowadays, due to its sensitivity to soil borne viruses and the neutral character of the produced wines. De Chaunac is a Seibel cross of unknown parentage.

Seyval Blanc is a Seyve-Villard hybrid of *V. vinifera, V. rupestris,* and *V. aestivalis lincecumii.* It is a cold hardy cultivar, which produces wine with a pomade fragrance and fairly bitter finish. Seyval Blanc is sensitive to bunch rot, but tolerant to many soil types.

Delaware, which is a cross between *V. labrusca, V. aestivalis,* and *V. vinifera,* is another important French–American hybrid In the past, it has been used in the production of sparkling wines. Delaware needs well drained soils to grow and tends to crack. It is susceptible to various fungal pathogens, such as *Phylloxera.* Duchess, which produces wine with a mild fruity aroma, is a cross between *Vitis labrusca* and *V. vinifera,* and is difficult to grow. Magnolia and Noble are self fertile strains of muscadine cultivars, which yield sweet bronze colored and dark red fruits, respectively.

13.2.3 American Hybrids

American hybrids are cultivars bred in North America, from hybrids between indigenous grapevines, and hybrids between these and *V. vinifera,* or both. Most important American hybrids are derived from *V. labrusca.* The major characteristics of American hybrids are low sugar content and high acidity, and abundant flavor, e.g., Niagara, Concord and Catawba. A method of reducing the intense labrusca flavor is early picking and cold fermentation. Norton and Cynthiana, derived from *Vitis aestivalis* (38), are grown in Arkansas and Missouri, and are resistant to indigenous diseases and pests. The group of American hybrids derived from *V. rotundifolia* have low sugar content, but are resistant to indigenous diseases, especially to Pierce's disease. American hybrids are extensively grown in Eastern North America, South America, Eastern Europe, and Asia, although their significance has decreased in North America.

Another group of hybrids are interspecific cultivars derived from crossing *V. vinifera* and species such as *V. amurensis, V. riparia, V. armata,* and *V. rotundifolia,.* Orion, Regent, and Phoenix are new hybrids developed in Germany, with winemaking properties equal or superior to most *V. vinifera.* Veeblanc was developed in Ontario from complex parentage and generates mild quality wine. Cayuga White from New York is a new hybrid with excellent winemaking attributes and resistance to fungal diseases. Its fragrance resembles apples and tropical fruits, with a rich mouth feel. The quality attributes of wine grapes can be improved by genetic engineering and clonal or somaclonal selection.

13.3 GENETIC ENGINEERING OF WINE GRAPES

The goal of grape breeding is to enhance agronomic properties of rootstock varieties and to improve winemaking attributes of fruit-bearing cultivars. There are a number of ways by which this can be accomplished, including simple cross and backcross between appropriate varieties. The improvement often involves the addition of novel qualities such as drought tolerance and disease resistance to established cultivars.

Standard breeding techniques are time consuming and expensive. Genetic engineering introduces the genes of choice without disrupting varietal characteristics of cultivars. The steps involved in genetic engineering include the isolation, amplification, and insertion of the target gene into the intended organism. An example is the insertion of a protein coat gene for the grapevine fanleaf virus (GFLV) into rootstock varieties in order to enhance their resistance (39). There are several laboratories around the world working on gene transfer technologies in grape vines. It has been reported that several transformed grape varieties are being evaluated in France (40) and the U.S. In Geneva, New York, a Thompson Seedless variety with an incorporated gene for resistance to tomato ringspot virus is being evaluated. Experiments in France include evaluation of rootstocks and Chardonnay carrying resistance genes to fanleaf virus; and another variety, Richter 1110, with an incorporated gene for resistance to chrome mosaic virus (CMV) (41). Attempts have also been made to introduce the chitinase gene into embryonic cultures of Merlot, Chardonnay, Pinot Noir, Concord, and Niagara varieties, through biolistic transformation. Increased chitinase activity may afford protection from fungal pathogens, such as *Botrytis*, which causes bunch rot, and *Uncinula necator*, which causes powdery mildew (42). These varieties are being evaluated for increased disease resistance. Superoxide dismutase is a key enzyme involved in the detoxification of active oxygen species and it is believed that a higher superoxide dismutase activity will confer added protection to grapevines during extreme cold conditions. A gene for superoxide dismutase was isolated from *Arabidopsis thaliana* and inserted into scion varieties in order to improve their cold resistance. A problem associated with these techniques is the uncertainty of the transfer and expression of traits under multigene control in another organism. Although genetic engineering requires isolated cells or tissues grown in tissue culture to develop into whole plants, the induction of embryos from grape tissue or cells is difficult. For transformed cells to be of value, they must be able to form embryos and differentiate into whole plants. The phenotypic variability of the progeny is another problem with propagating vines that originate from cells grown in tissue culture. The variability decreases with vine age or repeated propagation. Using buds from the fortieth node and onward from the stem apex, and grafting them onto desirable rootstocks, is a way to circumvent this problem (43). Progress has been made in both augmenting the proportion of embryonic tissue that develops into vines (44), and in understanding the origins of poor plantlet yield from grape cells (45).

13.3.1 Clonal Selection

Clones are forms of cultivars, derived vegetatively from a single parental plant, such that all derivatives are initially genetically identical (1). Clonal selection is a series of procedures designed to provide a premium stock of known characteristics. Basically, cuttings are multiplied and repeatedly assessed for their viticultural and fruit-bearing characteristics and their winemaking properties, as well as their resistance to systemic pathogens. Clonal selection is the primary means by which cultivar characteristics can be modified without significantly changing varietal attributes. Variations in monoclonal varieties consist of mutations accumulated since the origin of the cultivar which are reproduced unchanged in cuttings of the mutated clone, whereas in polyclonal varieties, variation occurs because the cultivar exists as a group of closely related strains. The difference between related clones may be phenotypical, such as color variants of Pinot Noir, Pinot Gris, and Pinot Blanc, or biotypical (almost genotypical), such as Sangiovese strains (46). Clones are grouped according to the combination of traits they possess. For example, among Pinot Noir cultivars, Pinot Fin, with trailing low yielding vines having small fruit clusters; Pinot Droit, having a high yielding vine with upright shoots; Pinot Fructifer, which has high yielding strains; and Mariafeld-type strains, with loose clustered, moderate yielding vines, show the variability in phenotypic characteristics (47). The main objective of clonal selection is the elimination of all systemic infectious organisms including pathogens. This is not always possible, as some systemic agents are not known to cause disease, and reinfection is always possible. The weakness of clonal selection is the resulting phenotypic variability, which is difficult to eliminate. In order to control the desirable variations, some breeders suggest planting several clones rather than just one (48).

Often, clonal instability is due to chimeric mutations, which occur in embryonic cells, and, as a result, cause vine tissues to differ from one another. Another important goal of clonal selection is the improvement in yield and grape quality. There is a negative association between yield and quality, but this can be overcome through rigorous viticultural practices such as more open canopy, increased plant density, and basal leaf removal. An exception to this rule is the Weinberg 29 clone of Riesling, which gives a high yield and shows a high level of sugar content in the fruits (49).

The assessment of quality is complex, time consuming, and costly. There is no unanimous criterion for quality. However, some parameters such as, ° Brix, pH, and acidity are used to assess maturity. Parameters such as the level of glycosyl glucose (G-G) in the must, the polyphenol content, and the color density and hue are commonly used as potential indicators of wine quality (50). The presence of specific flavor impact compounds such as 2-methyl-3-isobutylpyrazine in Cabernet Sauvignon is also another important characteristic. Unique varietal aromas can be integrated into clonal selection. Examples of this kind include the Muscat nuance of Chardonnay 77 and 809; the difference in wine making potential of Gamay 222 versus Gamay 509 in the production of Beaujolais cru vin de garde and vin nouveau (48). Because properties of a clone are dependent upon environmental conditions, some desirable traits may not be expressed, which requires vineyard growers to make their own assessment of a given clone.

13.3.2 Somaclonal Selection

Although clonal selection has reduced viral infection in grapes, more advances may come from incorporation of new genetic information through genetic engineering or induced mutations. Mutations can be created by exposing meristematic tissue or cells in tissue culture to mutagenic chemicals or radiation. Somaclonal selection enhances the expression and selection of clonal variation. It also involves the selective growth enhancement of cell lines. For example, isolation of transgenic vines or lines possessing tolerance for salinity or fungal toxins (51) is done by exposing cells to mutagenic agents, such as chemicals and radiation, during cell culture (39,52). However, the tolerance shown by tissue in culture may not always be expressed at the whole plant level (53).

13.3.3 Biotechnology in Viticulture

Biotechnology has been extensively applied in viticulture since the early work of breeders like F. Baco, B. Seyve, and A. Seibel. The main goal then was to develop cultivars with enhanced wine-producing characteristics of *Vitis vinifera* and the *Phylloxera* resistance of American species. Breeding has focussed on developing rootstocks resistant to drought, salt, lime, viruses, and nematodes. For example, *V. cinerea helleri* can donate resistance to lime induced chlorosis; *V. champinii* can supply tolerance to root knot nematodes; *V. vulpine* can provide drought tolerance under shallow soil conditions. Improvements have also been made in developing new pest and disease resistant fruit varieties. There is also a trend in producing grapevine varieties that produce fruits having increased nutritional value, such as antioxidant and vitamin contents, thereby improving the quality of the grapes and wine. Unfortunately, consumers are still sceptical of genetically modified organisms, which limits their acceptance. The benefits of such resistant varieties are their reduced production costs and use of pesticides. Their disadvantages however, are legal restrictions against interspecies crosses in the production of Appellation Contrôlée wines in most European countries. Recently 40 out of 60 Masters of Wine voted against any introduction of genetically modified (GM) vines or geneticallk modified organisms (GMOs) in any step of the wine making process (54). Despite these controversies, genetic engineering appears to be the most efficient way of successfully developing disease resistant cultivars.

Researchers are experimenting with the isolation and insertion of specific genes for resistance such as genes regulating the activities of chitinase, β-glucanase, and peroxidase isoenzymes.

13.4 GENETIC ENGINEERING OF YEAST FOR FERMENTATION

The two main organisms involved in fermentation are *Saccharomyces cerevisiae* and *Leuconostoc oenos*. Traditionally, indigenous yeasts conduct the fermentation, but induced fermentation with selected yeast strains is a standard procedure in most parts of the world (55). The reason for using specific yeasts is to avoid the production of off flavors sometimes associated with wild yeasts. New techniques such as mitochondrial DNA sequencing (56) and gene marker analysis (57) allow identification of strains responsible for fermentation. Regardless of yeast inoculation, red and white grape musts contain a significant amount of wild yeast, but their metabolic activity during induced fermentation is unknown. Thus, their significance in vinification is not clear. *Kloeckera apiculata* is the most frequently isolated wild yeast in grapes, believed to contribute greatly to the complexity of the wine. Other yeasts occasionally isolated are: *Brettanomyces, Candida, Debaryomyces, Hansenula, Kluyveromyces, Metschnikowia, Nadsonia, Pichia, Saccharomycodes,* and *Torulopsis* (58). *Saccharomyces cerevisiae* is the most important yeast species, because it may function as wine yeast, baker's yeast, distiller's yeast, and brewer's yeast. The original habitat of the yeast is uncertain (59), and only one healthy grape out of a thousand carries ideal wine yeasts (60). *Saccharomyces paradoxus* (*S. cerevisiae tetrasporus*), isolated from oak tree exudates, is assumed to be the ancestral form of *Saccharomyces cerevisiae*. Although *S. cerevisiae* is sometimes isolated from the intestinal track of fruit flies, the importance of insects in dispersal of *Saccharomyces* is not clear (59,61). *Saccharomyces* species such as *S. uvarum* and *S. bayanus* also can effectively conduct fermentations, and they are used in special applications; e.g., *S. uvarum* ferments well at temperatures down to 6°C and synthesizes desirable sensory components, whereas *S. bayanus* is well adapted for the production of sparkling wines and fine sherries. Other wild yeasts of sensory significance are *Candida stellata, Torulopsis delbruekii* and *Kloeckera apiculata. Candida stellata* persists in fermenting juice (62), and produces high concentrations of glycerol, increasing the mouth feel in wine. *Torulopsis delbruekii* positively influences sensory attributes of wine by producing low concentrations of acetic acid and succinic acid. *Kloeckera apiculata*, however, can produce up to 25 times the typical amount of acetic acid produced during induced fermentations (63), and is capable of producing above threshold amounts of 2-aminoacetophenone, giving a naphthalene like off odor to wines (64). *Kloeckera apiculata* may inhibit *S. cerevisiae* from completing fermentation (65). *Kloeckera apiculata* and *C stellata* normally produce only 4 to 10% alcohol, but can survive higher alcohol concentrations (66,67). Most of the other members of indigenous yeasts found on grapes are suppressed by low pH, high alcohol, oxygen deficiency, and sulphur dioxide concentrations. Most bacteria are inhibited by *S. cerevisiae*, except lactic acid bacteria. Commercially available strains of *S. cerevisiae* possess a wide range of winemaking characteristics such as low production of acetic acid, hydrogen sulphide (H$_2$S), and urea; the ability to produce fruit esters in abundance; and to enhance varietal flavorants. Yeasts can also be selected for their alcohol tolerance, fermentation speed, ability to restart stuck fermentation, production of specific wine styles (such as after carbonic maceration), resistance to killer factors, ability to ferment glucose or fructose selectively, fermentation at high pressure and at low temperature, and their ability to flocculate rapidly and completely after fermentation.

13.4.1 Ideal Yeast

Taking into account that each type of wine requires different types of yeast, it may be difficult to find an ideal type of yeast. However, the following are major characteristics of high quality wine yeasts:

1. Fermentation Speed: An ideal type of yeast should be able to initiate and complete the fermentation as quickly as possible after inoculation. The higher sugar content of the must, which creates an elevated osmotic pressure, is a limiting factor for most wine yeasts.

2. Alcohol tolerance: Yeast should be able to ferment all the sugars in the must in less than 3 weeks and be able to withstand up to 15% ethanol. Most fermentations stop at 1–2% sugar content creating conditions for spoilage. There are commercial strains of yeasts available that have the ability to restart stuck fermentations or to perform the second fermentation during sparkling wine production.

3. SO_2 Tolerance: Most of the yeasts are sensitive to SO_2, and sulphites are used as an antioxidant and a microbial inhibitor. The allowed concentration of SO_2 in the finished product is 70 parts per million in free state, and 350 parts per million in combined state, respectively. An ideal type of yeast should be able to dominate the fermentation even at this level of SO_2.

4. Cold tolerance: Cold tolerant yeasts are used especially for white wine production in order to keep the loss of the bouquet to a minimum. The fermentation temperature is often below 14°C.

5. Low foaming activity: Ideal yeasts should have a low foaming activity. High foaming yeasts spoil the quality of wine by producing off odors and off flavors.

6. Efficient conversion of sugar into alcohol: Ideal yeasts have the ability to convert sugar primarily into alcohol and produce low levels of other metabolites.

7. Production of desirable metabolites: Several secondary products of yeasts are components of the bouquet and flavor of the wine. The quality of a given wine is therefore dependent upon the production of the right amount of these metabolites. This is particularly important for those types of wines with low aging potential such as the Gamay Nouveau wines of Ontario, consumed within three months after production.

8. Low production of undesirable metabolites: The selection of yeasts is based on their ability to synthesize low levels of undesirable metabolites such as acetaldehyde, acetic acid, sulphur dioxide, hydrogen sulphide, and mercaptans.

9. Resistance to "Killer" yeasts: Most of the commercial yeasts are killer strains with the ability to ferment sugars completely and to resist toxins of contaminating killer yeasts. Killer yeasts are wild yeasts that possess the ability to synthesize and excrete proteins that kill other strain of yeasts. This is the reason why the yeast breeders breed yeasts containing the killer genes.

10. Flocculation: An ideal yeast strain should flocculate at the end of the fermentation, leaving the wine clear and requiring less rigorous filtration. Filtration negatively influences the complexity of wine.

11. Producing or reducing certain organic acids: Yeast strains are selected on the basis of their ability to deplete malic acid and other organic acids found in wine. This is very important taking into account the net difference in the acid content of grapes grown in hot and cool climate regions. Hot growing regions are ideal for the production of low acid grapes having high sugar content, whereas cool climate regions tend to produce high acid grapes.

13.4.2 Yeast Breeding and Wine Quality

The use of genetic engineering is very limited for improving wine yeasts, compared to yeasts used for other industrial fermentations. The reason for this is that more importance has been given to grape variety, fruit maturation, and fermentation temperature. Also, during genetic improvement, features controlled by one gene are easily influenced by other genes. For example, the expression of flocculation ability in yeasts is regulated by several genes, including epistatic genes and, possibly, cytoplasmic genetic factors (68). On the other hand, inactivating the gene that encodes sulphites reductase restricts the conversion of sulphites to sulphide (H_2S). Genetic improvement is attainable, although this may take time to achieve because many important enological attributes, such as the ability to ferment at low temperature and alcohol tolerance, are controlled by a combination of several genes. The consequences of changing the direction of a metabolic pathway on other characteristics of the yeasts are not very clear at present. For example increasing ester production could interfere with the alcohol tolerance ability of the strain (69). It is easier when the compound concerned is the end product of a pathway. There were no apparent undesirable side effects when genes for terpene synthesis were incorporated from a lab strain into a wine strain of *S. cerevisiae* (70). Rainieri and coworkers (71) found that thermotolerant *Saccharomyces cerevisiae* strains possess oenological potential and provide an important genetic resource for yeast improvement programs. In the search for a substitute for sulphite in winemaking, Suzzi and coworkers (72) selected a strain of *Saccharomyces cerevisiae* No 10278, able to form 30 to 80mg/L of sulphite, with an excellent stabilizing power. Techniques available for researchers for improving wine yeast qualities include selection and modification of the genetic makeup of the yeast. Simple selection is easy, if a selective medium is developed so that only cells containing the desirable trait can grow; otherwise, cells must be randomly isolated and individually studied for the presence of the desired trait. Modifying the genetic makeup involves procedures such as hybridization, mutagenesis, backcrossing, transformation, somatic fusion, and genetic engineering. To incorporate single traits into a desirable strain, backcross breeding is the preferred technique. Combined with a selection, backcross eliminates undesirable donor genes accidentally incorporated in the original cross.

Somatic fusion is used if a desirable feature missing in *S. cerevisiae* is found in another yeast species. This procedure requires enzymatic cell wall dissolution and mixing of protoplasts generated in the presence of polyethylene glycol, which enhances cell fusion. Instability is the major problem with somatic fusion, because fused cells often revert to one or the other of the original species.

On the other hand, the advantage of genetic transformation is that donor and recipient need not be closely related. The procedure involves bathing yeast protoplasts in a solution containing DNA from the donor organism. By this technique, the malolactic gene from *Lactococcus lactis* was transferred into *S. cerevisiae* along with the malate permease transport gene from *Schizosaccharomyces pombe* (73–76). Other genes of interest are pectic lyase to increase fruity flavor and β- (1-4)- endoglucanase to increase aroma. Boone and coworkers (77) integrated the yeast K1 killer toxin gene into the genome of laboratory and commercial yeasts by a gene replacement technology that generated recombinants containing only yeast DNA. The integration of the K1 killer gene into two K2 wine yeasts generated stable K1/K2 double killer strains, which have a wider spectrum of killing and a competitive advantage over other sensitive and killer strains of *Saccharomyces cerevisiae* in wine fermentations. Genetic engineering is useful if specific traits are desired, but not when multigenic characteristics are involved. However, the ethics of releasing genetically modified organisms that could enter the environment is questionable.

The main problem in breeding *S. cerevisiae* is its diploid character, which means each cell contain two copies of the genes. The diploid nature complicates improvement by masking the potentially desirable genes. More difficulties occur due to the low frequency of sexual reproduction, poor spore germination, and rapid return to the diploid form (75). Stability is the fundamental requirement for the yeast strain, which means that, for instance, flocculant strains should not loose their ability to flocculate.

13.4.3 Wine as a Functional Food

The beneficial effects of wine on human health may originate from their nonalcoholic components, such as flavanoids and phenols. Some phenolic compounds can be protective, whereas others have been found to be mutagenic at high doses in laboratory studies. Fazel and coworkers (78) found that quercetin can induce mutation during *in vitro* culture of animal cells, but is an anticarcinogen in whole animal dietary studies. This anomaly may be due to the differences in the concentrations of quercetin used and the low level of metal ions and free oxygen found in the animal body. Subbaramaiah and coworkers (79) observed that resveratrol, a member of the stilbene family found in wine, can inhibit the production of cyclooxygenase-2, thought to be important in carcinogenesis. As well, flavanols and flavones strongly reduce the action of the common dietary carcinogens such as the heterocyclic amines (80). In this regard, it is important to note that wines made from fruits such as cherry, blackberry and blueberry, and red grape show a very high complement of superoxide and hydroxyl radical scavenging ability due to the presence of several phenolic components (81). The flavanoids are also strong inhibitors of calcium second messenger function in animal systems, and the biological activities of wines may, to a large extent, result from this activity. Recently, during *in vitro* culture, the flavonoid wine components have been shown to be specifically cytostatic and cytotoxic to MCF-7 cells, which are estrogen receptor positive breast cancer cells. Under similar conditions, the normal human mammary epithelial cells were unaffected (82). Thus, by enhancing the components in the wine that afford health beneficial effects, the functional food value of the wine can be enhanced. Even though moderate wine consumption is believed to be beneficial to cardiovascular health, its effect on breast cancer development is still controversial. Several biotechnological approaches such as *in vitro* culture, genetic engineering of grapes for enhanced secondary metabolite biosynthetic pathways, and conditions of fermentation could ultimately enhance the functional food value of the wine.

13.5 CONCLUSIONS

It is possible to improve the quality attributes of wine using modern techniques such as selective breeding, hybridization, somatic fusion, transformation, and genetic engineering. They can be applied both to wine grapes and to yeasts. Although the safety of genetically modified food is an issue, the advantages of adapting these technologies are enormous; as they may help provide better yield, environmental protection and high quality wine products.

REFERENCES

1. Jackson, R.S. *Wine Science: Principles, Practices and Perception.* New York: Academic Press, 2000.
2. McGovern, P.E., D.L. Glusker, L.J. Exner, M.M. Voigt. Neolithic resinated wine. *Nature* 381:480–481, 1996.
3. Petrie, W.M.F. *Social Life in Ancient Egypt.* London: Methuen, 1923.
4. Zohary, D., M. Hopf. *Domestication of Plants in the Old World.* Oxford: Oxford University Press, 1988.

5. Nùñez, D.R., M.J. Walker. A review of paleobotanical findings of early Vitis in the Mediterranean and of the origins of cultivated grape-vines, with special references to prehistoric exploitation in the western Mediterranean. *Rev. Paleobot. Palynol.* 61:205–237, 1989.

6. Stevenson, A.C. Studies in the vegetational history of S.W. Spain, II: palynological investigations at Laguna de los Madres, Spain. *J. Biogeogr.* 12:293–314, 1985.

7. FAO. FAOSTAT database. http://apps1.fao.org/servlet/XteServlet?areas. September 2002.

8. Rimm, E.B., E.I. Giovannuchi, W.C. Willet, G.A. Colditz, A. Ascerio, B. Rosner, M.J. Stampfer. Prospecive study of alcohol consumption and risk of coronary disease in men. *Lancet* 338:464–468, 1991.

9. Soleas, G.J., E.P. Diamandis, D.M. Goldberg. Wine as a biological fluid: history, production and role in disease prevention. *J. Clin. Lab. Anal.* 11:287–313, 1997.

10. Lefort, P.L. Biometrical analysis of must aromagrams: application to grape breeding. In: *Proceedings of the 3rd International Symposium on Grape Breeding*, University of California, Davis, 1980, pp 120–129.

11. Fanizza, G. Multivariate analysis to estimate the genetic diversity of wine grapes (*Vitis vinifera*) for cross breeding in southern Italy. In: *Proceedings of the 3rd International Symposium on Grape Breeding*, University of California, Davis, 1980, pp 105–110.

12. Bowers, J.E., E.B. Bandman, C.P. Meredith. DNA fingerprint characterization of some wine grape cultivars. *Am. J. Enol. Vitic.* 44:266–274, 1993.

13. Bowers, J.E., G.S. Dangl, R. Vignani, C.P. Meredith. Isolation and characterization of new polymorphic simple sequence repeat loci in grape (*Vitis vinifera* L.). *Genome* 39:628–633, 1996.

14. Bowers, J.E., G.S. Dangl, C.P. Meredith. Development and characterization of additional microsatellite DNA markers for grape. *Am. J. Enol. Vitic.* 50(3):243–246, 1999.

15. Sefc, K.M., F. Regner, E. Turetschek, J. Glössl, H. Stcinkellner. Identification of microsatellite sequences in *Vitis riparia* and their applicability for genotyping of different *Vitis* species. *Genome* 42:1–7, 1999.

16. Thomas, M.R., N.S. Scott. Microsatellite repeats in grapevine reveal DNA polumorphisms, when analysed as sequence-tagged sites (STSs). *Theor. Appl. Genet.* 86:173–180, 1993.

17. Thomas, M.R., S. Matsumoto, P. Cairn, N.S. Scott. Repetitive DNA of grapevine: classes present and sequences suitable for cultivar identification. *Theor. Appl. Genet.* 86:173–180, 1993.

18. Thomas, M.R., N.S. Scott, R. Botta, J.M.H. Kijas. Sequence-tagged site markers in grapevine and citrus. *J. Jpn. Soc. Hortic. Sci.* 67:1189–1192, 1998.

19. Lefort, F., K.K.A. Roubelakis-Angelakis. Genetic comparison of Greek cultivars of *Vitis vinifera* L. by nuclear microsatellite profiling. *Am. J. Enol. Vitic.* 52:101–108, 2001.

20. Sefc, K.M., S. Guggenberger, F. Regner, C. Lexer, J. Glössl, H. Steinkellner. Genetic analysis of grape berries and raisins with microsatellite markers. *Vitis* 37:123–125, 1998.

21. Silvestroni, O, D. Di Pietro, C. Intrieri, R. Vignani, I. Filipetti, C. Del Casino, M. Scali, M. Cresti. Detection of genetic diversity among clones of cv. Fortana (*Vitis vinifera* L.) by microsatellite DNA polymorphism analysis. *Vitis* 36:147–150, 1997.

22. Vignani, R., J.E. Bowers, C.P. Meredith. Microsatellite DNA polymorphism analysis of clones of *Vitis vinifera* 'Sangiovese.' *Scientia. Hort.* 65:163–169, 1996.

23. Botta, R., N.S. Scott, I. Eynard, M.R. Thomas. Evaluation of microsatellite sequence-tagged site markers for characterizing *Vitis vinifera* cultivars. *Vitis* 34:99–102, 1995.

24. Wilson, E.K. So many grapes, so little time. *Chem. Eng. News* 79:37–39, 2001.

25. Cipriani, G., G. Frazza, E. Peterlunger, R. Testolin. Grapevine fingerprinting using microsatellite repeats. *Vitis* 33:211–215, 1994.

26. Grando, M.S., U. Malossini, I. Roncador, F. Mattivi. Parentage analysis and characterization of some Italian *Vitis vinifera* crosses. *Acta Hort.* 528:183–187, 2000.

27. Sefc, K.M., H. Steinkellner, H.W. Wagner, J. Glössl, F. Regner. Application of microsatellite markers for parentage studies in grapevine. *Vitis* 36:79–183, 1997.

28. Sefc, K.M., F. Regner, J. Glössl, H. Steinkellner. Genotyping of grapevine and rootstock using microsatellite markers. *Vitis* 37:15–20, 1998.

29. Sefc, K.M., H. Steinkellner, J. Glössl, S. Kampfer, F. Regner. Reconstruction of grapevine pedigree by microsatellite analysis. *Theor. Appl. Genet.* 97:227–231, 1998.

30. Thomas, M.R., P. Cairn, N.S. Scott. DNA typing of grapevines: a universal methodology and database for describing cultivars and evaluating genetic relatedness. *Plant Mol. Biol.* 25:939–949, 1994.

31. Alessandri, S., N. Vignozzi, A.M. Vignini. AmpeloCADs (Ampelographic Computer-Aided Digitizing System): an integrated system to digitize, file, and process biometrical data from *Vitis.* leaves. *Am. J. Enol. Vitic.* 47:257–267, 1996.

32. Patel, G.I., H.P. Olmo. Cytogenetics of Vitis, I: The hybrid *V. vinifera* x *V. rotundifolia. Am. J. Bot.* 42:141–159, 1955.

33. Okamoto, G., Y. Fujii, K. Hirano, A. Tai, A. Kobayashi. Pollen tube growth inhibitors from Pione grape pistils. *Am. J. Enol. Vitic.* 46:17–21, 1995.

34. Sefc, K.M., H. Steinkellner, H.W. Wagner, J. Glossl, F. Regner. Application of microsatellite markers to parentage studies in grapevine. *Vitis* 36:179–184, 1997.

35. Bowers, J.E., C.P. Meredith. The parentage of a classic wine grape, Cabernet Sauvignon. *Nat. Genet.* 16:84–87, 1997.

36. Regner, F., A. Stadlbauer, C. Eisenheld, H. Kaserer. Genetic relationships among Pinots and related Cultivars. *Am. J. Enol. Vitic.* 51:7–14, 2000.

37. Alleweldt, G. Disease resistant varieties. In: *Proceedings of the 8th Australian Wine Industry Technology Conference*, Stockley, C.S., et al., eds., 1993, pp 116–119.

38. Reisch, I.B., R.N. Goodman, M.H. Martens, N.F. Weeden. The relationships between Norton and Cynthians, red wine cultivars derived from *Vitis aestivalis. Am. J. Enol. Vitic.* 44:441–444, 1993.

39. Mauro, M.C., S. Toutain, B. Walter, L. Pink, L. Otten, P. Coutos-Thevenot, A. Deloire, P. Barvier. High efficiency regeneration of grapevine plants transformed with the GFLV coat protein gene. *Plant Sci.* 112:97–106, 1995.

40. Carbonneau, A. The early selection of grapevine rootstocks for resistance to drought conditions. *Am. J. Enol. Vitic.* 36:195–198, 1985.

41. Kikkert, J.R., B.I. Reisch. Genetic engineering of grapevines for improved disease resistance. In: *Grape Research News*, 7, Goffinet, M., ed., Geneva, New York: New York State Agricultural Experiment Station, 1996.

42 Kikkert, J.R., G.S. Ali, M.J. Striem, M.-H. Martens, P.G. Wallace, L. Molino, B.I. Reisch. Genetic engineering of grapevine (*Vitis* sp) for enhancement of disease resistance. *Acta Hort.* 447:273–279, 1997.

43. Grenan, S. Multiplication *in vitro* et charactéristiques juveniles de la vigne. *Bull. O.I.V.* 67:5–14, 1994.

44. Reustle, G., M. Harst, G. Alleweldt. Plant regeneration of grapevine (*Vitis* sp.) protoplast isolated from embryogenic tissue. *Plant Cell Rep.* 15:238–241, 1995.

45. Maes, O., P. Coutos-Thevenot, T. Jouennem, M. Boulay, J. Guern. Influence of extracellular proteins, proteases and protease inhibitors on grapevine somatic embryogenesis. *Plant Cell. Tissue Org. Cult.* 50:97–105, 1997.

46. Calō, A., A. Costacurta, G. Paludetti, M. Crespan, M. Giust, E. Egger, A. Grasselli, P. Storchi, D. Borsa, R. di Stephano. Characterization of biotypes of Sangiovese as a basis for clonal selection. In: *Proceedings of the International Symposium on Clonal Selection*, Rantz, J.M., ed., Davis, CA: American Society of Enology Viticulture, 1995, pp 99–104.

47. Wolpert, J.A. An overview of Pinot noir closes tested at UC, Davis. *Vineyard Winery Manage.* 21:18–21, 1995.

48. Boidron, R. Clonal selection in France: methods, organization, and use. In: *Proceedings of the International Symposium on Clonal selection*, Rantz, J.M., ed., Davis, CA: American Society of Enology Viticulture, 1995, pp 1–7.

49. Schöfflinger, H., G. Stellmach. Clone selection of grapevine varieties in Germany. *Fruit Var. J.* 50:235–247, 1996.

50. Somers, C. *The Wine Spectrum*. Adelaide, Australia: Winetitles, 1998.

51. Soulie, O., J.-P. Roustan, J. Fallot. Early *in vitro* selection of Eupypine-tolerant plantlets. Application to screening of *Vitis vinifera* cv. Ugni blanc somaclones. *Vitis* 32:243–244, 1993.

52. Kikkert, J.R., D. Hébért-Soulé., P.G. Wallace, M.J. Striem, B.I. Reisch. Transgenic plantlets of 'Chancellor' grapevine (*vitis* sp.) from biolistic transformation of embryonic cell suspensions. *Plant Cell Rep.* 15:311–316, 1996.

53. Lebrun, L., K. Rajasekaran, M.G. Mullins. Selection *in vitro* for NaCl-tolerance in *Vitis rupestris* Scheele. *Ann. Bot.* 56:733–739, 1985.

54. http://www.plant.uoguelph.ca/safefood/archives/agnet-archives.htm. Visited on July 29, 2002.

55. Barre, P., F. Vezinhet. Evolution towards fermentation with pure culture yeasts in wine making. *Microbiol. Sci.* 1:159–163, 1984.

56. Dubourdieu, D., A. Sokol, J.J. Zucca, P. Thalouarn, A. Dattee, M. Aigle. Identification des sources de levures isolées de vins par l'analyse de leur AND mitochondrial. *Connaiss. Vigne Vin.* 21:267–278, 1987.

57. Petering, J.E., P.A. Henschke, P. Langridge. The *Escherichia coli* β-glucuronidase gene as a marker for *Saccharomyces* yeast strain identification. *Am. J. Enol. Vitic.* 42:6–12, 1991.

58. Lafont-Lafourcade, S. Wine and Brandy. In: *Biotechnology Vol.5, Food and Feed Production with Microorganisms*, Reed, G., ed., Weinheim; Deerfield Beach, Florida; Basel: Verlag Chemie, 1983, pp 81–163. ISBN-0-89573-045-6.

59. Phaff, H.J. Ecology of yeast with actual and potential value in biotechnology. *Microb. Ecol.* 12:31–42, 1986.

60. Mortimer, R., M. Polsinelli. On the origin of wine yeast. *Res. Microbiol.* 150:199–204, 1999.

61. Wolf, E., I. Brenda. Qualität und resistenz, III. Das Futterwahlvermögen von *Drosophila melanogaster* gegenüber natürlichen Weinhefe-Arten und - Rassen. *Biol. Zentralbl.* 84:1–8, 1965.

62. Fleet, G.H., S. Lafon-Laforcade, P. Ribéreau-Gayon. Evolution of yeasts and lactic acid bacteria during fermentation and storage of Bordeaux wines. *Appl. Environ. Microbiol.* 48:1034–1038, 1984.

63. Ciani, M., F. Maccarelli. Oenological properties of non-*Saccharomyces* yeasts associated with winemaking. *World J. Microbiol. Biotechnol.* 14:199–203, 1998.

64. Sponholz, W.R., T. Hühn. Aging of wine: 1, 1, 6-Trimethyl-1, 2-dihycromaphthalene (TDN) and 2-aminoacetophenone. In: *Proceedings of the 4th International Symposium on Cool Climate Enology and Viticulture*, Henick-Kling, T., et al., eds., Geneva, New York: New York State Agricultural Experimental Station, VI:37–57, 1996.

65. Velázquez, J.B., E. Longo, C. Sieiro, J. Cansado, P. Calo, T.G. Villa. Improvement of the alcoholic fermentation of grape juice with mixed cultures of Saccharomyces cerevisiae wild strains; negative effects of *Kloeckera apiculata*. *World J. Microbiol. Biotechnol.* 7:485–489, 1991.

66. Gao, C., G.H. Fleet. The effects of temperature and pH on the ethanol tolerance of the wine yeasts, *Saccharomyces cerevisiae*, *Candida stellata* and *Kloeckera apiculata*. *J. Appl. Bacteriol.* 65:405–410, 1988.

67. Heard, G.M., G.H. Fleet. The effects of temperature and pH on the growth of yeast species during the fermentation of grape juice. *J. Appl. Bacteriol.* 65:23–28, 1988.

68. Teunissen, A.W.R.H., H.Y. Steensma. The dominant flocculation gene of *Saccharomyces cerevisiae* constitutes a new subtelomeric gene family. *Yeast* 11:1001–1013, 1995.

69. Guerzoni, M.E., R. Marchetti, P. Giudici. Modifications des composants aromatiques des vins obtenus par fermentation avec des mutants de *Saccharomyces cerevisiae*. *Bull. O.I.V.* 58:230–233, 1985.

70. Javelot, C., P. Girard, B. Colonna-Ceccaldi, B. Valdescu. Introduction of terpene production-ability to a wine strain of *Saccharomyces cerevisiae*. *J. Biotechnol.* 21:239–252, 1991.

71. Rainieri, S., C. Zambonelli, V. Tini, L. Castellari, P. Giudici. The enological traits of thermotolerant *Saccharomyces* strains. *Am. J. Enol. Vitic.* 49:319–324, 1998.

72. Suzzi, G., P. Romano, C. Zambonelli. *Saccharomyces* strain selection in minimizing SO2, requirement during vinification. *Am. J. Enol. Vitic.* 36:199–202, 1985.

73. Bony, M., F. Bidart, C. Camarasa, L. Ansanay, L. Dulau, P. Barre, S. Dequin, Metabolic analysis of *Saccharomyces cerevisiae* strains engineered for malolactic fermentation. *FEBS Lett.* 410:452–456, 1997.

74. Volschenk, H., M. Viljoen, J. Grobler, F. Bauer, A. Lonvaud-Funel, M. Denayrolles, R.E. Subden, H.J.J. Van Vuuren. Malolactic fermentation in grape musts by a genetically engineered strain of *Saccharomyces cerevisiae*. *Am. J. Enol. Vitic.* 48:193–197, 1997.

75. Dharmadhikari, M.R., K.L. Wilker. Deacidification of high malate must with *Schizosac-charomyces pombe*. *Am. J. Enol. Vitic.* 49:408–412, 1998.

76. Dequin, S., E. Baptista, P. Barre. Acidification of grape musts by *Saccharomyces cerevisiae* wine yeast strains, genetically engineered to produce lactic acid. *Am. J. Enol. Vitic.* 50:45–56, 1999.

77. Boone, C., A.M. Sdicu, J. Wagner, R. Degré, C. Sanchez, H. Bussey. Integration of the yeast K1 killer toxin gene into the genome of marked wine yeasts and its effects on vinification. *Am. J. Enol. Vitic.* 41:3742–3748, 1990.

78. Fazel, F., A. Rahman, I. Greensill, K. Ainley, S.M. Hasi, J.H. Parish. Strand scission in DNA by quercetin and Cu(II): identification of free radical intermediates and biological consequences of scission. *Carcinogenesis* 11:2005–2008, 1990.

79. Subbaramaiah, K., P. Michaluart, W.J. Chung, A.J. Danneberg. Resveratrol inhibits the expression of cyclooxygenase-2 in human mammarian and oral epithelial cells. *Pharmaceut. Biol.* 36:35–43, 1998.

80. Kazanawa, K., T. Yamashita, H. Ashida, G. Anno. Antimutagenecity of flavones and flavonols to heterocyclic amines by specific and strong inhibition of the cytochrome P450 1A family. *Biosci. Biotechnol. Biochem.* 62:970–977, 1998.

81. Pinhero, R.G., G. Paliyath. Antioxidant and calmodulin inhibitory activities of phenolic components in fruit wines and its biotechnological implications. *Food Biotechnol.* 15:179–192, 2001.

82. Hakimuddin, F., G. Paliyath, K. Meckling. Selective cytotoxicity of a red grape wine flavonoid fraction against MCF-7 cells. *Breast Canc. Res. Treatment.* 85: 65–79, 2004.

14

Biotechnology of Nonnutritive Sweeteners

Reena Randhir and Kalidas Shetty

CONTENTS

14.1   Introduction..327
14.2   Saccharin..329
14.3   Acesulfame-K 329
14.4   Sucralose 330
14.5   Aspartame 331
14.6   Neotame 333
14.7   Stevioside...333
14.8   Monellin...335
14.9   Brazzein 336
14.10 Thaumatin 336
14.11 Mabinlin...339
14.12 Conclusion 339
References...340

14.1 INTRODUCTION

Consumer preference for sweetness in a diet with reduced calories has created a market-place for sweeteners with few or no calories. Sweeteners are broadly classified as providing energy (nutritive) or not providing energy (nonnutritive). Nutritive sweeteners (e.g., sucrose, fructose) are generally recognized as safe (GRAS) by the Food and Drug Administration (FDA). The Code of Federal Regulations states that nonnutritive sweeteners are "substances having less than 2 percent of the calorie value of sucrose per equivalent unit of sweetening capacity" (1,2). By increasing the palatability of nutrient dense foods and beverages, sweeteners can promote health. Scientific evidence supports neither that intake of nutritive sweeteners by themselves increase the risk of obesity, nor that nutritive or nonnutritive sweeteners cause behavioral disorders. However, nutritive sweeteners increase risk of dental caries. High fructose intakes may cause hypertriglyceridemia and gastrointestinal symptoms in susceptible individuals (1,2).

Nonnutritive sweeteners are used to:

1. Expand food and beverage choices for those who want to control calorie, carbohydrate, or specific sugar intake
2. Assist weight control or reduction
3. Aid in the management of diabetes and hyperlipemia
4. Assist the control of dental caries
5. Enhance the usability of pharmaceuticals and cosmetics
6. Provide sweetness when sugar supply is limited
7. Assist the cost effective use of limited resources

These concerns are shaping the need for manufacturers and food and beverage processors to reconsider what additives they use in their products, and what alterations need to be made to meet consumer demand for something sweet, low calorie, and affordable (1,2). The search for noncarbohydrate nonnutritive sweeteners from natural sources has led to the discovery of many intensely sweet tasting substances. The characteristics of an ideal nonnutritive sweetener are:

1. A sweetness quality and profile identical to that of sucrose
2. Sensory and chemical stability under the relevant food processing and storage conditions
3. Compatibility with other food ingredients and stability toward other constituents in the food
4. Complete safety, shown as freedom from toxic, allergenic and other physiological properties
5. Complete freedom from metabolism in the body
6. High specific sweetness intensity (2)

Most sweet compounds, including all popular sweeteners, are small molecular weight compounds of different chemical nature, but there are also sweet macromolecules, both synthetic and natural, such as sweet proteins. Sweet molecules elicit their taste, in humans and other mammals, by interacting with the recently discovered T1R2–T1R3 taste receptors (3–5). The sequence of this protein indicates that it is a metabopromic 7 transmembrane G-protein coupled receptor with a high homology to the mGluR subtype 1 (4). The structure of the N-terminal part of the mGluR has been recently determined by x-ray diffraction and has been used as a template to build a homodimeric T1R3–T1R3 receptor model (4). It is very likely that small molecular weight sweet molecules occupy a pocket analogous to the glutamate pockets in the mGluR (5), possibly similar to the active site models predicted by indirect receptor mapping studies (6–8). These significant investigations have facilitated the understanding of sweetness perception in humans; and have also made it possible to modify genetically the sequence of the natural sweet proteins as necessary. The following review is a brief summary of the current nonnutritive sweeteners, namely Saccaharin, Acesulfame K, Sucralose, Aspartame, Neotame, Stevioside, Monellin, Brazzein, Thaumatin, and Manbinlin, which, although they are permitted food ingredients under different national statutory regulations, are not universally accepted. The evaluation of the sweetness of a given substance in relation to sucrose is made on a weight basis. Table 14.1 provides the relative sweetness of the nonnutritive sweeteners discussed in this chapter.

Table 14.1
Relative sweetness of nonnutritive sweeteners

Nonnutritive Sweetener	Approximate Sweetness (Sucrose = 1)
Saccharin	300
Acesulfame K	200
Sucralose	600
Aspartame	180–200
Neotame	8000
Stevioside	200
Monellin	1500–2000
Brazzein	2000
Thaumatin	1600
Mabinlin	100

14.2 SACCHARIN

Saccharin, commonly known as Sweet n Low™, was first discovered in 1879 by Remsen and Fahlberg and has been used commercially since then (9). It is 300 times as sweet as sucrose, produces no glycemic response in humans, synergizes the sweetening power of nutritive and nonnutritive sweeteners, and its sweetness is thermostable. It occurs as a white crystalline powder with a molecular formula of $C_7H_5NO_3S$ and molecular weight of 183.18 (Figure 14.1). It has been widely used to sweeten foods and beverages without calories or carbohydrates for over a century (10,11). It was widely used during the sugar shortages of the two world wars. Research also has shown that saccharin is beneficial to people with diabetes and obese people, and helps to reduce dental cavities (12). It is used in products such as soft drinks, tabletop sweeteners, baked goods, jams, chewing gum, canned fruit, candy, dessert toppings, and salad dressings. It is also used in cosmetic products, vitamins, and pharmaceuticals (10).

Although extensive research indicated that saccharin was safe for human consumption, there has been controversy over its safety. The basis for the controversy rests primarily on findings of bladder tumors in some male rats when fed high doses of sodium saccharin (13–18). Other saccharin research, however, indicates safety at human levels of consumption with no detectable metabolism using analytical techniques. The vast majority of the data on the biotransformation of saccharin demonstrate that this compound is excreted unchanged, predominantly in the urine of all lab animals tested (20–22). The production of saccharin is simple, inexpensive, and can be done in bulk. The production process uses the basic route described by Remsen and Fahlberg. Methyl anthranilate is diazotized by treatment with sodium nitrite and hydrochloric acid to form 2-carbomethoxybenzenediazonium chloride. Sulfonation of this compound produces 2-carbomethoxybenzenesulfinic acid, which is converted to 2-carbomethoxybenzenesulfonyl chloride with chlorine. Saccharin is synthesized by amidation of this sulfonychloride, followed by acidification. This is then treated with either sodium hydroxide or sodium bicarbonate to produce sodium saccharin (23).

14.3 ACESULFAME-K

Acesulfame-K, commonly known as Sunette™ or Sweet One™, was discovered by Clauss and Jensen, 1967 (24). It was approved by the FDA as tabletop sweetener and as an additive in a variety of desserts, confections, and alcoholic beverages. It is 200 times sweeter

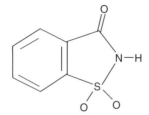

Figure 14.1 Chemical structure of saccharin.

than sucrose, is noncarcinogenic and produces no glycemic response (25). Its chemical name is potassium salt of 5,6-methyl-1,2,3-oxathiazin-4(3H)-one-2,2-dioxide and its chemical formula is $C_4H_4KNO_4S$ (Figure 14.2). Its sweetening power is not reduced by heating, and can synergize the sweetening power of other nutritive and nonnutritive sweeteners. It does not provide any energy, is not metabolized in the body, and is excreted unchanged (12,26,27). However, toxicology research performed on albino mice for the genotoxic and clastogenic potential of acesulfame-K revealed chromosome aberrations in the bone marrow cells. In view of this significant *in vivo* mammalian genotoxicity data, it is advised that acesulfame-K should be consumed with caution (28).

Synthesis of acesulfame-K starts with the reaction of acetoacetic acid tert-butyl ester with fluorosulfonyl isocyanate (29). Both compounds form the intermediate of α–N-fluorosulfonylcarbamoyl acetoacetic acid tert-butyl ester. This compound is unstable, and by releasing CO_2 and isobutene is converted to N-fluorosulfonyl acetoacetic acid amides (30). This reaction can be accelerated by heating the compound slightly. In the presense of potassium hydroxide it cyclizes to the dihydrooxathiazinone dioxide ring system by separating out fluorides. Because these are highly acidic compounds, salts of this ring compound are formed in reactions with KOH, NaOH or $Ca(OH)_2$ (30).

14.4 SUCRALOSE

Sucralose is also known by the brand name Splenda™ and is 600 times sweeter than sucrose. It is the only low calorie sweetener made from sugar, with a chemical formula of $C_{12}H_{19}C_{13}O_8$ and molecular weight of 397.64 (Figure 14.3). It was discovered in 1976 and scientific studies conducted over a 20 year period have conclusively determined that sucralose is safe for everyone to consume (31,32). It was approved by the FDA, and by the Joint FAO/WHO Expert Committee on Food Additives (JECFA) in 1990. Sucralose has been approved by prominent regulatory authorities throughout the world, and has been consumed by millions of people internationally since 1991.

It can be used in place of sugar to eliminate or reduce calories in a wide variety of products, including beverages, baked goods, desserts, dairy products, canned fruits, syrups, and condiments. Heating or baking does not reduce its sweetening power. It has no calories, and the body does not recognize it as a carbohydrate (33). Clinical research showed that it produced no glycemic response in the human body; approximately 15% of sucralose was passively absorbed in the body, and the majority was excreted unchanged (34–36). The small amount that was passively absorbed was not metabolized and was eliminated within 24 hours. The FDA concluded that it does not pose a carcinogenic, reproductive, or neurological risk to humans, and this is supported by clinical research (35,37). However a few controversial investigations indicate the potential hazards of sucralose on human health (38).

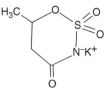

Figure 14.2 Chemical structure of Acesulfame-K

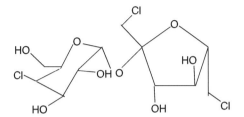

Figure 14.3 Chemical structure of Sucralose.

Production of sucralose starts with a cane sugar molecule, then three hydrogen-oxygen groups are substituted with three tightly bound chlorine atoms, which makes it inert. A high yielding bioorganic synthesis of sucralose (4,1′,6′-trichloro-4,1′,6′-trideoxy-galactosucrose) involves the chemical chlorination of raffinose to form a novel tetrachlororaffinose intermediate (6,4′,1″,6″-tetracyhloro-6,4′,1″,6″-tetradeoxygalactoraffinose; TCR) followed by the enzymatic hydrolysis of the α-1-6 glycosidic bond of the TCR to give sucralose and 6-chlorogalactose. The most active enzyme was produced by a strain of *Mortierella vinacea* and had a maximum rate of 118 μmol sucralose/g dry weight cells/hour, which was approximately 5% of the activity toward raffinose, and a Km of 5.8 mM toward TCR. The enzyme was used in the form of mycelial pellets in a continuous packed bed column reactor. Synthesis of raffinose was achieved from saturated aqueous solutions of galactose and sucrose using a selected α-galactosidase from *Aspergillus niger* (39).

14.5 ASPARTAME

Aspartame was discovered by accident by James Schlatter, a chemist at G.D Searle Co. in 1965, when he was testing an antiulcer drug. Its chemical name is N-L- aspartyl-l-phenyl-alanine-1-methyl ester and marketed as Nutrasweet™, Equal™, Spoonful™, and Equal Measure™. It is an odorless, white crystalline powder that has a sweet taste (40,41). It is widely used as a flavor enhancer, and to sweeten foods and beverages (42). It provides the same energy as any protein (4 calories per gram) because it is a combination of phenyl-alanine and aspartic acid (two amino acids), which is then combined with methanol (Figure 14.4). It is 180–200 times sweeter than sucrose, so the small amount needed to sweeten products does not actually contribute a significant number of calories (40,41). Clinical studies on the safety of aspartame have been widely conducted on animals and humans (43,44). The data demonstrated a substantial margin of safety for aspartame and its metabolites at reasonable levels of consumption. However, the product is required to display a warning label about the possibility of Phenylketonuria, a genetic disease where the body cannot produce the enzyme necessary to use the amino acid phenylalanine, which is one of the breakdown metabolites of aspartame (45,46).

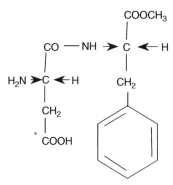

Figure 14.4 Chemical structure of Aspartame.

The two primary components of aspartame (phenylalanine and aspartic acid) are chiral, which means that they have two isomers that are mirror images that cannot be superimposed upon each other. This means that the final aspartame molecule will have two stereogenic centers. If the wrong enantiomers are used, the aspartame molecule will not have the correct shape to fit the binding site of the sweetness receptors on the tongue (47). In the synthesis of aspartame, the starting materials are a racemic mixture (equal quantities of both enantiomers) of phenylalanine and aspartic acid. Only the L-enantiomer of phenyl-alanine is used: this is separated from the racemate by reacting it with acetic anhydride and sodium hydroxide. If the product of this reaction is then treated with the enzyme porcine kidney acylase, and an organic extraction with H⁺ carried out, the L-enantiomer is found in the aqueous layer and the D-enantiomer remains in the organic layer (48,49).

Treatment of L-phenylalanine with methanol and hydrochloric acid esterifies the -CO₂H group, and this ester is then reacted with aspartic acid to give the final product. It is important that the amine group on aspartic acid be protected with carbobenzyloxy groups, and the acid group nearest the amine protected with benzyl groups, to prevent the L-phenylalanine reacting with these and giving unwanted byproducts. The acid group that is required to react is activated with Castro's reagent. Castro's reagent is displaced as L-phenylalanine is added, but the protective groups must be removed after the reaction. Carbobenzyloxy is removed by a reaction with hydrogen and platinum(IV) oxide with methanol and chloroform; benzyl is removed by reaction with hydrogen, palladium, and carbon, plus methanol and chloroform, completing the aspartame synthesis (48,49).

Another method for the synthesis of this high intensity sweetener is a very simple example of how proteases may be used in peptide synthesis. Most proteases show specific-ity in their cleavage sites, and may be used to synthesise specific peptide linkages (50–53). Aspartame is the dipeptide of L-aspartic acid with the methyl ester of L-phenylalanine (L-aspartyl-L-phenylalanyl-O-methyl ester). The chemical synthesis of aspartame requires protection of both the carboxyl group and the amino group of the L-aspartic acid. Even then, it produces aspartame in low yield and at high cost. If the carboxyl group is not pro-tected, a cost saving is achieved, but about 30% of the isomer is formed and must subse-quently be removed. When thermolysin is used to catalyze aspartame production, the regiospecificity of the enzyme eliminates the need to protect this carboxyl group, but the amino group must still be protected (usually by means of reaction with benzyl chlorofor-mate to form the benzyloxycarbonyl derivative, i.e., BOC-L-aspartic acid) to prevent the synthesis of poly-L-aspartic acid. More economical racemic amino acids can also be used, as only the desired isomer of aspartame will be formed.

If stoichiometric quantities of L-aspartic acid and L-phenylalanine methyl ester are reacted in the presence of thermolysin, an equilibrium reaction mixture is produced giving relatively small yields of aspartame. However, if two equivalents of the phenylalanine methyl ester are used, an insoluble addition complex forms in high yield at concentrations above 1 M. The loss of product from the liquid phase due to this precipitation greatly increases the overall yield of this process. Later, aspartame may be released from this adduct by simply altering the pH. The stereospecificity of the thermolysin determines that only the L-isomer of phenylalanine methyl ester reacts but the addition product is formed equally well from both the D- and L-isomers. This allows the use of racemic phenylalanine methyl ester, the L-isomer being converted to the aspartame derivative and the D-isomer forming the insoluble complex shifting the equilibrium to product formation. D-phenylalanine ethyl ester released from the addition complex may be isomerised enzymically to reform the racemic mixture. The BOC-aspartame may be deprotected by a simple hydrogenation process to form aspartame (51).

14.6 NEOTAME

Neotame is the FDA approved sweetener developed by Monsanto Chemical Corporation, and was discovered by Nofri and Tinti. It is reported to be approximately 8,000 times sweeter than sugar. Neotame has approximately 40 times the sweetness potency of aspartame (54). Its chemical name is N-[N-(3,3-dimethylbutyl)-L-aspartyl]-L-phenylalanine 1-methyl ester (Figure 14.5). It was designed to overcome some of the problems with aspartame. It is more thermostable and does not break down during processing, which is a major drawback of aspartame. It is suggested for use as a tabletop sweetener, to sweeten frozen desserts, chewing gum and candy, baked goods, fruit spreads, and readymade cereals (55,56).

Neotame is a derivative of aspartame dipeptide made of amino acids: aspartic acid and phenylalanine (54). It is quickly metabolized and fully eliminated through normal biological processes. The dimethylbutyl part of the molecule is added to block the action of peptidases which are enzymes that break the peptide bond between the aspartic acid and phenylalanine. This reduces the availability of phenylalanine, eliminating the need for a warning on labels directed at people with phenylketonuria, who cannot properly metabolize phenylalanine (57,58).

14.7 STEVIOSIDE

Stevioside is a nonnutritive sweetener extracted from the leaves of the plant *Stevia rebaudiana*, which belongs to the Asteraceae family and was rediscovered by Bertoni in 1888 (59–61) (Figure 14.6). It is native to Brazil, Venezuela, Colombia, and Paraguay where the native Guarani tribes have used *caa-ehe* (stevia) for over 1500 years to sweeten otherwise unpalatable medicinal drinks. It is a herbaceous perennial, which is normally used as a natural herbal sweetener (59). The glycosides are found between the veins of the leaves which can be up to 200 times sweeter than sucrose. The sweetness is mainly attributed to the two compounds, namely stevioside (3–10% of dry leaf weight) and rebaudioside A (1–3%) (62–66). Research conducted at the College of Pharmacy at the University of Illinois found that when a bacterium was exposed to stevioside, the DNA of the bacteria was genetically altered (67). Mutagenic effects of steviol, the aglycone of stevioside, were

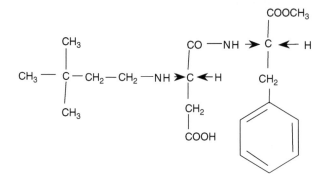

Figure 14.5 Chemical structure of Neotame.

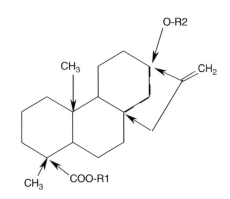

Figure 14.6 Chemical structure of Stevioside.

also reported in *Salmonella typhimurium* TM677. After metabolic activation, it was shown that so far unknown steviol metabolites caused mutations in *Salmonella typhimurium* TM677, including transitions, transversions, duplications, and deletions at the guanine phosphoribosyltransferase (*gpt*) gene (67,68). Stevia, or stevioside, has not been granted GRAS status by the FDA, but the Dietary Supplement Act of 1994 allows stevia to be sold in the US as a dietary supplement. However, it has been used as a sweetener in South America for centuries, and in Japan for over 30 years.

The other uses for the plant and its extracts are in weight loss programs, because of its ability to reduce the cravings for sweet and fatty foods. The plant has also been used to treat diseases such as diabetes, hypoglycemia, candidiasis, high blood pressure, skin abrasions, and for inhibiting growth and reproduction of bacterialike plaque (69). The advantages of stevioside as a dietary supplement for human subjects are manifold: it is stable, it is noncalorific, it supports good dental health by reducing the intake of sugar, and opens the possibility for safe use by diabetic and phenylketonuria patients, and by obese persons. Stevioside has a few advantages over artificial sweeteners, in that it is stable at high temperatures (100°C) and a wide pH range (3–9), and does not darken with cooking. It is suggested in the use of sweetening soft drinks, ice cream, cookies, pickles, chewing gum, tea, and skin care products.

The biosynthesis of stevioside is via the recently discovered 2-C-Methyl-D-erythritol-4-phosphate pathway (MEP) due to the *ent*-kaurene skeleton of stevioside (69). The genes of the enzymes catalyzing the first two steps of this pathway, 1-deoxy-D-xylulose-5-phosphate synthase (DXS) and 1-deoxy-D-xylulose-5-phosphate reductoisomerase

(DXR) were cloned using reverse transcriptase polymerase chain reaction (PCR). The DXS and DXR from *Stevia* both contain an N-terminal plastid targeting sequence and show high homology to other known plant DXS and DXR enzymes. Furthermore, it was demonstrated through heterologous expression in *Escherichia coli* that the cloned cDNAs encode these functional proteins (70–73).

14.8 MONELLIN

Monellin is a sweet tasting protein isolated from the tropical African fruit *Dioscoreophyllum cumminsii*. Inglett and May first reported the isolation of a sweet substance from the berries in 1967. The unusual protein possesses the interesting property of having a very high specificity for the sweet receptors in the tongue (74). Monellin is made up of two dissimilar polypeptide chains, which are noncovalently associated. It consists of two peptide chains, the A chain of 44 residues and the B chain of 50 residues (75–80). These two chains, A and B, are linked by noncovalent interactions (Figure 14.7). Single chain mutants of monellins in, which the two chains are covalently linked, retain all their sweetening power and have greatly increased thermal stability (81). The first single chain monellin is obtained by joining the C-terminal residue of the B chain directly to the N-terminal residue of the A chain. The second single chain monellin is obtained by linking B and A chains *via* the Gly-Phe dipeptide. It has a molecular weight of 11,000 and is approximately 1500–2000 times sweeter than sugar on a molar basis and several thousand times sweeter on a weight basis. It does not contain carbohydrates or modified amino acids. Monellin is extracted from the fruit by removing the skin, leaching the fruit into water, and concentrating the product by salting out with ammonium sulphate. It shows little promise as a commercial sweetener because of taste qualities, stability, and difficulties in propagating the plant (74).

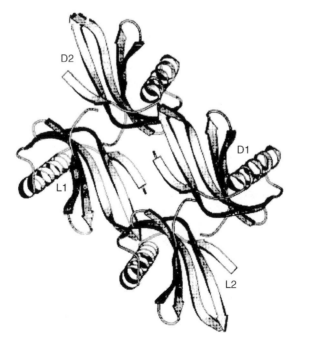

Figure 14.7   Schematic diagram of the spatial relationship of the two D-and two L-monellin molecules. Adapted from Hung L, M. Kohmura, Y. Ariyoshi, S. Kim. *J. Mol. Biol.* 285:311–321, 1999.

The heterologous expression of the monellin gene in the yeast *Candida utilis* has been investigated. A single chain monellin gene was expressed under the control of the glyceraldehyde-3-phosphate decarboxylase gene promoter from *C. utilis* (80). A promoter deficient marker gene allowed high copy number integration of vectors into either the rDNA locus or the URA3 gene locus. Monellin was produced at a high level, accounting for 50% of the soluble protein. No significant decrease in the production level of monellin was detected in transformants after 50 generations of nonselective growth (80,81).

14.9 BRAZZEIN

Brazzein is a sweet tasting protein found in the fruit of the African plant *Pentadiplandra brazzeana* (82). It contains no carbohydrate, and its structure bears no structural resemblance to sucrose, small molecule chemical sweeteners, or the other sweet tasting proteins monellin and thaumatin. Moreover, members of the class of sweet proteins contain no conserved stretches of amino acids (83). It is the smallest, most heat stable and pH stable member of the set of proteins known to have intrinsic sweetness. These properties make brazzein an ideal compound for investigating the chemical and structural requirements of a sweet tasting protein. It is 2,000 times sweeter than sucrose and has a molecular mass of 6,473 Da (83,84).

Brazzein is a single chain polypeptide of 54 amino acid residues with four intramolecular disulfide bonds, no free sulfhydryl group, and no carbohydrate (82). It is rich in lysine, but contains no methionine, threonine or tryptophan. It exists in two forms in the ripe fruit. The major form contains pyroglutamate (pGlu) at its N-terminus; the minor form is without the N-terminal pGlu (des-pGlu1). Taste comparisons of chemically synthesized brazzein and des-pGlu1 brazzein revealed that the latter protein has about twice the sweetness of the former (85). It is highly water soluble with an isoelectric point (*pI* = 5.4) lower than those of other sweet proteins (82). It is remarkably heat stable, and its sweet taste remains after incubation at 80°C for 4 h (86). Chemical modification studies suggested that the surface charge of the molecule is important and led to the conclusion that Arg, Lys, Tyr, His, Asp, and Glu are important for brazzein's sweetness. The structure of brazzein was determined by nuclear magnetic resonance (NMR) spectroscopy in solution at pH 5.2 and 22°C (86,87). The study revealed that brazzein contains one short α-helix (residues 21–29) and three strands of antiparallel β-sheet structures (strand I, residues 5–7; strand II, residues 44–50; strand III, residues 34–39) held together by four disulfide bonds (Figure 14.8). It is proposed that the small connecting loop containing His31 and the random coil loop around Arg43 are the possible determinants of the molecule's sweetness (88,89). Brazzein can be synthesized by the fluoren-9-yl-methoxycarbonyl solid phase method which produces a compound identical to natural brazzein as demonstrated by high performance liquid chromatography, mass spectroscopy, peptide mapping, and taste evaluation. The D-enantiomer of brazzein, which is a mirror image of brazzein, was also synthesized, but was devoid of any sweetness and was essentially tasteless (85).

14.10 THAUMATIN

Thaumatin is a sweet tasting protein isolated from the arils of *Thaumatococcus daniellii* (Benth), a plant native to tropical West Africa (90). It is approved for use in many countries and has application as both a flavor enhancer and a high intensity sweetener. The

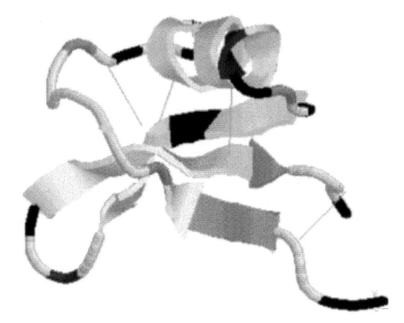

Figure 14.8 Diagram showing the three-dimensional backbone of brazzein. Adapted from Jin, Z., V. Danilova, P. Assadi, M. Fariba, D.J. Aceti, J.L. Markley, G. Hellekant. *FEBS Lett.* 544(3):33–37, 2003.

availability of thaumatin of plant origin is very limited (91), and it is notoriously difficult to produce thaumatin by recombinant DNA methods (92). Production has been attempted with *Escherichia coli* (93), *Bacillus subtilis* (94), *Streptomyces lividans* (95), *Saccharomyces cerevisiae* (96), and *Aspergillus oryzae* (96). A synthetic gene for thaumatin with fungal codon usage has been synthesized, but expression in *Aspergillus niger* gave poor yields (97). Expression could be limited by a weak promoter, copy number, insertion location, inefficient processing of the prepropeptides, or bottlenecks in protein traffic and translocation through the membrane systems of the protein secretory pathway (98,99). Efficient production of a heterologous protein is usually achieved by increasing gene dosage, although overloading of the secretory pathway may result in abnormal folding and protein degradation.

Naturally occurring thaumatin consists of six closely related proteins (I, II, III, a, b, and c), all with a molecular mass of 22 kDa (207 amino acids) (90). Neither protein contains bound carbohydrate or unusual amino acids. The proteins have an isoelectric point of 12 (90). The sweet taste of thaumatins can be detected at threshold amounts 1600 times less than that of sucrose on a weight basis. The three dimensional structure of thaumatin I has been determined at high resolution (99,100), revealing that the protein consists of three domains: an 11 strand, flattened ß-sandwich (1–53, 85–127 and 178–207, domain I); a small, disulfide rich region (54–84, domain III); and a large disulfide rich region (128–177, domain II) (Figure 14.9). The five lysine residues, modification of which affected sweetness, are separate and spread over a broad surface region on one side of the thaumatin I molecule. These lysine residues exist in thaumatin, but not in nonsweet thaumatinlike proteins, suggesting that these lysine residues are required for sweetness. These lysine residues may play an important role in sweetness through a multipoint interaction with a putative thaumatin receptor (100,101).

Thaumatin was also secreted by the methylotrophic transgenic yeast *Pichia pastoris.* The mature thaumatin II gene was directly cloned from Taq polymerase amplified PCR

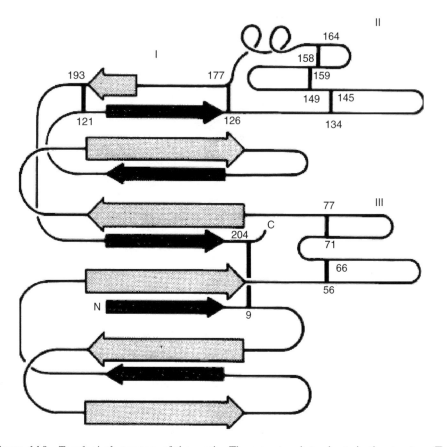

Figure 14.9 Topological structure of thaumatin. There are two beta sheets in the structure. The beta strands of the top sheet are indicated by broad arrows and those of the bottom sheet by narrow arrows. Also shown are the three domains of the protein and a crystallographic assignment of the disulphide bonds shown in black vertical bars. Figure adapted from: Vos, A.M., M. Hatada, H. Van der Wel, H. Krabbendam, A.F. Peerdeman, S.H. Kim. *Proc. Natl Acad Sci USA* 82(5):1406–1409, 1985.

products by using cloning methods, and fused the pPIC9K expression vector that contains *Saccharomyces cerevisiae* prepro alpha mating factor secretion signal (101,102). Several additional amino acid residues were introduced at both the N- and C-terminal ends by genetic modification to investigate the role of the terminal end region for elicitation of sweetness in the thaumatin molecule. The secondary and tertiary structures of purified recombinant thaumatin were almost identical to those of the plant thaumatin molecule. Recombinant thaumatin II elicited a sweet taste just as native plant thaumatin II; its threshold value of sweetness to humans was around 50 nM, which is the same as that of plant thaumatin II. These results demonstrated that the functional expression of thaumatin II attained by *Pichia pastoris* systems and that the N- and C-terminal regions of the thaumatin II molecule did not play an important role in eliciting the sweet taste of thaumatin (102).

Research indicated that obtaining of thaumatin producing strains by transformation with an expression cassette containing a synthetic thaumatin gene (with an optimized codon usage) and the inactivation of a specific protease resulted in a significant increase of extracellular thaumatin (99). An alternative method for reduction of expression of a particular gene is the use of antisense RNA. Although the technique is simple, the effectiveness of the method is influenced by many factors (103). This technique has been successfully used to silence the *creA* gene in *Aspergillus nidulans* (104). It was, therefore used to silence the *pepB* gene in *A. awamori* by the antisense RNA technique, as a first approach to elimination of the negative effect of the presence of aspergillopepsin B on thaumatin accumulation. Research revealed that significant amounts of antisense RNA of the *pepB* gene are formed by using a strong fungal promoter, but that the aspergillopepsin B is not completely removed from the broths. On the other hand, *pepB* gene disruption by replacement using the double marker selection procedure led to the complete loss of aspergillopepsin B and to a 100% increase in thaumatin accumulation under optimal fermentation conditions (105).

14.11 MABINLIN

A new sweet tasting protein named mabinlin II was extracted from the seeds of *Capparis masaikai*. It was purified by ammonium sulfate fractionation, carboxymethylcellulose to Sepharose ion exchange chromatography, and gel filtration. The sweetness of mabinlin II was unchanged by at least 48 h incubation at nearly boiling temperature (106). Purified mabinlin II thus obtained gave a single band having a molecular mass of 14 kDa on SDS/PAGE (Sodium Dodecyl Sulphate-Polyacrylamide Gel Electrophoresis). In the presence of dithiothreitol, mabinlin II gave two bands having molecular masses of 4.6 kDa and 5.2 kDa on SDS/PAGE. Two peptides (A chain and B chain) were separated from reduced and S-carboxamidomethylated mabinlin II by HPLC. The amino acid sequences of the A chain and B chain were determined by the automatic Edman-degradation method. The A chain and B chain consist of 33-amino acid and 72-amino acid residues, respectively. The A chain is mostly composed of hydrophilic amino acid residues and the B chain also contains many hydrophilic residues. High similarity was found between the amino acid sequences of mabinlin II and 2S seed storage proteins, especially 2S albumin AT2S3 in *Arabidopsis thaliana* (106,107).

14.12 CONCLUSION

The pursuit for the ideal sweetener continues. The ideal sweetener is expected to taste like sucrose, and be colorless, odorless, noncariogenic, and nontoxic; and to present a pleasant, untainted taste without a delayed onset of persistence in sweetness; but at the same time be noncaloric and low cost. The long term safety of the nonnutritive sweeteners, which is of prime importance to consumers, has yet to be thoroughly investigated. However, the wide variety of sweeteners available today enables the development of a much wider range of new, good tasting, low calorie products to meet consumer demand. A variety of low calorie sweeteners also provides products with increased stability, improved taste, lower production costs, and more choices for the consumer. The search for noncarbohydrate sweeteners from natural sources has led to the discovery of many intensely sweet tasting substances. The occurrence of sweet tasting proteins has provided a new approach to the potential treatment of diabetes, obesity, and other metabolic disorders. Such choices allow a person to control their caloric intake in a manner best suited for their health conditions.

REFERENCES

1. Nabors, L.B., R.C. Gelardi. *Alternative Sweeteners: Food Science and Technology.* New York: Marcel Dekker, 1985.
2. Calorie and noncalorie sweeteners: per capita consumption, 1970–1982. In: *Sugar and Sweetener Outlook and Situatuion Report*, Washington: U.S. Department of Agriculture, Sept. 1983, p 29.
3. Nelson, G., M.A. Hoon, J. Chandrashekar, Y. Zhang, N.J. Ryba, C.S. Zuker. Mammalian sweet taste receptors. *Cell* 106:381–390, 2001.
4. Margolskee, R.F. Molecular mechanisms of bitter and sweet taste transduction. *J. Biol. Chem.* 277:1–4, 2002.
5. Li, X., L. Staszewski, X. Xu, K. Durick, M. Zoller, E. Adler. Human receptors for sweet and umami taste. *Proc. Natl. Acad. Sci. USA* 99:4692–4696, 2002.
6. Shallenberger, R.S., T. Acree. Molecular theory of sweet taste. *Nature* 216:480–482, 1967.
7. Kier, L.B. A molecular theory of sweet taste. *J. Pharm. Sci.* 61:1394–1397, 1972.
8. Temussi, P.A., F. Lelj, T. Tancredi. Three-dimensional mapping of the sweet taste receptor site. *J. Med. Chem.* 21:1154–1158, 1978.
9. Kaufman, G.B., P.M. Priebe. The discovery of saccharin: a centennial retrospect. *Ambix* 25:191–207, 1879.
10. Smith, J.C., A. Sclafani. Saccharin as a sugar surrogate revisited. *Appetite* 38(2):155–160, 2002.
11. Leclercq, C., D. Berardi, M.R. Sorbillo, J. Lambe. Intake of saccharin, aspartame, acesulfame K and cyclamate in Italian teenagers: present levels and projections. *Food Additives Contaminants* 16(3):99–111, 1999.
12. Ilback, N.G., S. Jahrl, H. Enghardt-Barbieri, L. Busk. Estimated intake of the artificial sweeteners acesulfame-K, aspartame, cyclamate and saccharin in a group of Swedish diabetics. *Food Additives Contaminants* 20(2):99–114, 2003.
13. Turner, S.D., H. Tinwell, W. Piegorsch, P. Schmezer, J. Ashby. The male rat carcinogens limonene and sodium saccharin are not mutagenic to male big blue rats. *Mutagenesis* 16(4):329–332, 2001.
14. Kurokawa, Y., T. Umemura. Risk assessment on an artificial sweetener, saccharin. *J. Food Hyg. Soc. Jpn.* 37(5):341–342, 1996.
15. Arnold, D.L. Long term toxicity study of orthotoluenesulfanamide and sodium saccharin in the rat. *Toxicol. Appl. Pharmacol.* 52:113–152, 1980.
16. Smith, J.C., W. Castonguayt, D.F. Foster, L.M. Bloom. Detailed analysis of glucose and saccharin in the rat. *Physiol. Behav.* 24(1):173–176, 1980.
17. Arnold, D.L., C.A. Moodie, H.C. Grice, S.M. Charbonneau, B. Stavric, B.T. Collins, P.F. McGuire. Long-term toxicity of toluenesulfonamide and sodium saccharin in the rat. *Toxicol. Appl. Pharmacol.* 52(1):113–152, 1980.
18. Arnold, D.L., C.A. Moodie, H.C. Grice, S.M. Charbonneau, B. Stavric, B.T. Collins, P.F. McGuire, I.C. Munro. The effect of toluene sulfonamide and sodium saccharin on the urinary tract of neonatal rats. *Toxicol. Appl. Pharmacol.* 51(3):455–464, 1979.
19. Rao, T.K., D.R. Stoltz, J.L. Epler. Lack of enhancement of chemical mutagenesis by saccharin in the *Salmonella typhimurium* assay. *Arch. Toxicol.* 43(2):141–146, 1979.
20. Kessler, I., J.P. Clark. Saccharin, cyclamate and human bladder cancer: a case-control study: no evidence of an association. *JAMA* 240:349–355, 1978.
21. Bailey, C.J., J.M.E. Knapper, S.L. Turner, P.R. Flatt. Antihyperglycaemic effect of saccharin in diabetic ob/ob mice. *Br. J. Pharmacol.* 120(1):74–78, 1997.
22. Chappel, C. A review and biological risk assessment of sodium saccharin. *Reg. Toxicol. Pharmacol.* 15(3):253–270, 1992.
23. Walter, G.J., M.I. Mitchell. Saccharin. In: *Alternative Sweeteners: Food Science and Technology*, New York: Marcel Dekker 17:15–41, 1985.
24. Clauss, K., H. Jensen. Oxathiazinon dioxides: a new group of sweetening agents. *Angew. Chem.* 85:965, 1973.

25. Peck, A. Use of acesulfame K in light and sugar-free baked goods. *Cereal Foods World* 39(10):743–745, 1994.

26. Abe, Y., Y. Takeda, H. Ishiwata, T. Yamada. Purity and content of a sweetener, acesulfame potassium, and their test methods. *Shokuhin Eiseigaku Zasshi*. 41(4):274–279, 2000.

27. Suami, T., L. Hough, T. Machinami, T. Saito, K. Nakamura. Molecular mechanisms of sweet taste 8: saccharin, acesulfame-K, cyclamate and their derivatives. *Food Derivatives* 63(3):391–396, 1998.

28. Mukherjee, J., J. Chakrabarti. *In vivo* cytogenetic studies on mice exposed to acesulfame-K, a nonnutritive sweetener. *Food Chem. Toxicol.* 35(12):1177–1179, 1997.

29. Arpe, H.J. Acesulfame-K: a new noncaloric sweetener. In: *Health and Sugar Substitutes, Proceedings of the ERGOB conference, Geneva, 1978*, p 178.

30. Wolfhard, G., R. Lipinski. Acesulfame-K. In: *Alternative Sweeteners: Food Science and Technology*, New York: Marcel Dekker, 1985, pp 89–102.

31. Stroka, J., N. Dossi, E. Anklam. Determination of the artificial sweetener Sucralose® by capillary electrophoresis. *Food Additives Contaminants* 20(6):524–527, 2003.

32. Knight, I. The development and applications of sucralose, a new high-intensity sweetener. *Can. J. Physiol. Pharmacol.* 72(4):435–439, 1994.

33. Finn, J.P., G.H. Lord. Neurotoxicity studies on sucralose and its hydrolysis products with special reference to histopathologic and ultrastructural changes. *Food Chem. Toxicol.* 38(2): S7–S17, 2000.

34. Grice, H.C., L.A. Goldsmith. Sucralose: an overview of the toxicity data. *Food Chem. Toxicol.* 38(2):S1–S636, 2000.

35. Baird, I.M., R.J. Merritt, G. Hildick-Smith. Repeated dose study of sucralose tolerance in human subjects. *Food Chem. Toxicol.* 38(2):S123–S129, 2000.

36. Sims, J., A. Roberts, J.W. Daniel, A.G. Renwick. The metabolic fate of sucralose in rats. *Food Chem. Toxicol.* 38(2):S115–S121, 2000.

37. Mann, S.W., M.M. Yuschak, S.J.G. Amyes, P. Aughton, J.P. Finn. A carcinogenicity study of sucralose in the CD-1 mouse. *Food Chem. Toxicol.* 38(2):S91–S98, 2000.

38. Goldsmith, L.A. Acute and subchronic toxicity of sucralose. *Food Chem. Toxicol.* 38(2): S53–S69, 2000.

39. Bennett, C., J.S. Dordick, A.J. Hacking, P.S.J. Cheetham. Biocatalytic synthesis of disaccharide high-intensity sucralose via a tetrachlororaffinose intermediate. *Biotechnol. Bioeng.* 39(2):211–217, 1992.

40. Cloninger, M.R., R.E. Baldwin. L-Aspartyl-L-phenylalanine methyl esther (aspartame) as a sweetener. *J. Food. Sci.* 39:347–349, 1974.

41. Prat, L.L., J.M. Oppert, F. Bellisle, G.B. Guy. Sweet taste of aspartame and sucrose: effects on diet-induced thermogenesis. *Appetite* 34(3):245–251, 2000.

42. Fellows, J.W., S.W. Chang, W.H. Shazer. Stability of aspartame in fruit preparations used in yogurt. *J. Food Sci.* 56(3):689–691, 1991.

43. Janssen, C. M., C.A. van der Heijden. Aspartame: review of recent experimental and observational data. *Toxicology* 50:1–26, 1988.

44. Bianchi, R.G., E.T. Muir, D.L. Cook, E.F. Nutting. The biological properties of aspartame: actions involving the gastrointestinal system. *J. Envir. Pathol. Toxicol.* 93:355–362, 1980.

45. Oyama, Y., H. Sakai, T. Arata, Y. Okano, N. Akaike, K. Sakai, K. Noda. Cytotoxic effects of methanol, formaldehyde, and formate on dissociated rat thymocytes: a possibility of aspartame toxicity. *Cell Biol. Toxicol.* 18(1):43–50, 2002.

46. Ishii, H. Incidence of brain tumors in rats fed aspartame. *Toxicol. Lett.* 7:433–437, 1981.

47. Duerfahrt, T., S. Doekel, P.L.M. Quaedflieg, M.A. Marahiel. Construction of hybrid peptide synthetases for the production of alpha-L-aspartyl-L-phenylalanine, a precursor for the high-intensity sweetener aspartame. *Eur. J. Biochem.* 270(22):4555–4563, 2003.

48. Isono, Y., M. Nakajima. Enzymatic synthesis of aspartame precursor in solvent-free reaction system with salt hydrates. *Nippon Shokuhin Kagaku Kogaku Kaishi* 49(12):813–817, 2002.

49. Ahn, J.E., C. Kim, C.S. Shin. Enzymic synthesis of aspartame precursors from eutectic substrate mixtures. *Process. Biochem.* 37(3):279–285, 2001.

50. Garbow, J.R., J.J. Likos, S.A. Schroeder. Structure, dynamics, and stability of beta-cyclo-dextrin inclusion complexes of aspartame and neotame. *J. Agric. Food Chem.* 49(4):2053–2060, 2001.

51. Li, J.P., Z.Y. He. Preparing process of aspartame. *Zhongguo YiyaoGongye Zazhi* 31(3):106, 2000.

52. Karikas, G.A., K.H. Schulpis, G. Reclos, G. Kokotos. Measurement of molecular interaction of aspartame and its metabolites with DNA. *Clin. Biochem.* 31(5):405–407, 1998.

53. Nakaoka, H., Y. Miyajima, K. Morihara. Papain-catalyzed synthesis of aspartame precursor: a comparison with thermolysin. *J. Ferment. Bioeng.* 85(1):43–47, 1998.

54. Nofre, C., J.M. Tinti. Neotame: discovery, properties, utility. *Food Chem.* 69(3):245–257, 2000.

55. E.J. Munson, S.A. Schroeder, I. Prakash, D.J.W. Grant. Neotame anhydrate polymorphs II: quantitation and relative physical stability. *Pharm. Res.* 19(9):1259–1264, 2002.

56. Dong, Z., V.G. Young, A. Sheth, E.J. Munson, S.A. Schroeder, I. Prakash, D.J.W. Grant. Crystal structure of neotame anhydrate polymorph G. *Pharm. Res.* 19(10):1549–1553, 2002.

57. Mayhew, D.A., C.P. Comer, W.W. Stargel. Food consumption and body weight changes with neotame, a new sweetener with intense taste: differentiating effects of palatability from toxicity in dietary safety studies. *Regul. Toxicol. Pharmacol.* 38(2): 124–143, 2003.

58. Flamm, W.G., G.L. Blackburn, C.P. Comer, D.A. Mayhew, W.W. Stargel. Long-term food consumption and body weight changes in neotame safety studies are consistent with the allometric relationship observed for other sweeteners and during dietary restrictions. *Reg. Toxicol. Pharmacol.* 38(2):144–156, 2003.

59. Kroyer, G.T. The low calorie sweetener stevioside: stability and interaction with food ingredients. *Lebensmittel Wissenschaft Technologie* 32(8):509–512, 1999.

60. Adduci, J., D. Buddhasukh, B. Ternai. Improved isolation and putrification of stevioside. *J. Sci. Soc. Thailand* 13(3):179–183, 1987.

61. Ogawa, T., M. Nozaki, M. Masanao. Total synthesis of stevioside. *Tetrahedron* 36(18):2641–2648, 1980.

62. Geuns, J.M.C. Stevioside. *Phytochemistry* 64(5):913–921, 2003.

63. Brandle, J.E., N. Rosa. Heritability for yield, leaf-stem ratio and stevioside content estimated from a landrace cultivar of *Stevia rebaudiana. Can. J. Plant Sci.* 72:1263–1266, 1992.

64. Crammer, B., R. Ikan. Sweet glycosides from the stevia plant. *Chem. Br.* 22:915–916, 1986.

65. Metivier, J., A.M. Viana. The effect of long and short day length upon the growth of whole plants and the level of soluble proteins, sugars and stevioside in leaves of *Stevia rebaudiana* Bert. *J. Exp. Bot.* 30:1211–1222, 1979.

66. Soejarto, D.D., C.M. Compadre, P.J. Medon, S.K. Kamath, A.D. Kinghorn. Potential sweetening agents of plant origin, II: field search for sweet-tasting *Stevia* species. *Econ. Bot.* 37:71–79, 1983.

67. Matsui, M., K. Matsui, Y. Kawasaki, Y. Oda T. Noguchi, Y. Kitagawa, M. Sawada, M. Hayashi, T. Nohmi, K. Yoshihira, T. Ishidate, M. Sofuni. Evaluation of the genotoxicity of stevioside and steviol using six *in vitro* and one *in vivo* mutagenicity assays. *Mutagenesis* 11:573–579, 1996.

68. Matsui, M., T. Sofuni, T. Nohmi. Regionally-targeted mutagenesis by metabolically-activated steviol: DNA sequence analysis of steviol-induced mutants of guanine phosphoribosyltransferase (gpt) gene of *Salmonella typhimurium* TM677. *Mutagenesis* 11:565–572, 1996.

69. Gregersen, S., P.B. Jeppesen, J.J. Holst, K. Hermansen. Antihyperglycemic effects of stevioside in type 2 diabetic subjects. *Metab. Clin. Exp.* 53(1):73–76, 2004.

70. Totté, N., L. Charon, M. Rohmer, F. Compernolle, I. Baboeuf, J.M.C. Geuns. Biosynthesis of the diterpenoid steviol, an *ent*-kaurene derivative from *Stevia rebaudiana* Bertoni, via the methylerythritol phosphate pathway. *Tetrahedron Lett.* 41:6407–6410, 2000.

71. Kim, K.K., H. Yamashita, Y. Sawa, H. Shibata. A high activity of 3-hydroxy-3-methylglutaryl coenzyme A reductase in chloroplasts of *Stevia rebaudiana* Bertoni. *Biosc. Biotech. Biochem.* 60:685–686, 1996.

72. Brandle, J.E., A. Richman, A.K. Swanson, B.P. Chapman. Leaf ESTs from *Stevia rebaudiana*: a resource for gene discovery in diterpene synthesis. *Plant Mol. Biol.* 50:613–622, 2002.

73. Kim, K.K., Y. Sawa, H. Shibata. Hydroxylation of ent-kaurenoic acid to steviol in *Stevia rebaudiana* Bertoni: purification and partial characterization of the enzyme. *Arch. Biochem. Biophys.* 332: 223–230, 1996.

74. Ogata, C., M. Hatada, G. Tomlinson, W.C. Shin, S.H. Kim. Crystal structure of the intensely sweet protein monellin. *Nature* 328(6132):739–742, 1987.

75. Kotlovyi, V., W.L. Nichols, L.F. Ten Eyck. Protein structural alignment for detection of maximally conserved regions. *Biophys. Chem.* 105(2,3):595–608, 2003.

76. Sung, Y.H., H.D. Hong, C. Cheong, J.H. Kim, J.M. Cho, Y.R. Kim, W. Lee. Folding and stability of sweet protein single-chain monellin: an insight to protein engineering. *J. Biol. Chem.* 276(47):44229–44238, 2001.

77. Kim, S.H., C.H. Kang, R. Kim, J.M. Cho, Y.B. Lee, T.K. Lee. Redesigning a sweet protein: increased stability and renaturability. *Protein Eng.* 2:571–575, 1989.

78. Tomic, M.T., J.R. Somoza, D.E. Wemmer, Y.W. Park, J.M. Cho, S.H. Kim. 1H resonance assignments, secondary structure and general topology of single-chain monellin in solution as determined by 1H 2D-NMR. *J. Biomol. NMR* 2:557–572, 1985.

79. Hung, L., M. Kohmura, Y. Ariyoshi, S. Kim. Structural differences in D- and L-monellin in the crystals of racemic mixture. *J. Mol. Biol.* 285:311–321, 1999.

80. Kondo, K., M. Yutaka, H. Sone, K. Kobayashi, H. Iijima. High-level expression of a sweet protein, monellin, in the food yeast *Candida utilis*. *Nat. Biotechnol.* 15(5):453–457, 1997.

81. Kim, H., K. Lim. Large-scale purification of recombinant monellin from yeast. *J. Ferment. Bioeng.* 82(2):180–182, 1996.

82. Ming, D., G. Hellekant. A new high-potency thermostable sweet protein from *Pentadiplandra brazzeana* B. *FEBS Lett.* 355(1):106–108, 1994.

83. Jin, Z., V. Danilova, P. Assadi, M. Fariba, D.J. Aceti, J.L. Markley, G. Hellekant. Critical regions for the sweetness of brazzein. *FEBS Lett.* 544(1–3):33–37, 2003.

84. Ishikawa, K., M. Ota, Y. Ariyoshi, H. Sasaki, M. Tanokura, M. Ding, J. Caldwell, F. Abildgaad. Crystallization and preliminary x-ray analysis of brazzein, a new sweet protein. *Acta Crystallogr. D. Biol. Crystallogr.* 52:577–578, 1996.

85. Izawa, H., M. Ota, M. Kohmura, Y. Ariyoshi. Synthesis and characterization of the sweet protein brazzein. *Biopolymers* 39(1):95–101, 1996.

86. Caldwell, J.E., F. Abildgaard, Z. Dzakula, D. Ming, G. Hellekant, J.L. Markley. Solution structure of the thermostable sweet-tasting protein brazzein. *Nat. Struct. Biol.* 5(6):427–431, 1998.

87. Somoza, J.R., F. Jiang, L. Tong, C.H. Kang, J.M. Cho, S.H. Kim. Two crystal structures of a potently sweet protein: natural monellin at 2.75 A resolution and single-chain monellin at 1.7 A resolution. *J. Mol. Biol.* 234(2):390–404, 1993.

88. Assadi-Porter, F.M., D.J. Aceti, J.L. Markley. Sweetness determinant sites of brazzein, a small, heat-stable, sweet-tasting protein. *Arch. Biochem. Biophys.* 376(2):259–265, 2000.

89. Assadi-Porter, F.M., D.J. Aceti, H. Cheng, J.L. Markley. Efficient production of recombinant brazzein, a small, heat-stable, sweet-tasting protein of plant origin. *Arch. Biochem. Biophys.* 376:252–258, 2000.

90. Van der Wel, H., K. Loeve. Isolation and characterization of thaumatin I and II, the sweet-tasting proteins from *Thaumatococcus daniellii* Benth. *Eur. J. Biochem.* 31(2): 221–225, 1972.

91. Kong, J.Q., Q. Zhao, L. Gao, X.T. Qui, Q.J. Yang. Sweet protein thaumatin and its genetic engineering. *Yichuan* 25(2):232–236, 2003.

92. Zemanek, E., B.P. Wasserman. Issues and advances in the use of transgenic organisms for the production of thaumatin, the intensely sweet protein from *Thaumatococcus daniellii*. *Crit. Rev. Food Sci. Nutr.* 35:455–466, 1995.

93. Faus, I., C. Patiño, J.L. del Río, C. del Moral, H. Sisniega, V. Rubio. Expression of a synthetic gene encoding the sweet-tasting protein thaumatin in *Escherichia coli*. *Biochem. Biophys. Res. Commun.* 229:121–127, 1996.

94. Illingworth, C., G. Larson, G. Hellekant. Secretion of the sweet-tasting plant protein thaumatin by *Bacillus subtilis*. *Biotechnol. Lett.* 10:587–592, 1988.

95. Illingworth, C., G. Larson, G. Hellekant. Secretion of the sweet-tasting plant protein thaumatin by *Streptomyces lividans*. *J. Ind. Microbiol.* 4:37–42, 1989.

96. Edens, L., I. Born, A.M. Ledeboer, J. Maat, M.Y. Toonen, C. Visser, C.T. Verrips. Synthesis and processing of the plant protein thaumatin in yeast. *Cell* 37:629–633, 1984.

97. Gwynne, D.I., M. Devchand. Expression of foreign proteins in the genus *Aspergillus*. In: *Aspergillus, the Biology and Industrial Applications*, Bennet, J.W., M.A. Klich, eds., London: Butterworth, 1992, pp 203–214.

98. Faus, I., C. del Moral, N. Adroer, J.L. del Río, C. Patiño, H. Sisniega, C. Casas, J. Bladé, V. Rubio. Secretion of the sweet-tasting protein thaumatin by recombinant strains of *Aspergillus niger* var. *awamori*. *Appl. Microbiol. Biotechnol.* 49:393–398, 1998.

99. Verdoes, J.C., P.J. Punt, C.A.M.J.J. van den Hondel. Molecular genetic strain improvement for the over-production of fungal proteins by filamentous fungi. *Appl. Microbiol. Biotechnol.* 43:195–205, 1995.

100. De Vos, A.M., M. Hatada, H. Van der Wel, H. Krabbendam, A.F. Peerdeman, S.H. Kim. Three-dimensional structure of thaumatin I, an intensely sweet protein. *Proc. Natl. Acad. Sci. USA* 82:1406–1409, 1985.

101. Kaneko, R., N. Kitabatake. Structure–sweetness relationship in thaumatin: importance of lysine residues. *Chem. Senses* 26:167–177, 2001.

102. Masuda, T., S. Tamaki, R. Kaneko, R. Wada, Y. Fujita, A. Mehta, N. Kitabatake. Cloning, expression and characterization of recombinant sweet-protein thaumatin II using the methylotrophic yeast *Pichia pastoris*. *Biotechnol Bioeng*. 85(7):761–769, 2004.

103. Agrawal, S., E.R. Kandimalla. Antisense therapeutics: is it as simple as complementary base recognition? *Mol. Med. Today* 6:72–81, 2000.

104. Bautista, L.F., A. Aleksenko, M. Hentzer, A. Santerre-Henriksen, J. Nielsen. Antisense silencing of the *creA* gene in *Aspergillus nidulans*. *Appl. Environ. Microbiol.* 66:4579–4581, 2000.

105. Berka, R.M., M. Ward, L.J. Wilson, K.J. Hayenga, K.H. Kodama, L.P. Carlomagno, S.A. Thompson. Molecular cloning and deletion of the gene encoding aspergillopepsin A from *A. awamori*. *Gene* 86:153–162, 1990.

106. Liu, X., S. Maeda, Z. Hu, T. Aiuchi, K. Nakaya, Y. Kurihara. Purification, complete amino acid sequence and structural characterization of the heat-stable sweet protein, mabinlin II.

107. Nirasawa, S., Y. Masuda, K. Nakaya, Y. Kurihara. Cloning and sequencing of a cDNA encoding a heat-stable sweet protein, mabinlin II. *Gene* 181(1,2):225–227, 1996.

15

Biotechnological Approaches to Improve Nutritional Quality and Shelf Life of Fruits and Vegetables

Reena Grittle Pinhero and Gopinadhan Paliyath

CONTENTS

15.1 Introduction 346
15.2 Potatoes 346
    15.2.1 Factors Affecting Accumulation of Reducing Sugars in Potatoes..........347
    15.2.2 Low Temperature Sweetening in Potatoes...347
    15.2.3 Metabolism of Starch in Tuber 348
        15.2.3.1 Starch Synthesis 349
        15.2.3.2 Starch Degradation..350
    15.2.4 Starch–Sugar Balance 352
    15.2.5 Sucrose Metabolism...353
        15.2.5.1 Sucrose Degradation 355
    15.2.6 Glycolysis...356
    15.2.7 Oxidative Pentose Phosphate Pathway (PPP)..358
    15.2.8 Mitochondrial Respiration..358
    15.2.9 Compartmentation and Stress Induced Membrane Changes 359
        15.2.9.1 Lipid Composition...359
        15.2.9.2 Membrane Permeability 360
    15.2.10 Free Radicals and Antioxidant Enzymes 362
15.3 Tomato...363
    15.3.1 Role of Membrane in Shelf Life 363
    15.3.2 Phospholipase D Gene Family 364
        15.3.2.1 Role of PLD in Senescence and Chilling 364
        15.3.2.2 Regulation of PLD Activity 365
        15.3.2.3 Antisense Suppression of PLD Activity..................................365

        15.3.3 Lipoxygenase..366
                15.3.3.1 Role of Lipoxygenase in Membrane Deterioration 366
                15.3.3.2 Tomato Fruit Ripening and Lipoxygenases 368
                15.3.3.3 Regulation of LOX by Genetic Manipulation 368
        15.3.4 Cell Wall Metabolism and Fruit Softening 369
15.4 Conclusion...370
References...370

15.1   INTRODUCTION

Fruits and vegetables, which can be consumed as fresh and as processed products, are important ingredients of a healthy diet. They are valuable sources of vitamins, minerals, antioxidants, and fiber. The important quality factors of fruits and vegetables are their color, flavor, texture, and nutritive value. Consumers always prefer to buy fruits and vegetables of high quality. As used by the industry, quality is a concept involving measurable attributes: degree of purity, firmness, flavor, color, size, maturity, condition, or any other distinctive attributes or characteristics of the product (1). The qualities of the produce bought by the consumer are influenced by many factors, such as the cultivar, the environmental conditions affecting growth, cultural practices, exposure to pests and diseases, time of harvesting, and postharvest and storage conditions used. Today, with the advancement of technology in several areas, the only factor on which the grower has no control of is the environment of the field conditions. Heredity (the identity of the cultivar) plays a major role in determining the quality of fruits and vegetables, as evidenced by the various varietal differences in quality. Even though traditional crop breeding is still being used as one means of crop improvement, continuing advances in knowledge and technology have dramatically expanded the biotechnological tools available for genetic improvement and production of vegetables and fruits. The term biotechnology is broad, encompassing a wide range of disciplines in science. In this chapter, the focus will be on plant transformation, where genes are modified or transferred by molecular means, and the resulting improvements in fruit and vegetable quality. Because this chapter is focused on genetic engineering, we will be discussing only the parameters and mechanisms that affect nutritional quality and shelf life (other than pests and diseases), and which can be improved or modified by genetic engineering. The discussion will also be based on two important crops; potatoes and tomato.

15.2   POTATOES

Potatoes are ranked fourth in production of all agricultural commodities in the world and yield more dry matter and protein per hectare than the major cereal crops (2). They are consumed as fresh and processed products, and used as raw material for many industrial purposes such as starch extraction. Potato chips and french fries are two of the most popular processed potato products. The consumer preference for these products is influenced by the color and crispness of these products. The primary problem associated with potato processing is the nonenzymatic browning of the product that occurs under the high temperature conditions used during frying, when reducing sugar levels are high in the tissue, a phenomenon known as Maillard reaction (3). The reducing sugars, glucose and fructose, combine with the α-amino groups of amino acids at the high temperatures used in frying operations, resulting in darker and more bitter flavored french fries and chips that are unacceptable to the consumer. The ideal content of reducing sugars is 0.1% of the tuber fresh weight; 0.33% is the upper acceptable limit (4). A four year study was conducted to determine the compositional differences during low temperature storage between low sugar accumulating and high sugar accumulating cultivars in relation to potato chip processing quality (5). Pearson analysis of the above data showed that chip color was most closely correlated with reducing sugar concentration. Multiple regression analysis revealed that the relative contribution of each of the parameters studied, such as sucrose, reducing sugars, nitrogen, protein, ascorbic acid and dry matter content, varied greatly among cultivars and selections evaluated and from season to season (5).

15.2.1 Factors Affecting Accumulation of Reducing Sugars in Potatoes

The factors that affect the sweetening or breakdown of starch into sucrose and its component reducing sugars glucose and fructose are drought, excess nitrogen during growth, high temperature at harvest, handling, aging, identity of the cultivar, anaerobic conditions, and low temperature during post harvest storage (6).

15.2.2 Low Temperature Sweetening in Potatoes

Low temperature sweetening (LTS) in potato tubers is a phenomenon that occurs when tubers are stored at temperatures below 10°C in order to minimize respiration and sprouting. LTS results in the accumulation of starch breakdown products, primarily sucrose and the reducing sugars glucose and fructose (7), which cause Maillard browning during potato chip frying operations (3,8). Fry color of Russet Burbank and Shepody potatoes has been shown to be more closely associated with glucose concentration than with fructose, total reducing sugars, sucrose, or total sugars (9). In order to avoid Maillard browning, processing potatoes are generally stored at temperatures around 10°C; but at this storage temperature potato tubers will sprout. To prevent sprouting during storage, the processing tubers are treated with chemical sprout inhibitors. However, due to health and environmental concerns, there is increasing pressure to reduce the use of chemical sprout inhibitors. The only solution to avoid this problem is using cultivars that are resistant to LTS.

Low temperature storage of potato tubers has many advantages, such as control of sprout growth and senescent sweetening, reduction of physiological weight loss due to decreased respiration and losses associated with bacterial and fungal pathogens, and extended marketability. Low temperature storage has several advantages, but the main drawback associated with it is the accumulation of reducing sugars and the resulting browning of processed products such as chips and fries.

The mechanism responsible for the initiation and subsequent regulation of LTS in potato tubers has not been fully elucidated. Many theories have been proposed to explain LTS based on starch metabolism, sucrose metabolism, glycolysis and the oxidative pentose phosphate pathway (PPP), and mitochondrial respiration (10,11), as well as membrane instability, lipid peroxidation, and electrolyte leakage (12–15). It has been suggested that in mature, cold stored potato tubers, the glycolytic or respiratory capacity plays a key role in the ability of potatoes to regulate their sugar concentration (16). Although LTS has not been elucidated at the molecular level, many factors may play a role in it. For example, chilling may influence compartmentation and membrane permeability by altering the phase transition of lipids in the bilayer, resulting in the leakage of key ions such as inorganic phosphates. This can alter the activity and synthesis of key enzymes involved in the metabolic pathways, ultimately resulting in LTS (17).

Many theories have been postulated and documented to explain LTS at the level of carbohydrate metabolism in stored potato tubers (7,11,12,17,18). The mechanism is complex and may involve the interaction of several pathways of carbohydrate metabolism and the genes that regulate these pathways. This discussion focuses on a theoretical model of the mechanisms involved in LTS based on information available about the roles of the major tuber carbohydrate metabolic pathways as well as changes in membrane stability (Figure 15.1).

15.2.3 Metabolism of Starch in Tuber

Starch is the major component in the main crop plants of the world, as well as an important raw material for many industrial processes. Potato tubers contain 60–80% starch, of which sugars represent only a small fraction (up to 3% on a dry weight basis) (19). There is evidence that the principal event in LTS is the cold induced synthesis of sucrose (7,12). The carbon needed for the synthesis of sucrose and reducing sugars for LTS is generally, but not always, provided by a net breakdown of starch. An increase in potato tuber sugar content occurs early during cold storage; over two to three months at storage temperatures of 1–3°C, tubers can convert as much as 30% of their starch content (20). In mature King Edward tubers stored at 2°C, the sugar content increased from 0.3 to 2.5% in three months, with the initial appearance of sucrose followed by reducing sugars (19). Coffin et al. (21) found that sucrose content increased within two days of 5°C storage for both mature and immature tubers of four cultivars, while fructose and glucose content increased more slowly. Pollock and ap Rees (22) reported an increase in both sucrose and reducing sugar content within 5 days in tubers stored at 2°C, and after 20 days storage, the sugar content was approximately

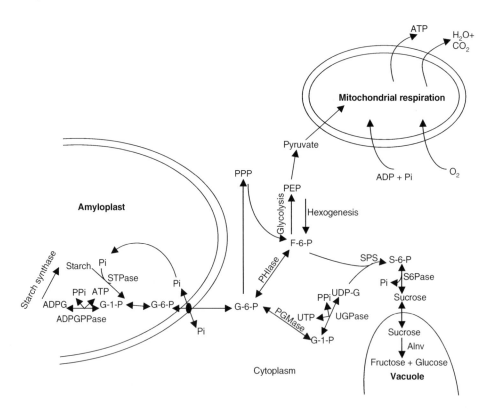

Figure 15.1 Starch-sugar interconversion in potato tubers. ADPG, ADP-glucose; ADPGPPase, ADP-glucose pyrophosphorylase; G-1-P, glucose-1-phosphate; G-6-P, glucose-6-phosphate; STPase, starch phosphorylase; F-6-P, fructose 6-phosphatase, SPS, sucrose-6-phosphate synthase; S-6-P, sucrose-6-phosphate; S6Pase, sucrose-6-P phosphatase; Ainv, acid invertase; PPP, pentose phosphate pathway; PEP, phosphoenolpyruvate (adapted from Sowokinos, 2001)

six times greater than at day 0. The sweetening response of tubers to low temperatures is fairly consistent, but is influenced by cultivar, locality, and conditions prior to cold storage. Isherwood (19) related energy requirements to possible biosynthetic pathways and concluded that sucrose was formed from starch when potato tubers were moved from 10° to 2°C and that starch was reformed when tubers were moved back from 2° to 10°C, although different metabolic pathways were involved. Reconditioning of tubers is sometimes used to improve chipping quality by decreasing the level of reducing sugars (20). After cold storage, potatoes are reconditioned at 18°C where sugar content decreases and starch content increases as sugars are resynthesized to starch. However, the response to reconditioning is neither consistent nor completely restorative, and tends to be cultivar dependent.

Preconditioning treatment has been used to lessen chilling injury in chilling sensitive plants. Storage at 10°C prior to cold storage can acclimatize tubers and lessen the LTS effect (8). Katahdin tubers preconditioned at 15.5°C for one to four weeks before 0°C storage did not show a change in sugar accumulation patterns or respiration rates (23). The use of intermittent warming (15.5°C for one week following 0°C for three weeks) decreased sugar levels and respiration rates to levels lower than those of tubers stored continuously at 0, 1 and 4.5°C, although sugar levels were not low enough for desirable chipping potatoes.

15.2.3.1 Starch Synthesis

Starch is synthesized in plastids (amyloplasts) upon tuber initiation, and both the number of starch grains and the grain size increase during tuber growth (Figure 15.1, Figure 15.2). Starch consists of two types of glucose polymers, the highly branched amylopectin, and relatively unbranched amylose. Potato starch is comprised of 21–25% amylose and 75–79% amylopectin (24). Starch is synthesized from ADPglucose by the concerted action of ADPglucose pyrophosphorylase (ADPGPase), starch synthase, and the starch branching enzymes (25). Following the conversion of sucrose into hexose phosphates in the cytosol, glucose-6-phosphate is transported into the amyloplast where it is converted into glucose-1-phosphate. A study involving antisense inhibition of plastidial phosphoglucomutase supported the theory that carbon from the cytosol was imported into potato tuber amyloplasts in the form of glucose-6-phosphate (26). Glucose-1-phosphate is subsequently converted to ADPglucose by ADPGPase. The starch synthases catalyze the polymerization of the glucose monomers into α-1,4-glucans using ADPglucose as a substrate, while the starch branching enzyme catalyzes the formation of the α-1,6-linkages of amylopectin (25).

ADPGPase is often referred to as a rate-limiting step in starch synthesis (25). It is subjected to allosteric activation by 3-P-glycerate and inhibition by inorganic phosphate (27). Strategies to alter the starch metabolism in tubers by genetic manipulation of ADPGPase may be helpful in reducing the accumulation of reducing sugars during LTS. It has been reported that transgenic tubers with an 80–90% reduction in ADPGPase activity have reduced starch content relative to wild type tubers (28,29). The reduction in ADPGPase activity resulted in a major reduction of carbon flux, with increased flux to sucrose and decreased flux to starch. Stark et al. (30) have reported that hexose accumulation was greatly reduced in cold stored tubers with overexpression of the mutated ADPGPase gene, glgC16, from *E. coli*. The glgC16 gene produces a mutant form of ADPGPase that is less responsive to allosteric effectors. It has been suggested that the observed decrease in hexose concentration could be due to the higher starch biosynthetic capacity of the transgenic tubers. Lorberth et al. (31) developed transgenic potatoes with decreased levels of R1 protein, a starch granule bound protein capable of introducing phosphate into starch-like glucans. By reducing the activity of the R1 protein using antisense technology, the phosphate content of starch was reduced, resulting in a starch that

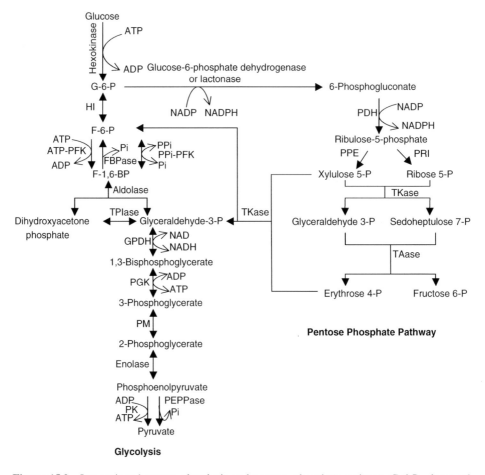

Figure 15.2 Interactions between glycolysis and pentose phosphate pathway. G-6-P, glucsoe 6-phosphate, HI, hexose phosphate isomerase, F-6-P, fructose 6-phosphate, ATP-PFK, ATP-dependent phosphofructokinase, FBPase, fructose 1,6-bisphosphatase, PPi-PFK, PPi-dependent phosphofruc-tose kinase, F 1,6-BP, fructose 1,6 bisphosphate, TPlase, triose phosphate isomerase; GPDH, glycer-aldehyde 3--phoshate dehydrogenase; PGK, phosphoglycerate kinase; PM, phosphoglycero mutase; PK, pyruvate kinase; PDH, 6-phosphogluconate dehydrogenase; PPE, phosphopentose epimerase; PRi, phoshoribo-isomerase; Tkase, transketolase; Taase, transaldolase.

was less susceptible to degradation at low temperatures relative to the starch of wild type tubers. It has been observed that after two months of storage at 4°C, the transgenic tubers contained up to ninefold lower concentrations of reducing sugars compared to the wild type. However, the commercial value of the modification of starch could not be assessed because the authors did not analyze the processing quality of transgenic tubers.

15.2.3.2   Starch Degradation

The differential response of potato cultivars to LTS may be the result of starch properties that affect the ability of enzymes to degrade it. The various starch properties ascribed are:

1.   Chemical modifications of glucose units by attachment of covalently bound phos-phate. Phosphate esters are attached to C3 or C6 glucosyl residues of amylopectin

in the larger interbranch chains and are absent around the branching points. This affects the cleavage sites and degradation product patterns (32).

2. Surface property alteration caused by the negative surface charge from surface phosphate, lipid, or protein can affect the properties of enzymes and other soluble compounds (33).

3. Association with starch-metabolizing enzymes such as endoamylase activity in cotton leaves (34) and starch synthase in potato tubers (35).

4. Physical characteristics of starch. Isolated starch grains from two cultivars differing in their sensitivity to LTS showed an increase in starch grain size over time with the disappearance of smaller starch grains while ND860-2, the resistant cultivar had a consistently smaller mean starch grain size (36,37). Higher levels of amylose in ND 860-2 were believed to be responsible for a more ordered crystallinity within the starch granule, decreased thermomechanical analysis swelling, increased resistance to gelatinization and decreased susceptibility to α-amylase hydrolysis.

In addition to the various properties of starch, which affect its degradation, other factors such as enzymes responsible for starch degradation during LTS have been studied. The pathway of starch breakdown during LTS is not well established. The degradation of starch is believed to occur in the amyloplast (38). The widespread distribution of α-glucan phosphorylase, α-amylase, β-amylase, and maltase (39) suggests that starch breakdown could be phosphoryltic, hydrolytic, or both. However it is assumed that starch breakdown in cold stored potato tubers is mainly conducted by starch phosphorylase, because sucrose is the first sugar to accumulate upon transfer of tubers to chilling temperatures (19). Amylase activities are too low at such cold temperature to catalyze the required rate of starch degradation (40), and no increase in either maltose or polymers of glucose larger than maltose, the common products of amylolytic starch degradation (41), have been observed during LTS (42).

Two types of potato phosphorylases are recognized based on glucan specificity, monomer size, and intracellular location. They are noninterconvertible proteins with different primary structures and different immunological properties (38). Type 1 isozyme, also known as type H, is localized in the cytoplasm, has a low affinity for maltodextrins, has a high affinity for branched polyglucans, and cross reacts with type H phosphorylase from potato leaves. Type 2 isozyme, also known as type L, is located in the amyloplast, has a high affinity for maltodextrins, has a low affinity for branched polyglucans, and cross reacts with type L enzyme from the leaf (38,43). Type L and type H isozymes do not cross react immunologically. The function of these isozymes in starch degradation and LTS is unknown. There are reports which suggest that starch breakdown during LTS is phosphorylitic. Kumar et al. (44) have demonstrated that although the activities of cytosolic and plastidic isozymes of starch phosphorylase were reduced by up to 70% in transgenic potatoes expressing antisense cDNA constructs of starch phosphorylase, this did not affect the accumulation of reducing sugars during 4°C storage.

Other studies suggest that starch breakdown in potato tubers during cold storage is not solely due to the activity of starch phosphorylase. Cochrane and coworkers (45), using a modified amylase assay, found that α- and β-amylases and α-glucosidase activities were much higher in tubers stored at the colder temperature (4°C) than those stored at 10°C, and in cultivars known to be more susceptible to LTS. It was considered inappropriate to correlate reducing sugar content and amylase activity, because the formation of reducing sugars is influenced by many other cold labile processes in the tuber. Reducing sugar content and the activities of α- and β-amylases and debranching enzymes were measured by Cottrell et al. (46) over 139 days in five cultivars of potato tuber stored at 4 and 10°C. The activities of these enzymes were always greater at 4 than at 10°C, but cultivars that accumulated high levels of reducing sugars did not always display the greatest level of hydrolytic enzyme activity (46). It has been reported that the onset of sugar accumulation in tubers during low temperature storage coincided with an increase in the activity of one specific isoform of amylase, the β-amylase in the cultivar Desirée (47–49). β-Amylase activity was present at low levels in tubers stored at 20°C, and increased from four- to fivefold within 10 days of storage at 3°C. However, no specific role has been established for this cold induced β-amylase in LTS.

15.2.3.2.1   Effect of Inorganic Phosphates   The intracellular compartmental-ization of Pi has been suggested to influence carbon partitioning in nonphotosynthetic potato tubers in a manner similar to its role in photosynthetic tissues. It has been observed that increased inorganic phosphate in the amyloplast shifted metabolic activities toward starch degradation rather than accumulation (Figure 15.1) (17). Increased Pi concentration inhibits ADPGPase and enhances starch breakdown by α-glucan phosphorylase. A high concentration of Pi was found in tuber amyloplasts (50), and Pi in cold stored tubers was later found to be cleaved off from starch (51). Higher levels of Pi were found in Russet Burbank potatoes stored at 5.5 than at 15.5°C (52). A highly significant correlation was found between the Pi content and the accumulation of reducing sugars. Amyloplasts were found to have high concentrations of Pi, citrate, Cl$^-$, and K$^+$. It was suggested that Pi leaks from the amyloplast to the cytoplasm during cold storage and induces higher sugar con-centration levels in tubers during LTS (53).

Another source of Pi in plant cells is the vacuole (54). The major portion of Pi is stored in potato tuber vacuoles where it is compartmentalized away from the cytoplasm. Loughman (55) examined the respiratory changes of potato tuber slices and found that the larger part of Pi in the cell was localized in the vacuole, and did not take part in the steady state metabolism of the cell. However, the Pi in the vacuole may become available for metabolism in the cytoplasm during cold storage when ionic pumps that utilize ATP in the tonoplast become unable to maintain ionic gradients (56). This scenario could happen by passive leakage of Pi into the cytoplasm or when the membrane becomes leaky due to changes in the properties of the lipid bilayer. Increased Pi in the cytoplasm could affect the metabolism by mobilizing carbon from the amyloplast into the cytoplasm in exchange for Pi transported into the amyloplast by the hexose phosphate translocator (Figure 15.1) (57). Inside the amyloplast stroma, Pi can serve as a substrate for the continued phosphorlysis of starch, via α-glucan phosphorylase, forming additional molecules of G-1-P. Elevated cytoplasmic levels of Pi initiated by leaky membranes, coupled with reduced levels of fructose 2,6,bisphosphate during cold stress, would direct carbons away from glycolysis and favor the buildup of hexose phosphates for gluconogenic reactions (17).

15.2.4   Starch–Sugar Balance

The close relationship between starch and sugar levels when potatoes are cooled from 10 to 2°C and then rewarmed from 2 to 10°C after an interval, has given rise to a misleading con-cept of a starch to sugar balance in which the overall change between the two compounds is seen as being reversible (19). All the available evidence suggests that sucrose is formed from starch by an irreversible pathway, and that starch is formed from sucrose by separate, but likewise irreversible, routes. The very close relationship between starch and sucrose in stored potatoes may be due to the fact that starch is the only possible source of carbon for sugar synthesis in the cold (7,11,12,17). There is a strong evidence to indicate that G-6-P is trans-ported into the amyloplast of potato tubers to support starch synthesis (26). Thus the pathways of starch and sugar biosynthesis compete for the same pool of precursors.

A net flux of carbon from starch synthesis into sucrose occurs in cold stored tubers as evidenced from the use of radiolabels in experiments (4,58). The coexistence of the pathways of sucrose synthesis and starch breakdown in stored tubers may be regulated by fine control mechanisms. In potato discs incubated with ^{14}C glucose at 3°C and 15°C, a large proportion of label is recovered in starch. At low temperature, in a cold sensitive, high sugar-accumulating cultivar, the ratio of ^{14}C recovered in sucrose to that recovered in starch increased (4,58), but was unaffected in a cold tolerant, low sugar-accumulating cultivar (4,58). This suggests that genotypic variation in the capacity to maintain an active starch synthesizing system may help in alleviating the rate of sucrose accumulation.

15.2.5 Sucrose Metabolism

Sucrose is the first sugar to accumulate during LTS. Its accumulation in potato tubers has been recorded within hours of their placement at LTS inducing temperatures, with the accumulation of reducing sugars occurring a few days later (19). Sucrose synthesis occurs in the cytoplasm of the tuber either by sucrose 6-P synthase (SPS), or by sucrose synthase (SS) and the hexose phosphates required for this are transported from the amyloplast via a phosphate translocator (Figure 15.1) (26,59).

Pressey (60) reported that SS activities decreased after harvest, and continued to do so under low temperature storage conditions. SPS activity also decreased if tubers were held at warm temperatures but rapidly increased when tubers were held at low temperatures. This observation indicates that SPS is the enzyme responsible for sucrose synthesis at low temperatures. Pollock and ap Rees (22) reported that sucrose synthesis during LTS is catalyzed by SPS and not by SS. This was also confirmed by ^{13}C NMR studies (61).

The increase in sucrose synthesis upon transferring the tubers to low temperature has been associated with the increased expression of an isoform of SPS (SPS-1b, 127 kDa) (62). The cold induced increase in the SPS-1b isoform was found to correlate well with the change in the kinetic properties of the enzyme. The major isoform found in tubers stored at room temperature is a 125-kDa protein (SPS-1a). Reconditioning of the tubers at 20°C resulted in the disappearance of the cold induced SPS isoform after 2–4 days (49). An increase in the total amount of SPS transcript was observed at low temperature in each of these studies. SPS from potato tubers has been shown to be subject to fine regulation by allosteric effectors and protein phosphorylation (63). Potato tuber SPS is allosterically activated by G-6-P and inhibited by protein phosphorylation.

Antisense technology has been used as an effective tool to investigate the roles of enzymes that lead to the production of sucrose, as well as reducing hexoses such as glucose and fructose in LTS. Many researchers investigating LTS mechanisms have used this technology to substantiate the role of enzymes in the carbohydrate metabolic pathway. For example, in experiments involving transgenic tubers where the SPS activity was reduced by 70–80% either by antisense or cosuppression, cold sweetening was reduced by inhibiting the increase of the cold induced isoform of SPS (64). The authors also observed that the V_{max} of SPS was 50 times higher than the net rate of sugar accumulation in wild type tubers, and found that SPS is strongly substrate limited, particularly for UDP-G (Figure 15.1). These results indicate that the rate of cold sweetening in wild type tubers is not strongly controlled by the overall SPS activity or the overall amount of SPS protein. Alterations in the kinetic properties of SPS during cold temperature storage were more effective in stimulating sucrose synthesis than changes in SPS expression. The observation that changes in the kinetic properties of potato tuber SPS coincide with the onset of sugar accumulation points to the fact that the fine regulation of SPS may be more important than coarse regulation in controlling the ability of a cultivar to sweeten during cold storage.

However, it should be noted that SPS may not be the only candidate that regulates sugar accumulation during LTS, because other factors that affect the availability of hexose phosphates, such as glycolysis and the pentose phosphate pathway, may have key roles to play (Figure 15.1, Figure 15.2).

UDP-glucose pyrophosphorylase (UGPase) is a cytosolic enzyme that catalyzes the formation of UDP-G, one of the substrates required for the synthesis of sucrose (Figure 15.1). Depending on the physiological state of the tubers (i.e, growth or post harvest storage), the UGPase reaction may be directed toward the synthesis or degradation of starch (10). During the process of cold sweetening, it has been suggested that UDP-G and PPi have regulatory roles in directing carbon flux into glycolysis, starch synthesis, hexose formation, or a combination of the three (17,65). The activity of UGPase has been correlated with the amount of glucose that tubers of different cultivars accumulate in cold storage (12), leading to the assumption that this enzyme might be a control point for low temperature sweetening, as it regulates the rate of SPS and sucroneogenesis by controlling the levels of UDP-G (17,66).

Genetic manipulation to down regulate the expression of UGPase in potato tubers has resulted in contrasting results based on the physiological stage of the tubers. In two separate experiments in which the UGPase activity was reduced by 30–50% compared to their wild types, the transgenic tubers accumulated lower levels of sucrose during storage relative to wild type tubers at 4°C and 12°C (67) and at 6°C and 10°C (68). It has been suggested that by limiting the rate of UDP-G synthesis, UGPase may exert control over the flux of carbon toward sucrose during the cold storage of tubers. These observations are supported by the results of Hill et al. (47) who observed that following the initiation of cold sweetening, the concentration of UDP-G changed in parallel with the concentration of sucrose.

In contrast to the above results, Zrenner et al. (69) observed that carbohydrate metabolism of growing tubers was not affected when the transgenic plants had a 96% reduction in UGPase activity as compared with the wild type plants. No significant changes were observed in the levels of fresh mass, dry mass, starch, hexose phosphates, or UDP-G at harvest relative to the wild type tubers. It was reported that 4–5% of UGPase activity was still in considerable excess compared to the activity of other glycolytic enzymes in the tuber, and the antisense construct may have to reduce UGPase to negligible levels in transgenic potatoes before any phenotypic differences are noticeable (70).

It should be noted that the flow of carbon is different based on the physiological state of the tuber. In the growing tuber, most of the incoming sucrose is used for the synthesis of starch, while in the stored tuber the hexose–phosphate produced from starch degradation is converted into sucrose. This explains the different responses obtained by Zrenner et al. (69), Spychalla et al. (67), and Borovkov et al. (68). In cold stored tubers, when the rate of starch breakdown exceeds the rates of glycolysis and respiration, the conversion of G-1-P to UDP-G is the only means of controlling the level of hexose phosphates. Hence it is possible that a significant effect of reduced UGPase activity may be observed only in tubers acting in the direction of sucrose synthesis, such as during post harvest storage.

Two UGPase alleles have been identified in potato tubers: UgpA and UgpB (67). In a survey conducted on a number of American and European cultivars and selections stored at 4°C, it was observed that a relationship existed between the allelic polymorphism of UGPase and the degree of sweetening. The genotypes that resist sweetening during cold storage have demonstrated a predominance of the allele UgpA; the genotypes susceptible to sweetening have a predominance of the allele UgpB (68).

In order to assess the role of UGPase in LTS, Sowokinos (13) cloned UGPase from 16 American potato cultivars and selections that have varying degrees of cold sweetening ability during storage at 3°C. It was observed that cultivars that were resistant to LTS possessed a UgpA: UgpB allelic ratio of 4:0 or 3:1. The cultivars demonstrating LTS revealed a ratio of 1:3 or 0:4 in favor of the UgpB allele. Sowokinos (13) also observed that the cold sensitive potato cultivars expressed up to three acidic isozymes of UGPase (UGP1, UGP2, UGP3) with UGP3 being the most abundant. In addition to the three isozymes present in the sensitive cultivars, the cold resistant cultivars possessed another two isozymes, UGP4 and UGP5 that were more basic in nature. Sowokinos (18) studied the physicochemical and catalytic properties of the purified UGP4 and UGP5 isozymes, and suggested that the differences in sugar accumulation between the cultivars and selections that are either sensitive or resistant to LTS may be partially due to the unique nature of expression and catalytic properties of the isoforms in resistant lines, including pH optimum, substrate affinities for G-1-P and UTP, V_{max}, and the magnitude of product inhibition with UDP-G. The overall effect of these differences in isozyme expression is that it may decrease the rate of UDP-G formation, resulting in a lower accumulation of reducing sugars in the cold resistant clones.

15.2.5.1 Sucrose Degradation

Sucrose plays a pivotal role in plant growth and development because of its function in translocation and storage, and the increasing evidence that sucrose (or some metabolite derived from it) may play a nonnutritive role as a regulator of cellular metabolism, possibly by acting at the level of gene expression (71). As mentioned earlier, sucrose is the first sugar to form during LTS, and the source of glucose and fructose accumulation appears to be the degradation of sucrose (72). Sucrose is broken down by two types of enzymes in plants. By invertase action, it is hydrolyzed into glucose and fructose; whereas by the action of SS, it is converted into UDP-G and fructose in the presence of UDP (73).

Potato tubers are known to possess both alkaline and acid invertases. Acid invertase is localized in vacuoles, whereas alkaline (neutral) invertase is localized in the cytoplasm (72,74). Acid invertase isoforms that are ionically bound to the cell wall have also been identified (74). Alkaline invertases are sucrose specific, while acid invertases cleave sucrose at the fructose residue but can also hydrolyze other β-fructose containing oligosaccharides such as raffinose and stachyose (74).

Based on several observations of sucrose synthase and acid invertase activities in developing, mature, and cold stored tubers, and given the fact that sucrose is stored mainly in the vacuole, it is believed that sucrose synthase is responsible for sucrose degradation in developing tubers, whereas acid invertase is the principal enzyme responsible for the breakdown of sucrose into hexoses during LTS (60,70,73,75–77). Based on the widely established inverse correlation between sucrose content and vacuolar acid invertase activity, it is strongly believed that sucrose is broken down by acid invertase in the vacuole and the resulting glucose and fructose are transported into the cytosol for the formation of hexose phosphate by hexoskinase (72,78–80). It has been reported that glucose concentrations are frequently higher than fructose concentrations in stored potato tubers (81). Zrenner et al. (82) evaluated the glucose to fructose ratio of 24 different cultivars and found that the ratios were between 1.1 and 1.6, which is a strong indicator that invertase is the key enzyme responsible for the conversion of sucrose to hexose.

Zrenner et al. (82) studied the effect of soluble acid invertase activity in relation to the hexose to sucrose ratio in 24 different potato cultivars and observed a strong correlation between the hexose to sucrose ratio and the extractable soluble invertase activity. They also isolated a cold inducible acid invertase cDNA from potatoes and developed transgenic potatoes expressing the invertase cDNA in an antisense orientation. The subsequent 12–58% reduction of acid invertase activity compared to the wild type tubers resulted in an accumulation of sucrose and a decrease in the concentration of hexoses. The hexose to sucrose ratio was found to decrease with decreasing invertase activities; however, the total amount of soluble sugars did not significantly change. Based on these observations, it was concluded that invertases do not control the total combined amount of glucose, fructose, and sucrose in cold stored potato tubers, but are involved in the regulation of the ratio of hexoses to sucrose (82). Greiner et al. (83) strongly inhibited the activity of cold induced vacuolar invertase in potato plants by repressing the activity, or by the expression of a putative vacuolar invertase inhibitor from tobacco (Nt-inh), in potato plants under the control of the CaMV 35S promoter. It was possible to decrease the cold induced hexose accumulation up to 75% without affecting tuber yield. Although the concentration of sugar produced during cold induced sweetening was decreased, the level of hexose accumulated was still in excess of what is commercially acceptable for the production of potato chips and fries. The observation that antisense expression of acid invertase did not control the total amount of soluble sugars in cold stored potato tubers (82) indicates that other factors in the carbohydrate metabolism may influence the regulation of the total amount of sugars accumulated, and that acid invertase could be only one of the enzymes involved in starch–sugar conversion.

Metabolism is more rigorously regulated by intracellular compartmentalization in plants than in animals (84), and compartmentalization of the pathways of carbohydrate catabolism is realized to be a distinct feature of plant respiration (70). It is now believed that a "futile" cycling (simultaneous synthesis and degradation) of sucrose functions continuously to allow plants to respond rapidly to demand for carbon (85). This metabolic cycle may also be involved in LTS. For instance, it has been observed that in the first two weeks of 4°C storage, the initial rates of sucrose accumulation corresponded closely with the estimated rates of sucrose synthesis (47). The rate of total soluble sugar accumulation decreased with increasing duration of cold storage. It is suggested that sugar accumulation decreased because the rate of recycling equalled the rate of synthesis.

15.2.6 Glycolysis

The effects of cold exposure on the metabolism of potato tubers indicate that cold induced sweetening may at least in part be due to differential sensitivity to low temperature of the enzymes in the glycolytic pathway (86). The available data suggest that phophofructokinase (PFK) and pyruvate kinase are more sensitive to cold than are the other enzymes involved in the metabolism of hexose-6-phosphates (Figure 15.2), and by lowering the temperature, divert the latter to sucrose (Figure 15.1) (22,86,87). Another glycolytic enzyme that has been studied in relation to LTS is fructose-1,6-bisphosphatase (FBPase). Plants possess ATP dependent (ATP-PFK) and PPi dependent (PPi-PFK) phosphofructokinases (88).

PPi-PFK is a cytosolic enzyme and experiments to examine the role of PPi-PFK during the aging of tissue slices from potato tubers (starch-storing tissue) and carrot roots (sucrose-storing tissue) showed that both vegetables showed the same pattern of changes of phosphorylated metabolites and fructose 2,6-bisphosphate. But, the consumption of PPi by tubers and the production of PPi in carrots indicated PPi-PFK control of the glycolytic flux in tubers and catalysis of the opposite reaction in carrot roots (89). PPi-PFK is activated by fructose-2,6-bisphosphate which does not affect ATP-PFK (90). The activity of PPi-PFK is often equal to or exceeds that of ATP-PFK (72,90). It has been reported that the activity of PPi-PFK was ten times that of ATP-PFK in developing potato tubers, and hence it is suggested that glycolysis may proceed regardless of the activity of ATP-PFK (91). The maximum activity of PPi-PFK has also been shown to be greatly reduced in tubers stored at a low temperature of 5°C due to a decrease in PPi-PFK affinity for fructose 2,6-bisphosphate, an increase in sensitivity to fructose 2,6-bisphosphate as an activator, and a decrease in fructose 2,6-bisphosphate concentration at decreasing temperature (92). By contrast, in another study, no evidence was found for a cold induced inhibition of PPi-PFK in tubers stored at 2°C and 8°C (93). Hence it was postulated that PPi-PFK contributes to LTS by regulating the PPi concentration below inhibitory levels, facilitating the formation of UDP-G and subsequent synthesis of sucrose (Figure 15.1 and Figure 15.2).

ATP-PFK has been implicated in the regulation of LTS (94,95). It is responsible for the irreversible ATP dependent conversion of fructose-6-phosphate to fructose-1,6-bisphosphate. Potato tubers have been reported to possess four isozymes of ATP-PFK (96). It has been reported that the temperature coefficient (Q_{10}) of three of the four isoforms were higher at 2–6°C than at 12–16°C, indicating the cold lability of these isoforms and their roles in the accumulation of hexose phosphate and sucrose synthesis in LTS (95). This result supports the suggestion of Bryce and Hill (97) that ATP-PFK dominates the control of glycolysis, and thereby respiration, in plants. However, the observation that respiration of potato tubers increases concomitantly with the initial increase in sugar concentration (19,98), and the fact that the conversion of fructose-6-phosphate is also catalyzed by PPi-PFK (Figure 15.2) indicate multiple regulatory controls in the biosynthesis of sugar phosphates.

Genetic manipulation of the two PFKs was carried out further to explore their roles in LTS. About 88–99% inhibition of PPi-PFK expression was obtained in stored tubers by antisense expression of PPi-PFK cDNA (99). Even though the transformation resulted in higher levels of hexose phosphates in transgenic tubers compared to their wild type tubers, no difference was observed between these tubers in the rates of sucrose and hexose accumulation, and the total amounts of sugars accumulated at 4°C. Besides, no change was observed in the maximum catalytic activities of ATP-PFK or other enzymes of glycolysis (pyruvate kinase) or sucrose breakdown (invertase and sucrose synthase) in the antisense tubers. This observation suggests that compensation occurs at the level of fine metabolic regulation rather than gene expression. The above results indicate that PPi-PFK may not control the rate of glycolysis at low temperatures, and that tubers possess excessive capacity to phosphorylate fructose-6-phosphate. The results are also not in agreement with the theory proposed by Claassen et al. (93), that PPi-PFK is involved in regulating the PPi concentration, as no evidence was observed to substantiate that the antisense and wild type tubers contained different PPi concentrations (99).

Expression of the *E. coli* pfkA gene in potato tubers resulted in a 14- to 21-fold increase in the maximum catalytic activity of ATP-PFK, without affecting the activities of other glycolytic enzymes (100). It was also found that no corresponding decrease in the concentration of hexose phosphate was observed, while the pool sizes of other glycolytic intermediates increased three- to eightfold. In another study, it was reported that a substantial increase in ATP-PFK activity did not affect the flux through glycolysis or a flux between glycolysis and the PPP (101). The above results suggest that ATP-PFK may not limit the rate of respiration of potato tubers. ATP-PFK is potently inhibited by phosphoenolpyruvate, and hence ATP-PFK activity may be dependent upon the activity of enzymes that metabolize phosphoenolpyruvate (PEP) such as pyruvate kinase and phosphoenolpyruvate phosphatase. The above contention is in agreement with the findings of Thomas et al. (102). Using metabolic control analysis (MCA) on tuber glycolysis, Thomas et al. (102) observed that ATP-PFK exerts little control over glycolytic flux, while far more control of flux resides in the dephosphorylation of PEP.

Fructose 1,6-bisphosphatase is localized in the plastids and in the cytosol. Cytosolic FBPase is involved in hexogenesis, converting fructose-1,6-bisphosphate to fructose-6-phosphate, which is used by SPS as one of the substrates for the production of sucrose-6-phosphate (Figure 15.1 and Figure 15.2). FBPase is potently inhibited by fructose-2,6-bisphosphate (90,103), a metabolite which is also a potent activator of PPI-PFK (104). In a study to investigate the role of FBPase in LTS, it was observed that there was a rapid increase in the levels of sucrose and reducing sugars in tubers stored at 2°C, but no change in FBPase activity, relative to 8°C storage.

In a study carried out to identify the regulatory steps in glycolysis, a decline in phosphoenolpyruvate and a rise in pyruvate were observed when potato tubers were stored under anoxic conditions (86). As this step is preceded by phosphofructokinase, pyruvate kinase cannot regulate glycolytic flux directly as it cannot control the entry of glucose-6-phosphate into glycolysis. However, pyruvate kinase could play a role in the regulation of the movement of carbon out of glycolysis and into the oxidative pentose phosphate pathway. It has been suggested that the cold lability of phosphofructokinase and pyruvate kinase could lead to a rapid reduction in hexose phosphate consumption, which could cause their diversion to sucrose (86).

The theory that the cold lability of enzymes in the glycolytic pathway diverts hexose 6-phosphate for sucrose production and thus to LTS cannot fully explain LTS, as it takes time for potatoes to sweeten fully (7). From the results of a study carried out by Marangoni et al. (16), by comparing LTS resistant (ND860-2) and LTS susceptible (Norchip) potato cultivars, it has been suggested that tubers with decreased invertase activity along with increased glycolytic or respiratory capacity, should be more tolerant to low temperature stress.

15.2.7   Oxidative Pentose Phosphate Pathway (PPP)

Although the PPP is usually depicted as being separate from glycolysis, the two pathways are intimately linked (Figure 15.2). They share the common intermediates glyceraldehyde 3-phosphate, fructose 6-phosphate and glucose 6-phosphate, and flow through either of the pathways will be determined by the metabolic needs of the cell. The main function of PPP is to generate NADPH for various biosynthetic reactions (105). It was proposed by Wagner et al. (106) that for low sugar accumulating cultivars, the PPP may provide a means of preventing the accumulation of high levels of sugars when tubers are stored below 10°C (by bypassing phosphofructokinase). However, no differences were observed in the specific activities of glucose 6-phosphate dehydrogenase and 6-phosphogluconate dehydrogenase among LTS resistant and LTS susceptible potato cultivars stored at 4°C and 12°C, respectively (107). It was observed that the LTS resistant cultivars exhibited higher activities of G6PDH and 6PGDH, relative to the LTS susceptible cultivars.

15.2.8   Mitochondrial Respiration

During storage of potato tubers below 5°C, in addition to changes in sugar accumulation patterns, respiration changes have also been observed (108). It has been reported that the cold resistant potato clone ND860-2 has shown a higher respiration rate throughout storage compared to the cold susceptible Norchip (109). Respiration decreases as storage temperature decreases, but at storage below 5°C, respiration is stimulated. There is a brief respiratory burst attributed to the combined effect of cyanide resistant (alternative pathway) and cytochrome mediated pathways (108), followed by a subsequent decrease in respiration rate to a new steady state (19,99).

It has been suggested that during chilling stress, an alternative oxidase pathway may play a protective role in the mitochondrion by preventing both an over reduction of the respiratory chain and the consequent production of reactive oxygen species that cause cellular damage (110). It has been suggested that the alternative pathway operates only during periods of high cellular energy charge, or when there is an imbalance between the supply of carbohydrates and the requirement for carbohydrates for structural growth, energy production, storage, and osmoregulation (111,112). There is also evidence which suggests that physical characteristics of the cellular membrane (i.e., mitochondrial membranes) may activate the alternative pathway (113). In a study carried out by Amir et al. (99) to study the relationship between respiration rate, sugar content, and ATP levels in cold stored tubers, an immediate decrease in respiration rate was observed upon storage at 4° C. The respiratory minimum was concomitant with an ATP maximum which is followed by a respiratory burst and a rapid decline in ATP content. This evidence suggests the presence of an active alternative pathway in cold stressed tubers. Expression of the alternative pathway is known to increase with decreasing temperatures (114). It has been suggested that sucrose formation could serve as an effective sink for excess ATP via the alternative pathway (113). In agreement with Solomos and Laties (113), it was observed that low O_2 levels, which inhibit the alternative pathway, were effective in suppressing sugar accumulation in tubers stored at 1°C (115) which suggests that LTS may be directly linked to the onset of cyanide resistant respiration.

15.2.9   Compartmentation and Stress Induced Membrane Changes

Membranes play an integral role in the response of plant tissues to chilling and freezing. It has been proposed that the thermotropic phase transition of membrane lipids might play an initiative role in the chilling sensitivity of plants (116–118). With further exposure to chilling, the phase separated biomembranes become incapable of maintaining ionic gradients and cellular metabolism becomes disrupted. The occurrence of phase separation as the initial event in chilling injury has been demonstrated in cyanobacterium *Anacystis nidulans* (119). It has been argued that such a phase separation would not occur in plant cells because they contain high levels of polyunsaturated fatty acids in their membranes. However, a positive correlation has been observed between chilling sensitivity of herbaceous plants and the level of saturated and transmonounsaturated molecular species of phosphatidylglycerol in thylakoid membranes (120,121).

It is likely that the regulation of starch breakdown and of sucrose synthesis is to some extent achieved by compartmentation. Therefore, it is possible that low temperature sweetening is, at least in part, due to effects of cooling on such compartmentation (7). In potatoes, studies have been performed on the effects of cold storage on lipid composition and membrane permeability (122,123), the associated biophysical changes of amyloplast membranes (124,125) and mitochondrial membranes (126), and lipid peroxidation (15,127–129). The results of these studies are described below.

15.2.9.1   Lipid Composition

Phospholipids and glycerolipids are the major potato lipids (130). It has been found that the combined proportion of polyunsaturated fatty acids (linoleic and linolenic) for all potato varieties examined consistently represents 70–76% of the total fatty acids, which help maintain membrane fluidity at lower temperatures (130).

In plants stressed by low temperature or other factors, the survival of the plant is based on the ability of the plant to maintain or reestablish membrane fluidity (131). Fatty acid desaturases play a central role in regulating the level of unsaturation of fatty acids in membrane lipids, which helps maintain membrane fluidity or refluidizes the membranes that have become rigid due to low temperature exposure (132). Bonnerot and Mazliak (133) reported cold induced oleyl-PC desaturase activity in microsomes from 16 h aged slices of potato tubers stored at 4°C for 3 months. Spychalla and Desborough (134) reported that the total amount of linoleic and linolenic acids remained constant, but the ratio of linolenic to linoleic increased over storage time for both tubers stored at 3°C and 9°C. Low temperature storage of potato tubers has been shown to increase the levels of

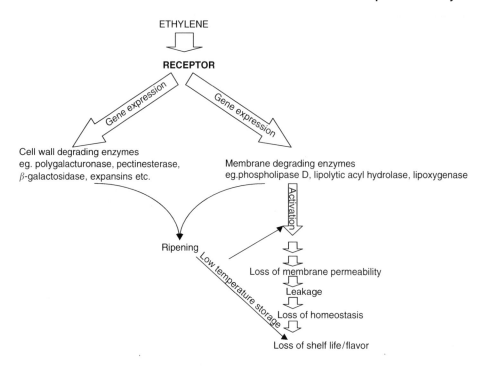

Figure 15.3 Schematic diagram illustrating the early events in the membrane and cell wall degradation during fruit ripening.

monogalactosyl diacylglycerol and digalactosyl diacylglycerol (135–137). Hence, it is possible that one or more of the low temperature induced changes in the lipid composition of the membrane may contribute to LTS or confer resistance to LTS.

15.2.9.2   Membrane Permeability

Several studies have reported that cold temperature damages membranes, resulting in the loss of compartmentation and homeostasis of the cell (138–140). As temperature decreases, membrane lipids undergo a phase transition from a liquid crystalline to a solid gel phase that results in cracks in the membrane, increases the membrane permeability and leakage of ions, and alters metabolism (141). The temperature range for the phase change in membranes is specific for each horticultural commodity and is a function of the heterogeneity of the lipid content, the ionic environment, and the presence of sterols and proteins (142).

From the earlier discussions, it is believed that LTS is caused by the effect of low temperature on many pathways of carbohydrate metabolism at the level of starch synthesis and breakdown, sucrose synthesis, hexogenesis, glycolysis, the PPP, and mitochondrial respiration. These pathways are also compartmentalized in the plant cell, and involve amyloplasts, the cytoplasm, vacuoles, and mitochondria. For the maintenance of homeostasis in normal cells, a tight control of the movement of the substrates or intermediates of these pathways is in place. However, when the plant, or a cell of the plant, experiences stress (such as low temperature stress as in the case of LTS), the normal metabolism of the cell will be lost, which might make the cell a candidate for readjustment at various levels. It is assumed that the changes caused by low temperature at the membrane level might be one of the factors contributing to LTS.

15.2.9.2.1   Tonoplast Membrane Permeability   The "leaky membrane theory" of LTS suggests that the cause of LTS may be a leaky tonoplast membrane that allows Pi to be leaked into the cytoplasm from the vacuole (17). A high concentration of Pi in cytoplasm is believed to mobilize carbon from the amyloplast to the cytoplasm, while cytoplasmic Pi participates with G-1-P in a reversible exchange across the amyloplast membrane. Increased Pi concentration in the amyloplast favors the α-glucan phosphorylase activated starch breakdown and inhibits ADPGPase mediated starch synthesis (17). In addition Pi affects fructose 2,6,-bisphosphate, phosphofructokinase, sucrose synthase, and UGPase (see section 15.2.3.2.1). The leaky membrane theory is been supported by subtle changes in fatty acid composition of potato membranes as well as by increases in electrical conductivity, which is an indicator of electrolyte leakage and membrane permeability. It has been reported that the relative change in electrical conductivity of four cultivars of potato paralleled the increase in sugar concentration when temperature was dropped from 20°C to 0°C (143). A difference in electrical conductivity was noticed among cultivars that had accumulated similar amounts of sugars. It was concluded that the increased electrical conductivity was not due to the increase in the accumulation of sugars, particularly because respiration rates were found to increase before the increase in sugar concentration and electrical conductivity.

Knowles and Knowles (122) studied the relationship between electrolyte leakage and degree of saturation of polar lipids in Russet Burbank seed tubers stored at 4°C, and observed an inverse linear relationship between the double bond index (DBI) and electrical conductivity ($r = -0.97$). The DBI revealed that the proportion of unsaturated fatty acids in membranes decreased over storage time with an accompanying increase in electrical conductivity. The authors concluded that the ability to increase membrane lipid unsaturation in storage could confer resistance to electrolyte leakage by maintaining the fluidity of the membranes. This result was supported by another study, in which it was observed that tubers stored at 3°C had greater increase in sugar content, total fatty acid saturation, and membrane permeability, as compared to tubers stored at 9°C (123). These studies suggest that high initial or high induced levels of lipid unsaturation could prevent increased membrane permeability during low temperature storage. It has been reported that cyanobacteria transformed with *des*A gene, which encodes a 12 acyl-lipid desaturase in *Synechocystis* PCC6803, did not show any significant changes in photosynthetic activity below 10°C, whereas in the wild type cells, the photosynthetic activity was decreased irreversibly (132).

15.2.9.2.2 Amyloplast Membrane Permeability It has been suggested that the low temperature induced defects in amyloplast membrane allow α-glucan phosphorylase from the cytoplasm to enter the amyloplast and degrade starch, resulting in the accumulation of sugars during LTS (144,145). Electron spin resonance study, used to examine the amyloplast membrane in potato tubers stored at 5.5 and 15.5°C, showed a strong relationship between membrane permeability and starch to sugar conversion (146). Studies using spin labeled probes revealed that at low temperatures, membranes exhibited decreased lipid fluidity. O'Donoghue et al. (125) observed that membrane lipid phase transitions were higher for Norchip (an LTS susceptible cultivar) than ND 860-2 (an LTS resistant cultivar) at both 4 and 12°C storage, and Norchip amyloplast membranes were more ordered at 4 than 12°C. The drop in double bond index (DBI) was 93% for Norchip while only 70% drop was observed for ND 860-2 due to loss of linoleic and linolenic acid. It was suggested that low temperature caused Norchip membranes to undergo deterioration to a greater extent than ND 860-2 membranes and this could have contributed to LTS. By contrast, it has been reported that the amyloplast membrane breaks down during senescence but remains relatively intact during LTS (147–149). Based on TEM examination of membranes from LTS resistant and LTS susceptible cultivars of potatoes stored at 5 and 10°C, Yada et al. (149) concluded that LTS is not the result of amyloplast membrane breakdown. However, it is likely that changes in membrane can take place at the molecular composition or organizational level, which can affect the permeability or transport properties, or both, and can contribute to LTS.

15.2.10 Free Radicals and Antioxidant Enzymes

Evidence from several lines of research suggests that a variety of toxic oxygen species such as superoxide radicals, hydrogen peroxide, and hydroxyl radicals are produced in plants exposed to various environmental stresses such as high and low temperatures, drought, light, and exposure to pollutants causing oxidative damage at the cellular level (150). Lipid peroxidation is considered to be one of the reasons for membrane deterioration during senescence and chilling injury (140), and results from the activity of lipoxygenase, resulting in the formation of lipid peroxides and free radicals (151). In a study to analyze the relationship between sugar accumulation and changes in membrane lipid composition associated with membrane permeability in early stages of LTS, sucrose cycling and accumulation were greatest for Norchip, a LTS susceptible cultivar at 4°C as compared to LTS tolerant ND 860-2 (15). No significant changes were observed in phospholipid, galactolipid, free sterol levels, or phospholipid to free sterol ratio. However, the double bond index obtained from the fatty acid profiles of the total lipid fraction decreased significantly (decreased unsaturation) for Norchip tubers at 4°C over time. Free fatty acid and diene conjugation values fluctuated and increased over time for both Norchip and ND 860-2 stored at 4°C and 12°C, with greater amplitude of fluctuations observed for Norchip stored at 4°C. From the results, it has been suggested that these effects may be due to the high levels of lipid acyl hydrolase and lipoxygenase found in potato tubers, and the observed peroxidation products could relate low temperature stress and the resultant LTS to chilling injury and drought stress (15). However, Fauconnier et al. (152) could not observe a correlation between cold sweetening and membrane permeability or lipid saturation status. They studied the effect of three storage conditions: at 4°C, at 20°C with sprout inhibitors, and at 20°C without sprout inhibitors, and observed that during storage at 20°C without sprout inhibitor, the increase in membrane permeability is inversely correlated to sucrose accumulation. It was also observed that lipoxygenase activity and gene expression are not correlated with the fatty acid composition of the membrane. It was also observed that the lipoxygenase activity and fatty acid hydroperoxide content are low in older tubers, irrespective of the storage conditions and the varieties. Spychalla and Desborough (134) studied the antioxidant potential of potato tubers stored at 3 and 9°C and observed that tubers stored at 3° C had higher superoxide dismutase activities than their 9°C counterparts and demonstrated time dependent increases in superoxide dismutase, catalase, and α-tocopherol during the 40 week storage period. They also observed that low sugar clones had significantly higher levels of superoxide dismutase and catalase than high sugar clones but significantly lower levels of α-tocopherol. The increased antioxidant responses could be due to increased free radical production as manifested by the higher levels of superoxide dismutase, catalase, and peroxidase activities in seed tubers stored at 4°C for 20 months as against those stored for 8 months (129).

From the above discussion, it can be concluded that LTS is not the result of a single cause. The sugar balance in potato is regulated by many intermediate carbohydrate metabolic pathways, which are subject to genetic and environmental control. It might be possible that the low temperature effects on enzymes involved in carbohydrate metabolism result in an imbalance in the normal metabolism combined with its effect on membrane fluidity, thus diverting or leaking the intermediates, or both, in the biosynthetic or respiratory pathway for sucrose and reducing sugar production. Hence, even though genetic engineering has great potential in manipulating or improving crop productivity and the quality of horticultural crops, because of the complexity of LTS, in depth research on various molecular and biochemical properties and their correlation to LTS need to be conducted before we can fully exploit that potential. The existence of cultivars resistant to LTS with several molecular, biochemical, or compositional characteristics, or combinations of these, might provide a better tool in understanding the mechanism that is responsible for resisting LTS in those cultivars. We can therefore be hopeful that the coordinated efforts of plant biochemists, molecular biologists and traditional plant breeders would help to better understand and control LTS, thus eliminating the use of chemical sprout inhibitors and their harmful safety issues.

15.3   TOMATO

Tomatoes rank second to potatoes in dollar value among all vegetables produced in the USA and in other parts of the world where they are grown (1). In terms of per capita consumption, processed tomato products lead all the other processed vegetables. The main factors in determining the postharvest deterioration of fruit and vegetable crops are the rate of softening of the fruit which influences quality, shelf life, wastage, infection by postharvest pathogens, and frequency of harvest, and which limit the duration of transportation and storage. Damage to the structure and function of the membrane affects the post harvest shelf life and quality of fruits, vegetables, and other food sources by causing leakage of ions from cellular storage compartments into the cytosol, thereby disrupting the homeostasis of the cell (153,154). A major problem faced by the fruit and vegetable fresh market and processing industry in the Northern Latitudes is the lack of a year round supply of high quality material. Even though cold storage can be used for long term storage of fruits and vegetables, in the case of tomatoes it is not effective due to the sensitivity of tomatoes to being chilled. Many factors affect the shelf life of tomato products, but our discussion will be based mainly on membrane changes during cold storage and the genetic manipulation to circumvent these factors.

15.3.1   Role of Membrane in Shelf Life

The development of fruit in many plants can be interpreted as following a two step process. During the first phase, the ovary or hypothalamus within the flower expands and develops into a full sized fruit. During the second phase, the full sized fruit undergoes ripening, a complex set of molecular and physiological changes in the fruit. The ripening process brings dramatic changes to the fruit: softening, biosynthesis of pigments, and increase in sugar content, flavor, and aroma. In climacteric fruits such as tomatoes, and many other fruits, ripening begins with increased respiration and ethylene biosynthesis (155). Fruit ripening can be considered as the beginning of senescence of the fruit (156).

Senescence can occur at various levels, from cellular to whole plant levels, and is regulated by genetic, hormonal, and environmental factors (140,157). The plant hormone ethylene plays a major role in the ripening and senescence processes, and extensive work has been conducted in the past two decades on the role of ethylene in fruit ripening and signal transduction. Genetic manipulation to increase the shelf life of fruits, especially tomatoes, has been extensively undertaken, resulting in several new transgenic varieties with improved storage and quality characteristics (158–164).

Senescence is characterized by membrane deterioration resulting from the catabolism of membrane lipids and proteins. The pathway of the catabolism of phospholipids has been elucidated from several senescing systems and involves the sequential action of enzymes that include phospholipase D (PLD, phosphatidyl choline hydrolase, EC 3.1.4.4, PLD), phosphatidate phosphatase (3-sn-phosphatidate phosphohydrolase, EC 3.1.3.4), lipolytic acyl hydrolase and lipoxygenase (linoleate, oxygen oxidoreductase, EC 1.13.11.12) (140). In most systems studied, the first step in the lipid catabolic pathway is the conversion of phospholipid to phosphatidic acid by PLD, even though phosphatidylinositol and its phosphorylated forms may be acted upon by both phospholipase C and PLD (165,166). Phosphatidic acid does not accumulate, as it is immediately converted to diacylglycerol by phosphatidate phosphatase. Diacylglycerol is deacylated by lipolytic acyl hydrolase, liberating free fatty acids. Among the free fatty acids, unsaturated fatty acids with 1–4 penta-diene systems (18:2 and 18:3) serve as substrates for lipoxygenase, resulting in the formation of fatty acid hydroperoxides. Fatty acid hydroperoxides undergo a variety of reactions by virtue of their active unstable structure, including the generation of free radicals. The free radicals damage the protein as well as the membrane, giving rise to the characteristics features of senescence. The above reactions are deemed autocatalytic, as the reaction products increasingly contribute to the formation of gel phase and nonbilayer lipid structures resulting in the destabilization of the membrane and eventually in the loss of homeostasis. Similar changes occur in the membrane in response to chilling injury, but also involve the effect of low temperature on the catalytic activity of enzymes involved as well as the effect on the phase transition temperature of the lipids (see section 15.2.9). Here we will be emphasizing the role of PLD and lipoxygenase in enhancing the shelf life of tomatoes during cold storage.

15.3.2 Phospholipase D Gene Family

Phospholipids provide the backbone for biomembranes and serve as rich sources of signaling messengers. Phospholipase D (PLD, EC 3.1.4.4) catalyzes the hydrolysis of structural phospholipids to generate phosphatidic acid and a free head group. PLD has been grouped into three classes based on their requirements for Ca^{2+} and lipids in *in vitro* assays: the conventional PLD that is most active at 20 to 100 mM levels of Ca^{2+}; the polyphosphoinositide (PI) dependent PLD that is most active at micromolar levels of Ca^{2+}; and the phosphatidylinositol (Ptdln) specific PLD that is Ca^{2+} independent (167). PLD has been cloned from a number of plants (168), animals (169), and fungi (170) and found to constitute a supergene family of many isoforms (171). The PLD isoforms from *Arabidopsis* have been divided into five groups: PLDα, β, γ, δ, and ε. (168). The PLDα gene product is responsible for the conventional PLD activity and differs physiologically from the PLDβ, and PLDγ isoforms based on Ca^{2+} requirements and pH (172). From tomatoes, three PLDα forms and two PLDβ classes have been cloned (173). Different expression patterns were observed in different plant tissues and organs for each PLD. In fruit, PLDα3 appeared to be transiently accumulated during early ripening, whereas PLDα2 was accumulated throughout fruit development and maturation (173).

15.3.2.1   Role of PLD in Senescence and Chilling

PLD activities have been observed in many cellular functions during seed germination, aging, various abiotic and biotic stresses, and senescence (140,167,174). In membranes of tomato fruit stored at low temperature, an accumulation of phospholipid catabolites occurs due to differential effects of reduced temperature on the activities of lipid degrading enzymes (175). Destabilization of membranes has been suggested as one of the causes of chilling injury (176). In ripening tomato fruits, rigidification of microsomal membranes has been reported to activate PLD and increase membrane catabolism (177). Increased PLD activities during and after chilling are suggested to result in chilling injury in maize (139) and cucumber fruits (178). In *Arabidopsis*, cold stress increased the expression of PLDα, but not PLDβ or PLDχ, implying that PLDα has a role in plant responses to low temperature stress (174). Recently the many roles of PLD in signal transduction have been reported (167). In tomatoes, PLDβ is suggested to have a role in signal transduction due to its low abundance, activation by micromolar concentration of Ca^{2+}, a concentration range that arises locally during signaling (173), whereas a metabolic role has been assigned for the various forms of PLDα. Our studies on the role of PLD during fruit ripening in cherry tomatoes showed that the soluble and membrane associated PLD activities increased during fruit development, which peaked at the mature green and orange stages (178a). It was reported by Jandus et al. (179) that PLD activity decreased slightly between the mature green and orange stages when the tomatoes ripened on the plants, but increased between the orange and red stages to values higher than at the mature green stage. However, when tomatoes were harvested at the mature green stage and left to ripen at room temperature, PLD activity decreased by about 40% between the mature green and red stages. It has been reported that phosphatidic acid increased as much as twofold while total phospholipids decreased about 20–25% during ripening of tomato pericarp (180,181). Treatment with lysophosphatidylethanolamine, which acts as a specific inhibitor of PLD activity (182) retarded senescence in tomato fruits and leaflets (183). From the observations on PLD activity at low temperature, ripening and senescence, it is conceivable that the quality of tomatoes, which is highly chilling sensitive, may be affected to a large degree by the modulation of PLD activity.

15.3.2.2   Regulation of PLD Activity

The activities of PLD are affected by a number of factors such as Ca^{2+} concentration, substrate lipid composition, pH changes, and mastoparan, a tetradecapeptide G-protein activator (167). Sequence analysis indicates that plant PLDs contain a Ca^{2+} to phospholipid binding fold, called the C2 domain at the N terminus. The C2 domains of PLDα and PLDβ have been demonstrated to bind Ca^{2+}, with PLDβ having a higher affinity for Ca^{2+}, whereas the Ca^{2+} requirement of PLDα is influenced by pH and substrate lipid composition (167). PLDα is active at near physiological, micromolar Ca^{2+} concentrations at an acidic pH of 4.5–5 in the presence of mixed lipid vesicles. PLDβ and PLDχ are optimally active under physiological micromolar concentration of Ca^{2+} concentrations at neutral pH and may play an active role in signal transduction. The relative distribution of PLD between the soluble and membrane fractions changes during development and in response to stress (184,185). It has been shown that Ca^{2+} binding increases the affinity of the C2 domains for membrane phospholipids (186). This shows that the C2 domain in PLD is responsible for mediating a Ca^{2+} dependent intracellular translocation between the cytosol and membranes. An increase in cytosolic Ca^{2+}, as well as a decrease in cytosolic pH, has been reported to occur in response to stress (187), which are favorable conditions for the activation of PLDα. PLDs associated with microsomal membranes are correlated with stress induced activation of PLD mediated hydrolysis (184,185). The increased association of preexisting PLD in the cell with membranes may represent a rapid and early step in PLD activation during stress responses (185).

15.3.2.3   Antisense Suppression of PLD Activity

In order to study the role of PLD in fruit ripening and senescence of fruits, we have developed transgenic tomatoes expressing antisense PLDα cDNA. The fruits from antisense Celebrity tomato (a fresh eating type) were smaller than the control fruits and showed a 30% decrease in PLD activity during development (178a). After storage for two weeks at room temperature, the control fruits developed wrinkles, indicative of senescence and dehydration, whereas the transgenic fruits appeared to be relatively normal. The transgenic fruits were also firmer, possessed a higher level of red pigmentation and increased level of soluble solids (178a). Transgenic Celebrity fruits showed a decrease in PLD expression as evidenced from Northern blot. Even though very few transcripts were detected at the mature green, orange and red stages in the antisense PLD celebrity fruits, PLD activity was present at these stages suggesting a very low turnover rate of PLD, and that PLD synthesized at young or intermediate stages remains functional even at the red stage (178a). These results suggest that for effective inhibition of PLD using antisense suppression, PLD expression has to be reduced at an early stage of fruit development using an appropriate fruit specific promoter. In our experiments we have used a constitutive promoter (CaMV 35S) for tomato transformation. The antisense Celebrity fruits also showed low levels of PLD activity during ripening, suggesting that the natural senescence process was retarded, which was translated into increased firmness in these fruits. This observation was contrary to earlier results in *Arabidopsis* where antisense suppression of PLDα resulted in retardation of ABA and ethylene promoted leaf senescence, without affecting the natural senescence of leaves (188). It has been reported that the phospholipid content of tomato fruit declines during ripening (180). This decrease in phospholipid content could be due to a high PLD activity. It was interesting to note that antisense suppression of PLDα in an ornamental cherry tomato cultivar, Microtom, did not show any significant reduction of PLD activity (178a). However, ethylene climacteric of the transgenic fruits was delayed by nearly six days, as compared to the control fruits. In Microtom, PLD activity declined during ripening in the control fruits, whereas transgenic fruits retained much higher levels of PLD activity. This may be related to the delayed climacteric in the transgenic fruits, indicating a slower rate of deterioration. *In situ* localization of PLD by immunolabeling followed by electron microscopy also supports this observation. These results suggest that fruits from different cultivars may differ in their pattern of senescence and the relative role of PLD may differ between fruits and leaves. It is unclear why PLD activity in transgenic Microtom was higher compared to control plants during ripening, as opposed to the observation in Celebrity. It has been reported that in PLDα suppressed *Arabidopsis*, the expression and activities of other PLD isoforms are not altered (189) which means that the other PLD members cannot compensate for the loss of PLDα.

15.3.3 Lipoxygenase

Lipoxygenases (LOX, EC 1.13.11.12) are a class of enzymes that catalyze the hydroperoxidation of *cis-cis*-1,4-pentadiene moieties in polyunsaturated fatty acids and that occur widely in both plant and animal kingdom. LOX pathways in higher plants and mammals are different in two main respects: in mammals the main LOX substrates are arachidonic and eicosapentaenoic acids whereas linoleate and α-linolenate are the most important LOX substrates in plants; and the hydroperoxide metabolizing enzymes are different in plants and mammals (190). Although LOX isoforms occur in most plant cells, the tissue specific expression level of LOX within a plant can vary substantially depending on developmental and environmental conditions (191). Plant lipoxygenases have been implicated as having a role in the loss of membrane integrity associated with senescence, flavor and odor formation, response to pest attack, and wounding (192). Products of the LOX pathway such as traumatin, jasmonic acid, oxylipins, and volatile aldehydes, are supposed to play a key role in signal transduction in response to wounding, as antimicrobial substances in host–pathogen interactions, as regulators of growth and development, and as aromatic compounds that affect food quality (191).

15.3.3.1 *Role of Lipoxygenase in Membrane Deterioration*

Lipid peroxidation is an inherent feature of senescence and generates a variety of activated oxygen species such as singlet oxygen and the alkoxy and peroxy radicals. Alkoxy and peroxy radicals are formed directly as decomposition products of lipid peroxides and singlet oxygen is formed through the interaction of lipid peroxy radicals (157). Lipid peroxidation is initiated either by the action of reactive oxygen species or enzymatically by the action of lipoxygenases (LOX). LOX appears to play an important role in the deterioration of membranes during senescence by initiating lipid peroxidation, and also by forming activated oxygen independently (193). Increasing data suggests that LOX is activated by various stresses such as wounding (194), water deficit (195), thermal stresses (196), and ozone stress (196). It is known that the primary site of action of various stresses is the biomembrane (151), which results in the liberation of free linoleic and linolenic acids (197). Some results also suggest the direct oxygenation of membrane lipids and biomembranes by LOX (195,198–200). The role of LOX in membrane deterioration is evident from its association with membrane in tomato fruits (192,201) and carnation petals (202) as a membrane bound LOX can attack the membrane lipids more readily than a soluble one. This is supported from the observation that LOX can oxygenate esterified polyenoic acids in complex lipids and biomembranes, in addition to free polyenoic fatty acids such as linoleic and linolenic acids (190). Yamauchi et al. (198) reported the oxygenation of dilinolenoyl monogalactosyldiacyl-glycerol in dipalmitoylphosphatidylcholine liposomes by a crude soybean LOX preparation containing all the three isozymes, thus demonstrating that plant LOX catalyzes the oxygenation of both free polyunsaturated fatty acids and monogalactolipids. Brash et al. (199) observed that soybean LOX-1 oxidizes fatty acid residues within phoshatidylcholine and other phospholipids such as phosphatidylinositol lipids and phosphatidylethanolamine. Similar results were reported by Kondo et al. (200) from their studies on LOX action in soybean seedlings and Maccarone et al. (195) on soybean LOX during water deficit. It has been proposed that the role of soybean cotyledon LOX during the early stages of seedling growth is the disruption of storage cell membranes, enhancing their permeability (203). These evidences corroborate the role of LOX in membrane deterioration. However, there is controversy regarding the role of LOX in senescence, as its activity increases during senescence in *Pisum sativum* foliage (204), whereas in soybean cotyledons, total LOX activity has been shown to decrease with advancing senescence (205). The occurrence of LOX activity in young and expanding tissues as well as the observation that soybean LOX is not induced in senescing tissues, argues against the role of LOX in senescence (191). However, soybean seedlings subjected to water stress showed an increase in specific activity of their major lipoxygenases, LOX-1 and LOX-2, which was paralleled by the increase of LOX content and mRNA, indicating that osmotic stress modulates the expression of LOX genes at the transcriptional level (195). Osmotic stress also increased the oxidative index of biomembranes by increasing the hydroperoxide content of the lipid ester fraction. Water deficit has been reported to impair cell membrane functioning (206). Based on the observed enhancement of both LOX activity and membrane oxidative state in response to water deficit, the authors suggested that their results corroborate the hypothesis of a role of LOX in plant membrane deterioration. Studies of the effects of thermal injury (heat shock and cold) and ozone treatment on LOX activity of soybean seedlings have shown that cold stress decreased the specific activities of LOX1 and LOX2, which is attributed to at least in part to a down regulation of gene expression at the translational level (196). Both heat shock and ozone treatment enhanced the LOX-specific activities, acting at the level of transcription of the genes. It is proposed that LOX-1 and LOX-2 are involved in the thermotolerance of soybeans and in the precocious aging induced by ozone. It has been suggested by the authors that cold, heat shock, and ozone can ultimately act on cell membranes. This corroborates the hypothesis for a major role of LOX in the control of membrane integrity. It has been observed that during ripening in tomato fruit pericarp, two distinct LOXs were identified based on their pH optima and their sensitivity to the LOX inhibitor nordihydroguaiaretic acid (207). Both these activities increased sharply during early ripening stages, and decreased after the fruit had ripened fully. It has been shown that the LOX require free fatty acids as their substrate and the timing and extent of peroxidative reactions initiated by LOX are determined by the availability of these substrates which are made available through the action of lipolytic acyl hydrolase (157,201). All these results suggest that a method of enhancing shelf life and quality of fruits could be by the regulation of LOX activity.

15.3.3.2 Tomato Fruit Ripening and Lipoxygenases

Studies on LOX genes during tomato fruit ripening using a low ethylene producing fruit containing an ACC oxidase (*ACO1*) sense suppressing transgene, and tomato fruit ripening mutants such as *Never-ripe* (*Nr*), and *ripening-inhibitor* (*rin*), have demonstrated that expression of three LOX genes *TomloxA*, *TomloxB* and *TomloxC* is regulated differentially during fruit ripening, and that ethylene and a separate developmental component are involved (208). The expression of *TomloxA* declines during ripening and this is delayed in the *ACO1* transgenic low ethylene and *Nr* fruit, indicating that this phenomenon is ethylene regulated. Transcript abundance also declines during *rin* fruit development indicating that developmental factors also influence the expression of *TomloxA*. *TomloxB* expression increases during fruit ripening, which is also stimulated by ethylene. TomloxC gene expression is up regulated in the presence of ethylene and during ripening. The principal substrates of LOX in tomato fruit are the linoleic and linolenic acids, and the action of the 13-lyase on the 13-fatty acid hydroperoxide products of these substrates results in the production of hexanal and hexenal respectively. Release of these aldehydes following disruption of the tissue results in the production of the typical aroma characteristic of fresh tomatoes (209). As the ripening process continues, the thylakoid membranes break down as the chloroplasts are transformed into chromoplasts. It has been suggested that LOX may be the trigger for the chloroplast to chromoplast transition (210), and the polyunsaturated fatty acids in the thylakoid may be acted upon by LOX, and subsequently by a lyase, to release hexanal and hexenal, which in turn influence the flavor and aroma characteristics of the fruit (209). It has been suggested that the various LOX genes in tomato fruit may aid either in the defense mechanisms early in unripe fruit, or flavor and aroma generation and seed dispersal mechanism in the later developmental stages (208).

15.3.3.3 Regulation of LOX by Genetic Manipulation

With the genetic tools available today, it is possible to identify the role or function of a particular gene or gene product either by over expressing or down regulating the gene. Overexpression of LOX2 gene from soybean embryos fused with the enhancer of alfalfa mosaic virus under the control of a duplicated CaMV 35S promoter in transgenic tobacco increased the fatty acid oxidative metabolism as evidenced by a 50–529% increase in C_6 aldehyde production (211). The impact on C_6 aldehyde formation was greater than the effect on production of fatty acid hydroperoxides, which is consistent with other studies indicating the greater involvement of soybean embryo LOX2 in generating C_6 aldehydes than that of other well characterized LOX isozymes. To evaluate the role of LOX in the onset of plant defense, transgenic tobacco plants expressing the antisense tobacco LOX cDNA were developed which showed strongly reduced elicitor and pathogen induced LOX activity (212). A linear relationship was observed between the extent of LOX suppression and the size of the lesion caused by the fungus, *Phytophthora parasitica*. The antisense plants also showed enhanced susceptibility toward the compatible fungus *Rhizoctonia solani*. The authors suggested that their results demonstrate the strong involvement of LOX in the establishment of incompatibility in plant–microorganism interactions, consistent with its role in the defense of host plants. Antisense tomato plants were developed with

TomloxA under the control of fruit specific promoter 2A11 and ripening specific promoter of polygalacturonase (213). Reduced levels of endogenous *TomloxA* and *TomloxB* mRNA (2–20% of wild type) were detected in transgenic fruit containing 2A11 promoter compared to nontransformed plants, whereas the level of mRNA for *TomloxC* was unaffected. LOX enzyme activity was also reduced in these transgenic plants. However, no significant changes were observed in flavor volatiles. The transgenic plants with PG promoter were less effective in reducing endogenous LOX mRNA levels. The authors concluded that either very low levels of LOX are sufficient for the generation of C_6 aldehydes and alcohols, or a specific isoform *TomloxC* in the absence of *TomloxA* and *TomloxB* is responsible for the production of these compounds. Transgenic tomato and tobacco plants were developed by transformation using the chimeric gene fusions of *TomloxA* and *TomloxB* promoter with β-glucuronidase (GUS) reporter gene. GUS activity in *tomloxA-gus* plants during seed germination peaked at day 5 and was enhanced by methyl jasmonate whereas no GUS activity was detected in *tomloxB-gus* seedlings (214). During fruit development, GUS expression in *tomloxA-gus* tobacco fruit increased 5 days after anthesis and peaked at 20 days after anthesis. In *tomloxB-gus* tobacco GUS activity increased at 10 days after anthesis and peaked at 20 days after anthesis. In *tomloxA-gus* tomato fruit, GUS activity was observed throughout fruit ripening, with highest expression at the orange stage, and the expression was localized to the outer pericarp during fruit ripening. In *tomloxB-gus* fruit, GUS activity was detected at the mature green stage, while expression was localized in the outer pericarp and columella. It has been shown that antisense transgenic potato plants with reduced levels of one specific 13-LOX isoform (LOX-H3) largely abolished the accumulation of proteinase inhibitors upon wounding, indicating that this LOX-H3 plays an important role in the regulation of wound induced gene expression (215).

The genetic manipulation studies on LOX show a specific role of LOX in various developmental processes and defense mechanisms. However, considering the presence of various isoforms in tomatoes and the roles they play, it is very important to characterize the specific role of each isoform before regulating its expression for a specific intent. For example, from the results of Beaudoni and Rothstein (214), it appears that antisense suppression of *TomloxA* may be helpful in enhancing the shelf life of the tomato fruit during storage without affecting the flavor of the fruit.

15.3.4   Cell Wall Metabolism and Fruit Softening

Tomato fruit ripening is a highly regulated developmental process requiring expression of a large number of gene products (216). Enzymes involved in the degradation of cell walls, complex carbohydrates, chlorophyll, and other macromolecules must be coordinately expressed with enzymes that make the fruit desirable nutritionally and aesthetically. Polyglalacturonases [PGs, poly ($1\rightarrow4$-α-D-galacturonide) glycanohydrolases] are enzymes that catalyze the hydrolytic cleavage of galacturonide linkages in the cell wall, and are the most widely studied among cell wall hydrolases. PG has been implicated as an important enzyme in fruit softening based on its appearance during ripening, corresponding to the increase in fruit softening. In a number of cultivars, there is a correlation between levels of PG activity and the extent of fruit softening; it degrades isolated fruit cell walls *in vitro* in a manner similar to that observed during ripening and several ripening mutants that have been described with delayed or decreased softening are deficient in PG activity (217). Results on the genetic manipulation of PG has suggested that PG activity alone is not sufficient to affect fruit softening (218), and other enzymes such as pectin-methylesterase, β-galactosidase, and expansins are involved in fruit softening. This discussion is not intended to cover this area, as it has been reviewed recently by Brummell and Harpster (218) and through another chapter in this book.

15.4 CONCLUSION

Fruit color, texture, nutritional value, and flavors are the most important parameters that affect the quality of fruits and vegetables. The post harvest storage conditions of fruits and vegetables also affect these parameters as well as the shelf life, and the type of fruit and vegetables determines the extent of the effect. In the case of potatoes and tomatoes, it is apparent from the above discussion that chilling induced changes affect the above parameters even though preharvest factors and cultivar identity also contribute to this effect. While damages to membranes are attributed to be responsible for these changes in both crops, in the case of potatoes, carbohydrate metabolism also plays a major role. Our understanding of the genes involved in the membrane deteriorative processes as well as carbohydrate metabolism have resulted in the alleviation of these effects to a certain extent. Considering the roles of PLD and LOX in the membrane deteriorative pathway, it might be possible to enhance the shelf life of tomato by generating a double transgenic plant with suppressed activities of PLD and LOX.

REFERENCES

1. Gould, W.A. *Tomato Production, Processing and Technology,* 3rd ed. Baltimore (ISBN 0-930027-18-3): CTI Publications Inc., 1992, pp 3–5, 253–254.
2. Duplessis, P.M., A.G. Marangoni, R.Y. Yada. A mechanism for low temperature induced sugar accumulation in stored potato tubers: the potential role of the alternative pathway and invertase. *Amer. Potato J.* 73:483–494, 1996.
3. Roe, M.A., R.M. Faulks, J.L. Belsten. Role of reducing sugars and amino acids in fry colour of chips from potatoes grown under different nitrogen regimes. *J. Sci. Food Agr.* 52:207–214. 1990.
4. Davies, H.V., R. Viola. Regulation of sugar accumulation in stored potato tubers. *Postharvest News Inf.* 3(5):97N–100N, 1992.
5. Blenkinsop, R.W., L.J. Copp, R.Y. Yada, A.G. Marangoni. Changes in compositional parameters of tubers of potato (*Solanum tuberosum*) during low-temperature storage and their relationship to chip processing quality. *J. Agric. Food Chem.* 50:4545–4553, 2002.
6. Wismer, W.V. Sugar accumulation and membrane related changes in two cultivars of potato tubers stored at low temperature. PhD dissertation, University of Guelph, Guelph, Ontario, 1995.
7. ap Rees, T., W.L. Dixon, C.J. Pollock, F. Franks. Low temperature sweetening of higher plants. In: *Recent Advances in the Biochemistry of Fruits and Vegetables,* Friend, J., M.J.C. Rhodes, eds., Academic Press: New York, 1981, pp 41–61.
8. Burton, W.G. The sugar balance in some British potato varieties during storage, II: the effects of tuber age, previous storage temperature, and intermittent refrigeration upon low-temperature sweetening. *Eur. Potato J.* 12:81–95, 1969.
9. Prichard, M.K., L.R. Adam. Relationships between fry color and sugar concentration in stored Russet Burbank and Shepody potatoes. *Amer. Potato J.* 71:59–68, 1994.
10. Sowokinos, J.R. Postharvest regulation of sucrose accumulation in transgenic potatoes: role and properties of potato tuber UDP-glucose pyrophosphorylase. In: *The Molecular and Cellular Biology of the Potato,* Belkap, W., W.D. Park, M.E. Vadya, eds., Biotechnology in Agriculture, Series 12, Wallingford, UK: CABI Publishing, 1994, pp 81–106.
11. Wismer, W.V., A.G. Marangoni, R.Y. Yada. Low temperature sweetening in roots and tubers. *Hort. Rev.* 27:203–231, 1995.
12. Sowokinos, J.R. Stress-induced alterations in carbohydrate metabolism. In: *The Molecular and Cellular Biology of the Potato,* Vayda, M.E., W.D. Park, eds., Wallingford, UK: CAB Int, 1990, pp 137–158.

13. Sowokinos, J.R. Allele and isozyme patterns of UDP-glucose pyrophosphorylase as a marker for cold-sweetening resistance in potatoes. *Amer. J. Potato Res.* 78:57–64, 2001.
14. Spychalla, J.P., S.L. Desborough. Fatty acids, membrane permeability, and sugars of stored potato tubers. *Plant Physiol.* 94:1207–1213, 1990.
15. Wismer, W.V., W.M. Worthing, R.Y. Yada, A.G. Marangoni. Membrane lipid dynamics and lipid peroxidation in the early stages of low-temperature sweetening in tubers of *Solanum tuberosum. Physiologia Plant.* 102:396–410, 1998.
16. A.G. Marangoni, P.M. Duplessis, R.Y. Yada. Kinetic model for carbon partitioning in *Solanum tuberosum* tubers stored at 2°C and the mechanism for low temperature stress-induced accumulation of reducing sugars. *Biophys. Chem.* 65:211–220, 1997.
17. Sowokinos, J.R. Effects of stress and senescence on carbon partitioning in stored potatoes. *Amer. Potato J.* 67:849–857, 1990.
18. Sowokinos, J.R. Biochemical and molecular control of cold-induced sweetening in potatoes. *Amer. J Potato Res.* 78:221–236, 2001.
19. Isherwood, F. Starch-sugar interconversion in *Solanum tuberosum. Phytochemistry* 12:2579–2591, 1973.
20. Salunkhe, D.K., B.B. Desai, J.K. Chavan. Potatoes. In: *Quality and Peservation of Vegetables,* Eskin, N.A.M., ed., Boca Raton, FL: CRC Press, 1989, pp 2–52.
21. Coffin, R.H., R.Y. Yada, K.L. Parkin, B. Grodzinski, D.W. Stanley. Effect of low temperature storage on sugar concentrations and chip colour of certain processing potato cultivars and selections. *J. Food Sci.* 52:639–645, 1987.
22. Pollock, C.J., T. ap Rees. Effect of cold on glucose metabolism by callus and tubers of *Solanum tuberosum. Phytochemistry* 14:1903-1906, 1975.
23. Hruschka, H.W., W.L. Smith Jr., J.E. Baker. Reducing chilling injury of potatoes by intermittent warming. *Amer. Potato J.* 46:38-53, 1969.
24. Van Es, A., K.J. Hartmans. Starch and sugars during tuberization, storage and sprouting. In: *The Storage of Potatoes,* Rastovski, A., A. van Es, eds., Wageningen: The Netherlands Centre for Agricultural Publishing and Documentation, 1981, pp 79–113.
25. Smith, A.M., K. Denyer, C. Martin. The synthesis of the starch granule. *Annu. Rev. Plant Physiol. Plant Mol. Biol.* 48:67–87, 1997.
26. Tauberger, E., A.R. Fernie, M. Emmermann, A. Renz, J. Kossmann, L. Willmitzer, R.N. Trethewey. Antisense inhibition of plastidial phosphoglucomutase provides compelling evidence that potato tuber amyloplasts import carbon from the cytosol in the form of glucose-6-phosphate. *Plant J.* 23:43–53, 2000.
27. Sowokinos, J.R., J. Preiss. Pyrophosphorylase in *Solanum tuberosum* L, III: purification, physical and catalytic properties of ADP-Glucose pyrophosphorylase in potatoes. *Plant Physiol.* 69:1459–1466, 1982.
28. Müller-Röber, B.T., U. Sonnewald, L. Willmitzer. Inhibition of ADP-glucose pyrophosphorylase leads to sugar storing tubers and influences tuber formation and expression of tuber storage protein genes. *EMBO J.* 11:1229–1238, 1992.
29. Sweetlove, L.J., B.T. Müller-Röber, L. Willmitzer, S.A. Hill. The contribution of adenosine 5'-diphosphoglucose pyrophosphorylase to the control of starch synthesis in potato tubers. *Planta* 209:330–337, 1999.
30. Stark, D.M., G.F. Barry, G.M. Kishore. Engineering plants for commercial products and applications. *Ann. New York Acad. Sci.* 792:26–37, 1996.
31. Lorberth, R., G. Ritte, L. Willmitzer, J. Kossmann. Inhibition of a starch-granule-bound protein leads to modified starch and repression of cold sweetening. *Nature Biotechnol.* 16:473–477, 1998.
32. Takeda, Y., S. Hizukuri, Y. Ozono, M. Suetake. Actions of porcine pancreatic and *Bacillus subtillis* α-amylase and *Aspergillus niger* glucoamylase on phosphorylase (1-4)-α-D-glucan. *Biochim. Biophys. Acta* 749:302–311, 1983.
33. Marsh, R.A., S.G. Waight. Starch degradation. In: *The Biochemistry of Plants: A Comprehensive Treatise,* Vol. 14, Preiss, J., ed., San Diego: Academic Press, 1982, pp 103–128.

34. Chang, C.W. Enzyme degradation of starch in cotton leaves. *Phytochemistry* 21:1263–1269, 1982.

35. Vos-Scheperkeuter, G.H., W. de Boer, R.G.F. Visser, W.J. Feenstra, B. Witholt. Identification of granule-bound starch synthase in potato tubers. *Plant Physiol.* 82:411-416, 1986.

36. Barichello, V., R.Y. Yada, R.H. Coffin, D.W. Stanley. Low temperature sweetening in susceptible and resistant potatoes: starch structure and composition. *J. Food Sci.* 55:1054–1059, 1990.

37. Barichello, V., R.Y. Yada, R.H. Coffin. Starch properties of various potato (*Solanum tuberosum* L) cultivars susceptible and resistant to low temperature sweetening. *J. Sci. Food Agr.* 56:385–397, 1991.

38. Steup, M. Starch degradation. In: *The Biochemistry of Plants. A Comprehensive Treatise,* Vol. 14: *Carbohydrates,* Preiss, J., ed., San Diego: Academic Press, 1988, pp 103–128.

39. ap Rees, T. Pathways of carbohydrate breakdown in plants. *Biochemistry* 11:89–127, 1974.

40. Morrell, S., T. ap Rees. Control of the hexose content of potato tubers. *Phytochemistry* 25:1073–1076, 1986.

41. Preiss, J., C. Levi. Starch biosynthesis and degradation. In: *The Biochemistry of Plants*, Vol. 2, Preiss, J., ed., New York: Academic Press, 1980 pp 371–423.

42. Zhou, D., T. Solomos. Effect of hypoxia on sugar accumulation, respiration, activities of amylase and starch phosphorylase, and induction of alternative oxidase and acid invertase during storage of potato tubers (*Solanum tuberosum* cv. Russet Burbank) at 1°C. *Physiologia Plant.* 104:255–265, 1998.

43. Nakano, K., H. Mori, T. Fukui. Molecular cloning of cDNA encoding potato amyloplast α-glucan phosphorylase and the structure of its transit peptide. *J. Biochem.* 106:691–695, 1989.

44. Kumar, G.N.M., L.O. Knowles, N. Fuller, N.R. Knowles. Starch phosphorylase activity correlates with senescent sweetening but not low temperature-induced sweetening in potato. *Plant Physiol.* 123(suppl):126, 2000.

45. Cochrane, M.P., C.M. Duffus, M.J. Allison, R.G. Mackay. Amylolytic activity in stored potato tubers. 2: the effect of low-temperature storage on the activities of α- and β-amylase and α-glucosidase in potato tubers. *Potato Res.* 34:333–341, 1991.

46. Cottrell, J.E., C.M. Duffus, L. Paterson, G.R. Mackay, M.J. Allison, H. Bain. The effect of storage temperature on reducing sugar concentration and the activities of three amylolytic enzymes in tubers of the cultivated potato, *Solanum tuberosum* L. *Potato Res.* 36:107–117, 1993.

47. Hill, L.M., R. Reimholz, R. Schröder, T.H. Nielsen, M. Stitt. The onset of sucrose accumulation in cold-stored potato tubers is caused by an increased rate of sucrose synthesis and coincides with low levels of hexose-phosphates, an activation of sucrose phosphate synthase and the appearance of a new form of amylase. *Plant Cell Env.* 19:1223–1237, 1996.

48. Neilson, T.H., U. Deiting, M. Stitt. A β-amylase in potato tuber is induced by storage at low temperature. *Plant Physiol.* 113:503–510, 1997.

49. Deiting, U., R. Zrenner, M. Stitt. Similar temperature requirement for sugar accumulation and for the induction of new forms of sucrose phosphate synthase and amylase in cold-stored potato tubers. *Plant Cell Env.* 21:127–138, 1998.

50. Schwimmer, S., A. Bevenue, W.J. Weston. Phosphorus components of the white potato. *Agr. Food Chem.* 3:257–260, 1955.

51. Samotus, B., S. Schwimmer. Effect of maturity and storage on distribution of phophorus among starch and other components of potato tuber. *Plant Physiol.* 37:519–522, 1962.

52. V.V. Shekhar, W.M. Iritani. Starch to sugar interconversion in *Solanum tuberosum* L, 1: influence of inorganic ions. *Amer. Potato J.* 55:345–350, 1978.

53. Isherwood, F.A., M.G.H. Kennedy. The composition of the expressed sap from cold stored potatoes. *Phytochemistry* 14:83–84, 1975.

54. Bielski, R.L. Phosphate pools, phosphate transport, and phosphate availability. *Annu. Rev. Plant Physiol.* 24:245–252, 1973.

55. Loughman, B.C. Uptake and utilization of phosphate associated with respiratory changes in potato tuber slices. *Plant Physiol.* 35:418–424, 1960.

56. Graham, D., B.D. Patterson. Responses of plants to low, nonfreezing temperatures: proteins, metabolism and acclimation. *Annu. Rev. Plant Physiol.* 33:347–372, 1982.

57. Borchert, S., H. Grosse, H.W. Heldt. Specific transport of inorganic phosphate, glucose-6-phosphate, dihydroxyacetone phosphate and 3-phosphoglycerate into amyloplasts from pea roots. *FEBS Lett.* 253:183–186, 1989.

58. Viola, R., H.V. Davies. Effect of temperature on pathways of carbohydrate metabolism in tubers of potato (*Solanum tuberosum* L). *Plant Sci.* 103:135–143, 1994.

59. Kammerer, B., K. Fischer, B. Hilpert, S. Schubert, M. Gutensohn, A. Weber, U.I. Flügge. Molecular characterization of a carbon transporter in plastids from heterotrophic tissue: the glucose 6-phosphate/phosphate antiporter. *Plant Cell* 10:105–117, 1998.

60. Pressey, R. Changes in sucrose synthetase and sucrose phosphate synthetase activities during storage of potatoes. *Amer. Potato J.* 47:245–251, 1970.

61. Viola, R., H.V. Davies, A.R. Chudek. Pathways of starch and sucrose biosynthesis in developing tubers of potato (*Solanum tuberosum* L) and seeds of faba bean (*Vicia faba* L): elucidation by ^{13}C NMR spectroscopy. *Planta* 183:202–208, 1991.

62. Reimholz, R., M. Geiger, V. Haake, U. Deiting, K.-P. Krause, U. Sonnewald, M. Stitt. Potato plants contain multiple forms of sucrose phosphate synthase, which differ in their tissue distributions, their levels during development, and their responses to low temperature. *Plant Cell Env.* 20:291–305, 1997.

63. Reimholz, R., P. Geigenberger, M. Stitt. Sucrose-phosphate synthase is regulated via metabolites and protein phosphorylation in potato tubers, in a manner analogous to the enzyme in leaves. *Planta* 192:480–488, 1994.

64. Krause, K.-P., L. Hill, R. Reimholz, T.H. Nielson, U. Sonnewald, M. Stitt. Sucrose metabolism in cold-stored potato tubers with decreased expression of sucrose phosphate synthase. *Plant Cell Env.* 21:285–299, 1998.

65. Jelitto, T., U. Sonnewald, L. Willmitzer, M. Hajirezeai, M. Stitt. Inorganic pyrophosphate content and metabolites in potato and tobacco plants expressing *E. coli* pyrophosphatase in their cytosol. *Planta* 188:238–244, 1992.

66. Sowokinos, J.R., J.P. Spychalla, S.L. Desborough. Pyrophosphorylases in *Solanum tuberosum*, IV: purification, tissue localization, and physicochemical properties of UDP-glucose pyrophosphorylase. *Plant Physiol.* 101:1073–1080, 1993.

67. Spychall, J.P., B.E. Scheffler, J.R. Sowokinos, M.W. Bevan. Cloning, antisense RNA inhibition, and the coordinated expression of UDP-glucose pyrophosphorylase with starch biosynthesis genes in potato tubers. *J. Plant Physiol.* 144:444–453, 1994.

68. Borovkov, A.Y., P.E. McClean, J.R. Sowokinos, S.H. Ruud, G.A. Secor. Effect of expression of UDP-glucose pyrophosphorylase ribozyme and antisense RNAs on the enzyme activity and carbohydrate composition of field-grown transgenic potato plants. *J. Plant Physiol.* 147:644–652, 1996.

69. Zrenner, R., L. Willmitzer, U. Sonnewald. Analysis of the expression of potato uridine diphosphate-glucose pyrophosphorylase and its inhibition by antisense RNA. *Planta* 190:247–252, 1993.

70. ap Rees, T., S. Morrell. Carbohydrate metabolism in developing potatoes. *Amer. Potato J.* 67:835–847, 1990.

71. Jang, J.C., J. Sheen. Sugar sensing in higher plants. *Plant Cell* 6:1665–1679, 1994.

72. ap Rees, T. Hexose phosphate metabolism by nonphotosynthetic tissues of higher plants. In: *The Biochemistry of Plants.* Vol. 14, Preiss, J., ed., New York: Academic Press, 1988, pp 1–33.

73. Ross, H.A., H.V. Davies. Sucrose metabolism in tubers of potato (*Solanum tuberosum* L): effects of sink removal and sucrose flux on sucrose-degrading enzymes. *Plant Physiol.* 98:287–293, 1992.

74. Sturm, A. Invertases: primary structures, functions and roles in plant development and sucrose partitioning. *Plant Physiol.* 121:1-7, 1999.

75. Pressey, R. Potato sucrose synthase: purification, properties and changes in activity associated with maturation. *Plant Physiol.* 44:759–764, 1969.

76. Pressey, R., R. Shaw. Effect of temperature on invertase, invertase inhibitor and sugars in potato tubers. *Plant Physiol.* 41:1657–1661, 1966.

77. Avigad, G. Sucrose and other disaccharides. In: *Encyclopedia of Plant Physiology,* Vol. 13A, Loewus, F.A., W. Tanner, eds., Berlin: Springer–Verlag, 1982, pp 217–317.

78. ap Rees, T. Pathways of carbohydrate breakdown in plants. *MTP Int. Rev. Sci. Plant Biochem.* 11:89–127, 1974.

79. Davies, H.V., R.A. Jefferies, L. Scobie. Hexose accumulation in cold-stored tubers of potato (*Solanum tuberosum* L): the effect of water stress. *J. Plant Physiol.* 134:471–475, 1989.

80. Richardson, D.I., H.V. Davies, H.A. Ross, G.R. Mackay. Invertase activity and its relationship to hexose accumulation in potato tubers. *J. Exp. Bot.* 41:95–99, 1990.

81. Davies, H.V., R. Viola. Control of sugar balance in potato tubers. In: *The Molecular and Cellular Biology of the Potato,* 2nd ed., Belknap, W.R., M.E. Vayda, W.D. Park, eds., Wallingford: CAB Int., 1994, pp 67–80.

82. Zrenner, R., K. Schüler, U. Sonnewald. Soluble acid invertase determines the hexose-to-sucrose ratio in cold-stored potato tubers. *Planta* 198:246–252, 1996.

83. Greiner, S., T. Rausch, U. Sonnewald, K. Herbers. Ectopic expression of a tobacco invertase inhibitor homolog prevents cold-induced sweetening of potato tubers. *Nat. Biotechnol.* 17:708–711, 1999.

84. Dennis, D.T., J.A. Miernyk. Compartmentation of nonphotosynthetic carbohydrate metabolism. *Annu. Rev. Plant Physiol.* 33:27–50, 1982.

85. Geigenberger, P., M. Stitt. A 'futile' cycle of sucrose synthesis and degradation is involved in regulating partitioning between sucrose, starch and respiration in cotyledons of germinating *Ricinus communis* L seedlings when phloem transport is inhibited. *Planta* 185:81–90, 1991.

86. Dixon, W.L., T. ap Rees. Identification of the regulatory steps in glycolysis in potato tubers. *Phytochemistry* 19:1297–1301, 1980.

87. Dixon, W.L., T. ap Rees. Carbohydrate metabolism during cold-induced sweetening of potato tubers. *Phytochemistry* 19:1653–1656, 1980.

88. Huber, S.C. Fructose 2,6-bisphosphate as a regulatory metabolite in plants. *Annu. Rev. Plant Physiol.* 37:233–246, 1986.

89. Hajirezaei, M., M. Stitt. Contrasting roles for pyrophosphate: fructose 6-phosphate phosphotransferase during aging of tissue slices from potato tubers and carrot storage tissues. *Plant Sci.* 77:177–183, 1991.

90. Stitt, M. Fructose 2,6,-bisphosphate as a regulatory molecule in plants. *Annu. Rev. Plant Physiol. Plant Mol. Biol.* 41:153–185, 1990.

91. Morrell, S., T. ap Rees. Sugar metabolism in developing tubers of *Solanum tuberosum.* *Phytochemistry* 25:1579–1585, 1986b.

92. Trevanion, S.J., N.J. Kruger. Effect of temperature on the kinetic properties of pyrophosphate: fructose 6-phosphate phosphotransferase from potato tuber. *J. Plant Physiol.* 137:753–759, 1991.

93. Claassen, P.A.M., M.A.W. Budde, H.J. de Ruyter, M.H. van Calker, A. van Es. Potential role of pyrophosphate: fructose-6-phosphate phosphotransferase in carbohydrate metabolism of cold stored tubers of *Solanum tuberosum* cv. Bintje. *Plant Physiol.* 95:1243–1249, 1991.

94. Bredemeijer, G.M.M., H.C.J. Burg, P.A.M. Claassen, W.J. Stiekema. Phosphofructokinase in relation to sugar accumulation in cold-stored potato tubers. *J. Plant Physiol.* 138:129–135, 1991.

95. Hammond, J.B.W., M.M. Burrell, N.J. Kruger. Effect of low temperature on the activity of phosphofructokinase from potato tubers. *Planta* 180:613–616, 1990.

96. Kruger, N.J., J.B.W. Hammond, M.M. Burrell. Molecular characterization of four forms of phosphofructokinose purified from potato tuber. *Arch. Biochem. Biophys.* 267:690–700, 1988.

97. Bryce, J.H., S.A. Hill. Energy production in plant cells. In: *Plant Biochemistry and Molecular Biology,* Lee, P.J., R.C. Leegood, eds., Chichester: Wiley Europe Ltd, 1993, pp 1–21.

98. Amir, J., V. Kahn, M. Unterman. Respiration, ATP level, and sugar accumulation in potato tubers during storage at 4°C. *Phytochemistry* 16:1495–1498, 1977.

99. Hajirezaei, M., U. Sonnewald, R. Viola, S. Carlisle, D. Dennis, M. Stitt. Transgenic potato plants with strongly decreased expression of pyrophosphate: fructose 6-phosphate phosphotransferase show no visible phenotype and only minor changes in metabolic fluxes in their tubers. *Planta* 192:16–30, 1994.

100. Burrell, M.M., P.J. Mooney, M. Blundy, D. Carter, F. Wilson, J. Green, K.S. Blundy, T. ap Rees. Genetic manipulation of 6-phosphofructokinose in potato tubers. *Planta* 194:95–101, 1994.

101. Thomas, S., P.J. Mooney, M.M. Burrell, D.A. Fell. Finite change analysis of glycolytic intermediates in tuber tissue of lines of transgenic potato (*Solanum tuberosum*) overexpressing phosphofructokinose. *Biochem. J.* 322:111–117, 1997.

102. Thomas, S., P.J. Mooney, M.M. Burrell, D.A. Fell. Metabolic control analysis of glycolysis in tuber tissue of potato (*Solanum tuberosum*): explanation for the low control of coefficient of phosphofructokinose over respiratory flux. *Biochem. J.* 322:119–127, 1997.

103. Stitt, M. Frucotse 2,6-bisphosphate and plant carbohydrate metabolism. *Plant Physiol.* 84:201–204, 1987.

104. Van Schaftingen, E., B. Lederer, R. Bartons, H.-G. Hers. A kinetic study of pyrophosphate: fructose-6-phosphate phosphotransferase from potato tubers. *Eur. J. Biochem.* 129:191–195, 1982.

105. D.T. Dennis, Y. Huang, F.B. Negm. Glycolysis, the pentose phosphate pathway and anaerobic respiration. In: *Plant metabolism*, 2nd ed., Dennis, D.T., D.H. Turpin, D.D. Lefebvre, D.B. Layzell, eds. UK: Addison Wesley Longman Ltd., 1997, pp 105–123.

106. Wagner, A.M., T.J.A. Kneppers, B.M. Kroon, L.H.W. van der Plas. Enzymes of the pentose phosphate pathway in callus-forming potato tuber discs grown at various temperatures. *Plant Sci.* 51:159–164, 1987.

107. Barichello, V., R.Y. Yada, R.H. Coffin, D.W. Stanley. Low temperature sweetening in susceptible and resistant potatoes: starch structure and composition. *J. Food Sci.* 55:1054–1059, 1990.

108. Sherman, M., E.E. Ewing. Temperature, cyanide, and oxygen effects on the respiration, chip color, sugars, and organic acids of stored tubers. *Amer. Potato J.* 59:165–178, 1982.

109. Barichello, V., R.Y. Yada, R.H. Coffin, D.W. Stanley. Respiratory enzyme activity in low temperature sweetening of susceptible and resistant potatoes. *J. Food. Sci.* 55:1060–1063, 1990.

110. Purvis, A.C., R.L. Shewfelt. Does the alternative pathway ameliorate chilling injury in sensitive plant tissues? *Physiologia Plant.* 88:712–718, 1993.

111. Lambers, H. Cyanide-resistant respiration: a non-phosphorylating electron transport pathway acting as an energy overflow. *Physiologia Plant.* 55:478–485, 1982.

112. Elthon, T.E., C.R. Stewart, C.A. McCoy, W.D. Bonner, Jr. Alternate respiratory path capacity in plant mitochondria: effect of growth, temperature, the electrochemical gradient, and assay pH. *Plant Physiol.* 80:378–383, 1986.

113. Solomos, T., G.G. Laties. The mechanism of ethylene and cyanide action in triggering the rise in respiration in potato tubers. *Plant Physiol.* 55:73–78, 1975.

114. Lance, C. Cyanide-insensitive respiration in fruits and vegetables. In: *Recent Advances in the Biochemistry of Fruit and Vegetables,* Friend, J., M.J.C. Rhodes, eds., New York: Academic Press, 1981, pp 63–87.

115. Sherman, M., E.E. Ewing. Effects of temperature and low oxygen atmospheres on respiration, chip color, sugars, and malate of stored potatoes. *J. Amer. Soc. Hort. Sci.* 108:129–133, 1983.

116. Lyons, J.M. Chilling injury in plants. *Annu. Rev. Plant Physiol.* 24:445–466, 1973.

117. Raison, J.K. The influence of temperature-induced phase changes on kinetics of respiratory and other membrane-associated enzymes. *J. Bioenerg.* 4:258-309, 1973.

118. Shewfelt, R.L. Response of plant membranes to chilling and freezing. In: *A biophysical approach to structure, development and senescence,* Leshem, Y.Y., ed., Dordrecht: Kluwer, 1992, pp 192–219.

119. Murata, N., I. Nishida. Lipids of blue-green algae (cyanobacteria). In: *The Biochemistry of Plants*, Vol. 9. Orlando: Academic Press, 1987, 315–347.

120. Murata, N. Molecular species composition of phosphadtidylglycerols from chilling-sensitive and chilling-resistant plants. *Plant Cell Physiol.* 24:81–86, 1983.

121. Murata, N., N. Sato, N. Takahashi, Y. Hamazaki. Compositions and positional distributions of fatty acids in phospholipids from leaves of chilling-sensitive and chilling-resistant plants. *Plant Cell Physiol.* 23:1071–1079, 1982.

122. Knowles, N.R., L.O. Knowles. Correlations between electrolyte leakage and degree of saturation of polar lipids from aged potato (*Solanum tuberosum* L) tuber tissue. *Ann. Bot.* 63:331–338, 1989.

123. Spychalla, J.P., S.L. Desborough. Fatty acids, membrane permeability, and sugars of stored potato tubers. *Plant Physiol.* 94:1207–1213, 1990.

124. Shekhar, V.C., W.M. Iritani, J. Magnuson. Starch-sugar interconversion in *Solanum tuberosum* L, 11: influence of membrane permeability and fluidity. *Amer. Potato J.* 56:225–234, 1979.

125. O'Donoghue, E.P., R.Y. Yada, A.G. Marangoni. Low temperature sweetening in potato tubers: the role of the amyloplast membrane. *J. Plant Physiol.* 145(3):335–341, 1995.

126. Gounaris, Y., J.R. Sowokinos. Two-dimensional analysis of mitochondrial proteins from potato cultivars resistant and sensitive to cold-induced sweetening. *J. Plant Physiol.* 140:611–616, 1992.

127. Berkeley, H.D., T. Galliard. Lipids of potato tubers: effect of growth and storage on lipid content of the potato tuber. *J. Sci. Food. Agr.* 25:861–867, 1974.

128. Lojkowska, E., M. Holubowska. Changes of the lipid catabolism in potato tubers from cultivars differing in susceptibility to autolysis during the storage. *Potato Res.* 32:463–470, 1989.

129. Kumar, G.N.M., N.R. Knowles. Changes in lipid peroxidation and lipolytic and free-radical scavenging enzyme activities during aging and sprouting of potato (*Solanum tuberosum*) seed tubers. *Plant Physiol.* 102:115–124, 1993.

130. Galliard, T. Lipid of potato tubers, 1: lipid and fatty acid composition of tubers from different varieties of potato. *J. Sci. Food. Agr.* 24:617–622, 1973.

131. Thompson, G.A., Jr., K.J. Elnspahr, S. Ho Cho, T.C. Peeler, M.B. Stephenson. Metabolic responses of plant cells to stress. In: *Biological Role of Plant Lipids*, Biacs, P.A., K. Gruiz, T. Kremmer, eds., London: Plenum Publishing, 1989, pp 497–504.

132. Nishida, I., N. Murata. Chilling sensitivity in plants and cyanobacteria: the crucial contribution of membrane lipids. *Annu. Rev. Plant Physiol. Plant Mol. Biol.* 47:541–568, 1996.

133. Bonnerot, C., P. Mazliak. Induction of the oleoly-phosphatidylcholine desaturase activity during the storage of plant organs: a comparison between potato and jerusalem artichoke tubers. *Plant Sci. Lett.* 35:5–10, 1984.

134. Spychalla, J.P., S.L. Desborough. Superoxide dismutase, catalase, and α-tocopherol content of stored potato tubers. *Plant Physiol.* 94:1214–1218, 1990.

135. Jarvis, M.C., J. Dalziel, H.J. Duncan. Variations in free sugars and lipids in different potato varieties during low-temperature storage. *J. Sci. Food Agr.* 25:1405–1409, 1974.

136. Galliard, T., H.D. Berkeley, J.A. Mathew. Lipids of potato tubers, effects of storage temperature on total, polar, and sterol lipid content and fatty acid composition of potato tubers. *J. Sci. Food Agr.* 26:1163–1170, 1975.

137. Lijenberg, C., A.S. Sandelius, E. Selstam. Effect of storage in darkness and in light on the content of membrane lipids of potato tubers. *Physiologia Plant.* 43:154–159, 1978.

138. Phan, C.-T. Temperature effects on metabolism. In: *Postharvest Physiology of Vegetables,* Welchmann, J., ed., New York: Marcel Dekker, 1987, pp 173–180.

139. Pinhero, R.G., G. Paliyath, R.Y. Yada, D.P. Murr. Modulation of phospholipase D and lipoxygenase activities during chilling: relation to chilling tolerance of maize seedlings. *Plant Physiol. Biochem.* 36:213–224, 1998.

140. Paliyath, G., M.J. Droillard. The mechanisms of membrane deterioration and disassembly during senescence. *Plant Physiol. Biochem.* 30:789–812, 1992.

141. Lyons, J.M., J.K. Raison. Oxidative activity of mitochondria isolated from plant tissues sensitive and resistant to chilling injury. *Plant Physiol.* 45:386–389, 1970.

142. Quinn, P.J., W.P. Williams. Plant lipids and their role in membrane function. *Prog. Biophys. Mol. Biol.* 34:109–173, 1978.

143. Workman, M., A. Cameron, J. Twomey. Influence of chilling on potato tuber respiration, sugar, o-dihydroxyphenolic content and membrane permeability, *Amer. Potato J.* 56:277–288, 1979.

144. Schneider, E.M., J.U. Becker, D. Volkmann. Biochemical properties of potato phosphorylase change with its intracellular localization as revealed by immunological methods. *Planta* 151:124–134, 1981.

145. Brisson, N., H. Giroux, M. Zollinger, A. Camirand, C. Simard. Maturation and subcellular compartmentation of potato starch phosphorylase. *Plant Cell* 1:559–566, 1989.

146. Shekhar, V.C., W.M. Iritani, J. Magnuson. Starch-sugar interconversion in *Solanum tuberosum* L, II: influence of membrane permeability and fluidity. *Amer. Potato J.* 56:225–234, 1979.

147. Isherwood, F. Mechanism of starch-sugar interconversion in *Solanum tuberosum*. *Phytochemistry* 15:33–41, 1976.

148. Sowokinos, J.R., P.H. Orr, J.A. Knoper, J.L. Varns. Influence of potato storage and handling stress on sugars, chip quality, and integrity of the starch (amyloplast) membrane. *Amer. Potato J.* 64:213–226, 1987.

149. Yada, R.Y., R.H. Coffin, K.W. Baker, M.J. Lsezkowiat. An electron microscopic examination of the amyloplast membranes from potato cultivar susceptible to low temperature sweetening. *Can. Inst. Food Sci. Technol. J.* 23:145–148, 1990.

150. Allen, R.D. Dissection of oxidative stress tolerance using transgenic plants. *Plant Physiol.* 107:1049–1054, 1995.

151. McKersie, B.D., Y.Y. Leshem. *Stress and Stress Coping in Cultivated Plants.* Dordrecht, Kluwer, 1994, pp. 79–100.

152. Fauconnier, M.L., J. Rojas-Beltra, D. Delcarte, F. Dejaeghere, M. Marlier, P. Du Jardin. Lipoxygenase pathway and membrane permeability and composition during storage of potato tubers (*Solanum tuberosum* L) in different conditions. *Plant Biol.* 4:77–85, 2002.

153. Shewfelt, R.L., R.E. McDonald, H.O. Hultin. Effect of phospholipid hydrolysis on lipid oxidation in flounder muscle microsomes. *J. Food Sci.* 46:1297–1301, 1981.

154. Stanley, D.W. Biological membrane deterioration and associated quality losses in food tissues. *Crit. Rev. Food Sci. Nutr.* 30:487–553, 1991.

155. Giovannoni, J. Molecular biology of fruit maturation and ripening. *Annu. Rev. Plant Physiol. Plant Mol. Biol.* 52:725–749, 2001.

156. Sacher, J.A. Studies of permeability, RNA and protein turnover during aging of fruit and leaf tissue. In: *Aspects of the Biology of Aging*, Woolhouse, H.W., ed., New York: Academic Press, 1967 pp 269–303.

157. J.H. Brown, G. Paliyath, J.E. Thompson. Physiological mechanisms of plant sensescence. In: *Plant Physiology: A Treatise, X*, Steward, F.C., R.G.S. Bidwell, eds., New York: Academic Press, 1991, pp 227–275.

158. Lanahan, M.B., H.C. Yen, J.J. Giovannoni, H.J. Klee. The *Never Ripe* mutation blocks ethylene perception in tomato. *Plant Cell* 6:521–530, 1994.

159. Lashbrook, C.C., D.M. Tieman, H.J. Klee. Differential regulation of the tomato ETR gene family throughout plant development. *Plant J.* 15:243–252, 1998.

160. Payton, S., R.G. Fray, S. Brown, D. Grierson. Ethylene receptor expression is regulated during fruit ripening, flower senescence and abscission. *Plant Mol. Biol.* 31:1227–1231, 1996.

161. Tieman, D.M., M.G. Taylor, J.A. Ciardi, H.J. Klee. The tomato ethylene receptors NR and LeETR4 are negative regulators of ethylene response and exhibit functional compensation within a multigene family. *Proc. Natl. Acad. Sci. USA* 97:5663–5668, 2000.

162. Wilkinson, J., M. Lanahan, H. Yen, J.J. Giovannoni, H.J. Klee. An ethylene-inducible component of signal transduction encoded by Never-ripe. *Science* 270:1807–1809, 1995.

163. Yen, H., S. Lee, S. Tanksley, M. Lanahan, H.J. Klee, J.J. Giovannoni. The tomato Never-ripe locus regulates ethylene-inducible gene expression and is linked to a homologue of the *Arabidopsis* ETR1 gene. *Plant Physiol.* 107:1343–1353, 1995.

164. Zhou, D., P. Kalaitzis, A.K. Mattoo, M.L. Tucker. The mRNA for an ETR1 homologue in tomato is constitutively expressed in vegetative and reproductive tissues. *Plant Mol. Biol.* 30:1331–38, 1996.

165. Leshem, Y.Y. *Plant Membranes: A Biophysical Approach to Structure, Development and Senescence.* London: Kluwer Academic Publishers, 1992, pp 174–188.

166. Paliyath, G., D.P. Murr, J.E. Thompson. Catabolism of phosphorylated phosphatidylinositols by carnation petal microsomal membranes enriched in plasmalemma and endoplasmic reticulum. *Physiol. Mol. Biol. Plants* 1:141–150, 1995.

167. Wang, X. Plant phospholipases. *Ann. Rev. Plant Physiol. Plant Mol. Biol.* 52:211-231, 2001.

168. Wang, X. Multiple forms of phospholipase D in plants: the gene family, catalytic and regulatory properties, and cellular functions. *Prog. Lipid Res.* 39:109–149, 2000.

169. Hammond, S.M., Y.M. Alshuller, T. Sung, S.A. Rudge, K. Rose. Human ADP-ribosylation factor-activated phosphatidylcholine-specific phospholipase D defines a new and highly conserved gene family. *J. Biol. Chem.* 270:29640–29643, 1995.

170. Rose, K., S.A. Rudge, M.A. Frohman, A.J. Morris, J. Engebrecht. Phospholipase D signaling is essential for meiosis. *Proc. Natl. Acad. Sci. USA* 92:12151–12155, 1995.

171. Liscovitch, M., M. Czarny, G. Fiucci , X. Tang. Phospholipase D: molecular and cell biology of a novel gene family. *Biochem. J.* 345:401–415, 2000.

172. Wang, X. The role of phospholipase D in signaling cascades. *Plant Physiol.* 120:645–651, 1999.

173. Laxalt, A.M., B. ter Riet, J.C. Verdonk, L. Parigi, W.I.L. Tameling, J. Vossen, M. Haring, A. Musgrave, T. Munnik. Characterization of five tomato phospholipase D. cDNAs: rapid and specific expression of *LePLDβ1* on elicitation with xylanase. *Plant J.* 26:237–247, 2001.

174. Wang, C., C. Zien, M. Afitlhile, R. Welti, D.F. Hildebrand, X. Wang. Involvement of phospholipase D in wound-induced accumulation of jasmonic acid in *Arabidopsis*. *Plant Cell* 12:2237–2246, 2000.

175. Todd, J.F., G. Paliyath, J.E. Thompson. Effect of chilling on the activities of lipid degrading enzymes in tomato fruit microsomal membranes. *Plant Physiol. Biochem.* 30:517–522, 1992.

176. Marangoni, A.G., D.W. Stanley. Phase transitions in microsomal membranes from chilling-sensitive and chilling-resistant tomato plants and fruit. *Phytochemistry* 28:2293–2301, 1989.

177. McCormac, D.J., J.F. Todd, G. Paliyath, J.E. Thompson. Modulation of bilayer fluidity affects lipid catabolism in microsomal membranes of tomato fruit. *Plant Physiol.* 31:1–8, 1993.

178. Parkin, K.L., S.J. Kuo. Chilling-induced lipid degradation in cucumber (*Cucumis sativa* L. cultivar Hybrid C) fruit. *Plant Physiol.* 90:1049–1056, 1989.

178a. Pinhero, R.G., K.C. Almquist, Z. Novotna, G. Paliyath. Developmental regulation of phospholipase D in tomato fruits. *Plant Physiol. Biochem.* 41:223–240, 2003.

179. Jandus, J., O. Valentovä, J. Kas, J. Daussant, C. Thévenot. Phospholipase D during tomato fruit ripening. *Plant Physiol. Biochem.* 35:123–128, 1997.

180. Güçlü, J., A. Paulin, P. Soudain. Changes in polar lipid during ripening and senescence of cherry tomato (*Lycopersicon esculentum*): relation to climateric and ethylene increases. *Physiol. Plant.* 77:413–419, 1989.

181. Whitaker, B.D. Lipid changes in mature-green tomato fruit during ripening, during chilling, and after rewarming subsequent to chilling. *J. Amer. Soc. Hort. Sci.* 119:994–999, 1994.

182. Ryu, S.B., B.H. Karlsson, M. Ozgen, J.P. Palta. Inhibition of phospholipase D by lysophosphatidylethanolamine, a lipid-derived senescence retardation. *Proc. Natl. Acad. Sci. USA* 94:12717–12721, 1997.

183. Farag, K.M., J.P. Palta. Use of lysophosphatidylethanolamine, a natural lipid, to retard tomato leaf and fruit senescence. *Physiologia Plant.* 87:515–521, 1993.

184. Ryu, S.B., X. Wang. Activation of phospholipase D and the possible mechanism of activation in wound-induced lipid hydrolysis in castor bean leaves. *Biochim. Biophys. Acta.* 1303:243–250, 1996.

185. Wang X., C. Wang, Y. Sang, L. Zheng, C. Qin. Determining functions of multiple phospholipase Ds in stress response of *Arabidopsis*. *Biochem. Soc. Trans.* 28:813–816, 2000.

186. Zheng, L., R. Krishnamoorthi, M. Zolkiewski, X. Wang. Distinct Ca^{2+} binding properties of novel C2 domains of plant phospholipase Dα and β. *J. Biol. Chem.* 275:19700–19706, 2000.

187. Zocchi, G., J.B. Hanson. Calcium influx into corn roots as a result of cold shock. *Plant Physiol.* 70:318–319, 1982.

188. Fan, L., S. Zheng, X. Wang. Antisense suppression of phospholipase D alpha retards abscisic acid- and ethylene-promoted senescence of postharvest *Arabidopsis* leaves. *Plant Cell* 9:2183–2196, 1997.

189. Fan, L., S. Zheng, D. Cui, X. Wang. Subcellular distribution and tissue expression of phospholipase Dα, Dβ, and Dχ in *Arabidopsis. Plant Physiol.* 119:1371–1378, 1999.

190. Grechkin, A. Recent developments in biochemistry of the plant lipoxygenase pathway. *Prog. Lipid Res.* 37:317–352, 1998.

191. Rosahl, S. Lipoxygenase in plants: their role in development and stress response. *Z. Naturforsch* 51:123–138, 1996.

192. Droillard, M.J., M.A. Rouet-Mayer, J.M. Bureau, C. Lauriere. Membrane-associated and soluble lipoxygenase isoforms in tomato pericarp: characterization and involvement in membrane alterations. *Plant Physiol.* 103:1211–1219, 1993.

193. Lynch, D.V., J.E. Thompson. Lipoxygenase-mediated production of superoxide anion in senescing plant tissue. *FEBS Lett.* 173:251–254, 1984.

194. Mauch, F., A. Kmecl, U. Schaffrath, S. Volrath, J. Gorlach, E. Ward, J. Ryals, R. Dudler. *Plant Physiol.* 114:1561–1566, 1997.

195. Maccarrone, M., G.A. Veldink, A.F. Agro, J.F.G. Vliegenthart. Modulation of soybean lipoxygenase expression and membrane oxidation by water deficit. *FEBS Lett.* 371:223–226, 1995.

196. Maccarrone, M., G.A. Veldink, J.F.G. Vliegenthart. Thermal injury and ozone stress affect soybean lipoxygenase expression. *FEBS Lett.* 309:225–230, 1992.

197. Conconi, A., M. Miquel, J.A. Browse, C.A. Ryan. Intracellular levels of free linolenic and linoleic acids increase in tomato leaves in response to wounding. *Plant Physiol.* 111:797–803, 1996.

198. Yamauchi, R., M. Kojima, K. Kato, Y. Ueno. Lipoxygenase-catalyzed oxygenation of monogalactosyldilinolenoylglycerol in dipalmitoylphosphatidylcholine liposomes. *Agr. Biol. Chem.* 49:2475–2477, 1985.

199. Brash, A.R., C.D. Ingram, T.M. Harris. Analysis of specific oxygenation reaction of soybean lipoxygeanse-1 with fatty acids esterified in phospholipids. *Biochemistry* 26:5465–5471, 1987.

200. Kondo, Y., Y. Kawai, T. Hayashi, M. Ohnishi, T. Miyazawa, S. Itoh, J. Mizutani. Lipoxygenase in soybean seedlings catalyzes the oxygenation of phospholipid and such activity changes after treatment with fungal elicitor. *Biochim. Biophys. Acta.* 1170:301–306, 1993.

201. Todd, J.F., G. Paliyath, J.E. Thompson. Characteristics of a membrane-associated lipoxygenase in tomato fruit. *Plant Physiol.* 94:1225-1232, 1990.

202. M.A. Rouet-Mayer, J.M. Bureau, C. Lauriere. Identification and characterization of lipoxygenase isoforms in senescing carnation petals. *Plant Physiol.* 98:971–978, 1992.

203. B.A. Vick, D.C. Zimmerman. Oxidative systems for modification of fatty acids: the lipoxygenase pathway. In: *The Biochemistry of Plants*, Vol. 9, Stumpf, P.K., E.E. Conn, eds., Orlando: Academic Press, 1987, pp 53–90.

204. Grossman, S., Y. Leshem. Lowering of endogenous lipoxygenase activity in *Pisum sativum* foliage by cytokinins as related to senescence. *Physiologia Plant.* 43:359–362, 1978.

205. Peterman, T.K., J.N. Siedow. Behaviour of lipoxygeanse during establishment, senescence and rejuvenation of soybean cotyledons. *Plant Physiol.* 78:690–695, 1985.

206. Navari-Izzo, F., N. Vangioni, M.F. Quartacci. Lipids of soybean and sunflower seedlings grown under drought conditions. *Phytochemistry* 29:2119–2123, 1990.

207. Ealing, P.M. Lipoxygenase activity in ripening tomato fruit pericarp tissue. *Phytochemistry* 36:547–552, 1994.

208. Griffiths, A., C. Barry, A.G. Alpuche-Soils, D. Grierson. Ethylene and developmental signals regulate expression of lipoxygenase genes during tomato fruit ripening. *J. Exp. Bot.* 50:793–798, 1999.

209. R.G. Buttery, R. Teranishi, R.A. Flath, L.C. Ling. In: *Flavor Chemistry: Trends and Developments,* Teranishi, R., R.G. Buttery, eds., Washington: American Chemical Society, 1989, pp 213.

210. Ferrie, B.J., N. Beaudoin, W. Burkhart, C.G. Bowsher, S.J. Rothstein. The cloning of two tomato lipoxygenase genes and their differential expression during tomato fruit ripening. *Plant Physiol.* 106:109–118, 1994.

211. Deng, W., W.S. Grayburn, T.R. Hamilton-Kemp, G.B. Collins. Expression of soybean-embryo lipoxygenase 2 in transgenic tobacco tisue. *Planta* 187:203–208, 1992.

212. Rancé, L., J. Fournier, M.T. Esquerré-Tugayé. The incompatible interaction between *Phytophthora parasitica* var. nicotianaerace and tobacco is suppressed in transgenic plants expressing antisense lipoxygenase sequences. *Proc. Natl. Acad. Sci. USA* 95:6554–6559, 1998.

213. Griffiths, A., S. Prestage, R. Linforth, J. Zhang, A. Taylor, D. Grierson. Fruit-specific lipoxygenase suppression in antisense-transgenic tomatoes. *Postharvest Biol. Tech.* 17:163-173, 1999.

214. Beaudoin, N., S.J. Rothstein. Developmental regulation of two tomato lipoxygenase promoters in transgenic tobacco and tomato. *Plant Mol. Biol.* 33:835–846, 1997.

215. Royo, J., J, Léon, G. Vancanneyt, J.P. Albar, S. Rosahl, F. Ortego, P. Castañera, J.J. Sánchez-Serrano. Antisense-mediated depletion of a potato lipoxygenase reduces wound induction of proteinase inhibitors and increases weight gain of insect pests. *Proc. Natl. Acad. Sci. USA* 96:1146–1151, 1999.

216. Grierson, D., A.A. Kader. Fruit ripening and quality. In: *The Tomato Crop: A Scientific Basis for Improvement,* Atherton, J.G., J. Rudich, eds., London: Chapman & Hall, 1986, pp 241–280.

217. Kramer, M., R.E. Sheehy, W.R. Hiatt. Progress towards the genetic engineering of tomato fruit softening. *TIBTECH* 7:191–193, 1989.

218. Brummell, D.A., M.H. Harpster. Cell wall metabolism in fruit softening and quality and its manipulation in transgenic plants. *Plant Mol. Biol.* 47:311–340, 2001.

16

Egg Yolk Antibody Farming for Passive Immunotherapy

Jennifer Kovacs-Nolan and Yoshinori Mine

CONTENTS

16.1 Introduction 382
16.2 Avian Egg Antibodies 382
    16.2.1 Avian Immune System 382
    16.2.2 Molecular Properties of IgY..383
        16.2.2.1 Structure of IgY 383
        16.2.2.2 Physicochemical Properties of IgY....................................383
    16.2.3 Advantages of IgY 385
16.3 Passive Immunization with IgY..386
    16.3.1 Immunotherapeutic Applications of IgY...386
        16.3.1.1 Rotavirus 386
        16.3.1.2 *Escherichia coli* 386
        16.3.1.3 *Salmonella species* 388
        16.3.1.4 *Pseudomonas aeruginosa* 389
        16.3.1.5 *Streptococcus mutans*..390
        16.3.1.6 *Helicobacter pylori* 390
        16.3.1.7 Coronavirus..390
        16.3.1.8 Fish Pathogens 391
        16.3.1.9 Others 391
16.4 Antibody Stability 391
    16.4.1 Physiology of the Gastrointestinal Tract...391
    16.4.2 Degradation and Stability of IgY 392
16.5 Biotechnology of Avian Egg Antibodies..393
References...393

16.1   INTRODUCTION

Immunoglobulins, or antibodies, are large molecules formed by the immune system, developed by higher organisms to combat the invasion of foreign substances (1). In both humans and animals, the administration of specific antibodies is an attractive approach to establishing protective immunity against viral and bacterial pathogens, and has been prompted by the need to find alternatives to antibiotics in response to the increasing number of antibiotic resistant organisms, as well as organisms that are not responsive to traditional antibiotic treatment (2). There are also individuals who are unable to mount an active immune response against pathogens, including infants, children with congenital or acquired immunodeficiency syndromes, and those rendered immunodeficient by chemotherapy, malnutrition, or aging (3,4), who would benefit significantly from effective passive immunization techniques.

Passive immunization, a technique in which preformed pathogen specific antibodies are administered orally to individuals to prevent, and in some cases treat, infectious diseases, may be one of the most valuable applications of antibodies (5). It is proposed that the antibodies may exert a sort of antimicrobial activity against pathogens, by binding, immobilizing, and consequently reducing or inhibiting the growth, replication, or colony forming abilities of pathogens (5).

Traditionally, polyclonal antibodies have been produced in mammals such as mice, rats, rabbits, and goats. The antibodies are then collected from the serum of the animal (6). However, these antibodies cannot be prepared on the industrial scale that is required for passive immunization techniques, because of the limitation in the amount of blood that can be collected at any one time. Chickens have attracted considerable attention as an alternative source of antibodies for the prevention and treatment of infectious diseases (2,7,8). The serum immunoglobulin of chickens, referred to as immunoglobulin IgY, is transferred in large quantities to the egg yolk in order to give acquired immunity to the developing embryo (9), and can be readily extracted from the egg yolk.

The large scale potential of IgY, as well as its source, a normal dietary component, makes it ideally suited for passive immunization applications. Egg yolk antibodies, therefore, have significant potential for food fortification and nutraceutical applications for the reduction of the morbidity and mortality associated with infection by human and animal pathogens.

The following chapter details the structural and physical characteristics of IgY, the numerous advantages and immunotherapeutic applications of IgY, and future prospects in IgY research.

16.2   AVIAN EGG ANTIBODIES

16.2.1   Avian Immune System

Immunoglobulin production in chickens is significantly different from that of mammals. While the rearrangement of immunoglobulin genes is an ongoing process in mammals, in chickens, it occurs in one single wave during embryogenesis, limiting the possible number of antibodies to the number of B-cell precursors (10,11), estimated to be around $2-3 \times 10^4$ cells (12). In contrast to mammalian systems, gene rearrangement also contributes little to chicken antibody diversity (13). However, despite the fact that chickens have an extremely limited number of immunoglobulin genes compared to mammals, they have developed other methods of producing a wide range of immune responses and diverse antibody molecules (13), enabling their use for the production of specific antibodies.

Three immunoglobulin classes have been shown to exist in the chicken: IgA, IgM, and IgY. The IgA and IgM are similar to mammalian IgA and IgM. Chicken IgY is the functional equivalent of IgG, the major serum antibody found in mammals, and makes up about 75% of the total antibody population (2). The serum concentrations of IgY, IgA, and IgM have been reported to be 5.0, 1.25, and 0.61 mg/mL, respectively (14). In mammals, the transfer of maternal antibodies can take place after birth, however, in the chicken, the maternal antibodies must be transferred to the developing embryo to give acquired immunity to the chick (5,7). Antibody, specifically IgA and IgM, is incorporated into the egg white during egg formation. Serum IgY is selectively transferred to the yolk via a receptor on the surface of the yolk membrane which is specific for IgY translocation (15,16). Morrison et al. (17) identified the regions within the IgY molecule which are important for its uptake and specific translocation into the yolk. Egg white contains IgA and IgM at concentrations of around 0.15 and 0.7 mg/mL, respectively, whereas the yolk may contain up to 25 mg/mL of IgY (18). Mammalian equivalents of IgE and IgD have not been identified in chickens (13).

16.2.2 Molecular Properties of IgY

16.2.2.1 Structure of IgY

The structure of IgY is significantly different from that of mammalian IgG (7), even though they share a similar function. IgY contains two heavy (H) and two light (L) chains and has a molecular mass of 180 kDa, larger than that of mammalian IgG (159 kDa). IgY also possesses a larger molecular weight H chain (68 kDa) as compared to that in mammals (50 kDa). The H chain of IgG consists of four domains: the variable domain (V_H) and three constant domains ($C\gamma1$, $C\gamma2$ and $C\gamma3$). The $C\gamma1$ domain is separated from $C\gamma2$ by a hinge region, which gives flexibility to the Fab fragments (the portion which contains the antigen binding activity). In contrast, the H chain of IgY does not have a hinge region, and possesses four constant domains (Cv1-Cv4) in addition to the variable domain (Figure 16.1). Sequence comparisons between IgG and IgY have shown that the $C\gamma2$ and $C\gamma3$ domains of IgG are closely related to the Cv3 and Cv4 domains of IgY; while the equivalent of the Cv2 domain is absent in the IgG chain, having been replaced by the hinge region (19). The content of β-sheet structure in the constant domains of IgY has been reported to be lower than that of IgG, and the flexibility between the Cv1 and Cv2 domains, corresponding to the hinge region of IgG, is less than that of IgG (20). Unlike IgG, IgY has two additional Cys residues, Cys331 and Cys 338, in the Cv2-Cv3 junction, which likely participate in the inter-v chain disulfide linkages (19).

Both IgG and IgY are known to contain Asn linked oligosaccharides. However, the structures of oligosaccharides in IgY differ from those of any mammalian IgG, containing unusual monoglucosylated oligomannose-type oligosaccharides with $Glc_1Man_{7-9}GlcNAc_2$ structure (21,22).

Furthermore, the isoelectric point of IgY is lower than that of IgG (23), and it has been suggested that IgY is a more hydrophobic molecule than IgG (24). It also does not have the ability to precipitate multivalent antigens, except at high salt concentrations (25), possibly due to the steric hindrance caused by the closely aligned Fab arms. High salt concentrations may serve to release the Fab arms, thereby permitting precipitation (19).

16.2.2.2 Physicochemical Properties of IgY

IgY and IgG display differences in their stability when subjected to pH, heat, and proteolytic enzymes. Although the stability of both immunoglobulins is similar when subjected to alkaline conditions, IgY demonstrated much less stability than rabbit IgG to acid denaturation. Shimizu et al. (20,26) found that the activity of IgY was decreased by incubating at pH 3.5 or lower, and completely lost at pH 3. The rabbit IgG antibodies did not demonstrate

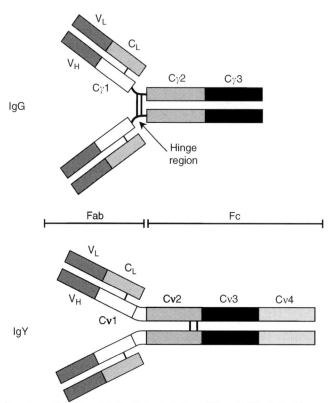

Figure 16.1 Structure of IgG and IgY. (Adapted from Warr, G.W., K.E. Magor, D.A. Higgins, *Immunol. Today* 16:392–398, 1995.)

a loss of activity until the pH was decreased to 2, and even then some activity still remained. Similar results were also observed by Hatta et al. (27), using IgY produced against human rotavirus. The denaturing effects of acidic conditions on IgY were also found to be reduced by the addition of sucrose (28) or sorbitol (29). IgY is also more sensitive to heating than the rabbit IgG. Shimizu et al. (20) found that the activity of IgY was decreased by heating for 15 minutes at 70°C or higher, whereas the activity of the IgG did not decrease until 75–80°C or higher. The addition of stabilizers, such as sucrose, maltose, or glycerol to IgY preparations has been found to reduce the extent of thermal denaturation when heated at high temperatures (28,30). The addition of sugars may also increase IgY stability under various processing conditions, including high pressure (5000 kg/cm^2) (28) and freeze drying (31). It has also been observed that that IgY is more susceptible to digestion with proteolytic enzymes such as pepsin, trypsin, and chymotrypsin than IgG (32), but this will be discussed in more detail later.

Because immunoglobulins are large, complex molecules, the structural features causing the stability differences of the two immunoglobulins are unknown. Shimizu et al. (20) predicted that the lower content of β-structure in IgY may indicate that the conformation of IgY is more disordered, and therefore less stable, than mammalian IgG. The lack of a hinge region in IgY may be another factor affecting molecular stability. The lower flexibility of the Cv1 and Cv2 domains of IgY, as compared to the hinge region of IgG, may cause the rapid inactivation of the antibody by various treatments, because the flexibility of the hinge region is considered to influence the overall properties of immunoglobulin molecules (33).

16.2.3   Advantages of IgY

The production of polyclonal antibodies in chickens provides many advantages over the traditional method of producing antibodies in mammals (Table 16.1). The method of producing antibodies in hens is much less invasive, involving only the collection of eggs rather than blood, and is therefore less stressful to the animal (34). Animal care costs are also lower for chickens than for mammals such as rabbits (2). In contrast to mammalian serum, egg yolk contains only a single class of antibody, IgY, which can be easily purified from the yolk by precipitation techniques (35). The phylogenetic distance between chickens and mammals allows the production of antibodies in chickens against highly conserved mammalian proteins. This would not be possible in mammals, as demonstrated by Knecht et al. (36), who produced and compared antibodies against human dihydroorotate dehydrogenase in both rabbits and chickens. In contrast with mammalian systems, much less antigen is required to produce an efficient immune response in chickens (37), and sustained high titers reduce the need for frequent injections (35). Chicken antibodies will also recognize different epitopes than mammalian antibodies, giving access to a different antibody repertoire than with mammalian antibodies (7). Hens therefore provide a more hygienic, cost efficient, convenient, and plentiful source of antibodies, as compared to the traditional method of obtaining antibodies from mammalian serum (2,35).

In terms of safety, eggs are normal dietary components, so no risk of side effects from IgY consumption would be expected (2). In contrast to IgG, IgY does not activate mammalian complement or interact with mammalian Fc receptors that could mediate inflammatory response in the gastrointestinal tract (2). Even though IgY has been detected at low levels in the serum of calves and piglets after antibodies were orally administered, no absorption of intact antibodies has been shown in humans (2), so the risk of allergic reaction to purified IgY preparations would be expected to be low.

Nakai et al. (38) estimated that the productivity of antibodies in hens as nearly 18 times greater than that in rabbits based on the weight of antibody produced per head, which means that the use of hens can effect an overall reduction in animal use for the same level of antibody production (2). Because of the high yolk IgY concentrations, over 100 mg of IgY can be obtained from one egg (39). A laying hen produces approximately 20 eggs per month, therefore over 2 g of IgY per month may be obtained from a single chicken (7). Once purified, IgY has been found to be stable for years at 4°C (40). The large scale feeding, and automated collection and separation of eggs is a common practice (41), as is

Table 16.1

Comparison of the production and characteristics of IgG and IgY. (Based on Schade, R., C. Pfister, R. Halatsch, P. Henklein. *ATLA* 19:403–419, 1991.)

	Mammals (IgG)	Chicken (IgY)
Source of antibody	Blood serum	Egg yolk
Antibody sampling	Bleeding (invasive)	Egg collection (noninvasive)
Antibody amount	200 mg/bleed (40 ml blood)	50–100 mg/egg (5–7 eggs/week)
Amount of antibody per year	1400 mg	40 000 mg
Amount of specific antibody	~5%	2–10%
Protein A/G binding	Yes	No
Interference with rheumatoid factors	Yes	No
Activation of mammalian complement	Yes	No

vaccination of chicken flocks to control infections (42), making the large scale production and harvesting of antibodies in chicken eggs for therapeutic purposes quite feasible.

16.3   PASSIVE IMMUNIZATION WITH IGY

16.3.1   Immunotherapeutic Applications of IgY

Antibiotics have been widely used to treat infections. This method of treatment is being threatened by the growing numbers of immunosuppressed individuals, the emergence of new pathogens (including those which do not respond to antibiotics, such as viral pathogens), and the increasing resistance of organisms to antimicrobial drugs (43,44). This has prompted much research into the administration of specific antibodies as an alternative to antibiotics and antimicrobial chemotherapy to treat infections. Passive immunization also offers benefits in applications for which conventional vaccines may be less suitable, including providing immediate protection, controlled duration of activity, as well as selectivity of target organism (45).

The potential applications of IgY for the prevention and treatment of infections caused by pathogenic bacteria and viruses have been studied at length (Table 16.2).

16.3.1.1   Rotavirus

Human rotavirus (HRV) has been identified as the major causative agent of acute infantile gastroenteritis (46), infecting up to 90% of children under the age of three, and resulting in more than 600 000 deaths annually (47). Human rotavirus causes a shortening and atrophy of the villi of the small intestine (46), resulting in decreased water absorption, leading to severe diarrhea and vomiting and eventually death due to dehydration (48,49). Yolken et al. (3) found that orally administered antibodies isolated from the eggs of chickens immunized with three different serotypes of rotavirus (mouse, human, and monkey) were capable of preventing rotavirus induced diarrhea in mice infected with murine rotavirus, whereas IgY isolated from the eggs of unimmunized chickens failed to prevent rotavirus infection. Using an HRV infection model in suckling mice, Hatta et al. (27) reported that antiHRV IgY decreased the incidence of rotavirus induced diarrhea in mice, both when administered before and after HRV challenge, suggesting its use for both therapeutic and prophylactic applications. Similarly, Ebina (50) also observed the prevention of HRV induced symptoms in mice using antiHRV IgY. The production of specific IgY against recombinant HRV coat protein VP8*, a cleavage product of the rotavirus spike protein VP4 has been reported (51). VP4 has been implicated in several important functions, including cell attachment and penetration, hemagglutination, neutralization, and virulence (52–54). VP8 has been found to play a significant role in viral infectivity and neutralization of the virus. Kovacs-Nolan et al. (51) immunized chickens with recombinant VP8*, and found that the resulting antiVP8* IgY exhibited significant neutralizing activity against the Wa strain of HRV in vitro, indicating that antiVP8* IgY may be used for the prevention and treatment of HRV infection. The oral administration of antiHRV egg yolk to children infected with HRV resulted in a modest improvement in symptoms, but further studies are required (55).

Neonatal calf diarrhea caused by bovine rotavirus (BRV) is a significant cause of mortality in cattle (56). The passive protection of calves against BRV infection, using antiBRV IgY, has also been demonstrated (57).

16.3.1.2   Escherichia coli

Diarrhea due to enterotoxigenic Escherichia coli (ETEC) is a major health problem in humans and animals. ETEC is the most common cause of enteric colibacillosis encountered

Table 16.2

Specific IgY against various bacterial and viral antigens.

Pathogen	Effect of IgY	Ref.
Rotavirus	Prevented murine rotavirus in mice	3
	Prevented HRV-induced gastroenteritis in mice	27, 50
	Prevented HRV infection *in vitro*, using IgY against recombinant subunit protein VP8	51
	Protected calves from BRV-induced diarrhea	57
	Improved HRV-associated symptoms in children	55
Escherichia coli	Prevented K88+, K99+, 987P+ ETEC infection in neonatal piglets	63, 65
	Inhibited adhesion of ETEC K88 to piglet intestinal mucosa, *in vitro*	64
	Inhibited attachment of F18+ *E. coli* to porcine intestinal mucosa *in vitro*, and reduced diarrheal cases in infected piglets	66
	Protected neonatal calves from fatal enteric colibacillosis	67
	Prevented diarrhea in rabbits challenged with ETEC	68
	Inhibited growth of *E. coli* and colonization of cells by EPEC, using IgY against BfpA	69
	Protection of chickens against APEC-associated respiratory disease	70
Salmonella spp.	Inhibited SE and ST growth *in vitro*	72
	Inhibited adhesion of SE to human intestinal cells *in vitro*	73
	Protected mice challenged with SE or ST from experimental salmonellosis, using IgY against OMP, LPS and Fla	75, 81
	Protected mice challenged with SE from experimental salmonellosis, using IgY against SEF14	82
	Prevented fatal salmonellosis in neonatal calves exposed to ST or S. Dublin	83
Pseudomonas aeruginosa	Prevented pathogensis of *P. aeruginosa in vitro*	85
	Inhibited *P. aeruginosa* adhesion to epithelial cells *in vitro*	86
	Prolonged time before *P. aeruginosa* re-colonization, and prevented chronic infection in CF patients	84, 87
	Anti-*P. aeruginosa* IgY stable in saliva for 8 hrs	88
Streptococcus mutans	Reduced caries development in rats	91
	Prevented *S. mutans* adhesion *in vitro*, and *in vivo* in humans	93
	Prevented *S. mutans* accumulation and reduction of caries in rats using IgY against *S. mutans* GBP	94
	Reduced caries incidence in desalivated rats, using IgY against Gtf	89
Helicobacter pylori	Inhibited growth of *H. pylori* and reduced binding to human gastric cells, *in vitro*	97, 98
	Decreased gastric mucosal injury and inflammation in a gerbil model of *H. pylori*-induced gastritis	97, 98
Bovine coronavirus	Protected neonatal calves from BCV-induced diarrhea	101
Yersinia ruckeri	Protected rainbow trout against *Y. ruckeri* infection	103

(Continued)

Table 16.2 (*Continued*)

Pathogen	Effect of IgY	Ref.
Edwardsiella tarda	Prevented Edwardsiellosis in Japanese eels infected with *E. tarda*	104
Staphylococcus aureus	Inhibited production of *Staphylococcus aureus* enterotoxin-A	85
IBDV	Protected chicks from infectious bursal disease virus	106
PEDV	Protected piglets against porcine epidemic diarrhea virus	107
Cryptosporidium	Prevented cryptosporidiosis in mice	108

in children in developing countries and travelers to these countries, accounting for one billion diarrheal episodes annually (58). It is also a problem in neonatal calves (59) and piglets (60), causing significant economic losses to the pig industry from both mortality and reduced growth rates, killing 1.5–2.0% of weaned pigs (61).

The strains of ETEC which are associated with intestinal colonization and cause severe diarrhea are the K88, K99, and 987P fimbrial adhesins (62). The production of IgY against these fimbrial antigens has been described and was found to inhibit the binding of *E. coli* K88+, K99+, and 987P+ strains to porcine epithelial cells (63) and to porcine intestinal mucus (64), *in vitro*. The antifimbrial IgY was also found to protect piglets against infection with each of the three strains of *E. coli* in a dose dependent manner in passive immunization trials (63). In another feeding study, 21 day old pigs were challenged with ETEC. AntiK88 IgY was administered to the piglets in milk three times a day for 2 days. Control piglets developed severe diarrhea within 12 hrs and 30% of the pigs died. In contrast, the pigs that were fed IgY exhibited no signs of diarrhea 24 or 48 hrs after treatment (65). Imberechts et al. (66) produced IgY against *E.coli* F18ac fimbriae, and found that the IgY inhibited attachment of F18ac+ *E. coli* to porcine intestinal mucosa *in vitro*, and decreased incidences of diarrhea and death in animals infected with F18ac+ *E. coli*.

The passive protective effect of antiETEC IgY against fatal enteric colibacillosis in neonatal calves has also been examined (67). Calves fed milk containing IgY had only transient diarrhea, 100% survival and good body weight gain. O'Farrelly et al. (68) reported the use of IgY for the prevention of diarrhea in rabbits through the oral administration of antiETEC IgY. The use of IgY against enteropathogenic *E. coli* (EPEC), another major cause of childhood diarrhea, has also been examined. de Almeida et al. (69) produced IgY against a recombinant EPEC protein (BfpA), one of the factors required for EPEC pathogenesis, and found that the antiBfpA antibodies were capable of inhibiting the colonization of cells by EPEC, as well as inhibiting *E. coli* growth *in vitro*. Finally, IgY has also been suggested for the passive protection of chickens against respiratory disease caused by avian pathogenic *E. coli* (APEC) (70).

Because anti*E. coli* IgY has been shown to inhibit binding of *E. coli* both *in vitro* and *in vivo*, the clinical application of IgY for the passive protection against *E. coli* related diarrhea in humans is now being examined.

16.3.1.3   *Salmonella species*

Salmonella species are a significant cause of food poisoning, and it is estimated that 2 to 4 million cases of salmonellosis occur in the U.S. annually (71). Symptoms include fever, abdominal pain, headache, malaise, lethargy, skin rash, constipation, and changes in mental state. However, infants, the elderly, and immunocompromised individuals may develop more severe symptoms. In these cases, the infection may spread from the intestines to the blood stream and to other sites in the body, and can cause death. *Salmonella enteritidis* (SE) and *Salmonella typhimurium* (ST), in particular, are the major agents of salmonellosis (71).

Lee et al. (72) found that IgY against SE and ST was capable of inhibiting the growth of SE and ST *in vitro* when added to culture medium. They also suggested that the IgY bound to antigens expressed on the *Salmonella* surface, resulting in structural alterations on the bacterial cell. AntiSE IgY was also found to inhibit the adhesion of SE to human intestinal cells *in vitro* (73).

Salmonella has several surface components which are virulence related, including outer membrane protein (OMP) (74,75), lipopolysaccharides (LPS) (76,77), flagella (Fla), and in some strains fimbrial antigens (78,79). OMP is involved in pathogenicity determination (80), and has been used successfully for vaccine applications in both active and passive immunization studies (74,75). Lipopolysaccharides have also been shown to be highly immunogenic, producing large amounts of LPS specific IgY, and have demonstrated potential application for the inhibition of *Salmonella* adhesion (76,77). The ability of antiOMP, antiLPS or antiflagella (Fla) IgY to passively protect mice against experimental salmonellosis was examined. In mice challenged with SE, administration of specific IgY resulted in a survival rate of 80% (OMP specific IgY), 47% (LPS specific IgY), and 60% (Fla specific IgY), compared to only 20% in control mice. In the case of ST, the survival rate was 40%, 30%, and 20% using OMP, LPS or Fla specific IgY, while none of the control mice survived (81). A novel fimbrial antigen, SEF14, produced primarily by SE and *S. dublin* strains, has been described (82). Mice challenged with SE and treated with antiSEF14 IgY had a survival rate of 77.8%, compared to a 32% survival rate in the control mice.

Salmonella infection in calves is also a worldwide problem, with ST and *S. dublin* accounting for most salmonellosis within the first 2 weeks after birth. The passive protective effect of IgY was investigated by orally inoculating calves with SE or *S. dublin*, and administering IgY against SE or *S. dublin* orally 3 times a day for 7 to 10 days after challenge. All control calves died within 7–10 days. Calves given a low titer dose of IgY displayed 60–70% mortality, and calves treated with a high titer dose of IgY had only fever and diarrhea (83).

16.3.1.4 *Pseudomonas aeruginosa*

Cystic fibrosis (CF) is the most common fatal genetic disease in Caucasians. Colonization of the airways of CF patients by *Pseudomonas aeruginosa* (PA) is the main cause of morbidity and mortality, and once a chronic infection has been established it is very difficult to eliminate, even with the use of antibiotics (84).

It has been shown that specific IgY is capable of preventing the pathogenesis of PA (85). Carlander et al. (86) found that IgY raised against PA was capable of inhibiting adhesion of PA to epithelial cells *in vitro*, but did not inhibit bacterial growth, suggesting that specific IgY might be capable of interfering with the bacterial infection process and prevent colonization in CF patients.

In vivo studies with CF patients have demonstrated that antiPA IgY, when administered orally to CF patients on a continuous basis in the form of a mouth rinse, is capable of preventing, or prolonging the time before, PA recolonization (84,87). Kollberg et al. (84) found that the antiPA IgY prevented recolonization in 40% of CF patients, compared to only 14% with no recolonization in the control (untreated) group. As well, none of the IgY treated patients became chronically colonized with PA, in contrast to recolonization in 24% of the control patients. These results indicate that antiPA IgY may reduce the need for antibiotic treatment of CF patients.

The stability of antiPA IgY in saliva from healthy individuals, following a mouth rinse with an aqueous IgY solution, was examined (88). Antibody activity in the saliva remained after 8 hours. After 24 hours, the antibody activity had decreased significantly, but was still detectable in some subjects, suggesting the use of specific IgY for the treatment of various local infections, such as the common cold, and tonsillitis.

16.3.1.5   Streptococcus mutans

The role of *Streptococcus mutans* in the development of dental caries is well recognized (89). The molecular pathogenesis of *S. mutans* associated dental caries involves a series of binding events that eventually lead to the accumulation of sufficient numbers of these cariogenic bacteria to cause disease (90). IgY against *S. mutans* MT8148 or cell associated glucosyltransferases (CA-Gtf), one of the major virulence factors in the pathogenesis of dental caries (89), was prepared and tested against dental caries (91,92). Consumption of a cariogenic diet which contained more than 2% IgY containing yolk powder resulted in significantly lower caries scores (91), and effective passive protection against the colonization of *S. mutans* in the oral cavity, in a rat model of dental caries. It has also been reported that mouth rinse containing IgY specific to *S. mutans* was effective in preventing the dental plaque of humans *in vitro* and *in vivo* (93). Krüger et al. (89) found that the incidence of caries in desalivated rats fed antiCA-Gtf IgY was decreased in severity and number, indicating that IgY may be effective as prophylaxis for high caries risk patients.

Glucan-binding proteins (GBPs) are believed to be involved in *S. mutans* biofilm development, and antibodies against GBP-B appear to have the potential to modulate infection and disease caused by *S. mutans* (94). Smith et al. (94) found that rats treated with antiGBP-B IgY displayed a decrease in *S. mutans* accumulation, as well as a decrease in the overall amount of dental caries, as compared to control rats.

16.3.1.6   Helicobacter pylori

Helicobacter pylori is the most common cause of gastritis and gastric ulcers, and has been implicated in the development of gastric carcinomas. Around 25–50% of the population carry *H. pylori*, and this number is higher in developing countries (95). Antibiotic therapy is often employed to treat *H. pylori* infection, but this therapy is often undesirable due to increasing medical costs, low eradication rate, and the development of antibiotic resistant strains (29,96).

Anti*H. pylori* IgY was shown by Shin et al. (97) to significantly inhibit the growth of *H. pylori, in vitro*, and inhibit urease activity (a critical virulence factor), as well as reduce the binding of *H. pylori* to human gastric cells. The oral administration of specific IgY decreased gastric mucosal injury and inflammation *in vivo*, using a gerbil model of *H. pylori* induced gastritis (97,98), suggesting the use of anti*H. pylori* antibodies for the treatment of *H. pylori* associated gastric diseases.

Immunodominant *H. pylori* proteins were also identified, and used to produce highly specific anti*H. pylori* and antiurease IgY, to address concerns that antibodies produced against whole cell *H. pylori* might cross react with other bacteria, including normal human flora (96,99).

16.3.1.7   Coronavirus

Bovine coronavirus (BCV) is an important cause of neonatal calf diarrhea and acute diarrhea in adult cattle (100). Ikemori et al. (101) examined the protective effect of antiBCV antibodies in egg yolk in calves challenged with BCV. They found that control calves which received no antibodies experienced severe diarrhea and died within 6 days after infection. Calves fed milk containing egg yolk, however, all survived and had positive weight gain. The results indicate that orally administered egg yolk antibodies are capable of passively protecting calves against BCV infection.

16.3.1.8   Fish Pathogens

Yersinia ruckeri is the causative agent of enteric redmouth disease, a systemic bacterial septicemia, in salmonid fish (102). The persistence of *Y. ruckeri* in carrier fish for long periods and the shedding of bacteria in feces can present a continuing source of infection. The clearance of *Y. ruckeri* from carrier fish by orally administered anti*Y. ruckeri* antibody would provide a cost effective alternative to slaughtering fish which pose a health risk. The passive immunization of rainbow trout (*Oncorhynchus mykiss*) against *Y. ruckeri* infection has been examined (103). Groups of rainbow trout fed anti*Y. ruckeri* IgY prior to being challenged with *Y. ruckeri* had lower mortalities compared with fish receiving normal feed. The group fed IgY also appeared to have a lower number of infected fish, based on organ and intestine culture. The number of IgY fed fish carrying *Y. ruckeri* in intestine samples also appeared to be lower than the normal feed controls, regardless of whether the feeding was given before or after the challenge. The oral administration of specific IgY against fish pathogens could provide an alternative to the use of antibiotics and chemotherapy for prevention of bacterial diseases in fish farms, and would provide a novel approach for the prevention of viral infections in fish, as no medicine has been reported to be effective.

  Edwardsiella tarda, another important fish pathogen, is spread by infection through the intestinal mucosa. Edwardsiellosis in Japanese eels is a serious problem for the eel farming industry, and egg yolk antibodies have been investigated for the prevention of this infectious disease, as treatment with antibiotics has been found to promote the growth of bacteria resistant strains (104). Eels were challenged with *E. tarda*, and anti*E. tarda* IgY was orally administered. Infected eels died within 15 days, whereas the eels given IgY survived without any symptoms of *E. tarda* infection (104,105).

16.3.1.9   Others

In addition, specific IgY has been shown to be effective at preventing and treating several other pathogens. Sugita-Konishi et al. (85) found that specific IgY was capable of inhibiting the production of *Staphylococcus aureus* enterotoxin-A *in vitro*. Its use has also been suggested for the passive protection of chicks against infectious bursal disease virus (IBDV) (106), for the protection against porcine epidemic virus (PEDV) in piglets (107), and for the prevention of cryptosporidiosis due to *Cryptosporidium* infection (108). Finally, IgY against *Bordetella bronchiseptica*, *Pasteurella multocida*, and *Actinobacillus pleuropneumoniae*, causative agents of swine respiratory diseases, has been proposed as an alternative method to control infectious respiratory diseases in pigs (109,110).

16.4   ANTIBODY STABILITY

16.4.1   Physiology of the Gastrointestinal Tract

The human gastrointestinal tract consists of the stomach, small intestine (which is divided into the duodenum, jejunum, and ileum), and the large intestine (111). Ingested proteins are subjected to denaturation by the acidic pH of the stomach, as well as to degradation by the proteolytic enzymes present in the stomach and small intestine (112).

  The pH of the stomach ranges from 1 to 3.5 due to secretion of hydrochloric acid by the parietal cells (112–114). In neonates, however, the gastric fluid has a pH near neutral, which rapidly decreases to a pH below 3 within the first few days after birth (2). In both infants and adults, the pH of the gastric fluid will rapidly rise to about 4 or 5 soon after food reaches the stomach (111), but will return to low pH, occurring much more slowly in infants than in adults (115). It is in the stomach that proteins are subjected to digestion by pepsin, which is secreted by chief cells (116). Pepsin is optimally active at pH 1.8 to 3.5 (111), but will start to lose activity around pH 4 and is completely inactivated in alkaline conditions (111,117). The digestive activity of pepsin is terminated when the gastric contents are mixed with alkaline pancreatic juice in the small intestine (116). The extent of gastric protein digestion is determined by the physical state of the ingested protein, the activity of the pepsin in the gastric juice, and the length of time it stays in the stomach (111). The transit time in the stomach has been found to vary considerably, and has been reported to range from 0.5 to 4.5 (112,118), with a median time of 1 to 1.5 hours (114,118).

Trypsin and chymotrypsin continue digestion in the small intestine, where the pH ranges from around pH 6 to 7.4 (113,114). Transit through the small intestine has been reported to take from 1 to 4 hours (119), but may be as high as 7.5 hours (114). The pH rises to around 7.5 to 8 in the large intestine, where the transit time can be as long as 17 hours (112,114).

Along with the duration of exposure in each of the segments of the gastrointestinal tract, the ratio of enzyme to protein will also determine to what extent a protein is degraded (112).

16.4.2 Degradation and Stability of IgY

Several investigations have been carried out *in vitro* to study the effects of gastrointestinal enzymes and pH on the structure and activity of IgY, in order to predict its fate when administered orally (112). Antibodies were tested with pepsin at pH 2 to 4, and trypsin and chymotrypsin at a pH around neutral, simulating passage through the stomach and small intestine.

IgY has been found to be relatively resistant to trypsin and chymotrypsin digestion, but sensitive to pepsin digestion (120). It has, however, been found to be more susceptible to pepsin, trypsin, and chymotrypsin digestion than IgG (26,32). Using IgY raised against HRV, Hatta et al. (27) demonstrated that almost all of the IgY activity was lost following digestion with pepsin, but activity remained even after 8 hours incubation with trypsin or chymotrypsin. They found that digestion of IgY with pepsin at pH 2 resulted in complete hydrolysis of the antibody molecule, leaving only small peptides, however IgY digested with pepsin at pH 4 retained some of its activity, with heavy (H) and light (L) chains still present, along with smaller peptides. Shimizu et al. (120) also obtained similar results following the digestion of IgY against *Escherichia coli* with gastrointestinal enzymes.

In vivo results have also yielded similar results. Ikemori et al. (121) described the feeding of anti*E.coli* IgY to calves, and found that the activity of the antibodies was decreased significantly following passage through the stomach. Similarly, the feeding of specific IgY, also against *E. coli*, was carried out by Yokoyama et al. (122) using pigs. The morphology and physiology of the gastrointestinal tract of pigs has been found to be similar to that of humans, and therefore pigs represent a good animal model to study digestion in humans (123). Using piglets of varying ages, from 10 hours (neonatal) to 28 days old, it was observed that the IgY was degraded upon passage through the gastrointestinal tract, with some intact IgY detected in the intestine of neonatal pigs but to a lesser extent in the older pigs, possibly reflecting the increasing gastrointestinal maturity of the pigs.

Methods of encapsulating IgY, to provide resistance to the proteolytic and denaturing conditions of the stomach have been examined. Shimizu et al. (124) encapsulated IgY using lecithin or cholesterol liposomes, and found that the stability of the antibodies to *in vitro* gastric conditions was increased. Encapsulation of IgY in a water in oil in water (W/O/W) emulsion was examined by Shimizu and Nakane (125), who found that the antibody activity was significantly reduced by encapsulation alone, suggesting that denaturation of the IgY was occurring at the oil to water interface. Ikemori et al. (121) treated anti*E. coli* IgY with hydroxypropyl methylcellulose phthalate, an enteric coating substance used for drugs, and examined its activity following passage through the gastrointestinal tract of calves. They found that although the activity of the encapsulated IgY was decreased, it was not decreased to the same extent as the nonencapsulated antibodies. This research would indicate that the successful encapsulation of IgY would be an effective means for protecting IgY under gastric conditions.

16.5 BIOTECHNOLOGY OF AVIAN EGG ANTIBODIES

As demonstrated, the chicken is a useful animal for the production of specific antibodies for immunotherapeutic purposes. The many advantageous characteristics of IgY also make it ideal for use in diagnostic applications, where monoclonal antibodies have traditionally been used. Hybridomas can be cloned and grown in large quantity for indefinite periods of time, secreting high concentrations of monoclonal antibodies. The production of monoclonal IgY has also been examined, generated through the fusion of spleen cells from immunized chickens with chicken B cells, to produce monoclonal IgY-secreting hybridomas (126–131). These cells are capable of providing a consistent supply of antibody with a single and known specificity and homogeneous structure (1). Several researchers have described the production of single chain fragment variable region (scFv) monoclonal IgY via recombinant DNA technology (132–134), in order to improve upon the low levels of antibodies produced in the chicken hybridoma systems (133). Using RNA extracted from chicken hybridoma cells, Kim et al. (134) were able to express recombinant monoclonal IgY in *E. coli* and reported the production of 5–6 mg of IgY per litre of culture, suggesting that the production of monoclonal IgY on a large scale may be possible. Lillehoj and Sasaki (128) and Kim et al. (134) produced monoclonal IgY against *Eimeria* species, an intracellular parasite responsible for avian coccidiosis, in order to study the avian immune response to this parasite and to aid in vaccine development, as it was thought that monoclonal antibodies from mice would not adequately reflect the avian immune response due to the differences in antibody repertoire. Given that it is possible to produce antibodies against a vast array of antigens and epitopes using chickens, antibodies against any number of bacterial, viral, or biological antigens could be produced, suggesting the significant potential of avian antibodies for further use in immunodiagnostics and identification of disease markers, immunotherapy, and the treatment and prevention of disease, as well as affinity purification methods. Research is being continued to develop potential methods of production and application of egg yolk antibodies. Romito et al. (135) suggested the immunization of chickens with naked DNA, to elicit antigen specific IgY. Using DNA rather than a protein antigen could eliminate the protein expression and purification steps, and would allow the production of antibodies against pathogenic or toxic antigens. Not only are chickens useful for the production of specific IgY, but Mohammed et al. (136) demonstrated the deposition of recombinant human antibodies into the egg yolk of transgenic chickens, suggesting an extension of the production of specific IgY in eggs.

REFERENCES

1. Janeway, C.A., P. Travers. Structure of the antibody molecule and immunoglobulin genes. In: *Immunology: The immune system in health and disease.* Janeway, C.A., P. Travers, eds., London: Current Biology Ltd., 1996.

2. Carlander, D., H. Kollberg, P.-E. Wejaker, A. Larsson. Peroral immunotherapy with yolk antibodies for the prevention and treatment of enteric infections. *Immunol. Res.* 21:1–6, 2000.

3. Yolken, R.H., F. Leister, S.B. Wee, R. Miskuff, S. Vonderfecht. Antibodies to rotavirus in chickens' eggs: a potential source of antiviral immunoglobulins suitable for human consumption. *Pediatrics* 81:291–295, 1988.

4. Hammarström, L. Passive immunity against rotavirus in infants. *Acta Paediatr. Suppl.* 430:127–132, 1999.

5. Sim, J.S., H.H. Sunwoo, E.N. Lee. Ovoglobulin IgY. In: *Natural Food Antimicrobial Systems.* Naidu, A.S., ed., New York: CRC Press, 2000, pp 227–252.

6. Schade, R., C. Staak, C. Hendrikson, M. Erhard, H. Hugl, G. Koch, A. Larsson, W. Pollmann, M. van Regenmortel, E. Rijke, H. Spielmann, H. Steinbush, D. Straughan. The production of avian (egg yolk) antibodies: IgY. *ATLA* 24:925–934, 1996.

7. Carlander, D., J. Stalberg, A. Larsson. Chicken antibodies: a clinical chemistry perspective. *Ups. J. Med. Sci.* 104:179–190, 1999.

8. Sim, J.S., S. Nakai, W. Guenter. *Egg Nutrition and Biotechnology.* Oxon, UK: CAB Publishing, 1999.

9. Janson, A.K., C.I. Smith, L. Hammarström. Biological properties of yolk immunoglobulins. *Adv. Exp. Med. Biol.* 371:685–690, 1995.

10. Reynaud, C.-A., V. Anquez, J.-C. Weill. Somatic hyperconversion diversifies the single V_H gene of the chicken with a high incidence in the D region. *Cell* 59:171–183, 1989.

11. Reynaud, C.-A., A. Dahan, V. Anquez, J.-C. Weill. The chicken D locus and its contribution to the immunoglobulin heavy chain repertoire. *Eur. J. Immunol.* 21:2661–2670, 1991.

12. Pink, J.R.L., O. Vainio, A.-M. Rijnbeek. Clones of B lymphocytes in individual follicles of the bursa of Fabricius. *Eur. J. Immunol.* 15:83–87, 1985.

13. Sharma, J.M. The structure and function of the avian immune system. *Acta Vet. Hung.* 45:229–238, 1997.

14. Leslie, G.A., L.N. Martin. Studies on the secretory immunologic system of fowl, 3: serum and secretory IgA of the chicken. *J. Immunol.* 110:1–9, 1973.

15. Loeken, M.R., T.F. Roth. Analysis of maternal IgG subpopulations which are transported into the chicken oocyte. *Immunology* 49:21–28, 1983.

16. Tressler, R.L., T.F. Roth. IgG receptors on the embryonic chick yolk sac. *J. Biol. Chem.* 262:15406–15412, 1987.

17. Morrison, S.L., M.S. Mohammed, L.A. Wims, R. Trinh, R. Etches. Sequences in antibody molecules important for receptor-mediated transport into the chicken egg yolk. *Mol. Immunol.* 38:619–625, 2002.

18. Rose, M.E., E. Orlans, N. Buttress. Immunoglobulin classes in the hen's eggs: their segregation in yolk and white. *Eur. J. Immunol.* 4:521–523, 1974.

19. Warr, G.W., K.E. Magor, D.A. Higgins. IgY: clues to the origins of modern antibodies. *Immunol. Today* 16:392–398, 1995.

20. Shimizu, M., H. Nagashima, K. Sano, K. Hashimoto, M. Ozeki, K. Tsuda, H. Hatta. Molecular stability of chicken and rabbit immunoglobulin G. *Biosci. Biotech. Biochem.* 56:270–274, 1992.

21. Ohta, M., J. Hamako, S. Yamamoto, H. Hatta, M. Kim, T. Yamamoto, S. Oka, T. Mizuochi, F. Matsuura. Structure of asparagine-linked oligosaccharides from hen egg yolk antibody (IgY). Occurrence of unusual glucosylated oligo-mannose type oligosaccharides in a mature glycoprotein. *Glycoconj. J.* 8:400–413, 1991.

22. Matsuura, F., M. Ohta, K. Murakami, Y. Matsuki. Structures of asparagines linked oligosaccharides of immunoglobulins (IgY) isolated from egg-yolk of Japanese quail. *Glycoconj. J.* 10:202–213, 1993.

23. Polson, A., M.B. von Wechmar, G. Fazakerley. Antibodies to proteins from yolk of immunized hens. *Immunol. Commun.* 9:495–514, 1980.

24. Davalos-Pantoja, L., J.L. Ortega-Vinuesa, D. Bastos-Gonzalez, R. Hidalgo-Alvarez. A comparative study between the adsorption of IgY and IgG on latex particles. *J. Biomater. Sci. Polym. Ed.* 11:657–673, 2000.

25. Hersh, R.T., A.A. Benedict. Aggregation of chicken gamma-G immunoglobulin in 1.5 M sodium chloride solution. *Biochim. Biophys. Acta* 115:242–244, 1966.

26. Shimizu, M., H. Nagashima, K. Hashimoto. Comparative studies on molecular stability of immunoglobulin G from different species. *Comp. Biochem. Physiol.* 106B:255–261, 1993.

27. Hatta, H., K. Tsuda, S. Akachi, M. Kim, T. Yamamoto, T. Ebina. Oral passive immunization effect of anti-human rotavirus IgY and its behaviour against proteolytic enzymes. *Biosci. Biotech. Biochem.* 57:1077–1081, 1993.

28. Shimizu, M., H. Nagashima, K. Hashimoto, T. Suzuki. Egg yolk antibody (IgY) stability in aqueous solution with high sugar concentrations. *J. Food Sci.* 59:763–766, 1994.

29. Lee, K.A., S.K. Chang, Y.J. Lee, J.H. Lee, N.S. Koo. Acid stability of anti-Helicobacter pylori IgY in aqueous polyol solution. *J. Biochem. Mol. Biol.* 35:488–493, 2002.

30. Chang, H.M., R.F. Ou-Yang, Y.T. Chen, C.C. Chen. Productivity and some properties of immunoglobulin specific against *Streptococcus mutans* serotype c in chicken egg yolk (IgY). *J. Agric. Food Chem.* 47:61–66, 1999.

31. Jaradat, Z.W., R.R. Marquardt. Studies on the stability of chicken IgY in different sugars, complex carbohydrates and food materials. *Food Agric. Immunol.* 12:263–272, 2000.

32. Otani, H., K. Matsumoto, A. Saeki, A. Hosono. Comparative studies on properties of hen egg yolk IgY and rabbit serum IgG antibodies. *Lebensm. Wiss. Technol.* 24:152–158, 1991.

33. Pilz, I., E. Schwarz, W. Palm. Small-angle x-ray studies of the human immunoglobulin molecule. *Eur. J. Biochem.* 75:195–199, 1977.

34. Schade, R., C. Pfister, R. Halatsch, P. Henklein. Polyclonal IgY antibodies from chicken egg yolk: an alternative to the production of mammalian IgG type antibodies in rabbits. *ATLA* 19:403–419, 1991.

35. Gassmann, M., P. Thommes, T. Weiser, U. Hubscher. Efficient production of chicken egg yolk antibodies against a conserved mammalian protein. *FASEB J.* 4:2528–2532, 1990.

36. Knecht, W., R. Kohler, M. Minet, M. Loffler. Anti-peptide immunoglobulins from rabbit and chicken eggs recognize recombinant human dihydroorotate dehydrogenase and a 44-kDa protein from rat liver mitochondria. *Eur. J. Biochem.* 236:609–613, 1996.

37. Larsson, A., D. Carlander, M. Wilhelmsson. Antibody response in laying hens with small amounts of antigen. *Food Agr. Immunol.* 10:29–36, 1998.

38. Nakai, S., E. Li-Chan, K.V. Lo. Separation of immunoglobulin from egg yolk. In: *Egg Uses and Processing Technologies: New Developments*, Sim, J.S., S. Nakai, eds.,Wallingford, UK: CAB International, 1994, pp 94–105.

39. Akita, E.M., S. Nakai. Immunoglobulins from egg yolk: isolation and purification. *J. Food Sci.* 57:629–634, 1992.

40. Larsson, A., R. Balow, T.L. Lindahl, P. Forsberg. Chicken antibodies: taking advantage of evolution: a review. *Poult. Sci.* 72:1807–1812, 1993.

41. Cotterill, O.J., L.E. McBee. Egg breaking. In: *Egg Science and Technology*, 4th ed., Stadelman, W.J., O.J. Cotterill, eds., New York: The Haworth Press, Inc., 1995, pp 231–263.

42. Sharma, J.M. Introduction to poultry vaccines and immunity. *Adv. Vet. Med.* 41:481–494, 1999.

43. Casadevall, A., M.D. Scharff. Return to the past: the case for antibody-based therapies in infectious diseases. *Clin. Infect. Dis.* 21:150–161, 1995.

44. Wierup, M. The control of microbial diseases in animals: alternatives to the use of antibodies. *Int. J. Antimicrob. Agents* 14:315–319, 2000.

45. Zeitlin, L., R.A. Cone, T.R. Moench, K.J. Whaley. Preventing infectious disease with passive immunization. *Microbes Infect.* 2:701–708, 2000.

46. Kapikian, A.Z., R.M. Chanock. Rotaviruses. In: *Fields Virology*, 3rd ed., Fields, B.N., D.N. Knipe, P.M. Howley, R.M. Chanock, J.L. Melnick, T.P. Monath, B. Rolzman, S.E. Strauss, eds., New York: Raven Press, 1996, pp 1657–1708.

47. Clark, H.F., R.I. Glass, P.A. Offit. Rotavirus vaccines. In: *Vaccines*, 3rd ed., S.A. Plotkin, W.A. Orentstein, eds., Philadelphia: W.B. Saunders Company, 1999, pp 987–1005.

48. Ludert, J.E., A.A. Krishnaney, J.W. Burns, P.T. Vo, H.B. Greenberg. Cleavage of rotavirus VP4 *in vivo*. *J. Gen. Virol.* 77:391–395, 1996.

49. Hochwald, C., L. Kivela. Rotavirus vaccine, live, oral, tetravalent (RotaShield). *Pediatr. Nurs.* 25:203–204, 1999.

50. Ebina, T. Prophylaxis of rotavirus gastroenteritis using immunoglobulin. *Arch. Virol. Suppl.* 12:217–223, 1996.

51. Kovacs-Nolan, J., E. Sasaki, D. Yoo, Y. Mine. Cloning and expression of human rotavirus spike protein, VP8*, in *Escherichia coli. Biochem. Biophys. Res. Commun.* 282:1183–1188, 2001.

52. Mackow, E.R., R.D. Shaw, S.M. Matsui, P.T. Vo, M.-N. Dang, HB Greenberg. The rhesus rotavirus gene encoding protein VP3: location of amino acids involved in homologous and heterologous rotavirus neutralization and identification of a putative fusion region. *Proc. Natl. Acad. Sci. USA* 85:645–649, 1988.

53. Both, G.W., A.R. Bellamy, D.B. Mitchell. Rotavirus protein structure and function. In: *Rotaviruses*. Ramig, R.F., ed., New York: Springer-Verlag, 1994, pp 67–106.

54. Desselberger, U., M.A. McCrae. The rotavirus genome. In: *Rotaviruses*. R.F. Ramig, ed., New York: Springer-Verlag, 1994, pp 31–66.

55. Sarker, S.A., T.H. Casswall, L.R. Juneja, E. Hoq, E. Hossain, G.J. Fuchs, L. Hammarström. Randomized, placebo-controlled, clinical trial of hyperimmunized chicken egg yolk immunoglobulin in children with rotavirus diarrhea. *J. Pediatr. Gastroenterol. Nutr.* 32:19–25, 2001.

56. Lee, J., L.A. Babiuk, R. Harland, E. Gibbons, Y. Elazhary, D. Yoo. Immunological response to recombinant VP8* subunit protein of bovine rotavirus in pregnant cattle. *J. Gen. Virol.* 76:2477–2483, 1995.

57. Kuroki, M., Y. Ohta, Y. Ikemori, R.C. Peralta, H. Yokoyama, Y. Kodama. Passive protection against bovine rotavirus in calves by specific immunoglobulins from chicken egg yolk. *Arch. Virol.* 138:143–148, 1994.

58. Sack, R.B. Antimicrobial prophylaxis of travellers' diarrhea: a selected summary. *Rev. Infect. Dis.* 8:S160–S166, 1986.

59. Moon, H.W., S.C. Whipp, S.M. Skartvedt. Etiologic diagnosis of diarrheal diseases of calves: frequency and methods for detecting enterotoxic and K99 antigen produced by *Escherichia coli. Am. J. Vet. Res.* 37:1025–1029, 1976.

60. Morris, J.A., W.J. Sojka. *Escherichia coli* as a pathogen in animals. In: *The Virulence of Escherichia* coli. Sussman, M., ed., London: Academic Press, Inc., 1985, pp 47–77.

61. Hampson, D.J. Postweaning *Escherichia coli* diarrhoea in pigs. In: *Escherichia coli in Domestic Animals and Humans.* Gyles, C.L., ed., Oxon, UK: CAB International, 1994, pp 171–191.

62. Parry, S.H., D.M. Rooke. Adhesins and colonization factors of *Escherichia coli.* In: *The Virulence of Escherichia coli.* Sussman, M., ed., London: Academic Press, Inc., 1985, pp 79–159.

63. Yokoyama, H., R.C. Peralta, R. Diaz, S. Sendo, Y. Ikemori, Y. Kodama. Passive protective effect of chicken egg yolk immunoglobulins against experimental enterotoxigenic *Escherichia coli* infection in neonatal piglets. *Infect. Immun.* 60:998–1007, 1992.

64. Jin, L.Z., K. Samuel, K. Baidoo, R.R. Marquardt, A.A. Frohlich. *In vitro* inhibition of adhesion of enterotoxigenic *Escherichia coli* K88 to piglet intestinal mucus by egg yolk antibodies. *FEMS Immunol. Med. Microbiol.* 21:313–321, 1998.

65. Marquardt, R.R., L.Z. Jin, J.W. Kim, L. Fang, A.A. Frohlich, S.K. Baidoo. Passive protective effect of egg yolk antibodies against enterotoxigenic *Escherichia coli* K88+ infection in neonatal and early weaned piglets. *FEMS Immunol. Med. Microbiol.* 23:283–288, 1999.

66. Imberechts, H., P. Deprez, E. Van Driessche, P. Pohl. Chicken egg yolk antibodies against F18ab fimbriae of *Escherichia coli* inhibit shedding of F18 positive *E. coli* by experimentally infected pigs. *Vet. Microbiol.* 54:329–341, 1997.

67. Ikemori, Y., M. Kuroki, R.C. Peralta, H. Yokoyama, Y. Kodama. Protection of neonatal calves against fatal enteric colibacillosis by administration of egg yolk powder from hens immunized with K99-piliated enterotoxigenic *Escherichia coli. Am. Vet. Res.* 53:2005–2008, 1992.

68. O'Farrelly, C., D. Branton, C.A. Wanke. Oral ingestion of egg yolk immunoglobulin from hens immunized with an enterotoxigenic *Escherichia coli* strain prevents diarrhea in rabbits challenged with the same strain. *Infect. Immun.* 60:2593–2597, 1992.

69. de Almeida, C.M., V.M. Quintana-Flores, E. Medina-Acosta, A. Schriefer, M. Barral-Netto, W Dias da Silva. Egg yolk anti-BfpA antibodies as a tool for recognizing and identifying enteropathogenic *Escherichia coli*. *Scand. J. Immunol.* 57:573–582, 2003.

70. Kariyawasam, S., B.N. Wilkie, C.L. Gyles. Resistance of broiler chickens to *Escherichia coli* respiratory tract infection induced by passively transferred egg-yolk antibodies. *Vet. Microbiol.* 98:273–284, 2004.

71. Bell, C., A. Kyriakides. *Salmonella: A Practical Approach to the Organism and its Control in Foods*. New York: Blackie Academic & Professional, 1998.

72. Lee, E.N., H.H. Sunwoo, K. Menninen, J.S. Sim. *In vitro* studies of chicken egg yolk antibodies (IgY) against *Salmonella enteritidis* and *salmonella typhimurium*. *Poult. Sci.* 81:632–641, 2002.

73. Sugita-Konishi, Y., M. Ogawa, S. Arai, S. Kumagai, S. Igimi, M. Shimizu. Blockade of *Salmonella enteritidis* passage across the basolateral barriers of human intestinal epithelial cells by specific antibody. *Microbiol. Immunol.* 44:473–479, 2000.

74. Isibasi, A., V. Ortiz, M. Vargas, J. Paniagua, C. Gonzales, J. Moreno, J. Kumate. Protection against *Salmonella typhi* infection in mice after immunization with outer membrane proteins isolated from *Salmonella typhi* 9, 12, d, Vi. *Infect. Immun.* 56:2953–2959, 1988.

75. Udhayakumar, V., V.R. Muthukkaruppan. Protective immunity induced by outer membrane proteins of *Salmonella typhimurium* in mice. *Infect. Immun.* 55:816–821, 1987.

76. Sunwoo, H.H., T. Nakano, T. Dixon, J.S. Sim. Immune responses in chicken against lipopoly-saccharides of *Escherichia coli* and *Salmonella typhimurium*. *Poultry Sci.* 75:342–345, 1996.

77. Mine, Y. Separation of *Salmonella enteritidis* from experimentally contaminated liquid eggs using a hen IgY immobilized immunomagnetic separation system. *J. Agric. Food Chem.* 45:3723–3727, 1997.

78. Thorns, C.J., M.G. Sojka, D. Chasey. Detection of a novel fimbrial structure on the surface of *Salmonella enteritidis* by using a monoclonal antibody. *J. Clin. Microbiol.* 28:2409–2414, 1990.

79. Thorns, C.J., M.G. Sojka, M. McLaren, M. Dibb-Fuller. Characterization of monoclonal antibodies against a fimbrial structure of *Salmonella enteritidis* and certain other serogroup D salmonellae and their application as serotyping reagents. *Res. Vet. Sci.* 53:300–308, 1992.

80. Galdiero, F., M.A. Tufano, M. Galdiero, S. Masiello, M.D. Rosa. Inflammatory effects of *Salmonella typhimurium* porins. *Infect. Immun.* 58:3183–3186, 1990.

81. Yokoyama, H., K. Umeda, R.C. Peralta, T. Hashi, F. Icatlo, M. Kuroki, Y. Ikemori, Y. Kodama. Oral passive immunization against experimental salmonellosis in mice using chicken egg yolk antibodies specific for *Salmonella enteritidis* and *S. typhimurium*. *Vaccine* 16:388–393, 1998.

82. Peralta, R.C., H. Yokoyama, Y. Ikemori, M. Kuroki, Y. Kodama. Passive immunization against experimental salmonellosis in mice by orally administered hen egg yolk antibodies specific for 14-kDa fimbriae of *Salmonella enteritidis*. *J. Med. Microbiol.* 41:29–35, 1994.

83. Yokoyama, H., R.C. Peralta, K. Umeda, T. Hashi, F.C. Icatlo, M. Kuroki, Y. Ikemori, Y. Kodama. Prevention of fatal salmonellosis in neonatal calves, using orally administered chicken egg yolk *Salmonella*-specific antibodies. *Am. J. Vet. Res.* 59:416–420, 1998.

84. Kollberg, H., D. Carlander, H. Olesen, P.-E. Wejaker, M. Johannesson, A. Larsson. Oral administration of specific yolk antibodies (IgY) may prevent *Pseudomonas aeruginosa* infections in patients with cystic fibrosis: a phase I feasibility study. *Pediatr. Pulmonol.* 35:433–440, 2003.

85. Sugita-Konishi, Y., K. Shibata, S.S. Yun, H.K. Yukiko, K. Yamaguchi, S. Kumagai. Immune functions of immunoglobulin Y isolated from egg yolk of hens immunized with various infectious bacteria. *Biosci. Biotech. Biochem.* 60:886–888, 1996.

86. Carlander, D.O., J. Sundstrom, A. Berglund, A. Larsson, B. Wretlind, H.O. Kollberg. Immunoglobulin Y (IgY): a new tool for the prophylaxis against *Pseudomonas aeruginosa* in cystic fibrosis patients. *Pediatr. Pulmonol.* 19:241 (Abstract), 1999.

87. Carlander, D., H. Kollberg, P.E. Wejaker, A. Larsson. Prevention of chronic *Pseudomonas aeruginosa* colonization by gargling with specific antibodies: a preliminary report.

In: *Egg Nutrition and Biotechnology.* Sim, J.S., S. Nakai, W. Guenter, eds., New York: CABI Publishing, 2000, pp 371–374.

88. Carlander, D., H. Kollberg, A. Larsson. Retention of specific yolk IgY in the human oral cavity. *BioDrugs* 16:433–437, 2002.

89. Krüger, C., S.K. Pearson, Y. Kodama, A. Vacca Smith, W.H. Bowen, L. Hammarström. The effects of egg-derived antibodies to glucosyltransferases on dental caries in rats. *Caries. Res.* 38:9–14, 2004.

90. Hamada, S., H.D. Slade. Biology, immunology, and cariogenicity of *Streptococcus mutans.* *Microbiol. Rev.* 44:331–384, 1980.

91. Otake, S., Y. Nishihara, M. Makimura, H. Hatta, M. Kim, T. Yamamoto, M. Hirasawa. Protection of rats against dental caries by passive immunization with hen egg yolk antibody (IgY). *J. Dent. Res.* 70:162–166, 1991.

92. Hamada, S., T. Horikoshi, T. Minami, S. Kawabata, J. Hiraoka, T. Fujiwara, T. Ooshima. Oral passive immunization against dental caries in rats by use of hen egg yolk antibodies specific for cell-associated glucosyltransferase of *Streptococcus mutans.* *Infect. Immun.* 59:4161–4167, 1991.

93. Hatta, H., K. Tsuda, M. Ozeki, M. Kim, T. Yamamoto, S. Otake, M. Hirosawa, J. Katz, N.K. Childers, S.M. Michalek. Passive immunization against dental plaque formation in humans: effect of a mouth rinse containing egg yolk antibodies (IgY) specific to *Streptococcus mutans.* *Caries. Res.* 31:268–274, 1997.

94. Smith, D.J., W.F. King, R. Godiska. Passive transfer of immunoglobulin Y to *Streptococcus mutans* glucan binding protein B can confer protection against experimental dental caries. *Infect. Immun.* 69:3135–3142, 2001.

95. Dunn, B.E., H. Cohen, M.J. Blaser. *Helicobacter pylori.* *Clin. Microbiol. Rev.* 10:720-741, 1997.

96. Shin, J.-H., S.-W. Nam, J.-T. Kim, J.-B. Yoon, W.-G. Bang, I.-H. Roe. Identification of immunodominant *Helicobacter pylori* proteins with reactivity to *H. pylori*-specific egg-yolk immunoglobulin. *J. Med. Microbiol.* 52:217–222, 2003.

97. Shin, J.-H., M. Yang, S.W. Nam, J.T. Kim, N.H. Myung, W.-G. Bang, I.H. Roe. Use of egg yolk-derived immunoglobulin as an alternative to antibiotic treatment for control of *Helicobacter pylori* infection. *Clin. Diagn. Lab. Immunol.* 9:1061–1066, 2002.

98. Roe, I.H., S.W. Nam, M.R. Yang, N.H. Myung, J.T. Kim, J.H. Shin. The promising effect of egg yolk antibody (immunoglobulin yolk) on the treatment of *Helicobacter pylori*-associated gastric diseases. *Korean J. Gastroenterol.* 39:260–268, 2002.

99. Shin, J.-H., I.-H. Roe, H.-G. Kim. Production of anti-*Helicobacter pylori* urease-specific immunoglobulin in egg yolk using an antigenic epitope of *H. pylori* urease. *J. Med. Microbiol.* 53:31–34, 2004.

100. Kapil, S., A.M. Trent, S.M. Goyal. Excretion and persistence of bovine coronavirus in neonatal calves. *Arch. Virol.* 115:127–132, 1990.

101. Ikemori, Y., M. Ohta, K. Umeda, F.C. Icatlo Jr, M. Kuroki, H. Yokoyama, Y. Kodama. Passive protection of neonatal calves against bovine coronavirus-induced diarrhea by administration of egg yolk or colostrum antibody powder. *Vet. Microbiol.* 58:105–111, 1997.

102. Stevenson, R.M.W., D. Flett, B.T. Raymond. Enteric redmouth (ERM) and other enterobacterial infections of fish. In: *Bacterial Diseases of Fish,* Inglis, V., R.J. Roberts, N.R. Bromage, eds., Oxford: Blackwell Scientific Publications, 1993, pp 80–105.

103. Lee, S.B., Y. Mine, R.M.W. Stevenson. Effects of hen egg yolk immunoglobulin in passive protection of rainbow trout against *Yersinia ruckeri.* *J. Agric. Food Chem.* 48:110–115, 2000.

104. Hatta, H., K. Mabe, M. Kim, T. Yamamoto, M.A. Gutierrez, T. Miyazaki. Prevention of fish disease using egg yolk antibody. In: *Egg Uses and Processing Technologies: New Developments.* Sim, J.S., S. Nakai, eds., Oxon, UK: CAB International, 1994, pp 241–249.

105. Gutierrez, M.A., T. Miyazaki, H. Hatta, M. Kim. Protective properties of egg yolk IgY containing anti-*Edwardsiella tarda* antibody against paracolo disease in the Japanese eel, *Anguilla japonica* Temminck & Schlegel. *J. Fish Dis.* 16:113–122, 1993.

106. Eterradossi, N., D. Toquin, H. Abbassi, G. Rivallan, J.P. Cotte, M. Guittet. Passive protection of specific pathogen free chicks against infectious bursal disease by in-ovo injection of semi-purified egg-yolk antiviral immunoglobulins. *J. Vet. Med.* B44:371–383, 1997.

107. Kweon, C.-H., B.-J. Kwon, S.-R. Woo, J.-M. Kim, G.-H. Woo, D.-H. Son, W. Hur, Y.-S. Lee. Immunoprophylactic effect of chicken egg yolk immunoglobulin (IgY) against porcine epidemic diarrhea virus (PEDV) in piglets. *J. Vet. Med. Sci.* 62:961–964, 2000.

108. Cama, V.A., C.R. Sterling. Hyperimmune hens as a novel source of anti-*Cryptosporidium* antibodies suitable for passive immune transfer. *J. Protozool.* 38:42S–43S, 1991.

109. Ling, Y.S., Y.J. Guo, J.D. Li, L.K. Yang, Y.X. Luo, S.X. Yu, L.Q. Zhen, S.B. Qiu, G.F. Zhu. Serum and egg yolk IgG antibody titers from laying chickens vaccinated with *Pasteurella multocida*. *Avian Dis.* 42:186–189, 1998.

110. Shin, N.R., I.S. Choi, J.M. Kim, W. Hur, H.S. Yoo. Effective methods for the production of immunoglobulin Y using immunogens of *Bordatella bronchiseptica*, *Pasteurella multocida* and *Actinobacillus pleuropneumoniae*. *J. Vet. Sci.* 3:47–57, 2002.

111. Granger, D.N., J.A. Barrowman, P.R. Kvietys. *Clinical Gastrointestinal Physiology*. Toronto: W.B. Saunders Company, 1985.

112. Reilly, R.M., R. Domingo, J. Sandhu. Oral delivery of antibodies; future pharmacokinetic trends. *Clin. Pharmacokinet.* 4:313–323, 1997.

113. Evans, D.F., G. Pye, R. Bramley, A.G. Clark, T.J. Dyson, J.D. Hardcastle. Measurement of gastrointestinal pH profiles in normal ambulant human subjects. *Gut* 29:1035–1041, 1988.

114. Fallingborg, J., L.A. Christensen, M. Ingeman-Nielsen, B.A. Jacobsen, K. Abildgaard, H.H. Rasmussen, S.N. Rasmussen. Measurement of gastrointestinal pH and regional transit times in normal children. *J. Pediatr. Gastroenterol. Nutr.* 11:211–214, 1990.

115. Koldovsky, O. Development of human gastrointestinal functions: interaction of changes in diet composition, hormonal maturation, and fetal genetic programming. *J. Am. Coll. Nutr.* 3:131–138, 1984.

116. Castro, G.A. Digestion and absorption of specific nutrients. In: *Gastrointestinal Physiology*, 2nd ed., Johnson, E.R., ed., Toronto: The C.V. Mosby Company, 1981, pp 123–140.

117. Magee, D.F., A.F. Dalley. *Digestion and the Structure and Function of the Gut*. Basel, Switzerland: Karger, 1986.

118. Madsen, J.L. Gastrointestinal transit measurements, a scintographic method. *Dan. Med. Bull.* 41:398–411, 1994.

119. Davis, S.S. Small intestine transit. In: *Drug Delivery to the Gastrointestinal Tract*, Hardy, J.G., S.S. Davis, C.G. Wilson, eds., Toronto: John Wiley & Sons, 1989, pp 49–61.

120. Shimizu, M., R.C. Fitzsimmons, S. Nakai. Anti-*E. coli* immunoglobulin Y isolated from egg yolk of immunized chickens as a potential food ingredient. *J. Food Sci.* 53:1360–1366, 1988.

121. Ikemori, Y., M. Ohta, K. Umeda, R.C. Peralta, M. Kuroki, H. Yokoyama, Y. Kodama. Passage of chicken egg yolk antibodies treated with hydroxypropyl methylcellulose phthalate in the gastrointestinal tract of calves. *J. Vet. Med. Sci.* 58:365–367, 1996.

122. Yokoyama, H., R.C. Peralta, S. Sendo, Y. Ikemori, Y. Kodama. Detection of passage and absorption of chicken egg yolk immunoglobulins in the gastrointestinal tract of pigs by use of enzyme-linked immunosorbent assay and fluorescent antibody testing. *Am. J. Vet. Res.* 54:867–872, 1993.

123. Miller, E.R., D.E. Ullrey. The pig as a model for human nutrition. *Ann. Rev. Nutr.* 7:361–382, 1987.

124. Shimizu, M., Y. Miwa, K. Hashimoto, A. Goto. Encapsulation of chicken egg yolk immunoglobulin G (IgY) by liposomes. *Biosci. Biotechnol. Biochem.* 57:1445–1449, 1993.

125. Shimizu, M., Y. Nakane. Encapsulation of biologically active proteins in a multiple emulsion. *Biosci. Biotechnol. Biochem.* 59:492–496, 1995.

126. Nishinaka, S., T. Suzuki, H. Matsuda, M. Murata. A new cell line for the production of chicken monoclonal antibody by hybridoma technology. *J. Immunol. Methods* 139:217–222, 1991.

127. Asaoka, H., S. Nishinaka, N. Wakamiya, H. Matsuda, M. Murata. Two chicken monoclonal antibodies specific for heterophil hanganutziu-deicher antigens. *Immunol. Lett.* 32:91–96, 1992.

128. Lillehoj, H.S., K. Sasaki. Development and characterization of chicken-chicken B cell hybridomas secreting monoclonal antibodies that detect sporozoite and merozoite antigens of *Eimeria. Poult. Sci.* 73:1685–1693, 1994.

129. Nishinaka, S., H. Akiba, M. Nakamura, K. Suzuki, T. Suzuki, K. Tsubokura, H. Horiuchi, S. Furusawa, H. Matsuda. Two chicken B cell lines resistant to ouabain for the production of chicken monoclonal antibodies. *J. Vet. Med. Sci.* 58:1053–1056, 1996.

130. Matsushita, K., H. Horiuchi, S. Furusawa, M. Horiuchi, M. Shinagawa, H. Matsuda. Chicken monoclonal antibodies against synthetic bovine prion protein peptides. *J. Vet. Med. Sci.* 60:777–779, 1998.

131. Matsuda, H., H. Mitsuda, N. Nakamura, S. Furusawa, S. Mohri, T. Kitamoto. A chicken monoclonal antibody with specificity for the N-terminal of human prion protein. *FEMS Immunol. Med. Microbiol.* 23:189–194, 1999.

132. Yamanaka, H.I., T. Inoue, O. Ikeda-Tanaka. Chicken monoclonal antibody isolated by a phage display system. *J. Immunol.* 157:1156–1162, 1996.

133. Nakamura, N., Y. Aoki, H. Horiuchi, S. Furusawa, H.I. Yamamoto, T. Kitamoto, H. Matsuda. Construction of recombinant monoclonal antibodies from a chicken hybridoma line secreting a specific antibody. *Cytotechnology* 32:191–198, 2000.

134. Kim, J.K., W. Min, H.S. Lillehoj, S. Kim, E.J. Shon, K.D. Song, J.Y. Han. Generation and characterization of recombinant ScFv antibodies detecting *Eimeria acervulina* surface antigens. *Hybridoma* 20:175–181, 2001.

135. Romito, M., G.J. Viljoen, D.H. Du Plessis. Eliciting antigen-specific egg yolk IgY with naked DNA. *Biotechniques* 31:670–677, 2001.

136. Mohammed, S.M., S. Morrison, L. Wims, K. Ryan Trinh, A.G. Wildeman, J. Bonsellar, R.J. Etches. Deposition of genetically engineered human antibodies into the egg yolk of hens. *Immunotechnology* 4:115–125, 1998.

17

Human Gut Microflora in Health and Disease: Focus on Prebiotics

Tamara Casci, Robert A. Rastall, and Glenn R. Gibson

CONTENTS

17.1 Introduction to Human Gut Microflora and Fermentation Properties 402
17.2 Gut Microflora in Health and Disease 404
    17.2.1 Acute Infections and Inflammatory Reactions 404
    17.2.2 Antibiotic Associated Diarrhea (AAD)..405
    17.2.3 Immune Stimulation...405
    17.2.4 Lactose Intolerance 406
    17.2.5 Colon Cancer...406
    17.2.6 Cholesterol Lowering Effect and Lipid Metabolism 408
17.3 Structure to Function Relationships in Prebiotic Oligosaccharides 408
    17.3.1 Fructooligosaccharides..409
        17.3.1.1 Fermentability of FOS 409
        17.3.1.2 FOS Interactions with Lipid Metabolism 411
        17.3.1.3 FOS Interactions with Mineral Absorption 411
        17.3.1.4 FOS Interactions with Cancer...411
    17.3.2 Galactooligosaccharides...412
    17.3.3 Isomaltooligosaccharides 413
    17.3.4 Lactulose 414
        17.3.4.1 Colon cancer 415
        17.3.4.2 Mineral absorption 415
        17.3.4.3 Constipation 415
    17.3.5 Soya Products..416
    17.3.6 Lactosucrose..418
    17.3.7 Xylooligosaccharides 418
17.4 Structure to Function Relationships in Other Oligosaccharides..........................419
    17.4.1 Lactitol..419

    17.4.2 Laevan Type Prebiotics 419
    17.4.3 Gentiooligosaccharides 420
    17.4.4 Glucose Based Oligosaccharides 421
17.5 Conclusions 421
References..422

17.1 INTRODUCTION TO HUMAN GUT MICROFLORA AND FERMENTATION PROPERTIES

In humans, the colon receives dietary material, which has already been digested in the upper gut, from the ileum, and the contents are then mixed and retained for 6–12 hours in the cecum and right colon. Thereafter, digesta are ejected and pass through the transverse to the left colon for storage, absorption of water, and eventual excretion (1). The typical retention time for the colon is about 60 h in U.K. subjects (2). Thus, the large gut is a tubular system, with nutrients flowing into the cecum, and bacteria, their metabolic products, and undigested food being excreted in feces. The major end products of fermentation in humans are the SCFA (short chain fatty acids), acetate, propionate, and butyrate (3); gases H_2 and CO_2 (4,5); ammonia (6); amines (7); phenols (8,9); and energy, which bacteria use for growth and maintenance of cellular function. Only about 5% of total SFCA are excreted in feces. *In vivo*, SCFA contribute about 10% of the daily energy requirements of the host. Most SCFA produced by gut bacteria are absorbed and metabolized in the body (10).

A range of substrates, of dietary origin or produced by the host, is available for fermentation by the colonic microflora. From the diet, resistant starch is the most quantitatively important (2) and is readily fermented by gut bacteria including *Bacteroides*, *Bifidobacterium* spp., and *Fusobacterium* spp. (11). Nonstarch polysaccharides form the next largest contribution to the fermentable substrates and include plant derived substrates like pectin, cellulose, hemicelluloses, and chemically related substances (2). Oligosaccharides like lactose, lactulose, raffinose, stachyose, and fructooligosaccharides also escape absorption by the small intestine and are fermented in the colon. Mucin glycoproteins, which are produced by goblet cells in the colonic epithelium, are predominant endogenous substances fermented in the colon (2) and can be metabolized by a range of microorganisms, most importantly *Bifidobacterium*, *Ruminococcus* spp., and some *Bacteroides* (12–14). Proteins and peptides originating in the diet, in pancreatic secretions or produced by bacteria, are available in the colon (2,6), although to a lesser extent than are carbohydrates.

The principal polysaccharide degrading bacteria are thought to be the Gram-negative anaerobes belonging to the genus *Bacteroides* (15). They are able to synthesize a wide range of cell associated polysaccharide depolymerases and glycosidases that allows them to grow upon various polysaccharides (15,16). The breakdown of highly polymerized materials in the gut is a cooperative activity, with enzymes from many different species participating in the process (2).

Gibson and Roberfroid (17) defined prebiotics as "nondigestible food ingredients that beneficially affect the host by selectively stimulating the growth and/or the activity of one or a limited number of bacteria normally resident in the colon, and thus attempt to improve host health."

Prebiotic substances, such as lactulose, fructooligosaccharides (including inulin), and galactooligosaccharides, selectively stimulate the growth of *Bifidobacterium* spp. in the colon (18). Isomaltooligosaccharides, glucooligosaccharides, lactitol, xylitol, and xylooligosaccharides (XOS) are candidate prebiotics, but the specific enzymes for the degradation of these molecules have not yet been fully evaluated, so the full explanatory mechanism for any purported prebiotic effect is not yet evident (19).

Fermentation of prebiotic oligosaccharides by probiotic bacteria in the gastrointestinal tract has been shown to have a number of possible health implications including alleviation of lactose maldigestion, increased resistance to gut invasion by pathogenic bacteria, stimulation of the immune system, protection against cancer, effect on mineral bioavailability, and lowering of lipid levels. Potential health benefits may include reduction of the risk of intestinal infectious diseases, cardiovascular disease, noninsulin dependent diabetes, obesity, osteoporosis, and cancer (18).

Zubillaga et al. (18) reported on the link between probiotics and certain health conditions including ulceration, diarrhea, *H. pylori* infection, lactose intolerance, cancer, and immune system function. It was demonstrated that *L. acidophilus* and *B. bifidum* can act as an "ecological" therapy for gastritis and duodenitis, and that high intakes of fermented milk products were associated with reduced risk of ulceration.

In the context of diarrhea, it has been found that a fermented product containing *L. acidophilus* could inhibit growth of *S. dysenteriae, S. typhosa,* and *E. coli.* The beneficial effect has been attributed to antimicrobial substances produced by *L. acidophilus, in vivo,* which neutralize the enterotoxins of *E. coli* (20).

Helicobacter pylori infection was found to be associated with the deficiency of *Lactobacillus* species in the stomach (18). Moreover, more than one *in vitro* study has shown the ability of *L. salivarius* to inhibit growth of *H. pylori* (21–23).

Lactobacillus acidophilus is associated with an improvement of *in vitro* lactose fermentation (24). However, this is also known to be true for conventional yogurts. *Lactobacillus delbruekii bulgaricus, S. thermophilus, L. acidophilus,* and *Bifidobacterium* are thought to be antimutagenic and anticarcinogenic (25), and are potentially useful for inhibiting cancer. However, the evidence is still unclear and contradictory (18).

Spanhaak et al. (26) showed that consumption of milk fermented with *L. casei* strain Shirota could modulate the composition and metabolic activities of the intestinal flora. This agrees with the findings of Schiffrin et al. (27) who showed that *L. acidophilus* Lal and *B. bifidus* Bb 12 did not modify lymphocyte subsets in humans.

The gut microflora exert a considerable influence on host biochemistry including: enzymatic activity of intestinal contents, SCFA production in the lumen, oxidation reduction potential of luminal contents, host physiology, host immunology, and modification of host synthesized molecules (28). The SCFA and, to a lesser extent, gases primarily provide a source of energy and growth substrates to other gut microorganisms, salvage dietary energy (60–70% of epithelial cell energy is derived from butyrate), affect colonic mucosal cell proliferation and differentiation, control luminal pH, improve mineral ion absorption, and regulate metabolic pathways (29).

In vitro and *in vivo* studies show that the intermediate and end products of fermentation formed by colonic bacteria depend partly upon the chemical composition of the polysaccharide substrate. For example, starch fermentation yields high levels of butyrate, whereas with a more oxidized substrate such as pectin, more acetate is produced (30,31). Cummings and Macfarlane (2) suggested that this may be explained by the fact that different bacteria take part in fermenting different polysaccharides. The relative rate of depolymerization of complex carbohydrates in the colon may influence the types of fermentation product that are generated (2). A contribution is also given by the physiology and anatomical architecture of the gut (32,33). Transit time of digesta through the colon strongly influences the activities of the gut microflora (2) and fecal ammonia concentrations (34). Best estimates indicate that the amount of NSP (nonstarch polysaccharides) available for fermentation from the U.K. diet is in the range of 8–18 g/d (2,35). In persons consuming western type diets, the amount of fermented carbohydrate, other than starch and NSP, is unlikely to be more than 10 g/d (2). Principal sugars escaping absorption in the small intestine are lactose, raffinose, and stachyose (2).

In humans, acetate, propionate, and butyrate account for approximately 85–95% of total SCFA in all regions of the colon. Other acidic products of fermentation are found in the large gut, such as the branched chain fatty acids, isobutyrate, 2-methylbutyrate, and isovalerate, which are products of amino acid fermentation; while other organic acids, including the electron sink products lactate and succinate, accumulate to a lesser degree (36). Further acidic, neutral, or basic products of the hind gut fermentation include phenols, indoles, amines, and ammonia (7), which are generated during the catabolism of amino acids.

The precise mechanisms whereby SCFA are absorbed in humans are not known with certainty; however, the process is concentration dependent, and SCFA transport is associated with the appearance of bicarbonate ions and stimulation of sodium absorption, and is independent of bulk water flow (3,37,38).

17.2 GUT MICROFLORA IN HEALTH AND DISEASE

17.2.1 Acute Infections and Inflammatory Reactions

Disturbances in the normal intestinal microbial community structure can result in a proliferation of pathogens. Acute inflammatory reactions cause diarrhea and sometimes vomiting, and can be associated with a number of bacteria and viruses including *E. coli*, *Campylobacter* species, *Vibrio cholerae, S. aureus, B. cereus, Clostridium perfringens*, *Salmonella* species, *Shigella* species, *Yersinia* species, and protozoa, especially *Giardia lamblia, Entamoeba histolytica,* and *Cryptosporidium parvum* (39,40). Bacteria can also be linked to more chronic diseases in the colon. For example, *C. difficile* has been targeted as the primary causative agent of pseudomembranous colitis.

The GI tract functions as a barrier against antigens from microorganisms and food (41). Among the possible mechanisms of probiotic therapy is promotion of a nonimmunological gut defense barrier, which includes the normalization of increased intestinal permeability and altered gut microecology. Another possible mechanism is improvement of the intestinal immunological barrier; particularly through immunoglobulin A responses and alleviation of gut inflammatory responses, which produce a stabilising effect. Many probiotic effects are mediated through immune regulation, particularly from the balance control of proinflammatory and antiinflammatory cytokines (41).

Research has shown that *Bifidobacterium* spp. can exert powerful antagonistic effects toward *E. coli* and other pathogens like *Campylobacter* and *Salmonella*. The inhibitory effect was variable in species of *Bifidobacterium* spp., with *B. infantis* and *B. longum* exerting the greatest effect (42).

Several open studies suggest a beneficial role of *L. rhamnosus* GG, *S. boulardii*, and *L. plantarum* LP299v during *C. difficile* related infections (43–50). Some studies have shown promising effects of probiotics on inflammatory bowel disease in animals. Intracolonic administration of *L. reuteri* R2LC to rats decreased the disease, whereas *L. reuteri* HLC was ineffective (51).

While there is likely to be a more chronic component to Crohn's disease, *Eubacterium, Peptostreptococcus, Pseudomonas, Bacteroides vulgatus,* mycobacteria, and *C. difficile* have all been associated with its onset (52). In an open study, a 10 day administration of *L. rhamnosus* GG to 14 children with active or inactive Crohn's disease resulted in an increase in immunoglobulin A secreting cells to β-lactoglobulin and casein, which indicated an interaction between the probiotic and the local immune system (53). Various epidemiological and animal studies, from numerous groups, have agreed that ulcerative colitis has a microbial etiology. In particular, sulphate reducing bacteria are also thought to play an important role (54).

Studies have shown a probiotic preparation to be as effective as mesalazine in reducing and preventing relapse in ulcerative colitis (55,56). *Helicobacter pylori* survives in the mucosal layer of the gastric epithelium, and is thought to be involved in the etiology of ulcers, type B gastritis, and stomach cancer. It secretes ammonia as part of its normal metabolism, thereby allowing it to survive at the acidic pH in the stomach (57). Attempts to eradicate *H. pylori, in vivo* with a probiotic have mainly failed (58).

17.2.2 Antibiotic Associated Diarrhea (AAD)

Diarrhea occurs in about 20% of patients who receive antibiotics. Antibiotic associated diarrhea (AAD) results from microbial imbalance that leads to a decrease in the fermentation capacity of the colon. Invasion with *C. difficile, Klebsiella oxytoca,* and *C. septicum* are causes of AAD (39,59).

Studies have shown that oral administration of *S. boulardii* can decrease the risk of AAD (50,59). Other studies have shown the efficacy of *S. boulardii* (60), *Enterococcus faecium* SF68, and *L. rhamnosus* GG in shortening the duration of AAD (59).

Several trials have suggested a preventive effect of some fermented products on the risk of diarrhea in children (61,62). Saavedra et al. (63) showed that feeding *B. bifidum* and *S. thermophilus* to infants significantly reduced the risk of diarrhea and the shedding of rotavirus.

17.2.3 Immune Stimulation

In a human trial, administration of probiotic yogurt gave an increase in the production of γ-interferon (64). In animal models, probiotics have been shown to stimulate the production of antibodies (local and systemic), enhance the activity of macrophages, increase γ-interferon levels, and increase the concentration of natural killer cells (65).

An enhancement in the circulating IgA antibody secreting cell response was observed in infants supplemented with a strain of *L. casei*, and this correlated with shortened duration of diarrhea in the study group when compared to a placebo group (66). Other studies have reported an enhancement in the nonspecific immune phagocytic activity of granulocyte populations in the blood of human volunteers after consumption of *L. acidophilus* and *B. bifidum* (67–69). It is possible that stimulation of intestinal IgA antibody responses induced by lactic acid bacteria (LAB) may be explained partly by an effect on phagocyte cell functions. Ingestion of yogurt has been reported to stimulate cytokine production, including γ-interferon, in human blood mononuclear cells (70).

It has been shown in experimental animals that the postnatal maturation of small intestinal brush border membranes was associated with increased food protein binding capacity (71). The capacity of antigens to bind to epithelial cells is related to the rate and route of antigen transfer and shown to influence the intensity of mucosal immune responses (72). Oral introduction of *Lactobacillus* species can enhance nonspecific host resistance to microbial pathogens, thereby facilitating the exclusion of pathogens in the gut (73,74). Several strains of live lactic acid bacteria have been shown *in vitro* to induce the release of the proinflammatory cytokines tumor necrosis factor α, and interleukin 6, reflecting stimulation of nonspecific immunity (75).

Oral introduction of *L. casei* and *L. bulgaricus* activates the production of macrophages (73) and administration of *L. casei* and *L. acidophilus* activates phagocytosis in mice (74). Enhanced phagocytosis by *L. acidophilus* Lal was also reported in humans (67).

Probiotic bacteria have been shown to modulate phagocytosis differently in healthy and allergic subjects: in healthy persons there was an immunostimulatory effect, whereas in allergic persons, down regulation of the inflammatory response was detected (76).

Consumption of the adhesive strains *L. johnsonii* Lj1 and *B. lactis* Bb12 has been shown to increase the phagocytosis of *E. coli, in vitro* (67). A combination of both strains, together with *S. thermophilus*, has also been shown to possess adjuvant activity when consumed in combination with an attenuated *Salmonella typhimurium*. The preparation caused a significant increase in a specific serum IgA to the *Salmonella* (77).

In order to influence the immune system, a probiotic microorganism must activate the lymphoid cells of the gut associated lymphoid tissue, which are diffusely distributed amongst epithelial cells and populate the lamina propria and submucosa (78). Some probiotic strains are clearly antagonistic against *Helicobacter pylori, in vitro* although as mentioned earlier, experiments have not been so successful *in vivo* (59). A significant reduction of urease activity *in vivo* (which reflects the activity of *H. pylori*) has been reported in patients treated with a supernatant of *L. johnsonii* La1 (79).

Lactic acid bacteria have been observed to produce antimicrobial substances. The most widely produced antimicrobial substances are organic acids, especially lactic and acetic acids. Hydrogen peroxide and carbon dioxide are also widely produced by LAB. If probiotic LAB are metabolically active during their passage throughout the intestinal tract, it is very likely that some of these substances will be produced. A reduction in fecal pH confirms this. The production of other antimicrobial components, diacetyl, reuterin, pyroglutamic acid, and especially bacteriocins, is not certain under *in vivo* conditions. To date, no studies have been performed to investigate the production or efficacy of bacteriocins in the intestine. Evidence exists that they are produced and active in the oral cavity (80). Mechanisms implicated in the protective effects of probiotics may therefore include the production of acids, hydrogen peroxide, or antimicrobial substances; competition for nutrients or adhesion receptors; antitoxin actions; and stimulation of the immune system (59).

17.2.4   Lactose Intolerance

Lactose intolerance is a problem for approximately 70% of the world's population and is linked to a low intestinal β-galactosidase activity. Inadequate hydrolysis of dietary lactose leads to abdominal discomfort (81). Some lactic acid bacteria, including *L. acidophilus* and *B. bifidum,* produce β-D-galactosidase, the action of which increases tolerance to dairy products (81). There is good evidence for the alleviation of symptoms of lactose intolerance by specific probiotics and, recently, immune changes related to milk hypersensitivity have been shown to be down regulated (76).

It is well established that persons with lactose intolerance experience improved digestion and tolerance of the lactose contained in yogurt than of that contained in milk (82). At least two mechanisms have been reported to explain this: digestion of lactose in the gut lumen by lactase contained in the yogurt bacteria, and slower intestinal delivery or transit time of yogurt compared to milk (82–85). Kim and Gilliland (81) suggested that improved lactose digestion was not caused by lactose hydrolysis prior to consumption, but rather by the action of the enzyme in the digestive tract.

17.2.5   Colon Cancer

Bacterial links to colonic cancer are also evident. The colon is the second most common site of tumor formation in humans, and several bacterial products are carcinogenic (86). Development of colon cancer consists of a sequence of events that, although incompletely understood, occurs in definable steps. First is an initiating step, in which a carcinogen, produced by metabolic activation of a precursor, produces an alteration in DNA. At present, it is believed that several mutations must occur for a tumor to develop. The postinitiation steps are much less clear, but usually involve changes in signal transduction pathways. The next clear step is an overgrowth in colonic crypts, which can be seen morphologically in an aberrant crypt. Aberrant crypts, which are considered preneoplastic structures, are enlarged and elevated relative to normal crypts, and have a serpentine growth pattern. Aberrant crypts may occur singly or as groups of aberrant crypts within a single focus. A certain small but unknown fraction of these aberrant crypts will progress to polyps and eventually to tumors (87).

The carcinogenic effect of endogenous toxic and genotoxic compounds is probably influenced through the activity of the bacterial enzymes NAD(P)H dehydrogenase, nitroreductase, β-glucuronidase, β-glucosidase, and 7-α-dehydroxylase. Harmful and beneficial bacteria commonly found in the intestine differ in their enzyme activities (88). *Bifidobacterium* spp. and lactobacilli have negligible activities of xenobiotic metabolising enzymes, unlike bacteroides, clostridia, and enterobacteriaceae.

Ingestion of viable probiotics and prebiotics is associated with anticarcinogenic effects, one mechanism of which is the detoxification of genotoxins in the gut (89). Because of the complexity of cancer initiation, progression, and exposure in the gut, many types of interactions may be envisaged. It has been shown (89) that short lived metabolite mixtures isolated from milk that was fermented with strains of *L. bulgaricus* and *S. thermophilus,* were more effective in deactivating aetiological risk factors of colon carcinogenesis than cellular components of microorganisms. The underlying mechanisms are not known. After ingestion of resistant starch, the gut flora induces the chemopreventive enzyme glutathione transferase π (EC 2.5.1.18) in the colon. Together, these factors lead to a reduced load of genotoxic agents in the gut and increased production of agents that deactivate toxic components. Butyrate is one such protective agent and is associated with a lowering of cancer risk (89). The role of butyrate in modulation of nucleic acid metabolism is of particular interest, especially its effects upon regulation of gene expression and cell growth. Butyrate can reversibly alter the *in vitro* properties of human colorectal cancer cell lines by prolonging doubling time and slowing down growth rates (90). Low concentrations of this SCFA reduce DNA synthesis and suppress proliferation in a variety of cell types (91–93).

Ammonia is mainly formed through the deamination of amino acids (6,94). It is rapidly absorbed from the gut and detoxified by urea formation in the liver. Low concentrations can alter the morphology and intermediate metabolism of colonic epithelial cells (95,96), affect DNA synthesis, and reduce their lifespan. Thus, a high ammonia concentration in the colonic lumen may select for neoplastic growth.

Phenols and indoles do not appear to be carcinogenic alone, but act as cocarcinogens (97). Increased production of these metabolites has been found to occur in a variety of disease states (9,98,99). Phenolic and indolic compounds are produced by deamination of the aromatic amino acids, tyrosine, phenylalanine, and tryptophan. Phenols are not found in urine of germ free animals (8).

In colon cancer, the evidence of probiotic efficacy is still not certain. In animal models it was shown that a dietary intake of *B. longum* significantly suppressed development of azoxymethane (AOM) induced aberrant cryptic foci formation in the colon (100). This study was also confirmed by another using *B. longum* and inulin (101). Kinouchi et al. (102) showed that *L. acidophilus* and *B. adolescentis* suppressed ileal ulcer formation in animal models. *Lactobacilus acidophilus* fed to healthy volunteers significantly decreased β-glucuronidase, nitroreductase, and axoreductase activities (103).

Oral administration of *L. casei* strain Shirota has been shown to have inhibitory properties upon chemically induced tumors in rats (104). Also, it may prolong survival during specific cancer treatments (105). The beneficial effect of probiotic based cancer therapy has been associated with their antimutagenic properties and ability to modulate immune parameters including T-cell, natural killer (NK) cell, and macrophage activities, which are important in hindering tumor development (27,76,104). Some, but not all, epidemiological studies suggest that consumption of fermented dairy products may have certain protective effects against large colon adenomas or cancer (106).

Dietary administration of lyophilized cultures of *B. longum* strongly suppressed AOM induced colon tumor development. The inhibition of colon carcinogenesis was associated with a decrease in colonic mucosal cell proliferation and colonic mucosal and tumor ODC (ornithine decarboxylase) and *ras*-p21 oncoprotein activities (107). *Ras* activation represents one of the earliest and most frequently occurring genetic alterations associated with human cancers, especially that of the colon (108). Elevated levels of *ras*-p21 have been correlated with increased cell proliferation, histological grade, nuclear anaplasia, and degree of undifferentiation (109). The major conclusion from animal data is that there appears to be a synergistic effect of consumption of probiotics and prebiotics on the attenuation of development of colon cancer. The effect is often not large, but it is possible that it could be beneficial, in combination with other methods to reduce risk (87).

17.2.6 Cholesterol Lowering Effect and Lipid Metabolism

Prebiotics in the form of fermented milk are thought to have cholesterol lowering properties in humans. However, studies have given equivocal findings. Even recent studies fail to provide convincing evidence that "live" fermented milk products have any cholesterol lowering efficacy in humans (110).

In animals, dietary FOS caused a suppression of hepatic triglyceride and VLDL (very low density lipoprotein) synthesis, resulting in marked reductions in triglycerides, and to a lesser extent cholesterol, levels (110). Evidence for similar effects in humans is sparse and more studies are needed, particularly with respect to effects on postprandial triglyceride concentrations (110).

Propionate metabolism has also been studied in ruminants, where it is a major gluconeogenic precursor (111). In rats, it has been shown to lower cholesterol (112,113) either by inhibition of hepatic cholesterol synthesis or by redistribution of cholesterol from plasma to the liver (113). In humans, however, no change was seen in total cholesterol levels, although HDL (high density lipoprotein) cholesterol increased when a group of females were fed sodium propionate (114).

In CHD (coronary heart disease) there are variable data and no firm conclusions can be drawn (65). Schaafsma et al. (115) found that daily feeding of 125 mL of a test probiotic milk significantly lowered LDL (low density lipoprotein) cholesterol levels and total serum cholesterol.

17.3 STRUCTURE TO FUNCTION RELATIONSHIPS IN PREBIOTIC OLIGOSACCHARIDES

Eight NDOs have been licensed as having FOSHU (Foods for Specified Health Use) status by the Ministry of Health and Welfare in Japan: GOS (galactooligosaccharides), FOS (fructooligosaccharides), lactulose, XOS (xylooligosaccharides), soybean oligosaccharides, raffinose, lactosucrose, and isomaltooligosaccharides (116). The efficacy of prebiotics toward promoting human health is strongly related to their chemical structure. This section of the review therefore elucidates the interactions between oligosaccharide structure and prebiotic effects.

17.3.1 Fructooligosaccharides

"Fructooligosaccharides" (FOS) is a common name for fructose oligomers that are mainly composed of 1-kestose (GF_2), nystose (GF_3), and 1^F- fructofuranosyl nystose (GF_4) in which fructosyl units (F) are bound at the β-2,1 position of sucrose (GF) (117). FOS are produced from sucrose through the transfructosylating action of either β-fructofuranosidases (β-FFase, EC 3.2.1.26) or β-D-fructosyltransferases (β-FTase, EC 2.4.1.9) (119).

17.3.1.1 Fermentability of FOS

In vitro experiments (anaerobic batch culture fermenters inoculated with human fecal bacteria and a carbohydrate substrate) on the comparative fermentation of chicory inulin and FOS showed that both inulin and FOS were rapidly and completely metabolized by the microbial flora in the fermenters. The relative rate of fermentation was similar for both substrates. A more detailed analysis however, revealed that the rate of degradation of oligomers with a degree of polymerization (DP < 10) was approximately twice that of molecules with a higher DP. Also, GF_n type and F_m type components of inulin hydrolysates disappeared from the culture medium at a similar rate, and the two types are known to have similar bifidogenic effects (119). There is no difference between FOS obtained by partial enzymatic hydrolysis (containing fructan chains ending with a glucose moiety as well as fructan chains ending with a fructose moiety) and FOS obtained by enzymatic synthesis from sucrose (120).

The efficiency of FOS over sucrose and other sugars as substrates for most strains of *Bifidobacterium* spp. is due to the intracellular 2,1 β-D fructan-fructanhydrolase (E.C 3.2.1.7) activity (119,121), which liberates monomeric fructose molecules that are then transported into the bacterial cell (122). The fructose moiety is then metabolized in the specific "bifidus" pathway (19). The enzyme also gives the genus a competitive advantage in the human gut (121). The β(2-1) bond linking the fructose moieties in the FOS chains is central in the nondigestibility as well as the bifidogenic properties of FOS (123).

Lactobacillus plantarum and *Lactobacillus rhamnosus* were found to be capable of metabolising only tri- and tetrasaccharide fractions, not pentasaccharides (124). This suggests that there may be specific transport systems in these organisms for trisaccharides and tetrasaccharides (125). In general, feeding of FOS increases *Bifidobacterium* spp. and *Lactobacillus* species, increases SCFA concentrations, and decreases *Clostridium* spp., *Fusobacterium* spp., *Bacteroides*, and pH (29,126–128). The intake doses of FOS that elicit a bifidogenic effect in human studies has ranged from 4 to 15 g/d (128,129). That *Bifidobacterium* species selectively ferment the fructans in preference to other carbohydrate sources such as starch, fructose, pectin, and polydextrose was confirmed in a volunteer trial using fructooligosaccharides (127) and by other human studies (130,131).

Using continuous chemostat cultures inoculated with fecal slurries, Gibson and Wang (132) showed that after six turnovers, chicory derived FOS, but not glucose, were able to selectively stimulate *Bifidobacterium* spp. growth; the number of these bacteria was almost three orders of magnitude higher than *Bacteroides*. With glucose as the substrate, *Bacteroides* were two orders of magnitude higher than *Bifidobacterium* spp. (119). *In vitro* studies showed that the majority of *Bifidobacterium* species utilized FOS and low polymerized inulins, although twelve out of the thirty strains tested were not stimulated. Incorporation of FOS into the diet stimulated the proliferation of fecal *Bifidobacterium* spp. by 1.6 log cfu/g under *in vivo* conditions, whereas *B. animalis* KSp4 was not effective (133).

Administration of *Bifidobacterium* spp. together with the prebiotic improved the bifidogenic effect by around 1.4 log cfu/g of feces. Supplementation had almost no effect upon some groups of gut microflora (133). This indicates that utilization of inulin by *Bifidobacterium* spp. depends on the degree of polymerization (DP) of fructooligomer chains and purity of the preparations. Among FOS, low DP inulin and high DP inulin, FOS was the most fermentable substrate and inulin the least.

It has also been suggested that fructans with higher DP are less rapidly fermented by *Bifidobacterium* (133). The results of this study indicate that the ability of *Bifidobacterium* spp. to metabolize FOS is a species dependent feature, and only to a small extent strain dependent. Muramatsu et al. (121) showed that the β-fructofuranosidase from *B. adolescentis* G1 has a unique substrate specificity to fructooligomers of DP 2-8 than to inulin and sucrose. All *L. acidophilus* strains tested fermented FOS, a result consistent with that recently reported by Sghir et al. (134). Such strains were: 33200 (ATCC), 837 (ATCC), DDS-1 (NC), and NCFM (NCSU), the latter two being widely promoted as probiotics. Three other commercial probiotic strains, *L. plantarum* MR240, *L. casei* MR191, and *L. casei* 685, also fermented FOS, whereas *Lactobacillus* strain GG was found to be a nonfermenter (despite being a well studied probiotic strain). Interestingly, most of the *L. bulgaricus* and *S. thermophilus* strains (the bacteria traditionally used in yogurt manufacture) were not FOS fermenters (124).

Perrin et al. (122) compared the physiological behavior of *B. infantis* ATCC 15697 growing on synthetic FOS or its components. The studies were carried out in batch cultures on semisynthetic media. Glucose, fructose, sucrose, and FOS as carbohydrate sources were compared. Glucose was the preferred substrate for growth and biomass production. Fructose was the best for lactate and acetate production. With FOS more metabolites were produced than with fructose. In a mixture of FOS, the shorter saccharides were used first [this has also been observed with other strains (135)] and fructose was released in the medium. Fructofuranosidase was inducible by fructose. The conclusion was that the glucose contained in FOS and sucrose might sustain growth and cell production while fructose may enable the production of major metabolites (122).

As a consequence of the metabolism of the FOS by fermentative bacteria, SCFA and lactic acid are produced. Both lead to a drop in the pH of the large intestine. This is beneficial for the organism as it constitutes an ideal medium for the development of the bifidogenic flora and, at the same time, limits the development of bacteria which are considered pathogens. The content of the metabolites produced by putrefactive bacteria decrease following the ingestion of FOS (136).

The addition of *L. acidophilus* 74-2 with FOS gave rise to an increase of *Bifidobacterium* spp. Major positive changes occurred in the production of organic acids: a strong upward trend was observed especially in the case of butyric and propionic acids. A noticeable increase of β-galactosidase activity was monitored while the activity of β-glucuronidase, generally considered undesirable, declined (137).

A variety of fermentation products influence the anatomy, physiology, and immunology of the host; and conversely the general constitution of the host affects the microbial community in the intestine (137). Gibson and Wang (132) have shown that *in vitro*, when *Bifidobacterium* spp. are given FOS as a substrate they secrete a peptide that is inhibitory to most of the common organisms causing acute diarrhea. In human healthy volunteer studies, oral dosing with 15 g of FOS daily led to *Bifidobacterium* spp. becoming the dominant species in the large bowel (127).

It was found that neither FOS nor inulin was, to a significant extent, fermented by *B. bifidum* (138). Bacteria other than the *Bifidobacterium* spp. can ferment FOS. These include *K. pneumoniae*, *S. aureus* and *epidermidis*, *E. faecalis* and *faecium*, *Bacteroides vulgatus, thetaiotamicron, ovatus* and *fragilis, L. acidophilus,* and *Clostridium* species, mainly *butyricum* (119). However, their ability to do so in mixed culture is repressed by *Bifidobacterium* spp. competition.

17.3.1.2   FOS Interactions with Lipid Metabolism

Evidence suggests that FOS acts as a soluble fiber and can be useful in increasing intestinal motility and transport, and in the reduction of high levels of plasma cholesterol (136). FOS can decrease triaglycerol in VLDL when given to rats. The triaglycerol lowering action of FOS is due to a reduction of *de novo* fatty acid synthesis in the liver through inhibition of all lipogenic enzymes (139).

Postprandial insulin and glucose concentrations are low in the serum of FOS fed animals and this could explain, at least partially, the metabolic effects of FOS. Some events occurring in the gastrointestinal tract after oligofructose feeding could be involved in the antilipogenic effects of this fructan: the production of propionate through fermentation, a modulation of the intestinal production of incretins, or modification of the availability of digestible carbohydrates. Recent studies have shown that the hypotriglyceridemic effect of fructans also occurs in humans (139).

The effects of inulin-type fructans on glycemia and insulinemia are not yet fully understood, and available data are often contradictory. Other nondigestible carbohydrates are known to modify the kinetics of absorption of carbohydrates, thus decreasing the incidence of glycemia and insulinemia (140,141).

17.3.1.3   FOS Interactions with Mineral Absorption

The production of a high concentration of short chain carboxylic acids in the colonic fermentation of nondigestible carbohydrates facilitates the absorption of minerals, Ca^{2+} and Mg^{2+} in particular (142). Roberfroid (143) argued that inulin-type fructans may improve mineral absorption and balance by an osmotic effect that transfers water into the large bowel, thus increasing the volume of fluid in which these minerals can dissolve. Also, acidification of the colon (as a consequence of carbohydrate fermentation) raises the concentration of ionized minerals, which favors passive diffusion. *In vivo* human studies have confirmed the positive effect of inulin and FOS upon absorption of calcium, but not of iron, magnesium, or zinc (144,145).

It has been observed that the effect of FOS on metabolic calcium balance, calcium content of bone, and prevention of ovariectomy induced loss of trabecular structure became more prominent when dietary calcium was high (146,147). It has been discussed that at least part of the stimulating effect of FOS on mineral absorption might be attributed to SCFA production (148).

17.3.1.4   FOS Interactions with Cancer

Prebiotics, in particular the chicory derived β(2-1) fructans, have been shown to exert cancer protective effects in animal models (149). A study was carried out to determine the effects of two chicory fructans: FOS (Raftilose P95, average DP 4) and long chain inulin (Raftiline HP, average DP 25) on apoptosis and bacterial metabolism in rats (149). Both substrates were shown to exert protective effects at an early stage in the onset of cancer.

It has been shown that inulin at dietary concentrations of 5–10% suppressed ACF (foci) in the rat colon (101,123,150).

NDOs may exert cancer protective effects at the cellular level following SCFA formation during fermentative bacterial metabolism. Butyrate, acetate, and propionate regulate colonic epithelial turnover and butyrate induces apoptosis in colon adenoma and cancer lines (151,152).

Rao et al. (150) and Rowland et al. (101) have reported protective effects of inulin at a postinitiation stage in the form of suppression of early preneoplastic lesions following treatment with a carcinogen. In this study, there was a trend toward increased colonic β-glucosidase activity and decreased ammonia concentration in the rats fed oligofructose or inulin. This was a positive response as ammonia is a known tumor promoter (96), so suppression of ammonia formation in the gut can be considered beneficial. β-glucuronidase is involved in DMH (1,2-dimethylhydrazine) and AOM metabolism, causing the release of MAM (methylazoxymethanol) from its biliary conjugate in the colon. Therefore, a decrease in β-glucuronidase may be protective (149). Large concentrations of butyrate can be produced as products of colonic fermentation by gut bacteria. Young (153) proposed that butyrate possessed antitumor properties, indicating the potential of probiotic strains such as *Lactobacillus* for preventing cancer. This study has shown a strong increase of β-galactosidase activity during administration of *L. acidophilus* 74-2 with FOS. After addition had stopped, the β-galactosidase activity again declined suggesting a causal link between the intake of a prebiotic product and increase of β-galactosidase activity, whereas β-glucuronidase activity declined (137).

17.3.2 Galactooligosaccharides

6' Transgalactosylated oligosaccharides (GOS, 154) are oligosaccharides produced by transgalactosylation of lactose using a β-galactosidase (β-Gal) (155). *In vitro,* GOS increased ATP, SCFA, and acetate concentrations, decreased pH and were 95% degraded from the fourth day of intake onward. They lead toward an increase in *Bifidobacterium* spp. species, while breath hydrogen excretion was reduced. However, methane excretion and levels of *Enterobacter* species are not affected (156).

GOS can be fermented by a number of intestinal bacteria including *Bifidobacterium* spp., *Bacteroides, Enterobacteria,* and *Lactobacillus* species (154,157,158).

A few human feeding studies investigating the prebiotic potential of GOS suggest that they may have bifidogenic potential (156,159). The bifidogenic nature of GOS has been related to a linkage specificity of the *Bifidobacterium* β-galactosidase, which cleaves β1-3 and β1-6 linkages, instead of β1-4 linkages (160).

Two probiotic strains, *Bifidobacterium lactis* DR10 and *Lactobacillus rhamnosus* DR20, were tested for their ability to utilize and grow upon GOS present in a commercial hydrolysed lactose milk powder (161). The results demonstrated that *B. lactis* DR10 preferentially utilized tri- and tetrasaccharides whereas *L. rhamnosus* DR20 preferred sugars with a lower degree of polymerization (di- and monosaccharides). Surprisingly, monosaccharides including glucose were the final sugars to be utilized by *B. lactis* DR10 (161). According to Rastall and Maitin (125) this suggests that *B. lactis* has a specific transport system for GOS that is not present in *L. rhamnosus*. A β-galactosidase that displays hydrolytic activity toward GOS but not lactose has been described in *B. adolescentis* (162). It has a preferential activity toward GOS with DP 2-3.

Fifty-four strains of lactic acid bacteria were studied for their ability to ferment lactose derived oligosaccharides. A good correlation was seen between the ability of a strain to utilize oligosaccharide and the presence of the lactose hydrolysing enzyme β-galactosidase (161). It has been suggested that GOS can be used more readily and selectively *in vitro* by *Bifidobacterium* spp. than other oligosaccharides such as lactulose and raffinose (163). However, the mechanism of utilization of this oligosaccharide by this group of bacteria is poorly understood. It is known that even though most *Bifidobacterium* spp. strains of human gut origin can readily use GOS, only a few strains from other genera, including *Lactobacillus* species, possess this ability (158,163). 4'-Galactosyllactose is selectively utilized by all the *Bifidobacterium* strains tested (157,158,164) compared with lactulose and raffinose whose specificity is less marked. Other studies also show that some strains of *Lactobacillus, Bacteroides* and *Clostridium* ferment GOS, and TD (transgalactosylated disaccharides) might even be better substrates for these bacteria (157,158). The bacterial species that ferment 4'- and 6'-GOS overlap, although some differences are found (116).

Physiological effects of GOS fermentation in the gut are:

1. Improvement of defecation. Deguchi et al. (165) showed that ingestion of 5g of GOS daily improved defecation frequency in a group of women.
2. Elimination of ammonia. In a human study, the ingestion of 4'-GOS at a dose of 3 g/d reduced the fecal ammonia concentration (166). The fact that other putrefactive products such as phenol, p-cresol, and indole were also decreased indicates that there was a dramatic change in metabolism upon ingestion of 4'-GOS.
3. Colon cancer prevention.
4. Stimulation of calcium absorption. Chonan et al. (167) found a stimulatory effect of GOS on calcium absorption and calcium content in femur and tibia with 0.5% calcium supplementation but not with 0.05%.

In a human volunteer trial, it was found that GOS increased *Bifidobacterium* spp. and *Lactobacillus* numbers while decreasing *Bacteroides* species and *Candida* species (159).

At a dose of 10 g/d, GOS given in the diet for 21 days decreased breath H_2 excretion and increased fecal *Bifidobacteria* without affecting *Enterobacteria* species or fecal pH (147). Similar results were found with inulin and lactose.

Van Laere et al. (162) reported on the effects of both reducing and nonreducing GOS comprising 2 to 8 residues on the growth of *B. adolescentis* DSM20083 and on the production of a novel β-galactosidase (β-GalII). In cells grown on GOS, in addition to the lactose degrading β-Gal (β-Gal I), another β-Gal (β-Gal II) was detected and it showed activity toward GOS but not toward lactose (162). During fermentation of GOS the di- and trisaccharides were first metabolized. β-Gal II was active toward β-galactosyl residues that were 1-4, 1-6, 1-3, and 1-1 linked, indicating its role in the metabolism of GOS by *B. adolescentis*, one of the predominant human fecal bacteria (162). This enzyme might allow *B. adolescentis* to utilize the oligosaccharides more efficiently than other microorganisms. Therefore, this GOS mixture, containing mainly higher molecular weight materials, might be an interesting prebiotic substrate. During the growth of *B. adolescentis* on GOS a large number of glycosidases are produced, including two arabinofuranohydrolases, which are involved in the degradation of arabinoxylooligosaccharides. This may offer an additional competitive advantage, since it allows the organism to scavenge the environment for a range of substrates and use the degradation products for growth (162). No endoglycanase activity could be detected in the cell extract, suggesting that *B. adolescentis* adopted a strategy aimed at utilizing polysaccharide degradation products generated by other microorganisms instead of taking part in the initial depolymerization stage of polysaccharides (162).

These results indicate that certain *Bifidobacterium* spp. together with prebiotics may be used for prophylaxis against opportunistic intestinal infections with antibiotic resistant pathogens (168).

17.3.3   Isomaltooligosaccharides

Isomaltooligosaccharides (IMO) are composed of glucose monomers linked by α1-6 glucosidic linkages. Isomalto-900, a commercial product, is produced by incubating α-amylase, pullulanase, and α-glucosidase with cornstarch. The major oligosaccharides in this mixture are isomaltose, isomaltotriose, and panose. IMO occur naturally in various fermented foods and sugars such as soybean sauce and honey (169).

IMO have been shown to be fermented by *Bifidobacterium* spp. and *Bacteroides fragilis*, but not by *E. coli* and other bacterial populations (169). IMO are efficient prebiotics in that they stimulate a lactic microflora as well as allow elevated production of butyrate, which is thought to be a desirable metabolite in the gut (65). Furthermore, the IMO mixture consisting of a high degree of branched oligosaccharides (due to α-GTase) is expected to be particularly effective at promoting *Bifidobacterium* spp. growth in the human intestine as growth of *Bifidobacterium* spp. is known to be enhanced in proportion to the degree of glucosidic polymerization of the IMO components (170). Kaneko et al. (171) reported that the IMO mixture containing trisaccharides and tetrasaccharides was more effective in enhancing the microflora than the IMO mixture of disaccharides.

Olano-Martin et al. (172) looked into the fermentability of dextran and novel oligodextrans (I, II, and III) by gut bacteria. In anaerobic batch culture, dextran and oligodextran increased the number of *Bifidobacterium* spp. and butyrate concentration. Using a three stage continuous culture cascade system, it was shown that a low molecular weight oligodextran (IV) was the best substrate for *Bifidobacterium* spp. and lactobacilli. Moreover, dextran and oligodextrans stimulated butyrate production, which suggests that oligodextrans indeed have potential as modulatory agents for the gut microflora.

17.3.4   Lactulose

Lactulose (Galβ1,4 Fru) is formed by heating lactose syrup to caramelization (173). It was discovered by Montgomery and Hudson in 1930 (174) and its growth promoting effect for *Bifidobacterium* species was determined by Petuely (175).

It is difficult to crystallize in water solution, however, a new crystal form of lactulose was recently reported (173,176). It has an energy value of 2.2 and 2.0 kcal/g for miniature pigs and rats, respectively (177). The medical applications of lactulose include the treatment of hyperammonemia, hepatic encephalopathy, and chronic constipation. Lactulose is also thought to prevent renal (178) and liver failures, is expected to lower incidence of colorectal cancer, shows a therapeutic effect in salmonella carrier state, has an antiendoxin effect, and contributes toward a decreased response in diabetics (179). Lactulose prevents postsurgery complications of obstructive jaundice, activates the immune system, prevents contracting infections, including urinary tract infection and respiratory tract infections, and prevents the recurrence of colorectal adenomas (179). In a human study, total bacterial counts were not changed by administration of lactulose (180); however, the *Bifidobacterium* count increased significantly, and lectinase positive *Clostridium* spp., including *C. perfringens*, and *Bacteroides,* decreased significantly. As a result, the ratio of *Bifidobacterium* in total bacterial counts increased from 8.3% before intake to 47.4% after intake. Indole and phenol, metabolites of intestinal bacteria, decreased significantly during intake, as did ammonia. Enzymatic activities of β-glucuronidase, nitroreductase, and azoreductase in wet feces decreased during intake. Fecal pH decreased and fecal moisture content increased significantly during intake (179). Lactulose promoted the growth of a range of lactic acid bacteria, although their growth was more rapid on lactose, particularly for *B. bifidum* (181). When lactulose was incubated with various intestinally derived strains, *B. bifidum, L. casei,* and *Lactobacillus* species utilized the disaccharide well with no gas production, while *C. perfringens* metabolized lactulose with a large amount of gas production (182).

Lactulose has been known to exert an antiendotoxin effect. Liehr and Heine (183) reported that endotoxin mediated galactosamine hepatitis in rats, as induced by intravenous injection of galactosamine, developed necrosis of the hepatocytes, and the inflammatory reaction of the liver tissue was virtually prevented by oral (184) and intravenous (184) administration of lactulose beforehand. Iwasaki et al. (185) reported that when lactulose was administered to hepatic cirrhosis and hepatic cancer patients, improvement of endotoxemia was observed in 50% of the patients.

Chronic portal systemic encephalopathy is a metabolic disorder of the central nervous system and is characterized by psychiatric and neurological disturbances that may progress from mild mental aberration to coma. It is manifest when ammonia, which is produced in the intestine as a metabolite by intestinal putrefactive bacteria, escapes detoxication by hepatic function and is transported to the brain where it acts as a toxin. Lactulose reduces the ammonia level in blood and alleviates symptoms, because lactulose regulates intestinal activity, stimulates the growth of *Bifidobacterium* spp., and suppresses intestinal pH and therefore ammonia production. This lowers ammonia absorption in the intestine.

Organic acids and SCFA produced in the colon migrate to the blood and are metabolized in the liver. In one study, the production of C_4 and C_6 fatty acids was completely inhibited by lactulose in an *in vitro* incubation system. It was suggested that lactulose thus detoxified the profile of SCFA produced in the presence of proteins (186). Genovese et al. (187) observed that lactulose improved blood glucose response to an oral glucose test in noninsulin dependent diabetic patients by preadministration of lactulose. Also, lactulose activated the cell mediated immune system depressed during liver cirrhosis (188). A significant reduction in the number of patients contracting urinary infections and respiratory infections using lactulose therapy has been found (189).

Onishi et al. (190) studied the effect of lactulose on ammonia metabolism during dynamic exercise. Increases in blood ammonia in graded exercise were suppressed, and oxygen intake at moderate exercise was increased significantly through preadministration of lactulose. It was thus suggested that lactulose has some benefits when used in combination with exercise.

17.3.4.1 *Colon cancer*

In the context of colon carcinoma, lactulose administration to healthy volunteers lowered fecal concentrations of secondary bile salts (191,192). However Rooney et al. (193) did not confirm this observation. Roncucci et al. (194) reported that lactulose decreased the recurrence rate of colon adenomas. Lactulose suppressed DNA damage in the colon mucosa of rats treated with DMH (1,2 dimethylhydrazine)(195). Roncucci et al. (194) evaluated the effect of lactulose on the recurrence rate of adenomatous polyps. It was found that administration of lactulose reduced the recurrence rate of adenomas in the large bowel.

17.3.4.2 *Mineral absorption*

Lactulose significantly increased calcium absorption but only if the diet contained >0.3% Ca (196). Magnesium absorption from the cecum was affected by inulin but not by the background amount of dietary calcium (197). Suzuki et al. (198) observed that, in rats, lactulose intake promoted the absorption or retention in the body of Ca, Mg, Zn, Cu, and Fe. Igarashi and Ezawa (199) studied the effect of lactulose on bone strength. The results suggested that lactulose promoted calcium absorption in osteoporosis and increased the strength of bones in rats suffering osteoporosis.

17.3.4.3 Constipation

Lactulose has gained popularity in the treatment of constipation and chronic hepatic encephalopathy (147,200–203). Lactulose prevents constipation through its hyperosmotic effects, mainly based on unabsorbed SCFA and cations associated with it (204). Its effects are manifest through activation of intestinal peristaltic movement and intestinal acidification and osmotic reaction. However, this could be true of all prebiotics and may have more to do with all NDO structures than lactulose specifically (205).

17.3.5 Soya products

The primary oligosaccharides contained in soybeans are raffinose (Galα1,6Gluα1,2βFru) and stachyose (Galα1,6Galα1,6Gluα1,2βFru). They are growth promoters of *B. infantis,* *B. longum,* and *B. adolescentis* but do not stimulate *E. coli, E. faecalis,* or *L. acidophilus* (173,206). Also, raffinose effects a relatively short generation time of *B. infantis,* independent of the substrate upon which the organism has been previously cultured. In a study by Fooks et al. (65), all *Bifidobacterium* species tested fermented this carbohydrate with the exception of *B. bifidum,* while *L. salivarius, B. fragilis,* and *Mitsuokella multiacida* metabolized the substrate to a lesser degree. In another study, it was found that addition of a low concentration [0.1% (w/v)] of SOR (refined soybean oligosaccharides) to a two stage continuous culture of fecal bacteria (207) resulted in a threefold increase in the number of bacteria in the total bacterial count.

De Boever et al. (208) examined the survival of *L. reuteri* when challenged with glycodeoxycholic acid (GDCA), deoxycholic acid (DCA), and soygerm powder. Moreover the impact of *L. reuteri* on the bioavailability of isoflavones present in soygerm powder was examined. When 4 g/l soygerm powder was added, the *Lactobacillus* strain survived the bile salt burden better and the membrane damage in the haemolysis test decreased. The *Lactobacillus* strain cleaved β-glycosidic isoflavones during fermentation of milk supplemented with soygerm powder. Interactions between the bacteria and soygerm powder suggest that combining both in fermented milk could exhibit advantageous synbiotic effects. De Boever et al. (209) have shown that consumption of a soygerm powder may beneficially modulate the colonic fermentation. The soygerm powder contained constituents with putative health benefits such as β-glycosidic isoflavones. They can play an important role in the health benefits from soya food, such as lower incidence of estrogen related cancers, cardiovascular diseases, and osteoporosis (210). Saponins are also present in soygerm powder and these compounds may be responsible for the lowering of the serum cholesterol (211). In this study, soygerm powder and a *Lactobacillus reuteri* strain were combined. Because viability is essential for *L. reuteri* to exert its probiotic effects, the ability of soygerm powder to enhance the survival of *L. reuteri* when the latter was exposed to bile salts was investigated. Free isoflavones are only released from soygerm powder in the colon. A second objective was therefore to determine whether *L. reuteri* deconjugated β-glycosidic isoflavones during the fermentation of milk supplemented with soygerm powder, thereby increasing their bioavailability. Soygerm powder contains putative health promoting estrogenic isoflavones. These molecules occur as biologically inactive β-glycosides, which are poorly resorbed by humans. Isoflavones can only be absorbed in a biologically active form after cleavage of their β-glycosidic linkage. It is generally accepted that this does not occur until the compounds reach the colonic microbiota, which produce β-glycosidase (EC 3.2.1.21) (212). Day et al. (213), found that lactase phlorizin hydrolase (EC 3.2.1.108), a β-glycosidase present in the brush border of the mammalian small intestine, was capable of hydrolysing isoflavone glycosides thereby making the free isoflavones available for resorption. Nevertheless, a substantial loss of isoflavone glycosides into the large intestine can be expected because of a limited residence of food compounds in the small intestine and a possible decrease of lactase phlorizin hydrolase levels in adulthood (214). To take advantage of the putative beneficial isoflavones, Izumi et al. (215) suggested that a constant plasma concentration of isoflavones was needed. Therefore one should try to maximize the resorption of free isoflavones by consuming soya foods already containing free isoflavones or by releasing free isoflavones in the small intestine. The enzyme profile of the fermentations indicated that the presence of high levels of β-galactosidase released the isoflavones from the soygerm powder.

De Boever et al. (209) used the *in vitro* model SHIME (simulation of the human intestinal microbial ecosystem) to study the effect of a soygerm powder rich in β-glycosidic phytoestrogenic isoflavones on the fermentation pattern of the colon microbiota and to determine to what extent the latter metabolized the conjugated phytoestrogens. The addition resulted in an overall increase in bacterial marker populations (*Enterobacteriaceae*, coliforms, *Lactobacillus* species, *Staphylococcus* species and *Clostridium* species), with a significant increase in the *Lactobacillus* population. The SCFA concentration increased about 30% during the supplementation period; this was due mainly to a significant increase of acetic and propionic acids. Gas analyses revealed that the methane concentration increased significantly. Ammonia and sulphide concentrations were not influenced by soygerm supplementation. Use of an electronic nose apparatus indicated that odor concentrations decreased significantly during the treatment period. The β-glycosidic bonds of the phytoestrogenic isoflavones were cleaved under the conditions prevailing in the large intestine. The increased bacterial fermentation after addition of the soygerm powder was paralleled by substantial metabolism of the free isoflavones resulting in recovery of only 12–17% of the supplemented isoflavones (209).

Isoflavones have been shown to compete effectively with endogenous mammalian estrogens in binding to the estrogen receptor of mammalian cells preventing estrogen stimulated growth of cancerous cells (216,217). Microbiota play an important role in the generation of biologically active isoflavones, but also inactivate these bioactive compounds after further bacterial fermentation, resulting in a loss of their purported beneficial effects.

Much of the research concerning the health promoting effects of soya products has focused on the putative role of one dietary ingredient, the isoflavones. Soya is an important source of many other nutrients, however, including dietary fibre, oligosaccharides (which could classify as prebiotics), proteins, trace minerals, and vitamins, which could influence the host wellbeing (218). Soya products typically contain high concentrations of the α-glycosidic GOS, raffinose, and stachyose. These sugars are not absorbed in the upper part of the gastrointestinal tract or hydrolysed by human digestive enzymes (219). Upon delivery to the colon, the sugars are fermented by the colonic microbiota possessing α-galactosidase activity. Administration of soybean oligosaccharides to an *in vitro* culture of human gut microbiota has been reported to be bifidogenic (220). Therefore, Gibson and Roberfroid (17) suggested a potential prebiotic role for soybean oligosaccharides. Studies fully addressing this hypothesis have not been published.

Soygerm powder is marketed mainly because it contains a large amount of isoflavones, through which the product could relieve menopausal problems, reduce osteoporosis, improve blood cholesterol levels, and lower the risk of certain types of cancer (221,222). The oligosaccharides present in the soygerm powder consist mainly of indigestible raffinose (13 g/kg) and stachyose (87 g/kg), oligosaccharides which, therefore, serve as substrates for bacterial fermentation (219). During the treatment period, the concentration of all bacterial groups increased and the greatest increase was observed for *Lactobacillus* species. An increase in lactic acid bacteria is often driven by a decrease in luminal pH because the capacity to thrive in low pH environments gives these bacteria a selective advantage over other species (134). The pH reduction in a microniche could select for lactic acid producing bacteria in this environment, possibly excluding long term colonization by invasive pathogens (132). During the supplementation period, the fermentative capacity of the microbial community was increased, and this led to a general rise in SCFA. The largest increase was observed in vessel 3, indicating a high level of carbohydrate fermentation. A rise in SCFA concentration is a positive property because these acids, especially butyric acid, are the main energy source for colonocytes and influence colonic function by stimulating water and sodium absorption and modulating motility (223). Furthermore, butyric acid induces differentiation, stimulates apoptosis of cancerous cells *in vitro,* and thus arrests the development of cancer (224). During supplementation of soygerm powder, no effect on ammonia and hydrogen sulphide concentration was observed. Ammonia can alter the morphology and intermediary metabolism of intestinal cells, increase DNA synthesis, and promote tumorigenesis (225). Hydrogen sulphide selectively inhibits butyric acid oxidation in colonocytes and this may play a pathogenic role in inflammatory bowel diseases such as colitis. Hence, increases in the concentration of these compounds are considered to be potentially harmful to the host. Quantitative gas analyses revealed that methane concentration increased significantly in the last two colonic compartments during the supplementation period. Production of methane by the colonic microbiota is important because it permits a more complete fermentation, in that the removal of H_2 is energetically more favorable (226).

Phenolic and indolic compounds have been linked to a variety of disease states in humans, including initiation of cancer, malabsorption, and anaemia (227). Data revealed that during addition of the soygerm powder, the amount of odor significantly decreased. It is the first study in which the odor produced by the gut microbiota was monitored (209). This indicates that consumption of soygerm powder may influence the gut microbiota in a beneficial manner.

17.3.6 Lactosucrose

Lactosucrose is produced from a mixture of lactose and sucrose using the enzyme β-fructosidase. The fructosyl residue is transferred from the sucrose to the 1-position of the glucose moiety in the lactose, producing a nonreducing oligosaccharide (116) according to the following reaction (228):

$$Gal\beta 1 \rightarrow 4Glc + Glc\alpha 1 \leftrightarrow 2\beta Fru \rightarrow Gal\beta 1 \rightarrow 4Glc\alpha 1 \leftrightarrow 2\beta Fru + Glc$$

β-fructosidase is an artificial oligosugar that is indigestible and therefore low in calories. It enhances the growth of intestinal *Bifidobacteria*. Lactosucrose plays a major role in the refinement of sugars in the food industry (229).

In one study, lactosucrose was used by *B. infantis, B. longum,* and *B. adolescentis,* but not by *L. acidophilus, E. coli* or *E. faecalis* (206). *Bifidobacteria* species grew at a higher rate on lactosucrose than on raffinose (Galα-1,6Gluα-1,2βFru), and the former was utilized by a smaller number of *Enterobacteriaceae* strains than glucose, lactose or raffinose. Fujita et al. (230) found lactosucrose to be similar to FOS in its effects on intestinal bacteria in that it promoted *Bifidobacterium* spp. but few of the other organisms tested.

17.3.7 Xylooligosaccharides

Xylooligosaccharides (XOS) are chains of xylose molecules linked by β-1,4 bonds and mainly consist of xylobiose, xylotriose, and xylotetraose (231), and are found naturally in bamboo shoots, fruits, vegetables, milk, and honey (232). Xylose containing oligosaccharides can be XOS or WBO (wheat bran oligosaccharides). They are produced enzymatically by hydrolysis of xylan from birch wood (233), oats (234), or corncobs (235) by a xylanase from *Trichoderma* (233). WBO are prepared from wheat bran hemicellulose, consisting mainly of branched xylose containing and arabinose containing oligosaccharides such as arabinosylxylobiose, arabinosylxylotriose, arabinosylxylotetraose, and diarabinosylxylotetraose (236). XOS are poorly utilized by *Bifidobacterium* strains, but xylobiose was reported to be bifidogenic (231,237). Only a few *Bifidobacterium* spp. can ferment xylan (238,239). Few studies have been conducted on XOS. Okazaki et al. (237) carried out a small trial in which five male subjects were fed 1 g/d and five were fed 2 g/day XOS for three weeks. It was found that a dose of 1 g/d XOS was sufficient to elicit a bifidogenic response. In a study by Yamada et al. (236), although WBO could not stimulate all the *Bifidobacterium* strains tested, the substrate was concluded to be more selective than FOS or SOE (soybean oligosaccharide extract). A pure culture study by Jaskari et al. (234) concluded raffinose to be superior to xylooligomers and FOS in increasing *Bifidobacterium* populations without increasing potential pathogens. Van Laere et al. (240) concluded that FOS promoted *Lactobacillus* and other intestinal bacteria more than XOS. Hopkins et al. (231) concluded that overall GOS, Oligomate, and FOS were better bifidogenic substrates than XOS, inulin, and pyrodextrins.

17.4 STRUCTURE TO FUNCTION RELATIONSHIPS IN OTHER OLIGOSACCHARIDES

17.4.1 Lactitol

Lactitol (4-0-[β-galactopyranosyl]-D-glucitol) is a sweet tasting sugar alcohol derived from lactose by reduction of the glucose part of the disaccharide (241).

The fermentation of lactitol encourages the growth of saccharolytic bacteria and decreases the amount of proteolytic bacteria, which are responsible for the production of ammonia, carcinogenic compounds, and endotoxins (241). Most ingested lactitol reaches the colon because a suitable β-galactosidase is missing in the small intestine. Cell free extracts of *L. acidophilus* and *B. longum* showed high fermentation activity for both lactulose and lactitol. However, cell free extracts of *E. coli* and *B. fragilis* showed low activity with lactulose and lactitol indicating that they can aid the growth of saccharolytic bacteria in the colon (241). A 10% addition of lactitol to bacterial cultures reduces pH and gas formation (241). Growth of *Bifidobacterium* and *Lactobacillus* species was induced by lactitol (241). The addition of lactitol at 1, 2, 5, or 10% concentration caused inhibition of proteolytic, toxin producing strains and increased saccharolytic bacterial levels to an extent proportional to the lactitol concentration (241). It also caused a decline in pH and inhibition of NH_3 production (241). Lactitol is mainly fermented by anaerobic organisms (241). Scevola et al. (242) also described a reduction in growth of *E. coli* with lactitol. One possible explanation for this effect may be blockage of the receptor sites of the lectins by which the bacteria adhere to cell walls (241). Patients ingesting lactitol had a higher SCFA concentration, which resulted in a pH reduction. This resulted in increased solubility of minerals in the colon (241). Lactitol is used in the treatment of hepatic encephalopathy and other liver diseases because it aids the growth of saccharolytic bacteria such as *Bifidobacterium* and *Lactobacillus* species.

Changing the microbial balance in favor of saccharolytic bacteria inhibits production of endotoxins in the colon and thereby lowers the amount absorbed in blood plasma (241). A marked decrease in the fecal cholesterol concentration and in the concentration of its metabolites, and in total fecal bile acid concentrations of humans has been observed upon lactitol addition to a normal diet (241). These results indicate that saccharolytic microflora were stimulated by lactitol, which has a positive effect on cholesterol.

17.4.2   Laevan Type Prebiotics

Laevan type exopolysaccharide (EPS) from *Lactobacillus sanfranciscensis*, laevan, inulin, and FOS were studied for their prebiotic properties (243). An enrichment of *Bifidobacterium* species was found for EPS and inulin but not for laevan or FOS. The bifidogenic effect of EPS was confirmed using a selective growth medium. The use of EPS and FOS resulted in enhanced growth of *Eubacterium biforme* and *Clostridium perfringens,* respectively (243). Numbers of *Bifidobacterium* spp. increased exclusively with EPS whereas those of *Lactobacillus* decreased concomitantly. With EPS, the coliform counts increased to a reduced extent in comparison to a control without added carbohydrate. Growth of *Clostridium* was stimulated not only by the two EPS but also by laevan (243). In another study, Korakli et al. (238) investigated the metabolism by *Bifidobacterium* spp. of EPS produced by *L. sanfranciscensis*. The prebiotic properties of bread, and metabolism by *Bifidobacterium* spp. of water soluble polysaccharides (WSP) from wheat and rye were also investigated (238). Cereal products are the most important staple food throughout the world. Cereal grains are predominantly composed of starch; nonstarch polysaccharides composed of glucose (β-glucan), fructose (polyfructan), xylose, and arabinose (arabinoglycan) are present (244). Some of these polysaccharides, like starch, are partially digested, and others are believed to be dietary fibre, such as arabinoxylan. In addition to polyfructan, wheat and rye flours contain kestose, nystose, and other FOS of the inulin type (233). Comparison of an intestinal microflora incubated in media with EPS and without a carbohydrate source showed that EPS favored the growth of *Bifidobacterium* spp. by more than two log values and the growth of *Clostridia* by less than one log (243). EPS of *L. sanfranciscensis* TMW1.392 is composed predominantly of fructans and is therefore probably not degraded under the conditions prevailing in the stomach or small intestine. Thus, EPS meets some requirements for its use as a prebiotic (239). Polysaccharides composed of glucose in WSP isolated from rye sourdough fermented with *L. sanfranciscensis* and from control rye dough could not be fermented by *B. adolescentis*. This can probably be attributed to the enzymatic degradation of a starch fraction during dough fermentation. Furthermore, polyfructan in flour was completely degraded by cereal enzymes (238).

In conclusion, EPS produced by *L. sanfranciscensis* TMW1.392 during sourdough fermentation was metabolized by *Bifidobacterium* spp. species. The WSP in wheat and rye were also degraded by *Bifidobacterium* spp. However, polyfructan and the starch fraction, which possess a bifidogenic effect, were degraded by cereal enzymes during dough fermentation while the EPS was retained. The stability of EPS should enable it to withstand the baking process. This suggests that EPS will improve the nutritional properties of sourdough fermented products (238).

17.4.3   Gentiooligosaccharides

Gentiooligosaccharides (GeOS) are glucose polymers of the form Gluβ1-6[Gluβ1-6]$_n$ where n=1-5. These are not hydrolyzed in the stomach or small intestine and should therefore reach the colon intact (235). Rycroft et al. (245) investigated the fermentation properties of GeOS as compared to FOS and maltodextrin in mixed fecal culture. Gentiooligosaccharides gave the largest significant increases in *Bifidobacterium,*

Lactobacillus, and total bacterial numbers during *in vitro* batch culture incubations. However, FOS appeared to be a more selective prebiotic as it did not significantly stimulate growth of bacterial groups which were not probiotic in nature. Gentiooligosaccharides and maltodextrin produced the highest levels of SCFA. Lowest gas production was seen with GeOS and highest with FOS (245). Gentiooligosaccharides possessed bifidogenic activity *in vitro,* which was maintained over the 24-hour fermentation period (245). Large amounts of lactate were produced from GeOS and maltodextrin metabolism (21.60±7.26 mmol/L and 21.86± 10.58 mmol/L respectively compared to 11.44±3.21 mmol/L in FOS). Berggren et al. (246) had previously observed that barley β-glucans generated greater SCFA than FOS when fed to rats. This was also the case in this study, where GeOS were studied, not β-glucans. Highest butyrate production was recorded on FOS in accordance with previous batch fermentation studies (247,248), human trials (127), and rat studies (249,250). FOS also produced the lowest levels of propionate in the fermenters. Lowest gas production in the fermentations was detected with GeOS, whereas FOS produced the highest volume. This is an undesirable characteristic in a prebiotic (245).

17.4.4   Glucose Based Oligosaccharides

Glucose based oligosaccharides include cellooligosaccharides (β1,4), maltooligosaccharides (α1,4), gentiooligosaccharides (β1,6), isomaltooligosaccharides (α1,6) oligosaccharides, and those with mixed linkages such as glucooligosaccharides (α1,2, α1,4 and α1,6 linked), maltodextrin based oligosaccharides, and oligolaminarans.

Yazawa et al. (251) found that oligosaccharides of DP 3-5 from dextran hydrolysate promoted *B. infantis* but not *E. coli, E. faecalis* or *L. acidophilus.* Kohmoto et al. (169) observed that panose, isomaltose, isomaltotriose, and IMO900P were utilized as well as raffinose by all the *Bifidobacterium* spp. tested except *B. Bifidum,* which gave no growth on any of the substrates. *Bacteroides* species utilized all the sugars tested, but fewer *Clostridium* species grew on the IMO than the raffinose. Overall IMO900P appeared at least as selective, if not more so, for *Bifidobacterium* spp. than raffinose. IMO (NC) mixtures of different molecular weight distributions were incubated with fecal bacteria in batch cultures and complex gut models (172). The lower molecular weight products showed similar prebiotic activity to FOS with respect to increases in *Bifidobacterium* spp., decreases in *Bacteroides* and *Clostridium* spp., and the production of lactate and acetate. The higher molecular weight products generated higher butyrate concentrations. In the gut model, a validated three stage compound continuous culture system that represents the conditions typical of the human gastrointestinal tract (252), high molecular weight products stimulated *Bifidobacterium* spp. and *Lactobacillus* strains, while stimulating butyrate production, which persisted throughout the model and was greatest in vessel 3 (representing the distal colon). Once again this suggests that DP affected the prebiotic activity and fermentation of a molecule. In this case, a molecule with a low DP seemed to be more bifidogenic *in vitro.* IMO900P clearly had prebiotic potential *in vivo* and the minimum effective dose was 13–15 g/d IMO900 (253). The IMO3 fraction (trisaccharide) had a greater prebiotic effect than IMO2 (171). This may have been due to the greater hydrolysis of IMO2 than IMO3 by isomaltase in the jejunum so more substrate was available for the colonic microflora when IMO3 was administered.

In a limited pure culture study (254), while most *Bacteroides* strains, particularly *B. thetaiomicron,* degraded α-GOS, eight *Bifidobacterium* strains did not do so to the same extent, and three could not degrade it at all. Two *C. butyricum* strains degraded around 50% whereas most *Lactobacillus* strains hydrolysed less than 10% α-GOS, which appears not to have a prebiotic effect. IMO appeared to be prebiotic. Differences among glucooligosaccharides were due to the DP or linkage types of the molecules, suggesting that a low DP and α- rather than β- linkages favor fermentation and prebiotic activity.

17.5 CONCLUSIONS

Increasing consumer awareness for health and nutritional issues makes the market for prebiotics very promising; however, because of the high product costs and the still fragmented scientific backup of some of the health claims, this is still a disorganized area. Structure to function relationship studies, together with the development of cheaper manufacturing techniques, and more understanding of the physiological mechanism by which prebiotics bring about improved health, provide one answer to establishing the functional oligosaccharide market.

This chapter has reviewed research and conclusions on the complex effects of oligosaccharide structure and size on the human microflora. A large comparative study, as reported by Djouzi and Andrieux (250), involved feeding 4% (w/w) FOS, TOS (transgalactooligosaccharides), and α-GOS diets to rats for four weeks. TOS and FOS led to desirable changes in microflora composition but had little effect on enzyme activities. α-GOS did not affect microflora composition but favorably altered the metabolism. The desirable changes with α-GOS were reduced gas production and increased glycosidase activity. GOS was superior to FOS in the greater *Bifidobacterium* population and the increased β-galactosidase activity when GOS was administered. Galactooligosaccharides, however, gave higher hydrogen and methane excretions. Inulin had a similar effect to FOS although higher doses of the former were needed (127,131). This suggests that a lower molecular weight gives a higher prebiotic activity, in accordance with pure culture studies (251). Transgalactosylated disaccharides were not as effective prebiotics as TOS and Oligomate, as the former needed a higher dose (157,255,256). Therefore, longer chain GOS were more effective than their disaccharide counterparts. Although these findings are contradictory, they suggest that neither disaccharides nor polysaccharides are the most effective prebiotics, and that oligosaccharides of DP>3 would be most likely prebiotic candidates.

In human trials with soybean oligosaccharides (257,258) less SOE than raffinose was needed to gain a prebiotic effect, and since 3 times as much stachyose (tetrasaccharide) than raffinose is present in SOE, it is likely that stachyose has the best prebiotic activity of the two (220). In the case of IMO, a low DP was more bifidogenic *in vitro* fermentation studies (172). α-galactooligosacchride showed some prebiotic activity, MDO (maltodextrin based oligosaccharide) behaved like a polysaccharide, and IMO behaved like a prebiotic. So a low DP and α- rather than β- linkages favor fermentation and prebiotic activity (172,247,248,254). The complexity of structure to function studies is also due to population variation which accounts for many inconsistencies between studies, and the observation that very often the bifidogenic effect is dependent on the original level of the *Bifidobacterium* strains present (259). For this reason, there is a great need for an extensive comparative *in vitro* study of all prebiotics available, involving the same parameters and in the same controlled conditions, so as to allow a comparison of the same amounts. This will help to better inform the most efficacious prebiotics.

REFERENCES

1. Wiggins, H.S. Gastroenterological functions of dietary fibre. In: *Dietary Fibre*, Birch, G.G., K.J. Parker, eds., London: Applied Science, 1983, pp 205–219.

2. Cummings, J.H., G.T. Macfarlane. The control and consequences of bacterial fermentation in the human colon. *J. Appl. Bacteriol.* 70:443–459, 1991.
3. Cummings, J.H. Short chain fatty acids in the human colon. *Gut* 22:763–779, 1981.
4. Wolin, M.J., T.L. Miller. Carbohydrate fermentation. In: *Human Intestinal Microflora in Health and Disease*, Hentges, D.J., ed., London: Academic Press, 1983, pp 147–165.
5. Allison, C., G.T. Macfarlane. Effect of nitrate upon methane production by slurries of human faecal bacteria. *J. Gen. Microbiol.* 134:1397–1405, 1988.
6. Macfarlane, G.T., J.H. Cummings, C. Allison. Protein degradation by human intestinal bacteria. *J. Gen. Microbiol.* 132:1647–1656, 1986.
7. Drasar, B.S., M.J. Hill. *Human Intestinal Flora*. London: Academic Press, 1974.
8. Bakke, O.M., T. Midtvedt. Influence of germfree status on the excretion of simple phenols of possible significance in tumor promotion. *Experientia* 26:519–519, 1970.
9. E. Bone, A. Tamm, M. Hill. The production of urinary phenols by gut bacteria and their possible role in the causation of large bowel cancer. *Am. J. Clin. Nutr.* 29:1448–1454, 1976.
10. Macfarlane, G.T., G.R. Gibson. Carbohydrate fermentation, energy transduction and gas metabolism in the human large intestine. In: *Gastrointestinal Microbiology*, Vol. 1, Mackie, R.I., B.H. White, eds., London: Chapman & Hall, 1997, pp 269–318.
11. Macfarlane, G.T., H.N. Englyst. Starch utilisation by the human large intestinal microflora. *J. Appl. Bacteriol.* 60:195–201, 1986.
12. Hoskins, L.C., E.T. Boulding. Mucin degradation in human colon ecosystems. *J. Clin. Invest.* 67:163–172, 1981.
13. Roberton, A.M., R.A. Stanley. *In vitro* utilization of mucin by *Bacteroides fragilis. Appl. Environ. Microbiol.* 43:325–330, 1982.
14. Hoskins, L.C., M. Agustines, W.B. McKee, E.T. Boulding, M. Kriaris, G. Niedermeyer. Mucin degradation in human colon ecosystems. Isolation and properties of fecal strains that degrade ABH blood group antigens and oligosaccharides from mucin glycoproteons. *J. Clin. Invest.* 75:944–953, 1985.
15. Salyers, A.A., J.A.Z. Leedle. Carbohydrate utilization in the human colon. In: *Human Intestinal Microflora in Health and Disease,* Hentges, D.J., ed., London: Academic Press, 1983, pp 129–146.
16. Macfarlane, G.T., S. Hay, S. Macfarlane, G.R. Gibson. Effect of different carbohydrates on growth, polysaccharidase and glycosidase production by *Bacteroides ovatus*, in batch and continuous culture. *J. Appl. Bacteriol.* 68:179–187, 1990.
17. Gibson, G.R., M.B. Roberfroid. Dietary modulation of the human colonic microbiota: introducing the concept of prebiotics. *J. Nutr.* 125:1401–1412, 1995.
18. Zubillaga, M., R. Weill, E. Postaire, C. Goldman, R. Caro, J. Boccio. Effect of probiotics and functional foods and their use in different diseases. *Nutr. Res.* 21:569–579, 2001.
19. Gibson, G.R., R. Fuller. Aspects of *in vitro* and *in vivo* research approaches directed toward identifying probiotics and prebiotics for human use. *J. Nutr.* 130:391S–395S, 2000.
20. Rani, B., N. Khetarpaul. Probiotic fermented food mixture: possible applications in clinical anti-diarrhoea usage. *Nutr. Health* 12:97–105, 1998.
21. Aiba, Y., N. Suzuki, A.M. Kabir, A. Takagi, Y. Koga. Lactic acid-mediated supression of *Helicobacter pylori* by the oral administration of *Lactobacillus salivarius* as a probiotic in a gnotobiotic murine model. *Am. J. Gastroenterol.* 93:2097–2101, 1998.
22. Bazhenov, L.G., V.M. Bondarenko, E.A. Lykova. The antagonostic action of lactobacilli on *Helicobacter pylori. Zh. Mikrob. Epid. Immun.* 3:89–91, 1997.
23. Kabir, A.M., Y. Aiba, A. Takagi, S. Kamiya, T. Miwa, Y. Koga. Prevention of *Helicobacter pylori* infection by lactobacilli in a gnotobiotic murine model. *Gut* 41:49–55, 1997.
24. Jiang, T., D.A. Savaiano. *In vitro* lactose fermentation by human colonic bacteria is modified by *Lactobacillus acidophilus* supplementation. *J. Nutr.* 127:1489–1495, 1997.
25. Pool-Zobel, B.L., R. Munzner, W.H. Holzapfel. Antigenotoxic properties of lactic acid bacteria in the *S. typhimurium* mutagenicity assay. *Nutr. Cancer* 23:261–270, 1993.

26. Spanhaak, S., R. Hanevaar, G. Schaafsma. The effect of consumption of fermented by *Lactobacillus casei* strain Shirota on the intestinal microflora and immune parameters in humans. *Eur. J. Clin. Nutr.* 52:899–907, 1998.

27. Schiffrin, E.J., D. Brassart, A.L. Servin, F. Rochat, A. Donnet-Hughes. Immune modulation of blood leukocytes in humans by lactic acid bacteria: criteria for strain selection. *Am. J. Clin. Nutr.* 66:515S–520S, 1997.

28. Tannock, G.W. *Normal Microflora: An Introduction to Microbes Inhabiting the Human Body*. London: Chapman & Hall, 1995.

29. O' Sullivan, M.G. Metabolism of bifidogenic factors by gut flora: an overview. *Bull. IDF* 313:23–30, 1996.

30. Englyst, H.N., S. Hay, G.T. Macfarlane. Polysaccharide breakdown by mixed populations of human faecal bacteria. *FEMS Microbiol. Ecol.* 95:163–171, 1987.

31. Scheppach, W., C. Fabian, M. Sachs, H. Kasper. The effect of starch malabsorption on fecal short chain fatty acid excretion in man. *Scand. J. Gastroenterol.* 23:755–759, 1988.

32. Wiggings, H.S., J.H. Cummings. Evidence for the mixing of residue in the human gut. *Gut* 17:1007–1011, 1976.

33. Macfarlane, G.T., J.H. Cummings, S. Macfarlane, G.R. Gibson. Influence of retention time on degradation of pancreatic enzymes by human colonic bacteria grown in a 3-stage continuous culture system. *J. Appl. Bacteriol.* 67:521–527, 1989.

34. Cummings, J.H., M.J. Hill, E.S. Bone, W.J. Branch, D.J.A. Jenkins. The effect of meat protein and dietary fibre on colonic function and metabolism, part II: bacterial metabolites in feces and urine. *Am. J. Clin. Nutr.* 32:2094–2101, 1979.

35. Bingham, S.A., S. Pett, K.C. Day. NSP intake of a representative sample of British adults. *J. Hum. Nutr. Diet.* 3:339–344, 1990.

36. Cummings, J.H., E.W. Pomare, W.J. Branch, C.P.E. Naylor, G.T. Macfarlane. Short chain fatty acids in human large intestine, portal, hepatic and venous blood. *Gut* 28:1221–1227, 1987.

37. Argenzio, R.A., M. Miller, W. Von Engelhardt. Effect of volatile fatty acids on water and ion absorption from the goat colon. *Am. J. Physiol.* 229:997–1002, 1975.

38. Argenzio, R.A., S.C. Whipp. Interrelationship of sodium, chloride, bicarbonate and acetate transport by the colon of the pig. *J. Physiol.* 295:315–381, 1979.

39. Salminen, S., C. Bouley, M.-C. Boutron-Ruault, J.H. Cummings, A. Franck, G.R. Gibson, E. Isolauri, M.-C. Moreau, M. Roberfroid, I. Rowland. Functional food science and gastrointestinal physiology and function. *Br. J. Nutr.* 80(1):S147–S171, 1998.

40. Macfarlane, G.T., G.R. Gibson. Bacterial infections and diarrhoea. In: *Human Colonic Bacteria: Role in Nutrition, Physiology, and Pathology*, Gibson, G.R., G.T. Macfarlane, eds., Boca Raton, FL: CRC Press, 1995, pp 201–226.

41. Isolauri, E., Y. Sutas, P. Kankaanpaa, H. Arvilommi, S. Salminen. Probiotics: effects on immunity. *Am. J. Clin. Nutr.* 73:444S–50S, 2001.

42. Ziemer, C.J., G.R. Gibson. An overview of probiotics, prebiotics and synbiotics in the functional food concept: perspectives and future strategies. *Int. Dairy J.* 8:473–479, 1998.

43. Gorbach, S.L., T.W. Chang, B. Goldin. Successful treatment of relapsing *Clostridium difficile* colitis with *Lactobacillus* GG. *Lancet* 2:1519–1519, 1987.

44. Kimmey, M.B., G.W. Elmer, C.M. Surawics, L. McFarland. Prevention of further recurrence of *Clostridium difficile* colitis with *Saccharomyces boulardii*. *Dig. Dis. Sci.* 35:897–901, 1990.

45. Buts, J.P., G. Corthier, M. Delmee. *Saccharomyces boulardii* for *Clostridium difficile* associated enteropathies in infants. *J. Pediatr. Gastroenterol. Nutr.* 16:419–425, 1993.

46. Surawics, C.M., L. McFarland, G.W. Elmer, J. Chinn. Treatment of recurrent *Clostridium difficile* colitis with vancomycin and *Saccharomyces boulardii*. *Am. J. Gastroenterol.* 84:1285–1287, 1989.

47. Biller, J.A., A.J. Katz, A.F. Flores, T.M. Buie, S.L. Gorbach. Treatment of recurrent *Clostridium difficile* colitis with *Lactobacillus* GG. *J. Pediatr. Gastroenterol. Nutr.* 21:224–226, 1995.

48. Bennet, R.G., S.L. Gorbach, B.R. Goldin, T.-W. Chang, B.E. Laughon, W.B. Greenough III, J.G. Bartlett. Treatment of relapsing *C. difficile* diarrhea with *Lactobacillus GG. Nutr. Today* 31:35S–38S, 1996.

49. Levy, J. Experience with live *Lactobacillus plantarum* 299v: a promising adjunct in the management of recurrent *Clostridium difficile* infection. *Gastroenterology* 112:A379, 1997.

50. McFarland, L.V., C.M. Surawics, R.N. Greenberg, R. Fekety, G.W. Elmer, K.A. Moyer, S.A. Melcher, K.E. Bowen, J.L. Cox, Z. Noorani. A randomized placebo-controlled trial of *Saccharomyces boulardii* in combination with standard antibiotics for *Clostridium difficile*. *JAMA* 271:1913–1918, 1994.

51. Fabia, R., A. Ar'Rajab, M.L. Johansson, R. Willen, R. Anderson. The effect of exogeneous administration of *Lactobacillus reuteri* R2LC and oat fibre on acetic acid-induced colitis in the rat. *Scand. J. Gastroenterol.* 111:334–344, 1993.

52. Gibson, G.R., G.T. Macfarlane. Intestinal bacteria and disease. In: *Human Health: Contribution of Microorganisms,* Gibson, S.A.W., ed., London: Spinger-Verlag, 1994, pp 53–62.

53. Malin, M., H. Suomalainen, M. Saxelin, E. Isolauri. Promotion of IgA immune response in patients with Crohn's disease by oral bacteriotherapy with *Lactobacillus GG. Ann. Nutr. Metab.* 40:137–145, 1996.

54. Gibson, G.R., J.H. Cummings, G.T. Macfarlane. Growth and activities of sulphate-reducing bacteria in gut contents of healthy subjects and patients with ulcerative colitis. *FEMS Microbiol. Ecol.* 86:103–112, 1991.

55. Kruis, W., E. Schutz, P. Fric, B. Fixa, G. Judmaier, M. Stolte. Double-blind comparison of an oral *Escherichia coli* preparation and mesalazine in mantaining remission of ulcerative colitis. *Aliment. Pharmacol. Ther.* 11:853–858, 1997.

56. Rembacken, B.J., A.M. Snelling, P.M. Hawkey, D.M. Chalmers, A.T. Axon. Non-pathogenic *Escherichia coli* versus mesalazine for the treatment of ulcerative colitis: a randomised trial. *Lancet* 354:635–639, 1999.

57. Rolfe, R.D. Colonisation resistance. In: *Gastrointestinal Microbiology. Vol 2. Gastrointestinal microbes and host interactions,* Mackie, R.I., White, B.A., R.E. Isaacson, eds., New York: Chapman and Hall, 1997.

58. Bazzoli, F., R.M. Zagari, S. Fossi, M.C. Morelli, P. Pozzato, M. Ventrucci, G. Mazzella, D. Festi, E. Roda. *In vivo Helicobacter pylori* clearance failure with *Lactobacillus acidophilus*. *Gastroenterology* 102:(A38), 1992.

59. Marteau, P.R., M. de Vrese, C.J. Cellier, J. Schrezenmeir. Protection from gastrointestinal diseases with the use of probiotics. *Am. J. Clin. Nutr.* 73:430S–436S, 2001.

60. Adams, J., C. Barret, A. Barret-Bellet, E. Benedetti, A. Calendini, P. Daschen. Essais cliniques controlès en double insu de l'ultra-levure lyophilisèe. Etude multicentrique par 25 mèdecins de 388 cas. *Gazette Med. France* 84:2072–2078, 1977.

61. Marteau, P., P. Pochart, Y. Bouhnik, J.C. Rambaud. Fate and effects of some transiting microorganisms in the human gastrointestinal tract. *World Rev. Nutr. Diet.* 74:1–21, 1993.

62. Gibson, G.R., J.M. Saavedra, S. McFarlane, G.T. McFarlane. Probiotics and intestinal infections. In: *Probiotics 2: Applications and Practical Aspects,* Fuller, R., ed., New York: Chapman & Hall,1997, pp 10–38.

63. Saavedra, J.M., N.A. Bauman, I. Oung, J.A. Perman, R.H. Yolken. Feeding of *Bifidobacterium bifidum* and *Streptococcus thermophilus* to infants in hospital for prevention of diarrhea and shedding of rotavirus. *Lancet* 344:1046–1049, 1994.

64. Halpern, G.M., K.J. Vruwink, J. Van der Water, C.L. Keen, M.E. Gershwin. Influence of long-term yogurt consumption in young adults. *Int. J. Immunother.* 7:205–210, 1991.

65. Fooks, L.J., R. Fuller, G.R. Gibson. Prebiotics, probiotics and human gut microbiology. *Int. Dairy J.* 9:53–61, 1999.

66. Kaila, M., E. Isolauri, E. Soppi, V. Virtanen, S. Laine, H. Arvilommi. Enhancement of the circulating antibody secreting cell response in human diarrhea by a human *Lactobacillus* strain. *Pediatr. Res.* 32:141–144, 1992.

67. Schiffrin, E.J., F. Rochat, H. Link-Amster, J.M. Aeschlimann, A. Donnet-Hughes. Immunomodulation of human blood cells following the ingestion of lactic acid bacteria. *J. Dairy Sci.* 78:491–497, 1994.

68. Marteau, P., M. Minekus, R. Havenaar, J.H.J. Huis in't Veld. Survival of lactic acid bacteria in a dynamic model of the stomach and small intestine: validation and the effects of bile. *J. Dairy Sci.* 80:1031–1037, 1997.

69. Marteau, P., J.P. Vaerman, J.P. Dehennin, S. Bord, D. Brassart, P. Pochart, J.F. Desjeux, J.C. Rambaud. Effects of intrajejunal perfusion and chronic ingestion of *Lactobacillus johnsonii* strain La1 on serum concentrations and jejunal secretions of immunoglobulins and serum propteins in healthy humans. *Gastroenterol. Clin. Biol.* 21:293–298, 1997.

70. Solis-Pereira, B., D. Lemonnier. Induction of human cytokines by bacteria used in dairy foods. *Nutr. Res.* 13:1127–1140, 1996.

71. Bolte, G., M. Knauss, I. Metzdorf, M. Stern. Postnatal maturation of rat small intestinal brush border membranes correlated with increase in food protein binding capacity. *Dig. Dis. Sci.* 43:148–155, 1998.

72. Van der Heijden, P.J., A.T.J. Bianchi, M. Dol, J.W. Pals, W. Stok, B.A. Bokhout. Manipulation of intestinal immune responses against ovalbumin by cholera toxin and its B subunit in mice. *Immunology* 72:89–93, 1991.

73. G. Perdigon, M.E. de Macias, S. Alvarez, G. Oliver, A.A. de Ruiz Holgado. Effect of perorally administered lactobacilli on macrophage activation in mice. *Infect. Immun.* 53:404–410, 1986.

74. Perdigon, G., M.E. de Macias, S. Alvarez, G. Oliver, A.P. de Ruiz Holgado. Systemic augmentation of the immune response in mice by feeding fermented milks with *Lactobacillus casei* and *Lactobacillus acidophilus*. *Immunology* 63:17–23, 1998.

75. Miettinen, M., J. Vuopio-Varkila, K. Varkila. Production of human tumor necrosis factor alpha, interleukin-6, and interleukin-10 is induced by lactic acid bacteria. *Infect. Immun.* 64:5403–5405, 1996.

76. Pelto, L., E. Isolauri, E.M. Lilius, J. Nuutila, S. Salminen. Probiotic bacteria down-regulate the milk-induced inflammatory response in milk-hypersensitive subjects but have an immunostimulatory effect in healthy subjects. *Clin. Exp. Allergy* 28:1471–1479, 1998.

77. Link-Amster, H., F. Rochat, K.Y. Saudan, O. Mignot, J.M. Aeschlimann. Modulation of a specific humoral immunoresponse and changes in intestinal flora mediated through fermented milk intake. *FEMS Immunol. Med. Microbiol.* 10:55–64, 1994.

78. Madara, L.J. The chameleon within: improving antigen delivery. *Science* 277:910–911, 1997.

79. Michetti, P., G. Dorta, P.H. Wiesel, D. Brassart, E. Verdu, M. Herranz, C. Felley, N. Porta, M. Rouvet, A.L. Blum, I. Corthesy-Theulaz. Effect of whey-based culture supernatant of *Lactobacillus acidophilus* (johnsonii) La1 on *Helicobacter pylori* infection in humans. *Digestion* 60:203–209, 1999.

80. Ouwehand, A.C., P.V. Kirjavainen, C. Shortt, S. Salminen. Probiotics: mechanisms and established effects. *Int. Dairy J.* 9:43–52, 1999.

81. Kim, H.S., S. Gilliland. *L. acidophilus* as a dietary adjunct for milk to aid lactose digestion in humans. *J. Dairy Sci.* 66:959–966, 1983.

82. De Vrese, M., A. Stegelmann, B. Richter, S. Fenselau, C. Laue, J. Schrezenmeir. Probiotics-compensation for lactase insufficiency. *Am. J. Clin. Nutr.* 73:421S–429S, 2001.

83. Marteau, P., B. Flourie, P. Pochart, C. Chastang, J.F. Desjeux, J.C. Rambaud. Effect of the microbial lactase activity in yogurt on the intestinal absorption of lactose: an *in vivo* study in lactase-deficient humans. *Br. J. Nutr.* 64:71–79, 1990.

84. Mahè, S., P. Marteau, J.F. Huneau, F. Thuilier, D. Tomè. Intestinal nitrogen and electrolyte movements following fermented milk ingestion in human. *Br. J. Nutr.* 71:169–180, 1994.

85. Lin, M., C.L. Yen, S.H. Chen. Management of lactose maldigestion by consuming milk containing lactobacilli. *Dig. Dis. Sci.* 43:133–137, 1998.

86. Rowland, I.R. Toxicology of the colon-role of the intestinal microflora. In: *Human Colonic Bacteria: Role in Nutrition, Physiology, and Pathology,* Gibson, G.R., G.T. Macfarlane, eds., Boca Raton, FL: CRC Press, 1995, pp 115–174.

87. Brady, L.J., D.D. Gallaher, F.F. Busta. The role of probiotic cultures in the prevention of colon cancer. *J. Nutr.* 130:410S–414S, 2000.

88. Mital, B.K., S.K. Garg. Anticarcinogenic, hypocholesterolemic, and antagonistic activities of *Lactobacillus acidophilus. Crit. Rev. Microbiol.* 21:175–214, 1995.

89. Wollowski, I., G. Rechkemmer, B. Pool-Zobel. Protective role of probiotics and prebiotics in colon cancer. *Am. J. Clin. Nutr.* 73:451S–455S, 2001.

90. Sakata, T. Stimulatory effect of short-chain fatty acids on epithelial cell proliferation in the rat intestine: a possible explanation for trophic effects of fermentable fibre, gut microbes and luminal trophic factors. *Br. J. Nutr.* 58:95–103, 1987.

91. Leder, A., P. Leder. Butyric acid, a potent inducer of erythroid differentiation in cultural erythroleukemic cells. *Cell* 5:319–322, 1975.

92. Hagopian, H.K., M.G. Riggs, L.A. Swartz, V.M. Ingram. Effect of *n*-butyrate on DNA synthesis in chick fibroplast and Hela cells. *Cell* 12:855–860, 1977.

93. Borenfreund, E., E. Schmid, A. Bendich, W.S. Franke. Constitutive aggregates of intermediate-sized filaments of the vimentin and cytokeratin type in cultured hepatoma cells and their dispersal by butyrate. *Exp. Cell Res.* 127:215–235, 1980.

94. Wrong, O.M., A.J. Vince, J.C. Waterlow. The contribution of endogenous urea to fecal ammonia in man, determined by ^{15}N labelling of plasma urea. *Clin. Sci.* 68:193–199, 1985.

95. Visek, W.J. Effects of urea hydrolysis on cell life-span and metabolism. *Fed. Proc.* 31:1178–1193, 1972.

96. Visek, W.K. Diet and cell growth modulation by ammonia. *Am. J. Clin. Nutr.* 31:S216–S220, 1978.

97. Dunning, W.T., M.R. Curtis, M.E. Mann. The effect of added dietary tryptophane on the occurrence of 2-acetylaminofluorene-induced liver and bladder cancer in rats. *Cancer Res.* 10:454–459, 1950.

98. Bryan, G.T. The role of urinary tryptophan metabolites in the etiology of bladder cancer. *Am. J. Clin. Nutr.* 24:841–847, 1971.

99. Duran, M., D. Ketting, P.K. DeBrec, C. van der Heiden, S.K. Wadman. Gas chromatographic analysis of urinary phenols in patients with gastrointestinal disorders and normals. *Clin. Chim. Acta* 45:341–347, 1973.

100. Kulkarni, N., B.S. Reddy. Inhibitory effect of *Bifidobacterium longum* cultures on the azoxymethane-induced aberrant crypt foci formation and fecal bacterial β-glucuronidase. *Proc. Soc. Exp. Biol. Med.* 207:278–283, 1994.

101. Rowland, I.R., C.J. Rumney, J.T. Coutts, L.C. Lievense. Effect of *Bifidobacterium longum* and inulin in gut bacterial metabolism and carcinogen-induced aberrant crypt foci in rats. *Carcinogenesis* 19:281–285, 1998.

102. Kinouchi, T., K. Kataoka, S. Ruo Bing, H. Nakayama, M. Uejima, K. Shimono, T. Kuwahara, S. Akimoto, I. Hiraoka, Y. Onishi. Culture supernatants of *Lactobacillus acidophilus* and *Bifidobacterium adolescentis* repress ileal ulcer formation in rats treated with a non-steroidal anti-inflammatory drug by suppressing unbalanced growth of aerobic bacteria and lipid peroxidation. *Microbiol. Immunol.* 42:347–355, 1998.

103. Goldin, B.R., S.L. Gorbach. Alterations of the intestinal microflora by diet, oral antibiotics, and *Lactobacillus*: decreased production of free amines from aromatic nitro compounds, azo dyes and glucorinides. *J. Natl. Cancer Inst.* 73:689–695, 1984.

104. Kato, I., K. Endo, T. Yokokura. Effects of oral administration of *Lactobacillus casei* on antitumor responses induced by tumor resection in mice. *Int. J. Immunopharmacol.* 16:29–36, 1994.

105. Okawa, T., H. Niibe, T. Arai, K. Sekiba, K. Noda, S. Takeuchi, S. Hashomoto, N. Ogawa. Effect of LC9018 combined with radiation therapy on carcinoma of the uterine cervix: a phase III, multicenter, randomized, controlled study. *Cancer* 72:1949–1954, 1993.

106. Rafter, J.J. The role of lactic acid bacteria in colon cancer prevention. *Scand. J. Gastroenterol.* 30:497–502, 1995.

107. Reddy, B.S. Prevention of colon cancer by pre- and probiotics: evidence from laboratory studies. *Br. J. Nutr.* 80(2):S219–S223, 1998.

108. Barbacid, M. *Ras* oncogenes: their role in neoplasia. *Eur. J. Clin. Invest.* 20:225–235, 1990.

109. Kotsinas, A., D.A. Spandidos, P. Romanowski, A.H. Wyllie. Relative expression of wild-type and activated Ki-*ras*2 oncogene in colorectal carcinomas. *Int. J. Oncol.* 3:841–845, 1993.

110. Taylor, G.R.J., C.M. Williams. Effect of probiotics and prebiotics on blood lipids. *Br. J. Nutr.* 80(2):S225–S230, 1998.

111. Bergman, D.N., W.E. Roe, K. Kon. Quantitative aspects of propionate metabolism and glucogenesis in sheep. *Am. J. Physiol.* 211:793–799, 1966.

112. Chen, W.-J.L., J.W. Anderson, D. Jennings. Propionate may mediate the hypocholesterolemic effects of certain soluble plant fibres in cholesterol-fed rats. *Proc. Soc. Exp. Biol. Med.* 175:215–218, 1984.

113. Illman, R.J., D.L. Topping, G.H. McIntosh, R.P. Trimble, G.B. Storer, M.N. Taylor, B.-Q. Cheng. Hypocholesterolaemic effects of dietary propionate studies in whole animals and perfused rat liver. *Ann. Nutr. Metab.* 32:97–107, 1988.

114. Venter, C.S., H.H. Vorster, J.H. Cummings. Effects of dietary propionate on carbohydrate and lipid metabolism in man. *Am. J. Gastroenterol.* 85:549–552, 1990.

115. Schaafsma, G., W.J.A. Meuling, W. van Dokkum, C. Bouley. Effects of a milk product, fermented by *Lactobacillus acidophilus* and with fructo-oligosaccharides added, on blood lipids in male volunteers. *Eur. J. Clin. Nutr.* 52:436–440, 1998.

116. Sako, T., K. Matsumoto, R. Tanaka. Recent progress on research and applications of non-digestible galacto-oligosaccharides. *Int. Dairy J.* 9:69–80, 1999.

117. Yun, J.W. Fructooligosaccharides: occurrence, preparation, and application. *Enzyme Microb. Technol.* 19(2):107–117, 1996.

118. Hidaka, H., M. Hirayama, N. Sumi. A fructooligosaccharide-producing enzyme from *Aspergillus niger* ATCC 20611. *Agric. Biol. Chem.* 52:1181–1187, 1988.

119. Roberfroid, M.B., J.A.E. Van Loo, G.R. Gibson. The bifidogenic nature of chicory inulin and its hydrolysis products. *J. Nutr.* 128(1):11–19, 1998.

120. Crittenden, R.G., M.J. Playne. Production, properties and applications of food-grade oligo-saccharides. *Trends Food Sci. Technol.* 7:353–361, 1996.

121. Muramatsu, K., S. Onodera, S. Kikuchi, N. Shiomi. Substrate specificity and subsite affinities of β-fructofuranosidase from *Bifidobacterium adolescentis* G1. *Biosci. Biotech. Biochem.* 58:1642–1645, 1994.

122. Perrin, S., M. Warchol, J.P. Grill, F. Schneider. Fermentations of fructo-oligosaccharides and their components by *Bifidobacterium infantis* ATCC 15697 on batch culture in semi-synthetic medium. *J. Appl. Microbiol.* 90:859–865, 2001.

123. Rao, V.A. The prebiotic properties of oligofructose at low intake levels. *Nutr. Res.* 21:843–848, 2001.

124. Kaplan, H., R.W. Hutkins. Fermentation of fructooligosaccharides by lactic acid bacteria and *Bifidobacteria*. *Appl. Environ. Microbiol.* 66:2682–2684, 2000.

125. Rastall, R.A., V. Maitin. Prebiotics and synbiotics: towards the next generation. *Curr. Opin. Biotechnol.* 13:490–496, 2002.

126. Fuller, R., G.R. Gibson. Modification of the intestinal microflora using probiotics and pre-biotics. *Gastroenterology* 32(222):28–31, 1997.

127. Gibson, G.R., E.B. Beatty, X. Wang, J.H. Cummings. Selective stimulation of *Bifidobacteria* in the human colon by oligofructose and inulin. *Gastroenterology* 108:975–982, 1995.

128. Gibson, G.R., A. Willems, S. Reading, M.D. Collins. Fermentation of non-digestible oligo-saccharides by human colonic bacteria. *Proc. Nutr. Soc.* 55:899–912, 1996.

129. Roberfroid, M.B. Functional effects of food components and the gastrointestinal system: chicory fructooligosaccharides. *Nutr. Rev.* 54:S38–S42, 1996.

130. Buddington, R.K., C.H. Williams, S.C. Chen, S.A. Witherley. Dietary supplementation of neosugar alters the fecal flora and decreases activities of some reductive enzymes in human subjects. *Am. J. Clin. Nutr.* 63:709–716, 1996.

131. Kleesen, B., B. Sykura, H.J. Zunft, M. Blaut. Effects of inulin and lactose on fecal microflora, microbial activity and bowel habit in elderly constipated persons. *Am. J. Clin. Nutr.* 65:1397–1402, 1997.

132. Gibson, G.R., X. Wang. Regulatory effects of *Bifidobacteria* on the growth of other colonic bacteria. *J. Appl. Bacteriol.* 77:412–420, 1994.

133. Bielecka, M., E. Biedrzycka, A. Majkowska. Selection of probiotics and prebiotics for synbiotics and confirmation of their *in vivo* effectiveness. *Food Res. Int.* 35:125–131, 2002.

134. Sghir, A., J.M. Clow, R.I. Mackie. I: Continuous culture selection of *Bifidobacteria* and lactobacilli from human faecal samples using fructooligosaccharide as selective substrate. *J. Appl. Microbiol.* 85:769–777, 1988.

135. Hartemink, R, M.C.J. Quataert, K.M.J. Van Laere, M.J.R. Nout, F.M. Rombouts. Degradation and fermentation of fructo-oligosaccharides by oral streptococci. *J. Appl. Bacteriol.* 79:551–557, 1995.

136. Losada, M.A., T. Olleros. Towards a healthier diet for the colon: the influences of fructooligosaccharides and lactobacilli on intestinal health. *Nutr. Res.* 22:71–84, 2002.

137. Gmeiner, M., W. Kneifel, K.D. Kulbe, R. Wouters, P. De Boever, L. Nollet, W. Verstraete. Influence of a synbiotic mixture consisting of *Lactobacillus acidophilus* 74-2 and a fructooligosaccharide preparation on the microbial ecology sustained in a simulation of the human intestinal microbial ecosystem (SHIME reactor). *Appl. Microbiol. Biotechnol.* 53:219–223, 2000.

138. Wang, X. Comparative aspects of carbohydrate fermentation by colonic bacteria. PhD dissertation, University of Cambridge, Cambridge, U.K., 1993.

139. Delzenne, N.M., N. Kok. Effects of fructans-type prebiotics on lipid metabolism. *Am. J. Clin. Nutr.* 73:456S–458S, 2001.

140. Stanley, J.C., E.A. Newsholme. The effect of dietary guar gum on the activities of some key enzymes of carbohydrate and lipid metabolism in mouse liver. *Br. J. Nutr.* 53:215–222, 1985.

141. Leclère, C.J., M. Champ, J. Boillot, G. Guille, G. Lecannu, C. Molis, F. Bornet, M. Krempf, J. Delortlaval, J.P. Galmichem. Role of viscous guar gums in lowering the glycemic response after a solid meal. *Am. J. Clin. Nutr.* 59:914–921, 1994.

142. Roberfroid, M.B. Prebiotics and probiotics: are they functional foods? *Am. J. Clin. Nutr.* 71:1682S–1687S, 2000.

143. Ellegard, L., H. Andersson, I. Bosaeus. Inulin and oligofructose do not influence the absorption of cholesterol or the excretion of cholesterol, Ca, Mg, Zn, Fe, or bile acids but increases energy excretion in ileostomy subjects. *Eur. J. Clin. Nutr.* 51:1–5, 1997.

144. Coudray, C., J. Bellanger, C. Castiglia-Delavaud, C. Remesy, M. Vermorel, Y. Rayssignuier. Effect of soluble and partly soluble dietary fibres supplementation on absorption and balance of calcium, magnesium, iron and zinc in healthy young men. *Eur. J. Clin. Nutr.* 51:375–380, 1997.

145. Van den Heuvel, E.G.H.M., T. Muys, W. van Dokkum, G. Schaafsma. Oligofructose stimulates calcium absorption in adolescents. *Am. J. Clin. Nutr.* 69:544–548, 1999.

146. Scholz-Ahrens, K.E., J. van Loo, J. Schrezenmeir. Effect of oligofructose on bone mineralization in ovariectomized rats is affected by dietary calcium. *Am. J. Clin. Nutr.* 73:498S–498S, 2001.

147. Scholz-Ahrens, K.E., G. Schaafsma, E.G.H.M. van den Heuvel, J. Schrezenmeir. Effect of prebiotics on mineral metabolism. *Am. J. Clin. Nutr.* 73:459S–464S, 2001.

148. Ohta, A., M. Ohtsuki, S. Baba, T. Adachi, T. Sakata, E.I. Sagaguchi. Calcium and magnesium absorption from the colon and rectum are increased in rats fed fructooligosaccharides. *J. Nutr.* 125:2417–2424, 1995.

149. Hughes, R., I.R. Rowland. Stimulation of apoptosis by two prebiotic chicory fructans in the rat colon. *Carcinogenesis* 22(1):43–47, 2001.

150. Rao, C.V., D. Chou, B. Simi, H. Ku, B.S. Reddy. Prevention of colonic aberrant crypt foci and modulation of large bowel microbial activity by dietary coffee fiber, inulin and pectin. *Carcinogenesis* 10:1815–1819, 1998.

151. Hague, A., D.J.E. Elder, D.J. Hicks, C. Paraskeva. Apoptosis in colorectal tumor cells: induction by the short chain fatty acids butyrate, propionate and acetate and by the bile salt deoxycholate. *Int. J. Cancer* 60:400–406, 1995.

152. Hague, A., C. Paraskeva. The short chain fatty acid butyrate induces apoptosis in colorectal tumor cell lines. *Eur. J. Cancer Prev.* 4:359–364, 1995.

153. Young, G. Prevention of colon cancer: role of short chain fatty acids produced by intestinal flora. *Asia Pac. J. Clin. Nutr.* 5:44–47, 1996.

154. Matsumoto, K., Y. Kobayashi, N. Tamura, T. Watanabe, T. Kan. Production of galactooligosaccharides with β-galactosidase. *Denpun Kagaku* 36:123–130, 1989.

155. Van der Meer, R., I.M.J. Bovee-Oudenhoven. Dietary modulation of intestinal bacterial infections. *Int. Dairy J.* 8:481–486, 1998.

156. Bouhnik, Y., B. Flourie, L. D'Agay-Abensour, P. Pochart, G. Gramet, M. Durand, C.-J. Rambaud. Administration of transgalacto-oligosaccharides increases fecal *Bifidobacteria* and modifies colonic fermentation metabolism in healthy humans. *J. Nutr.* 127:444–448, 1997.

157. Tanaka, R., H. Takayama, M. Morotomi, T. Kuroshima, S. Ueyama, K. Matsumoto, A. Kuroda, M. Mutai. Effects of administration of TOS and *Bifidobacteria breve* 4006 on the human faecal flora. *Bifidob. Microfl.* 2:17–24, 1983.

158. Ishikawa, F., H. Takayama, K. Matsumoto, M. Ito, O. Chonan, Y. Deguchi, H. Kituchi-Hakayawa, M. Watanuki. Effects of β 1-4 linked galactooligosaccharides on human faecal microflora. *Bifidus* 9:5–18, 1995.

159. Ito, M., Y. Deguchi, K. Matsumoto, M. Kimura, N. Onodera, T. Yajima. Influence of galacto-oligosaccharides on human faecal microflora. *J. Nutr. Sci. Vitaminol.* 39:635–640, 1993.

160. Dumortier, V., C. Brassart, S. Bouquelet. Purification and properties of a β-galactosidase from *Bifidobacterium bifidum* exhibiting a transgalactosylation reaction. *Biotechnol. Appl. Biochem.* 19:341–354, 1994.

161. Gopal, P.K., P.A. Sullivan, J.B. Smart. Utilisation of galacto-oligosaccharides as selective substrates for growth by lactic acid bacteria including *Bifidobacterium lactis* DR10 and *Lactobacillus rhamnosus* DR20. *Int. Dairy J.* 11:19–25, 2001.

162. Van Laere, K.M.J., T. Abee, H.A. Schopls, G. Beldman, A.G.J. Voragen. Characterization of a novel β-galactosidase from *Bifidobacterium adolescentis* DSM 20083 active towards transgalactooligosaccharides. *Appl. Environ. Microbiol.* 66(4):1379–1384, 2001.

163. Tanaka, R., K. Matsumoto. Recent progress on probiotics in Japan, including galacto-oligosaccharides. *Bull. Int. Dairy Fed.* 336:21–27, 1998.

164. Ohtsuka, K., Y. Benno, A. Endo, H. Ueda, O. Ozawa, T. Ulchida, T. Mitsuoka. Effects of 4'galactosyllactose on human intestinal microflora. *Bifidus* 2:143–149, 1989.

165. Deguchi, Y., K. Matsumoto, T. Ito, M. Watanuki. Effects of β1-4 galactooligosaccharides administration on defecation of healthy volunteers with constipation tendency. *Jap. J. Nutr.* 55:13–22, 1997.

166. Tamai, S., K. Ohtsuka, O. Ozawa, T. Ulchida. Effect of a small amount of galactooligosaccharides on fecal *Bifidobacterium*. *J. Jap. Soc. Nutr. Food Sci.* 45:456–460, 1992.

167. Chonan, O., H. Takahashi, H. Yasui, M. Watanuki. Effects of beta 1→4 linked galactooligosaccharides on use of magnesium and calcification of the kidney and heart in rats fed excess dietary phosphorus and calcium. *Biosci. Biotech. Biochem.* 60:1735–1737, 1996.

168. Asahara, T., K. Nomoto, K. Shimizu, M. Watanuki, R. Tanaka. Increased resistance of mice to *Salmonella enterica serovar Typhimurium* infection by synbiotic administration of *Bifidobacteria* and transgalactosylated oligosaccharides. *J. Appl. Microbiol.* 91:985–996, 2001.

169. Kohmoto, T., F. Fukui, H. Takaku, Y. Machida, M. Arai, T. Mitsuoka. Effect of isomalto-oligosaccharides on human fecal flora. *Bifidob. Microfl.* 7:61–69, 1988.

170. Lee, H.-S., J.-H. Auh, H.-G. Yoon, M.-J. Kim, J.-H. Park, S.-S. Hong, M.-H. Kang, T.-J. Kim, T.-W. Moon, J.-W. Kim, K.-H. Park. Cooperative action of α-glucanotransferase and

maltogenic amylase for an improved process of isomaltooligosaccharide (IMO) production. *J. Agric. Food Chem.* 50:2812–2817, 2002.

171. Kaneko, T., Kohmoto, T., H. Kikuchi, M. Shiota, H. Iino, T. Mitsuoka. Effects of isomalto-oligosaccharides with different degrees of polymerisation on human faecal *Bifidobacteria*. *Biosci. Biotech. Biochem.* 58:2288–2290, 1994.

172. Olano-Martin, E., K.C. Mountzouris, G.R. Gibson, R.A. Rastall. *In vitro* fermentability of dextran, oligodextran and maltodextrin by human gut bacteria. *Br. J. Nutr.* 83:247–255, 2000.

173. Tamura, Y., T. Mizota, S. Shimamura, M. Tomita. Lactulose and its application to the food and pharmaceutical industries. *Bull. Int. Dairy Fed.* 289:43–53, 1993.

174. Montgomery, E.M., C.S. Hudson. Relations between rotatory power and structure in the sugar group: XXVII. synthesis of a new disaccharide ketose. *J. Am. Chem. Soc.* 52:2101–2106, 1930.

175. Petuely, F. Bifidusflora bei Flaschenkindern durch bifidogene Substanzen (Bifidusfacktor). *Z. Kinderheilkd.* 79:174–179, 1957.

176. Jeffrey, G.A., D. Huang, P.E. Pfeffer, R.L. Dudley, K.B. Hicks, E. Nitsch. Crystal structure and n.m.r. analysis of lactulose trihydrate. *Carbohydr. Res.* 266:29–42, 1992.

177. Bird, S.P., D. Hewitt, M.I. Gurr. Energy values of lactitol and lactulose as determined with miniature pigs and growing rats. *J. Sci. Food Agric.* 51:233–246, 1990.

178. Matsuda, Y., Y. Iematsu, T. Shibamoto, S. Oyama. Effect of lactulose on acute renal failure in rabbits. *J. Med.* 27:49–54, 1992.

179. Mizota, T. Lactulose as a growth promoting factor for *Bifidobacterium* and its physiological aspects. *Bull. Int. Dairy Fed.* 313:43–48, 1996.

180. Terada, A., H. Hara, M. Kataoka, T. Mitsuoka. Effect of lactulose on the composition and metabolic activity of the human faecal flora. *Microbiol. Ecol. Health Dis.* 5:43–50, 1992.

181. Smart, J.B. Transferase reactions of β-galactosidases: new product opportunities. *Bull. Int. Dairy Fed.* 289:16–22, 1993.

182. Sahota, S.S., P.M. Bramley, I.S. Menzies. The fermentation of lactulose by colonic bacteria. *J. Gen. Microbiol.* 128:319–325, 1982.

183. Liehr, H., W.D. Heine. Treatment of endotoxemia in galactosamine hepatitis by lactulose administered intravenously. *Hepato-Gastroenterology* 28:296–298, 1981.

184. Liehr, H., G. Emglish, U. Rasenack. Lactulose: a drug with antiendotoxin effect. *Hepato-Gastroenterology* 27:356–360, 1980.

185. Iwasaki, M., I. Maruyama, N. Ikejiri, M. Abe, T. Maeda, H. Nagata, H. Abe, K. Tanigawa. Liver disease and endotoxin (II): endotoxin in severe liver lesion. *Jap. J. Gastroenterol.* 77:386–393, 1980.

186. Mortensen, P.B., H.S. Rasmussen, K. Holtug. Lactulose detoxifies *in vitro* short-chain fatty acid production in colonic contents induced by blood: implications for hepatic coma. *Gastroenterology* 94:750–754, 1988.

187. Genovese, S., G. Riccardi, A.A. Rivellese. Lactulose improves blood glucose response to an oral glucose test in non-insulin dependent diabetic patients. *Diabetes Nutr. Metab.* 5:295–297, 1992.

188. Vendemiale, G., G. Palasciano, F. Cirelli, M. Altamura, A. De Vincentis, E. Altomare. Crystalline lactulose in the therapy of hepatic cirrhosis: evaluation of clinical and immunological parameters: preliminary results. *Drug Res.* 42(II):969–972, 1992.

189. Fulton, J.D. Infection limitation with lactulose therapy. *J. Clin. Exp. Gerontol.* 10(3&4):117–124, 1988.

190. Onishi, S., H. Yamazaki, C. Obayashi, Y. Suzuki. Effect of lactulose on ammonia metabolism during exercise. *Resp. Circ.* 38:693–697, 1990.

191. Nagengast, F.M., M.P. Hectors, W.A. Buys, J.H. van Tongeren. Inhibition of secondary bile acid formation in the large intestine by lactulose in healthy subjects of two different age groups. *Eur. J. Clin. Invest.* 18:56–61, 1988.

192. Owen, R.W. Faecal steroids and colorectal carcinogenesis. *Scand. J. Gastroenterol.* 32(222):76–82, 1997.

193. Rooney, P.S., L.M. Hunt, P.A. Clarke, K.A. Gifford, J.D. Hardcastle, N.C. Armitage. Wheat fibre, lactulose and rectal mucosal proliferation in individuals with a family history of colorectal cancer. *Br. J. Surg.* 81:1792–1794, 1994.

194. Roncucci, L., P.D. Donato, L. Carati, A. Ferrari, M. Perini, G. Bertoni, G. Bedogni, B. Paris, F. Svanoni, M. Girola, M.P. Leon. Antioxidant vitamins or lactulose for the prevention of the recurrence of colorectal adenomas. *Dis. Colon Rectum* 36:227–234, 1993.

195. Rowland, I.R., C.A. Bearne, R. Fischer, B.L. Pool-Zobel. Effect of lactulose on DNA damage induced by DMH in the colon of human flora-associated rats. *Nutr. Cancer Prev.* 26:37–47, 1996.

196. Brommage, R., C. Binacua, S. Antille, A.-L. Carriè. Intestinal calcium absorption in rats is stimulated by dietary lactulose and other resistant sugars. *J. Nutr.* 123:2186–2194, 1993.

197. Rèmèsy, C., M.-A. Levrat, L. Gamet, C. Demignè. Cecal fermentations in rats fed oligosaccharides (inulin) are modulated by dietary calcium level. *Am. J. Physiol.* 264:G855–G862, 1993.

198. Suzuki, K., M. Uehara, Y. Endo, S. Goto. Effect of lactose, lactulose and sorbitol on iron, zinc and copper utilization in rats. *J. Jap. Soc Nutr. Food Sci.* 39:217–219, 1986.

199. Igarashi, C., I. Ezawa. Effects of whey calcium and lactulose on the strength of bone in ovariectomized osteoporosis model rats. *Pharmacometrics* 42:245–253, 1991.

200. Van Loo, J., J. Cummings, N. Delzenne, H. Englyst, A. Franck, M. Hopkins, N. Kok, G. Macfarlane, D. Newton, M. Quigley, M. Roberfroid, T. van Vliet, E. van den Heuvel. Functional food properties of non-digestible oligosaccharides: a consensus report from the ENDO project (DGXII AIRII-CT94-1095). *Br. J. Nutr.* 81:121–132, 1999.

201. Wesselius de Casparis, A., S. Braadbaat, G.E. Bergh-Bolken, M. Mimica. Treatment of chronic constipation with lactulose syrup: results of a double-blind study. *Gut* 9:84–86, 1968.

202. Attar, A., M. Lemann, A. Ferguson, M. Halphen, M.C. Boutron, B. Flourie, E. Alix, M. Salmeron, F. Guillemot, S. Chaussade, A.M. Menard, J. Moreau, G. Naudin, M. Barthet. Comparison of a low dose polyethylene glycol electrolyte solution with lactulose for treatment of chronic constipation. *Gut* 44:226–230, 1999.

203. Clausen, M.R., P.B. Mortensen. Disaccharides and colonic flora. Clinical consequences. *Drugs* 53:930–942, 1997.

204. Teuri, U., R. Korpela. Galacto-oligosaccharides relieve constipation in elderly people. *Ann. Nutr. Metab.* 42:319–327, 1998.

205. Marteau, P. Prebiotics and probiotics for gastrointestinal health. *Clin. Nutr.* 20(1):41–45, 2001.

206. Minami, Y., K. Yazawa, Z. Tamura, T. Tanaka, T. Yamamoto. Selectivity of utilisation of galactosyl-oligosaccharides by *Bifidobacteria*. *Chem. Pharm Bull.* 31:1688–1691, 1983.

207. Saito, Y., T. Takano, I.R. Rowland. Effects of soybean oligosaccharides on the human gut microflora in *in vitro* culture. *Microbiol. Ecol. Health Dis.* 5:105–110, 1992.

208. De Boever, P., R. Wouters, W. Verstraete. Combined use of *Lactobacillus reuteri* and soygerm powder as food supplement. *Lett. Appl. Microbiol.* 33:420–424, 2001.

209. De Boever, P., B. Deplancke, W. Verstraete. Fermentation by gut microbiota cultured in a simulator of the human intestinal microbial ecosystem is improved by supplementing a soygerm powder. *J. Nutr.* 130:2599–2606, 2000.

210. Messina, M.J. Legumes and soybeans: overview of their nutritional profiles and health effects. *Am. J. Clin. Nutr.* 70:439S–450S, 1999.

211. Setchell, K.D.R. Phytoestrogens: the biochemistry, physiology, and implications for human health of soy isoflavones. *Am. J. Clin. Nutr.* 68:1333S–1346S, 1998.

212. King, R.A., D.B. Bursill. Plasma and urinary kinetics of the isoflavones daidzein and genistein after a single soy meal in humans. *Am. J. Clin. Nutr.* 67:867–872, 1998.

213. Day, A.J., F.J. Canada, J.C. Diaz, P.A. Kroon, R. Mclauchlan, C. Faulds, G.W. Plumb, M.R.A. Morgan, G. Williamson. Dietary flavonoid and isoflavone glycosides are hydrolysed by the lactase site of lactase phlorizin hydrolase. *FEBS Lett.* 468:166–170, 2000.

214. Harvey, C.B., Y. Wang, L.A. Highes, D.M. Swallow, W.P. Thurrell, V.R. Sams, R. Barton, S. Lanzomiller, M. Sarner. Studies on the expression of intestinal lactase in different individuals. *Gut* 36:28–33, 1995.

215. Izumi, T., M.K. Piskula, S. Osawa, A. Obata, K. Tobe, M. Saito, S. Kataoka, Y. Kubota, M. Kikuchi. Soy isoflavone aglycones are absorbed faster and in higher amounts than their glycosides in humans. *J. Nutr.* 130:1695–1699, 2000.

216. Brzezinski, A., A. Debi. Phytoestrogens: the "natural" selective estrogen receptor modulators? *Eur. J. Obstet. Gynaecol. Reprod. Biol.* 85:47–51, 1999.

217. Molteni, A., L. Brizio-Molteni, V. Persky. *In vitro* hormonal effects of soybeans isoflavones. *J. Nutr.* 125:751S-756S, 1995.

218. Slavin, J.L., M.C. Martini, D.R. Jacobs, L. Marquart. Plausible mechanims for the protectiveness of whole grains. *Am. J. Clin. Nutr.* 70:459S–463S, 1999.

219. Suarez, F.L., J. Springfield, J.K. Furne, T.T. Lohrmann, P.S. Kerr, M.D. Levitt. Gas production in humans ingesting a soybean flour derived from beans naturally low in oligosaccharides. *Am. J. Clin. Nutr.* 69:135–139, 1999.

220. Hayakawa, K., J. Mizutani, K. Wada, T. Masai, I. Yoshihara, T. Mitsuoka. Effect of soybean oligosaccharides on human fecal flora. *Microb. Ecol. Health Dis.* 3:293–303, 1990.

221. S. Watanabe, M. Yamaguchi, T. Sobue, T. Takahashi, T. Miura, Y. Arai, W. Mazur, K. Wahala, H. Aldercreutz. Pharmakokinetics of soybean isoflavones in plasma, urine and feces of men after ingestion of 60 g baked soybean powder (kinako). *J. Nutr.* 128:1710–1715, 1998.

222. Zhang, Y., G.J. Wang, T.T. Song, P.A. Murphy, S. Hendrich. Urinary disposition of the soybean isoflavones daidzein, genistein and glycitein differs among humans with moderate isoflavone degradation activity. *J. Nutr.* 129:957–962, 1999.

223. Cherbut, C., A.C. Aubè, H.M. Blottière, J.P. Glamiche. Effects of short-chain fatty acids on gastrointestinal motility. *Scand. J. Gastroenterol.* 32:52–57, 1997.

224. Scheppach, W., H.P. Bartram, F. Richter. Role of short-chain fatty acids in the prevention of colorectal cancer. *Eur. J. Cancer* 31(7-8):1077–1080, 1995.

225. Ichikawa, H., T. Sakata. Stimulation of epithelial cell proliferation of isolated distal colon of rats by continuous colonic infusion of ammonia or short-chain fatty acids is nonadditive. *J. Nutr.* 128:843–847, 1998.

226. Christl, S.U., P.R. Murgatroyd, G.R. Gibson, J.H. Cummings. Production, metabolism and excretion of hydrogen in the large intestine. *Gastroenterology* 102:1269–1277, 1992.

227. Macfarlane, G.T., S. Macfarlane. Human colonic microbiota: ecology, physiology and metabolic potential of intestinal bacteria. *Scand. J. Gastroenterol.* 32:3–9, 1997.

228. Gibson, G.R., P.B. Ottaway, R.A. Rastall. *Prebiotics*. Oxford: Chandos Publishing, 2000.

229. Pilgrim, A., M. Kawase, M. Ohashi, K. Fujita, K. Murakami, K. Hashimoto. Reaction kinetics and modeling of the enzyme-catalysed production of lactosucrose using β-fructofuranosidase from *Arthrobacter* sp. K-1. *Biosci. Biotech. Biochem.* 65(4):758–765, 2001.

230. Fujita, K., K. Hara, S. Sakai, T. Miyake, M. Yamashita, Y. Tsunetomi, T. Mitsuoka. Effect of 4^G-β-D-galactosylsucrose (lactosucrose) on intestinal flora and its digestibility in humans. *Denpun Kagaku* 38:249–255, 1991.

231. Hopkins, M.J., J.H. Cummings, G.T. Macfarlane. Interspecies difference in maximum specific growth rates and cell yields of *Bifidobacteria* cultured on oligosaccharides and other simple carbohydrate sources. *J. Appl. Microbiol.* 85:381–386, 1998.

232. Vasquez, M.J., J.L. Alonso, H. Dominguez, J.C. Parajo. Xylooligosaccharides: manufacture and applications. *Trends Food Sci. Technol.* 11:387–393, 2000.

233. Campbell, J.M., L.L. Bauer, G.C. Fahey, A.J.C.L. Hogarth, B.W. Wolf, D.E. Humter. Selected fructooligosaccharide (1-kestose, nystose, and 1^F-β-fructofuranosylnystose) composition of foods and feeds. *J. Agric. Food Chem.* 45:3076–3082, 1997.

234. Jaskari, J., P. Kontula, A. Siitonen, H. Jousimies-Somer, T. Mattila-Sandholm, K. Poutanen. Oat β-glucan and xylan hydrolysates as selective substrates for *Bifidobacterium* and *Lactobacillus* strains. *Appl. Microbiol. Biotech.* 49:175–181, 1998.

235. Playne, M.J., R. Crittenden. Commercially available oligosaccharides. *Bull. Int. Dairy Fed.* 313:10–12, 1996.

236. Yamada, H., K. Itoh, Y. Morishita, H. Taniguchi. Structure and properties of oligosaccharides from wheat bran. *Cereal Food World* 38:490–492, 1993.

237. Okazaki, M., S. Fujikawa, N. Matsumoto. Effects of xylooligosaccharide on growth of *Bifidobacteria. J. Jap. Soc. Nutr. Food Sci.* 43:395–401, 1990.

238. Korakli, M., M.G. Ganzie, R.F. Vogel. Metabolism by *Bifidobacteria* and lactic acid bacteria on polysaccharides from wheat and rye, and exopolysaccharides produced by *Lactobacillus sanfranciscensis. J. Appl. Microbiol.* 92:958–965, 2002.

239. Crociani, F., A. Alessandrini, M.M. Mucci, B. Biavati. Degradation of complex carbohydrates by *Bifidobacterium* spp. *Int. J. Food Microbiol.* 24:199–210, 1994.

240. Van Laere, K.M.J., R. Hartemink, M. Bosveld, H.A. Schols, A.G.J. Voragen. Fermentation of plant cell wall derived polysaccharides and their corresponding oligosaccharides by intestinal bacteria. *J. Agric. Food Chem.* 48:1644–1652, 2000.

241. Kummel, K.F., S. Brokx. Lactitol as a functional prebiotic. *Cereal Food World* 46(9):424–429, 2001.

242. Scevola, D., G. Bottari, L. Oberto, V. Monzillo. Intestinal bacterial toxins and alcohol liver damage. Effects of lactitol, a synthetic disaccharide. *Clin. Dietol.* 20:297–297, 1993.

243. Dal Bello, F., J. Walter, C. Hertel, W.P. Hammes. *In vitro* study of prebiotic properties of levan-type exopolysaccharides from lactobacilli and non-digestible carbohydrates using denaturing gradient gel electrophoresis. *Syst. Appl. Microbiol.* 24:232–237, 2001.

244. Belitz, H.D., W. Grosch. *Food Chemistry.* Berlin: Springer-Verlag, 1999.

245. Rycroft, C.E., M.R. Jones, G.R. Gibson, R.A. Rastall. Fermentation properties of gentiooligosaccharides. *Lett. Appl. Microbiol.* 32:156–161, 2001.

246. Berggren, A.M., I.M.E. Bjorck, E.M.G.L. Nyman, B.O. Eggum. Short-chain fatty acid content and pH in cecum of rats given various sources of carbohydrates. *J. Sci. Food Agric.* 63:397–406, 1993.

247. Michel, C., C. Benard, M. Lahaye, D. Formaglio, B. Kaeffer, B. Quemener, S. Berot, J.C. Yvin, H.M. Blottiere, C. Cherbut. Algal oligosaccharides as functional foods: *in vitro* study of their cellular and fermentative effects. *Sci. Aliment.* 19:311–332, 1999.

248. Flickinger, E.A., B.W. Wolf, K.A. Garleb, J. Chow, G.J. Leyer, P.W. Johns, G.C. Fahey. Glucose-based oligosaccharides exhibit different *in vitro* fermentation patterns and affect *in vivo* apparent nutrient digestibility and microbial populations in dogs. *J. Nutr.* 130:1267–1273, 2000.

249. Campbell, J.M., G.C. Fahey Jr., B.W. Wolf. Selected indigestible oligosaccharides affect large bowel mass, cecal and fecal short-chain fatty acids, pH and microflora in rats. *J. Nutr.* 127:130–136, 1997.

250. Djouzi, Z., C. Andrieux. Compared effects of three oligosaccharides on metabolism of intestinal microflora of rats inoculated with a human fecal flora. *Br. J. Nutr.* 78:313–324, 1997.

251. Yazawa, K., K. Imai, Z. Tamura. Oligosaccharides and polysaccharides specifically utilisable by *Bifidobacteria. Chem. Pharm. Bull.* 26:3306–3311, 1978.

252. Macfarlane, G.T., S. Macfarlane, G.R. Gibson. Validation of a three-stage compound continuous culture system for investigating the effect of retention time on the ecology and metabolism of bacteria in the human colon. *Microb. Ecol.* 35:180–187, 1998.

253. Kohmoto, T., F. Fukui, H. Takaku, T. Mitsuoka. Dose-response test of isomalto-oligosaccharides for increasing faecal *Bifidobacteria. Agric. Biol. Chem.* 55:2157–2159, 1991.

254. Djouzi, Z., C. Andrieux, V. Pelenc, S. Somarriba, F. Popot, F. Paul, P. Monsan, O. Szylit. Degradation and fermentation of α-gluco-oligosaccharides by bacterial strains from human colon: *in vitro* and *in vivo* studies in gnotobiotic rats. *J. Appl. Bacteriol.* 79:117–127, 1995.

255. Ito, M., Y. Deguchi, A. Miyamori, K. Matsumoto, H. Kikuchi, K. Matsumoto, Y. Kobayashi, T. Yajima, T. Kan. Effects of administration of galactooligosaccharides on the human fecal microflora, stool weight and abdominal sensation. *Microb. Ecol. Health Dis.* 3:285–292, 1990.

18

Immunomodulating Effects of Lactic Acid Bacteria

Hanne Risager Christensen and Hanne Frøkiær

CONTENTS

18.1 Introduction 436
18.2 The Gut Mucosa – A Unique Compartment of the Immune System....................437
    18.2.1 Organization of GALT 437
    18.2.2 Immune Responses of GALT and Peripheral Immune System 439
        18.2.2.1 Nonspecific Response Mechanisms...440
        18.2.2.2 Specific Response Mechanisms – Immunity vs. Tolerance 440
18.3 Influence of the Gut Flora on the Immune Function 444
    18.3.1 The Gut Flora 444
    18.3.2 Interactions Between the Gut Flora and the Host Immune System..........446
18.4 Immunomodulatory Mechanisms of LAB 447
    18.4.1 Methods Used for Assessing Immunomodulatory Effect of LAB............448
    18.4.2 Effects of LAB on Nonimmunologic Mucosa Barrier Functions..............450
    18.4.3 Effects on Phagocytic Cells 450
    18.4.4 Effects on NK Cell Activity..451
    18.4.5 Interactions Between LAB and Intestinal Epithelial Cells......................452
    18.4.6 Effects on Cytokine Production 452
    18.4.7 Effects on Lymphocyte Proliferation 453
    18.4.8 Effects on Specific Immune Responses – Adjuvant Capacity of LAB 454
    18.4.9 Enhancement of IgA Production 455
18.5 Effects of LAB on Diseases...455
    18.5.1 Effects of LAB on Intestinal Infections...456
    18.5.2 Effects on Inflammatory Bowel Disease 457
    18.5.3 Antiallergy Properties of LAB 458
    18.5.4 Antitumor Effect of LAB 459
    18.5.5 Impact of Viability and Dose on the Effects of LAB 460

18.6 Concluding Remarks and Future Perspectives 461
Acknowledgment 461
References..461

18.1 INTRODUCTION

For decades, it has been believed that consumption of fermented dairy products is beneficial for health. In 1907, Eli Metchnikoff, the Nobel Prize winner and director of the Pasteur
Institute, postulated in his book *The Prolongation of Life* that eating yogurt would extend
the life span due to the health-improving influence of the lactic acid bacteria (LAB)
involved in yogurt fermentation on the gut flora. Since that time, many studies have been
conducted on gut-associated microorganisms, of which LAB are important members (1–
3). These studies have led to the growing recognition today that the microflora colonizing
the gastrointestinal tract of man and animal indeed play an important and essential role for
the health of the host.

The gut flora comprises a highly evolved and extremely complex microbial ecosystem of hundreds of different microbial species. This metabolically highly active society is
in close proximity with the gut epithelia through which it provides an array of signals and
products that affect homeostasis of various physiologic processes of the body. Of these
processes, the immune function constitutes a very important part; stimulating signals from
the gut bacteria are critical for the development and maintenance of the mucosal and systemic immune compartments (4,5). A corollary, which has collected increasing evidence
of support, is that the panel of stimulating signals bestowed by the gut flora is dynami-
cally affected by the microbial composition, with some strains exerting stronger bioactivities than others (6–8). The gut flora are composed of a group of resident (autochthonous)
bacterial species, and a variable group of transient (allochthonous) species that temporar-
ily fills in functional gaps of the flora. In view of these facts, the concept of manipulating
the composition of the gut flora in attempt to improve health of the host has emerged. That
is, an increase in numbers but also potency of specific bacterial groups possessing particular health-promoting properties is desirable. One approach much investigated for this
purpose is oral supplementation; e.g., via functional food products, with so called "probiotic" microorganisms that supply the gut flora with niches of beneficial functionality. The
word probiotic is derived from the Greek meaning "for life." The definition of probiotics
has changed over time; in 1989, Fuller defined probiotics as, "a live microbial food supplement which beneficially affects the host by improving the intestinal microbial balance"
(9). However, to acknowledge emerging scientific data on proven effects, a broadening of
the definition was recently proposed: "Probiotics are microbial cell preparations or components of microbial cells that have a beneficial effect on the health and well-being of the
host" (10). In addition to the original concept of probiotics as acting solely via improvements of imbalances of the gut flora as a whole, this newly proposed definition embraces
the bioactive effects of microbial compounds, not necessarily within the viable cell, but
which are active also as nonviable whole cell or even as fractionated preparations. This
broader concept gains ground in the recognition that supplementation of microbial preparations apart from viable cultures in some instances exerts a health promoting effect.
Moreover, probiotic effects, e.g., on immune parameters, are not necessarily mediated via
a preceding intervention with the gut flora balance (6,11,12).

The probiotic microorganisms currently used are predominantly, though not exclusively, lactic acid bacteria (LAB) such as lactobacilli or bifidobacteria (13). LAB are major
components of the gut flora (1–3). Research has revealed these bacteria are far from just

inert commensals, but may constitute one of the most significant microbial populations providing immunoregulatory signals (7,14–16).

In this chapter, we will discuss the current knowledge of how the gut flora as a whole affects the immune system, and then focus on the immunomodulating effects of LAB present either as a natural part of the gut flora or supplemented as probiotics. However, to provide a basis for understanding the functional influence conferred by the gut flora on the immune system, we will begin by giving a view of the organization and function of the intestinal immune system and its affiliation to the peripheral immune system.

18.2 THE GUT MUCOSA – A UNIQUE COMPARTMENT OF THE IMMUNE SYSTEM

The gut mucosal membranes form the interface between the highly controlled internal microenvironment of the body (basolateral side) and the external environment (gut lumen, apical side). To allow efficient absorption of nutrients, the surface area of the mucosal membranes is enormous, reaching around 400 m^2, which exceeds by far that of the skin (~2 m^2) (17). Along this huge area, the gut mucosa continually experiences a massive exposure to both pathogenic and harmless agents derived from the ingested food as well as the gut flora.

The immune compartment present in the gut mucosa — the gut associated lymphoid tissue (GALT) — is recognized as the largest immune organ of the body and plays a crucial role for keeping the body healthy. GALT constitutes a major part in the gut barrier function against pathogens and other harmful antigens entering the body through the gut. Equally important but oppositely directed, GALT is important in down regulating responses toward ingested harmless food antigens to permit entrance of nutritional food antigens into the body – a mechanism designated oral tolerance (18). In keeping with all this, GALT is not only responsible for local immune responses but also for systemic responses via primed immune cells that exit GALT and enter into peripheral circulation. As already touched on, appropriate development and balance maintenance of these rather distinct functions of GALT is affected by the microenvironment of the gut, most importantly the gut flora, including LAB, which is the core subject of the present chapter.

18.2.1 Organization of GALT

GALT is compartmentalized into inductive sites and effector sites; i.e., sites where priming and dissemination, respectively, of an immune response primarily takes place (17). This division is far from absolute, but is, however, very useful in understanding GALT functioning. GALT comprises the highly organized structures of the Peyer's patches and the mesenteric lymph nodes, and the more diffuse tissues consisting of a large number of lymphoid cells scattered throughout the lamina propria and intestinal epithelium (intraepithelial lymphocytes) (Figure 18.1).

Peyer's patches and the mesenteric lymph nodes are thought to be the main inductive sites and are found in all mammals. Peyer's patches are located primarily on the antimesenteric side of the small intestine. They contain a number of B cell rich zones with germinal centers called follicles, and, adjacent to these in the parafollicular areas, there are T cell rich zones. The subepithelial dome area overlying the follicles contains many antigen presenting cells, primarily dendritic cells. This area is extremely important for antigen processing and presentation for the initiation of an immune response such as IgA production. The dome area is covered by a follicle associated epithelium facing the gut lumen. In addition to conventional enterocytes, the follicle associated epithelium contains the very specialized microfold (M) cells. M cells lack microvilli, exhibit reduced enzymatic activity, and are adept at uptake and transport of intact luminal antigens (such as soluble food antigens, microorganisms, and viruses) into the Peyer's patch for further immunologic processing. M cells are particularly efficient in taking up particulate material such as microorganisms; in fact, particulate material seems to be almost exclusively taken up by M cells, making these cells the only portal for such antigens. Lymphoid structures similar to Peyer's patches are present also in the large intestine, particularly in the appendix and cecum. They consist only of isolated lymphoid follicles but are believed to function in a manner similar to Peyer's patches.

The mesenteric lymph nodes serve as secondary lymph nodes (Figure 18.1). Located in the center of the mesenteric tissue, they assemble all of the lymphatic vessels draining the gut mucosa as afferent (ingoing) lymphatic vessels into the nodes. Their efferent (outgoing) lymphatic vessels lead the lymph fluid to peripheral circulation via the thoracic duct. The mesenteric lymph nodes, therefore, seem to act as a crossover point between the peripheral and mucosal immune response.

The main effector site of intestinal immune responses is the lamina propria, which is the inner part of the intestinal villi (Figure 18.1). The lamina propria contains a large and heterogeneous group of scattered immune cells, including antigen presenting cells (dendritic cells and macrophages) and T and B cells highly differentiated into mature effector

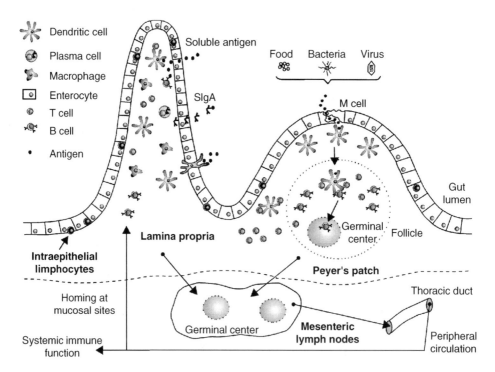

Figure 18.1 Schematic depiction of the gut-associated lymphoid tissue (GALT). The Peyer's patches and mesenteric lymph nodes are organized lymphoid tissues, while scattered immune cells are abundant in lamina propria and inside the epithelium (intraepithelial lymphocytes). Luminal antigen can enter either via M cells (primarily particulate antigen) overlying the Peyer's patch, through the enterocytes (soluble antigen) or potentially via capture by traversing dendritic cells. Immune cells travel from Peyer's patches and lamina propria to mesenteric lymph nodes from where they can enter the blood stream via the thoracic duct. Upon peripheral circulation, some immune cells, particularly IgA+ B cells, home at effector sites primarily lamina propria.

cells, as will be discussed later. Many of the cells are primed in the Peyer's patches from where they migrate out and traffic through peripheral circulation to the lamina propria to exert immune responses. The most predominant type of response of the lamina propria is IgA production by plasma cells (differentiated B cells). Despite being a major effector site, certain immune responses may be induced in the lamina propria as well. Likewise, certain effector functions may disseminate in the Peyer's patches and mesenteric lymph nodes.

The important compartment of the intraepithelial lymphocytes can neither be grouped exclusively into an inductive or an effector site. The intraepithelial lymphocytes represent one of the largest lymphoid populations in the body and consist mainly of CD8$^+$ (cytotoxic) and CD4$^+$ (helper) T cells with different T cell receptor (TCR) phenotypes ($\alpha\beta$ TCR or $\gamma\delta$ TCR). The proximity of the intraepithelial lymphocytes to the enterocytes and the gut lumen makes them one of the first cellular mucosal immune components to respond to luminal antigens. They perform an effector function by exerting cytolytic activity, playing a role in the barrier function against pathogens; and an inductive function, as they seem to play a role in the induction of oral tolerance toward food antigens. Moreover, they appear very important in regulation of immune homeostasis including regulation of IgA production in the lamina propria.

As a second point of entry besides M cells, antigens can enter GALT through a population of intraepithelial dendritic cells (Figure 18.1) (19). Dendritic cells can take up and process antigens, and *in vitro* models suggest that dendritic cells can extend their dendrites between the enterocytes and into the lumen without disrupting the tight junctions that bind the enterocytes together (20).

Finally, it is important for the overview of GALT to mention the enterocytes. Besides their principal role as cells absorbing degraded luminal nutrients, enterocytes are also capable of transporting soluble intact antigens and thus make up a third point of antigen entry (21). Also, an increasing amount of data shows that these cells participate significantly in the initiation and regulation of immune responses via interaction with GALT, especially the intraepithelial cells (22). Enterocytes moreover, serve to transport IgA into the gut lumen.

18.2.2 Immune Responses of GALT and Peripheral Immune System

In GALT, the various immunocompetent cells such as epithelial cells, dendritic cells, T cells, B cells, and macrophages form a unique homeostatic immune network. During initiation of an immune response, the functionally distinct cells of GALT act coordinately to provide stimuli and share common regulatory elements. These processes are characterized by the production of a highly controlled set of molecules like cytokines, chemokines, and costimulatory molecules that alert the host to breach in the mucosal barrier and focus the immune response at the site of infection. Cytokines are soluble intercellular signaling molecules of principal importance for immune cell communication. They act to stimulate or suppress cell functions; and the chemokines, which constitute a cytokine subgroup, act specifically to attract immune cells.

The gastrointestinal surfaces, like the rest of the body, are protected against pathogens by both antigen nonspecific and specific mechanisms. The specific mechanisms implement immune cells committed specifically to act against a given antigen. The nonspecific mechanisms act without proceeding antigen specific commitment and include both nonimmune mechanisms as well as mechanisms exerted by immune cells. Thus, the nonspecific protective mechanisms are constantly in place to take action instantaneously, and therefore serve as the first line of defense. In contrast, the specific immune response takes time to establish, but afterward exerts a much more precisely targeted and effective response, especially toward reencountered antigens.

The gut flora and probiotic LAB affect both nonspecific and specific immune mechanisms. However, up till now most of the knowledge about the immune modulating effects of LAB was limited to only a few components of these immune mechanisms. To understand the relevance of such effects, an outline of the pertinent mechanisms of the nonspecific and specific immune system, particularly in the gut, is given in what follows.

18.2.2.1 Nonspecific Response Mechanisms

Most macromolecules are prevented from direct contact with the epithelial surface by mechanisms belonging to the nonimmune part of the nonspecific defense. The enterocytes have a folded apical surface of rigid structures (microvilli) that are coated with a thick layer of membrane anchored, negatively charged, mucin-like glycoproteins called glycocalyx. This layer forms a diffusion barrier and a highly degradative microenvironment. Bactericidal compounds such as lactoferrin, peroxidase, lysozyme, and defensins are produced by Paneth cells also situated in the epithelium. Another important nonimmune defense mechanism is the capability of the gut flora to compete with pathogens for nutrients and adhesion receptors and, moreover, to produce antimicrobial substances and acids that inhibit pathogenic colonization (23). These mechanisms are collectively called colonization resistance.

Interaction of microorganisms, such as pathogens, with epithelial cells, including enterocytes, triggers the epithelial cells to secrete factors, like cytokines and growth factors, that activate nonspecific mechanisms of the immune defense. The production of chemokines such as interleukin-8 (IL-8), monocyte chemoattractant protein-1 (MCP-1) and proinflammatory cytokines like IL-1β and tumor necrosis factor-α (TNF-α) by epithelial cells, stimulates activation and migration of phagocytic cells such as macrophages and neutrophils to the site of contact (infection) (24). Phagocytic cells bind and internalize antigens (a process called phagocytosis) through a repertoire of receptors, e.g., the antibody receptors (FcγRI, FcγRII, FcγRIII, FcαR and FcεRI), complement receptors (CR1 and CR3), and mannose receptor (MR). When phagocytosed, the microorganisms are killed by bactericidal compounds such as lysozyme or reactive oxygen species (ROS) — a process called the oxidative burst. Motile dendritic cells in Peyer's patches, lamina propria or within the epithelium are also capable of phagocytosing antigen, although much less efficiently. Upon stimulation by most pathogens, secretion of the cytokine IL-12 by macrophages and dendritic cells enhances cytotoxicity of natural killer (NK) cells. NK cells are another important part of the nonspecific immune system.

In addition to the importance for the initiation of the nonspecific immune defense, dendritic cells are of particular importance for the induction of a specific immune response. In this way, dendritic cells create a link between the nonspecific and the specific immune systems (25).

18.2.2.2 Specific Response Mechanisms – Immunity vs. Tolerance

An impressive feature of the protective mechanisms of GALT is the capacity to discriminate between harmless antigens of the diet or resident microflora, and antigens from potential pathogens. The outcome of exposure to foreign antigens can involve three types of antigen-specific mechanisms: mucosally induced tolerance, also termed oral tolerance; induction of mucosal IgA response; and induction of systemic response (18). It is essential for normal life that the intestinal immune system be exquisitely regulated to mobilize the right type of response to any encountered antigen; i.e., induction of tolerance versus immunity.

The mechanisms regulating the distinction between these responses in the intestine are far from completely understood. It is clear, however, that they depend on establishment and maintenance of homeostasis among the immunocompetent cells. It is increasingly believed that under well established homeostasis, antigenic molecules such as pure soluble antigens encountered by GALT are not intrinsically immunogenic but rather induce tolerance as the default response. In contrast to such "naked" antigens, antigens that are endowed with "danger" signals, such as microorganisms bearing pathogenic structures, drive the immune response away from the default response and instead induce active immunity (26). During imbalances in immune homeostasis, the immune system may, however, become either hyporesponsive to such danger signals and hence increase the propensity for development of infections or cancer; or it may become hyperresponsive, resulting in misinterpretations of weak signals from harmless antigens as being danger signals, which is the case for organ specific autoimmune diseases (reactions against "self" antigens) and hypersensitivity reactions such as allergy (e.g., reactions against food antigens).

When GALT, or other lymphoid tissues, encounter foreign antigens, via transport through M cells or enterocytes, the antigen is processed by professional antigen-presenting cells (macrophages or dendritic cells) (24). By binding to surface major histocompatibily molecules (MHC), antigen fractions are presented to T cells. T cells, primarily T helper (Th) cells, that by chance have a receptor (TCR) specific to the antigen, bind to the antigen-presenting cells. During the subsequent intimate cell–cell interaction involving the MHC-antigen-TCR complex, costimulatory molecules, and cytokines, the T cells become activated and polarized to exert specific effector functions.

Activation and polarization of naïve Th cells are crucial for the type of immune response generated. Dendritic cells are the principal stimulators of naïve Th cells and, thus, the gatekeepers of an immune response (27). They perceive stimulating signals from microorganisms (pathogens and gut flora) through a panel of surface receptors (pattern recognition receptors) such as Toll-like receptors that recognize conserved structural patterns on the microorganisms; including lipopolysaccharide on Gram-negative bacteria, peptidoglycans on Gram-positive bacteria, and specific immunostimulatory DNA sequences (e.g., unmethylated CpG), carbohydrates, and glycolipids, occurring in many microbes (28). The dendritic cells, moreover, perceive signals such as cytokines coming from the epithelium and resident effector cells. Depending on the specific combination and timing of perceived signals, the dendritic cells mature and respond differentially whereby they control the type of immune response to be initiated (29–31). Dendritic cells accomplish this by driving differential polarization of naïve Th cells into active mature effector cells, producing different sets of cytokines that direct different types of immune responses as depicted in Figure 18.2. Three functionally discrete groups of T cells characterized by different cytokine profiles exist. Type 1 Th cells (Th1 cells) are characterized by secretion of IL-2 and IFN-γ, support cell-mediated immunity (cytotoxic T lymphocyte activation) and are immunopathologically associated with autoimmune diseases. In the gut, diseases like Crohn's disease are associated with a deleterious Th1 cell-driven response toward certain species of the gut flora. Th1 cells are induced in response to IL-12 produced by dendritic cells.

Type 2 Th cells (Th2 cells) produce IL-4, IL-5, IL-9, and IL-13 and support antibody production, and are immunopathologically associated with allergy such as food allergy, where a humoral IgE response is generated. A primary factor produced by dendritic cells for the initiation of Th2 cells is not identified, but absence of IL-12 and early presence of IL-4 (not produced by the dendritic cells) is important (32).

In contrast to Th1 and Th2 cells, mediating immunity, the third subpopulation of T cells, termed regulatory T (T_R) cells, are specialized in down regulating (suppressing) immune responses. T_R cells include two subtypes: Th3 and Tr1 cells, both of which are important effector cells in down regulating Th1 and Th2 cell activity and therefore play a central role in tolerance mechanisms and maintenance of immunologic homeostasis (33,34).

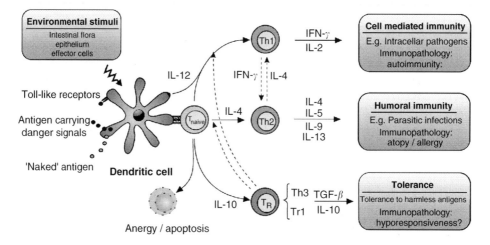

Figure 18.2 Schematic representation of polarized patterns of cytokine production and effector functions of differentiated T helper (Th) cells upon activation by dendritic cells. See text for further explanation.

T_R cells exert their immunoregulatory effect through expression of inhibitory cell surface molecules, or by secretion of immunoregulatory cytokines, most importantly TGF-β and IL-10, or both. Another response suppressive mechanism of the immune system is elimination of potential reactive (i.e., antigen specific) T cells either by "paralyzing" them (anergy) or by inducing apoptosis (deletion).

Many mechanistic details in the regulation of priming (Th1 or Th2) versus tolerance (T_R) are still incompletely understood. Induction of T_R cells or specific T cell anergy or deletion are the major mechanisms involved in the induction of oral tolerance, which depends on the dose of ingested antigen (35). An important prerequisite for tolerance induction is a preserved integrity of the mucosal barrier to avoid excessive antigen leakage into the mucosa (36). Even so, uptake and processing of intact soluble antigen by enterocytes are increasingly believed to play a major role in the induction of oral tolerance. This includes both presentation by the enterocytes themselves (MHC or CD1d molecules) to intraepithelial or lamina propria lymphocytes as well as basolateral releases of exosomes containing MHC-antigen complexes to be taken up by other cells, such as dendritic cells, present locally or at distant sites. Shortly after intake of antigen, antigen-loaded dendritic cells appear in Peyer's patches, lamina propria, and mesenteric lymph nodes, and whether tolerance or priming is the eventual outcome, antigen-specific activation of T cells takes place both in GALT and in the periphery soon after ingestion of antigen (37). Therefore, dissemination of activating or activated immune cells, or antigen bound to MHC or not, from inductive sites of GALT into peripheral tissues as well as into mucosal effector sites, creates an important link between GALT and the peripheral immune system.

It has become evident that enterocytes are not only capable of processing antigens, but, in fact, although they are not immune cells, they can respond actively to contact with luminal antigens, particular luminal bacteria such as LAB or pathogens (38). Enterocytes, like immune cells, express Toll-like receptors, through which they are stimulated by the gut flora and pathogens. The enterocytes respond by secreting a wide range of factors that affect intestinal immune cells, including macrophages, dendritic cells, and intraepithelial lymphocytes, which in turn influence the enterocytes by altering their functionality (39,40). In this way, the enterocytes, representing a predominant cell population, are intimately involved in the immunologic homeostasis and defense against pathogens, and also regulation of IgA production (41). As discussed later, several recent studies have focused on how the gut flora and LAB affect cellular interactions of enterocytes and intraepithelial cells and the resultant signal generation, important for regulation of tolerance vs. immunity. Impact on these interactions might be a key gate for the immunomodulating effect of LAB.

A last important hallmark of mucosal immunity is the secretory IgA (SIgA) response, which is neither a Th1, Th2, nor a T_R response. IgA is overwhelmingly the most important immunoglobulin in the intestine and other mucosal surfaces. Peyer's patches and mesenteric lymph nodes are important inductive sites for B cells to be committed for IgA production (Figure 18.1) (42). Antigen-loaded dendritic cells in Peyer's patches stimulate activation of B cells both through direct interaction with the B cells and via Th cell activation (IL-4-producing Th2 cells and TGF-β-producing Th3 cells). T cells colocalize with germinal center B cells, which are subsequently undergoing isotype switching from IgM to IgA. This occurs either in the Peyer's patch follicles or via cell migration to follicles of the mesenteric lymph nodes. Antigen-specific activated T and B cells then leave the inductive sites and migrate via the thoracic duct into peripheral circulation. Migrating cells finally home to mucosal effector sites, e.g., intestinal lamina propria or other mucosal membranes, where B cells, upon antigenic restimulation, complete their differentiation into IgA secreting plasma cells.

Production of IgA in lamina propria depends on the presence of IL-5, IL-6, IL-10 and TGF-β produced by effector T cells (43). Plasma cells synthesize IgA as a dimer of two IgA molecules interconnected by a J chain (44). Dimeric IgA is a ligand for the polymeric Ig receptor located on the basolateral surface of enterocytes. Here, bound IgA is taken up and transported across the cytoplasm to be released into the gut lumen by cleavage of the poly-Ig receptor molecule, leaving a so called secretory piece still bound to the IgA molecule. Due to the secretory piece, SIgA is resistant to the highly proteolytic environment in the gut.

The presence of SIgA in the gut lumen prevents infection of epithelial cells by coating pathogenic microorganism or virus and so forestalls binding to the cell surface. SIgA can likewise prevent absorption of antigen. By mediating transport of antigen across the

Table 18.1

Characteristics for Classical Immune Responses

Type of Response		Characteristics
Nonspecific responses		
Nonimmune:	Glycocalyx layer	Diffusion barrier
	Bactericidal compounds	Lysozyme, lactoferrin, defensins
	Gut flora	Competition/antimicrobial
Immune:	Phagocytosis	Neutrophils, macrophages, dendritic cells, IL-8[*], ROS
Specific responses		
Cell-mediated response		Th1, cytotoxic T cells, IL-1, IL-2, IL-12, IFN-γ, TNF-α, ROS
Systemic antibody response		Th2 and B cells, IgG, IgE, IL-4, IL-5, IL-9, IL-13
Tolerance		T_R (Th3, Tr1) cells suppressing Th1/Th2 cells, IL-10, TGF-β
Mucosal IgA response		(Th2), Th3 and B cells, IgA, TGF-β, IL-5, IL-6

[*] Abbreviations: IL: interleukin, IFN: interferon, TNF: tumor necrosis factor, ROS: reactive oxygen species, TGF: transforming growth factor

epithelium, SIgA can also neutralize, and carry back into the lumen, antigen that has leaked through the epithelial barrier. In this way, SIgA prevents pathogenic antigens from coming into contact with the body – a mechanism called immune exclusion. Due to the huge area of the gastrointestinal tract, great amounts of SIgA are necessary in performing its task. In man, approximately 3 grams of SIgA are delivered each day into the intestinal lumen; and 80–90% of all immunoglobulin-producing B cells (plasma cells) of the body, in fact, reside in the mucosa and exocrine glands (42,44).

Essential characteristics for these various immune responses, which are often subjects for measurement in evaluating immune modulation, are summarized in Table 18.1.

18.3 INFLUENCE OF THE GUT FLORA ON THE IMMUNE FUNCTION

It is well documented that the intestinal ecosystem features dynamic and reciprocal interactions among its microflora, the epithelium, and the immune system. Because LAB are a part of the gut flora, either naturally or by supplementation, an understanding of the mechanisms whereby LAB compel their immune modulating effects on the host requires an outline of what position LAB occupy in the gut flora and how the gut flora as a whole interact with the immune system of its host.

18.3.1 The Gut Flora

In man, greater than 75% of the fecal output is composed of bacterial cells derived from more than 400 different strains, totaling approximately 10^{14} viable bacteria, of which only about 40% are culturable (2). This enormous number of bacterial cells exceeds by a factor of 10 the number of eukaryotic cells constituting the human body. The diversity and numerical importance of the bacteria of the gastrointestinal tract vary in the different gut sections, and the

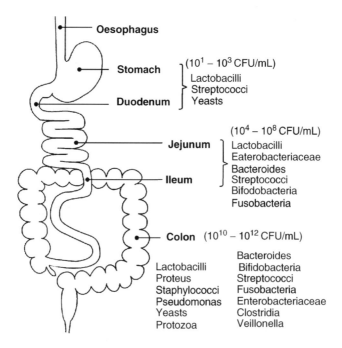

Figure 18.3 Microbial colonization of the human gut (175).

resident LAB are no exception (Figure 18.3). In adults, species of lactobacilli and streptococci as well as some other Gram-positive bacteria vastly dominate the flora of the sparsely populated stomach and duodenum. This is due to their ability to adhere to the epithelium and to resist acid, bile salts, and pancreatic secretions (45). At the distal part of the small intestine (jejunum and ileum), the microbial concentration and diversity increases, and here bifidobacteria species also enter the scene. In the colon which harbors a very diverse population of mostly strict anaerobic species, the levels of bifidobacteria reach 10^{10}–10^{11} cells/g whereas lactobacilli and enterococi species number only 10^7–10^8 cells/g (46). In total, the population of bifidobacteria and lactobacilli in the gut of adults represent no more than 1–5% and <1%, respectively, of the fecal flora when analyzed by molecular analyses.

The predominant flora of adults is remarkably stable, whereas in children it is less stable, and in infancy highly unstable (46). In infancy, vast changes occur during development of the flora. Immediately after birth, the sterile gut of the fetus is rapidly colonized, and levels up to 10^9 bacteria/g feces are reached within less than 48 hours. The composition of this initial flora represents a span of microbes originating from the environment, e.g., from the mother, but gives no indications of those likely to be retained in the tract (47). Subsequently, a selection of bacterial strains develops consistently along with maturation of the intestinal mucosa and under strong influence of both the microbial exposure and diet. Bifidobacteria constitute a major part of the flora of infants, ranging from 60 to 90%, whereas the values for Lactobacilli resemble those of adults (46). However, the flora varies greatly with neonatal feeding. In breast-fed infants, Bifidobacteria become dominant and the minor flora consist mainly of lactobacilli and streptococci. In contrast, formula-fed infants have similar amounts of bifidobacteria and bacteroides (~40%) and a minor flora frequently containing staphylococci, *Escherichia coli*, and clostridia (47). Upon weaning, the microbial profile becomes adult-like for both breast-fed and formula-fed with a significant increase of bacteroides.

Pioneer bacteria entering the gut in infancy can favor their own growth and inhibit growth of later introduced bacteria by inducing specific and lasting glycosylation of the glycocalyx as well as by modulating enterocyte gene expression. The initial colonization, therefore, widely shapes the final composition of the stable gut flora in adult life. As will be discussed later, increasing evidence shows that the development and maintenance of immunologic homeostasis is influenced by the composition of the gut flora, particularly the content of LAB. For example, distinct differences in the gut flora regarding bifidobacteria persist between atopic and nonatopic 2-year-olds (48). Thus, intake from birth or early infancy of probiotics may induce long term and enhanced efficacy of immune functions as well as other physiological conditions due to colonization as an integral part of the stable gut flora (autochthonous flora); contrasting probiotics ingested by older children and adults getting access only transiently (allochthonous flora) if at all, and thus demanding chronic administration of high concentrations.

In the gut flora of adults, the composition of the bifidobacteria population tends to be generally stable, while the composition of the lactobacilli tends to fluctuate more with only some persisting strains (46,49). Therefore, at least for the lactobacilli population, there seems to be a bacterial niche prone to alterations at a higher level than just bacteria in transit, e.g., through dietary changes or bacterial supplementation with probiotic lactobacilli.

Newly introduced harmless bacteria such as LAB stimulate the immune system including tolerance mechanisms and production of IgA. The produced IgA mediates immune exclusion whereby further contact between the bacteria and the immune system is attenuated as discussed in a previous section (50). Therefore, it has been suggested that for the allochthonous part of the flora, a high turnover of different strains of bacteria through the diet may participate in the continuous stimulation of GALT as a part of maintaining homeostasis.

18.3.2   Interactions Between the Gut Flora and the Host
Immune System

In the elucidation of the impact of the gut flora on the immune system, germ-free (gnotobi-otic) animals (pigs, mice, or rats) are very useful. Germ-free animals are obtained by deliver-ing the pups through Caesarian section and hand-rearing them in sterile isolators. By comparing germ-free and conventional animals (animals with normal gut flora), or by delib-erately colonizing the intestine of germ-free animals with defined microorganisms, it is pos-sible to study the interaction between the gut flora, or gut flora components, and the host immune system. With this approach, species of LAB have often been investigated (51–53). A general observation from such studies is, however, that colonization with a single or a few gut flora-derived microbes is less effective in modifying GALT than complete conventional-ization of the animals, revealing the importance of diverse bacterial stimuli.

The immune system of newborn conventional animals and humans matures along with colonization of the intestine. Such maturation does not occur in germ-free animals, leaving GALT development rudimentary and characterized by underdeveloped Peyer's patches and mesenteric lymph nodes lacking follicular germinal centers, reduced epithe-lial cell turnover, and low densities of immune cells both in gut mucosa and in peripheral lymph nodes (54). Upon microbial colonization, GALT and the systemic immune com-partment undergo marked changes, of which the chief structural and functional effects on the immune system are listed in Table 18.2.

IgA production is one function highly influenced by the presence of intestinal microbes. Gut colonization instigates maturation of Peyer's patches, and consequently the numbers of IgA-producing plasma cells in the lamina propria increases (50,55). Within the compartment of intraepithelial lymphocytes, the number and cytolytic potential of $\alpha\beta$ T cells increases after microbial stimulation while the population of $\gamma\delta$ T cells is unaffected by the gut flora (56). As mentioned earlier, gut flora microbes can mediate phenotypic alterations of enterocytes affecting bacterial adhesion (expression of glycoconjugates) but

Table 18.2

Effect of Gut Flora on Structure and Function of the Immune System Observed in Studies of Germ-Free vs. Conventional Animals.

Compartment	Structural or Functional Change
Gut mucosa	Development of Peyer's patches
	Expansion of IgA-producing cells in lamina propria
	Expansion of intraepithelial T cells ($\alpha\beta$ TCR[†])
	Acquisition of cytotoxicity of T cells ($\alpha\beta$ TCR)
	Induction of MHC molecules on enterocytes
	Induction of glycoconjugates in enterocytes
	Enhanced cytokine production
Systemic	Increased number of immune cells in lymph nodes
	Enhanced cell-mediated immune response
	Enhanced cytokine production capacity of peritoneal cells
	Enhanced oxidative burst of macrophages and neutrophils
	Development of the capability to induce and maintain oral tolerance in antibody response

[†] Abbreviations: TCR: T cell receptor, MHC: major histocompatibility complex
Source: References 176 and 177.

also immune regulation; upon bacterial colonization, the expression of MHC class II molecules on enterocytes is provoked gradually (57). Thus, in contrast to the dietary antigens (ingested also by the germ-free mice), bacterial compounds are capable of stimulating the MHC class II expression, important for immunologic presentation of exogenous antigens. MHC class II expression is regulated by cytokines, and intestinal bacteria also greatly affect cytokine production in gut cells, which, in turn, conceivably affects the MHC class II expression (12).

The functional impact of diverse cellular changes may affect different immune functions. Besides enhancing the IgA production as previously mentioned, a major function of the gut flora is the impact on the host's capability to induce oral tolerance. Intake of dietary antigens by germ-free animals elicits an IgE antibody response along with increased levels of IL-4. This aberrant Th2 response, not induced in conventional mice, can, however, be corrected by reconstituting the gut flora of germ-free animals with *Bifidobacterium infantis* (a predominant gut bacteria of infants) at the neonatal stage, but not at a later age, signifying the importance of gut flora including LAB in the maturation of the neonatal immune system (51).

The neonate immune system of humans is immature with a poorly developed GALT. Intraepithelial lymphocytes are not expanded, and germinal centers are not formed, until sometime after birth, again reflecting the dependency of microbial stimuli (4,58). Newborns are generally Th2-skewed and have a reduced capability to induce Th1-mediated responses, and have poor IgA and IgG production capacity. The gut flora is believed to be a major Th2-counterbalancing ($Th1/T_R$-driving) force in the postnatal establishment of immune homeostasis (4). Many observations support the notion that most mucosal immune cells are competent even before birth, but need to undergo an activation process initiated by environmental signals, most significantly originating from the gut flora. Studies in animals (59) and in humans (60) have identified deficient antigen-presenting cell function as the chief explanation for immunologic immaturity; in neonates dendritic cells and macrophages are unable to deliver adequate costimulatory signals to activate naïve T cells. Thus, the effect of the gut flora, in particular the LAB dominating the flora of infants, may exert an effect on antigen-presenting cells, such as dendritic cells, either directly, or indirectly through interactions with other cells, e.g., enterocytes and intraepithelial lymphocytes, which, in turn, affect the antigen-presenting cells. As will be discussed in far more detail later, many LAB possess Th1- and T_R-inducing capacities and, indeed, interact with the gut epithelium (61,62). Such mechanisms of LAB may be crucial for the establishment and maintenance of immunologic homeostasis, and currently these phenomena have led to focused research into the immunomodulatory mechanisms of LAB as described in what follows.

18.4 IMMUNOMODULATORY MECHANISMS OF LAB

Numerous immunomodulating effects of LAB are observed in cellular mechanistic *in vitro* studies. Common to most of them is the absence of a mechanistic link to the immunomodulating effects observed *in vivo* and, one step further, to the observed "end point" effects on diseases or their prevention. This fact, in addition to the observation that a great deal of variation occurs among the different LAB even at the strain level, has caused probiotic research to be extremely multifarious and sometimes confusing. However, as more and more data accumulate some common characteristics come to light. In this section, we will go over some of the most prevailing immune mechanistic effects found for LAB as evident from either *in vitro* cell studies or *in vivo* studies in experimental animal models

Table 18.3

Summary of Immune Mechanistic Modulatory Effects of LAB[†]

Effect	Characteristics
Increased barrier integrity (Reduced permeability)	Reduced bacterial translocation and Ag leakage
Enhanced colonization resistance	Competitive LAB (adhesion sites, nutrients)
	Altered intestinal surface glycosylation (altered adhesion)
	LAB metabolites (antimicrobial substances, lactate)
Improved phagocytic cell function	Enhanced phagocytic and bactericidal capacity
	Reduced pathogen-induced apoptosis
Improved NK cell activity	Increased tumoricidal activity
Cytokine regulation	Enhanced production of pro-type 1 response cytokines
	Enhanced homeostasis (Th1/Th2/T_R cell balance)
Epithelial interaction	Counter-balance proinflammatory signals from G[−] bacteria
	Stimulation of TGF-β production in enterocytes
Lymphocyte proliferation	Altered mitogen-induced T and B cell proliferation
Adjuvant capacity	Increased systemic and secretory Ag-specific Ab response
Secretory response	Nonspecific increase of IgA (incl. IgA[+] B cells in the gut)
	Enhanced Ag-specific IgA (e.g., against rotavirus)

[†] Abbreviations: LAB: lactic acid bacteria, Ag: antigen, Ab: antibody, G[−]: Gram-negative

Table 18.4

Commonly Used Methods for Studying the Effects of Probiotic Bacteria on the Immune System

Method
Phagocytic capacity
Tumoricidal activity (NK cells)
Cell proliferation
Antigen-specific stimulation (primed cells)
Polyclonal mitogen stimulation
Cytokine determination
Supernatants from cell cultures
Serum
Cellular surface marker expression
Antibody determination
Antigen-specific
Isotype profile (mice: Th1: IgG2a, Th2: IgG1+IgE)

or humans. These mechanisms are summarized in Table 18.3. In the subsequent section we will go into the effect of LAB on diseases and the impact of viable vs. nonviable bacteria and effective dose.

18.4.1   Methods Used for Assessing Immunomodulatory Effect of LAB

Several methods have been employed for studying immune mechanisms as well as for evaluating clinical studies of immunomodulating effects of LAB, ranging from classical immune methods to sophisticated protocols for examining more unique parameters.

Table 18.4 summarizes the most commonly used immune methods, which will be discussed in brief in what follows.

Many different types of cells or cell mixtures are being investigated. When human cells are used for evaluating immune functions, whole blood or blood cell subgroups, such as peripheral blood mononuclear cells (PBMC) or polymorphonuclear cells are routinely used. Carcinoma cell lines of human enterocytes (Caco-2 and HT-29) are often used to study gut epithelial functions, sometimes in combination with PBMC to reflect the interaction between immune cells of GALT and enterocytes. With mice and rats, which are by far the most frequently used experimental animals to study immune functions, cells from the spleen, peritoneum, Peyer's patches, mesenteric lymph nodes, and lamina propria are widely used. Also, tissue samples for histological studies as well as secondary cultures (e.g., dendritic cells derived from bone marrow) and cell lines (macrophages, T, and B cells) are frequently used.

For the evaluation of nonspecific immune functions, cellular phagocytosis and NK cell activity are often determined, as these functions are important in combating microbial infection and growth of neoplastic cells. Phagocytosis is typically measured by cellular *in vitro* uptake of pathogenic bacteria like *E. coli* or *Staphylococcus aureus* following coculture of cell samples containing phagocytic cells such as whole blood, PBMC, or peritoneal lavages, and pathogens. Enumeration is either done by direct microscopic count or by flow cytometric analysis using fluorescence labeled bacteria. In humans, NK cell activity is normally measured as killing of dye-labeled or ^{51}Cr-labeled human tumor cells following coculture with PBMC. Phagocytic capacity and NK cell activity have been used to evaluate human clinical intervention studies.

Lymphocyte proliferation is a common method for measuring cellular response capacity toward *in vitro* stimulation either with antigen using *in vivo*-primed cells (immunization and vaccination), or nonspecifically with polyclonal mitogens. The most common mitogens applied to test lymphocyte proliferation are the T cell mitogens concanavalin A and phytohemagglutinin, and the B cell mitogens lipopolysaccharide and pokeweed mitogen. Although a commonly used method, the biological significance of mitogen stimulation is often questioned because of the broad nonspecific nature of the stimulation. Detection of cellular incorporation of [^{3}H]thymidine into DNA is commonly used for estimating changes in cell number.

Production of cytokines and other immune regulatory compounds are also methods frequently used in the assessment of immune mechanistic effects. Cytokines can be measured in supernatants of cell cultures or in whole blood samples with or without preceding stimulation with antigen or mitogen. Because many cytokines function at a focal level, e.g., in the synapse of interacting cells, and are often present only briefly and in low amounts, their detection is hampered and the biological significance of assessing their level in many *in vitro* experimental settings remains to be established. Cytokines are typically measured by enzyme-linked immunosorbent assay (ELISA) but can also be measured at the mRNA level.

Other measurements have been used to determine the effects of probiotics on immune status. For example, the number and type of immune cells such as blood or intestinal B and T cells can be determined by the use of fluorescence labeled antibodies directed against cell surface markers using flow cytometric analysis. Moreover, qualitative and quantitative determination of antibody production with or without preceding priming is used. Specific antibody responses to vaccination antigens or to bacterial antigens are used as an index for the adjuvant capacity of LAB; i.e., their capacity to potentiate a response to an antigen. The profile of the types of antibody produced indicates the type of response initiated, e.g., IgE in an allergic response.

18.4.2   Effects of LAB on Nonimmunologic Mucosa Barrier Functions

Many effects of LAB are very likely to be mediated through epithelial cell functions. For the nonimmunologic part, the intestinal epithelium provides a physical protective barrier, whose integrity is differentially influenced by various microbial signals. The integrity of the epithelial barrier is dependent on a tight cellular structure of enterocytes joined together by tight junctions. Defects in the barrier function can occur during invasion by pathogens, food intolerance reactions, or exposure to chemicals or radiation. Leakage through tight junctions or direct cell disruption results in increased permeability and consequently translocation of microbes and dietary antigens, which deteriorate the condition and further compromise barrier function (63). Factors such as proinflammatory cytokines, secreted by enterocytes and intraepithelial lymphocytes during contact with certain pathogenic bacteria, affect the integrity of the tight junctions and thus intestinal permeability, resulting in elevated transport of fluid and electrolytes eventually leading to diarrhea. Due to the same molecular mechanisms, beneficial microorganisms of the gut flora, including certain LAB, seem instead to enhance the integrity of the enterocytes tight junctions. For example, administration of the probiotic bacteria *L. rhamnosus* GG or *L. plantarum* 299v has been found to reverse increased intestinal permeability caused by cow's milk in suckling rats, or by *E. coli* infection in rats, respectively (64,65). Likewise, the probiotic product mixture VSL#3, which contains *B. longum*, *B. infantis*, *B. breve*, *L. acidophilus*, *L. casei*, *L. delbruekii* ssp. *bulgaricus*, *L. plantarum*, and *S. salivarius* ssp. *thermophilus*, was found to normalize colonic barrier integrity in parallel with a reduction of the proinflammatory cytokines TNF-α and IFN-γ in a colitis mouse model, where the permeability was otherwise increased by conventional gut colonization compared to germ-free animals (66).

Besides down regulation of proinflammatory factors, epithelial integrity could also be improved by LAB enhancing enterocyte proliferation and, thus increasing cellular turnover of injured and dysfunctional enterocytes. Such an effect has been seen in rats supplemented with *L. casei* (67). Alternatively or in addition, improved permeability could be mediated by LAB through pathogen exclusion due to enhanced resistance to pathogen colonization, which is another nonimmune defense mechanism promoted by certain LAB. Intestinal LAB such as strains of *Lactobacillus* are known to possess antimicrobial activities against pathogenic bacteria such as production of bactericidal peptides (68), and have been found to inhibit adhesion of pathogens to enterocytes *in vitro* (69). Bifidobacteria and some *Lactobacillus* strains, including many probiotics, adhere to enterocytes and are in this way believed to compete with pathogens for adhesion sites (70). The ability of certain bacteria to regulate surface expression of glycosylated compounds that mediate attachment of pathogens to the epithelial surface is another possible explanation for the improving effects of LAB on resistance to pathogen colonization (71). The capability of LAB to up regulate mucin production may also improve colonization resistance. *L. plantarum* 299v has been found *in vitro* to inhibit adhesion of *E. coli* to enterocytes and to up regulate mucin mRNA expression in the cells (72). Mucin up regulation may aid clearance of pathogens while also increasing attachment of LAB, as certain probiotic LAB are known to adhere to intestinal mucus (70). Much still needs to be confirmed and further clarified about the effect of LAB and the gut flora on the nonimmune related defense mechanisms.

18.4.3   Effects on Phagocytic Cells

Probiotic LAB possess the capability to enhance nonspecific immune functions of phagocytic cells (neutrophils and monocytes or macrophages). Although the exact mechanisms involved are largely unknown, the effect seems evident both from *in vitro* and *in vivo* studies, showing, however, a great variation among strains.

Increased phagocytic capacity is one important effect found for certain probiotic LAB: Ingestion of strains such as *L. johnsonii* La1, *L. rhamnosus* HN001, *B. lactis* Bb-12, and *B. lactis* HN019 significantly increases the phagocytic capacity of blood leukocytes or peritoneal cells (mice only), as shown in mice (73) and humans, including the elderly (6,74–77). In humans, this effect was shown to be stronger in individuals with poor phagocytic capacity prior to treatment (78). In contrast, ingestion of *L. casei* Shirota by humans was found not to influence the phagocytic capacity (79). Enhanced phagocytic capacity might be mediated through LAB-induced up regulation of phagocytic receptors, as observed for *L. rhamnosus* GG given to healthy humans, who showed an increased expression of CR1, CR3, FcγRI, and FcαR in neutrophils (80).

Increased phagocytic capacity mediated by probiotics is one mechanism that potentially accounts for enhanced resistance against pathogen infections. For example, mice challenged with pathogens like *Salmonella typhimurium* or *E. coli* 0157:H7 showed increased resistance to infection in conjunction with enhanced phagocytic capacity when fed probiotic LAB (*L. rhamnosus* HN001 or *B. lactis* HN019) (81,82).

Certain LAB are known to affect components involved in bactericidal activity of phagocytes, which is another mechanism potentially explaining increased resistance to infection. LAB enhance macrophage enzymatic activity when fed to mice (*L. casei* and *L. acidophilus*) (83) and have been found to enhance the oxidative burst (production of ROS) of phagocytic cells in humans (*L. johnsonii* La1) (14). Moreover, when exposed *in vitro* to macrophages, many strains of LAB readily induce production of nitric oxide and hydrogen peroxide, although the physiological relevance of such an *in vitro* effect remains unclear (84).

Pathogens such as *Salmonella typhimurium* have developed mechanisms to disseminate into deep tissues like the intestinal mucosa by inducing apoptosis (programmed cell death) of phagocytic and other immune cells. Recently, LAB strains (*L. delbruekii* ssp. *bulgaricus* and *S. thermophilus*) were found to inhibit such apoptosis induction and in so doing possibly enhance the capability of the phagocytic cells to kill the pathogen (85). This effect might, however, be mediated indirectly by an increased bactericidal capacity of the phagocytes that instigate killing of the pathogen before apoptosis is induced. Inhibition of pathogen-induced apoptosis of immune cells is yet another mechanism through which LAB might improve resistance to infection.

18.4.4 Effects on NK Cell Activity

NK cells are another group of cells playing an important role in the nonspecific immune defense against tumor cells, and viral and bacterial infections. A vast majority of NK cells reside in the liver, but they are also found at mucosal sites (86). Many studies, involving both animals and humans, have indicated that ingestion of certain LAB significantly enhances NK cell activity of PBMC observed as an increased tumoricidal activity (77,78,87,88). In some studies, a concomitant increase in the relative proportion of NK cells (cells expressing the CD56 surface molecule) among the isolated PBMC was observed, which could at least partly account for the observed increased tumoricidal activity. The increase in NK cell activity has been found to be positively correlated with age, and thus to benefit particularly elderly people of whom a high proportion have decreased NK cell activity (78). However, a direct link between LAB-induced increased NK cell activity and human health has yet to be documented.

NK cells are activated as an early event in an immune response by IL-12 produced by monocytes, macrophages, and dendritic cells. Activated NK cells produce IFN-γ that functions in the nonspecific immune response to further activate nonspecific immune mechanisms such as phagocytosis. NK cells themselves are also activated by IFN-γ.

Moreover, IFN-γ, like IL-12, is important in the promotion of specific Th1-polarized immune responses. The population of NK cells contained in PBMC is activated indirectly by LAB to produce IFN-γ through the effect of LAB on monocytes, which involves induction of IL-12 and costimulatory molecules (89). As will be discussed in what follows, many LAB induce IL-12 production in monocytes, macrophages, and dendritic cells (90–92), which potentially mediates the enhancing effect of LAB on NK cell activity.

18.4.5 Interactions Between LAB and Intestinal Epithelial Cells

Intestinal epithelial cells are endowed with several important immunoregulatory functions, as we have discussed, including antigen presentation and production of inflammatory and regulatory cytokines. Contact with pathogenic microorganisms such as *Salmonella typhimurium*, *E. coli* or *Helicobacter pylori*, activates enterocytes to secrete cytokines that induce migration of inflammatory cells like neutrophils into the mucosa. Such cytokines include the neutrophil attractant IL-8, and the proinflammatory cytokines IL-1β and TNF-α. Gram-positive versus Gram-negative bacteria appear to exert rather different effects on enterocytes (93). Gram-negative bacteria like *E. coli*, or *E. coli*-derived lipopolysaccharides, activate enterocytes to produce IL-8 and TNF-α through binding to the pattern recognition receptor CD14 (a cofactor for binding of several bacterial components to a Toll-like receptor). In contrast, Gram-positive bacteria including *L. johnsonii* La1 and *L. acidophilus* La10, as well as the cell wall glycolipid lipoteichoic acid derived thereof, inhibit the proinflammatory effect of *E. coli* (94). In line with this, *L. plantarum* 299v has been shown to inhibit *E. coli*-induced neutrophil transepithelial migration (95).

Migration of intraepithelial lymphocytes into the epithelial compartment depends on expression of adhesion molecules (αEβ7 integrin) on their surface, and these molecules are up regulated by TGF-β (96). In contrast to Gram-negative bacteria inducing inflammatory cytokines in epithelial cells, the probiotic *L. johnsonii* La1 was shown to induce TGF-β, potentially affecting the number of intraepithelial lymphocytes, and immunologic homeostasis, through the suppressive effect of TGF-β (61).

These observations suggest that LAB and other Gram-positive bacteria of the gut flora posses a capability to counterbalance the proinflammatory response of Gram-negative bacteria of the flora and so participate in the maintenance of immunologic homeostasis. The interaction of bacteria with enterocytes was found to be dependent on the presence of other immune cells, predominantly Th cells, and, in turn, to affect the activity of the immune cells, such as inducing immunosuppressive antigen-presenting cells (40,97). This provides evidence that luminal bacterial signaling is transduced to the host by a network of cross-talking immune cells in the gut, in which the gut epithelial cells participate considerably.

18.4.6 Effects on Cytokine Production

A large number of studies has clearly shown that LAB can influence cytokine production in immunocompetent cells of animals and humans. From *in vitro* studies with different types of immune cells such as human PBMC, murine spleen cells, dendritic cells or macrophage cell lines, a pattern for the type of cytokines typically induced following exposure to LAB has emerged: many lactobacilli, as well as other Gram-positive bacteria, are often capable of inducing Th1 response-related cytokines; IL-12, IL-18, IFN-γ, and TNF-α, and furthermore often induce IL-6 and IL-10 (87,90–92,98–100). In contrast to this, Gram-negative bacteria, e.g., nonpathogenic *E. coli*, are weak type 1 cytokine inducers but preferentially induce IL-10 (98,101,102). IFN-γ production by T cell or NK cells is induced through induction of the early response cytokine IL-12 in antigen presenting cells (monocytes, macrophages, or dendritic cells) (99,101,103). A large variation in the capacities to induce IL-12 occurs amongst species and strains of LAB; *L. casei* and *L. paracasei*, have, for instance, been repeatedly reported to include some of the strongest IL-12 inducing strains (91,92,99,104). IL-10 is, like IL-12, induced at different levels by different LAB, and strong IL-12-inducers appear often also to be strong IL-10 inducers. Induction of cytokines occurs in a dose dependent manner. In addition, the ratio between the induced cytokine levels varies not only with the strain of LAB but also with the concentration and duration of exposure, which adds further complexity to the phenomenon of bacterial signaling *in vivo* (92,102).

In addition to the capability of many LAB to be Th1-response potentiators *in vitro*, it has recently been found for a strain of *L. paracasei* (NCC2461), that, in addition to being a strong IL-10 and IL-12-inducer, this strain induced the development of a T_R-like Th cell population producing TGF-β and IL-10 in an *in vitro* system (104). Extension of such effect to *in vivo* conditions could explain some of the effects observed for LAB. In support of this, a recent *in vivo* study in mice showed that T_R cells with specificity toward the gut flora appeared to be dominant in the immune homeostasis of GALT (105).

In vivo cytokine production upon ingestion of LAB has been studied in both humans and animals. The results of these studies vary greatly and seem to depend on the LAB used and the experimental setting. As with the *in vitro* studies, animal feeding studies tend to show that some LAB strains favor Th1 responses. For example, *L. casei* CRL431 and *L. casei* Shirota fed to allergy-primed mice increased serum levels of IFN-γ and reduced IL-4 (106,107). However, in a study with healthy mice fed different LAB, no changes were observed for IFN-γ, TNF-α, or IL-6 (108). In another study, feeding yogurt to mice reduced cytokine expression, especially TNF-α (109). Moreover, in a study based on counting local cytokine-producing cells in the gut villi of healthy mice fed a range of lactobacilli, no changes were observed for IFN-γ, IL-4, or IL-10 (12). Yet in this study, significant induction of TNF-α and IL-2 was observed for some strains (*L. reuteri* and *L. brevis*), clearly showing a great variation between species and strains of LAB in their capacity to modulate the immune function at the site of delivery. Such local responses, however, may not necessarily be reflected systemically.

In human studies, administration of *L. rhamnosus* in combination with *L. reuteri* to patients with atopic disease did not alter IL-2, IL-4, IL-10, or IFN-γ production by PBMC, despite convalescence (110). However, in a similar study with atopic individuals ingesting *L. rhamnosus* GG, a transient up regulation of IL-10 was evident (111). Likewise, in patients suffering from intestinal inflammation, probiotics were shown to up regulate the tissue levels of IL-10 (112). In healthy individuals, intake of *L. brevis* ssp. *coagulans* was shown to increase a virus-induced IFN-α (an important cytokine in nonspecific cell mediated immunity) response. This effect followed a dose dependent pattern, with those subjects with initially lowest levels displaying the greatest increase (113). To date no studies have shown results on the effect of LAB on IL-12 production in humans. IL-12 and as other cytokines function focally between adjacent cells, which hampers factual *in vivo* assessment of changes of these cytokines.

On the whole, probiotic LAB are potentially capable of either improving homeostasis by inducing antiinflammatory cytokines like IL-10, or supporting Th1 responses by enhancing cytokines like IL-12. The pattern of LAB-induced modulation of cytokine production *in vivo* is likely to depend on the condition of the host; i.e., reflecting either immunologic imbalance due to disease or allergy, or proper immunologic balance when healthy.

18.4.7 Effects on Lymphocyte Proliferation

Lymphocyte proliferation responses to mitogens are used to examine T and B cell function upon treatment with immunosuppressive or immunoenhancing agents. The immunomodulating effect of ingested LAB on T and B cell proliferative capacity depends upon the species and strain of LAB involved. In some studies, ingestion of LAB such as *L. rhamnosus* HN001, *L. acidophilus* HN017, and *B. lactis* HN019 by mice was found to augment T and B cell function as evidenced by elevated proliferation response to concanavalin A, phytohe-magglutinin, or lipopolysaccharide stimulation (73,114). However, in similar settings, strains of *L. casei, L. gasseri,* and *L. rhamnosus* inhibited mitogen-induced T and B cell prolifera-tion, while *L. acidophilus* was without influence (115). Ingestion of LAB has, in some instances, been found to alter the proportion of T and B cells in lymphoid tissues, which may indirectly affect the proliferation response. Nevertheless, in cases with unchanged relative T cell populations, LAB were still found to alter the proliferation response (116). Cells from spleen, mesenteric lymph nodes and Peyer's patches have all exhibited altered proliferation response, showing, as for other analyses, that LAB exert immunomodulatory effects both at sites remote from and adjacent to the point of delivery. No explicit physiologic interpretation of altered proliferation response tested *in vitro* can be made. LAB inhibiting lymphocyte proliferation might be useful in the management of immune hyperresponsive diseases like allergy, autoimmunity, or intestinal inflammation, whereas LAB that enhance proliferation may be of general benefit to healthy individuals to improve the immune defense.

18.4.8   Effects on Specific Immune Responses – Adjuvant Capacity of LAB

A primary mechanism by which probiotic LAB mediate immune enhancement is through an adjuvant effect. An adjuvant is a compound that enhances the immune response toward a specific antigen when administered along with the antigen. Some examples of adjuvant capac-ity of certain LAB include studies where *L. rhamnosus, L. acidophilus, L. fermentum,* or *B. lactis* fed to healthy mice, significantly enhanced the formation of serum antibodies toward oral or parenterally administered antigens like cholera toxin, tetanus toxoid, or ovalbumin (12,73,117). Moreover, some LAB have been found to potentiate the antibody response against *E. coli* when cocolonized in germ-free animals (52,118). However, in this setting it was found that a large proportion of the antibody induced toward the LAB themselves cross reacted with the *E. coli*, which at least could partly account for the apparent adjuvant effect of LAB on the immune response against pathogens.

The LAB adjuvant effect has also been seen at the cytokine level. The *in vitro* antigen stimulated production of cytokines such as IFN-γ (Th1), IL-4, and IL-5 (Th2) in spleen and peritoneal cells of animals primed with ovalbumin and the Th2-driving adjuvant Al(OH)3 was enhanced in animals fed *L. rhamnosus*, which indicates an adjuvant effect on both Th1 and Th2 cytokines (119). Feeding *L. casei* Shirota likewise enhanced Th1 cytokine produc-tion, but reduced Th2 cytokine production and, moreover, reduced the formation of ovalbu-min-specific IgE (106). Hence, in addition to exhibiting an adjuvant capacity, different LAB strains seem to modulate cytokine regulation differentially, possibly resulting in the polariza-tion of the response. Accordingly, a study in mice showed that intake of *L. acidophilus, L. delbruekii* ssp. *bulgaricus,* and *L. casei* all enhanced the IgG1 response (Th2-associated isotype) toward parenterally injected ovalbumin, whereas *L. acidophilus* was the only strain that also enhanced the IgG2a response (Th1-associated isotype), and was also the only strain inducing detectable IL-12-producing cells in lamina propria (117).

Whether such modulations of Th1 and Th2 cytokine production during an antigen-specific reaction are beneficial to the host still remains to be established. It is, however, hypothesized that mechanisms of LAB mediating suppression of Th2 responses, either by favoring Th1 responses or by enhancing tolerance induction mechanisms (in general improving Th1/Th2/T_R-homeostasis), may be effective in the antiallergy effect observed for certain probiotic LAB (62).

Adjuvant effects of LAB are also observed in human studies. For example, enhanced specific antibody responses to rotavirus are reported when patients who are infected with, or vaccinated against, rotavirus, ingest *L. rhamnosus* GG (120,121). However, in a study with subjects receiving attenuated *Salmonella typhi* as an oral vaccine, ingestion of *L. rhamnosus* GG did not enhance antibody formation, indicating significant variations between different experimental settings (122).

In many vaccination regimes and in the defense against pathogenic infections, a secretory IgA response is important. Like systemic antibody responses already discussed, certain LAB enhance secretory IgA responses, which is discussed in what follows.

18.4.9   Enhancement of IgA Production

Augmentation of IgA production following ingestion of LAB is one of the better documented effects of probiotic LAB. LAB enhance IgA production in both a specific and a nonspecific manner. From studies in mice, oral administration of a wide range of lactobacilli and bifidobacteria species are found to increase the number of IgA^+ B cells in the gut mucosa as well as total IgA production (16,53,123,124). Likewise, in humans, ingestion of LAB, such as *L. rhamnosus* GG or *L. johnsonii* La1, is found to increase the total IgA level in blood and feces (15,120,125).

On an antigen-specific level, many species of LAB also promote secretory IgA production as observed both in mice and humans. In mice, several studies show that along with an increased protection against pathogens, probiotic LAB increases pathogen-specific IgA (114,126). Moreover, oral administration of LAB is shown to enhance IgA against dietary antigens like the milk protein β-lactoglobulin or toxins (16,123,127). Such effects are also found in humans, showing that ingestion of *L. johnsonii* La1, bifidobacteria or *L. rhamnosus* GG increases the specific IgA formation to attenuated *Salmonella typhi* (125), poliovirus vaccinations (15), and dietary proteins (in patients with gut inflammation) (128), respectively. Increased rotavirus-specific IgA response in rotavirus patients given probiotic LAB such as *L. rhamnosus* GG has been demonstrated in several studies and is believed to be a contributing mechanism in the observed protection against reinfection with rotavirus (120,129,130).

The cytokine IL-6 is important for the maintenance of an IgA response; Th2 cytokines (IL-4 and IL-5) are more important during response induction (131). Thus, the ability of many LAB to induce IL-6 but not Th2 cytokines, as discussed in a previous section, may be linked to the capability of LAB to act as an adjuvant for IgA production.

18.5   EFFECTS OF LAB ON DISEASES

Scientific evidence is now being accumulated to support both therapeutic and prophylactic properties of probiotic LAB and fermented products. Many beneficial health effects have been claimed for probiotics, especially concerning their potential to prevent or ameliorate intestinal diseases. There are, however, fundamental differences between various LAB, and the clinical effects vary greatly, with only a few probiotic strains showing efficacy in randomized placebo-controlled clinical studies. The probiotic organism that has received by far the most clinical attention to date is *L. rhamnosus* GG. Several other organisms have, however, also been thoroughly studied, of which the most important and their effects are listed in Table 18.5. Most clinical studies have been performed using viable bacteria. However, in some instances, nonviable bacterial preparations have been shown also to be effective, which will be discussed at the end of this section.

Table 18.5

Examples of Probiotics with Reported Immunomodulating Effects in Humans

Bacteria	Effects	Ref.
L. rhamnosus GG	Prevention and treatment of diarrhea, vaccine adjuvant, antiallergy, increased IgA	120,121,128,135,161,162
L. johnsonii La1	Vaccine adjuvant, enhanced phagocytosis, antiinfection (*Helicobacter*), increased IgA	6,14,125,178
L. plantarum 299v	Antiinflammatory bowel disease	145
L. rhamnosus HN001	Enhanced phagocytosis and NK cell activity, antiinfection	77,81,82,179
B. lactis HN019	Enhanced phagocytosis and NK cell activity, antiinfection	76,114,132,180
L. reuteri DSM12246	Treatment of diarrhea, antiallergy	110,136
VSL#3 (mixture of LAB)	Antiinflammatory bowel disease, prevention of diarrhea	148,181,182
L. casei Shirota	Treatment of diarrhea, anticancer, antiallergy, antiinfection, enhanced NK cell activity	88,104,106,168,183,184
L. casei CRL431	Treatment of diarrhea, antiinfection, increased IgA	185–188
B. lactis Bb-12	Antiallergy, prevention of diarrhea vaccine adjuvant, increased IgA, enhanced phagocytosis	6,15,134,161

18.5.1   Effects of LAB on Intestinal Infections

Some LAB have proved effective in preventing gastrointestinal infections and in the recovery from infectious diarrhea due to miscellaneous causes including traveler's diarrhea and rotavirus.

Acute diarrhea caused by gastrointestinal infections is common throughout the world, especially among children. One aspect of the effect of LAB against infectious diseases in the gut is due to enhanced resistance to colonization, e.g., by production of antibacterial metabolites, and, thus, to a lesser extend related to a direct effect on the immune system. However, combating pathogenic infections has also been shown to be due to stimulation of the immune defense, particularly by increasing phagocytic capacity and the secretory antibody (IgA) response, as discussed previously. Studies in mice have demonstrated that in parallel with such immune stimulating effects, dietary supplementation with *L. rhamnosus* HN001 or *B. lactis* HN019 reduces the severity of infections with the pathogens *E. coli* O157:H7 and *Salmonella typhimurium* (81,82,114,132). In contrast, in a clinical study with healthy volunteers challenged with enterotoxigenic *E. coli*, administration of *L. acidophilus* and *L. bulgaricus* did not reduce the attack rate, incubation period, or duration of diarrhea (133).

Nevertheless, in studies concerning children with acute infectious diarrhea (viral or bacterial), probiotics such as *L. reuteri* DSM 12249, *L. rhamnosus* GG, or *B. bifidum* effectively ameliorated the diarrhea in both hospitalized and nonhospitalized patients (134–136). Probiotics have been found to be particularly effective against rotaviral infections. The effect of *L. rhamnosus* GG in the treatment of rotavirus diarrhea in infants and children is one of the best documented probiotic effects in human subjects (137). Mechanisms including normalization of gut flora and gut permeability, as well as enhancement of mucosal immune response, have been reported in several studies; e.g., clearance of virus after primary infection and subsequent protection against reinfection (seen for at least 1 year) correlates well with the production of mucosal and serum IgA, which are enhanced by probiotics (44).

Antibiotic-associated diarrhea is another form of acute diarrhea that occurs frequently. It is caused by a disruption of the ecosystem of the gut flora resulting in decreased colonization resistance and altered fermentation capacity (microbial imbalance). Under these conditions *Clostridium difficile* or *Klebsiella oxytoca* may colonize the gut and induce colitis. *L. rhamnosus* GG has shown to be effective in preventing antibiotic-associated diarrhea in children while being ineffective in adult patients. The probiotic microorganisms *Enterococcus faecium* SF68 and the yeast *Saccharomyces boulardii*, have also shown prophylactic effects in antibiotic-associated diarrhea (138).

Traveler's diarrhea is another common form of acute diarrhea. In several studies, probiotic LAB demonstrated no effect in preventing traveler's diarrhea, and data to support a prophylactic effect is limited. In one study, however, the incidence of diarrhea in tourists going to Egypt was reduced considerably in subjects given capsules of *S. thermophilus, L. bulgaricus, L. acidophilus,* and *B. bifidum*, indicating that the effect of probiotics in traveler's diarrhea not only depends upon the probiotic strain or strains, but also on the specific conditions, such as the type of microbial exposure, which can vary at different destinations (139).

Colonization of the gastric mucosa with *Helicobacter pylori* is associated with gastritis, ulcers, and some malignancies. Several probiotic LAB have demonstrated anti-*Helicobacter* activity in *in vitro* studies. *In vivo*, LAB strains including *L. johnsonii* La1 have also been found to exhibit anti-*Helicobacter pylori* activity by suppressing this organism, but do not appear to eradicate it (140).

In summary, probiotic LAB seem to possess the potential to be used in the management of several infectious diseases. In particular, probiotic bacteriotherapy of children and infants, to reduce the incidence and shorten the duration of acute infectious diarrhea, might be effective. This fact may be related to the ability of the bifidobacteria predominating in the flora of breast-fed infants to promote protection against pathogenic infections (141).

18.5.2 Effects on Inflammatory Bowel Disease

Inflammatory bowel disease (IBD), including ulcerative colitis, Crohn's disease, and pouchitis (inflammation of the ileal pouch arising after surgical resection of colon), refers to intestinal disorders of unknown cause that are characterized by chronic or recurrent intestinal inflammation (142). It appears that IBD, particularly Crohn's disease, develops due to genetically influenced dysregulation of mucosal immunity against components of the intestinal flora. Cytokine imbalance due to dysregulated Th cells appears to play an important role in the pathogenesis: Crohn's disease is mediated predominantly by Th1 response, whereas ulcerative colitis is usually associated with a Th2 response.

Some probiotics seem to be very efficient in the treatment of IBD. As discussed earlier in this chapter, IL-10 is a key regulatory cytokine for tolerance to luminal antigens, including the gut flora. IL-10 gene-deficient mice respond to their gut flora by developing spontaneous colitis resembling Crohn's disease and are, therefore, an effective experimental model for IBD (143). A gut flora deficient in LAB is a characteristic of IL-10 gene-deficient mice, and administration of probiotic LAB, such as *L. reuteri* and *L. plantarum* 299v, has been shown to be an effective treatment of these mice (144,145). As also mentioned earlier, bacterial immunostimulatory DNA sequences, especially unmethylated CpG motifs, are known to activate epithelial cells through interaction with Toll-like receptors (28). In different IBD mouse models, treatment with immunostimulatory DNA inhibited induction of colonic proinflammatory cytokines and chemokines (146). Likewise, genomic unmethylated DNA from the probiotic mixture VSL#3 ameliorated colitis in a similar model, whereas methylated DNA was ineffective, clearly demonstrating the importance of immunostimulatory DNA in probiotics effective against IBD and perhaps against other diseases involving an immune imbalance as well (147).

In humans, a number of clinical studies have now demonstrated an efficacy of probiotics against ulcerative colitis, Crohn's disease, or pouchitis. The VSL#3 product (*B. longum, B. infantis, B. breve, L. acidophilus, L. casei, L. delbruekii* ssp. *bulgaricus, L. plantarum,* and *S. salivarius* ssp. *thermophilus*) has proved very efficient in prophylaxis and reduction of recurrence of chronic relapsing pouchitis (148–150). Significantly, in IBD patients treated with VSL#3, the tissue levels of IL-10 were increased revealing an immunomodulatory mechanism elicited by the probiotics (112).

Other nonLAB microorganisms such as *E. coli* Nissle 1917 and *Saccharomyces boulardii* have also proved efficient in the treatment of IBD (151). Intense research is ongoing to elucidate further the efficacy of probiotic LAB in the treatment and prevention of IBD.

18.5.3 Antiallergy Properties of LAB

The prevalence of allergic diseases has progressively increased in Western industrialized countries (152). Allergy and atopy, featuring IgE production, are closely associated with a Th2 cell-driven response. Atopic eczema due to food allergy represents the earliest manifestations of allergic disease, starting from infancy.

A so called "hygiene hypothesis" was originally formulated by Strachan in 1989 (153), stating a possible explanation for the increasing prevalence of allergy. In essence, this hypothesis stated that the lower microbial exposure of the immune system due to increased hygiene causes an underdeveloped Th1 polarization (newborns are Th2-skewed) and, thus, an imbalance in the Th1/Th2 homeostasis, favoring the Th2-driven allergic responses. While this original hypothesis mainly implied the consequence of increased hygiene to be a reduction in the number of Th1-driving infections known to reduce allergic disorders, a revised hygiene hypothesis was proposed by Wold in 1998 (154), saying that the overly hygienic lifestyle in modern Western countries, rather than limiting pathogenic infections, has altered the normal intestinal colonization pattern in infancy leading to a failure to induce and maintain oral tolerance of innocuous antigens.

Indeed, there is data accumulating showing a clear tendency of atopic subjects to have an aberrant composition of the gut flora in comparison with nonatopic subjects: infants in whom atopy develops tend to have fewer bifidobacteria and more clostridia (8,155,156). Recent findings have, moreover, indicated a correlation between allergic disease and the composition of the intestinal bifidobacteria population, pointing toward allergic infants having a more adult-like bifidobacteria flora with high levels of *B. adolescentis* whereas healthy infants have high levels of *B. bifidum* (157,158). Interestingly, the adhesion capacity to human intestinal cells of the isolated fecal bifidobacteria from healthy infants has been found to be significantly higher than for allergic infants. Furthermore, when testing the capacity to induce cytokines in a macrophage cell line, bifidobacteria from allergic infants vs. healthy infants were found to induce more proinflammatory cytokines such as IL-1β, IL-12, and TNF-α, whilst inducing much less IL-10 with many strains not inducing IL-10 at all. This provides preliminary indications that although bifidobacteria are Th1 cytokine inducers and putatively counteract Th2 responses, the incapability of inducing the response suppressive IL-10 in combination with the TNF-α-inducing capacity (TNF-α is increased during allergic inflammation in the gut) of the gut flora derived bifidobacteria of allergic infants possibly plays a decisive role in allergic disease etiology (159).

A corollary of the documented immunologic repercussions to aberrant gut flora, as well as the general finding of LAB to improve immune homeostasis, is the investigation of the effect of supplementing LAB on allergy. As mentioned earlier, oral administration of *L. casei* Shirota to mice prior to Th2 response priming reduced IgE and IL-4 levels while increasing the Th1 cytokines IL-2 and IFN-γ (106). Likewise, in a mouse model of atopy, administration of *L. casei* CRL431 was shown to down regulate IgE and IL-4 synthesis by increasing IFN-γ. This was, however, only effective when administered before sensitization, indicating the impact of timing (107).

In humans, a number of studies have indicated the efficacy of *Lactobacillus* administration in the treatment and prevention of atopic eczema. In this regard, *L. rhamnosus* GG is the most investigated probiotic LAB strain and has proved efficient in several settings. In studies of infants with atopic eczema, supplementation with *L. rhamnosus* GG has been found to improve clinical symptoms, alleviate intestinal inflammation and enhance IL-10 production significantly (111,160,161). One of these studies also included *B. lactis* Bb-12, which demonstrated this strain to be equally efficient in reducing the severity of eczema (161). In a study involving older children (1–13 years) with chronic atopic eczema, oral treatment with a combination of *L. rhamnosus* 19070-2 and *L. reuteri* DSM12246, decreased the severity of eczema in more than half of the treated subjects (110). These studies substantiate the efficacy of some probiotics in the treatment of manifested allergy.

Clinical research has also aimed at the efficacy of probiotics to prevent development of allergy. The efficiency of *L. rhamnosus* GG to prevent the occurrence of atopic eczema in infants from atopic families has been studied (162). The bacteria were given to mothers during pregnancy, and mothers who were breast feeding continued to take the bacteria after delivery; otherwise the children were given bacteria for 6 months. At 2 years of age, the frequency of atopic eczema in the probiotic group was reduced by half (from 46% to 23%), demonstrating a potential of this probiotic LAB in prophylaxis of allergy. In line with this, administration of *L. rhamnosus* GG to the pregnant and lactating mothers increased the level of TGF-β in breast milk and reduced the risk of the children developing atopic eczema during the first two years of life (163).

There are indications that using probiotics in the management of allergy is most if not solely effective in allergies occurring in infancy or early childhood. For example, *L. rhamnosus* GG failed to prevent both birch pollen allergy and apple allergy in young adults (164). Supplementation of infants via formulas fortified with probiotics or supplementation of mothers during pregnancy and lactation are definite future perspectives for the applicability of probiotics. Research is ongoing, working to identify new probiotics and their mechanisms of action and optimal protocol of administration, especially regarding timing.

18.5.4   Antitumor Effect of LAB

Colorectal cancer is one of the most important causes of death from cancer in Western countries. Epidemiological studies show a diet associated risk, in particular of the low fiber "Western diet"; and many studies confirm the involvement of intestinal microflora in the onset of colon cancer. In animals, several studies have shown that bacteria of the bacteroides and clostridia genera increase the incidence and growth of colonic tumors, whereas other genera such as lactobacilli and bifidobacteria prevent tumorigenesis (165). Likewise, in a human study, high risk of colon cancer was found to be associated with presence of species of *Bacteroides*, whereas low risk was associated with presence of species of *Lactobacillus* (166). Thus, much attention has focused on decreasing cancer through diet alterations, in particular increasing intake of fiber and probiotics both intervening with the gut flora.

No direct experimental evidence has, yet, been established that consumption of LAB suppresses cancer in man; however, a vast amount of indirect evidence exists from laboratory experiments using experimental animal models (167). From such studies, several non-immunologic mechanisms of LAB have been suggested, such as decrease of fecal enzymes that may be involved in carcinogen formation, elimination of carcinogens, and production of antitumorigenic compounds. However, antitumor effects of LAB in experimental animals seem also to be mediated through immunologic mechanisms. As discussed in the previous section, LAB increase specific and nonspecific immune mechanisms, potentially protecting against tumor development. In different models of tumor bearing mice, injection of heat-killed *L. casei* Shirota (termed LC9018) to the site of tumor growth showed a potent inhibition of tumor growth and increased survival (106,168). This antitumor effect was mediated through production of the tumor-suppressing type 1 cytokines IL-12, IFN-γ, and most importantly TNF-α, as injection of neutralizing anticytokine antibodies abolished the antitumor effect of the bacteria. Studies on the effect of yogurt (*L. delbrueekii* ssp. *bulgaricus* and *S. thermophilus*) showed that consumption of yogurt by mice inhibited chemically induced colorectal cancer in conjunction with an increase in IgA$^+$ B cells in the colon, and with IFN-γ and TNF-α production (169). In summary, these and other studies suggest that certain LAB impart an antitumor effect through modulation of the host's immune system, specifically cell-mediated immunity. However, such antitumor effects of LAB in humans remain to be determined.

18.5.5   Impact of Viability and Dose on the Effects of LAB

Besides strain-related primary selection criteria (adhesion, acid, and bile tolerance, not discussed in this chapter), some practical aspects have an impact on the immunomodulating effects of probiotics; most importantly the consequences of viable vs. nonviable bacterial preparations, and effective doses. Probiotics, as defined in the introduction, include also non-viable bacteria and bacterial components. However, research has shown that some effects demand the bacteria be alive whereas other effects are independent of viability. Effects exerted via prolonged epithelial adherence or (transient) colonization logically require live bacteria although there are indications that nonviable bacteria can, in fact, adhere to epithelial cells (170). A variety of immunomodulating activities, seem to be attributable to specific cell wall or intracellular bacterial components, and are exhibited by nonviable bacteria and even by the isolated components of the cell (100,147).

Only a few studies have focused specifically on testing the effects of viable vs. nonviable bacteria. Viable and heat-killed LAB were equally efficient in enhancing phagocytic capacity in mice, whereas only the viable bacteria enhanced IgA production (171,172). Likewise, only viable *L. rhamnosus* GG enhanced IgA production in patients with acute rotavirus diarrhea, while both viable and nonviable probiotics shortened the diarrhea (130). In an attempt to assess the efficacy of viable vs. heat-killed *L. rhamnosus* GG on atopic disease, the heat-killed preparation induced diarrhea and the study could not be completed, opening the question of safety of ingestion of large amounts of heat-killed bacteria (173). These and other studies clearly indicate that the importance of viable vs. nonviable bacteria depends on the effect in question, although viable probiotics tend to have more health effects than nonviable (113,174). The method of inactivation, e.g., inactivation by natural exhaustion in stored fermented products, which introduces cellular changes; inactivation by heat killing experimentally or due to pasteurization, which highly denatures cellular compounds; or inactivation by UV or gamma irradiation, which spares protein denaturation but affects DNA, may also play a role.

Another important factor involving the effects of probiotic LAB is effective dose. Typically, probiotic LAB are only effective if the dosage is sufficiently high. Accordingly, the minimum daily dose of viable *L. johnsonii* La1, which significantly increases phago-cytic capacity and oxidative burst in humans, is 10^9 bacteria (14). In a study in mice, the minimum daily dose for a significant increase in blood cell phagocytic capacity was 10^7 with a clear dose-response effect up to 10^{11} bacteria per day (172). Most clinical studies apply doses at 10^9–10^{11} viable bacteria/day.

18.6 CONCLUDING REMARKS AND FUTURE PERSPECTIVES

Probiotics are just one aspect of the emerging field of functional foods, and have recently received renewed enthusiasm and far more rigorous scientific pursuit. Well documented beneficial effects on gastrointestinal disturbances are mediated by probiotics. The research worldwide is now focusing on the development of target specific probiotic products with well characterized microorganisms selected for their specific health enhancing properties. For this purpose, more research into the intestinal flora and its interaction with the host immune function is a prerequisite to enhance the scientific soundness of identifying novel probiotics. Currently, many questions remain to be answered as to how probiotic LAB exert their ostensibly wide spectrum of immunomodulating effects. It appears reasonable to assume that the multifactorial effects of LAB stem from merely one or a few primary effects, yet targeted at some early stages of the immune response and thus causing altera-tions in a variety of mechanisms along the response cascade. Accordingly, increasing amounts of data indicate that LAB greatly affect epithelial cells and antigen-presenting cells such as the dendritic cells important for the initiation and polarization of an immune response. Effects on these cells disseminate like ripples in a pond, potentially modulating multiple immune functions regulated by these cells. Equally important, such primary effects of probiotics can also evoke diversified outcomes depending on the immune status of the host; i.e., healthy, malfunctioning, or diseased.

Like many other areas of biology, regulation of the immune system is extremely com-plex. Metaphorically, the phenomenon of immune regulation could frankly be compared with jumping in a waterbed: when jumping down the bed, other parts of the bed inevitably go up, with some areas of the bed moving more vigorously than others; until stabilization, many areas move alternately up and down at different amplitudes and in an almost unpredictable pattern very closely linked to even small disparities in factors related to both the "jumper" and the bed — the "jumper" being the probiotic and the bed being the consumer.

ACKNOWLEDGMENT

The authors wish to thank Dr. Charlotte Madsen for critically reading through the manuscript.

REFERENCES

1. Molin, G., B. Jeppsson, M.L. Johansson, S. Ahrne, S. Nobaek, M. Stahl, S. Bengmark. Numerical taxonomy of *Lactobacillus* spp. associated with healthy and diseased mucosa of the human intestines. *J. Appl. Bacteriol.* 74:314–323, 1993.
2. Conway, P.L. Development of intestinal microbiota. In: *Gastrointestinal microbiology*, Mackie, R.I., B.A. White, R.E. Isaacson, eds., New York: Chapman & Hall, 1997, pp 3–38.
3. Ahrne, S., S. Nobaek, B. Jeppsson, I. Adlerberth, A.E. Wold, G. Molin. The normal *Lactobacillus* flora of healthy human rectal and oral mucosa. *J. Appl. Microbiol.* 85:88–94, 1998.

4. Bjorksten, B. The intrauterine and postnatal environments. *J. Allergy. Clin. Immunol.* 104:1119–1127, 1999.

5. Holt, P.G., C.A. Jones. The development of the immune system during pregnancy and early life. *Allergy* 55:688–697, 2000.

6. Schiffrin, E.J., D. Brassart, A.L. Servin, F. Rochat, A. Donnet-Hughes. Immune modulation of blood leukocytes in humans by lactic acid bacteria: criteria for strain selection. *Am. J. Clin. Nutr.* 66:515S–520S, 1997.

7. Blum, S., Y. Delneste, S. Alvarez, D. Haller, P.F. Perez, C.H. Bode, W.P. Hammes, A.M.A. Pfeifer, E.J. Schiffrin. Interactions between commensal bacteria and mucosal immunocompetent cells. *Int. Dairy J.* 9:63–68, 1999.

8. Kirjavainen, P.V., E. Apostolou, T. Arvola, S.J. Salminen, G.R. Gibson, E. Isolauri. Characterizing the composition of intestinal microflora as a prospective treatment target in infant allergic disease. *FEMS Immunol. Med. Microbiol.* 32:1–7, 2001.

9. Fuller, R. Probiotics in man and animals. *J. Appl. Bacteriol.* 66:365–378, 1989.

10. Salminen, S., A. Ouwehand, Y. Benno, Y.K. Lee. Probiotics: how should they be defined? *Trends Food Sci. Tech.* 10:107–110, 1999.

11. Hamann, L., V. El Samalouti, A.J. Ulmer, H.D. Flad, E.T. Rietschel. Components of gut bacteria as immunomodulators. *Int. J. Food. Microbiol.* 41:141–154, 1998.

12. Maassen, C.B.M., C. Holten-Neelen, F. Balk, J.H. Bak-Glashouwer, R.J. Leer, J.D. Laman, W.J.A. Boersma, E. Claassen. Strain-dependent induction of cytokine profiles in the gut by orally administered *Lactobacillus* strains. *Vaccine* 18:2613–2623, 2000.

13. Ouwehand, A.C., S. Salminen, E. Isolauri. Probiotics: an overview of beneficial effects. *Antonie Van Leeuwenhoek* 82:279–289, 2002.

14. Donnet-Hughes, A., F. Rochat, P. Serrant, J.M. Aeschlimann, E.J. Schiffrin. Modulation of nonspecific mechanisms of defense by lactic acid bacteria: effective dose. *J. Dairy Sci.* 82:863–869, 1999.

15. Fukushima, Y., Y. Kawata, H. Hara, A. Terada, T. Mitsuoka. Effect of a probiotic formula on intestinal immunoglobulin A production in healthy children. *Int. J. Food Microbiol.* 42:39–44, 1998.

16. Takahashi, T., E. Nakagawa, T. Nara, T. Yajima, T. Kuwata. Effects of orally ingested *Bifidobacterium longum* on the mucosal IgA response of mice to dietary antigens. *Biosci. Biotechnol. Biochem.* 62:10–15, 1998.

17. Mowat, A.M., J.L. Viney. The anatomical basis of intestinal immunity. *Immunol. Rev.* 156:145–166, 1997.

18. Strobel, S., A.M. Mowat. Immune responses to dietary antigens: oral tolerance. *Immunol. Today* 19:173–181, 1998.

19. Kelsall, B.L., W. Strober. Dendritic cells of the gastrointestinal tract. *Springer Semin. Immunopathol.* 18:409–420, 1997.

20. Rescigno, M., M. Urbano, B. Valzasina, M. Francolini, G. Rotta, R. Bonasio, F. Granucci, J.P. Kraehenbuhl, P. Ricciardi-Castagnoli. Dendritic cells express tight junction proteins and penetrate gut epithelial monolayers to sample bacteria. *Nat. Immunol.* 2:361–367, 2001.

21. Dearman, R.J., H. Caddick, D.A. Basketter, I. Kimber. Divergent antibody isotype responses induced in mice by systemic exposure to proteins: a comparison of ovalbumin with bovine serum albumin. *Food Chem. Toxicol.* 38:351–360, 2000.

22. Perdue, M.H. Mucosal immunity and inflammation, III: the mucosal antigen barrier: cross talk with mucosal cytokines. *Am. J. Physiol.* 277:G1–G5, 1999.

23. Kraehenbuhl, J.P., E. Pringault, M.R. Neutra. Intestinal epithelia and barrier functions. *Aliment. Pharmacol. Ther.* 11:3–8, 1997.

24. Kelsall, B., W. Strober. Gut-associated lymphoid tissue: antigen handling and T-lymphocyte responses. In: *Mucosal Immunology*, Ogra, P.L., J. Mestecky, M.E. Lamm, W. Strober, J. Bienenstock, J.R. McGhee, eds., San Diego: Academic Press, 1999, pp 293–317.

25. Palucka, K., J. Banchereau. Dendritic cells: a link between innate and adaptive immunity. *J. Clin. Immunol.* 19:12–25, 1999.

26. Strober, W., R.L. Coffman. Tolerance and immunity in the mucosal immune system: introduction. *Res. Immunol.* 148:489–490, 1997.

27. Banchereau, J., R.M. Steinman. Dendritic cells and the control of immunity. *Nature* 392:245–252, 1998.

28. Underhill, D.M., A. Ozinsky. Toll-like receptors: key mediators of microbe detection. *Curr. Opin. Immunol.* 14:103–110, 2002.

29. Vieira, P.L., E.C. de Jong, E.A. Wierenga, M.L. Kapsenberg, P. Kalinski. Development of Th1-inducing capacity in myeloid dendritic cells requires environmental instruction. *J. Immunol.* 164:4507–4512, 2000.

30. de Jong, E.C., P.L. Vieira, P. Kalinski, J.H.N. Schuitemaker, Y. Tanaka, E.A. Wierenga, M. Yazdanbakhsh, M.L. Kapsenberg. Microbial compounds selectively induce Th1 cell-promoting or Th2 cell-promoting dendritic cells *in vitro* with diverse Th cell-polarizing signals. *J. Immunol.* 168:1704–1709, 2002.

31. Shortman, K., Y.J. Liu. Mouse and human dendritic cell subtypes. *Nat. Rev. Immunol.* 2:151–161, 2002.

32. Banchereau, J., F. Briere, C. Caux, J. Davoust, S. Lebecque, Y.J. Liu, B. Pulendran, K. Palucka. Immunobiology of dendritic cells. *Annu. Rev. Immunol.* 18:767–811, 2000.

33. Roncarolo, M.G., R. Bacchetta, C. Bordignon, S. Narula, M.K. Levings. Type 1 T regulatory cells. *Immunol. Rev.* 182:68–79, 2001.

34. Singh, B., S. Read, C. Asseman, V. Malmstrom, C. Mottet, L.A. Stephens, R. Stepankova, H. Tlaskalova, F. Powrie. Control of intestinal inflammation by regulatory T cells. *Immunol. Rev.* 182:190–200, 2001.

35. Friedman, A., H.L. Weiner. Induction of anergy or active suppression following oral tolerance is determined by antigen dosage. *Proc. Natl. Acad. Sci. USA* 91:6688-6692, 1994.

36. Brandtzaeg, P. Current understanding of gastrointestinal immunoregulation and its relation to food allergy. *Ann. NY Acad. Sci.* 964:13–45, 2002.

37. Smith, K.M., J.M. Davidson, P. Garside. T-cell activation occurs simultaneously in local and peripheral lymphoid tissue following oral administration of a range of doses of immunogenic or tolerogenic antigen although tolerized T cells display a defect in cell division. *Immunology* 106:144–158, 2002.

38. Jung, H.C., L. Eckmann, S.K. Yang, A. Panja, J. Fierer, E. Morzyckawroblewska, M.F. Kagnoff. A distinct array of proinflammatory cytokines is expressed in human colon epithelial cells in response to bacterial invasion. *J. Clin. Invest.* 95:55–65, 1995.

39. Eckmann, L., H.C. Jung, C. Schurermaly, A. Panja, E. Morzyckawroblewska, M.F. Kagnoff. Differential cytokine expression by human intestinal epithelial cell lines regulated of interleukin-8. *Gastroenterology* 105:1689–1697, 1993.

40. Haller, D., C. Bode, W.P. Hammes, A.M.A. Pfeifer, E.J. Schiffrin, S. Blum. Non-pathogenic bacteria elicit a differential cytokine response by intestinal epithelial cell/leucocyte co-cultures. *Gut* 47:79–87, 2000.

41. Fujihashi, K., M.N. Kweon, H. Kiyono, J.L. VanCott, F.W. vanGinkel, M. Yamamoto, J.R. McGhee. A T cell/B cell epithelial cell Internet for mucosal inflammation and immunity. *Springer Semin. Immunopathol.* 18:477–494, 1997.

42. Brandtzaeg, P., I.N. Farstad, F.E. Johansen, H.C. Morton, I.N. Norderhaug, T. Yamanaka. The B-cell system of human mucosae and exocrine glands. *Immunol. Rev.* 171:45–87, 1999.

43. Kett, K., K. Baklien, A. Bakken, J.G. Kral, O. Fausa, P. Brandtzaeg. Intestinal B cell isotype response in relation to local bacterial load: evidence for immunoglobulin A subclass adaptation. *Gastroenterology* 109:819–825, 1995.

44. Macpherson, A.J., L. Hunziker, K. McCoy, A. Lamarre. IgA responses in the intestinal mucosa against pathogenic and non-pathogenic microorganisms. *Microb. Infect.* 3:1021–1035, 2001.

45. Dunne, C. Adaptation of bacteria to the intestinal niche: Probiotics and gut disorder. *Inflamm. Bowel Dis.* 7:136–145, 2001.

46. Vaughan, E.E., M.C. de Vries, E.G. Zoetendal, K. Ben Amor, A.D.L. Akkermans, W.M. de Vos. The intestinal LABs. *Antonie Van Leeuwenhoek* 82:341–352, 2002.

47. Harmsen, H.J.M., A.C.M Wildeboer-Veloo, G.C. Raangs, A.A. Wagendorp, N. Klijn, J.G. Bindels, G.W. Welling. Analysis of intestinal flora development in breast-fed and formula-fed infants by using molecular identification and detection methods. *J. Pediatr. Gastroenterol. Nutr.* 30:61–67, 2000.

48. Bjorksten, B., P. Naaber, E. Sepp, M. Mikelsaar. The intestinal microflora in allergic Estonian and Swedish 2-year-old children. *Clin. Exp. Allerg.* 29:342–346, 1999.

49. Tannock, G.W., K. Munro, H.J.M Harmsen, G.W. Welling, J. Smart, P.K. Gopal. Analysis of the fecal microflora of human subjects consuming a probiotic product containing *Lactobacillus rhamnosus* DR20. *Appl. Environ. Microbiol.* 66:2578–2588, 2000.

50. Shroff, K.E., K. Meslin, J.J. Cebra. Commensal enteric bacteria engender a self-limiting humoral mucosal immune response while permanently colonizing the gut. *Infect. Immun.* 63:3904–3913, 1995.

51. Sudo, N., S. Sawamura, K. Tanaka, Y. Aiba, C. Kubo, Y. Koga. The requirement of intestinal bacterial flora for the development of an IgE production system fully susceptible to oral tolerance induction. *J. Immunol.* 159:1739–1745, 1997.

52. Herias, M.V., C. Hessle, E. Telemo, T. Midtvedt, L.A. Hanson, A.E. Wold. Immunomodulatory effects of *Lactobacillus plantarum* colonizing the intestine of gnotobiotic rats. *Clin. Exp. Immunol.* 116:283–290, 1999.

53. Ibnou-Zekri, N., S. Blum, E. Schiffrin, T. von der Weid. Divergent patterns of colonization and immune response elicited from two intestinal *Lactobacillus* strains that display similar properties *in vitro. Infect. Immun.* 71:428–436, 2003.

54. Guarner, F., J.R. Malagelada. Gut flora in health and disease. *Lancet* 361:512–519, 2003.

55. Crabbe, P.A., H. Bazin, H. Eyssen, J.F. Heremans. Normal microbial flora as major stimulus for proliferation of plasma cells synthesizing IgA in gut-germ-free intestinal tract. *Int. Arch. Allerg. Appl. Immunol.* 34:362–367, 1968.

56. AbreuMartin, M.T., S.R. Targan. Regulation of immune responses of the intestinal mucosa. *Crit. Rev. Immunol.* 16:277–309, 1996.

57. Matsumoto, S., H. Setoyama, Y. Umesaki. Differential induction of major histocompatibility complex molecules on mouse intestine by bacterial colonization. *Gastroenterology* 103:1777–1782, 1992.

58. Brandtzaeg, P. Development and basic mechanisms of human gut immunity. *Nutr. Rev.* 56: S5–S18, 1998.

59. Ridge, J.P., E.J. Fuchs, P. Matzinger. Neonatal tolerance revisited: turning on newborn T cells with dendritic cells. *Science* 271:1723–1726, 1996.

60. Hunt, D.W.C., H.I. Huppertz, H.J. Jiang, R.E. Petty. Studies of human cord-blood dendritic cells: evidence for functional immaturity. *Blood* 84:4333–4343, 1994.

61. Blum, S., D. Haller, A. Pfeifer, E.J. Schiffrin. Probiotics and immune response. *Clin. Rev. Allerg. Immunol.* 22:287–309, 2002.

62. Cross, M.L. Immunoregulation by probiotic *Lactobacilli*: pro-Th1 signals and their relevance to human health. *Clin. Appl. Immunol. Rev.* 3:115–125, 2002.

63. Lewis, S.A., J.R. Berg, T.J. Kleine. Modulation of epithelial permeability by extracellular macromolecules. *Physiol. Rev.* 75:561–589, 1995.

64. Isolauri, E., H. Majamaa, T. Arvola, I. Rantala, E. Virtanen, H. Arvilommi. *Lactobacillus casei* strain GG reverses increased intestinal permeability induced by cow milk in suckling rats. *Gastroenterology* 105:1643–1650, 1993.

65. Mangell, P., P. Nejdfors, M. Wang, S. Ahrne, B. Westrom, H. Thorlacius, B. Jeppsson. *Lactobacillus plantarum* 299v inhibits *Escherichia coli*-induced intestinal permeability. *Dig. Dis. Sci.* 47:511–516, 2002.

66. Madsen, K., A. Cornish, P. Soper, C. McKaigney, H. Jijon, C. Yachimec, J. Doyle, L. Jewell, C. De Simone. Probiotic bacteria enhance murine and human intestinal epithelial barrier function. *Gastroenterology* 121:580–591, 2001.

67. Ichikawa, H., T. Kuroiwa, A. Inagaki, R. Shineha, T. Nishihira, S. Satomi, T. Sakata. Probiotic bacteria stimulate gut epithelial cell proliferation in rat. *Dig. Dis. Sci.* 44:2119–2123, 1999.

68. Jacobsen, C.N., N.V. Rosenfeldt, A.E. Hayford, P.L. Moller, K.F. Michaelsen, A. Paerregaard, B. Sandstrom, M. Tvede, M. Jakobsen. Screening of probiotic activities of forty-seven strains of *Lactobacillus* spp. by *in vitro* techniques and evaluation of the colonization ability of five selected strains in humans. *Appl. Environ. Microbiol.* 65:4949–4956, 1999.

69. Bernet, M.F., D. Brassart, J.R. Neeser, A.L. Servin. *Lactobacillus acidophilus* La1 binds to cultured human intestinal cell lines and inhibits cell attachment and cell invasion by enterovirulent bacteria. *Gut* 35:483–489, 1994.

70. Ouwehand, A.C., P.V. Kirjavainen, M.M. Gronlund, E. Isolauri, S.J. Salminen. Adhesion of probiotic micro-organisms to intestinal mucus. *Int. Dairy J.* 9:623–630, 1999.

71. Bry, L., P.G. Falk, T. Midtvedt, J.I. Gordon. A model of host-microbial interactions in an open mammalian ecosystem. *Science* 273:1380–1383, 1996.

72. Mack, D.R., S. Michail, S. Wei, L. McDougall, M.A. Hollingsworth. Probiotics inhibit enteropathogenic *E. coli* adherence *in vitro* by inducing intestinal mucin gene expression. *Am. J. Physiol.* 276:G941–G950, 1999.

73. Gill, H.S., K.J. Rutherfurd, J. Prasad, P.K. Gopal. Enhancement of natural and acquired immunity by *Lactobacillus rhamnosus* (HN001), *Lactobacillus acidophilus* (HN017) and *Bifidobacterium lactis* (HN019). *Br. J. Nutr.* 83:167–176, 2000.

74. Chiang, B.L., Y.H. Sheih, L.H. Wang, C.K. Liao, H.S. Gill. Enhancing immunity by dietary consumption of a probiotic lactic acid bacterium (*Bifidobacterium lactis* HN019): optimization and definition of cellular immune responses. *Eur. J. Clin. Nutr.* 54:849–855, 2000.

75. Gill, H.S., K.J. Rutherfurd. Probiotic supplementation to enhance natural immunity in the elderly: effects of a newly characterized immunostimulatory strain *Lactobacillus rhamnosus* HN001 (DR20 (TM)) on leucocyte phagocytosis. *Nutr. Res.* 21:183–189, 2001.

76. Gill, H.S., K.J. Rutherfurd, M.L. Cross, P.K. Gopal. Enhancement of immunity in the elderly by dietary supplementation with the probiotic *Bifidobacterium lactis* HN019. *Am. J. Clin. Nutr.* 74:833–839, 2001.

77. Sheih, Y.H., B.L. Chiang, L.H. Wang, C.K. Liao, H.S. Gill. Systemic immunity-enhancing effects in healthy subjects following dietary consumption of the lactic acid bacterium *Lactobacillus rhamnosus* HN001. *J. Am. Coll. Nutr.* 20:149–156, 2001.

78. Gill, H.S., K.J. Rutherfurd, M.L. Cross. Dietary probiotic supplementation enhances natural killer cell activity in the elderly: an investigation of age-related immunological changes. *J. Clin. Immunol.* 21:264–271, 2001.

79. Spanhaak, S., R. Havenaar, G. Schaafsma. The effect of consumption of milk fermented by *Lactobacillus casei* strain *Shirota* on the intestinal microflora and immune parameters in humans. *Eur. J. Clin. Nutr.* 52:899–907, 1998.

80. Pelto, L., E. Isolauri, E.M. Lilius, J. Nuutila, S. Salminen. Probiotic bacteria down-regulate the milk-induced inflammatory response in milk-hypersensitive subjects but have an immunostimulatory effect in healthy subjects. *Clin. Exp. Allerg.* 28:1474–1479, 1998.

81. Gill, H.S., Q. Shu, H. Lin, K.J. Rutherfurd, M.L. Cross. Protection against translocating *Salmonella typhimurium* infection in mice by feeding the immuno-enhancing probiotic *Lactobacillus rhamnosus* strain HN001. *Med. Microbiol. Immunol.* 190:97–104, 2001.

82. Shu, Q., H.S. Gill. Immune protection mediated by the probiotic *Lactobacillus rhamnosus* HN001 (DR20 (TM)) against *Escherichia coli* O157: H7 infection in mice. *FEMS Immunol. Med. Microbiol.* 34:59–64, 2002.

83. Perdigon, G., M.E.N. Demacias, S. Alvarez, M. Medici, G. Oliver, A.P.D. Holgado. Effect of a mixture of *Lactobacillus casei* and *Lactobacillus acidophilus* administered orally on the immune system in mice. *J. Food Prot.* 49:986–987, 1986.

84. Park, S.Y., G.E. Ji, Y.T. Ko, H.K. Jung, Z. Ustunol, J.J. Pestka. Potentiation of hydrogen peroxide, nitric oxide, and cytokine production in RAW 264.7 macrophage cells exposed to human and commercial isolates of *Bifidobacterium*. *Int. J. Food. Microbiol.* 46: 231–241, 1999.

85. Valdez, J.C., M. Rachid, N. Gobbato, G. Perdigon. Lactic acid bacteria induce apoptosis inhibition in *Salmonella typhimurium* infected macrophages. *Food. Agric. Immunol.* 13:189–197, 2001.

86. Pang, G., A. Buret, R.T. Batey, Q.Y. Chen, L. Couch, A. Cripps, R. Clancy. Morphological phenotypic and functional characteristics of a pure population of CD56+ CD16− CD3− large granular lymphocytes generated from human duodenal mucosa. *Immunology* 79: 498–505, 1993.

87. Desimone, C., B.B. Salvadori, R. Negri, M. Ferrazzi, L. Baldinelli, R. Vesely. The adjuvant effect of yogurt on production of gamma-interferon by Con-A-stimulated human peripheral blood lymphocytes. *Nutr. Rep. Int.* 33:419–433, 1986.

88. Nagao, F., M. Nakayama, T. Muto, K. Okumura. Effects of a fennented milk drink containing *Lactobacillus casei* strain *Shirota* on the immune system in healthy human subjects. *Biosci. Biotechnol. Biochem.* 64:2706–2708, 2000.

89. Haller, D., P. Serrant, D. Granato, E.J. Schiffrin, S. Blum. Activation of human NK cells by *Staphylococci* and *Lactobacilli* requires cell contact-dependent costimulation by autologous monocytes. *Clin. Diagn. Lab. Immunol.* 9:649–657, 2002.

90. Miettinen, M., S. Matikainen, J. Vuopio-Varkila, J. Pirhonen, K. Varkila, M. Kurimoto, I. Julkunen. *Lactobacilli* and *Streptococci* induce interleukin-12 (IL-12), IL-18, and gamma interferon production in human peripheral blood mononuclear cells. *Infect. Immun.* 66:6058–6062, 1998.

91. Hessle, C., L.A. Hanson, A.E. Wold. *Lactobacilli* from human gastrointestinal mucosa are strong stimulators of IL-12 production. *Clin. Exp. Immunol.* 116:276–282, 1999.

92. Christensen, H.R., H. Frokiaer, J.J. Pestka. *Lactobacilli* differentially modulate expression of cytokines and maturation surface markers in murine dendritic cells. *J. Immunol.* 168:171–178, 2002.

93. Lammers, K.M., U. Helwig, E. Swennen, F. Rizzello, A. Venturi, E. Caramelli, M.A. Kamm, P. Brigidi, P. Gionchetti, M. Campieri. Effect of Probiotic strains on interleukin 8 production by HT29/19A cells. *Am. J. Gastroenterol.* 97:1182–1186, 2002.

94. Vidal, K., A. Donnet-Hughes, D. Granato. Lipoteichoic acids from *Lactobacillus johnsonii* strain La1 and *Lactobacillus acidophilus* strain La10 antagonize the responsiveness of human intestinal epithelial HT29 cells to lipopolysaccharide and gram-negative bacteria. *Infect. Immun.* 70:2057–2064, 2002.

95. Michail, S., F. Abernathy. *Lactobacillus plantarum* inhibits the intestinal epithelial migration of neutrophils induced by enteropathogenic *Escherichia coli*. *J. Pediatr. Gastroenterol. Nutr.* 36:385–391, 2003.

96. Cepek, K.L., S.K. Shaw, C.M. Parker, G.J. Russel, J.S. Morrow, D.L. Rimm, M.B. Brenner. Adhesion between epithelial cells and T lymphocytes mediated by cadherin and the alpha(E)beta(7) integrin. *Nature* 372:190–193, 1994.

97. Haller, D., P. Serrant, G. Peruisseau, C. Bode, W.R. Hammes, E. Schiffrin, S. Blum. IL-10 producing CD14(low) monocytes inhibit lymphocyte- dependent activation of intestinal epithelial cells by commensal bacteria. *Microbiol. Immunol.* 46:195–205, 2002.

98. Pereyra, B.S., D. Lemonnier. Induction of human cytokines by bacteria used in dairy foods. *Nutr. Res.* 13:1127–1140, 1993.

99. Kato, I., K. Tanaka, T. Yokokura. Lactic acid bacterium potently induces the production of interleukin-12 and interferon-gamma by mouse splenocytes. *Int. J. Immunopharmacol.* 21:121–131, 1999.

100. Tejada-Simon, M.V., J.J. Pestka. Proinflammatory cytokine and nitric oxide induction in murine macrophages by cell wall and cytoplasmic extracts of lactic acid bacteria. *J. Food. Prot.* 62:1435–1444, 1999.

101. Haller, D., S. Blum, C. Bode, W.P. Hammes, E.J. Schiffrin. Activation of human peripheral blood mononuclear cells by nonpathogenic bacteria *in vitro*: evidence of NK cells as primary targets. *Infect. Immun.* 68:752–759, 2000.

102. Hessle, C., B. Andersson, A.E. Wold. Gram-positive bacteria are potent inducers of monocytic interleukin-12 (IL-12) while gram-negative bacteria preferentially stimulate IL-10 production. *Infect. Immun.* 68:3581–3586, 2000.

103. Shida, K., K. Makino, A. Morishita, K. Takamizawa, S. Hachimura, A. Ametani, T. Sato, Y. Kumagai, S. Habu, S. Kaminogawa. *Lactobacillus casei* inhibits antigen-induced IgE secretion through regulation of cytokine production in murine splenocyte cultures. *Int. Arch. Allerg. Immunol.* 115:278–287, 1998.

104. von der Weid, T., C. Bulliard, E.J. Schiffrin. Induction by a lactic acid bacterium of a population of CD4(+) T cells with low proliferative capacity that produce transforming growth factor beta and interleukin-10. *Clin. Diagn. Lab. Immunol.* 8:695–701, 2001.

105. Cong, Y., C.T. Weaver, A. Lazenby, C.O. Elson. Bacterial-reactive T regulatory cells inhibit pathogenic immune responses to the enteric flora. *J. Immunol.* 169:6112–6119, 2002.

106. Matsuzaki, T., J. Chin. Modulating immune responses with probiotic bacteria. *Immunol. Cell. Biol.* 78:67–73, 2000.

107. De Petrino, S.F., M.E.B. Bonet, O. Meson, G. Perdigon. The effect of *lactobacillus casei* on an experimental model of atopy. *Food Agric. Immunol.* 14:181–189, 2002.

108. Tejada-Simon, M.V., Z. Ustunol, J.J. Pestka. Effects of lactic acid bacteria ingestion of basal cytokine mRNA and immunoglobulin levels in the mouse. *J. Food Prot.* 62:287–291, 1999.

109. Ha, C.L., J.H. Lee, H.R. Zhou, Z. Ustunol, J.J. Pestka. Effects of yogurt ingestion on mucosal and systemic cytokine gene expression in the mouse. *J. Food Prot.* 62:181–188, 1999.

110. Rosenfeldt, V., E. Benfeldt, S.D. Nielsen, K.F. Michaelsen, D.L. Jeppesen, N.H. Valerius, A. Paerregaard. Effect of probiotic *Lactobacillus* strains in children with atopic dermatitis. *J. Allerg. Clin. Immunol.* 111:389–395, 2003.

111. Pessi, T., Y. Sutas, M. Hurme, E. Isolauri. Interleukin-10 generation in atopic children following oral *Lactobacillus rhamnosus* GG. *Clin. Exp. Allerg.* 30:1804–1808, 2000.

112. Ulisse, S., P. Gionchetti, S. D'Alo, F.P. Russo, I. Pesce, G. Ricci, F. Rizzello, U. Helwig, M.G. Cifone, M. Campieri, C. De Simone. Expression of cytokines, inducible nitric oxide synthase, and matrix metalloproteinases in pouchitis: effects of probiotic treatment. *Am. J. Gastroenterol.* 96:2691–2699, 2001.

113. Kishi, A., K. Uno, Y. Matsubara, C. Okuda, T. Kishida. Effect of the oral administration of *Lactobacillus brevis* subsp. *coagulans* on interferon-alpha producing capacity in humans. *J. Am. Coll. Nutr.* 15:408–412, 1996.

114. Shu, Q., H. Lin, K.J. Rutherfurd, S.G. Fenwick, J. Prasad, P.K. Gopal, H.S. Gill. Dietary *Bifidobacterium lactis* (HN019) enhances resistance to oral *Salmonella typhimurium* infection in mice. *Microbiol. Immunol.* 44:213–222, 2000.

115. Kirjavainen, P.V., H.S. El-Nezami, S.J. Salminen, J.T. Ahokas, P.F. Wright. The effect of orally administered viable probiotic and dairy *Lactobacilli* on mouse lymphocyte proliferation. *FEMS Immunol. Med. Microbiol.* 26:131–135, 1999.

116. Novakovic, S., I. Boldogh. *In vitro* TNF-alpha production and *in vivo* alteration of TNF-alpha RNA in mouse peritoneal macrophages after treatment with different bacterial derived agents. *Cancer Lett.* 81:99–109, 1994.

117. Perdigon, G., C.M. Galdeano, J.C. Valdez, M. Medici. Interaction of lactic acid bacteria with the gut immune system. *Eur. J. Clin. Nutr.* 56:S21–S26, 2002.

118. Herias, M.V., T. Midtvedt, L.A. Hanson, A.E. Wold. Increased antibody production against gut-colonizing *Escherichia coli* in the presence of the anaerobic bacterium *Peptostreptococcus. Scan. J. Immunol.* 48:277–282, 1998.

119. Cross, M.L., R.R. Mortensen, J. Kudsk, H.S. Gill. Dietary intake of *Lactobacillus rhamnosus* HN001 enhances production of both Th1 and Th2 cytokines in antigen-primed mice. *Med. Microbiol. Immunol.* 191:49–53, 2002.

120. Kaila, M., E. Isolauri, E. Soppi, E. Virtanen, S. Laine, H. Arvilommi. Enhancement of the circulating antibody secreting cell response in human diarrhea by a human *Lactobacillus* strain. *Pediatr. Res.* 32:141–144, 1992.

121. Isolauri, E., J. Joensuu, H. Suomalainen, M. Luomala, T. Vesikari. Improved immunogenicity of oral DxRRV reassortant rotavirus vaccine by *Lactobacillus casei* GG. *Vaccine* 13:310–312, 1995.

122. Fang, H., T. Elina, A. Heikki, S. Seppo. Modulation of humoral immune response through probiotic intake. *FEMS Immunol. Med. Microbiol.* 29:47–52, 2000.

123. Fukushima, Y., Y. Kawata, K. Mizumachi, J. Kurisaki, T. Mitsuoka. Effect of *Bifidobacteria* feeding on fecal flora and production of immunoglobulins in lactating mouse. *Int. J. Food Microbiol.* 46:193–197, 1999.

124. Perdigon, G., E. Vintini, S. Alvarez, M. Medina, M. Medici. Study of the possible mechanisms involved in the mucosal immune system activation by lactic acid bacteria. *J. Dairy Sci.* 82:1108–1114, 1999.

125. Link-Amster, H., F. Rochat, K.Y. Saudan, O. Mignot, J.M. Aeschlimann. Modulation of a specific humoral immune response and changes in intestinal flora mediated through fermented milk intake. *FEMS Immunol. Med. Microbiol.* 10:55–63, 1994.

126. Marteau, P., P. Seksik, R. Jian. Probiotics and health: new facts and ideas. *Curr. Opin. Biotechnol.* 13:486–489, 2002.

127. Tejada-Simon, M.V., J.H. Lee, Z. Ustunol, J.J. Pestka. Ingestion of yogurt containing *Lactobacillus acidophilus* and *Bifidobacterium* to potentiate immunoglobulin A responses to cholera toxin in mice. *J. Dairy. Sci.* 82:649–660, 1999.

128. Malin, M., H. Suomalainen, M. Saxelin, E. Isolauri. Promotion of IgA immune response in patients with Crohn's disease by oral bacteriotherapy with *Lactobacillus* GG. *Ann. Nutr. Metab.* 40:137–145, 1996.

129. Majamaa, H., E. Isolauri, M. Saxelin, T. Vesikari. Lactic acid bacteria in the treatment of acute rotavirus gastroenteritis. *J. Pediatr. Gastroenterol. Nutr.* 20:333–338, 1995.

130. Kaila, M., E. Isolauri, M. Saxelin, H. Arvilommi, T. Vesikari. Viable versus inactivated *Lactobacillus* strain GG in acute rotavirus diarrhea. *Arch. Dis. Child.* 72:51–53, 1995.

131. Husband, A.J., D.R. Kramer, S. Bao, R.M. Sutherland, K.W. Beagley. Regulation of mucosal IgA responses *in vivo*: cytokines and adjuvants. *Vet. Immunol. Immunopathol.* 54:179–186, 1996.

132. Shu, Q., H.S. Gill. A dietary probiotic (*Bifidobacterium lactis* HN019) reduces the severity of *Escherichia coli* O157: H7 infection in mice. *Med. Microbiol. Immunol.* 189:147–152, 2001.

133. Clements, M.L., M.M. Levine, R.E. Black, R.M. Robinsbrowne, L.A. Cisneros, G.L. Drusano, C.F. Lanata, A.J. Saah. *Lactobacillus* prophylaxis for diarrhea due to enterotoxigenic *Escherichia coli. Antimicrob. Agents Chemother.* 20:104–108, 1981.

134. Saavedra, J.M., N.A. Bauman, I. Oung, J.A. Perman, R.H. Yolken. Feeding of *Bifidobacterium bifidum* and *Streptococcus thermophilus* to infants in hospital for prevention of diarrhea and shedding of rotavirus. *Lancet* 344:1046–1049, 1994.

135. Guandalini, S., L. Pensabene, M. Abu Zikri, J.A. Dias, L.G. Casali, H. Hoekstra, S. Kolacek, K. Massar, D. Micetic-Turk, A. Papadopoulou, J.S. de Sousa, B. Sandhu, H. Szajewska, Z. Weizman. *Lactobacillus* GG administered in oral rehydration solution to children with acute diarrhea: a multicenter European trial. *J. Pediatr. Gastroenterol. Nutr.* 30:54–60, 2000.

136. Rosenfeldt, V., K.F. Michaelsen, M. Jakobsen, C.N. Larsen, P.L. Moller, M. Tvede, H. Weyrehter, N.H. Valerius, A. Paerregaard. Effect of probiotic *Lactobacillus* strains on acute diarrhea in a cohort of nonhospitalized children attending day-care centers. *Pediatr. Infect. Dis. J.* 21:417–419, 2002.

137. Szajewska, H., J.Z. Mrukowicz. Probiotics in the treatment and prevention of acute infectious diarrhea in infants and children: a systematic review of published randomized, double-blind, placebo-controlled trials. *J. Pediatr. Gastroenterol. Nutr.* 33:S17–S25, 2001.

138. Bergogne-Berezin, E. Treatment and prevention of antibiotic associated diarrhea. *Int. J. Antimicrob. Agents* 16:521-526, 2000.

139. Black, F.T., P.L. Andersen, J. Ørskov, F. Ørskov, K. Gaarslev, S. Laulund. Prophylactic efficacy of *Lactobacilli* on traveler's diarrhea. *Travel Med.* 7:333–335, 1989.

140. Michetti, P. *Lactobacilli* for the management of *Helicobacter pylori. Nutrition* 17:268–269, 2001.

141. Rubaltelli, F.F., R. Biadaioli, P. Pecile, P. Nicoletti. Intestinal flora in breast- and bottle-fed infants. *J. Perinat. Med.* 26:186–191, 1998.

142. Sartor, R.B. Pathogenesis and immune mechanisms of chronic inflammatory bowel diseases. *Am. J. Gastroenterol.* 92:S5–S11, 1997.

143. Madsen, K.L., D. Malfair, D. Gray, J.S. Doyle, L.D. Jewell, R.N. Fedorak. Interleukin-10 gene-deficient mice develop a primary intestinal permeability defect in response to enteric microflora. *Inflamm. Bowel Dis.* 5:262–270, 1999.

144. Madsen, K.L., J.S. Doyle, L.D. Jewell, M.M. Tavernini, R.N. Fedorak. *Lactobacillus* species prevents colitis in interleukin 10 gene-deficient mice. *Gastroenterology* 116:1107–1114, 1999.

145. Schultz, M., C. Veltkamp, L.A. Dieleman, W.B. Grenther, P.B. Wyrick, S.L. Tonkonogy, R.B. Sartor. *Lactobacillus plantarum* 299V in the treatment and prevention of spontaneous colitis in interleukin-10-deficient mice. *Inflamm. Bowel Dis.* 8:71–80, 2002.

146. Rachmilewitz, D., F. Karmeli, K. Takabayashi, T. Hayashi, L. Leider-Trejo, J.D. Lee, L.M. Leoni, E. Raz. Immunostimulatory DNA ameliorates experimental and spontaneous murine colitis. *Gastroenterology* 122:1428–1441, 2002.

147. Rachmilewitz, D., F. Karmeli, K. Takabayashi, E. Raz. Amelioration of experimental colitis by probiotics is due to the immunostimulatory effects of its DNA (abstr). *Gastroenterology* 122(1):T1004, 2002.

148. Gionchetti, P., F. Rizzello, A. Venturi, P. Brigidi, D. Matteuzzi, G. Bazzocchi, G. Poggioli, M. Miglioli, M. Campieri. Oral bacteriotherapy as maintenance treatment in patients with chronic pouchitis: a double-blind, placebo-controlled trial. *Gastroenterology* 119:305–309, 2000.

149. Gionchetti, P., F. Rizzello, A. Venturi, U. Helwig, C. Amadini, K.M. Lammers, F. Ugolini, G. Poggioli, M. Campieri. Prophylaxis of pouchitis onset with probiotic therapy: A double-blind, placebo controlled trial (abstr). *Gastroenterology* 118(1,2):1214, 2000.

150. Mimura, T., F. Rizzello, S. Schreiber, I.C. Talbot, R.J. Nicholls, P. Gionchetti, M. Campieri, M.A. Kamm. Once daily high dose probiotic therapy maintains remission and improved quality of life in patients with recurrent or refractory pouchitis: a randomised, placebo-controlled, double-blind trial (abstr). *Gastroenterology* 122(1):667, 2002.

151. Kruis, W., P. Fric, M.S. Stolte. Maintenance of remission in ulcerative colitis is equally effective with *Escherichia coli* Nissle 1917 and with standard mesalamine (abstr). *Gastroenterology* 120(1):680, 2001.

152. Bjorksten, B., D. Dumitrascu, T. Foucard, N. Khetsuriani, R. Khaitov, M. Leja, G. Lis, J. Pekkanen, A. Priftanji, M.A. Riikjarv. Prevalence of childhood asthma, rhinitis and eczema in Scandinavia and Eastern Europe. *Eur. Respir. J.* 12:432–437, 1998.

153. Strachan, D.P., Hay fever, hygiene, and household size. *Br. Med. J.* 299:1259–1260, 1989.

154. Wold, A.E. The hygiene hypothesis revised: is the rising frequency of allergy due to changes in the intestinal flora? *Allergy* 53:20–25, 1998.

155. Bottcher, M.F., E.K. Nordin, A. Sandin, T. Midtvedt, B. Bjorksten. Microflora-associated characteristics in faeces from allergic and nonallergic infants. *Clin. Exp. Allergy.* 30:1590–1596, 2000.

156. Kalliomaki, M., P. Kirjavainen, E. Eerola, P. Kero, S. Salminen, E. Isolauri. Distinct patterns of neonatal gut microflora in infants in whom atopy was and was not developing. *J. Allerg. Clin. Immunol.* 107:129–134, 2001.

157. He, F., A.C. Ouwehand, E. Isolauri, H. Hashimoto, Y. Benno, S. Salminen. Comparison of mucosal adhesion and species identification of *Bifidobacteria* isolated from healthy and allergic infants. *FEMS Immunol. Med. Microbiol.* 30:43–47, 2001.

158. Ouwehand, A.C., E. Isolauri, F. He, H. Hashimoto, Y. Benno, S. Salminen. Differences in *Bifidobacterium* flora composition in allergic and healthy infants. *J. Allerg. Clin. Immunol.* 108:144–145, 2001.

159. He, F., H. Morita, H. Hashimoto, M. Hosoda, J.I. Kurisaki, A.C. Ouwehand, E. Isolauri, Y. Benno, S. Salminen. Intestinal *Bifidobacterium* species induce varying cytokine production. *J. Allerg. Clin. Immunol.* 109:1035–1036, 2002.

160. Majamaa, H., E. Isolauri. Probiotics: a novel approach in the management of food allergy. *J. Allerg. Clin. Immunol.* 99:179–185, 1997.

161. Isolauri, E., T. Arvola, Y. Sutas, E. Moilanen, S. Salminen. Probiotics in the management of atopic eczema. *Clin. Exp. Allergy.* 30:1604–1610, 2000.

162. Kalliomaki, M., S. Salminen, H. Arvilommi, P. Kero, P. Koskinen, E. Isolauri. Probiotics in primary prevention of atopic disease: a randomised placebo-controlled trial. *Lancet* 357:1076–1079, 2001.

163. Rautava, S., M. Kalliomaki, E. Isolauri. Probiotics during pregnancy and breastfeeding might confer immunomodulatory protection against atopic disease in the infant. *J. Allerg. Clin. Immunol.* 109:119–121, 2002.

164. Herlin, T., S. Haahtela, T. Haahtela. No effect of oral treatment with an intestinal bacterial strain, *Lactobacillus rhamnosus* (ATCC 53103), on birch-pollen allergy: a placebo-controlled double-blind study. *Allergy* 57:243–246, 2002.

165. Horie, H., K. Kanazawa, M. Okada, S. Narushima, K. Itoh, A. Terada. Effects of intestinal bacteria on the development of colonic neoplasm: an experimental study. *Eur. J. Cancer Prevent.* 8:237–245, 1999.

166. Moore, W.E.C., L.H. Moore. Intestinal floras of populations that have a high risk of colon cancer. *Appl. Environ. Microbiol.* 61:3202–3207, 2003.

167. J. Rafter. Lactic acid bacteria and cancer: mechanistic perspective. *Br. J. Nutr.* 88:S89–S94, 2002.

168. Takahashi, T., A. Kushiro, K. Nomoto, K. Uchida, M. Morotomi, T. Yokokura, H. Akaza. Antitumor effects of the intravesical instillation of heat killed cells of the *Lactobacillus casei* strain *Shirota* on the murine orthotopic bladder tumor MBT-2. *J. Urol.* 166:2506–2511, 2001.

169. Perdigon, G., A.D. de LeBlanc, J. Valdez, M. Rachid. Role of yoghurt in the prevention of colon cancer. *Eur. J. Clin. Nutr.* 56:S65–S68, 2002.

170. Coconnier, M.H., M.F. Bernet, G. Chauviere, A.L. Servin. Adhering heat-killed human *Lactobacillus acidophilus,* strain LB, inhibits the process of pathogenicity of diarrhoeagenic bacteria in cultured human intestinal cells. *J. Diarrhoeal Dis. Res.* 11:235–242, 1993.

171. Neumann, E., M.A.P. Oliveira, C.M. Cabral, L.N. Moura, J.R. Nicoli, E.C. Vieira, D.C. Cara, G.I. Podoprigora, L.Q. Vieira. Monoassociation with *Lactobacillus acidophilus* UFV-H2b20 stimulates the immune defense mechanisms of germfree mice. *Braz. J. Med. Biol. Res.* 31:1565–1573, 1998.

172. Gill, H.S., K.J. Rutherfurd. Viability and dose-response studies on the effects of the immunoenhancing lactic acid bacterium *Lactobacillus rhamnosus* in mice. *Br. J. Nutr.* 86:285–289, 2001.

173. Kirjavainen, P.V., S.J. Salminen, E. Isolauri. Probiotic bacteria in the management of atopic disease: Underscoring the importance of viability. *J. Pediatr. Gastroenterol. Nutr.* 36:223–227, 2003.

174. Rayes, N., D. Seehofer, S. Hansen, K. Boucsein, A.R. Muller, S. Serke, S. Bengmark, P. Neuhaus. Early enteral supply of *Lactobacillus* and fiber versus selective bowel decontamination: a controlled trial in liver transplant recipients. *Transplantation* 74:123–128, 2002.

175. Holzapfel, W.H., P. Haberer, J. Snel, U. Schillinger, F.H.F. Huis in't Veld, V. Overview of gut flora and probiotics. *Int. J. Food. Microbiol.* 41:85–101, 1998.

176. Umesaki, Y., H. Setoyama. Structure of the intestinal flora responsible for development of the gut immune system in a rodent model. *Microbes Infect.* 2:1343–1351, 2000.

177. Elwood, C.M., O.A. Garden. Gastrointestinal immunity in health and disease. *Vet. Clin. N. Am. Small Anim. Pract.* 29:471–487, 1999.

178. Michetti, P., G. Dorta, P.H. Wiesel, D. Brassart, E. Verdu, M. Herranz, C. Felley, N. Porta, M. Rouvet, A.L. Blum, I. Corthesy-Theulaz. Effect of whey-based culture supernatant of *Lactobacillus acidophilus (johnsonii)* La1 on *Helicobacter pylori* infection in humans. *Digestion* 60:203–209, 1999.

179. Gill, H.S., M.L. Cross, K.J. Rutherfurd, P.K. Gopal. Dietary probiotic supplementation to enhance cellular immunity in the elderly. *Br. J. Biomed. Sci.* 58:94–96, 2001.

180. Arunachalam, K., H.S. Gill, R.K. Chandra. Enhancement of natural immune function by dietary consumption of *Bifidobacterium lactis* (HN019). *Eur. J. Clin. Nutr.* 54:263–267, 2000.

181. Delia, P., G. Sansotta, V. Donato, G. Messina, P. Frosina, S. Pergolizzi, C. De Renzis, G. Famularo. Prevention of radiation-induced diarrhea with the use of VSL#3, a new high-potency probiotic preparation. *Am. J. Gastroenterol.* 97:2150–2152, 2002.

182. Brigidi, P., B. Vitali, E. Swennen, G. Bazzocchi, D. Matteuzzi. Effects of probiotic administration upon the composition and enzymatic activity of human fecal microbiota in patients with irritable bowel syndrome or functional diarrhea. *Res. Microbiol.* 152:735–741, 2001.

183. de Waard, R., E. Claassen, G.C.A.M. Bokken, B. Buiting, J. Garssen, J.G. Vos. Enhanced immunological memory responses to *Listeria monocytogenes* in rodents, as measured by delayed-type hypersensitivity (DTH), adoptive transfer of DTH, and protective immunity, following *Lactobacillus casei Shirota* ingestion. *Clin. Diagn. Lab. Immunol.* 10:59–65, 2003.

184. Hori, T., J. Kiyoshima, H. Yasui. Effect of an oral administration of *Lactobacillus casei* strain *Shirota* on the natural killer activity of blood mononuclear cells in aged mice. *Biosci. Biotechnol. Biochem.* 67:420–422, 2003.

185. Perdigon, G., S. Alvarez, A.P.D. Holgado. Immunoadjuvant activity of oral *Lactobacillus casei*: influence of dose on the secretory immune response and protective capacity in intestinal infections. *J. Dairy Res.* 58:485–496, 1991.

186. Perdigon, G., M.E.N. Demacias, S. Alvarez, G. Oliver, A.A.P.D. Holgado. Prevention of gastrointestinal *infection using immunobiological methods with milk fermented with Lactobacillus* casei and *Lactobacillus acidophilus. J. Dairy Res.* 57:255–264, 1990.

187. Gonzalez, S., G. Albarracin, M.L.R. Pesce, M. Male, M.C. Apella, A.A.P.D. Holgado, G. Oliver. Prevention of infantile diarrhoea by fermented milk. *Microbiol. Aliment. Nutr.* 8:349–354, 1990.

188. Gonzalez, S.N., R. Cardozo, M.C. Apella, G. Oliver. Biotherapeutic role of fermented milk. *Biotherapy* 8:129–134, 1995.

19

Enzymatic Synthesis of Oligosaccharides: Progress and Recent Trends

V. Maitin and R. A. Rastall

CONTENTS

19.1 Oligosaccharides 474
    19.1.1 Biological Importance of Oligosaccharides.............................474
    19.1.2 Oligosaccharides as Drugs and Functional Foods 474
19.2 Chemical Synthesis 475
    19.2.1 Enzymatic Synthesis 476
    19.2.2 Glycosyltransferases...476
    19.2.3 Glycosidases...479
19.3 Equilibrium Controlled Synthesis 480
    19.3.1 Kinetically Controlled Synthesis..481
19.4 Influence of Reaction Components and Conditions on Yield and Selectivity 482
    19.4.1 Donor:Acceptor Sugar Ratio..482
    19.4.2 Temperature..484
    19.4.3 Enzyme Origin 484
    19.4.4 Use of Organic Solvents 485
    19.4.5 Substrate Concentration 486
    19.4.6 pH..487
    19.4.7 Time...487
    19.4.8 Enzyme Immobilization 488
19.5 Recombinant and Engineered Glycosidases in Oligosaccharide Synthesis 489
    19.5.1 Mutant Glycosidases 489
        19.5.1.1 Glycosynthases 489
        19.5.1.2 Mutant Enzymes other than Glycosynthases............................491
        19.5.1.3 Mutant Glycosidases Created by Directed Evolution 491
19.6 Conclusions 492
References...492

19.1 OLIGOSACCHARIDES

Carbohydrates play an important role in numerous biological processes. However, until a few decades ago, the chemistry and biology of carbohydrates was a Cinderella field (1), an area with significant activity but lacking the glamor of genomes and proteins. In recent years, concomitant with the rapid establishment of the field of glycobiology based on a range of synthetic and analytical methods, the potential of carbohydrates in the maintenance of health and as therapeutic agents has been realized (2,3). Oligosaccharides consist of two to ten monosaccharide residues linked via glycosidic linkages; these can be liberated by depolymerization (4,5). Oligosaccharides composed of monosaccharides only are termed simple or true oligosaccharides and those linked to nonsaccharides such as peptides and lipids are termed conjugate oligosaccharides.

19.1.1 Biological Importance of Oligosaccharides

According to Varki (6), oligosaccharides either mediate specific recognition events or provide modulation of biological processes. The former activity includes the role of cell surface oligosaccharides and their conjugates as recognition sites for bacteria, viruses, toxins, antibodies, and hormones. The latter activity refers to their functions in cell recognition, cell growth, cell differentiation, cell adhesion, signal transduction, development, regulation, and other intercellular communications. Several workers seem to agree that the major function of oligosaccharides is in molecular recognition (2,3,7). The involvement of oligosaccharides in a wide range of specific recognition mechanisms results from their incredible structural diversity. This diversity arises from the fact that they can be highly branched, and the monomeric units may be connected to one another by many different linkage types.

19.1.2 Oligosaccharides as Drugs and Functional Foods

It has been established that many pathogens use carbohydrate binding proteins, or lectins, to attach to cells and initiate disease (8,9). Specific membrane bound oligosaccharides serve as receptors for the attachment of the pathogen onto the cell. One of the natural defense mechanisms of the body against infection involves decoy oligosaccharides present in the mucosal lining of epithelial cells and in saliva, tears, urine, sweat, and breast milk. This is especially significant in breast milk, where the presence of many oligosaccharides at high concentration (up to millimolar levels) protects infants from many infections (9,10). An invading pathogen binds to these decoy oligosaccharides instead of the host cell and the host is thus protected from infection. This is the basic concept behind use of oligosaccharides as antiinfective agents. Because these oligosaccharides prevent the adhesion of the pathogen to the host cell, the term antiadhesives is used to describe this class of carbohydrate drugs. They also have a possible antiinflammatory application against tumors or virally infected cells. Many pharmaceutical companies are now involved in research and development of carbohydrate drugs (2,11).

Two types of antiinfectives are under investigation: receptor oligosaccharides, for use against respiratory and gastrointestinal disease; and sialic acid analogues for use against influenza (10) and cholera. It is important that the antiadhesives used are safe, nontoxic, and nonimmunogenic. Short chain oligosaccharides, for example, human oligosaccharides about 1kDa in size are promising antiadhesive agents. This is because of their high water solubility, and resistance to heat and pH, making their delivery and administration more effective. Another important factor in their efficacy is the strength of binding to the bacterial lectins. For this reason, polyvalent oligosaccharides have been found to be much more effective than monovalent ones because they form a stable structure by binding to more than one adhesin (9). Several glycomimetics, such as glycodendrimers (12) and mixed type glycoclusters (13) have emerged as valuable tools in this regard. The concept of antiinfective therapy is an exciting one, considering its potential commercial impact and wide application. However, at present, it is in its developmental stages. There are practical limitations to its widespread use because of factors such as the lack of animal models for clinical trials, availability of appropriate oligosaccharides, and limited knowledge of the host surface glycoconjugates (9,10).

Apart from their role in recognition, there has been recent interest in the use of oligosaccharides as functional food ingredients to promote growth of probiotic or beneficial bacteria in the human gut (14). Oligosaccharides provide a combination of suitable physicochemical and physiological properties, which make them very good candidates for use as food ingredients. The 1991 legislation for Foods for Specified Health Use (FOSHU) of the Japanese government comprised 223 items, of which more than half incorporate oligosaccharides as the functional components (15). Food grade oligosaccharides are manufactured using simple methods, and avoiding expensive purification steps, in order to keep the prices competitive. As a result, they are generally impure products containing residual feedstock carbohydrate and monomer sugars (11). Pharmaceutical applications in most cases require pure compounds of a defined chemical structure, and synthesis of these oligosaccharides is complex and expensive.

The biological importance of oligosaccharides coupled with their potential as prophylactic agents and functional food ingredients has generated considerable research interest in this area about various aspects of their structure, function, and uses. The limiting factors in this research and application of the results are the limited quantity and range of available oligosaccharides, and their very high prices. It is therefore important to develop oligosaccharide synthesis methods that are efficient, cost effective, and easy to scale up, in order to increase the repertoire of oligosaccharides available.

There are two methods of oligosaccharide synthesis: chemical and enzymatic.

19.2 CHEMICAL SYNTHESIS

Traditionally, oligosaccharides are synthesized by chemical methods. Though well developed, chemical methods of synthesis are complicated and involve a number of steps. Because carbohydrates have several hydroxyl groups of similar reactivity, there are numerous protection and deprotection steps required in order to achieve regioselectivity (16). In spite of this, a mixture of isomeric oligosaccharides is often produced. Moreover, the number of steps increases with the size of the oligosaccharide (up to 7 steps needed for disaccharide synthesis and more than ten for a trisaccharide), making the synthesis of even simple oligosaccharides a lengthy process. The method needs expensive chemicals, yields are low, and scale up is difficult. For an overview of chemical oligosaccharide synthesis, the reader is referred to exhaustive reviews by Boons (17) and Davis (18). Several oligosaccharides of biological importance have been synthesized using chemical methods, including oligosaccharides corresponding to the capsular polysaccharide of *Cryptococcus neoformans* and outer cell wall of *Moraxella catarrhalis* (19), a high mannose nonasaccharide corresponding to part of the glycoprotein gp120 of the viral coat of HIV-1(20), myoinositol containing compounds (21), glycopeptide containing tumor associated carbohydrate antigens (22), α-1,2-linked disaccharide derivatives, (23) and recently, mixed type glycocluster oligosaccharide mimetics containing sugar derivatives from different sugar series (galactoside, mannoside and fucoside) (13) with potential as antiadhesives. Chemical methods have also been reported for synthesis of sugar nucleotides needed for enzymatic synthesis reactions using Leloir-transferases (24). New strategies to improve the chemical synthesis process are being developed and are summarized by Bartolozzi and Seeberger (25). A significant achievement has been that of programmable one pot synthesis. Based on difference in reactivity between glycosyl donors, a database, search engine, and computer program called Optimer was developed (26). The program enables the user to select from a list of the best reagent combinations for the synthesis of an oligosaccharide of interest, along with predicted yields (27). The program has been successfully employed for controlled synthesis of both linear and branched trisaccharides and tetrasaccharides, allowing the possibility of creating a library of oligosaccharides. To facilitate the synthesis of some linkages, which are difficult to synthesize by chemical methods, chemists are now using chemoenzymatic synthesis, wherein an enzyme is used to synthesize part of the oligosaccharide structure, in combination with chemical methods. One such example is the synthesis of β-mannosyl linkage of N-glycans; Watt et al. (28) used β-mannosyltransferase for the synthesis of the core of N-linked oligosaccharides.

In spite of these advancements, the area of chemical synthesis remains a challenge because of its technical complexity and absence of general procedures applicable to synthesis of a range of oligosaccharides, which makes their adaptation to industrial scale next to impossible. Their application is currently limited to synthesis of some bioactive oligosaccharides for research purposes; large scale manufacture for food and drug use remains impractical.

19.2.1 Enzymatic Synthesis

Because most of the complex heterooligosaccharides in nature are produced by enzymes, it was thought that it must be possible to use the synthetic potential of enzymes *in vitro* to bring about oligosaccharide manufacture (29). In recent years, enzymatic synthesis of oligosaccharides has become the method of choice for most researchers. It helps to overcome two major problems posed by chemical synthesis: selective protection and deprotection is not necessary, and the configuration of the newly formed anomeric centres can be controlled well. As a result, a mixture of isomers is not obtained. Other advantages include lack of byproducts, wider range of reaction conditions and regiospecificity and stereospecificity of the enzymes. Thus enzymes can be used to synthesize oligosaccharides by single step reactions and to synthesize novel oligosaccharides. Production of the enzymes by fermentation and immobilization to enable reuse can reduce the cost of the enzymes, making the process very economical.

Two main types of enzymes are used to catalyze oligosaccharide synthesis: glycosyltransferases (biosynthetic) and glycosidases (hydrolytic).

19.2.2 Glycosyltransferases

Glycosyltransferases are type II membrane bound glycoproteins consisting of a short N-terminal cytoplasmic domain, a transmembrane domain that anchors it to the cell membrane, and the C-terminal catalytic domain (30). They are currently classified into 60 families according to the sequence based classification of Coutinho and Henrissat (31). Glycosyltransferases catalyze the stereospecific and regiospecific transfer of a monosaccharide from a donor substrate (glycosyl nucleotide) to an acceptor substrate. They are classified according to the sugar transferred from the donor to the acceptor and by the acceptor specificity. For example, β 1, 4-galactosyltransferase (GalT) from bovine milk catalyzes the transfer of a galactose (Gal) unit from UDP-Gal to the 4-OH group of terminal N-acetylglucosamine (GlcNAc) β -R acceptors generating Gal β 1-4GlcNAc β -R (N-acetyllactosamine-R) structures. A completely different enzyme, β 1,3-GalT is required for synthesis of isomeric Gal β 1-3 GlcNAc β -R structures from the same substrates (32). The regiospecificity and stereospecificity, the high selectivity for the acceptor substrate, and the high yields that can be achieved are attractive features of these catalysts (16). The reaction can be represented by the following general equation:

$$\text{Nucleotide sugar} + \text{R-OH} \rightarrow \text{R-O-sugar} + \text{nucleotide} \qquad (19.1)$$

where R is a free saccharide or a saccharide linked to a protein or lipid.

In vivo, the synthesis occurs via three fundamental steps: activation, transfer, and modification (33). In the first stage of the sequence, a monosaccharide is transformed into a sugar-1-phosphate (sugar-1-P) by a kinase enzyme. This then reacts with a nucleoside triphosphate (NTP) in a reaction catalyzed by nucleoside transferase, and forms a chemically activated nucleoside diphosphate sugar (NDP-sugar). The sugar unit is then transferred to a sugar acceptor by glycosyltransferases Each one is specific for a certain donor and acceptor and for the linkage position of the new glycosidic bond (29). Owing to this high specificity, they have been used in synthesis of oligosaccharides *in vitro*. However, such high specificity also means that a wide range of glycosyltransferases is required to enable the synthesis of a variety of desired oligosaccharides *in vitro*. At the moment, not many glycosyltransferases are commercially available [nine, as reported by Bastida et al. (34), of which only β-1,4-galactosyltransferase is available in amounts higher than 1 unit]. They are difficult to isolate and purify from their natural sources because they are usually membrane bound and present in low concentrations. Several glycosyltransferase genes, including those coding for glucosyltransferase (35,36), sialyltransferase (37,38), fucosyltransferase (34,39), mannosyltransferase (40,41), and galactosyltransferase (42–45) have been cloned, but many more need to be cloned and probably modified to make the enzymes more widely available and overcome problems of instability. In practice, the range of routinely used glycosyltransferases at the moment is mainly limited to β-1,4-galactosyltransferase, α-2,3-and α-2,6-sialyltransferases and α-1,3/4-fucosyltransferases (46,47).

Besides the availability of the actual enzyme, the availability of the sugar nucleotide donors was also a problem until recently, as they were very expensive and unstable. Eight sugar nucleotides: UDP-Glc, UDP-Gal, GDP-Fuc, CMP-NeuAc, UDP-glucuronic acid, GDP-Man, UDP-GlcNAc, and UDP-GalNAc act as donor substrates for most mammalian glycosyltransferases. With the exception of UDP-glucuronic acid, they are now commercially available, though still expensive. Because the initial report for regeneration of UDP-Glc and UDP-Gal by Wong et al. (48), numerous regeneration methods have been developed for all of these sugar nucleotides, which not only reduce their cost but also overcome the problem of product inhibition resulting from the released nucleoside monophosphates or diphosphates in the reaction (Figure 19.1). Representative papers include regeneration of UDP-Glu, UDP-Gal, CMP- NeuAc (49), and UDP-Gal, UDP-2-deoxyGal, UDP-Galactosamine (50) using multienzyme systems, regeneration of GDP-Man (40), and a simplified pathway for regeneration of UDP-Gal (51). Gram scale synthesis has been reported for UDP-Gal (52), and yields up to multikilogram levels have been achieved for some oligosaccharides such as SLe^x (32).

It must be realized, however, that the enzymes that are involved in these pathways are not recovered and recycled, which is essential for the process to be feasible at industrial scales. The obvious biotechnological solution was enzyme immobilization (53,54). Nishiguchi et al. (55) carried out the synthesis of a trisaccharide derivative Neu-5Acα (2→6) Galβ(1→4)GlcNAcβ-O-(CH$_2$)$_6$-NH$_2$ using recombinant β1,4-galactosyltransferase and α2,6-sialyltransferase immobilized by coupling reactions with activated Sepharose. The oligosaccharide was built up by glycosylation on a water soluble primer having GlcNAc residues and an α–chymotrypsin sensitive linker, for the release of final product by α-chymotrypsin treatment.

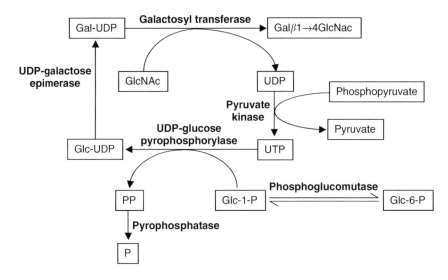

Figure 19.1 Multienzyme synthesis of *N*-acetyllactosamine using galactosyltransferase with cofactor regeneration

Several workers have also investigated the feasibility of using whole cells of recombinant bacteria as catalysts, as this approach obviates the need for enzyme isolation and in many cases tedious purification. Herrmann et al. (56) were able to synthesize disaccharides Manα1, 2Man, Manα1, 2ManOMe, and two dimannosylated glycopeptides using whole recombinant *E. coli* cells expressing cloned α-1,2-mannosyltransferase reported earlier (40). The group of Ozaki has produced a series of papers reporting the use of coupled recombinant or metabolically engineered bacterial strains: *Corynebacterium ammoniagenes* for NTP (re)generation, recombinant *E. coli* with NDP-sugar generation and regeneration cycle expression and recombinant *E. coli* with glycosyltransferase (57): for large scale production of globotriose (188 g/L) (58), N-acetyllactosamine (107 g/L) (59), 3'-sialyllactose (33 g/L), (60) and sialyl-T$_n$ epitope (45 g/L) (61). Chen et al. (62) have developed a superbug, an engineered *E. coli* strain containing a plasmid harboring an artificial gene cluster containing all the necessary genes for sugar nucleotide generation and regeneration as well as oligosaccharide accumulation for the synthesis of an α-Gal epitope. Another engineered *E. coli* strain (lacZ⁻) has been used by Priem et al. (63) for the production of human milk oligosaccharides by efficient conversion from lactose. The lacZ⁻ mutant accumulated lactose when grown on glycerol. The accumulated lactose acted as an acceptor for the heterologous β1,3-N-acetylglucosaminyltransferase enzyme expressed from a *Neisseria meningitidis* gene. The resultant trisaccharide acted as acceptor for the coexpressed *N. meningitidis* β1,4-galactosyltransferase enzyme. This system produced lacto-N-neotetraose and lacto-N-neohexaose at over 5 g/L yield. Similarly, a nanA⁻ *E. coli* was able to produce sialyllactose.

Another particularly elegant regeneration methodology is sugar nucleotide regeneration beads or superbeads (57,64,65). The superbeads technology involves overexpression of individual enzymes along the sugar nucleotide biosynthetic pathway in a His$_6$-tagged form followed by their coimmobilization onto nickel nitrilotriacetate agarose beads. The beads can be employed for synthesis of specific oligosaccharide sequences by combination with a variety of glycosyltransferases. UDP-Gal regenerating superbeads were developed as a model system and in combination with β1,4-GalT and/or α 1,3/1,4-GalT, 100 mg-1.03 g scale synthesis of biologically important oligosaccharides was achieved (66). The first generation superbeads required four enzymes and the reactions were performed in stirred batch mode whereas the second generation superbeads (65) involve seven enzymes and the reactions performed by circulation of the reaction mixture through the superbeads in a packed bed column reactor configuration, proving to be more efficient.

The aforementioned advancements over the last decade have transformed the status of Leloir glycosyltransferases in oligosaccharide synthesis by alleviating the problems associated with them earlier: of low availability of enzymes and sugar nucleotide donors, product inhibition, and high reagent costs. These developments, together with their inherent regioselectivity and stereoselectivity should significantly increase their applicability.

19.2.3 Glycosidases

As the name suggests, the natural function of glycosidases is hydrolytic cleavage of glycosidic bonds. Glycosidases occur widely in nature; in viruses, microorganisms, and plant and animal cells (16), and consequently are more available than glycosyltransferases. They are also more stable, act on easily available substrates, and do not need cofactors.

Their main drawback is that if nonlinkage specific enzymes are used, the regioselectivity tends to be low, which results in the formation of a product which is a mixture of several different linkages (usually 1→6, 1→4, 1→3 and 1→2), posing problems of purification. Linkage specific glycosidases, however, generally produce products with very high regioselectivity. In spite of this, keeping in mind their availability, relative economy of use and simplicity of the reaction, glycosidases have become established as a popular choice for oligosaccharide synthesis. While the high specificity of glycosyltransferases can allow precise sequential construction of desired oligosaccharides, the wide acceptor specificity of glycosidases makes them useful for synthesis of novel oligosaccharides of unknown and potentially exploitable biological activities.

Glycosidases can be divided into two groups: the exoglycosidases, which cleave glycosidic bonds at the nonreducing end of the oligosaccharide, and the endoglycosidases, which cleave internal glycosidic bonds (33). In most instances, synthesis has been carried out using retaining exoglycosidases. According to Coutinho and Henrissat (31), glycosidases are currently classified into 90 families based on sequence similarities. Based on mechanism (67–71), they are classified as retaining or inverting glycosidases depending on whether the hydrolysis proceeds by net retention or inversion of the stereochemistry of the anomeric center. In either case, the reaction is acid base catalyzed through an oxocarbenium ionlike transition state, and involves two carboxylic groups at the active site. Inverting glycosidases operate via direct displacement of the leaving group by water. The two carboxylic groups are suitably positioned at the active site such that one provides base catalytic assistance to the attack of water, while the other provides acid catalytic assistance to cleavage of the glycosidic bond. Retaining glycosidases use a double displacement mechanism involving the formation of a covalent glycosyl enzyme intermediate. Unlike the inverting enzymes, where one residue functions as a general acid and the other as the general base, in retaining enzymes one residue serves both functions, acting as acid catalyst for the glycosylation step and base catalyst for the deglycosylation step. The second carboxylic group acts as a nucleophile and a leaving group. A significant difference between the two enzyme classes is the separation between the carboxylic acid side chains, which is approximately 11 Å in inverting enzymes and 5.5 Å in retaining enzymes.

Detailed studies have been carried out on the mechanism of retaining glycosidases using several strategies, details of which can be found in reviews by Withers (68,71). These include intermediate trapping by (1) modification of the substrate through fluorination of its 2-hydroxyl or 5-hydroxyl group by fluorosugar reagents which has led to the identification of the catalytic nucleophiles of several families of retaining glycosidases; and

(2) mutation of the enzymes to decrease the turnover of the intermediates and enable determination of their three dimensional structure by crystallographic analysis. Crystallographic studies have also been performed on the enzyme itself (72) at various stages of reaction, providing snapshots of the free enzyme, enzyme substrate complex, covalent intermediate, and product complex.

The mechanism of efficient acid base catalysis by a single residue in retaining glycosidases was investigated using *Bacillus circulans* xylanase as a model. This was achieved by measuring the pK_a values of the two relevant carboxylic acid residues in the active site by ^{13}C NMR (73). While the pK_a of the catalytic residue (Glu172) was 6.7 in the first step, it dropped to 4.2 in the intermediate, allowing it to act as the base catalyst in the second step. Mutagenesis studies attribute the higher pK_a of 6.7 to the negative charge on the other residue Glu 78; in the absence of this charge, the pK_a drops to the lower value of 4.2. During the reaction, this shift in pK_a occurs when the charge on the catalytic nucleophile is removed upon ester formation. The carboxylic acid residue in question is thus able to operate as an acid or base catalyst depending on the charge on the second residue. This has been suggested as a general mechanism for retaining glycosidases.

There are two well established approaches for oligosaccharide synthesis using glycosidases: equilibrium controlled synthesis and kinetically controlled synthesis.

19.3 EQUILIBRIUM CONTROLLED SYNTHESIS

The equilibrium approach is based on the reversal of the catabolic role of enzymes in a thermodynamically controlled reaction (74). It involves a simple reversion of the glycoside hydrolysis reaction by combining a free monosaccharide and nucleophile by direct coupling (condensation). This is also referred to as direct glycosylation or reverse hydrolysis. The equilibrium constant of the reaction favors hydrolysis over glycoside formation and the yields therefore are usually low. In order to shift the reaction equilibrium toward product formation, various approaches have been used (33,46,74,75). The most common way is to use high concentrations of the products of the forward reaction (e.g., monosaccharides) and reduced concentrations of the reactants. This is achieved by incubating the enzyme in a highly concentrated solution of monosaccharide, of the order of 70–80% (w/w), that is, in conditions of lowered water activity (74). Second, to accelerate the otherwise low reaction rate, higher reaction temperatures of 50–60°C are used (74). Increased yields result from controlled and selective removal of the product oligosaccharides from the reaction mixture over the course of the reaction using molecular traps such as activated carbon (76,77). This makes reaction yields comparable to those achieved through transglycosylation. The greatest advantage of this method is the simplicity of the reaction procedure. Both homooligosaccharides and heterooligosaccharides can be synthesized using the equilibrium approach.

Mannooligosaccharides, both homo- and hetero-, have been obtained in good yields by reversal of the α-mannosidase reaction. There is interest in their synthesis because not only are they present in the high mannose type sugar chains of glycoproteins, but they have potential as antiadhesives. Different linkage isomers show varying degrees of inhibition at varying concentrations (78–80).The exceptionally high solubility of mannose in water has enabled synthesis to be carried out in mannose concentrations up to 85% (w/w) and maximum total yields of up to 70% mannooligosaccharide have been achieved using α-mannosidase from Jack bean (75,81). The same enzyme has been used to synthesize heteromannooligosaccharides by cocondensation with a range of acceptor sugars, in addition to homooligosaccharides (82,83). In general, the yields were higher at 70% total sugar concentration than at 50%, except in the case of lactulose, where the yield at 50% was more than double than at 70% total sugar. Higher yields were obtained for heteromannooligosaccharides using reverse hydrolysis compared to kinetic syntheses. A range of linkages, $\alpha(1\rightarrow1)$, $\alpha(1\rightarrow2)$, $\alpha(1\rightarrow3)$, $\alpha(1\rightarrow4)$, and $\alpha(1\rightarrow6)$, were obtained and the product spectrum showed a predominance of lower oligosaccharides, mainly disaccharides, trisaccharides, and tetrasaccharides in the case of homooligosaccharides. Partially purified α-mannosidases from almond meal and limpets were employed by Singh et al. (84) for synthesis of $\alpha(1\rightarrow2)$ and $\alpha(1\rightarrow3)$ linked mannose disaccharides in ratios of 65:35 and 57:43 respectively. Fungal α-mannosidase from *Aspergillus niger* (85) produced $\alpha(1\rightarrow2)$, $\alpha(1\rightarrow3)$ and $\alpha(1\rightarrow6)$ linked disaccharides and trisaccharides, but no $\alpha(1\rightarrow1)$ or $\alpha(1\rightarrow4)$ linkages were detected. The yield of individual regioisomers in equilibrium synthesis is correlated with the standard free energy. In the synthesis of oligosaccharides using α-glucosidase, the standard free energy of the $\alpha(1\rightarrow6)$ disaccharide was the lowest, followed by $\alpha(1\rightarrow2)$, $\alpha(1\rightarrow3)$, $\alpha(1\rightarrow4)$, and $\alpha(1\rightarrow1)$. The standard free energy was in inverse correlation to the oligosaccharide yields (76). Regioselective synthesis of $\alpha(1\rightarrow2)$ mannobiose and mannotriose was achieved (86) using specific 1,2-α-mannosidase from *Aspergillus phoenicis*. A novel α-1,6- mannosidase isolated from the same organism (87) resulted in synthesis of $\alpha(1\rightarrow6)$-linked mannobiose and mannotriose with absolute regioselectivity.

Other fungal enzymes used in synthesis by this approach include glucoamylase from *Aspergillus niger* (88) and β-glucanase from *Penicillium emersonii* (89). While homooligosaccharides were obtained in moderate yields (14–16%), the yields of heterooligosaccharides were rather low compared to the mannosidase enzymes. Using purified *A. niger* glucoamylase (90), a thorough investigation was carried out of its ability to form condensation products from aqueous mixtures of individual carbohydrates (arabinose, fructose, galactose, myoinositol, lyxose, mannose, ribose, and xylose) or their mixtures with glucose. Heterodisaccharides were produced with each of the eight sugars in combination with glucose. Interestingly, the enzyme was able to condense not only glucose but also galactose and mannose individually into disaccharides.

Cellulases (cellobiohydrolase I and endoglucanase I) from recombinant *Trichoderma reesei* (91), α-amylase from *Bacillus licheniformis* (92) and α-glucosidase from *Bacillus stearothermophilus* (93) provide some instances of use of bacterial glycosidases in reverse hydrolysis. Cello bio hydrolase (CBH) I gave yields in the order of 40% in the best case experimental conditions. Homooligosaccharides and heterooligosaccharides from DP 1-10 were obtained using α-amylase; however the products in this case as well in the case of CBH I were not characterized. The *B. stearothermophilus* α-glucosidase was able to produce isomaltose (51%), nigerose (25%), maltose (14%), and kojibiose (10%) in 50% glucose solution, and heterooligosaccharides were obtained both with mannose and xylose, with quantity being dependent on glucose:saccharide acceptor ratios.

Recently (94), a range of glycosidases were screened for their ability to synthesize thioglucosides, which in the context of glycobiology are useful as specific enzyme inhibitors and as glycosyl donors in oligosaccharide synthesis. Of the enzymes tested, β-glucosidase from almonds showed the highest activity in the reverse hydrolysis reaction and high yields of 1-propanethioglucoside were obtained (68% and 41%), based on 1-propanethiol and glucose respectively.

19.3.1 Kinetically Controlled Synthesis

This approach, also known as transglycosylation, employs an activated glycoside donor to form an active intermediate in high concentrations. This intermediate is then trapped by a second nucleophile (other than water) to yield a new glycoside (Figure 19.2) (33,46). The donor glycosides include oligosaccharides, substituted aryl glycosides and glycosyl fluorides.

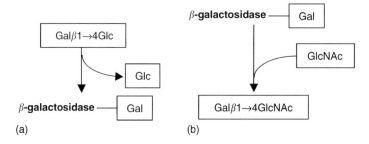

Figure 19.2 Synthesis of *N*-acetyllactosamine using β-galactosidase from *Bacillus circulans*. (a) Activation to form a galactosyl-enzyme complex. (b) Galactosyl transfer to *N*-acetyl glucosamine acting as an acceptor in competition with water.

The reaction is efficiently controlled kinetically as the product is a potential substrate of the glycosidase enzyme (95). As a result, higher product yields using lower enzyme concentration can be obtained. However, the donor glycoside is used up during the reaction in greater than stoichiometric quantities and cannot be reused. This drives up the cost of synthesis.

Table 19.1 summarizes some recent glycosides synthesized using the kinetic approach.

19.4   INFLUENCE OF REACTION COMPONENTS AND CONDITIONS ON YIELD AND SELECTIVITY

19.4.1   Donor:Acceptor Sugar Ratio

In equilibrium controlled heterooligosaccharide synthesis, the proportions can be regulated by the ratio of the individual substrates in the reaction mixture. Usually, heterooligosaccharides are obtained exclusively at higher ratios of the saccharide acceptor to donor, but yields are invariably lower. Using glucoamylase from *A. niger*, heterooligosaccharides were produced exclusively at a fucose:glucose ratio of 85:15, with a yield of 1.6%; highest yield of 3% was at 65:35 ratio but comprised only 42% heterooligosaccharides (88). The same trend was observed in the case of α-glucosidase from *B. stearothermophilus* (93); at a xylose:glucose ratio of 90:10, heterodisaccharides were the major products, but their final concentration was less than half (22mM) of that obtained at a ratio of 50:50 (53mM).

In kinetic synthesis, the donor and acceptor determine the structure of the glycoside synthesized. Although the same glycoside can be synthesized with several different donor and acceptor combinations, the yields vary depending on their nature. For instance (Table 19.1), while α-L-Fuc-(1→3)-α-GlcNAc-OAll and α-L-Fuc-(1→3)-β-GlcNAc-OAll could be synthesized in 34% and 25% yields respectively using α-L-Fuc-O-NP as donor and α/β-GlcNAc-OAll as acceptors, the yield was only 8% for the former when α-L-Fuc-F was the donor instead, and the second glycoside could not be isolated (96). The low yield was attributed to the increased hydrolysis rate of the fluoride donor.

The donor:acceptor ratios need to be optimized for best yields. Equimolar ratios gave higher yields compared to ratios of 1:0.5 and 1:0.2, in a study of galactooligosaccharide synthesis using β-galactosidases from three different sources (97); presumably due to higher transglycosylation rates and reduced secondary hydrolysis at 1:1 ratios. The nature of the glycosyl acceptor also influences the regiospecificity of the transfer. For instance, addition of α-Gal-O-p-NO$_2$Ph to α-Gal-O-Me gives predominantly 1→6 linked product, while the same addition to β-Gal-O-Me gives predominantly the 1→3 linked product (33).

Table 19.1
Synthesis of oligosaccharides using the kinetic approach

Enzyme	Source	Glycosyl Donor	Glycosyl Acceptor	Main Glycoside Product	Yield (%)	Ref.
N-acetylhexosaminidase	A. oryzae	p-NP- β-GlcNAc	GalNAc	β-GlcNAc-(1→6)-GalNAc	26	151
		p-NP- β-GalNAc	GalNAc	β-GalNAc-(1→6)-GalNAc	38	151
β-galactosidase	B. circulans	p-NP- β-Gal	β-GlcNAc-(1→6)-GalNAc	β-Gal-(1→4)- β-Gal(1→4)- β-GlcNAc-(1→6)-GalNAc	48	151
β-galactosidase	B. circulans	O-pNP-β-Gal	OAll- α/ β-GlcNAc	β-Gal-(1→4)-α/β-GlcNAc-OAll	α:66, β:30	152
			OAll- α/ β-Gal	β-Gal-(1→4)-α/β-Gal-OAll	α:55, β:49	
			Ph-S- β-Gal	β-Gal-(1→4)- β-Gal-S-Ph	63	
			O-pNP-β-Gal	β-Gal-(1→4)- β-Gal-O-pNP	23	
				β-Gal-(1→4)- β-Gal-(1→4)- β-Gal-O-pNP	10	
β-galactosidase	Porcine liver	o-NP-β-Gal	p-NP-α/β-Gal/GlcNAc/GalNAc	β-Gal-(1→6)-α/β-Gal-pNP	α:63.9, β:21.2	104
				β-Gal-(1→6)-α/β-GalNAc-pNP	α:78.7, β:28.1	
				β-Gal-(1→6)-α/β-GlcNAc-pNP	α:83.8, β:25.7	
β-galactosidase	B. circulans	o-NP-β-Gal	p-NP-α/β-Gal/GlcNAc/GalNAc	β-Gal-(1→3)-α-Gal-pNP	9,30	104
				β-Gal-(1→6)-α/β-Gal-pNP	α:22.8, β:11.7	
				β-Gal-(1→3)-α/β-GalNAc-pNP	α:75.9, β:20.9	
				β-Gal-(1→6)-α/β-GalNAc-pNP	α:3.20, β:18.6	
				β-Gal-(1→3)-α/β-GlcNAc-pNP	α:79.3, β:29.5	
				β-Gal-(1→6)-α/β-GlcNAc-pNP	α:0.30, β:6.0	
β-galactosidase	A. oryzae	o-NP-β-Gal	Xylose	Galactosyl-xylose	21	111
			Ethyleneglycol	Galactosyl-ethyleneglycol	70	
α-L- fucosidase	P. multicolor	α-L-Fuc-O-NP(1)	α-GlcNAc-OMe(3)	α-L-Fuc-(1→3)-α-GlcNAc-OMe (1+3)	29	96
		α-L-Fuc-F(2)	α/β-GlcNAc-OAll(α:4, β:5)	α-L-Fuc-(1→3)-α/β-GlcNAc-OAll (α:1+4, β:1+5)		
			GlcNAc(6)	α-L-Fuc-(1→3)-α-GlcNAc-OMe (2+3)	α:34, β: 25	
			Glc(7)	α-L-Fuc-(1→3)-α/β-GlcNAc-OAll (α:2+4, β:2+5)	13	
				α-L-Fuc-(1→3)-GlcNAc(2+6)	α:8, β: -7	
				α-L-Fuc-(1→3)-Glc (2+7)	34	

19.4.2 Temperature

The equilibrium synthesis reactions are typically carried out at temperatures of 50–60°C and kinetic synthesis reactions at 37°C, although temperatures ranging from room temperature to 45°C have been employed, depending on the enzyme source and thermal stability. The high substrate concentrations used in reverse hydrolysis exert a protective effect on the enzyme, thus allowing use of higher temperatures.

Statistical analysis was performed for effect of temperature (20, 30, 40, and 50°C) on lactose conversion by β-galactosidase from *B. circulans* with the aim of developing a kinetic model which could aid reactor design for the process (98). Reaction rate as well as yield increased at higher temperatures.

A recent study employing dextransucrase has reported surprising results where not only yield but also selectivity improved considerably with decreasing reaction temperature, best results being obtained in reactions in the frozen state (99). The yield of leucrose (0.1M sucrose as substrate) increased from 10% at 25°C to 65% at −4°C. Similarly, in reactions containing sucrose with isomaltulose as acceptor, overall yields (trisaccharide, tetrasaccharide, and pentasaccharide) at −4 to −10°C were 80–86%, double the yield of reactions in the 5–25°C range.

Temperature has been reported to have a bearing on the selectivity of β-glycosidases for donors differing in their glycon parts, as observed from an investigation with β-glucosidases and galactosidases from almond, a mesophilic (*K. fragilis*), and three thermophilic organisms (*C. saccharolyticum, S. solfataricus, P. furiosus*) (100). Selectivity, calculated as the specificity constant (V_{max}/K_m) or V_{max} ratio of glucoside to galactoside donor (p-NP or phenyl-β-glucoside/galactoside), decreased with increasing temperature in case of almond, *K. fragilis* and *C. saccharolyticum,* and remained constant for the other two thermophilic enzymes.

Klepsiella fragilis β-galactosidase became a β-glucosidase when the temperature was raised from 25 to 50°C.

The range of temperatures employed in enzymatic oligosaccharide synthesis combined with the high sugar concentrations has been reported in some instances (101,102) to limit optimum oligosaccharide yields owing to glycation induced enzyme inactivation. We have provided possible evidence for this (103) during synthesis of mannooligosaccharides by reverse hydrolysis. Inhibitors of the Maillard reaction were partially able to alleviate these effects resulting in reduced loss of enzyme activity and increased oligosaccharide yield.

19.4.3 Enzyme Origin

Regioselectivity can vary with different enzymes, and careful selection of the enzyme can effect the formation of desired linkage. A major objective of researchers in this field is to build up a library of glycosidases capable of selectively catalysing the formation of any desired glycosidic bond (47).

Same donors and acceptors lead to a range of glycosides varying in linkage as well as yield (Table 19.1), when β-galactosidases from two different sources were used (porcine liver and *Bacillus circulans*). Only (1→6) linked products were obtained with the former, while the bacterial enzyme gave rise to (1→3) linkages as well (104). β-glucosidases and galactosidases from almond, *K. fragilis, C. saccharolyticum, S. solfataricus,* and *P. furiosus* differed in their substrate specificity and donor selectivity and in the extent to which it was influenced by temperature (100). Another study investigated β-galactosidases from *B. circulans, A. oryzae, K. lactis, and K. fragilis* (98). There were obvious differences with respect to the amount, size, and type of oligosaccharides produced by these enzymes. *Bacillus circulans* enzyme produced the highest yield and variety of oligosaccharides, and also largest saccharides (up to pentasaccharides). The *A. oryzae* enzyme also produced higher oligosaccharides but in smaller quantities. The enzymes from *Kluyveromyces* mainly produced trisaccharides, with negligible amounts of higher saccharides. Novel galactooligosaccharides with distinct spectra were synthesized with β-galactosidase from five different species of *Bifidobacterium* (105). Higher yields (48%) of β-galactooligosaccarides were obtained with thermostable β-galactosidase from *Sulfolobus solfataricus* (97) compared to those from *A. oryzae* (36%) and *E. coli* (32%). Maximum oligosaccharide yields were obtained fastest with the *A. oryzae* enzyme, were highest with the *S. solfataricus* enzyme, and lowest with the *E. coli* enzyme.

Partially purified α-mannosidases from almond meal and limpets, when incubated with a high concentration of mannose in reverse synthesis reaction, gave rise to the same (1→2)- and (1→3)-mannose disaccharides, but in different ratios of 65:35 and 57:43 respectively (84). Taking advantage of the differences in oligosaccharide yields and specificities of mannosidases from different sources, we have recently devised a simple method of regioselective synthesis of α (1→3)-mannobiose (106). A commercial α-mannosidase from almond was highly efficient in reverse hydrolysis, and oligosaccharide yields of 45–50% were achieved. The products were a mixture of disaccharides (30.75%, w/w), trisaccharides (12.26%, w/w) and tetrasaccharides (1.89%, w/w) with 1→2, 1→3, and 1→6 isomers. α-1,2-linkage specific mannosidase from *P. citrinum* (106) and α-1,6-linkage specific mannosidase from *Aspergillus phoenicis* (87) were used in combination to hydrolyse the respective linkages from the mixture of isomers, resulting in α-(1→3)-mannobiose in 86.4% purity.

19.4.4 Use of Organic Solvents

Many workers have experimented with the use of nonaqueous media and organic cosolvents in enzymatic synthesis. While in reverse hydrolysis the aim is to reduce the water activity of the reaction systems and drive the reaction toward synthesis, in transglycosylation the aim is to reduce the extent of competing hydrolysis.

The feasibility of this approach depends to an extent on the enzyme source (107), as many enzymes undergo inactivation in such conditions and reduced yields or decreased reaction rates are observed. The effect of the solvent on enzyme activity under reaction conditions should be studied, and higher enzyme concentrations are generally required than in buffer. Using o-nitrophenyl galactoside as donor and 3-O-methyl Glc or GlcNAc as acceptor, β-galactosidase from *E. coli* synthesized 3-O-methyl *allo*lactose, Gal β(1→6) Gal and N-acetyl *allo*lactosamine in the presence of different concentrations of diglyme at a much reduced rate (55–144 h reaction time), compared to reaction time of only 6 h in aqueous buffer, and there was no improvement in yields. Conversely, with β-galactosidases from *K. fragilis* and *A. oryzae*, synthesis occurred only in the presence of organic solvents (trimethyl or ethyl phosphate, or tetraglyme in >60% v/v concentration) but the yields were not high. In aqueous buffer, these enzymes hydrolysed the donor and no transfer to acceptor was achieved.

Laroute and Willemot (108) investigated the effect of 66 different organic solvents on glucoamylase from *Rhizopus oryzae* and β-galactosidase from *Aspergillus flavus*. Both enzymes demonstrated similar activities in the range of solvents tested, residual activities after 24 h ranging from 0 (methanol, ethylene glycol, dimethyl sulfoxide, and amines) to 100% (1,4-butanediol, 1,5-pentanediol, water, and tributyl phosphate), being 65–100% for most ethers, alcohols, and esters. The exception was acetates, in which glucoamylase was denatured but β-galactosidase retained 60–70% activity. A protective effect against thermal denaturation was correlated with increasing carbon chain length and lower hydroxyl content. No clear correlation could be established between enzyme stability and hydrophobicity of the medium. In the case of ethers, higher half lives were obtained with water immiscible solvents. In the context of oligosaccharide synthesis, the study suggests that the study of enzyme stability is not sufficient to determine if a solvent would be suitable in synthesis, as synthesis failed to occur with glucoamylase in reaction mixtures containing tetrahydrofuran and tetrahydrofuranfurfuryl alcohol, though they did not cause enzyme denaturation. Most solvents allowed synthesis to occur and best yields were obtained with diethylene glycol diethyl ether (37%), 1-octanol (33%), and 1-hexanol (31%). Another study with glucoamylases in aqueous ether mixtures (109) showed that although the enzyme was stable, the final oligosaccharide concentration was lower compared to aqueous media due to limited solubility of glucose in organic rich media. Neither the product specificity nor the relative order of initial rates of disaccharide formation by glucoamylase was modified in aqueous ethers. Glucoamylase was considered unsuitable for oligosaccharide synthesis by the authors owing to its broad specificity.

The hydrophobicity of the solvent had a bearing on the size of oligosaccharides synthesized with thermophilic β-glycosidases from *S. solfataricus* and *P. furiosus* (110), synthesis of larger saccharides being favored in presence of a more hydrophobic solvent. Initial reaction rate increased as a function of log P. Toluene and nonane were found to be the best as organic phases, resulting in maximum yields equivalent to the aqueous solution.

Cosolvents DMF and acetone at 50% (v/v) decreased the yields of galactosyl and xylose compared to those in the absence of cosolvents, (111) but the rate of transglycosylation was doubled in the presence of 50% (v/v) acetone.

Improved synthesis of galactooligosaccharides was obtained from lactose by *A. oryzae* β-galactosidase in sodium bis (2-ethylhexyl) sulfosuccinate (AOT) isooctane reverse micelles (112). At 45% (w/v) lactose, a maximum yield of 42.5% (w/w) GOS was obtained in the AOT isooctane solution compared to 31% (w/w) in aqueous solution. The improved yield is a result of inhibition of the hydrolytic activity of the enzyme due to decreased water activity and increased lactose concentration in the core of the micelles.

Attempts to increase oligosaccharide synthesis using α-amylase from *B. licheniformis* in the presence of various organic solvents (ethanol, methanol, N-propanol, propanediol, N-butanol, dioxane, and dimethyl sulfoxide) were unsuccessful (92) and this could not be attributed to the loss of enzyme activity as the enzyme was stable and even stimulated in the presence of low concentrations of some solvents.

19.4.5 Substrate Concentration

As in the case of organic solvents, high substrate concentration drives the equilibrium reactions toward synthesis and decreases the hydrolytic side reaction in transglycosylation. High substrate concentrations also protect the enzyme against thermal denaturation. Substrate concentration is a major determinant of synthesis yields.

In reverse hydrolysis reaction using α-mannosidase from Jack bean, the ratio of initial reaction rate in 83% and 40% mannose was approximately two (81). Maximum yield was obtained at 70% mannose using α-mannosidases from Jack bean and *A. phoenicis* (83); reduced yields were obtained with the Jack bean enzyme at 85% mannose; and synthesis was completely inhibited at 80% mannose in the case of *A. phoenicis*. Synthesis using β-glucanase from *P. emersonii* gave maximum yield of 14–16% at 60% glucose concentration and yield fell sharply at 70% glucose. The initial lactose conversion rate and oligosaccharide production rate were independent of the initial lactose concentration during synthesis of oligosaccharides by thermostable β-glucosidase from *Pyrococcus furiosus* (108). However, the relative oligosaccharide yield increased slightly. Increase in concentration of xylose did not influence the rate of o-nitrophenyl-β-galactose (ONPG) conversion significantly in β-galactosidase catalysed synthesis of galactosyl xylose (106) but yield increased to 21% from 12% with increase in xylose concentration from 0.05M to 2.7M. In a study of oligosaccharide synthesis using β-galactosidases from four different sources (97), a higher lactose concentration increased yields in all the cases. Similarly, an increase in the maltose concentration (15–50%) led to an increase in the maximum trisaccharide (131– 498 mM) and tetrasaccharide (39–283 mM) concentrations using α-glucosidase from B. *stearothermophilus* and trisaccharide concentration (24–217 mM) for the same enzyme from brewer's yeast (109).

19.4.6   pH

pH of reaction has not been investigated as a variable in many synthesis studies. In most cases, the pH was chosen on the basis of the enzyme,e and variation of pH did not exert a significant influence on yields. In kinetic syntheses, pH can affect the proportion of hydrolysis and transglycosylation.

The influence of pH on the stability of Jack bean alpha mannosidase was studied (81) by measuring its residual activity in 83% mannose solution at pH 4.5, 5.5, and 6.5 after 24 h at 75°C and 80°C. At 75°C, the enzyme was completely stable at all pH values, whereas at 80°C, the residual activities were 35%, 65%, and 71% at pH 4.5, 5.5, and 6.5 respectively. At 75°C, the initial reaction rate at pH 4.5 and 5.5 was 0.42M disaccharides formed per hour and 0.23 M at pH 6.5. At 55°C, however, pH (4.5 or 5.5) did not influence reaction rates.

A pH to temperature correlation was also seen while optimization of galactooligosaccharide production from lactose using β-glycosidases from S. *solfataricus* and P. *furiosus* (110); optimal pH for synthesis differed with temperature. Optimal transglycosylation was obtained at pH 8.0 at 22°C and pH 5.0 at 90°C. This difference could be due to variable influence of pH on protein structure at different temperatures.

No effect of pH was found on the reaction rate between a pH range of 5.0–7.0 using P. *furiosus* β-glucosidase (113), reduced reaction rate was observed at pH 4.0. Relative yield was highest at pH 6.0 in the range tested.

19.4.7   Time

In reverse hydrolysis, the reaction time is the time taken to attain equilibrium, which is usually the point at which no increase in oligosaccharide yield is observed for a period of two or more days and is the point of maximum yield. This is influenced by many factors such as substrate concentration, enzyme source, activity and stability, reaction temperature, and pH. The time scales in equilibrium synthesis are generally in the order of 1–2 weeks although shorter times have been reported for mannooligosaccharide synthesis using Jack bean alpha mannosidase (81) and β-glucanase (89).

The time of reaction is more crucial in the case of transglycosylation than equilibrium synthesis and needs to be carefully monitored to prevent decreased yields due to secondary hydrolysis of products.

Using α-glucosidase from B. *stearothermophilus*, trisaccharides were rapidly synthesized up to a value of 27% on a molar basis from maltose within two days and their concentration declined thereafter (114). Tetrasaccharides reached their maximum concentration in 5 days and did not undergo appreciable hydrolysis. Only trisaccharides were formed by α-glucosidase from yeast, but they were stable over the time period of reaction (10 days). A time dependent modulation of product spectrum was found in the study; with α-(1→4) linked trisaccharides synthesized exclusively by the bacterial enzyme at 2 and 6 h intervals and a low proportion (<5%) of (1→3) and (1→6) linkages appearing after 20 h of reaction. All tetrasaccharides were α-(1→4) linked. The yeast glucosidase also produced mainly α-(1→4) linkages at 6 h, with 12% α-(1→3). The percentage of α-(1→3) linkage increased to 34% after 20 h accompanied by 7% α-(1→6).

19.4.8 Enzyme Immobilization

Synthesis using immobilized glycosidases drives the cost of synthesis down by enabling reuse of the catalyst. The yields are generally lower compared to synthesis using the free enzyme, but were improved in some cases. Immobilization is also able to modulate product spectrum, primarily due to steric effects on the immobilized enzyme. The reactor configuration and mode of operation (batch or continuous) also affects yield and spectrum.

A tetrasaccharide OS-1 (gal/glu 3:1) and a trisaccharide OS-2 (gal/glu 2:1) were synthesized in 32–35% total yield by free and immobilized β-galactosidase from *Thermus aquaticus* YT-1 (115). The enzyme was immobilized by cocrosslinking with bovine serum albumin followed by entrapment in agarose beads of about 2.0 mm diameter. The immobilized enzyme gave significantly higher lactose conversion rates, higher yields, and produced more OS-2 and less OS-1 than the free enzyme.

Using β-galactosidase from *Bullera singularis* immobilized by adsorption onto Chitopearl resin, galactooligosaccharides were produced in 55% yields with a productivity of 4.4 g/L/h from 100 g/L lactose solution in a continuous process in a packed bed reactor (116). In comparison, β-galactosidase from *A. oryzae* when immobilized on glutaraldehyde treated chitosan beads and used in a plug reactor (117) produced only 18% yield from 100 g/L lactose. Although the thermal stability of the immobilized enzyme was higher, yield of enzyme activity and galactooligosaccharides was lower than the free enzyme. High productivities of 80 and 106 g/L/h with corresponding yields of 21 and 26% (w/w) have been reported for the *A. oryzae* β-galactosidase immobilized on cotton cloth and operated in a plug flow reactor (118), from 200 g/L and 400 g/L lactose respectively.

A dramatic rise in yield (573%, oligomers mg/ml.mg of enzyme) was reported (119) for glucooligosaccharide synthesis using immobilized β-glucosidase from almond compared to the free enzyme at 7.5 M (1350 g/L) glucose concentration. Immobilization led to a decrease in hydrolytic activity and increase in the synthetic activity of the enzyme. It has been suggested that this might result from creation of a hydrophobic microenvironment in the region of the active site.

Immobilized α-glucosidase was employed in a column system comprising an immobilized α-glucosidase column and an activated carbon column sequentially (77). The disaccharides formed in the enzyme column were adsorbed in the carbon column preferentially, expelling glucose and making it available for recycling into the enzyme column. In this manner, formation of disaccharides occured by repeated condensation. This system resulted in a higher yield of α-(1→4) linked disaccharide than α-(1→6), which was different from the batch system. This was due to lower energy barrier of the α-(1→4) linked disaccharide enzyme activation intermediate, resulting in its faster formation and immediate adsorption in the activated carbon column.

The method of immobilization influenced the regioselectivity of linkage specific α-mannosidase from *A. phoenicis* (120). The enzyme immobilized by entrapment in alginate beads gave rise predominantly to α-(1→2) product, as in case of the free enzyme, but the predominant product was Man α-(1→6) Man in the cases of enzyme immobilized on China clay and DE-52. Additionally, immobilization on China clay gave rise to α-(1→3) linked disaccharide. The loss of regioselectivity has been correlated with the likely conformational freedom of the immobilized enzyme as the covalently crosslinked enzyme displayed the lowest regioselectivity followed by the noncovalently modified enzyme tightly bound to an ion exchange resin, and then by the enzyme entrapped in alginate beads. Total oligosaccharide yield was over 10% lower than that obtained with the free enzyme. Mannose at 80% concentration was used for synthesis with immobilized enzyme, but was inhibitory in synthesis with free enzyme.

19.5   RECOMBINANT AND ENGINEERED GLYCOSIDASES IN OLIGOSACCHARIDE SYNTHESIS

Recombinant glycosidases

The cloning of an increasing number of glycosidases is improving their availability, and many of these are being investigated for their synthetic potential.

Yield of cloned α-galactosidase from *Bifidobacterium adolescentis* expressed in *E. coli* (121) was 100 times higher than the native enzyme and could be purified by a single anion exchange chromatography step. The enzyme exhibited transglycosylating activity and α-galactooligosaccharides were synthesized from melibiose and stachyose.

Purified recombinant β-glycosidase from *Thermus thermophilus,* overexpressed in *E. coli,* had an optimum temperature of 88°C and performed transglycosylation at high temperature with yields of over 63% in transfucosylation reactions. The enzyme catalysed the hydrolysis of β-D-galactoside, β-D-glucoside, and β-D-fucoside. The enzyme specificity in decreasing order toward different linkages was β (1→3) (100%)> β (1→2) (71%)> β (1→4) (40%)> β (1→6) (10%). Potential application of the glycosidase was suggested in the synthesis of fucosyl adducts and fucosyl sugars.

β-glucosidase II from *Pichia etchellsii* expressed in *E. coli* (122) was able to carry out glucooligosaccharide synthesis both by reverse hydrolysis (glucobiose and triose, at 18 and 6 mmol/L from 167 mmol/L glucose) and transglycosylation (glucotriose and pentose at 4.5 and 2 mmol/L from 79 mmol/L cellobiose). High conversion of 25% was observed in transglycosylation reaction with sophorose. The enzyme exhibited low regioselectivity and β (1→2), β (1→3), and β (1→6)-linked glucobioses were obtained in equilibrium synthesis.

19.5.1   Mutant Glycosidases

19.5.1.1   Glycosynthases

Glycosynthases (123–126) are engineered mutant glycosidases constructed by mutation of the catalytic nucleophile of a retaining glycosidase to a small nonnucleophilic residue. These mutant enzymes are able to efficiently synthesize oligosaccharides, but lack hydrolytic activity due to their inability to form the glycosyl enzyme intermediate. Their hydrolytic inactivity obviates the main problem of secondary hydrolysis in transglycosylation and oligosaccharide synthesis using glycosynthases resulted in high yields of over 95% in some instances. The first reported glycosynthase was a Glu358Ala mutant of *Agrobacterium* species β-glucosidase (Abg) (123). The k_{cat} value was 10^7 times lower than the wild-type enzyme, but was restored 10^5-fold by addition of small anions such as azide or formate (127). In the case of azide, the reaction intermediate was identified as α-glucosyl azide indicating that the reaction proceeded via the direct attack of azide. This implies that the mutant underwent a shift in mechanism from retaining to inverting. The transfer of donor sugar to an acceptor is catalysed only from activated donors with a good leaving group such as fluoride and an opposite anomeric configuration to the normal substrate of the wild-type enzyme. The glycosyl fluoride substrate mimics the glycosyl intermediate and the mechanism of transfer is similar to that occurring during transglycosylation. Formate functions differently (124,128), acting as an assistant nucleophile in the formation of the intermediate from an activated β-substrate, termed as a retaining glycosynthase mechanism. During synthesis using Abg Glu 358 Ala, α-glucosyl fluoride proved to be a better donor, resulting in longer oligosaccharides and higher yields, than α-galactosyl fluoride, which catalysed only a single galactosyl transfer to give disaccharides. Synthesis of several valuable substrates and inhibitors of cellulases including those unattainable by the wild-type enzyme was achieved using this glycosynthase in gram scales.

The efficiency of the glycosynthase was increased dramatically with a 24-fold improvement in rate constant when serine instead of alanine was used to substitute the catalytic nucleophile, i.e., in an E358S mutant (129). The increased reaction rates resulted in higher yields. Also, the enzyme was able to glycosylate some glycosides like PNP-GlcNAc which were weak acceptors in the case of E358A. A specific stabilizing interaction such as hydrogen bonding between the Ser hydroxyl group and the anomeric fluorine of the α-glycosyl fluoride has been suggested as the reason for the increased glycosylation ability of Abg E358S. From a 20:1 culture, 3.5 g of the pure mutant enzyme could be obtained, which affords the possibility of its application in large scale oligosaccharide synthesis.

A β-mannosynthase constructed by Glu519Ser mutation of recombinant β-mannosidase from *Cellulomonas fimi* overexpressed in *E. coli* has been used for synthesis of β-mannosides using the donor α-D-mannosyl fluoride and various acceptors (130). Highest yield of 99% was obtained using PNP β-D-cellobioside as acceptor. The products contained a mixture of β-(1→3) and β-(1→4) linkages. The Glu519Ala mutant was both a poor hydrolase and mannosynthase and only 8% yields of disaccharides and trisaccharides were obtained.

Other glycosynthases include E387G mutant of *Sulfolobus solfataricus* β-glucosidase and those derived from endoglycosidases: E134A mutant of *B. licheniformis* glucanase and E197A mutant of *Humicola insolens* cellulase. The details of these enzymes are tabulated and discussed in reviews by Jakeman and Withers (125) and Williams and Withers (126). The two endoglycosynthases have been used in tandem in a one pot synthesis (131) of hexasaccharide substrates of 1,3/1,4- β-glucanases in 70–80% yields, useful in their kinetic and structural studies.

All of these glycosynthases give rise to β-(1→3) or β-(1→4) linked products. Jakeman and Withers (132) have generated a glycosynthase (Glu537Ser) from *E. coli* β-galactosidase capable of forming Gal-β- (1→6)-linkages. Product yields of 40–63% were obtained by reactions between α-D-galactosyl-fluoride (α-Gal-F) and β-D-glucopyranoside (pNPGlc), pNP β-cellobioside, and phenyl β-D-glucoside. A second mutation generated close to the enzyme's active site resulted in the double mutant Glu537Ser/Gly794Asp which was much more active and resulted in improved product yields of 80–85%. X-ray crystallography studies showed that the introduced Asp lies within the active site and might facilitate transglycosylation by interacting with the acceptor sugar.

A new endoglycosynthase has been generated by E231G, E231S, or E231A mutation of barley (1,3)-β-D-glucan endohydrolase (133). E231G exhibited the highest catalytic efficiency. The mutant enzymes had no activity on laminarin but activity could be restored partially with formate. The glycosynthase was able to bring about autocondensation of α-laminaribiosyl fluoride and heterocondensation of α-laminaribiosyl fluoride and 4'-nitrophenyl β-D-glucopyranoside to form polymeric products: linear (1,3)-β-D-glucans of DP 30−34.

The Abg E358S and E358G mutants of *Agrobacterium* species β-glucosidase have been used for solid phase oligosaccharide synthesis (134). A galactose moiety was transferred by these enzymes from α-Gal-F with >90% efficiency to acceptors linked to PEGA resin via a backbone amide linker (BAL) and the recovery of products was high.

The increasing repertoire of glycosynthases with a broadening donor and acceptor range, combined with their application in methodologies such as solid phase synthesis suitable for large scale use, make them attractive catalysts for oligosaccharide synthesis.

19.5.1.2   Mutant Enzymes other than Glycosynthases

The performance of β-glucosidase CelB from hyperthermophilic *Pyrococcus furiosus* in oligosaccharide synthesis was improved by two active site mutations (135): a phenylalanine to tyrosine (F426Y), and methionine to lysine (M424K). F426Y improved oligosaccharide yield from 40 to 45% compared to the wild type enzyme, and M424K increased the pH optimum of transglycosylation. The improvement in transglycosylation capacity and oligosaccharide yield was especially significant at low lactose concentrations of 10 to 20%. The F426Y/M424K double mutant gave 40% yield at 10% lactose, compared to only 18% for the wild-type enzyme. It also displayed a higher ratio of tetrasaccharides to trisaccharides. The increased yields were due to an increase in the transglycosylation:hydrolysis ratio. The high yields at low lactose concentrations are desirable characteristics for an industrial process.

β-galactosidase BIF3 from *Bifidobacterium bifidum* has a molecular mass of 188kDa and consists of a signal peptide, an N-terminal β-galactosidase region and a C-terminal galactose binding motif. The enzyme was converted from a hydrolytic to an efficient transgalactosylating enzyme by deletion of about 580 amino acid residues from the C-terminal including a putative galactose binding motif by deletion mutagenesis and termed a transgalactosylase (136). The truncated enzyme expressed in *E. coli* displayed a 9:1 ratio of trangalactosylating to hydrolytic activity during galactooligosaccharide synthesis over a range of lactose concentrations from 10 to 40% and yields of 37–44% were obtained.

19.5.1.3   Mutant Glycosidases Created by Directed Evolution

Directed evolution has been revolutionary in its approach to creating enhanced biocatalysts in that no structural information is required for the catalyst whose improvement is sought. The main requirements for successful directed evolution (137) are: the functional expression of the enzyme in a suitable microbial host, the availability of a screen (or selection) sensitive to the desired properties, and identifying a workable evolution strategy.

The techniques generally used for directed evolution were oligonucleotide directed mutagenesis and error prone PCR until the invention of a method called DNA shuffling by Stemmer (138). Since then, this has been the most widely used method for a range of enzymes. The method was further optimized by Zhao and Arnold (139) for improving the fidelity of recombination. Other methods of directed evolution include staggered extension process (140), random elongation mutagenesis (141), a phage display based method (142), cycling mutagenesis (143), and heteroduplex recombination (144).

Some examples of properties of biocatalysts improved by directed evolution can be found in Affholter and Arnold (145). Applications fall into a few major categories: improving function in nonnatural or extreme environments (where activities or stabilities are low), improving activity toward a new substrate, tuning specificity, and increasing functional expression in a heterologous host.

All of these general categories apply in the context of glycosidase catalysed synthesis of oligosaccharides and a few examples are available in recent literature, including improvement of regioselectivity.

Directed evolution of *E. coli* β-galactosidase carried out by DNA shuffling and colony screening in a plate based assay using chromogenic fucose substrates led to a 10- to 20-fold increase in K_{cat}/K_m for fucose substrates relative to the native enzyme (146). Thirteen base changes resulting in six amino acid changes were responsible for the change from galactosidase to fucosidase specificity.

Using staggered extension process, it has been attempted (147) to evolve the minor 1, 3 regioselectivity of α-galactosidase from *Bacillus stearothermophilus,* which is of greater biological and medical significance than the major 1, 6 selectivity. The initial screening of the mutants was based on their inability to grow on melibiose (1, 6-linked). Just one generation of random mutagenesis and one recombination of the best mutants led to the loss of 1, 6 hydrolytic and synthetic activity to a large extent. The mutant enzymes had preference for 1, 3 linkage ($R_{1,3/1,6}$=6.9, compared to 0.3 for the wild type), however no improvement in yield was obtained and it remained constant at about 5%. The increase in 1, 3 selectivity is thus primarily due to decrease in 1, 6 regioselectivity rather than an improvement in the 1, 3 regioselectivity.

The genes coding for β-glucosidase Cel B from *Pyrococcus furiosus* and β-glycosidase LacS from *Sulfolobus solfataricus* were subjected to evolution (148) by DNA family shuffling. Three rounds of screening led to isolation of two improved hybrids with 1.5-fold–3.5-fold and 3.5-fold–8.6-fold increases in lactose hydrolysis rates respectively. The study is an example of successful DNA shuffling between sequences of enzymes with limited homology.

Devising a suitable and efficient screen is, in fact, the most important step of a directed evolution experiment, and it is the first law of directed evolution that "You get what you screen for" (149). On this principle, a screening method has been developed by Mayer et al. (150) for identification of functional synthetic enzymes for oligosaccharide synthesis, in this case glycosynthases. The method was a plate based coupled enzyme screen employing two plasmids coexpressed in the same cell: one coding for the screening enzyme (releases a chromogenic product only from glycosynthase derived product and not the acceptor) and the other for the glycosynthase. The screen was used to detect glycosynthase activity in a library of random mutants generated by saturation mutagenesis of the catalytic nucleophile in *Agrobacterium* species β-glucosidase (Abg). In addition to the glycosynthases generated from this enzyme in previous studies (123,129), two new glycosynthases, E358G and E358C, were discovered using this assay. The E358G had a K_{cat}/K_m which was more than double than that of the previous best mutant, E358S.

19.6   CONCLUSIONS

It is evident that the use of biotechnology to manufacture oligosaccharides of food and medical importance has developed rapidly. We now have a range of synthetic approaches which can be employed to bring about the synthesis of a wide range of biologically important structures. The development of glycosynthases and the Superbead technology will increase the availability of complex mammalian oligosaccharides and facilitate drug development based on glycoconjugates. The use of glycosidases as synthetic catalysts is also maturing and we can now manufacture multigram quantities of several bioactive oligosaccharides without the need for derivatization. This approach is also a promising route to novel oligosaccharide structures. The application of directed evolution techniques to glycosidases can be expected to increase the range of glycosidases with desired specificities. Genetic engineering techniques can also be seen as a means for scaling up the manufacture of glycosidases from plant and mammalian sources and should, at least in principle, make them more readily available.

Developments in carbohydrate biotechnology are already facilitating efforts to understand the roles of complex carbohydrates in biological systems and they will also enable the exploitation of such knowledge for practical ends.

REFERENCES

1. Hurtley, S., R. Service, P. Szuromi. Cinderella's coach is ready. *Science* 291:2337–2337, 2001.
2. McAuliffe, J.C., O. Hindsgaul. Carbohydrate drugs: an ongoing challenge. *Chem. Ind.* 5:170–174, 1997.

3. Sharon, N., H. Lis. Carbohydrates in cell recognition. *Sci. Am.* 268:82–89, 1993.

4. El Khadem, S.H. Structure of oligosaccharides. In: *Carbohydrate Chemistry: Monosaccharides and their Oligomers.* San Diego: Academic Press, 1988, pp 191–222.

5. Pazur, J.H. Oligosaccharides. In: *The Carbohydrates: Chemistry and Biochemistry,* 2nd ed., Vol. IIA. Pigman, W., D. Horton, eds., New York: Academic Press, 1970, pp 69–137.

6. Varki, A. Biological roles of oligosaccharides: all the theories are correct. *Glycobiology* 3:97–130, 1993.

7. Dwek, R.A. Glycobiology: toward understanding the function of sugars. *Chem. Rev.* 96:683–720, 1996.

8. Ofek, I., N. Sharon. Adhesins as lectins: specificity and role in infection. *Curr. Top. Microbiol. Immunol.* 151:91–113, 1990.

9. Zopf, D., S. Roth. Oligosaccharides anti-infective agents. *Lancet* 347:1017–1021, 1996.

10. Karlsson, K.-A. Meaning and therapeutic potential of microbial recognition of host glyco-conjugates. *Mol. Microbiol.* 29:1–11, 1998.

11. Playne, M.J., Glycoscience: oligosaccharides as drugs, functional foods, and receptors in the gut. *Aust. Biotechnol.* 12:35–37, 2002.

12. Röckendorf, N., T.K. Lindhorst. Glycodendrimers. *Top. Curr. Chem.* 217:201–238, 2001.

13. Patel, A., T.K. Lindhorst. Synthesis of "mixed type" oligosaccharide mimetics based on a carbohydrate scaffold. *Eur. J. Org. Chem.* 1:79–86, 2002.

14. Crittenden, R.G., M.J. Playne. Production, properties and application of food-grade oligo-saccharides. *Trends Food Sci. Tech.* 7:353–361, 1996.

15. Nakakuki, T. Present status and future of functional oligosaccharide development in Japan. *Pure Appl. Chem.* 74:1245–1251, 2002.

16. Nilsson, K.G.I. Enzymatic synthesis of oligosaccharides. *Trends Biotech.* 6:256–264, 1988.

17. Boons, G.J. Strategies in oligosaccharide synthesis. *Tetrahedron* 52:1095–1121, 1996.

18. Davis, B.G. Recent developments in oligosaccharide synthesis. *J. Chem. Soc. Perkin Trans.* 1(14):2137–2160, 2000.

19. Ekelöf, K., P.J. Garegg, L. Olsson, S. Oscarson. Towards the 21st century: the emerging importance of oligosaccharide synthesis. *Pure Appl. Chem.* 69:1847–1852, 1997.

20. Grice, P., S.V. Ley, J. Pietruszka, H.M.I. Osborn, H.W.M. Priepke, S.L. Warriner. A new strategy for oligosaccharide assembly exploiting cyclohexane-1,2-diacetal methodology: an efficient synthesis of a high mannose type nonasaccharide. *Chem. Eur. J.* 3:431–440, 1997.

21. Garegg, P.J., P. Konradsson, D. Lezdins, S. Oscarson, R. Katinka, L. Öhberg. Synthesis of oligosaccharides of biological importance. *Pure Appl. Chem.* 70:293–298, 1998.

22. Allen, J.R., C.R. Harris, S.J. Danishefsky. Pursuit of optimal carbohydrate-based anticancer vaccines: preparation of a multiantigenic unimolecular glycopeptide containing the Tn, MBr1, and Lewis (y) antigens. *J. Am. Chem. Soc.* 123:1890–1897, 2001.

23. Wang, C.-C., J.-C. Lee, S.-Y. Luo, H.-F. Fan, C.-L. Pai, W.-C. Yang, L.-D. Lu, S.-C. Hung. Synthesis of biologically potent α-1,2-linked disaccharide derivatives via regioselective one-pot protection-glycosylation. *Angew Chem. Int. Ed.* 41:2360–2362, 2002.

24. Arlt, M., O. Hidsgaul. Rapid chemical synthesis of sugar-nucleotides in a form suitable for enzymatic oligosaccharide synthesis. *J. Org. Chem.* 60:14–15, 1995.

25. Bartolozzi, A., P.H. Seeberger. New approaches to the chemical synthesis of bioactive oli-gosaccharides. *Curr. Opin. Struct. Biol.* 11:587–592, 2001.

26. Ye, X.-S., C.-H. Wong. Anomeric reactivity-based one-pot oligosaccharide synthesis: a rapid route to oligosaccharide libraries. *J. Org. Chem.* 65:2410–2431, 2000.

27. Koeller, K.M., C.-H. Wong. Complex carbohydrate synthesis tools for glycobiologists: enzyme-based approach and programmable one-pot strategies. *Glycobiology* 10:1157–1169, 2000.

28. Watt, G.M., L. Revers, M.C. Webberley, I.B.H. Wilson, S.L. Flitsch. The chemoenzy-matic synthesis of the core trisaccharide of N-linked oligosaccharides using a recombinant β-mannosyltransferase. *Carbohydr. Res.* 305:533–541, 1997.

29. Theim, J. Applications of enzymes in synthetic carbohydrate chemistry. *FEMS Microbiol. Rev.* 16:193–211, 1995.

30. Öhrlein, R. Glycosyltransferase-catalysed synthesis of non-natural oligosaccharides. *Top. Curr. Chem.* 200:227–254, 1999.

31. Coutinho, P.M., B. Henrissat. Carbohydrate-active enzymes server at URL: http://afmb.cnrs-mrs.fr/CAZY/index.html, 1999.

32. Palcic, M.M. Biocatalytic synthesis of oligosaccharides. *Curr. Opin. Biotechnol.* 10:616–624, 1999.

33. Toone, E.J., E.S. Simon, M.D. Bednarski, G.M. Whitesides. Enzyme-catalysed synthesis of carbohydrates. *Tetrahedron* 45:5365–5422, 1989.

34. Bastida, A., A. Fernandez-Mayoralas, R.G. Arrayas, F. Iradier, J.C. Carretero, E. Garcia-Junceda. Heterologous over-expression of alpha-1,6-fucosyltransferase from Rhizobium sp.: application to the synthesis of the trisaccharide beta-D-GlcNAc(1 -> 4)-(alpha-L-Fuc-(1 -> 6))-D-GlcNAc, study of the acceptor specificity and evaluation of polyhydroxylated indolizidines as inhibitors. *Chem. Eur. J.* 7:2390–2397, 2001.

35. Creeger, E.S., L.I. Rothfield. Cloning genes for bacterial glycosyltransferases. *Methods Enzymol.* 83:326–331, 1982.

36. Russel, R.R.B., M.L. Gilpin, H. Mukasa, G. Dougan. Characterization of glucosyltransferase expressed from a *Streptococcus sobrinus* gene cloned in *Escherichia coli*. *J. Gen. Microbiol.* 133:935–944, 1987.

37. Weinstein, J., E.U. Lee, K. McEntee, P.H. Lai, J.C. Paulson. Primary structure of beta-galctosidase alpha-2,6-sialyltransferase- conversion of membrane-bound enzyme to soluble forms by cleavage of the NH2-terminal signal anchor. *J. Biol. Chem.* 262:17735–17743, 1987.

38. Wen, D.X., B.D. Livingston, K.F. Medzihradszky, S. Kelm, A.L. Burlingame, J.C. Paulson. Primary structure of Gal-beta-1,3(4)GlcNAc alpha-2,3-sialyltransferase determined by mass-spectrometry sequence-analysis and molecular-cloning-evidence for a protein motif in the sialyltransferase gene family. *J. Biol. Chem.* 267:21011–21019, 1992.

39. Ernst, L.K., V.P. Rajan, R.D. Larsen, M.M. Ruff, J.B. Lowe. Stable expression of blood group-H determinants and GDP-L-fucose-beta-D-galactosidase 2-alpha-L-fucosyltransferase in mouse cells after transfection with human DNA. *J. Biol. Chem.* 264:3436–3447, 1989.

40. Wang, P., G.-J. Shen, Y.-F. Wang, Y. Ichikawa, C.-H. Wong. Enzymes in oligosaccharide synthesis: active domain overproduction, specificity study, and synthetic use of an α-1,2-mannosyltransferase with regeneration of GDP-Man. *J. Org. Chem.* 58:3985–3990, 1993.

41. Revers, L., R.M. Bill, I.B.H. Wilson, G.M. Watt, S.L. Flitsch. Development of recombinant, immobilised beta-1,4-mannosyltransferase for use as an efficient tool in the chemo-enzymatic synthesis of N-linked oligosaccharides. *Biochim. Biophys. Acta Gen. Subjects* 1428:88–98, 1999.

42. Nakazawa, K., K. Furukawa, H. Narimatsu, A. Kobata. Kinetic study of human beta-1, 4-galactosyltransferase expressed in *Escherichia-coli*. *J. Biochem.* 113:747–753, 1993.

43. Malissard, M., L. Borsig, S. DiMarco, M.G. Grutter, U. Kragl, C. Wandrey, E.G. Berger. Recombinant soluble beta-1,4-galactosyltransferases expressed in *Saccharomyces cerevisiae*: purification, characterization and comparison with human enzyme. *Eur. J. Biochem.* 239:340–348, 1996.

44. Ju, T.Z., K. Brewer, A. D'Souza, R.D. Cummings, W.M. Canfield. Cloning and expression of human core beta 1,3- galactosyltransferase. *J. Biol. Chem.* 277:178–186, 2002.

45. Kudo, T., T. Iwai, T. Kubota, H. Iwasaki, Y. Takayma, T. Hiruma, N. Inaba, Y. Zhang, M. Gotoh, A. Togayachi, H. Narimatsu. Molecular cloning and characterization of a novel UDP-Gal: GalNAc alpha peptide beta 1,3-galactosyltransferase (ClGal-T2), an enzyme synthesizing a core 1 structure of O-glycan. *J. Biol. Chem.* 277:47724–47731, 2002.

46. Ichikawa, Y., G.C. Look, C.-H. Wong. Enzyme-catalyzed oligosaccharide synthesis. *Anal. Biochem.* 202:215–238, 1992.

47. Crout, D.H.G., G. Vic. Glycosidases and glycosyl transferases in glycoside and oligosaccharide synthesis. *Curr. Opin. Chem. Biol.* 2:98–111, 1998.

48. Wong, C.-H., S.L. Haynie, G.M. Whitesides. Enzyme-catalysed synthesis of N-acetyllactosamine with *in situ* regeneration of uridine 5'-diphosphate glucose and uridine 5'- diphosphate galactose. *J. Org. Chem.* 47:5416–5418, 1982.

49. Ichikawa, Y., J.L.C. Liu, G.J. Shen, C.H. Wong. A highly efficient multienzyme system for the one-step synthesis of a sialyl trisaccharide-insitu generation of sialic-acid and N-acetyllactosamine coupled with regeneration of UDP-Glucose, UDP-Galactose, and CMP-Sialic acid. *J. Am. Chem. Soc.* 113:6300–6302, 1991.

50. Wong, C.-H., R. Wang, Y. Ichikawa. Regeneration of sugar-nucleotide for enzymatic oligosaccharide synthesis: use of Gal-1-phosphate uridyltransferase in the regeneration of UDP-galactose, UDP-2-deoxygalactose, and UDP-galactosamine. *J. Org. Chem.* 57:4343–4344, 1992.

51. Bulter, T., L. Elling. Enzymatic synthesis of nucleotide sugars. *Glycoconjugate J.* 16:147–159, 1999.

52. Bulter, T., L. Elling. Enzymatic synthesis of UDP-galactose on a gram scale. *J. Mol. Catal. B Enzym.* 8:281–284, 2000.

53. Thiem, J., W. Treder. Synthesis of the trisaccharide Neu-5-Ac-Alpha(2-)6)Gal-Beta (1-)4)GlcNAc by the use of immobilized enzymes. *Angew. Chem. Int. Ed. Engl.* 25:1096–1097, 1986.

54. Lubineau, A., K. BassetCarpentier, C. Auge. Porcine liver (2->3)-alpha-sialyltransferase: substrate specificity studies and application of the immobilized enzyme to the synthesis of various sialylated oligosaccharide sequences. *Carbohydr. Res.* 300:161–167, 1997.

55. Nishiguchi, S., K. Yamada, Y. Fuji, S. Shibatani, A. Toda, S.-I. Nishimura. Highly efficient oligosaccharide synthesis on water-soluble polymeric primers by recombinant glycosyltransferases immobilized on solid supports. *Chem. Commun.* 19:1944–1945, 2001.

56. Herrmann, G.F., P. Wang, G.-J. Shen, C.-H. Wong. Recombinant whole cells as catalysts for the enzymatic synthesis of oligosaccharides and glycopeptides. *Angew. Chem. Int. Ed. Engl.* 33:1241–1242, 1994.

57. Nahalka, J., Z. Liu, X. Chen, P.G. Wang. Superbeads: immobilization in "sweet" chemistry. *Chem. Eur. J.* 9:372–377, 2003.

58. Koizumi, S., T. Endo, K. Tabata, A. Ozaki. Large-scale production of UDP-galactose and globotriose by coupling metabolically engineered bacteria. *Nat. Biotechnol.* 16:847–850, 1998.

59. Endo, T., S. Koizumi, K. Tabata, S. Kakita, A. Ozaki. Large-scale production of N-acetyllactosamine through bacterial coupling. *Carbohydr. Res.* 316:179–183, 1999.

60. Endo, T., S. Koizumi, K. Tabata, A. Ozaki. Large-scale production of CMP-NeuAc and sialylated oligosaccharides through bacterial coupling. *Appl. Microbiol. Biotechnol.* 53:257–261, 2000.

61. Endo, T., S. Koizumi, K. Tabata, S. Kakita, A. Ozaki. Large-scale production of the carbohydrate portion of the sialyl-Tn epitope, alpha-Neup5Ac-(2 -> 6)-D-GalpNAc, through bacterial coupling. *Carbohydrate Res.* 330:439–443, 2001.

62. Chen, X., Z.Y. Liu, J.B. Zhang, W. Zhang, P. Kowal, P.G. Wang. Reassembled biosynthetic pathway for large-scale carbohydrate synthesis: alpha-Gal epitope producing "superbug". *Chem. Biochem.* 3:47–53, 2002.

63. Priem, B., M. Gilbert, W.W. Wakarchuk, A. Heyraud, E. Samain. A new fermentation process allows large-scale production of human-milk oligosaccharides by metabolically engineered bacteria. *Glycobiology* 12:235–240, 2002.

64. Chen, X., J. Fang, J. Zhang, Z. Liu, J. Shao, P. Kowal, P. Andreana, P.G. Wang. Sugar nucleotide regeneration beads (Superbeads): a versatile tool for the practical synthesis of oligosaccharides. *J. Am. Chem. Soc.* 123:2081–2082, 2001.

65. Liu, Z.Y., J.B. Zhang, X. Chen, P.G. Wang. Combined biosynthetic pathway for *de novo* production of UDP-galactose: catalysis with multiple enzymes immobilized on agarose beads. *Chem. Biochem.* 3:348–355, 2002.

66. Zhang, J., B. Wu, Z. Liu, P. Kowal, X. Chen, J. Shao, P.G. Wang. Large-scale synthesis of carbohydrates for pharmaceutical development. *Curr. Org. Chem.* 5:1169–1176, 2001.

67. McCarter, J.D., S.G. Withers. Mechanisms of enzymatic glycoside hydrolysis. *Curr. Opin. Struct. Biol.* 4:885–897, 1994.

68. Withers, S.G. 1998 Hoffman La Roche Award Lecture: understanding and exploiting glycosidases. *Can. J. Chem.* 77:1–11, 1999.

69. Rye, C.S., S.G. Withers. Glycosidase mechanisms. *Curr. Opin. Chem. Biol.* 4:573–580, 2000.

70. Withers, S.G. Mechanisms of glycosyl transferases and hydrolases. *Carbohydr. Polym.* 44:325–337, 2001.

71. Vasella, A., G.J. Davies, M. Bohm. Glycosidase mechanisms. *Curr. Opin. Chem. Biol.* 6:619–629, 2002.

72. Davies, G.J., L. Mackenzie, A. Varrot, M. Dauter, A.M. Brzozowski, M. Schulein, S.G. Withers. Snapshots along an enzymatic reaction coordinate: analysis of a retaining beta-glycoside hydrolase. *Biochemistry* 37:11707–11713, 1998.

73. McIntosh, L.P., G. Hand, P.E. Johnson, M.D. Joshi, M. Korner, L.A. Plesniak, L. Ziser, W.W. Wakarchuk, S.G. Withers. The pK(a) of the general acid/base carboxyl group of a glycosidase cycle during catalysis: A C-13-NMR study of *Bacillus circulans* xylanase. *Biochemistry* 35:9958–9966, 1996.

74. Rastall, R.A., C. Bucke. Enzymatic synthesis of oligosaccharides. *Biotechnol. Genet. Eng. Rev.* 10:253–281, 1992.

75. Johansson, E., L. Hedbys, P.-O. Larsson, K. Mosbach, A. Gunnarsson, S. Svensson. Synthesis of mannose oligosaccharides via reversal of the α-mannosidase reaction. *Biotechnol. Lett.* 8:421–424, 1986.

76. Ajisaka, K., Y. Yamamoto. Control of regioselectivity in the enzymatic syntheses of oligosaccharides using glycosides. *Trends Glycosci. Glycotechnol.* 14:1–11, 2002.

77. Ajisaka, K., H. Nishida, H. Fujimoto. Use of an activated carbon column for the synthesis of disaccharides by use of a reverse hydrolysis activity of beta-galactosidase. *Biotechnol. Lett.* 9:387–392, 1987.

78. Firon, N., I. Ofek, N. Sharon. Carbohydrate specificity of the surface lectins of *Escherichia coli*, *Klebsiella pneumoniae*, and *Salmonella typhimurium*. *Carbohydr. Res.* 120:235–249, 1983.

79. Sharon, N. Bacterial lectins, cell-cell recognition and infectious disease. *FEBS Lett.* 217:145–157, 1987.

80. Pan, Y.T., B. Xu, K. Rice, S. Smith, R. Jackson, A.D. Elbein. Specificity of the high-mannose recognition site between *Enterobacter cloacae* pili adhesin and HT-29 cell membranes. *Infect. Immun.* 65:4199–4206, 1997.

81. Johansson, E., L. Hedbys, K. Mosbach, P.-O. Larsson, A. Gunnarsson, S. Svensson. Studies of the reversed α-mannosidase reaction in high concentrations of mannose. *Enzyme Microb. Technol.* 11:347–352, 1989.

82. Rastall, R.A., N.H. Rees, R. Wait, M.W. Adlard, C. Bucke. α-mannosidase-catalysed synthesis of novel manno-,lyxo-,and heteromanno-oligosaccharides: a comparison of kinetically and thermodynamically mediated approaches. *Enzyme Microb. Technol.* 14:53–57, 1992.

83. Suwasono, S., R.A. Rastall. Enzymatic synthesis of manno- and heteromanno-oligosaccharides using α-mannosidases: a comparative study of linkage-specific and non-linkage-specific enzymes. *J. Chem. Technol. Biotechnol.* 73:37–42, 1998.

84. Singh, S., M. Scigelova, D.H.G. Crout. Glycosidase-catalysed synthesis of mannobioses by the reverse hydrolysis activity of alpha-mannosidase: partial purification of alpha-mannosidases from almond meal, limpets and *Aspergillus niger*. *Tetrahedron Asymmetry* 11:223–229, 2000.

85. Ajisaka, K., I. Matsuo, M. Isomura, H. Fujimoto, M. Shirakabe, M. Okawa. Enzymatic synthesis of mannobioses and mannotrioses by reverse hydrolysis using α-mannosidase from *Aspergillus niger*. *Carbohydr. Res.* 270:123–130, 1995.

86. Suwasono, S., R.A. Rastall. A highly regioselective synthesis of mannobiose and mannotriose by reverse hydrolysis using specific 1,2-α-mannosidase from *Aspergillus phoenicis*. *Biotechnol. Lett.* 18:851–856, 1996.

87. Athanasopoulos, V.I., K. Niranjan, R.A. Rastall. Regioselective synthesis of mannobiose and mannotriose by reverse hydrolysis using a novel 1,6-α-D-mannosidase from *Aspergillus phoenicis*. *J. Mol. Catal. B Enzym.* 27:215–219, 2004.

88. Rastall, R.A., M.W. Adlard, C. Bucke. Synthesis of hetero-oligosaccharides by glucoamylase in reverse. *Biotechnol. Lett.* 13:501–504, 1991.

89. Rastall, R.A., S.F. Pikett, M.W. Adlard, C. Bucke. Synthesis of oligosaccharides by reversal of a fungal β-glucanase. *Biotechnol. Lett.* 14:373–378, 1992.

90. Pestlin, S., D. Prinz, J.N. Starr, P.J. Reilly. Kinetics and equilibria of condensation reactions between monosaccharide pairs catalyzed by *Aspergillus niger* glucoamylase. *Biotechnol. Bioeng.* 56:9–22, 1997.

91. Gama, F.M., M. Mota. Cellulases for oligosaccharide synthesis: a preliminary study. *Carbohydr. Polym.* 37:279–281, 1998.

92. Chitradon, L., P. Mahakhan, C. Bucke. Oligosaccharide synthesis by reversed catalysis using α-amylase from *Bacillus licheniformis*. *J. Mol. Catal. B Enzym.* 10:273–280, 2000.

93. Malá, Š., B. Králová. Heterooligosaccharide synthesis catalyzed by α-glucosidase from *Bacillus stearothermophilus*. *J. Mol. Catal. B Enzym.* 10:617–621, 2000.

94. Meulenbeld, G.H., B.M. Roode, S. Hartmans. Enzymatic synthesis of thioglucosides using almond β-glucosidse. *Biocatal. Biotrans.* 20:251–256, 2002.

95. Monsan, P., F. Paul. Enzymatic synthesis of oligosaccharies. *FEMS Microb. Rev.* 16:187–192, 1995.

96. Farkas, E., J. Thiem, K. Ajisaka. Enzymatic synthesis of fucose-containing disaccharides employing the partially purified α-L-fucosidase from *Penicillium multicolor*. *Carbohydr. Res.* 328:293–299, 2000.

97. Reuter, S., A.R. Nygaard, W. Zimmermann. β-galactooligosaccharide synthesis with β-galactosidases from *Sulfolobus solfataricus, Aspergillus oryzae,* and *Escherichia coli*. *Enzyme Microb. Technol.* 25:509–516, 1999.

98. Boon, M.A., A.E.M. Janssen, K. van't Riet. Effect of temperature and enzyme origin on the enzymatic synthesis of oligosaccharides. *Enzyme Microb. Technol.* 26:271–281, 2000.

99. Daum, B., K. Buchholz. High yield and high selectivity of reactions in the frozen state: the acceptor reaction of dextransucrase. *Biocatal. Biotrans.* 20:15–21, 2002.

100. Hansson, T., P. Adlercreutz. The temperature influences the ratio of glucosidase and galactosidase activities of β-glycosidases. *Biotechnol. Lett.* 24:1465–1471, 2002.

101. Bruins, M.E., E.W. Van Hellemond, A.E.M. Janssen, R.M. Boom. Maillard reactions and increased enzyme inactivation during oligosaccharide synthesis by a hyperthermophilic glycosidase. *Biotechnol. Bioeng.* 81:546–552, 2003.

102. Bruins, M.E., A.J.H. Thewessen, A.E.M. Janssen, R.M. Boom. Enzyme inactivation due to Maillard reactions during oligosaccharide synthesis by a hyperthermophilic glyclosidase: influence of enzyme immobilization. *J. Mol. Catal. B Enzym.* 21:31–34, 2003.

103. Maitin, V., R.A. Rastall. Enzyme glycation influences product yields during oligosaccharide synthesis by reverse hydrolysis. *J. Mol. Catal. B Enzym.* 30:195–202, 2004.

104. Zeng, X., R. Yoshino, T. Murata, K. Ajisaka, T. Usui. Regioselective synthesis of p-nitrophenyl glycosides of β-D-galactopyranosyl-disaccharides by transglycosylation with β-D-galactosidases. *Carbohydr. Res.* 325:120–131, 2000.

105. Rabiu, B.A., A.J. Jay, G.R. Gibson, R.A. Rastall. Synthesis and fermentation properties of novel galacto-oligosaccharides by β-galactosidases from *Bifidobacterium* species. *Appl. Environ. Microbiol.* 67:2526–2530, 2001.

106. Maitin, V., V. Athanasopoulos, R.A. Rastall. Synthesis of FimH receptor-active mannooligosaccharides by reverse hydrolysis using α-mannosidases from *Penicillium citrinum, Aspergillus phoenicis* and almond. *Appl. Microbiol. Biotechnol.* 63:666–671, 2004.

107. Finch, P., J.H. Yoon. The effects of organic solvents on the synthesis of galactose disaccharides using β-galactosidases. *Carbohydr. Res.* 303:339–345, 1997.

108. Laroute, V., R.-M. Willemot. Effect of organic solvents on stability of two glycosidases and on glucoamylase-catalysed oligosaccharide synthesis. *Enzyme Microb. Technol.* 14:528–534, 1992.

109. Cantarella, L., Z.L. Nikolov, P.J. Reilly. Disaccharide production by glucoamylase in aqueous ether mixtures. *Enzyme Microb. Technol.* 16:383–387, 1994.

110. Hansson, T., P. Adlercreutz. Optimization of galactooligosaccharide production from lactose using β-glycosidases from hyperthermophiles. *Food Biotechnol.* 15:79–97, 2001.

111. Giacomini, C., G. Irazoqui, P. Gonzalez, F. Batista-Viera, B.M. Brena. Enzymatic synthesis of galactosyl-xylose by *Aspergillus oryzae* β-galactosidase. *J. Mol. Catal. B Enzym.* 19,20:159–165, 2002.

112. Chen, S.-X., D.-Z. Wei, Z.-H. Hu. Synthesis of galacto-oligosaccharides in AOT/isooctane reverse micelles by β-galactosidase. *J. Mol. Catal. B Enzym.* 16:109–114, 2001.

113. Boon, M.A., J. van der Oost, W.M. de Vos, A.E.M. Janssen, K. van't Riet. Synthesis of oligosaccharides catalysed by thermostable β-glucosidase from *Pyrococcus furiosus*. *Appl. Biochem. Biotechnol.* 75:269–278, 1998.

114. Malá, Š., H. Dvořáková, R. Hrabal, B. Králová. Towards regioselective synthesis of oligosaccharides by the use of α-glucosidases with different substrate specificity. *Carbohydr. Res.* 322:209–218, 1999.

115. Berger, J.L., B.H. Lee, C. Lacroix. Oligosaccharides synthesis by free and immobilized β-galactosidases from *Thermus aquaticus* YT-1. *Biotechnol. Lett.* 17:1077–1080, 1995.

116. H.-J. Shin, J.-M. Park, J.-W. Yang. Continuous production of galacto-oligosaccharides from lactose by *Bullera singularis* β-galactosidase immobilized in chitosan beads. *Process Biochem.* 33:787–792, 1998.

117. Sheu, D.-C., S.-Y. Li, K.-J. Duan, C.W. Chen. Production of galactooligosaccharides by β-galactosidase immobilized on glutaraldehyde-treated chitosan beads. *Biotechnol. Techniques* 12:273–276, 1998.

118. Albayrak, N., S.-T. Yang. Production of galacto-oligosaccharides from lactose by *Aspergillus oryzae* β-galactosidase immobilized on cotton cloth. *Biotechnol. Bioeng.* 77:8–19, 2002.

119. Ravet, C., D. Thomas, M.D. Legoy. Gluco-oligosaccharide synthesis by free and immobilized β-glucosidase. *Biotechnol. Bioeng.* 42:303–308, 1993.

120. Suwasono, S., R.A. Rastall. Synthesis of oligosaccharides using immobilized 1,2-α-mannosidase from *Aspergillus phoenicis*: immobilization dependent modulation of product spectrum. *Biotechnol. Lett.* 20:15–17, 1998.

121. Van den Broek, L.A.M., J. Ton, J.C. Verdoes, K.M.J. Van Laere, A.G.J. Voragen, G. Beldman. Synthesis of α-galacto-oligosaccharides by a cloned α-galactosidase from *Bifidobacterium adolescentis*. *Biotechnol. Lett.* 21:441–445, 1999.

122. Bhatia, Y., S. Mishra, V.S. Bisaria. Biosynthetic activity of recombinant *Escherichia coli*-expressed *Pichia etchellsii* β-glucosidase II. *Appl. Biochem. Biotechnol.* 102,103:367–379, 2002.

123. Mackenzie, L.F., Q. Wang, R.A.J. Warren, S.G. Withers. Glycosynthases: mutant glycosidases for oligosaccharide synthesis. *J. Am. Chem. Soc.* 120:5583–5584, 1998.

124. Moracci, M., A. Trincone, M. Rossi. Glycosynthases: new enzymes for oligosaccharide synthesis. *J. Mol. Catal. B Enzym.* 155–163, 2001.

125. Jakeman, D.L., S.G. Withers. Glycosynthases: new tools for oligosaccharide synthesis. *Trends Glycosci. Glycotechnol.* 14:13–25, 2002.

126. Williams, S.J., S.G. Withers. Glycosynthases: mutant glycosidases for glycoside synthesis. *Aust. J. Chem.* 55:3–12, 2002.

127. Wang, Q., R.W. Graham, D. Trimbur, R.A.J. Warren, S.G. Withers. Changing enzymatic reaction mechanisms by mutagenesis: conversion of a retaining glucosidase to an inverting enzyme. *J. Am. Chem. Soc.* 116:11594–11595, 1994.

128. Moracci, M., A. Trincone, B. Cobuzzi-Ponzano, G. Perugino, M. Ciaramella, M. Rossi. Enzymatic synthesis of oligosaccharides by two glycosyl hydrolases of *Sulfolobus solfataricus*. *Extremophiles* 5:145–152, 2001.

129. Mayer, C., D.L. Zechel, S.P. Reid, R.A.J. Warren, S.G. Withers. The E358S mutant of *Agrobacterium* sp. β-glucosidase is a greatly improved glycosynthase. *FEBS Lett.* 466:40–44, 2000.

130. Nashiru, O., D.L. Zechel, D. Stoll, T. Mohammadzadeh, R.A.J. Warren, S.G. Withers. β-mannosynthase: synthesis of β-mannosides with a mutant β-mannosidase. *Angew. Chem. Int. Ed.* 40:417–420, 2001.

131. Faijes, M., J.K. Fairweather, D. Hugues, A. Planas. Oligosaccharide synthesis by coupled endo-glycosynthases of different specificity: a straightforward preparation of two mixed-linkage hexasaccharide substrates of 1,3/1,4- β-glucanases. *Chem. Eur. J.* 7:4651–4655, 2001.

132. Jakeman, D.L., S.G. Withers. On expanding the repertoire of glycosynthases: mutant β-galactosidases forming β-(1,6)-linkages. *Can. J. Chem.* 80:866–870, 2002.

133. Hrmova, M., T. Imai, S.J. Rutten, J.K. Fairweather, L. Pelosi, V. Bulone, H. Driguez, G.B. Fincher. Mutated barley (1,3)-β-D-glucan endohydrolase synthesize crystalline (1,3)-β-D-glucans. *J. Biol. Chem.* 277:30102–30111, 2002.

134. Tolborg, J.F., L. Peterson, K.J. Jensen, C. Mayer, D.L. Jakeman, R.A.J. Warren, S.G. Withers. Solid-phase oligosaccharide and glycopeptide synthesis using glycosynthases. *J. Org. Chem.* 67:4143–4149, 2002.

135. Hansson, T., T. Kaper, J. van der Oost, W.M. de Vos, P. Adlercreutz. Improved oligosaccharide synthesis by protein engineering of β-glucosidase CelB from hyperthermophilic *Pyrococcus furiosus*. *Biotechnol. Bioeng.* 73:203–210, 2001.

136. Jorgensen, F., O.C. Hansen, P. Stougaard. High-efficiency synthesis of oligosaccharides with a truncated β-galactosidase from *Bifidobacterium bifidum*. *Appl. Microbiol. Biotechnol.* 57:647–652, 2001.

137. Kuchner, O., F.H. Arnold. Directed evolution of enzyme catalysts. *Trends Biotech.* 15:523–530, 1997.

138. Stemmer, W.P.C. DNA shuffling by random fragmentation and reassembly: *in vitro* recombination for molecular evolution. *Proc. Natl. Acad. Sci. USA* 91:10747–10751, 1994.

139. Zhao, H., F.H. Arnold. Optimization of DNA shuffling for high-fidelity recombination. *Nucleic Acids Res.* 25:1307–1308, 1997.

140. Zhao, H., L. Giver, Z. Shao, J.A. Affholter, F.H. Arnold. Molecular evolution by staggered extension process (StEP) *in vitro* recombination. *Nat. Biotechnol.* 16:258–261, 1998.

141. Matsuura, T., K. Miyai, S. Trakulnaleamsai, T. Yomo, Y. Shima, S. Miki, K. Yamamoto, I. Urabe. Evolutionary molecular engineering by random elongation mutagenesis. *Nat. Biotechnol.* 17:58–61, 1999.

142. Pedersen, H., S. Holder, D.P. Sutherlin, U. Schwitter, D.S. King, P.G. Schultz. A method for directed evolution and functional cloning of enzymes. *Proc. Natl. Acad. Sci. USA* 95:10523–10528, 1998.

143. Buchholz, F., P.-O. Angrand, A.F. Stewart. Improved properties of FLP recombinase evolved by cycling mutagenesis. *Nat. Biotechnol.* 16:657–662, 1998.

144. Volkov, A.A., Z. Shao, F.H. Arnold. Recombination and chimeragenesis by *in vitro* hetero-duplex formation and *in vivo* repair. *Nucleic Acids Res.* 27, e18:1–6, 1999.

145. Affholter, J., F. Arnold. Engineering a revolution. *Chem. Br.* 35:48–51, 1999.

146. Zhang, J.-H., G. Dawes, W.P.C. Stemmer. Directed evolution of a fucosidase from a galactosidase by DNA shuffling and screening. *Proc. Natl. Acad. Sci. USA* 94:4504–4509, 1997.

147. Dion, M., A. Nisole, P. Spangenberg, C. Andre, A. Glottin-Fleury, R. Mattes, C. Tellier, C. Rabiller. Modulation of the regioselectivity of a *Bacillus* α-galactosidase by directed evolution. *Glycoconjugate J.* 18:215–223, 2001.

148. Kaper, T., S.J.J. Brouns, A.C.M. Geerling, W.M. De Vos, J. Van der Oost. DNA family shuffling of hyperthermostable b-glycosidases. *Biochem. J.* 368:461–470, 2002.

149. Arnold, F.H. Unnatural selection: molecular sex for fun and profit. *Eng. Sci.* 1/2:41–50, 1999.

150. Mayer, C., D.L. Jakeman, M. Mah, G. Karjala, L. Gal, R.A.J. Warren, S.G. Withers. Directed evolution of new glycosynthases from *Agrobacterium* β-glucosidase: a general screen to detect enzymes for oligosaccharide synthesis. *Chem. Biol.* 8:437–443, 2001.

151. Singh, S., M. Michaela Scigelova, G. Vic, D.H.G. Crout. Glycosidase-catalysed oligosaccharide synthesis of di-, tri- and tetra-saccharides using the N-acetylhexosaminidase from *Aspergillus oryzae* and β-galactosidase from *Bacillus circulans*. *J. Chem. Soc. Perkin Trans.* 1(16):1921–1926, 1996.

152. Farkas, E., J. Thiem. Enzymatic synthesis of galactose-containing disaccharides employing β-galactosidase from *Bacillus circulans*. *Eur. J. Org. Chem.* 11:3073–3077, 1999.

153. Ichikawa, Y. Glycozymes: tools for oligosaccharide synthesis and targets for drug development. *Trends Glycosci. Glycotechnol.* 9:S47–S59, 1997.

20

Metabolic Engineering of Bacteria for Food Ingredients

Ramon Gonzalez

CONTENTS

20.1 Introduction 502
20.2 Amino Acids..503
    20.2.1 Modification of Central Metabolic Pathways 503
    20.2.2 Modification of Biosynthetic Pathways.......................................505
    20.2.3 Modification of Transport Systems..507
    20.2.4 Use of Analytical Tools in the ME Toolbox 507
    20.2.5 Use of LAB in the Production of Amino Acids............................508
20.3 Multifunctional Organic Acids...508
    20.3.1 Succinic Acid..508
    20.3.2 Pyruvic Acid...510
    20.3.3 Lactic Acid 511
    20.3.4 Acetic Acid 511
20.4 Vitamins..512
    20.4.1 Riboflavin (Vitamin B2) 513
    20.4.2 Folate (Vitamin B11)..513
    20.4.3 L-Ascorbic Acid (Vitamin C) 513
20.5 Carbohydrates...513
20.6 Bacteriocins..514
20.7 Low Calorie Sugars..515
20.8 Aroma Compounds 515
20.9 Future Trends and Potentials..516
References..517

20.1 INTRODUCTION

Metabolic engineering (ME; see Table 20.1 for a complete list of acronyms used in this chapter) has been defined as the directed improvement of product formation or cellular properties through the modification of specific biochemical reactions or the introduction of new ones with the use of recombinant DNA technology (1,2). Therefore, the *analysis* and *modification* of metabolic pathways is of central importance to ME. The *analytical* part uses experimental and modeling techniques [e.g., labeling experiments, Metabolic Flux Analysis (MFA)], which allow the systematic study of cellular responses (in terms of metabolic fluxes) to genetic and environmental perturbations. This facilitates a rational design of metabolic *modifications*, which are implemented using recombinant DNA technology. Both fields, the *analysis* and the *modification* of metabolic pathways, will be covered in this chapter.

Because of the central role of fermentation in food production and preparation, microorganisms are constantly encountered in these processes. ME can rationally improve the properties of these microorganisms and their efficiency for producing different products. Bacteria occupy a central place among these microorganisms, as either the bacteria *per se* (e.g., as starter cultures and probiotics) or products synthesized by bacteria (i.e., used as biocatalyst in the synthesis of amino acids, organic acids, vitamins, and carbohydrates). This illustrates the tremendous potential of ME for improving the bacteria used by the food industry.

Food additives produced by bacteria can be incorporated into foods as nutritional supplements, flavor enhancers, texturizers, acidulants, preservatives, emulsifiers, surfactants, thickeners, or functional food ingredients. A functional food (also referred to as a nutraceutical or pharmaceutical food) is defined as a food that can beneficially affect one

Table 20.1

Acronyms

Acronym	Definition
AcCoA	Acetyl Coenzyme A
AlaDH	Alanine Dehydrogenase
EPS	Exopolysaccharides
GDH	Glutamate Dehydrogenase
GRAS	Generally Regarded As Safe
LAB	Lactic Acid Bacteria
LAC	Lactic Acid
LDH	Lactate Dehydrogenase
ME	Metabolic Engineering
MFA	Metabolic Flux Analysis
PDH	Pyruvate Dehydrogenase
PEP	Phosphoenolpyruvate
PPC	phosphoenolpyruvate carboxylase
PPP	Pentose Phosphate Pathway
PTS	Phosphotransferase System
PYC	Pyruvate Carboxylase
PYR	Pyruvate
TCA	Tricarboxylic Acid
2,5-DKG	2,5-diketo-D-gluconic Acid
2-KLG	2-keto-L-gulonic Acid

or more targeted bodily functions beyond adequate nutritional effects in a way that improves health and well-being and reduces the risk of disease (3). Functional food ingredients include low calorie sugars, polysaccharides, and vitamins.

Applying ME principles and tools to the production of food ingredients by bacteria has resulted in the efficient production of both native and totally novel products by several cultures, including strains of lactic acid bacteria (LAB), *Escherichia coli*, *Bacillus subtilis*, and *Corynebacterium glutamicum*. These microorganisms have been selected based on a variety of criteria including, availability of genetic tools for cloning, expression, disruption, or replacement of genes (i.e., sequenced genome, promoters, plasmids and cloning vectors, gene transfer methods, and gene expression systems); capacity to metabolize a wide spectrum of carbon sources at high fluxes; rapid growth on inexpensive carbon and nitrogen sources; GRAS (generally regarded as safe) status; resistance to bacteriophage attack; well established physiological knowledge; efficient transport of the final product to the extracellular medium and tolerance to its high levels; and use in large-scale fermentations and production at an industrial level. The examples in this chapter (summarized in Table 20.2) will cover the analysis and modification of central metabolic pathways, biosynthetic pathways, and transport systems involved in producing amino acids, organic acids, vitamins, carbohydrates, bacteriocins, low calorie sugars, and aroma compounds. It is worth noting that many of these products are commercially produced.

20.2 AMINO ACIDS

Amino acids have a great variety of current and potential uses as food, pharmaceuticals, and animal feed. Their main application field is food, where about 50% of the product is applied (4). They can be used as nutritional supplements, flavor enhancers, sweeteners, and in pre- and postoperative nutrition therapy (5). This section will discuss using ME to produce some of these amino acids. Recent progress in this field can be divided into three categories (6): (1) modification of central metabolic pathways, (2) modification of biosynthetic pathways, and (3) modification of transport systems. Figures 20.1, 20.2, and 20.3 summarize some of the examples discussed below.

20.2.1 Modification of Central Metabolic Pathways

From reasoning based on metabolic pathways structure, rerouting a carbon source to produce a desired amino acid should start by increasing the availability of precursor metabolites, energy, and reducing equivalents used in its synthesis. Central metabolic pathways meet these criteria, and therefore engineering central metabolism is essential for the efficient production of amino acids. The analytical tools included in the ME toolbox have played an essential role

Table 20.2

Examples of bacteria that have been engineered for the production of food ingredients

Microorganism(s)	Product (Reference)
Corynebacterium glutamicum	L-tryptophan (16), L-phenylalanine (11,12), L-tyrosine (14), L-threonine (15), L-isoleucine (13), and L-histidine (10)
Lactic acid bacteria	L-alanine (21), Vitamins (50), Exopolysaccharides (57), Bacteriocins (61), and Aroma Compounds (67)
Bacillus subtilis	Vitamins (48) and Lactic Acid (44)
Escherichia coli	Succinate (23,24), Pyruvate (33), Lactate (41,42), and Acetate (45)

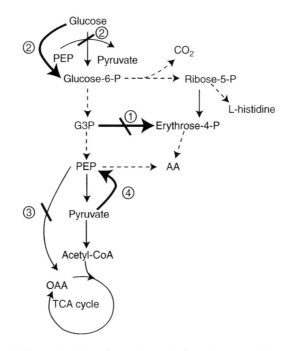

Figure 20.1 Metabolic engineering of central metabolic pathways to increase the synthesis of histidine and aromatic amino acids. Solid and dashed lines represent single and multiple steps, respectively. Solid bars over the arrows represent blocked enzymes, while thick arrows represent amplified enzymes. The following strategies are illustrated. Increasing the production of histidine by increasing the availability of Ribose-5-P, 1- Transketolase-deficient strains. Increasing the production of AA by increasing the supply of erythrose-4-P, 1- Overexpression of transketolases in AA producer. Increasing the production of AA by increasing the availability of PEP, 2- Inactivation of PEP-dependent PTS system for the transport of glucose and amplification of sugar-phosphorylating kinase gene; 3- Inactivation of PEP carboxylase; 4- Amplification of PEP synthase. Abbreviations, AA, aromatic amino acids; G3P, glyceraldehyde-3-P; and PEP, phosphoenolpyruvate.

in elucidating the function of different central pathways and suggesting useful strategies for redirecting carbon flow toward the biosynthesis of amino acids. For example, it has been shown that the pentose phosphate pathway (PPP) supports higher fluxes during the production of L-lysine compared to the production of L-glutamic acid in *C. glutamicum* (7,8). This was attributed to the higher requirements of reducing power (NADPH) in the production of L-lysine. Another example is improving aromatic amino acids and L-histidine production in *C. glutamicum* by increasing the availability of their precursor metabolites, erythrose 4-phosphate and ribose 5-phosphate, respectively, as well as NADPH. This can be done by modifying the flux through the PPP, either by increasing the activity of transketolase (and providing more erythrose 4-phosphate for aromatic amino acids biosynthesis) or by decreasing the activity of transketolase (and providing more ribose 5-phosphate for L-histidine biosynthesis) as shown in Figure 20.1, strategy 1 (6). Both approaches have produced *C. glutamicum* strains with an increased capacity for making aromatic amino acids (9) and L-histidine (10). Figure 20.1 shows additional examples of engineering central metabolic pathways to increase the availability of precursor metabolites used to synthesize aromatic amino acids in *C. glutamicum*. In general, these strategies are based on increasing the availability of erythrose 4-phosphate (strategy 1, Figure 20.1), phosphoenolpyruvate (strategies 2, 3, and 4, Figure 20.1), and ribose-5-P (strategy 1,

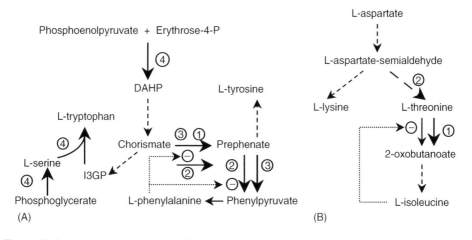

Figure 20.2 Metabolic engineering of biosynthetic pathways to increase the synthesis of aromatic amino acids, threonine, and isoleucine in *C. glutamicum*. See Figure 20.1 for explanation of different types of lines/arrows. In some cases, feedback inhibition has been represented using round dotted lines. Examples illustrated here include amplification of a gene encoding a rate-limiting step, introduction of a heterologous enzyme subjected to a different regulatory mechanism, and redirection of metabolic flux in a branch point (Ikeda (6)). (A) Increasing the synthesis of aromatic amino acids, 1 - Overexpression of chorismate mutase increased L-phenylalanine production; 2- Overexpression of mutated (insensitive to L-phenylalanine) chorismate mutase–prephenate dehydratase from *E. coli* increased the production of L-phenylalanine; 3- Simultaneous amplification of chorismate mutase and prephenate dehydratase resulted in increased production of L-tyrosine and L-phenylalanine; 4- Coexpression of two enzymes catalyzing the initial steps in the biosynthesis of aromatic amino acids (3-deoxy-D-arabino-heptulosonate 7-phosphate synthase) and L-serine (3-phosphoglycerate dehydrogenase) together with tryptophan-biosynthetic enzymes increased tryptophan production. (B) Increasing the synthesis of isoleucine and threonine, 1- Expression of L-isoleucine-insensitive *E. coli* threonine dehydratase (catabolic) enhanced isoleucine production; 2- Amplification of a threonine biosynthetic operon resulted in increased production of threonine in a lysine-producing *C. glutamicum* strain. Abbreviations, DAHP, 3-deoxy-D-arabino-heptulosonate-7-phosphate; I3GP, indole-3-glycerol-phosphate.

Figure 20.1) by inactivating the enzymes involved in their consumption and/or amplifying the enzymes involved in their production (for a review of these strategies, see Reference 6).

20.2.2 Modification of Biosynthetic Pathways

After engineering central metabolic pathways, a sufficient supply of precursor metabolites, energy, and reducing power is ensured, and efforts then need to be focused on engineering biosynthetic pathways that convert precursor metabolites into amino acids. Several strategies have been used to achieve this goal (6), and some of them are illustrated in Figure 20.2. For example, the gene that encodes a rate-limiting enzyme can be amplified, resulting in the release of a bottleneck. Ozaki et al. (11) and Ikeda et al. (12) used this strategy to improve the production of L-phenylalanine in *C. glutamicum*. Overexpression of the gene that encodes chorismate mutase in *C. glutamicum* K38 resulted in a 50% increase in the yield of L-phenylalanine (11). On the other hand, introducing heterologous enzymes subject to different regulatory mechanisms can also result in the release of a bottleneck. For example, overexpressing a mutated (insensitive to L-phenylalanine) *E. coli* gene that encoded the bifunctional enzyme chorismate mutase–prephenate dehydratase in *C. glutamicum* KY10694 led to a 35% increase in the production of L-phenylalanine (12). In addition, expressing the *E. coli* catabolic threonine dehydratase (insensitive to L-isoleucine) in *C. glutamicum*

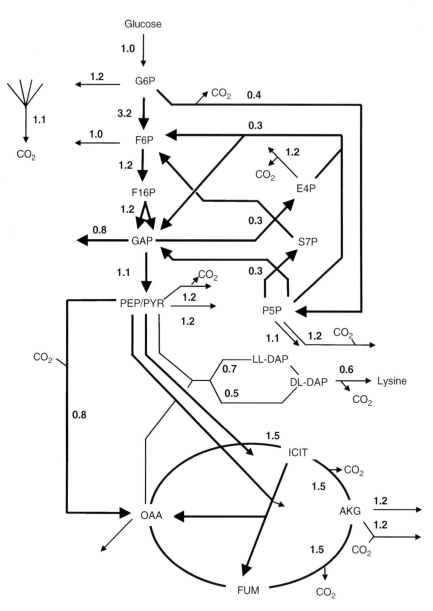

Figure 20.3 Labeling experiments-based MFA of two isogenic glutamate dehydrogenase mutants (homologous, NADPH-dependent, and heterologous, NADH-dependent) of the lysine producer strain *C. glutamicum* MH20-22B. Flux values were converted to flux ratios and expressed as (NADH-dependent mutant)/(NADPH-dependent mutant). Numbers near the thick lines give the estimated net fluxes while those near the thin arrows give the measured fluxes required for biomass synthesis. Adapted from Marx et al. (20). Abbreviations, AKG, α-ketoglutarate; DL-DAP, DL-diaminopimelate; E4P, erythrose-4-P; FUM, fumarate; F16P, fructose-1-6-bisphosphate; F6P, fructose-6-phosphate; GAP, glyceraldehyde-3-phosphate; G6P, glucose-6-phosphate; ICIT, isocitrate; LL-DAP, LL-diaminopimelate; OAA, oxaloacetate; PEP, phosphoenolpyruvate; PYR, pyruvate; P5P, pentose-5-phosphate; S7P, sedoheptulose-7-phosphate.

resulted in increased isoleucine production (13). A second strategy could be the redirection of metabolic flux in a branch point. Ikeda and Katsumata (14) engineered a tryptophan-producing mutant of *C. glutamicum* to produce L-tyrosine or L-phenylalanine in abundance

(26 and 28 g/L, respectively) by overexpressing the branch-point enzymes (chorismate mutase and prephenate dehydratase), catalyzing the conversion of the common intermediate chorismate into tyrosine and phenylalanine. Using a similar approach, Katsumata et al. (15) produced threonine using a lysine-producing *C. glutamicum* strain, by amplifying a threonine biosynthetic operon. Other strategies could include introducing heterologous enzymes that use different cofactors than those used by the native enzyme as well as amplifying the enzyme that catalyzes the steps linking central metabolism and the biosynthetic pathway (6). Ikeda et al. (16) achieved a 61% increase in tryptophan yield (50 g/L of tryptophan) in a tryptophan-producing *C. glutamicum* KY10894 by coexpressing two enzymes catalyzing the initial steps in the biosynthesis of aromatic amino acids (3-deoxy-D-arabino-heptulosonate 7-phosphate synthase) and serine (3-phosphoglycerate dehydrogenase) together with tryptophan biosynthetic enzymes.

20.2.3 Modification of Transport Systems

By altering amino acid transport systems, one could expect to decrease their intracellular concentration and avoid feedback inhibition. The following two examples illustrate that show the significance of this strategy. During the production of L-tryptophan by *C. glutamicum*, the accumulation of this product in the extracellular medium resulted in a backflow into the cells, which produced severe feedback inhibition in the biosynthesis of tryptophan (6). Ikeda and Katsumata (17) solved this problem by creating mutants with lower levels of tryptophan uptake, which resulted in an accumulation of 10 to 20% more tryptophan than in their parent. Another example of transport engineering is the increase in the fermentation yield of cysteine by overexpressing multidrug efflux genes (*mar* genes) in a cysteine-producing strain of *E. coli* (6).

20.2.4 Use of Analytical Tools in the ME Toolbox

In this section, I will give some examples of using MFA [See reference (2) for a detailed description of MFA] to improve the production of specific amino acids, including the two amino acids with the largest production volume worldwide (L-glutamic acid with 800,000 tons/year and L-lysine with 600,000 tons/year).

Glutamate. As with other amino acids, analysis and modification of central metabolic and biosynthetic pathways in glutamate-producing *C. glutamicum* strains have contributed to their improvement. MFA, including estimation of fluxes using labeling experiments, has elucidated the relative contribution of different pathways (e.g., Embden-Meyerhof and hexose monophosphate) under various physiological conditions and genetic backgrounds [(18) and references therein]. For example, MFA has established the relationship between the decline of oxoglutarate dehydrogenase complex and the flux distribution at the metabolic branch point of 2-oxoglutarate glutamate. This suggests that metabolic flux through anaplerotic pathways could be limiting the production of glutamate. Many of these findings have either been verified by genetic approaches or have led to a rational modification of metabolic pathways to improve the production of glutamate [(18) and references therein].

L-lysine. L-lysine production is another example of the application of labeling experiments-based MFA. Very comprehensive approaches have been used to assess all major fluxes in the central metabolism of *C. glutamicum*, which reveal patterns that can be used for designing ME strategies. In a comparison of six metabolic patterns, several metabolic fluxes that depend significantly on the physiological state of the cells were identified [(19) and references therein]. These included the coordinated flux through PPP and the tricarboxylic acid (TCA) cycle, the high capacity for the reoxidation of NADPH, and futile cycles between C3-compounds of glycolysis and C4-compounds of the TCA cycle. As an example, Figure 20.3 shows a comparison of the metabolic flux distribution between two isogenic glutamate dehydrogenase mutants (homologous, NADPH-dependent, and heterologous, NADH-dependent) of the L-lysine producer strain *C. glutamicum* MH20-22B (20). Metabolic flux patterns revealed that the PPP flux was high only for a high demand of NADPH and a low TCA cycle flux. The heterologous, NADH-dependent, glutamate dehydrogenase mutant required more NADH, resulting in an increased TCA cycle flux and then more NADPH supplied by the isocitrate dehydrogenase step in the TCA cycle. This led to a decreased flux of the PPP due to lower NADPH requirements (TCA cycle already produced part of it). The inverse is true for the homologous, NADPH-dependent, glutamate dehydrogenase mutant. There is a higher NADPH requirement, the PPP flux is higher, and the TCA cycle flux is lower.

20.2.5 Use of LAB in the Production of Amino Acids

LAB also has been used to produce different amino acids. *Lactococcus lactis* (21) has been engineered to produce L-alanine as the only end product of fermentation (more than 99%). Rerouting the carbon flux toward alanine was achieved by expressing *Bacillus sphaericus* alanine dehydrogenase (AlaDH) in lactate dehydrogenase (LDH) deficient strains. Finally, stereospecific production (> 99%) of L-alanine was achieved by disrupting the gene encoding alanine racemase.

20.3 MULTIFUNCTIONAL ORGANIC ACIDS

The ability to produce organic acids via fermentation is of great interest to the food industry due to the acids' widespread use as acidulants, food preservatives, beverage ingredients, sweeteners, and flavor enhancers.

Many bacteria, such as *E. coli*, carry out mixed acid fermentation of sugars (e.g., glucose) in which the principal products are formate (or CO_2 and H_2), acetate, lactate, succinate, and ethanol (Figure 20.4) (22). Under these conditions, anaerobic fermentative pathways have two main functions: to produce energy in an anaerobic environment via substrate level phosphorylation, and to provide a source of regeneration of NAD+, closing the redox balance. The aforementioned products can be organized in three groups according to the redox properties of the homofermentative pathway used in their synthesis: (1) net regeneration of redox equivalent by the homofermentative pathway (e.g., lactate, fumarate, and malate), (2) net production of redox equivalents by the homofermentative pathway (e. g., acetate and pyruvate), and (3) net consumption of redox equivalent by the homofermentative pathway (e.g., succinate).

In this section, I will present some examples of ME of *E. coli* for the homofermentative production of some of these organic acids, including succinate, pyruvate, lactate, and acetate. In general, all strategies have been focused on altering the carbon flux ratio at three branch points, phosphoenolpyruvate (PEP), pyruvate (PYR), and acetyl coenzyme A (AcCoA).

20.3.1 Succinic Acid

The production of succinate by several bacterial strains, including the obligate anaerobe *Anaerobiospirullum succiniciproducens*, *Actinobacillus* sp., and *Enterococcus* sp. have been reported. However, ME efforts have focused on engineering *E. coli* to produce succinate as the main fermentation product. In what follows I will summarize those results, which are also schematically represented in Figure 20.20.

E. coli wild-type strains produce succinate only as a minor product of fermentation. Genetic manipulations are required to increase succinate production and reduce byproduct formation. For example, increasing flux at the first step in the succinate branch by

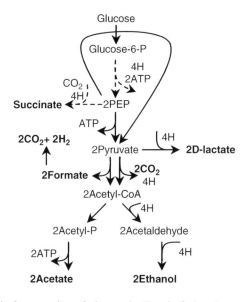

Figure 20.4 Anaerobic fermentation of glucose in *E. coli*. Only relevant central metabolic pathways have been included and it has been considered that glucose is transported into the cell only by the PEP-dependent PTS. Not all steps and metabolites are shown. Solid and broken lines represent single and multiple steps, respectively. Primary fermentation products are indicated in bold. Abbreviations, PEP, phosphoenolpyruvate.

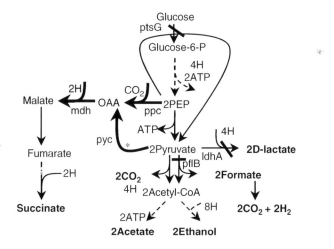

Figure 20.5 Metabolic engineering of *E. coli* to produce succinate. See Figure 20.1 for explanation of different types of lines/arrows. Genes involved in engineered steps are indicated by italics. Primary fermentation products are indicated in bold. Strategies include the blocking (genes *ldhA*, *pflB*, and *ptsG*) and amplification (genes *ppc* and *mdh*) of homologous enzymes as well as the introduction of heterologous enzymes (*R. etli pyc* gene). The asterisk (*) indicates that this step does not exist in *E. coli*. Abbreviations, *ldhA*, gene encoding D-lactate dehydrogenase; *mdh*, gene encoding malate dehydrogenase; PEP, phosphoenolpyruvate; *pflB*, gene encoding pyruvate-formate lyase; *ppc*, gene encoding PEP carboxylase; *ptsG*, gene encoding the enzyme IICB^{Glc} of the PTS system; *pyc*, gene encoding *R. etli* pyruvate carboxylase.

overexpressing PEP carboxylase results in an increase in succinate percentage yield (molar basis) from 12 to 45% (23). PEP is also a required cosubstrate for glucose transport via the phosphotransferase system (PTS) in wild-type *E. coli*. Thus, another approach is to direct pyruvate to the succinate branch. This is achieved by transforming a wild-type *E. coli* strain with plasmid pTrc99A-*pyc*, which expresses *Rhizobium etli* pyruvate carboxylase. This strategy results in an increase in both succinate percentage yield (17%) and productivity (0.17 g/L/h) (24). In order to prevent the accumulation of other undesired products, mutations in genes involved in producing lactate (*ldhA*, encoding lactate dehydrogenase) and formate (*pflB*, encoding pyruvate–formate lyase) were introduced, obtaining the strain *E. coli* NZN111, that grew poorly on glucose under anaerobic conditions. When the gene encoding malic enzyme from *Ascaris suum* was transformed into NZN111, succinate percentage yield and productivity were further increased to 39% and 0.29 g/L/h, respectively (25,26). Expression of *mdh* gene encoding malate dehydrogenase also resulted in improved production of succinate by *E. coli* NZN111 (27).

Donnelly et al. (28) reported an unknown spontaneous chromosomal mutation in NZN111, which permitted anaerobic growth on glucose, and this strain was designated as AFP111. When AFP111 was grown anaerobically under 5% H_2 and 95% CO_2, a succinate percentage yield of 70% and a succinate–acetate molar ratio of 1.97 were obtained. Further improvements of succinate production with this strain included dual phase fermentation (aerobic growth for biomass generation follow by anaerobic growth for succinate production) that resulted in a succinate percentage yield of 99% and a productivity of 0.87 g/L/h (29). AFP111 mutation was mapped to the *ptsG* gene, encoding an enzyme of the PTS (30).

Gokarn et al. (31) found that the expression of *R. etli* pyruvate carboxylase (PYC) in *E. coli* during anaerobic glucose metabolism caused a 2.7-fold increase in succinate concentration, making it the major product by mass. The increase came mainly at the expense of lactate formation. However, in a mutant lacking lactate dehydrogenase activity, expression of PYC resulted in only a 1.7-fold increase in succinate concentration. An accumulation of pyruvate and NADH, metabolites that affect the interconversion of the active and inactive forms of the enzyme pyruvate formate-lyase, may have caused the decreased enhancement of succinate. The same group (32) had previously shown that the presence of the *R. etli pyc* gene in *E. coli* (JCL1242/pTrc99A-pyc) restored the succinate producing ability of *E. coli ppc* null mutants (JCL1242), with PYC competing favorably with both pyruvate formate lyase and lactate dehydrogenase. Flux calculations indicated that during anaerobic metabolism the *pyc*(+) strain had a 34% greater specific glucose consumption rate, a 37% greater specific rate of ATP formation, and a 6% greater specific growth rate than the *ppc*(+) strain. The results demonstrate that when phosphoenolpyruvate carboxylase (PPC) or PYC are expressed, the metabolic network adapts by altering the flux to lactate and the molar ratio of ethanol to acetate formation.

20.3.2 Pyruvic Acid

Several bacterial strains have been used to produce pyruvic acid via the fermentation of different sugars. These include strains of *Escherichia*, *Pseudomonas*, *Enterococcus*, *Acinetobacter*, and *Corynebacterium* (33). The most successful processes (highest yield and concentrations, and shortest fermentation times) are those involving *E. coli* strains. The main strategy used has been limiting pyruvate consumption by pyruvate dehydrogenase complex (PDH) under aerobic conditions. Considering that lipoic acid is a cofactor of PDH, a screen by Yokota et al. (34) identified a lipoic acid auxotroph, *E. coli* W1485*lip2*, which accumulated 25.5 g/L pyruvate from 50 g/L glucose at 32 hours. Pyruvate production was further improved by introducing a mutation in F_1-ATPase, obtaining strain TBLA-1 (35). Strain TBLA-1 exhibited a higher capacity for producing pyruvate, more than 30 g/L of pyruvate were produced from 50 g/L of glucose in only 24 hours. The growth of TBLA-1 decreased to 67% of that in the parent strain, due to lower energy production (TBLA-1 is an F_1-ATPase-defective mutant). The enhanced pyruvate productivity in strain TBLA-1 was thought to be linked to increased activities of some glycolytic enzymes (36). Although there was an increase in glycolytic flux due to a decrease in ATP production (i.e., an F_1-ATPase mutation), the physiological mechanism mediating these changes was not identified. Recently, it was shown that the glycolytic flux in *E. coli* is controlled by the demand for ATP (37). By increasing the ATP hydrolysis, the glycolytic flux was increased by approximately 70%, indicating that glycolytic flux is mainly ($> 75\%$) controlled by reactions hydrolyzing ATP. In light of these results, it is clear that increased pyruvate production in strain TBLA-1 when compared to its wild type *E. coli* W1485*lip2* is due to lower ATP production in TBLA-1, which in turn is driving glycolysis and resulting in higher glycolytic fluxes.

20.3.3 Lactic Acid

Many microorganisms produce D-lactic acid (D-LAC), and some LABs, such as *Lactobacillus bulgaricus*, produce highly pure D-LAC (38). L-LAC has also been produced using other LABs, such as *Lactobacillus helveticus*, *Lactobacillus amylophilus*, and *Lactobacillus delbruekii* (39). Since LABs have complex nutritional requirements and low growth rates, and exhibit incomplete or negligible pentose utilization (40), other bacterial strains have been engineered to produce optically pure D- or L-LAC including *E. coli* (41,42), *Rhizopus orizae* (43), and *B. subtilis* (44). Chang et al. (41) engineered *E. coli* to produce optically pure D- or L-LAC. A *pta* mutant of *E. coli* RR1, which was deficient in the phosphotransacetylase of the Pta-AckA pathway, was found to metabolize glucose to D-LAC and to produce a small amount of succinate byproduct under anaerobic conditions. An additional mutation in *ppc* (encoding PPC) made the mutant produce D-LAC like a homofermentative LAB. In order to produce L-LAC, a nonindigenous fermentation product, an L-lactate dehydrogenase gene from *Lactobacillus casei* was introduced into a *pta ldhA* strain, which lacked phosphotransacetylase and D-LAC dehydrogenase. This recombinant strain was able to metabolize glucose to L-LAC as the major fermentation product, and produced up to 45 g/L of L-LAC. Zhou et al. (42) constructed derivatives of *E. coli* W3110 capable of producing D-LAC in a mineral salts medium. They eliminated competing pathways by chromosomal inactivation of genes encoding fumarate reductase (*frdABCD*), alcohol/aldehyde dehydrogenase (*adhE*), and pyruvate formate lyase (*pflB*). D-LAC production by these new strains approached the theoretical maximum yield of two molecules per glucose molecule and a chemical purity of D-LAC of ~98% with respect to soluble organic compounds. The cell yield and LAC productivity were increased by a further mutation in the acetate kinase gene (*ackA*). The aforementioned ME strategies used to engineer homolactic pathways in *E. coli* are summarized in Figure 20.6.

20.3.4 Acetic Acid

Acetic acid obtained through fermentation is mainly used in the food market (vinegar, meat preservative). Microorganisms currently used in its production are *Saccharomyces cerevisiae*, *Acetobacter aceti*, and *Clostridia* species. *E. coli* W3110 was recently engineered to produce acetic acid from glucose (45). The resulting strain (TC36) converted 60 g/L of glucose into 34 g/L of acetate in 18 h. Strain TC36 was constructed by sequentially assembling deletions that inactivated oxidative phosphorylation ($\Delta atpFH$), disrupted the cyclic function of the tricarboxylic acid pathway ($\Delta sucA$), and eliminated native fermentation pathways ($\Delta focA$-*pflB*, $\Delta frdBC$, $\Delta ldhA$, and $\Delta adhE$). These mutations minimized the loss of substrate carbon and the oxygen requirement for redox balance. Although TC36 produces only four ATPs per glucose, this strain grows well in a mineral salts medium and

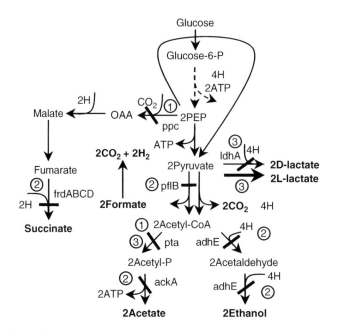

Figure 20.6 Metabolic engineering of *E. coli* to produce D- and L-lactate. See Figure 20.1 for explanation of different types of lines/arrows. Genes involved in engineered steps are indicated by italics. Primary fermentation products are indicated in bold. 1- Engineering a homo-D-lactic pathway by introducing mutations in *pta* and *ppc* genes; 2- Engineering a homo-D-lactic pathway by introducing mutations in *frdBC*, *ackA*, *adhE*, and *pflB* genes; 3- Engineering a homo-L-lactic pathway by introducing mutations in *pta* and *ldhA* genes and expressing an L-lactate dehydrogenase gene from *L. casei*. Abbreviations, *adhE*, gene encoding acetaldehyde/alcohol dehydrogenase; *ackA*, gene encoding acetate kinase; *frdABCD*, operon encoding fumarate reductase; *ldhA*, gene encoding lactate dehydrogenase; PEP, phosphoenolpyruvate; *pflB*, gene encoding pyruvate–formate lyase; *ppc*, gene encoding PEP carboxylase; *pta*, gene encoding phosphotransacetylase.

has no auxotrophic requirement. Glycolytic flux in TC36 (0.3 μmol·min^{-1}·mg^{-1} protein) was twice that of the parent. Higher flux was attributed to a deletion of membrane-coupling subunits in $(F_1F_0)H^+$-ATP synthase that inactivated ATP synthesis while retaining cytoplasmic F_1-ATPase activity. The oxygen requirements of this system are energy intensive, so in terms of commercial production of acetate it would be highly desirable to have an anaerobic system able to efficiently produce acetic acid. Our laboratory is currently engineering *E. coli* to produce acetate as the main fermentation product under anaerobic conditions (data not published). The general strategy is based on first eliminating competing pathways (by inactivating correspondent genes) in order to redirect the carbon flux toward the synthesis of acetate, and then transforming the engineered pathway into a viable alternative for cell growth (by introducing modifications that will balance consumption and production of NADH).

20.4 VITAMINS

Vitamins are nutraceuticals or functional foods. They serve as the cofactors of many of the enzymes involved in several metabolic reactions, and so are essential components in the human diet. In this section I will give some examples of using ME to produce three vitamins (riboflavin, folate, and ascorbic acid) in bacterial strains.

20.4.1 Riboflavin (Vitamin B2)

Both *B. subtilis* and *Lac. lactics* have been engineered to produce riboflavin. In *B. subtilis* overexpression of the *rib* genes (involved in the synthesis of riboflavin), which are organized in a cluster (46), is achieved by replacing the two regulated promoters with constitutive ones derived from a phage (47). Although multiple copies of this four gene construction were inserted at two different sites in the genome, a separate overexpression of *ribA* was necessary to reach maximum productivity (48). The *ribA* gene encodes a bifunctional protein with dihydroxy–butanone phosphate synthetase activity at the N-terminal, and GTP cyclohydrolase II activity at the C-terminal. This showed that the initial steps of riboflavin synthesis limited productivity in the last published stage of strain improvement. A metabolic analysis of the wild-type and riboflavin-producing *B. subtilis* strains found that the flux of the oxidative branch of the pentose phosphate pathway was increased in the production strain (49). Therefore, this branch may become rate limiting and new ME strategies may be needed for future strain improvement. *Lac. lactis* is also being engineered to produce riboflavin by overexpressing the gene (*ribA*) encoding the enzyme (GTP cyclohydrolase) catalyzing the first reaction for its synthesis from GTP. This enzyme had been previously reported to limit flux (48) and overexpression of *ribA* resulted in a three-fold increase in riboflavin production (50).

20.4.2 Folate (Vitamin B11)

Lac. lactis has been engineered to produce folate (50). Strategies include the expression of several genes involved in the biosynthesis of folate from GTP, either individually or in combination, and the expression of heterologous γ-glutamyl hydrolase (from humans and rats). These approaches have resulted in three- to six-fold increases in the production of folate and an alteration in the folate spatial distribution (i.e., a shift from mainly intracellular to extracellular accumulation).

20.4.3 L-Ascorbic Acid (Vitamin C)

Almost all biological processes for synthesizing L-ascorbic acid end with the synthesis of 2-keto-L-gulonic acid (2-KLG), which is later converted into ascorbic acid by conventional chemical processing technology (i.e., esterification and lactonization). Therefore, efforts have been focused on engineering strains capable of efficiently producing 2-KLG from different sugars. Two strains of *Gluconobacter oxydans* were engineered to convert sorbitol into 2-KLG. Genes encoding the enzymes sorbitol dehydrogenase and sorbose dehydrogenase were cloned from *G. oxydans* T-100 into *G. oxydans* G624, allowing the production of 2-KLG in three steps (51). An *Erwinia herbicola* has been engineered to produce 2-KLG from glucose in a single fermentation step (52). *E. herbicola* naturally produces 2,5-diketo-D-gluconic acid (2,5-DKG) via glucose oxidation, but lacks the enzyme 2,5-DKG reductase, which can convert 2,5-DKG into 2-KLG. Therefore, *E. herbicola* was engineered by expressing the gene encoding 2,5-DKG reductase from *Corynebacterium* sp., resulting in a recombinant strain capable of converting glucose into 2-KLG in a single fermentation step. This process has been further simplified and 2-KLG concentrations > 120 g/L have been achieved (53).

20.5 CARBOHYDRATES

Several nondigestible polysaccharides have a positive effect on the development of beneficial intestinal microflora, and therefore are known by their function as prebiotics (i.e., nondigestible food ingredients that provide specific nutrients to beneficial bacteria hosted in the colon). Among them are fructo-oligosacharides, galacto-oligosacharides, gluco-oligosacharides, transgalacto-oligosacharides, isomalto-oligosacharides and xylo-oligosacharides (50). LABs are a good source of nondigestible polysaccharides, mainly exopolysacharides (EPS), which are also recognized by their contribution to the texture, mouth feel, taste perception, and stability of the final food product (54). Based on the available genetic information on genes required for EPSs biosynthesis in LABs, heterologous production of EPSs has been achieved in a LAB strain that did not have the capacity to produce them (55). A similar experiment has shown that EPSs with different compositions can be produced from expression of the same cluster of genes due to either selectivity in the export and polymerization system or limitation in the supply of precursors (56). Van Kranenburg et al. (57) have shown that EPS production in *Lac. lactis* can be increased by overexpressing the *epsD* gene, encoding a priming glucosyltransferase. I will give two other examples of using ME strategies to improve EPS production and modify EPS composition in *Lac. lactis*. Looijesteijn et al. (58) showed that overexpression of the *fbp* gene encoding fructose bisphosphatase resulted in increased *Lac. lactis* growth and EPS synthesis using fructose as a sugar source. In addition, Boels et al. (59) studied the UDP-glucose pyrophosphorylase (*galU*) and UDP-galactose epimerase (*galE*) genes of *Lac. lactis* MG1363 to investigate their involvement in the biosynthesis of UDP-glucose and UDP-galactose, which are precursors of glucose- and galactose-containing EPS in *Lac. lactis*. Overexpression of both genes increased the intracellular pools of UDP-glucose and UDP-galactose. These examples clearly illustrate the role ME plays (and will play in the future) in the bacterial production of polysaccharides.

20.6 BACTERIOCINS

One of the most important properties of a starter culture is its capacity to synthesize antimicrobial metabolites, including bacteriocins, which, in turn, can help improve the quality of fermented products (i.e., control of pathogens, extension of shelf life, and improvement of sensory qualities). For example, ability of LAB to act as a preservative in foods is partially attributed to the production of bacteriocins [for a review of bacteriocins produced by LAB with potentials as biopreservatives in foods see O'Sullivan et al. (60)]. Bacteriocins are a diverse group of ribosomally synthesized antimicrobial proteins or peptides, some of which undergo posttranslational modifications (Rodriguez et al. (61)). Bacteriocins can be incorporated into food after they are produced in an individual process or by using live cultures that produce bacteriocins *in situ* in the food. The second option seems to be more cost effective and acceptable (i.e., most strains have a GRAS status). A potential problem associated with the use of bacteriocins in food preservation is the development of resistant populations of pathogenic bacteria. ME plays a central role in solving this problem by enabling the creation of multiple bacteriocin producers (62,63). Horn et al. (64) took advantage of the similarities in the production and secretory systems of pediocin PA-1, from *Pediococcus acidilactici*, and lactococcin A, from *Lac. lactis* bv. diacetylactis WM4. By using the lactococcin A secretory system, they engineered a lactoccocus strain for expressing and secreting pediocin PA-1. This heterologous expression system has been used to coproduce enterocin-A and pediocin PA-1 (62), and pediocin PA-1 and nisin (63). Recently, O'Sullivan et al. (60) have examined the coproduction of lacticin 3147 and lacticin 481. Rodriguez et al. (61) recently published an excellent review of the heterologous production of bacteriocins by LAB. In the future, ME analytical tools (e.g., MFA) could play an important role in improving the heterologous production of bacteriocin in bacteria.

This assertion is based on its contribution to the study of the synthesis of heterologous proteins in yeast (65).

20.7 LOW CALORIE SUGARS

Low calorie sugars, like mannitol, sorbitol, tagatose, and trehalose, contribute to weight loss, which is one reason there is such a high consumer demand for them. Tagatose is a low calorie sweetener that is poorly degraded by the human body. It is considered both a prebiotic and an antiplaque agent (www.tagatose.dk). Recently, *Lac. lactis* has been engineered to produce tagatose by disrupting the *lacC* and/or *lacD* genes, which resulted in the production of either tagatose 6-phosphate or tagatose 1,6-diphosphate (66). Further genetic modifications resulted in the production of tagatose 1,6-diphosphate as the sole end product. Current efforts are focused on the dephosphorylation of tagatose 1,6-diphosphate and excretion of the final product, tagatose.

20.8 AROMA COMPOUNDS

Diacetyl is an important flavor compound in many dairy products like butter, buttermilk, and cheeses (it provides a buttery flavor). *Lac. lactis* has been engineered to synthesize diacetyl as the main fermentation product (67). Efficient diacetyl production resulted from inactivation of the *aldB* gene (encoding α-acetolactate decarboxylase) and overproduction of NADH-oxidase activity as illustrated in Figure 20.7. This approach resulted in the conversion of 80% of the carbon source into diacetyl. Another important group of aroma compounds is obtained from enzymatic degradation of amino acids including isovalerate, isobutyrate, phenylacetate, phenylacetaldehyde, and indole. Rijnen et al. (68) have shown that overexpression of a heterologous glutamate dehydrogenase (GDH) gene in *Lac. lactis*

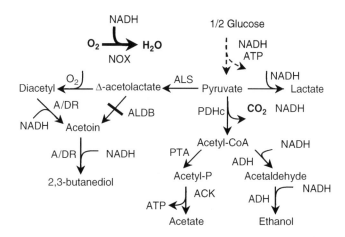

Figure 20.7   Metabolic engineering of *Lac. lactis* to produce diacetyl (ALDB-deficient and NOX overexpressed). Adapted from Hugenholtz et al. (67). See Figure 20.1 for explanation of different types of lines/arrows. Enzymes involved in reactions of interest are indicated. Abbreviations, ACK, acetate kinase; ADH, aldehyde/alcohol dehydrogenase; A/DR, acetoin/diacetyl reductase; ALDB, α-acetolactate decarboxylase; ALS, α-acetolactate synthase; NOX, NADH oxidase; PDHc, pyruvate dehydrogenase complex; PTA, phosphotransacetylase.

increased the conversion of amino acids into aroma compounds. Therefore, instead of adding α-ketoglutarate to cheese, a GDH-producing lactococcal strain could be used to enhance amino acid degradation in aroma compounds.

20.9 FUTURE TRENDS AND POTENTIALS

In the foreseeable future, ME will continue to play a central role in the use of bacteria to produce food ingredients. One potential area for the application of ME is the removal of undesirable sugars present in many foods. A typical example is lactose, which is present in liquid dairy products. Many people are lactose-intolerant and cannot consume products in which lactose is present. Engineering LAB to obtain starter cultures with elevated β-galactosidase activities has been proposed as a way to eliminate lactose (50). Raffinose is another example. Due to the absence of α-galactosidase, it cannot be degraded intestinally by humans. Consuming it causes intestinal disturbances. Engineering LAB to produce high levels of α-galactosidase and using it as starter cultures for removing raffinose has been proposed as a good solution to this problem (50). As a first step, the gene *melA* encoding α-galactosidase from *Lactobacillus plantarum* has been cloned (69) and is being expressed in *Lac. lactis* (50).

Energy metabolism is beginning to attract attention as a new target for ME, especially for engineering central metabolic pathways and transport systems. It was recently shown that an increase in ATP hydrolysis results in an increase in the glycolytic flux of *E. coli* (~70%), indicating that glycolytic flux is mainly (>75%) controlled by reactions hydrolyzing ATP (37). Therefore, one could engineer energy metabolism in order to improve the synthesis of products, such as organic acids, for which the glycolytic flux is a determinant (i.e., the introduction of an independent ATP "sink" would increase the glycolytic flux and the synthesis of the organic acid). One could even think that a similar approach will also be valid for synthesizing secondary metabolites whose precursors are provided by the glycolytic pathway. A second approach points toward a completely different direction, the creation of an energy surplus (in contrast to the energy "sink" created in the previous approach). For example, let us consider that the excretion of a certain product P (e.g., an amino acid) is an energy-requiring process. If a condition of energy surplus is created, cells will use any energy consuming process to dissipate the excess of energy, and therefore the secretion of product P will be stimulated.

The existence of sequenced genomes for several bacteria involved in producing food ingredients (e.g., *Lac. lactis, C. glutamicum, B. subtilis, E. coli*) will dramatically impact ME opportunities. Once genome sequences are available, technologies such as cDNA microarrays can be used to simultaneously monitor and study the expression levels of individual genes at genomic scale (known as the field of transcriptomics). This technology has already played an important role in understanding bacterial metabolism under a great variety of conditions, and in the improvement of numerous bacterial strains (70–74). However, the main contributions from the area of functional genomics are still to come and will result from the combined use of technologies that permit large-scale study of cell functioning and its interaction with the environment (i.e., combined analysis using genomic (DNA), transcriptomic (RNA), proteomic (protein), metabolomic (metabolites), and fluxomic (fluxes/enzymes) information). The era of functional genomics will contribute to an improved understanding of the molecular mechanisms underlying the relationships between microorganism and environment and in that way will establish the link between genotype and phenotype. Thus, in the future, functional genomics is expected to play a large role in the development of bacterial strains for the food industry.

REFERENCES

1. Bailey, J.E. Toward a science of metabolic engineering. *Science* 252:1668–1675, 1991.
2. Stephanopoulos, G., A.A. Aristidou, J. Nielsen. *Metabolic Engineering: Principles and Methodologies*, 1st ed. San Diego, CA: Academic Press, 1998.
3. Mollet, B., I. Rowland. Functional foods: at the frontier between food and pharma. *Curr. Opin. Biotech.* 13:483–485, 2002.
4. Mueller, U., S. Huebner. Economic aspects of amino acids production. *Adv. Biochem. Engin./Biotechnol.* 79:137–170, 2003.
5. Leuchtenberger, W. Amino acids: technical production and use. In: *Biotechnology*, Rehm, H.J., G. Reed, eds., Weinheim, Germany: VCH, 1996, pp 465–502.
6. Ikeda, M. Amino acid production processes. *Adv. Biochem. Engin./Biotechnol.* 79:1–35, 2003.
7. Vallino, J.J., G. Stephanopoulos. Metabolic flux distributions in *Corynebacterium glutamicum* during growth and lysine overproduction. *Biotechnol. Bioeng.* 41:633–646, 1993.
8. Marx, A., A.A. de Graaf, W. Wiechert, L. Eggeling, H. Sahm. Determination of the fluxes in the central metabolism of *Corynebacterium glutamicum* by nuclear magnetic resonance spectroscopy combined with metabolite balancing. *Biotechnol. Bioeng.* 49:111–129, 1996.
9. Ikeda, M., R. Katsumata. Hyperproduction of tryptophan by *Corynebacterium glutamicum* with the modified pentose phosphate pathway. *Appl. Environ. Microbiol.* 65:2497–2502, 1999.
10. Ikeda, M., K. Okamoto, T. Nakano, N. Kamada. Production of metabolite biosynthesized through phosphoribosylpyrophosphoric acid. Japan Patent 12,014,396 A, 2000.
11. Ozaki, A., R. Katsumata, T. Oka, A. Furuya. Cloning of the genes concerned in phenylalanine biosynthesis in *Corynebacterium glutamicum* and its application to breeding of a phenylalanine producing strain. *Agric. Biol. Chem.* 49:2925–2930, 1985.
12. Ikeda, M., A. Ozaki, R. Katsumata. Phenylalanine production by metabolically engineered *Corynebacterium glutamicum* with the pheA gene of *Escherichia coli*. *App. Microb. Biotechnol.* 39:18–323, 1993.
13. Guillouet, S., A.A. Rodal, G. An, P.A. Lessard, A.J. Sinskey. Expression of the *Escherichia coli* catabolic threonine dehydratase in *Corynebacterium glutamicum* and its effect on isoleucine production. *Appl. Environ. Microbiol.* 65:3100–3107, 1999.
14. Ikeda, M., R. Katsumata. Metabolic engineering to produce tyrosine or phenylalanine in a tryptophan producing *Corynebacterium glutamicum* strain. *App. Environ. Microbiol.* 58:781–785, 1992.
15. Katsumata, R., T. Mizukami, Y. Kikuchi, K. Kino. Threonine production by the lysine producing strain of *Corynebacterium glutamicum* with amplified threonine biosynthetic operon. In: *Genetics of Industrial Microorganisms. Proceedings of the Fifth International Symposium on the Genetics of Industrial Microorganisms*. Alacevic, M., D. Hranueli, Z. Toman, eds., Karlovac, Yugoslavia: Ognjen Prica Printing Works, 1987, pp 217–250.
16. Ikeda, M., K. Nakanishi, K. Kino, R Katsumata. Fermentative production of tryptophan by a stable recombinant strain of *Corynebacterium glutamicum* with a modified serine-biosynthethic pathway. *Biosci. Biotechnol. Biochem.* 58:674–678, 1994.
17. Ikeda, M., R. Katsumata. Tryptophan production by transport mutants of *Corynebacterium glutamicum*. *Biosci. Biotechnol. Biochem.* 59:1600–1602, 1995.
18. Kimura, E. Metabolic Engineering of glutamate production. *Adv. Biochem. Engin./ Biotechnol.* 79:37–57, 2003.
19. Pfefferle, W., B. Möckel, B. Bathe, A. Marx. Biotechnological manufacture of lysine. *Adv. Biochem. Engin./Biotechnol.* 79:59–112, 2003.
20. Marx, A., B.J. Eikmanns, H. Sahm, A.A. de Graaf, L. Eggeling. Response of the central metabolism in *Corynebacterium glutamicum* to the use of an NADH-dependent glutamate dehydrogenase. *Metabolic Eng.* 1:35–48, 1999.
21. Hols, P., M. Kleerebezem, A.N. Schanck, T. Ferain, J. Hugenholtz, J. Delcour, W.M. de Vos. Conversion of *Lactococcus lactis* from homolactic to homoalanine fermentation through metabolic engineering. *Nat. Biotechnol.* 17:588–592, 1999.

22. Clark, D.P., The fermentation pathways of *Escherichia coli. FEMS Microbiol. Rev.* 63:223–234, 1989.

23. Millard, C.S., Y.P. Chao, J.C. Liao, M.I. Donnelly. Enhanced production of succinic acid by overexpression of phosphoenolpyruvate carboxylase in *Escherichia coli. Appl. Environ. Microbiol.* 62:1808–1810, 1996.

24. Gokarn, R.R., E. Altman, M.A. Eiteman. Metabolic analysis of *Escherichia coli* glucose fermentation in presence of pyruvate carboxylase. *Biotechnol. Lett.* 20:795–798, 1998.

25. Stols, L., M.I. Donnelly. Production of succinic acid through overexpression of NAD^+-dependent malic enzyme in an *Escherichia coli* mutant. *Appl. Environ. Microbiol.* 63:2695–2701, 1997.

26. Stols, L., G. Kulkarni, B.G. Harris, M.I. Donnelly. Expression of *Ascaris suum* malic enzyme in a mutant *Escherichia coli* allows production of succinic acid from glucose. *Appl. Biochem. Biotechnol.* 63–65:153–158, 1997.

27. Boernke, W.E., C.S. Millard, P.W. Stevens, S.N. Kakar, F.J. Stevens, M.I. Donnelly. Stringency of substrate specificity of *Escherichia coli* malate dehydrogenase. *Arch. Biochem.* 322:43–52, 1995.

28. Donnelly, M.I., C.S. Millard, D.P. Clark, M.J. Chen, J.W. Rathke. A novel fermentation pathway in an *Escherichia coli* mutant producing succinic acid, acetic acid, and ethanol. *Appl. Biochem. Biotechnol.* 70–72:187–198, 1998.

29. Nghiem, N.P., M. Donnelly, C.S. Millard, L. Stols. Method for the production of dicarboxylic acids. U.S. patent 5,869,301, 1999.

30. Chatterjee, R.C., S. Millard, K. Champion, D.P. Clark, M.I. Donnelly. Mutation of the *ptsG* gene results in increased production of succinate in fermentation of glucose by *Escherichia coli. Appl. Environ. Microbiol.* 67:148–154, 2001.

31. Gokarn, R.R., J.D. Evans, J.R. Walker, S.A. Martin, M.A. Eiteman, E. Altman. The physiological effects and metabolic alterations caused by the expression of *Rhizobium etli* pyruvate carboxylase in *Escherichia coli. Appl. Microbiol. Biotechnol.* 56:188–195, 2001.

32. Gokarn, R.R., M.A. Eiteman, E. Altman. Metabolic analysis of *Escherichia coli* in the presence and absence of the carboxylating enzymes phosphoenolpyruvate carboxylase and pyruvate carboxylase. *Appl. Environ. Microbiol.* 66:1844–1850, 2000.

33. Li,Y., J. Chen, S.-Y. Lun. Biotechnological production of pyruvic acid. *Appl. Microbiol. Biotechnol.* 57:451–459, 2001.

34. Yokota, A., H. Shimizu, Y. Terasawa, N. Takaoka. Pyruvic acid production by a lipoic acid auxotroph of *Escherichia coli* W1485. *Appl. Microbiol. Biotechnol.* 41:638–643, 1994.

35. Yokota, A., Y. Terasawa, N. Takaoka, H. Shimizu, F. Tomita. Pyruvate production by an F_1-ATPase-defective mutant of *Escherichia coli. Biosci. Biotechnol. Biochem.* 58:2164–2167, 1994.

36. Yokota, A., M. Henmi, N. Takaoka, C. Hayashi, Y. Takezawa, Y. Fukumori, F. Tomita. Enhancement of glucose metabolism in a pyruvic acid-hyperproducing *Escherichia coli* mutant defective in F_1-ATPase activity. *J. Ferment. Bioeng.* 83:132–138, 1997.

37. Koebmann, B.J., H.V. Westerhoff, J.L. Snoep, D. Nilsson, P.R. Jensen. The glycolytic flux in *Escherichia coli* is controlled by the demand for ATP. *J. Bacteriol.* 184:3909–3916, 2002.

38. Benthin, S., J. Villadsen. Production of optically pure D-lactate by Lactobacillus bulgaricus and purification by crystallization and liquid/liquid extraction. *Appl. Microbiol. Biotechnol.* 42:826–829, 1995.

39. Vickroy, T.B., Lactic acid. In: *Comprehensive Biotechnology: The Principles, Applications, and Regulations of Biotechnology in Industry, Agriculture and Medicine*, Vol. 2. Moo-Young, M., ed., New York: Pergamon Press, 1985, pp 761–776.

40. Stanier, R.Y., J.L. Ingraham, M.L. Wheelis, P.R. Painter. *The Microbial World*, 5th ed., Englewood Cliffs, NJ: Prentice-Hall, 1986, pp 495–504.

41. Chang, D.E., H.C. Jung, J.S. Rhee, J.G. Pan. Homofermentative production of D- or L-lactate in metabolically engineered *Escherichia coli* RR1. *Appl. Environ. Microbiol.* 65:1384–1389, 1999.

42. Zhou, S., T.B. Causey, A. Hasona, K.T. Shanmugam, L.O. Ingram. Production of optically pure D-lactic acid in mineral salts medium by metabolically engineered *Escherichia coli* W3110. *Appl. Environ. Microbiol.* 69:399–407, 2002.

43. Skory, C.D. Isolation and expression of lactate dehydrogenase genes from *Rhizopus oryzae*. *Appl. Environ. Microbiol.* 66:2343–2348, 2000.

44. Ohara, H., H. Okuyama, S. Sawa, Y. Fujii, K. Hiyama. Development of industrial production of high molecular weight poly-L-lactate from renewable resources. *Nippon Kagaku Kaishi* 6:323–331, 2001.

45. Causey, T.B., S. Zhou, K.T. Shanmugam, L.O. Ingram. Inaugural article: engineering the metabolism of *Escherichia coli* W3110 for the conversion of sugar to redox-neutral and oxidized products: homoacetate production. *Proc. Natl. Acad. Sci. USA* 100:825–832, 2003.

46. Mironov, V.N., A.S. Kraev, M.L. Chikindas, B.K. Chernov, A.I. Stepanov, K.G. Skriabin. Functional organization of the riboflavin biosynthesis operon from *Bacillus subtilis* SHgw. *Mol. Gen. Genet.* 242:201–208, 1994.

47. Perkins, J.B., A. Sloma, T. Hermann, K. Theriault, E. Zachgo, T. Erdenberger, N. Hannett, N.P. Chatterjee, V. Williams, G.A. Rufo, R. Hatch, J. Pero. Genetic engineering of *Bacillus subtilis* for the commercial production of riboflavin. *J. Ind. Microbiol. Biotechnol.* 22:8–18, 1999.

48. Humbelin, M., V. Griesser, T. Keller, W. Schurter, M. Haiker, H.-P. Hohmann, H. Ritz, G. Richter, A. Bacher, A.P.G.M. van Loon. GTP cyclohydrolase II and 3,4-dihydroxy-2-butanone 4-phosphate synthase are the rate-limiting enzymes in riboflavin synthesis of an industrial *Bacillus subtilis* strain used for riboflavin production. *J. Ind. Microbiol. Biotechnol.* 22:1–7, 1999.

49. Sauer, U., V. Hatzimankatis, H.P. Hohmann, M. Manneberg, A.P.G.M. van Loon, J.E. Bailey. Physiology and metabolic fluxes of the wild-type and riboflavin-producing *Bacillus subtilis*. *Appl. Environ. Microbiol.* 62:3687–3696, 1996.

50. Hugenholtz, J., W. Sybesma, M.N. Groot, W. Wisselink, V. Ladero, K. Burgess, D. van Sinderen, J. Piard, G. Eggink, E.J. Smid, G. Savoy, F. Sesma, T. Jansen, P. Hols, M. Kleerebezem. Metabolic engineering of lactic acid bacteria for the production of nutraceuticals. *Antonie van Leeuwenhoek* 82:217–235, 2002.

51. Saito, Y., Y. Ishii, H. Hayashi, Y. Imao, T. Akashi, K. Yoshikawa, Y. Noguchi, S. Soeda, M. Yoshida, M. Niwa, J. Hosoda, K. Shimomura. Cloning of genes coding for L-sorbose and L-sorbosone dehydrogenases from *Gluconobacter oxydans* and microbial production of 2-keto-L-gulonate, a precursor of L-ascorbic acid, in a recombinant *G. oxydans* strain. *Appl. Environ. Microbiol.* 63:454–460, 1997.

52. Anderson, S., C.B. Marks, R. Lazarus, J. Miller, S. Stafford, J. Seymour, D. Light, W. Rastetter, D. Estell. Production of 2-keto-L-gulonate, an intermediate in L-Ascorbate synthesis by a genetically modified *Erwinia herbicola*. *Science* 230:144–149, 1985.

53. Chotani, G., T. Dodge, A. Hsu, M. Kumar, R. LaDuca, D. Trimbur, W. Weyler, K. Sanford. The commercial production of chemicals using pathway engineering. *Biochim. Biophys. Acta* 154:3434–455, 2000.

54. Jolly, J., J.F.V. Sebastien, P. Duboc, J.-R. Neeser. Exploiting exopolysaccharides from lactic acid bacteria. *Antonie van Leeuwenhoek* 82:367–374, 2002.

55. Germond, J.E., M. Delley, N. D'Amico, S.J. Vincent. Heterologous expression and characterization of the exopolysaccharide from *Streptococcus thermophilus* Sfi39. *Eur. J. Biochem.* 268:5149–56, 2001.

56. Stingele, F., J.R. Neeser, B. Mollet. Identification and characterization of the eps (exopolysaccharide) gene cluster from *Streptococcus thermophilus* Sfi6. *J. Bacteriol.* 178:1680–90, 1996.

57. van Kranenburg, R., I.C. Boels, M. Kleerebezem, W.M. de Vos. Genetics and engineering of microbial exopolysaccharides for food: approaches for the production of existing and novel polysaccharides. *Curr. Opin. Biotechnol.* 10:498–504, 1999.

58. Looijesteijn, P.J., I.C. Boels, M. Kleerebezem, J. Hugenholtz. Regulation of exopolysaccharide production by *Lactococcus lactis* subsp. cremoris by the sugar source. *Appl. Environ. Microbiol.* 65:5003–5008, 1999.

59. Boels, I.C., R. van Kranenburg, J. Hugenholtz, M. Kleerebezem, W.M. de Vos. Sugar catabolism and its impact on the biosynthesis and engineering of exopolysaccharide production in lactic acid bacteria. *Int. Dairy J.* 11:723–732, 2001.

60. O'Sullivan, L., R.P. Ross, C. Hill. Potential of bacteriocin-producing lactic acid bacteria for improvements in food safety and quality. *Biochimie* 84:593–604, 2002.

61. Rodriguez, J.M., M.I. Martinez, N. Horn, H.M. Dodd. Heterologous production of bacteriocins by lactic acid bacteria. *Int. J. Food Microbiol.* 80:101–116, 2002.

62. Martinez, J.M., J. Kok, J.W. Sanders, P.E. Hernandez. Heterologous coproduction of Enterocin A and Pediocin PA-1 by *Lactococcus lactis*: detection by specific peptide-directed antibodies. *Appl. Environ. Microbiol.* 66:3543–3549, 2000.

63. Horn, N., M.I. Martinez, J.M. Martinez, P.E. Hernandez, M.J. Gasson, J.M. Rodriguez, H. Dodd. Enhanced production of pediocin PA-1 and coproduction of nisin and pediocin PA-1 by *Lactococcus lactis*. *Appl. Environ. Microbiol.* 65:4443–4450, 1999.

64. Horn, N., M.I. Martinez, J.M. Martinez, P.E. Hernandez, M.J. Gasson, J.M. Rodriguez, H. Dodd. Production of pediocin PA-1 by *Lactococcus lactis* using the lactococcin A secretory apparatus. *Appl. Environ. Microbiol.* 64:818–823, 1998.

65. Gonzalez, R., B.A. Andrews, J. Molitor, J.A. Asenjo. Metabolic analysis of the synthesis of high levels of intracellular human SOD in *S. cerevisiae* rhSOD 2060 411 SGA122. *Biotechnol. Bioeng.* 82:152–169, 2003.

66. Hugenholtz, J., E.J. Smid. Nutraceutical production with food-grade microorganisms. *Curr. Opin. Biotechnol.* 13:497–507, 2002.

67. Hugenholtz, J., M. Kleerebezem, M. Starrenburg, J. Delcour, W. de Vos, P. Hols. *Lactococcus lactis* as a cell factory for high-level diacetyl production. *Appl. Environ. Microbiol.* 66:4112–4114, 2000.

68. Rijnen, L., P. Courtin, J.C. Gripon, M. Yvon. Expression of a heterologous glutamate dehydrogenase gene in *Lactococcus lactis* highly improves the conversion of amino acids to aroma compounds. *Appl. Environ. Microbiol.* 66:1354–1359, 2000.

69. Silvestroni, A., C. Connes, F. Sesma, G.S. De Giori, J.C. Piard. Characterization of the melA locus for alpha-galactosidase in *Lactobacillus plantarum*. *Appl. Environ. Microbiol.* 68:5464–5471, 2002.

70. Rhodius, V., T.K. Van Dyk, C. Gross, R.A. LaRossa. Impact of genomic technologies on studies of bacterial gene expression. *Ann. Rev. Microbiol.* 56: 599–624, 2002.

71. Tao, H., R. Gonzalez, A. Martinez, M. Rodriguez, L.O. Ingram, J.F. Preston, K.T. Shanmugam. Engineering a homo-ethanol pathway in *Escherichia coli*: increased glycolytic flux and levels of expression of glycolytic genes during xylose fermentation. *J. Bacteriol.* 183:2979–2988, 2001.

72. Gonzalez, R., H. Tao, S.W. York, K.T. Shanmugam, L.O. Ingram. Global gene expression differences associated with changes in glycolytic flux and growth rate in *Escherichia coli* during the fermentation of glucose and xylose. *Biotechnol. Prog.* 18:6–20, 2002.

73. Gonzalez, R., H. Tao, J.E. Purvis, S.W. York, K.T. Shanmugam, L.O. Ingram. Gene array-based identification of changes that contribute to ethanol tolerance in ethanologenic *Escherichia coli*: comparison of KO11 (parent) to LY01 (resistant mutant). *Biotechnol. Prog.* 19:612–623, 2003.

74. Dharmadi, Y., R. Gonzalez DNA Microarrays: experimental issues, data analysis, and application to bacterial systems. *Biotechnol. Prog.* 20:1309–1324, 2004.

21

Technologies Used for Microbial Production of Food Ingredients

Anthony L. Pometto III and Ali Demirci

CONTENTS

21.1 Introduction 521
21.2 Microorganism Selection and Development...522
21.3 Culture Media and Upstream Components...523
21.4 Bioreactor Monitoring Systems and Design..525
21.5 Fermentation Types Employed: The Actual Production Process........................528
21.6 Novel Bioreactors Design 529
21.7 Future Research...530
References..531

21.1 INTRODUCTION

The goal of this chapter is to supplement the industrial microbiology components in the first edition of *Food Biotechnology* by Knorr (1) and to present an updated overview of various technologies currently under investigation and employed by the food industry for the production of microbial food ingredients. For the detailed description of industrial microbiology principals, Demain and Davies *Manual of Industrial Microbiology and Biotechnology*, 2nd edition (2) is highly recommended.

For the food industry, the bottom line is cost. They are selling commodity products (food) with various levels of preconsumer processing. The goal is always to produce the most nutritious and safe product at the lowest possible cost.

For centuries microorganisms have been employed for the production of fermented food products (i.e., cheese, soy sauce, sauerkraut, wine, and bread). The consumption of some live microbial cultures (probiotics) has proven to provide a health benefit to humans and animals (3,4). Some foods containing live cultures are yogurt, buttermilk, and acidophilus milk. These microbial fermented food products also have an extended shelf-life compared to the perishable starting raw material. Thus, microorganisms not only provide a nutritional benefit to humans but act to extend the shelf-life of the food supply.

Microorganisms employed by the food industry include bacteria, yeasts, and molds. These microorganisms have several morphological and physiological differences. Morphologically bacteria are small and difficult to remove, yeasts are larger and will sometime settle out of solution, whereas molds are filamentous and are typically removed by filtration. Physiologically they differ in pH preferences (yeasts and molds prefer a lower pH than bacteria), nutrient requirements (different concentrations and types of nitrogen and other trace elements), growth rates (bacteria grow much faster than yeasts and molds), and more. Thus, different culture media, fermentation methods, and product recovery methods are required depending on the microbial system being cultured.

21.2 MICROORGANISM SELECTION AND DEVELOPMENT

Microorganisms are the biocatalysts that produce and maintain a host of enzymatic pathways that are used to produce the food component of interest. The characteristics of a good industrial microorganism for the production of food ingredients are (1) it must be effective in producing large quantities of a single product, (2) it can be efficiently isolated and purified, (3) it is easy to maintain and cultivate, (4) it is genetically stable, (5) it grows best in an inexpensive culture medium, and (6) it is safe for human consumption. The first step is to isolate the hardiest starter culture possible, then to begin strain improvements via classical mutagenesis or genetic engineering.

A classic example would be the production of L-phenylalanine for the artificial sweetener aspartame (NutraSweet® J.W. Childs Equity Partners II L.P.), which is a dipeptide of L-phenylalanine and L-aspartic acid. When NutraSweet first entered the market in 1981, the L-phenylalanine supply became the bottleneck for production. L-Phenylalanine, L-tyrosine, and L-tryptophan are produced via the shikimic acid pathway in all organisms. To develop a bacterium which over produced L-phenylalanine, first classical chemical mutagenesis of an L-tyrosine auxotroph of *Corynebacterium glutamicum* was employed using -phenylalanine analog resistance in an effort to reduce end product inhibition, and L-tyrosine production (5). Analogs such as *p*-aminophenylalanine, *p*- and *m*-fluorophenylalanine, and β-2-thienylalanine were incorporated into the cellular protein thus poisoning the cell. To combat this poison, surviving mutants must overproduce L-phenylalanine, thus neutralizing the toxic effects of the analogs. This process was repeated several times with mutants resistant to increasing concentrations of analogs. The final analog resistant bacterium selected by Hagino and Nakayama (5) produced 9.5 g/L of L-phenylalanine.

The over producing bacterium was then transformed with plasmids containing L-phenylalanine analog resistant chorismate mutase and prephenate dehydratase genes (6). These are two key enzymes in the shikimic acid pathway for L-phenylalanine production. Except for constitutive enzymes, most enzymes in the cell have a short half life in the cell. Thus, an increase in key enzyme concentrations and residence time in the cell will also increase production. Ozaki et al. (6) transformates produced 19.0 g/L of -Lphenylalanine, thus, illustrating how classical mutagenesis and molecular genetic techniques are employed to further enhance production of some desired metabolites for the food industry.

Another alternative method was whole bioconversion developed by Yamada et al. (7) which produced L-phenylalanine from trans-cinnamic acid via L-phenylalanine ammonia lyase (PAL) reversal in *Rhodotorula glutinis*. In the presence of concentrated ammonium hydroxide, the PAL reversal demonstrated a 70% conversion yield which produced 17.5 g/L of L-phenylalanine. By utilizing a whole cell bioconversion process, no enzyme purification step was needed and the enzyme proved to be more stable within the yeast under the harsh conditions employed. This process was used to produce some of the initial L-phenylalanine used for the production of NutraSweet. Eventually, however, production by the genetically engineered bacterium exceeded the levels in fed-batch fermentation, which did not involve caustic chemicals, and thus, became the method of choice. L-Phenylalanine purification is performed by ion exchange chromatography for all methods.

The genetic stability of cultures requires minimum culture transfers and long term storage capabilities. Fermentation media are inoculated from working cultures which are produced every few months from master cultures depending on the microorganism. The most common procedures for long term storage are freeze drying ($< -18°C$) and ultra-low temperature storage ($-70 - -80°C$). Freeze drying requires a cryoprotectant, such as sterile skim milk, followed by freeze drying and vial sealing under a vacuum (2). Sealing under nitrogen gas can also help to stabilize the culture and extend the shelf life. Ultra-low temperature storage is in a rich culture medium with 20% sterile glycerol. Some cultures are sensitive to freeze drying, thus, ultra-low is the most common method employed today, because of long-term culture viability. The risk is loss of electrical power and refrigeration problems.

Suspended cell cultures or spore suspensions are used as inocula for these industrial scale fermentations. Purity is constantly checked until inoculation. For suspended cell inoculation the sequence employed would be culture slant, shake-flask culture, benchtop fermentor, pilot-scale fermentation, then into full scale fermentation. Many fungal fermentations, such as citric acid and soy sauce fermentations, required a suspension of viable fungal spores as the inoculum. These spore suspensions are generated on large agar trays, and then are aseptically transferred into culture bottles suspended in sterile water or saline (2).

Microbial systems frequently constitute efficient mechanisms for the production of nutritionally important ingredients at a relatively low cost, for example, the production of selenomethione in yeast. A slight modification in the yeasts culture medium will force the yeast to substitute the sulfur group in methionine and cysteine with a selenium in standard fed-batch fermentation (8,9). To identify this medium change, Demirci and Pometto (8) developed a gradient delivery unit producing a gradient of sodium selenite or sodium selenate in a continuous bioreactor (Figure 21.1). It has been shown that selenium has several health benefits (10) including a cancer-protective effect (11), and a profound effect on the survival of HIV-infected patients (12).

Furthermore, microbial systems are ideal for the production of essential micronutrients such as amino acids, vitamins, and enzymes, and bulk ingredients such as organic acids and alcohols, whole cell flavor enhancers, and polysaccharides.

21.3 CULTURE MEDIA AND UPSTREAM COMPONENTS

The ideal culture medium will use inexpensive components to supply their complex nutrient requirements. Miller and Churchill (13) provide an excellent summary of inexpensive media components and their makeup. These ingredients are crop, animal, marine or yeast based components. The culture medium alone can represent 30 to 70% of the fermentation production costs. Slight changes in medium micronutrients can have a major impact on the fermentation (14,15). Thus, the food industry demands a consistent product from suppliers of these complex components. Failure to provide a consistent product will eliminate the commercial use. What decides the culture medium makeup? Essentially, it is the nutritional requirements of the microorganism of choice and its ability to biosynthesize essential elements such as amino acids, vitamins, lipids, and carbohydrates. For example, bacteria and yeast are high in protein (40–50%), whereas molds are not (10–25%). Yeast

Continuous fermentation with gradient system

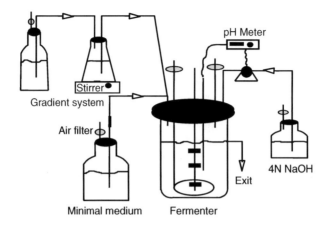

Figure 21.1 Continuous bioreactor gradient delivery system design used for developing the production protocols for selenomethionine production in yeast.

can generally grow in a minimal medium, whereas lactic acid bacteria, essential for the fermented dairy products, require a host of micronutrients to grow.

Each microorganism has different micronutrient requirements. Typically the exact micronutrient which benefits the fermentation that is being supplied by the complex nutrient employed is unknown. Generally it involves specific amino acids, vitamins, trace elements, and lipids. Furthermore, the concentration ratio of carbohydrate to nitrogen and phosphorous has a dramatic impact on microbial growth. Carbon–nitrogen imbalances can result in the production of other byproducts such as extracellular polysaccharide, and fermentation end products such as ethanol. In the case of ethanol production in yeast, excess glucose in the presence of oxygen will direct yeast to produce ethanol. This is called the Crabtree effect (16,17). Yeast cells typically consist of 48% carbon and 8% nitrogen on a dry weight basis. When the C:N ratio is 10:1, yeasts grow aerobically consuming little substrate while producing maximum cell mass, CO_2 and H_2O, but when the C:N ratio is 50:1 yeasts grow anaerobically, consuming much substrate while producing little cell mass, and much CO_2 and ethanol. This difference in yeast growth is also called the Pasteur effect (18).

Why the increase in industrial microbiology fermentation processes over the past 20 years? It is the result of the corn syrup sweetener industry and computer technology. Corn is 70% starch, and when dried to <19%, it can be stored for more than two years. Annually 10 billion bushels of corn are produced in the Unites States of America. Thus, the liquefaction of corn starch to glucose syrups for the production of high fructose corn sweeteners represents a consistent, low cost supply of substrate for most industrial microbiology fermentations. Not only is glucose produced, but customized substrates can be also produced. For example, corn syrups containing 19% dextrose, 14% maltose, 12% maltotriose, and 55% higher saccharides are used to control microbial growth rates and biological heat production in many fermentations. Thus, glucose is the platform chemical used for the microbial production of organic acids, amino acids, vitamins, and more. In fact some food grade fermentation facilities have located adjacent to a corn sweetener facility to permit glucose syrup to be piped directly to their fermentors. Glucose corn syrup is also shipped via truck or rail cars as liquid or dried product.

Furthermore, the advent of computer process controls of industrial scale fermentors has removed many of the fears associated with commercial scale fermentation by providing reliable and easy to operate dissolved oxygen, pH, foam, temperature, and sterilization controls of the process. Sirakaya et al. (19) described fermentation software to monitor and control the utilization fermentation process.

21.4 BIOREACTOR MONITORING SYSTEMS AND DESIGN

The stirred tank bioreactor design is the most common fermentor and consists of agitator, baffles, aeration sparger for aerobic culture growth, sterilizable monitoring probes for pH, dissolved oxygen, temperature, and antifoam, filling and draining ports, and often culture medium sterilization capabilities in the reactor tank. Reactor agitation is essential for temperature control, pH adjustments, oxygen absorption into the liquid medium, overall culture health, and mixing of any required additions of substrate, and nutrients. Typical commercial reactor working volumes for food grade ingredients are 5,000 to 40,000 gallons.

Fermentation health requires real time monitoring system. Microbial growth can be monitored via several methods. The most common method is indirect measurement of biomass by absorbance of the fermentation broth at 620 nm by using a spectrophotometer. The measured absorbance values can then be used to estimate biomass concentrations by using a standard curve. Standard curves are developed by collecting fresh log phase cells via centrifugation, washing cells with water or 0.1 M ammonium acetate pH 7.0 buffers, then serially diluting the pellet (20). Absorbance for each dilution is then determined spectrophotometrically at 620 nm. The actual dry weight biomass (g/l) is determined for each dilution via direct biomass measurement after oven drying of each dilution in preweighed boats at 70°C for 24–36 hr. This needs to be performed in at least replicates of three. By washing biomass with water, any influence of culture medium on dry weights can be eliminated or minimized. Finally, a standard curve can be constructed by plotting absorbance versus actual dry biomass weight (g/l). This method allows for quick, reliable, and easy conversion of absorbance to dry biomass (g/L).

For determining microbial health, viable cell counts can be rapidly performed by using an EPICS XL-MCL flow cytometer (Beckman-Coulter, Miami, FL) in conjunction with the Live/Dead BacLight™ bacterial viability test kit (Molecular Probes, Eugene, OR) (21). BacLight uses a mixture of Syto 9 fluorescence, which is measured as a log FL1 (525 nm) signal, and the propidium iodide fluorescence which is measured as a log FL4 (675 nm) signal. A two color histogram is collected with gating on the bacteria only population from the two parameter light scatter distribution and is used for the analysis of green only (live bacteria), red only (dead bacteria), and both colors (stressed bacteria).

Substrate consumption and product formation rates can be monitored by high pressure liquid chromatography (HPLC) or by membrane bound enzymes biosensors, which requires 20 and 1 min to run, respectively. HPLC analysis is time consuming, but the concentration of multiple metabolites can be monitor simultaneously. HPLC does not provide real time feedback on the health of the fermentation, and it has long sample preparation and run times. In contrast, membrane bound enzyme biosensors such as YSI 2700 select analyzer (Yellow Springs Instruments, Yellow Springs, OH) can analyze a sample in 1 min. However, these units are restricted by the availability of substrate specific oxidases which generate H_2O_2, the measurable product by their electrode. Some compounds currently measurable are glucose, ethanol, maltose, lactic acid, and lysine. Sample preparation is simply filtration (0.45 :m) and dilution with water if the value falls outside the instrument window.

Organic acid production can be continuously monitored via alkali addition rates for pH control. Alkali consumption can be easily monitored by feeding alkali solution from a sterile burette (22). Microbial respiration for aerobic cultures can be continuously monitored via dissolved oxygen probe or the concentration of CO_2 in the exit gas, which can be monitored via off gas analyzers or simply via alkali (4 N NaOH) traps followed by pH titration. However, more technologies are needed to acquire real time measurements of microbial growth, ensuring optimal fermentation time, and product formation in the shortest time possible.

For rapid analysis of biomass, substrate and product concentrations without any sample preparation, Fourier transform mid infrared (FT-MIR) spectroscopy has been successfully utilized for lactic acid (23) and ethanol (24) fermentation. Calibration models have been developed by using principle least square (PLS) and principle component regression (PCR) on suitable spectral wavenumber regions. This calibration model is then used for unknowns. The advantage of this system is that not only no sample preparation required, but it also provides analysis for substrate, product and biomass at the same time. This method can be used for online, real time analysis for monitoring and process control purposes.

For aerobic culture fermentation, house air is generally supplied under pressure. Oxygen transfer into the culture medium depends on the air bubble residence time in the culture medium and bubble size. The smaller the air bubbles the greater the O_2 transfer. Thus, all stirred tank reactors have aeration spargers that generate bubbles right beneath the first agitator blade. The exiting air bubble collides with the standard flat Rashton turbine agitator blade which strikes the bubbles hard and fast as they leave the sparger generating smaller air bubbles for improved oxygen transfer. Many commercial fermentations then follow up the agitator shaft with a series of down draft marine agitator blades, which look like a motor boat propeller (Figure 21.2). This series of down draft marine blades push the air bubble back down as it migrates up the reactor. This increases gas bubble residence time in the liquid medium before exiting out the top. In some fermentations, the rate limiting

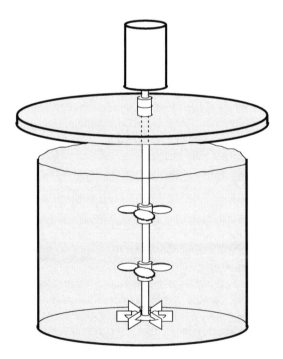

Figure 21.2 Diagram of stirred tank reactor with Rashton turbine agitator blade at the reactor bottom to break up air bubbles followed by a series of down draft marine blades to increase bubble residence time in the culture medium.

substrate will be oxygen. Thus, along with the series of down draft marine agitator blades a supply of pure oxygen may be needed. Pilot scale recombinant *Pichia* fermentations often require pure oxygen supplements to ensure optimal growth.

It is obvious that each microorganism differs in oxygen uptake rate (OUR). Therefore, oxygen transfer rates (OTR) (the rate of oxygen transfer from bubble to fermentation medium) must be equal to or higher than OUR. Maximum OTR can be calculated by using the following equation:

$$\text{OTR (mg O}_2/\text{min)} = k_L a \, C_L^* \, V_L$$

where C_L^* = Equilibrium dissolved oxygen level (mg O_2/L), $k_L a$ = Mass transfer coefficient (min^{-1}), and V_L = Liquid volume in tank (L).

Thus, $k_L a$ is a critical parameter affecting OTR and the desired dissolved oxygen concentration in a fermentation broth during fermentation, and $k_L a$ can be determined experimentally. Briefly, the percentage dissolved oxygen (%DO) level is reduced to almost zero by sparging nitrogen into the culture medium. Then, air or oxygen gas is sparged at the desired temperature, agitation, and aeration conditions. Percent DO values are recorded over time until %DO reaches to saturation level. After converting %DO values into dissolved oxygen concentrations (C_L), plotting ln ($C_L^* - C_L$) versus time gives a straight line with a negative slope which is equal to "$-k_L a$". By knowing $k_L a$, one can calculate OTR with the given aeration and agitation. If OTR is less than OUR, some changes can be implemented to improve OTR, such as increasing aeration rate or agitation. Even utilization of various types of propellers can be compared by calculating OTR under each condition.

Agitation and aeration of stirred tank fermentors may also generate foam, which is the entrapment of gas in lipid, polysaccharide, or protein matrix (25). If not controlled, a foam buildup can literally empty the reactor. Mechanical foam breakers are like giant fans which physically break the foam and blow it back down. Mechanical foam breakers have their limitations and ultimately food grade antifoams are employed to control culture broth foam. A list of common antifoams can be found in Hall et al. (25).

Modification to this basic bioreactor design occurs from specific microbial needs. For example, if your production microorganism is sensitive to agitator shear then an air lift bioreactor is employed. Air lift bioreactors consists of a central or an external draft tube whereby air bubbles passing up these tubes generate convection mixing of the medium (26). Many fungal fermentations require the production of fungal balls for maximum product formation. These fungal balls are very sensitive to the shear caused by the agitator blades. For example, *Aspergillus niger* citric acid fermentation requires a defined medium with specific concentrations of trace elements (i.e., copper, manganese, magnesium, iron, zinc, and molybdenum). The bioreactor is typically lined with glass to prevent the addition of any stray trace elements. A fungal spore suspension is used as an inoculum, and mycelium, for optimal production rates, consists of very small solid pellets, or spheres which require a bioreactor with minimum shear. Throughout the entire fermentation period, the minimum dissolved oxygen concentration of 20–25% of saturation is required (27). Thus, tall air lift fermentors are the bioreactors of choice for these types of fermentations.

Biological heat can also be a problem. Bacteria, yeast, and molds will generate different levels of biological heat because their growth rates are so different. All bioreactors require some kind of jacketed cooling and heating system. Also, the time of year and location of the facility will also dictate the level of cooling required. Biological heat is directly related to growth rates. The faster the growth rate the more heat generated. Thus, a rate limiting substrate can be used to control microbial growth. One example is the use of a substrate containing mono, di, oligo, and poly saccharides. For example, a liquefied corn starch described above containing 19% dextrose, 14% maltose, 12% maltotriose, and 55%

higher saccharides is commonly used to control microbial growth rates. Enzymatic hydrolyses of the di, tri, and oligo saccharides will dictate the level of available glucose in the fermentor, and thus, control microbial growth rates.

21.5 FERMENTATION TYPES EMPLOYED: THE ACTUAL PRODUCTION PROCESS

The work horse of the industry is batch and fed-batch fermentation. Batch fermentations are closed fermentations (28). The fermentation sequence starts with medium sterilization, reactor inoculation [1 to 2% (v/v) typically], incubation for complete microbial growth cycle with lag, log, and stationary phases, fermentation termination, draining the reactor for product recovery down stream, cleanup of the reactor, and starting over. Percentage yield is calculated by taking the concentration of the product formed (g/l) divided by the concentration of substrate consumed (g/l) times 100%, whereas productivity is a measure of the product formation rates. It is calculated by dividing the product concentration (g/l) by the fermentation time (hours); thus, it is presented in g/l/hr.

In fed-batch fermentation, additional carbohydrate is supplied to the batch fermentation during the run (28). High carbohydrate concentrations in the initial culture medium are toxic to many microorganisms. Thus, an optimal carbohydrate concentration is employed initially, which permits maximum culture growth to late log phase. When the carbohydrate concentration is reduced to almost zero, additional sterile carbohydrate is injected into the bioreactor to bring the carbohydrate concentration back to the starting level. When this is consumed, the process is repeated until end product inhibition forces the whole bioconversion to stop. Ideally, at the end of the fermentation you will have a product concentration which is three to four times greater than single batch fermentation with no residual carbohydrate. This will generate the highest yield possible. Also, due to increased end product accumulation with each carbohydrate addition, culture production rates will decrease. Thus, the decision as to how many fed-batch fermentations to perform before harvest is based on the desired final product concentration and the optimal fermentor use time. For example, in lactic acid fermentations a final product concentration >120 g/l is desired to enhance product recovery (29). For *Lactobacillus casei* this concentration can only be achieved via fed batch fermentation for a total fermentor run time of eight days.

Continuous fermentations are open fermentations, whereby fresh medium is continually added to the bioreactor, while spent culture medium, cells, and product are continually leaving (28). This fermentation is desired by the industry, because the reactor volume is 10 to 100 times smaller than batch fermentations, a steady stream of fermentation product is produced which will optimize downstream processes, bioconversion rates are always at maximum, operation costs are less, and the system can be fully automated and computer controlled to the point where only two operators are needed to manage the fermentation each day. However, it requires a continuous supply of sterilized or pasteurized culture medium, dilution rates are linked to microbial growth rates and the operational speed of downstream recovery process. Startup is slow, so any facility shut downs have an impact on production, and you are constantly fighting contamination (30). Thus, not all fermentations can be operated this way. The best example is ethanol production for gasohol, which is commonly a continuous fermentation with a four bioreactor train with increasing working volumes in each bioreactor. Thus, the dilution rate is decreasing in each bioreactor over the course of the fermentation. This dilution gradient in the train is critical, because as ethanol builds up in the culture medium, the yeast growth rate slows. Specific growth rate equals dilution rate. Finally a holding tank at the end is used to ensure

Figure 21.3 Example of *Lactobacillus casei* biofilm development on PCS tubes mounted on the agitator shaft for repeat batch fermentation. PCS blend employed contained 50% (w/w) polypropylene, 35% (w/w) ground soybean hulls, 5% (w/w) bovine albumin, 5% (w/w) Ardamine Z yeast extract, and 5% (w/w) soybean flour.

complete bioconversion of any residual sugars to ethanol. The CO_2 is collected and concentrated, then sold as another valuable byproduct. Fermented beverages (i.e., wine and beer) are still performed in batch.

For some continuous fermentations, an increased concentration of biomass in the reactor is required. This can be achieved by cell recycling, immobilized cell, and biofilm fermentations. Cell recycling reactors employ a filtration unit that allows for the constant bleeding of culture supernatant while retaining biomass (26). However, filtration unit fouling is a problem and must be constantly monitored. This type of operation has found use in the treatment of food processing from some starchy food waste streams (26).

21.6 NOVEL BIOREACTOR DESIGNS

One of the most common forms of immobilized cell bioreactor is entrapment, where high concentrations of cells are trapped in a polymer matrix such as alginate and κ-carrageenin (28,31). Thus, a high cell density is continuously retained in the fermentor while substrate is continuously converted to product. This higher concentration of biocatalysts in the reactor results in higher productivities and yields. The disadvantages of this method are migration of substrate through the matrix to the cell and the migration of product out, potentially high concentrations of product around the cells causing end product inhibition, cell leakage from the polymer matrix due to cell growth, and bead swelling and disintegration over time causing the whole fermentation to be stopped, cleaned, and restarted.

Biofilms are a natural form of cell immobilization in which microorganisms are attached to a solid surface (32). In this bioreactor, cells are continually growing, and sloughing off. Thus, the reactor is a mixture of immobilized and suspended cells. Continuous biofilm fermentations are truly open immobilized cell bioreactors (30). Their operation is equivalent to a suspended cell continuous fermentation with the added advantage of increase biomass concentrations in the bioreactor. Biofilms are typically resistant to harsh conditions, and can tolerate changes in the fermentation feed and conditions. However, not all microorganisms form biofilms. Filamentous microorganisms such as fungi and actinomycetes are natural biofilm formers. For nonfilamentous bacteria to form a biofilm, an extracellular polysaccharide needs to be generated by the bacterium (32).

Some bacteria will form biofilms on any surface such as metal, plastic, and glass. However, certain bacteria, such as lactobacilli, require something to stimulate this biofilm development (33). Plastic composite support (PCS) developed at Iowa State University has proven to stimulate biofilm development of *Lactobacillus casei* (22,34,36), *Zymomonas mobilis* (37,38), *Saccharomyces cerevisiae* (37,38,39), and *Actinobacillus succinogenes* (40). PCS is a high temperature extruded material consisting of at least 50% polypropylene, plus ground soybean hulls, bovine albumin and various culture micronutrients. Soybean hulls keep the extruded product porous due to the release of steam as the PCS leaves the extruder die. Bovine albumin performs as a natural plasticer which protects the temperature sensitive micronutrients. Micronutrients are selected based on the specific cultural requirements for amino acids, vitamins, and lipids. Monosaccharides are avoided due to poor PCS production. For example, the PCS blend for lactobacilli contains 50% (w/w) polypropylene, 35% (w/w) ground dried soybean hulls, 5% (w/w) bovine albumin, 5% (w/w) yeast extract, 5% (w/w) soybean flour, and mineral salts (35). PCS have been evaluated in batch (22), fed-batch (29), and continuous (30) lactic acid fermentations (Figure 21.3). In every application the percentage yields and productivity rates were significantly higher than suspended cell lactic acid fermentations. Furthermore, repeat batch fermentations have operated for more than 1.5 years with virtually no change in percentage yields and productivities. This longevity is attributed to the fact that once a biofilm has established on these customized materials, it will continue to grow as a biofilm. This is supported by the fact that a PCS biofilm reactor washed with concentrated ammonium hydroxide, rinsed with mineral salts solution, and then reinoculated with a fresh culture and medium will reestablish itself overnight. Commercially, the quick vinegar process is the most common biofilm process in current operation which uses wood chips for supports and *Acetobacter aceti* for production (27).

Solid substrate fermentation is when a substrate such as soybeans is ground, inoculated with *Aspergillus oryzae*, then incubated for three days for the production of soy sauce (27). It is a simple fermentation process and commonly used for aerobic fermentation due to its large surface area. Thus, oxygen concentration is high without using any mechanical forced air systems. Solid substrate fermentations require large areas or incubation space. A temperature controlled environment, intermittent monitoring for contamination and quality of starting material is essential for success.

21.7 FUTURE RESEARCH

Research is still needed for isolation of new microbial strains with improved production efficiencies and higher yields. More real time measurements are needed for culture conditions and metabolite formation. Recovery will continue to be the key factor associated with final product purity and cost. As an industry we cannot rely solely on genetic engineering as our method of improving current fermentations. As we have illustrated, there are many other techniques which can be employed to improve productivity and yield, including new inexpensive medium ingredients, more continuous fermentation processes, and new exotic microbial reservoirs in nature and in the food industry.

REFERENCES

1. Knorr, D. *Food Biotechnology*. New York: Marcel Dekker, 1987.
2. Demain, A.L., J.E. Davies. *Manual of Industrial Microbiology and Biotechnology*, 2nd ed. Washington: ASM Press, 1999.
3. Hoover, D.G., L.R. Steenson. *Bacteriocins of Lactic Acid Bacteria*. San Diego: Academic Press, Inc., 1993.
4. Tannock, G.W. *Probiotics, A Critical Review*. Norfolk, England: Horizon Scientific Press, 1999.
5. Hagino, H., K. Nakayama. L-Phenylalanine production by analog-resistant mutants of *Corynebacterium glutamicum*. *J. Agric. Biol. Chem.* 38:157–161, 1974.
6. Ozaki, A., R. Katsumata, T. Oka, A. Furuya. Cloning of the genes concerned in phenylalanine biosynthesis in *Corynebacterium glutamicum* and its application to breeding of a phenylalanine producing strain. *J. Agric. Biol. Chem.* 49:2925–2930, 1985.
7. Yamada, S., K. Nabe, N. Izuo, K. Nakamichi, I. Chibata. Production of L-phenylalanine from trans-cinnamic acid with *Rhodotorula glutinis* containing L-phenylalanine ammonia lyase activity. *Appl. Environ. Microbiol.* 42:773–778, 1981.
8. Demirci, A., A.L. Pometto III. Production of organically bound selenium yeast by continuous fermentation. *J. Agric. Food Chem.* 47:2491–2495, 1999.
9. Demirci, A., A.L. Pometto III, D. Cox. Enhanced Organically Bound Selenium Yeast Production by Fed-Batch Fermentation. *J. Agric. Food Chem.* 47:2496–2500, 1999.
10. Burk, R.F. *Selenium in Biology and Human Health*. Heidelburg: Springer-Verlag, 1994.
11. Combs Jr., G.F. Selenium as a cancer-protective agent. In: *The bulletin of Selenium-Tellurium Development Association*. February, 1997, pp 1–4.
12. Bologna, R., F. Indacochea, G. Shor-Posner, E. Mantero-Atienza, M. Grazziutii, M. Sotomayor, M.A. Fletcher, C. Cabrejos, G.B. Scott, M.K. Baum. Selenium and immunity in HIV-1 infected pediatric patients. *J. Nutr. Immunol.* 3:41–49, 1994.
13. Miller, T.L., B.W. Churchill. Substrates for large-scale fermentations. In: *Manual of Industrial Microbiology and Biotechnology,* Demain, A.L., N.A. Solomon, eds., Washington: American Society for Microbiology, 1986, pp 122–136.
14. Demirci, A., A.L. Pometto III, B. Lee, P. Hinz. Media evaluation in repeated batch lactic acid fermentation with *Lactobacillus plantarum* and *Lactobacillus casei* subsp. *rhomnosus*. *J. Agric. Food Chem.* 46:4771–4774, 1998.
15. Lee, B., A.L. Pometto III, A. Demirci, P. Hinz. Media evaluation in microbial fermentations for enzyme production. *J. Agric. Food Chem.* 46:4775–4778, 1998.
16. De Deken, R.H. The Crabtree-effect: a regulatory system in yeast. *J. Gen. Microbiol.* 44:149–156, 1966.
17. Postma, E., C. Verduyn, W.A. Scheffers, J.P. van Dijken. Enzymic analysis of the Crabtee effect in glucose-limited chemostat cultures of *Saccharomyces cerevisiae. Appl. Environ. Microbiol.* 55:468–477, 1989.
18. Gottschalk, G. *Bacterial Metabolism*, 2nd edition. New York: Springer-Verlag, 1986.
19. Sirakaya, B., J. Gerber, A. Demirci. Fermentation software for bioprocessing: Monitoring growth conditions without programming expertise. *Gene. Eng. News* 21(13):40,71, 2001.
20. Demirci, A., A.L. Pometto III. Enhanced production of D(-)-lactic acid by mutants of *Lactobacillus delbrueckii* ATCC 9649. *J. Indus. Microbiol.* 11:23–28, 1992.
21. Demirci, A., A.L. Pometto III, K.R. Harkins. Rapid Screening of solvents and carrier compounds for lactic acid recovery by emulsion liquid extraction and toxicity on *Lactobacillus casei* (ATCC 11443). *Bioseparation* 7:297–308, 1999.

22. Ho, L.K.G., A.L. Pometto III, P.N. Hinz, J.S. Dickson, A. Demirci. Ingredients selection for plastic composite-supports used in L(+)-lactic acid biofilm fermentation by *Lactobacillus casei* subsp. *rhamnosus*. *Appl. Environ. Microbiol.* 63:2516–2523, 1997.

23. Sivakesava, S., J. Irudayaraj, A. Demirci. Simultaneous determination of multiple components in *Lactobacillus casei* fermentation system using FT-MIR, NIR and FT-Raman spectroscopy. *Process. Biochem.* 37:371–378, 2001a.

24. Sivakesava, S., J. Irudayaraj, A. Demirci. Monitoring a bioprocess for ethanol production using FT-MIR and FT-Raman spectroscopy. *J. Ind. Microbiol. Biotechnol.* 26:185–190, 2001b.

25. Hall, M.J., S.D. Dickinson, R. Pritchard, J.I. Evans. Foams and foam control in fermentation processes, *Prog. Ind. Microbiol.* 12:171–234, 1973.

26. Jin, B., J. van Leeuwen, H.W. Doelle. The influence of geometry on hydrodynamic and mass transfer characteristics in a new external airlift reactor for the cultivation of filamentous fungi. *World J. Microbiol. Biotech.* 15(1):83–90, 1999.

27. Crueger, W., A. Crueger. *A Textbook of Industrial Microbiology*, 2nd Ed. Massachusetts: Sinauer Associates, Inc., 1990.

28. Bjurstrom, E.E. Fermentation systems and processes. In: *Food Biotechnology,* Knorr, D., ed., New York: Marcel Dekker, 1987, pp 193–222.

29. Velázquez, A.C., A.L. Pometto III, K.L.G. Ho, A. Demirci. Evaluation of plastic composite-supports in repeated-fed-batch biofilm lactic acid fermentation by *Lactobacillus casei*. *J. Appl. Biotech. Microbiol.* 55:434–441, 2001.

30. Cotton, J.C., A.L. Pometto III, J. Gvozdenovic-Jeremic. Continuous lactic acid fermentation using a plastic composite support biofilm reactor. *J. Appl. Biotech. Microbiol.* 57:626–630, 2001.

31. Demain, A.L., N.A. Solomon. *Manual of Industrial Microbiology and Biotechnology.* American Society for Microbiology, 1986.

32. Characklis, W.G. Biofilm processes,. In: *Biofilms,* Charackis, W.G., K.C. Marshall eds., New York: Wiley-Interscience, 1990, p 195–207.

33. Demirci, A., A.L. Pometto III, K.E. Johnson. Lactic acid production in a mixed culture biofilm reactor. *Appl. Environ. Microbiol.* 59:203–207, 1993.

34. Demirci, A., A.L. Pometto III. Repeated-batch fermentation in biofilm reactors with plastic-composite supports for lactic acid production. *Appl. Microbiol. Biotechnol.* 43:585–589, 1995

35. Ho, K.L.G., A.L. Pometto III, P.N. Hinz. Optimization of L (+)-lactic acid production by ring/disc plastic composite-supports through repeated-batch biofilm fermentation. *Appl. Environ. Microbiol.* 63:2533–2542, 1997.

36. Ho, K.L.G., A.L. Pometto III, P.N. Hinz, A. Demirci. Nutrients leaching and end product accumulation in plastic composite-supports for L(+)-lactic acid biofilm fermentation. *Appl. Environ. Microbiol.* 63:2524–2532, 1997.

37. Kunduru, M.R., A.L. Pometto III. Evaluation of plastic composite-supports for enhanced ethanol production in biofilm reactors. *J. Ind. Microbiol.* 16:241–248. 1996.

38. Kunduru, M.R., A.L. Pometto III. Continuous ethanol production by *Zymomonas mobilis* and *Saccharomyces cerevisiae* in biofilm reactors. *J. Ind. Microbiol.* 16:249–256, 1996.

39. Demirci, A., A.L. Pometto III, K.L.G. Ho. Ethanol production by *Saccharomyces cerevisiae* in biofilm reactors. *J. Ind. Microbiol.* 19:299–304, 1997.

40. S.E. Urbance, A.L. Pometto III, A.A. DiSpirito, A. Demirci. Medium evaluation and plastic composite support ingredient selection for biofilm formation and succinic acid production by *Actinobacillus succinogenes*. *Food Biotechnol.* 17:53–65, 2003.

22

Production of Carotenoids by Gene Combination in *Escherichia coli*

Gerhard Sandmann

CONTENTS

22.1 Carotenoids: Properties, Commercial Aspects, and
Biological Function in Human Health 533
22.2 *Escherichia coli* as a Carotenoid Production System 535
22.3 Principles of Carotenoid Biosynthesis..536
22.4 Production of Carotenoids in *Escherichia coli*: Achievements and Perspectives..... 538
References..542

22.1 CAROTENOIDS: PROPERTIES, COMMERCIAL ASPECTS, AND BIOLOGICAL FUNCTION IN HUMAN HEALTH

Carotenoids are water-soluble natural pigments of 30–50 carbon atoms. Even shorter structures called apocarotenoids result from their oxidative cleavage. Carotenoids are synthesized de novo by archaea, bacteria, fungi, and higher and lower plants. Animals are supplied with carotenoids from their food and are able to further modify the chemical structure. More than 600 carotenoids are known as intermediates or end products of different biosynthetic branches in various organisms. Most carotenoids consist of 40 carbon atoms; possess an acyclic chain; or carry β-ionone, ε-ionone, or aromatic end groups. Acyclic carotenoids are often modified at C-1 and C-2, e.g., by 1-HO, 1-CH$_3$O, or 2-keto groups. Typical substitutions of a β-ionone end are 3-hydroxy, 4-keto, and 5,6-epoxy moieties. The 3-hydroxy group can participate in glycosilation or formation of fatty acid esters. The most prominent feature of a carotenoid molecule is the polyene chain. Delocalization of the Π-electrons is responsible for their color and their antioxidative potential. Carotenoids can interact with radical chain reactions and are capable of energy dissipation from photosensitizers as heat. In photosynthesis, carotenoids function as light-harvesting antenna, transferring light energy to chlorophyll. More details on the structure and function of carotenoids can be found in References 1 and 2.

The commercially most important carotenoids are astaxanthin and β-carotene (structures 22.1 and 22.2, Figure 22.1) (3). Most of the carotenoids on the market is produced by chemical synthesis. A chemical process for the synthesis of lycopene (structure 22.3, Figure 22.1) has recently been established. The industrial use of carotenoids involves their application in nutrient supplementation, for pharmaceutical purposes, as food colorants, and in animal feeds. β-Carotene or other carotenoids with an unsubstituted β-ionone group are essential in human nutrition as a provitamin A. Taken up by the body, they are then metabolized to vitamin A, an integral component of the visual process. Lutein and zeaxanthin (structures 4 and 5, Figure 22.1) are constituents of the yellow eye spot, the macula lutea. There, both carotenoids act as protectants, preventing the retina from photodamage (4). Other carotenoids have a less defined role in human health. Evidence is accumulating that various carotenoids stimulate the immune system and play an important role in the prevention of degenerative diseases and cancer (5).

In addition to chemical synthesis, some biological systems for carotenoids are in use (6). Astaxanthin is accumulated in the green alga *Haematococcus pluvialis* or in the fungus

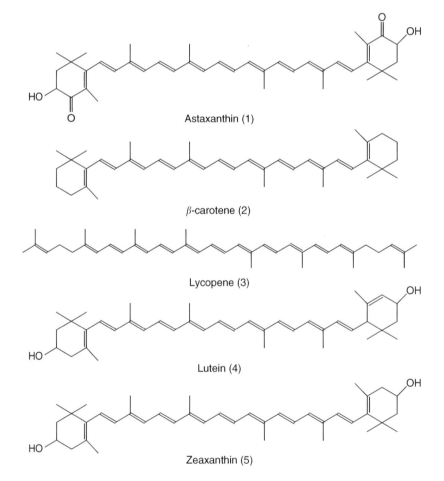

Astaxanthin (1)

β-carotene (2)

Lycopene (3)

Lutein (4)

Zeaxanthin (5)

Figure 22.1 Chemical structures of carotenoids of commercial interest or with an effect on human health.

Phaffia rhodozyma (synonymous to *Xanthophyllomyces dendrorhous*). β-Carotene is produced by cultivation of the green alga *Dunaliella salina* or the fungus *Blakeslea trispora.* Palm kernel oil is another source for β-carotene. Natural sources of other carotenoids are flowers of *Tagetes erecta* for lutein (as fatty acid esters), red fruit of *Capsicum annuum* for capsanthin, and capsorubin or tomato for lycopene.

In recent years, several transgenic microorganisms have been explored as production hosts for carotenoids. These were the fungi *Saccharomyces cerevisiae* and *Candida utilis* as well as the bacteria *Zymomonas mobilis, Agrobacterium tumefaciens,* and *Escherichia coli (E. coli)* (7–9). Among them, *E. coli* is the most advanced production system.

22.2 *ESCHERICHIA COLI* AS A CAROTENOID PRODUCTION SYSTEM

Escherichia coli was the first host for functional expression of carotenogenic genes in order to analyze the function of the corresponding enzymes (10,11). Later, this bacterium was very useful in the heterologous overexpression of individual carotenogenic enzymes as functional proteins for their purification and biochemical characterization (12).

Escherichia coli is a very convenient host for heterologous carotenoid production. Because of its fast and easy cultivation in substantial quantity, it can be transformed simultaneously with several plasmids as long as they belong to different incompatibility groups, i.e., possessing different origins of replication. For their stable maintenance in the cells, it is essential that each plasmid carries a different antibiotic resistance. Only when this selection pressure is maintained spontaneous plasmid loss prevented. Several useful plasmids for cotransformation are available. Convenient vectors are pUC-related plasmids with the pMB1 origin of replication and ampicillin resistance, pACYC184 with a p15A origin of replication and chloramphenicol resistance, pRK404 with an RK2 origin of replication and tetracycline resistance, and pBBR1MCS2 with a SC101 origin of replication and kanamycin resistance. These plasmids can be used for expression of individual genes or groups of genes which mediate the formation of certain basic carotenoid structures. For example, one plasmid may carry the genes necessary to obtain certain carotenoid intermediates and others the genes for systematic modifications of the structure (13,14). It should be pointed out that *E. coli* is noncarotenogenic which increases the flexibility to build up a desired carotenoid synthesis pathway. However, *E. coli* has to cope with the drain of prenyl pyrophosphates when carotenogenesis is established. Other plasmids may be used to overexpress certain genes to enhance the *E. coli* metabolic capacity for the supply of carotenoid precursors. Details on the choice of plasmids and growth conditions for carotenoid producing *E. coli* transformants are given in a recent publication (15).

In the majority of bacteria, including *E. coli*, formation of prenyl pyrophosphates, which are the precursors of carotenoids, proceed via a reaction sequence which is different from the mevalonate pathway typically found in fungi and animals (16). It is referred to either as the 1-deoxyxylulose-5-phosphate pathway, because this is the first intermediate, or as the 2-C-methyl-D-erythritol-4-phosphate pathway, because this is the first product which is converted to prenyl pyrophosphate exclusively. To date, most of the reaction steps have been elucidated. Initially, a C_2 unit from pyruvate is condensed to glyceraldehyde by a thiamin-dependent 1-deoxyxylulose-5-phosphate synthase (Figure 22.2). Then, the product is converted to 2-C-methyl-D-erythrol-4-phosphate by 1-deoxyxylulose-5-phosphate reductoisomerase. The following reactions involve the formation of 4-diphosphocytidyl-2-C-methylerythrol in a CTP-dependent reaction and phosphorylation at position 2 by adenosine triphosphate. Subsequently, cytidine-5-phosphate is released and via 2-C-methyl-D-erythritol-2,4-cyclodiphosphate 1-HO-2-methyl-2-butenyl-4-diphosphate is formed.

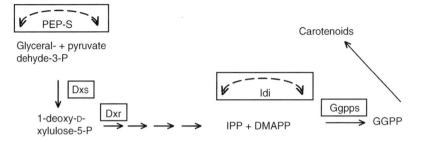

Figure 22.2 Metabolic engineering of the *Escherichia coli* deoxyxylulose-5-phosphate pathway and prenyl pyrophosphate formation for optimum precursor supply. Overexpression of the indicated enzymes either relieve a limitation of the metabolic flow or provide a more favorable balance for two-substrate reactions. PEP-S, phosphoenolpyruvate synthase; Dxs, deoxyxylulose-5-phosphate synthase; Dxr, deoxyxylulose-5-phosphate reducto isomerase; Idi, isopentenyl pyrophosphate isomerase; Gggpps, geranylgeranyl pyrophosphate synthase.

This product may be the branching point for independent routes to isopentenyl pyrophosphate (IPP) and dimethylallyl pyrophosphate (DMAPP). However, the details of the final steps to IPP and DMAPP are not fully understood yet. See Reference 17 for the newest insight into later reactions of this novel pathway.

Precursor supply for carotenogenesis can be increased in *E. coli* by overexpression of limiting enzymes of the deoxyxylulose-5-phosphate pathway and subsequent reactions. Supply of prenyl pyrophosphates and subsequent carotenogenesis was stimulated by transformation with the genes encoding 1-deoxy-D-xylulose-5-phosphate synthase, 1-deoxy-D-xylulose-5-phosphate reductoisomerase, and IPP isomerase under a strong promoter (Figure 22.2). Because geranylgeranyl pyrophosphate is the direct substrate for the formation of the first carotenoid in the pathway and its level is comparably low in *E. coli*, high expression levels of geranylgeranyl pyrophosphate synthase are very important for carotenogenesis (18). Another bottleneck for carotenoid biosynthesis in *E. coli* was relieved in the pathway by overexpressing the gene which encodes phosphoenolpyruvate synthase, a pyruvate consuming enzyme, indicating that the pools of glyceraldehyde 3-phosphate and pyruvate, which both are substrates of 1-deoxy-D-xylulose 5-phosphate synthase, have to be more balanced in the direction of glyceraldehyde 3-phosphate (19). By overexpression of different combinations of the limiting enzymes mentioned above, concentrations of various carotenoids to a final yield of 1.5 mg/g dry weight could be reached (20).

22.3 PRINCIPLES OF CAROTENOID BIOSYNTHESIS

C_{40} carotenoids are synthesized by head-to-head condensation of two molecules of C_{20} geranylgeranyl pyrophosphate catalyzed by the enzyme phytoene synthase (Figure 22.3A). A very similar condensation reaction from C_{15} farnesyl pyrophosphate leads to the formation of C_{30} carotenoids. In the next steps, the conjugated double bond system of C_{30} and C_{40} carotenes is sequentially extended, each time integrating one of the isolated double bonds. Different phytoene desaturases exist with respect to the catalyzed number of desaturation steps. The enzyme Pds from plants inserts only two double bonds symmetrically at positions 11 and 11′ (Figure 22.3B). The bacterial phytoene desaturase from *Rhodobacter capsulatus*, CrtI$_{Rc}$, carries out a 3-step desaturation with an additional

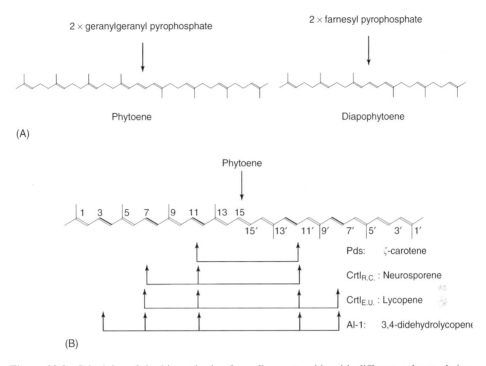

Figure 22.3 Principles of the biosynthesis of acyclic carotenoids with different polyene chains. (A) Establishment of the basic C_{30} and C_{40} carbon chain. (B) Extension of the conjugated double bond system. Double bonds in bold are introduced by the individual phytoene desaturases as indicated by the arrows yielding the specified carotenes.

double bond at position 7. The most abundant bacterial phytoene desaturase like $CrtI_{Eu}$ catalyzes a 4-step desaturation with additional double bonds at position 7 and 7′. The enzyme Al-1 from the fungus *Neurospora crassa* catalyzes a 5-step desaturation with an additional double bond at position 3. These individual desaturation reactions lead to the formation of the products ξ-carotene, neurosporene, lycopene, and 3,4-didehydrolycopene, respectively.

Lycopene can be modified at the terminal double bonds by addition of water, resulting in 1-HO derivatives which can be methylated to 1-CH_3O carotenoids (Figure 22.4). Desaturation at position 3,4 requires the presence of the 1-HO or 1-CH_3O group. These types of carotenoids are typical for photosynthetic bacteria. Finally, a 2-keto group can be introduced under aerobic growth. Another addition to the 1,2 double bond of lycopene is involved in the synthesis of C_{45} and C_{50} carotenoids. This chain elongation at C-2 proceeds via addition of one or two dimethyl allyl cation. Then, the molecule is stabilized by abstraction of a proton from C-17 and C17′.

Cyclization of lycopene ends to ionone groups involves protonation of the 1,2 double bond and the addition of the resulting carbo cation to the 5,6 double bond (Figure 22.4). Stabilization occurred by proton abstraction from either C-6 or C-4 yields β- or ε-rings, respectively. The possible individual substitutions of a β-ionone end group are summarized in Figure 22.4. A cyclic carotenoid may carry a 3-HO, 4-keto, and/or 5,6-epoxy moiety. It should be pointed out that only the combinations of 3-HO and 4-keto, or 3-HO and 5,6-epoxy are known. Details on the carotenogenic pathway can be found in many reviews (1,21,22).

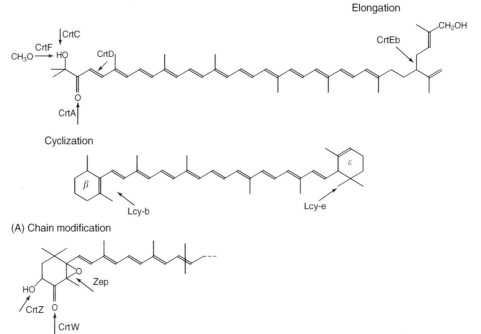

Figure 22.4   Reactions modifying the lycopene molecule (A) or a β-ionone ring (B).

22.4 PRODUCTION OF CAROTENOIDS IN *ESCHERICHIA COLI*: ACHIEVEMENTS AND PERSPECTIVES

Escherichia coli was the first noncarotenogenic organism into which carotenoid biosynthesis was engineered (10,11). Carotenogenic cells were obtained when gene clusters from *Erwinia* species was functionally analyzed. The expression before and after deletions of individual genes lead to the accumulation of the end product zeaxanthin diglucoside and all intermediates like phytoene, lycopene, β-carotene, and zeaxanthin. Many different carotenoids have been produced since, by using genes from many different organisms. Depending on the structure and the amount of carotenoids formed, pigmentation of *E. coli* can be very strong ranging in color from yellow to orange and red.

Genes used for heterologous carotenoid production in *E. coli* originated not only from bacteria but also from fungi, algae, and higher plants. An example is given in Figure 22.5 that demonstrates how genes from all theses organisms were combined in *E. coli,* interacting simultaneously in carotenogenesis. Initially, formation of lycopene is catalyzed by three enzymes encoded by bacterial genes, crtE, crtB, and crtI. Then, a fungal monocyclase Al-2 takes over, forming γ-carotene. Into this molecule, a plant hydroxylase Bhy introduces a 3-hydroxy group. Due to the functionality of all these enzymes in the foreign host *E. coli*, the final product 3-hydroxy-γ-carotene is obtained. The production of this carotenoid is also an example of the synthesis of a carotenoid which is not formed in any of the organisms from which these genes originated. The synthesis of novel carotenoids by combining genes from different organisms only works because the substrate specificity of the enzyme is such that it does not need to recognize the entire substrate molecule, but only certain regions of the molecule which are suitable for conversion. This combinatorial approach can result in the formation of totally novel carotenoids. For example, 1-hydroxylated acyclic

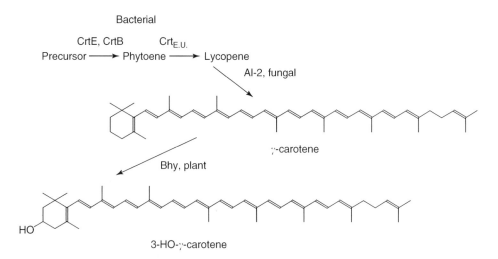

Bacterial

CrtE, CrtB   Crt$_{E.U.}$

Precursor ⟶ Phytoene ⟶ Lycopene

AI-2, fungal

γ-carotene

Bhy, plant

HO

3-HO-γ-carotene

Figure 22.5   Example of heterologous production of the carotenoid 3-HO-γ-carotene by combination of genes from bacteria, fungi and plants.

structures with up to 13 conjugated double bonds were designed with an antioxidative potential which was superior to other related carotenoids (23).

The production of many other carotenoids was successful in the factory *E. coli*. Their synthesis was assembled in a modular way by transformation with the appropriate genes which encode the enzymes responsible for the individual catalytic steps. For this approach, it is important to know the substrate and product specificity of the encoded enzymes. Sometimes, the products of highly homologous genes do not catalyze the same reaction. The 3,4-carotene desaturases CrtD$_{Rg}$ and CrtD$_{Rs}$ from the photosynthetic purple bacteria *Rubrivivax gelatinosus* and *Rhodobacter sphaeroides*, respectively; both enzymes catalyze the conversion of 1-HO-neurosporene and 1-HO-lycopene. However, only the enzyme from *R. gelatinosus* is able to desaturate 1-HO-3′,4′-didehydrolycopene as indicated in Figure 22.6A (24).

A list of genes for the synthesis of different carotenoid structures together with the necessary information on the catalytic function of their products can be found in Reference 25. Table 22.1 comprises the individual carotenoids which can be synthesized in *E. coli*. They include carotenoids of different chain lengths, cyclic and acyclic structures with shorter or longer conjugated double bond systems, unsubstituted hydrocarbons or hydroxy, and keto and epoxy derivatives. All these carotenoids can be produced in *E. coli* in quantities of up to 1–1.5 mg/g dry weight. Many of the carotenoids in Table 22.1 are found in suitable organisms only as biosynthetic intermediates which accumulate only in trace amounts, making it very difficult to extract and purify them. Others have not been detected in biological material before. These carotenoids are marked in bold face. Experimental details for the heterologous production of carotenoids in *E. coli* were recently published (15). Several of the double bonds in a carotenoid molecule can be formed in a cis or trans configuration. For some of the carotenoids produced in *E. coli*, the composition of these geometrical isomers has been determined (26). In general, the all-trans isomer is dominating. In the case of β-carotene, small quantities of 9-, 13-cis, and 9,13,13′-tetracis isomers were also formed. The formation of the major lycopene isomers is determined by the type of gene used to transform *E. coli*. With crtI related genes, mainly all-trans and smaller amounts of 5-, 9- and 13-cis isomers accumulate.

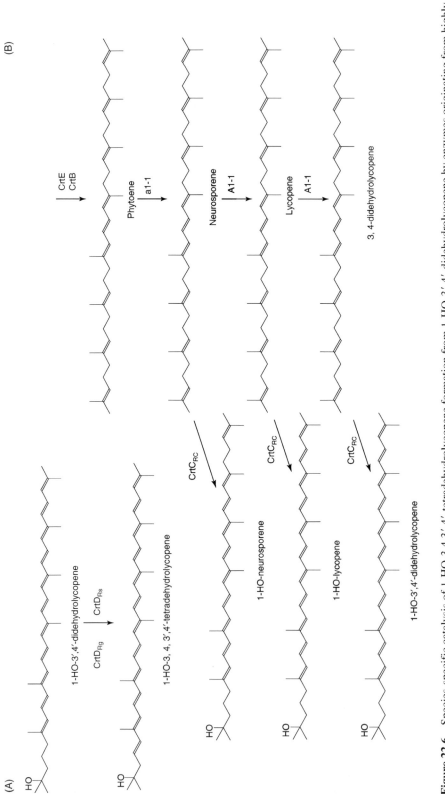

Figure 22.6 Species-specific catalysis of 1-HO-3,4,3′,4′-tetradehydrolycopene formation from 1-HO-3′,4′-didehydrolycopene by enzymes originating from highly homologous genes (A) and an example of intermediate flow into side reactions in the pathway leading to 1-HO-3′,4′-didehydrolycopene (B).

Table 22.1

Gene combinations and production of carotenoids in *E. coli*.

Erwinia:	crtE, crtB,	+ crtI$_{Eu}$, crtY, crtX
		Phytoene, Lycopene, β-Carotene, Zeaxanthin,
		Zeaxanthinglucosid
	+ Rhodobacter crtI$_{Rc}$, crtC + Erwinia crtY	
		1′,3-(HO)$_2$-γ Carotene§,
	+ Rhodobacter crtI$_{Rc}$ + Erwinia crtY	
		7,8-Dihydro-β-carotene*, 7,8,7′,8′-Tetrahydro-β-carotene*
	+ Rhodobacter/Rubrivivax crtI$_{Rc}$, crtD, crtC	
		1-HO-Lycopene; 1,1′-(HO)$_2$-Lycopene, 1′-HO-γ-Carotene
		Neurosporene, 1-HO-Neurosporene, Demethylsperoidene
		1-HO-3,4-Didehydrolycopene§, 1,1′-(HO)$_2$-3,4-
		Didehydrolycopene§
		1,1′-(HO)$_2$-3,4,3′,4′-Tetradehydrolycopene§
		7,8-Dihydrozeaxanthin, 3-HO-β-Zeacarotin
	+ Neurospora al-1 + Rhodobacter/Rubrivivax crtC, crtD	
		3,4-Didehydrolycopene, **1-HO-3′,4′-Didehydrolycopene,§**
		1-HO-3,4,3′,4′-Tetradehydrolycopene,§
	+ Erwinia crtI$_{Eu}$, crtY + Synechocystis crtO	
		Echinenone
	+ Erwinia crtI$_{Eu}$, crtY + Haematococcus bkt	
		Canthaxanthin
	+ Synechococcus crtP	
		ζ-Carotene
	+ Erwinia crtI$_{Eu}$ + Corynebacterium crtEb	
		Flavuxanthin, Nonaflavuxanthin
		Staphylococcus crtM
		Diapophytoene
	+ Staphylococcus crtN	
		Diapo-ζ-carotene, Diaponeurosporene
	+ Staphylococcus crtN + Anabaena crtQa	
		Diapolycopene

Note: Carotenoids in bold face have not been found before in biological material; § indicates identification by NMR; others by HPLC, absorbance, and mass spectroscopy.

Source: From Sandmann, G., *Chem. Biochem.*, 3, 629–635, 2002. With permission.

When the crtI gene is replaced by another phytoene desaturase gene crtP together with the ζ-carotene desaturase gene crtQb, prolycopene (=7,9,7′,9′-tetracis) dominates and all-trans lycopene is minor.

In some cases, the end product of an established carotenoid pathway may be much lower than expected. Instead, a mixture of structurally related compound is obtained. This may be the case when the enzymes react, not one after the other in a fixed sequence, but due to their broad substrate specificity in a metabolic web. When, e.g., the desaturase Al-1 and the 1,2-hydratase CrtC interact, not only the end product of the desaturation reaction 3,4-didehydrolycopene but also the intermediates neurosporene and lycopene are modified to the corresponding 1-HO derivatives by the hydratase (Figure 22.6B).

The major advantage of *E. coli* for carotenoid production is the versatility for the generation of very diverse structures. However, compared to the biological carotenoid production systems like *Dunaliella, Haematococcus, Blakeslea,* and *Phaffia* mentioned before, the yields are lower. Metabolic engineering of the *E. coli* terpenoid biosynthetic reactions led to a considerable increase of carotenoid production and there are indications that precursor supply is not limiting in these engineered strains. The major problem for *E. coli* seems to be carotenoid storage. All the lipophilic carotenoids are sequestered in the cell membranes. Therefore, future activities may aim to extend the carotenoid storage capacity by genetic modification of the density of membranes in *E. coli* cells or by establishing plastoglobuli-like structures.

REFERENCES

1. Goodwin, T.W. *The Biochemistry of the Carotenoids: Vol. I Plants,* 2nd ed. London: Chapman and Hall, 1980.
2. Britton, G. Structure and properties of carotenoids in relation to function. *FASEB J.* 9:1551–1558, 1995.
3. Nonomura, A.M. Industrial biosynthesis of carotenoids, in *Carotenoid Chemistry and Biology,* Krinsky, N.I., M.M. Mathews-Roth, R.F. Taylor, eds. New York: Plenum Press, 1989, pp 365–375.
4. Landrum, J.T., R.A. Bone. Carotenoids in the retina and lens: possible acute and chronic effects on human visible performance. *Arch. Biochem. Biophys.,* 385:28–40, 2001.
5. Mayne, S.T. Beta-carotene, carotenoids, and disease prevention in humans. *FASEB J.* 10:690–701, 1996.
6. Nelis, H.J., A.P. de Leenheer. Microbial sources of carotenoid pigments used in foods and feeds. *J. Appl. Bacteriol.* 70:181–191, 1991.
7. Misawa, N., S. Yamano, H. Ikenaga. Production of β-carotene in *Zymomonas mobilis* and *Agrobacterium tumefaciens* by introduction of the biosynthesis genes from *Erwinia uredovora. Appl. Environ. Microbiol.* 57:1874–1849, 1991.
8. Yamano, S., T. Ishii, N. Nakagawa, H. Ikenaga, N. Misawa. Metabolic engineering for production of β-carotene and lycopene in *Saccharomyces cerevisiae. Biosci. Biotechnol. Biochem.* 58:1112–1114, 1994.
9. Miura, Y., K. Kondo, T. Saito, H. Shimada, P.D. Fraser, N. Misawa. Production of the carotenoids lycopene, β-carotene, and astaxanthin in the food yeast *Candida utilis. Appl. Environ. Microbiol.* 64:1226–1229, 1998.
10. Misawa, N., M. Nakagawa, K. Kobayashi, S. Yamano, K. Nakamura, K. Harashima. Elucidation of the *Erwinia uredovora* carotenoid biosynthetic pathway by functional analysis of gene products expressed in *Escherichia coli. J. Bacteriol.* 172:6704–6712, 1990.
11. Schnurr, G., A. Schmidt, G. Sandmann. Mapping of a carotenogenic gene cluster from *Erwinia herbicola* and functional identification of its six genes. *FEMS Microbiol. Lett.* 78:157–162, 1991.
12. Sandmann, G. High level expression of carotenogenic genes for enzyme purification and biochemical characterization. *Pure Appl. Chem.* 69:2163–2168, 1997.
13. Sandmann, G., M. Albrecht, G. Schnurr, P. Knörzer, P. Böger. The biotechnological potential and design of novel carotenoids by gene combination in *Escherichia coli. Trends Biotechnol.* 17:233–237, 1999.
14. Schmidt-Dannert, C. Engineering novel carotenoids in microorganisms. *Curr. Opin. Biotechnol.* 11:255–261, 2000.
15. Sandmann, G. Combinatorial biosynthesis of novel carotenoids in *Escherichia coli.* In: *E. coli Gene Expression Protocols, Methods in Molecular Biology,* Vol. 205, Vaillantcourt, P.E., ed., Totowa, NJ: Humana Press, 2003, pp 303–314.
16. Lichtenthaler, H.K. The 1-deoxy-D-xylulose-5-phosphate pathway of isoprenoid biosynthesis in plants. *Annu. Rev. Plant Physiol. Plant Mol. Biol.* 50:47–65, 1999.

17. Rhodich, F., S. Hecht, K. Gärtner, P. Adam, C. Krieger, S. Amslinger, D. Arigoni, A. Bacher, W. Eisenreich. Studies on the non-mevalonate terpene biosynthetic pathway: metabolic role of IspH (LytB) protein. *Proc. Natl. Acad. Sci. USA* 99:1158–1163, 2002.

18. Ruther, A., N. Misawa, P. Böger, G. Sandmann. Production of zeaxanthin in *Escherichia coli* transformed with different carotenogenic plasmids. *Appl. Microbiol. Biotechnol.* 48:62–167, 1997.

19. Farmer, W.R., J.C. Liao. Precursor balancing for metabolic engineering of lycopene production in *Escherichia coli*. *Biotechnol. Prog.* 17:57–83, 2001.

20. Albrecht, M., N. Misawa, G. Sandmann. Metabolic engineering of the terpenoid biosynthetic pathway of *Escherichia coli* for production of the carotenoids β-carotene and zeaxanthin. *Biotechnol. Lett.* 21:791–795, 1999.

21. Cunningham, F.X., E. Gantt. Genes and enzymes of carotenoid biosynthesis in plants. *Annu. Rev. Plant Physiol. Mol. Biol.* 49:557–583, 1998.

22. Sandmann, G. Carotenoid biosynthesis and biotechnological application. *Arch. Biochem. Biophys.* 385:4–12, 2001.

23. Albrecht, M., S. Takaichi, S. Steiger, Z.Y. Wang, G. Sandmann. Novel hydroxycarotenoids with improved antioxidative properties produced by gene combination in *Escherichia coli*. *Nat. Biotechnol.* 18:843–846, 2000.

24. Steiger, S., S. Takaichi, G. Sandmann. Heterologous production of two novel acyclic carotenoids, 1,1'-dihydroxy-3,4-didehydrolycopene and 1-hydroxy-3,4,3',4' tetrahydrolycopene by combination of the crtC and crtD genes from *Rhodobacter* and *Rubrivivax*. *J. Biotechnol.* 97:51–58, 2002.

25. Sandmann, G. Combinatorial biosynthesis of carotenoids in a heterologous host: a powerful approach for the biosynthesis of novel structures. *Chem. Biochem.* 3:629–635, 2002.

26. Breitenbach, J., G. Braun, S. Steiger, G. Sandmann. Chromatographic performance on a C_{30}-bonded stationary phase of mono hydroxycarotenoids with variable chain length or degree of desaturation and of lycopene isomers synthesized by different carotene desaturases. *J. Chromatogr.* 936:59–69, 2001.

23

Production of Amino Acids: Physiological and Genetic Approaches

Reinhard Krämer

CONTENTS

23.1 Introduction 545
23.2 Physiological, Biochemical, and Genetic Background of Amino Acid Production ...548
23.3 Physiological, Biochemical, and Genetic Tools used to Improve Amino Acid
      Production 554
23.4 Amino Acid Production: Technical Aspects 558
23.5 Selected Examples of Amino Acids..559
      23.5.1 Glutamate 559
      23.5.2 L-phenylalanine 564
      23.5.3 L-tryptophan 565
      23.5.4 L-lysine 565
      23.5.5 Threonine..568
      23.5.6 Other Amino Acids...569
23.6 Future Developments and Perspectives 569
References..575

23.1 INTRODUCTION

Amino acids are the basic constituents of cellular proteins, as well as important nutrients for all living cells. Their importance is closely related to their nutritional value, in particular for those amino acids which are essential for man and animals (lysine, threonine, methionine, tryptophan, phenylalanine, valine, leucine, isoleucine, and, to some extent, arginine and histidine). Amino acids are used for a variety of purposes, depending on their taste, nutritional value, other physiological activities, and chemical properties. Besides their application in human nutrition (flavor enhancers, sweeteners), they are used as feed supplements (animal nutrition), as well as for cosmetics, in the pharmaceutical industry (infusion solutions, dietary food), and as building blocks for chemical synthesis (Table 23.1).

The use of amino acids in food consumes nearly 40% of total amino acid production. Major products used in the food industry are L-glutamate (monosodium glutamate, MSG), L-aspartate, L-phenylalanine, glycine, and L-cysteine. These amino acids are produced by a variety of companies in East Asia, the USA, and Europe (1).

The history of amino acid production is, to a large part, the history of the major producing organism, namely *Corynebacterium glutamicum*. In 1908, K. Ikeda discovered that sodium glutamate was responsible for the specific flavoring component of kelp, a traditional taste enhancing material in Japanese cooking. For a long time, sodium glutamate was commercially isolated from vegetable proteins, because chemical synthesis of glutamate leads to the formation of racemate (racemic acid ester). The sodium salt of the D-isomer, however, is tasteless. In 1957 a screen for bacteria with the capacity to excrete amino acids, in K. Kinoshita and S. Udaka were successful in finding a Gram-positive soil bacterium that was auxotrophic for the vitamin biotin and that spontaneously excreted glutamate under conditions of biotin limitation (2). Since then, a number of bacteria have been isolated that were also characterized by the peculiar capacity of excreting glutamate. Later, they were all reclassified as *C. glutamicum* (3). *Corynebacterium glutamicum*, a soil bacterium, belongs to the group of mycolic acid containing actinomycetes, together with *Nocardia* and *Mycobacteria*. After the successful introduction of *C. glutamicum* for commercial production of large quantities of amino acids, other bacteria, namely *Escherichia coli*, have also successfully been used for this purpose. Interestingly, a bacterium closely related to *C. glutamicum*, namely *Corynebacterium ammoniagenes*, is commercially used for the production of large quantities of nucleotides, another important flavor enhancer in food biotechnology. Recently, the importance of glutamate as a flavor enhancer has been explained by the discovery of special receptor proteins in the human tongue responsible for sensing the taste of glutamate (4).

The majority of amino acids used in the food industry are used as flavor enhancers (mainly L-glutamate, but to some extent also glycine and L-phenylalanine), as sweeteners (L-aspartate and L-phenylalanine, the building blocks for the sweetener aspartame), and for

Table 23.1

Currrent production and application of amino acids. (Data are estimated values for the year 2003)

Amino Acid	Estimated Production (t/y)	Preferred Production Method	Major Use
L-Glutamate	1.100.000	Fermentation	Flavor enhancer
L-Lysine	550.000	Fermentation	Feed additive
D,L-Methionine	550.000	Chemical synthesis	Feed additive
L-Threonine	50.000	Fermentation	Feed additive
Glycine	22.000	Chemical synthesis	Food additive
L-Aspartate	10.000	Enzymatic catalysis	Sweetener (Aspartame)
L-Phenylalanine	10.000	Fermentation	Sweetener (Aspartame)
L-Cysteine	3.000	Enzymatic method	Food additive
L-Arginine	1.000	Fermentation	Pharmaceutical
L-Tryptophan	500	Fermentation	Pharmaceutical
L-Valine	500	Fermentation	Pharmaceutical
L-Leucine	500	Fermentation	Pharmaceutical

a variety of purposes as food additives (e.g., cysteine as a flour additive). In addition, essential amino acids are used as ingredients in dietary food.

In addition to the importance of amino acids as food supplements, there is a huge market for amino acids as additives to animal feed (Table 23.1). The basic goal is to improve the value of animal feed, which typically lacks the essential amino acids lysine, methionine, threonine, and tryptophan as well as other essential amino acids to a minor extent (5,6). The economic driving force for manufacturing and adding amino acids to animal feed is twofold: improvement of the quality of feed leads to a reduction of the quantity needed, which leads to cost reduction; and an enormous reduction of animal waste (manure), because the individual animal will take in as much food as necessary to obtain a sufficient amount of the most limiting nutrients, which, in general, are essential amino acids. The second aspect is thus not only of economic, but also of high ecologic impact. Some aspects of the production of amino acids which are used as feed supplements will also be included in this chapter, because a number of highly relevant and interesting scientific developments in terms of physiology, biochemistry, and molecular approaches have been made in connection with these amino acids, particularly lysine. The results and tools developed in this field will certainly also find their application in the production of amino acids as food additives.

Amino acids which are used in bulk quantities in food and feed applications can be divided into four categories: (1) Those that are mainly synthesized by chemical procedures leading to racemic mixtures. The main example for this class is methionine, because animals (and humans) are able to convert the D-form of this amino acid into the L-form. The achiral amino acid glycine can also be produced by chemical synthesis. (2) Those that are synthesized by enzymatic synthesis, frequently using the enzymes of immobilized bacterial cells. A typical product manufactured by this method is aspartate, for which the enzyme aspartase is used, either from *E. coli* cells immobilized with organic polymers or from *Pseudomonas putida* cells fixed in the natural polymer carrageenan (6). (3) Those that are produced by gene tailored organisms. Glutamate, which is produced by *C. glutamicum*, represents the oldest biotechnological product in this field, and is at the same time the amino acid produced in the largest quantities. (4) The rest of the amino acids, namely lysine, threonine, and phenylalanine, which are produced by using fermentative techniques and microorganisms. The processes for their production substantially differ from those developed for glutamate. In addition, more sophisticated chemical procedures are used in some cases, which lead to stereospecific synthesis of particular amino acids. Most probably, these processes will be replaced in the future by biotechnological techniques based on the use of engineered microorganisms, if the market volume for these amino acids becomes large enough to call for the development of appropriate processes.

The organisms used for biotechnological processes of amino acid production are mainly *C. glutamicum* (glutamate, lysine, aromatic amino acids) and *E. coli* (threonine, aromatic amino acids). A number of other organisms have also been used, e.g., *Serratia marcescens*, a relative to *E. coli*, and others (5,6), these processes, however, are not of high economic impact so far. There is a clear tendency to use only the two major organisms (*C. glutamicum* and *E. coli*) as biotechnological workhorses and, on the basis of the wealth of knowledge on the genetic, biochemical, and in particular physiological level, to adapt their extraordinary capacities to the needs of producing the desired metabolite.

This chapter will have two foci. First, the physiological basis of amino acid excretion in bacteria, a phenomenon that does not seem to be very "physiological" at first view, will be discussed. Second, genetic and biochemical strategies that have been used to improve and to optimize the production of amino acids in a biotechnological scale will be described.

A detailed understanding of the background of amino acid production in terms of physiology, biochemistry, and genetics has proven to be absolutely essential for applying rational strategies to the optimization of amino acid production by metabolic design.

23.2 PHYSIOLOGICAL, BIOCHEMICAL, AND GENETIC BACKGROUND OF AMINO ACID PRODUCTION

It may be argued from a phenomenological point of view that amino acid producing strains can simply be classified into three different categories: wild-type strains able to excrete particular amino acids under specific culture conditions; regulatory mutants from which feedback control of amino acid biosynthesis has been removed; and (3) genetically modified strains in which the biosynthetic capacity has been amplified. These categories may also be combined, as well. This does not, however, fully explain the complexity of the metabolic background behind the peculiar capability to excrete amino acids.

A rational approach to amino acid production is based on a detailed understanding of the particular physiological reasons leading to amino acid excretion. As a matter of fact, amino acid excretion was for a long time interpreted as not being a physiological event, and this perception is still commonly accepted: "The secretion of amino acids serves no purpose for a microbe (7)." Amino acid excretion was thought to result from the impact of biotechnological techniques on the metabolic situation and the integrity of bacterial cells. On the contrary, we now know that amino acid production in most cases has to be rationalized in terms of exploiting and optimizing the physiological property of active amino acid excretion that is already naturally present in many, if not all bacterial organisms (8,9).

Consequently, discussion of the physiological, biochemical, and genetic background of amino acid production necessarily should start by raising and answering the following question. Why would microorganisms, which are generally thought to apply economic strategies in order to survive and to outcompete their neighbors, waste precious compounds such as amino acids, which have to be synthesized at high cost of carbon, nitrogen, and energy? In other words: what is the physiological meaning of the presence of amino acid excretion systems in microorganisms?

Before discussing details of cellular metabolism, one should be aware of a basic misunderstanding. When describing bacterial energy metabolism, F. Harold has conclusively stated: "It must be concluded that the bacterial solution to energy surplus is in general to waste it (10)." Under conditions of carbon and energy surplus, the major survival strategy of bacteria is not the optimization of metabolism in terms of economy and efficiency but in terms of metabolism or growth speed.

On a more detailed level, there are actually at least two plausible reasons that may cause amino acid excretion under physiological conditions (Figure 23.1). They will be explained predominantly in *C. glutamicum*, because of its widespread use in amino acid production. As a soil bacterium, *C. glutamicum* uses all available materials from its natural surroundings as carbon and energy sources. Peptides, derived from decomposing organic matter, are certainly an important source for these needs, and, like most bacteria, *C. glutamicum* harbors efficient uptake systems for peptides. On the other hand, this organism is not very well equipped with efficient degradation systems for various amino acids, e.g., lysine, arginine, branched chain, and aromatic amino acids. If peptides are used as carbon and energy sources, amino acids close to the central metabolism, e.g., glutamate, aspartate, or alanine, are efficiently metabolized. Those amino acids that are not efficiently metabolized, however, will accumulate and will lead to problems due to high internal concentrations. This has been demonstrated by the addition of peptides that contain lysine,

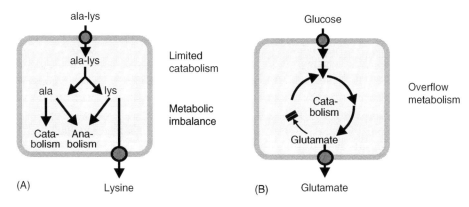

Figure 23.1   Physiological concepts of amino acid excretion. Two major concepts explaining excretion of amino acids based on physiological reasons. (A) Amino acids which are accumulated because of limited catabolic capacity or of a metabolically imbalanced situation are excreted (examples L-lysine in *C. glutamicum*, L-threonine in *E. coli*). (B) Amino acid excretion occurs under conditions of metabolic overflow, i.e., when a metabolic limitation occurs in the presence of a surplus of carbon, nitrogen and energy (example L-glutamate in *C. glutamicum*).

threonine, or isoleucine, as substrates, which leads to high cytoplasmic accumulation (11–13). Even more convincing, it was shown that deletion of the specific lysine excretion system in *C. glutamicum* may cause internal lysine concentrations of more than 1 M as a result of peptide feeding (14). This experiment is actually the most convincing explanation for the physiological necessity of amino acid excretion systems, at least in *C. glutamicum*. This explanation has been designated "metabolic imbalance." In a broader sense, this concept may hold for any imbalanced situation in the cell that can be corrected via export activity. A recent, very interesting example is documented by the discovery of a novel cysteine export system in *E. coli*, the function of which seems to be maintaining redox balance in connection with cytochrome biosynthesis (15).

In addition to lysine, this concept holds for a number of amino acids in *C. glutamicum*, such as threonine and branched chain amino acids, but is not applicable, of course, for glutamate, a metabolite too closely related to the central metabolism. Furthermore, this concept does not easily explain amino acid excretion in other organisms, for example *E. coli*, which is able to degrade those amino acids. Nevertheless, even in those cases it may be an obvious advantage for the bacterial cell to be able to effectively control the steady state concentration of cytoplasmic pools of some amino acids which are not closely related to the central metabolism by efflux, and not only by degradation, in times of a surplus of carbon sources.

In the case of glutamate excretion in *C. glutamicum*, however, a completely different explanation which is based on more general metabolic situations, has been found to apply. It has been studied in a number of microorganisms besides *C. glutamicum*, in particular *E. coli*, and may thus hold for several other cases also (17). If there happens to be an unlimited supply of carbon and nitrogen as well as of energy, and if at the same time some other metabolic limitation occurs which leads to decreased growth or even growth arrest, bacterial cells in general decide not to limit substrate uptake, but to continue the central metabolism at high speed and to waste the surplus of resources instead of switching down their metabolic activity to save nutrients. This situation is called "overflow metabolism" (16), and it is typical for glutamate overproduction in *C. glutamicum* (11). Acetate overproduction in *E. coli*, an unwanted side reaction frequently observed in high cell density fermentation and under oxygen limitation, is caused by the same factors. Limitations leading to overflow mechanisms can naturally occur for different reasons or can be artificially introduced in various ways, including lack of essential metal ions or cofactors. This holds for glutamate production in *C. glutamicum*, where limitation may occur due to lack of the essential cofactor biotin. This type of metabolic condition, leading to overproduction of metabolites in general and amino acids in particular, has also been used for optimization of amino acid production. In principle, introducing auxotrophies for amino acids other than those excreted in production strains leads to controlled growth limitation, and thus may enhance amino acid overproduction. As a matter of fact, glutamate production in *C. glutamicum* seems to be more complicated than the general scenario described above. As will be explained in more detail in section 23.5, in addition to growth limitation, glutamate excretion in *C. glutamicum* is somehow related to alterations in the organism's cell wall, as well as to changes in the activity of cytoplasmic enzymes.

In a number of amino acid production processes, however, these two concepts do not explain the actual situation during fermentation. Under conditions when the flux through central and peripheral pathways is strongly increased due to mutations and engineered enzymatic steps, the pressure for amino acid excretion is much higher as compared to normal physiological situations. Consequently, the basic concepts discussed above still hold as an explanation for the presence of effective excretion transport systems, but they are not applicable to the actual metabolic situation.

Nevertheless, given that amino acids have been shown not to passively cross the permeability barrier of the plasma membrane under production conditions (11,17), it is accepted that most, if not all, amino acid production processes necessarily depend on the presence of active export systems for two reasons. In addition to the trivial argument that most amino acids simply cannot cross the phospholipid bilayer without the help of transport proteins, it is also important to keep the internal amino acid concentration sufficiently low by active extrusion in order to avoid undesired feedback inhibition. This can be achieved only by energy dependent, active export systems, which guarantee that in spite of high rates of biosynthesis as well as increasing external product concentrations, the cytoplasmic amino acid concentration will remain at a relatively low level, compatible with normal physiological situations.

These considerations are also relevant for another situation, which is interesting both from a fundamental and an applied point of view, namely the question of whether amino acid production is coupled to cell growth or not. Whereas this seems to be clear for the basic physiological concepts discussed above, it is no longer obvious for typical production strains. Amino acid excretion as explained by metabolic imbalance may certainly be coupled to growth, but the concept of metabolic overflow at least requires limitation of growth if not growth arrest. On the other hand, with the exception of glutamate production, in typical production strains amino acid excretion is the result of the pressure of an engineered flux increase in the direction of the desired product, and amino acid export just enlists the available efflux systems. Consequently, whether production is coupled to growth or not is more likely to be governed by technical aspects, such as whether the production phase is specifically turned on during fermentation (e.g., threonine in *E. coli*) or not (e.g., lysine in *C. glutamicum*).

For bulk amino acid production, only a small number of microorganisms are used, namely *C. glutamicum* (glutamate, lysine, and aromatic amino acids), *E. coli* (threonine, aromatic amino acids, and cysteine), and, to some extent, *Serratia marcescens* (arginine and histidine). In view of this fact and on the basis of the available models explaining the physiological background of amino acid production, the question arises whether these organisms may have special capacities with respect to amino acid excretion. *Escherichia. coli* is on this list precisely because it is the major source of physiological and biochemical knowledge available. There are already genetic tools at hand that directed the choice of this organism for its use in amino acid production. The same argument holds true for *C. glutamicum* at another level, given that this organism is presumably the best studied bacterium with respect to technical fermentation.

On the other hand, it has frequently been argued that those bacteria would be particularly suitable for the purpose of metabolite production that have a less tightly regulated central and peripheral metabolism. This argument was put forward for *E. coli* and *S. marcescens*, the latter being identified as less tightly regulated. The success of *E. coli* in amino acid production, however, clearly argues against this interpretation. In the same line, *C. glutamicum* is designated as having a less well organized and regulated central and peripheral metabolism as than *E. coli*. There are, for example, three aspartokinase isoenzymes found in *E. coli* at the entrance to the biosynthesis pathway of the aspartate family of amino acids which are separately inhibited by lysine, threonine, and isoleucine, whereas *C. glutamicum* has only one enzyme, which is feedback inhibited in a concerted manner by lysine and threonine. It is, however, not obvious why concerted feedback control should in principle be inferior to control of individual enzymes. Moreover, in the meantime *C. glutamicum* was found to have a number of highly elaborated regulation networks, for example with respect to nitrogen control (9,18), or for the cellular response to osmotic stress (19). *Corynebacterium glutamicum* seems to be especially well equipped in some metabolic control regions, particularly the anaplerotic pathways. A somewhat loose pathway regulation, however, seems in fact to be present in *C. glutamicum* at least in some cases. It has been shown, for example, that simple addition of methionine, which, after being taken up and inhibiting its own biosynthesis, leads to overflow in the direction to lysine biosynthesis, which subsequently causes lysine excretion (20). The argument that less tightly regulated species are more favorable production organisms in general is also contradicted by the fact that *E. coli* seems to be clearly better suited for threonine production than *C. glutamicum*, in spite of the fact that its biosynthesis pathway is much better regulated.

Taken together, the reason for choosing a particular production organism does not seem to be directly derived from its physiological or genetic suitability for this purpose. The choice is obviously driven by the amount of genetic and physiological knowledge available, and by the recognition of a particular organism to be productive (*C. glutamicum*) or easy to handle (*E. coli*).

Nevertheless, *C. glutamicum* and related species seem to be an exception, at least in two aspects: Its physiological capacity to excrete large amounts of glutamic acid already under natural conditions seems to be exceptional, and the peculiar situation that some treatments affecting the state of the cell wall or plasma membrane trigger an increase in the propensity to excrete amino acids has not been reported for any other bacterium. It is important to note that this particular property does not seem to be restricted to glutamate excretion only.

With respect to the central and peripheral metabolism in amino acid producing bacteria, there are a number of specific strategies for metabolic design, which are relevant for metabolite overproduction in general and for amino acid production in particular. Schematically, the metabolic network related to metabolite production can be divided into several individual components (Figure 23.2). Basically, we can discriminate four different major steps being involved: substrate uptake, central metabolic pathways, amino acid biosynthesis, and amino acid excretion. This simple view has to be extended by further important considerations, in particular specific (local) regulatory mechanisms both on the level of gene expression and enzyme activity. Furthermore, the integration into global regulatory networks such as carbon control and nitrogen control has to be considered, as well as the importance of cellular metabolic balances like energy balance and redox balance.

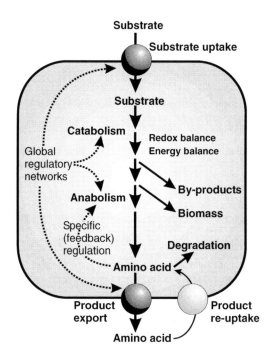

Figure 23.2 Basic steps involved in amino acid production. Summary of major metabolic and regulatory events important for amino acid production by bacteria (see text).

Moreover, specific situations have to be taken into consideration, such as amino acid degradation pathways or product reuptake by amino acid uptake systems.

Most of these aspects will be discussed in detail in section 23.5, in connection with processes used for the production of individual amino acids. This holds for regulatory concepts, as well as for more specific questions like redox balance in the case of lysine, or degradation activity in the case of threonine. The general pathway architecture, i.e., particular reactions as well as general regulation patterns of the major pathways of substrate degradation (e.g., glycolysis, pentose phosphate pathway), central metabolism (e.g., tricarboxylic acid cycle, glyoxylate cycle, anaplerotic reactions), and the common anabolic pathways of amino acid biosynthesis are in principle well known and studied in detail. The special situation of massive overproduction of an anabolic end product, i.e., an amino acid, which is largely different from the usual state under growth conditions, makes it necessary to focus in detail on particular parts of the process. A rational strategy for metabolic design seems to be relatively easy in the case of the final steps, i.e., the anabolic reaction sequences directly leading to the particular amino acid, where a straightforward deregulation or overexpression strategy seems promising. It is, however, obviously not that simple in the case of catabolic and central metabolic pathways involved in the supply of precursor molecules of amino acid biosynthesis, because these pathways also have to fulfil many other purposes in the cell, which may be in conflict with the basic biotechnological goal. This holds in particular for aspects of biomass production (growth), of energy consumption (efficiency), and the balance between desired and undesired pathways (yield).

Many studies have been done on the importance of central metabolic pathways in *C. glutamicum* for the efficient synthesis of products such as lysine or glutamate (21). In these studies, the core significance of anaplerotic reactions within this metabolic node has been demonstrated, making the set of anaplerotic reactions a particularly fascinating target of metabolic engineering. Because of the central position of these reactions, the results obtained are certainly of high relevance for the production of any amino acid the precursors of which are provided by reactions or side reactions of the citric acid cycle. Two precursor metabolites oxaloacetate and pyruvate, which serve as precursors for about 35% of cell biomass and metabolic products, are generated in the anaplerotic node. Several research groups have shown that altogether at least 5 individual enzymes are present in this complex metabolic process. The glyoxylate cycle also fulfils anaplerotic functions (Figure 23.3). By disruption of the gene encoding the phospho*enol*pyruvate carboxylase (22) and by detailed flux analysis using ^{13}C NMR techniques (23,24), it was shown that the pyruvate carboxyl-ase plays a major role in *C. glutamicum*. Altogether, a surprisingly complicated metabolic pattern is found: pyruvate dehydrogenase providing acetyl-CoA for the citric acid cycle; pyruvate carboxylase together with phospho*enol*pyruvate carboxylase supplying oxaloac-etate, as well as oxaloacetate decarboxylase, phospho*enol*pyruvate carboxykinase, and the malic enzyme. In addition, the glyoxylate cycle provides C4 precursor compounds. This may not be complete, given that the presence of a pyruvate oxidase was also revealed in the genome of *C. glutamicum*. It may be concluded that the anaplerotic node with its complex scenario of partly counteractive reactions is one of the "secrets" of *C. glutamicum* that explains its exceptional metabolic flexibility and versatility. In addition to obviously being an interesting target for rational metabolic design, it is also an instructive example, in which it has been shown that analysis of the significance of individual enzymes in these pro-cesses is not possible by in vitro enzyme tests, nor by genetic studies, but only by applying refined methods of metabolic flux analysis (25–28).

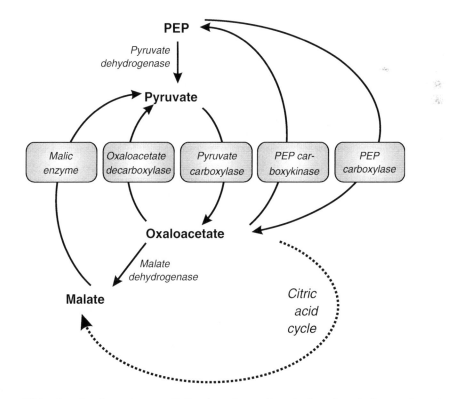

Figure 23.3 Anaplerotic pathways of *C. glutamicum*. Complexity of anabolic reactions in *C. glutamicum*. The figure represents a minimum setup of reactions present (see text).

23.3 PHYSIOLOGICAL, BIOCHEMICAL, AND GENETIC TOOLS USED TO IMPROVE AMINO ACID PRODUCTION

In the past decades, classical approaches including many rounds of mutation, selection, and screening, in some cases combined with molecular biology techniques such as threonine production, have led to impressively efficient strains of *C. glutamicum* and *E. coli* that are used in amino acid production. After introduction of modern genetic tools and sophisticated methods for pathway analysis, the era of rational metabolic design was begun (29,30). Now additional aspects have come into scientific focus, namely substrate uptake, precursor synthesis by central pathways, further steps within the anabolic pathway, redox and energy balance, and last but not least, active product excretion. At the beginning, straightforward molecular approaches such as direct deletion or overexpression of particular genes were, in general, not as successful as conservative methods of strain breeding. This is most likely due to the fact that optimum productivity with respect to a single anabolic end product, i.e., an amino acid, is based on fine tuning of the whole cellular network of pathways and regulation circuits. In order to render modern approaches of metabolic design more efficient than classical procedures, detailed knowledge of the metabolic web, on all possible levels of understanding, is needed. This will include everything from pathway architecture and regulation of the individual enzymes involved, to the sum of carbon, nitrogen, energy and redox fluxes including the construction of the related regulatory circuits. In the genomic and postgenomic era, the full set of information on the level of the genome, the transcriptome, the proteome, and the metabolome becomes more and more available. In the near future it will be fully complemented by dynamic information on metabolic pools and fluxes (the fluxome).

In addition to approaches of optimizing biosynthesis and excretion of amino acids based on the pathway architecture, which are already present in the respective production organisms, metabolic design certainly offers the possibility of introducing new reactions and pathways not so far found in the particular bacterium under study.

As targets for improving a producer organism, all of the different compounds of the metabolic network responsible for amino acid biosynthesis and excretion have to be considered. The vast majority of efforts have been spent on strategies to modify the terminal anabolic pathways (5,31). The most direct approach, which has been applied in all processes except glutamate production, is the amplification of the biosynthetic flux through a reaction which has high control strength on the whole pathway, frequently called a "bottleneck." This may be achieved by deregulation, i.e., release from feedback control by the respective end products of the pathway, a typical result of classical strategies in which high concentrations of substrate analogues were added, giving rise to inhibition insensitive mutant forms of originally feedback inhibited enzymes. Another strategy involves increased or decreased synthesis of enzymes at the corresponding metabolic branch points. An instructive example is lysine production, where a decrease in the dihydrodipicolinate synthase (DapA) increased the flux to lysine. A similar example is redirection of metabolic flux from tryptophan to tyrosine or phenylalanine in *C. glutamicum* (31).

Such strategies can be applied to all other aspects of the metabolic network involved in amino acid production (Figure 23.1). Particularly interesting and successful are the strategies directed toward influencing the connections of anabolic pathways to the central metabolism (precursor supply), decreasing degradation activity, and engineering of transport systems (avoiding futile cycles). Some aspects have not yet been fully exploited, such as optimization of substrate uptake and product excretion, and some fields are not really well understood to a level sufficient to be approached by a rational metabolic design, e.g., global regulatory mechanisms, and energy balance. These will briefly be discussed in section 23.6.

Biochemical and genetic information was successfully used in the case of precursor supply, most prominently concerning anaplerotic reactions. After the set of reactions within the anaplerotic node had been elucidated in *C. glutamicum* (3), the significance of phospho*enol*pyruvate carboxykinase (32) and, in particular, of pyruvate carboxylase, for amino acid production was elucidated. The importance of the latter enzyme for lysine and glutamate production was demonstrated by deletion and overexpression of the corresponding gene (33,34). Further examples of the successful optimization of precursor supply, threonine and aromatic amino acid production in *E. coli* and *C. glutamicum,* are described in section 23.5.

Another obvious strategy involves the inactivation of amino acid degradation pathways that are interfering with high product yield. This is not that important in *C. glutamicum* in view of the limited catabolic capacity of this organism, but it is very important in *E. coli*. An example of this strategy is threonine production in *E. coli.*

Although certainly significant for bacterial metabolism under production conditions, energy and redox balance has not yet been studied in great detail. Because of the obvious need of a sufficient supply of redox equivalents (NADPH) in the case of lysine production, the relation of metabolic flux through glycolysis versus the pentose phosphate pathway has been analyzed using ^{13}C NMR or MS flux analysis (section 23.5). Direct approaches by introducing NADH dependent instead of NADPH dependent enzymes have not yet been successful.

Transport reactions are important in amino acid production and are thus candidates for optimization in three different ways. First, highly active nutrient uptake is a prerequisite for efficient synthesis of amino acids, which, second, have to be actively secreted in order to be accumulated in the surrounding medium. Third, the latter process may be counteracted by the activity of amino acid uptake systems. Sugar uptake via phospho*enol*pyruvate dependent phosphotransferase systems (PTS) has been very well studied in *E. coli* (35), but unfortunately little information is available for *C. glutamicum* (36–38). PTS systems are highly interesting in connection with amino acid production not only because they are responsible for sugar uptake, but also due to their impact on regulatory networks in the cell with respect to control of carbon metabolism, including various transport systems. Amino acid excretion was not taken into consideration as a relevant aspect of amino acid production for a long time. First on the biochemical level (8), and in recent years by successful molecular identification (39), the discovery of an increasing number of specific and energy dependent amino acid exporters has opened a new world of transport reactions and, at the same time, targets for optimization strategies (Table 23.2). Several new families of transporters are known now: the lysE family (40), the RhtB family, the ThrE family, and the LIV-E family (39). All these families have a number of members found in the genomes of a variety of bacteria. It is interesting that all carriers related to amino acid excretion identified so far belong, with one exception, to the class of secondary carriers, i.e., they are coupled to the electrochemical potential as the driving force. So far, only a single ABC-type, primary carrier has been detected (Table 23.2). It should be mentioned, however, that glutamate excretion was found to depend on the cytoplasmic ATP concentration (41). The first members known were *LysE*, the lysine exporter in *C. glutamicum* (14), *RhtB* and *RhtC* from *E. coli* which are involved in threonine production by this organism (42), *ThrE*, the threonine exporter in *C. glutamicum* (43), and *BrnFE*, the leucine, isoleucine, and valine exporter in *C. glutamicum* (44). Further examples of newly discovered amino acid export systems are the two exporters involved in cysteine and O-acetylserine efflux in *E. coli*, which belong to completely different transporter families from those described above (45,46). The surprisingly large number of homologues found in the genomes of numerous bacterial species indicates that, in clear contrast to the previous perception, amino acid excretion seems to be a common event in the prokaryotic world.

Table 23.2

Amino acid excretion systems identified on the molecular level

Organism	Gene	Transporter Family[1]	Substrate	No. of Putative Transmembrane Segments
C. glutamicum	*lysE*	LysE	L-lysine, L-arginine	5–6
E. coli	*rhtB, rhtC*	RhtB	L-threonine[2], L-homoserine[2], L-homoserine lactone[2]	5–7
E. coli	*ydeD*	PecM	O-acetyl-L-serine, L-cysteine, L-glutamine, L-asparagine	9–10
C. glutamicum	*thrE*	ThrE	L-threonine	10
C. glutamicum	*brnFE*	LIV-E	L-isoleucine, L-leucine, L-valine	7 and 4[3]
E. coli	*yfiK*	RhtB	O-acetyl-L-serine, L-cysteine	5–7
E. coli	*cydDC*	ABC-type[4]	L-cysteine	

[1] The transporter family within the class of secondary transporters is indicated.

[2] Only resistance to the substances listed but no transport was measured.

[3] BrnFE is a transport system consisting of two different subunits.

[4] In contrast to all other transporters listed in Table 23.2, CydDC is an ABC-type transporter. For references, see text.

It has been demonstrated that amino acid uptake systems may be of significance for amino acid production in two different ways. Tryptophan production by *C. glutamicum* was found to be improved when the uptake system for aromatic amino acids was decreased in its activity. A similar situation was found for threonine production. The advantage of decreasing the activity of amino acid uptake for production may be twofold: a possible energy consuming futile cycle of energy dependent uptake and export can be avoided, and feedback inhibition as well as repression by high internal amino acid concentrations will not occur. For glutamate excretion, it has been shown that the risk of energy consuming futile cycling seems to be avoided by alternative regulation of uptake and excretion systems (47).

In addition to the list of targets for optimization, based on the dissection of metabolic and regulatory events, there are a number of additional aspects to consider. A highly interesting point which frequently is discussed in connection with strain optimization is the presence of auxotrophies for amino acids or related compounds in production strains. Amino acid auxotrophies in strains obtained by classical breeding techniques are in general interpreted as causing release from feedback control or decreasing alternative branches in anabolic metabolism, a conclusive example being lysine production strains (5,6). This explanation, however, is not always conclusive, because frequently auxotrophies for amino acids other than those related to the biosynthesis pathway of the amino acid excreted are found in producer strains. It should be considered that growth limitation may have an effect on productivity as has been discussed in connection with models for overflow metabolism. The disadvantage of the presence of auxotrophies in production strains, i.e., that fermentation media need substantial and defined supplementation with the respective compounds, have been overcome by the introduction of "leaky" strains. A leaky biosynthetic pathway is still functional, but its activity is reduced to a level where it can provide only limited amounts of the required metabolite. Examples are strains with reduced activity of citrate synthase (48) or homoserine dehydrogenase (49). Consequently, the respective amino acid as the end product of the leaky pathway will become growth limiting (metabolic overflow situation) and the cytoplasmic concentrations will be very low (release of feedback control).

A particularly interesting property in this respect is the importance of cell wall integrity for amino acid production in *C. glutamicum*. As discussed in section 23.5 in detail, interfering with cell wall integrity of *C. glutamicum* using a variety of methods causes this organism to excrete glutamate. It is necessary to emphasize, however, that this turns out to be a more general situation, obviously not restricted to glutamate excretion only. It has been shown that modification of the lipid composition (50) as well as the content of mycolic acids in the cell wall (51), also leads to an increase in the excretion of other amino acids. This exceptional property of *C. glutamicum*, however, has not yet been exploited in a rational way, because the molecular basis of these events is still not understood.

A chapter on optimization strategies for amino acid producing strains necessarily should include a brief overview on the modern methods used for this purpose. When developing metabolic concepts of amino acid overproduction, a number of tools from different scientific disciplines have been used and integrated in recent years. These approaches have, of course, been successfully employed also for purposes of rational metabolic design. They include molecular tools on the level of functional genomics, i.e., identification, modification, and controlled expression of genes coding for enzymes involved in amino acid synthesis pathways, in pathways responsible for precursors and supplementary building blocks, and in pathways leading to side products branching off the respective biosynthesis pathway, as well as modulation of the expression of genes coding for regulatory proteins. Analysis of the transcriptome as well as the proteome provides a more detailed understanding of regulatory networks working in the cell. Substantial and successful efforts have been made in *C. glutamicum* in this direction in recent years. With the methods of metabolome analysis, a completely new level of insight into the cell is starting to become available, in particular if its dynamics, i.e., the metabolic fluxes in the cell (fluxome) are also determined. A reasonably well characterized metabolome and the pattern of metabolic fluxes would be a solid basis for the rational understanding of amino acid production and metabolic design. We are, however, far from this situation, because a basic inventory of metabolites (metabolome) of not a single bacterial cell is available.

The genome of *C. glutamicum* has been sequenced independently several times (30), and a number of global studies on the transcriptome are now available (30,52–56). At the same time, proteome analysis has successfully been adapted to *C. glutamicum*, too (57–59), and has recently been extended to phosphoproteome analysis (60).

The metabolome approach, i.e., focusing on metabolic pools and their dynamics, has been applied on a more conceptual level already long before serious efforts at unravelling the metabolome of the cell and the metabolic fluxes have been started. This refers to attempts of identifying and quantifying limiting metabolic steps or "bottlenecks," which was a highly influential concept in the field of amino acid production. Earlier perceptions, which suggested that there would exist one major and dominating "bottleneck" controlling the flux through a particular pathway, were soon proven to be not fully correct or not very useful in application. Certainly, the first entrance step into an amino acid biosynthesis pathway is in general a true bottleneck, because it controls the flux through the consecutive anabolic pathway. In general, however, it is not sufficient to simply deregulate the first control point in order to make possible effective overproduction of a particular metabolic end product. The instructive work of Niederberger et al. (61) may serve as an example, where it has been shown by detailed analysis of aromatic amino acid biosynthesis in yeast that flux control is, in general, distributed among a series of enzymes, and is not confined to a single enzymatic step. There have been a large number of approaches to understand metabolic fluxes on the basis of theoretical considerations as well as experimental measurements (25). Well known, among others, are the concepts of control analysis (62,63) as well as metabolic balancing techniques (64,65). The latter have been extensively applied to studying fluxes in *C. glutamicum* under conditions of amino acid production (21,66–68). These concepts have certainly contributed to understanding the complex situation of inter-related fluxes in cells. The basic problem of most of these methods, however, lies in the fact that in general the complete and exact set of data on metabolite pools, and in particular on their dynamic behavior, is not available to a sufficient extent. Only after application of the sophisticated techniques of nuclear magnetic resonance spectrometry (NMR) (21,26,69) or mass spectrometry (28,67–70) to the analysis of fluxes in *C. glutamicum* was a better understanding of the dynamic aspects of bacterial metabolism provided. In particular, the novel combination of metabolite balancing and MALDI-TOF mass spectrometry seems to be a very useful tool in unravelling dynamic fluxes in bacterial cells (71). For a full understanding, however, a detailed analysis of the metabolome, in addition to the major metabolites which also have been identified in the studies mentioned above, is necessary.

23.4 AMINO ACID PRODUCTION: TECHNICAL ASPECTS

Since the discovery of *C. glutamicum* as a bacterium with the ability to produce large amounts of glutamate directly from sugar and ammonia, new processes of amino acid production were developed, and a new industry started manufacturing these products by using fermentative techniques. In view of the main goal of this article, namely to understand the physiological, biochemical, and genetic background of amino acid production, it seems valuable to have a brief look at the processes applied for amino acid fermentation. Among many techniques used namely extraction, chemical synthesis, enzymatic synthesis, and fermentation, only the latter, which, except for the production of methionine, is used for all large scale bulk amino acids, is discussed here. It is furthermore obvious that amino acid fermentation is just one step in the whole technical process, followed by downstream processing and wastewater treatment, which are also not discussed here.

Fermentation techniques used can be divided into batch and fed batch cultures, and continuous culture. Industrial amino acid fermentation is mostly performed using the former two procedures. Figure 23.4 shows a typical lysine fermentation procedure with *C. glutamicum*. In general, fed batch processes, in which the substrate is not only provided at the beginning of the process but is added repeatedly during fermentation, lead to higher productivity as compared to batch processes because of several reasons. By lowering the initial concentration of nutrients (sugar), the lag time can be shortened and growth as well as yield can be increased. This is mainly due to osmotic stress caused by high nutrient concentrations (72). Frequently, auxotrophic strains are used. In that case, excess supply of the required nutrient at the beginning of the fermentation may result in a decreased production yield, because of feedback effects caused by the nutrient. Very high growth rates under unlimited supplies of substrate may lead to oxygen limitation in large fermentors, which results in a decreased yield and an increased formation of byproducts, mainly organic acids.

Continuous culture has a number of advantages, as well as disadvantages, in the case of amino acid production (31,72). The production phase can be controlled much better, and studies have been published where the yields were significantly higher as compared to the

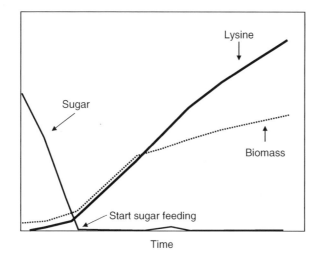

Figure 23.4   Technical l-lysine production process with *C. glutamicum*. Starch hydrolysate or sucrose is used as carbon source and ammonium sulfate as nitrogen source. After consumption of the initial amount of sugar, substrates are added continuously and lysine accumulates up to 170 g/l. The figure was kindly provided by Dr. L. Eggeling, Research Center Jülich.

same process in fed batch mode (31). As a matter of fact, it has been reported for glutamate production, for example, that the volumetric productivity with 5–8 g L^{-1} h^{-1} of continuous cultivation was at least twice as high as in fed batch fermentations (73). On the other hand, continuous fermentation is technically very demanding, as it requires continuous feeding of sterilized fresh media and sterile air. Furthermore, bacteria often undergo spontaneous mutations which, under the metabolic pressure of amino acid fermentation, may lead to reduced productivity during continuous culture for a long period of time.

It is worth noting that continuous culture is in principle only possible if amino acid excretion occurs during growth. There are several processes in which the production phase is clearly [L-threonine fermentation by *E. coli* (74)] or, to a significant extent (glutamate production by *C. glutamicum*), separated from the growth phase. Because cells are constantly kept in the growth phase in continuous culture, this procedure would not easily be applicable in these cases. Apart from that, continuous culture is a much better tool for studying microbial physiology in response to environmental conditions than batch culture. Consequently, the information obtained from continuous culture experiments is useful for optimizing conditions and feeding strategies in fed batch processes.

23.5 SELECTED EXAMPLES OF AMINO ACIDS

Besides those amino acids which are currently of core interest in the food industry, e.g., glutamate, phenylalanine, aspartate, and cysteine, a number of other amino acids are used for different purposes, namely as feed additives and for the pharmaceutical and cosmetic industry. Some of the latter will be briefly discussed here, because they represent instructive examples of production process development, the physiological and genetic background of which is reasonably well studied.

23.5.1 Glutamate
By far the most important amino acid in food biotechnology, and the first amino acid to be produced by fermentation, is glutamate (monosodium glutamate, MSG). In spite of this fact,

and in spite of numerous scientific investigations, it is still less well understood from the physiological and genetic point of view as are, for example, lysine or threonine production.

The historical background of glutamate production, as well as some of its fundamental aspects with respect to the importance of central metabolism and cell wall properties has already been mentioned. For the production of L-glutamate, only *C. glutamicum* is used, although *E. coli* and other organisms are also able to excrete glutamate. Currently, more than 1,000,000 tons of glutamate are produced annually.

The physiological and biochemical background of glutamate production in *C. glutamicum* is highly interesting, and differs significantly from that which is presumably responsible for the excretion of other amino acids as well as from amino acid excretion in other bacteria. L-glutamate is closely related to the central metabolism, synthesized from oxoglutarate by the NADPH dependent glutamate dehydrogenase, which, at the same time, represents the major reaction of nitrogen assimilation under conditions of nitrogen excess in the environment. Under nitrogen limiting conditions, the ATP dependent glutamate synthase glutamate oxoglutarate glutamine aminotransferase system (GS/GOGAT) is predominantly used. The first attempt to explain the property of *C. glutamicum* being an exceptionally good glutamate producer was based on the observation that the activity of the oxoglutarate dehydrogenase (OHDC), i.e., the citric acid cycle reaction following the branching point to glutamate, was surprisingly low. The straightforward explanation of the carbon flow being redirected to glutamate because of the insufficient activity of a major enzyme, thus leading to massive glutamate over synthesis and ultimately to glutamate production, appears convincing.

However, it turned out not to be true, at least in its simple meaning. Later on, it was revealed that *C. glutamicum* in fact possesses significant OHDC activity, which was found to be strongly reduced under conditions which lead to glutamate excretion (75). It has been proven by many investigations studying carbon flux under glutamate production conditions that in fact a flux redirection from oxoglutarate to glutamate is observed under glutamate production conditions (76–78). Moreover, disruption of the OHDC in *C. glutamicum* led to glutamate production comparable to that observed under biotin limited conditions (79). Unfortunately, the reason for the observed reduction of stability and activity of OHDC is not understood so far. It should be mentioned here that another concept has recently been put forward to explain the observation of a reduced flux through the second part of the citric acid cycle, based on the fact that Gram-positive bacteria similar to *C. glutamicum* may need energy input into the succinate dehydrogenase (SDH) reaction. The SDH reaction, which would then depend on the electrochemical proton potential, might possibly not function properly under conditions where the membrane surrounding it is somehow altered, as is suggested after treatments leading to glutamate excretion (80).

For these considerations and their possible consequences in understanding the physiological and biochemical background of glutamate excretion, a critical evaluation of the methods used for inducing glutamate production by *C. glutamicum* is necessary (79). Several different kinds of treatments lead to glutamate excretion. Based on the original observations, glutamate was found to be excreted under conditions of biotin limitation (2). *Corynebacterium glutamicum* requires biotin for growth. Biotin is a cofactor of several enzymes, most importantly of pyruvate carboxylase and acetyl-CoA carboxylase, which is essential for fatty acid synthesis. Glutamate overproduction can also be induced by addition of some kind of detergents, e.g., polyoxyethylene sorbitane monopalmitate (Tween 40) or polyoxyethylene sorbitane monostearate (Tween 60). The reason for induction of glutamate excretion is definitely not simply an increase in membrane permeability, because closely related detergents like the monolaurate (Tween 20) or monooleate esters (Tween 80) are not effective. Consequently, it is assumed that the detergents used exert a regulatory effect rather than a direct increase in permeability. Addition of β-lactam antibiotics (penicillin), which are known to interact in cell wall biosynthesis, leads to glutamate excretion. Strains which are auxotrophic for precursors required in phospholipid biosynthesis, such as glycerol or fatty acids, are able to excrete glutamate into the surrounding medium. The apparently simple method of applying increased temperature also leads to increased glutamate excretion, in particular in the case of temperature sensitive mutants (81). Recently, it was shown that direct changes in the architecture of the cell wall, brought about by disrupting enzymes involved in the synthesis of cell wall components in *C. glutamicum*, such as fatty acids (50) or trehalose (51), lead to increased glutamate excretion, too. In addition to these procedures, also general membrane acting factors, such as addition of local anaesthetics or application of osmotic stress, may induce glutamate excretion in *C. glutamicum* (82,84).

The cell wall of *C. glutamicum* has a very complex structure, consisting of an inner layer composed of cross linked structures of arabinogalactan, peptidoglycan, lipomannan, and lipoarabinogalactan (Figure 23.5). In the middle part, a mycolic acid bilayer structure is present, which, as a true bilayer, is an exception among Gram-positive bacteria, shared only with the other members of the mycolata (*Mycobacteria* and *Nocardia*). The outer layer of the cell wall, finally, is mainly composed of polysaccharides also containing free lipids like trehalose mono- and dimycolates (85,86). It seems to be obvious that all treatments and conditions, inducing glutamate excretion, although they appear to be very diverse, may influence synthesis and composition of the plasma membrane or the cell wall of *C. glutamicum*. Based on this fundamental interpretation, the basic explanation for the trigger being responsible for glutamate excretion was for a long time the concept of an increased leakiness of the membrane or cell wall of *C. glutamicum* (79). In 1989, however, it was shown that glutamate excretion is not due to increased leakiness but, on the contrary, mediated by a specific, energy dependent transport system (8,87). This export system was shown to be responsible for glutamate excretion under conditions of a metabolic overflow situation. Unfortunately, in contrast to lysine, threonine, and isoleucine, the gene or genes encoding this excretion system were not identified until now. The fact that glutamate excretion has been experimentally demonstrated to depend on the presence of a specific export system, however, does not directly help to understand the influence of factors triggering glutamate export or the involvement of the cell wall in this process. This situation led to a revival of the "barrier hypothesis" (82) in which the permeability properties of the cell wall were directly made responsible for the ability to effectively excrete glutamate. All these concepts have to take into account that glutamate uptake systems are located in the plasma membrane, which, under conditions where no glutamate excretion can be observed, take up glutamate with high activity (41,88). Consequently, this requires efficient permeation of glutamate through the cell wall barrier. Thus, glutamate uptake has to take place at the same time that the barrier concept argues for a strictly limited permeability in the direction of efflux, making this concept unlikely for thermodynamic reasons. Furthermore, porin proteins have been identified in the *C. glutamicum* cell wall, providing permeability for hydrophilic solutes (85,89–91).

Concerning a direct involvement of the plasma membrane in glutamate efflux, it has been shown that an increase in passive glutamate permeability is not responsible for glutamate excretion (87). The concept that the observed increase in efflux is correlated with a change in the lipid composition also has to be abandoned, at least in the strict sense that membrane alteration is an essential prerequisite for glutamate export activity (92). It has recently been shown that *C. glutamicum* strains in which the OHDC has been disrupted become efficient glutamate producers without having a changed membrane composition (79). Another concept, related to changes of the physical state of the membrane triggering glutamate excretion, was based on the assumption of a changed viscosity of the cell wall

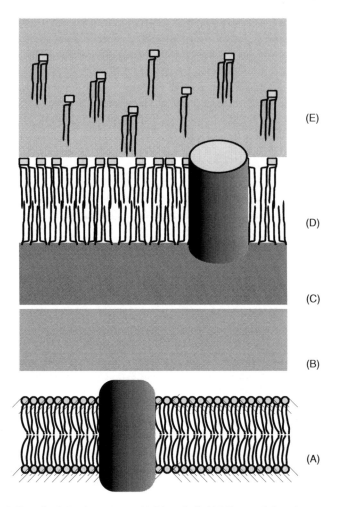

(E)

(D)

(C)

(B)

(A)

Figure 23.5 Cell wall of *C. glutamicum*. A) Phospholipid bilayer of the plasma membrane with embedded membrane proteins, e.g., amino acid transport systems. B) Peptidoglycan layer. C) Arabinogalactan layer. D) Mycolic acid bilayer of the cell wall. The inner leaflet consists mainly of bound mycolic acids covalently linked to the arabinogalactan layer, whereas free mycolates make up the outer leaflet. Porin channels are embedded in the mycolic acid layer. E) The outer layer mainly composed of polysaccharides also contains trehalose mono- and dimycolates, and in addition, proteins. Drawn after (85).

or plasma membrane under these conditions (93), which, however, was contradicted by results obtained later (82). Nevertheless, the fact remains that the capacity of glutamate excretion seems somehow to be correlated to the state of the membrane or cell wall.

Recently a gene was identified (*dtsR*), which has structural similarity to genes of acetyl-CoA carboxylases, the deletion of which led to fatty acid auxotrophy, a decrease in ODHC activity, and glutamate excretion even in the presence of biotin. Consequently, *dtsR* was proposed to be the real target in detergent treatment as well as under biotin limitation. The future development of this line of research seems to be highly promising.

A possible conceptual way out of this dilemma is based on a hypothesis derived from experiments on the contribution of cell wall biosynthesis to cell division (94, M. Wachi, personal communication). In this concept, the obvious impact on cell wall integrity is thought to trigger a regulatory signal transduced to the cytoplasm via the plasma membrane in *C. glutamicum*, leading to unknown mechanisms resulting in down regulation or inactivation of the oxoglutarate dehydrogenase, and possibly to activation of the glutamate excretion system.

In addition to the impact of cell wall properties on glutamate efflux, particular properties of the central metabolism, particularly the availability of precursor molecules, such as oxaloacetate, in *C. glutamicum* are also of significance for glutamate production. Consequently, both understanding and possible manipulation of the anaplerotic node in *C. glutamicum*, responsible for the supply of oxaloacetate, especially the activity of pyruvate carboxylase, play a pivotal role for effective glutamate production (21,79). The relative contribution of several enzymes being in principle able to contribute to oxaloacetate (the essential precursor compound) supply, namely phospho*enol*pyruvate carboxylase (PEPC), phospho*enol*pyruvate carboxykinase (PEPCk), pyruvate carboxylase (PC), malic enzyme (ME), and the glyoxylate pathway, has been estimated in a number of articles. Most investigations show that PC may possibly be responsible for the greatest contribution to the oxaloacetate pool (24,95,96), whereas the authors of another study using high temperature triggered glutamate excretion came to the conclusion that PEPC may be the most important enzyme (81,96). This result is highly relevant, in particular because PC contains biotin and the latter study was carried out under conditions of biotin limitation.

Flux analysis using NMR spectrometry has turned out to be very helpful in interpreting the metabolic changes related to conditions of active glutamate excretion. The carbon flux distribution within the central metabolism has been evaluated (Figure 23.6). The relative contribution of glycolysis in comparison to the pentose phosphate pathway was found to be increased under conditions of glutamate excretion (37,97), and the bidirectional carbon flux between pyruvate and oxaloacetate was decreased significantly (97). It should be mentioned

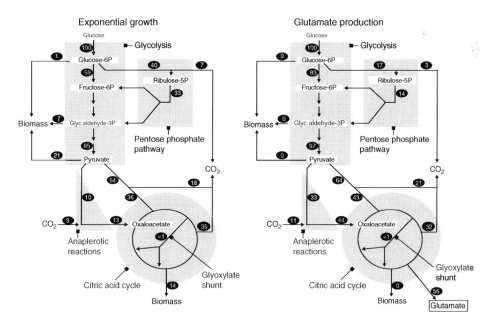

Figure 23.6 Metabolic flux distribution in *C. glutamicum* under conditions of growth and glutamate production. The major metabolic pathways starting from the substrate glucose and leading to biomass, CO_2 and glutamate are represented. The numbers give relative values of metabolic fluxes (see text). Adapted from (112).

that an opposite effect was found for lysine production in both cases. Also, massive flux redirection occurs with respect to anaplerotic reactions and biomass formation (Figure 23.6).

23.5.2 L-phenylalanine

Phenylalanine production is particularly important because of its use in the synthesis of the low calorie sweetener aspartame (methyl ester of the dipeptide L-aspartyl-L-phenylalanine), which is 150–200 times sweeter than sucrose. Besides fermentation, phenylalanine is also produced using chemical synthesis. For microbial production of L-phenylalanine mainly *E. coli* and *C. glutamicum* (and related species) are used (98). Strains have been obtained which are able to produce up to 50g/l of L-phenylalanine (99).

The biosynthetic pathway of aromatic amino acids is divided into a common part leading to chorismate from which the three terminal pathways to L-phenylalanine, L-tyrosine and L-tryptophan branch off (100). This pathway is efficiently regulated in *E. coli* by several different mechanisms both on the level of gene expression and protein activity (Figure 23.7).

Development of phenylalanine production in *E. coli* was based on several strategies to increase the carbon flow of aromatic amino acid biosynthesis. (31,101). The availability of an essential precursor molecule, erythrose-4-phosphate has been increased by modifying the expression of enzymes of the pentose phosphate pathway, namely transketolase (102) and transaldolase. PEP is a key intermediate in the synthesis of aromatic amino acids as well as in many other metabolic reactions. Consequently, efforts have been made to improve the supply of this compound either by direct means (31,101), e.g., by inactivating enzymes that compete for PEP, increasing the flux to PEP, and reducing the use of PEP for sugar uptake via the PTS, or by modulating general networks of carbon control in *E. coli* (103). In addition to increasing the availability of precursors, the specific

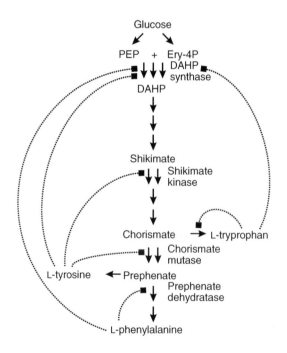

Figure 23.7 Biosynthesis pathways of aromatic amino acids. Simplified version of L-phenylalanine synthesis in *E. coli* including pathways to L-tryptophane and L-tyrosine. The dotted lines indicate regulation mechanisms, both feedback and transcriptional control are active at the indicated points.

pathways responsible for aromatic amino acid biosynthesis have also been successfully engineered. Alleviation of feedback inhibition of 3-deoxy-d-arabino-heptulosonate 7-phosphate synthase (DAHP) synthase isoenzymes by phenylalanine and tyrosine has been achieved, previously through application of appropriate amino acid analogues, and presently by site specific mutation. The genes encoding the enzymes chorismate mutase and prephenate dehydratase have been over expressed to increase the flux toward phenylalanine (101). In general, phenylalanine producers are tyrosine auxotrophs. Consequently, enzymes of the common pathway of aromatic amino acid synthesis are no longer feedback inhibited by this amino acid; moreover, tyrosine accumulation as a result of the shared pathway is prevented. A further beneficial effect is the possibility of a controlled growth limitation during the production phase by appropriate tyrosine feeding.

23.5.3 L-tryptophan

L-tryptophan is an amino acid with increasing potential commercial interest. It is currently mainly used as feed additive, despite its value as an essential amino acid, because of a number of casualties of eosinophilia myalgia syndrome (EMS) in humans which occurred due to the consumption of impure L-tryptophan manufactured by fermentation. L-tryptophan is produced by strains of *E. coli* (104) and *C. glutamicum* (105). In order to obtain effective tryptophan producing organisms, similar alterations of precursor pathways and of the biosynthetic route as described above for phenylalanine production have been carried out. In *E. coli*, it was not necessary to block the route to phenylalanine and tyrosine within the common pathway to aromatic amino acids, which would have led to auxotrophy for these two amino acids (106). Overexpression of the gene encoding the anthranilate synthase (*trpE*), which is the first specific enzyme in the tryptophan synthesis pathway, was sufficient to divert carbon flux to L-tryptophan. In addition, the gene encoding tryptophanase (*trnA*), the enzyme responsible for tryptophan degradation in *E. coli*, was removed in order to optimize tryptophan production (107).

Tryptophan production in *C. glutamicum* is an example where it has been shown that product uptake or reuptake may, in fact, be important in the production process. Engineering a decrease in tryptophan uptake led to a significant enhancement of tryptophan production (108). It was argued that removal of tryptophan uptake results in lower internal tryptophan concentrations and thus reduced feedback effects. Most interestingly, it was shown that tryptophan accumulation by the mutant lacking L-tryptophan uptake was improved mainly in the final part of the fermentation, when high external tryptophan concentrations were accumulated at the outside (31), thus illustrating the advantage of this strategy under conditions of a high backpressure from the external product.

23.5.4 L-lysine

Although lysine is used predominantly as a feed and not as a food additive, it is the best studied example of an amino acid production process in terms of molecular and biochemical aspects. Lysine biosynthesis and production have been intensively investigated with respect to the anabolic pathways involved, pathway regulation, and excretion. Consequently, it may serve here as a paradigm for understanding the tools available and the approaches used in the production of these kinds of amino acids. This holds true for several aspects which seem to be similar in the production of a number of these metabolites, namely pathway fine tuning and regulation, as well as metabolic connection to related pathways; the significance of central pathways for precursor supplementation, in particular the importance of the anaplerotic node; the impact of the cytoplasmic redox and energy balance; the demonstration of how helpful flux analysis may turn out to be in understanding amino acid biosynthesis pathways within the complex cellular metabolic network; and the fundamental demonstration of how important active and energy dependent export systems may be in the course of amino acid production.

Lysine is solely produced by various strains of *C. glutamicum* in fermentative processes, and is used predominantly as a feed additive. Large scale production started in 1958 in Japan. Worldwide production is about 550,000 tons of L-lysine annually (72).

Lysine, like methionine, threonine and isoleucine, is a member of the aspartate family of amino acids. In *C. glutamicum*, the biosynthetic pathway is split into two branches in its lower part (Figure 23.7). The first enzyme of this pathway, aspartokinase, represents the major control point of carbon flow. It is inhibited in *C. glutamicum* by its end products lysine and threonine; consequently, all lysine producing strains contain a deregulated form of this enzyme, which, in general, carries a single amino acid exchange at position 279 of the α-subunit (109). The following were identified as being important in the flux control of this pathway (110,111): (1) deregulation of the aspartokinase (*LysC*); (2) increase of the activity of a feedback resistant aspartokinase, typically by overexpression; (3) increase in the activity of the dihydrodipicolinate synthase (*DapA*), the enzyme situated right after the branching point to the other members of the aspartate family of amino acids; (4) availability of oxaloacetate as precursor for aspartate; (5) availability of NADPH as reducing equivalent: redox balance; and (6) high activity of a specific, energy dependent lysine excretion system *(LysE)*.

Another interesting point about lysine biosynthesis in *C. glutamicum* is that the pathway is split at the level of piperideine-2,6-dicarboxylate (Figure 23.8). In contrast, *E. coli* possesses only the succinylase variant, and some *Bacillus* species only the dehydrogenase variant. One reason for this redundancy is the fact that D,L-diaminopimelate is an essential precursor for cell wall biosynthesis. Only flux analysis using NMR spectrometry revealed that the two pathways are used to highly different extents under differing environmental conditions (112). Because of a low affinity toward its substrate ammonium, the dehydrogenase cannot contribute significantly to lysine biosynthesis at low ammonium concentrations. At high concentrations, however, this pathway is responsible for about 75% of total flux to lysine. If either the succinylase or the dehydrogenase variant is inactivated, lysine production is reduced to about 40%. Consequently, both pathways are functionally combined to ensure the proper supply of D,L-diaminopimelate for the cell.

Last but not least, lysine is an ideal example to demonstrate the importance of the presence of an appropriate excretion system for amino acid production. By 1991, lysine export was shown to depend on the presence of a specific excretion system (11,113). Lysine export is coupled to the electrochemical proton potential, which provides the driving force for its active extrusion from the cytoplasm. Moreover, this system was the first amino acid export system, the encoding gene of which had been identified (14). *LysE*, the lysine excretion carrier, represents a relatively small membrane protein of 25.4 kDa with 5-6 transmembrane domains, presumably being present as a dimer in its active state. If this transport system is deleted, no further export of lysine is possible. Upon addition of lysine containing peptides, the internal lysine concentration reaches toxic values of up to 1 M in this mutant strain of *C. glutamicum*, thus providing an elegant proof for the significance of amino acid excretion carriers in cases of internal amino acid accumulation. *LysE* was the first member of a growing family of transport proteins involved in metabolite export. In addition to the lysine exporter *LysE*, a corresponding transcription factor, named *LysG*, was identified and later characterized (14).

Lysine production is one of the best studied examples of flux analysis in bacteria. Several techniques have been employed for the elucidation of flux distribution in *C. glutamicum* under conditions of lysine production in comparison to normal growth conditions (25). Moreover, methods of flux analysis using ^{13}C-NMR spectrometry

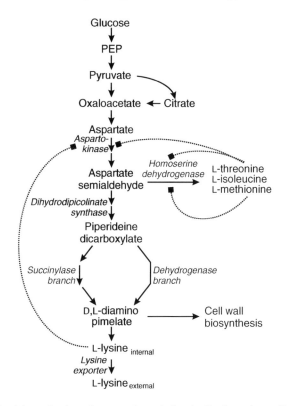

Figure 23.8 Lysine biosynthesis pathway and regulation in *C. glutamicum*. Simplified version of the biosynthesis pathway of the aspartate family of amino acids branches at aspartate semialdehyde to lysine, and to methionine, threonine and isoleucine, respectively. Some reactions of precursor supply are also indicated. The lysine anabolic pathway is split at the level of piperideine-2,6-dicarboxylate into the succinylase branch (4 enzymes, supply of succinyl-CoA and glutamate) and the dehydrogenase branch (1 enzyme, supply of NADPH and NH_4). NADPH is further needed in the steps leading to aspartate semialdehyde and piperideine-2,6-dicarboxylate. Pyruvate is added at the dipicolinate synthase. D,L-diaminopimelate is a precursor of cell wall biosynthesis. Feedback control mechanisms are indicated by dotted lines.

(21,69,95,97,114) as well as MALDI-TOF mass spectrometry (28,67–71) led to a detailed description of the activity of particular enzymes of central and peripheral pathways during lysine production. This made quantification of anaplerotic fluxes, of glycolysis in relation to the pentose phosphate pathway, and the individual share of the two branches of the split lysine pathway possible. Furthermore, particular modifications applied during metabolic engineering approaches could be controlled with respect to their impact on individual fluxes within the reaction network related to lysine production.

An interesting application of this approach concerns the redox balance under production conditions. For several steps within lysine biosynthesis, reducing power in the form of NADPH has to be provided. By using ^{13}C-NMR analysis, the supply of NADPH by the pentose phosphate pathway and by isocitrate dehydrogenase could be calculated. The share of the carbon flux through the pentose phosphate pathway (NADPH generating), in the total flux through the sum of glycolysis (NADH generating), and in the pentose phosphate pathway was found to be significantly increased under lysine production due to the demand of NADPH (68,97). A comparison of this flow and the total flow of NADPH in *C. glutamicum* under conditions of effective lysine production indicated that a limitation by the availability of NADPH is likely. In a recent study in which various lysine producing strains with different lysine yields were systematically compared, an interesting close correlation of lysine yield and the extent of flux through the pentose phosphate pathway as well as the extent of anaplerotic flux was observed (71). The increase of nicotinamide nucleotide transhydrogenase activity in bacteria has also been demonstrated as a successful approach to overcome problems with respect to the redox balance due to a limitation in the NADPH supply (125).

23.5.5 Threonine

Besides lysine, threonine is another member of the aspartate family of amino acids. Like the essential amino acid lysine, threonine is mainly used as a feed additive and its production will thus be described here only briefly. After glutamate and lysine, threonine ranks third in the production volume with about 20,000 tons annually. In contrast to the previous two amino acids, threonine is produced using *E. coli* strains (116). However, studies to improve the application of *C. glutamicum* for threonine production have been reported (117,118). In addition to its commercial significance, threonine is a highly instructive example when reviewing systematic strategies to improve amino acid production. Above and beyond the common features essential for amino acid producing strains, as already discussed for lysine, development of threonine producers is interesting because of two major points. Development of threonine producing strains resembles a straightforward approach in combining classical strain breeding techniques with recombinant DNA technology. It is an example where a broad combination of metabolic changes finally led to the desired result. Among these changes is the introduction of the capacity to use alternate carbon sources as well as the abolishment of amino acid degradation pathways.

In the late 1960s an *E. coli* strain was described which produced threonine because of the presence of a mutant aspartokinase resistant to feedback inhibition by threonine, which was achieved by selection using the threonine analog α-amino-β-hydroxyvaleric acid (119). In the years 1977–1992, a highly productive *E. coli* strain was developed by the group of Debabov (116). The biosynthetic pathway leading to threonine was systematically optimized by several rounds of mutation and selection as well as genetic engineering. In addition to pathway engineering, the production strain was equipped with the capacity to use sucrose as a carbon source, in contrast to wild-type *E. coli* strains which cannot use sucrose. A further significant improvement of threonine production was achieved by reducing threonine degradation. For this purpose, the major threonine degrading enzyme, threonine dehydratase, was inactivated. At the end, a strain producing up to 100 g/L of threonine within 36 h of fermentation was obtained with practically no other products in significant amounts (116).

Threonine is also an example where knowledge on substrate transport in general and on specific excretion systems in particular is available. In *E. coli*, a number of threonine uptake systems are present, the correct contribution of which is not yet understood completely. Interestingly, it has been shown that threonine producer strains have acquired a significantly reduced capacity for threonine uptake in the course of strain construction (74,130). Consequently, expression of the *tdcC* gene encoding an L-threonine uptake system led to decreased threonine production (31). This is another example demonstrating the significance of avoiding product reuptake for effective production strains.

Altogether three genes are assumed to encode for threonine export systems in *E. coli*, namely *rhtA, rhtB* and *rhtC* (42), which all belong to a large family of carrier proteins the function of which has not yet been elucidated. The participation of some of these carriers in threonine export, however, has been questioned (120). Although it seems clear that at least one of these transport proteins contributes to threonine export, the situation does not seem to be fully clear yet (39).

Threonine export has been characterized on the molecular level in the case of *C. glutamicum* (43). Also in this case, the recently identified exporter protein *ThrE* again represents the first member of a new transporter family. In the course of the years, there has been considerable effort to optimize *C. glutamicum* with respect to threonine production also (116,117,118), however, the production capacity of these strains never reached that of good *E. coli* strains, due to a relatively low export capacity, as had been found on the biochemical level before (12,121). In agreement with this explanation, in a recent publication it was shown that both reduction of the activity of the major threonine degradation pathway (serine hydroxymethyltransferase, *glyA*) and overexpression of the threonine exporter *ThrE* led in fact to an increase in threonine productivity (118).

23.5.6 Other Amino Acids

A number of amino acids, which are of significant relevance for the food industry, e.g., L-aspartic acid (besides L-phenylalanine necessary for synthesis of aspartame), cysteine, glycine, and alanine, are still produced by chemical synthesis, by enzymatic methods, or by using immobilized cells (6). In recent years, new developments have been described which may lead to introduction of fermentation procedures for the production of at least some of these amino acids.

A new and promising example in this direction was recently reported for cysteine, an amino acid used as a flour additive, which is produced mainly by extraction or enzymatic methods because of the lack of effective producer strains. Cysteine biosynthesis is in general tightly regulated because of its dependence on sulphur metabolism as well as its potential danger of toxic or unwanted side products. Based on recombinant *E. coli* strains, a process leading to significant cysteine production has recently been developed (122). In the development of a cysteine producing strain, care had to be taken not to simply over express the respective genes encoding cysteine biosynthesis enzymes, but to guarantee a coordinated increase in the flux via this end product, in order to avoid accumulation of intermediates which potentially lead to toxic side products. Interestingly, in the effort to optimize cysteine production in *E. coli* strains, two specific exporters related to cysteine efflux were discovered. First, the gene *ydeD* was identified which after over expression significantly increased the efflux of thiazolidine derivatives of cysteine (45). These derivatives are formed intracellularly by the reaction of cysteine with ketones or aldehydes such as pyruvate or glyoxylate. They can be cleaved again in the external medium. The successful screening strategy was based on the property of genes to augment cysteine production in an industrial producer strain. The same strategy led recently to the identification of a further gene, *yfiK*, the product of which had a similar effect on efflux of cysteine in *E. coli* (46). These two transport proteins are both able to excrete O-acetylserine and cysteine, as well as derivatives of it (*YdeD*). Interestingly, a further, completely different cysteine exporter was recently identified in *E. coli*, obviously related to redox homeostasis during cytochrome biosynthesis (15).

23.6 FUTURE DEVELOPMENTS AND PERSPECTIVES

Classical procedures of strain breeding by mutation combined with selection and screening procedures have proven in the past to be powerful tools for providing efficient amino acid producing strains. With easy access to modern molecular techniques, strain improvement on the basis of detailed biochemical knowledge becomes more and more important. The obvious strategy applied previously was directed toward identifying and overcoming metabolic bottleneck reactions. This strategy, however, was frequently not as successful as expected, in particular when it was applied to strains which already had a substantial capacity of amino acid production. Production strains which have evolved through many rounds of mutagenesis and selection are already fine tuned in their metabolic setup for increased production to a significant extent. Genetic manipulation that has strong impact on metabolism, as is normally the case when deleting or strongly over expressing particular genes, frequently leads to disruption of the subtle metabolic balance already obtained in these strains.

In the era of rational metabolic design, we need a combination of data and knowledge from many different fields for understanding bacterial metabolism, to design and to further optimize amino acid producing strains effectively. Process optimization needs an increase in systemic knowledge of how a bacterial cell functions in all relevant aspects. This strategy asks for more direct and accurate methods to analyze and monitor the current status of bacterial cells during fermentation. In the following, a variety of tools presently available for analysis and understanding of the complex network of bacterial metabolism will be reviewed.

The availability of a large number of bacterial genomes, including those of biotechnological relevance such as *E. coli* and *C. glutamicum*, fostered the direct application of genomic information for the design of production strains (30). Genome annotation was helpful for defining the inventory of enzymes and pathways with putative biotechnological significance. Post genome approaches, i.e., transcriptome and proteome analysis, are currently applied with increasing frequency. Several examples have been mentioned previously, such as manipulations within the anaplerotic node of reactions in *C. glutamicum*. A recent, very instructive example is represented by the design of a new production strain utilizing a novel procedure, designated "genome breeding" by the authors (31,123). Based on comparative genome analysis of wild-type and production strains, a set of mutations was identified by which these strains differ. Based on physiological and biochemical knowledge, mutations which were supposed to be relevant for amino acid production were selected and introduced step by step into the wild type strain. The response of these manipulations on amino acid excretion finally led to a production strain with high productivity, but a minimal set of known mutations. In the example mentioned (123) the authors succeeded in creating a lysine producing strain carrying single amino acid replacements in just three enzymes, namely aspartokinase, homoserine dehydrogenase, and pyruvate kinase. The strain obtained showed extremely high lysine productivity, and, due to the fact that growth and glucose consumption was almost as fast as in the wild type, the fermentation period was shortened to nearly half of that normally required. Further developments are to be expected with a systematic application of this approach, exploiting the large variety of available useful mutants which have originated during the long history of classical strain breeding. The aim of this strategy is to combine beneficial genetic properties identified in the genomes of various mutants into newly engineered producing strains (30). Future efforts in this direction necessarily need more input from biochemistry, given that the function of a large number of potentially interesting gene products, in particular transport proteins, are presently not known.

The availability of the genome is the basis for analyzing the transcriptome of a cell, i.e., the sum of mRNA synthesized at a given time. The information obtained with DNA array techniques may be extremely relevant in the near future for strain improvement in two respects. It is the basis for identifying regulons and modulons, sets of genes which are regulated in a concerted manner under particular metabolic states or conditions of environmental stress. This information can then be used to understand connections within local and global regulatory networks, which are relevant to the biosynthetic pathway under study. Instructive examples in connection with central carbon metabolism were recently reported.

DNA microarray analysis led to the finding of the reciprocal regulation of two genes encoding two different glyceraldehyde-3-phosphate dehydrogenase enzymes with different cofactor specificity, in dependence on the flux direction through the central metabolic pathways (54). A combination of microarray analysis and transcriptional analysis of specific genes revealed a surprising complexity of the regulon for acetate utilization in *C. glutamicum* (134).

Furthermore, DNA arrays are in principle a promising future tool for direct process monitoring. Efforts are underway in many labs to establish systems in which DNA array techniques and information on metabolic patterns is combined, in order to be able to follow the metabolic state of a cell suspension in a fermentor as closely as possible to an online situation. If this kind of analytic information were available, the performance of a bacterial culture in a fermenter, in particular in a prolonged fermentation mode, could be significantly improved. Fermentation may be run at the edge of the mass transfer capacity of the fermentor given precise monitoring of the physiological state of the cell (72). This is still far from being realistic due to technical reasons, mainly with respect to the time scale between drawing a sample and obtaining the transcriptome pattern, as well as the routine cost. This kind of monitoring system would be highly attractive for the purpose of keeping bacteria constantly at a metabolic state of optimal performance.

Proteome data will certainly be helpful, too, in the future, in particular if they are combined with information on posttranslational modification of the synthesized proteins, e.g., phosphorylation (60). Posttranslational modifications, which are not accessible by DNA array techniques, are potentially of high interest and may play an important role for regulation networks, in particular in central carbon metabolism. So far, the major effort is still directed toward a complete inventory of the proteins synthesized, mainly due to the fact that some classes of proteins, among them membrane proteins and low abundance proteins such as transcription factors, are still not detectable by 2D gel electrophoresis (57–59). The same holds for another developing tool, namely the investigation of protein to protein interactions, which may become a significant help in unravelling regulatory networks based on direct protein crosstalk.

Certainly a tool of utmost relevance will be metabolome analysis, based on gas chromatography (GC)- or liquid chromatography-mass spectrometry (LC-MS) techniques. As a matter of fact, it is not widely recognized that we still do not have an idea of the total inventory of even the major metabolites in any single bacterial cell. It is obvious that this information would be extremely helpful for understanding the differences between certain metabolic conditions, for example, during fermentation, or for identifying metabolic changes in production strains as compared to the wild type. Moreover, metabolome data represents the basis for sophisticated metabolic flux analysis.

Several examples of the value of metabolic flux analysis combined with metabolite balancing have been described in this chapter. With the introduction of NMR and, in particular, MS techniques into this field, a complete analysis of the dynamic situation of cell metabolism (the fluxome) became, in principle, accessible for routine studies. The evaluation of the significance and value of particular changes in the metabolism of bacterial strains under different metabolic situations seems to be an essential basis for a rational metabolic design. Flux analysis is furthermore a prerequisite for understanding the dynamic situation of cell metabolism including all regulatory connections. Last but not least, it is obvious that the information obtained by various levels of analyses mentioned here can be integrated into a systems biology approach only with the help of powerful bioinformatics tools.

Reaching the aim of an increase in the systemic understanding of cell metabolism, however, does not only require global approaches, but certainly depends to a large extent on additional information from biochemistry and physiology. An instructive example mentioned above is the identification and analysis of specific export proteins to be responsible for amino acid excretion. Furthermore, the architecture of complex regulatory networks cannot be fully elucidated without the help of biochemical techniques.

Although not the topic of this chapter, further development in fermentation techniques is still a major option for improvement of amino acid production, based on detailed knowledge of biochemistry and physiology of the bacterial cell. A potentially promising example concerns the type of fermentation procedure to be used. It has been discussed before that continuous fermentation would in principle provide an interesting method with several advantages, which, however, due to some drawbacks is not yet used for amino acid production. Continuous fermentation needs coupling of growth and amino acid production, which is in fact frequently observed for lysine production by *C. glutamicum*. In a number of procedures, growth and production are artificially uncoupled to some extent, as in for threonine production by *E. coli*. Unfortunately, it is not yet well understood in terms of metabolic, energetic, and regulatory reasons how and to what extent amino acid excretion can be coupled to growth under optimum production conditions.

In the preceding part of this chapter, tools available for future developments in amino acid production have been listed. However, it seems interesting to identify targets for these efforts which seem to be attractive on the basis of current knowledge. To a large extent, this would be *C. glutamicum*, because the wealth of biochemical information available is certainly more impressive for *E. coli*.

In spite of the fact that central and peripheral metabolism in *C. glutamicum* has been studied to a significant extent, this does not hold for sugar metabolism and, in particular, for sugar uptake. Although the flux through glycolysis has been analysed, the regulation of important enzymes, such as fructose bisphosphate kinase, is not yet known. Interestingly, two genes encoding isoenzymes of each the fructose bisphosphate kinase and of the glyceraldehyde-3-phosphate dehydrogenase have been found in the genome of *C. glutamicum* (72). It is well known from *E. coli* that the phospho*enol*pyruvate dependent phosphotransferase systems (PTS) are, besides their significance in sugar uptake, important for regulating other uptake systems and pathways of central carbon metabolism. In several cases, as in for glutamate production, it is not yet known how the cell actually regulates the activity of amino acid excretion systems in dependence of the availability of the carbon and energy supply. It seems to be a plausible hypothesis that sugar uptake via PTS may be involved in this kind of regulation. Moreover, when the carbon flow to amino acid end products in the course of process optimization is further increased, at some point substrate uptake may become limiting in the overall pathway to the product. A further aspect of engineering PTS systems for sugar uptake may be noted here. Efforts have been made to bypass this uptake system in *E. coli* in connection with production of aromatic amino acids, in order to increase the cellular content of phospho*enol*pyruvate, which is both a precursor of aromatic amino acids and the energy donor of the PTS system. In fact, this manipulation resulted in a higher yield of phenylalanine (101), representing another example of transport design, different from that mentioned in connection with reuptake of substrate (tryptophan), as well as in connection with excretion systems.

Certainly a core field of interest in future development will be regulatory networks, including response to stress conditions. Whereas regulation of particular biosynthetic pathways seems to be quite well understood, both in *E. coli* and *C. glutamicum*, this is not true to the same extent for global regulatory mechanism, including the response of the cell to limitation of carbon (carbon control) or nitrogen sources (nitrogen control). Both control systems are well studied in *E. coli* (135), and the latter has also been elucidated in *C. glutamicum* (18,136). Unfortunately, this does not hold for the important network of carbon control in this bacterium. Knowledge on this topic would be furthermore attractive for investigating suspected regulatory interactions between the signal transduction mechanisms of carbon and nitrogen control, which would certainly be essential for a full understanding of the integration of amino acid metabolism in the cellular network. The wealth of regulatory mechanisms as elucidated in *E. coli* to a significant extent by numerous studies is unfortunately not available in *C. glutamicum*. This refers to the variety of sigma factors, transcription factors, two component systems, attenuation mechanisms, RNA stability, posttranslational modifications, enzyme stability, and degradation. A better knowledge of these mechanisms and their significance for amino acid synthesis would also be the basis for the design of appropriate DNA arrays for controlling the status of fermentation processes.

These regulation networks are also relevant for cellular mechanisms of stress response. Bacterial cells are known to possess sophisticated strategies for stress response (heat, osmotic stress, nutrient limitation, and many others) which involve coordinated regulation on the level of gene expression and protein activity. A well studied example is the sigma B regulon in the Gram-positive *Bacillus subtilis* (127). In case of *C. glutamicum*, only the response to osmotic stress (19,128), nitrogen limitation (18,126), and, to some extent, phosphate limitation (129) have been studied in detail so far. Because enforcement of high yield production in fermentors is certainly a kind of stress in terms of growth limitation, pathway rearrangement, concentration of nutrients, and products, both specific and general stress responses will be active under these conditions and need to be further elucidated.

Another field of significant importance, which is still neglected to some extent, is energy metabolism. This is of particular interest for glutamate production in *C. glutamicum*, which seems to be related to a metabolic overflow situation including mechanisms wasting metabolic energy such as futile cycling (130). Information is available on architecture and organization of the respiratory chain and related redox enzymes in *C. glutamicum* (131–136), whereas the role of the core energy converter ATPase and that of possible futile cycles under conditions of metabolic overflow (glutamate production) is not yet clear. Interesting data have recently been published indicating a close relation of ATPase activity and efficiency of glutamate excretion in *C. glutamicum* (137).

In view of an integrative pattern of reactions being involved in amino acid production (Figure 23.1), the presence, activity and regulation of amino acid excretion systems seems also to be highly relevant. Despite the fact that the importance of active amino acid excretion has been elegantly proven in the case of lysine (14,138), increasing knowledge of these systems and their availability for constructing recombinant producing strains has so far not successfully been integrated into application, with just one exception. The possibility has to be taken into consideration, of course, that, at least in the case of lysine excretion, the activity of the exporter may not be limiting even in producer strains. As has been mentioned above, overexpression of a gene coding for a secondary carrier protein has significantly improved export of cysteine derivatives in engineered *E. coli* strains (45). It has recently been reported that for threonine excretion in a recombinant *C. glutamicum* strain, overexpression of the gene encoding the threonine exporter in fact improved threonine accumulation in the medium (118). The observed positive effect may be due to the fact that the capacity of the amino acid export systems present, and is very limited in both cases reported, which was in fact shown for threonine in *C. glutamicum* (12,121).

In any case, it would be of utmost interest to identify the export system responsible for excretion of glutamate in *C. glutamicum*, which has so far been characterized only at the biochemical level (87,139). Furthermore, the observation that many more products can, in principle, also be observed to be excreted under particular physiological conditions by *C. glutamicum*, *E. coli* and a number of other bacteria, strongly argues for the presence of many more export systems. Because the discovery of some of these systems has already led to the definition of completely new families of transport proteins, more "surprises" of this kind have to be expected.

A truly special property of *C. glutamicum* is the dependence of amino acid excretion on the integrity of the cell wall or plasma membrane. Although in general only discussed in connection with glutamate export, it is important to note that the excretion of other amino acids, such as lysine and proline is also found to be increased when the cell wall is affected . Several different hypotheses for explaining the contribution of the cell wall to amino acid excretion have already been discussed in section 23.5. If a likely explanation were correct, that of a regulatory connection between activity of cell wall synthesis and enzymes of the central metabolism in *C. glutamicum*, it would add an interesting target for further optimization of amino acid excretion for application of molecular and biochemical techniques. Recent publications indicate significant progress toward an improved understanding of cell wall biosynthesis in *C. glutamicum* (140–142).

Finally, after discussing possible tools and targets for future optimization strategies, it is worth briefly discussing more general aims in amino acid production, other than the specific strategies of increasing metabolic fluxes toward the desired end product. In the case of low price products, such as glutamate, lysine, or threonine, future efforts will concentrate on strategies to increase the yield of amino acid production processes. Current yields (g amino acid / g sugar) are 0.45–0.55 for glutamate, 0.4–0.5 for lysine and threonine, and 0.2–0.25 for phenylalanine (31). It is very difficult to calculate true maximum yields, because some additional factors such as ATP stoichiometry and efficiency of energy metabolism are not exactly known. It is obvious, however, that the enormous heat production by bacterial cultures during industrial fermentation due to the high respiration activity argues for a partially uncoupled energy metabolism. Whether this is an intrinsic property of respiratory chain activity and energy metabolism under conditions of amino acid production, or whether it is due to unknown futile cycles is not clear. Consequently, it may in principle be an interesting future target for improvement of efficiency.

Besides improving already available processes in *C. glutamicum* and *E. coli*, efforts on the design of new products or on new processes for available products are a constant challenge for research in amino acid production. The example of L-cysteine production has been mentioned above, where a new process in *E. coli* characterized by novel aspects of strain construction has been reported recently. It may also be possible and attractive in the future to develop a fermentation process for methionine, the only amino acid which is still chemically produced in racemic form in bulk quantities.

For research in microbial biotechnology, as well as for application in industrial processes, a broad basis of physiological, genetic, and biochemical knowledge would be highly attractive. In spite of all classical and modern tools available, we are still far from a complete understanding even of a single bacterial cell in terms of systems biology. The methods applied so far are increasingly well suited for a descriptive understanding; however, they are clearly not detailed enough to be used for predictive statements. Consequently, the final goal of this multidisciplinary work should aim at the "virtual bacterial cell," an aim which is certainly still far from reality, both in the case of *E. coli* and *C. glutamicum*. Due to the combined effort of many groups working on a multilevel analysis of these microorganisms, however, it seems highly probable that a bacterial cell will be the first cell to be understood on that level.

The introduction of techniques of amino acid production by using microorganisms has led to an enormous increase in the beneficial use of these compounds. The application of the molecular tools available will certainly help to overcome remaining limitations, provided these tools can be applied. In that sense, further improvement of amino acid production in the food industry will only be economically relevant if application of genetically engineered bacterial producer strains will not be restricted to an unnecessary extent.

REFERENCES

1. Hodgson, J. The changing bulk biocatalyst market. *Biotechnology* 12:789–790, 1994.
2. Kinoshita, K., S. Udaka, M. Shimono. Studies on the amino acid fermentation, I: production of L-glutamic acid by various microorganisms. *J. Gen. Appl. Microbiol.* 7:193–205, 1957.
3. Liebl, W., M. Ehrmann, W. Ludwig, K.H. Schleifer. Transfer of *Brevibacterium divaricatum* DSM 20297T, "*Brevibacterium flavum*" DSM 20411, "*Brevibacterium lactofermentum*" DSM 20412 and DSM 1412, and *Corynebacterium glutamicum* and their distination by rRNA gene restriction patterns. *Int. J. Syst. Bacteriol.* 41 :225–235, 1991.
4. Nelson, G., J. Chandrashekar, M.A. Hoon, L.X. Feng, G. Zhao, N.J.P. Ryba, C.S. Zuker. An amino-acid taste receptor. *Nature* 416:199–202, 2002.
5. Aida, K., I. Chibata, K. Nakayama, K. Takinami, H. Yamada. *Biotechnology of Amino Acid Production*, Vol. 24. *Progr. Ind. Microbiol.* Amsterdam: Elsevier, 1986.
6. Leuchtenberger, W. Amino acids: technical production and use. In: Biotechnology, Vol. 6, Products of Primary Metabolism, Rehm, H.J. and G., Reed, eds., Weinheim, Germany: VCH Publishers, 1996, pp 455–502.
7. Johnson-Green, P. *Introduction to Food Biotechnology.* Boca Raton: CRC Press, 2002.
8. Kramer, R. Genetic and physiological approaches for the production of amino acids. *J. Biotechnol.* 45:1–21, 1996.
9. Burkovski, A., R. Kramer. Bacterial amino acid transport proteins: occurrence, functions, and significance for biotechnological applications. *Appl. Microbiol. Biotechn.* 58:265–274, 2002.
10. Harold, F. *The Vital Force: A Study of Bioenergetics.* New York: Freeman, 1986, p 163.
11. Broer, S., R. Kramer. Lysine excretion by *Corynebacterium glutamicum*, 1: identification of a specific secretion carrier system. *Eur. J. Biochem.* 202:131–135, 1991.
12. Palmieri, L., D. Berns, R. Kramer, M. Eikmanns. Threonine diffusion and threonine transport in *Corynebacterium glutamicum* and their role in threonine production. *Arch. Microbiol.* 165:48–54, 1996.
13. Hermann, T., R. Kramer. Mechanism and regulation of isoleucine excretion in *Corynebacterium glutamicum. Appl. Environ. Microbiol.* 32:3238–3244, 1996.
14. Vrljic, M., H. Sahm, L. Eggeling. A new type of transporter with a new type of cellular function: L-lysine export from *Corynebacterium glutamicum. Mol. Microbiol.* 22:815–826, 1996.
15. Pittman, M.S., H. Corker, G.H. Wu, M.B. Binet, A.J.G. Moir, R.K. Poole. Cysteine is exported from the *Escherichia coli* cytoplasm by CydDC, an ATP-binding cassette-type transporter required for cytochrome assembly. *J. Biol. Chem.* 277:49841–49849, 2002.
16. Tempest, D.W., O.M. Neijssel. Physiological and energetic aspects of bacterial metabolite overproduction. *FEMS Microbiol. Lett.* 100:169–176, 1992.
17. Kramer, R. Analysis and modeling of substrate uptake and product release by prokaryotic and eukaryotic cells in. In: *Advances in Biochemical Engineering/Biotechnology*, Vol. 54, Scheper, T., ed., Heidelberg: Springer-Verlag, 1996, p 31–74.
18. Burkovski, A. I do it my way: regulation of ammonium uptake and ammonium assimilation in *Corynebacterium glutamicum. Arch. Microbiol.* 179:83–88, 2003.
19. Morbach, S., R. Kramer. Impact of transport processes in the osmotic response of *Corynebacterium glutamicum. J. Biotechnol.* 104:69–75, 2003.
20. Vrljic, M., W. Kronemeyer, H. Sahm, L. Eggeling. Unbalance of L-lysine flux in *Corynebacterium glutamicum* and its use for the isolation of excretion-defective mutants. *J. Bacteriol.* 177:4021–4027, 1995.
21. De Graaf, A.A., L. Eggeling, H. Sahm. Metabolic Engineering for L-Lysine Production by *Corynebacterium glutamicum.* In: *Advances in Biochemical Engineering/Biotechnology,* Vol. 73, Scheper, T. ed., Heidelberg: Springer-Verlag, 2001, p 9–29.
22. Peters-Wendisch, P.G., B.J. Eikmanns, G. Thierbach, B. Bachmann, H. Sahm. Phosphoenolpyruvate carboxylase in *Corynebacterium glutamicum* is dispensable for growth and lysine production. *FEMS Microbiol. Lett.* 112:269–274, 1993.

23. Peters-Wendisch, P.G., V.F. Wendisch, A.A. de Graaf, B.J. Eikmanns, H. Sahm. C-3-carboxylation as an anaplerotic reaction in phosphoenolpyruvate carboxylase-deficient *Corynebacterium glutamicum*. *Arch. Microbiol.* 165:387–396, 1996.

24. Petersen, S., A.A. de Graaf, L. Eggeling, M. Mollney, W. Wiechert, H. Sahm. *In vivo* quantification of parallel and bidirectional fluxes in the anaplerosis of *Corynebacterium glutamicum*. *J. Biol. Chem.* 275:35932–35941, 2000.

25. Eggeling, L., H. Sahm, A.A. de Graaf. Quantifying and directing metabolic flux: application to amino acid overproduction. In: *Advances in Biochemical Engineering/Biotechnology*, Vol. 54, Scheper, T., ed., 1996, p 1–30.

26. Wiechert, W., A.A. de Graaf. *In vivo* stationary flux analysis by ^{13}C labeling experiments. In: *Advances in Biochemical Engineering/Biotechnology*, Vol. 54, Scheper, T., ed., 1996, p 109–154.

27. Wiechert, W., M. Mollney, S. Petersen, A.A. de Graaf. A universal framework for ^{13}C metabolic flux analysis. *Metab. Eng.* 3:265–283, 2001.

28. Kromer, J.O., O. Sorgenfrei, K. Klopprogge, E. Heinzle, C. Wittmann. In-depth profiling of lysine-producing *Corynebacterium glutamicum* by combined analysis of transcriptome, metabolome, and fluxome. *J. Bacteriol.* 186:1769–1784, 2004.

29. Kirchner, O., A. Tauch. Tools for genetic engineering in the amino acid-producing bacterium *Corynebacterium glutamicum*. *J. Biotechnol.* 104(1–3)287–299, 2003.

30. Ikeda, M., S.Nakagawa. The *Corynebacterium glutamicum* genome: features and impacts on biotechnological processes. *Appl. Microbiol. Biotechnol.* 62:99–109, 2003.

31. Ikeda, M. amino acid production processes. In: *Advances in Biochemical Engineering/Biotechnology*, vol. 79, Scheper, T. ed., Heidelberg: Springer-Verlag, 2003, p 1–35.

32. Riedel, C., D. Rittmann, P. Dangel, B. Mockel, S. Petersen, H. Sahm, B.J. Eikmanns. Characterization of the Phosphoenolpyruvate carboxykinase gene from *Corynebacterium glutamicum* and significance of the enzyme for growth and amino acid production. *J. Mol. Microbiol. Biotechnol.* 3:573–583, 2001.

33. Peters-Wendisch, P., B. Schiel, V.F. Wendisch, E. Katsoulidis, B. Mockel, H. Sahm, B.J. Eikmanns. Pyruvate carboxylase is a major bottleneck for glutamate and lysine production by *Corynebacterium glutamicum*. *J. Mol. Microbiol. Biotechnol.* 3:295–300, 2001.

34. Koffas, M.A.G., G.Y. Jung, J.C. Aon, G. Stephanopoulos. Effect of pyruvate carboxylase overexpression on the physiology of *Corynebacterium glutamicum*. *Appl. Environ. Microbiol.* 68:5422–5428, 2002.

35. Postma, P.W., J.W. Lengeler, G.R. Jacobson. Phosphoenolpyruvate: carbohydrate phosphotransferase systems of bacteria. *Microbiol. Rev.* 57:543–594, 1993.

36. Lee, J.K., M.H. Sung, K.H. Yoon, J.H. Yu, T.K. Oh. Nucleotide sequence of the gene encoding the *Corynebacterium glutamicum* mannose enzyme II and analyses of the deduced protein sequence. *FEMS Microbiol. Lett.* 119:137–145, 1994.

37. Dominguez, H., C. Rollin, A. Guyanvarch, J.L. Guerquin-Kern, M. Cocain-Bousquet, N.D. Lindley. Carbon-flux distribution in the central metabolic of *Corynebacterium glutamicum* during growth on fructose, *Eur. J. Biochem.* 254:96–102, 1998.

38. Parche, S., A. Burkovski, G.A. Sprenger, B. Weil, R. Kramer, F. Titgemeyer. *Corynebacterium glutamicum*: a dissection of the PTS. *J. Mol. Microbiol. Biotechnol.* 3:423–428, 2001.

39. Eggeling, L., H. Sahm. New ubiquitours translocators: Amino acid export by *Corynebacterium glutamicum* and *Escherichia coli*. *Arch. Microbiol.* 180:155–160, 2003.

40. Yen, M.R., Y.H. Tseng, P. Simic, H. Sahm, L. Eggeling, M.H. Saier, Jr. The ubiquitous ThrE family of putative transmembrane amino acid efflux transporters. *Res. Microbiol.* 153:19–25, 2002.

41. Kramer, R., C. Lambert. Uptake of glutamate in *Corynebacterium glutamicum*, 2: evidence for a primary active transport system. *Eur. J. Biochem.* 194:937–944, 1990.

42. Aleshin, V.V., N.P. Zakataeva, V.A. Livshits. A new family of amino-acid-efflux proteins. *Trends Bioch. Sci.* 24:133–135, 1999.

43. Simic, P., H. Sahm, H. Eggeling. L-threonine export: use of peptides to identify a new translocator from *Corynebacterium glutamicum*. *J. Bacteriol.* 183:5317–5324, 2001.

44. Kennerknecht, N., H. Sahm, M.R. Yen, M. Patek, M.H. Saier, Jr., L. Eggeling. Export of L-isoleucine from *Corynebacterium glutamicum*: a two-gene-encoded member of a new translocator family. *J. Bacteriol.* 184:3947–3956, 2002.

45. Dassler, T., T. Maier, C. Winterhalter, A. Bock. Identification of a major facilitator protein from *Escherichia coli* involved in efflux of metabolites of the cysteine pathway. *Mol. Microbiol.* 36:1101–1112, 2000.

46. Franke, I., A. Resch, T. Dassler, T. Maier, A. Bock. YfiK from *Escherichia coli* promotes export of *O*-Acetylserine and Cysteine. *J. Bacteriol.* 185:1161–1166, 2003.

47. Kramer, R., C. Hoischen. Futile cycling caused by the simultaneous presence of separate transport systems for uptake and secretion of amino acids in *Corynebacterium glutamicum*. In: *Biothermokinetics*, Westerhoff, H, ed., Intercept Publ., 1994, pp 19–26.

48. Shiio, I., H. Ozaki, K. Ujigawa-Takeda. Production of aspartic-acid and lysine by citrate synthase mutants of *Brevibacterium-flavium*. *Agric. Biol. Chem.* 46:101–110, 1982.

49. Shiio, I., K.J. Sano. Microbial production of L-lysine, 2: production by mutants sensitive to threonine or methionine. *Gen. Appl. Microbiol.* 15:267–275, 1969.

50. Nampoothiri, K.M., C. Hoischen, B. Bathe, B. Mockel, W. Pfefferle, K. Krumbach, H. Sahm, L. Eggeling. Expression of genes of lipid synthesis and altered lipid composition modulates L-glutamate efflux of *Corynebacterium glutamicum*. *Appl. Microbiol. Biotechnol.* 58:89–96, 2002.

51. Wolf, A., R. Kramer, S. Morbach. Three pathways for trehalose metabolism in *Corynebacterium glutamicum* ATCC13032 and their significance in response to osmotic stress. *Mol. Microbiol.* 49:1119–1134, 2003.

52. Loos, A., C. Glanemann, L.B. Willis, X.M. O'Brien, P.A. Lessard, R. Gerstmeir, S. Guillouet, A.J. Sinskey. Development and validation of corynebacterium DNA microarrays. *Appl. Environ. Microbiol.* 67:2310–2318, 2001.

53. Muffler, A., S. Bettermann, M. Haushalter, A. Horlein, U. Neveling, M. Schramm, O. Sorgenfrei. Genome-wide transcription profiling of *Corynebacterium glutamicum* after heat shock and during growth on acetate and glucose. *J. Biotechnol.* 98:255–268, 2002.

54. Hayashi, M., H. Mizoguchi, N. Shiraishi, M. Obayashi, S. Nakagawa, J. Imai, S. Watanabe, T. Ota, M. Ikeda. Transcriptome analysis of acetate metabolism in *Corynebacterium glutamicum* using a newly developed metabolic array. *Biosci. Biotechnol. Biochem.* 66:1337–1344, 2002.

55. Wendisch, V.F. Genome-wide expression analysis in *Corynebacterium glutamicum* using DNA microarrays. *J. Biotechnol.* 104(1-3):273–285, 2003.

56. Huser, A.T., A. Becker, I. Brune, M. Dondrup, J. Kalinowski, J. Plassmeier, A. Puhler, I. Wiegrabe, A. Tauch. Development of a *Corynebacterium glutamicum* DNA microarray and validation by genome-wide expression profiling during growth with propionate as carbon source. *J. Biotechnol.* 106:269–286, 2003.

57. Hermann, T., G.Wersch, E.M. Uhlemann, R. Schmid, A. Burkovski. Mapping and identification of *Corynebacterium glutamicum* proteins by two-dimensional gel electrophoresis and microsequencing. *Electrophoresis* 19:3217–3221, 1998.

58. Hermann, T., W. Pfefferle, C. Baumann, E. Busker, S. Schaffer, M. Bott, H. Sahm, N. Dusch, J. Kalinowski, A. Puhler, A.K. Bendt, R. Kramer, A. Burkovski. Proteome analysis of *Corynebacterium glutamicum*. *Electrophoresis* 22:1712–1723, 2001.

59. Schaffer, S., B.Weil, V.D. Nguyen, G. Dongmann, K. Gunther, M. Nickolaus, T. Hermann, M. Bott. A high-resolution reference map for cytoplasmic and membrane-associated proteins of *Corynebacterium glutamicum*. *Electrophoresis* 22:4404–4422, 2001.

60. Bendt, A.K., A. Burkovski, S. Schaffer, M. Bott, M. Farwick, T. Hermann. Towards a phosphoproteome map of *Corynebacterium glutamicum*. *Proteomics* 3:1637–1646, 2003.

61. Niederberger, P., R. Prasad, G. Miozzari, H. Kacser. A strategy for increasing an *in vivo* flux by genetic manipulations: the trpyptophan system of yeast. *Biochem. J.* 287:473–479, 1992.

62. Kacser, H., J.A. Burns. The control of flux. *Symp. Soc. Exp. Biol.* 27:65–104, 1973.

63. Heinrich, R., T.A. Rapoport. A linear steady-state treatment of enzymatic chains: critique of the crossover theorem and a general procedure to identify interaction sites with an effector. *Eur. J. Biochem.* 42:97–105, 1974.

64. Stephanopoulos, G., J.J.Vallino. Network rigidity and metabolic engineering in metabolite overproduction. *Science* 252:1675–1681, 1991.

65. Vallino, J.J., G. Stephanopoulos. Metabolic flux distributions in *Corynebacterium glutamicum* during growth and lysine overproduction. *Biotechnol. Bioeng.* 41:633–646, 1993.

66. Marx, A., A.A. de Graaf, W. Wiechert, L. Eggeling, H. Sahm. Determination of the fluxes in the central metabolism of *Corynebacterium glutamicum* by nuclear magnetic resonance spectroscopy combined with metabolite balancing. *Biotechnol. Bioeng.* 49:111–129, 1996.

67. Wittmann, C., E. Heinzle. Modeling and experimental design for metabolic flux analysis of lysine-producing *Corynebacteria* by mass spectrometry. *Metab. Eng.* 3:173–191, 2001.

68. Wittmann, C., E. Heinzle. Application of MALDI-TOF MS to lysine-producing *Corynebacterium glutamicum*: a novel approach for metabolic flux analysis. *Eur. J. Biochem.* 268:2441–2455, 2001.

69. Drysch, A., M. El Massoudi, W. Wiechert, A.A. de Graaf, R. Takors. Serial flux mapping of *Corynebacterium glutamicum* during fed-batch L-lysine producing using the sensor reactor approach. *Biotechnol. Bioeng.* 5:497–505, 2004.

70. Kiefer, P., E. Heinzle, O. Zelder, C. Wittmann. Comparative metabolic flux analysis of lysine-producing *Corynebacterium glutamicum* cultured on glucose and fructose. *Appl. Environ. Microbiol.* 70:229–239, 2004.

71. Wittmann, C., E. Heinzle. Genealogy profiling through strain improvement by using metabolic network analysis: metabolic flux genealogy of several generations of lysine-producing corynebacteria. *Appl. Environ. Microbiol.* 68:5843–5859, 2002.

72. Pfefferle, W., B. Mockel, B. Bathe, A. Marx. Biotechnological manufacture of Lysine. In: *Advances in Biochemical Engineering/Biotechnology*, Vol. 79, Scheper, T. ed., Heidelberg: Springer-Verlag, 2003, pp 59–112.

73. Koyoma, Y., T. Ishii, Y. Kawahara, Y. Koyoma, E. Shimizu, T. Yoshioka. EU Patent Application EP0844308 A2 980527, 1998.

74. Okamoto, K., K. Kino M. Ikeda. Hyperproduction of L-threonine by an *Escherichia coli* mutant with impaired L-threonine uptake. *Biosci. Biotechnol. Biochem.* 61:1877–1882, 1997.

75. Kawahara, Y., K. Takahashi-Fuke, E. Shimizu, T. Nakamatsu, S. Nakamori. Relationship between the Glutamate production and the activity of 2-Oxoglutarate dehydrogenase in *Brevibacterium lactofermentum*. *Biosci. Biotech. Biochem.* 61:1109–1112, 1997.

76. Kimura, E., C. Abe, Y. Kawahara, T. Nakamatsu. Molecular cloning of a novel gene, *dtsR*, which rescues the detergent sensitivity of a mutant derived from *Brevibacterium lactofermentum*. *Biosci. Biotech. Biochem.* 60:1565–1570, 1996.

77. Kimura, E., C. Abe, Y. Kawahara, T. Nakamatsu, H. Tokuda. A *dtsR* gene-disrupted mutant of *Brevibacterium lactofermentum* requires fatty acids for growth and efficiently produces L-glutamate in the presence of an excess of biotin. *Biochem. Biophys. Res. Comm.* 234:157–161, 1997.

78. Kimura, E., C. Yagoshi, Y. Kawahara, T. Ohsumi, T. Nakamatsu. Glutamate overproduction in *Corynebacterium glutamicum* triggered by a decrease in the level of a complex comprising DtsR and a biotin-containing subunit. *Biosci. Biotechnol. Biochem.* 63:1274–1278, 1999.

79. Kimura, E. Metabolic engineering of Glutamate production. In: *Advances in Biochemical Engineering/Biotechnology*, vol. 79, Scheper, T. ed., Heidelburg: Springer-Verlag, 2003, p. 37–57.

80. Schirawski, J., G. Unden. Menaquinone-dependent succinate dehydrogenase of bacteria catalyzes reversed electron transport driven by the proton potential. *Eur. J. Biochem.* 257:210–215, 1998.

81. Delaunay, S., P. Gourdon, P. Lapujade, E. Mailly, E. Oriol, J.M. Engasser, N.D. Lindley, J.L. Goergen. *Enz. Microbial. Technol.* 25:762–768, 1999.

82. Lambert, C., A. Erdmann, M. Eikmanns, R. Kramer. Triggering Glutamate excretion in *Corynebacterium glutamicum* by modulating the membrane state with local anesthetics and osmotic gradients. *Appl. Environ. Microb.* 61:4334–4342, 1995.

83. Eggeling, L., H.Sahm. The cell wall barrier of *Corynebacterium glutamicum* and amino acid efflux. *J. Biosci. Bioeng.* 92:201–213, 2001.

84. Gourdon, P., M. Raherimandimby, H. Dominguez, M. Cocaign-Bousquet, N.D. Lindley. Osmotic stress, glucose transport capacity and consequences for glutamate overproduction in *Corynebacterium glutamicum*. *J. Biotechnol.* 104:77–85, 2003.

85. Puech, V., M. Chami, A. Lemassu, M.A. Laneelle, B. Schiffler, P. Gounon, N. Bayan, R. Benz, M. Daffé. Structure of the cell envelope of *corynebacteria*: importance of the non-covalently bound lipids in the formation of the cell wall permeability barrier and fracture plane. *Microbiology* 147:1365–1382, 2000.

86. Bayan, N., C. Houssin, M. Chami, G. Leblon. Mycomembrane and S-layer: two important structures of *Corynebacterium glutamicum* cell envelope with promising biotechnology applications. *J. Biotechnol.* 104:55–67, 2003.

87. Hoischen, C., R. Kramer. Evidence for an efflux carrier system involved in the secretion of glutamate by *Corynebacterium glutamicum*. *Arch. Microbiol.* 151:342–347, 1989.

88. Kronemeyer, W., N. Peekhaus, R. Kramer, H. Sahm, L. Eggeling. Structure of the *gluABCD* cluster encoding the glutamate uptake system of *Corynebacterium glutamicum*. *J. Bacteriol.* 177:1152–1158, 1995.

89. Lichtinger, T., A. Burkovski, M. Niederweis, R. Kramer, R. Benz. Biochemical and biophysical characterization of the cell wall porin of *Corynebacterium glutamicum*: the channel is formed by a low molecular mass polypeptide. *Biochemistry* 37:15024–15032, 1998.

90. Costa-Riu, N., A. Burkovski, R. Kramer, R. Benz. PorA represents the major cell wall channel of the gram-positive bacterium *Corynebacterium glutamicu*. *J. Bacteriol.* 185:4779–4786, 2003.

91. Costa-Riu, N., E. Maier, A. Burkovski, R. Kramer, F. Lottspeich, R. Benz. Identification of an anion-specific channel in the cell wall of the Gram-positive bacterium *Corynebacterium glutamicum*. *Mol. Microbiol.* 50:1295–1308, 2003.

92. Hoischen, C., R. Kramer. Membrane alteration is necessary but not sufficient for effective glutamate secretion in *Corynebacterium glutamicum*. *J. Bacteriol.* 172:3409–3416, 1990.

93. Duperray, F., D. Jezequel, A. Ghazi, L. Letellier, E. Shechter. Excretion of glutamate from *Corynebacterium glutamicum* triggered by amine surfactans. *Biochim. Biophys. Acta* 1103:250–258, 1992.

94. Kijima, N., D. Goyla, A. Takada, M. Wachi, K. Nagai. Induction of only limited elongation instead of filamentation by inhibition of cell division in *Corynebacterium glutamicum*. *Appl. Microbiol. Biotechnol.* 50:227–232, 1998.

95. Park, S.M., A.J. Sinskey, G. Stephanopoulos. Metabolic and physiological studies of *Corynebacterium glutamicum* mutants. *Biotechnol. Bioeng.* 55:864–879, 1997.

96. Delaunay, S., D. Uy, M.F. Baucher, J.M. Engasser, A. Guyonvarch, J.L. Goergen. Importance of phosphoenolpyruvate carboxylase of *Corynebacterium glutamicum* during the temperature triggered glutamic acid fermentation. *Metabol. Eng.* 1:334–343, 1999.

97. Marx, A., K. Striegel, A.A. de Graaf, H. Sahm, L. Eggeling. Response of the central metabolism of *Corynebacterium glutamicum* to different flux burdens. *Biotechnol. Bioeng.* 56:168–180, 1997.

98. De Boer, L., L. Dijkhuizen. Microbial and enzymatic processes for L-phenylalanine oproduction. *Adv. Biochem. Eng. Biotechnol.* 41:1–27, 1990.

99. Backmann, K., M.J. O'Connor, A. Maruya, E. Rudd, D. McKay, R. Balakrishnan, M. Radjai, V. DiPasquantonio, D. Shoda, R. Hatch, K. Venkatasubramanian. Genetic engineering of metabolic pathways applied to the production of phenylalanine. *Ann. N. Y. Acad. Sci.* 589:16–24, 1990.

100. Pittard, A.J. Biosynthesis of aromatic amino acids. In: *Escherichia coli and Salmonella, Cellular* and Molecular Biology, Neidhardt, F.C., et al., eds., Washington DC: *Amer. Soc. Microbiol.*, 1996, pp 458–484.

101. Bongaerts, J., M. Kramer, U. Muller, L. Raeven, M. Wubbolts. Metabolic engineering for microbial production of aromatic amino acids and derived compounds. *Metab. Eng.* 3:289–300, 2001.

102. Ikeda, M., R. Katsumata. Hyperproduction of tryptophan by *Corynebacterium glutamicum* with the modified pentose phosphate pathway. *Appl. Environ. Microbiol.* 65:2497–2502, 1999.

103. Tatarko, M., T.Romeo. Disruption of a global regulatory gene to enhance central carbon flux into phenylalanine biosynthesis in *Escherichia coli. Curr. Microbiol.* 43:26–32, 2001.

104. Berry, A. Improving production of aromatic compounds in *Escherichia coli* by metabolic engineering. *Trends Biotechnol.* 14:250–256, 1996.

105. Katsumata, R., M. Ikeda. Hyperproduction of tryptophan in *Corynebacterium-glutamicum* by pathway engineering. *Biotechnology* 11:921–925, 1993.

106. LaDuca, R. J., A. Berry, G. Chotani, T. C. Dodge, G. Gosset, F. Valle, J.C. Liao, J. Yong-Xiao, S. D. Power. Metabolic pathway engineering of aromatic compounds. In: *Manual of Industrial Microbiology and Biotechnology*, Demain, A.L., J.E. Davies, eds., Washington DC: *Amer. Soc. Microbiol.*, 1999, pp 605–615.

107. Aiba, S., H. Tsunekawa, T. Imanaka. New approach to tryptophan production by *Escherichia coli*: genetic manipulation of composite plasmids *in vitro. Appl. Environ. Microbiol.* 43:289–297, 1982.

108. Ikeda, M., R. Katsumata. Tryptophan production by transport mutants of *Corynebacterium glutamicum. Biosci. Biotechnol. Biochem.* 59:1600–1602, 1995.

109. Kalinowski, J., J. Cremer, B. Bachmann, L. Eggeling, H. Sahm, A. Puhler. Genetic and biochemical analysis of the aspartokinase from *Corynebacterium glutamicum. Mol. Microbiol.* 5:1197–1204, 1991.

110. Follettie, M.T., O.P. Peoples, C. Agoropoulou, A.J. Sinskey. Gene structure and expression of the *Corynebacterium glutamicum*: molecular cloning, nucleotide sequence, and expression, *J. Bacteriol.* 175:4096–4103, 1993.

111. Jetten, M.S.M., A.J. Sinskey. Recent advances in the physiology and genetics of amino acid-producing bacteria. *Crit. Rev. Biotechnol.* 15:73–103, 1995.

112. Sonntag, K., L. Eggeling, A.A. de Graaf, H. Sahm. Flux partitioning in the split pathway of lysine synthesis in *Corynebacterium glutamicum*. Quantification by 13C- and 1H-NMR spectroscopy. *Eur. J. Biochem.* 213:1321–1325, 1993.

113. Broer, S., R.Kramer. Lysine excretion by *Corynebacterium glutamicum*, 2: Energetics and mechanism of the transport system. *Eur. J. Biochem.* 202:137–143, 1991.

114. Marx, A., B.J. Eikmanns, H. Sahm, A.A. de Graaf, L. Eggeling. Response of the central metabolism in *Corynebacterium glutamicum* to the use of an NADH-dependent glutamate dehydrogenase. *Metabol. Eng.* 1:35–48,1999.

115. Kojima, H., K. Totsuka. Patent Application WO9511985, 1995.

116. Debabov, V.G. The threonine story. In: *Advances in Biochemical Engineering/Biotechnology,* Vol. 79, Scheper, T. ed., Heidelberg: Springer-Verlag, 2003, p 113–136.

117. Ishida, M., K. Sato, K. Hashiguchi, H. Ito, H. Enei, S. Nakamori. High fermentative production of L-threonine from acetate by a *Brevibacterium flavum* stabilized strain transformed with a recombinant plasmid carrying the *Escherichia coli thr* operon. *Biosci. Biotechnol. Biochem.* 57:1755–1756, 1993.

118. Simic, P., J. Willuhn, H. Sahm, L. Eggeling. Identification of *glyA* (encoding Serine Hydr oxymethyltransferase) and its use together with the exporter ThrE to increase L-Threonine accumulation by *Corynebacterium glutamicum. Appl. Environ. Microbiol.* 68:3321–3327, 2002.

119. Shiio, I., S. Nakamori. Microbial production of L-threonine, I: production by *Escherichia coli* mutant resistant to alpha-amino-beta-hydroxyvaleric acid. *Agric. Biol. Chem.* 33:1152–1161, 1969.

120. Kruse, D., R. Kramer, L. Eggeling, M. Rieping, W. Pfefferle, J.H. Tchieu, Y.J. Chung, M.H. Saier, A.Burkovski. Influence of threonine exporters on threonine production in *Escherichia coli. Appl. Microbiol. Biotechnol.* 59:205–210, 2002.

121. Reinscheid, D.J., W. Kronemeyer, L. Eggeling, B.J. Eikmanns, H. Sahm. Stable expression of HOM-1-THRB in *Corynebacterium glutamicum* and its effect on the carbon flux to threonine and related amino acids. *Appl. Environ. Microbiol.* 60:126–132, 1994.

122. Leinfelder, W., C. Winterhalter. Japan Patent 11 056 381 A, 1999.

123. Ohnishi, J., S. Mitsuhashi, M. Hayashi, S. Ando, H. Yokoi, K. Ochiai, M. Ikeda. A novel methodology employing *Corynebacterium glutamicum* genome information to generate a new L-Lysine-producing mutant. *Appl. Microbiol. Biotechnol.* 58:217–223, 2002.

124. Gerstmeir, R., V.F. Wendisch, S. Schnicke, H. Ruan, M. Farwick, D. Reinscheid, B.J. Eikmanns. Acetate metabolism and its regulation in *Corynebacterium glutamicum. J. Biotechnol.* 104:99–122, 2003.

125. Merrick, M., R.A. Edwards. Nitrogen control in bacteria. *Microbiol. Rev.* 59:604–622, 1995.

126. Burkovski, A. Ammonium assimilation and nitrogen control in *Corynebacterium glutamicum* and its relatives: an example for new regulatory mechanisms in actinomycetes. *FEMS Microbiol. Rev.* 27:617–628, 2003.

127. Hecker, M., U. Volker. General stress response of *Bacillus subtilis* and other bacteria. *Adv. Microb. Physiol.* 44:35–91, 2001.

128. Ronsch, H., R. Kramer, S. Morbach. Impact of osmotic stress on volume regulation, cytoplasmic solute composition and lysine production in *Corynebacterium glutamicum* MH20-22B. *J. Biotechnol.* 104:87–97, 2003.

129. Ishige, T., M. Krause, M. Bott, V.F. Wendisch, H. Sahm. The phosphate starvation stimulon of *Corynebacterium glutamicum* determined by DNA microarray analyses. *J. Bacteriol.* 185:4519–4529, 2003.

130. Russell, J.B., G.M. Cook. Energetics of bacterial growth: balance of anabolic and catabolic reactions. *Microbiol. Rev.* 59:48–62, 1995.

131. Matsushita, K., T. Yamamoto, H. Toyoma, O. Adachi. NADPH oxidase system as a superoxide-generating cyanide-resistant pathway in the respiratory chain of *Corynebacterium glutamicum. Biosci. Biotechnol. Biochem.* 62:1968–1977, 1998.

132. Matsushita, K., A. Otofuji, M. Iwahashi, H. Toyama, O. Adachi. NADH dehydrogenase of *Corynebacterium glutamicum*: purification of an NADH dehydrogenase II homolog able to oxidize NADPH. *FEMS Microbiol. Lett.* 204:271–276, 2001.

133. Kusumoto, K., M. Sakiyama, J. Sakamoto, S. Noguchi, N. Sone. Menaquinol oxidase activity and primary structure of cytochrome bd from the amino acid fermenting bacterium *Corynebacterium glutamicum. Arch. Microbiol.* 173:390–397, 2000.

134. Molenaar, D., M.E. van der Rest, A. Drysch, R. Yucel. Functions of the membrane-associated and cytoplasmic malate dehydrogenases in the citric acid cycle of *Corynebacterium glutamicum. J. Bacteriol.* 182:6884–6891, 2000.

135. Bott, M. A. Niebisch. The respiratory chain of *Corynebacterium glutamicum. J. Biotechnol.* 104, 129–153, 2003.

136. Niebisch, A., M. Bott. Purification of a cytochrome bc-aa3 supercomplex with quinol oxidase activity from *Corynebacterium glutamicum*. Identification of a fourth subunity of çytochrome aa3 oxidase and mutational analysis of diheme cytochrome c1. *J. Biol. Chem.* 278:4339–4346, 2003.

137. Sekine, H., T. Shimada, C. Hayashi, A. Ishiguro, F. Tomita, A. Yokota. H+-ATPase defect in *Corynebacterium glutamicum* abolishes glutamic acid production with enhancement of glucose consumption rate. *Appl. Microbiol. Biotechnol.* 57:534–540, 2001.

138. Broer, S., L. Eggeling, R. Kramer. Strains of *Corynebacterium glutamicum* with different lysine productivities may have different lysine excretion systems. *Appl. Environ. Microbiol.* 59:316–321, 1993.

139. Gutmann, M., C. Hoischen, R. Kramer. Carrier-mediated glutamate secretion by *Corynebacterium glutamicum* under biotin limitation. *Biochim. Biophys. Acta.* 1112:115–123, 1992.

140. Brand, S., K. Niehaus, A. Puhler, J. Kalinowski. Identification and functional analyis of six mycolyltransferase genes of Corynbebacterium glutamicum ATCC 13032. *Arch. Microbiol.* 180:33–44, 2003.

141. De Sousa-D'Auria, C., R. Kacem, V. Puech, M. Tropis, G. Leblon, C. Houssin, M. Daffé. New insights into the biogenesis of the cell envelope of corynebacteria: identification and functional characterization of five new mycolyltransferase genes in *Corynbebacterium glutamicum*. *FEMS Microbiol. Lett.* 15:35–44, 2003.

142. Portevin, D., C. De Sousa-D'Auria, C. Houssin, C. Grimaldi, M. Chami, M. Daffé, C. Guilhot. A polyketide synthase catalyzes the last condensation step of mycolic acid biosynthesis in mycobacteria and related organisms. *Proc. Natl. Acad. Sci. USA* 101:314–319, 2004.

24

Biotechnology of Microbial Polysaccharides in Food

Ian W. Sutherland

CONTENTS

24.1 Introduction 584
24.2 Natural Occurrence of Microbial Polysaccharides in Foods 584
     24.2.1 Microbial Polysaccharides Present in Food as
           Products of Microbial Food Components 584
     24.2.2 Physical Properties 588
     24.2.3 Production and Synthesis 588
24.3 Microbial Polysaccharides Incorporated as Food Additives.................................589
     24.3.1 Xanthan 589
           24.3.1.1 Industrial Substrates..590
           24.3.1.2 Biosynthesis of Xanthan 592
           24.3.1.3 Physical Properties of Xanthan..594
           24.3.1.4 Food Usage of Xanthan 595
           24.3.1.5 Two Are Better Than One – Synergistic Gels
                   Involving Xanthan 597
     24.3.2 Gellan 597
           24.3.2.1 Structure...598
           24.3.2.2 Physical Properties...598
           24.3.2.3 Food Applications 600
     24.3.3 Curdlan 600
24.4 Exopolysaccharides as a Source of Flavor Components 601
24.5 Gazing Into the Crystal Ball:
     The Future for Microbial Polysaccharides in Food 602
24.6 Legislative Acceptability of Microbial Polysaccharides.......................................603
References...605

24.1 INTRODUCTION

Polysaccharides are incorporated into foods to alter the rheological properties of the water present and thus change the texture of the product. Most of the polysaccharides used are employed because of their ability to thicken substances or to cause gel formation. They are also used to stabilize foams and emulsions, suspend particulate materials and inhibit or decrease syneresis, as well as increase water retention. Advantage is also taken of the ability of some mixtures of polysaccharides to exhibit synergistic gelling — basically for the two polymers to yield a gel at concentrations of each which will not in themselves form gels. Associated with these readily measurable physical properties are others such as "mouth feel," which are more difficult to define but which also show some correlation with physical properties. In addition, polysaccharides are incorporated because of their capacity to control the texture of foods and to prevent or reduce ice crystal formation in frozen foods; they may also influence the appearance, color, and flavor of prepared foodstuffs. It must also be remembered that many foodstuffs already contain animal or plant polysaccharides such as hyaluronic acid, starch, or pectin. Thus addition of any further polysaccharide or polysaccharides will in all probability involve interactions with these components as well as with proteins, lipids, and other food components. Currently the polysaccharides added to food are sourced from bacteria, algae, and plants, and are exemplified by xanthan and gellan, alginate and carrageenan, and pectin and starch respectively.

The use of microbial polysaccharides in food is also governed by a number of factors unrelated to their physical properties. This includes the observation that humans generally do not metabolize microbial polysaccharides. They are nutritionally inert and nontoxic. Currently there is considerable consumer concern over the use of "food additives," including hydrocolloids. Most regulatory bodies review their lists of permitted food additives at regular intervals, when they may amend the regulations. This may also lead to improved or altered specifications. On the other hand, many of the microbial species, which are currently used in food preparations, synthesize and excrete polysaccharides as well as other products.

Very many bacteria, yeasts, and fungi can produce polysaccharides. While much interest in these polymers is due to their role in infection or adhesion, some of these polymers have proved to be useful industrial products. Dextran was the first microbial polysaccharide to be commercialized and to receive approval for food use. Although it is no longer used for this purpose, several other microbial polymers are now commercial products with a variety of uses. Use of these microbial exopolysaccharides (EPS) in foods is more limited. Only two bacterial polysaccharides are currently extensively employed by the food industry, except in Japan where polymers of this type are regarded as natural products. It must also be remembered that although microbial polysaccharides may be incorporated into foods as food additives, many other EPS are integral products of microorganisms which are involved in the preparation of the food. A number of natural foods contain polysaccharide-producing microorganisms and considerable interest has currently been shown in the lactic acid bacteria (LAB) which are widely used in fermented milk products and other fermented foods. This review will consider both microbial polysaccharides as food additives and those EPS which are present as normal products of microorganisms used in fermented foods.

24.2 NATURAL OCCURRENCE OF MICROBIAL POLYSACCHARIDES IN FOODS

24.2.1 Microbial Polysaccharides Present in Food as Products of Microbial Food Components

A number of the microorganisms that are used in the preparation of foodstuffs are capable of synthesizing extracellular polysaccharides. This is especially true of the lactic acid bacteria (*Streptococcus*, *Lactobacillus* and *Lactococcus* spp.), and a number of structural studies have recently elucidated the nature of some of these polysaccharides. These and other reports can be found in the reviews of De Vuyst and Degeest (1) and Laws et al. (2). Less is yet known about the relationship between structure and physical properties than is the case for Gram-negative bacterial products such as xanthan and gellan (3). As well as the deliberate addition of microbial polysaccharides to food products to obtain specific properties, there are a number of bacterial fermentations in which polysaccharide is produced and is needed to yield a specific type of product. An example of this can be found in certain types of fermented milk product such as yogurts. In some of these, the production of polysaccharide during bacterial growth is claimed to enhance the product, particularly in respect of the body and texture of the product and in its smoothness and mouth feel. This is particularly true of countries such as the Netherlands and France in which the addition of plant or animal stabilizers is prohibited. The polysaccharide from *Streptococcus salivarus thermophilus* is used in this way. Unfortunately, the production of an apparently neutral glycan by these bacteria is unstable and the use of the bacteria tends to lead to a lack of uniformity in the product. Polysaccharide-producing strains of *Lactobacillus delbruekii bulgaricus* are also used for this purpose, strains being available which synthesize from 14 to 400 mg/litre of a culture of viscous polysaccharide in which galactose is the major component. There are also some fermented milk products in which the production of a thick, gel-like texture results from the exopolysaccharide synthesized by the bacteria used. These are traditional products from parts of Finland, which do not have widespread sales elsewhere. Thus, the importance of polysaccharides from mesophilic and thermophilic LAB is recognized in the production and rheological characteristics of fermented milks, but problems have been caused by the instability of products and requirement of many of these bacteria to grow in complex media (4). The textural and other improvements are made by these polysaccharides, even though the actual yields are low. As pointed out by Ruas-Madiedo et al. (5) some of the EPS from LAB may also possess health-promoting effects such as antitumor and immune-modulating activities in addition to possible prebiotic activity and cholesterol-lowering activity (6).

As LAB are recognized food microorganisms, increasing interest has recently been shown in the EPS synthesized by them. It is also accepted that at present EPS derived from these bacteria could not yield significant commercial products (7). Now that a range of polysaccharides from LAB has been studied, it is clear that no common pattern of structures exists, although certain monosaccharides including D-glucose, D-galactose, and L-rhamnose are commonly found (1). While all the heteropolysaccharides are composed of regular repeat units, some carry phosphate groups in addition to their monosaccharide components, while others lack the inorganic residues. Most of the polysaccharides contain the sugars common to other bacterial EPS and as yet few of the rarer monosaccharides have been reported. Among these are galactofuranose found in the pentasaccharide repeat units of the *Lactobacillus rhamnosus* strains C83 EPS (Figure 24.1) (8) and KL37c (9). This monosaccharide is also present in the exopolysaccharide produced by *Streptococcus thermophilus* EU20 (10) as well as *S. thermophilus* strain SY 89 in which it is found together with galactopyranose (Figure 24.2) (11). The majority of EPS from LAB are composed of a small range of common sugars very frequently including D-glucose, D-galactose, and L-rhamnose (2). These are mainly neutral hexoses or methylpentoses, but may also include the corresponding uronic acids and *N*-acetylaminosugars (12). Although L-fucose is common among the EPS of Gram negative bacteria, L-rhamnose is the methylpentose most frequently found in EPS from LAB. Many of the EPS are polyanionic due to the presence of either uronic acids or phosphate groups. The most common uronic acid is D-glucuronic acid, but D-galacturonic acid is also frequently found. The structures are repeat units of 3–7 monosaccharides and although the polymers are essentially linear macromolecules, they

→3)-α-D-Glcp-(1→2)-β-D-Galf -(1→6)-α-D-Galp-(1→6)-α-D-Glcp-(1→3)-β-D-Galf-(1→

(8,9)

Figure 24.1   The pentasaccharide repeat unit of the *Lactobacillus rhamnosus* strain C83 EPS (8).

$$→3)\text{-}α\text{-D-Glcp-}(1→3)\text{-}β\text{-D-Glcp -}(1→3)\text{-}β\text{-D-Galf-}(1→$$
$$_1↑^6$$
$$β\text{-D-Galp-1}$$

(11)

Figure 24.2   The tetrasaccharide repeat unit of *Streptococcus thermophilus* strain SY89 containing equimolar amounts of galactopyranose and galactofuranose (11).

$$β\text{-D-Glcp} \quad CH_3.COO.$$
$$_6↓^1 \qquad ^2↓$$
$$→ 4)\text{-}β\text{-D-Glcp-}(1→ 3)\text{-}β\text{-L- Rhap-}(1→$$
$$_1↑^3$$
$$α\text{-L-Rhap (4}←\text{sn-glycerol 3-phosphate)}$$

(13)

Figure 24.3   The triply branched repeat unit of the EPS of *Lactobacillus sake* 0-1 (13).

$$→2)\text{-}α\text{-L-Rhap-}(1→2)\text{-}α\text{-D-Galp-}(1→3)\text{-}α\text{-D-Glcp-}(1→3)\text{-}α\text{-D-Galp-}(1→3)\text{-}α\text{-L-Rhap-}(1→$$
$$↑_4$$
$$β\text{-D-Galp-}(1→4)\text{-}β\text{-D-Glcp-1}$$

(14)

Figure 24.4   The heptasaccharide repeat of *Lactococcus lactis cremoris* B39 (14).

also very commonly carry branches. Examples of this can be seen in the triply branched repeat unit of *Lactobacillus sake* 0-1 (Figure 24.3) (13) or the heptasaccharide repeat of *Lactococcus lactis cremoris* B39 (Figure 24.4) (14). The branches may be on different monosaccharides, or on the same monosaccharide as is the case for the EPS from *Lactobacillus lactis cremoris* SBT 0495 in which a single D-galactose carries both a rhamnose and a galactose-1-phosphate (15). Strains of the same species may yield several different EPS chemotypes. An isolate of *Lactobacillus delbruekii bulgaricus* from kefir grains yielded a polysaccharide with almost equimolar amounts of D-glucose and D-galactose (16), whereas the EPS from another strain of the same subspecies had a very different structure as can be seen in Figure 24.5 (17). Another strain possessed a heptasaccharide repeating unit in which the ratio of D-galactose: D-glucose: L-rhamnose was 5:1:1 with the structure shown in Figure 24.6 (18). Strain 291 differed again in yielding a pentasaccharide repeat unit containing only glucose and galactose (Figure 24.7) (19).

In addition to monosaccharides, bacterial EPS frequently carry other substituents, which may either be organic or inorganic in nature. The most common organic substituents are either *O*-acetyl groups or pyruvate ketals. The pyruvate ketals are most commonly found attached to D-galactose, D-glucose, or D-mannose residues. It is fairly unusual to find both pyruvate and acetate attached to the same monosaccharide residue. As more structures from Gram-positive species are studied, a number of these have been found to contain pyruvate. One such is EPS from *Lactobacillus rhamnosus* RW-959M in which a galactose

→2)-α-L-Rhap-(1→4)-α-D-Glcp-(1→3)-β-L-Rhap-(1→4)-β-D-Glcp-1→4)-α-D-Glcp-(1→
$$\begin{array}{c} 3 \\ \uparrow \\ 1 \\ \text{α-L-Rhap} \end{array}$$

(17).

Figure 24.5 The structure of the EPS from *Lactobacillus delbrueckii bulgaricus* EU23 (17).

→2)-α-D-Galp-(1→3)-β-D-Glcp-(1→3)-β-D-Galp-(1→4)-α-D-Galp-(1→
$$\begin{array}{c} 1\uparrow3 \qquad\qquad\qquad\qquad 1\uparrow4 \qquad\qquad 1\uparrow3 \\ \text{β-D-Galp} \qquad\qquad\qquad \text{β-D-Galp} \qquad \text{α-L-Rhap} \end{array}$$

(18)

Figure 24.6 The heptasaccharide structure of the EPS from *Lactobacillus delbrueckii bulgaricus* rr (18).

→4)-β-D-Glcp-(1→4)-α-D-Glcp-(1→4)-β-D-Galp-(1→
$$\begin{array}{c} 1\uparrow6 \\ \text{β-D-Galp-(1→4)-β-D-Glcp} \end{array}$$

(19)

Figure 24.7 The structure of the EPS from *Lactobacillus delbrueckii bulgaricus* 291 (19).

residue carried a 4,6-linked ketal (20). In a number of EPS the acyl groups are present in nonstoichiometric amounts. Thus some xanthans contain 0.3 moles of pyruvate per repeat unit. Phosphate groups occur widely among EPS from Gram-positive species including some of the industrially important *Streptococcus* and *Lactobacillus* species, which are employed in the fermentation of milk, yogurt, and cheeses. Quite a few of the extracellular polysaccharides from LAB are neutral polymers as can be seen in Figures 2 and 5. Although some of the LAB can produce homopolymers composed of a single monosaccharide, the majority are heteropolysaccharides. Among the homopolysaccharides synthesized by LAB are glucans and fructans, several of which are produced only when the bacteria are grown in the presence of sucrose. One example is *Lactobacillus reuteri* LB121 which yielded both a fructan and a glucan with masses of 3.5 and 150 kDa respectively (21). Unusually, one strain of *L. lactis lactis* (H414) yielded a galactan. An enzyme from the *L. reuteri* 121 strain was later used to produce a high mass (>10^7) fructan with β(2→1) linkages (22). This was thus the first report of bacterial inulin synthesis in a GRAS listed bacterial species (the only other observation of bacterial inulin formation related to a non-GRAS listed species). The strain was also unusual in being capable of forming a highly branched α-D-glucan (23).

It should also be remembered that a number of bacterial strains might have the capacity to form more than one type of polysaccharide, although it is unusual for two or more EPS to be secreted simultaneously. An example of this was *Streptococcus thermophilus* LY03 which yielded two EPS of the same composition, D-glucose and D-galactose in the same molar ratio of 1:4, but differing in mass (24). In contrast to this, strain S3 formed a polymer composed of D-galactose and L-rhamnose in the ratio 2:1 and was also partially acetylated on a D-galactofuranosyl side chain residue (25). The hexasaccharide repeat unit contained three D-galactopyranosyl residues. One of the simplest structures identified to date is the neutral trisaccharide repeat unit from *Lactobacillus* spp. G-77

→6)-α-D-Glcp-(1→6)-α-D-Glcp-(1→
$_1\uparrow 2$
α-D-Glcp

(26)

Figure 24.8   The neutral trisaccharide repeat unit from *Lactobacillus* spp. G-77 (26).

(Figure 24.8) (26). This bacterial species was capable of forming two EPS, the other being a 2-substituted (1→3)-β-D-glucan, a structure that was also found in the EPS from *Pediococcus damnosus* (27).

Among the most unusual structures determined so far is that for an EPS from *Streptococcus thermophilus* 8S (28). This proved to possess a pentasaccharide repeat unit containing D-galactose: D-glucose: D-ribose: N-acetyl-D-galactosamine in the molar ratio 2:1:1:1. In addition an open chain nonionic acid was identified. A substituent of this type has not been reported from other bacterial EPS. Some high mass LAB polymers appear to possess more complex structures with nonstoichiometric amounts of their components. Thus the EPS from *Lactobacillus helveticus* TY1-2 was formed from heptasaccharide repeat units with the molar ratio glucose: galactose: glucosamine 3.0:2.8:0.9 (29). The actual structure was in the form of a trisaccharide backbone with the aminosugar residue carrying both a monosaccharide (a D-galactosyl residue) and a trisaccharide side chain.

24.2.2 Physical Properties

Rheological studies on the EPS from *Propionobacterium acidi-propiionici* DSM4900 revealed shear thinning behavior and suggested that it formed an entangled polymer solution (30), while a polymer from *Streptococcus thermophilus* SY containing rhamnose, glucose, and galactose, was claimed to yield pseudoplastic solutions and have a viscosity higher than xanthan (31). Another LAB EPS, from *L. saké*, had an estimated mass of 4–9 x 10^6 Da and was also claimed to have a higher intrinsic viscosity than xanthan (32). Studies have also been made on the effects of chemical modifications to the EPS from a small number of LAB. Tuinier et al. (33) found that removal of side chain D-galactosyl residues from the EPS of *L. lactis cremoris* polymers increased chain stiffness and thickening efficiency. Deacetylation of the EPS from strain B891 left the chain stiffness unaltered.

In one of the few attempts to model the structure of EPS from *Lactobacillus helveticus*, Faber et al. (34) noted that the EPS from strain 766 exhibited a flexible twisted secondary structure. In their tertiary structure, the polysaccharide chains tended to adopt a random coil conformation.

24.2.3 Production and Synthesis

Yields of most of the EPS from LAB are generally low. However *L. reuteri* LB121 was claimed to form 10g. l^{-1} of two homopolysaccharides (21). Various studies in the laboratory have utilized whey-based media to permit the isolation and characterization of the polysaccharides products from LAB. Thus Ricciardi et al. (31) noted that synthesis of the polymer in *S. thermophilus* was growth associated and obtained yields of 152mg. l^{-1}. Attempts have also been made to improve EPS synthesis through identification and augmentation of key enzymes involved in the earlier stages of carbohydrate metabolism. However this has not always proved successful. In *S. thermophilus*, over expression of *galU* (UDP-glucose pyrophosphorylase) enhanced enzyme activity but not EPS synthesis (35). Only when *galU* and *pgmA* (phosphoglucomutase) were both over expressed were higher yields of EPS obtained. It has also been recognized that in *Lactococcus lactis*, there was competition between EPS synthesis and bacterial growth, with glucose-6-phosphate as the key intermediate metabolite (36). This study indicated possible mechanism by which EPS production might be enhanced through genetic manipulation of the EPS bio-synthetic and glycolysis pathways. However a later report from the same researchers (37) revealed that although over expression of the *galU* gene greatly increased sugar nucleotide (UDP-glucose and UDP-galactose) levels, EPS production was not significantly different from the wild type.

24.3 MICROBIAL POLYSACCHARIDES INCORPORATED AS FOOD ADDITIVES

In the US, polysaccharides which are to be used as food additives are subject to Generally Regarded As Safe (GRAS) regulations and must be approved by the USDA. Currently only two microbial polysaccharide products have such approval – xanthan and gellan, although curdlan may also be under evaluation. Previously dextran was also an approved food additive, but it is not currently used in food manufacture. In Japan, a wider view is taken and microbial EPS are regarded as natural products. One thus finds the bacterial product curdlan being used in various foodstuffs, while pullulan (from the fungus *Aureobasidium pullulans*) is also acceptable. In the case of pullulan, the polysaccharide also has potential value as a food packaging material, a role which is also strictly controlled.

24.3.1 Xanthan

Xanthan, a product from the plant pathogenic bacterium *Xanthomonas campestris*, is typically produced copiously as an extracellular slime by the various pathovars of *Xanthomonas campestris* as well as by some other *Xanthomonas* species. Acid hydrolysates of xanthan contain D-glucose, D-mannose, and D-glucuronic acid in the molar ratio 2:2:1. Other components have been reported in polysaccharides from bacterial isolates described as *Xanthomonas* species, but the strains involved have generally been poorly characterized. The primary structure of the polysaccharide is a pentasaccharide repeat unit, effectively a cellulose chain to which trisaccharide side chains are attached at the C-3 position on alternate D-glucosyl residues (38,39) (Figure 24.9). The polysaccharides from several other *Xanthomonas* species seem to share the same composition as that from pathovars of *X. campestris*. Depending on the bacterial strain and on the physiological conditions for bacterial growth, the polysaccharide may carry varying amounts of *O*-acetyl groups on the

$$\to [-4-\beta\text{-D-Glc}p-(1\to4)-\beta\text{-D-Glc}p-1-]\to$$
$$_1\uparrow3$$
$$\alpha\text{-D-Man}p\text{-O.CO.CH}_3$$
$$_1\uparrow2$$
$$\alpha\text{-D-Glc}p\text{A}$$
$$_1\uparrow4$$
$$\beta\text{-D-Man}p = \text{Pyr}$$

Typically the internal α-D-mannosyl residue is fully acetylated but only c. 30% of the β-mannosyl termini are ketalated.

(38)

Figure 24.9 The structure of xanthan from *Xanthomonas campestris* (38).

C-6 position of the internal α-ᴅ-mannosyl residue and of 4,6-carboxyethylidene (pyruvate ketal) on the side chain terminal β-ᴅ-mannosyl residue respectively. Material from some strains (not used for food) carries two acetyl groups on the internal mannosyl residue, and there is some evidence to suggest that certain xanthan preparations may even contain in excess of 2 moles of acetate per repeat unit. However, the material used as a food additive derives from a standard strain (frequently designated NRRL B-1459) and generally contains pyruvate and acetate in approximate molar ratios of 0.3 and 1.0 per pentasaccharide repeat unit.

   X. campestris will grow and produce EPS on a wide range of carbon substrates including amino acids, citric acid cycle intermediates, and carbohydrates. Either ammonium salts or amino acids can be used as nitrogen sources. Various ions are needed for bacterial growth and polysaccharide synthesis. Limitation of any of the ions required for substrate uptake or for precursor or polymer synthesis can affect the yield and properties of the EPS. Xanthan formation by *X. campestris* resembles many other bacterial-EPS-producing systems in that polymer production is favored by a high ratio of carbon source/limiting nutrients such as nitrogen. Typically, media for laboratory synthesis of xanthan contain 0.1–0.2% ammonium salt and 2–3% glucose or sucrose. Xanthan can even be produced in fairly good yield when the bacteria are grown in a simple synthetic medium composed of glucose, ammonium sulphate, and salts, but production is improved in the presence of organic nitrogen sources. The quality and the final yield of xanthan may be enhanced by the addition of small amounts of organic acids or of citric acid cycle intermediates such as α-ketoglutaric acid (40). Oxygen is required for growth and for xanthan production, and as the culture viscosity increases as xanthan is formed, oxygen may rapidly become limiting. For satisfactory cultivation in a fermenter, vigorous aeration was essential and fermentation vessels had to be designed to ensure minimal dead space, which would otherwise lead to stagnant areas of culture. The xanthan production process is complicated by mass transfer reactions in the high viscosity broths. Oxygen solubility decreased with increasing xanthan concentration and the diffusion constant of oxygen in dilute solutions of the polysaccharide was reduced relative to water (41). The presence of polysaccharide in the fermentation broth may also affect the availability of carbon substrates such as glucose and sucrose by interaction with these carbohydrates. Consequently, fed-batch culture is usually preferred. Nakajima et al. (42) suggested that in the design of the fermentation vessel, the volume exposed to high shear is of critical importance. The specific rate of xanthan production depended on the volume of the high shear region.

   Xanthan is produced throughout growth of *X. campestris* and in the stationary phase. The specific rate of xanthan synthesis was closely related to the bacterial growth rate in batch culture; maximal during exponential growth and minimal during the stationary phase. In this respect, xanthan production resembled the synthesis of several other bacterial exopolysaccharides. Differences could be seen in the viscosity and the acylation of the xanthan synthesized in nutrient limited media during various phases of the growth cycle (43,44). Acetyl CoA and phosphoenolpyruvate may not be readily available during certain stages of growth to permit the complete acylation of the xanthan repeating units. Consequently, the xanthan synthesized in batch culture represents the products of all phases of growth and is possibly a mixture of several molecular types with varying degrees of acylation and varying mass.

24.3.1.1 Industrial Substrates

In the laboratory, pure substrates such as glucose or sucrose are used, whereas in industrial production different substrates are employed. The substrates must be cheap, plentiful carbon sources, which frequently include starch, starch hydrolysates, corn syrup, molasses, glucose, and sucrose (derived from either sugar beet or from sugar cane). It is imperative that they also be acceptable for food use. Optimal synthesis of xanthan requires a balance between utilizable carbon and nitrogen sources. Care must therefore be taken to obtain consistency in yield and product quality when using substrates which may contain nitrogen in addition to carbohydrate. The nutritional versatility of *X. campestris* is clearly a major factor in favor of its use for commercial xanthan production. However, the quality of the xanthan produced from different substrates may vary considerably — the molecular weight and hence the rheological characteristics of xanthan synthesized from glucose or starch may well differ from that formed when whey or other proteinaceous material is employed. The nitrogen sources used for industrial production of xanthan may include yeast hydrolysates, soybean meal, cottonseed flour, distillers' solubles, or casein hydrolysates. In batch culture for industrial production of the polysaccharide, there must be careful control of pH and of the aeration rate. Adequate oxygen transfer may be difficult to achieve unless the fermentation vessel has been carefully designed to ensure that mixing is optimal. To minimize this problem, fed-batch processes may be used. Even so, the conditions used for production and processing must be carefully standardized to ensure that product yield and quality are consistent. In the laboratory, the polymer can be readily prepared in high yield, and then separated from the bacterial cells by high speed centrifugation or by precipitation with quaternary ammonium compounds. As an industrial product, it is manufactured in stirred tank fermenters on a large scale by batch or fed-batch fermentation by a number of commercial companies; normally fermentation proceeds for 3 days at 30°C. In this respect, Linton (45) has claimed that "exopolysaccharide production is a very efficient process" and calculations have shown that the conversion efficiency of substrate to xanthan is very high. After completion of fermentation, the product is subjected to heat treatment to eliminate viable bacteria and to destroy hydrolytic enzymes such as cellulases, amylases, pectinases, and proteases. This treatment also enhanced the rheological properties of xanthan in solution (46). Recovery of the polysaccharide from industrial cultures requires removal or destruction of the cells followed by precipitation with the polar organic solvent isopropanol. To reduce the cost of the process, the solvents are later recovered for reuse by distillation. The fibrous precipitate is dried, milled, and sieved to give material of different mesh sizes. Bacterial cells are difficult to remove from the highly viscous culture broth, although pasteurization may lead to some autolysis and degradation of cell material as well as improving the subsequent separation of cells from the polymer. Polysaccharide recovery in the presence of polar solvents may be also be increased or accelerated by the addition of electrolytes. Subsequent purification can be achieved by a variety of techniques including fractional precipitation and chromatography. Dilute solutions of xanthan may also be subjected to clarification by filtration. The purified xanthan material can then be subjected to various analytical procedures to verify the composition, and to standard techniques for the determination of polysaccharide structure. The procedures used in the production and recovery of xanthan have been reviewed by Garcia-Ochoa et al. (47) and Galindo (46).

Although it is not degraded in the human or animal body, xanthan is biodegradable. It is a substrate for a range of xanthanases (enzymes with endo-1,4-β-D-glucanase activity cleaving the main chain of the polysaccharide) and xanthan lyases. Most crude enzyme preparations contain at least two different types of enzyme activity, although there may also be associated glycosidases (Figure 24.10). Substrates are randomly cleaved to yield oligosaccharides of different sizes; there is accompanying rapid loss of solution viscosity. Most commercial cellulase preparations lack activity against xanthan unless the polysaccharide is in dilute, ion free solution (48,49). The second type of enzyme found associated with the bacterial preparations is a xanthan lyase (4, 5 transeliminase) which cleaves the β-D-mannosyl-D-glucuronic acid linkage of the trisaccharide side chains (50). The activity of the xanthan

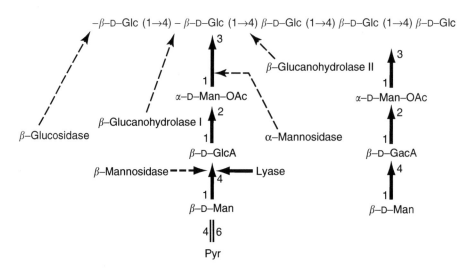

Figure 24.10 Enzymatic degradation of xanthan.

lyase was not greatly influenced by the presence or absence of pyruvate groups on the terminal mannose or of ester-linked acetyl groups on the internal mannose. The enzymes that degrade xanthan are useful laboratory tools for the determination of some of the structurally subtle, acyl modifications found on many xanthan molecules. The fragments generated are more suitable than undegraded polysaccharide for structural studies using fast atom bombardment spectroscopy. They have also proved useful in studies of monoclonal antibody specificity (51). Both the monoclonal antibodies and the enzymes could be used for highly specific quantitation of xanthan in processed foods, as has been suggested recently by Ruijssenaars et al. (52,53), or for the determination of xanthan purity.

24.3.1.2 Biosynthesis of Xanthan

Because of its industrial applications and the very unusual properties of the polysaccharide in aqueous solution, xanthan has been the subject of a large number of structural and physico-chemical studies. These have revealed much information on structure to function relationships in microbial polysaccharides. The biosynthesis of xanthan has also been extensively examined. It shares a common mode of biosynthesis with other bacterial polysaccharides composed of regular repeating units, which are found as cell wall components or extracellular products. Membrane-bound enzymes utilize various activated carbohydrate donors in a tightly regulated sequence to form the polysaccharide on an acceptor molecule. The oligosaccharide repeat units of xanthan are produced by the sequential addition of monosaccharides from the energy rich sugar nucleotides (usually nucleoside diphosphate sugars) to a C_{55} isoprenoid lipid acceptor molecule. At the same time, acyl adornments are added from appropriate activated donors. Thus, the xanthan backbone is formed by the sequential addition of D-glucose-1-phosphate and D-glucose respectively from two moles of UDP-D-glucose. Thereafter, D-mannose and D-glucuronic acid are added from GDP-mannose and UDP-glucuronic acid respectively. Each step requires a specific enzyme and a specific substrate. Absence of the enzyme (or the substrate) inhibits synthesis of the polysaccharide. Depending on the strain used and the physiological conditions under which the bacteria have been grown, and hence on the exact structure of the polymer formed, *O*-acetyl groups are transferred from acetyl CoA to the internal mannose residue, and pyruvate, from phosphoenolpyruvate (PEP), is added to the terminal mannose. This sequence of reactions

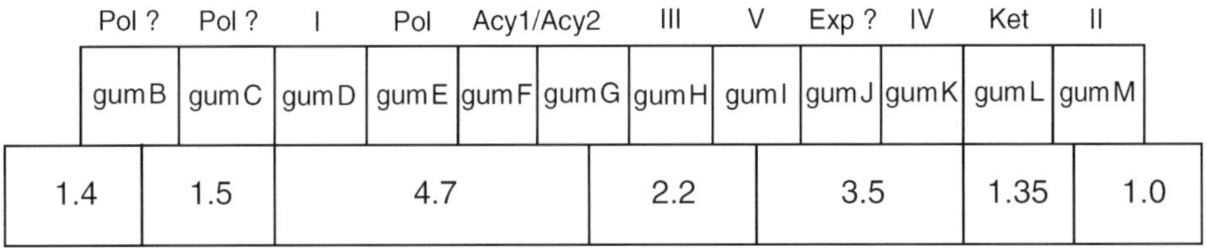

Pol ?	Pol ?	I	Pol	Acy1/Acy2	III	V	Exp ?	IV	Ket	II
gum B	gum C	gum D	gum E	gum F	gum G	gum H	gum I	gum J	gum K	gum L	gum M

1.4	1.5	4.7	2.2	3.5	1.35	1.0

The restriction map of Xanthomonas campestris. The BamHI restriction map (bottom) indicates the order and the approximate size (in kilobases) of the fragments of the 16 kb of DNA of the xanthan gene cluster. The genetic map (centre) indicates the twelve separate xanthan genes, designated gum B to gum M. The biochemical functions (top) indicate the enzymic activity identified for each gene: I–V, transferases I–V; Acy, acetylases (1 and 2); Ket, ketalase; Pol, polymerase; Pol?, possible involvement with polymerisation or export; Exp, post-polymerisation processing or export.

Figure 24.11 Genetic map of the genome segment of *X. campestris* involved in polysaccharide production (54,55).

was elegantly demonstrated by Ielpi and his colleagues (54,55) and is indicated in Figure 24.11. After the pentasaccharide repeat unit has been formed, oligomers are produced by transfer to other lipid intermediates, gradually increasing the size of the carbohydrate chain. Although the structure of the repeating unit is determined by the sequential transfer of the different monosaccharides and acyl groups from their respective donors by highly specific sugar *transferases*, the *polymerase* enzyme responsible for polymerization of the pentasaccharides into a macromolecule of $M_r > 10^6$ is now known to be less specific. This is clear from the ability of mutants to form xanthans defective in their trisaccharide side chains. Such a lack of absolute specificity might account for the natural evolution of bacterial exopolysaccharides such as xanthan, and for the existence of families of structurally related but nonidentical microbial polysaccharides. The final stages of exopolysaccharide secretion from the cytoplasmic membrane passage across the periplasm and outer membrane and finally excretion into the extracellular environment — are less well defined than the earlier biosynthetic steps. The process undoubtedly requires an energy source and may well be analogous to the export of lipopolysaccharide to the outer membrane, in which ATP provides the energy. The genetic system controlling xanthan production in *X. campestris* was elucidated by Vanderslice et al. (56,57). A linear sequence of twelve genes responsible for xanthan production was found. This included seven gene products needed for monosaccharide transfer and for acylation at the lipid intermediate level to form the completely acylated repeat unit (Figure 24.11; Table 24.1). Interestingly, two acetyl transferases were found, one of which appears to function when the pyruvate transferase is defective. The region of the chromosome controlling xanthan synthesis comprised 13 kb. However, other genes, responsible for less well defined functions, may yet be identified in *X. campestris*. The later stages of xanthan biosynthesis presumably include features common to those exopolysaccharide-producing bacteria which have been studied in detail. Thus, in *E. coli*, a region of the chromosome extending to 11.6 kb codes for five proteins, expression of which appears to be necessary for exopolysaccharide production. These proteins *may* form part of a multicomponent system responsible for exporting the polymer from the site of polymerization at the cytoplasmic membrane, to the exterior of the bacterial cell. One of the proteins involved appeared to have attributes common to other ATP-utilizing systems and it may be concerned with excretion (58).

Regulation of xanthan synthesis is not yet well defined. Many of the enzymes needed for the formation of precursors not specifically associated with exopolysaccharide production appear to be under separate control. In *X. campestris*, there are no significant

Table 24.1

Genes Involved in Xanthan Synthesis

Gene products of genes involved in non-specific precursor synthesis:	
UDP-D-glucose pyrophosphorylase	
Gene products of genes involved in specific precursor synthesis:	
GDP-D-mannose pyrophosphorylase	
UDP-D-glucose dehydrogenase	
Gene products of genes involved specifically in synthesis:	
D-Glucose-1-phosphate transferase	(*gumD*)
D-Glucosyl transferase	(*gumM*)
D-Mannosyl transferase 1	(*gumH*)
D-Glucuronosyl transferase	(*gumK*)
D-Mannosyl transferase 2	(*gumI*)
Ketalase	(*gumL*)
Acetylase 1 acetylates the internal mannose	(*gumF*)
Acetylase 2 acetylates the external mannose	(*gumG*)
Polymerase	
(Polysaccharide excretion mechanism)	(*gumB, gumC, gumE*)

competing demands for substrate as any carbohydrate supplied to the bacteria is metabolized to provide energy, or converted either to essential cell components, wall material, or exopolysaccharide with high efficiency. There are effectively no intracellular storage polymers such as glycogen competing with exopolysaccharide in their synthetic demands on substrate. Xanthan is also the only polysaccharide synthesized by *X. campestris*, unlike several other bacterial species where two or even three extracellular polymers, together with intracellular oligosaccharides, may be produced.

24.3.1.3 Physical Properties of Xanthan

The unusual structure of xanthan as a substituted cellulose confers a number of physical properties to the polymer which are utilized in the food (and other) industries. It is stable at both acid and alkaline pH and forms a pseudoplastic dispersion in water. Relatively low polysaccharide concentrations produce highly viscous solutions and the viscosity does not change greatly on raising the temperature. The solutions are compatible with many other food ingredients and give good flavor release. Currently an estimated 20,000–30,000 metric tons are produced per annum for food and nonfood uses by a number of manufacturers in the US, UK, and elsewhere in the EU, as well as other countries such as China.

   The techniques applied to determine the physical attributes of the exopolysaccharide, together with a thorough knowledge of its chemical structure, have enabled the assembly of much information on structure to function relationships pertaining to xanthan and have also provided an excellent model for other microbial exopolysaccharides. Values for the persistence length obtained using techniques such as light scattering and viscometry in dilute solution range from 110–150 nm. A wide range of values proposed for the molecular weight of xanthan has resulted partly from the application of different techniques and also from the tendency of the polysaccharide strands to aggregate in solution. The most consistent results have been obtained using light scattering techniques (Table 24.2). Shatwell (59) noted a range of 0.9–1.6×10^6 Da for four xanthan preparations with differing acyl substitution from different pathovars grown under standard cultural conditions. Other estimated results ranged between the extremes of 1.1 and 47×10^6 Da. Clearly the very high values must be

Table 24.2

Molecular Mass of Xanthan

Strain	Molecular Weight ($\times 10^6$)
X. campestris pv. *campestris* 646	0.9–1.2
X. campestris pv. *phaseoli* 1128	1.27
X. campestris pv. *phaseoli* 556	1.48
X. campestris pv. *oryzae* PXO61	1.60

Molecular Weights of Xanthans obtained by Static Light Scattering Measurements
Results of Shatwell et al. (59).

of dubious accuracy. Measurements of contour length similarly produced a wide range of estimates (4–120×10^6). X-ray fiber diffraction studies on xanthan indicated that it formed aligned helices in which there was poor lateral packing. This is consistent with an absence of crystallization and of gelation. The polysaccharide adopted a right handed helical conformation with fivefold symmetry in which the helix pitch was 4.7nm (60). Increase of the temperature of aqueous solutions of xanthan eventually produced a partial melting of the double helix. Xanthan is generally considered to adopt an ordered double helical conformation when in solution at lower temperatures. This conformation causes the apparent stiffness of the polysaccharide at low temperature or in high ionic strength. This stiffness was higher than that estimated for most other macromolecules and is comparable to that of the fungal β-D-glucans which exist as triple helices (61). The conformation was markedly affected by the presence of ions; salt increased the melting temperature. This conformational change, the shift from order to disorder, could be displaced to higher temperatures by the presence of increasing amounts of ions (Figure 24.12) (62). Acyl groups also influence the conformational changes, but light scattering experiments indicated that the inherent stiffness of the polysaccharide molecule is not greatly affected by the pattern of acylation. The presence of acetyl groups on the xanthan structure had a stabilizing effect on the ordered form of the polysaccharide, whereas pyruvate groups had the opposite effect. This could be attributed to intramolecular electrostatic repulsion between pyruvate groups. Conversely, the apolar interactions of the methyl groups of the acetate affected stabilization.

Aqueous solutions of xanthan are highly pseudoplastic whether or not acetyl and ketal substituents are present. The absence of the acyl groups does not normally affect the solubility of xanthans. The viscosity, the degree of pseudoplasticity, and the value for the transition from soft gel to pseudoplastic behavior are all directly related to the polysaccharide concentration. The effects resulting from the addition of salts to the solution also depend on polymer concentration. At ca 0.3% (w/v) xanthan, salts had practically no measurable effect, but at higher polymer concentrations increased viscosity could be observed. This was presumably due to alteration of the macromolecular association. There was probably no significant difference between the viscosity and the degree of acylation. Acetyl and pyruvate contents appeared to have no influence on dilute solution viscosity of xanthan or on the intrinsic viscosity at a given molecular weight. The viscosity of truncated xanthans in which the terminal mannosyl residue is missing was much lower than the wild type material. Surprisingly, the polytrimer, in which the terminal disaccharide is missing, was more viscous than the original xanthan (57,63).

24.3.1.4 Food Usage of Xanthan

Most industrial applications of xanthan derive from its ability to dissolve in hot or cold water to yield high viscosity, pseudoplastic solutions, even at low polysaccharide concentrations. Many of the industrial applications of xanthan were originally in the food industry. It

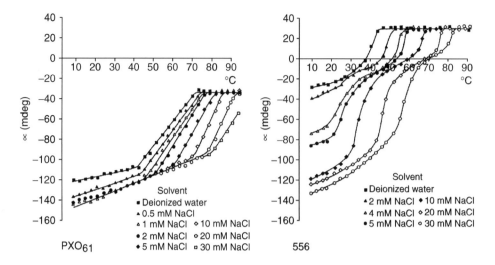

Figure 24.12 Effect of salts on the transition of xanthan from the ordered to the disordered state. Results for two different xanthans (61 - acetylated, non-pyruvylated; 556 - non-acetylated, pyruvylated) monitored by optical rotation measurement. Units mdeg (62).

received approval for food use from the US FDA in 1969 and about 60% of the xanthan currently produced is probably food grade polymer. In many foodstuffs, the useful attributes xanthan include rapid flavor release, good mouth feel and compatibility with other food ingredients including proteins, lipids, and polysaccharides. Salad dressings, relishes, and yogurts to which xanthan is added are of low pH. Xanthan, which is stable over a wide range of pH values, is thus very well suited to such applications. Polysaccharides are added to foods to provide suitable rheological properties during processing and a final appearance which will enhance consumer appeal during purchase and consumption.

Xanthan is incorporated into foods to alter the rheological properties of the water present and thus change the texture of the product. Associated with these readily measurable properties, are others such as mouth feel which are more difficult to define but which also show some correlation with physical properties. Mouth feel has been related to viscosity and, in particular, to nonNewtonian behavior. This also relates to the masking effect of viscosity on the intensity of taste. There is also a specific relationship between the polysaccharide and flavors present in any food. Thus xanthan provides good perception of sweetness and flavor when compared with some nonmicrobial polysaccharides such as gum guar or carboxymethylcellulose. Xanthan is also used for its capacity to control the texture of foods and to prevent or reduce ice crystal formation in frozen foods. As it is compatible with many of the other polysaccharides traditionally used in food, a wide range of textures can be obtained.

Many foods are essentially colloids in which there are complex interactions between the various ingredients. Emulsions of oil in water or water in oil form an important group of food products in which oils or lipids, water, and other ingredients are processed to form sauces, spreads, etc. Some of the emulsion-containing systems are dried and then reconstituted, as in packaged instant soups, desserts, and sauces. Whether used after the initial processing or after reconstitution with water, such emulsions may need to be stabilized with polysaccharides. In this role, xanthan has found many applications in dry mix food products. It can be dispersed in cold or hot water to provide thixotrophic dispersions which can be subsequently heated or refrigerated. The xanthan prevents the constituents from reverting to their original separate phases and ensures the long term stability required for food products.

Some of the foodstuffs into which xanthan is incorporated are of relatively low pH. Xanthan, with its high stability over a wide pH range, is well suited to such applications (64).

Many polysaccharides are employed to provide gelation of foodstuffs. Starch is widely used because of its low cost and its compatibility with many food ingredients, but is not suitable for all food applications. Xanthan on its own only forms very weak gels, but when mixed with various plant galactomannans or glucomannans, and then heated and cooled, it forms thermoreversible gels (*vide infra*).

24.3.1.5 Two Are Better Than One – Synergistic Gels Involving Xanthan

Many food products are formulated using a mixture of two or more polysaccharides to obtain the desired properties after processing (65). The aim may be to reduce the overall content of additive polysaccharides or to form a structured product. Aqueous solutions of xanthan, when mixed with certain plant gluco- or galactomannans, can be induced to form gels, whereas neither component will form a gel on its own. The mixed solution must be heated above the transition temperature of the xanthan in order to denature the xanthan helix and then allowed to cool. X-ray fiber diffraction studies on the mixed gels (66) reveal unique fiber patterns. Previously, gelation of the polymer mixtures had been attributed to the interaction between the helix of the microbial polysaccharide and unsubstituted regions of the plant glucomannan backbone. These results indicated that intermolecular binding could be due to cocrystallization resulting from the stereochemical compatibility between the cellulosic and mannan backbones. Acetyl groups on the xanthan molecules played an important role in inhibiting gelation of the polysaccharide mixtures and increasing the polymer concentrations required (59,62,67).

The most widely used mixture in food manufacture is xanthan and locust bean gum (another widely used food polysaccharide from seeds of the Mediterranean leguminous tree *Ceratonia siliqua*). As pointed out by Copetti et al. (68) concentrated solutions of either xanthan or locust bean gum (LBG) reveal very different properties – xanthan behaves as a weak gel network while LBG represents a hyperentangled macromolecular solution. LBG solutions revealed shear thinning behavior and a Newtonian region at low shear rates, whereas the mixed systems behaved as weak gels. The maximum synergistic effect was observed in a ratio of xanthan: LBG of 1:1. The rheological properties of the mixed gels also depended on the ratio of D-mannose: D-galactose in the LBG (69). LBG with higher mannose–galactose ratios, when mixed with xanthan, showed strongly increased synergism when temperatures were > 60°C. The melting points observed also ranged from 40–53°C, again increasing with increased mannose in the LBG. The xanthan-LBG network was formed from xanthan supermolecular strands and was unaffected by the heat treatment or the LBG fraction as addition of LBG failed to influence the xanthan structure. Although these gels are opaque, this is still satisfactory for many food applications where clarity is not required. *Synergistic gels* of this type are employed in a range of foods including spreads and cream cheeses. (Table 24.3).

Mixtures of xanthan with various types of starch have also been examined (70). The pasting peak viscosity of potato starch was reduced in the presence of xanthan, an effect thought to be due to repulsion of the negatively charged bacterial polysaccharide by phosphate groups associated with the starch. The opposite effect, increased viscosity, was seen with xanthan/maize starch mixtures.

24.3.2 Gellan

The polysaccharide (gellan® or gelrite®) was isolated from a bacterial strain initially designated *Auromonas* (*Pseudomonas*) *elodea* but now named *Sphingomonas paucimobilis*

Table 24.3
Food Uses of Synergistic Gels of Xanthan and Glycomannans

Application	Products
Gel formation and stabilization	Cheese and cream cheese
	Dessert gels
	Ice cream
	Milk shakes; milk drinks
	Puddings and pie fillings
Viscosity control	Chocolate drinks
	Cottage cheese dressings
	Instant soups
	Milk shakes
	Ice cream

(71,72). It is an intriguing new polysaccharide of microbial origin with a range of potential food applications, and is the only bacterial polysaccharide other than xanthan to be currently GRAS listed. Like xanthan, it is produced by aerobic submerged fermentation. After production is complete, the viscous broth is pasteurized then treated with mild alkali to remove acyl groups. It is then hot filtered to remove bacterial cell debris. This polysaccharide is an excellent gelling agent providing clear, brittle gels at much lower concentration than agar. Recently a mutant yielding enhanced production of gellan has been reported (73). As well as food use, gellan has been widely employed as a gelling agent in plant biotechnology under the trade name of gelrite or phytogel; it can also be used in place of agar in the formulation of bacterial culture media. Other *Sphingomonas* isolates have yielded a family of polysaccharides which are closely related structurally. Although none form gels, they do yield highly viscous aqueous solutions and have been proposed for various industrial, nonfood uses.

24.3.2.1 Structure

The structure of gellan is a repeating linear tetrasaccharide composed of D-glucose, D-glucuronic acid, and L-rhamnose in a ratio of 2:1:1, but devoid of any side chains (74). In the native material, *O*-acetyl and glyceryl residues are present on the D-glucosyl residue adjacent to D-glucuronic acid (75) (Figure 24.13). As a family of polysaccharides with closely similar carbohydrate structures has now been identified, this has permitted examination of the influence of various substituents and side chains on the physical properties of the polysaccharides (76). One polysaccharide is an exact analogue of gellan, in which the L-rhamnosyl residue has been replaced by L-mannose (77). Most of these polymers possess nonstoichiometric amounts of *O*-acetyl groups, the position of which still remains uncertain in some preparations. None have so far been identified as having food applications.

24.3.2.2 Physical Properties

The commercial gellan product has a mass of ~0.5×10^6 Da. The deacylated polysaccharide forms gels with properties common in some respects to those formed with agar, alginate, or carrageenan. Gelation was dependent on ionic strength as well as on the nature of the cation. The gelation temperature increased with increasing polysaccharide concentration (78). In texture, the gels most resembled those formed with agar or κ-carrageenan. In common with agar gels, marked melting/setting hysteresis was observed. The native structure of the gellan

-3-β- D-Glc$p$-(1→4)-β-D-Glc$p$A-(1→4)-β- D-Glc$p$-(1→4)-α- L-Rha$p$-(1-

(74,75)

Figure 24.13   The structure of gellan (74,75).

polymer that contains O-glycerate and ~6% acetate formed weak, elastic, thermoreversible gels. From x-ray fiber diffraction studies, a threefold helical structure with an axial repeat of 2.82 nm has been suggested. The glyceryl and acetyl groups inhibited crystallization of localized regions of the gellan chains (79). Rheological data suggested that in solution, the polysaccharide adopted a locally rigid conformation. The changes observed on deacetylation of the native polysaccharide, resulting in both increased crystallinity and gel stiffness, were interpreted as suggesting that longer sections of gellan contribute to the junction zones formed (80). When the polysaccharide was deacylated, extensive intermolecular association occurred and produced strong, brittle gels closely resembling those obtained from agar. In the crystalline structure, the two left handed threefold helical chains are arranged parallel to one another as an intertwined duplex in which interchain hydrogen bonds stabilize the structure (81). Computer models indicate that an almost fully extended polysaccharide chain of 3_2 double helices packs antiparallel, providing good compatibility with measured x-ray intensities (82). A range of gel textures can be produced through control of the degree of acylation of the polymer. The gels also resembled agar in that they were rigid, brittle, and thermoreversible. They could be reautoclaved and were stable to most commercial enzymes. The polymer behaved as a stiff coil when dissolved in dimethyl sulfoxide (DMSO) but preserved its helical secondary structure in aqueous solutions. The difference between melting and setting temperatures was 45–60°C (64), the gelling temperature increasing with the cation concentration. Unlike the more highly charged macromolecules alginate or pectin, there is marked lack of specificity among the alkaline earth cations, and the temperatures at the midpoint of the sigmoidal transitions (T_m) increased only moderately with ionic strength (83). Gelling and melting temperatures increased when either the gellan or the salt concentration was increased within the concentration range 0.3–2.0% (84), but the values were dependent on the cation used. This was ascribed to the increased number of junction zones produced, and decreased rotational freedom, causing higher heat resistance and increased elastic modulus. The mechanical properties of the gel directly related to the ionic selectivity, with monovalent ions in the order K > Na > Li. Grasdalen and Smidsrød (85) also concluded that gelation of the *A. elodea* subsp. *paucimobilis* polysaccharide occurred in two steps — chain ordering and chain association. Increasing the ion or polysaccharide concentration in the medium raised the gelling and melting point of the polysaccharide gels. The presence of the cations increased the number of junction zones and decreased the rotational freedom. The structures were rendered more heat resistant and the elastic modulus of the gel increased (84). It was also suggested that there was a well defined binding site for Ca^{2+} (86). The O-acetyl groups on the native polysaccharide had only a weak effect on the aggregation of the gellan molecules, whereas the L-glyceryl residues were detrimental to crystal packing (81). The potassium salt possessed a crystalline structure in which the L-glyceryl groups prevented the coordinated interactions of ions and carboxylate groups which are required for strong gelation (87). Once deacylated, the coordination of calcium between double helices was so strong that molecular aggregation occurred at very low cation concentrations. The backbone conformation of gellan on reexamination proved to comprise two left handed, threefold helical chains organized in parallel. In aqueous solution, these produced an intertwined duplex in which the chains were displaced half a pitch with respect to each other (88,89).

24.3.2.3 Food Applications

Gellan forms clear, brittle, thermoreversible gels, which give excellent flavor release. The significance of this property in various food applications was indicated by Owen (90). Gellan can also be used in combination with other polysaccharides used by the food industry, although increased concentrations reduced gel strengths. It does, however, induce synergism when blended with gelatine or with gum Arabic, with gel strength increased by 40–60% (64). As pointed out by Gibson (91), blending gellan with other gums or mixtures of gums could produce a wide range of textures ranging from hard and brittle to soft and elastic. The exact texture obtained would depend on the proportion of the different gelling agents. Thus, blending with agar or κ-carrageenan, each of which also form brittle gels, reduced gel strength but retained the brittle texture. The very clear gels formed by low levels of gellan can be expected to have considerable aesthetic as well as practical appeal. Mixtures of gellan and gelatine also provide a range of textures (92). It was suggested that gellan has potential food use where a highly gelled structure or mouth feel was required (Table 24.4) (64).

24.3.3 Curdlan

Different customs exist in different countries. Unlike people in North America and Europe, people in Japan regard microbial polysaccharides as natural products which can be added to food without specific regulatory control. There is thus an upward trend in the use of microbial exopolysaccharides such as curdlan in Japan. Market shortfalls in some of the natural (plant or algal) polymers might lead to increased use of microbial products in their place. This has been exemplified by the fall in the supply of gum arabic in 1985 to 25% of the 1984 level and other shortfalls in the supplies of locust bean gum and guar gum.

Table 24.4

Uses of Gellan

a) Food Application	Function	Concentration Required
Jellies	Gelling agent in desserts.	0.15–0.2%
Jams	In low calorie spreads and fillings	0.12–0.3
Confectionery		0.8–1.0
Processed foods	Gelling in fruit, vegetables, meats	0.2–0.3%
Processed meats	Texture modification	0.1–1%
Icings	Coating agent	0.05–0.12%
Pie fillings	Texturiser	0.25–0.35%

b) Potential Food Applications of Gellan	
Dairy products	Ice cream, milkshakes, yogurts
Fabricated foods	Fabricated fruits, meats.
Icings and frostings	Bakery icings, frostings
Jams and jellies	Low calorie jams, jellies, fillings
Pie fillings and puddings	Instant desserts, jellies
Water-based gels	Aspics
Pet foods	Gelled pet foods, meat chunks

Gibson (91)

Gels can be formed by several neutral bacterial polysaccharides. These include the linear β-D-glucan homopolymer, curdlan, which is formed by a number of bacterial species related to the Gram negative bacterium *Agrobacterium radiobacter* and *Rhizobium* (93) as well as *Cellulomonas flavigena* (94,95). Curdlan was approved for food use in Japan, Korea, and Taiwan. It is a β-(1→3)-linked D-glucan which is insoluble in water, however, when aqueous solutions are heated it forms a gel. It is unusual in that it may either form a thermoreversible (low set) gel with some similarity to agar or, on heating to higher temperatures (greater than 80°C), yield a high set gel which is not normally thermoreversible (96). In the presence of sugars, salt or starch, the gel strength may be lowered but it is unaffected by freezing and thawing, although syneresis is observed. Curdlan is probably unique among biological gums in that a gel is formed when the polysaccharide suspension is heated, and increased heating may even strengthen the gel. The gel thus formed is stable over a wide pH range and is stable to freezing and thawing. From x-ray crystallographic studies, it has been suggested that curdlan forms a triple helical crystalline structure (97). When added to food as a gelling agent, curdlan yields gels intermediate between the elasticity of gelatine and the brittleness of agar (98). Applications of curdlan in food can be seen in Table 24.5.

The different types of gel formed by curdlan under different conditions vary in their specific properties (99), some showing pseudocrystallinity. Differences in syneresis were also noted. When examined by electron microscopy, considerable differences were revealed (100). Some formed fibril units while in others assemblies of these were seen culminating in net-like structures. Curdlan resembles many microbial homopolysaccharides in that it is biodegradable through the action of specific enzymes such as that from a *Bacillus circulans* strain which yields laminaribiose as the sole hydrolysis product (101). However, some of the gel forms are extremely resistant to most β-D-glucanases.

24.4 EXOPOLYSACCHARIDES AS A SOURCE OF FLAVOR COMPONENTS

As many microbial exopolysaccharides contain appreciable amounts of 6-deoxysugars, it has been suggested that the polymers might be used as sources of these sugars (102). The 6-deoxyhexoses in turn, could then be used as intermediates in the synthesis of furaneol and its derivatives. These compounds can be used as flavoring agents for the food industry. Furaneol yields a powerful caramel-like flavor, but when modified by the addition of various short chain fatty acid esters gives a range of either meat or fruit flavors. Furaneol is fairly expensive (~$1–200/kg), and it has been proposed that this cost could be reduced through the use of polysaccharide derived deoxysugars. If this were to be achieved economically, it would require release of the sugars from the polymers by enzymic treatment.

Table 24.5

Food Applications of Curdlan

Food Application	Function	Concentration Required
Jellies	Gelling agent	1–5%
Processed foods	Gelling agent	1–10%
Processed meats	Texture modification	0.1–1%
Sauces	Improved viscosity	0.2–0.7%
Freeze-dried foods	Improved rehydration	0.5–1%

Nakao (96)

Using immobilized enzymes or cells and microbial polysaccharides containing a high percentage of deoxysugars this might be feasible; however, enzymes capable of degrading the appropriate exopolysaccharides to their component sugars have not as yet been described.

24.5 GAZING INTO THE CRYSTAL BALL: THE FUTURE FOR MICROBIAL POLYSACCHARIDES IN FOOD

In the hope that LAB EPS with improved physical properties can be developed, or that strains can be produced with higher yields of EPS, various groups have recently studied the biosynthesis of these polymers. Normal yields of EPS are usually in the range 100–250mg litre^{-1}. (12) Yields are much lower than those obtained for xanthan or gellan. In *Lactococcus lactis* strain NIZO B40, the genes for EPS synthesis were encoded on a 40kb plasmid (103). A 12 kb region contained 14 genes with products having sequence homologies to other gene products involved in EPS, LPS, and teichoic acid biosynthesis. As with production of xanthan and similar EPS, synthesis involved an isoprenoid lipid carrier and genes responsible for transfer of sugars from sugar nucleotides to this carrier were identified. Comparison of the *eps* gene clusters of *S. thermophilus* and *L. lactis* also revealed considerable homology between genes with similar putative functions (103). It also proved possible to introduce the *eps* gene cluster from *Streptococcus thermophilus* Sfi6 into *Lactococcus lactis* MG1363 to produce an altered polysaccharide (104), while eps-encoding genes carried on plasmids from *L. lactis* could be transferred to other strains of the same bacterial species (105). Portions of the gene clusters in LAB revealed high conservation and some similarity to the genes required for capsular EPS synthesis in other Gram positive Streptococci including *S. pneumoniae* and *S. agalactiae*. It has been suggested that successful genetic engineering and targeting of specific genes leading to alteration of the repeat units might beneficially change the physical properties of the polysaccharides. However, low yields might then present problems as was found with EPS from gram negative bacteria including some xanthan mutants (56,57).

The role of specific enzymes involved in carbohydrate metabolism and EPS synthesis was also studied. In *S. thermophilus*, strain differences were observed in the role of UDP-glucose pyrophosphorylase in polysaccharide production (106). In this bacterial species, the amount of EPS produced in synthetic media depended on the carbohydrate source, with highest yields resulting from a combination of lactose and glucose (24). Attempts have also been made to produce EPS from LAB using defined media but at least for *L. delbruekii bulgaricus* various amino acids and vitamins were necessary for both growth and EPS production (107). As pointed out by Jolly et al. (7), a better understanding of structure function relationships of the EPS from LAB is still required, and this must be allied to studies of the interactions with other components present in the food matrix (108). If after achieving such an understanding, polysaccharide structures can be manipulated to yield specific properties, the question remains whether such products would be legislatively and publicly acceptable.

An alternative approach to alter the physical properties of polysaccharides has made use of glycosidases and other enzymes to remove terminal sugars from side chains. Thus, van Casteren et al. (14) employed an enzyme from *Aspergillus aciduleatus* to remove a terminal β-D-galactosyl residue from the disaccharide side chains in a *L. lactis cremoris* polymer (Figure 24.4). Another possible source of novel EPS may be yeasts, several of which already find applications in food either directly or as sources of pigments and flavors. A few species, including *Rhodotorula glutinis* formed highly viscous polyanionic EPS (109) in good yield. The polysaccharide product from strain KCTC7989 formed pseudoplastic solutions in water. There is also the possibility that polysaccharides from suitable nonfood microorganism might be designed to provide very specific physical properties. One candidate for this approach might be acetan from *Acetobacter xylinum*. This polysaccharide resembles xanthan both in its cellulosic backbone and in some of its physical properties (110,111). Mutants with altered chemical structure and differing physical properties have been isolated (112).

24.6 LEGISLATIVE ACCEPTABILITY OF MICROBIAL POLYSACCHARIDES

Before new microbial polysaccharides can be permitted for use as food additives, they must be submitted to a process of approval. In addition to the technological justification for the inclusion of the polymer in foodstuffs, evidence has to be provided of the need to use the polysaccharide and of its safety in use. The producer must also demonstrate that the new additive will benefit the consumer. This covers various categories:

1. The presentation of the foodstuff
2. The need to keep the food wholesome until eaten
3. The extension of dietary choice
4. The need for nutritional supplement
5. Convenience of purchasing, packaging, storage, preparation, and use
6. Economic advantage including longer shelf life or reduced price

Not all these factors will be appropriate in the case of microbial polysaccharides. Any company intending to produce or market the polymer must first embark on an extensive research program. In the U.K., evidence is then presented to the Food Advisory Committee in which there are representatives of consumer and enforcement bodies, the food industry, medicine, and academia. This advises the relevant Government Departments (Health; DEFRA). If the Food Advisory Committee considers that the need has been proved, safety is then examined and is assessed by the Committee on Toxicity of Chemicals in Food, Consumer Products, and Environment. Tests include determination of the LD_{50} in rats, feeding trials to animals through two generations of reproduction, and teratology testing. Tests for dust inhalation must also be performed. The costs for a development programm of this type may well be over $10 million and 3–6 years may elapse before approval is given. In addition to official approval, there must be consumer acceptance. A major factor in favor of microbial polysaccharides is their lack of digestion by humans and animals. They can thus be used as constituents of low calorie diets as well as other processed foodstuffs. After the Food Advisory Committee has made its recommendation, the polysaccharide still requires approval from the EU if the foodstuffs in which it will be incorporated will be marketed in other Member Countries of the EU. In the U.K., the schedule of "Emulsifiers and Stabilizers in Food Regulations" is promulgated by DEFRA (previously the Ministry of Agriculture, Fisheries, and Food). The EU has attempted to harmonize food additive regulations within its member countries through a "Directive of the Council of Ministers on Emulsifiers, Stabilizers, Thickeners, and Gelling Agents for Food Use." This provides a list of the agents that are fully accepted for use in member states. The compounds are designated in Annex 1 of the EU Directive 80/597/EEC with an appropriate serial number (E400=alginic acid, E406=agar, E415=xanthan, etc). Further details include the foodstuffs in which the use of the emulsifiers is permitted, together with criteria for purity and labeling. The attempted standardization of labeling procedures provides details of trade name, manufacture, designated number, etc. In practice many processed foods are clearly labeled with their contents and consumers can see for themselves if xanthan (or gellan) is one of these.

The World Health Organization (WHO) and Food and Agricultural Organization (FAO), joint expert committee on food additives has the responsibility of proposing Acceptable Daily Intakes. Thus an ADI of 0–10mg/kgm body weight has been established for xanthan. The Committee divided food additives into three categories:

1. Fit for use in food
2. Need to be evaluated
3. Should not be used in foods

After evaluation, detailed specifications for the food additives include identity and purity. Typically, WHO provides data on biological aspects such as ingestion, calorific availability, and digestibility, and on toxicology. Toxicological evaluation includes short term and long term evaluation using a range of different animals. A two year study using rats failed to show any carcinogenic or other toxic effects which could be attributed to xanthan. Tests on human volunteers may also be included; again in the case of xanthan no adverse effects were found although this polysaccharide could effect a slow but significant weight loss in individuals suffering from obesity (113). In the USA, xanthan is permitted as a food additive under regulations controlled by the Food and Drug Administration (FDA). It is on the list of substances GRAS list, being approved in the Federal Register for use as a stabilizer, emulsifier, foam enhancer, thickening, suspending, and bodying agent. Use is permitted in sauces, gravies, and coatings applied to meat and poultry products. It may also be added to cheeses and cheese products, certain milk and cream products, food dressings, table syrups, and frozen desserts. A further use, also requiring FDA approval, is as a component of paper and paperboard intended for use in food packaging and likely to be in contact with the food. Although dextran did receive approval for food use, there is apparently no current food usage of this EPS. Curdlan, which might have a number of food applications, is used in Japan as a natural material, but is not currently permitted in the USA. The other microbial exopolysaccharide which is currently GRAS listed as an approved food additive is gellan.

The definition of EPS, for legislative purposes and for patents, is not particularly satisfactory. They are generally identified by the molar ratio or the composition of their carbohydrate and noncarbohydrate components. More detailed information on structure may be provided, together with properties such as optical rotation, viscosity, physical appearance and ash content. Xanthan is currently described in the National Formulary as: "…a high molecular weight polysaccharide gum produced by a pure culture fermentation of a carbohydrate with *X. campestris*, then purified by recovery with isopropyl alcohol, dried, and milled. It contains D-glucose and D-mannose as the dominant hexose units, along with D-glucuronic acid, and is prepared as the sodium, potassium or calcium salt. It yields not less than 4.2% and not more than 5% carbon dioxide, calculated on the dried basis, corresponding to not less than 91% and not more than 108% xanthan gum. When employed as a food additive, xanthan must conform to the definition provided in the National Formulary (UK) of "…a high molecular weight polysaccharide gum produced by a pure culture fermentation of a carbohydrate with *X. campestris*…". "The food grade polysaccharide must also meet the specifications listed in the Food Chemicals Codex (UK). These include absence of *E. coli* and *Salmonella,* and limits on arsenic (<3ppm), heavy metals (<30 ppm) and isopropanol (<750 ppm). If more exopolysaccharides are proposed for food usage, some may have similar composition and improved methods of definition will be needed. In foods such as yogurt, when addition of xanthan is permitted, it may be difficult using current methodology to distinguish it from polymers synthesized by the LAB present. It is thus surprising that few methods have been developed for the identification and quantification of the EPS once present in food. With the discovery of enzymes specifically acting on the exopolysaccharides, these could well be used to provide highly specific assay procedures for the polymers once they have been incorporated into foods containing other carbohydrate-containing materials. A study by Craston et al. (114) indicated that the gellan lyase from heterologous bacteria (a small pink-pigmented pseudomonad) might prove useful in specific determination of gellan in processed foods. A similar enzyme has been isolated and characterized by Hashimoto et al. (115,116). Their use would have considerable advantages over the tradition assay methods which fail to distinguish between different polysaccharides or carbohydrate-containing material, present in foods. Similarly, pyruvated-mannose-specific xanthan lyase lyases studied by Ruijssenaars et al. (52,53) could be used to quantify xanthan in food as well as determining its quality. Alternatively, techniques based on monoclonal antibodies prepared against xanthan such as those developed by Haaheim and Sutherland (51) could be used. Other enzymes degrading microbial polysaccharides in food, including some of those from LAB, have also been identified but it remains unclear whether they would be as suitable as the xanthan and gellan lyases for quantification of the polymers in foodstuffs (115).

REFERENCES

1. De Vuyst, L., B. Degeest. Heteropolysaccharides from lactic acid bacteria. *FEMS Microbiol. Rev.* 23:153–177, 1999.
2. Laws, A.P., Y. Gu, V.M. Marshall. Biosynthesis, characterization and design of bacterial exopolysaccharides from lactic acid bacteria. *Biotechnol. Adv.* 19:597–625, 2000.
3. Sutherland, I.W. Structure function relationships in microbial exopolysaccharides. *Biotechnol. Adv.* 12:393–448, 1995.
4. Cerning, J. Exocellular polysaccharides produced by lactic acid bacteria. *FEMS Microbiol. Rev.* 87:113–130, 1990.
5. Ruas-Madiedo, P., J. Hugenholtz, P. Zoon. An overview of the functionality of exopolysaccharides produced by lactic acid bacteria. *Int. Dairy J.* 12:163–171, 2002.
6. Nakajima, H., Y. Suzuki, H. Kaizu, T. Hirota. Cholesterol-lowering activity of ropy fermented milk. *J. Food Sci.* 57:1327–1329, 1992.
7. Jolly, L., S.J.F. Vincent, P. Duboc, J.-R. Neeser. Exploiting exopolysaccharides from lactic acid bacteria. *Antonie van Leeuwenhoek* 82:367–374, 2002.
8. Vanhaverbeke, C., C. Bosso, P. Colin-Morel, C. Gey, L. Gamar-Nourani, K. Blondeau, J.M. Simonet, A. Heyraud. Structure of an extracellular polysaccharide produced by *Lactobacillus rhamnosus* strain C83. *Carbohydr. Res.* 314:211–220, 1998.
9. Lipinski, T., C. Jones, X. Lemercinier, A. Korzeniowska-Kowal, M. Strus, J. Rybka, A. Gamian, P.B. Heczko. Structural analysis of the *Lactobacillus rhamnosus* strain KL37C exopolysaccharide. *Carbohydr. Res.* 338:605–609, 2003.
10. Marshall, V.M., H. Dunn, M. Elvin, N. McLay, Y. Gu, A.P. Laws. Structural characterisation of the exopolysaccharide produced by *Streptococcus thermophilus* EU20. *Carbohydr. Res.* 331:413–422, 2001.
11. Marshall, V.M., A.P. Laws, Y. Gu, F. Levander, P. Rådstrom, L. De Vuyst, B. Degeest, F. Vaningelen, H. Dunn, M. Elvin. Exopolysaccharide-producing strains of thermophilic lactic acid bacteria cluster into groups according to their EPS structure. *Lett. Appl. Microbiol.* 32:433–437, 2001.

12. Ricciardi, A., F. Clementi. Exopolysaccharide from lactic acid bacteria: structure, production and tecnological applications. *Ital. J. Food Sci.* 12:23–45, 2000.

13. Robijn, G.W., D.J.C. van den Berg, H. Haas, J.P. Kamerling, J.F.G. Vliegenhart. Determination of the structure of the exopolysaccharide produced by *Lactobacillus saké* O-1. *Carbohydr. Res.* 276:117–136, 1995.

14. Van Casteren, W.H.M., C. Dijkema, H.A. Schols, G. Beldman, A.G.J. Voragen. Structural characterisation and enzymic modification of the exopolysaccharide produced by *Lactococcus lactis* subsp. *cremoris* B39. *Carbohydr. Res.* 324:170–181, 2000.

15. Nakajima, H., T. Hirota, T. Toba, T. Itoh, S. Adachi. Structure of the extracellular polysaccharide from slime-forming *Lactococcus lactis* subsp. *cremoris* SBT 0495. *Carbohydr. Res.* 224:245–253, 1992.

16. Frengova, G.I., E.D. Simova, D.M. Beshkova, Z.I. Simov. Exopolysaccharide produced by lactic acid bacteria of kefir grains. *Z. Naturforsch.(C.)* 57:805–810, 2002.

17. Harding, L.P., V.M. Marshall, M. Elvin, Y. Gu, A.P. Laws. Structural characterisation of a perdeuteriomethylated exopolysaccharide by nmr spectroscopy: characterisation of the novel exopolysaccharide produced by *Lactobacillus delbrueckii* subsp. *bulgaricus* EU23. *Carbohydr. Res.* 338:61–67, 2003.

18. Gruter, M., B.R. Leeflang, J. Kuiper, J.P. Kamerling, J.F.G. Vliegenhart. Structural characterisation of the exopolysaccharide produced by *Lactobacillus. Carbohydr. Res.* 239:209–226, 1993.

19. Faber, E.J., J.P. Kamerling, J.F.G. Vliegenhart. Structure of the extracellular polysaccharide produced by *Lactobacillus delbrueckii* subsp. bulgaricus 291. *Carbohydr. Res.* 331:183–194, 2001.

20. Van Casteren, M.R., C. Pau-Roblot, A. Begin, D. Roy. Structure determination of the exopolysaccharide produced by *Lactobacillus rhamnosus* strains RW-9595M. *Biochem. J.* 363:2–17, 2002.

21. Van Geel-Schutten, G.H., E.J. Faber, E. Smit, K. Bonting, M.R. Smith, B. Ten Brink, J.P. Kamerling, J.F.G. Vliegenhart, L. Dijkhuizen. Biochemical and structural characterization of the glucan and fructan exopolysaccharides synthesized by the *Lactobacillus reuteri* wild-type strain and by mutant strains. *Appl. Environ. Microbiol.* 65:3008–3014, 1999.

22. Van Hijum, S.A.F.T., G.H. Van Geel-Schutten, H. Rahaoui, M.J.E.C. van der Maarel, L. Dijkhuizen. Characterization of a novel fructosyltransferase from *Lactobacillus reuteri* that synthesizes high-molecular-weight inulin and inulin oligosaccharides. *Appl. Environ. Microbiol.* 68:4390–4398, 2002.

23. Kralj, S., G.H. Van Geel-Schutten, H. Rahaoui, R.J. Leer, E.J. Faber, M.J.E.C. van der Maarel, L. Dijkhuizen. Molecular characterization of a novel glucosyltransferase from *Lactobacillus reuteri* strain 121 synthesizing a unique, highly branched glucan with α-(1→4) and α-(1→6) glucosidic bonds. *Appl. Environ. Microbiol.* 68:4283–4291, 2002.

24. Degeest, B., L. De Vuyst. Correlation of activities of the enzymes α-phosphoglucomutase, UDP-galactose 4-epimerase and UDP-glucose pyrophosphorylase with exopolysaccharide biosynthesis by *Streptococcus thermophilus* LY03. *Appl. Env. Microbiol.* 66:3519–3527, 2000.

25. Faber, E.J., M.J. van den Haak, J.P. Kamerling, J.F.G. Vliegenhart. Structure of the exopolysaccharide produced by *Streptococcus thermophilus* S3. *Carbohydr. Res.* 331:173–182, 2001.

26. Duenas-Chasco, M.T., M.A. Rodriguez-Carvajal, P. Tejero-Mateo, J.L. Espartero, A. Irastorza-Iribas, A.M. Gil-Serrano. Structural analysis of the exopolysaccharides produced by *Lactobacillus* spp. G-77. *Carbohydr. Res.* 307:125–133, 1998.

27. Duenas-Chasco, M.T., M.A. Rodriguez-Carvajal, P.T. Mateo, G. Franco-Rodriguez, J.L. Espartero, A. Irastorza-Iribas, A.M. Gil-Serrano. Structural analysis of the exopolysaccharide produced by *Pediococcus damnosus* 2.6. *Carbohydr. Res.* 303:453–458, 1997.

28. Faber, E.J., D.J. van Haaster, J.P. Kamerling, J.F.G. Vliegenhart. Characterization of the exopolysaccharide produced by *Streptococcus thermophilus* 8S containing an open-chain nononic acid. *Europ. J. Biochem.* 269:5590–5598, 2002.

29. Yamamoto, Y., S. Murosaki, R. Yamauchi. Structural studies on an exocellular polysaccharide produced by *Lactobacillus helveticus* TY1-2. *Carbohydr. Res.* 261:67–78, 1994.

30. Gorret, N., C.M.G.C. Renard, M.H. Famelart, J.L. Maubois, J.L. Doublier. Rheological characterization of the exopolysaccharide produced by *Propionobacterium acidi-propionici* on milk microfiltrate. *Carbohydr. Polym.* 51:149–158, 2003.

31. Ricciardi, A., E. Parente, M.A. Crudele, F. Zanetti, G. Scolari, I. Mannazzu. Exopolysaccharide production by *Streptococcus thermophilus* SY: production and preliminary characterization of the polymer. *J. Appl. Microbiol.* 92:297–306, 2002.

32. van den Berg, D.J.C., G.W. Robijn, A.C. Janssen, M.L.F. Giuseppin, R. Vreeker, J.P. Kamerling, J.F.G. Vliegenhart, A.T. Ledeboer, C.T. Verrips. Production of a novel extracellular polysaccharide by *Lactobacillus saké* O-! and characterization of the polysaccharide. *Appl. Environ. Microbiol.* 61:2840–2844, 1995.

33. Tuinier, R., W.H.M. van Casteren, E. Looijesteijn, H.A. Schols, A.G.V. Voragen, P. Zoon. Effects of structural modifications on some physical characteristics of exopolysaccharides from *Lactococcus lactis*. *Biopolymers* 59:160–166, 2001.

34. Faber, E.J., J.A. van Kuik, J.P. Kamerling, J.F.G. Vliegenhart. Modeling of the structure in aqueous solution of the exopolysaccharide produced by *Lactobacillus helveticus* 766. *Biopolymers* 63:66–76, 2002b.

35. Levander, F., M. Svensson, P. Rådstrom. Enhanced exopolysaccharide production by metabolic engineering of *Streptococcus thermophilus*. *Appl. Env. Microbiol.* 68:784–790, 2002.

36. Ramos, A., I.C. Boels, W.M. de Vos, H. Santos. Relationship between glycolysis and exopolysaccharide biosynthesis in *Lactococcus lactis*. *Appl. Env. Microbiol.* 67, 33–41, 2001.

37. Boels, I.C., A. Ramos, M. Kleerebezem, W.M. de Vos. Functional analysis of the *Lactococcus lactis galU* and *galE* genes and their impact on sugar nucleotide and exopolysaccharide synthesis. *Appl. Env. Microbiol.* 67:3033–3040, 2001.

38. Jansson, P-E., L. Kenne, B. Lindberg. Structure of the extracellular polysaccharide from *Xanthomonas campestris*. *Carbohydr. Res.* 45:275–282, 1975.

39. Melton, L.D., Mindt, L., Rees, D.A., Sanderson, G.R. Covalent structure of the extracellular polysaccharide from *Xanthomonas campestris*. *Carbohydr. Res.* 46:245–257, 1976.

40. Souw, P. and Demain, A.L. Nutritional studies on xanthan production by *Xanthomonas campestris*. *Appl. Environ. Microbiol.* 37:1186–1192, 1979.

41. Ho, C.S., L.-K. Ju, R.F. Baddour. The anomaly of oxygen diffusion in aqueous xanthan solutions. *Biotechnol. Bioeng.* 32:8–17, 1988.

42. Nakajima, S., H. Funahashi, T. Yoshida. Xanthan gum production in a fermentor with twin impellors. *J. Ferment. Bioengin.* 70:392–397, 1990.

43. Tait, M.I., I.W. Sutherland, A.J. Clarke-Sturman. Effect of growth conditions on the production, composition and viscosity of *Xanthomonas campestris* exopolysaccharide. *J. Gen. Microbiol.* 132:1483–1492, 1986.

44. Casas, J.A., V.E. Santos, F. GarciaOchoa. Xanthan gum production under several operational conditions: molecular structure and rheological properties. *Enzyme Microb. Technol.* 26:282–291, 2000.

45. Linton, J.D. The relationship between metabolite production and the growth efficiency of the producing organism. *FEMS Microbiol. Rev.* 75:1–18, 1990.

46. Galindo, E. Aspects of the process for xanthan production. *Trans. I. Chem. E.* 72:227–237, 1994.

47. Garcia-Ochoa, F., V.E. Santos, J.A. Casas, E. Gómez. Xanthan gum: production, recovery and properties. *Biotechnol. Adv.* 18:549–579, 2002.

48. Rinaudo, M., M. Milas. Enzymic hydrolysis of the bacterial polysaccharide xanthan by cellulase. *Intern. J. Biol. Macromol.* 2:45–48, 1980.

49. Sutherland, I.W. Hydrolysis of unordered xanthan in solution by fungal cellulases. *Carbohydr. Res.* 131:93–104, 1984.

50. Sutherland, I.W. Xanthan lyases: novel enzymes found in various bacterial species. *J. Gen. Microbiol.* 133:3129–3134, 1987.

51. Haaheim, L., I.W. Sutherland. Monoclonal antibodies reacting with the exopolysaccharide from *Xanthomonas campestris*. *J. Gen. Microbiol.* 135:605–612, 1989.

52. Ruijssenaars, H.J., J.A.M. Debont, S. Hartmans. A pyruvated mannose-specific xanthan lyase involved in xanthan degradation by *Paenibacillus alginolyticus* XL-1. *Appl. Environ. Microbiol.* 65:2446–2452, 1999.

53. Ruijssenaars, H.J., S. Hartmans, J.C. Verdoes. A novel gene encoding xanthan lyase of *Paenobacillus alginolyticus* strain XL-1. *Appl. Environ. Microbiol.* 66:3945–3950, 2000.

54. Ielpi, L., R. Couso, M. Dankert. Lipid-linked intermediates in the biosynthesis of xanthan gum. *FEBS Lett.*, 130:253–256, 1981.

55. Ielpi, L., R.O. Couso, M. Dankert. Xanthan gum biosynthesis acetylation occurs at the prenyl-phospho-sugar stage. *Biochem. Int.*, 6:323–333, 1983.

56. Vanderslice, R.W., D.H. Doherty, M. Capage, M.R. Betlach, R.A. Hassler, N.M. Henderson, J. Ryan-Graniero, M. Tecklenberg. Genetic engineering of polysaccharide structure in *Xanthomonas campestris*. In: *Biomedical and Biotechnicalological Advances in Industrial Polysaccharides*, Crescenzi, V., I.C.M. Dea, S. Paoletti, S. Stivala, I.W. Sutherland, eds., New York: Gordon and Breach, 1989, pp 145–156.

57. Betlach, M.R., M.A. Capage, D.H. Doherty, R.A. Hassler, N.M. Henderson, R.W. Vanderslice, J.D. Marelli, M.B. Ward. Genetically engineered polymers: manipulation of xanthan biosynthesis. In: *Industrial Polysaccharides*, Yalpani, M., ed., Amsterdam: Elsevier, 1987, pp 145–156.

58. Whitfield, C., M.A. Valvano. Biosynthesis and expression of cell-surface polysaccharides in Gram-negative bacteria. *Adv. Microb. Physiol.* 35:135–246, 1993.

59. Shatwell, K.P. The influence of acetyl and pyruvic acid substituents on the solution and interaction properties of xanthan. Ph.D. Thesis, Edinburgh University, 1989.

60. Moorhouse, R., M.D. Walkinshaw, S. Arnott. Xanthan gum - molecular conformation and interactions. *ACS Symp.* 45:90–102, 1977.

61. Norisuye, T., T. Yanaki, H. Fujita. Triple helix of a *Schizophyllum commune* polysaccharide in aqueous solution. *J. Polym. Sci.* 18:547–548, 1980.

62. Shatwell, K.P., I.W. Sutherland, I.C.M. Dea, S.B. Ross-Murphy. The influence of acetyl and pyruvate substituents on the solution properties of xanthan polysaccharide. *Carbohydr. Res.* 206:87–103, 1990.

63. Tait, M.I., I.W. Sutherland. Synthesis and properties of a mutant type of xanthan. *J. Appl. Bacteriol.* 66:457–460, 1989.

64. Kang, K.S., D.J. Pettit. Xanthan, gellan, welan and rhamsan. In: *Industrial Gums*, Whistler, R., J.N. BeMiller, eds., New York: Academic Press, 1993, pp 341–397.

65. Morris, V.J. and Wilde, P.J. Interactions of food biopolymers. *Curr. Opin. Colloid Interface Sci.* 2:567–572, 1997.

66. Morris, V.J., M. Miles. Effect of natural modifications on the functional properties of extracellular bacterial polysacharides. *Int. J. Biol. Macromol.* 8:342–348, 1986.

67. Shatwell, K.P., I.W. Sutherland, S.B. Ross-Murphy, I.C.M. Dea. Influence of the acetyl substituent on the interaction of xanthan with plant polysaccharides, I: xanthan-locust bean gum systems. *Carbohydr. Polym.* 14:29–51, 1991.

68. Copetti, G., M. Grassi, R. Lapasin, S. Pricl. Synergistic gelation of xanthan gum with locust bean gum: a rheological investigation. *Glycoconjugate J.* 14:951–961, 1997.

69. Lundin, L., A. Hermansson. Supermolecular aspects of xanthan-locust bean gum gels based on rheology and electron microscopy. *Carbohydr. Polym.* 26:129–140, 1995.

70. Shi, X., J.N. BeMiller. Effects of food gums on viscosities of starch suspensions during pasting. *Carbohydr. Polym.* 50:7–18, 2002.

71. Pollock, T.J. Gellan related polysaccharides and the genus *Sphingomonas*. *J. Gen. Microbiol.* 139:1939–1945, 1993.

72. Kang, K.S., G.T. Veeder, P.J. Mirrasoul, T. Kaneko, I.W. Cottrell. Agar-like polysaccharide produced by a *Pseudomonas* species: production and basic properties. *Appl. Environ. Microbiol.* 43:1086–1091, 1982.

73. West, T.P. Isolation of a mutant strain of *Pseudomonas* sp. ATCC31461 exhibiting elevated polysaccharide production. *J. Ind. Microbiol. Biotechnol.* 29:185–188, 2002.

74. Jansson, P.-E., B. Lindberg, P.A. Sandford. Structural studies of gellan gum, an extracellular polysaccharide elaborated by *Pseudomonas elodea*. *Carbohydr. Res.* 124:135–139, 1983.

75. Kuo, M., A.J. Mort, A. Dell. Identification and location of L-glycerate, an unusual acyl substituent in gellan gum. *Carbohydr. Res.* 156:173–187, 1986.

76. Cairns, P., M.J. Miles, V.J. Morris. X-ray fibre diffraction studies of members of the gellan family of polysaccharides. *Carbohydr. Polym.* 14:367–371, 1991.

77. O'Neill, M.A., A.G. Darvill, P. Albersheim, K.J. Chou. Structural analysis of an acidic polysaccharide secreted by *Xanthobacter* sp. (ATCC53272). *Carbohydr. Res.* 206:289–296, 1990.

78. Nakamura, K., K. Harada, Y. Tanaka. Viscoelastic properties of aqueous gellan solutions: the effects of concentration on gelation. *Food Hydrocol.* 7:435–447, 1993.

79. Carroll, V., G.R. Chilvers, D. Franklin, M.J. Miles, V.J. Morris. Rheology and microstructure of solutions of the microbial polysaccharide from *Pseudomonas elodea*. *Carbohydr. Res.* 114:181–191 1983.

80. Carroll, V., M.J. Miles, V.J. Morris. Fibre diffraction studies of the extracellular polysaccharide from Pseudomonas elodea. *Intern. J. Biol. Macromol.* 4:432–433, 1982.

81. Chandrasekaran, R., V.G. Thailambal. The influence of calcium ions, acetate and L-glycerate groups on the gellan double helix. *Carbohydr. Polym.* 12:431–442, 1990.

82. Upstill, C., E.D.T. Atkins, P.T. Attwool. Helical conformations of gellan gum. *Int. J. Biol. Macromol.* 8:275–288, 1986.

83. Crescenzi, V., M. Dentini, I.C.M. Dea. The influence of side-chains on the dilute solution properties of three structurally related bacterial anionic polysaccharides. *Carbohydr. Res.* 160:283–302, 1987.

84. Moritaka, H., K. Nishinari, N. Nakahama, H. Fukuba. Effects of potassium chloride and sodium chloride on the thermal properties of gellan gum gels. *Biosci. Biotech. Biochem.* 56:595–599, 1992.

85. Grasdalen, H., O. Smidsrød. Gelation of gellan gum. *Carbohydr. Polym.* 7:371–394, 1988.

86. Larwood, V.L., B.J. Howlin, G.A. Webb. Solvation effects on the conformational behaviour of gellan and calcium ion binding to gellan double helices. *J. Mol. Modeling* 2:175–182, 1996.

87. Chandrasekaran, R., A. Radha, V.G. Thailambal. Roles of potassium ions, acetyl and L-glyceryl groups in the native gellan double helix. *Carbohydr. Res.* 224:1–17 1992.

88. Chandrasekaran, R., A. Radha. Molecular architectures and functional-properties of gellan gum and related polysaccharides. *Trends Food Sci. Technol.* 6:143–148, 1995.

89. Chandrasekaran, R., L.C. Puigjaner, K.L. Joyce, S. Arnott. Cation interactions in gellan: an X-ray study of the potassium salt. *Carbohydr. Res.* 181:23–40, 1988.

90. Owen, G. Gellan quick-setting gum systems. In: *Gums and Stabilisers for the Food Industry* 4, Phillips, G.O., D.J. Wedlock, P.A. Williams, eds., Oxford: IRL Press, 1989, pp 173–182.

91. Gibson, W. Gellan. In: *Thickening and Gelling Agents for Food*, Imeson, K., ed., Glasgow: Blackie & Son, 1992, pp 227–250.

92. Wolf, C.L., W.M. LaVelle, R.C. Clark. Gellan gum/gelatine blends. U.S. Patent 4,876,105, 1989.

93. Harada, T., S. Sato, A. Harada. Curdlan. *Bull. Kobe Univ.* 20:143–164, 1987.

94. Buller, C.S. K.C. Voepel. Production and purification of an extracellular polyglucan produced by *Cellulomonas flavigena* strain KU. *J. Ind. Microbiol.* 5:139–146, 1990.

95. Kenyon, W.J., C.S. Buller. Structural analysis of the curdlan-like exopolysaccharide produced by *Cellulomonas flavigena*. *J. Ind. Microb. Biotech.* 29:200–203, 2002.

96. Nakao, Y. Properties and food applications of curdlan. *Agro-Food-Industry Hi-Tech* 8:12–15, 1997.

97. Chuah, C.T., A. Sarko, Y. Deslandes, R.H. Marchessault. Triple helical crystalline structure of curdlan and paramylon hydrates. *Macromolecules* 16:1375–1382, 1983.

98. Kimura, H., S. Moritaka, A. Misaki. Curdlan. *J. Food Sci.* 43, 200–203, 1973.

99. Kanzawa, Y., A. Harada, A. Koreeda, T. Harada, K. Okuyama. Difference of molecular association in two types of curdlan gel. *Carbohydr. Polym.* 10:299–313, 1989.

100. Harada, T., Y. Kanzawa, K. Kanenaga, A. Koreeda, A. Harada. Electron microscopic studies on the ultrastructure of curdlan and other polysaccharides used in foods. *Food Struct.* 10:1–18, 1991.

101. Kanzawa, Y., T. Kurasawa, Y. Kanegae, A. Harada, T. Harada. Purification and properties of a new exo-(1→3)-β-D-glucanase from *Bacillus circulans* YK9 capable of hydrolysing resistant curdlan with formation of only laminaribiose. *Microbiology* 140:637–642, 1994.

102. Graber, M., A. Morin, F. Duchiron, P.F. Monsan. Microbial polysaccharides containing 6-deoxysugars. *Enzyme Microb. Technol.* 10:198–205, 1988.

103. van Kranenburg, R., J.D. Marugg, I.I. Van Swam, N.J. Willem, W.M. de Vos. Molecular characterization of the plasmid-encoded *eps* gene cluster essential for exopolysaccharide biosynthesis in *Lactococcus lactis*. *Mol. Microbiol.* 24:387–397, 1997.

104. Stingele, F., S.J.F. Vincent, E.J. Faber, J.W. Newell, J.P. Kamerling, J.-R. Neeser. Introduction of the exopolysaccharide gene cluster from *Streptococcus thermophilus* Sfi6 into *Lactococcus lactis* MG1363: production and characterization of an altered polysaccharide. *Mol. Microbiol.* 32:1287–1295, 1999.

105. van Kranenburg, R., I.I. Vos, M.K. van Swam, W.M. de Vos. Functional analysis of glycosyltransferase genes from *Lactococcus lactis* and other gram positive cocci: complementation, expression and diversity. *J. Bacteriol.* 181:6347–6353, 1999.

106. Escalante, A., C. Wacherrodarte, M. Garciagaribay, A. Farres. Enzymes involved in carbohydrate metabolism and their role on exopolysaccharide production in *Streptococcus thermophilus*. *J. Appl. Microbiol.* 84:108–114, 1998.

107. Grobben, G.J., I. Chinjoe, V.A. Kitzen, I.C. Boels, F. Boer, J. Sikkema, M.R. Smith, J.A.M. Debont. Enhancement of exopolysaccharide production by *Lactobacillus delbrueckii* subsp. *bulgaricus* NCFB 2772 with a simplified defined medium. *Appl. Environ. Microbiol.* 64:1333–1337, 1998.

108. Duboc, P., B. Mollet. Applications of exopolysaccharides in the dairy industry. *Int. Dairy J.* 11:759–768, 2001.

109. Cho, D.H., H.J. Chae, E.Y. Kim. Synthesis and characterization of a novel extracellular polysaccharide by *Rhodotorula glutinis*. *Appl. Chem. Biotech.* 95:183–193, 2001.

110. Couso, R.O., L. Ielpi, M.A. Dankert. A xanthan gum-like polysaccharide from *Acetobacter xylinum*. *J. Gen. Microbiol.* 133:2133–2135, 1987.

111. Morris, V.J., G.J. Brownsey, P. Cairns, G.R. Chilvers, M.J. Miles. Molecular origins of acetan solution properties. *Intern. J. Biol. Macromol.* 11:326–328, 1989.

112. MacCormick, C.A., J.E. Harris, A.P. Gunning, V.J. Morris. Characterization of a variant of the polysaccharide acetan produced by a mutant of *Acetobacter xylinum* strain CR1/4. *J. Appl. Bacteriol.* 74:196–199, 1993.

113. WHO, 1987.

114. Craston, D.H., P. Farnell, J.M. Francis, S. Gabriac, W. Matthews, M. Saeed, I.W. Sutherland. Determination of gellan gum by capillary electrophoresis and Ce-Ms. *Food Chem.* 73:103–110, 2001.

115. Hashimoto, W., K. Maesaka, N. Sato, S. Kimura, K. Yamamoto, H. Kumagai, K. Murata. Microbial system for polysaccharide depolymerization: enzymatic route for gellan depolymerization by *Bacillus* sp. GL1. *Arch. Biochem. Biophys.* 339:17–23, 1997.

116. Hashimoto, W., E. Kobayashi, H. Nankai, N. Sato, T. Miya, S. Kawai, K. Murata. Unsaturated glucuronyl hydrolase of *Bacillus* sp GL1: novel enzyme prerequisite for metabolism of unsaturated oligosaccharides produced by polysaccharide lyases. *Arch. Biochem. Biophys.* 368:367–374, 1999.

117. Ruijssenaars, H.J., F. Stingele, S. Hartmans. Biodegradability of food-associated extracellular polysaccharides. *Curr. Microbiol.* 40:194–199, 2000.

118. Robijn, G.W., R.C. Gallego, D.J.C. van den Berg, H. Haas, J.P. Kamerling, J.F.G. Vliegenhart. Structural characterization of the exopolysaccharide produced by *Lactobacillus acidophilus* LMG9433 *Carbohydr. Res.* 288:203–218, 1996.

119. Robijn, G.W., A. Imberty, D.J.C. van den Berg, A.M. Ledeboer, J.P. Kamerling, J.F.G. Vliegenhart, S. Perez. Predicting helical structures of the exopolysaccharide produced by *Lactobacillus sake* O-1. *Carbohydr. Res.* 288:57–74, 1996.

25

Solid-State Bioprocessing for Functional Food Ingredients and Food Waste Remediation

Kalidas Shetty

CONTENTS

25.1 Introduction 611
25.2 Modern Applications 612
25.3 SSB for Food Waste Bioprocessing and Applications..................................612
25.4 Complexity of SSB Systems and Need for "Systems Biology" Approaches........612
25.5 SSB in Traditional Food Products and Implications.....................................613
25.6 Approach for Development of Functional Foods and Ingredients Using SSB 614
    25.6.1 Soybeans...614
    25.6.2 Fava Beans and Other Legumes 615
    25.6.3 Cranberry..616
    25.6.4 Enzymes in Phenolic Mobilization and Metabolic Biology of PLPPP.....617
25.7 Extension of PLPPP Strategies for Liquid Yeast Fermentation and
    Lactic Acid Bacteria..617
25.8 Concluding Perspectives 619
References..621

25.1 INTRODUCTION

Solid-state bioprocessing (SSB) is an aerobic microbial metabolic process and is generally referred to as the process in which 1) microbial growth and product formation occur, 2) on suitable solid substrates, 3) in the near absence of free water or in low water that is absorbed within the solid substrate matrix (1–3).

Therefore, referring to this process as a solid-state fermentation (SSF) is a misnomer. Fermentation generally refers to growth with no oxygen, as in alcoholic or lactic acid fermentation. Solid-state bioprocessing deals with the utilization of water-insoluble materials for microbial growth and metabolite production (1). However, suitable materials are those that absorb well and have sufficient moisture to support aerobic growth (1). The SSB system has been widely used in Asia for over 2000 years for products like Koji for soy sauce and miso, as well as tempeh, and is being increasingly modernized and applied to production of other industrial products, including enzymes and pharmaceuticals (1). In the west, an SSB-like process is used in meat preservation, bread making, and cheese and mushroom production (4). This chapter will be limited to the discussion of certain unique applications of SSB for development of ingredients for functional foods, and some aspects of solid waste remediation for food processing wastes. A more detailed conceptual discussion of SSB-like process is covered elsewhere in the book.

25.2 MODERN APPLICATIONS

Since the late nineteenth and early twentieth century, SSB has been used for the production of many enzymes by microorganisms (4,5). It was also extended to the production of important metabolites, such as gluconic acid, citric acid, and gallic acid (4,5). Later, in the mid twentieth century, SSB was extended to the production of antibiotics and steroid transformation. Today it has many applications, especially the production of value added products from agricultural and food processing byproducts, which otherwise have very low value. Select examples of such applications, where SSB is required for initial aerobic breakdown, include the production of fertilizer and soil amendments (6,7), animal feed in the form of single cell proteins (8–12), mushrooms (13–15), ethanol (16–19), methane (bio) gas (20,21), citric acid (22–25), butanol (26), flavor and aroma volatiles (27), and enzymes (28–30). In terms of food applications, there is continuous contemporary improvement in the use of pure cultures and in metabolic processing for bread yeasts (31), cheese (32) and fermented sausage (33,34), Koji production for soy products (35), and tempeh (36,37), along with means of enhancing nutritional quality with such products as carotenoids (38), reducing toxins (39,40), and moderating antinutritional factors (41).

25.3 SSB FOR FOOD WASTE BIOPROCESSING AND APPLICATIONS

Solid-state bioprocessing (SSB) has been used for composting and sewage treatment since early in the twentieth century, because the early phases of waste breakdown requires the decomposition of polymeric substrates by aerobic processes (1–3). This concept has been extended to producing fertilizer and soil amendments (6,7) and to mushroom production (14,15). Another development of a value added product is the conversion of fruit processing wastes to fungal inoculants for agricultural and environmental applications (42–45). With suitable combinations of fruit processing wastes and fishery wastes, beneficial microorganism rich amendments can be developed for reclamation of marginal soils, and to provide proper addendum for efficient composting at landfills and the waste remediation of pollutants.

25.4 COMPLEXITY OF SSB SYSTEMS AND NEED FOR "SYSTEMS BIOLOGY" APPROACHES

Solid-state bioprocessing (SSB) systems, in almost all cases, involve complex microecosystems that have multiple components and variables; such complex systems require a more integrated "Systems Biology" approach. Systems Biology as described by Hood (46) is the study of the relationships of all elements in a biological system, rather than of each element one at a time. In the case of SSB systems, this can be extended to the consideration of a "microecosystem" with multiple substrates and a succession of microbial processes and metabolic interconversions. Further extending the concept of Systems Biology, Kitano (47) has looked at Systems Biology as being similar to understanding the overall workings of a machine or modern mechanical device, as a car or airplane. According to Kitano (47), to understand biology at the systems level one must examine the structure and dynamics of cellular and organismal function, rather than the characteristics of isolated parts. He elegantly provides four key properties from which systems level understanding of biological systems can be derived; these are:

1. Understanding systems structures such as genes, enzymes, and pathways
2. Understanding the dynamics of the pathways and interrelationships
3. Understanding control mechanisms that determine function
4. Understanding of control points

Once these four key properties are understood, one can design or redesign the overall function, the basic structure, and the functional relationships of each biological system (47). Various elements of this Systems Biology perspective, based on parallels to the workings of modern mechanical devices, have merits. However, the complexity and diversity of the process of designing and redesigning biological systems and biological and microbiological ecosystems for human applications require a more comprehensive perspective of evolutionary relationships across the mosaic of prokaryotes and eukaryotes, and the eco-evolutionary development and interrelationships that span two or three billion years. In particular, the perspective will have to start from fundamental energy production and the management of various biological systems individually and in an ecosystem. Further, the variations of energy and redox management from oxygen rich environments to low oxygen and anaerobic environments in complex ecosystems, such as are found in several SSB systems, have to be tackled before we can move on. From fundamental insights into metabolic critical points in energy and redox management across biological systems, we have a better framework to link and connect genetic, signal, enzyme, protein, and related metabolic critical points that drive the overall biological systems and the multiple biological systems in an ecosystem. This energy and redox management based Systems Biology rationale is a suitable way to determine critical control points (CCP) in SSB systems involving complex biological substrates and microbial systems, in order to better design and fine tune SSB systems for any applications.

25.5 SSB IN TRADITIONAL FOOD PRODUCTS AND IMPLICATIONS

To develop an energy and redox management based Systems Biology in SSB systems for food product design and functional food applications, it would make sound and logical sense to go back to over 2000 years of empirical SSB systems, and understand how they have been established in our modern food and cultural mosaic. In this book these concepts have been highlighted by insights into select traditional SSB systems from human development in Africa, China, India, and the Mediterranean. Examples of improvements in bioprocessing in these traditional SSB systems that have come with the knowledge of modern biology, especially microbiology, nutrition, the metabolic biology of energy management, and molecular biology, can provide the framework to adapt more integrated Systems Biology approaches.

25.6 APPROACH FOR DEVELOPMENT OF FUNCTIONAL FOODS AND INGREDIENTS USING SSB

Knowledge of traditional SSB systems can provide the principles, safe biological systems, and framework to develop functional foods (improvement of conventional foods with added health benefits) and ingredients. In line with this thinking, and based on the rationale of incorporating a more integrated Systems Biology approach, several innovative SSB systems have been developed for harnessing the potential of phenolic functional ingredients from safe and established dietary sources. In this chapter discussion will focus on the cultivation of soybeans, fava beans, and cranberries.

In these approaches, SSB systems have been adapted, over a period of more than 2000 years, to enrich phenolic metabolites from soybean, fava bean and cranberry crops for enhanced health benefits. In the case of cranberries, there is cross cultural integration of a traditional American Indian medicinal berry with Asian SSB based food processing using modern metabolic biology principles. In the case of soybeans and fava beans, we have used the traditional Asian tempeh SSB system and adapted it to enriching the beans with functional phenolics that have diverse health application potential. These strategies offer unlimited potential to capture the phytochemical potential of diverse dietary botanicals with well developed traditional microbial processing systems, using modern microbiology, metabolic biology, molecular biology, nutritional biochemistry, and analytical methods.

25.6.1 Soybeans

Epidemiological research has associated high soy intake with a lower risk for certain types of cancer (48,49). Soybean is a rich source of phenolic antioxidants with isoflavonoids being major components. The chief isoflavonoid found in soybean is genistein. Research has shown the chemopreventive properties of purified and synthetic genistein (50–54). However, more recent research has shown that fermented soymilk, developd through bioprocessing, performed better at reducing the incidence of mammary tumor risk than a similar mixture of its constituent isoflavonoids, suggesting that the food background may play a positive role in the chemopreventive actions of soy, in addition to the resident isoflavonoids like genistein (55). As fermented soymilk is rich in phenolic aglycones, which are more active and more readily taken up than their β-glycosides, increasing the free phenolic content of soy-based food through microbial bioprocessing may positively affect its medicinal and nutritional value (56–59). Thus, SSB of soybean substrates by a dietary fungus such as *Rhizopus oligosporus* or *Lentinus edodes* has been developed to mobilize free phenolic antioxidants (60–62). This concept was later extended to mobilization of phenolic antioxidants in soy kefir milk fermentation (63).

When SSB by *Rhizopus oligosporus* of whole soybean was undertaken based on the tempeh bioprocessing model, it was found that while total soluble phenolic content increased 120–135% in the extracts, increased antioxidant activity based on a free radical scavenging assay was limited to the early growth period (60). Higher antioxidant activity was linked to increased glucosidase and glucuronidase activity. Overall results based on phenolic antioxidant and enzyme activities suggested a possible involvement of lignin remobilization and degradation activities, potentially as a part of a detoxification pathway by *Rhiozpus oligosporus* (60).

Phenolic mobilization by SSB using *R. oligosporus* was further investigated, using defatted soybean powder. In this SSB system, phenolic content increased 41% in water extracts and 255% in ethanolic extracts after 10 days of growth (61). In this system beta-glucosidase was the main carbohydrate-cleaving enzyme that is required for efficient phenolic mobilization from defatted powdered soybean (61). This tempeh SSB model was extended to *Lentinus edodes*, which is a slow growing fungus used in shiitake mushroom (fruiting phase of *L. edodes*) production that grows only in the vegetative mycelial phase in soybean (62). It was determined that *L. edodes* SSB was effective in mobilizing phenolic antioxidants from defatted soybean powder (62). Postprocessing phenolic content had increased 90% to 11.3 mg/gram dry weight in water extract and 232% to 5.8 mg/gram dry weight in ethanolic extracts. Furthermore, phenolic antioxidant mobilization by *L. edodes* was related to the activities of the carbohydrate-cleaving enzymes alpha-amylase and beta-glucosidase, as well as laccase, an enzyme involved in lignin biodegradation (62).

Further, in another study, phenolic mobilization by *L. edodes* was associated with manganese peroxidase and laccase (64). Taken together, these results so far indicate that adapting traditional tempeh fermentation using the same food grade fungus *R. oligosporus,* or exploiting the slower growing vegetative mycelial organism of Shiitake mushrooms, *L. edodes*, phenolic ingredients can be mobilized from soybean whole beans or defatted powders. The functional relevance of these phenolic metabolites were confirmed when it was clear that the phenolic mobilization by *R. oligosporus* on day 2 of SSB growth, that coincided with an increased laccase activity, also had the highest phenolic antioxidant activity and antimicrobial activity against ulcer bacteria *Helicobacter pylori* (65) and the food borne bacterial pathogen *Listeria monocyotogenes* (McCue et al., unpublished results). Results also indicate that phenolic mobilization by *L.edodes* vegetative mycelium enhanced DNA protection using a plasmid model system and had antihypertension potential (McCue et al., unpublished results).

25.6.2 Fava Beans and Other Legumes

Dopamine has several key physiological and biochemical roles in human health and well-being. The role of dopamine as a hormone for urinary and vasodilatory function has been postulated (66). The renal production of dopamine has been shown to be an important factor in the control of blood pressure (67). Further, the importance of dopamine for controlling normal motor functions is evidenced by the abnormality of the basal ganglia in Parkinson's disease, where marked loss of these dopamine linked neurons has been observed (68). A recent interesting article has proposed a general theory that attributes the origins of human intelligence to an expansion of dopaminergic systems in human cognition (69). According to Previc (69), "Dopamine is postulated to be the key neurotransmitter regulating six predominantly left hemispheric cognitive skills to human language and thought such as motor planning, working memory, cognitive flexibility, abstract reasoning, temporal analysis /sequencing and generativity."

L-Tyrosine is considered to be the initial precursor in the biosynthesis of dopamine via L-DOPA (68). Natural L-DOPA is found in significant quantities in fava beans. Studies have shown an antiParkinson effect of fava bean consumption (70–72). There is documentation of a substantial increase in plasma L-DOPA levels following consumption of fava beans, which correlated with substantial improvement in motor performance of patients (70). In a recent clinical case study, fava bean consumption prolonged on periods in patients with Parkinson's disease who had on–off fluctuations (73). This prolonged on period was not observed when higher levels (800–1000 mg) of pure, synthetic L-DOPA was given (73), suggesting that L-DOPA in the fava bean background is significant in the positive clinical benefits observed.

L-DOPA has been identified in seedlings, pods, and beans of the fava bean plant (74). A substantial clinical improvement in 6 patients with Parkinson's disease was observed after ingestion of 250 g of the fava bean fruit that included pods and seeds (70). A separate study has shown that the amount of L-DOPA in seedlings is up to 20 times higher than in seeds (75). Verad et al. (74) suggested that germinated seedlings have other advantages, such as ease of use and fewer side effects (i.e., flatulence, cramping). While a previous study (70) reported that consumption of boiled pods and seeds of the fava bean provided substantial improvement in the debilitating effects of Parkinson's disease, a later study (74) showed that consumption of germinated seedlings was even better, and that high plasma L-DOPA levels consistent with substantial clinical improvements were observed. The study is important, because germinated fava bean sprouts could be grown and used any time of the year and may be particularly useful for lower income populations. According to these studies, germinated sprouts of the fava bean offer an excellent and complimentary dietary approach to the management of Parkinson's disease, along with other prescribed treatments. In addition, we have explored the use of SSB fava bean systems using *Rhizopus oligosporus*, which offers more potential for enhanced functionality from phenolic enrichment by recruiting the complex metabolic interactions of fungal colonization of fava beans.

A fava bean SSB system was adopted from the soybean tempeh model using *Rhizopus oligosprous* (76). The L-DOPA content in the SSB fungal grown fava bean increased significantly, to approximately twice that of the control, accompanied by moderate phenolic linked antioxidant activity and high fungal superoxide dismutase (SOD) activity during the early stages of growth (76). This indicated that L-DOPA can be mobilized and formed from fava bean substrates by fungal SSB and this also contributed to the antioxidant functionality of such extracts. A high SOD activity during early and later growth stages indicates likely oxidation stress of the initial fungal colonization and in late stages is likely due to nutrient depletion, when levels of free soluble phenolics are high (76). High levels of soluble phenolics during late stages correlated to an increase in beta-glucosidase activity, indicating that the enzyme plays a role in the likely release of phenolic alycones and the high antioxidant activity of late stages. Therefore use of SSB for fava bean using *R. oligosporus* can improve antioxidant functionality and Parkinson's relevant L-DOPA content (76).

25.6.3 Cranberry

Pioneering investigation has shown that phenolic glycosides in cranberries can be hydrolyzed by enzyme glycosidases using food grade fungus *Lentinus edodes* during SSB (77). Further, preliminary results have shown that during the SSB there is a strong correlation between high antioxidant activity of the fruit extracts and glycosidase activity (78). Based on these preliminary results, the phenolic antioxidant content and activity is being investigated in fruit extracts and pomace byproducts of the cranberry using *R. oligosporus* and *L. edodes* (78,79). In this investigation, the food grade fungi *L. edodes* and *R. oligosporus* are used to mobilize the release of functional phenolic antioxidants linked to beta-glucosidase during SSB of cranberry fruit pulp extracts and pomace byproducts. This helps enhance the value of cranberries and other fruits as a source of food grade health ingredients that can be used and reconstituted in a wide array of processed food. Specifically, this approach has allowed the development of high ellagic acid enriched extracts, with enhanced antimicrobial activity against food borne pathogens (80,81) and ulcer associated *Helicobacter pylori* (82). This process not only enhances the economic value of fruits and fruit ingredients, it could provide an abundant source of health relevant phenolic antioxidants with nutraceutical value. This value added SSB approach has substantial implications for improving the functional value of food products and could be extended to a wide array of foods and food byproducts. We have expanded on these studies to regulate flavonoids and diphenyls in other important fruits, such as apples, raspberries, strawberries, and grapes.

25.6.4 Enzymes in Phenolic Mobilization and Metabolic Biology of PLPPP

Phenolic mobilization from soybean substrates has been linked to the activity of carbohydrate-cleaving enzymes alpha amylase and beta-glucosidase (62), and to lignin degrading enzymes, such as laccase and peroxidase (64). Based on the low phenolics and the high antioxidant activity linked to peak laccase activity in early phase of *R. oligosporus* growth, it has been suggested that polymeric forms of phenolic may have more functional activity. It is likely that the polymeric form is the form with the most potent health benefits, as seen in anti*Helicobacter pylori* inhibition studies (65). These studies are now being extended to inhibition studies in breast cancer cell culture models (McCue et al., unpublished results). Based on carbohydrate mobilization and likely polymeric phenolic forms produced at early stages, it is likely that a growing fungus has to adjust to these conditions metabolically.

In this chapter a model is proposed that as free soluble sugars are enhanced during fungal growth and as polymeric phenolic forms are enhanced, the fungus likely switches to an alternative stress induced pathway. The polymeric phenolic forms may stack on the outer membrane of the fungal mycelia and some degraded phenolics may be have the ability to both enter the cytosol and donate protons to the cytosol through a localized hyper-acidification and transport through H+-transporters (Figure 25.1). When these combinations of events occur the fungus switches to an alternative proline linked pentose–phosphate pathway (PLPPP).

By switching to a praline linked oxidative phosphorylation for ATP synthesis (away from a NADH based one that requires full operation of the tricarboxylic acid cycle), the fungus avoids entering a futile cycle, and at the same time supports the pentose phosphate pathway to supply the reductant NADPH for anabolic pathways. In this way, NADPH can not only meet the needs of proline, which can enter the mitochondria through proline dehydrogenase, but also meet the needs of an antioxidant enzyme response through superoxide dismuatase (SOD) and catalase (CAT). This is also likely linked to the glutathione redox flux (83). This PLPPP critical control point is hypothesized to be important in the adaptation of all fungal systems, including yeast based bioprocessing (both SSB and liquid). It may be especially important under high oxygen conditions, when oxygen pressure coupled to soluble sugar uptake is high. It could also be closely tied to the functional activities of phenolics from substrates on which these microorganisms grow. This concept is being investigated for fungal and yeast growth in the bioprocessing of soybeans, fava beans, cranberries, apples and black tea, and could be extended to any substrates where phenolic mobilization from botanical or other food substrates are involved.

25.7 EXTENSION OF PLPPP STRATEGIES FOR LIQUID YEAST FERMENTATION AND LACTIC ACID BACTERIA

Investigation has confirmed the involvement of lignin-degrading enzymes in phenolic antioxidant mobilization during yogurt production from soymilk by active probiotic kefir cultures (63). Kefir cultures contain a consortium of lactic acid bacteria and a yeast,

Saccharomyces fragilis (63). This approach was based on the rationale that fermented soymilk is rich in phenolic aglycones, which are more active and more readily taken up than their β-glycosides. Therefore, increasing the free phenolic content of soy based food through microbial bioprocessing may positively affect its medicinal and nutritional value (56–59). During the investigation, total soluble phenolic content and free radical scavenging antioxidant activity was measured every 8 h for 48 h. Further, the activity of several enzymes (β-glucosidase, laccase, peroxidase) associated with the microbial degradation of polymeric phenolics and lignin and previously linked to phenolic mobilization from soybean during solid-state bioprocessing by dietary fungi (62,64) were also investigated. Soluble phenolic content increased with kefir culture time and was strongly correlated to total peroxidase and laccase activity. However, phenolic content dropped sharply at 48 h. Antioxidant activity increased with kefir culture time and was strongly correlated to decreased soluble phenolic content over the same time period.

This research has important implications for the optimization of functional phytochemicals in commercial soymilk based yogurts, which can be targeted for disease chemoprevention strategies. This soymilk model can be extended to mobilization of phenolics in other liquid fermentation systems such as those for producing wine, beer, apple cider, buttermilk and any other process involving fermented yeast and acidic fermentative bacteria that utilizes an initial aerobic and subsequent low or no oxygen fermentation systems. In such systems the adjustment to initial oxygen stress and subsequent switch to low oxygen or no oxygen involves adjustment to ethanol or organic acids that are produced on the growing liquid substrates.

In the case of yeast, the fungal PLPPP model (Figure 25.1) described earlier, with modifications based on oxygen stress, may be involved. In the case of acid-producing fermentative bacterial systems which are prokaryotes, the fundamental energy metabolism and redox management takes place between the outer plasma membrane (like an independent mitochondria) and cytosol. It is conceivable that the acid-producing bacteria have evolved a high proton management cytosolic dehydrogenase system that can take the inward proton flux and protons generated from new lactic acid production under reduced oxygen and couple it to plasma membrane driven oxidative phosphorylations or substrate level phosphorylations linked to organic acid. It would make sense for this dehydrogenase linked energy and reductant management to occur through the same proline linked pentose–phosphate pathway (PLPPP), with the adjustment within the simpler prokaryotic coupling between the plasma membrane and cytosol (Figure 25.2). The ultimate efficiency and need for such a control point is the same as for mitochondrial linked systems in yeast or filamentous fungi. In these systems, energy metabolism through proline and reductant needs for anabolism through NADPH from the pentose phosphate pathway can be provided during the initial crucial switch from high oxygen conditions to low and no oxygen conditions, when substrate level phosphorylation linked to alcohol or organic acids predominate.

It is also conceivable that organic acid-producing prokaryotes are more versatile in this redox switch, and therefore could have evolved a coupled mechanism to better survive high oxygen, high stress, and eventually high organic acids under much reduced oxygen. This rationale could be used in developing antimicrobial systems against pathogenic bacteria, where combinations of exogenous acids and hyperacidifying, proton-donating phenolic metabolites could be recruited to inhibit or kill pathogenic bacteria (83). Using the same logic, organic acid and phytochemical enriched SSB and liquid fermentation systems could be effective ways to design antimicrobials against pathogenic bacteria for improving food safety, as seen in many traditional fermented foods. In such a situation critical points at the level of energy and reductant management between the plasma membrane and cytosol become important, whether involving our hypothetical PLPPP or an alternative model.

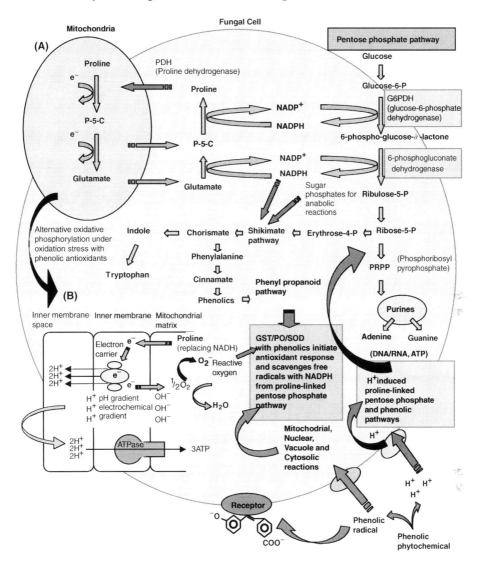

Figure 25.1 Extension of plant proline linked pentose–phosphate pathway model for the effect of external phenolic phytochemicals in yeast and fungal systems during solid-state bioprocessing under aerobic exposure when oxidation stress is maximum and sugar uptake in rapid. (Abbreviations: P5C; pyrroline-5-carboxylate, GST;Glutathione-s-transferase, PO;peroxidase, SOD;superoxide dismutase)

25.8 CONCLUDING PERSPECTIVES

More than 2000 years of microbial based bioprocessing (aerobic and fermentative–anaerobic) of foods and beverages in diverse food cultures around the world has provided us with many complex food substrate and microbial systems for developing functional foods and ingredients. In addition, value added food waste remediation systems have been developed. These food substrates and microbial systems can be further developed for human health benefits and environmental management using the Systems Biology approach of modern biology. These complex systems involve complex consortia of microbial ecosystems and metabolic biology acting on multiple food substrates. To understand these complex systems,

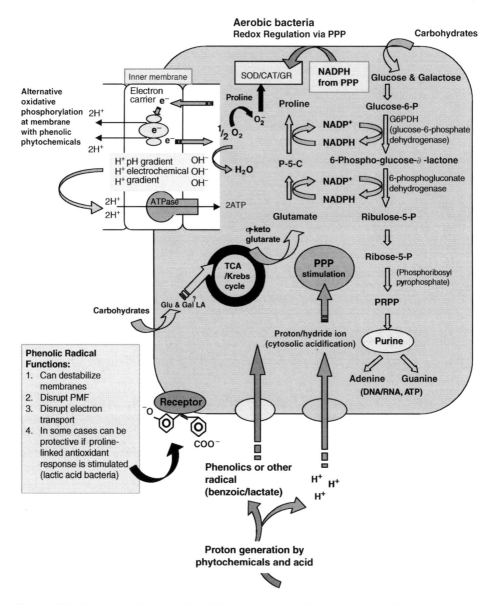

Figure 25.2 Extension of plant proline linked pentose–phosphate pathway model for the effect of external phenolic phytochemicals in prokaryotic oxygen utilizing bacterial systems, where released phenolics from fermenting substrates could contribute to a redox response and tolerance to acidity, unlike nonacid tolerant bacteria, which are inhibited by phenolics (Abbreviations: P5C;pyrroline-5-carboxylate, SOD;superoxide dismutase, CAT;catalase, GR;glutathione reductase, PPP;pentose phosphate pathway; Krebs {TCA-tricarboxylic acid} cycle), Glu; Glucose, Gal; Galactose, LA;lactic acid)

a more integrated Systems Biology approach that first requires understanding of critical control points in energy and redox management across individual organisms to consortia is essential. An understanding of ecoevolutionary relationships of energy and redox metabolic biology across prokaryotes to eukaryotes from oxygen rich to low oxygen and to fermentative environments is also required. This energy- and redox-linked metabolic biology based platform can help us integrate exciting information from the genetic, signal, and enzyme–protein fields. Such an approach gives us a better understanding of the critical control points that define the functioning of organisms and ecosystems. The understanding of such control points can be utilized to better design functional food and ingredients using SSB systems.

Among many possible critical energy and redox-modulating pathways, we have focused on the relevance of the proline linked pentose–phosphate pathway (PLPPP) in SSB systems. Under the diverse stresses of high or low oxygen environments, organisms likely use PLPPP as an alternative pathway, with proline as a reductant, for oxidative phosphorylation at outer plasma membrane in prokaryotes and mitochondria in eukaryotic yeast and fungi. At the same time the pentose–phosphate pathway is utilized for redox management in the cytosol, to drive reductants such as NADPH for anabolism, including antioxidant enzyme response pathways involving superoxide dismutase and catalase. The phenolics mobilized from food substrates will, in many ways, define the behavior of the microbial energy and redox responses. The metabolic and cellular biology of these inter-relationships can be harnessed to maximize enhancement and optimization of functional foods and ingredients.

REFERENCES

1. Zheng, Z., K. Shetty. Solid-state fermentation and value-added utilization of fruit and veg-etable processing by-products. In: *Wiley Encyclopedia of Food Science and Technology,* 2nd ed., F.J. Francis, ed., New York: Wiley Interscience, 1999, pp 2165–2174.
2. Cannel, E., M. Moo-Young. Solid-state fermentation systems. *Process. Biochem.* 4:2–7, 1980.
3. Mudgett, R.E. Solid-state fermentations. In: *Manual of Industrial Microbiology and Biotechnology,* Demain, A.L., N.A. Solomon, eds., Washington, D.C.: American Society of Microbiology, 1986, pp 66–83.
4. Aidoo, K.E., R. Hendry, B.J.B Wood. Solid substrate fermentation. *Adv. Appl. Microbiol.,* 28:201–237, 1982.
5. Pandey, A. Recent progress in developments in solid-state fermentation. *Process. Biochem.* 27:109–117, 1992.
6. Logsdon, G. Pomace is a grape resource. *Biocyle* 33:40–41, 1992.
7. Hang, Y.D. Improvement of the nutritional value of apple pomace by fermentation. *Nutr. Rep. Int.* 38:207–211, 1988.
8. Rahmat, H., R.A. Hodge, G.J. Manderson. Solid-substrate fermentation of *Kloeckera apiculate* and *Candida utilis* on apple pomace to produce an improved stock-feed. *World J. Microbiol. Biotechnol.* 11:168–170, 1995.
9. Menezes, T.J.B. Protein enrichment in citrus wastes by solid-state fermentation. *Process. Biochem.* 24:167–171, 1989.
10. Chung, S.L., S.P. Meyer. Bioprotein from banana wastes. *Dev. Indust. Microbiol.* 20:723–731, 1979.
11. Enwefa, O. Biomass production from banana skins. *Appl. Microbiol. Biotechnol.* 36:283–284, 1991.
12. Fellows, P.J., J.T. Worgan. Growth of *Saccharromycopsis fibuliger* and *Candida utilis* in mixed culture in apple processing wastes. *Enzyme Microbiol. Technol.* 9:434–437, 1987.
13. Moyson, E., H. Verachtert. Growth of higher fungi on wheat straw and their impact on the digestability of the substrate. *Appl. Microbiol. Biotechnol.* 36:421–424, 1991.
14. Upadhyay, R.C., H.S. Sidhi. Apple pomace: a good substrate for the cultivation of edible mushrooms. *Curr. Sci.* 57:1189–1190, 1988.
15. Worrall, J.J., C.S. Yang. Shiitake and oyster mushroom production on apple pomace and sawdust. *Hort. Sci.* 27:1131–1133, 1992.
16. Hang, Y.D. Production of alcohol from apple pomace. *Appl. Environ. Microbiol.* 42:1128–1129, 1981.

17. Sandhu, D.K., V.K. Joshi. Solid-state fermentation of apple pomace for concomitant production of ethanol and animal feed. *J. Sci. Ind. Res. India* 56:86–90, 1997.

18. Saha, B.C., B.S. Dien, R.J. Bothast. Fuel ethanol production from corn fiber: current status and technical prospect. *Appl. Biochem. Biotechnol.* 70:115–125, 1998.

19. Roukas, T. Solid-state fermentation of carob pods for ethanol production. *Appl. Microbiol. Biotechnol.* 41:296–301, 1994.

20. Jewell, W.J., R.J. Cummings. Apple pomace energy and solids recovery. *J. Food Sci.* 48:407–410, 1984.

21. Knol, W., M.M. Van der Most, J. De Waart. Biogas production by anaerobic digestion of fruit and vegetable wastes. *J. Food Sci. Agric.* 29:822–827, 1978.

22. Hang, Y.D., E.E. Woodams. Apple pomace: a potential substrate for citric acid production by *Aspergillus niger. Biotechnol. Lett.* 6:763–764, 1984.

23. Tran, C.T., D.A. Mitchell. Pineapple waste: a novel substrate for citric acid production by solid-state fermentation. *Biotechnol. Lett.* 17:1107–1110, 1995.

24. Hang, Y.D., E.E. Woodams. Production of citric acid from corncobs by *Aspergillus niger. Bioresource Technol.* 64:251–253, 1998.

25. Roukas, T. Citric acid production from carob pod by solid-state fermentation. *Enzyme Microbial. Technol.* 24:54–59, 1999.

26. Voget, C.E., C.F. Mognone, R.J. Ertola. Butanol production from apple pomace. *Biotechnol. Lett.* 7:43–46, 1985.

27. Bramorski, A. Production of volatile compounds by edible *Rhizopus oryzae* during solid-state cultivation of tropical agro-industrial substrates. *Biotechnol. Lett.* 20:359–362, 1998.

28. Hours, R.A., C.E. Voget, R.J. Ertola. Apple pomace as raw material for pectinase production in solid-state fermentation. *Biol. Wastes* 23:221–228, 1988.

29. Rombouts, F.M., W. Pilnik. Enzymes in fruit and vegetable juice technology. *Process. Biochem.* 13:9–13, 1978.

30. Krishna, C., M. Chandrasekaran. Banana waste as substrate for alpha-amylase production by *Bacillus subtilis* (CBTK 106) under solid-state fermentation. *Appl. Microbiol. Biotechnol.* 46:106–111, 1996.

31. Reed, G. Production of baker's yeast. In: *Prescott and Dunn's Industrial Microbiology*, 4th ed., Reed, G. ed., Westport, CT: AVI, 1982, pp 593–633.

32. Pederson, C.S. *Microbiology of Food Fermentations,* 2nd ed., Westport, CT: AVI, 1979.

33. Selgas, M.D. Potential technological interest of *Mucor* strain to be used in dry fermented sausage production. *Food Res. Int.* 28:77–82, 1995.

34. Lucke, F.K. Fermented meat products. *Food Res. Int.* 27:299–307, 1994.

35. Wood, B.J. Technology transfer and indigenous fermented foods. *Food Res. Int.* 27:269–280, 1994.

36. Campbell-Platt, G. Fermented foods: a world perspective. *Food Res. Int.* 27:252–257, 1994.

37. Keuth, S., B. Bisping. Vitamin B12 production by *Citrobacter freundii* or *Klebsiella pneumoniae* during tempeh fermentation and proof of enterotoxin absence by PCR. *Appl. Environ. Microbiol.* 60:1495–1499, 1994.

38. Han, J.R. Sclerotia growth and carotenoid production of *Penicillium* sp. PT95 during solid-state fermentation of corn meal. *Biotechnol. Lett.* 20:1063–1065, 1998.

39. Steinkraus, K.H. Nutritional significance of fermented foods. *Food Res. Int.* 27:259–267, 1994.

40. Van-Veen, A.G., D.C.W. Graham, K.H. Steinkraus. Fermented peanut presscake. *Cereal Sci. Today* 13:96–99, 1968.

41. Reddy, N.R. Reduction of antinutritional and toxic compounds in plant foods by fermentation. *Food Res. Int.* 27:281–290, 1994.

42. Zheng, Z., K. Shetty. Cranberry processing waste for solid-state fungal inoculant production. *Process. Biochem.* 33:323–329, 1998.

43. Zheng, Z., K. Shetty. Solid state production of beneficial fungi on apple processing waste using glucosamine as the indicator of growth. *J. Agric. Food Chem.* 46:783–787, 1998.

44. Zheng, Z., K. Shetty. Effect of apple pomace-based *Trichoderma* inoculants on seedling vigor in pea (*Pisum sativum*) germinated in potting soil. *Process. Biochem.* 34:731–735, 1999.

45. Zheng. Z., K. Shetty. Enhancement of pea (*Pisum sativum*) seedling vigor and associated phenolic content by extracts of apple pomace fermented with *Trichoderma* spp. *Process. Biochem.* 36:79–84, 2000.

46. Hood, L. Systems biology: integrating technology, biology and computation. *Mech. Ageing Dev.* 124:9–16, 2003.

47. Kitano, H. Systems biology: a brief overview. *Sci.* 295:1662–1664, 2002.

48. McCue, P., K. Shetty. Health benefits of soy isoflavonoids and strategies for enhancement: a review. *Crit. Rev. Food Sci. Nutr.* 44:1–7, 2004.

49. Dai, Q., A.A. Franke, F. Jin, X.O. Shu, J.R. Hebert, L.J. Custer, J. Cheng, Y.T. Gao, W. Zheng. Urinary excretion of phytoestrogens and risk of breast cancer among Chinese women in Shanghai. *Cancer Epid. Biomarkers Prev.* 11:815–821, 2002.

50. Darbon, J.M., M. Penary, N. Escalas, F. Casagrande, F. Goubin-Gramatica, C. Baudouin, B. Ducommun. Distinct Chk2 activation pathways are triggered by genistein and DNA-damaging agents in human melanoma cells. *J. Biol. Chem.* 275:15363–15369, 2000.

51. Lamartiniere, CA. Protection against breast cancer with genistein: a component of soy. *Am. J. Clin. Nutr.* 71:1705S–1707S, 2000.

52. Xu, J., G. Loo. Differential effects of genistein on molecular markers related to apoptosis in two phenotypically dissimilar breast cancer cell lines. *J. Cell. Biochem.* 82:78–88, 2001.

53. Lamartiniere, C.A., M.S. Cotroneo, W.A. Fritz, J. Wang, R. Mentor-Marcel, E. Elgavish. Genistein chemoprevention: timing and mechanisms of action in murine mammary and prostate. J. Nutr. 132:552S–558S, 2002.

54. Tanos, V., A. Brzezinski, O. Drize, N. Strauss, T. Peretz. Synergistic inhibitory effects of genistein and tamoxifen on human dysplastic and malignant epithelial breast cancer cells *in vitro. Eur. J. Obstet. Gyn. Reprod. Biol.* 102:188–194, 2002.

55. Ohta, T., S. Nakatsugi, K. Watanabe, T. Kawamori, F. Ishikawa, M. Morotomi, S. Sugie, T. Toda, T. Sugimura, K. Wakabayashi. Inhibitory effects of *Bifidobacterium*-fermented soy milk on 2-amino-1-methyl-6-phenylimidazo[4,5-*b*]pyridine-induced rat mammary carcinogenesis, with a partial contribution of its component isoflavones. *Carcinogenesis* 25:937–941, 2000.

56. Izumi, T., M.K. Piskula, S. Osawa, A. Obata, K. Tobe, M. Saito, S. Kataoka, Y. Kubota, M. Kikuchi,. Soy isoflavone aglycones are absorbed faster and in higher amounts than their glucosides in humans. *J. Nutr.* 130:1695–1699, 2000.

57. Rao, M., G. Muralikrishna. Evaluation of the antioxidant properties of free and bound phenolic acids from native and malted finger millet (Ragi, Eleusine coracana Indaf-15). *J. Agric. Food Chem.* 50:889–892, 2002.

58. Setchell, K., N. Brown, L. Zimmer-Nechemias, W. Brashear, B. Wolfe, A. Kirschner, J. Heubi. Evidence for lack of absorption of soy isoflavone glycosides in humans, supporting the crucial role of intestinal metabolism for bioavailability. *Am. J. Clin. Nutr.* 76:447–453, 2002.

59. Yuan, L., C. Wagatsuma, M. Yoshida, T. Miura, T. Mukoda, H. Fujii, B. Sun, J.H. Kim, Y.J. Surh. Inhibition of human breast cancer growth by GCP™ (genistein combined polysaccharide) in xenogenetic athymic mice: involvement of genistein biotransformation by α-glucuronidase from tumor tissues. *Mutat. Res.* 523–524:55–62, 2003.

60. McCue, P., K. Shetty. Role of carbohydrate-cleaving enzymes in phenolic antioxidant mobilization from whole soybean fermented with *Rhizopus oligosporus. Food Biotechnol.* 17:27–37, 2003.

61. McCue, P, A. Horii, K. Shetty. Solid-state bioconversion of phenolic antioxidants from defatted soybean powders by *Rhizopus oligosporus*: role of carbohydrate cleaving enzymes. *J. Food Biochem.* 27:501–514.

62. McCue, P, A. Horii, K. Shetty, Mobilization of phenolic antioxidants from defatted soybean powders by *Lentinus edodes* during solid-state bioprocessing is associated with enhanced production of laccase. *Innovative Food Sci. Emerging Technol.* 5:385–392, 2004.

63. McCue, P., K. Shetty. Phenolic antioxidant mobilization during yogurt production from soymilk. *Process. Biochem.* 40:1791–1797, 2005.

64. McCue, P., K. Shetty. A model for the involvement of lignin degradation enzymes in phenolic antioxidant mobilization from whole soybean during solid-state bioprocessing by *Lentinus edodes*. *Process. Biochem.* 40:1143–1150, 2005.

65. McCue P., Y.-T. Lin, R.G. Labbe, K. Shetty. Sprouting and solid-state bioprocessing by *Rhizopus oligosporus* increase the *in vitro* antibacterial activity of aqueous soybean extracts against *Helicobacter pylori*. *Food Biotechnol.* 18:229–249, 2004.

66. Vered, Y., I. Grosskopf, D. Palevitch, A. Harsat, G. Charach, M.S. Weintraub, E. Graff. The influence of *Vicia faba* (broad bean) seedlings on urinary sodium excretion. *Plant. Med.* 63:237–240, 1997.

67. Missale, C., S.R. Nash, S.W. Brown, M. Jaber, M.G. Caron. Dopamine Receptors: From Structure to Function. Physiological Reviews, 78:189–225, 1998.

68. Elsworth, J.D., R.H. Roth. Dopamine synthesis, uptake, metabolism, and receptors: relevance to gene therapy of Parkinson's disease. *Exper. Neurol.* 144:4–9, 1997.

69. Previc, F.H. Dopamine and the origins of human intelligence. Brain Cognit. 41:299–350, 1999.

70. Rabey, J.M., Y. Vered, H. Shabtai, E. Graff, A.D. Korczyn. Improvement of Parkinson's features correlate with high plasma levodopa values after broad bean (*Vicia faba*) consumption. *J. Neurol. Neurosurg. Psych.* 55:725–727, 1992.

71. Kempster, P.A., Z. Bogetic, J.W. Secombe, H.D. Martin, N.D.H. Balazss, M.L. Wahlqvist. Motor effects of broad beans (*Vicia faba*) in Parkinson's disease: single dose studies. *Asia Pac. J. Clin. Nutr.* 2:85–89, 1993.

72. Spengos, M., D. Vassilopoulos. Improvement of Parkinson's diseases after *Vicia faba* consumption. In: *Book of Abstracts, 9th International Symposium on Parkinson's Disease, 1988, p 46.*

73. Apaydin, H., S. Ertan, S. Ozekmekci. Broad bean (Vicia faba): a natural source of L-DOPA: prolongs "on" periods in patients with Parkinson's disease who have "on-off" fluctuations. *Movement Disorders* 15:164–166, 2000.

74. Vered, Y., J.M. Rabey, D. Palevitch, I. Grosskopf, A. Harsat, A. Yanowski, H. Shabtai, E. Graff. Bioavailability of levodopa after consumption of *Vicia faba* seedlings by Parkinsonian patients and control subjects. *Clin. Neuropharmacol.* 17:138–146, 1994.

75. Wong, K.P., B. Geklim. L-DOPA in the seedlings of *Vicia faba*: its identification, quantification and metabolism. *Biogen. Amines* 8:167–173, 1992.

76. Randhir, R., D. Vattem, K. Shetty. Solid-state bioconversion of fava bean by *Rhizopus oligosporus* for enrichment of phenolic antioxidants and L-DOPA. *Innovative Food Sci. Emerging Technol.* 5:235–244, 2004.

77. Zheng, Z., K. Shetty. Solid-state bioconversion of phenolics from cranberry pomace and role of *Lentinus edodes* beta-glucosidase. *J. Agric. Food Chem.* 48:895–900, 2000.

78. Vattem, D.A., K. Shetty. Solid-state production of phenolic antioxidants from cranberry pomace by *Rhizopus oligosporus*. *Food Biotechnol.* 16:189–210, 2002.

79. Vattem, D.A., K. Shetty. Ellagic acid production and phenolic antioxidant activity in cranberry pomace mediated by *Lentinus edodes* using solid-state system. *Process. Biochem.* 39:367–379, 2003.

80. Vattem, D.A., Y.T. Lin, R.G. Labbe, K. Shetty. Phenolic antioxidant mobilization in cranberry pomace by solid-state bioprocessing using food grade fungus *Lentinus edodes* and effect on antimicrobial activity against select food-borne pathogens. *Innovative Food Sci. Emerging Technol.* 5:81–91, 2004.

81. Vattem, D.A., Y.T. Lin, R.G. Labbe, K. Shetty. Antimicrobial activity against select food-borne pathogens by phenolic antioxidants enriched cranberry pomace by solid-state bioprocessing using food-grade fungus *Rhizopus oligosporus*. *Process. Biochem.* 39:1939–1946, 2004.

82. Vattem, D.A., Y.T. Lin, K. Shetty. Enrichment of phenolic antioxidants and anti-*Helicobacter pylori* properties of cranberry pomace by solid-state bioprocessing. *Food Biotechnol.* 19:51–68, 2005.

83. Shetty, K., M.L. Wahlqvist. A model for the role of proline-linked pentose phosphate pathway in phenolic phytochemical biosynthesis and mechanism of action for human health and environmental applications: a review. *Asia Pac. J. Clin. Nutr.* 13:1–24, 2004.

Index

A

AA, *see* Arachidonic acid

L-AA, *see* L-Ascorbic acid

AAD, *see* Antibiotic associated diarrhea

Aberrant crypts, colon cancer and, 407

Abiotic stress, plant adaptation to, 210

ABTS, *see* 2,2′-Azinobis-(3-ethylbenzthiazoline-6-sulfonic acid)

Acceptable Daily Intakes, 604

Acesulfame-K, 329, 331

Acetate–malonate pathway, 2, 152, 211

Acetic acid, 511

Acetobacter xylinum, 603

N-Acetylaminosugars, 585

Acetylation, 68

Acetyl-CoA carboxylases, 562

Acidophilus milk, 521

Actinobacillus pleuropneumoniae, 391

Activated oxygen species, lipid peroxidation and, 366

Acute diarrhea, LAB and, 456

ADA, *see* American Dietetic Association

Adenylate cyclase, Forskolin induced activation of, 189

Adjuvant, definition of, 454

ADP-glucose pyrophosphorylase (AGPase), 56, 349

Aerobic culture fermentation, 526

Aeromonas hydrophila, tannins and, 286, 295

AFLP, *see* Amplified fragment length polymorphisms

AGPase, *see* ADP-glucose pyrophosphorylase

Agrobacterium

    radiobacter, 601

    species β-glucosidase, mutant of, 489

    tumefaciens, 123

AhR, *see* Aryl hydrocarbon receptor

Alanine, production of, 569

Aldehydes, reaction of cysteine with, 569

Alkaline invertases, potato tuber, 355

Allergic disorders, hygiene hypothesis and, 458

Almond, α-mannosidase from, 485

Alzheimer's disease, 98

    F2-isoprostanes and, 237

    ROS and, 161

American cranberry, 156

American Dietetic Association (ADA), 8

American hybrid wines, 316

Amino acid(s)

    active product excretion, 554

    aspartate, 566

    auxotrophies, 556

    biosynthetic pathways, 505

    categories of bulk quantities, 547

    current production and application of, 546

    deamination of, 407

    degradation pathways, inactivation of, 55

    deregulation, 554

    energy and redox balance, 555

    excretion

        metabolic imbalance, 550

        physiological concepts of, 549

        systems, 556

    export systems, 555

    fermentation, 558

    identity

        β-conglycinin, 29

        glycinin, 31

    metabolic engineering and, 503

    overproduction, 550, 557

    precursor synthesis by central pathways, 554

    producing strains, categories of, 548

    soybean bile acid binding stability and, 43

    strain breeding by mutation, 569

    substrate uptake, 554

    sulfur containing, 36

    transport systems, altering of, 507

    uptake systems, 556

    uses of, 545, 546

Amino acids, production of, 545–582

    basic steps involved in, 552

    cell wall integrity and, 557

    challenge for research in, 574

    continuous culture, 558

    examples of amino acids, 559–569

        glutamate, 559–564

        L-lysine, 565–568

        other amino acids, 569

        L-phenylalanine, 564–565

threonine, 568–569
L-tryptophan, 565
fermentation and, 572
future developments, 569–574
LAB and, 508
organisms used for biotechnological processes of,
        547
physiological, biochemical, and genetic
        background of amino acid production,
        548–553
physiological, biochemical, and genetic tools
        used to improve amino acid production,
        554–558
strategies to increase, 574
technical aspects, 558–559
transport reactions in, 555
Ammonia metabolism, lactulose and, 415
Amphiregulin (AR), 262
Amplified fragment length polymorphisms (AFLP),
        312
Amylase, debranching enzymes, 58
β-Amylase, potato tubers and, 352
Amylopectin
    chain length, 71
    cluster models of, 54
    double helices, 65
    elongated glucan chains within, 57
    gelation of mixtures of amylose and, 67
    molecules, organization of, 55
    proportion of in starches, 53
    retrogradation, 65
    structure, changes in, 61, 63
    synthesis, SS and, 58
Amyloplasts, 56, 352
Amylose
    destabilization, 66
    free starch, 68, 69
    gelation, 66, 67
    retrogradation, 66
    rice, 61
amylose extender mutation, 62, 70, 71
Anacystis nidulans, 359
Anchusa officinalis, 190
Anemia, 113
Angiotensin converting enzyme inhibitors, 16
Animal feed, amino acids as additives to, 547
Anthocyanins, 136, 164
Anthranilate synthase, genes encoding, 565
Antiallergenic drugs, 268
Antibiotic associated diarrhea (AAD), 405, 457
Antibody stability, 391
Anticytokine antibodies, 460
Antifungal agent, 10
Antigen(s)
    naked, 441
    -presenting cells, 441, 447
        immunosuppressive, 452
        LAB and, 461
    self, 441

vaccination, antibody responses to, 449
Antimicrobial functionality, 302
Antimicrobial phenolics, 11, 210
Antinutrients, 108, 113
Antioxidant(s)
    assays, 232
    defense
        phenolic phytochemicals in nutritional
            management of, 161
        response, 161, 162
        systems, 159
    definition of, 211
    enzymes, potato, 362
    food preservation and, 9–10
    homeostasis, 167
    lipid-soluble, 124
    phenolic, 171, 286
    plant phenolics as, 211
    protection factor (APF), 235
    responsive element (ARE), 166
    rich foods, 212
    synthetic, 234
    vitamin C as, 125
Antioxidant functionality, biochemical markers for,
        229–251
    antioxidant assays, 232–235
        2,2'-azinobis-(3-ethylbenzthiazoline-6-
            sulfonic acid), 233
        based on free radical neutralization, 232–233
        β-carotene oxidation assay, 235
        ferric reducing ability of plasma, 234
        oxygen radical absorbance capacity assay,
            234–235
    direct measurement of reactive oxygen species,
        231–232
    DNA oxidation products, 237–243
        catalase, 241
        glutathione peroxidase, 240
        glutathione reduced and oxidized, 238–240
        glutathione-S-transferases, 240
        myeloperoxidase, 242–243
        NADPH oxidoreductase, 242
        8-oxo-7,8-dihydro-2'-deoxyguanosine, 237–
            238
        superoxide dismutase, 240–241
        xanthine oxidase and uric acid, 241–242
    lipid oxidation products, 235–237
        F2 isoprostanes, 237
        LDL resistance to oxidation, 236–237
        malondialdehyde, 236
    plant phenolic dependent peroxidases, 243
APF, *see* Antioxidant protection factor
Apocarotenoids, 533
Apoptosis induction, LAB inhibition of, 451
AR, *see* Amphiregulin
Arabidopsis
    cold stress in, 364
    thaliana, 126, 317, 339
Arabinogalactan, 561

Arabinosylxylobiose, 418
Arabinoxylan, 420
Arabinoxylooligosaccharides, degradation of, 413
Arachidonic acid (AA), 237, 256, 266
ARE, *see* Antioxidant responsive element
Arginine, 545
Aroma compounds, 515
Aromatase inhibitors, 260
Aromatic amino acids, biosynthetic pathway of, 564
Aryl hydrocarbon receptor (AhR), 255, 261
Ascorbic acid, oxidation linked diseases and, 161
L-Ascorbic acid (L-AA), 125, 126, 513
Aspartame, 331, 332, 564
Aspartate, 566, 568
Aspartokinase, 566
Aspergillus
    aciduleatus, 602
    flavus, organic solvents and, 485
    niger, 331
        citric acid fermentation, 527
        fungal α-mannosidase from, 481
    oryzae, 337
Astaxanthin, 299
Atherosclerosis, F2-isoprostanes and, 237
Atopic eczema, 458, 459
ATP
    dependent phophofructokinase (ATP-PFK), 356
    sink, 516
    stoichiometry, amino acid production and, 574
    synthesis, NADH linked, 169
ATP-PFK, *see* ATP dependent phophofructokinase
Aureobasidium pullulans, 589
Auxins, 213
Avian immune system, 382
2,2′-Azinobis-(3-ethylbenzthiazoline-6-sulfonic acid) (ABTS), 233
Azoxymethane (AOM), 407
Aztec marigold, antimutagenicity activity of carotenoids from, 120

B

Bacillus
    cereus
        acute inflammatory reactions and, 404
        sage extract and, 286
    circulans
        enzyme, oligosaccharides produced by, 484
        β-galactosidase from, 482
        xylanase, 480
    stearothermophilus
        α-galactosidase from, 491
        trisaccharide synthesis and, 487
    subtilis, 337, 573
        oxalate decarboxylase from, 33
        strains, riboflavin-producing, 513
Backbone amide linker (BAL), 490
Baco Noir, 315

Bacteria, *see also* Lactic acid bacteria, immunomodulating effects of
    changes in metabolism of, 571
    example of flux analysis in, 566
    food additives produced by, 502
    FOS and, 410
    glutamate and, 546
    Gram-negative, 441, 452
    Gram-positive, 441, 445, 452, 561
    interaction of with enterocytes, 452
    phagocytosis and, 405
    pioneer, 445
    probiotic, 448
    spontaneous mutations of, 559
    strategies for stress response, 573
    viability, 460
Bacteria, metabolic engineering of, 501–520
    amino acids, 503–508
        modification of biosynthetic pathways, 505–507
        modification of central metabolic pathways, 503–505
        modification of transport systems, 507
        use of analytical tools in ME toolbox, 507–508
        use of LAB in production of amino acids, 508
    aroma compounds, 515–516
    bacteriocins, 514–515
    carbohydrates, 513–514
    future trends, 516
    low calorie sugars, 515
    multifunctional organic acids, 508–512
        acetic acid, 511–512
        lactic acid, 511
        pyruvic acid, 510–511
        succinic acid, 508–510
    vitamins, 512–513
        L-ascorbic acid, 513
        folate, 513
        riboflavin, 513
Bacterial energy metabolism, 548
Bacterial fermentations, yogurt and, 585
Bacterial glycosidases, reverse hydrolysis and, 481
Bacterial pathogens, phenolic antimicrobials from plants for control of, 285–309
    dietary soybean approach for control of *H. pylori*, 298–301
        phenolic response of soybean extracts inhibiting *H. pylori* urease, 301
        solid-state bioprocessing of soybean using *Rhizopus oligosporus*, 301
        sprouting of soybean and phenolic antioxidant stimulation methods, 299–301
    filling gaps in strategy with screening of elite phenolic phytochemical-producing lines, 291–298
        control of *H. pylori* by phenolic antimicrobials from plants, 296–297

current treatment strategies for *H. pylori*, 297–298

dietary phenolic phytochemical strategies for control of *H. pylori*, 298

novel model for mechanism of action, 292–296

phenolic antimicrobials from plants, 286–287

potential of *Lamiaceae* as source of phenolic antimicrobials, 288–291

Listeria monocytogenes, 289–291

Staphylococcus aureus, 288–289

recent progress, 301–302

Bactericidal peptides, *Lactobacillus* and, 450

Bacteriocins, 514

Bacteroides, 402, 459

Baicalein, 269

Baker's yeast, 319

BAL, *see* Backbone amide linker

Barbera, 314

Barley starches, waxy, 70

Barrier hypothesis, 561

Batch fermentations, 528

BCV, *see* Bovine coronavirus

Bean ferritin rice, iron content of, 113

Benzoic acid derivatives, 2, 152–153

Berry(ies)

extracts, free radical scavenging activity of, 156

phenolic phytochemicals from, 153

Bifidobacteria, 436, 445

allergic disease and, 458

enterocytes and, 450

Bifidobacterium, 402

bifidum, β-galactosidase from, 491

colon cancer and, 407

FOS and, 409

IMO fermentation by, 414

infantis, 447

lactis, GOS and, 412

Bile acid binding ability, soybean protein, 42

Bioactive ingredients, food as vehicle for, 9

Bioavailability of micronutrients, 108

Bioconversion, definition of, 108

Biological heat, 527

Biosynthetic pathway, leaky, 557

Biotin, fatty acid synthesis and, 560

Biphenyls, 2, 153, 154

Biradical, 157

Blakeslea trispora, 535

Blindness, vitamin A deficiency and, 122

Blood clotting, calcium and, 116

Bordetella bronchiseptica, 391

Bottleneck, biosynthetic flux, 554, 557

Bovine coronavirus (BCV), 390

Brain damage, greatest single cause of preventable, 99

Brazzein, 336, 337

Breast cancer chemoprevention, phytochemicals and, 253–283

background, 254–255

molecular pathways of breast cancer chemoprevention, 258–269

arachidonic acid metabolism, 266–269

CYP1A1 and aryl hydrocarbon receptor, 261–262

estrogen signaling, 259–261

growth factor signaling, 262–266

p53, 258–259

strategies for phytochemical enrichment for chemoprevention, 269–272

clonal screening of phytochemicals from heterogeneous genetic background, 270

consistency of phytochemicals, 269

liquid fermentation, 272

solid-state bioprocessing, 271

stress and elicitor induced sprouting, 270–271

susceptibility and chemoprevention, 255–257

Brewer's yeast, 319

Bullera singularis, β-galactosidase from, 488

Butyrate

nucleic acid metabolism and, 407

starch fermentation and, 403

C

Cabernet Sauvignon, 314

Caffeic acid, 3, 261

Calcium

absorption, lactulose and, 415

cellular homeostasis and, 168

homeostasis, impairment of, 170

osteoporosis and, 116

Campylobacter species, acute inflammatory reactions and, 404

Cancer, *see also specific types*

carotenoid-rich foods and, 119

cell(s)

butyric acid and, 418

estrogen stimulated growth of, 417

proliferation, genistein and, 135

chemoprevention, soybean isoflavonoids and, 135

flavonoids and, 156

FOS interactions with, 411

predictive factor for, 98

risk, soy intake and, 269

role for phenolics in, 10

ROS and, 2, 161

vitamin C and, 125–126

Candida

species, 164

utilis, 336

Canola oil, 85

Carbohydrate(s), 513

biotechnology, 492

metabolic pathway, role of enzymes in, 352

metabolism, potato tuber, 347

zinc and, 116

Carboxylic acid residues, 480

Carcinogen formation, fecal enzymes and, 460
Cardiovascular diseases (CVDs), 2, 4, 125, 163, 288
α-Carotene, 119
β-Carotene, 119
ζ-Carotene desaturase (ZDS), 120
Carotenogenesis, precursor supply for, 536
Carotenoid(s)
    application of in food industry, 117
    biosynthesis
        bottleneck for, 536
        pathway, 121
    composition of, 533
    as food colorants, 119
    groups of, 120
    metabolic engineering, focus of research on, 120
    oxidation linked diseases and, 161
Carotenoids, production of by gene combination in *Escherichia coli*, 533–543
    achievements and perspectives, 538–542
    Escherichia coli as carotenoid production system, 535–536
    principles of carotenoid biosynthesis, 536–537
    properties, commercial aspects, and biological function of carotenoids, 533–535
κ-Carrageenan, 598
Castro's reagent, 332
Catalase (CAT), 3, 160, 211, 230, 240, 617
Cataract formation, vitamin C and, 125–126
Catechin, 3
CBH, *see* Cello bio hydrolase
CCP, *see* Critical control point
Celebrity fruits
    antisense, 366
    transgenic, 365
Cello bio hydrolase (CBH), 481
Cell redox status, 159
Cellular hyperacidification, 302
Cellulomonas fimi, β-mannosidase from, 490
Ceratonia siliqua, 597
Cereal
    grain(s)
        composition of, 420
        main component of, 52
        seeds, protein content of, 26
Cerebrovascular disease, ROS and, 161
CF, *see* Cystic fibrosis
Chambourcin, 315
Chardonnay, 313
CHD, *see* Coronary heart disease
Cheese(es)
    microbe laced, 172
    species employed in fermentation of, 587
Chemiluminiscent dyes, common, 235
Chemopreventive agents, plant derived compounds as, 254, 257
Chicken antibodies, 385
Chinese herbal medicine, 298
Chloroplast to chromoplast transition, 368
Chorismate mutase, genes encoding, 565

Chrome mosaic virus (CMV), 317
Chromophore oxidation, chemiluminescence assays based on, 242
Chronic portal systemic encephalopathy, 415
Citric acid cycle, 553, 590
CLA, *see* Conjugated linoleic acid
Clark-type electrode, 241
Clonal profile antimicrobial efficacy, 292
Clonal selection, wine cultivars and, 317
Clostridium
    difficile, 457
    perfringens, 404, 420
CMV, *see* Chrome mosaic virus
Cobra venom factor, 188, 270
Code of Federal Regulations, statement on nonnutritive sweeteners, 327
Cold stress, phenylpropanoid compounds and, 11
Cold tolerant yeasts, 320
Coleus blumei, 190
Colitis, probiotics and, 458
Collagen biosynthesis, L-AA and, 125
Colon cancer
    bacterial links to, 406
    lactulose and, 415
    prevention, GOS and, 413
Colonization resistance, 440
Colorectal cancer, LAB and, 459
Commercial fortification of foods, 109
Common aromatic biosynthetic pathway, 213
Concanavalin A, 449
β-Conglycinin, 27
    amino acid identity, 29
    emulsifying ability, 38
    structural features of, 34
    three dimensional structure of, 30
    X-ray crystallography of, 29
Conjugated linoleic acid (CLA), 86, 91
Constipation, lactulose and, 415
Corn starch, liquefaction of to glucose syrups, 524
Coronary heart disease (CHD), 408
    carotenoid-rich foods and, 119
    flavonoids and, 156
Coronavirus, 390
Corynebacterium
    ammoniagenes, 546
    glutamicum, 503, 522, 546, 549
        anaplerotic pathways of, 553
        ATPase and, 573
        cell wall structure of, 561, 562
        glutamate excretion and, 549
        lysine production by, 572
        metabolic flux distribution in, 563
        OHDC activity of, 560
        production of lysine by, 566
        special property of, 574
        stress response of, 573
COX, *see* Cyclooxygenases
COX-2
    inhibitors, 16

overexpression, 267
resveratrol and, 322
suppression, genistein and, 135
upregulation, 257
Crabtree effect, 524
Cranberries, functional phytochemicals from,
        151–185
    chemical nature and biosynthesis of phenolic
            phytochemicals, 152–171
        biological function of phenolic
                phytochemicals, 162–171
        oxidative stress mediated pathogenesis, 157–
                162
        phenolic phytochemicals from berries, 153–
                157
    cranberry synergies with functional
            phytochemicals and other fruit extracts,
            173–174
    innovative strategies to enrich fruits with phenolic
            antioxidants, 171–172
    phenolic phytochemicals, 152
    solid-state bioprocessing, 172–173
Cranberry
    extract(s)
        α-amylase inhibition activity of, 174
        anticancer properties of, 165
    pomace, 156
    powder, antimutagenic functionality of, 174
    SSB, 616
Critical control point (CCP), 18, 218, 221, 302
    antimicrobial action, 295
    NADPH and, 201
    PLPPP and, 192
    rosmarinic acid synthesis, 187
    Systems Biology, 613
Crohn's disease, 404, 441, 457
Crop improvement, means of, 346
Cryptosporidium parvum, acute inflammatory
        reactions and, 404
β-Cryptoxanthin, 119, 122
Cupin superfamily, 33
Curcuma longa, cancer chemopreventive properties
        for, 10
Curcumin, 262
Curdlan, 600
    food applications of, 601
    permission to use, 604
    polysaccharide suspension of, 601
CVDs, *see* Cardiovascular diseases
Cyanide resistant respiration, potato tubers and, 359
Cyclooxygenases (COX), 256
Cysteine
    exporter, 569
    production of, 569
Cystic fibrosis (CF), 389
Cytochrome P450 monooxygenases, 191
Cytokine(s)
    description of, 439
    expression, yogurt and, 453

intestinal permeability and, 450
production
    immune mechanistic effects and, 449
    LAB and, 452
proinflammatory, 450, 452
regulation, LAB adjuvant effect and, 454
TGF-β, 265
Cytosolic phenolic radical proton linked redox cycle,
        293
Cytotoxic oxygen species, 158

D

DA, *see* Dopamine
DAHP, *see* 3-Deoxy-d-arabino-heptulosonate 7-
        phosphate synthase
Dairy products
    flavor compound in, 515
    intolerance to, 109
DAN, *see* Diaminonapthalene
DBI, *see* Double bond index
DCA, *see* Deoxycholic acid
DCIS, *see* Ductal carcinoma *in situ*
Debranching enzymes, 58, 63
De Chaunac, 316
Decoy oligosaccharides, 474
Decoy sugars, 164
Degree of polymerization (DP), 66, 409
Dehydroascorbate, 125
Delaware, 316
Dementia, ROS and, 161
3-Deoxy-d-arabino-heptulosonate 7-phosphate
        synthase (DAHP), 565
Deoxycholic acid (DCA), 416
Dextran, fermentability of, 414
Dextrins, beta-limit, 72
DHA, *see* Docosohexaenoic acid
Diabetes
    F2-isoprostanes and, 237
    management, phytochemicals for, 18
    ROS and, 2, 161
    type II, 4, 163
Diacetyl, 515
Diacylglycerol, deacylated, 364
Diaminonapthalene (DAN), 236
D,L-Diaminopimelate, 566
Diarrhea
    acute infectious, 456
    acute inflammatory reactions and, 404
    antibiotic associated, 405, 457
    E. coli and, 386
    L. acidophilus and, 403
    probiotics and, 456
    rotavirus, IgA production in patients with, 460
    traveler's, 457
Dienoic conjugated fatty acids, 92
Dietary antioxidants, 161
Differential Scanning Calorimetry (DSC), 44, 65, 68

Digestive processes, 108
Dihydrodipicolinate synthase, 566
Dimethylallyl pyrophosphate (DMAPP), 536
7,12-Dimethylbenz[a]anthracene (DMBA), 165, 262
Dioscoreophyllum cumminsii, 335
Dipalmitoylphosphatidylcholine liposomes, 367
1,1-Diphenyl-2-picrylhydrazyl (DPPH), 15, 271
Diphenylpropanes, 153
Directed evolution
    experiment, most important step of, 494
    first law of, 492
    mutant glycosidases created by, 491
    techniques use for, 492
Direct glycosylation, 480
Disease
    prevention, antioxidants and, 212
    related stress, 98
DisPE, *see* Disproportionating enzyme
Disproportionating enzyme (DisPE), 56
Distiller's yeast, 319
DMAPP, *see* Dimethylallyl pyrophosphate
DMBA, *see* 7,12-Dimethylbenz[a]anthracene
DNA
    array techniques
        fermentation and, 571, 573
        genome availability and, 570
    damage
        antioxidants and, 257
        induced carcinogenesis, 174
    maintenance reactions, micronutrients and, 98
    oxidation, 237, 238
    replication, zinc and, 116
    shuffling, 491
    topoisomerases, inhibitor of, 166
Docosohexaenoic acid (DHA), 89
Docosopentaenoic acid (DPA), 91
Domain classes, reclassification of enzymes into, 60
Donor:acceptor sugar ratio, 482
L-DOPA and phenolic antioxidants, enhancement of
    in fava bean, 209–227
    biosynthesis of phenolic metabolites and L-
        DOPA, 213–215
    implications, 222–223
    linking L-DOPA synthesis to pentose phosphate
        and phenylpropanoid pathways, 218–
        220
    Parkinsonian syndrome, 215–218
        L-DOPA from fava bean, 217–218
        L-DOPA from natural sources, 217
    plant phenolics in human health and as
        antioxidants, 211–212
    recent progress on elicitation linked bioprocessing
        to enhance L-DOPA and phenolics in
        fava bean sprouts, 220–222
    recent progress on solid-state bioprocessing of
        fava bean to enhance L-DOPA and
        phenolics using food grade fungal
        systems, 222
    role of phenolic secondary metabolites, 209–210

Dopamine (DA), 215
    pathway, 216
    release, 215
    renal production of, 615
Double bond index (DBI), 361
DP, *see* Degree of polymerization
DPA, *see* Docosopentaenoic acid
DPPH, *see* 1,1-Diphenyl-2-picrylhydrazyl
Drugs, oligosaccharides as, 474
DSC, *see* Differential Scanning Calorimetry
Ductal carcinoma *in situ* (DCIS), 254
Duodenitis, ecological therapy for, 403

E

ECL, *see* External chain length
ECM, *see* Extracellular matrix
EGCG, *see* Epigallocatechin-3-gallate
EGF, *see* Epidermal growth factor
EGFR, *see* Epidermal growth factor receptor
Egg yolk antibody farming for passive
        immunotherapy, 381–400
    antibody stability, 391–393
        degradation and stability of IgY, 392–393
        physiology of gastrointestinal tract, 391–392
    avian egg antibodies, 382–386
        advantages of IgY, 385–386
        avian immune system, 382–383
        molecular properties of IgY, 383–384
    biotechnology of avian egg antibodies, 393
    passive immunization with IgY, 386–391
        coronavirus, 390–391
        Escherichia coli, 386–388
        fish pathogens, 391
        Helicobacter pylori, 390
        others, 391
        pseudomonas aeruginosa, 389–390
        rotavirus, 386
        Salmonella species, 388–389
        Streptococcus mutans, 390
Eicosanoids, immune response, 89
Eicosopentaenoic acid (EPA), 89, 90, 91
Electrolyte leakage, potato tubers and, 347
Electron sink products, 404
Electron spin resonance spectroscopy (ESR), 232
Electron transport chain (ETC), 159, 170
Eleostearic acid, anticarcinogenic effects of, 92
ELISA, *see* Enzyme-linked immunosorbent assay
Ellagic acid, 10, 165, 169, 259
Ellagitannins, 165
Emulsifying ability, soybean proteins, 38
Emulsion-containing systems, 596
Endoglycosynthases, one pot synthesis of
        hexasaccharide substrates, 490
Entamoeba histolytica, acute inflammatory reactions
        and, 404
Enterococcus faecium, 457

Enterocytes
    Bifidobacteria and, 450
    cellular turnover of injured, 450
    GALT and, 439
Enteropathogenic *E. coli* (EPEC), 388
Enterotoxigenic *Escherichia coli* (ETEC), 386, 388
Environmental stresses, phenolic metabolites and,
        152
Enzyme(s)
    complex, polyketide synthase, 90
    immobilization, 488
    -linked immunosorbent assay (ELISA), 449
    reclassification of, 60
    regioselectivity of, 484
    retaining, 479
    synthesis
        organic solvents and, 485
        pH of reaction in, 487
Eosinophilia myalgia syndrome, 565
EPA, *see* Eicosopentaenoic acid
EPEC, *see* Enteropathogenic *E. coli*
Epidermal growth factor (EGF), 255
Epidermal growth factor receptor (EGFR), 256
Epigallocatechin gallate, 298
Epigallocatechin-3-gallate (EGCG), 263, 266
EPS, *see* Exopolysaccharides
Equal™, 331
Equal Measure™, 311
ER, *see* Estrogen receptor
ERK, *see* Extracellular signal regulated kinase
Erwinia uredovora, 122, 123
Erythrose-4-phosphate, 564
Escherichia coli, 172, 337, 550, *see also* Carotenoids,
        production of by gene combination in
        Escherichia coli
    acute inflammatory reactions and, 404
    adhesion of to enterocytes, 450
    amino acids and, 546
    Bifidobacterium spp. effects on, 404
    cranberry phenolics and, 163
    cranberry pomace and, 165
    cysteine export in, 573
    cysteine production in, 569
    -derived lipopolysaccharides, 452
    enteropathogenic, 388
    enterotoxigenic, 386, 388, 403
    formula-fed infants and, 445
    infections, prevention of, 167
    inhibition of growth of, 114
    pfkA gene, expression of in potato tubers, 357
    phagocytosis measurement and, 449
    phenylalanine production in, 564
    soybean and, 29
    tannins and, 286
    threonine production using, 568
    tryptophan degradation in, 565
ESR, *see* Electron spin resonance spectroscopy
Esterified fatty acids, 92

Estrogen
    biosynthetic pathway, 255
    phytochemicals, 260
    receptor (ER), 134, 259
        isoflavones and, 417
        signaling, 259
    regulation, 259
    signaling, 259, 260
ETC, *see* Electron transport chain
ETEC, *see* Enterotoxigenic *Escherichia coli*
Ethylene biosynthesis, tomato, 363
Ethyl methanesulphonate, 61
Eubacterium biforme, 420
Exopolysaccharides (EPS), 419, 514, 584
    common sugars in, 585
    definition of, 604
    Lactobacillus delbrueckii bulgaricus, 587
    Lactobacillus helveticus, 588
    Lactobacillus sake, 586
    Pediococcus damnosus, 588
    physical attributes of, 594
    production, process of, 591
    Propionobacterium acidi-propiionici, 588
    Streptococcus thermophilus, 588
External chain length (ECL), 66
Extracellular matrix (ECM), 168
Extracellular signal regulated kinase (ERK), 268
Extracted starch, nonfood industrial uses of, 52

F

FAO, *see* Food and Agricultural Organization
Fast atom bombardment spectroscopy, 592
Fat-rich foods, oxidative rancidity of, 125
Fat-soluble vitamins, chemical structures of, 119
Fatty acid(s)
    composition, potato membrane, 361
    esterified, 92
    ratio of omega-6 to omega-3, 85
    synthesis, biotin and, 560
Fava bean, *see* L-DOPA and phenolic antioxidants,
        enhancement of in fava bean
FBPase, *see* Fructose-1,6-bisphosphatase
Fed-batch fermentation, 528
Fermentation(s)
    aerobic culture, 526
    batch, 528
    description of, 611–612
    end products, 524
    genetic engineering of yeast for, 319
    microorganisms and, 502
    processes, examples, 172
    solid-state, 137, 611
    solid substrate, 530
    spore suspensions used for, 523
Fermented food products, use of microorganisms for
        production of, 521
Fermentor, most common, 525

Ferric reducing ability of plasma (FRAP), 234
Ferritin, 113
Ferulic acid, 3
FGF, *see* Fibroblast growth factor
Fibroblast growth factor (FGF), 262
Fish
    oils, 86
    pathogens, 391
F2-isoprostanes, 237
Flavonoid(s), 10, 153
    acetate–malonate pathway and, 211
    as AhR agonists, 261
    families of, 2
    UV irradiation and, 210
    wine components, 322
Flavonols, 152
Flavor enhancers, use of amino acids as, 546
Fluorescent dyes, common, 235
Folate, 513
Food(s)
    additive(s)
        consumer concern over use of, 584
        legislative acceptability of, 603
        use of dextran as, 589
    allergy, atopic eczema and, 458
    antioxidant rich, 212
    biotechnology, most important amino acid in, 559
    commercial fortification of, 109
    diversification, 109
    effectiveness, validation of, 7
    functional, 502
        health care and, 6
        oligosaccharides as, 474
        phytochemicals and, 4
        sales of, 9
        ultimate success of, 6
        wine as, 322
    genetically modified, 136
    grade oligosaccharides, 475
    ingredients, *see* Microbial production of food
            ingredients, technologies used for
    microbial fermentation of, 172
    pharmaceutical, 502
    pigment, carotenoids as, 117
    poisoning, *Salmonella*, 388
    polysaccharides added to, 596
    potential medicinal effects of, 4
    preservation, antioxidants and, 9–10
    preservatives, *Lamiaceae* family, 288
    proteins
        bioactive peptides isolated from, 42
        properties of, 26
    refined, exposure of to light, 124
    synergistic supplementation of, 174
    U.S. regulation of, 8
    waste remediation systems, 619, *see also* Solid-
            state bioprocessing for functional food
            ingredients and food waste remediation
Food Advisory Committee, 603

Food and Agricultural Organization (FAO), 604
Foods for Specified Health Use (FOSHU), 4, 408,
        475
Fortification, commercial, 109
FOS, *see* Fructooligosaccharides
FOSHU, *see* Foods for Specified Health Use
Fourier transform mid infrared (FT-MIR)
        spectroscopy, 526
FRAP, *see* Ferric reducing ability of plasma
Free radical(s), 197–198
    neutralization, antioxidant assays based on, 232
    oxidative stress and, 162
Freeze drying, 523
French–American hybrid wines, 315
French paradox, 4, 163
Frozen foods
    reduced ice crystal formation in, 596
    use of polysaccharides in, 584
Fructooligosaccharides (FOS), 402, 408
    bacteria fermenting, 410
    Bifidobacterium and, 409
    efficiency of, 409
    fermentability of, 409
    interactions with cancer, 411
    interactions with lipid metabolism, 411
    interactions with mineral absorption, 411
Fructose-1,6-bisphosphatase (FBPase), 356, 357
Fruit(s)
    beverages, polyphenolic compounds in, 153–156
    Celebrity, 365, 366
    chilling injury, 364
    juice blends, 174
    most abundant phenolic compound in, 152
    ripening
        cell wall degradation during, 360
        stages, LOX and, 367
        tomato, 363
    softening, cell wall metabolism and, 369
    wastes, solid-state bioprocessing and, 172
Fruits and vegetables, improvement of nutritional
            quality and shelf life of, 345–380
    potatoes, 346–363
        compartmentation and stress induced
                membrane changes, 359–361
        factors affecting accumulation of reducing
                sugars in potatoes, 347
        free radicals and antioxidant enzymes, 362–
                363
        glycolysis, 356–358
        low temperature sweetening in potatoes,
                347–348
        metabolism of starch in tuber, 348–352
        mitochondrial respiration, 358–359
        oxidative pentose phosphate pathway, 358
        starch–sugar balance, 352–353
        sucrose metabolism, 353–356
    tomato, 363–369
        cell wall metabolism and fruit softening, 369
        lipoxygenase, 366–369

phospholipase D gene family, 364–366

role of membrane in shelf life, 363–364

Frying oils, 86

FT-MIR spectroscopy, *see* Fourier transform mid infrared spectroscopy

Functional food(s), 502

    efficacy, validation of, 7

    health care and, 5

    oligosaccharides as, 474

    phytochemicals and, 4

    sales of, 9

    ultimate success of, 6

    wine as, 322

Fungal bioconversion system, 222

Fungal detoxification, 173

Fungus

    bioprocessing of plant based foods by dietary, 137

    food grade, 175

    hydrolases produced by, 172

Furaneol, 601

G

Galactooligosaccharides (GOS), 402, 408

    bacteria fermenting, 412

    fermentation in, physiological effects of, 413

    hydrogen excretion, 422

    SCFA concentrations with, 412

D-Galacturonic acid, 125, 585

Gallic acid, 3

GALT, *see* Gut associated lymphoid tissue

Gamay Noir à jus Blanc, 314

Garcinia mangostana, seed oil of, 88

Gas chromatography (GC), 571

Gastric ulcers, *Helicobacter pylori* and, 390

Gastritis, ecological therapy for, 403

Gastrointestinal disease, *Ocimum sanctum* and, 211

Gastrointestinal tract, physiology of, 391

GBPs, *see* Glucan-binding proteins

GBSS, *see* Granule bound starch synthase

GC, *see* Gas chromatography

GDCA, *see* Glycodeoxycholic acid

GDH, *see* Glutamate dehydrogenase

Gelatinization, 64

Gellan, 597

    food applications, 600

    gelling and melting temperatures, 599

    physical properties of, 598

    polysaccharide production and, 585

    structure of, 598, 599

    uses of, 598, 600

Generally Regarded As Safe (GRAS), 173, 327, 503, 589, 598

Genetically modified (GM) foods, 136

Genetically modified organisms (GMOs), 318

Genistein, 135, 139, 260

Genome breeding, 570

Gentiooligosaccharides (GeOS), 420

GeOS, *see* Gentiooligosaccharides

Geranylgeranyl pyrophosphate (GGPP), 120

Gewürztraminer, 313

GFLV, *see* Grapevine fanleaf virus

G-G, *see* Glycosyl glucose

GGPP, *see* Geranylgeranyl pyrophosphate

Giardia lamblia, acute inflammatory reactions and, 404

Gibberellins biosynthetic pathway, 121

Gleditsia sinensis, 266

Glucan-binding proteins (GBPs), 390

Gluconobacter oxydans, 513

Glucooligosaccharides, 402

Glucose

    based oligosaccharides, 421

    corn syrup, 524

    glycoside formation and, 171

Glucose-1-phosphate, conversion of to ADPglucose, 349

Glucose-6-phosphate dehydrogenase, 218, 358

β-Glucuronidase reporter gene, 369

Glutamate, 507, 559

    dehydrogenase (GDH), 515

    efflux, plasma membrane and, 561

    excretion

        Corynebacterium glutamicum, 549

        membrane triggering, 561

    overproduction, 560

    production, background, 560

Glutaredoxins (GRx), 159

Glutathione (GSH), 159, 166, 211, 230

    disulfide (GSSG), 160

    metabolism, 239

    peroxidase, 240

    S-transferase (GST), 3, 166, 230, 240

Glyceraldehyde-3-phosphate dehydrogenase enzymes, 571

Glycerolipids, potato, 359

Glycine, production of, 569

Glycine max, 10

Glycinin, 27

    amino acid identity, 31

    crystallization of purified, 29

    disulfide bonds, 40

    emulsifying ability, 39

    improvement in physicochemical functions of, 39

    processes of, 28

    restriction enzyme sites, 40

    structural features of, 35

Glycocalyx, 440

Glycodeoxycholic acid (GDCA), 416

Glycolysis

    ATP-PFK and, 357

    potato tuber, 356

Glycones, 135

Glycosidases
    application of directed evolution techniques to,
        492
    groups, 479
    mutant, 489, 491
    natural function of, 479
    recombinant, 489
    retaining, 479, 480
Glycoside(s), 2, 153
    formation, glucose and, 171
    hydrolysis reaction, reversion of, 480
    plant phenolics as, 210
Glycosyl glucose (G-G), 318
Glycosyltransferases, 476, 479
Glycosynthase efficiency, 490
GM foods, *see* Genetically modified foods
GMOs, *see* Genetically modified organisms
GOS, *see* Galactooligosaccharides
Gosyuyu, 298
GPX, *see* Guaiacol peroxidase
Graciano, 315
Gram-positive bacteria, 445
Granule bound starch synthase (GBSS), 56, 57, 60
Grape(s)
    growing regions, world, 312
    reduced viral infection in, 318
    seed extracts
        α-amylase inhibition activity of, 174
        antioxidant activity of, 156
Grapevine fanleaf virus (GFLV), 316
GRAS, *see* Generally Regarded As Safe
Green revolution, plant breeding efforts during, 109
Green tea, 266
Growth factor signaling, 262, 256
Growth rings, 53
GRx, *see* Glutaredoxins
GSH, *see* Glutathione
GSSG, *see* Glutathione disulfide
GSSH, *see* Oxidized glutathione
GST, *see* Glutathione S-transferase
Guaiacol peroxidase (GPX), 14, 214, 270
GuLO, *see* L-Gulono-γ-lactone oxidase
L-Gulono-γ-lactone oxidase (GuLO), 126
Gums, blending of gellan with, 600
Gut associated lymphoid tissue (GALT), 437
    immune homeostasis of, 453
    immune responses of, 439
    immunity vs. tolerance, 440
    nonspecific response mechanisms, 440
    organization of, 437
    schematic depiction of, 438
Gut flora, *see also* Human gut microflora in health
        and disease
    composition of, 436
    influence of on immune function, 444
    interactions between host immune system and,
        446
Gut mucosa, immune system and, 437

H

Haematococcus pluvialis, 534
HDL, *see* High density lipoprotein
Heat shock proteins, 210
Helicobacter pylori, 164, 172
    control of by phenolic antimicrobials, 296
    cranberry pomace and, 165
    current treatment strategies, 297
    dietary soybean approach for control of, 298
    enterocytes and, 452
    gastric ulcers and, 390
    gastritis and, 457
    infection(s)
        prevention of, 167
        vitamin C and, 126
    Lactobacillus deficiency and, 403
    pathogenesis mediated by, 161, 241, 297
    peptic ulcer caused by, 175
    RA and, 189
    strategies for control of, 298
    ulcer associated, 271, 616
    urease, soybean extracts and, 301
Heme iron, 113
Hemoglobin depletion bioassay, 114
Hepatocyte growth factor (HGF), 262
Heptasaccharide repeating unit, 586
Herbal medicines, key market for, 9
Herbs
    phytochemical consistency and, 287
    strategies for isolating clonal dietary, 15
Heterodisaccharides, production of, 481
Heteropolysaccharides, phosphate groups in, 585
HETEs, *see* Hydroxyeicosatetraenonic acids
Hexose accumulation tuber, 356
HGF, *see* Hepatocyte growth factor
Hidden hunger, 98, 109
High amylose starch, 70
High density lipoprotein (HDL), 408
High oleic soybean oil (HOS), 87
High pressure liquid chromatography (HPLC), 236,
        525
Hippuric acid, 164
Histidine, 545
HIV-infected patients, selenium and, 523
Hizukuri cluster model, 56
HMEC, *see* Human mammary epithelial cells
Hormone replacement therapy (HRT), 135
HOS, *see* High oleic soybean oil
HPLC, *see* High pressure liquid chromatography
HRT, *see* Hormone replacement therapy
HRV, *see* Human rotavirus
Human gut microflora in health and disease, 401–434
    fermentation properties, 402–404
    gut microflora, 404–408
        acute infections and inflammatory reactions,
            404–405
        antibiotic associated diarrhea, 405

cholesterol lowering effect and lipid metabolism, 408
colon cancer, 406–408
immune stimulation, 405–406
lactose intolerance, 406
structure to function relationships in other oligosaccharides, 419–421
gentiooligosaccharides, 420–421
glucose based oligosaccharides, 421
lactitol, 419
laevan type prebiotics, 419–420
structure to function relationships in prebiotic oligosaccharides, 408–419
fructooligosaccharides, 409–412
galactooligosaccharides, 412–413
isomaltooligosaccharides, 413–414
lactosucrose, 418
lactulose, 414–416
soya products, 416–418
xylooligosaccharides, 418–419
Human mammary epithelial cells (HMEC), 267
Human rotavirus (HRV), 386
Humicola insolens cellulase, mutant of, 490
Hydrocolloids, 584
Hydrogenated oils, 85, 87
Hydroxybenzene, 152
p-Hydroxybenzoic acid, 164
Hydroxycinnamic acid, 294
Hydroxyeicosatetraenonic acids (HETEs), 268
Hydroxypropylation, 68
Hygiene hypothesis, 458
Hypericum perforatum, 269
Hypertension management, 16
Hypervariable region, 32
Hypoglycemia, 334
Hyptis verticillata, 188

I

IBD, *see* Inflammatory bowel disease
IBDV, *see* Infectious bursal disease virus
Ideal sweetener, pursuit for, 339
IFN-γ, 450
NK cells and, 451–452
T cell production of, 452
IFS, *see* Isoflavone synthase
IgA production
intestinal microbes and, 446
LAB enhancement of, 455
IGF, *see* Insulin-like growth factor
IGFBP, *see* Insulin-like growth factor binding protein
IGFR, *see* Insulin-like growth factor receptor
IgY
advantages of, 385
characteristics of, 385
degradation of, 392
passive immunization with, 386
physicochemical properties of, 383

-secreting hybridomas, 393
stability of, 392
structure of, 383
IL-8, *see* Interleukin-8
Immobilized cell bioreactor, 529
Immune exclusion, 444
Immune function, influence of gut flora on, 444
Immune homeostasis, postnatal establishment of, 447
Immune response(s)
characteristics of classical, 443
eicosanoids and, 89
Immune system
activation, lactulose and, 414
avian, 382
gut flora and, 446
methods for studying effects of probiotic bacteria on, 448
neonate, 447
probiotics and, 406
regulation, LAB and, 461
Immunoglobulin classes, 382
IMO, *see* Isomaltooligosaccharides
Indica rice line, 113
Indole-3-carbinol, 262
Inducible nitric oxide synthase (iNOS), 173
Industrial amino acid fermentation, 558
Industrial scale fermentors, 524
Infant formula, TAG-enriched, 89
Infectious bursal disease virus (IBDV), 391
Inflammatory bowel disease (IBD), 404, 457
Inflammatory diseases, phenolic antimicrobials and, 288
Inorganic phosphates, potato tubers and, 352
iNOS, *see* Inducible nitric oxide synthase
Inositol triphosphate kinase pathway, 263
Insulin-like growth factor (IGF), 256
binding protein (IGFBP), 263
receptor (IGFR), 263
γ-Interferon, yogurt administration and, 405
Interleukin-8 (IL-8), 440, 452
Intermediate starch, 54
component, characteristics of, 55
definition of, 55
International Union for pure and applied chemists (IUPAC), 152
Intestinal IgA antibody responses, LAB and, 405
Intestinal infections, effects of LAB on, 456
Inulin, 402
starch and, 73
-type fructans, 411
Iodine deficiency, 99
IPP, *see* Isopentenyl pyrophosphate
Iron
-binding glycoprotein, 114
compounds, water-soluble, 109
deficiency, 112
heme, 113
losses, 112
nonheme, 113

regulatory mechanisms, 112
store protein, natural, 113
supplementation, 109
ISA, *see* Isoamylase
Isoamylase (ISA), 56, 58
Isocitrate dehydrogenase, 567
Isoflavone(s)
estrogen receptor and, 417
synthase (IFS), 136
Isoflavonoids, 2, 10, 269
biological activities of, 135
health-promoting effects of, 135
metabolism of, 137
Isoleucine, 545, 551
Isomaltooligosaccharides (IMO), 402, 413, 421
Isopentenyl pyrophosphate (IPP), 536
IUPAC, *see* International Union for pure and applied chemists

J

Jack bean alpha mannosidase, 487
Japanese cooking, traditional taste enhancing material in, 546
JNK, *see* c-Jun N-terminal kinase
c-Jun N-terminal kinase (JNK), 258

K

Kaempferol, 259
Katahdin tubers, 349
Kefir cultures, 617
Kelp, 546
Ketones, reaction of cysteine with, 569
Killer yeasts, 320
King Edward tubers, sugar content in, 348
Klebsiella
fragilis β-galactosidase, 484
oxytoca, 405, 457
Kloeckera apiculata, 319
Koji production, 612
Krebs/TCA cycle, 11, 12–13

L

LAB, *see* Lactic acid bacteria
Labeling procedures, attempted standardization of, 604
β-Lactam antibiotics, 561
Lactate dehydrogenase (LDH), 508
Lactic acid bacteria (LAB), 295–296, 436, 584, 620
adjuvant effect, 454
amino acid production and, 508
gene clusters in, 602
health-promoting effects of EPS from, 585

hydrogen peroxide produced by, 406
immune mechanistic modulatory effects of, 448
intestinal IgA antibody responses induced by, 405
lactose derived oligosaccharides and, 412
metabolic engineering and, 503
phagocytic capacity of, 451
PLPPP strategies, 617
polysaccharides and, 584
starter cultures obtained with, 516
therapeutic properties of, 455
Lactic acid bacteria, immunomodulating effects of, 435–471
effects of LAB on diseases, 455–461
antiallergy properties of lab, 458–459
antitumor effect, 459–460
impact of viability and dose on effects of LAB, 460–461
inflammatory bowel disease, 457–458
intestinal infections, 456–457
future perspectives, 461
gut mucosa, 437–444
immune responses of GALT and peripheral immune system, 439–444
organization of GALT, 437–439
immunomodulatory mechanisms, 447–455
effects on cytokine production, 452–453
effects on lymphocyte proliferation, 453–454
effects on NK cell activity, 451–452
effects on nonimmunologic mucosa barrier functions, 450
effects on phagocytic cells, 450–451
effects on specific immune responses, 454–454
enhancement of IgA production, 455
interactions between LAB and intestinal epithelial cells, 452
methods used for assessing immunomodulatory effect of LAB, 448–449
influence of gut flora on immune function, 444–447
gut flora, 444–445
interactions between gut flora and host immune system, 446–447
Lactitol, 402, 419
Lactobacilli, 436, 445
Lactobacillus
acidophilus, 407, 452, 454
antimicrobial activities of, 450
bulgaricus, 511
casei fermentation, 528, 529
colon cancer risk and, 459
deficiency, 403
delbrueckii bulgaricus
EPS structure of, 587
polysaccharide-producing strains of, 585
helveticus, EPS of, 588
plantarum, 409, 516
rhamnosus, 409, 453, 454, 456

GOS and, 412
        pentasaccharide repeat unit of, 586
        treatment of atopic eczema with, 459
    sake, triply branched repeat unit of EPS of, 586
    spp., neutral trisaccharide repeat unit from, 587,
        588
Lactococcus lactis, 602
Lactoferrin, 114
Lactose intolerance, 406
Lactosucrose, 418
Lactulose, 408
    antiendotoxin effect of, 414
    colon cancer and, 415
    constipation and, 415
    formation of, 414
    liver cirrhosis and, 415
    medical applications of, 414
    mineral absorption and, 415
Laevan type prebiotics, 419
Latent TGF binding proteins (LTBPs), 265
LBG, *see* Locust bean gum
LC-MS, *see* Liquid chromatography-mass
        spectrometry
LDH, *see* Lactate dehydrogenase
LDL, *see* Low density lipoprotein
Leaky membrane theory, LTS, 360
Legume(s)
    phenolic phytochemicals in, 2
    seeds, protein content of, 26
    sprouting, 270
    substrates, dietary fungal bioprocessing of, 271
Lentinus edodes, 173, 616
Letrozole, 260
Leucine, 545
Leuconostoc oenos, fermentation, 319
Leukotrienes (LTs), 268
Lignin, 214, 272
Linoleic acid, 85
Linolenic acids
    liberation of, 367
    tomato fruit, 368
Lipid(s)
    free radicals and, 162
    metabolism
        FOS interactions with, 411
        prebiotics and, 408
    oxidation, 124, 235
    peroxidation
        activated oxygen species and, 366
        NADPH dependent, 166
        potato membrane deterioration and, 362
    -soluble vitamins, good sources of, 99
    zinc and, 116
Lipoarabinogalactan, 561
Lipomannan, 561
Lipopolysaccharides (LPS), 389
Lipoxygenase (LOX), 256, 366
    activity, pathogen induced, 368
    chloroplast to chromoplast transition and, 368

    inhibitor, 367
    membrane deterioration and, 366, 370
    pathways, products of, 366
    regulation by genetic manipulation, 368
    role in plant membrane deterioration, 367
    tomato fruit ripening and, 368
Liquid chromatography-mass spectrometry (LC-MS),
        571
Listeria monocytogenes, 172, 615
    control of by oregano clonal extracts, 289
    cranberry pomace and, 165
    RA and, 189
    tannins and, 286
Live cultures, foods containing, 521
Liver
    carbon tetrachloride toxicity and, 163
    cirrhosis, lactulose and, 415
Locust bean gum (LBG), 597
Low amylose starch, 69
Low calorie sugars, 515
Low density lipoprotein (LDL), 136, 163, 211, 408
Low temperature sweetening (LTS), 347
    cause of, 360
    differential response of potato cultivars to, 350
    leaky membrane theory, 360
    regulation, ATP-PFK in, 357
    resistant cultivars, 358
    starch degradation during, 351
    sucrose synthesis and, 348
    susceptible cultivar, 361
LOX, *see* Lipoxygenase
LPS, *see* Lipopolysaccharides
LTBPs, *see* Latent TGF binding proteins
LTs, *see* Leukotrienes
LTS, *see* Low temperature sweetening
Lucigenin, 242
Lunasin, 26, 42
    antimitotic activity of, 45
    production of, 46
Lung cancer, flavonoids and, 156
Lutein, 119, 123, 534
Lycopene, 119
    cyclization of, 537
    principal dietary sources of, 121
β-Lycopene cyclase gene, 122
Lymph nodes, LAB and, 438
Lymphocyte proliferation, 449, 453
Lysine, 545, 551
    biosynthesis pathway, 567
    production of, 566
L-Lysine production, 504, 507
Lysophosphatidic acid acyltransferase, 88–89

M

Mabinlin, 339
Macular degeneration, carotenoid-rich foods and, 119

Maillard reaction, 346, 347, 484
Maize
    endosperm mutant, 62
    starch(es)
        high amylose, 71
        mutants, 59
Major histocompatibility molecules (MHC), 441
MALDI-TOF mass spectrometry, 567
Malic enzyme, 563
Malondialdehyde (MDA), 236
Maltodextrin based oligosaccharide (MDO), 422
Maltooligosaccharides, 57
Mammary gland development, 254
Manganese superoxide dismutase, 138
Mannooligosaccharides, 480
Mannose
    receptor (MR), 440
    sensitive fimbriae, 164
β-Mannosidase, 490
α-1,2-Mannosyltransferase, 478
β-Mannosyltransferase, 476
Marechal Foch, 315
Margarine, 87
Marigold varieties, pigmentation of, 119
Matrix metalloproteinase (MMP), 259
MCA, *see* Metabolic control analysis
MCP-1, *see* Monocyte chemoattractant protein-1
MDA, *see* Malondialdehyde
MDO, *see* Maltodextrin based oligosaccharide
ME, *see* Metabolic engineering
Meal supplements, 9
Melanin pigment production, 293
Melanoid polymers, 286
Membrane integrity, role of LOX in control of, 367
Mental retardation, greatest single cause of
        preventable, 99
Mentha spicata, 187
Merlot, 314
Metabolic control analysis (MCA), 357
Metabolic engineering (ME), 502
Metabolic Flux Analysis (MFA), 502
Metabolic harmony, 98
Metabolic homeostasis, 97
Metabolic imbalance, amino acids and, 549
Metabolome analysis, 571
Methionine, 545, 574
N-Methyl-N′-nitro-N-nitrosoguanidine (MNNG),
        174
Methylpentose, 585
MFA, *see* Metabolic Flux Analysis
MHC, *see* Major histocompatibility molecules
Microbe laced cheeses, 172
Microbial polysaccharides, biotechnology of,
        583–610
    exopolysaccharides as source of flavor
        components, 601–602
    future for microbial polysaccharides in food,
        602–603

    legislative acceptability of microbial
        polysaccharides, 603–605
    microbial polysaccharides incorporated as food
        additives, 589–601
        curdlan, 600–601
        gellan, 597–600
        xanthan, 589–597
    natural occurrence of microbial polysaccharides,
        584–589
        physical properties, 588
        production and synthesis, 588–589
        products of microbial food components,
        584–588
Microbial production of food ingredients,
        technologies used for, 521–532
    bioreactor monitoring systems and design, 525–
        528
    culture media and upstream components, 523–
        525
    fermentation types employed, 528–529
    future research, 530–531
    microorganism selection and development, 522–
        523
    novel bioreactor designs, 529–530
Microecosystem, SSB, 613
Micronutrient(s), 97, *see also* Minerals and
        vitamins, molecular biotechnology for
        nutraceutical enrichment
    antinutritional substances reducing, 110–111
    bioavailability of, 108
    deficiencies, antinutritional factors and, 112
    malnutrition
        approaches to reduce, 108
        treatment of symptoms of, 109
Microorganisms, food industry, 522
Microsporum species, 164
Microtom, PLD activity in, 366
Microwave radiation, 137
Milk
    products, fermented, 408, 585
    species employed in fermentation of, 587
Mineral(s)
    absorption
        FOS interactions with, 411
        lactulose and, 415
    cheap sources of, 112
    dietary allowances per day, 100–102
Minerals and vitamins, molecular biotechnology for
        nutraceutical enrichment, 97–132
    alleviating nutritional disorders by nutraceuticals,
        97–108
    micronutrient bioavailability and reduction of
        micronutrient malnutrition, 108–112
    minerals, 112–117
        calcium, 116
        iron, 112–116
        phytic acid, 116–117
        zinc, 116
    vitamins and nutraceuticals, 117–128

carotenoids as food pigment and provitamin
A, 117–123
vitamin C, 125–128
vitamin E, 124–125
Mitochondrial respiration, potato tuber, 358
MMP, *see* Matrix metalloproteinase
MNNG, *see* N-Methyl-N′-nitro-N-nitrosoguanidine
Momordica charantia, 92
Monellin, 335
Monoclonal antibody specificity, xanthan and, 592
Monocyte chemoattractant protein-1 (MCP-1), 440
Monooxygenase CYP1A1, 255, 261
Monosodium glutamate (MSG), 546, 559
Monostarch phosphorylation, 68
Montelukast, 268
Moraxella catarrhalis, 475
Moritella species, polyketide synthase and, 90
Mouth feel, 584
MPO, *see* Myeloperoxidase
MR, *see* Mannose receptor
MSG, *see* Monosodium glutamate
Mucin
glycoproteins, 402
up regulation, 450
Mucuna pruriens, 217
Muller-Thurgau variety, 313
Multiple pathway enzymes, modification of, 63
Muscat Blanc, 314
Muscle contraction, calcium and, 116
Mushroom(s)
growth substrate for, 172
production, 612
Mutant glycosidases, 489, 491
Myeloperoxidase (MPO), 158, 242
Myrcetin, 10

N

NADH
cellular needs for, 175
synthesis, deregulation of, 11
NADPH, 621
carbohydrates generating, 218
cellular demand for, 169
critical control point for generating, 218
cytosolic proton/hydride ion flux and, 199
demand for proline synthesis, 12
dependent enzymes, 555
equivalent: redox balance and, 566
L-lysine production and, 504
oxidase (NOX), 231, 242
pentose phosphate pathway and, 567
PLPPP and, 223
PPP and, 358
proline biosynthesis and, 139
recycling, 296

reoxidation of, 507
SOD and, 617
Naked antigens, 441
Natural killer (NK) cells, 407–408, 440
immune functions and, 449, 451
measurement of activity, 449
NBMA, *see* N-Nitrosobenzylmethylamine
Nebbiolo, 315
Neisseria meningitidis, 478
Neotame, 333, 334
Neurodegenerative diseases
ROS and, 161
vitamin C and, 125–126
Neurospora crassa, 537
Neutrophil attractant, cytokines and, 452
Nitric oxide (NO), 157, 158
o-Nitrophenyl-β-galactose (ONPG), 486
N-Nitrosobenzylmethylamine (NBMA), 165
n-Nitroso methyl urea (NMU), 260
NK cells, *see* Natural killer cells
NMR, *see* Nuclear magnetic resonance spectrometry
NMU, *see* n-Nitroso methyl urea
NO, *see* Nitric oxide
Nonenzymatic browning, potatoes and, 346
Nonheme iron, 113
Noninsulin-dependent diabetes mellitus, 123
Nonnutritive sweeteners, biotechnology of, 327–344
acesulfame-K, 329–30
aspartame, 331–333
brazzein, 336
mabinlin, 339
monellin, 335–336
neotame, 333
saccharin, 329
stevioside, 333–335
sucralose, 330–331
thaumatin, 336–339
Nonstarch polysaccharides, 402
Nonsteroidal antiinflammatory drugs (NSAIDs), 267
Nopal cactus, 123
Norchip
amyloplast membranes, 361
sucrose cycling for, 362
Nordihydroguaiaretic acid, 367
NOX, *see* NADPH oxidase
NSAIDs, *see* Nonsteroidal antiinflammatory drugs
NTP, *see* Nucleoside triphosphate
Nuclear magnetic resonance spectrometry (NMR),
558, 563
Nucleic acid(s)
free radicals and, 162
metabolism, butyrate and, 407
zinc and, 116
Nucleoside triphosphate (NTP), 477
Nutraceutical enrichment, *see* Minerals and
vitamins, molecular biotechnology for
nutraceutical enrichment
Nutrasweet™, 331, 522, 523

O

Obesity
    health problems related to, 26
    nonnutritive sweeteners and, 327
Ocimum sanctum, 187, 211
Octenylsuccination, 68
ODC, *see* Ornithine decarboxylase
OHDC, *see* Oxoglutarate dehydrogenase
Oils, plant, genetic modification of, 85–95
    fatty acids for food ingredients, 89–92
    frying oils, 86–87
    hydrogenated oils, 85
    oils for baking and confectionary applications,
        87–89
Oleic acid, 87
Oligosaccharide(s)
    biological importance of, 475
    food grade, 475
    glucose based, 421
    glycosidase catalysed synthesis of, 491
    one pot synthesis, 476
    simple, 474
    synthesis
        glucoamylase and, 486
        kinetic approach, 483
        Leloir glycosyltransferases in, 479
Oligosaccharides, enzymatic synthesis of, 473–500
    biological importance of oligosaccharides, 474
    chemical synthesis, 475–480
        enzymatic synthesis, 476
        glycosidases, 479–480
        glycosyltransferases, 476–479
    equilibrium controlled synthesis, 480–482
    influence of reaction components and conditions
        on yield and selectivity, 482–488
        donor:acceptor sugar ratio, 482–483
        enzyme immobilization, 488
        enzyme origin, 484–485
        pH, 487
        substrate concentration, 486–487
        temperature, 484
        time, 487
        use of organic solvents, 485–486
    oligosaccharides as drugs and functional foods,
        474–475
    recombinant and engineered glycosidases in
        oligosaccharide synthesis, 489–492
        glycosynthases, 489–490
        mutant enzymes other than glycosynthases,
        491
        mutant glycosidases created by directed
        evolution, 491–492
Omega-3 fatty acids, ratio of omega-6 to, 85
OMP, *see* Outer membrane protein
Oncorhynchus mykiss, 391
ONPG, *see* o-Nitrophenyl-β-galactose
Open reading frames (ORFs), 90
Optimer program, 476

ORAC assay, *see* Oxygen radical absorbance capacity
    assay
Oral tolerance, 440
Oregano
    clonal extracts, *Listeria monocytogenes* and, 289
    essential oil of, 286–287
    phenolics, 194, 271
ORFs, *see* Open reading frames
Organic acid(s)
    multifunctional, 508
    production, 172, 525
Origanum vulgare, 15, 187, 291
Ornithine decarboxylase (ODC), 165, 408
Orthosiphon aristatus, 188
Oryza sativa, 123
OSI, *see* Oxidative stability index
Osteoporosis, calcium and, 116
OTR, *see* Oxygen transfer rates
OUR, *see* Oxygen uptake rate
Outer membrane protein (OMP), 389
Ovalbumin/antiovalbumin mediated passive
    cutaneous anaphylaxis, 188
Ovarian hormones, 259
Overflow metabolism, 549
Oxaloacetate, 553
Oxidation
    LDL and, 211, 236
    linked diseases, 161
        diet and, 199, 210
        fruit juice blends and, 174
        fruit and vegetable management of, 231
        ROS and, 2
    linked immune dysfunction, 197
Oxidative burst, 440, 451
Oxidative damage, smoking and, 126
Oxidative rancidity, 125
Oxidative stability index (OSI), 86
Oxidative stress
    -associated diseases, 126
    biological adaptive responses to manage, 160
    cause of, 230
    chronic diseases and, 162
    control of, 138
    excessive ROS formation and, 235
    free radical mediated, 157
    linked disease conditions, 199
    plant peroxidases and, 243
    protein disulfides and, 159
Oxidized glutathione (GSSH), 159
Oxoglutarate dehydrogenase (OHDC), 560
Oxygen
    radical absorbance capacity (ORAC) assay, 234
    transfer rates (OTR), 527
    uptake rate (OUR), 527
Ozone, plants exposed to, 10, 210

P

PAL, *see* Phenylalanine ammonia lyase
Palm oil, 85
Paneth cells, 440
Pangium edule, 218
Parellada, 313
Parkinson's disease (PD), 215
  neuronal cell model system, 217
  ON-OFF fluctuations, 217
  ROS and, 161
  SSB and, 616
Parkinson's management, L-DOPA for, 16
Partial hydrogenation, frying oils, 86
Passive immunization, 382, 386
Passive immunotherapy, *see* Egg yolk antibody
      farming for passive immunotherapy
Pasteur effect, 524
Pasteurella multocida, 391
Pattern recognition receptors, 441
PBMC, *see* Peripheral blood mononuclear cells
PC, *see* Pyruvate carboxylase
PCBs, *see* Polychlorinated biphenyls
PD, *see* Parkinson's disease
PDGF, *see* Platelet derived growth factor
PDS, *see* Phytoene desaturase
Pearson vs. Shalala dietary supplement claims case, 7
Pediococcus
  acidilactici, 514
  damnosus, EPS of, 588
PEDV, *see* Porcine epidemic virus
PEITC, *see* Phenethyl isothiocyanate
Penicillin, 561
Penicillium emersonii, 481
Pentose phosphate pathway (PPP), 11, 167, 213, 347
  coupling of proline biosynthesis with, 169
  deregulation of, 193
  interactions between glycolysis and, 350
  L-lysine and, 504
  NADPH and, 567
  overexpressed, 194
  oxidative, 358
  pyruvate kinase and, 358
  stimulation, 14, 168, 169, 175
PEP, *see* Phosphoenolpyruvate
PEPCk, *see* Phosphoenolpyruvate carboxykinase
Peptidoglycan, 561
Perilla frutescens, 187, 189, 270
Peripheral blood mononuclear cells (PBMC), 449, 451
Peroxidases, phenolic dependent, 243
Peyer's patches, 437, 438, 440, 449, 454
PFK, *see* Phophofructokinase
PG, *see* Polyglalacturonase
PGE$_2$, *see* Prostaglandin E$_2$
PGs, *see* Prostaglandins
Phaffia rhodozyma, 534–535
Phagocytes, bactericidal activity of, 451

Phagocytic receptors, LAB-induced up regulation
      of, 451
Phagocytosis, 43–44, 440
Pharmaceutical food, 502
Phenethyl isothiocyanate (PEITC), 259
Phenolic acids, 2
Phenolic antioxidants, *see* L-DOPA and phenolic
      antioxidants, enhancement of in fava
      bean
Phenolic compounds, major classes of, 153
Phenolic dependent peroxidases, 243
Phenolic phytochemicals
  antibacterial properties of, 163
  anticarcinogenic properties of, 3
  antioxidant properties of, 3, 163
  biological function of, 162
  biosynthesis of, 155
  categories of, 162
  cellular homeostasis and, 162
  current proposed mechanisms of action of, 175
  free radical scavenging activity, 167
  gene expression and, 162
  membrane interaction, 169
  membrane stacking, 170
  modulation of gene expression by, 169
  most abundant group of, 153
  potential anticarcinogenic functions of, 163
  rich sources of, 152
  sources of, 1, 9
  synthesis of, 2
  tissue culture based screening of, 291
  weak dissociation of, 169–170
Phenolic phytochemicals, clonal screening and sprout
      based bioprocessing of, 1–23
  biological function of phenolic phytochemicals,
      2–4
  clonal screening of phenolic synthesis in herb
      shoot culture systems, 11–14
  phenolic phytochemical ingredients and benefits,
      1–2
  phenolic synthesis in seed sprouts, 14–15
  phytochemicals and functional food, 4–9
  relevance of phenolic antioxidants, 9–11
  strategies and implications, 15–16
Phenolics, *see also* Plant phenolics
  antimicrobial, 11, 210
  membrane permeability, 295
  oxidative polymerization of, 243
  synthesis, stimulation of, 137
Phenylalanine, 545
  ammonia lyase (PAL), 213, 522
  RA biosynthesis and, 190, 191
Phenylketonuria, 331
Phenylpropanoid(s), 213–214
  –acetate–malonate pathway, 2
  biosynthesis, 14, 15
PHO, *see* Phosphorylase
Phophofructokinase (PFK), 356
Phosphatidyl choline hydrolase, 363

Phosphatidylglycerol, 359
Phosphoenolpyruvate (PEP), 213, 357, 592
  carboxykinase (PEPCk), 563
  carboxylase, 510, 563
  sugar uptake and, 555, 572
  xanthan and, 590
Phospholipase D (PLD), 363, 364
  activity
    antisense suppression of, 365
    inhibitor of, 365
    Microtom, 366
    regulation of, 365
  expression, transgenic Celebrity fruits, 365
  membrane deteriorative pathway and, 370
  role in senescence and chilling, 364
Phospholipids, potato, 359
Phosphorylase (PHO), 56
Phosphotransferase system (PTS), 510, 555, 572
Photobacter profundum, polyketide synthase and, 90
Photo oxidative damage, prevention of, 119
β-Phycoerythrin, 234
Phytic acid, 113, 116, 117
Phytoalexins, 11, 210, 262
Phytochemicals, *see* Phenolic phytochemicals,
    clonal screening and sprout based
    bioprocessing of
Phytoene
  desaturase (PDS), 120
  synthase (PSY), 120, 121
Phytoestrogens, 134
Phytohemagglutinin, 449
Phytophthora parasitica, 368
Pichia etchellsii, β-glucosidase from, 489
Pimpinella anisum, 10
p53-induced genes, activation of, 258
Pinot Noir, 314
Pisum sativum, 14, 270, 367
PKC, *see* Protein kinase C
PKS enzyme complex, *see* Polyketide synthase
    enzyme complex
Plant(s)
  defense
    role of LOX in, 368
    system, elicitor mediated activation of, 137
  lipoxygenases, membrane integrity and, 366
  membranes, chilling sensitivity and, 359
  –microorganism interactions, LOX involvement
    in, 368
  oils, *see* Oils, plant, genetic modification of
  peroxidases, oxidative stress and, 243
  phenolics
    esterified form of, 210
    polymers formed from, 210
    synthesis of, 213
  primary products in, 209
  respiration, distinct feature of, 356
  starches, *see* Starches, plant, genetic modification
    of

stress response, secondary metabolites in, 209
transformation, 346
Plastic composite support, biofilm development and,
    530
Platelet derived growth factor (PDGF), 262
PLD, *see* Phospholipase D
PLPPP, *see* Proline linked pentose–phosphate
    pathway
PMF, *see* Proton motive force
Poliovirus vaccinations, 455
Polychlorinated biphenyls (PCBs), 261
Polyglalacturonase (PG), 369
Polyketide synthase (PKS) enzyme complex, 90
Polyphenoloxidase (PPO), 210
Polyphenols, 2, 152
Polysaccharide(s), *see also* Microbial
        polysaccharides, biotechnology of
  gel formation using, 584
  nonstarch, 402, 420
  production, role of UDP-glucose
        pyrophosphorylase in, 602
  recovery, 591
  water soluble, 420
Polyunsaturated fatty acids (PUFAs), 86, 89, 124
Porcine epidemic virus (PEDV), 391
Potato(es)
  amylose free mutant of, 70
  consumer preference for, 346
  GBSSI antisense lines of, 60
  lipids, major, 359
  membrane(s)
    amyloplast membrane permeability, 361
    changes in fatty acid composition of, 361
    deterioration, lipid peroxidation and, 362
    permeability, 359, 360
    tonoplast membrane permeability, 360
  phosphate content in, 72
  phosphorylases, glucan specificity of, 351
  processing, primary problem associated with, 346
  starch, high amylose, 63
  starch–sugar balance, 352
  sugar balance in, 362
  tuber(s)
    alkaline and acid invertases in, 355
    β-amylase activity, 352
    inorganic phosphates and, 352
    low temperature storage of, 347
    oleyl-PC desaturase activity in, 359
    reconditioning of, 349
    starch breakdown in, 351
    starch degradation, 350
    starch-sugar interconversion in, 348
    starch synthesis, 349
    UGPase alleles identified in, 354
Pouchitis, probiotics and, 458
PPI, *see* Proton pump inhibitors
PPi dependent phophofructokinase (PPi-PFK), 356
PPi-PFK, *see* PPi dependent phophofructokinase
PPO, *see* Polyphenoloxidase

PPP, *see* Pentose phosphate pathway
Prebiotic(s)
    anticarcinogenic effects, 407
    cholesterol lowering effect of, 408
    definition of, 402
    laevan type, 419
    market for, 421
    substances, 402
Pregnancy, breast cancer risk and, 255
Prephenate
    aminotransferase, 191
    dehydratase, genes encoding, 565
Proanthocyanidins, 2, 152
Probiotic(s)
    bacteria, methods for studying effects of on
        immune system, 448
    based cancer therapy, 408–409
    definition of, 436, 460
    diarrhea and, 456
    effects of on inflammatory bowel disease, 404
    IBD treatment with, 457, 458
    immunomodulating effects of, 456
    microorganisms, 436
Proglycinin, modified, 41
Prokaryotes, 292
Proline
    biosynthesis, 139, 169
    dehydrogenase, 139
Proline linked pentose–phosphate pathway (PLPPP),
        11, 15, 139, 302
    alternative, 617
    clonal tissue culture systems and, 200
    cytosolic proton linked modulation of, 296
    L-DOPA regulation and, 218
    metabolic biology of, 617
    model, 195, 197, 219, 220, 618, 619
    phenolic antimicrobial activity model, 293, 294
    phenolic synthesis driven through, 221
    proton linked redox cycle and, 293
    RA biosynthesis and, 192, 193
    SSB and, 621
    strategies, liquid yeast fermentation and, 617
    stress response and, 138
Propionate metabolism, 408
Propionobacterium acidi-propiionici, 588
Prostaglandin E_2 (PGE$_2$), 266–267
Prostaglandins (PGs), 266
Prostate cancer
    genistein and, 140
    soy food consumption and, 134
Protein(s)
    digestibility, 26, 37
    engineering, 27
        improvement in physicochemical functions
            by, 38
        nutritional quality and, 36
        physicochemical functions and, 42
    free radicals and, 162
    glucan-binding, 390

    heat shock, 210
    iron store, 113
    kinase C (PKC), 166
    physiological property of, 26
    stability, RA and, 200
    tumor suppressor, 255, 262
    zinc and, 116
Proteus mirabilis, 164
Proton motive force (PMF), 293, 295
Proton pump inhibitors (PPI), 297
Protozoa, acute inflammatory reactions and, 404
Provitamin A carotenoid, most potent, 119
Prunella vulgaris, 188
Pseudomonas
    aeruginosa, cystic fibrosis and, 389
    elodea, 221, 300
    putida, 547
PSY, *see* Phytoene synthase
Ptilota filicina, isomerase genes cloned from, 91
PTS, *see* Phosphotransferase system
PU, *see* Pullulanase
Pubertal hormones, 254
PUFAs, *see* Polyunsaturated fatty acids
Pullulanase (PU), 56, 58
Pyrococcus furiosus, β-glucosidase from, 492
Pyruvate, 553
    carboxylase (PC), 563
    dehydrogenase complex, 510

Q

Quercetin, 3, 10
Quinone: NADPH oxidoreductase, 166
Quinones, 286, 293

R

RA, *see* Rosmarinic acid
Radiation, cytogenetic damage induced by, 166
Raffinose, 516
Raisins, microsatellite typing of, 312
Raloxifene, 260
RBD vegetable oil, *see* Refined, bleached, deodorized
        vegetable oil
RDAs, *see* U.S. Recommended Dietary Allowances
RE, *see* Reducing equivalents
Reactive nitrogen species (RNS), 158
Reactive oxygen species (ROS), 2, 157, 440
    antioxidant systems and, 159
    detection, quantitative assays, 232
    direct measurement of, 231
    DNA oxidation and, 211, 238
    excessive generation of, 161
    formation, 159, 230
    generation, 138
    imbalance in, 161

tumor cell mitochondria and, 140
increase, phagocyte derived, 139
major physiological source of, 211
mediated pathogenicity, 158
oxidation-linked diseases and, 2
as signaling molecules, 159
superoxide anion and, 157
Receptor oligosaccharides, 474
Recombinant DNA technology, 502
Recombinant glycosidases, 489
Reducing equivalents (RE), 13
Refined, bleached, deodorized (RBD) vegetable oil,
86
Refined foods, exposure of to light, 124
Regulatory T cells, 441
Relishes, xanthan added to, 596
Restriction fragment length polymorphisms (RFLP),
312
Resveratrol, 3, 262, 268
COX-2 and, 322
wine and, 173
Retaining enzymes, 479, 480
Retinal photodamage, prevention of, 534
Retinoic acid, 264
Retrogradation, 64
Reverse hydrolysis, 480
aim of, 485
glucooligosaccharide synthesis by, 489
reaction time in, 487
substrate concentrations used in, 484
synthesis of mannooligosaccharides by, 484
use of bacterial glycosidases in, 481
RFLP, *see* Restriction fragment length
polymorphisms
Rheumatoid arthritis, 162, 198
Rhizoctonia solani, 368
Rhizopus
oligosporus, 172, 173, 222, 614
oryzae glucoamylase, effect of organic solvents
on, 485
Rhodobacter
capsulatus, 536
sphaeroides, 539
Rhodotorula glutinis, 522, 602
Riboflavin, 513
Rice
amylose produced from, 61
crops, lactoferrin-rich, 114
grains, iron levels of, 114
starches, low amylose, 70
Wx alleles of, 61
Riesling, 313
RNS, *see* Reactive nitrogen species
ROS, *see* Reactive oxygen species
Rosmarinic acid (RA), 2, 12, 187, 188
cobra venom factor and, 270
source of, 269
Rosmarinic acid biosynthesis and mechanism of
action, 187–207

biosynthesis in cell cultures, 190–194
biosynthesis and generation of high RA clonal
lines, 192
pathways associated with biosynthesis, 190–
191
role of proline linked pentose–phosphate
pathway in biosynthesis in clonal
systems, 192–194
mechanism of action through stimulation of host
antioxidant response, 194–200
sources and functional pharmacological effects,
188–189
Rosmarinus officinalis, 10, 187
Rotaviral infections, probiotics and, 456
Rotavirus, 386
diarrhea, IgA production in patients with, 460
-specific IgA response, 455
Rubrivivax gelatinosus, 539
Russet Burbank potatoes, fry color of, 347

S

Saccharin, 329, 330
Saccharomyces
boulardii, IBD treatment with, 458
cerevisiae, 337
diploid character of, 321
fermentation, 319
killer strains of, 321
strains, thermotolerant, 321
fragilis, 617–618
Sage extract, *Bacillus cereus* and, 286
Salad dressings, xanthan added to, 596
Salmonella
enterica, 33
enteritidis (SE), 389
food poisoning, 388
species, acute inflammatory reactions and, 404
typhimurium, 174, 451
enterocytes and, 452
LAB effects on, 456
Salvia cavaleriei, 188
Sausages, fermented, 172
Sauvignon Blanc, 313
SBE, *see* Starch branching enzymes
SCFA, *see* Short chain fatty acids
Schizosaccharomyces pombe, malate permease
transport gene from, 321
Scopoletin, 10
Scutellaria baicalensis, 268
SDH, *see* Succinate dehydrogenase
SE, *see* Salmonella enteritidis
Secretory IgA (SIgA), 443
Seed(s)
protein content of, 26
sprouts, phenolic synthesis in, 14
storage proteins, 27

Selective estrogen receptor modulators (SERM), 135, 260
Self antigens, 441
SERM, *see* Selective estrogen receptor modulators
Serratia marcescens, 550, 551
Seyval Blanc, 316
SFC, *see* Solid fat content
Shepody potatoes, fry color of, 347
Shewanella species, polyketide synthase and, 90
Shigella species, acute inflammatory reactions and, 404
Shikimate
    metabolites, biosynthesis of, 194
    pathway, 2, 213, 270
SHIME model, 417
Shiraz, 315
Short chain fatty acids (SCFA), 402, 403
    absorption, 404
    concentration, lactitol and, 419
    gentiooligosaccharides and, 420
    metabolism of in liver, 415
    production, 410
    transport, 404
Shortening, 87
Sialic acid analogues, 474
SIgA, *see* Secretory IgA
Signaling messengers, rich sources of, 364
Simple oligosaccharides, 474
Skin tumor formation, inhibition of, 3
Smoking, oxidative damage and, 126
SOD, *see* Superoxide dismutase
SOE, *see* Soybean oligosaccharide extract
Solid fat content (SFC), 85–86
Solid-state bioprocessing (SSB), 171, 172, 175, 271, 611
    filamentous fungi and, 302
    fungal based, 222
    phenolic mobilization by, 615
    Rhizopus oligosporous and, 301
Solid-state bioprocessing for functional food ingredients and food waste remediation, 611–624
    approach for development of functional foods and ingredients, 614–617
        cranberry, 616–617
        enzymes in phenolic mobilization and metabolic biology of PLPPP, 617
        fava beans and other legumes, 615–616
        soybeans, 614–615
    complexity of SSB systems and need for systems biology approaches, 612–613
    extension of PLPPP strategies for liquid yeast fermentation and lactic acid bacteria, 617–619
    modern applications, 612
    SSB for food waste bioprocessing and applications, 612
    traditional food products and implications, 613–614

Solid-state fermentation (SSF), 137, 611
Solid substrate fermentation, 530
Sorbitol, 384
Soy
    foods, prostate cancer and, 134
    products
        Koji production for, 612
        liquid fermentation, 272
        lunasin in, 46
Soybean(s)
    Bowman-Birk protease inhibitor, 26
    ferritin gene, 113
    genetic modulation of, 136
    LOX, 367
    oil, 85
    oligosaccharide extract (SOE), 419, 422
    peptide, 26
    phenolic mobilization from, 618
    primary oligosaccharides in, 416
    protein(s), *see also* Soybean proteins, molecular design of
        disordered regions of, 30
        phagocytosis stimulating peptide from, 43
    sprouting, 299
    substrates, SSB of, 614
Soybean isoflavonoids, potential health benefits of, 133–149
    consumption of soybean and reduced incidence of disease, 134–139
        isoflavone enrichment of soybean, 136–137
        major bioactivities of soybean isoflavonoids, 135–136
        model mechanism for action of soybean isoflavonoids against cancer, 137–138
        soybean isoflavonoids, 134–135
        stress response and proline linked pentose–phosphate pathway, 138–139
    hypothetical model for cancer chemopreventive action of soybean isoflavone genistein, 139–141
    implications, 141–142
Soybean proteins, molecular design of, 25–50
    future trends, 46
    improvement and addition of physiological properties of soybean proteins, 42–46
        addition of phagocytosis stimulating activity, 43–45
        antimitotic activity of lunasin, 45–46
        fortifying of bile acid binding ability, 42–43
    improvement of nutritional quality by protein engineering, 36–38
        increasing digestibility, 37–38
        increasing sulfur containing amino acids, 36–37
    improvement in physicochemical functions by protein engineering, 38–42
        improvement in physicochemical functions of glycinin, 39–42

structure to physicochemical function
  relationships of soybean proteins, 38–39
soybean proteins, 27–28
  β-conglycinin, 27
  glycinin, 27–28
structural features of soybean proteins, 34–36
  β-conglycinin, 34–35
  glycinin, 35–36
three dimensional structures of soybean proteins,
  28–34
  crystallization, 28–29
  cupin superfamily, 33–34
  X-ray crystallography, 29–32
Specialty chemical, world's most important
  manufactured, 126
Specialty oil market, 86
Spoonful™, 331
Sports bars, 9
Sprout based bioprocessing, *see* Phenolic
  phytochemicals, clonal screening and
  sprout based bioprocessing of
SPS, *see* Sucrose 6-P synthase
SS, *see* Starch synthase
SSB, *see* Solid-state bioprocessing
SSF, *see* Solid-state fermentation
STAGs, *see* Structured triacylglycerols
Staphylococcus aureus
  acute inflammatory reactions and, 404
  control of by phenolic phytochemicals, 288
  inhibition of, 286
  phagocytosis measurement and, 449
Staple food, diet formed mainly of, 99
Starch(es)
  applications of, 68
  biosynthesis, pathway of, 57
  branching enzymes (SBE), 56, 58, 349
    activity, modifications in, 62
    deficiencies in, 63
  chemical classification of, 53
  crystalline order in, 55
  degradation, 350, 354
  extracted, nonfood industrial uses of, 52
  fermentation, butyrate and, 403
  functionality, 65
  glucan polymers of, 64
  granule(s)
    appearance of, 53
    heating of, 64
    size, mechanisms influencing, 72
  Hizukuri cluster model, 56
  incorporation of inulins with, 73
  intermediate, 54
  metabolism of in tuber, 348
  -metabolizing enzymes, 351
  modified, chromatography of, 66
  phosphate content of, 349–350
  phosphorylase, 351
  physical characteristics of, 351
  resistant, 402

synthase (SS), 57, 58, 59, 349
  activity, modifications in, 61
  mutant, first reported, 61
synthesis, rate-limiting step in, 349
tapioca, 72
waxy barley, 70
Starches, plant, genetic modification of, 51–84
  applications, 68–73
    amylopectin chain length, 71–72
    amylose free starch, 68–69
    high amylose starch, 70–71
    low amylose starch, 69–70
    novel starches, 72–73
  functionality, 64–68
    amylopectin retrogradation, 65–66
    amylose gelation, 66–67
    gelation of mixtures of amylose and
      amylopectin, 67–68
  modifications, 59–64
    DeBE activity, 63
    GBSSI activity, 60–61
    multiple pathway enzymes, 63–64
    SBE activity, 62–63
    SS activity, 61–62
  structure, 52–56
  synthesis, 56–59
    branching enzyme, 58
    debranching enzyme, 58–59
    granule bound starch synthase, 57
    soluble starch synthase, 58
Starvation, diseased cell and, 141
Stearic acid, 88
Stevioside, 333, 334
Stizolobium deeringianum, 217
St. John's wort, 269
Stomach cancer
  flavonoids and, 156
  Helicobacter pylori and, 296
Strawberry, D-galacturonic acid reductase gene
  expression from, 125
Streptococci, 445
Streptococcus
  faecalis, tannins and, 286, 295
  mutans, 164, 390
  salivarus thermophilus, polysaccharide from, 585
  thermophilus, 602
    EPS of, 587, 588
    tetrasaccharide repeat unit of, 586
Streptomyces lividans, 337
Stress response, cellular mechanisms of, 573
Structured triacylglycerols (STAGs), 89
Succinate
  –acetate molar ratio, 510
  dehydrogenase (SDH), 560
Succinic acid, 508
Sucralose, 330, 331
Sucrose
  breakdown of starch into, 347
  cold induced synthesis of, 348

enzymes leading to production of, 352
    functions, futile cycling of, 356
    metabolism, potato tuber, 353
    synthase, 352
Sucrose 6-P synthase (SPS), 352
Sugar(s)
    alcohol, fermentation of, 419
    decoy, 164
    low calorie, 515
    nucleotide regeneration beads, 478
    ratio, donor:acceptor, 482
    transferases, 593
Sugary-2 locus, 61
Sulfolobus solfataricus 485, 490
Sulfur containing amino acids, 36
Sunette™, 329
Sunflower oil, 85, 87
Superbead technology, 478, 492
Superbug, 478
Superoxide anion, 157
Superoxide dismutase (SOD), 3, 15, 211
    forms of, 240
    gene, *Arabidopsis thaliana*, 317
    importance of, 160
    manganese, 138
    NADPH and, 617
    nutrient depletion and, 222
    stimulation, 221
Superyeasts, 312
Sweet compounds, 328
Sweet n Low™, 329
Sweet One™, 329
Swine respiratory diseases, causative agents of, 391
Systems Biology, 613, 614, 619

T

TAGs, *see* Triacylglycerols
Tamoxifen, 260, 264, 265
Tannins, 2, 154, 286
    condensed, 171
    hydrolyzed, 173
Tapioca starch, 72
TBA, *see* Thiobarbituric acid
TCA cycle, *see* Tricarboxylic acid cycle
TCDD, *see* 2,3,7,8-Tetrachlorodibenzo-p-dioxin
T cell(s)
    antigen receptor mediated signaling, 189
    antigen-specific activation of, 442
    function, lymphocyte proliferation and, 453
    IFN-γ production by, 452
    proliferation, RA and, 189
    receptor (TCR), 439
    regulatory, 441
TCR, *see* T cell receptor
α-TE, *see* α-Tocopherol
Teichoic acid biosynthesis, 602
Tempeh, 134, 302, 612

Tempranillo, 315
Terminal end buds, 254
2,3,7,8-Tetrachlorodibenzo-p-dioxin (TCDD), 166
Tetrasaccharides, 487
TGF-α, *see* Transforming growth factor alpha
TGF-β, *see* Transforming growth factor beta
Thaumatin, 336, 338
Thermus
    aquaticus, β-galactosidase from, 488
    thermophilus, β-glycosidase from, 489
Thin layer chromatography (TLC), 238
Thiobarbituric acid (TBA), 236
Thioglucosides, 481
Thioredoxin (Trx), 160
Threonine, 545, 551, 568
    degrading enzyme, 568
    export, 569
Thromboxanes (TXs), 266
Thylakoid membranes, phosphatidylglycerol in, 359
Thymus vulgaris, 10, 187, 188, 291
TLC, *see* Thin layer chromatography
TMT, *see* Tocopherol methyltransferases
TNF-α, *see* Tumor necrosis factor-α
Tocopherol(s)
    methyltransferases (TMT), 125
    oxidation linked diseases and, 161
α-Tocopherol (α-TE), 124
Toll-like receptors, 441, 442, 457
Tomato(es)
    carotenogenesis in, 120
    chilling injury, 364
    fruit ripening
        cell wall metabolism and, 369
        LOX and, 368
        mutants, 368
    lack of year round supply of, 363
    membrane
        deterioration, 363
        shelf life and, 363
    plants, antisense, 368–369
    post harvest shelf life, 363
    processed tomato products, 363
    ripening, 363
Touriga National, 315
Toxic oxygen species, 362
Traminer, 313
Transaldolase, 564
Transcriptomics, 516
Transforming growth factor alpha (TGF-α), 262
Transforming growth factor beta (TGF-β), 262, 264, 452
trans free oils, 87
Transgalactosylating enzyme, 491
Transgenic Celebrity fruits, PLD expression in, 365
Transglycosylation, 481, 486
Transketolase, 504, 564
Traveler's diarrhea, 457
Trehalose, 561
Triacylglycerols (TAGs), 89

Tricarboxylic acid (TCA) cycle, 169, 507, 508, 617
Trichosanthes kirilowii, 92
Trx, *see* Thioredoxin
Trycophyton species, 164
Tryptophan, 545
  biosynthetic enzymes, 507
  production, 556
Tumor
  angiogenesis, 266
  cell death, genistein and, 141
  cell growth inhibition, cranberry extracts and, 165
  development
    probiotic based cancer therapy and, 407–408
    soy and, 260
  necrosis factor-α (TNF-α), 440, 450, 458
  suppression, eleostearic acid and, 92
  suppressor protein, 255, 262
TXs, *see* Thromboxanes
Tyrosine auxotrophs, 565

U

UDP-Gal regenerating superbeads, 478
UDP-glucose pyrophosphorylase (UGPase), 354, 588
UGPase, *see* UDP-glucose pyrophosphorylase
Ulcerative colitis, probiotics and, 458
Uncinula necator, 317
Urinary tract infections (UTIs), 163, 414
U.S. Recommended Dietary Allowances (RDAs),
      98, 112
U.S. statutory drug definitions, 8
UTIs, *see* Urinary tract infections

V

Vaccination antigens, antibody responses to, 449
Vaccinium
  angustifolium, 156
  macrocarpon, 156
  myrtillus, 156
Valine, 545
Vascular endothelial growth factor (VEGF), 164,
      256, 266
Vegetable(s), *see also* Fruits and vegetables,
      improvement of nutritional quality and
      shelf life of
  oil
    major crops used for, 85
    RBD, 86
  phenolic phytochemicals in, 152
VEGF, *see* Vascular endothelial growth factor
Vibrio
  cholerae, acute inflammatory reactions and, 404
  parahemolyticus, 172
    cranberry pomace and, 165
    inhibition of, 286

Vicia faba, ʟ-DOPA glycoside from, 217
Vidal Blanc, 315
Vigna radiata, 14
Virtual bacterial cell, 574
Vitamin(s), 512, *see also* Minerals and vitamins,
      molecular biotechnology for
      nutraceutical
  definition of, 117
  dietary allowances per day, 103–107
  fat-soluble, 119
  water-soluble, 118
Vitamin A
  compound, most potent, 120
  deficiency, worldwide, 122
Vitamin C, 125, 513
Vitamin E
  biosynthesis pathway, 121
  supplementation, 124
Viticulture, biotechnology in, 318
Vitis vinifera cultivars
    red cultivars, 314
    white cultivars, 313

W

Water deficit, cell membrane functioning and, 367
Water soluble polysaccharides (WSP), 420
Water-soluble vitamins, chemical structures of, 118
WAXS, *see* Wide angle x-ray scattering
Waxy barley starches, 70
Waxy rice starch, 68–69
WBO, *see* Wheat bran oligosaccharides
Western diet, LAB and, 459
Wheat bran oligosaccharides (WBO), 418
WHO, *see* World Health Organization
Wide angle x-ray scattering (WAXS), 64
Wild blueberry extract, α-amylase inhibition activity
      of, 174
Wild yeast, isolation of in grapes, 319
Wine(s)
  Chilean red, 298
  consumption, world, 312
  grapes, genetic engineering of, 316
  major phenolic phytochemicals in, 152
  phenolic components of, 322
  polyphenols, synergistic interactions between,
      173
  resveratrol and, 173
  yeast, 319
Wine industry, biotechnology in, 311–326
  genetic engineering of wine grapes, 316–319
    biotechnology in viticulture, 318–319
    clonal selection, 317–318
    somaclonal selection, 318
  genetic engineering of yeast for fermentation,
      319–322
    ideal yeast, 320
    wine as functional food, 322

yeast breeding and wine quality, 321–322
grape cultivars and wine types, 312–316
American hybrids, 316
French–American hybrids, 315–316
Vitis vinifera cultivars, 313–315
Winemaking, history of, 311
World Health Organization (WHO), 604
World population, growth of, 108
Wound healing
vitamin C and, 125
zinc and, 116
Wounding, phenylpropanoid compounds and, 11
WSP, *see* Water soluble polysaccharides

X

Xanthan
aqueous solutions of, 595
biodegradability of, 591
biosynthesis of, 592
concentration, oxygen solubility and, 590
food usage of, 595
gum, 14, 271
-LBG network, 597
lyase, 591–592
media for laboratory synthesis of, 590
molecular mass of, 595
monoclonal antibody and, 592, 605
National Formulary description of, 604
oligosaccharide repeat units of, 592
permitted uses of, 604
physical properties of, 594
polysaccharide production and, 585
production, 589, 591
synergistic gels involving, 597, 598
synthesis
genes involved in, 594
regulation of, 593
Xanthine oxidase, 159, 240
Xanthine oxidoreductase (XOR), 241
Xanthomonas campestris, 589, 593
Xanthophylls, 120, 122
Xenobiotic(s)
detoxification of, 3, 163
response element, 262
XOR, *see* Xanthine oxidoreductase

XOS, *see* Xylooligosaccharides
Xylitol, 402
Xylooligosaccharides (XOS), 402, 408, 418

Y

Yeast(s)
alcohol tolerance of, 319, 320
amino acid biosynthesis in, 558
breeding, wine quality and, 321
cells, composition of, 524
cold tolerance of, 320
genetic engineering of for fermentation, 319
glucan, 271
ideal, 320
inoculation, 319
isolated, 319
killer, 320
Yersinia
ruckeri, 391
species, acute inflammatory reactions and, 404
Yogurt(s)
bacterial fermentations in, 585
commercial soymilk based, 272
cytokine expression and, 453
γ-interferon and, 405
lactic acid bacteria and, 436
manufacture, bacteria used in, 410
species employed in fermentation of, 587
xanthan added to, 596

Z

Zataria multiflora, 190
ZDS, *see* ζ-Carotene desaturase
Zeaxanthin, 123, 534, 538
Zileuton, 268
Zinc
deficiency, 99
-dependent enzymes, 116
DNA replication and, 116
supplementation, 109
Zinfandel, 314
Zingiber cassumunar, 10
Zymomonas mobilis, 535